D1252676

HANDBOOK OF STRATA-BOUND AND STRATIFORM ORE DEPOSITS

Volume 9
REGIONAL STUDIES AND SPECIFIC DEPOSITS

HANDBOOK OF STRATA-BOUND AND STRATIFORM ORE DEPOSITS

Edited by
K.H. WOLF

PART I

1. Classifications and Historical Studies
2. Geochemical Studies
3. Supergene and Surficial Ore Deposits; Textures and Fabrics
4. Tectonics and Metamorphism
   Indexes Volumes 1–4

PART II

5. Regional Studies
6. Cu, Zn, Pb, and Ag Deposits
7. Au, U, Fe, Mn, Hg, Sb, W, and P Deposits
   Indexes Volumes 5–7

PART III

8. General Studies
9. Regional Studies and Specific Deposits
10. Bibliography and Ore Occurrence Data
    Indexes Volumes 8–10

ELSEVIER SCIENTIFIC PUBLISHING COMPANY
Amsterdam – Oxford – New York 1981

HANDBOOK OF STRATA-BOUND AND STRATIFORM ORE DEPOSITS

PART III

Edited by

K.H. WOLF

Volume 9

REGIONAL STUDIES AND SPECIFIC DEPOSITS

ELSEVIER SCIENTIFIC PUBLISHING COMPANY
Amsterdam – Oxford – New York 1981

ELSEVIER SCIENTIFIC PUBLISHING COMPANY
335 Jan van Galenstraat
P.O. Box 211, 1000 AE Amsterdam, The Netherlands

Distributors for the United States and Canada:

ELSEVIER/NORTH-HOLLAND INC.
52, Vanderbilt Avenue
New York, N.Y. 10017

Library of Congress Cataloging in Publication Data (Revised)
Main entry under title:

Handbook of strata-bound and stratiform ore deposits.

Includes bibliographical references and indexes.
CONTENTS: I. Principles and general studies, v. 1. Classifications and historical studies.-- v. 2. Geochemical studies.--v. 3. Supergene and surficial ore deposits : textures and fabrics.-- [etc.]--v. 9. Regional studies and specific deposits.
1. Ore-deposits. 2. Geology. I. Wolf, Karl H.
QE390.H36 553 77-461887

ISBN: 0-444-41824-5 (vol. 9)

Printed in The Netherlands.

LIST OF CONTRIBUTORS TO THIS VOLUME

A.C. BROWN
Department of Mineral Engineering, Ecole Polytechnique, Montreal, Que., Canada

U. FÖRSTNER
Institut für Sedimentforschung, Universität Heidelberg, Heidelberg, German Federal Republic

G.P. GLASBY
New Zealand Oceanographic Institute, Department of Scientific and Industrial Research, Wellington North, New Zealand

D.I. GROVES
Department of Geology, The University of Western Australia, Nedlands, W.A., Australia

W.W. HANNAK
Hannover, German Federal Republic

T. HOPWOOD
North Adelaide, S.A., Australia

D.R. HUDSON
C.S.I.R.O. Division of Mineralogy, Floreat Park, W.A., Australia

J.H. JOHNSTON
Department of Chemistry, Victoria University of Wellington, Wellington, New Zealand

M.M. KIMBERLEY
Department of Geology and Erindale College, University of Toronto, Mississauga, Ont., Canada

K.E. KNEDLER
Department of Chemistry, Victoria University of Wellington, Wellington, New Zealand

W. KREBS
Institut für Geologie und Paläontologie, Technische Universität Braunschweig, Braunschweig, German Federal Republic

J.R. KYLE
Department of Geological Sciences, The University of Texas at Austin, Austin, Texas, U.S.A.

D.E. LARGE
Institut für Geologie und Paläontologie, Technische Universität Braunschweig, Braunschweig, German Federal Republic

B. LEHMANN
Freie Universität Berlin, Fachbereich Geowissenschaften, Institut für Angewandte Geologie, Berlin, German Federal Republic

M.A. MEYLAN
Department of Geological Sciences, University of Wisconsin–Milwaukee, Milwaukee, Wisconsin, U.S.A.

H.-J. SCHNEIDER
Freie Universität Berlin, Fachbereich Geowissenschaften, Institut für Angewandte Geologie, Berlin, German Federal Republic

CONTENTS

Chapter 8. SEDIMENT-HOSTED SUBMARINE EXHALATIVE LEAD–ZINC DEPOSITS – A REVIEW OF THEIR GEOLOGICAL CHARACTERISTICS AND GENESIS
by D.E. Large

Chapter 9. THE GEOLOGY OF THE MEGGEN ORE DEPOSIT
by W. Krebs

Chapter 1

THE TIMING OF MINERALIZATION IN STRATIFORM COPPER DEPOSITS

A.C. BROWN

INTRODUCTION

Numerous detailed studies of stratiform [1] copper deposits reported in recent years have rejected the classic igneous-hydrothermal epithermal/telethermal origin for such mineralization, and have instead suggested that an origin intimately related to the processes of sedimentation would be much more suited to the observed character of these deposits. Overwhelming features, such as the widespread continuity of mineralization within a sedimentary basin and the distribution of mineralization within narrow stratigraphic limits, are commonly cited as evidence of sedimentary origins. However, such features do not exclude the possibility of diagenetic additions of ore metals to particularly favorable stratigraphic horizons, and in fact recent studies indicate repeatedly that one or more diagenetic events preceded the deposition of ore metals. For example, the deposition of iron sulfides, and in some cases sulfate-bearing minerals, is commonly observed to have occurred before the introduction of ore-stage metals.

The deposition of iron sulfides is of course a common phenomenon in anaerobic marine muds in which sulfate from overlying sea water is reduced by bacteria (cf. Chapter 6, Vol. 2, this Handbook) to form dissolved sulfide species which in turn react with sulfophile metals to precipitate primitive sulfide phases. Normally iron is the most readily available metal, and the initial solid phases are iron sulfides which transform during early diagenesis into relatively stable minerals such as fine-grained pyrite and pyrrhotite (Berner, 1971).

If on the other hand a more sulfophile metal such as copper, was present in this early diagenetic environment, it would certainly react with the bacterially generated sulfide before the ubiquitous iron, and the sediment would accumulate copper-bearing sulfides in proportion to the copper and reduced sulfur available. The investigations described below indicate instead that copper and its associated metals (e.g., lead, zinc, cadmium, cobalt) form ore minerals within the host sediments by reaction (replacement) with one or more prior sulfur-bearing minerals, which had already accumulated within the strata

---

[1] For an explanation of the usage of terms employed in this chapter, see the Appendix.

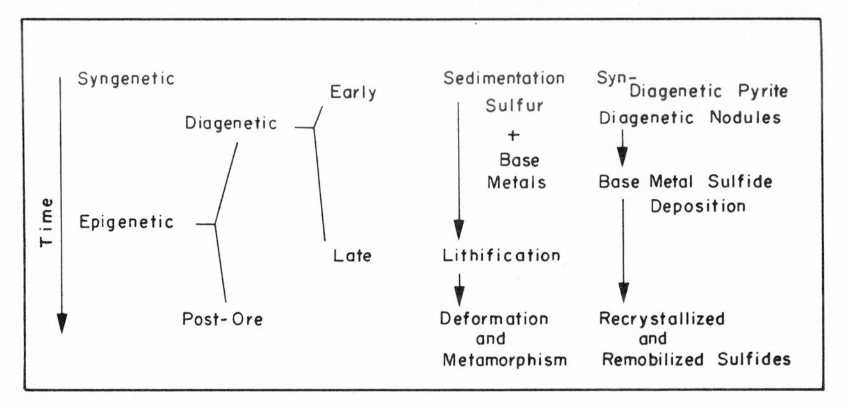

Fig. 1. The relative timing of syngenetic and epigenetic processes responsible for the formation of stratiform copper deposits as discussed in this chapter. (Modified after Brown, 1978.)

during sedimentation and subsequent early diagenesis. Consequently, the base metal mineralization is temporally separated from the sedimentary event by at least one early stage of diagenesis, such as the formation of primitive or crystalline pyrite.

The relative timing of these successive mineralization stages as envisaged in this paper is illustrated schematically in Fig. 1. The data from which these concepts have developed are drawn from a number of cited observations on many of the world's most prominent stratiform copper deposits: e.g., the Copperbelt of Central Africa, the Kupferschiefer of Europe, and the White Pine and Creta ores of the United States (see indexes in Volumes 4 and 7 of this Handbook for chapters discussing some of these deposits).

The intent of this analysis is to emphasize the post-sedimentary controls over the ultimate form and dimensions of such stratiform deposits, which in turn must influence one's concept of exploration targets in sedimentary terrain. While the targets are certainly large and attractive, knowledge of local ore controls, including diagenetic features, should definitely be applied to optimize the results in any exploration program.

CRITERIA FOR POST-SEDIMENTARY MINERALIZATION AT WHITE PINE, MICHIGAN

One of the most overwhelming arguments for a syngenetic, sedimentary origin for the White Pine mineralization (Fig. 2) is the widespread and uniform distribution of disseminated copper sulfides along bedding and within narrowly defined stratigraphic limits at the base of the Nonesuch Shale. Fig. 3 illustrates the typical stratigraphic section in question, and shows the restriction of ore zones principally to two fine-grained, organic-rich sedimentary units, the Parting Shale and the Upper Shale [1].

---

[1] The term "shale" may be misleading because even the most shaly beds are composed to a large extent of silt. Some beds are characterized by thin laminae grading upward from silt to shale (e.g., beds 23 and 43); others consist of massive silt with (e.g., beds 27 and 47) or without (e.g., beds 24, 26, 44 and 46) shaly partings (Ensign et al., 1968; Wiese, 1973).

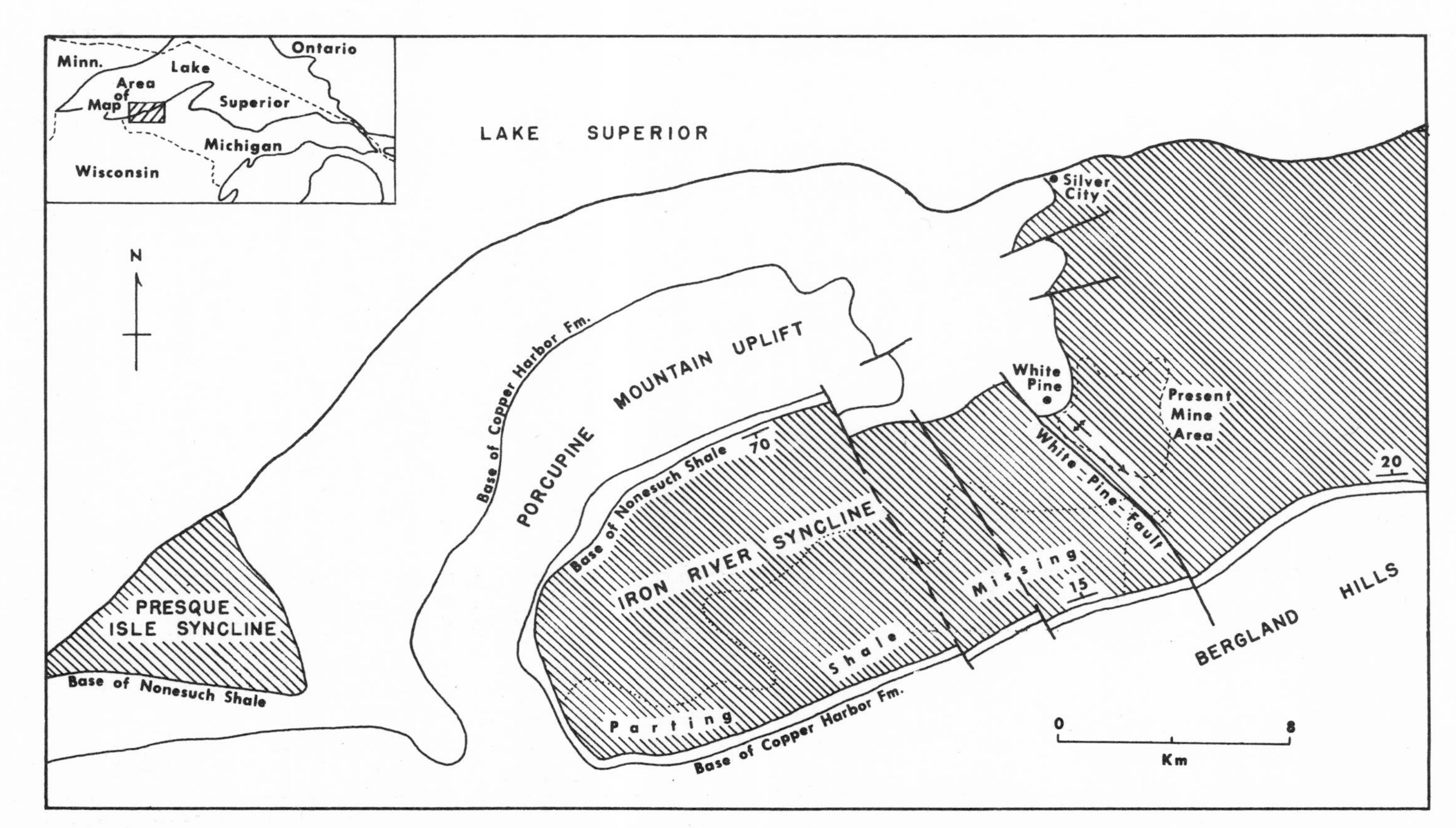

Fig. 2. Index map showing the major stratigraphic and structural features of the White Pine area, Michigan. (Modified after White and Wright, 1966, *Econ. Geol.*, 61: 1171–1190; and Brown, 1971, *Econ. Geol.*, 66: p. 544.)

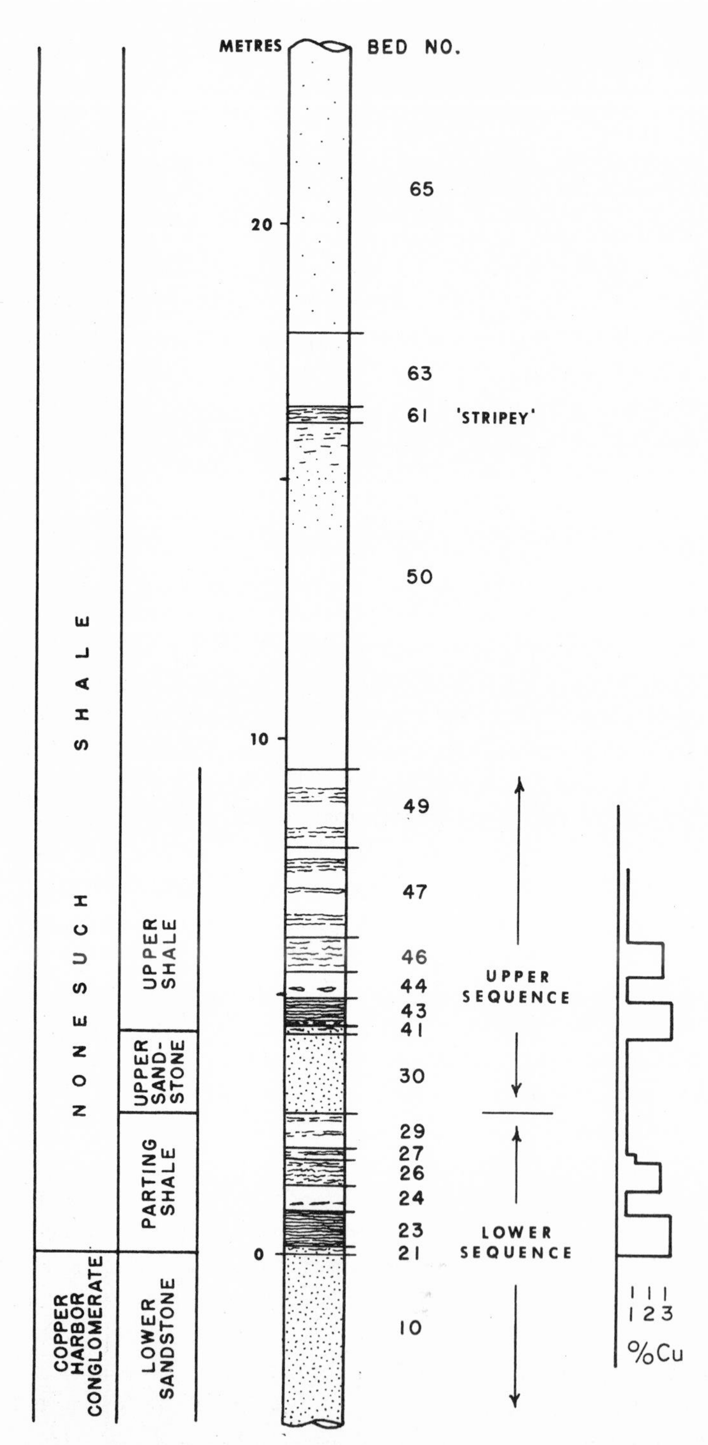

Fig. 3. Stratigraphic column through the basal Nonesuch Shale and underlying Copper Harbor Conglomerate at White Pine. (Modified after Brown, 1968.)

Early observers also noted a close spatial association between high-grade mineralization (especially native copper which may occur in important amounts in sandstone units such as the Lower Sandstone of the uppermost Copper Harbor Formation; see Hamilton, 1967) and the White Pine fault. From this relationship, it was once postulated that part, if not all, of the White Pine mineralization was structurally controlled and hence epigenetic (see Appendix) in origin.

The following paragraphs demonstrate that both of these concepts are untenable in the light of more detailed observations, and that most probably the copper mineralization at White Pine took place after sedimentation and before structural deformation of the lithified Nonesuch Shale. Jensen (see Wiese, 1973) and Burnie et al. (1972) have shown that the sulfur of both the mineralized and pyritic zones is characterized by a wide spread in $\delta^{34}S$ values consistent with a biogenic origin for sulfur throughout the lower Nonesuch. Consequently, a diplogenetic (see Appendix) origin with syn-diagenetic iron sulfides replaced by copper during diagenesis seems to be the most satisfactory explanation at this time.

Configuration of the mineralized zone

Whereas the exploited ore horizons at White Pine are very closely defined by stratigraphic limits, the complete envelope of copper mineralization (including sparsely mineralized zones, which unfortunately are often erroneously labelled as "sterile zones" and grouped with pyritic strata) is gently transgressive to bedding, as indicated by detailed mapping of the upper limit of mineralization (fringe) across the Nonesuch basins (Fig. 4).

Furthermore, studies of drill core and underground sections show that the cupriferous zone consists of one continuous mineralized zone superimposed on basal Nonesuch beds with a single undulating, blanket-like fringe surface that does not oscillate laterally with facies variations in the host-rock sediments, as would be expected if the mineralization were syngenetic and copper deposition were controlled by lateral oscillations of the shoreline during Nonesuch sedimentation (see Garlick, 1961, 1976). For example, the White Pine mineralization blankets two recognized transgressive cycles (Lower Sandstone–Parting Shale and Upper Sandstone–Upper Shale) in the basal Nonesuch (Fig. 3) without any relationship to these obviously important oscillations in the sedimentary environment.

According to a simple statistical analysis, a minor facies variation at the Copper Harbor–Nonesuch contact did, however, play a very significant role in defining the upper limit of copper mineralization across the Nonesuch basins. The lowest beds (Nos. 21 and 23) of the Nonesuch Formation contain minor amounts of copper where sandy, and major concentrations of copper where thick, fine-grained and rich in organic matter. Variations in the amount of copper in these basal beds have been examined statistically and found to correlate significantly (e.g., $R = -0.92$ for $N = 11$; $R = -0.88$ for $N = 36$) and inversely with the height of the fringe surface within the Nonesuch section. This observa-

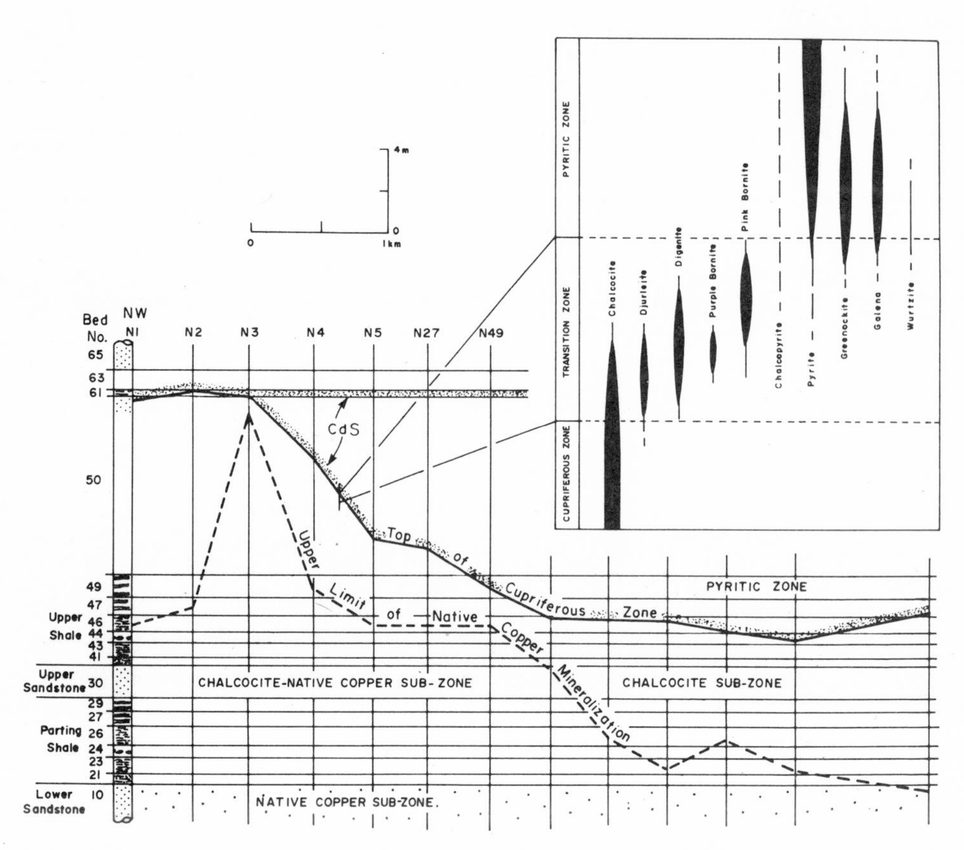

Fig. 4. Sketch showing the undulating configuration of the top of the cupriferous zone (fringe) in the basal Nonesuch Shale, and the occurrence of anomalous greenockite (CdS) concentrations at the fringe surface and along the No. 61 horizon. The inset shows the typical mineralogic sub-zones recognized within the transition between the cupriferous and pyritic zones. (After Brown, 1968; and Brown, 1971, *Econ. Geol.,* 66: p. 533.)

tion indicates that little copper remained for mineralization of higher stratigraphic units in areas where a large proportion of the total introduced copper was precipitated in the Nos. 21 and 23 beds at the base of the Nonesuch Formation. This interpretation becomes particularly pertinent to the origin of the Nonesuch mineralization when it is realized that the observed correlation is logical only if copper did in fact enter the Nonesuch through those basal beds. In a syngenetic model, there would be no reason to expect a relationship between the position of the top of the cupriferous zone and the amount of mineralization in beds commonly several metres to tens of metres below.

A study of the undulating fringe surface across the White Pine fault indicates rather clearly that this surface was displaced by the fault. This observation leads to the conclusion that the ore was not emplaced under the control of major structural features, and led Ensign et al. (1968) to suggest that the occurrence of high-grade mineralization in the neighborhood of the White Pine fault is probably coincidental. Also, mapping on a regional scale within the Nonesuch basin has since revealed the existence of several other major faults which do not appear to exhibit high-grade mineralization halos.

Mineralogy

The addition of copper to initially pyritic sediments as suggested above is supported as well by several mineralogic features. For example, disseminated fine-grained pyrite is replaced in a systematic step-by-step manner by chalcopyrite, bornite, digenite and finally chalcocite along the upper limit of the cupriferous zone. Hence it appears that the fringe surface defines the ultimate position of a mineralization front which advanced upward through the basal Nonesuch beds. Within the cupriferous zone proper, this replacement process is suggested by the occurrence of chalcocite nodules surrounded by halos of dusty hematite; presumably the iron released during replacement of initially pyritic nodules by chalcocite was redeposited in the immediate surroundings as an iron oxide.

The presence of greenockite (CdS), and to a lesser extent galena and wurtzite, in anomalous concentrations directly above the cupriferous zone is also readily explicable if copper entered the Nonesuch strata from below. Cd, Pb and Zn are less sulfophile than copper, but more sulfophile than iron and, if these three metals occurred as trace elements either in the ore solution or the host rock, they would be continually swept upward by an advancing cupriferous mineralization front and redeposited at the base of the pyritic zone by reaction with iron sulfide. This concept has been verified in the laboratory using chromatographic column techniques (Ritchie, 1964; Brown, 1974).

Other pertinent studies might be cited in support of a post-sedimentary origin for the White Pine mineralization. For example, Burnie et al. (1972) found that the sulfur isotopic ratios for the pyritic zone resemble closely those of the mineralized zone, and concluded that the sulfur of the entire basal Nonesuch was probably distributed initially as one common iron sulfide, which was subsequently replaced by copper sulfides in the present cupriferous zone.

Brown (1968, 1970, 1971), D.E. White (1968) and W.S. White (1971) have also examined the possibility of satisfactory short and long range transport models and ultimate sources of copper to complement their concepts of a diagenetic replacement process at White Pine. They conclude that reasonable sources and mechanisms could have existed (see a later section of this chapter).

It would seem more appropriate now, however, to examine other stratiform base-metal deposits and cite criteria for their post-sedimentary origins.

SEAL LAKE, LABRADOR

The sub-economic copper mineralization at Seal Lake, Labrador, shows remarkable similarities to that of White Pine. The two districts occur in geologically similar rift basins filled with volcanics and sediments of comparable lithologies and age. However, the Seal Lake basin has suffered intense isoclinal deformation and thrusting from the adjacent Grenville orogenic activity.

Detailed mapping, core logging and microscopic studies (Gandhi and Brown, 1975) have shown that, while the most obvious mineralization at Seal Lake is now fracture-controlled, the initial pre-deformation occurrence of sulfides was clearly limited at the local stratigraphic level to a narrow interval of widespread dark-grey sulfide-rich shales overlying quartzites and red-beds.

The discovery of mineralization in an area of minor deformation permitted observation of mineralization interpreted to represent the pre-deformation character of the Seal Lake mineralization. Here, the sulfides were found to exhibit an ordered array of copper, copper–iron and then iron sulfides, as described for White Pine above (Fig. 5). Furthermore, textures such as pseudomorphs after pyrite indicate that the copper-rich sulfides replace iron-rich sulfides as witnessed at White Pine. A minor occurrence of sphalerite was also found stratigraphically above the cupriferous horizon in one drill core.

Due to the complex geology of the Seal Lake region, it is not possible to reconstruct details of plausible mineralization models as has been done at White Pine, but by analogy of geologic settings and mineralogic features, it seems certain that a post-sedimentary, pre-deformation origin would apply to the Seal Lake mineralization as well.

CRETA, SOUTH-CENTRAL U.S.A.

A persistent, organic-rich cupriferous shale some 20 cm thick within flat-lying Permian strata of Texas, Oklahoma and Kansas has been described by Johnson (1974), Smith (1976; Vol. 6 of this Handbook) and others as a marine-to-brackish water sediment which paralleled an adjacent continental landmass to the southeast. The widespread distribution of copper (grades up to 5% along a total distance of approximately 200 km) within this narrow stratigraphic interval argues strongly in favor of a syn-sedimentary origin of mineralization.

However, mineralogic studies by Gann and Hagni (1974) and Hagni and Gann (1976) indicate that pyrite has invariably been replaced by copper-bearing sulfides (principally chalcocite and digenite) and never the reverse. In addition, they report that the Creta pyrite occurs as several textural varieties (colloform, framboidal and several crystalline forms) which are presumably syn-diagenetic in origin, but which may represent multiple substages of pyrite deposition within that earliest diagenetic time interval. The subsequent replacement of these different pyrites by copper sulfides necessarily

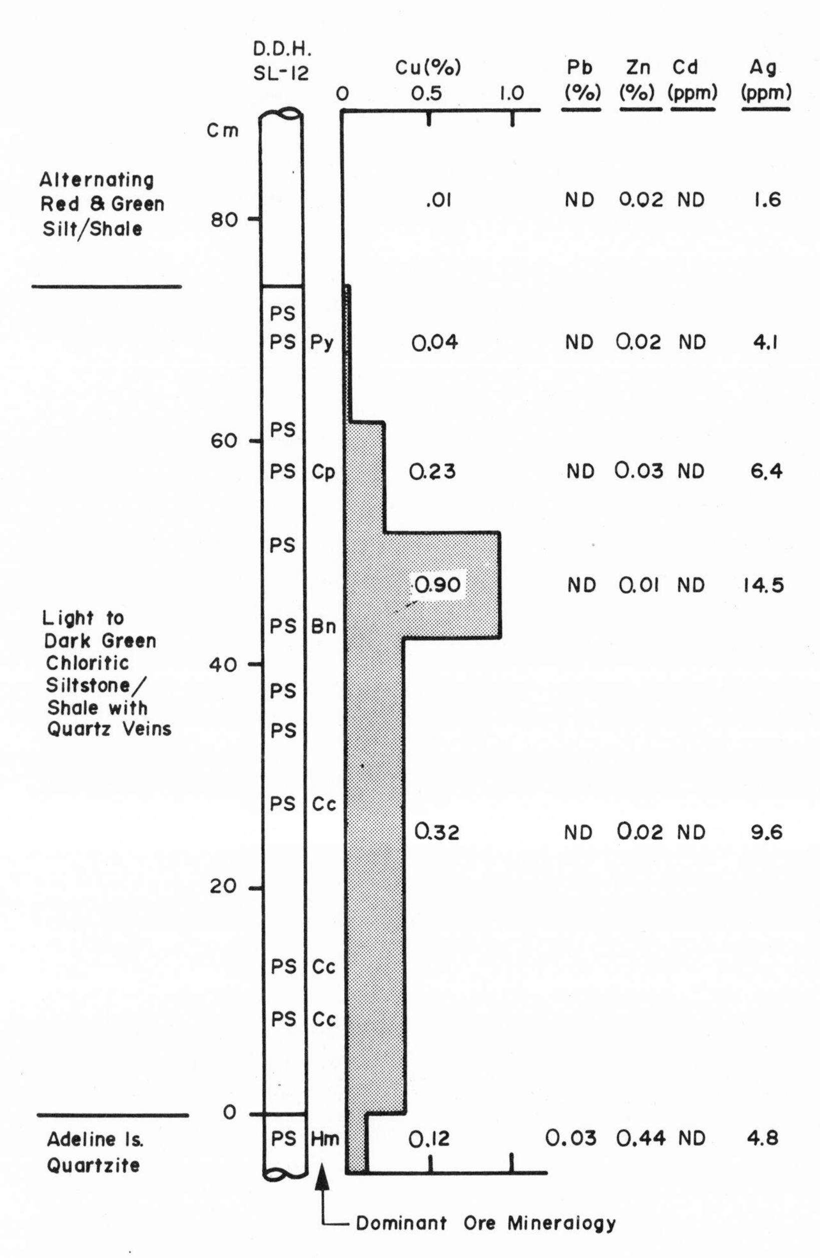

Fig. 5. Stratigraphic column with indicated ore mineralogy and metal distribution at Seal Lake, Labrador. (After Gandhi and Brown, 1975, *Econ. Geol.*, 70: 145–163, with permission.)

postpones the mineralization event until some time following those very early pyrite-forming stages, and hence requires a post-sedimentary (see Appendix) origin for the Creta copper deposits.

THE KUPFERSCHIEFER, NORTH-CENTRAL EUROPE

The Kupferschiefer has long been cited as a classic sedimentary copper deposit extending from England to Poland (see Chapter 7, Vol. 6, this Handbook). However, the occurrence of copper is restricted to only about 0.2% of the Permian Zechstein basin, and especially restricted to areas of marine sediments overlying margins of molasse basins within the Variscan fold belt (Rentzsch, 1974; Rentzsch et al., 1976). The very extensive reserves of copper discovered in Poland since the last world war (Krason, 1967; Kirkham, 1975; Mining Magazine, 1977), together with the famous ores of Germany, comprise a major copperbelt in north-central Europe.

Many authors (e.g., Schouten, 1937; Rentzsch and Knitzschke, 1968) have reported that pyrite of presumed syngenetic origin has been replaced by copper sulfides in the Kupferschiefer. Also, the common copper–iron sulfides as well as lead and zinc sulfides are typically zoned laterally and vertically, especially relative to underlying hematitic beds and overlying distant pyritic beds. These pertinent relationships are illustrated in Fig. 6 which is a more informative version from Rentzsch (1974) of a similar illustration by Kautzsch (1962).

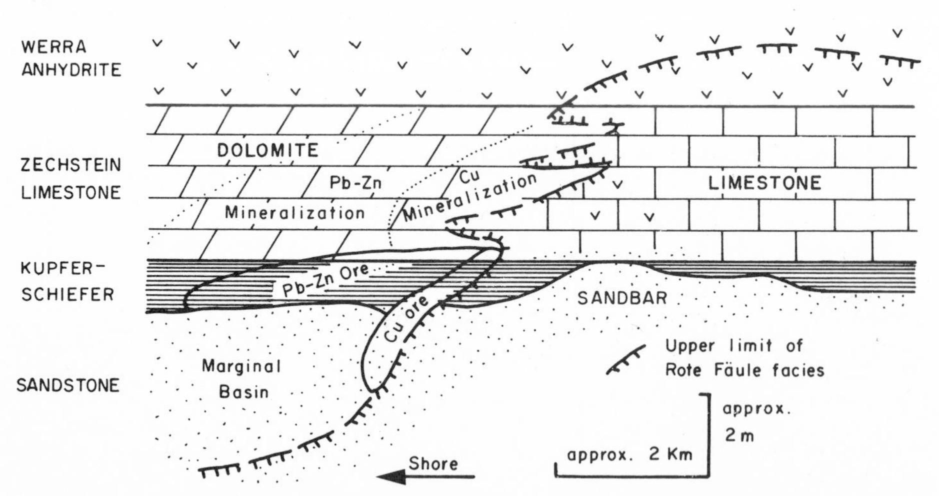

Fig. 6. Stratigraphic profile through the basal Zechstein showing the hematitic Rote Fäule alteration facies which transgresses bedding at gentle angles (note vertical exaggeration) above offshore sandbars, and showing the zoning of base metal sulfides in unaltered organic-bearing strata above and laterally adjacent to the Rote Fäule facies. (After Rentzsch, 1974.)

The Kupferschiefer horizon is underlain by continental red-beds that grade laterally from coarse- to fine-grained sediments toward the axis of the Zechstein basin. With marine transgression, some red-bed material was reworked to form shallow off-shore marine sandbars. The Kupferschiefer itself was then deposited as a 30–60 cm thick, black laminated marine shale which in general is characterized by abundant organic matter and fine-grained disseminated pyrite. The black shale grades rapidly upward into the Zechstein Limestone and then into the Werra Anhydrite unit, which signals the evolution of the Zechstein embayment to a restricted marine basin environment. The Zechstein carbonate unit is normally dolomitic on the shoreward side of the sandbars.

A zone of diagenetic reddening, known as the Rote Fäule facies, transgresses this stratigraphic section at low angles (note the vertical exaggeration of Fig. 6). Copper mineralization (predominantly chalcocite) lies directly above the Rote Fäule, and grades upward and laterally (leftward in Fig. 6) through intermediate copper–iron sulfides into the pyritic zone. Lead–zinc mineralization overlies the cupriferous zone. This zoning is strikingly similar to that observed at White Pine and is of course a simple reflection of the relative solubilities of these metals in a sulfide-bearing environment.

However, Garlick (1961, 1976; Chapter 6, Vol. 6, this Handbook) attributes such transgressive zoning of metals in the Zambian copperbelt to lateral zoning of metal deposition during onlap–offlap sedimentation: copper added to the marine basin by fluvial water would precipitate near-shore, while iron precipitated off-shore. Nevertheless, this zoning is also explicable by precipitation from epigenetically introduced ore solutions, with metals deposited according to the relative solubilities of the metal sulfides. In the case of the Kupferschiefer mineralization, the Garlick model is in fact refuted by the zonation in closed rings around areas of Rote Fäule facies which have been identified as off-shore features (Rentzsch, 1974). Furthermore, the spatial control of mineralization by the Rote Fäule facies, which is in turn a superimposed diagenetic feature, clearly places the timing of mineralization of the Kupferschiefer within the diagenetic realm.

CENTRAL AFRICAN COPPERBELTS

The extensive copper province of Central Africa has commonly been discussed under two headings: the Zambian (formerly Northern Rhodesian) copperbelt, and the Shaban (formerly Katangan) copperbelt, but it is generally recognized now as a single district composed of a predominantly detrital-facies environment in Zambia and a carbonate-facies environment in Shaba. The observations selected for illustration in this chapter are limited to one example of recent studies carried out in each sub-province.

The Zambian copperbelt

Annels (1974) reported that anhydrite nodules are conspicuously absent from high-grade ore zones at the Nkana North Limb deposit, and that ore is characterized instead by nodules composed of carbonate and quartz. This inverse correlation between anhydrite and mineralization suggests to Annels that anhydrite may have contributed sulfur for diagenetic reduction by bacterial or inorganic means. With the destruction of anhydrite, the nodular structure is conserved by an aggregate of carbonate and quartz, while the sulfur is reduced and redeposited in the form of sulfides. Under bacterial reduction, the metabolic release of CO_2 is incorporated into the carbonate; on the other hand, an inorganic transformation according to a reaction of the type:

$$CaSO_4 + (C + 4\,H) \rightarrow H_2S + CaCO_3 + H_2O$$

would have more appeal as the sediment temperature rose under burial to conditions unsuitable for bacterial activity.

The above concepts are supported by further textural observations. Annels noted the occurrence of sulfide encrustations on carbonate–quartz nodules, and in addition he observed that those sulfides consisted initially of pyrite which in mineralized zones was replaced in part or whole by copper-bearing sulfides. Thus the introduction of copper did not take place until after the diagenetic transformation of anhydrite, which itself would normally be considered to have formed as nodules during some still earlier diagenetic time.

Of course, the above features concern only a minor portion of the sulfide content of a deposit such as Nkana North Limb, and moreover, it could be dangerous to project these findings to the entire copperbelt. But the relationships described illustrate the capabilities of persistent detailed studies to establish reasonable successions of sedimentary and diagenetic events involved in the ultimate mineralization of sediments, and in this case to indicate clearly that copper entered the host rock after several early diagenetic transformations. This approach is certainly reinforced by the success of similar textural studies which have aided in determining the timing of mineralization in the Shaban copperbelt as described below.

The Shaban copperbelt

The Kamoto ores of Shaba in Zaire also bear evidence of mineralization following one or more prior diagenetic events. Bartholomé et al. (1971, 1972) have shown, for example, that: (1) the only sulfide enclosed by authigenic quartz is pyrite (Fig. 7A); (2) the only sulfide included in diagenetic magnesite nodules is pyrite (Fig. 7B); (3) copper sulfides replaced early syn-diagenetic pyrite throughout the most of the ore zones, but small remnants of pyrite are preserved within porphyroblasts of carrollite ($CuCO_2S_4$) in banded continuity with banded disseminations of copper-bearing sulfides in the adjacent host rock (Fig. 7C); and (4) the influx of cobalt followed an initial stage of non-cobaltiferous pyrite deposition as demonstrated by cobalt-free cores within grains (Fig. 7D). Each of these independent features points to the same conclusion: a diagenetic

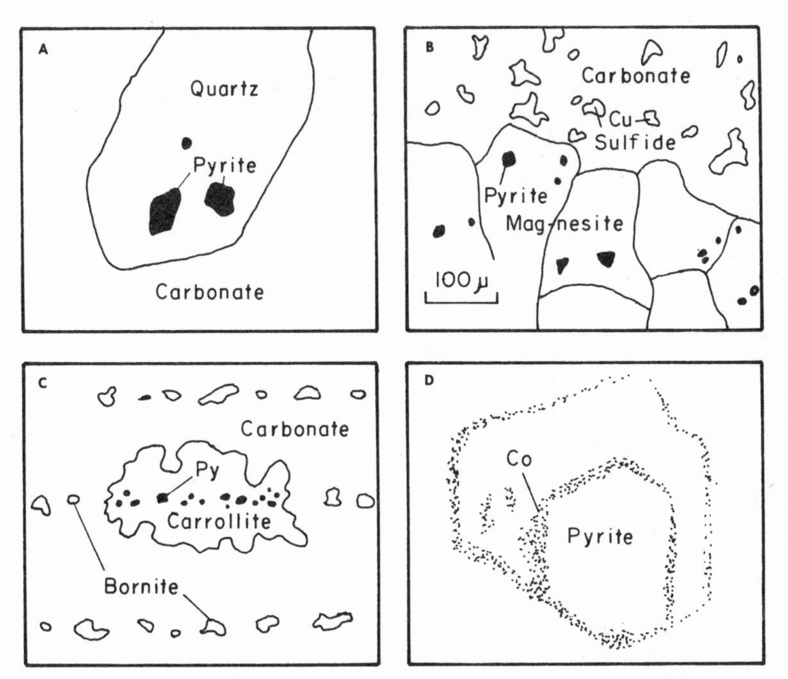

Fig. 7. Textures illustrating post-syndiagenetic pyrite mineralization in the Kamoto ores, Shaba. (After Bartholomé et al., 1971, 1972.) A. Pyrite without copper-bearing sulfides included in authigenic quartz. B. Pyrite without copper-bearing sulfides included in diagenetic magnesite nodules. C. Pyrite preserved from later replacement by copper-bearing sulfides by inclusion in carrollite porphyroblasts. D. Cobalt-free cores of early pyrite overgrown by later cobalt-rich pyrite. Cobalt distributions determined by scanning electron microprobe.

event involving pyrite alone preceded the deposition of copper. To be more precise, copper could not have been present even in the interstitial fluid during these diagenetic processes, since it would have precipitated preferentially before any iron sulfide. Moreover, the Kamoto ores exhibit once more that fundamental textural evidence of copper mineralization after initial iron sulfide enrichment: the replacement of pyrite by copper-rich sulfides (Bartholomé, 1962, 1963).

In 1974 and 1976, Bartholomé and colleagues (Bartholomé, 1974; Dimanche, 1974; Dimanche and Bartholomé, 1976) introduced still another approach to the analysis of timing of stratiform copper mineralization. It had been noticed earlier (Bartholomé, 1962) that rutile and other titania-bearing minerals occur in common abundance adjacent to sulfide grains. If one considers the diagenetic transformation of initial detrital ilmenite grains, these associations assume important genetic significance. The unstable ilmenite grain first liberates iron and leaves a residue of titanium oxide. The iron may then follow one of three possible routes as defined by the Eh–pH diagram of Fig. 8: (1) migrate elsewhere if the interstitial fluid is sufficiently acid; (2) precipitate in situ as pyrite if conditions are reducing, not too acid, and a source of sulfur is available to deposit this very

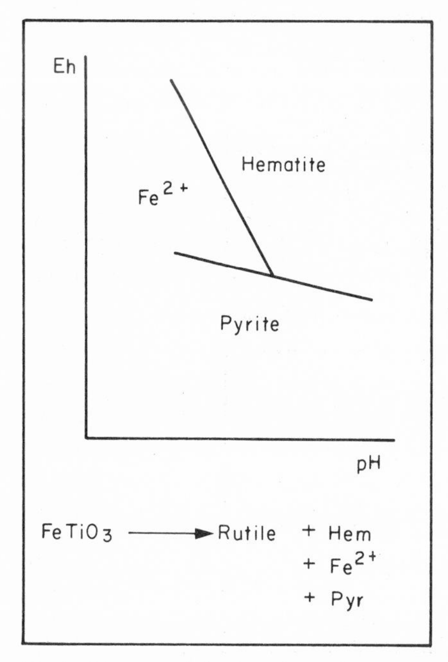

Fig. 8. Three possible alterations of detrital ilmenite depending on the Eh–pH conditions of the interstitial solution. (After Brown, 1978.)

minor quantity of diagenetic-stage pyrite; or (3) precipitate in situ as hematite if conditions are oxidizing, and neutral to alkaline. In the case of stratiform copper mineralization, there is normally an abundance of organic matter to assure reducing conditions and perhaps also supply the necessary organic or biogenic sulfide to precipitate pyrite. At the same time, the titania content of the ilmenite is redeposited as stable TiO_2 in the form of the ilmenite grains and permits identification of the diagenetic transformations described above. Bartholomé was able to recognize these textural features in the Kamoto ores and furthermore observed that the pyrite so formed was *subsequently* replaced in part or whole by copper sulfides. Once again we see that a minor, but nevertheless genetically significant diagenetic process, occurred in the host rock before the introduction of copper. Bartholomé has also observed the same phenomenon in other ores such as White Pine, and obviously the technique described should be applied more extensively elsewhere [1].

[1] See Adams et al. (1974), Morton (1974) and Reynolds et al. (1977) for examples of ilmenite alteration applied to the genesis of uranium ores in clastic sediments.

OTHER DEPOSITS

An exhaustive survey of stratiform copper deposits showing evidence of post-sedimentary mineralization would include many more well-known occurrences. This compilation terminates with only a brief mention of three additional deposits of this type.

Morton et al. (1973), Harrison (1974), Kirkham (1974) and others have recently described the many occurrences of stratiform copper mineralization in the Belt Supergroup rocks of Late Precambrian age in the northwestern U.S.A. and adjacent Canada. In a detailed textural analysis of these showings, Trammell (1975) observed that copper-bearing sulfides (chalcocite, digenite, bornite and chalcopyrite) replaced initial disseminated pyrite, and concluded, with other pertinent geologic data, that mineralization probably took place during compaction and diagenesis of the Belt sediments.

Anhaeusser and Button (1972) report that the Witvlei stratiform copper–silver deposit of South West Africa is characterized by replacement of pyrite and progressive enrichment in copper by chalcopyrite, bornite, digenite, chalcocite and native copper. They observe many similarities of the Witvlei mineralization to deposits described above, such as the Kupferschiefer and White Pine deposits, and suggest that copper was remobilized from underlying volcanic strata which are known to contain small copper showings.

Finally, the initial descriptions of the Upper Proterozoic Redstone copperbelt mineralization of the Northwest Territories of Canada (Coates, 1964; Watson and Mustard, 1973; Kirkham, 1974; Ruelle, 1978; and others) indicate that this extensive occurrence of stratiform copper originated by replacement of biogenic iron sulfides. The host rock is interpreted by Ruelle as a supratidal flat (sabkha) horizon which trapped copper liberated from underlying clastic sediments during the process of red-bed formation. The resultant deposits closely resemble those of Zaire, with carbonates and red-beds (including sabkha facies) hosting mineralization along a strike-length of some 250 km.

COMPLEMENTARY SOURCES AND TRANSPORT MECHANISMS

In the above review of evidences for post-sedimentary mineralization, only the process of deposition has been discussed in any detail. Presuming those arguments worthy of further examination, it is reasonable to ask where the ore metals might have come from, and how they were transported to the postulated sites of deposition in post-sedimentary time.

It is not the intent of the following to provide complete details of plausible sources and transport mechanisms for the deposits described above – such information is quite unique to each occurrence and is generally discussed in a thorough manner by one or more authors already mentioned. It is possible, however, to summarize those concepts briefly below.

Mineralization models compatible with post-sedimentary emplacement of ore metals are generally conceived after study of the enclosing lithologies (Wolf, 1976) and possible large-scale structural or tectonic features in the sedimentary basin. Visualization of basin characteristics during and shortly after deposition of the ore-bearing sediment is particularly pertinent.

Without rejecting the possibility entirely, classic magmatic (telethermal) concepts with ore solutions guided by structural channels, etc., can usually be set aside as incompatible with ore deposition principally during early diagenetic time when the fine-grained host sediment was still reasonably soft and permeable, and not yet sufficiently consolidated and lithified to support fractures [1].

However, magmatic activity, especially of volcanic affinity, should not be neglected, for in many cases there may be evidence of volcanic or high-level intrusions penecontemporaneous with the host strata, and ore fluids may have been generated either directly from unseen magmas at depth or indirectly by remobilization under the influx of magmatic heat. At White Pine and Seal Lake, for example, mineralization occurs in sediments closely overlying immense thicknesses of rift-valley tholeiitic basalts. In the case of White Pine, this volcanic pile is known to include a large felsic dome stratigraphically and spatially beneath the mine area, and it is tempting to suggest that ore fluids escaped locally into the Copper Harbor Conglomerate from a buried volcanic vent beneath the Nonesuch Shale.

On the other hand, there are numerous other plausible sources of metal aside from latent magmatic activity. In general, the alternatives rely on some mechanism of remobilization of metals already distributed in one or more units of the enclosing strata or basement:

(1) From the ore horizon itself: remobilization of trace amounts of metals normally found in euxinic sediments.

(2) From adjacent red-beds: metals may have been absorbed on matrix clays, or on amorphous iron hydroxides, which subsequently released the metals during crystallization to the usual hematitic pigment of red-beds. Some red-beds (e.g., White Pine) contain high-grade copper zones which, while possibly representing deep native-copper facies of the overlying stratiform sulfide mineralization, may also have contributed copper to the stratiform ore zone by remobilization.

(3) From deep-seated strata and basement rocks: normal crustal rocks contain suf-

---

[1] At White Pine, there is evidence (Brown, 1971) of minor fracture-controlled mineralization which locally disturbs the usually smooth fringe surface and which may represent the last gasp of the main ore-stage mineralization front which advanced upward into the Nonesuch Shale; and there is evidence (Carpenter, 1963) of a later minor and paragenetically distinct fracture-controlled mineralization. The principal ore-stage stratiform mineralization disseminated within bedding is considered to have preceded those mineralizations and to have been emplaced in unfractured soft sediments.

ficient traces of metal to account for overlying stratiform concentrations, and regoliths at the basement contact may carry additional enrichments. Ore-grade concentrations in the basement would add further potential to a basement source, but unless observed in close spatial association with nearby stratiform mineralization, this proposal would be quite conjectural. However, at White Pine, the underlying Portage Lake volcanic series is well-known some 100 km to the northeast for its native copper ores which undoubtedly are connected genetically with White Pine to form a common copper–silver province. If, as suggested by Jolly (1974), the ores of the native copper district owe their existence to remobilization from basalts during low-grade metamorphism, the White Pine ore fluids may also have been generated in a similar manner.

Recent geologic literature contains descriptions of numerous specific models in which stratiform copper mineralization is assumed to be post-sedimentary in origin, and compatible means of metal transport and sources of metal have been proposed. Brown (1968, 1971) does not suggest any preference for sources of the White Pine copper, but postulates the existence of a chloride-rich brine with some 10^{-3} m/l dissolved copper within the vast Copper Harbor Formation beneath the Nonesuch Shale. Based on that reservoir of ore solutions, calculations show that mineralization could take place within approximately $3 \cdot 10^5$ to $3 \cdot 10^6$ years using infiltration and/or diffusion modes of transport within the basal Nonesuch. Conceivably, an influx of ore solution into the Nonesuch could have been the predominant process during early diagenesis when shale permeabilities were high, and diffusion could have played a more prominent role as permeabilities decreased. The possibility of salt-sieving effects (Hanshaw, 1962) during mineralization has not been seriously evaluated for the White Pine situation, and could well have been important.

D.E. White (1968) accepted a post-sedimentary origin for the White Pine mineralization and proposed that the upper limit of the cupriferous zones marks the position of an essentially horizontal density-controlled interface between a heavy saline ore solution below and normal low-density meteoric water above. Undulations in the fringe surface relative to stratigraphy would then be explained by undulations in the attitude of the Nonesuch beds. This gravity-stratification model is attractive in that it is consistent with the observed layering of brines in present-day oil field basins, and it also provides a mechanism for recycling of spent brines to depths where they might once again pick up metals for further mineralization. Circulation of brines may be triggered by abnormal heat sources such as latent volcanic heat beneath the Nonesuch.

Brown (1970) appreciates the advantages of White's model, but disputes the correlation of the fringe surface with the upper limit of a circulating brine cell. For example, the correlation of the fringe position with the amount of copper in the basal Nonesuch bed is not explained by White's model. A density-controlled interface at some unknown distance above the fringe surface would, however, seem very logical.

Brown (1974) discovered as well that CdS concentrations could be elevated beyond the apparent cupriferous front, if ore solutions were pulsed momentarily upward. This phenomenon could explain the observation of consistently stratiform disseminations of greenockite in pyritic portions of the No. 61 bed some 20 m above the base of the Nonesuch (see Fig. 4). Apparently the abrupt pulse of copper-rich solution can effectively mobilize any Cd, Pb or Zn accumulated ahead of the usual mineralization front and carry such metals upward a distance equivalent to the pulse of ore solution. In the case of the Nonesuch stratigraphy, the No. 61 bed would probably intervene in this exceptional transport process: it is the first major sulfide- and carbonate-rich bed to be encountered above the usual ore zones and would probably screen out these metals, perhaps through neutralization of the ore fluid. At the same time, the amount of copper in the temporary pulse of ore solution would be too little to leave any significant copper mineralization above the normal background values of the pyritic zone. The cause of this pulsation of the ore fluid is unknown, but may have been a fluid pressure differential arising out of minor tectonic adjustments in the sedimentary basin.

Other deposits have been interpreted in terms of a sabkha mineralization model described by Renfro (1974). In these cases, evaporitic sediments of supratidal flat origin have been recognized overlying terrestrial red-beds, and it is proposed that trace amounts of ore metals carried in the slightly acid pore fluids of the terrestrial sediments are brought toward the surface beneath active tidal flats during high rates of evaporation in a semi-arid climate. An H_2S-charged algal mat horizon directly underlying the supratidal sediment would precipitate copper and other associated metals which migrate seaward and upward across the red-bed–marine sediment interface.

This sabkha model explains the post-sedimentary, early diagenetic origin of stratiform copper mineralization to a very encouraging degree at Creta (see Smith, 1976; Vol. 6 of this Handbook), and has attractive potentials for several other deposits as suggested by Renfro, Ruelle and others. The Shaban copperbelt, for example, is characterized by a transgression from red continental conglomerates to greyish marine tidal flat sediments (Bartholomé et al., 1972); Renfro (1974) would interpret the latter as coastal sabkha sediments. The lack of anhydrite, but high degree of diagenetic magnesium and silica metasomatism at Kamoto (Bartholomé, 1974), suggests a modified sabkha mode might be applicable, but thus far no definitive studies of this possibility have been made.

In many other deposits, an association of stratiform copper mineralization with transgressions from continental red-beds to marine shales and evaporites has been noted (Davidson, 1966), and with further study, additional sabkha affiliations may be disclosed. For the purposes of this paper, the sabkha model represents only one more possible mode of mineralization consistent with evidence of post-sedimentary copper deposition, and only serves to increase the possibility of a reasonable explanation for the origin of stratiform deposits found by independent means to have been deposited after sedimentation.

CONCLUDING REMARKS

The illustrations and discussions above have been presented with one immediate aim – to demonstrate evidence for and the feasibility of stratiform copper mineralization in post-sedimentary time. The fact that these evidences are repetitious and are drawn from many well-known and carefully studied deposits reinforces the arguments presented. The fact that many of these evidences have only come to light in recent years despite many prior studies emphasizes our need for continued detailed investigation.

As long as future research continues to support the conclusions of this review paper, a second and undoubtedly more important aim will be served – exploration for stratiform copper ores can benefit from application of such post-sedimentary models. For example, shorelines may continue to act as regional guides within a generally favorable sedimentary basin, but local offshore features, sulfide zoning with replacement textures, aquifer pinch-outs, algal mat distributions, etc., may guide the exploration geologist more precisely to the location of ore-grade mineralization, and to different locations within a generally favorable ore horizon than would be conceived following syn-sedimentary concepts. Consequently, this academic exercise aimed at diagnosing the genesis of stratiform copper ores has very significant economic consequences, and it is hoped that it will prove its usefulness in future mineral exploration programs for this important class of ore deposit.

ACKNOWLEDGEMENTS

Without the careful observations and interpretations published by the numerous authors mentioned, this synthesis would not have been possible. In addition, the writer would like to acknowledge the enormous help received through the numerous comments and constructive criticism of many colleagues. Financial support for these continuing studies into the genesis of stratiform copper ores has been received largely from a National Research Council of Canada operating grant.

APPENDIX

Usage of some descriptive and genetic terms in this chapter

Stratiform copper deposits of the type discussed in this chapter are now generally considered to have formed either during sedimentation or during subsequent diagenesis of the host sediments. Unfortunately there is not always consistent usage among geologists of the terms intended to identify specific periods within the sedimentary and diagenetic history of such sediments. Without resorting to authoritative definitions, etc. (the reader might well consult appropriate references), the author points out various common usages of terms in the brief list below, and indicates his preferences. The relative timing implied by some of the terms in this paper is illustrated in Fig. 1. The reader is also referred to Bartholomé (1974), Wolf (1976) and Samama (1976) for further comments on concepts pertinent to the timing of stratiform copper mineralization.

Sedimentary: A very common and attractive descriptive term for prominently well-bedded and laterally extensive stratiform mineralization; applied genetically, it signals an origin contemporaneous with sedimentation of the host strata, an interpretation which may not be justified. It is sometimes used in the descriptive sense (with or without that specification) for mineralization determined to be post-sedimentary!

Post-sedimentary: A term which clearly excludes a sedimentary origin, but lacks further temporal significance (e.g., Brown, 1978). It is essentially synonymous with "epigenetic", but stresses an opposition to a possible "sedimentary" timing of mineralization. It may be preferable to "diagenetic" in that the latter could have unwanted genetic significance as well as desired temporal connotations. However, the term admittedly suffers from lack of precision within the "epigenetic" time interval and might benefit, albeit awkwardly, from combination with additional expressions such as "pre-deformation", "pre-lithification", etc. where justified.

Epigenetic: Appropriate for mineralization later in age than the enclosing host rock, hence applicable to mineralization following sedimentation, and hence synonymous with "post-sedimentary". Historically it bears an unfortunate connotation of magmatic hydrothermal mineralization with structural controls – this restriction should be persistently and carefully avoided.

Syngenetic and syn-sedimentary: Strictly speaking, these terms are appropriate for mineralization during sedimentation (a generally accepted concept for such ores as volcanogenic massive sulfide deposits and sedimentary iron ores), but unfortunately they are commonly applied as well to early diagenetic mineralization.

Syn-diagenetic: An appropriate term for "syngenetic" pyrite in euxenic sediments since it acknowledges the production of biogenic sulfide in the diagenetic zone beneath the water–sediment interface. Some authors (e.g., Heyl, 1968) use the lengthier expression "syngenetic and early diagenetic".

Diagenetic: Defines mineralization within an interval immediately following sedimentation and, structural effects excluded, extending through to possible metamorphism. Also employed to imply mineralization not only during but *due to* some diagenetic process.

Diplogenetic: A very neglected but useful expression which designates an origin due to two genetic processes. For example, the sulfur of chalcocite may have been derived by replacement of syn-diagenetic biogenic pyrite, and consequently pre-dates the deposition of copper during somewhat later diagenesis.

Stratiform: A term with specific literal meaning and wide usage. Regrettably, for the deposit-type discussed in this paper, it also permits confusion with other types of stratiform copper mineralization (e.g., volcanogenic massive sulfide deposits). Also, many of our so-called stratiform deposits exhibit peneconcordant behavior (e.g., White Pine, Kupferschiefer).

To date the writer has not encountered a universally satisfying term for the group of deposits in question here. Evidences point to a diplogenetic origin involving syn-diagenetic, biogenic sulfide precipitated normally as iron sulfide; and a post-sedimentary, most probably early diagenetic, introduction of base metals deposited by reaction with the earlier iron sulfide. They are stratiform for the most part, but some deposits transect bedding in a peneconcordant fashion. They are sediment-hosted, but not necessarily non-volcanogenic. Would not communications be simpler if only they could be shown to be sedimentary in origin!

REFERENCES

Adams, S.S., Curtis, H.S. and Hafen, P.L., 1974. Alteration of detrital magnetite-ilmenite in continental sandstones of the Morrisson Formation, New Mexico. In: *Formation of Uranium Ore Deposits.* Int. At. Energy Agency, Vienna, pp. 219–253.

Anhaeusser, C.R. and Button, A., 1972. A petrographic and mineragraphic study of the copper-bearing formation in the Witvlei area, South West Africa. *Univ. Witwatersrand, Econ. Geol. Res. Unit, Inf. Circ.,* 66: 39 pp.

Annels, A.E., 1974. Some aspects of the stratiform ore deposits of the Zambian copperbelt and their genetic significance. In: P. Bartholomé (Editor), *Gisements Stratiformes et Provinces Cuprifères.* Soc. Géol. Belg., Liège, pp. 235–254.

Bartholomé, P., 1962. *Les minerais cupro-cobaltifères de Kamoto (Katanga-Ouest). I. Pétrographie, II. Paragenèse.* Stud. Univ. "Louvanium", Kinshasa, 40 pp. and 24 pp.

Bartholomé, P., 1963. Sur la zonalité dans les gisements du copperbelt de l'Afrique centrale. *IAGOD symposium – Problems of Postmagmatic Ore Deposition,* 1: 317–321.

Bartholomé, P., 1974. On the diagenetic formation of ores in sedimentary beds, with special reference to the Kamoto ore deposits. In: P. Bartholomé (Editor), *Gisements Stratiformes et Provinces Cuprifères.* Soc. Géol. Belg., Liège, pp. 203–214.

Bartholomé, P., Katekesha, F. and Lopez-Ruiz, J., 1971. Cobalt zoning in microscopic pyrite from Kamoto, Republic of the Congo (Kinshasa). *Miner. Deposita,* 6: 167–176.

Bartholomé, P., Eviand, P., Katekesha, F., Lopez-Ruiz, J. and Ngongo, M., 1972. Diagenetic ore-forming processes at Kamoto, Katanga, Republic of the Congo. In: G.C. Amstutz and A.J. Bernard (Editors), *Ores in Sediments.* Springer, Berlin, pp. 21–41.

Berner, R.A., 1971. *Principles of Chemical Sedimentology.* McGraw-Hill, New York, N.Y., 240 pp.

Brown, A.C., 1968. *Zoning in the White Pine Copper Deposit, Ontonagon County, Michigan.* Thesis, Univ. Mich., Ann Arbor, 199 pp.

Brown, A.C., 1970. Environments of generation of some base-metal ore deposits (Discussion). *Econ. Geol.,* 65: 60–61.

Brown, A.C., 1971. Zoning in the White Pine copper deposit, Ontonagon County, Michigan. *Econ. Geol.,* 66: 543–573.

Brown, A.C., 1974. An epigenetic origin for stratiform Cd–Pb–Zn sulfides in the lower Nonesuch Shale, White Pine, Michigan. *Econ. Geol.,* 69: 271–274.

Brown, A.C., 1978. Stratiform copper deposits – Evidence for their post-sedimentary origin. *Minerals Sci. Eng.,* 10(3): 172–181.

Burnie, S.W., Schwarcz, H.P. and Crocket, J.H., 1972. A sulfur isotopic study of the White Pine mine, Michigan. *Econ. Geol.,* 67: 895–914.

Carpenter, R.H., 1963. Some vein-wall rock relationships in the White Pine mine, Ontonagon County, Michigan. *Econ. Geol.,* 58: 643–666.

Coates, J.A., 1964. *The Redstone Bedded Copper Deposit and Discussion of the Origin of Red Bed Copper Deposits.* Thesis, Univ. B.C., 75 pp. (unpubl).

Davidson, C.F., 1966. Some genetic relationships between ore deposits and evaporites. *Inst. Min. Metall., Trans.,* 75: B216–B225.

Dimanche, F., 1974. Paragenèses des sulfures de cuivre dans les gisements de Shaba (Zaire). I. Kipushi, II. Kamoto. In: P. Bartholomé (Editor), *Gisements Stratiformes et Provinces Cuprifères.* Soc. Géol. Belg., Liège, pp. 185–203.

Dimanche, F. and Bartholomé, P., 1976. The alteration of ilmenite in sediments. *Miner. Sci. Eng.,* 8: 187–201.

Ensign, C.O., Jr., White, W.S., Wright, J.C., Patrick, J.L., Leone, R.J., Hathaway, D.J., Trammell, J.W., Fritts, J.J. and Wright, T.L., 1968. Copper deposits in the Nonesuch Shale, White Pine, Michigan. In: J.D. Ridge (Editor), *Ore Deposits in the United States 1933/1967.* Am. Inst. Min. Metall., Graton-Sales Vol., 1: 460–488.

Fleischer, V.D., Garlick, W.G. and Haldane, R., 1976. Geology of the Zambian copperbelt. In: K.H. Wolf (Editor), *Handbook of Strata-bound and Stratiform Ore Deposits,* 6. Elsevier, Amsterdam, pp. 223–352.

Gandhi, S.S. and Brown, A.C., 1975. Cupriferous shales of the Adeline Island Formation, Seal Lake Group, Labrador. *Econ. Geol.,* 70: 145–163.

Gann, D.E. and Hagni, R.D., 1974. Ore microscopy of copper ore at the Creta mine, southern Oklahoma. *Geol. Soc. Am., South-Central Meet., Abstr.,* p. 104.

Garlick, W.G., 1961. The syngenetic theory. In: F. Mendelson (Editor), *The Geology of the Northern Rhodesian Copperbelt.* MacDonald and Co., London, 523 pp.

Garlick, W.G., 1976. Genesis of the ore shale deposits; Genesis of the arenaceous ore deposits; The syngenetic explanation; and Syngenesis versus epigenesis. In: K.H. Wolf (Editor), *Handbook of Strata-bound and Stratiform Ore Deposits, 6.* Elsevier, Amsterdam, pp. 323–352.

Hagni, R.D. and Gann, D.E., 1976. Character of copper ore in the Prewitt Shale (Permian) at Creta mine, southwestern Oklahoma. *Geol. Soc. Am., Annu. Meet., Denver, Abstr.*, p. 899.

Hamilton, S.K., 1967. Copper mineralization in the upper part of the Copper Harbor Conglomerate at White Pine, Michigan. *Econ. Geol.*, 62: 885–904.

Hanshaw, B.B., 1962. *Membrane Properties of Compacted Clays.* Thesis, Harvard Univ., 113 pp., unpubl.

Harrison, J.E., 1974. Copper mineralization in miogeosynclinal clastics of the Belt Supergroup, Northwestern United States. In: P. Bartholomé (Editor), *Gisements Stratiformes et Provinces Cuprifères.* Soc. Géol. Belg., Liège, pp. 353–366.

Heyl, A.V., 1968. Minor epigenetic, diagenetic and syngenetic sulfide, fluorite, and barite occurrences in the Central United States. *Econ. Geol.*, 63: 585–594.

Johnson, K.S., 1974. Permian copper shales of southwestern United States. In: P. Bartholomé (Editor), *Gisements Stratiformes et Provinces Cuprifères.* Soc. Géol. Belg., Liège, pp. 383–393.

Jolly, W.T., 1974. Behavior of Cu, Zn, and Ni during prehnite-pumpellyite rank metamorphism of the Keweenawan basalts, Northern Michigan. *Econ. Geol.*, 69: 1118–1125.

Jung, W. and Knitzschke, G., 1976. Kupferschiefer in the German Democratic Republic (GDR) with special reference to the Kupferschiefer deposit in the southeastern Harz foreland. In: K.H. Wolf (Editor), *Handbook of Strata-bound and Stratiform Ore Deposits, 6.* Elsevier, Amsterdam, pp. 253–406.

Kautzsch, E., 1962. General discussions. In: J. Lombard and P. Nicolini (Editors), *Symposium on Stratiform Copper Deposits in Africa.* Assoc. Afr. Geol. Surv., Paris, pp. 200–202.

Kirkham, R.V., 1974. A synopsis of Canadian stratiform copper deposits in sedimentary sequences. In: P. Bartholomé (Editor), *Gisements Stratiformes et Provinces Cuprifères.* Soc. Géol. Belg., Liège, pp. 367–382.

Kirkham, R.V., 1975. Visit to Kupferschiefer copper deposits in Poland provides data for comparing Canadian occurrences. *North. Miner,* Nov. 27, pp. A9–A11.

Krason, J., 1967. Quelques résultats des recherches sur le Permien Polonais. *Bull. Soc. Géol. Fr.*, 9: 701–713.

Mining Magazine, 1977. Polish copper. *Mining Journal Ltd.*, November 1977, pp. 500–506.

Morton, R.D., 1974. Sandstone-type uranium deposits in the Proterozoic strata of northwestern Canada. In: *Formation of Uranium Ore Deposits.* Int. At. Energy Agency, Vienna, pp. 255–273.

Morton, R., Goble, E. and Goble, R.J., 1973. Sulfide deposits associated with Precambrian Belt-Purcell strata in Alberta and British Columbia, Canada. *Belt Symposium*, 1: 159–179.

Renfro, A.R., 1974. Genesis of evaporite-associated stratiform metalliferous deposits – a sabkha process. *Econ. Geol.*, 69: 33–45.

Rentzsch, J., 1974. The "Kupferschiefer" in comparison with the deposits of the Zambian copperbelt. In: P. Bartholomé (Editor), *Gisements Stratiformes et Provinces Cuprifères.* Soc. Géol. Belg., Liège, pp. 235–254.

Rentzsch, J. and Knitzschke, G., 1968. Die Erzmineralparagenesen des Kupferschiefers und ihre regionale Verbreitung. *Freiberg. Forschungsh. C,* 231: 189–211.

Rentzsch, J., Schirmer, B., Röllig, G. and Tischendorf, G., 1976. On the metal source of non-ferrous mineralizations in the Zechstein basement (Kupferschiefer type). In: *The Current Metallogenic Problems of Central Europe.* Geological Institute, Warsaw, pp. 171–188.

Reynolds, R.L., Goldhaber, M.B. and Grauch, R.I., 1977. Uranium Associated with iron-titanium oxide minerals and their alteration products in a South Texas roll-type deposit. In: *Short Papers of the U.S. Geological Survey Uranium–Thorium Symposium.* U.S. Geol. Surv. Circ., 753: 37–39.

Ritchie, A.S., 1964. *Chromatography in Geology.* Elsevier, New York, N.Y., 185 pp.

Ruelle, J.C., 1978. Depositional environment and genesis of stratiform copper deposits of the Redstone copper belt, MacKenzie Mountains, N.W.T. *Geol. Soc. Am., Abstr. with Progr.*, 10(7): 482–483.

Samama, J.C., 1976. Comparative review of the genesis of the copper-lead sandstone-type deposits. In: K.H. Wolf (Editor), *Handbook of Strata-bound and Stratiform Ore Deposits, 6*. Elsevier, Amsterdam, pp. 1–20.

Schouten, C., 1937. *Metasomatische Probleme.* Amsterdam.

Smith, G.E., 1976. Sabkha and tidal-flat facies control of stratiform copper deposits in north Texas. In: K.H. Wolf (Editor), *Handbook of Strata-bound and Stratiform Ore Deposits, 6.* Elsevier, Amsterdam, pp. 407–446.

Trammell, J.W., 1975. *Strata-bound Copper Mineralization in the Empire Formation and Ravalli Group, Belt Supergroup, Northwest Montana.* Thesis, Univ. Washington, 97 pp.

Watson, I.M. and Mustard, D.K., 1973. The Redstone bedded copper deposit (abstr.) *Symp. Sed. Geol. Min. Dept. Can. Cordillera.* Geol. Assoc. Can., pp. 20–22.

White, D.E., 1968. Environments of generation of some base-metal ore deposits. *Econ. Geol.*, 63: 301–335.

White, W.S., 1971. A paleohydrologic model for mineralization of the White Pine copper deposit, northern Michigan. *Econ. Geol.*, 66: 1–13.

White, W.S. and Wright, J.C., 1966. Sulfide-mineral zoning in the basal Nonesuch Shale, northern Michigan. *Econ. Geol.*, 61: 1171–1190.

Wiese, R.G., Jr., 1973. Mineralogy and geochemistry of the Parting Shale, White Pine, Michigan. *Econ. Geol.*, 68: 317–331.

Wolf, K.H., 1976. Ore genesis influenced by compaction. In: G.V. Chilingarian and K.H. Wolf (Editors), *Compaction of Coarse-Grained Sediments, II.* Elsevier, Amsterdam, pp. 475–675.

Chapter 2

OOLITIC IRON FORMATIONS

MICHAEL M. KIMBERLEY

INTRODUCTION

Iron formations of all ages and textures may be classified according to the sedimentary environment in which they formed (Kimberley, 1978, 1979b). Most chert-poor oolitic iron formations appear to have resulted from sedimentation within continental seas which were locally open to an ocean, but which were largely surrounded by exposed land. Examples of this type of iron formation include the Minette and Clinton deposits. The stratigraphic record of these "oolitic-inland-sea" iron formations is reviewed herein and a Tertiary example in Colombia, South America, is described in detail. A genetic hypothesis for this and other iron formations, which are at least partly oolitic, is presented and defended petrographically and geochemically.

The term "iron formation" is used here much as it was by Leith (1903), as an abbreviation of "iron-bearing formation", a lithostratigraphic unit dominantly composed of iron-rich chemical sedimentary rock. No special chert content, texture, or sedimentary structure is implied for the chemical sedimentary rock, here called ironstone, but the minimum iron content is 15%. Edwards (1958) has adopted similar usage of terms. Stratigraphic limits of an iron formation which contains non-ironstone interbeds are the borders of its uppermost and lowermost ironstone beds. Some iron formations are parts of formally designated formations.

Many geological environments have generated iron formations (Kimberley, 1978). These include: (1) landlocked seas, e.g., Oligocene iron formations north of the Aral Sea, U.S.S.R. (Davidson, 1961; Strakhov, 1969); (2) shallow extensions of the world ocean largely surrounded by exposed land, e.g., the Pliocene iron formation under and adjacent to the Sea of Azov, U.S.S.R. (Markevich, 1960; Sokolova, 1964); (3) shallow seas on extensive, terrigenous-sediment-poor continental shelves or oceanic platforms, e.g., the Lower Proterozoic (Precambrian X) iron formation of Labrador and adjacent Quebec, Canada (Gross, 1968); (4) shallow volcanic platforms, e.g., the Devonian Lahn-Dill iron formations of Germany (Bottke, 1965; Quade's Chapter 6, Vol.7 in this Handbook series); and (5) deep oceans, e.g., the turbidite-associated Archean Santa Claus iron formation of Western Australia (Dunbar and McCall, 1971).

The first two environments listed above, jointly here termed the inland-sea environment, have typically produced chert-poor, oolitic iron formations, the conceptual "ironstone" of James (1966). These beds of oolitic and/or pisolitic ironstone are invariably intercalated with shallow-water sandstone, shale, terrigenous-sediment-rich limestone, and/or oolitic limestone. They and subordinate interbeds are therefore termed sandy, clayey, and oolitic, shallow-inland-sea iron formation, abbreviated to the acronym, SCOS-IF. At present, SCOS-IF is mostly termed "oolitic ironstone" (e.g., Stanton, 1972). However, cherty banded iron formations are commonly also oolitic (Dimroth, 1976) and, outside of North America, are commonly also called "ironstone" (e.g., Dunbar and McCall, 1971). The world-wide volume of cherty banded ferriferous oolite (Dimroth, 1976) may well exceed the volume of chert-poor ferriferous oolite and, consequently, oolitic texture is a poor criterion for differentiation. Reliance on banding for differentiation (e.g., Cloud, 1976) is also fallible because of voluminous non-banded, non-oolitic, cherty Precambrian (Goodwin, 1962) and Phanerozoic ironstone (Bottke, 1965). Paleoenvironmental classifications, e.g., SCOS-IF, are more comprehensive and informative, and are equally applicable independent of age (Kimberley, 1978).

SCOS iron formations are rarely more than 30 m thick or more than 150 km in extent. The predominant texture is oolitic, but a significant proportion of the ironstone which has formed in the inland-sea environment is structureless and fine-grained (Hallimond, 1925), i.e., femicrite (Dimroth, 1976). The predominant ferriferous mineral is either goethite, hematite, chamosite, or siderite, rarely magnetite or pyrite. Both iron and phosphorus are commonly concentrated about six times mean crustal abundance, i.e., about 34% Fe and 0.6% P (Taylor, 1964). Manganese, vanadium, and arsenic are similarly concentrated and vary proportionately with iron (Aldinger, 1957a, b; Harder, 1964a, b; Kolbe, 1970). Banding is generally indistinct or absent in deposits of all ages. Textures and sedimentary structures indicate highly agitated, very shallow-water sedimentation; e.g., intraclasts of penecontemporaneously lithified oolite are commonly surrounded by individual ooids. Normally, a significant proportion of the ooid nuclei are fragments of broken ooids, and associated clastic quartz grains are more angular than in most quartzose sandstone. Where fossil fragments are present, at least some are partially to completely ferruginized. The most common sedimentary cycle involving oolitic ironstone has repeating sandstone–ironstone–shale units, from bottom to top. Volcanic rocks and bedded chert are generally not associated with SCOS-IF.

Wherever paleogeographic reconstruction is possible, inland-sea sedimentation is deduced for all chert-poor, oolitic iron formations. Where reconstruction is impossible, inland-sea sedimentation generally may be inferred by a combination of close stratigraphic proximity to fluvial sandstone, paucity of terrigenous-sediment-free carbonate, and widespread evidence of desiccation and turbulence.

SCOS-IF ironstone is normally distinct from glauconitic sandstone because of higher iron content and lesser abundance of clastic grains, besides the presence of oolitic or pisolitic texture and scarcity of glauconite. Distinct ironstone with micritic texture also occurs in beds of a few centimeters in thickness within coal measures, as in the Carboniferous of Scotland (MacGregor et al., 1920) and Permian of South Africa (Wagner, 1928). Only SCOS-IF is discussed comprehensively herein. Its chronological development is reviewed in the first section of this paper and a description of Tertiary SCOS iron formations in Colombia and Venezuela is given in the second section. A genetic model for SCOS-IF as an early diagenetic replacement of calcareous oolite is presented in the third section.

CHRONOLOGICAL DEVELOPMENT OF OOLITIC-INLAND-SEA IRON FORMATION

Precambrian oolitic-inland-sea iron formation

Oolitic-inland-sea iron formation (SCOS-IF) ranges in age from Early Proterozoic (Precambrian X), about 2200 m.y. B.P., to Pliocene, about 4.5 m.y. B.P. The oldest SCOS-IF occurs in South Africa and Western Australia. The Lower Proterozoic Timeball Hill Formation of South Africa is a SCOS-IF-bearing sequence of carbonaceous shale and cross-bedded quartzite, interpreted by Eriksson (1973) to be largely deltaic. Like all subsequent SCOS iron formations, those within and correlative with the Timeball Hill Formation are only a few meters thick and are dominantly composed of chert-poor, oolitic and pisolitic ironstone (Wagner, 1928). Individual ooids are mostly composed of chamosite, hematite, and goethite, all locally replaced by porphyroblastic magnetite (Schweigart, 1965). Ooids within the locally ripple-marked ironstone commonly have fragments of broken ooids as ooid nuclei. Intraclasts of oolite occur, particularly in the correlative Daspoort SCOS-IF. Lower Proterozoic oolitic ironstone has also been reported from the Turee Creek Formation of Western Australia, a shale sequence with subordinate orthoquartzite (Button, 1975, 1976b; cf. Chapter 7, Vol.5 in this Handbook). Like the Timeball Hill Formation, the Turee Creek lies above and is separated by an unconformity from a thick cherty iron formation. These SCOS iron formations do not differ in any known way from several slightly metamorphosed Phanerozoic SCOS iron formations and are therefore evidence against Petranek's (1964a, p.51) suggestion that "... conditions stemming from the progressive and irreversible evolution of the Earth ..." were partly responsible for Ordovician SCOS-IF.

The next-younger SCOS iron formations, about 1400 m.y. old, are the roughly correlative Constance Range and Roper River deposits of northern Australia, 20 and 50 m thick, respectively (Trendall, 1973a; Edwards, 1958). As for all iron-formation thicknesses subsequently listed herein, these are maxima and include subordinate non-ironstone interbeds. Beds of ripple-marked, oolitic ironstone composed of siderite, chamosite, hematite, and magnetite are enclosed here within an orthoquartzite–shale sequence (Harms, 1965). Chert is present, but occurs as a late-diagenetic replacement of siderite cement (Cochrane and Edwards, 1960). A barred basin is interpreted by Harms (1965) on the basis of mud cracks, ripple marks, cross-bedding, and intraformational [1] conglomerate in associated detrital sedimentary rocks.

[1] Editor's note: Although a generally accepted *genetic* petrologic (and loosely petrographically

Precambrian SCOS iron formations are much fewer in number than Phanerozoic deposits. This may represent either an evolutionary increase in occurrence of the necessary chemical environment within epicontinental basins or just the presence of a greater volume of inland-sea sequences of Phanerozoic versus Precambrian age. Support may be found for the latter interpretation. Although fluvial conglomerate is now well documented from the Archean (Turner and Walker, 1973), no Archean inland-sea sequence has yet been reported. Most Lower Proterozoic volcano-sedimentary sequences display lateral continuities of a few hundred kilometers and stratigraphic thicknesses of hundreds of meters, commonly including interbedded dolostone, quartzite, and red and black ferriferous shale, present beneath a thick cherty iron formation which is overlain by black shale or argillite (Gross, 1973). These dimensions and the world-wide similarity of lithostratigraphic sequences are here considered indicative of continental-shelf sedimentation, rather than inland-sea sedimentation. An inland sea would not likely remain free of detrital sediment long enough for accumulation of an extensive cherty iron formation hundreds of meters thick.

Paleozoic oolitic-inland-sea iron formation

SCOS iron formations of Late Cambrian or Early Ordovician age occur in the Bliss Sandstone of New Mexico. The Bliss grades upward from a conglomerate developed on Precambrian basement to cross-bedded sandstone with predominantly subangular grains (Kelly, 1951). Beds of oolitic, hematite–chamosite ironstone locally exceed 5 m in thickness and are interbedded with ferriferous sandstone and limestone, including oolitic limestone (Kelly, 1951). The Lower Ordovician Wabana SCOS iron formations of Newfoundland are cross-bedded, ripple-marked, and enclosed in a thick sandstone–shale sequence which locally displays raindrop impressions (Hayes, 1915, 1929). Middle Lower Ordovician SCOS iron formations of Czechoslovakia are enclosed in tuffaceous shale transgressive onto Precambrian basement. Areal variations in bed thickness and current indicators record a topographically variable basin margin, which at least partially enclosed the areas of oolite sedimentation (Berg et al., 1942a, b; Skocek, 1963a, b; Petranek, 1964b). Paleogeographic reconstruction of a contemporary SCOS iron formation-bearing sequence in Thuringia, East Germany, reveals a basin about 100 km in diameter which received clastic sediment from over half its perimeter (Deubel et al., 1942). Upper Lower Ordovician deposits of Normandy, France, are interbedded with metamorphosed trilobite-bearing shales. Oolitic siderite–chamosite–hematite ironstone here displays porphyroblastic

---

applied) expression, "intra-formational" is slowly being recognized as a misnomer if used as, e.g., in the present case, where it *wrongly* suggests the presence of *internally* formed (i.e., *intra-*) components originating *within* a formation or lithologic unit (e.g., intraformational solution or collapse or cataclastic breccia). To allow a clear distinction, in the present case the phrase "intraenvironmental congomerate" (i.e., formed within the primary sedimentary milieux, in which also the rounding of the pebbles may have occurred), would be more accurate.

magnetite much like the Lower Proterozoic South African ironstone (Cayeux, 1909; Hoenes and Tröger, 1945).

Upper Cambrian to Middle Ordovician SCOS iron formations in Wales were deposited near coastlines and invariably covered by silicate mud (Pulfrey, 1933). Oolitic and pisolitic chamosite–thuringite ironstone displays partial replacement by siderite, magnetite, pyrite, and secondary chlorite (Hallimond, 1925, 1951; Pulfrey, 1933). Upper Upper Ordovician iron formations near Mayville, Wisconsin, formed in a shallow sea between the young Appalachians and the interior craton, as did the Lower Middle Silurian Clinton deposits, closer to the then-more-deeply-eroded northern Appalachians. Hematite predominates in oolitic ironstone of both the Mayville (Thwaites, 1914; Hawley and Beavan, 1934) and Clinton iron formations (Alling, 1947; Schoen, 1962; Hunter, 1970). Lower Silurian hematitic iron formations near Birmingham, Alabama, formed in the same inland sea and were interbedded with sandstone, siltstone, and cross-bedded limestone conglomerate (Burchard, 1910; Burchard and Butts, 1910; Bearce, 1973). Lower Silurian, magnetite-rich SCOS iron formations in northern Spain occur in quartzite–ironstone–slate cycles, in ascending sequence (Rechenberg, 1956).

Lower Devonian, metamorphosed, magnetite-rich SCOS iron formations up to 3 m thick are interbedded with slate in Nova Scotia (Wright, 1975, p.77). A 2 m-thick Devonian SCOS-IF in Arizona occurs between underlying silty limestone and overlying shale. Where the bed of oolitic hematite–chamosite ironstone is missing, either a poorly sorted, hematite-cemented sandstone is found or hematite nodules within the uppermost silty limestone (Willden, 1960, 1961). Devonian beds of oolitic chamosite–siderite ironstone, each 1–20 m thick, are interbedded with black phyllite in Yugoslavia (Latal, 1952; Page, 1958). Thin Middle Upper Devonian deposits of Belgium (Cayeux, 1909) are enclosed in a shale unit which thickens southward but, like associated Devonian strata, uniformly displays evidence of only shallow-water sedimentation (Rutten, 1969). Middle and Upper Devonian beds of oolitic hematite ironstone up to 3 m thick are intercalated with shale and sandstone in the Tuyun district of China (Ikonnikov, 1975).

No Carboniferous SCOS iron formations are known, and only the Late Permian is represented, by the Desert Basin deposit of northwestern Australia. Beds of oolitic ironstone up to 30 m thick are enclosed here in ferruginous sandstone lenses which are correlative with highly fossiliferous sandstone (Edwards, 1958). The Triassic also lacks known SCOS-IF.

Mesozoic oolitic-inland-sea iron formation

The Jurassic of northwestern Europe represents the epitome of epicontinental-sea sedimentation (Gignoux, 1950) and the most intricately-connected of these shallow seas generated abundant SCOS-IF (Karrenberg et al., 1942). Subordinate SCOS-IF formed contemporaneously elsewhere.

Lower Lower Jurassic SCOS-IF occurs in Germany, France, and Iraq. The German iron formation is the thickest at 19 m, but almost half of this is interbedded shale (Simon, 1969). The Iraqi iron formation, correlative with shale, consists of pisolitic, intraclastic and laminated, as well as oolitic, ironstone (Skocek et al., 1971). The French deposit is only about a meter thick and is associated with ferruginous limestone (Cayeux, 1922, p.36).

Middle Lower Jurassic SCOS-IF occurs in Germany and England. The Oberbank-Echte iron formation of Germany, 7 m thick, overlies oolitic marl and underlies oolitic limestone (Finkenwirth and Simon, 1969; Schellmann, 1969). Its epicontinental sea was bounded about 35 km to the east, 85 km to the west-southwest and 125 km to the north-west; several other SCOS iron formations resulted from sedimentation in this sea (Berg and Hoffmann, 1942). The Marlstone, Cleveland, and Frodingham iron formations of England, from 5 to 9 m thick, respectively overlie ferriferous conglomerate, shale, and limestone. In turn, they are overlain by calcareous conglomerate, ferriferous shale and ferriferous claystone. In all three iron formations, a large proportion of the ooid nuclei are ooid fragments; most detrital quartz grains are subangular to angular, and fossils have been partially ferruginized. The Marlstone, with calcite–chamosite ooids, is locally gradational to oolitic limestone (Hallimond, 1925; Whitehead et al., 1952; Edmonds et al., 1965). The Cleveland also contains calcite-bearing ferriferous ooids (Sorby, 1857; Lamplugh et al., 1920). The Frodingham is cross-bedded, as is the Marlstone, and locally pisolitic (Davies and Dixie, 1951; Whitehead et al., 1952).

Upper Lower Jurassic SCOS-IF is well known from the Lorraine and other districts of France and also occurs in Germany, England, Scotland, and the U.S.S.R. The Lorraine ironstone beds are up to 5 m thick and are interbedded with sandstone, shale, and sandy, ferriferous, and/or oolitic limestone (Cayeux, 1922; Bichelonne and Angot, 1939; Bubenicek, 1971). Goethite, chamosite, and siderite predominate over hematite in the oolitic ironstone, whereas hematite is more abundant in the contemporary Avelas deposit of southern France (Cayeux, 1922). A 9 m-thick goethitic ironstone bed near Ringsheim, West Germany, is underlain by shale and overlain by argillaceous limestone (Urban, 1966). Both the 12 m-thick Northampton and 4 m-thick Rosedale SCOS iron formations of England are predominantly chamositic and overlain by oolitic limestone. The Northampton is underlain by sandy limestone (Lamplugh et al., 1920; Taylor, 1949, 1951) and the Rosedale by argillaceous sandstone (Hallimond, 1925; Hemingway, 1951). The thin, chamositic Raasay iron formation of Scotland is enclosed in shale (MacGregor et al., 1920) as is the thick Malka iron formation of the U.S.S.R. (Sokolova, 1964; Timofeeva, 1966; Timofeeva and Balashov, 1972).

Most Middle Jurassic SCOS iron formations occur in Switzerland and adjacent France. The 4 m-thick Lower Middle Jurassic Doubs SCOS-IF of France is enclosed in oolitic limestone (Cayeux, 1922). Three Upper Middle Jurassic Swiss deposits, Erzegg, Windgaelle, and Chamoson, are 2–3 m thick and associated with shale or, locally, argillaceous limestone (Deverin, 1945). All three are principally composed of chamosite; Chamoson being the type locality for the mineral.

Lower Upper Jurassic deposits are more widespread in Europe. The 18 m-thick Gifhorn SCOS-IF of West Germany (Kolbe and Simon, 1969) is associated with calcareous shale and impure limestone, like the thin Marsannay-le-Bois deposit of France (Cayeux, 1922). The 5 m-thick Westbury SCOS-IF of England is enclosed in shale. The Westbury, Northampton, and Avelas deposits all contain calcite-bearing ferriferous ooids like those of the Marlstone SCOS-IF.

In the Early Cretaceous, SCOS iron formations were still forming in France and England. The Lower Lower Cretaceous Vassy deposit of France, only 2 m thick, is enclosed in ferriferous shale (Cayeux, 1922). It contains non-marine pelecypods and represents the oldest known landlocked-sea subtype of SCOS-IF. The lack of older known deposits partly may reflect lesser precision in paleontological differentiation between the two environments in older sequences, but probably mostly reflects generally brief preservation of landlocked-sea sedimentary units. Vassy ironstone is typical SCOS-IF ironstone, with ooid fragments commonly forming ooid nuclei, subangular detrital quartz grains, and intraclasts of oolite. Goethite predominates over both siderite and chamosite (Cayeux, 1922).

The Middle Lower Cretaceous Claxby iron formation of England contains marine pelecypods and ammonites. It is associated with glauconitic sandstone as is the Upper Lower Cretaceous Seend deposit of England (Lamplugh et al., 1920; Hallimond, 1925). Lower Cretaceous SCOS-IF of Syria, up to 10 m thick, is pisolitic and rich in both plant remains and angular quartz (El Sharkawi et al., 1976). Ironstone overlies sandstone and underlies interbedded claystone–sandstone. By contrast, the 2 m-thick Upper Lower Cretaceous Ramin SCOS-IF of nearby Israel is oolitic and enclosed in ferruginous limestone (Boscovitz-Rohrlich et al., 1963).

The transition from Early to Late Cretaceous in northwestern Europe was marked by a profound deepening of basins, general loss of epicontinental seas (Gignoux, 1950), and termination of SCOS-IF sedimentation. However, SCOS iron formations increasingly formed elsewhere. In western Canada, the 9 m-thick Peace River deposit is enclosed in Upper Cretaceous pyritic shale (Mellon, 1962; Bertram and Mellon, 1973; Petruk, 1977). The 11 and 12 m-thick Bahariya and Aswan SCOS iron formations of Egypt are contained in Upper Cretaceous sandstone (Nassim, 1950; Nakhla and Shehata, 1967) and a 15 m-thick Upper Upper Cretaceous iron formation overlies argillaceous sandstone in Nigeria (Jones, 1955, 1965; Adeleye, 1973). Goethite is more abundant than chamosite in oolitic-pisolitic ironstone of all four; at Aswan hematite exceeds goethite.

Tertiary oolitic-inland-sea iron formation

Tertiary examples are similarly widespread. Deposits in eastern Colombia and northwestern Venezuela are described by Kimberley (in press) and in the next section of this paper. The Eocene Djebel Ank SCOS-IF of Tunisia is 8 m thick, between shale and overlying argillaceous limestone (Solignac, 1930; Nicolini, 1967). The Middle Eocene Weches iron formation of Texas reaches 46 m in thickness, including interbedded sandstone and mudrock (Sellards and Baker, 1934; Eckel, 1938). Middle Eocene iron formations of Louisiana are much thinner (Jones, 1969). The Oligocene Shumaysi SCOS-IF of Arabia, up to 14 m thick, lies between sandstone and overlying shale. Shumaysi oolitic-pisolitic ironstone contains the usual subangular quartz grains, ooid fragments as ooid nuclei and, like most Tertiary SCOS-IF, locally abundant fossil wood (Al-Shanti, 1966).

The example of landlocked-sea SCOS-IF listed in the introduction is the Middle Oligocene Kutan–Bulak deposit of the U.S.S.R., enclosed in fluvial sandstone north of the present Aral Sea. Oolitic-intraclastic ironstone containing non-marine pelecypods occurs here in beds up to 15 m thick (Davidson, 1961; Strakhov, 1969). The youngest major SCOS-IF, the Pliocene Kerch deposit, also occurs in the southern U.S.S.R., under and adjacent to the Sea of Azov (Markevich, 1960; Strakhov, 1969). Although less than 5 m.y. old, it is one of the thickest at over 16 m and most extensive at about 150 km (Sokolova, 1964; Berggren and Van Couvering, 1974). It is more manganiferous than average SCOS-IF, but displays the same ferriferous minerals, textures, and sedimentary structures as do the older unmetamorphosed deposits. The Kerch iron formation provides a clear indication that the potential for extensive SCOS-IF sedimentation continues to exist in epicontinental basins.

TERTIARY OOLITIC-INLAND-SEA IRON FORMATION OF COLOMBIA AND VENEZUELA

Introduction

Sandy, clayey, and oolitic, shallow-inland-sea iron formation (SCOS-IF) of Eocene and Miocene age occurs in four areas of northwestern South America (i.e., Sabanalarga, Paz de Rio, Cúcuta, and Lagunillas) spread over a distance of 650 km along a trend of N14E (Fig. 1). In the Middle Cretaceous, this region formed a northward-sloping continental shelf on which marine limestone and shale accumulated (Young et al., 1956). Commencing in the Maastrichtian, uplift occurred along the present Caribbean coast, resulting in alternating fluvial environments and epicontinental sea-lacustrine environments through the Tertiary in the interior.

In the Paz de Rio–Sabanalarga region of Colombia, Maastrichtian coal swamps were followed in the Paleocene by fluvial sandstone which became increasingly interbedded upward with red-mottled, light gray mudstone and minor coal. Uplift of the Central Cordillera of the Colombian Andes in the Middle Eocene brought locally deep erosion of the upper mudstone–sandstone beds and deposition of fluvial conglomerate and sandstone, which became similarly increasingly interbedded upward with red-mottled light gray mudstone (Irving, 1975). In the Late Eocene, a landlocked or epicontinental marine sea transgressed bleached and mottled coastal sand and, with numerous subordinate fluvial regressions, accumulated over a kilometer of sediment before the Miocene.

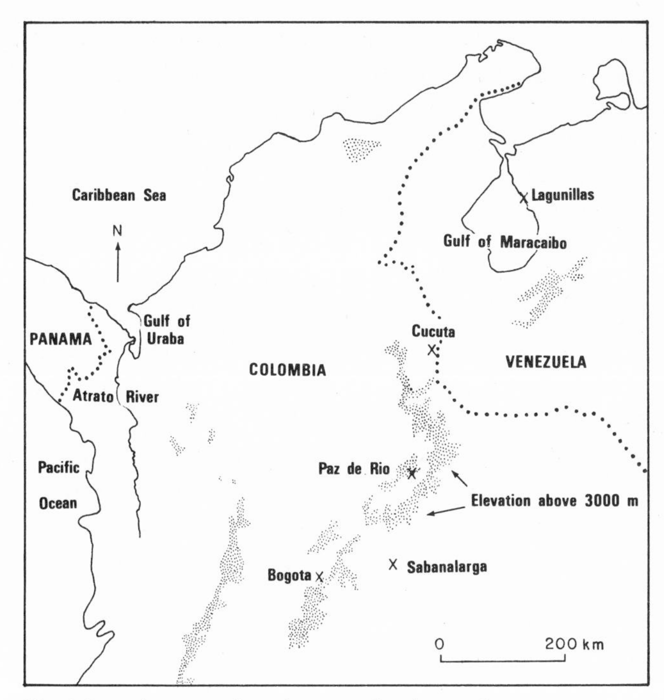

Fig. 1. Location map of northwestern South America.

At least two, possibly correlative, oolitic iron formations, up to 8 m thick or more, formed locally during the first transgressive–regressive cycle.

In the Lagunillas area of Venezuela, Lower or Middle Miocene marine oolite transgressed over "... a thin series of bleached and mottled clays, sands, lignites and black carbonaceous sticky clays which probably originated as swamp muck" (Hedberg, 1928; Sutton, 1946, p.1698). Existing literature and oil-company core logs describe this ironstone as dark green oolitic glauconite. However, H.D. Hedberg (personal communication, 1972) notes that this mineral identification was based only on visual inspection and weathered samples provided by him texturally resemble typical chamositic-goethitic oolite. Only goethite and quartz were detected in the weathered samples by X ray powder diffractometry. Thickness of the SCOS-IF is about 30 m.

Near Cúcuta, Colombia (Fig. 1), thin beds of goethitic and chamositic ironstone are interbedded with deltaic mudstone and sandstone of Late (?) Miocene age (James and Van Houten, 1979). Oolite and ooid-bearing sandstone formed during transgressions caused by decreases in detrital sedimentation. Intervening regressions ended with soil formation on deltaic mudflats (James and Van Houten, 1979).

Stratigraphy of the Paz de Rio area, Colombia

The only commercially exploited oolitic iron formation in northwestern South America is mined near Paz de Rio, Colombia (6°11′N, 72°43′W; Figs. 1 and 2). The Paz de Rio SCOS iron formation varies from 0.5 m to over 8 m in thickness and is found near the base of the 1400 m-thick Concentracion Formation (Fig.3) in all outcrops of basal beds except one (Alvarado and Sarmiento-Soto, 1944). The Concentracion and underlying Picacho Formations extend for a maximum of 57 km in a topographically-depressed outcrop–belt trending N30E within the Eastern Cordillera (Cordillera Oriental) of the Colombian Andes (Fig.1; Irving, 1971).

Type sections of the Picacho and Concentracion formations occur near and along the Soapaga River (Figs. 2 and 3). In the lower 50 m of the Lower Picacho, fluvial, cross-bedded sandstone and conglomerate are equally represented. In the Lower Picacho, mud-

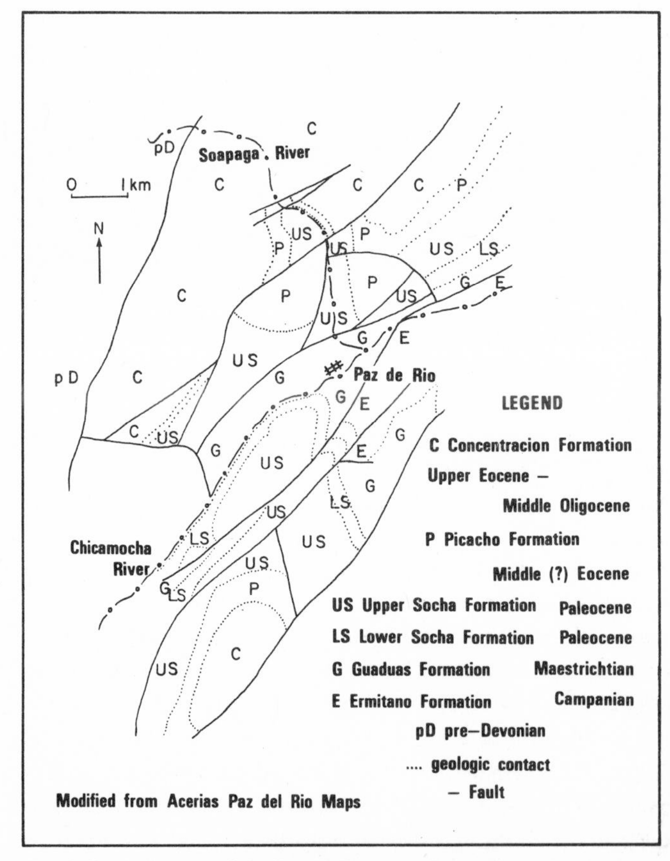

Fig. 2. Geologic map of the Paz de Rio area, Colombia. (After Kimberley, in press.)

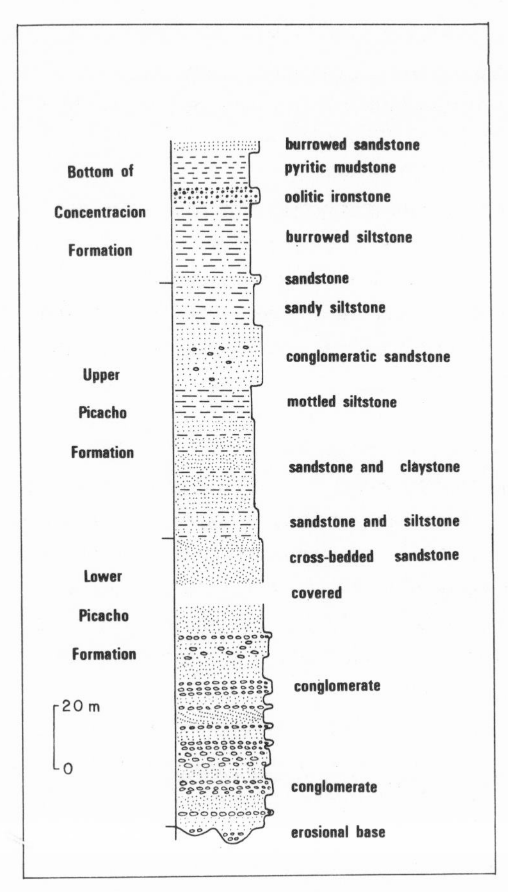

Fig. 3. Stratigraphic section of the Picacho Formation and basal Concentracion Formation along the Soapaga River, Paz de Rio area, Colombia.

rock is present only as abundant mudstone clasts in the fluvial conglomerate. The base of the 80 m-thick Upper Picacho is the base of the lowest mudrock bed, generally red-mottled, light to medium gray siltstone which is interbedded with sandstone. Fossil tree leaves and wood fragments increase in abundance upward through the Upper Picacho and the uppermost siltstone contains abundant limonitic veins. A conformably overlying sandstone bed, basal Concentracion Formation sandstone, contains abundant fossil logs, more highly compressed than those within overlying ironstone. Above 3 m, the basal sandstone grades to a 22 m-thick bed of siltstone which is red-mottled only in the lower portion and decreases in sand upward. Pellet-filled burrows are common in this probable beach–lagoon sequence, which grades up to equally burrowed, oolitic ironstone of an oolitic sand bar or sand belt (Ball, 1967). The oolitic ironstone, 5 m thick, grades up to basinal mudstone which is carbonaceous and pyritic.

Overlying beds of the 1400 m-thick Concentracion Formation record repeated shallow-water transgression and regression. About 700 m stratigraphically above the Paz de Rio iron formation, thin beds of oolitic ironstone and ooid-bearing sandstone are interbedded with dark gray claystone and minor lignite. Siderite spherules are common in mudrocks throughout the Concentracion Formation. Persistence of a water body during accumulation of 1400 m of shallow-water sediment is suggestive of an epicontinental marine or landlocked sea, rather than a freshwater lake. The occurrence of flaser bedding in a few Concentracion beds indicates some tidal-flat sedimentation (Reineck and Wunderlich, 1968). However, shelly fauna are extremely rare and the only shells found to date resemble a contemporary non-marine Colombian assemblage (Pilsbry and Olsson, 1928; Berry and Walthall, 1961).

General features of the Paz de Rio iron formation

The Paz de Rio iron formation is underlain by massive, light olive gray (5Y6/1) to dark greenish gray (5GY4/1) sandstone or siltstone and overlain by black, pyritic mudstone. Sandstone beneath the iron formation locally displays scours, but elsewhere the contact is gradational and the sandstone displays burrows filled with chamositic ooids (Plate I, B). Locally, irregular hematitic veins less than 1 mm wide permeate subjacent sandstone and gradually decrease in abundance downward. The upper contact is generally gradational through mudstone with sideritic nodules.

The iron formation is normally just one bed of oolitic ironstone, but two beds separated by sandstone occur in a fault block 2 km west of Paz de Rio (Fig. 2). Where just ironstone, both the uppermost and lowermost portions of the iron formation are consistently poorest in ferric minerals and range from dark greenish gray (5G4/1) to brownish black (5YR2/1), whereas the central and predominant portion of the bed ranges

---

PLATE I

Photographs and photomicrographs of ironstone and adjacent rocks in the Paz de Rio area, Colombia. (After Kimberley, in press.)

A. Well-preserved tree seeds from basal oolitic ironstone, Some ooids are still attached.

B. Sandstone containing dark burrows filled with chamositic ooids. The sandstone directly underlies basal chamositic ironstone. Length of bar is 10 mm.

C. Hematite–goethite ooid partially replaced by sparry siderite and surrounded by siderite and subordinate chamosite. Ooid core is an ooid fragment. Transmitted and reflected light. Crossed nicols. Length of bar is 0.1 mm.

D. Deformed pyritic ooid containing chert (dark), surrounded by silicate mud (dark) and disseminated pyrite, from gradational ironstone–mudstone overlying uppermost ironstone. Reflected light. Length of bar is 0.1 mm.

E. Sideritic ooid with patchy remnant layers of hematite–goethite. Ooid had intersecting cracks before sideritization and hematite–goethite is preferentially preserved in one quadrant. Siderite growth in this quadrant has deformed hematite–goethite layers. Crossed nicols. Length of bar is 0.1 mm.

F. Intraclast of gradational limestone–ironstone in basal ironstone, surrounded by chamositic ooids, quartz sand, quartz granules, and small sandstone pebbles. Length of bar is 10 mm.

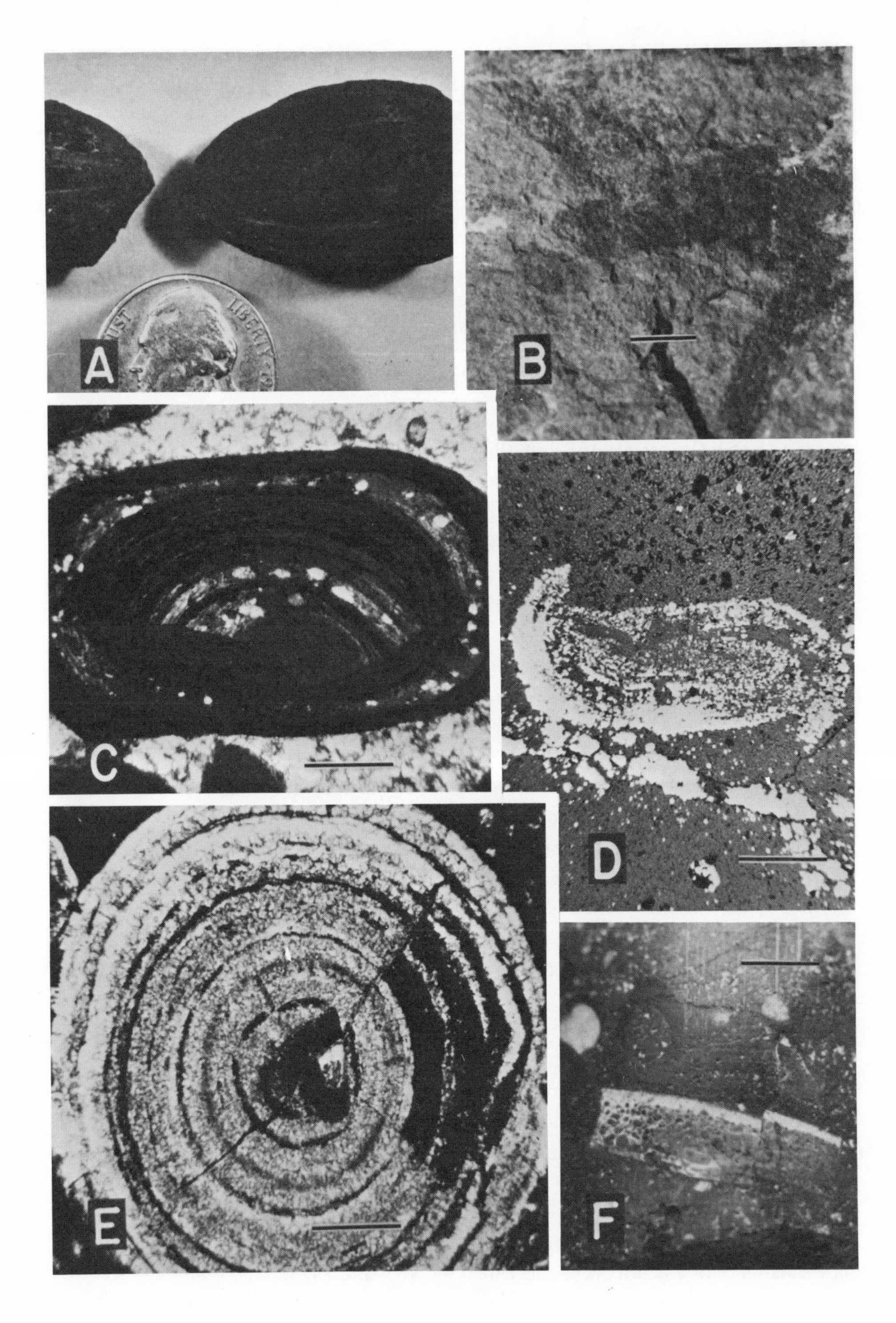
A
LIBERTY
B
C
D
E
F

from dark reddish brown (10R3/4) to moderate red (5R4/6). The proportion of quartz sand decreases markedly upward through the lower third of the iron formation and increases very slightly through the upper third.

Trace fossils and fossil wood are the only evidences of life in the iron formation and adjacent beds. However, the iron formation is generally highly burrowed, and fossil logs and tree seeds together constitute up to 5% of the basal ironstone (Plate I, A). The fossil wood displays excellently preserved carbonaceous cell walls lined with hematite and cells filled with siderite, except where replaced by pyrite. All portions of the iron formation display irregular burrows a few millimeters in diameter. Some burrows are lined with hematite and some are filled with the only quartz sand present in a thin section.

Mineralogy of the Paz de Rio iron formation

The oolitic ironstone has been studied petrographically for Acerias Paz del Rio by Rau (1963) and Gruner (1946), but they did little X-ray diffractometry and markedly disagree on whether hematite or goethite, respectively, is the predominant ferric mineral. The two minerals normally occur finely intergrown and are jointly here termed "hematite–goethite". Detailed X-ray powder diffractometry has shown that hematite and goethite are nearly equally abundant in about half the red oolite and that hematite predominantes or is exclusively detected in the other half (Kimberley, 1974).

Siderite is the only carbonate detected by diffractometry and occurs prominently in all the uppermost and lowermost dark ironstone and less prominently in central red ironstone. Chamosite occurs more evenly in both marginal dark and central red ironstone. Chlorapatite is sporadically present throughout and quartz peaks decrease upward through the lower third with the decrease in quartz sand. Only quartz and kaolinite are detected in overlying mudstone. No other minerals were identified and the only unidentified peak, at 2.34 Å, probably represents a complex phosphate.

Chemical compositions of minerals were determined qualitatively by wave-length profiling with an electron-probe microanalyzer, model EMX-SM of Applied Research Laboratories. A LiF scanner was used with a wavelength range of 1.2–3.6 Å (Kimberley, 1974). All siderite was found to contain substantial calcium and manganese. No other manganiferous phase was discovered. Apatite was the only other calcareous phase detected and its chlorine content was proven. Chamosite was the only aluminous silicate encountered and it exhibited minor potassium. Goethitic and hematitic areas consistently displayed both a little aluminum and a little silicon. This could be dissseminated chamosite or disseminated quartz in aluminous goethite and hematite (Janot et al., 1971). Pyrite was found throughout all dark ironstone and, within interstitial hydrocarbon, either native sulfur or some non-ferriferous sulfide appeared locally.

Petrography of the Paz de Rio iron formation

Eighty-three polished thin sections were studied from three sets of core through the iron formation and adjacent beds, boreholes 98, 165, and 173 of Acerias Paz del Rio. All of the ironstone displays ooids, although not all of the ironstone is oolite. Following Bathurst (1975), oolite is defined to be sediment or sedimentary rock dominantly composed of ooids. Like the ooids of all other SCOS iron formations examined by the author, those of Paz de Rio closely resemble Recent aragonitic ooids and are unlike any other chemical sedimentary grains (Rohrlich, 1974). Like aragonitic ooids, chamositic ooids of Paz de Rio and other SCOS iron formations display an extinction cross in cross-polarized light due to tangential crystal orientation (Berg, 1942). No radial structure was found like that within most ancient calcareous ooids. The ooids are also distinct from spherules in bauxite, because the latter locally display differences of a few orders of magnitude in diameters of adjacent concentrically layered spherules. Another unmistakable difference is that bauxite spherules commonly grow within older spherules and are enveloped within earlier-formed concentric layering (Bardossy and Mack, 1967). This was not observed at Paz de Rio or in any other SCOS iron formation and, consequently, Schellmann's (1969) suggestion that ferriferous oolitic structure forms diagenetically may be rejected. One-fifth to one-third of the ooids in all parts of the Paz de Rio iron formation contain fragments of ooids as cores. Ooid fragments without secondary growth are rare.

Hematite, goethite, siderite, chamosite, and pyrite are the only ferriferous minerals observed in Paz de Rio polished thin sections. Magnetite constituted a few percent of the ironstone in a mined-out area. All observed ferriferous minerals are present within ooids as well as interstitially, and all ferriferous minerals occur in both the central red and marginal dark ironstone. However, pyrite is rare in red ironstone. The red ironstone is mostly hematite and goethite with subordinate siderite and chamosite, whereas the dark ironstone is largely siderite with subordinate chamosite and hematite–goethite.

The central red ironstone contains relatively undeformed, well-sorted ooids averaging 0.7–0.9 mm in thin-section diameter and ranging up to 1.3 mm. An average thin-section diameter of 0.8 mm would correspond, for uniform spheres, to a true diameter of 1.0 mm (Krumbein, 1935). Most ooids contain very thin layers of hematite, goethite, intergrown hematite–goethite, and chamosite. Some ooids are partially replaced by siderite (Plate I, C and E), the interstices between ooids being filled with intergrown siderite–hematite–goethite and subordinate quartz silt. Locally, growth of siderite has deformed hematite–goethite oolitic layers, a feature unobserved in chamositic layers. All chamositic layers are exceedingly thin, but hematite–goethite layers are commonly patchy, thus suggesting at least a partial origin by oxidation of chamosite. Some veinlets transecting ooids contain hematite–goethite, but not chamosite. Authigenic chert and apatite partially replace hematite–goethite as well as chamosite within ooids. Pyrite constitutes 0.03% or less of red ironstone in polished thin sections whereas gypsum, an oxidation product, is apparent in some hand samples.

The basal dark ironstone is locally poorly sorted, with adjacent ooids ranging from 0.1 to 1.8 mm in diameter and quartz grains ranging from 0.15 to 1.6 mm. Wood fragments and interstitial hydrocarbon are most abundant in areas of poor sorting. Most of the basal ironstone is non-oolitic siderite, with subordinate detrital quartz and minor goethite. The ooids are mostly chamosite. Some siderite has rimmed ooids and partially replaced quartz grains. Quartz generally ranges from 10 to 50% of the rock, but thin interbeds of ferriferous sandstone occur locally. Coarse sand grains are well rounded, whereas grains less than 0.8 mm in diameter are subangular to angular and non-spherical. The principal associated heavy mineral is zircon. Angular quartz grains form nuclei within about 10% of the ooids and also occur within some oolitic sheaths. Patches of replacement pyrite constitute 0.5–1% of the rock. Ooids and quartz grains have been concentrated around burrows and within central portions of burrows. The burrows average 7–10 mm in diameter and are rimmed locally by hematite–goethite. Chamositic ooids have been highly compacted, except where enclosed by siderite, unlike little-deformed hematitic ooids in red ironstone.

Both lower and upper dark ironstone are gradational to central red ironstone. Siderite and subordinate chamosite are the predominant ferriferous minerals in the upper as in the lower ironstone, but the rocks differ in most other respects. Quartz grains constitute less than 1% of the upper ironstone and ooid sorting is fairly good, ooid diameters averaging about 0.8 mm and ranging to 1.4 mm. Most ooids contain chamosite largely to completely replaced by sparry siderite and apatite. Even some quartz cores of ooids are partially replaced by siderite. Residual chamosite commonly appears to be as finely layered as aragonite in the freshest Recent ooids. Hematite and goethite are generally absent, whereas pyrite constitutes 0.5–5% of the rock, in irregular patches and as replacements of ooids in which oolitic structure is preserved (Plate I, D). Oolitic portions of the rock grade through portions with ooid-sized spheres devoid of internal structure, i.e. pellets, to micritic, finely intermixed chamosite, siderite, and apatite. Locally, there are lenticular, interlaminated oolitic and micritic bands parallel to bedding, both types averaging 5 mm in thickness. Non-oolitic bands contain thin laminae of chamosite between thick laminae of siderite. In oolitic bands, ooids of sparry siderite are surrounded by finely intergrown chamosite and apatite. An intraclast composed largely of calcite was found in one core of upper dark ironstone (Plate I, F). No other calcite has been found in the Paz de Rio SCOS-IF. Dark ironstone and associated sandstone locally contain irregular concretions of chert several decimeters long.

Chemistry of the Paz de Rio iron formation

Stratigraphic variability in chemical composition within the central red and lower dark portions of the iron formation is revealed in several thousand analyses performed by Acerias Paz del Rio, all made available for this study. Three of the most complete sets of analyses, two from opposite ends of the region of ore bodies, 10.5 km apart (Tables I and III), and one from the center of the region (Table II), exhibit the upward decrease in SiO_2 content found in all intervening areas, largely due to upward decrease in detrital quartz. The interval of most marked SiO_2-decrease normally corresponds to the transition from basal dark to central red ironstone. Other chemical trends are less consistent,

TABLE I

Chemical variation in tunnel 11, El Banco ore body, Paz de Rio iron formation [1]

Sample interval from top of red ironstone (m)	SiO_2	Al_2O_3	Fe	Mn	P	S	CaO	MgO	Loss on calcination
0–1	7.22	6.50	49.71	0.19	1.23	0.05	0.70	0.18	11.01
1–2	12.55	7.04	44.84	0.10	1.40	0.05	1.35	0.45	11.00
2–3	13.55	5.24	44.14	0.04	1.02	0.03	1.36	0.27	13.45
3–4	8.95	6.55	48.79	0.04	1.16	0.01	0.90	0.09	11.05
4–5	12.64	6.57	43.91	0.14	1.33	0.01	1.36	0.58	12.23
5–6	13.59	5.06	44.84	0.25	0.98	0.04	0.91	0.09	12.81
6–7	32.54	5.00	33.11	0.28	1.05	0.03	0.76	0.09	11.14
7–7.5	47.70	5.09	25.21	0.35	0.98	0.06	0.45	0.09	7.74

[1] All samples were dried at 110°C before analysis. Loss on calcination is the weight loss on heating at 1000°C for 75 minutes. All values are in weight percent. (After Kimberley, in press.)

but fairly common through the iron formation, e.g. the increases in manganese and sulfur in the basal relative to central ironstone. These elements generally have a range of up to 0.5% Mn and 0.15% S. Aluminum content is remarkably constant, rarely lying outside the range of 4.0–7.5% Al_2O_3 in either red or dark ironstone. Phosphorus also displays a fairly narrow range, from 0.7 to 1.4% P, except where aluminum is scarce (Table II).

TABLE II

Chemical variation in drill hole 162, Salitre ore body, Paz de Rio iron formation [1]

Sample interval from top of red ironstone (m)	SiO_2	Al_2O_3	Fe	Mn	P	S	CaO	MgO	Loss on calcination
0.00–0.50	7.09	5.32	45.85	0.30	1.19	0.05	2.86	1.00	14.64
0.50–1.00	5.18	4.30	50.04	0.18	1.01	0.01	1.61	0.95	12.31
1.00–1.50	5.59	4.69	50.04	0.24	1.01	0.02	1.47	0.53	12.27
1.50–2.00	5.21	4.68	49.26	0.18	0.97	0.06	1.32	0.79	13.51
2.00–2.50	7.92	5.72	44.54	0.31	1.37	0.06	3.15	0.90	14.23
2.50–3.00	7.52	5.49	48.83	0.19	0.90	0.07	1.76	0.21	12.08
3.00–3.50	5.65	5.55	49.02	0.21	0.99	0.08	2.21	0.84	11.98
3.50–3.75	9.89	5.83	45.23	0.14	1.04	0.09	2.65	0.53	12.24
3.75–4.00	10.72	6.33	41.63	0.26	1.53	0.06	6.03	1,37	11.74
4.00–4.15	19.98	5.85	37.52	0.42	0.41	0.08	1.62	0.53	15.24
4.15–4.40	27.45	3.47	33.67	0.39	0.25	0.09	1.18	0.63	17.46
4.40–4.65	26.50	3.67	34.18	0.39	0.29	0.10	1.18	0.74	16.80
4.65–4.75	29.85	4.41	31.35	0.33	0.29	0.14	1.18	0.63	16.30

[1] All samples were dried at 110°C before analysis. Loss on calcination is the weight loss on heating at 1000°C for 75 minutes. All values are in weight percent. (After Kimberley, in press.)

TABLE III

Chemical variation in drill hole 119, Sativa ore body, Paz de Rio iron formation [1]

Sample interval from top of red ironstone (m)	SiO_2	Al_2O_3	Fe	Mn	P	S	CaO	MgO	Loss on calcination
0–1	5.40	5.60	49.42	0.23	1.30	0.11	1.20	0.86	12.84
1–2	5.63	5.54	48.56	0.31	1.88	0.08	2.25	0.43	11.97
2–3	6.48	5.72	49.34	0.17	1.15	0.08	1.35	0.54	11.92
3–4	15.07	5.43	45.72	0.16	1.02	0.09	1.80	0.43	9.20
4–5	26.40	4.36	36.64	0.25	0.99	0.09	1.80	0.65	10.39
5–<6	39.24	4.99	28.97	0.25	0.70	0.09	1.13	0.86	9.74

[1] All samples were dried at 110°C before analysis. Loss on calcination is the weight loss on heating at 1000°C for 75 minutes. All values are in weight percent. (After Kimberley, in press.)

Magnesium fluctuates from 0.02 to 1.4% MgO and is irregularly distributed. Calcium is commonly correlative with phosphorus because of their coexistence in apatite; local antipathetic relationships may be due to incorporation of calcium in siderite, as found by electron-probe microanalysis.

Sedimentary environment of the Paz de Rio iron formation

The Paz de Rio iron formation is interpreted to have been an oolitic sand bar which transgressed a sandy lagoon and beach and was itself transgressed by basinal mud, like the transgressive Jurassic SCOS-IF sequences of Europe (Brockamp, 1942). The poorly sorted sand and wood fragments of the basal dark ironstone would have been trapped in the highly burrowed lagoonal sediment along with ooids eroded and transported landward from the sand bar. Upper dark ironstone would represent ooids eroded and transported seaward, increasingly mixed basinward with carbonaceous mud. A subsequent regression lacking oolite is represented by a gradation from mudstone to burrowed sandstone, about 10 m stratigraphically above the iron formation.

A similar alluvium–beach–basin sedimentary sequence, but lacking oolite, was found around and under a modern epicontinental sea, the Gulf of Urabá, in northwestern Colombia (Fig.1). A major river, the Atrato, is building a bird-foot delta into this protected Caribbean gulf. Large lakes connected to channels of the Atrato are continually filtering fluvial mud (Vann, 1959) and forming beds comparable to mudstone beds in the Upper Picacho Formation. The delta is well forested and decay products from buried plant matter induce dissolution of iron from surrounding mud, hence limonitic mottling upon oxidation. The shelly and tree seed-rich beach is thoroughly burrowed by crabs and is narrow because the tidal range is less than 50 cm. Core-sampling in the gulf reveals areas of wood accumulation within about a kilometer of shore, partly due to break-up and water-logging of the abundant vegetation rafts supplied by the Atrato River. Average grains size in cores regularly diminishes away from shore and color darkens. Sediments comparable to most Concentration Formation rock types were found by coring in and around the Gulf of Urabá.

Sabanalarga oolitic-inland-sea iron formation

An iron formation which may well be correlative with the Paz de Rio SCOS-IF occurs 128 km south of Paz de Rio, 4 km east of Sabanalarga, Boyacá. The two iron formations are separated by about 70 km of mostly uplifted Cretaceous rocks. Possibly contemporaneous, "... dark gray conglomerate containing black oölites" occurs 50 km farther south-southwest of Sabanalarga (Segovia, 1967, p.1021).

The Sabanalarga SCOS-IF reaches 4 m in thickness and extends for a maximum of 15 km along strike, grading laterally to sandstone. Conformably underlying the iron formation is a rock unit like the Picacho Formation in the Paz de Rio area, i.e. interbedded conglomerate and sandstone that fines upward through 120 m and finally grades through a few meters of interbedded sandstone and mottled mudstone (Camacho et al., 1969). Conformably overlying the iron formation is a thick Concentracion-like unit of interbedded mudstone and sandstone.

The freshest outcrop of the iron formation is in Quinchalera Creek where almost 3 m of oolitic sandstone and ironstone overlie scoured fine sandstone and underlie dark gray shale. Scours up to 3 cm deep are filled with oolitic coarse sandstone which grades upward to finer-grained, sandy chamositic oolite. The scoured fine sandstone is grayish yellow green (5GY7/2) and becomes red-mottled more than 1 m below the oolite. This red mottling was also apparent in cores drilled 1 km south of Quinchalera Creek. Along the creek, dark gray shale overlying the iron formation becomes sandier and flaser-bedded upward through 30 m, but this thickness has been locally augmented because the shale bed is normally 10 m elsewhere and rarely 18 m thick. Fossil leaves are extremely abundant in the upper gray shale. The unit overlying the shale varies laterally within a few meters from 1 to 3.4 m in thickness and contains cross-bedded, medium-grained sandstone in beds 2–30 cm thick, interbedded with 0.5–3 cm of siltstone. Trace fossils cover the sandstone bedding planes, but shelly fossils are absent. A massive sandstone bed over 10 m thick caps the sequence. As at Paz de Rio, oolitic and quartzose transgressive sand belt has been covered by mud, followed by a regression.

Thin sections of fresh core samples reveal a gradual upward increase in chamositic ooids relative to subangular and subrounded quartz sand. The proportions of goethite and siderite relative to chamosite also increase upward. As in the Paz de Rio SCOS-IF, the lowermost, purely chamositic ooids are the most highly deformed but, where undeformed, display extremely delicate oolitic layering. The proportion of ooids with ooid fragments as nuclei is also similar, one-fifth to one-third, and the range is uniform throughout. Although goethite and siderite increase upward, both remain subordinate to chamosite. Goethite appears to be an oxidation product of selected oolitic layers, whereas siderite occurs mainly as rhombs cross-cutting all primary structures. Apatite is present as interstitial radiating fibers. The upper portion of the iron formation was weathered where drilled and the lowermost oxidation affected chamositic ooids, but not interstitial chamosite. One chemical analysis of the ironstone is 30% Fe, 0.64% P, 0.09% S, and 36% insolubles, mostly SiO_2 (Camacho et al., 1969).

PROPOSED ORIGIN OF OOLITIC-INLAND-SEA IRON FORMATIONS

Statement of the problem

Genetic modelling for SCOS-IF is well constrained because, as James (1966, p.48) notes, "... with ages as young as Pliocene, significant atmospheric or biospheric modification seems an unlikely possibility ...". Nonetheless, there is a remarkable lack of consensus regarding the processes of iron mobilization, transportation (Table IV), and

TABLE IV

Proposed modes of all or much iron transportation to areas of SCOS-IF concentration

Proposed mode(s)	Adherent(s)
Fluvial solution or undifferentiated fluvial solution and/or colloidal suspension	Adeleye (1973), Bubenicek (1971), Hunter (1970), Strakhov (1969), Taylor (1969), Uthe (1969), Harder (1964b), Skocek (1963a), Cochrane and Edwards (1960), Castaño and Garrels (1950), Landergren (1948), Alling (1947), Hallimond (1925), Cayeux (1922), Hayes (1915), Newland (1910)
Fluvial colloidal suspension	Millot (1970), Correns (1969), Petranek (1964), Caillère and Kraut (1953), Berg (1944)
Ferric oxides and hydroxides adsorbed on fluvial clays	Schoen (1962), Carroll (1958), Deverin (1945), Hallimond (1925)
Upwelling marine solution	Gruss and Thienhaus (1969), Braun (1964), Finkenwirth (1964), Borchert (1960)
Sedimentation of overlying mud	this paper, Hummel (1922), Sorby (1857)
Ascension of hydrothermal solution followed by subaqueous dispersion	Gross (1965), Schweigart (1965), Villain (1902)
Ascending pore water from underlying marine or transgressed coastal sediment	Kolbe (1970), Aldinger (1957b)
Groundwater from peripheral areas	James (1966)

precipitation (Table V) to form SCOS-IF. This paper presents a modification of the first genetic model which was based upon thin-section study (Sorby, 1857) and critically compares it with subsequent hypotheses. The presentation follows that of Kimberley (1979a).

The most genetically-important sedimentary features of SCOS-IF are considered to be the combination of petrographic evidence for an early-formed ferrous silicate and evidence in sedimentary structures and textures for an extremely shallow, agitated, and presumably oxygenated sedimentary environment. Chamosite, ferrous aluminous iron serpentine, occurs in varying proportions throughout most facies of most well-described SCOS iron formations, and the majority of SCOS-IF petrographers have interpreted it to have been the first-formed, or one of the first-formed, ferriferous minerals (e.g., Tegengren, 1921; Davies and Dixie, 1951; Harms, 1965; Brookfield, 1971; Kimberley, 1974). Chamosite is unstable, however, in the presence of free oxygen (Curtis and Spears, 1968).

In the afore-mentioned Eocene Paz de Rio SCOS-IF of Colombia, chamosite is the most delicately layered mineral in the ooids (Plate II, A). The regularity of internal structure appears to be as good as that of any Recent aragonitic ooids and better than that of any known Pleistocene ooids. In doubly polarized light, chamositic ooids display an extinction cross (Plate II, A) as do aragonitic ooids, because of tangential orientation of crystallites (Berg, 1942). Layering in hematite-goethite ooids is almost as regular (Plate II, B); individual layers in Paz de Rio ooids display slight variations in thickness unobserved in chamositic ooids. The chamositic ooids, therefore, could not have formed by alteration of hematite-goethite ooids but the latter could be oxidation products of the former.

Oolitic texture, like that characteristic of SCOS-IF, presently forms in less than 2 m of water depth (Bathurst, 1975). In general, a significant proportion of SCOS-IF ooids contain ooid fragments as nuclei; ooid fragmentation was probably preceded by exposure and desiccation (Adeleye, 1975; Halley, 1977). Intraclasts of ironstone within ironstone are locally abundant (Cayeux, 1922). The Paz de Rio iron formation overlies fluvial sedimentary rocks by only 25 m and has been thoroughly burrowed (Kimberley, 1974). Chamosite generally exhibits no oxidation near or within burrows, but a few burrows are lined with hematite. Fresh chamositic ooids locally fill burrows into sandy siltstone underlying the Paz de Rio iron formation.

Recognition of this "*ferrous mineralogy–oxygenated environment*" paradox is not new. Hallimond (1951) attempted to resolve the problem by suggesting that the characteristically shallow-water features of SCOS-IF formed in deep water. Harder (1964a) believed that chamosite has formed diagenetically from other ferriferous minerals. The former proposal has not been supported, however, by any other SCOS-IF investigator and the latter has not been defended petrographically.

TABLE V

Proposed modes of all or much SCOS-IF iron concentration

Proposed mode(s)	Adherent(s)
Inorganic oxidation in a water body	Uthe (1969), Curtis and Spears (1968), James (1966), Finkenwirth (1964), Skocek (1963a), Borchert (1960), Cochrane and Edwards (1960), Castaño and Garrels (1950), Alling (1947), Smyth (1892)
Increase in electrolytes by mixing of fresh with marine water	Hunter (1970), Correns (1969), Uthe (1969), Harder (1964a), Petranek (1964b), Skocek (1963a), Alling (1947), Berg (1944)
Direct or indirect precipitation by bacteria	Hallimond (1925), Cayeux (1922), Harder (1919), Newland (1910)
Cooling, degassing and/or mixing with seawater of hydrothermal solutions	Gross (1965), Schweigart (1965), Villain (1902)
Selective removal of non-ferriferous minerals by currents due to density and grain-shape differences	Brookfield (1971), Bubenicek (1971), Skocek (1963a)
Physical separation of iron-rich solutions and/or colloidal suspensions from fluvial detritus caught in a "clastic trap"	Adeleye (1973), Castaño and Garrels (1950)
Subaerial weathering of iron-rich rock, e.g., peridotite, or lateritic peneplanation, followed by marine sedimentation of the highly ferriferous residue	Milllot (1970), Erhart (1967)
Electrolytic and acid-base precipitation during dissolution of aragonitic sediment by descending leachate from closely overlying mud being leached by organic acids	this paper
Submarine dissolution by decomposing organic matter and nearby precipitation	Hummel (1922)
Increase in pH and/or Eh of groundwater or marine pore water on mixing with normal seawater	James (1966), Aldinger (1957b)
Separation of loosely-bound iron from fluvial clays during transportation through anoxygenic coastal water	Deverin (1945)
Supersaturation due to evaporation in a closed or restricted basin	Schoen (1962)
Early diagenetic replacement of aragonitic sediment by solutions from nearby, dissolving glauconite	Brown (1914)

Outline of the genetic model

The genetic model presented herein is an elaboration of that offered by Henry Clifton Sorby (1857), based on a study of the Jurassic Cleveland Hill SCOS-IF of England. Sorby (1857, p.460) wrote:

> "The general conclusion that I therefore draw from these facts is, that, at first, the Cleveland Hill Ironstone was a kind of oolitic limestone, interstratified with ordinary clays containing a large amount of the oxides of iron, and also organic matter, which, by their mutual re-action, gave rise to a solution of bicarbonate of iron – that this solution percolated through the limestone, and, removing a large part of the carbonate of lime by solution, left in its place carbonate of iron; and not that the rock has formed as a simple deposit at the bottom of the sea."

Expanding on Sorby's hypothesis, it is here proposed that during a regression closely following sedimentation of aragonitic oolite, weathering of deltaic mud produced ferriferous leachate which permeated underlying oolite and ferruginized aragonite and highly magnesian calcite (Fig.4). Given that oolite presently forms in extremely shallow water (Bathurst, 1975), little terrigenous sedimentation would be required to cover an extensive bed of oolite with organic-rich mud. Organic-rich waters are generally ferriferous because of reducing conditions and mobilization of iron by organic acids (Gruner, 1922). Filtered organic-rich groundwater may contain more than 10^8 times the thermodynamically-predicted concentration of iron (Shapiro, 1964). However, it is uncertain whether aqueous iron occurs mostly as a colloidal suspension of ferric hydroxide dispersed by dissolved organic acids (Shapiro, 1964) or as Fe(II)-organic complexes (Theis and Singer, 1972). In both sets of experiments, iron was resistant to precipitation over a wide range of Eh and pH. Shapiro (1964) found the most effective inducement for precipitation to be an increase in electrolytes through the addition of salt, particularly divalent and higher valency salts. Increase in ionic strength generally induces colloid coagulation (e.g., Stumm and Morgan, 1970). It is therefore proposed that the principal concentrating process for SCOS-IF metals has been electrolytic precipitation due to dissolution of aragonite and highly magnesian calcite, aided by a related increase in pH. Fresh fluvial mud would continually be brought to the area of eluviation and weathered mud removed.

Evaluation of the genetic model

This eluviation-replacement hypothesis may be elucidated and evaluated through calculation of mass-transfer rates necessary to account for the youngest and perhaps most voluminous SCOS-IF, namely the Pliocene Kerch iron formation which rims and underlies the southwestern part of the Sea of Azov, U.S.S.R., north of the Black Sea. This SCOS-IF easily provides the best constraint on sedimentation rate of voluminous ironstone, having formed within a few hundred thousand years about 4.5 m.y. ago (Sokolova, 1964; Berggren and Van Couvering, 1974). The age of this deposit is less than the residence time of

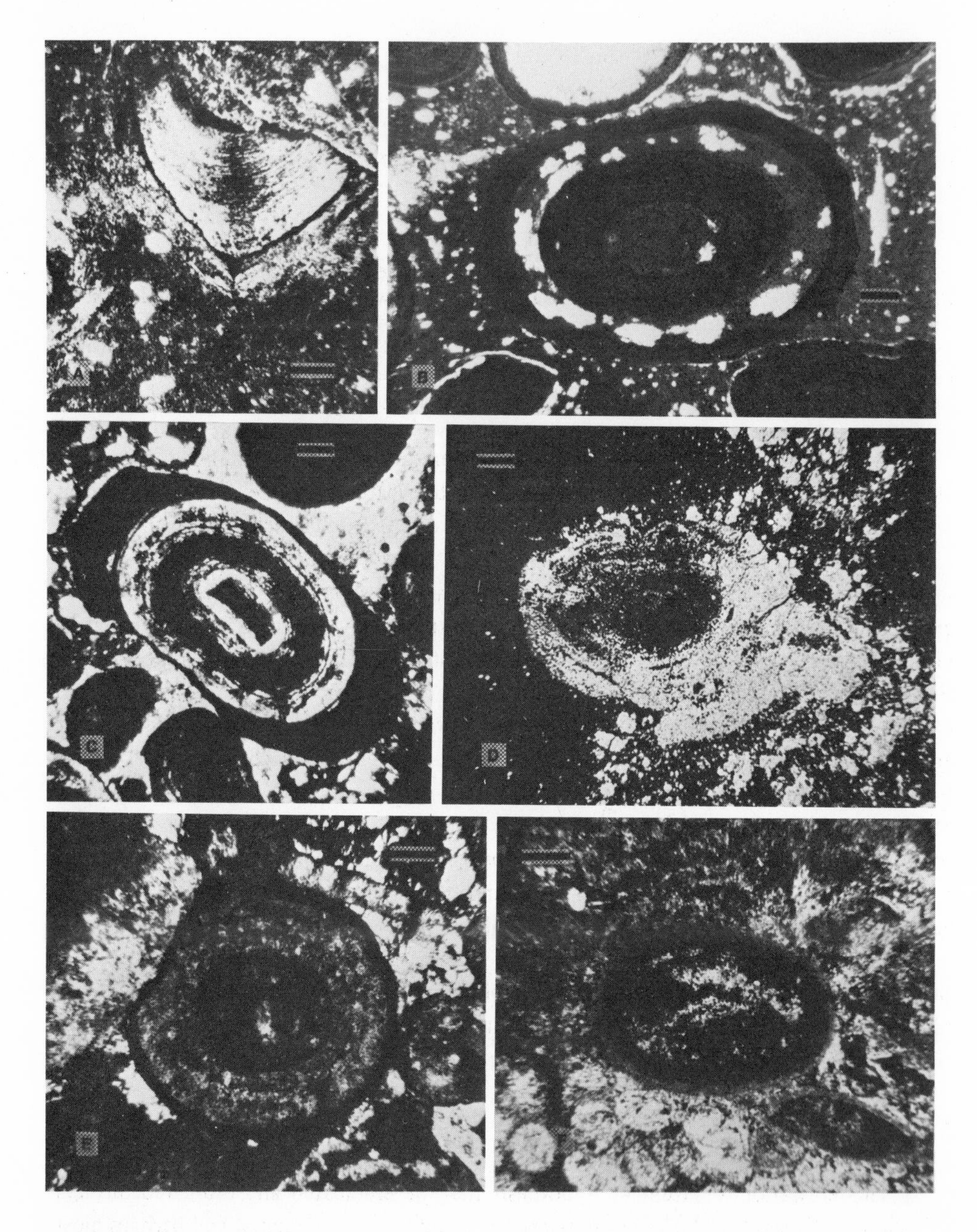

PLATE II

Photomicrographs of Eocene and Jurassic oolite. All scale bars are 0.1 mm. (After Kimberley, 1979a.) A. Chamositic ooid fragment displaying partial extinction in doubly polarized light due to preferred orientation of crystallites. Ooid fragment is rimmed by pyrite and surrounded by finely intergrown chamosite, siderite, pyrite, authigenic silica, and apatite. Eocene Paz de Rio iron formation, Colombia.

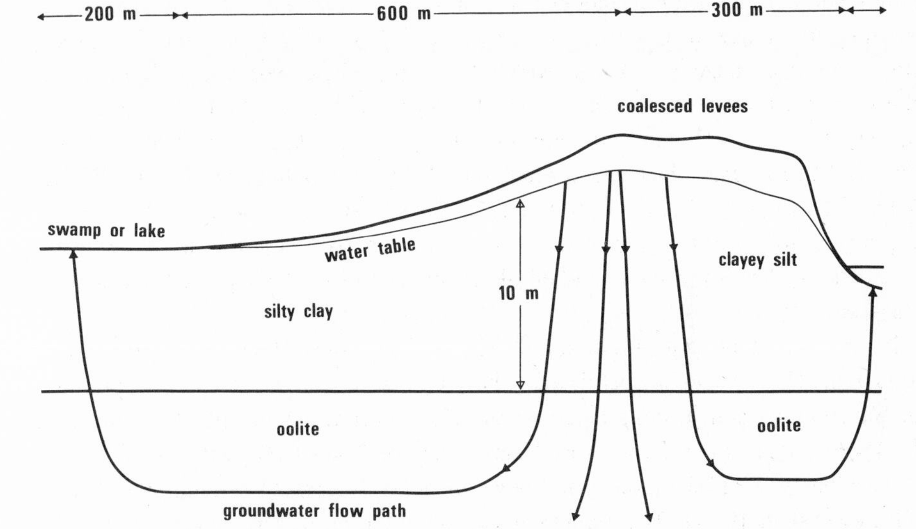

Fig. 4. Hypothetical cross-section of a delta covering a bed of aragonitic oolite, illustrating groundwater flow paths to a distributary and to an interdistributary swamp or lake. The oolite grades upward to transgressive inland-sea mud which grades in turn to regressive deltaic mud. (After Kimberley, 1979a.)

most major elements in the ocean (Garrels and Perry, 1974) and comparable to the time required for complete removal of atmospheric oxygen by weathering, given no oxygen replenishment (Holland, 1973). Referring to the Kerch SCOS-IF, James (1966, p.48) concluded that, "... the range of environmental conditions in the Pliocene could not have differed greatly from those of the present except in relative importance." Rates of

---

B. Hematite–goethite ooid partially replaced by siderite. A hematitic core (dark) with two patches of siderite (white) is enveloped by interlayered hematite–goethite, followed by purely hematitic layers preferentially replaced by siderite, and finally interlayered hematite–goethite. Interstices are filled with finely intergrown hematite, chamosite, siderite, and apatite. An apatite blade (white) occurs above the scale bar. Combined plane polarized transmitted and reflected light. Paz de Rio iron formation.

C. Hematitic ooid with preferential replacement of certain layers by siderite. Ooid core is an ooid fragment. Doubly polarized transmitted light. Paz de Rio iron formation.

D. Chamositic ooid largely replaced by pyrite (white) and a little authigenic silica (dark like chamosite), surrounded by intergrown pyrite, chamosite, and minor authigenic silica. Plane polarized reflected light. Paz de Rio iron formation.

E. Calcite (white) – goethite (dark) ooid. Goethite presumably replaced aragonite and the remaining aragonite altered to calcite. Plane polarized transmitted light. Jurassic Marlstone iron formation, England.

F. Hematite (dark) – dolomite (white)·ooid in stromatolitic dolostone. Plane polarized light. Eocene Green River Formation, Wyoming.

iron supply and concentration in modern environments therefore may be compared with the minimum rate required to account for observed iron concentration in the Kerch iron formation. Average thickness of the Kerch SCOS-IF, where best exposed, is 7.4 m (Sokolova, 1964). Illustrated estimates of areal extent inferred from limited outcrop and drilling vary from $5 \cdot 10^3$ km^2 (Strakhov, 1969) to $11 \cdot 10^3$ km^2 (Markevich, 1960). Assuming a 7.4 m average thickness throughout, the corresponding range in volume is 37–81 km^3 and the range in mass of iron is about 4 to $9 \cdot 10^{13}$ kg. The mass of economically-exploitable Kerch iron has been estimated to exceed 10^{12} kg (Percival, 1952). The range in mass of chemically concentrated aluminum is about 6 to $13 \cdot 10^{12}$ kg, largely within chamosite.

The rate of deltaic sediment supply to the envisaged area of soil formation above aragonitic oolite (Fig.4) may be postulated to have been comparable to that of the present Danube River which annually discharges about 10^{11} kg of sediment into the Black Sea (Smith, 1966). Assuming average crustal abundance of iron in this sediment (Taylor, 1964), leaching of 8–17% of the iron above an oolite bed could produce the Kerch SCOS-IF in less than 10^5 yr. The maximum possible duration of this process is not limited to the maximum duration of oolitic sedimentation, but to the total duration of ferriferous groundwater permeation since sediment burial about 4.5 m.y. ago.

Another potentially limiting factor is the rate of groundwater supply of eluviated iron to underlying oolite. This may be evaluated in a hydrological model (Fig.4) based on general relief and sediment distribution of deltas, including ancient deltas (e.g., Donaldson et al., 1970). In this model, groundwater flows from an area of coalesced levees toward a distributary in one direction and toward an interdistributary swamp or lake in the other. Groundwater first descends through about 10 m of mud to an oolite bed. Half of the lateral flow occurs in the oolite and the other half in all sediment beneath the oolite. Downward groundwater flow into the oolite is assumed to be constant along each of the two directions of lateral flow. The levee-distributary side would be composed largely of clayey silt and the levee-swamp side largely of silty clay (e.g., Kolb and Van Lopik, 1966). Groundwater flow rates would differ accordingly.

A head loss of 3 m within the oolite is assumed for a lateral flow path of 300 m to the edge of the distributary (Fig.4). Water pressure would decrease parabolically because of the assumption of constant influx. A conservative estimate of the average permeability of the oolitic sediment would be 10 darcys. Robinson (1967) measured 3–9 darcys in partially cemented oolite with a high proportion of unconnected porosity. The thickness of oolite is taken to be that of the Kerch SCOS-IF, namely 7.4 m. By integration of Darcy's Law (e.g., DeWiest, 1965) and substitution of these values, an influx of 3 m/yr is calculated along the 300 m. Similar calculation for a hypothetical 600 m flow path from levee edge to swamp or lake, assuming 2 m of head loss, indicates a constant influx of 0.5 m/yr. Downward flow averaged over the entire delta surface of Fig. 4 is 1 m/yr.

Downward influx would come partly from rainfall and partly from seasonal flooding. Annual water-level fluctuation in the lower Volga ranged from 3 to 15 m before flood control (Encyclopaedia Britannica, 1974). Annual water-level fluctuations of about 9 m are observed in the Atrato River basin of Colombia (Vann, 1959), a basin filling with sediment similar to that associated with the Paz de Rio SCOS-IF (Kimberley, 1974). Rainfall on the Atrato delta is 1.84 m/yr (Vann, 1959), three times the continental average (Garrels and Mackenzie, 1971).

Downward flow rates of 0.5 and 3 m/yr correspond to reasonable permeabilities for uncompacted silty clay and clayey silt, assuming a constant head loss of 1 m between the water table and the oolite. Darcy's Law permeabilities for flow through 10 m are respectively 10^{-2} and 10^{-1} darcy, approximately. Schoeller (1962, p.140) lists permeability measurements for silty clay and clayey silt ranging from 10^{-7} to 1 darcy, with the typical value for silt given as 10^{-1} darcy. Golder and Gass (1962) estimate the permeability of silt to vary mostly between 10^{-1} and 1 darcy. Sandy clay soil in Virginia also displays this range in permeability (Lambe, 1955). Silt is the dominant grain size in all thick mudrock beds that the author has found overlying iron formations. Pure clay has a much lower permeability and local interdistributary lenses of pure clay could divert descending groundwater. However, this would have little effect on total groundwater supply to some underlying oolite bed. The permeability of mud directly underlying the distributary channel would be relatively high due to grain-contact disruption by constant upward flow.

Given the widespread mottling of modern deltaic sediment (e.g., Donaldson et al., 1970), substantial dissolution of iron apparently is a general subsurface feature of deltas. In the present groundwater-flow model, the aqueous iron content reached within descending groundwater is assumed to be that reached within anoxygenic lacustrine sediment, 20 ppm Fe (Emerson, 1976). Values up to 43 ppm have been recorded in certain swamp waters (Moore, 1910). The Kerch SCOS-IF contains about 1.1 g Fe cm^{-3} (Sokolova, 1964). A 7.4 m-thick bed of this oolite would have received sufficient iron for complete ferruginization from 1 m/yr of such descending groundwater within about $4 \cdot 10^5$ yr. Distributary migration would ensure areally-uniform ferruginization within this time span.

Evidence for the genetic model

Oolitic inland-sea (SCOS) iron formations are transgressive and overlain by mudrock, argillaceous sandstone, or argillaceous limestone (Brockamp, 1942; Kimberley, 1979b). At Paz de Rio, oolite transgressed a sandy lagoon rich in logs and tree seeds and was covered by basinal mud. Cell structure of wood locally abundant within the ironstone is remarkably well preserved, indicative of very early mineralization (St. John, 1927). A vertical gradation from mudstone to burrowed sandstone about 10 m above the iron formation records a subsequent regression. Soil formation during this regression is presumed to have produced descending ferriferous groundwater, which became sufficiently anoxygenic to precipitate chamosite. Sustained anoxygenic conditions at Paz de Rio are evidenced by partial replacement of chamosite by siderite (Plate II, B and C), as in most other SCOS iron formations. The tendency for some secondary ferriferous minerals to preserve oolitic layering is exhibited by pyrite at Paz de Rio (Plate II, D) and elsewhere (Hayes, 1915; Kalliokoski, 1966). Preservation of oolitic layering during replacement is probably controlled by organic-rich layers which act like templates, as they apparently do during replacement of aragonitic ooids by calcite (Shearman et al., 1970).

The most desirable evidence for the eluviation-replacement model is partial replacement of calcareous oolite, a feature unobserved at Paz de Rio but apparent elsewhere. The extensive Jurassic Korallenoolith Formation of West Germany displays a complete gradation from oolitic limestone to oolitic ironstone (Freitag, 1970; Thienhaus, 1957; Berg et al., 1942b). Berg (1944, p.46) illustrates SCOS-IF ooids from Germany which are partly composed of calcite and partly of chamosite. Maubeuge (1972, p.470) has discovered unferruginized oolitic limestone within the Lorraine SCOS-IF and Cayeux (1922, p.464) found primary calcareous ooids in the Avelas SCOS-IF of southern France. Gradational oolitic limestone–ironstone also occurs in the Jurassic of England (Edmonds et al., 1965), illustrated here by a calcareous ooid collected from the Marlstone SCOS-IF which is partially replaced by goethite (Plate II, E). The Jurassic Raasay SCOS-IF of Scotland is correlative with "... greenish ferruginous oolitic limestone ..." (MacGregor et al., 1920, p.201) and SCOS-IF is interbedded with oolitic limestone in the Upper Cambrian Bliss Sandstone of New Mexico (Kelly, 1951).

Other samples of partially ferruginized oolitic limestone and dolostone in inland-sea environments include oolite in the Eocene Green River Formation of the western U.S.A. Bradley (1929, p.221) reports that the iron is concentrated within ooids relative to the matrix and that ooids display only concentric, no radial, structure. The author has found both characteristics to be typical of oolitic ironstone. A ferriferous Green River Formation ooid is illustrated in Plate II, F. Calcitic-dolomitic oolite within the Triassic Buntsandstein Formation of Germany contains glauconite, pyrite, and apatite (Langbein, 1975). Sedimentary rocks associated with weakly ferruginized Buntsandstein carbonate units, which include the type algal stromatolite (Kalkowsky, 1908), differ from those associated with SCOS-IF by being poor in silt and clay, the size range of potential detrital iron source for weathering-leachate groundwater.

The lack of previous recognition by SCOS-IF investigators of partial ferruginization of calcareous oolite may be attributed to the misconception that oolite is initially calcitic. A paragenetic sequence like aragonite–chamosite–calcite therefore would have been interpreted to be simply chamosite–calcite, given better preservation of oolitic layering in chamositic than in calcitic portions. Hallimond (1925, p.10) observed that in Great Britain, "No trace of original calcite pisoliths or ooliths was found in the ironstones." Taylor (1949, p. 81) shared this error and Caillère and Kraut (1953) have repeated Hallimond's observation for the Lorraine SCOS-IF of France. Recent ooids are dominantly composed of aragonite; aragonite unreplaced during ferruginization would eventually convert to thermodynamically stable, magnesium-poor calcite (Berner, 1971). Delicate sedimentary structures are usually not well preserved during this conversion (Shearman et al., 1970; Rohrlich, 1974) and the more regular arrangement of chamosite, hematite, and goethite in some partially calcitic ooids may be attributed to later calcitization of unferruginized aragonite.

Hallimond (1925, p.11) was well aware of calcitic ooids in the Marlstone SCOS-IF but proposed "... the formation of calcite ooliths from chamosite ooliths ...". Much of the Marlstone calcite does seem to replace chamosite, but pockets of apparently originally calcareous ooids also occur (Plate II, E). In some Marlstone ooids, delicate oolitic layers grade laterally from pure chamosite to pure calcite well within the ooids (Kimberley, 1974).

Several other characteristics of SCOS iron formations and Quaternary sediments support the present elaboration of Sorby's (1857, 1906) genetic hypothesis. Sorby's (1857) observation that the present seafloor lacks sediment comparable to oolitic ironstone remains valid. The only claim of discovery of modern marine ferriferous oolite, by Pratje (1930) in the North Sea, has been rejected (James, 1966). The textural similarities between ironstone and limestone noted by Sorby (1857) and Cayeux (1922) occur locally in all well-described fossiliferous ironstone.

Macroscopic characteristics supporting the eluviation-replacement hypothesis include the typical upward transition from oolitic ironstone to argillaceous sedimentary rock and the typical lack of such a vertical transition above oolitic limestone (Kimberley, 1974). Purely calcareous sequences provide no evidence of iron having been available for diagenetic ferruginization. The general distribution of SCOS-IF near or just within latitudinal belts of contemporaneous salt sedimentation (Brockamp, 1942) is consistent with the modern distribution of epicontinental calcareous oolite. The hypothesis is not dependent upon marine sedimentation of oolite and could account for the non-marine Oligocene Kutan-Bulak SCOS-IF (Strakhov, 1969) as a diagenetic replacement of calcareous oolite, like that presently covering 3000 km^2 in the nearby nonmarine Aral Sea (Dickey, 1968).

SCOS-IF is generally thinner than mudrock beds which accumulated during the same time interval further offshore (Cayeux, 1922, p.639; Petranek, 1964b, p.145; Hallam, 1966; Hunter 1970). This is consistent with the slow subsidence envisaged to occur in the area of oolitic sedimentation during a subsequent regression. Slow subsidence would induce continued erosion of mud fluvially transported onto the oolite and subjected to partial leaching. The hypothesis also explains the paleogeographic relationship of oolitic ironstone to oolitic limestone in Jurassic shale–sandstone–(limestone or ironstone) cyclothems in England. Oolitic limestone repeatedly formed in the more southerly cyclothems far from shore, whereas oolitic ironstone formed in the nearshore environments of Yorkshire (Hemingway, 1951).

Supportive microscopic characteristics include the commonly equal or greater size of dense ferriferous ooids relative to quartz grains outside of the ooids. This represents hydrodynamic disequilibrium, but porous aragonitic ooids would have been approximately in settling-equilibrium with the quartz. Recent aragonitic ooids are lighter than skeletal carbonate grains of the same size (Newell and Rigby, 1957, p.57). Pyritic ooids in the Ordovician Wabana SCOS-IF are similar in size to adjacent detrital quartz grains, a characteristic cited by Kalliokoski (1966, p. 878) as evidence for diagenetic pyritization.

Previous arguments against related genetic models

Many objections have been raised against previous calcareous-replacement hypotheses. The common occurrence of purely calcitic fossils surrounded by completely ferriferous ooids was considered by Hallimond (1925, p. 90) to be conclusive evidence against preferential replacement of ooids. However, Sorby (1857) found a correlation between degree of ferruginization and kind of fossil, and implied that ooids had been particularly susceptible to ferruginization like certain fossils, because of their initial mineralogy. Cayeux (1922, p.913) reported that the major types of fossils in SCOS-IF of France were locally ferruginized by all common ferriferous minerals, but that internal structures of Echinodermata fossils were consistently much better preserved by ferriferous minerals than those of Mollusca. This difference is probably related to the high skeletal porosity, 40%–50%, and high-magnesian-calcite mineralogy of the Echinodermata (Deverin, 1945; Macurda and Meyer, 1975) versus low porosity and magnesium-poor calcite mineralogy of Mollusca (Chave, 1954). Calcite with more than 8.5% $MgCO_3$ is less stable than aragonite in seawater (Berner, 1975). Highly magnesian Echinodermata fossils would have been similarly more reactive to groundwater and would have been replaced rapidly, a necessary condition for excellent preservation (Schopf, 1975). The author has found that fossils which were originally aragonite or highly magnesian calcite, are generally more ferruginized and more commonly preserved as impressions than fossils which consisted originally of magnesium-poor calcite. However, more work is needed to test this observation statistically.

The fact that ooids are typically more completely ferruginized than associated Mollusca of presumed similar initial mineralogy is attributed to relatively high connected porosity, hence permeability of the oolitic ultra-structure (Loreau and Purser, 1973). Two investigators have suggested that any replacement would have progressed inward from the perimeters of ooids (Smyth, 1892; Wagner, 1928); however, electron microscopy has shown that aragonitic ooids are porous aggregates of crystallites (Loreau and Purser, 1973) which could transmit pore fluids freely. The observation that a large proportion of ooids in all paleo-environmental types of oolitic iron formations have fragments as nuclei (Dimroth and Chauvel, 1973) is a more serious objection, because ooid fragments form only a very small proportion of Recent ooid nuclei. Beds of calcareous oolite with a large proportion of fragmental nuclei do occur in several evaporite-bearing sequences and fragmentation seems to be related to regression, exposure, and desiccation (Halley, 1977). Major regression following oolitic sedimentation is an essential feature of the genetic hypothesis supported herein, and hence minor regressions sufficient to result in desiccation and fragmentation would be expected.

The claim of specialized fauna in some SCOS-IF of England relative to contemporary limestone (Arkell, 1936) has been discredited by subsequent investigation (Brookfield, 1973), and the 20% smaller size of bryozoa in Silurian Clinton ironstone than in Clinton limestone (Alling, 1947) may be related to better sorting of shallow-water oolitic sediment which became ironstone. Phanerozoic SCOS-IF is generally highly fossiliferous and primary sedimentation must have occurred in water bodies with sub-toxic metal contents (Cayeux, 1922).

Intraclasts, ripped-up clasts of oolite, commonly occur in oolitic ironstone. Cayeux (1922, p.926) reported a different paragenetic history for the intraclasts and the rest of Lorraine ironstone, whereas Bichelonne and Angot (1939, p.463), studying the same SCOS-IF, questioned this interpretation. In general, mineralogical differences between ooids and interstitial areas, or among different ooids, do not prove that ferruginization took place in separate localities prior to mixing. Chemically distinct micro-environments, caused by slight differences in permeability or organic-matter content, could have profound effects on ferriferous mineral development.

The common interlayering within SCOS-IF ooids of ferrous and ferric minerals, e.g., chamosite and hematite, has also been cited in opposition to a replacement origin. In reference to the Ordovician Wabana SCOS-IF, Gruner (1922, p.416) wrote: "It would be difficult to believe that one concentric layer could have remained in one state of oxidation while an adjacent one changed to some other state, unless there was a conspicuous original difference." However, the principal petrographer of Wabana SCOS-IF, A.O. Hayes (1929, p.690), was later unconvinced and suggested that, ". . . the possibility of oxidation of the chamosite as a source for the large amount of hematite needs study." The greater regularity of chamositic than hematitic or goethitic oolitic layers has been cited herein as possible evidence for such a partial oxidation of Paz de Rio ooids, and there is general agreement that chamosite was the first-formed ferriferous mineral in the widespread Jurassic SCOS-IF of Britain (e.g., Davies and Dixie, 1951; Brookfield, 1971). Tegengren (1921, p.9) found a similar relationship in Carboniferous SCOS-IF of China and Harms (1965) in the Middle Proterozoic (Precambrian Y) Constance Range SCOS-IF of Australia.

The agreement by petrographers does not mitigate against Gruner's (1922) logic. To propose differential oxidation of microns-thick oolitic layers, one must demonstrate differences between them. Shearman et al. (1970) have done this by showing that Recent aragonitic ooids contain variable amounts of organic matter in their layers and that this organic matter may survive alteration of aragonite. Organic-poor chamositic layers would have been the first oxidized. It is unlikely, however, that all ferriferous minerals in SCOS-IF are alteration products of chamosite. Other ferriferous minerals have probably replaced aragonite or filled pores contemporaneous with or subsequent to the precipitation of chamosite.

The existence of large-scale mineralogical and facies variations has also been considered to contradict replacement (Pettijohn, 1975, p.425). Consistent lateral facies variations are rare in individual SCOS iron formations, the best exception between the Ordovician Sarka SCOS-IF of Czechoslovakia (Petranek, 1964b). From paleoshore to basin, the Sarka SCOS-IF is gradational through more than 11 km from oolitic hematite ironstone, through siderite-clay ironstone with inclusions of oolitic hematite ironstone, to pure siderite-clay ironstone. Oolitic chamosite ironstone is a subordinate lithology of greater abundance basinward. The iron formation has a nearshore maximum of 20 m in a single ironstone bed, but this splits at about 3 km basinward into two ironstone beds which grade to a total of less than 5 m in thickness. As in other SCOS-IF, subordinate chamosite occurs in nearly all hematite ironstone and vice versa (Skocek, 1963a, b). Although the proportions of ferrous and ferric minerals vary widely, the common assemblage of hematite-chamosite-siderite would, at equilibrium, define the same chemical environment. However, paragenetic relationships are apparent (Skocek, 1963a, b; Petranek, 1964b), as in other SCOS-IF, and chemical equilibrium beyond the scale of a few microns is unlikely. Hematite, goethite, chamosite, siderite, pyrite, and carbonaceous matter commonly coexist in individual thin sections of Paz de Rio SCOS-IF (Kimberley, 1974).

Within the proposed genetic model, large-scale mineralogical variation would be related to control of paleoslope on biogenic weathering-leaching reactions and on water-table fluctuation; e.g., landward portions of the ferruginized calcareous beds would be most susceptible to subsequent oxidation above the water table, as presently occurring in the Kerch SCOS-IF (Sokolova, 1964). Variation in ironstone grain size would be partially related to water depth during aragonite sedimentation.

Vertical mineralogical variation in laterally homogeneous SCOS-IF may be correlated to paleoslope by deduction of transgressive or regressive sedimentation. The Eocene Paz de Rio SCOS-IF is interpreted to have formed by transgression of an oolitic sand bar over a well-burrowed lagoon with abundant logs and tree seeds (Kimberley, in press). In turn, basinward carbonaceous silicate mud transgressed over the oolite. Aragonite mud, scattered ooids, and carbonaceous matter which accumulated shoreward and basinward of the oolitic bar are now represented by the stratigraphically upper and lower, greenish gray to brownish black siderite–quartz ironstone which is marginal to the central oolitic hematite–goethite ironstone. This mineralogical distribution is attributed to a higher content of organic matter and lesser permeability of marginal sediment during diagenetic replacement.

The role of siderite in SCOS-IF deserves special attention because two early authors considered it to be normally the first-formed ferriferous mineral (Sorby, 1857; Cayeux, 1909, 1922), whereas the majority of other petrographers have reported that it typically replaces chamosite (e.g., Hayes, 1915; Hallimond, 1925; Wagner, 1928; Taylor, 1949; Kimberley, 1974). In the Sarka and Peace River iron formations, siderite is reported to be largely a diagenetic cement (Skocek, 1963b, p.102; Mellon, 1962). Siderite is generally ubiquitous in SCOS-IF, with the exception of Clinton iron formations (Schoen,

1962), and must indicate some commonly attained diagenetic state, presumably abundant aqueous CO_2 (Stumm and Morgan, 1970, p.343). The common correlation between siderite and allogenic clay content corroborates this because CO_2-generating organic matter preferentially accumulates with silt and clay (Pettijohn, 1975, p.443).

The only other serious objection previously offered to a replacement hypothesis involves local occurrences of iron-poor limestone closely overlying ironstone, presumed to be incompatible with ferruginization by descending groundwater (Villain, 1902; Newland, 1910). In the genetic model proposed here, the leached sediment is largely or entirely eroded, and succeeding calcareous sediment deposited directly on ironstone may remain calcareous unless covered by argillaceous sediment subject to leaching. Newland (1910) also argues that the shale above ironstone would have been impermeable to groundwater. However, the pre-compaction permeability of surficial mud is substantial (Lambe, 1955) and may be augmented by bioturbation.

OTHER GENETIC MODELS FOR OOLITIC-INLAND-SEA IRON FORMATIONS

Constraint on sedimentation rate

Extremely diverse genetic hypotheses have been offered to explain sandy, clayey, and oolitic shallow-inland-sea iron formation (SCOS-IF) since the pioneering work of Sorby (1857). Sorby's hypothesis is incompatible with all of these and must be rejected if any subsequent hypothesis is more tenable. The various modes of iron supply (Table IV) and concentration (Table V) may be evaluated most simply against constraints imposed by the size and duration of sedimentation of the Kerch iron formation (Sokolova, 1964). This SCOS-IF is restricted within the Pliocene Epoch to the Middle Cimmerian Stage which lasted about $9 \cdot 10^5$ yr (Berggren and Van Couvering, 1974). Actual duration of oolitic sedimentation was probably much less than this, because an overlying bed of Middle Cimmerian mudstone is almost as thick as the iron formation (Putzer, 1943) and because shallow-water sedimentation is characteristically intermittent (Newell, 1972).

Hypothesis of marine iron supply

The first hypothesis to be considered is that of Borchert (1960) who suggested an offshore marine source for SCOS-IF iron. A marine source seems unlikely because the minimum masses of Kerch iron and aluminum estimated herein are respectively 8.5 and 5 times total marine masses, based on Turekian's (1968) data. However, the Black Sea–Sea of Azov area probably represents the largest modern analogue of Borchert's (1960, 1965) envisaged iron-concentrating paleoenvironment and his hypothesis therefore deserves special attention. Borchert (1960, p.272) proposed that iron could be copiously dissolved in a CO_2-rich marine zone between an overlying O_2-rich and underlying H_2S-

rich zone. The dissolved iron content of the Black Sea water indeed displays a five-fold increase through a few tens of meters between O_2-rich and H_2S-rich zones, but the maximum iron content does not exceed 50 ppb (Brewer and Spencer, 1974). Water which becomes anoxygenic enters the Black Sea through the Bosporous at a rate of about 190 km^3/yr (Brewer and Spencer, 1974). If all entering water were to reach the maximum observed dissolved iron content, upwell onto the Sea of Azov shelf, and precipitate all the iron, it would take over $4 \cdot 10^6$ years to produce the Kerch iron formation, more than four times the duration of the Middle Cimmerian Stage and nearly the age of the deposit. (See figs. 91 and 92 in Wolf's Chapter 1, Vol.8 in this Handbook.)

Hypotheses of fluvial iron supply

A more potent genetic process seems to be required. The most-cited process of iron supply, i.e., fluvial solution and/or colloidal suspension (Table IV), may readily be studied in modern settings. Coonley et al. (1971) showed that precipitation of all dissolved and finely colloidal iron into the detrital sediment of New Jersey's Mullica River would result in ironstone. However, they found that iron flocculated before reaching the open ocean and that the floccules were swept out to the Atlantic shelf, so that none of the Mullica-derived sediments are anomalously ferriferous. The author conducted a coring program around the mouth of a more protected major tropical river, the Atrato of Colombia, which discharges into the Gulf of Urabá. The only evidence for chemical concentration of iron found there is indirect. Dark green clay in the brackish gulf contains locally abundant, minute gypsum crystals with a δS^{34} of –6‰ (Kreuger Enterprises Laboratory). This isotopic ratio is suggestive of an origin by oxidation of diagenetic ferrous sulfide (Thode et al., 1961; Wall and Reed, 1976). Possible iron concentration seaward of a "clastic trap" (Table V) is improbable in the Gulf of Urabá, because the Atrato River continuously supplies terrigenous-sediment-rich vegetation rafts which disperse and break up throughout the gulf.

Authigenic, ferrous iron-rich, 7 Å-basal-spacing clay has been discovered in other tropical deltaic sediment by Porrenga (1965, 1967a, b). Porrenga's choice of the name "chamosite" for this poorly organized clay is somewhat misleading because it typically has a greater magnesium content than chamosite, the abundant iron serpentine in SCOS-IF. The disordered ferriferous clay described by Porrenga has not precipitated directly from seawater, as presumed for SCOS-IF chamosite by most authors, but has formed diagenetically within tropical and non-tropical sediment rich in decaying fecal matter (Porrenga, 1967a; Giresse, 1969; Rohrlich et al., 1969).

Despite the lack of modern marine concentration of fluvial iron (cf. Förstner's Chapter 5, this volume), fluvial supply deserves quantitative testing for the Kerch area. Putzer (1943) postulated fluvial iron supply from exposed portions of the Kerch peninsula. However, the following calculation reveals that even the Volga River, the greatest river in Europe, could barely supply sufficient iron to form the Kerch SCOS-IF. Annual discharge of the Volga is $2.5 \cdot 10^{14}$ l/yr (Encyclopaedia Britannica, 1974). The dissolved iron content is unknown, but the average for analyzed European rivers is 0.8 ppm (Livingstone, 1963). This is twice the world average (Gibbs, 1972) and the lesser polluted Volga may well be less ferriferous than the other European rivers. These values would include substantial particulate ferric hydroxide, even if water samples were filtered (Stumm and Morgan, 1970, p.548; Shapiro, 1964), and so would include much of the colloidal iron invoked by several authors (Table IV). Accepting the average European value as a maximum, the Volga could annually contribute $2 \cdot 10^8$ kg of dissolved and finely colloidal iron in all chemical species. At this rate, over $2 \cdot 10^5$ yr would be required to form the Kerch iron formation, given concentration of all precipitated iron. However, ferriferous floccules are presently removed by moderate currents which are probably weaker than those which formed the ubiquitous oolitic texture in Kerch ironstone (Coonley et al., 1971). Any reasonable assumption of partial separation would lead to prohibitively long calculated durations of chemical sedimentation. Postulation of fluvial supply by a more tropical river would not alter the conclusion; the Amazon averages only 0.03 ppm Fe (Gibbs, 1972), two orders of magnitude less than that assumed by Gruner (1922, p.455) in a similar calculation for the Biwabik iron formation.

Fluvial aluminum supply may be modelled by postulating a Volga-sized discharge with the median dissolved aluminum content of major North American rivers (Durum and Haffty, 1963). Over 10^5 yr would be required to form observed chamosite in the Kerch SCOS-IF. Given the geochemical dissimilarities between aluminum and iron, it is unlikely that perfect concentration of both would occur together. Moreover, the paucity of terrigenous sediment in much of the Kerch iron formation (Sokolova, 1964) is inconsistent with the widespread dispersal of terrigenous sediment by vegetation rafts discharged from modern rivers. Given the local abundance of fossil wood (barite-replaced) in the Kerch iron formation (Sokolova, 1964), nearby coasts must have been well forested. The hypothesis of direct fluvial supply of SCOS-IF iron is therefore rejected.

Hypotheses of lateral and upward groundwater supply of iron

Another possible mode of iron supply is seaward groundwater flow from the periphery of the area of iron concentration (Table IV). This may be tested for the Kerch deposit by assuming a high iron content for groundwater and by employing Darcy's Law to calculate the minimum time required to supply the observed iron. Strakhov's (1969) estimate of the total iron-formation perimeter, 710 km, is used to determine the minimum required time. In the extreme, ferriferous groundwater could have flowed basinward through the entire perimeter. Oolite would have been the most permeable sediment in the iron formation-bearing suite and it may be postulated that half of the ferriferous groundwater reached the basin center by flowing through the oolite and the other half flowed through all other sediment. The iron formation is underlain by mudstone and argillaceous limestone which would have had low permeabilities.

The permeability of oolite assumed in a previous section, based on measurements by Robinson (1967), is 10 darcys (cf. also Chilingar and Wolf, 1975). Given consistent textural indications in the Kerch SCOS-IF of very shallow-water sedimentation, total relief across the basin was probably less than 30 m. A much larger area of Pleistocene calcareous oolite sedimentation, the Bahaman Andros platform, has less than 12 m total relief (Newell and Rigby, 1957; Bathurst, 1975) except near a few faults. The maximum distance from the closest basinal perimeter is 17 km on Strakhov's (1969) map and 47 km on Markevich's (1960). Assuming the smaller distance, a 30 m head loss, and a permeability of 10 darcys, average groundwater flow rate may be approximated with Darcy's Law to be 5.4 m/yr. Dissolved iron content could be as high as that reached within anoxygenic lacustrine sediment, about 20 ppm (Emerson, 1976). Assuming 20 ppm Fe, a 710 km perimeter, a 7.4 m thickness of oolite, and a 5.4 m/yr flow rate, more than $3.5 \cdot 10^7$ yr would be required to transfer inward the observed mass of iron. The hypothesis of peripherally-supplied groundwater, therefore, also must be rejected.

Miscellaneous hypotheses

The other hypotheses for iron concentration at the sediment–water interface (Tables IV and V) are similarly dismissable. As previously shown, several world oceans would have to evaporate to produce the Kerch SCOS-IF. The hypothesis of marine concentration of ferric oxides and hydroxides adsorbed on fluvial clays by chemical reduction in anoxygenic areas is subject to the same limitations as that of Borchert (1960). Iron supply in vertically ascending pore fluids from compacting sediment may be modelled for the limiting case of 20 ppm Fe, an extreme value for sub-soil groundwater (Hem, 1960). The minimum thickness of the ascending column of such water required to form the Kerch SCOS-IF would be 400 km, an unreasonable amount. The suggestion that the oolitic texture of SCOS-IF ironstone may form diagenetically as in spherulitic bauxite (Schellmann, 1969) is inconsistent with the observed sorting of SCOS-IF ooids and the lack of any evidence of oolitic growth within and truncating oolitic layers.

Proposed modern lacustrine ferriferous oolite

Lacustrine concentrations of iron in Lake Malawi and Lake Chad of Africa have been presented as Recent analogues of oolitic iron formations. Müller and Förstner [1] (1973) propose that hydrothermal solutions rise through as much as several kilometers of Lake Malawi sediment, leach iron, and deposit the locally observed few centimeters to tens of centimeters of ferriferous sediment at the top of the sedimentary pile. The dominant ferriferous mineral is nontronite which is exceedingly rare in iron formations, except for the Cretaceous Peace River SCOS-IF (Petruk, 1977; Petruk et al., 1977). Ferriferous spherules with thin concentric superficial layers, found at only one Lake Malawi locality, were termed "oolites". However, the illustrated internal structure of

[1] See also Förstner's Chapter 4, this volume.

these spherules is not as regular as that of typical ooids in SCOS-IF. Moreover, the hypothesis of extensive subaqueous dispersion of hydrothermal solutions has been discredited by Strakhov and Nesterova (1968) who showed that iron and vanadium are not dispersed from volcanic centers in the Kurile Arc. There is no reported evidence of hydrothermal leaching beneath any SCOS-IF.

Lemoalle and Dupont (1973, p.177) suggest that their study of ferriferous spherules in Lake Chad ". . . should be helpful when assessing the chemical conditions of the formation of iron and silica oolite beds." Taking their data for the total volume of "oolitic" ferriferous sediment and mass of contained iron, one may calculate an average iron content of 0.405 g cm^{-3}. Assuming 30% porosity and solid-sediment density of 2.7 g cm^{-3}, the bulk sediment would average 21% Fe. The only substantially ferriferous mineral discovered was goethite which contains up to 63% Fe. Therefore, the montmorillonite reported to be the other principal mineral in the "oolitic" sediment must be rather abundant. Since this sediment contains little matrix, the montmorillonite must occur principally as "oolite" nuclei. The author knows of no iron formation in which even a small proportion of the ooids have nuclei of montmorillonite and concludes that the Lake Chad ferriferous deposit is distinct from SCOS-IF.

RELEVANT QUATERNARY CARBONATE FERRUGINIZATION

The possibility of early diagenetic ferruginization has been confirmed by the recent discovery of partially ferruginized Pleistocene oolite on northern Andros Island, Bahamas (Kimberley, 1975). Most rock exposures of the Bahaman and Cay Sal platforms, totalling about 155,000 km^2, are Pleistocene oolitic limestone which was originally largely composed of aragonite (Newell and Rigby, 1957). On parts of northern Andros, this was covered by probable volcanic ash which accumulated in depressions and weathered to a soil. Where sufficiently thick, this soil is stratified into upper red and lower dark brown horizons. Chemical reduction within the lower soil horizon has mobilized iron, which has been carried downward to react with directly underlying, calcareous oolite. The resulting red oolite is well-indurated and has taken up about 15% SiO_2, 10% Al_2O_3, and 1–2% Fe (X-ray fluorescence analysis). Most of the alumina may occur in eluviated clays, but at least 1/3 of the iron was chemically concentrated in a reducing environment. By determining the variation of magnetic properties with temperature, G.W. Pearce (personal communication, 1975) has shown 2/3 of the iron in the oolite to be within a paramagnetic silicate and/or carbonate and 1/3 within "protomagnetite". The original oolite would have been quite pure; Recent Bahaman oolite contains only 95 to 500 ppm Fe (Milliman, 1974, p.46).

Ferruginization of Pleistocene oolite has been found at several localities in northern Andros, at the northeastern corner, called Morgan's Bluff, on the northwestern coast at Red Bay, 20 km away, and about halfway in between, along the Red Bay road. At the

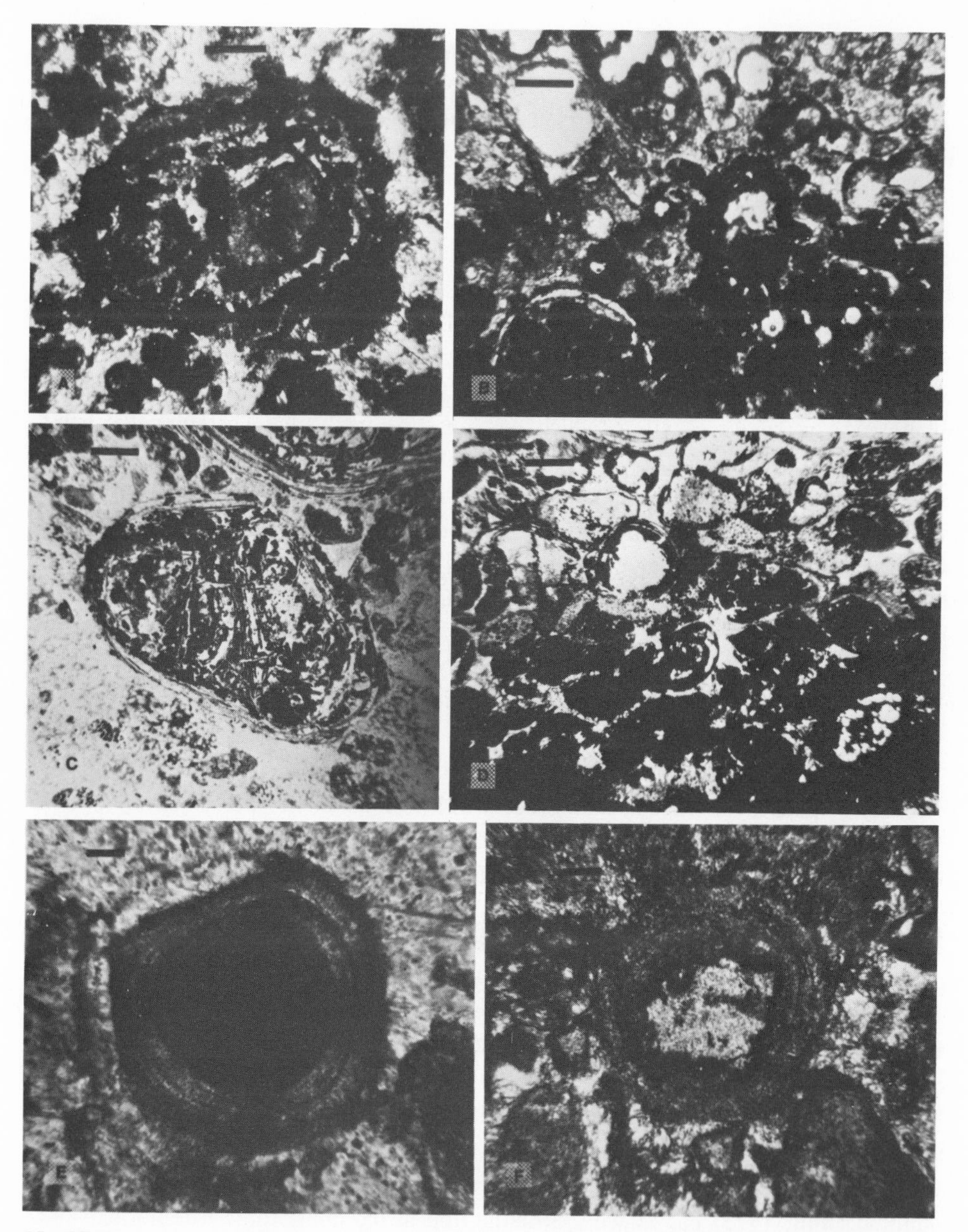

PLATE III

Photomicrographs of Quaternary and Precambrian oolite. (After Kimberley, 1979a.)

A. Quaternary compound ooid composed of calcite plus amorphous silica (white) and "protomagnetite" plus amorphous red material (dark). Doubly polarized light. Scale bar is 0.1 mm. Red Bay, Andros Island, Bahamas.

B. Quaternary banded oolite. Upper (white) band is calcite plus finely intergrown amorphous silica. Scale bar occurs within the largest of several holes in the section. Lower (dark) band is "protomag-

last locality, the oolite is markedly banded in red and white, resembling banded cherty ironstone. White bands contain roughly equal amounts of calcite and aragonite. The only major diagenetic addition to the white rock is silica, about 15%. Red bands contain no X-ray detectable aragonite; they are dominantly calcite with 15% amorphous silica. The microscopic texture of the rock is similar to that of banded ironstone in which oolitic texture is partially preserved in both iron-rich and iron-poor bands (Plate III, B and D). Individual ferruginized ooids are rather similar to those of any oolitic ironstone (Plate III, A and C). The banding and free silica content of this ferriferous oolite are not characteristic, however, of inland-sea ironstone but of ironstone which has formed on a platform like the Andros platform (Kimberley, 1978).

Weakly ferruginized, ten- to thirty-thousand-year-old aragonitic ooids have been found on the Atlantic continental shelf off the southeastern United States (Terlecky, 1967). Nine samples of these brown ooids averaged 0.62% Fe and ranged up to 1.4% Fe. Similar brown aragonitic ooids with 0.5% Fe have been reported by Davies (1970, p.134) from Shark Bay, Western Australia. Davies (1970, p.166) interprets the partial ferruginization of aragonite to have occurred during reducing diagenesis. Other Recent ferruginization includes the pseudomorphic replacement by poorly crystalline chamosite of foraminiferal tests off the coast of Gabon (Giresse, 1969).

In southern Bali, Indonesia, flat-lying Miocene reefal limestone interbedded with volcaniclastics is superficially iron stained by leachate from the weathering volcaniclastics (W.H. Nelson, personal communication, 1975). This limestone is calcitic, however, and no replacement of calcite can be detected in stained samples supplied to the author by D. Kadar (1973). Alteration of aragonite to calcite before burial could account for the restriction of the only known calcite to the only known intraclast in the Paz de Rio SCOS-IF. Prolonged exposure of aragonitic sediment to fresh water results in cementation and alteration to chemically-stable, magnesium-poor calcite (Bathurst, 1975).

---

netite" plus amorphous red material, calcite, and amorphous silica. Ooids are preferentially ferruginized around the fossil shell. Plane polarized light. Scale bar is 0.3 mm. North-central Andros Island, Bahamas.

C. Lower Proterozoic (Precambrian X) compound ooid composed of recrystallized chert (white) and hematite (dark) surrounded by fine-grained recrystallized chert (light stippled) and siderite (dark stippled). Plane polarized light. Scale bar is 0.1 mm. Sokoman iron formation, Quebec.

D. Lower Proterozoic banded oolite. Upper band is siderite (light stippled) and recrystallized chert (white). Lower band is partially oxidized greenalite (dark) and siderite. Plane polarized light. Scale bar is 0.3 mm. Gunflint iron formation, Ontario.

E. Calcitic ooid displaying extinction cross in doubly polarized light. Scale bar is 0.02 mm. Upper limestone member of Gunflint Formation at Hillcrest Park, Ontario.

F. Ooid within partially ferruginized Archean (Precambrian W) oolitic limestone. Volcanic ash nucleus is rimmed by hematite (dark) and enveloped by calcite finely interlayered with an unidentified green iron silicate (dark). Microprobe wavelength profiling reveals both substantial iron and aluminum in the silicate. Plane polarized light. Scale bar is 0.2 mm. Base of Outerring iron formation, Eastring Lake, Northwest Territories, Canada.

POSSIBLE IMPLICATIONS OF THE GENETIC MODEL

All known lines of evidence are consistent with an origin of oolitic inland-sea (SCOS) iron formations by early diagenetic replacement of calcareous sediment. There are several implications that this may have for other iron-formation types. Iron formations formed on shallow volcanic platforms (SVOP-IF) and in metazoan-poor, extensive, chemical-sediment-rich, shallow seas (MECS-IF) both locally display oolitic texture but are otherwise quite different from SCOS iron formations and more closely resemble each other (James, 1966). An hypothesis related to that proposed for SCOS-IF which may account for the thick cherty iron formations is that large flat platforms (like the Bahaman Andros platform) were normally emergent and received pyroclastics which weathered and were leached of iron and silicon, and that occasional rapid marine transgressions produced beds of aragonite which became ferruginized and silicified by the leachate (Kimberley, 1975). Weathered pyroclastics would have been eroded preferentially over silicified aragonite by extensive sheet-wash, like that presently flowing over much of the Upper Pleistocene Miami Oolite in southern Florida (White, 1970).

Unlike the genetic process proposed here for SCOS-IF, the foregoing mechanism has been found to be presently producing ferriferous rock, the Quaternary ferriferous rock described from northern Andros Island (Plate III, A and D). Beds of comparable ferriferous rock with up to 5% Fe occur within a Pliocene–Pleistocene dolomitic sequence on San Salvador, Bahamas, 400 km away (Supko, 1977). This dolomitic sequence resembles the Lower Proterozoic (Precambrian X) Malmani Dolomite of South Africa, with its beds of "iron-formation precursor" (Button, 1976a, b). The total area of very shallow and slightly exposed aragonitic sediment in the Bahaman-Cay Sal region is about 1.5 times Trendall and Blockley's (1970) estimate of original extent of the voluminous Hamersley MECS-IF (Newell and Rigby, 1957). Although substantial surficial ferruginization has only yet been found on northern Andros, all of this region may be underlain by ferriferous beds as on San Salvador.

Several characteristics of MECS-IF and SVOP-IF are consistent with the eluviation-replacement hypothesis. An evaporitic paleoenvironment suitable for aragonite sedimentation is recorded in the Lower Proterozoic Sokoman MECS-IF of Quebec by gypsum pseudomorphs and length-slow chalcedony (Chauvel and Dimroth, 1974; Dimroth, 1976). Calcareous ooids have been discovered in the slightly ferriferous Upper Limestone Member above a MECS iron formation in the Lower Proterozoic Gunflint Formation of Ontario (Plate III, E), and oolite at the base of the Archean (Precambrian W) Outerring SVOP-IF in the Northwest Territories of Canada is slightly ferriferous limestone (Plate III, F) with irregular veins and patches of ironstone. Although locally abundant, oolitic texture does not predominate in any cherty iron formation and most of the primary sediment is interpreted to have been aragonite mud. Fossil blue-green algae beneath (Cloud and Licari, 1972) and within Lower Proterozoic MECS iron formations (Barghoorn and Tyler, 1965) most likely indicate an oxygenic paleoenvironment in which any ferrous mineral would have been unstable.

Ferruginization by fresh groundwater under tropical weathering conditions is supported by Knauth's (1973, p.202) study of $\delta^{18}O$ and δD in Gunflint MECS-IF chert and is consistent with similarly low values of $\delta^{18}O$ found in other MECS iron formations (Perry and Tan, 1972; Becker and Clayton, 1976). Pyroclastics have been reported from MECS iron formations in South Africa, Western Australia, and elsewhere (La Berge, 1966a, b). Some MECS iron formations enclose extensive beds of tuffaceous mudrock (Moorhouse, 1960) and ash forms ooid cores in at least one SVOP-IF (Plate III, F). However, most cherty ironstone contains little or no tuff. This is here attributed to a high sedimentation rate of shallow-water aragonite during relatively brief periods of submergence. During emergence, pyroclastics would readily weather and the resultant clay minerals would be eroded off the carbonate platform.

Tropical weathering of pyroclastics on carbonate platforms normally releases silicon and concentrates aluminum and phosphorus, as exemplified by basaltic-ash soils which overlie Pleistocene limestone on Niue Island, a volcanic island in the South Pacific (Schofield, 1959; cf. also Chapter 3, Vol.3, in this Handbook). An average partial analysis of this soil is 0.7% SiO_2, 36.6% Al_2O_3, and 2.2% P_2O_5 (Birrell et al., 1939). Differences in weathering processes or pyroclastic supply between hypothesized successive calcareous sedimentation-replacement cycles would result in stratigraphic compositional differences in the ironstone or chert produced. Banding may be attributed to rhythmic concentration of carbonaceous matter in the primary chemical sediment, due to biological cycles and/or storm-layering, much like the similarly extensive bituminous laminae in Permian anhydrite of Texas and New Mexico (Udden, 1924; Trendall, 1973b). Distribution of minor carbonaceous matter could have controlled distribution of ferriferous minerals relative to chert during early diagenetic replacement. Non-ferriferous chert is also commonly banded (Sargent, 1929), including the shallow-water Devonian novaculite of Texas (Folk, 1973, p.717). The generally antithetic relationship between banding and abundance of metazoan fossils in ironstone may be related to bioturbation (Tyler and Twenhofel, 1952, p. 138).

Rare-earth element distributions recently determined for the Sokoman MECS-IF indicate an areally uniform chemical environment during precipitation of the ferrous silicate-siderite facies (Fryer, 1977b). These rocks display oolitic and intraclastic texture, indicative of extremely shallow-water sedimentation, and so interpretation of the observed cerium anomaly as evidence for contemporaneous oxidation of seawater, presumably by an oxygenic atmosphere, is inconsistent with the proposal of primary sedimentation of ferrous minerals (Fryer, 1977b). Uniform rare-earth-element distributions would be produced by areally uniform weathering of compositionally uniform volcanic ash, as on the virtually flat Andros platform. Rare-earth-element distributions differ substantially among cherty iron formations (Fryer, 1977a; Graf, 1977; Shimizu and Masuda, 1977) and these differences are more readily attributed to different weathering regimes or different pyroclastic compositions than to major variations in oceanic composition.

Five elements chemically concentrated in SCOS-IF (Si, Al, Mn, P, and V) are also individually concentrated in certain beds of iron-poor oolite which may be genetically related to SCOS-IF, e.g., Upper Cambrian oolitic chert in Pennsylvania (Choquette, 1955), kaolinitic oolite which directly overlies chamositic oolite of the Northampton SCOS-IF in Britain (Taylor, 1949), manganiferous oolite with minor associated oolitic ironstone in the voluminous Nikopol and Chiatura manganese deposits of the U.S.S.R. (Varentsov, 1964; Sokolova, 1964; Strakhov, 1969), oolitic phosphorite of the Permian Phosphoria Formation in and around Idaho (Mansfield, 1927; Sheldon, 1963), and Neogene vanadiferous oolitic limestone in the U.S.S.R. (Xolodov, 1973). Mineralization of all of these types of oolite may have involved early diagenetic processes. If so, Sorby's (1957) pioneering concept of diagenetic mineralization may prove to be as fundamental as his other pioneering concepts (Summerson, 1976).

REFERENCES

Adeleye, D.R., 1973. Origin of ironstones, an example from the middle Niger Valley, Nigeria. *J. Sediment. Petrol.*, 43: 709–727.

Adeleye, D.R., 1975. Derivation of fragmentary oolites and pistolites from desiccation cracks. *J. Sediment. Petrol.*, 45: 794–798.

Aldinger, H., 1957a. Zusammenfassung der Ergebnisse der wissenschaftlichen Sitzungen. *Z. Dtsch. Geol. Ges.*, 109: 2–6.

Aldinger, H., 1957b. Zur Entstehung der Eisenoolithe im schwäbischen Jura. *Z. Dtsch. Geol. Ges.*, 109: 7–9.

Alling, H.L., 1947. Diagenesis of the Clinton hematite ores of New York. *Bull. Geol. Soc. Am.*, 58: 991–1018.

Al-Shanti, A.M.S., 1966. Oolitic iron ore deposits in Wadi Fatima between Jeddah and Mecca, Saudi Arabia. *Saudi Arabia, Minist. Pet. Miner. Resour., Bull.*, 2: 51 pp.

Alvarado, B. and Sarmiento-Soto, R., 1944. Yacimientos de hierro de Paz de Rio (Boyacá). *Colomb. Serv. Geol. Nac., Inf.*, 468: 139 pp.

Arkell, W.J., 1936. The Corallian rocks of Dorset, pt. 1, The coast. *Proc. Dorset Nat. Hist. Archaeol. Soc.*, 57: 59–93.

Ball, M.M., 1967. Carbonate sand bodies of Florida and the Bahamas. *J. Sediment. Petrol.*, 37: 556–591.

Bardossy, G. and Mack, E., 1967. Zur Kenntnis der Bauxite des Parnass-Kiona-Gebirges. *Miner. Deposita*, 2: 334–348.

Barghoorn, E.S. and Tyler, S.A., 1965. Microorganisms from the Gunflint chert. *Science*, 147: 563–577.

Bathurst, R.G.C., 1975. *Carbonate Sediments and their Diagenesis.* Elsevier, Amsterdam, 2nd ed., 658 pp.

Bearce, D.N., 1973. Origin of conglomerates in Silurian Red Mountain Formation of central Alabama; their paleogeographic and tectonic significance. *Bull. Am. Assoc. Pet. Geol.*, 57: 688–701.

Becker, R.H. and Clayton, R.N., 1976. Oxygen isotope study of a Precambrian banded iron-formation, Hamersley Range, Western Australia. *Geochim. Cosmochim. Acta*, 40: 1153–1165.

Berg, G., 1942. Die Ausfällung des Eisens. In: B. Brockamp (Editor), *Zur Entstehung deutscher Eisenerzlagerstätten. Arch. Lagerstättenforsch.*, 75: 178–180.

Berg, G., 1944. Vergleichende Petrographie oolithischer Eisenerze. *Arch. Lagerstättenforsch.*, 76: 128 pp.

Berg, G. and Hoffman, K., 1942. Zur Paläogeographie und Entstehung der Eisenerze in den Lias-Schichten. In: B. Brockamp (Editor), *Zur Entstehung deutscher Eisenerzlagerstätten. Arch. Lagerstättenforsch.*, 75: 61–69.

Berg, G., Dahlgrun, F. and Martini, H.J., 1942a. Die Erze des böhmischen Untersilurs. In: B. Brockamp (Editor), *Zur Entstehung deutscher Eisenerzlagerstätten. Arch. Lagerstättenforsch.*, 75: 150–155.

Berg, G., Seitz, O. and Teichmuller, R., 1942b. Die Eisenerze im Korallenoolith von Braunschweig. In: B. Brockamp (Editor), *Zur Entstehung deutscher Eisenerzlagerstätten. Arch. Lagerstättenforsch.*, 75: 71–78.

Berggren, W.A. and Van Couvering, J.A., 1974. The Late Neogene: biostratigraphy, geochronology and paleoclimatology of the last 15 million years in marine and continental sequences. *Paleogeogr., Paleoclimatol., Paleoecol.*, 16: 1–216.

Berner, R.A., 1971. *Principles of Chemical Sedimentology.* McGraw-Hill, New York, N.Y., 240 pp.

Berner, R.A., 1975. The role of magnesium in the crystal growth of calcite and aragonite from sea water. *Geochim. Cosmochim. Acta,* 39: 489–504.

Berry, D.W. and Walthall, B.H., 1961. Paipa–Belencito–Paz de Rio, Department of Boyacá. *Colomb. Soc. Pet. Geol. Geophys., 3rd Field Conf.*, 40 pp.

Bertram, E.F. and Mellon, G.B., 1973. Peace river iron deposits. *Alberta Res. Inf. Ser.*, 75: 53 pp.

Bichelonne, J. and Angot, P., 1939. *Le Bassin ferrifère de Lorraine: Nancy–Strasbourg.* Berger-Levrault, 483 pp.

Birrell, K.S., Seelye, F.T. and Grange, L.I., 1939. Chromium in soils of Western Samoa and Niue Island. *N.Z.J. Sci. Technol., Sect. A,* 21: 91–95.

Borchert, H., 1960. Genesis of marine sedimentary iron ores. *Inst. Min. Metall. Trans.*, 69: 261–279.

Borchert, H., 1965. Formation of marine sedimentary iron ores. In: J.P. Riley and G. Kirow (Editors), *Chemical Oceanography, 2.* Academic Press, New York, N.Y., pp. 159–204.

Boscovitz-Rohrlich, V., Mitzmager, A. and Mizrahi, Y., 1963. Structure and benefication of a low grade iron ore. *Min. Mag. (London),* 108: 325–331.

Bottke, H., 1965. Die exhalativ-sedimentären devonischen Roteisensteinlagerstätten des Ostsauerlandes. *Geol. Jahrb. Beih.,* 3: 147 pp.

Bradley, W.H., 1929. Algae reefs and oolites of the Green River Formation. *U.S. Geol. Surv. Prof. Pap.*, 154-G: 201–223.

Braun, H., 1964. Zur Entstehung der marin-sedimentären Eisenerze. *Clausthaler Hefte Lagerstättenkd. Geochem. Miner. Rohst.*, 2: 133 pp.

Brewer, P.G. and Spencer, D.W., 1974. Distribution of some trace elements in Black Sea and their flux between dissolved and particulate phases. In: E.T. Ross and D.A. Ross (Editors), *The Black Sea – Geology Chemistry, and Biology. Mem. Am. Assoc. Pet. Geol.*, 20: 137–143.

Brockamp, B., 1942. Die paläogeographische Stellung der Eisenablagerungen. In: B. Brockamp (Editor), *Zur Entstehung deutscher Eisenerzlagerstätten. Arch. Lagerstättenforsch.*, 75: 181–186.

Brookfield, M., 1971. An alternative to the "clastic trap" interpretation of oolitic ironstone facies. *Geol. Mag.*, 108: 137–143.

Brookfield, M., 1973. The paleoenvironment of the Abbotsbury Ironstone (Upper Jurassic) of Dorset. *Paleontology,* 16: 261–274.

Brown, T.C., 1914. Origin of oolites and the oolitic texture in rock. *Bull. Geol. Soc. Am.*, 25: 745–780.

Bubenicek, L., 1971. Géologie du gisement de fer de Lorraine. *Bull. Cent. Rech. Pau, Soc. Natl. Pét. Aquitaine,* 5: 223–320.

Burchard, E.F., 1910. The Clinton iron-ore deposits in Alabama. *Trans. Am. Inst. Min. Eng.*, 40: 75–133.

Burchard, E.F. and Butts, C., 1910. Iron ores, fuels, and fluxes of the Birmingham district, Alabama. *U.S. Geol. Surv. Bull.*, 400: 204 pp.

Button, A., 1975. The Gondwanaland Precambrian project. *Univ. Witwatersrand, Econ. Geol. Res. Unit, Annu. Rep.*, 16: 37–54.

Button, A., 1976a. Iron-formation as an end member in carbonate sedimentary cycles in the Transvaal Supergroup, South Africa. *Econ. Geol.,* 71: 193–201.

Button, A., 1976b. Transvaal and Hamersley Basins – review of basin development and mineral deposits. *Miner. Sci. Eng.,* 8: 262–293.

Caillère, S. and Kraut, F., 1953. Considerations sur la genèse des minerais de fer oolithiques lorrains. In: F. Blondel (Editor), *La Genèse des Gîtes de Fer. Proc. 19th Int. Geol. Congr., Sect.,* 10: 101–117.

Camacho, R., Ulloa-Melo, C. and Pacheco, A., 1969. Yacimiento de hierro oolitico de Sabanalarga (Boyacá). *Inst. Nac. Invest. Geol. Mineras, (Colomb.), Tech. Pap.,* 15: 23 pp.

Carroll, D., 1958. Role of clay minerals in the transportation of iron. *Geochim. Cosmochim. Acta,* 14: 1–27.

Castaño, J.R. and Garrels, R.M., 1950. Experiments on the deposition of iron with special reference to the Clinton iron ore deposits. *Econ. Geol.,* 45: 755–770.

Cayeux, L., 1909. *Les Minerais de Fer Oolithique de France, 1. Minerais de Fer Primaires. Etudes des Gîtes Minéraux de la France,* Paris, Serv. Carte Géol. Impr. Natl., 344 pp.

Cayeux, L., 1922. *Les Minerais de Fer Oolithique de France, 2. Minerais de Fer Secondaires. Etudes des Gîtes Minéraux de la France,* Paris, Serv. Carte Géol. Impr. Natl., 1052 pp.

Chauvel, J.-J. and Dimroth, E., 1974. Facies types and depositional environment of the Sokoman Iron Formation, central Labrador trough, Quebec, Canada. *J. Sediment. Petrol.,* 44: 299–327.

Chave, D.E., 1954. Aspects of the biogeochemistry of magnesium: calcareous marine organisms. *J. Geol.,* 62: 266–283.

Chilingar, G.V. and Wolf, K.H. (Editors), 1975. Compaction of Coarse-Grained Sediments, Vol. I and II. Elsevier, Amsterdam, I: 552 pp., II: 808 pp.

Choquette, P.W., 1955. A petrographic study of the "State College" siliceous oolite. *J. Geol.,* 63: 337–347.

Cloud, P.E. and Licari, G.R., 1972. Ultrastructure and geologic relations of some two-aeon old Nostocacean algae from Northeastern Minnesota. *Am. J. Sci.,* 272: 138–149.

Cloud, P., 1976. Beginnings of biospheric evolution and their biogeochemical consequences. *Paleobiology,* 2: 351–387.

Cochrane, G.W. and Edwards, A.B., 1960. The Roper River oolitic ironstone formations. *Commonw. Sci. Ind. Res. Organ., Aust., Mineragraph. Invest., Tech. Pap.,* 1: 28 pp.

Coonley, L.S. Jr., Baker, E.B. and Holland, H.D., 1971. Iron in the Mullica River and in Great Bay, New Jersey. *Chem. Geol.,* 7: 51–63.

Correns, C.W., 1969. *Introduction to Mineralogy.* Springer, New York, N.Y., 2nd ed., 484 pp.

Curtis, C.D. and Spears, D.A., 1968. The formation of sedimentary iron minerals. *Econ. Geol.,* 63: 257–270.

Davidson, C.F., 1961. Oolitic ironstones of fresh-water origin. *Min. Mag. (London),* 104: 158–159.

Davies, G.R., 1970. Carbonate bank sedimentation, eastern Shark Bay, Western Australia. In: B.W. Logan et al. (Editors), *Carbonate Sedimentation and Environments, Shark Bay, Western Australia. Mem. Am. Assoc. Petrol. Geol.,* 13: 85–168.

Davies, W. and Dixie, R.J.M., 1951. Recent work on the Frodingham ironstone. *Proc. Yorksh. Geol. Soc.,* 28: 85–96.

Deubel, F., Berg, G. and v. Gaertner, H.R., 1942. Die Erze des thüringischen Untersilurs. In: B. Brockamp (Editor), *Zur Entstehung deutscher Eisenerzlagerstätten. Arch. Lagerstättenforsch.,* 75: 140–150.

Deverin, L., 1945. Etude pétrographique des minerais de fer oolithitques du Dogger des Alpes suisses. *Beitr. Geol. Schweiz, Geotech. Ser., Part* 13, 2: 115 pp.

De Wiest, R.J.M., 1965. *Geohydrology.* Wiley, New York, N.Y., 366 pp.

Dickey, P.A., 1968. Contemporary nonmarine sedimentation in Soviet Central Asia. *Bull. Am. Assoc. Pet. Geol.,* 52: 2396–2421.

Dimroth, E., 1976. Aspects of the sedimentary petrology of cherty iron-formation. In: K.H. Wolf (Editor), *Handbook of Strata-bound and Stratiform Ore Deposits,* 7. Elsevier, Amsterdam, pp. 203–254.

Dimroth, E. and Chauvel, J.J., 1973. Petrography of the Sokoman iron formation in part of the central Labrador trough, Quebec Canada. *Geol. Soc. Am. Bull.,* 84: 111–134.

Donaldson, A.C., Martin, R.H. and Kanes, W.H., 1970. Holocene Guadalupe delta of Texas Gulf Coast. In: J.P. Morgan (Editor), *Deltaic Sedimentation Modern and Ancient. Soc. Econ. Paleontol. Mineral. Spec. Publ.*, 15: 107–137.

Dunbar, G.J. and McCall, G.J.H., 1971. Archean turbidites and banded ironstone of the Mt. Belches area (Western Australia). *Sediment. Geol.*, 5: 92–113.

Durum, W.H. and Haffty, J., 1963. Implications of the minor element content of some major streams of the world. *Geochim. Cosmochim. Acta*, 27: 1–11.

Eckel, E.B., 1938. The brown iron ores of eastern Texas. *U.S. Geol. Surv. Bull.*, 902: 157 pp.

Edmonds, E.A., Poole, E.G. and Wilson, V., 1965. Geology of the country around Banbury and Edge Hill. *G.B. Geol. Surv. Mem.*, 137 pp.

Edwards, A.B., 1958. Oolitic iron formations in Northern Australia. *Geol. Rundsch.* 47: 668–682.

El Sharkawi, M.A., Mahfouz, S. and El Dallal, M.M.N., 1976. The pisolitic ironstone of Gdeidet Yabous and Naba Barada localities, Zebdani District, Syria. *Chem. Erde*, 35: 241–250.

Emerson, S., 1976. Early diagenesis in anaerobic lake sediments: chemical equilibria in interstitial waters. *Geochim. Cosmochim. Acta*, 40: 925–934.

Encyclopaedia Britannica, 1974, 15th ed., Vol. 19, Univ. Chicago, Chicago, Ill., 1180 pp.

Erhart, H., 1967. *La genèse des sols en tant que phénomène géologique.* Masson, Paris, 2nd ed., 177 pp.

Eriksson, K.A., 1973. The Timeball Hill Formation – a fossil delta. *J. Sediment. Petrol.*, 43: 1046–1053.

Finkenwirth, A., 1964. Das Eisenerz des Lias gamma am Kahlberg Bei Echte und der Weissjura in Süd-Hannover. *Geol. Jahrb. Beih.*, 56: 131 pp.

Finkenwirth, A. and Simon, P., 1969. Das Eisenerzlager des Lias γ der Grube Echte. In: H. Bottke et al. (Editors), *Sammelwerk Deutsche Eisenerzlagerstätten, II-1. Geol. Jahrb. Beih.*, 79: 58–84.

Folk, R.L., 1973. Evidence for peritidal deposition of Devonian Caballos novaculite, Marathon Basin, Texas. *Bull. Am. Assoc. Pet. Geol.*, 57: 702–725.

Freitag, K.-P., 1970. Feinstratigraphische und petrographische Untersuchungen im erzführenden Korallenoolith (Unter Malm) des westlichen Wesergebirges (Nordwestdeutschland). *Clausthaler Hefte Lagerstättenkd. Geochem. Miner. Rohst.*, 9: 185–214.

Fryer, B.J., 1977a. Rare earth evidence in iron-formations for changing Precambrian oxidation states. *Geochim. Cosmochim. Acta*, 41: 361–367.

Fryer, B.J., 1977b. Trace element geochemistry of the Sokoman Iron Formation. *Can. J. Earth Sci.*, 14: 1598–1610.

Garrels, R.M. and Mackenzie, F.T., 1971. *Evolution of Sedimentary Rocks.* Norton, New York, N.Y., 397 pp.

Garrels, R.M. and Perry, E.A. Jr., 1974. The cycling of carbon, sulfur, and oxygen through geologic time. In: D. Goldberg (Editor), *The Sea, 5.* Wiley-Interscience, New York, N.Y., pp. 303–336.

Gibbs, R.J., 1972. Water chemistry of the Amazon river. *Geochim. Cosmochim. Acta*, 36: 1061–1066.

Gignoux, M., 1950. *Géologie Stratigraphique.* Masson, Paris, 4th ed., 735 pp.

Giresse, P., 1969. Etude des différents grains ferrugineux authigènes des sédiments sous-marins au large du delta de l'Ogooué (Gabon). *Sci. Terre*, 14: 27–62.

Golder, H.Q. and Gass, A.A., 1962. Field tests for determining permeability of soil strata. In: *Field Testing of Soils. Am. Soc. Test. Mater., Spec. Tech. Publ.*, 322: 29–45.

Goodwin, A.M., 1962. Structure, stratigraphy, and origin of iron formations, Michipicoten area, Algoma District, Ontario, Canada. *Geol. Soc. Am. Bull.*, 73: 561–586.

Graf, J.L. Jr., 1977. Rare earth elements as hydrothermal tracers during the formation of massive sulfide deposits in volcanic rocks. *Econ. Geol.*, 72: 527–548.

Gross, G.A., 1965. Geology of iron deposits in Canada, Vol. 1, General geology and evaluation of iron deposits. *Can. Geol. Surv., Econ. Geol. Rep.*, 22: 181 pp.

Gross, G.A., 1968. Geology of iron deposits in Canada, 3. Iron ranges of the Labrador geosyncline. *Can. Geol. Surv., Econ. Geol. Rep.*, 22: 179 pp.

Gross, G.A., 1973. The depositional environment of principal types of Precambrian iron-formations. In: *Genesis of Precambrian Iron and Manganese Deposits. UNESCO Earth Sci.*, 9: 15–21.

Gruner, J.W., 1922. The origin of sedimentary iron formations: the Biwabik Formation of the Mesabi Range. *Econ. Geol.*, 17: 407–460.

Gruner, J.W., 1946. An investigation and report on the mineralogy, chemistry, and origin of the oolitic iron ores of the Paz del Rio district, Colombia, South America, with recommendations for mining and concentration. *Report to Acerias Paz del Rio, Colombia*, 37 pp. (Unpublished)

Gruss, H. and Thienhaus, R., 1969. Paläogeographie und Entstehung der Eisenerze des Ober-Aalenium (Dogger beta) Nordwestdeutschlands. In: H. Bottke et al. (Editors), *Sammelwerk Deutsche Eisenerzlagerstätten, II.1. Beih. Geol. Jahrb.*, 79: 167–172.

Hallam, A., 1966. Depositional environment of British Liassic ironstones considered in the context of their facies relationships. *Nature*, 209: 1306–1309.

Halley, R.B., 1977. Ooid fabric and fracture in the Great Salt Lake and the geologic record. *J. Sediment. Petrol.*, 47: 1099–1120.

Hallimond, A.F., 1925. Iron ores: bedded ores of England and Wales, petrography and chemistry. *G.B. Geol. Surv. Mem., Spec. Rep. Miner. Resour. G.B.*, 29: 139 pp.

Hallimond, A.F., 1951. Problems of the sedimentary iron ores. *Proc. Yorks. Geol. Soc.*, 28: 61–66.

Harder, E.C., 1919. Iron-depositing bacteria and their geological relations. *U.S. Geol. Surv. Prof. Pap.*, 113: 89 pp.

Harder, H., 1964a. On the diagenetic origin of bertierin (chamositic) iron ores. In: *Genetic Problems of Ores. 22nd Int. Geol. Congr. New Delhi*, 5: 193–198.

Harder, H., 1964b. The use of trace elements in distinguishing different types of marine sedimentary iron ores. In: *Genetic Problems of Ores, 22nd Int. Geol. Congr., New Delhi*, 5: 551–556.

Harms, J.E., 1965. Iron ore deposits of Constance Range. In: J. McAndrew (Editor), *Geology of Australian Ore Deposits, 1. Aust. Inst. Min. Metall., Melbourne*, 2nd ed., pp. 264–269.

Hawley, J.E. and Beavan, A.P., 1934. Mineralogy and genesis of the Mayville iron ore of Wisconsin. *Am. Mineral.*, 19: 493–514.

Hayes, A.O., 1915. Wabana iron ore of Newfoundland. *Can. Geol. Surv. Mem.*, 78: 163 pp.

Hayes, A.O., 1929. Further studies of the origin of the Wabana iron ore of Newfoundland. *Econ. Geol.*, 24: 687–690.

Hedberg, H.D., 1928. Some aspects of sedimentary petrography in relation to stratigraphy in the Bolivar Coast fields of the Maracaibo Basin, Venezuela. *J. Paleontol.*, 2: 32–42.

Hem, J.D., 1960. Restraints on dissolved ferrous iron imposed by bicarbonate redox potential, and pH. *U.S. Geol. Surv. Water-Supply Pap.*, 1459-B: 33–35.

Hemingway, J.E., 1951. Cyclic sedimentation and the deposition of ironstone in the Yorkshire Lias. *Proc. Yorks., Geol. Soc.*, 28: 67–74.

Hoenes, D. and Tröger, E., 1945. Lagerstätten oolithischer Eisenerze in Nordwestfrankreich. *Neues Jahrb. Mineral. Abh.*, 79(A): 192–257.

Holland, H.D., 1973. Ocean water, nutrients and atmospheric oxygen. *Proc. Symp. Hydrogeochemistry and Biogeochemistry, 1.* Clarke, Washington, D.C., pp. 68–81.

Hummel, K., 1922. Die Entstehung eisenreicher Gesteine durch Halmyrolyse (= submarine Gesteinszersetzung). *Geol. Rundsch.*, 13: 40–81 and 97–136.

Hunter, R.E., 1970. Facies of iron sedimentation in the Clinton Group. In: G.W. Fischer et al. (Editors), *Studies of Appalachian geology, Central and Southern.* Wiley-Interscience, New York, N.Y., pp. 101–121.

Ikonnikov, A.B., 1975. Mineral resources of China. *Geol. Soc. Am. Microform Publ.*, 2: 555 pp.

Irving, E.M., 1971. La evolucion estructural de los Andes mas septentrionales de Colombia. *Bol. Geol.* (Colombia), 19(2): 89 pp.

Irving, E.M., 1975. Structural evolution of the northernmost Andes, Colombia. *U.S. Geol. Surv. Prof. Pap.*, 846: 47 pp.

James, H.E. Jr. and Van Houten, F.B., 1979. Miocene goethitic and chamositic oolites, northeastern Colombia. *Sedimentology*, 26: 125–133.

James, H.L., 1966. Chemistry of the iron-rich sedimentary rocks. In: M. Fleischer (Editor), *Data of Geochemistry. U.S. Geol. Surv. Prof. Pap.*, 440-W: 61 pp (6th ed.).

Janot, C., Gibert, H., de Gramont, X. and Biais, R., 1971. Etude des substitutions Al-Fe dans les roches latéritiques. *Bull. Soc. Fr. Miner. Cristallogr.*, 94: 367–380.

Jones, H.A., 1955. *The Occurrence of Oolitic Ironstones in Nigeria: their Origin, Geologic History, and Petrology.* Thesis, Oxford Univ., 206 pp.

Jones, H.A., 1965. Ferruginous oolites and pisolites. *J. Sediment. Petrol.*, 35: 838–845.

Jones, H.L., 1969. *Petrography, Mineralogy, and Geochemistry of the Chamositic Iron Ores of North-central Louisiana.* Thesis, Univ. of Oklahoma, 196 pp. (unpublished)

Julivert, M., 1970. Cover and basement tectonics in the Cordillera Oriental of Colombia, South America, and a comparison with some other folded chains. *Geol. Soc. Am. Bull.*, 81: 3623–3646.

Kadar, D., 1973. Notes on the age of the limestones of the southern peninsula, Bali Island. *Indones. Geol. Surv., Tech. Publ. Paleontol. Ser.*, 5: 13–15.

Kalkowsky, E., 1908. Oolith und stromatolith im norddeutschen Buntsandstein: *Z.D. Geol. Ges.*, 60: 60–125.

Kalliokoski, J., 1966. Diagenetic pyritization in three sedimentary rocks. *Econ. Geol.*, 61: 872–885.

Karrenberg, H., Berg, G., Aldinger, H. and Frank, M., 1942. Die Eisenoolith von West- und Südwestdeutschland. In: B. Brockamp (Editor), *Zur Entstehung deutscher Eisenerzlagerstätten. Arch. Lagerstättenforsch.*, 75: 78–110.

Kelly, V.C., 1951. Oolitic iron deposits of New Mexico. *Bull. Am. Assoc. Pet. Geol.* 35: 2199–2228.

Kimberley, M.M., 1974. *Origin of Iron Ore by Diagenetic Replacement of Calcareous Oolite,* Thesis, Princeton Univ., Vol. 1, 345 pp., Vol. 2: 386 pp.

Kimberley, M.M., 1975. Proposal of iron formation origin by cycles of aragonite sedimentation, cover by volcanic ash or terrigenous mud, weathering, organic acid leaching of mud, acid-base aragonite replacement, and mud erosion: a Quaternary analogue (abstr.). *Geol. Soc. Am., Abstr. Progr.*, 7: 1146–1147.

Kimberley, M.M., 1978. Paleoenvironmental classification of iron formations. *Econ. Geol.*, 73: 215–229.

Kimberley, M.M., 1979a. Origin of oolitic iron formations. *J. Sediment. Petrol.* 49: 111–131.

Kimberley, M.M., 1979b. Geochemical distinctions among environmental types of iron formations. *Chem. Geol.*, 25: 185–212.

Kimberley, M.M., in press. The Paz de Rio oolitic inland-sea iron formation. *Econ. Geol.*, 75.

Knauth, L.P., 1973. *Oxygen and Hydrogen Isotope Ratios in Cherts and Related Rocks.* Thesis. Calif. Inst. Technol., 369 pp.

Kolb, C.R. and Van Lopik, J.R., 1966. Depositional environments of the Mississippi River deltaic plain – southeastern Louisiana. In: M.L. Shirley (Editor), *Deltas in their Geologic Framework.* Houston Geol. Soc., Houston, pp. 17–61.

Kolbe, H., 1970. Zur Entstehung und Charakteristick mesozoischer marinsedimentärer Eisenerze in östlichen Niedersachsen. *Clausthaler Hefte Lagerstättenkd. Geochem. der Mineral. Rohst.*, 9: 161–184.

Kolbe, H. and Simon, P., 1969. Die Eisenerze im Mittleren und Oberen Korallenoolith des Gifhorner Troges. In: H. Bottke et al. (Editors), *Sammelwerk Deutsche Eisenerzlagerstätten,* II.1. *Beih. Geol. Jahrb., Hannover,* 79: 256–338.

Krumbein, W.C., 1935. Thin section mechanical analysis of indurated sediments. *J. Geol.*, 43: 482–496.

La Berge, G.L., 1966a. Altered pyroclastic rocks in iron-formation in the Hamersley Range, Western Australia. *Econ. Geol.*, 61: 147–161.

La Berge, G.L., 1966b. Altered pyroclastic rocks in South African iron-formation. *Econ. Geol.*, 61: 572–581.

Lambe, T.W., 1955. The permeability of fine-grained soils. In: *Symposium on Permeability of Soils. Am. Soc. Testing Materials, Spec. Tech. Publ.* 163: 56–67.

Lamplugh, G.W., Wedd, C.B. and Pringle, J., 1920. Bedded ores of the Lias, Oolites, and later formations in England. *Iron Ores,* 12, *G.B. Geol. Surv. Mem.*, 240 pp.

Landergren, S., 1948. On the geochemistry of Swedish iron ores and associated rocks. *Sver. Geol. Unders., Ser. C,* 496: 182 pp.

Langbein, R., 1975. Petrologische Analyse der Sedimentations Bedingungen im höheren Mittleren Buntsandstein (smD-S) der DDR. *Chem. Erde,* 34: 85–100.

Latal, E., 1952. Die Eisenerzlagerstätten Jugoslaviens. *Symposium sur les gisements de fer du monde, 19th Int. Geol. Congr., Algiers,* 2: 529–563.

Leith, C.K., 1903. The Mesabi iron-bearing district of Minnesota. *U.S. Geol. Surv. Monogr.,* 43: 316 pp.

Lemoalle, J. and Dupont, B., 1973. Iron-bearing oolites and the present conditions of iron sedimentation in Lake Chad (Africa). In: G.C. Amstutz and A.J. Bernard (Editors), *Ores in Sediments. Springer, Berlin*: pp. 167–178.

Livingstone, D.A., 1963. Chemical composition of rivers and lakes. In: M. Fleischer (Editor), *Data of Geochemistry. U.S. Geol. Surv. Prof. Pap.,* 440-G: 64 pp. (6th ed.).

Loreau, J.-P. and Purser, B.H., 1973. Distribution and ultrastructure of Holocene ooids in the Persian Gulf. In: B.H. Purser (Editor), *The Persian Gulf.* Springer, New York, N.Y., pp. 279–341.

MacGregor, M., Lee, G.W. and Wilson, G.V., 1920. The iron ores of Scotland. *Iron Ores, 11. G.B. Geol. Surv. Mem.,* 236 pp.

Macurda, D.B. Jr. and Meyer, D.L., 1975. The microstructure of the crinoid endoskeleton. *Univ. Kansas Paleontol. Contrib., Pap.,* 74: 22 pp.

Mansfield, G.R., 1927. Geography, geology, and mineral resources of part of southeastern Idaho. *U.S. Geol. Surv. Prof. Pap.,* 152: 453 pp.

Markevich, V.P., 1960. The concept of facies. *Int. Geol. Rev.* 2: 367–379.

Maubeuge, P.-L., 1972. *Etudes stratigraphiques sur la formation ferrifère de Lorraine et ses morts-terrains.* Inter-impression-E, Paris, 487 pp.

Mellon, G.B., 1962. Petrology of Upper Cretaceous oolitic iron-rich rocks from northern Alberta. *Econ. Geol.,* 57: 921–940.

Milliman, J.D., 1974. *Recent Sedimentary Carbonates, Part 1, Marine Carbonates.* Springer, New York, N.Y., 375 pp.

Millot, G., 1970. *Geology of Clays.* Springer, New York, N.Y., 429 pp.

Moore, E.S., 1910. The occurrence and origin of some bog iron deposits in the district of Thunder Bay, Ontario. *Econ. Geol.,* 5: 528–538.

Moorhouse, W.W., 1960. Gunflint iron range in the vicinity of Port Arthur. *Ontario Dep. Mines, Annu. Rep.,* 69(7): 1–40.

Müller, G. and Förstner, U., 1973. Recent iron ore formation in Lake Malawi, Africa. *Miner. Deposita,* 8: 278–290.

Nakhla, F.M. and Shehata, M.R.N., 1967. Contributions to the mineralogy and geochemistry of some iron-ore deposits in Egypt (U.A.R.). *Miner. Deposita,* 2: 357–371.

Nassim, G.L., 1950. The oolithic hematite deposits of Egypt. *Econ. Geol.,* 45: 578–581.

Newell, N.D., 1972. Stratigraphic gaps and chronostratigraphy. *Proc. 24th Int. Geol. Congr.* Sect. 7: 198–203.

Newell, N.D. and Rigby, J.K., 1957. Geological studies on the Great Bahama Bank. In: R.J. Le Blanc and J.A. Breeding (Editors), *Regional Aspects of Carbonate Deposition. Soc. Econ. Paleontol. Mineral., Spec. Publ.,* 5: 15–72.

Newland, D.H., 1910. The Clinton iron-ore deposits in New York State. *Am. Inst. Min. Eng. Trans.,* 40: 165–183.

Nicolini, P., 1967. Remarques comparatives sur quelques éléments sédimentologiques et paléogeographiques liés aux gisements de fer oolithiques du Djebel Ank (Tunisie) et de Lorraine (France). *Miner. Deposita,* 2: 95–101.

Page, B.M., 1958. Chamositic iron ore deposits near Tajmiste, Western Macedonia, Yugoslavia. *Econ. Geol.,* 54: 1–21.

Percival, F.G., 1952. Les ressources mondiales en minerai de fer. In: F. Blondel and L. Marvier (Editors), *Symposium sur les Gisements de Fer du Monde, 1. Proc. 19th Int. Geol. Congr.,* pp. 17–31.

Perry, E.C. Jr. and Tan, F.C., 1972. Significance of oxygen and carbon isotope variations in early Precambrian cherts and carbonate rocks of southern Africa. *Geol. Soc. Am. Bull.*, 83: 647–664.

Petranek, J., 1964a. Ordovician – a major epoch in iron ore deposition. *Proc. 22nd Int. Geol. Congr.*, 15: 51–57.

Petranek, J., 1964b. Ordovician sedimentary iron ores in Ejpovice (in Czechoslovakian). *Sb. Geol. Ved Geol. (Collect. Geol. Sci.)*, 2: 39–153.

Petruk, W., 1977. Mineralogical characteristics of an oolitic iron deposit in the Peace River district, Alberta. *Can. Mineral.*, 15: 3–13.

Petruk, W., Farrell, D.M., Laufer, E.E., Tremblay, R.J. and Manning, P.G., 1977. Nontronite and ferruginous opal from the Peace River iron deposit in Alberta, Canada. *Can. Mineral.*, 15: 14–21.

Pettijohn, F.J., 1975. *Sedimentary Rocks.* Harper and Row, New York, N.Y., 3rd ed., 628 pp.

Pilsbry, H.A. and Olsson, A.A., 1928. Tertiary fresh-water mollusks of the Magdalena embayment, Colombia. *Acad. Nat. Sci. Phila., Proc.*, 87: 7–39.

Porrenga, D.H., 1965. Chamosite in Recent sediments of the Niger and Orinoco deltas. *Geol. Mijnbouw*, 44: 400–403.

Porrenga, D.H., 1967a. *Clay Mineralogy and Geochemistry of Recent Marine Sediments in Tropical Areas.* Stolk-Dordt, Amsterdam, 145 pp.

Porrenga, D.H., 1967b. Glauconite and chamosite as depth indicators in the marine environment. *Mar. Geol.*, 5: 495–501.

Pratje, O., 1930. Rezente marine Eisen-Ooide aus der Nordsee. *Zentralbl. Mineral. Geol. Palaeontol., Abt. B*: 289–294.

Pulfrey, W., 1933. The iron-ore oolites and pisolites of North Wales. *Q.J. Geol. Soc. London*, 89: 401–430.

Putzer, H., 1943. Die oolithischen Brauneisenerz-Lagerstätten der Kertsch-Halbinsel. *Z. Angew. Mineral.*, 4: 363–378.

Rau, E.L., 1963. Beneficiation of Acerias Paz del Rio, S.A., iron ore. *Project Rep. No. 210536, Colorado School of Mines Research Foundation*, 65 pp. (unpublished).

Rechenberg, H.P., 1956. Die Eisenerzlagerstätte "Vivaldi" bei Ponferrada, Léon, Spanien. *Neues Jahrb. Mineral. Abh.*, 89: 111–136.

Reineck, H.-E. and Wunderlich, F., 1968. Classification and origin of flaser and lenticular bedding. *Sedimentology*, 11: 99–104.

Reyes, I., 1966. Geologia de la region de Paz Vieja. *Intern. Rep., Acerias Paz del Rio S.A.*, Colombia, 32 pp. (unpublished)

Robinson, R.B., 1967. Diagenesis and porosity development in Recent and Pleistocene oolites from southern Florida and the Bahamas. *J. Sediment. Petrol.*, 37: 355–364.

Rohrlich, V., 1974. Microstructure and microchemistry of iron ooliths. *Miner. Deposita*, 9: 133–142.

Rohrlich, V., Price, N.B. and Calvert, S.E., 1969. Chamosite in the Recent sediments of Loch Etive, Scotland. *J. Sediment Petrol.*, 39: 624–631.

Rutten, M.G., 1969. *The Geology of Western Europe.* Elsevier, Amsterdam, 520 pp.

St. John, R.N., 1927. Replacement vs. impregnation in petrified wood. *Econ. Geol.*, 22: 729–739.

Sargent, H.C., 1929. Further studies in chert – I. *Geol. Mag.*, 66: 399–413.

Sarmiento-Soto, R., 1946. *Geology and Iron Ore Resources of the Paz de Rio Region, Boyacá, Colombia.* Thesis, Univ. of California at Los Angeles, 181 pp.

Schellmann, W., 1969. Die Bildungsbedingungen sedimentärer Chamosit- und Hamatit-Eisenerze, am Beispiel der Lagerstätte Echte. *Neues Jahrb. Mineral. Abh.*, 111: 1–31.

Schoen, R., 1962. *Petrology of Iron-Bearing Rocks of the Clinton Group in New York State.* Thesis, Harvard University, 151 pp.

Schofield, J.C., 1959. The geology and hydrology of Niue Island, South Pacific. *N.Z. Geol. Surv. Bull.*, 62: 28 pp.

Schopf, J.M., 1975. Modes of fossil preservation. *Rev. Palaeobot. Palynol.*, 20: 27–53.

Schweigart, H., 1965. Genesis of the iron ores of the Pretoria Series, South Africa. *Econ. Geol.,* 60: 269–298.

Segovia, A., 1967. Geology of Plancha L-12, Colombia, South America: a reconnaissance. *Geol. Soc. Am. Bull.,* 78: 1007–1028.

Sellards, E.H. and Baker, C.L., 1934. The geology of Texas, Vol. 2, Structural and economic geology. *Univ. Tex. Bull.,* 3401: 884 pp.

Shapiro, J., 1964. Effect of yellow organic acids on iron and other metals in water. *J. Am. Water Works Assoc.,* 56: 1062–1082.

Shearman, D.J., Twyman, J. and Karimi, M.Z., 1970. The genesis and diagenesis of oolites. *Proc. Geol. Assoc. (London),* 81: 561–575.

Sheldon, R.P., 1963. Physical stratigraphy and mineral resources of Permian rocks in western Wyoming. *U.S. Geol. Surv. Prof. Pap.,* 313-B: 49–273.

Shimizu, H. and Masuda, A., 1977. Cerium in chert as an indication of marine environment of its formation. *Nature,* 266: 346–348.

Simon, P., 1969. Die Lias-Eisenerze der Grube Friderike. In: H. Bottke et al. (Editors), Sammelwerk Deutsche Eisenerzlagerstätten, II-1. *Beih. Geol. Jahrb., Hannover,* 79: 40–58.

Skocek, V., 1963a. Oolitic iron ores from the regions of Raca and Bechlova. *Sb. Geol. Ved Geol. (Collec. Geol. Sci.),* 1: 31–63 (in Czechoslovakian).

Skocek, V., 1963b. Petrographic composition and genesis of iron ore in the Brezina region. *Rozpr. Cesk. Akad. Ved (Proc. Czech. Acad. Sci.),* 73(4): 109 pp. (in Czechoslovakian).

Skocek, V., Al-Qaraghuli and Saadallah, A.A., 1971. Composition and sedimentary structures of iron ores from the Wadi Husainiya area, Iraq. *Econ. Geol.,* 66: 995–1004.

Smith, A.E. Jr., 1966. Modern deltas: comparison maps. In: M.L. Shirley (Editor), *Deltas in their Geologic Framework.* Houston Geol. Soc., Houston, Texas, pp. 233–251.

Smyth, C.H. Jr., 1892. One the Clinton iron ore. *Am. J. Sci.,* 43, (258): 487–496.

Sokolova, E.I., 1964. Physicochemical investigations of sedimentary iron and manganese ores and associated rocks: Jerusalem. *Isr. Progr. Sci. Transl.,* 220 pp.

Solignac, M., 1930. Les caractères minéralogiques du minerai de fer oolithique du djebel el Ank (Tunisie méridionale). *C.R. Acad. Sci.,* 191: 107–109.

Sorby, H.C., 1857. On the origin of the Cleveland Hill ironstone. *Geol. Polytech. Soc. West Riding of Yorks. Proc.,* 3: 457–461.

Sorby, H.C., 1906. The origin of the Cleveland ironstone. *Naturalist,* October, 1906: 354–357.

Stanton, R.L., 1972. *Ore Petrology.* McGraw-Hill, New York, N.Y., 713 pp.

Strakhov, N.M., 1969. In: S.I. Tomkieff and J.E. Hemingway (Editors), *Principles of Lithogenesis, 2.* Oliver and Boyd, Edinburgh, 609 pp.

Strakhov, N.M. and Nesterova, I.L., 1968. Effects of volcanism on the geochemistry of marine deposits in the Sea of Okhotsk. *Geochem. Int.,* 5: 644–666.

Stumm, V. and Morgan, J.J., 1970. *Aquatic Chemistry, an Introduction emphasizing Chemical Equilibria in Natural Waters.* Wiley-Interscience, New York, N.Y., 583 pp.

Summerson, C.H. (Editor), 1976. *Sorby on Sedimentology: Geological Milestones, 1.* Rosenstiel Sch. Mar. Atmos. Sci., Univ. Miami, 225 pp.

Supko, P.R., 1977. Subsurface dolomites, San Salvador, Bahamas. *J. Sediment. Petrol.,* 47: 1063–1077.

Sutton, F.A., 1946. Geology of Maracaibo Basin, Venezuela. *Bull. Am. Assoc. Pet. Geol.,* 30: 1621–1741.

Taylor, J.H., 1949. Petrology of the Northampton Sand Ironstone formation. *G.B. Geol. Surv. Mem.,* 111 pp.

Taylor, J.H., 1951. Sedimentation problems of the Northampton Sand Ironstone. In: The Constitution and origin of sedimentary iron ores. *Proc. Yorks. Geol. Soc.,* 28: 74–85.

Taylor, J.H., 1969. Sedimentary ores of iron and manganese and their origin. *15th Inter-University Geol. Congr. Proc.,* Univ. of Leicester, England, pp. 171–186.

Taylor, S.R., 1964. Abundance of chemical elements in the continental crust: a new table. *Geochim. Cosmochim. Acta,* 28: 1273–1285.

Tegengren, F.R., 1921. The iron ores and iron industry of China, Part 2. *China Geol. Surv. Mem. Peking, Ser. A,* 2: 180 pp.

Terlecky, P.M. Jr., 1967. *The Nature and Distribution of Oolites on the Atlantic Continental Shelf of the Southeastern United States.* Thesis, Duke Univ., Durham, N.C., 46 pp.

Theis, T.L. and Singer, P.C., 1972. The stabilization of ferrous iron by organic compounds in natural waters. In: *Trace Metals and Metal-Organic Interactions in Natural Waters.* Ann Arbor Science, Ann Arbor, Mich., pp. 303–320.

Thienhaus, R., 1957. Zur Palaeogeographie der Korallenoolitherze des Wesergebirges. *Z. Dtsch. Geol. Ges.,* 109: 49–62.

Thode, H.G., Monster, J. and Dunford, H.B., 1961. Sulphur isotope geochemistry. *Geochim. Cosmochim. Acta,* 25: 159–174.

Thwaites, F.T., 1914. Recent discoveries of "Clinton" iron ore in eastern Wisconsin. *U.S. Geol. Surv. Bull.,* 540: 338–342.

Timofeeva, Z.V., 1966. Some features of lithology and geochemistry of iron-rich ore gangues and ores in the Bechasyl Plateau (northern Caucasus). *Litol. Polezn. Iskop. (J. Lithol. Mineral.)* 1966(1): 33–48 (in Russian).

Timofeeva, Z.V. and Balashov, Yu.A., 1972. The distribution of rare earth elements in oolitic iron ores of the northern Caucasus. *Litol. Polezn. Iskop. (J. Lithol. Mineral.)* 1972(3): 128–135 (in Russian).

Trask, P.D., 1939. Organic content of Recent marine sediments. In: P.D. Trask (Editor), *Recent Marine Sediments.* Am. Assoc. Pet. Geol., Tulsa, Okla., pp. 428–453.

Trendall, A.F., 1973a. Precambrian iron-formations of Australia. *Econ. Geol.,* 68: 1023–1034.

Trendall, A.F., 1973b. Iron-formations of Hamersley Group of Western Australia: type examples of varved Precambrian evaporites. In: *Genesis of Precambrian Iron and Manganese Deposits.* UNESCO Earth Sci., 9: 247–270.

Trendall, A.F. and Blockley, J.G., 1970. The iron formations of the Precambrian Hamersley Group, Western Australia, with special reference to the associated crocidolite. *West. Aust. Geol. Surv. Bull.,* 119: 366 pp.

Turekian, K.K., 1968. *Oceans.* Prentice-Hall, Englewood Cliffs. N.J., 120 pp.

Turner, C.C. and Walker, R.G., 1973. Sedimentology, stratigraphy, and crustal evolution of the Archean greenstone belt near Sioux Lookout, Ontario. *Can. J. Earth Sci.,* 10: 817–845.

Tyler, S.A. and Twenhofel, W.H., 1952. Sedimentation and stratigraphy of the Huronian of Upper Michigan. *Am. J. Sci.,* 250: 1–27 and 118–151.

Udden, J.A., 1924. Laminated anhydrite in Texas. *Bull. Geol. Soc. Am.,* 35: 347–354.

Urban, H., 1966. Bildungsbedingungen und Faziesverhältnisse der marinsedimentären Eisenerzlagerstätten am Kahlenberg bei Ringsheim/Baden. *Jahrb. Geol. Landesamt Baden-Würtemberg,* 8: 125–267.

Uthe, R., 1969. Geochemical considerations of Clinton iron ore deposition. *Compass Sigma Gamma Epsilon,* 46: 169–180.

Van der Hammen, T., 1954. El desarrollo de la flora colombiana en les periodos geologicos. I. Maestrichtiano hasta Terciario mas inferior. *Bol. Geol. (Colomb.),* 2: 49–106.

Van der Hammen, T., 1957. Estratigrafia palinologica de la Sabana de Bogota (Cordillera Oriental de Colombia). *Bol. Geol. (Colomb.),* 5: 191–203.

Van der Hammen, T., 1958. Estratigrafia del Terciario y Maestrichtiano continentales y tectogenesis de los Andes colombianos. *Bol. Geol. (Colomb.),* 6: 67–128.

Van Houten, F.B., 1967. Cenozoic oolitic iron ore, Paz de Rio, Boyacá, Colombia. *Econ. Geol.,* 62: 992–997.

Van Houten, F.B., 1972. Iron and clay in tropical savanna alluvium, northern Colombia: a contribution to the origin of red beds. *Geol. Soc. Am. Bull.,* 83: 2761–2772.

Vann, J.H., 1959. Landform–vegetation relationships in the Atrato delta. *Ann. Assoc. Am. Geogr.,* 49: 345–360.

Varentsov, I.M., 1964. *Sedimentary Manganese Ores.* Elsevier, Amsterdam, 119 pp.

Villain, F., 1902. *Le gisement de minerai de fer oolithique de la Lorraine.* Dunod, Paris, 179 pp.

Wagner, P.A., 1928. The iron deposits of the Union of South Africa. *S. Afr. Geol. Surv. Mem.*, 26: 268 pp.

Wall, B. and Reed, W.E., 1976. Sulfur isotopic evidence for the depositional environment of the Lower Tulare Formation, San Joaquin Valley, California (abstr.). *Geol. Soc. Am. Abstr. Progr.*, 8: 1158–1159.

White, W.A., 1970. The geomorphology of the Florida Peninsula. *Dep. Natl. Res., State of Florida. Geol. Bull.*, 51: 164 pp.

Whitehead, T.H., Anderson, W., Wilson, V., Wray, D.A. and Dunham, K.C., 1952. The Liassic ironstones. *G.B. Geol. Surv. Mem.*, 211 pp.

Willden, R., 1960. Sedimentary iron-formation in the Devonian Martin Formation, Christmas Quadrangle, Arizona. *U.S. Geol. Surv. Prof. Pap.*, 400-B: 21–23.

Willden, R., 1961. Composition of the iron-formation of Devonian age in the Christmas quadrangle, Arizona. *U.S. Geol. Surv. Prof. Pap.*, 424-D: 304–306.

Wright, J.D., 1975. Iron deposits of Nova Scotia. *N.S. Dep. Mines, Econ. Geol., Ser.*, 75-1: 154 pp.

Xolodov, V.N., 1973. *Sedimentary Ore Genesis and Metallogenesis of Vanadium.* Nauka, Moscow, 279 pp.

Young, G.A., Bellizzia, A., Renz, H.H., Johnson, F.W., Robie, R.H. and Vall., J.M., 1956. Geologia de las cuencas sedimentarias de Venezuela y de sus campos petroliferos. *20th Int. Geol. Congr., Mexico City*, 4: 161–322 pp.

Chapter 3

METALLIFEROUS DEEP-SEA SEDIMENTS

M.A. MEYLAN, G.P. GLASBY, K.E. KNEDLER and J.H. JOHNSTON

INTRODUCTION

Background

Metalliferous sediments are unconsolidated accumulations of variable proportions of hydrothermal, detrital, hydrogenous, and biogenous material (Heath and Dymond, 1977) in which the transition metal content is elevated above that of "normal" sediments. In the marine environment, there is a complete range of pelagic and hemipelagic sediment compositions between what would be considered non-metalliferous and those sediments that would be termed "metalliferous". Metal-rich marine sediments have been recovered from a number of sites, most prominently on the East Pacific Rise (EPR) and Mid-Atlantic Ridge (MAR), in the Bauer Basin and Red Sea. Because iron and manganese tend to be the principal metal constituents of metalliferous sediments, these deposits have also been designated deep-sea iron deposits, ferruginous sediments, or ferromanganoan sediments.

The first deep-sea metalliferous sediment sample was recovered during the HMS *Challenger* traverse of the southeastern Pacific (globigerina ooze from Stn 293, 39°04′S, 105°05′W, 3820 m depth). The sample contained 42% Fe (carbonate-free basis) and displayed an Fe/Al ratio of 21 – both values within the range of sediments later called metalliferous (Heath and Dymond, 1977). Murray and Renard (1891) did not recognize the unusual nature of this sample, and subsequent workers such as Revelle (1944) and El Wakeel and Riley (1961) did not distinguish some significantly metal-enriched pelagic sediment samples from their less metalliferous counterparts. These earlier workers did state, however, that the manganese, and even red clay itself, could be derived from the subaqueous decomposition of volcanic debris (cf. Chester and Aston, 1976).

Credit for the discovery of deep-sea metalliferous sediments as such should therefore go to K. Boström (e.g., Boström and Peterson, 1966), who pointed out that East Pacific Rise crest sediments from equatorial latitudes were enriched in Fe, Mn, Cu, Cr, Ni and Pb. The existence of widespread basal metal-rich sediments in deep ocean basins has been demonstrated by Deep Sea Drilling Project (D.S.D.P.) coring (e.g., Maxwell et al., 1970; Von der Borch and Rex, 1970). Metalliferous sediments from the axial deeps of the Red Sea were first reported by Miller et al. (1966).

Metal source and composition

The ubiquity and significance of deep-sea metalliferous sediments is now widely recognized. Depending upon the major source of the constituent metals, marine iron–manganese deposits may be classified as follows (Bonatti et al., 1972):

(1) hydrogenous – slow precipitation of metals from seawater,

(2) diagenetic – remobilization of metals from the underlying sediment column,

(3) hydrothermal – transportation of metals out of underlying sediment or basement rock by volcanically heated fluids, and

(4) halmyrolytic – low-temperature leaching of basaltic pyroclastics by seawater.

Manganese nodules are formed primarily by hydrogenous processes, with significant diagenetic input, while hydrothermal processes seem to be chiefly responsible for the formation of metalliferous sediments.

There is considerable overlap in composition between metalliferous sediments containing a hydrothermal component and those pelagic clays most enriched in authigenic components (which generally occur where sedimentation rates are lowest). This is confirmed by the work of Murray and Renard (1891) and Glasby et al. (1979), who have noted the presence of abyssal red clays containing 19.9% Fe, 2.1% Mn (Stn 281) and 10.3% Fe, 0.9% Mn (Stn G997), respectively, in the Southwestern Pacific Basin which, from the general geochemistry of the sediments in the basin, are clearly not metalliferous as defined in terms of hydrothermal origin. Because of this apparent overlap in composition, there has been no attempt in the literature to define metalliferous sediments in terms of minimum concentrations of iron and manganese.

Although absolute transition metal content cannot be used to unambiguously define whether or not a sediment is metalliferous, use of metal ratios has produced more meaningful comparisons, illustrating the generally Fe-rich, Al-poor nature of metalliferous sediments. Included are such ratios as Fe/Ti and Al/(Al + Fe + Mn) (Boström et al., 1973) (Fig.1a), Fe + Mn/Co + Ni and Fe + Mn/Al (Boström, 1973), an Mn–Fe–Al triangular diagram (Dymond et al., 1977), and the Mn–Fe–(Ni + Co + Cu) triangular diagram of Bonatti et al. (1972) (Fig.1b), which was used to relate elements of manganese nodules to their sources.

Economic significance and related ores

Certain marine metalliferous sediment deposits may prove economically feasible to mine, particularly those of the Red Sea (Bischoff and Manheim, 1969; Bäcker and Post, 1978). Marine metalliferous sediments have also received attention as modern analogues of sediments found above basalt in ancient ophiolite sequences, such as the ochers and

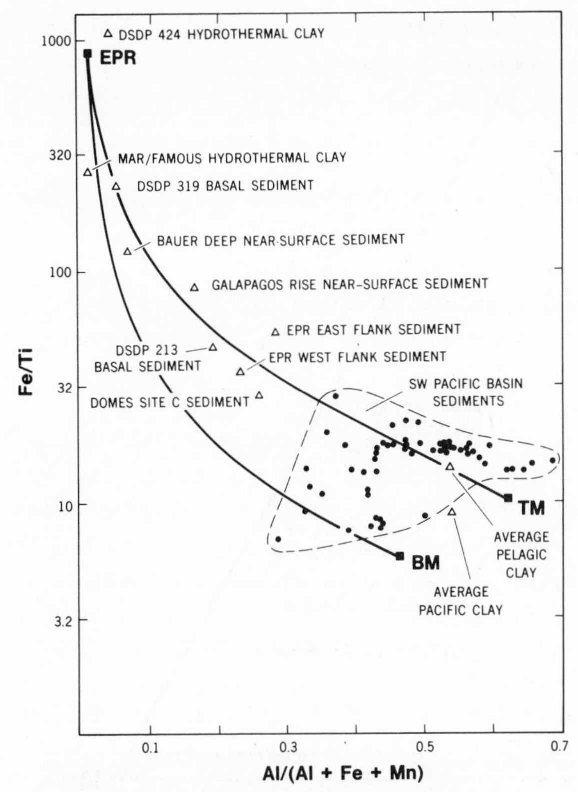

Fig.1a. Diagram of Fe/Ti vs. Al/(Al + Fe + Mn) ratios for marine sediments. *EPR* = East Pacific Rise hydrothermal end-member sediment composition; *TM* = terrestrial matter end-member sediment composition; *BM* = biological matter end-member sediment composition. (After Boström et al., 1973.) SW Pacific Basin sediments data from Meylan et al., in prep. For source of data and additional information regarding samples indicated by triangles, refer to Table X, Deposit Numbers 1, 2, 13, 29, 33, 38, 45, 78, and 89.

umbers of the Troodos Massif (Robertson, 1975; Robertson and Fleet, 1976). In addition to the Cyprus occurrences, metal-rich sediments presumably resulting from hydrothermal activity at ancient spreading centres have been noted in the Apennines of northern Italy (Bonatti et al., 1976a), in Newfoundland (Upadhyay and Strong, 1973; Duke and Hutchinson, 1974), and elsewhere (Bäcker, 1973; Anonymous, 1977).

Fossil manganese deposits in eugeosynclinal sequences, derived from submarine volcanic exhalations, are also known (Stanaway et al., 1978). (For details on the styles of mineralization characteristic of different types of geosynclines, see Chapter 1 by Evans in Vol. 4 of this Handbook).

Stratiform massive Fe–Cu sulphide deposits are found in pillow basalts immediately underlying the sediments of ophiolite sequences, but a detailed discussion of this type of hydrothermal metal enrichment is beyond the scope of this paper. Sillitoe (1972)

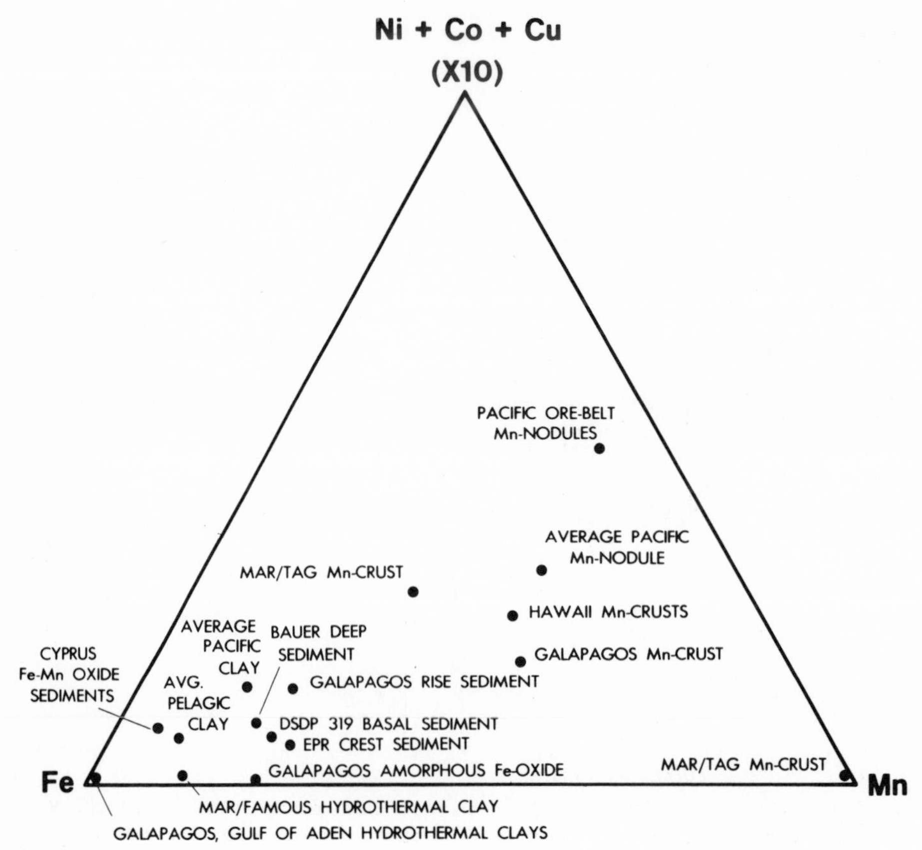

Fig.1b. Ternary diagram of percent Fe vs. Mn vs. (Ni + Co + Cu) × 10. (After Bonatti et al., 1972.) Average Pacific Mn-nodule composition from Hekinian et al. (1978). Generalized Clarion–Clipperton Fracture Zone area ("Pacific ore belt") Mn-nodule composition from unpublished data. Average Hawaiian Archipelago Mn-crust composition from Frank et al. (1976). For source of data and additional information regarding other samples, refer to Table X, Deposit Numbers 1, 2, 10, 12, 13, 23, 29, 33, 35, 36, 38, 45, 92, and 103.

reviewed the subject of ophiolitic massive sulphides in terms of their formation at spreading centres, and predicted that such deposits would eventually be located in a modern oceanic setting. Indeed, Francheteau et al. (1979) recently found massive sulphides on the East Pacific Rise.

This paper will focus on the occurrence, composition and origin of marine hydrothermal and hydrogenous metal-rich sediments, particularly those that can be categorized as epigenetic-hydrothermal and syngenetic-exhalative. Excellent reviews of the broader subject of marine metallogenesis have been provided by Bonatti (1975, 1978), Emery and Skinner (1977), and Rona (1977, 1978) (see also McMurtry, in press). In this Handbook series, the topic of metalliferous sediments has been introduced by Schott (Chapter 6,

Vol. 3) and Degens and Ross (Chapter 4, Vol. 4), the latter authors including a detailed discussion of the Red Sea deposits. Coverage of the Red Sea deposits in this paper will be limited to a general overview, along with the most recent published information. Marine manganese nodules have been reviewed by Glasby and Read (Chapter 7, Vol. 7).

PRINCIPAL TYPES OF DEEP-SEA METAL-ENRICHED SEDIMENTS

Carbonaceous sediments

Carbonaceous muds or oozes (fine terrigenous material with an admixture of biogenic debris) represent a type of sedimentary deposit that may contain unusually high concentrations of metals such as V, Cr, Zn, Ni, Mo, U, and S, in addition to being potential source rocks for oil (Cruickshank, 1974). Carbonaceous sediments typically occur in thicknesses up to a few tens of metres in shallow basins on the continental shelves of western Africa, South America and elsewhere, often associated with phosphate deposits. An example is the diatomaceous ooze with remarkable enrichments in heavy metals which was discovered in bays of the Gulf of Tajura (part of the Gulf of Aden) (Amann et al., 1973). Sediment containing up to 300 ppm U, 30–50 ppm Mo and 200–400 ppm V was recovered. Carbonaceous mud is presently forming mostly in shelf areas with upwelling nutrient-rich water where biogenic and geochemical processes concentrate the metals.

D.S.D.P. drilling has revealed the existence of pyrite- and siderite-bearing black muds and shales, formed in reducing environments, at or near the base of many Atlantic Ocean sediment cores (Fig.2). Most of the black shales are Cretaceous in age, and they apparently were deposited in shallow to deep water under restricted circulation or stagnant conditions at a time when abundant terrestrial organic detritus was being carried into the narrow, newly formed Atlantic Ocean (e.g., Scientific Party, D.S.D.P. Leg 48, 1976; but cf. Fischer et al., 1977). The dark, organic-rich units (carbonaceous clays and sapropels) are often interbedded with units deposited under more oxidizing conditions, such as the dolomitic silts and clays at Site 138, and micritic limestone at Site 330. Metallic sulphides in the black shales may have as their source hydrothermal exhalations from the Mid-Atlantic Ridge spreading belt (Emery and Skinner, 1977).

The Phosphoria Formation of the western United States, the Chattanooga Shale of the mid-continent of the United States (Vine and Tourtelet, 1970), and the Upper Permian Kupferschiefer bituminous marl of Germany (see Vaughn, Chapter 10, Vol. 2; and Jung and Knitzschke, Chapter 7, Vol. 6, in this series) appear to be fossil occurrences of metal-rich carbonaceous muds which accumulated in shelf areas.

Metalliferous sediments

More widespread on the ocean floor than organic-rich metal-bearing muds are metalliferous sediments, primarily formed as a result of hydrothermal phenomena and aug-

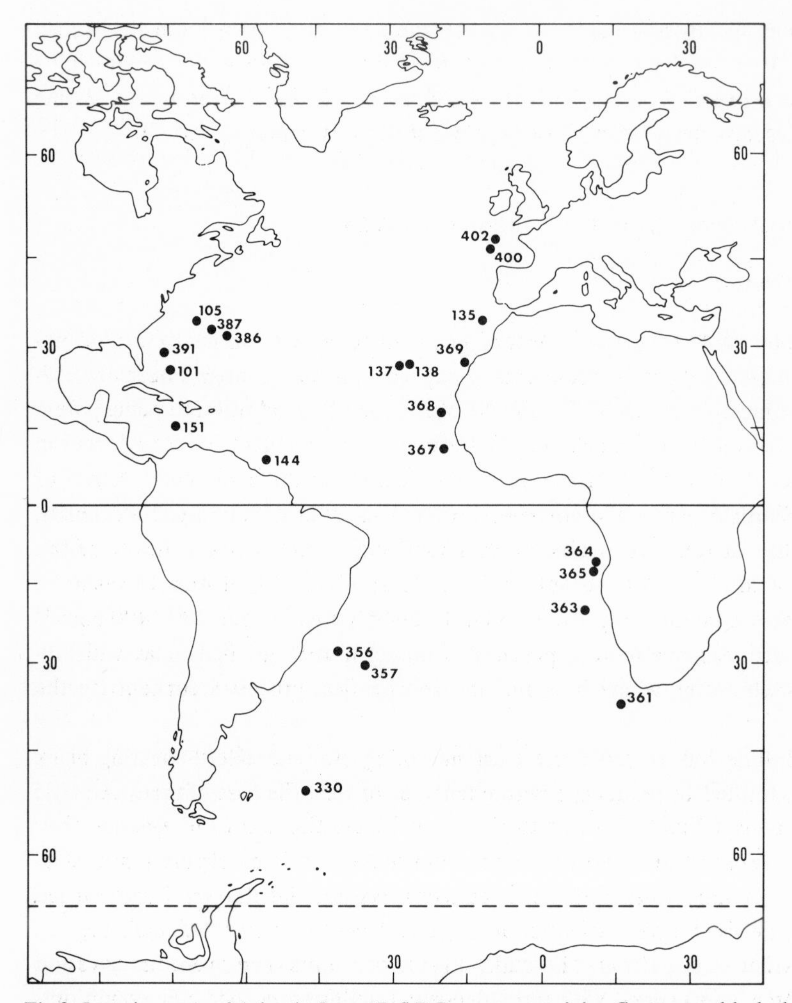

Fig.2. Location map of Atlantic Ocean D.S.D.P. cores containing Cretaceous black shales.

mented in metal content by authigenic or diagenetic processes. The important metal-bearing phases are metal sulphides, iron smectite, amorphous and poorly crystalline iron oxides, and ferromanganese oxides. These phases are mixed with varying proportions of calcareous and siliceous tests and volcanoclastic debris, as well as with other minerals formed *in situ* and with terrestrially derived detritus, the latter usually making up only a small fraction of the sediment.

Metalliferous sediments are known to occur on the crests of mid-ocean ridges, as basal deposits beneath pelagic sediments extending away from the ridge crests, in isolated basins, in fracture zones, and in deposits formed around volcanic islands. Cronan (1976a) has also suggested that they may be formed in marginal basins behind island arcs, such as the Lau Basin (see, however, Bertine, 1974; Bertine and Keene, 1975; Robinson, 1977; Knedler et al., in prep.), and a hydrothermal origin for metals in the lakes of the East African Rift system has been reported (Degens and Kulbicki, 1973; Müller and Förstner, 1973; Degens and Ross, Chapter 4, Vol. 4 and Förstner, Chapter 4, this volume, in this seris). The importance of plate tectonic processes in controlling the locations of various types of mineral deposits found on the sea floor has been discussed by Blissenbach (1972) and Blissenbach and Fellerer (1973).

NATURE OF METAL-BEARING PHASES

Iron oxide phases

Physical appearance. Hydrothermal precipitates from exhalations on the East Pacific Rise were originally described by Boström and Peterson (1966), who noted that abundant brown amorphous metal oxides constituted a separate, very finely disseminated or loosely aggregated material. The finely disseminated material imparts a bright orange-brown colour to the supernatant liquid of crestal sediment suspensions and does not settle for many weeks. Cronan and Garrett (1973) similarly described the non-biogenic constituents of basal metalliferous sediments in the eastern Pacific as consisting of translucent grains and globules of amorphous yellowish to dark reddish-brown iron oxides ranging from 1 to 25 μm in diameter, being dominantly about 5 μm (Von der Borch and Rex, 1970). Somewhat larger globules (100 μm) were found in sediments from the manganese nodule belt of the North Pacific described by Bischoff and Rosenbauer (1977). Scanning electron micrographs of the ferruginous spherules reveal the surface textures to be variable, but commonly consisting of a felted texture of rod-like forms, about 0.1 μm across and 1 μm in length, reminiscent of sheaths of iron-depositing bacteria (Von der Borch et al., 1971); other surfaces have a more crystalline appearance.

These amorphous metallic globules were given the name "RSO's" by Yeats and Hart et al. (1976); this is an abbreviation for "red-brown semi-opaque oxides." RSO's from the Nazca Plate were extensively studied by Bass (1976), who found them in all but one sample from D.S.D.P. Leg 34 cores. Variations in the size, colour, relief and transparency of the RSO's were observed between the particles from Sites 319 and 320 which, together with the higher density of RSO's from the older sediments from Site 319, suggested an ageing process. However, the observation that the particles from the basal sediment at Site 321 are almost entirely less dense than calcite argues against such an ageing process, favouring instead a sorting process where density and other properties vary with distance of the depositional site from the spreading centre (Bass, 1976). Bass also found that the RSO content varied roughly with goethite content and could be readily distinguished from the other main iron-containing mineral, smectite, by the lighter colour, extremely fine grain size and birefringence of the latter. Interestingly, Bass concluded that the study of crystalline components of metalliferous sediments had been over-emphasized in past studies by investigators who perhaps spuriously implied a genetic relation of such components to the largely amorphous RSO's.

Basal iron oxide-rich facies from the Wharton Basin (Pimm, 1974a, b), the Central Indian and Carlsberg Ridges (Cronan et al., 1974); and from the Madagascar Basin in the southwest Indian Ocean (Warner and Gieskes, 1974; Leclaire, 1974) are largely X-ray amorphous yellow-brown globules of varying shape. Goethite and some haematite have been identified in the basal sediments from the Wharton Basin.

The goethite-amorphous facies of the Red Sea contains yellow to brown irregular or roughly spherical (up to 50 μm diameter) aggregates (Herman and Rosenberg, 1969), 1–30 μm spherules (Bischoff, 1969a) or yellowish-red and brown spheroidal aggregates (1–10 μm) of amorphous iron hydroxides (Miller et al., 1966; Baturin et al., 1969).

Mineralogy. The similar appearance of these metalliferous sediments is striking, particularly when one considers their wide distribution both in locality and in age. Their amorphous character is frequently reported, although more sophisticated X-ray diffraction techniques (Dasch et al., 1971; Dymond et al., 1973) have revealed a complex mineralogy with smectite, δ-MnO_2, goethite and todorokite as the dominant phases. The iron oxide minerals which have been detected by X-ray diffraction and other means in metalliferous sediments include goethite, haematite, lepidocrocite, maghemite, magnetite and ferrihydrite (Table I).

Magnetic susceptibility and Mössbauer spectroscopic studies of metalliferous sediments are not commonly reported in the literature, yet are particularly useful in elucidating iron mineralogy. Strangway et al. (1969) discussed the magnetic properties of minerals from the Red Sea thermal brines, and with additional information on thermal behaviour, were able to identify goethite together with trace amounts of maghemite, siderite and lepidocrocite. More specifically, the goethite probably falls within the size range exhibiting superparamagnetic behaviour at room temperature, around 100 Å.

Mössbauer studies of metalliferous sediments from the Bauer Deep have been reported by Bagin et al. (1975), who found that superparamagnetic goethite (140–300 Å grains) was the predominant colloidal mineral in these sediments. Thermomagnetic studies of the sediments suggested the presence of less than 0.1% maghemite. Bagin et al. concluded that the maghemite probably formed by oxidation of finely dispersed magnetite. This is consistent with the observation that magnetic properties in the northern parts of the Bauer Deep were determined by paramagnetic and superparamagnetic minerals, while those in the southern parts were determined by magnetic minerals which reflected the influence of Easter Island.

Goethite is the product of hydrolysis of most solutions of iron (III) salts, except chloride, and will form directly from solutions by hydrolysis, or from amorphous hydroxide if its solubility has not been considerably reduced by ageing (Schwertmann, 1966). Schwertmann also discussed the influence of silica, phosphate and organic com-

TABLE I

Non-detrital minerals found in metalliferous sediments

Iron oxides	
Goethite	α-FeOOH
Haematite	α-Fe_2O_3
Lepidocrocite	γ-FeOOH
Maghemite	γ-Fe_2O_3
Magnetite	Fe_3O_4
Ferrihydrite	$2.5Fe_2O_3 \cdot 4.5H_2O$
"Limonite"	$FeOOH \cdot nH_2O$
Iron smectite	
Nontronite	$(Fe, Al)Si_2O_5(OH) \cdot nH_2O$
Fe-montmorillonite	$(Fe, Al)Si_2O_5(OH) \cdot nH_2O$
Sulphides	
Pyrite	FeS_2
Sphalerite	ZnS
Pyrrhotite	$Fe_{1-x}S$
Chalcopyrite	$CuFeS_2$
Marcasite	FeS_2
Marmatite	(Zn, Fe)S
Greigite	Fe_3S_4
Mackinawite	FeS
Manganese oxides	
Todorokite	$(Mn, Ba, Ca, Mg)Mn_3O_7 \cdot H_2O$
Birnessite	$(Na_{0.7}Ca_{0.3})Mn_7O_{14} \cdot 2.8H_2O$
δ-MnO_2	δ-MnO_2
Psilomelane	$BaMn_9O_{16}(OH)_4$
Manganite	γ-MnOOH
Groutite	α-MnOOH
Ranciéte	$(Ca, Mn)Mn_4O_9 \cdot 3H_2O$
Woodruffite(?)	$(Zn, Mn)_2Mn_5O_{12} \cdot 4H_2O$
Carbonates	
Calcite	$CaCO_3$
Aragonite	$CaCO_3$
Magnesian Calcite	$CaMg(CO_3)_2$
Dolomite	$MgCa(CO_3)_2$
Siderite	$FeCO_3$
Manganosiderite	$(Mn, Fe)CO_3$
Rhodochrosite	$MnCO_3$
Other minerals	
Barite	$BaSO_4$
Anhydrite	$CaSO_4$
Gypsum	$CaSO_4 \cdot 2H_2O$
Halite	NaCl
Vivianite	$Fe_3(PO_4)_2 \cdot 8H_2O$
Glauconite	$K_2(Mg, Fe)_2Al_6(Si_4O_{10})_3(OH)_{12}$
Palygorskite	$Mg_5Si_8O_{20}(OH)_2 \cdot 8H_2O$
Sepiolite	$Mg_2Si_3O_8 \cdot 2H_2O$
Phillipsite	$(K_2, Na_2, Ca)(Al_2Si_4O_{12}) \cdot 4.5H_2O$

(continued)

TABLE I (continued)

Clinoptilolite	$(Na, K, Ca)_{2-3}Al_3(Al, Si)_2Si_{13}O_{36} \cdot 12H_2O$
Analcite	$NaAl(SiO_3)_2 \cdot H_2O$
Garnet (Hydrogrossularite)	$Ca_3Al_2Si_3O_{12}(OH)_x$
Opal	$SiO_2 \cdot nH_2O$
Cristobalite	SiO_2
Tridymite	SiO_2
Native Copper	Cu

Compiled from numerous sources, especially relevant volumes of *Initial Reports of Deep Sea Drilling Project.*

pounds in the formation of goethite (see also Schwertmann and Taylor, 1972a, b). Nucleation in iron (III) solution and hydroxide gels has been studied by Atkinson et al. (1968, 1977).

There is a tendency for goethite to be reported in the older or basal metalliferous sediments which are now far removed from the spreading centre at which it is thought they were formed. This presumably results from the ageing of the amorphous iron-oxide phases.

Haematite is a common constituent of basal metalliferous sediments. For example, 11.7% haematite was reported in a reddish brown zeolitic clay above basalt at D.S.D.P. Site 9A (Peterson et al., 1970). At Site 105, also located in the northwestern Atlantic, haematite is mentioned as "common" in a basal hard reddish brown clayey limestone (Hollister and Ewing et al., 1972). Red scales of hematitic and less crystalline masses of limonitic material have been found in an iron oxide sediment at Site 213 in the Indian Ocean (Von der Borch and Sclater et al., 1974), and a haematite- and calcite-veined hyaloclastite was recovered from the lowermost samples of Site 307 in the northwestern Pacific (Larson and Moberly et al., 1975). Haematite is generally considered to be indicative of hydrothermal activity.

Magnetite found in marine sediments is usually believed to be a detrital volcanic mineral, but may also have a hydrothermal origin. Hackett and Bischoff (1973) reported a magnetite facies in the Atlantis II Deep. There, the magnetite coexists with haematite. Because the proportion of magnetite relative to haematite increases down the core sample, Hackett and Bischoff postulated a diagenetic transformation of well-crystallized haematite to well-crystallized magnetite. Magnetite has occasionally been detected in basal metalliferous sediments, such as the iron-rich laminations in brown clay at Site 248 in the Mozambique Basin (Simpson and Schlich et al., 1974).

Chukhrov (1973) and Chukhrov et al. (1973, 1974, 1977a, b) have made extensive investigations into the nature and occurrence of iron oxide hydroxides in sediments. Ferrihydrite, equivalent to the brown X-ray amorphous hydrous iron oxide of Towe and Bradley (1967), was first described by Chukhrov and has been found associated with goethite in sediments from Santorini (previously described by Zelenov, 1964). Elsewhere, the presence of ferrihydrite has been attributed to the rapid oxidation of Fe^{2+} in sediments from the Atlantis II Deep (identified on the basis of transformations reported by Bischoff, 1969b), in the Cheleken precipitates of thermal fields on land and in the ferruginous precipitates of the thermal fields of the Kurile Islands (Chukhrov, 1973; Chukhrov et al., 1974). The colloidal iron oxides from the East Pacific Rise were thought to be ferrihydrite because X-ray powder patterns were similar to those for haematite, the latter being the spontaneous transformation product of ferrihydrite under the appropriate conditions.

Ferrihydrite can, however, be transformed into goethite under highly acid or alkaline conditions and by the presence of solutions of Fe^{2+} salts (Chukhrov et al., 1974, 1975, 1977a). Both transformations are retarded if the ferrihydrite contains a fine silica admixture (Chukhrov et al., 1975). Schwertmann and Fischer (1973) proposed that ferrihydrite is a source of dissolved iron for the crystallization of goethite from solution.

Iron smectite phase

Smectite appears to have an ubiquitous association with metalliferous sediments, and in the Red Sea constitutes a facies immediately overlying the amorphous-goethite facies (Bischoff, 1969a). Bischoff (1972) used Mössbauer spectroscopy in conjunction with chemical and X-ray diffraction data to characterize the smectite mineral from the Red Sea geothermal system as having a composition intermediate between true dioctahedral nontronite and a trioctahedral hypothetical end member, in which ferrous iron occupies all octahedral sites.

Smectite is also a constituent of the mixed amorphous iron–manganese oxide-detrital facies of basal sediment in the eastern Pacific (Von der Borch and Rex, 1970). Smectite, recently characterized as an iron-rich, aluminum-poor nontronite (Dymond and Eklund, 1978), is the major component of Bauer Deep sediment (Dasch et al., 1971; Sayles and Bischoff, 1973; Sayles et al., 1975).

Dymond et al. (1973), on the basis of isotopic analyses, concluded that the smectite in East Pacific and Bauer Deep samples which they investigated was authigenic. The absence of a continental source for iron-rich montmorillonite of the Bauer Deep, observations of smectite pseudomorphs of radiolaria, and the chemical similarity between pore fluid and bottom water, led Sayles and Bischoff (1973) to postulate that at least part of the smectite originated from diagenetic processes which must have occurred near the seawater–sediment interface. Sayles et al. (1975) favoured an authigenic origin for the montmorillonite whereby hydrothermal solutions and seawater interacted at low temperatures.

More recently, Heath and Dymond (1977) have put forward a model depositional regime for the northwestern Nazca Plate, which elegantly explains the change in distribution of iron between the principal sedimentary components of this region. Essentially, it involves seawater leaching of fresh-formed basalt, liberating iron (and other elements) which form an amorphous hydroxide floc upon contact with cold, oxygen-bearing seawater. A small fraction of the iron may react with silica to form iron-rich smectite. These fine-grained sediments are readily transported by deep water crossing the rise into the Bauer Deep and Central Basin. As the sediment is moved, the surface-active hydroxide reacts with biogenically deposited silica to form more iron-rich smectite. Dymond and Eklund (1978) have further elaborated upon this model, which also accounts for the chemical composition of ferromanganese nodules and crusts (Lyle et al., 1977).

Corliss et al. (1978) attribute the formation of nontronite in the hydrothermal mounds near the Galapagos Rift to a change from reducing conditions within the mounds where Fe and Si (leached from basalt and sediments) are deposited as a nearly aluminum-free nontronite, to oxidizing conditions at the bottom–water interface, where silica-rich iron oxides and manganese oxides form the exposed surface of the mounds. The critical difference between these models appears to be the necessary redox conditions. The model of Heath and Dymond (1977) is a special case, where, at positive Eh values ferric oxyhydroxides are precipitated at the crest and subsequently react with biogenic silica to form smectite.

Hekinian et al. (1978) proposed two models to account for the green clay-rich material from the Galapagos Spreading Centre, one of which does not involve fluids interacting with sediments, but instead, emanation directly into seawater above the mounds followed by precipitation of the clay-rich material. The relatively unaltered nature of the basalt immediately underlying the hydrothermal deposits is compatible with this model. The chemical and mineralogical zonation of the hydrothermal deposits from Transform Fault "A" of the FAMOUS area suggests that the method of emplacement of the deposits in both areas was similar (Hoffert et al., 1978).

Sulphide phases

The most common sulphide in metal-enriched marine sediments is pyrite. It is ubiquitous, usually in minor amounts, in carbonaceous sediments, forming by reaction of dissolved H_2S and detrital iron minerals. Pyrite can also precipitate from hydrothermal fluids. It is associated with sphalerite, which is the dominant mineral, and lesser amounts of chalcopyrite, X-ray amorphous iron monosulphide, marcasite and marmatite in a distinct sulphide facies of the Red Sea geothermal area (Bischoff, 1969a; Stephens and Wittkop, 1969).

Bischoff proposed a sequence of processes to account for the formation of iron-bearing minerals of Red Sea brines. Here, precipitation of the iron sulphide is probably effected by cooling of the brine; this releases the metal from the chloride complex, which then is precipitated by sulphur, probably derived by bacterial reduction of sulphate. Further cooling and partial oxidation conditions would account for the precipitation of montmorillonite characterized by high ferrous iron content. Ferric hydroxides would be formed under completely oxidizing conditions and with time transformed to the more stable goethite modification. Local temperature anomalies could account for subsequent formation of haematite (and magnetite; see Hackett and Bischoff, 1973) or, alternatively, diagenesis involving bacterial activity would produce pyrrhotite.

A pyrite-bearing aragonitic non-fossiliferous limestone has been reported from basal sediments cored above alkali basalt beneath the Aleutian Abyssal Plain (Natland, 1973). Associated with the limestone are iron-rich clays and clay-free goethite-bearing calcareous ironstone, together constituting sediment of facies diversity similar to that of Red Sea sediments.

Massive sulphide deposits have recently been discovered at the axis of the East Pacific Rise (Francheteau et al., 1979). Chemical analyses of these sediments indicate two principal modes: one zinc-rich (23–28.7% Zn) and the other iron-rich (19.9–42.7% Fe). Preservation of the deposits could have been effected either by anoxic bottom waters or by rapid burial. Associated fragile ochre-coloured material made up of amorphous iron oxides is considered by these authors to result from the oxidation of the iron sulphide.

(See Vaughn, Chapter 10, Vol. 2 of this series, for a discussion of the mineralogy and geochemistry of sedimentary sulphides.)

Manganese oxide phase

Manganese can substitute for iron in the X-ray amorphous oxides which are dispersed in metalliferous sediments, and it can also form separate oxide minerals such as todorokite, birnessite and δ-MnO_2. These minerals may occur as finely disseminated grains in sediment, in the form of manganese-rich micronodules, or as crusts on submarine rocks, in addition to the more familiar manganese nodule form. Microprobe work on Bauer Deep sediments by Dymond and Eklund (1978) revealed that an iron-rich microlaminate component (in the form of micronodules) consists of alternating submicrometre-width layers of todorokite and iron hydroxide. The manganese minerals have been collectively referred to as manganese hydroxyoxides (e.g., Dymond et al., 1973; Rona, 1978).

Other phases

A number of other minerals of hydrothermal or hydrogenous origin are often detected in metalliferous sediments. Siderite has been reported in older sediments from the Atlantic Ocean (Peterson et al., 1970; Rex, 1970) and also as a major constituent of Recent sediment from Santorini (Butuzova, 1966; Bonatti et al., 1972; Puchelt et al., 1973). Peterson et al. (1970) attributed the frequent association of siderite with rhodocrosite and cristobalite to hydrothermal alteration of tuffs (see also Rex, 1970). Hydrothermal barite crystals have been found disseminated in East Pacific Rise sediments (Arrhenius and Bonatti, 1965) and in opal-cemented hyaloclastite from the Lau Basin (Bertine and Keene, 1975).

Size fractionation and chemical leaching of the iron phases

The distribution of iron in size-fractionated metalliferous sediments has been investigated by Dymond et al. (1973) and Sayles et al. (1975). Dymond et al. (1973) found that the <2 μm fraction of sediments from the central North Pacific is slightly enriched in iron relative to the 2–20 μm fraction, whereas the data of Sayles et al. (1975) indicate a much stronger fractionation in sediments from the Bauer Deep. Here, the level of iron in the colloidal fraction is 30–60% above the concentrations found in the >2 μm size fraction. Generally, the amount of dispersed colloidal iron is greatest in the finest fraction of marine sediments (Skornyakova, 1965, and references therein; Halbach et al., 1979).

Selective chemical leaching procedures have been useful in determining the particular association of iron, whether with manganese oxide, iron oxide or silicate phases, especially where fine particle size precludes mineralogical determinations by the more common (and less time consuming) X-ray diffraction method. The most frequently used method is that based on the Chester and Hughes (1967) leaching techniques. This essentially involves a three-step treatment with (1) dilute acetic acid to determine carbonate, loosely held ions and interstitial water evaporates, (2) mixed acid-reducing agent (acetic acid and hydroxylamine hydrochloride) which takes into solution the above and also ferromanganese oxides, and (3) hot hydrochloric acid which solubilizes most minerals, including iron oxide minerals, but not the most resistant silicates and aluminosilicates (see Cronan, 1976b).

Cronan and Garrett (1973) and Cronan (1976b) found that iron in basal metalliferous sediments from the eastern Pacific is concentrated in the HCl-soluble fraction, indicating that most of this element is not associated with carbonates or ferromanganese oxide phases. Sayles and Bischoff (1973) used the mixed acid-reducing agent on ferromanganoan sediments from the equatorial east Pacific, but advocated caution in interpreting the results of such analyses, because the method dissolves colloidal iron oxides as well as ferromanganese compounds, and probably also the montmorillonite. They concluded that 60–80% of the total iron must reside in the smectite. Similar results were obtained for sediments from the Bauer Deep by Sayles et al. (1975) and Heath and Dymond (1977). In the latter study, the selectivity of various leach solutions was investigated, and this resulted in the use of ammonium oxalate–oxalic acid buffer which removes amorphous ferric hydroxides and poorly crystalline ferromanganese oxyhydroxides, but not crystalline goethite. These authors found a systematic increase in oxalate-insoluble iron from northern East Pacific Rise to Bauer Deep to Central Basin samples which reflects a mineralogical transformation from a poorly crystallized iron hydroxide to an iron-rich smectite. Significantly different leach results were obtained for Bauer Deep sediments by Bagin et al. (1975), who concluded that most of the iron was associated with well-crystallized oxides and hydroxides. The iron silicate (montmorillonite) was considered to play a subordinate role and, further, no major differences in the mode of occurrence of iron between metalliferous sediments from the East Pacific Rise and the Bauer Deep were reported by these authors.

MARINE DISTRIBUTION AND NATURE OF SURFACE AND NEAR-SURFACE DEPOSITS

Metalliferous sediments have been found throughout the world ocean, except in the Arctic Ocean (Fig.3). Principal characteristics and compositional aspects of the major deposits are reviewed in this section.

East Pacific Rise

Iron–manganese oxides were first reported as an important constituent of East Pacific Rise sediments by El Wakeel and Riley (1961), who analyzed a chocolate-coloured ooze, described as a calcareous–manganiferous mud (sample 15) and found it to contain 13.9% Fe_2O_3, 4.8% MnO, and 70.6% $CaCO_3$. Subsequently, more detailed studies of EPR sediments were performed by Bonatti and Joensuu (1966), Boström and Peterson (1966, 1969) and Boström et al. (1969), amongst others. Boström has undoubtedly been the most active worker in this area (Boström, 1967, 1970a, 1973; Boström and D.E. Fisher, 1969; Boström and Valdes, 1969; Boström et al., 1971, 1973, 1974a, b, 1976; D.E. Fisher and Boström, 1969; Rydell et al., 1974; Veeh and Boström, 1971).

Bonatti and Joensuu (1966) found several types of metalliferous deposits on the upper flanks and crest of a seamount located near the EPR crest in the South Pacific (cf. Bonatti, 1967). A dredge haul consisted primarily of red-yellow fragments of a friable, powdery rock with about 30% iron in the form of poorly crystalline goethite. Also recovered were a few pebbles of black porous manganese oxides, and one small fragment of relatively fresh basalt, a part covered by a crust of manganese oxides less than 1 mm thick. Another seamount yielded massive basalt and hyaloclastics consisting of basaltic glass grains up to 1 mm in size, cemented in some samples by a red ferruginous matrix, and in others by a black matrix rich in manganese.

Most of the East Pacific Rise crest is, however, blanketed with calcareous ooze in which the metal components are admixed. At 6°N and 12–16°S, the crestal sediments contain abundant, brown, X-ray amorphous metal oxide precipitates (Boström and Peterson, 1966). The precipitates are extremely fine and do not represent coatings or impregnations of other mineral grains. Where abundant, the precipitate stains globigerina ooze a dark brown. The metal oxides are rich in iron and manganese and poor in aluminum and titanium, and are known to occur in the ooze in significant quantities to a depth of at least 6 m (Boström and Peterson, 1969). Iron–manganese micronodules and grain coatings play only a minor role in concentrating metals in crestal metalliferous sediments (Dymond et al., 1973).

Bischoff and Sayles (1972) noted a difference in the major component of the dilute acetic acid-insoluble (carbonate-free) fraction between the upper and lower parts of crestal and flank sediment cores. In the upper few meters, an amorphous silicate dominates, characterized by high water content (about 80% by weight) and a dark brown colour; it is microscopically isotropic and colloidal, and X-ray amorphous. In the deeper

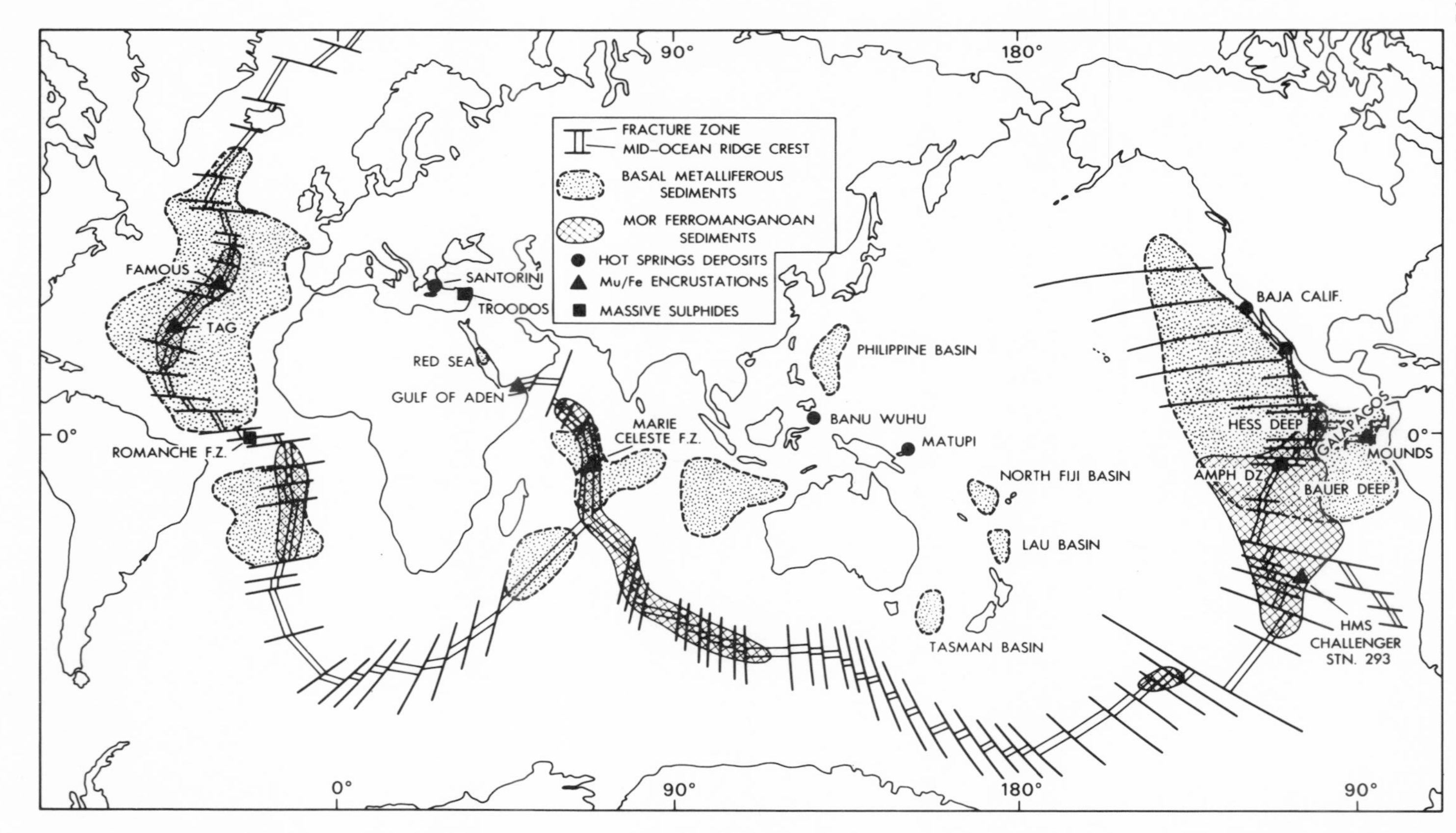

Fig. 3. Map of general location of marine metalliferous deposits. Modified from Rona (1978); MOR ferromanganoan sediment distribution after Boström et al. (1969), among others. Basal metalliferous sediment distribution from D.S.D.P. drilling.

parts of cores, however, nontronitic montmorillonite of varying degrees of crystallinity predominates. Since the crystallinity appeared to increase with depth in some cores, Bischoff and Sayles suggested the possibility that amorphous silicate material crystallizes to montmorillonite with time. However, Dymond et al. (1973) found that reducing acid-soluble goethite from several areas of the Pacific showed little evidence of recrystallization with age from the near-colloidal original precipitate.

In an excellent analysis of the factors controlling the formation of metalliferous sediments from the EPR-Bauer Deep region, Heath and Dymond (1977) have shown that analytical data for the sediment can be described in terms of a mixture of hydrothermal, detrital, hydrogenous, and biogenous material, and that the principal factors influencing the distribution of each element include the relative input of each of the above four sources, the lateral transport of sediments across the area and the transformation of the unstable metalliferous hydroxides into more stable smectite and ferromanganese oxyhydroxides.

Lopez and Corliss (1976) examined the chemistry of various components in a sediment core collected east of the East Pacific Rise. Ni, Co, Sb, W and Ce were strongly concentrated in manganese micronodules. Arsenic was concentrated in both micronodules and phillipsite. Rare earth elements (other than cerium) were concentrated in fish debris, and some phase other than barite also appeared to contain barium. As pointed out by Heath and Dymond (1977), these data and others emphasize the polygenetic origin of sediments in the southeastern Pacific.

Iron in East Pacific Rise sediments

In an attempt to elucidate further the nature of iron in metalliferous sediments from the East Pacific Rise, six crestal sediment samples (Fig.4 and Table II) were investigated using mineralogical, chemical and Mössbauer methods. The samples were collected

TABLE II

Sample identification

Sample core number	Sampling interval (cm)	Latitude	Longitude	Water depth (m)
SH1526	26– 29	8°55.30'S	108°20.05'W	3252
SH1529	21– 25	9°07.69'S	108°42.25'W	3798
SH1529	106–110	9°07.69'S	108°42.25'W	3798
SH1529	139–143	9°07.69'S	108°42.25'W	3798
SH1531	28– 31	9°05.22'S	108°55.44'W	3832
SH1531	115–117	9°05.22'S	108°55.44'W	3832

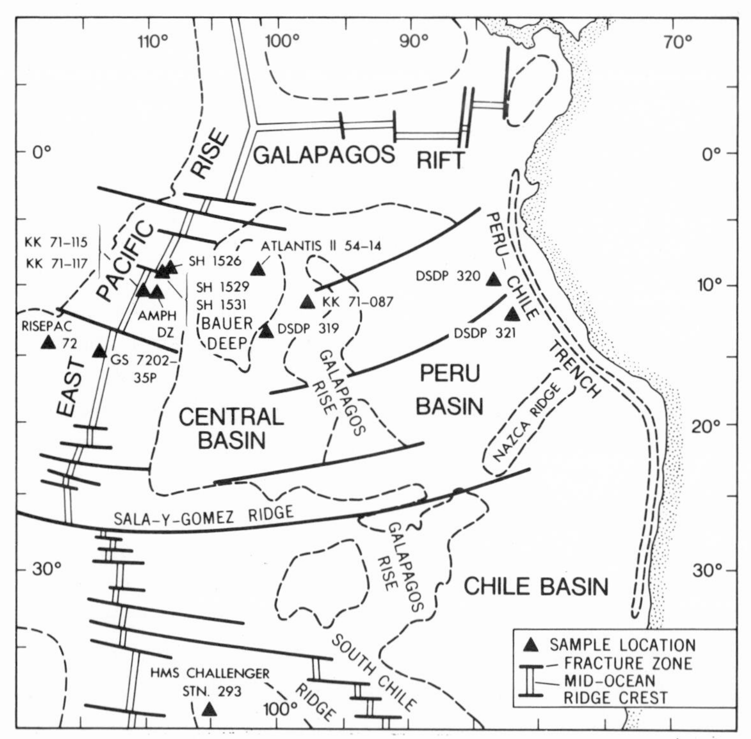

Fig.4. Map of southeastern Pacific Ocean metalliferous deposit locations. Bathymetry schematic; sea-floor tectonic pattern after Dymond and Veeh (1975). Analytical data for samples from specific deposit locations are given in Tables III, IV, X–XII.

in 1976 aboard RRS *Shackleton* by gravity corer, as part of a geochemical study of the area.

Previous investigation of the precise nature and mode of occurrence of iron in metalliferous sediments has been impaired by the extremely fine particle size of the precipitate, and consequently some of the more common geochemical investigative methods are unsuitable. Chemical analyses of metalliferous sediments are well documented in the literature, as are mineralogical studies of the crystalline (detectable by X-ray diffraction) phases. In an attempt to determine the character of iron in these sediment, selective chemical leaching techniques have been used to determine "mobile" and silicate-associated iron. These have proved particularly useful in calculating the partitioning of elements between the principal sedimentary components. Magnetic susceptibility studies are less common and ^{57}Fe Mössbauer spectroscopic data scarce. The latter method is, however, particularly appropriate to the study of iron in such sediments as it is applicable to X-ray amorphous substances and suffers no interference effects from other non-iron-containing constituents.

Experimental. X-ray powder diffraction studies of air-dried, unwashed sediments were carried out according to the procedure of Johnston and Glasby (1978) using a scan speed of 0.125° 2 Θ/min. Further characterization of the untreated sediment crystalline components was attempted by step scanning over selected angular intervals at 0.05° 2 Θ angular increments and counting for 100 seconds at each increment to increase the peak/background ratio and hence increase the sensitivity of this method.

In addition, the carbonate-free fractions of the sediments were investigated. Such fractions were obtained by leaching with 0.2*M* acetic acid buffered to pH 4.8 and subsequently washed and air dried. X-ray diffraction patterns of each sample, mounted in such a way as to minimize preferred orientation effects, were recorded before and after glyceration.

The major chemical compositions of the sediments were determined by X-ray fluorescence spectroscopy (XRF) using fused glass discs, according to the method of Norrish and Hutton (1969), and an automated Siemens SRS-1 spectrometer. Trace-element (except barium) analyses were carried out on pressed powder samples, and trace barium analyses on glass discs (Kennedy and Roser, in prep.).

^{57}Fe Mössbauer spectra were obtained at room temperature for four air-dried sediments [SH1526 (26–29 cm), SH1529 (21–25 cm, 106–110 cm and 139–143 cm)] and for the acetic acid-leached fractions of SH1529 (21–25 cm and 139–143 cm). The experimental procedure has been previously described by Johnston and Glasby (1978).

Two Lorentzian peaks were computer fitted to the experimental spectra using a modified version of the χ^2 non-linear regression minimization procedure of Stone (1967) (see also Mroczek, 1977). Evaluation of the acceptability of the fit was determined from chemical and statistical considerations. In order for the fit to be chemically acceptable the widths and areas of the peaks constituting the quadruple doublet had to be similar. For statistical acceptability, the χ^2 value had to lie between the 1% and 99% points on the χ^2 probability distribution curve. In addition, the residual deviations plot had to be statistically featureless.

X-ray diffraction analysis results and discussion. Calcite is the principal mineral in all six sediment samples. Minor amounts of plagioclase feldspar, mostly oligoclase–andesine varieties, occur in four samples, with greatest abundance in SH1531 (115–117 cm). Plagioclase was not detected in SH1526 (26–29 cm), and was observed only in the leached fraction of SH1529 (21–25 cm). Alkali feldspar (probably sanidine) was tentatively identified in the latter sample, and trace amounts were found in all other samples.

Smectite is the other major constituent of the non-carbonate fraction, and is most abundant in SH1529 (106–110 cm and 139–143 cm) and SH1531 (115–117 cm). Phillipsite was positively identified only in SH1529 (21–25 cm), and tokorokite was tentatively identified in this same sample by the presence of peaks at 9.7 Å and 4.82 Å. Step scanning procedures permitted the tentative identification of barite as a trace component in all six sediments.

The expansion of the basal spacing from 14.3 Å to approximately 17.5 Å upon glyceration, and the (060) spacings which range from 1.50 Å to 1.508 Å for all cases, suggest that the smectite is predominantly of dioctahedral character, best fitting montmorillonite (MacEwan, 1961) and similar to an iron-montmorillonite from south of the Gulf of California (Butuzova et al., 1976). Infrared spectra of the leached sediments support the above assignment. SH1531 (21–29 cm and 115–117 cm) also have diffraction peaks around 1.53 Å, suggesting that a trioctahedral clay component may be an additional constituent of these two sediments.

Peterson and Goldberg (1962), in their study of feldspar in sediments from the South Pacific, found that bytownite was the dominant plagioclase in all specimens taken from the East Pacific Rise north of Easter Island to about 13°S, and that farther north only small quantities, insufficient for identification, were found in the fine 4–8 μm fraction. South of Easter Island, oligoclase–andesine varieties of plagioclase and anorthoclase alkali feldspar with lesser amounts of labradorite–andesine and sanidine feldspars were found in sediments from the crest and flanks of the Rise. Labradorite has also been reported as the hydrochloric acid-insoluble residue of sediments from the crest and flanks of the East Pacific Rise 16–17°S (Bender et al., 1971). The finding in this study that oligoclase–andesine varieties are the dominant feldspars in five samples, occurring together with lesser amounts of alkali feldspar, is generally in accordance with the above work. One might have expected more calcic plagioclase because of its association with basaltic volcanism, but the absence of feldspars in SH1526 (26–29 cm) and their paucity in SH1529 (21–25 cm) may indicate that localized sources for these minerals are most important (see Peterson and Goldberg, 1962).

Montmorillonite is the other dominant crystalline component of the sediments investigated here, and has been found to be the dominant clay mineral in Cenozoic deep-sea sediments from the equatorial Pacific, formed by the alteration of volcanic debris (Heath, 1969). However, the carbonate-free fractions of sediments from active ridges are depleted in Al and Si, indicating that the absolute quantities of clay minerals are small (Boström et al., 1969). Iron-rich montmorillonite has been reported associated with many occurrences of metalliferous sediments, for example, from the North Pacific (Dymond et al., 1973) and DOMES Site C (Bischoff and Rosenbauer, 1977), although Marchig (1978) and Glasby et al. (1979) have disputed the metalliferous nature of the latter; the East Pacific Rise (Bender et al., 1971; Field et al., 1976); the Bauer Deep (Dasch et al., 1971; Sayles and Bischoff, 1973; Sayles et al., 1975); the iron-montmorillonite facies of the Red Sea (Bischoff, 1969a); the FAMOUS area of the Mid-Atlantic Ridge (Zemmels et al., 1977) and the sediments in the vicinity of the Carlsberg and Indian Ridges (Cronan et al., 1974). In two of the East Pacific Rise sediments investigated here, both di- and tri-octahedral clays probably are present. Such a mixture of smectites has been reported for Red Sea sediments (Stoffers and Ross, 1974).

Simple calculations using analytical data for SH1526 (26–29 cm), chosen because it has negligible feldspar, and assuming that all the silicon is bound in a smectite phase of composition similar to that reported for the Bauer Deep nontronite (Dymond and Eklund, 1978), i.e. Fe/Si = 0.85, indicate that approximately 40% of the iron must be present in non-silicate minerals. However, no crystalline oxides or oxide hydroxides of iron were detected using X-ray diffraction techniques in either the bulk or leached sediment. Therefore, it is inferred that the iron is present in an amorphous state. If a correction is made for opaline silica using the method outlined in Piper (1973) and Boström et al. (1973), the percentage of iron in a non-silicate phase increases to 84%. This correction procedure does not give Al/Si weight ratios for the opaline-free fraction (0.38) similar to those of the Bauer Deep nontronite (0.05), suggesting that the assumptions made may not be valid in this case. Element partitioning calculations by Heath and Dymond (1977) indicate that iron percentages in the hydroxide, smectite and aluminosilicate components of sediments from the East Pacific Rise are 81%, 18% and 1%, respectively, whereas in Bauer Deep sediments these become 44%, 40% and 16%, respectively. This suggests that, although montmorillonite is the major crystalline iron-containing mineral in the six sediments investigated here, it probably accounts for only a minor fraction of the total iron and amorphous or poorly crystalline components must be considered.

Chemical analysis results and discussion. The chemical composition of six sediment samples analyzed for this study is presented in Table III, together with selected analyses of other metalliferous sediments. The latter include two sediments (KK71-115 GC-015 and KK71-117 FFC-177) from a locality similar to those studied in this work; sediment from the crest of the EPR south of the area investigated (Rise Pac 72); Bauer Deep sediments (KK71-087 FFC-130, D.S.D.P. 319-18 and D.S.D.P. 319-19), and also Mid-Atlantic Ridge average sediment. Choice was made on the basis of similar calcium content in order to facilitate comparison. Note that the two D.S.D.P. 319 sediments were probably originally deposited near the now-extinct Galapagos Rise.

Fe, Ti, Al, Si, Mg, Cu and Zn increase with depth in cores SH1529 and SH1531. The highest concentrations of aluminum and silicon occur in SH1531 (115–117 cm), which also has the most feldspar, as suggested by the respective peak intensities in the X-ray diffraction patterns. The low Fe/Si and high Al/(Fe + Mn + Al) ratios also suggest that this sample has a greater non-iron-containing aluminosilicate content.

Feldspar and smectite probably account for most of the aluminum and silicon. Iron is largely distributed between the montmorillonite and the amorphous iron oxyhydroxide, the former accounting for some of the magnesium, potassium and sodium (not present as sea-salt). Comparison of major and trace element covariance using data which are not on a carbonate-free basis is not very meaningful, however, particularly where variations are small, as for example, in phosphorus content.

High strontium values are related to the high calcite content of the sediments. Barium is also significantly greater than in non-carbonate sediment (Cronan, 1976b) and probably present in the mineral barite (Goldberg and Arrhenius, 1958). Maximum concentrations of barium in marine sediments are found to occur over parts of the East

TABLE III

Chemical composition of East Pacific Rise metalliferrous sediments *1

Sample	Major elements (%)									
	Fe	Mn	Al	Si	Ti	P	Ca	Mg	K	Na
SH1526 (26–29 cm)	2.79	0.74	0.20	2.03	0.03	0.12	34.11	1.54	0.03	0.39
SH1529 (21–25 cm)	2.94	0.72	0.36	2.26	0.03	0.12	33.79	1.76	0.05	0.50
SH1529 (106–110 cm)	3.68	0.49	0.53	4.07	0.05	0.10	31.51	1.86	0.07	0.51
SH1529 (139–143 cm)	5.15	0.94	0.57	5.14	0.05	0.12	29.56	2.03	0.07	0.32
SH1531 (28–31 cm)	3.11	0.39	0.76	4.04	0.07	0.11	32.38	1.83	0.04	0.55
SH1531 (115–117 cm)	5.15	0.76	1.49	7.35	0.13	0.11	26.06	2.23	0.14	0.88
Estimated precision (%) *4	1	<9	2	1	3	3	1	2	>2	4
KK71–115 GC–015 *6	7.09	2.58	0.07	2.09			27.39 *5			
KK71–117 FFC–177 *7	5.65	2.01	0.06	2.06			29.19 *5			
Rise Pac 72 *8	2.1	0.5	0.21	1.34	0.005		34.44 *5			
KK71–087 FFC–130 *9	2.23	0.89	0.98	4.45			28.79 *5			
DSDP319–18 *10	2.63	0.77	0.08	0.49			35.00			
DSDP319–19 *11	1.64	0.44	0.11	0.78			34.80			
Mid-Atlantic Ridge near 45°N *12	1.51	0.08	1.10				32.45 *5			

*1 Element concentrations are given on absolute, not carbonate-free, basis.
*2 Loss on ignition: sediment (previously dried overnight at 105°C) ignited at 1000°C for one hour.
*3 Calculated as the total of the major oxides plus trace elements.
*4 From Kennedy and Roser (in prep.).
*5 Ca calculated as Ca from $CaCO_3$.
*6,7 East Pacific Rise sediment from 10°33'S, 110°36'W and 10°49'W, respectively. Recalculated from data in Heath and Dymond (1977).
*8 East Pacific Rise crestal sediment from 14°18'S, 117°34'W and 30–70 cm depth in core. From Boström and Peterson (1969).

Pacific Rise (Arrhenius and Bonatti, 1965) where high accumulation rates for this element have also been reported (Boström, 1973). Assuming that the sediments from the area investigated contain approximately 80% calcium carbonate by weight (Boström,

race Elements (ppm)							LOI (%) *2	Total (%) *3
r	Ba	Cu	Ni	Zn	Zr	Pb		
182	938	108	36	63	54	24	39.48	100.52
148	1275	102	51	53	52	28	38.29	100.48
206	1978	122	38	64	64	29	35.88	100.08
104	1469	177	53	91	61	28	33.96	100.46
148	1720	87	33	49	63	29	36.59	101.42
077	1985	166	77	84	66	34	30.38	99.62
1	1	4	10	4	15	12		
	1340	359	133	147				
	1460	283	112	122				
		100	55	75				
		240	220	78				
	580	127	63	52				
	527	112	42	44				

[9] Bauer deep sediment from 11°33′S, 97°27′W; 5–10 cm interval sampled. Recalculated from data in Heath nd Dymond (1977).
[10,11] Bauer deep sediment from 13°01.04′S, 101°31.46′W and 101.58 m, 109.92 m depths below ocean loor. Probably represents sediment which had formed near the now extinct Galapagos Rise (see Dymond et l. 1977). Recalculated from data in Dymond et al. (1976).
*12 Average of surface sediments from Mid-Atlantic Ridge near 45°N. Recalculated from data in Cronan 1972a).

1973), barite would account for less than 2% on a carbonate-free basis. It is therefore difficult to detect by conventional X-ray diffraction methods. Values reported in this study are higher than those for sediments over the entire East Pacific Rise (Boström,

1976), and previous studies (reviewed by Cronan, 1974) indicate that Ba is associated with sediments containing an abundance of biological remains or ferromanganese or iron oxide phases. Zirconium levels are higher than those reported for the EPR and the element has been found to be concentrated in this region (Boström et al., 1974a; Boström, 1976). Lead values are relatively uniform and compare favourably with those reported for Bauer Deep sediment (Dasch et al., 1971) after allowance has been made for the carbonate content.

It can be seen from Table III that sediments studied here are chemically similar to other metalliferous sediments. They tend to have lower Mn and higher Al, Si and Ba than other East Pacific Rise sediments, but are significantly more enriched in iron and manganese than Mid-Atlantic Ridge sediments. They have a major element composition similar to metalliferous sediment from the Bauer Deep (Heath and Dymond, 1977) but significantly lower Cu and Ni.

Elemental ratios have been used to characterize metalliferous sediments and infer genesis (Boström and Peterson, 1969; Boström et al., 1969; Boström, 1970a, 1973). Representative ratios have been calculated for the sediments studied here (Table IV). The values reflect slight deficiences in manganese, and excess aluminum relative to other metalliferous sediments, but the sediments are notably enriched in Fe and Mn and depleted in Al compared to the average pelagic clay. McArthur and Elderfield (1977),

TABLE IV

Elemental ratios for eastern Pacific metalliferous sediments

Sample	$\frac{100Fe}{Fe + Mn + Al}$	$\frac{100Mn}{Fe + Mn + Al}$	$\frac{100Al}{Fe + Mn + Al}$	$\frac{Fe + Mn}{Al}$	Fe/Si
SH1526 (26–29 cm)	74.8	19.8	5.4	17.6	1.37
SH1529 (21–25 cm)	73.1	17.9	8.9	10.1	1.30
SH1529 (106–110 cm)	78.3	10.4	11.2	7.9	0.90
SH1529 (139–143 cm)	77.3	14.1	8.6	10.7	1.00
SH1531 (28–31 cm)	73.0	9.2	17.8	4.6	0.77
SH1531 (115–117 cm)	69.6	10.3	20.1	4.0	0.70
GS7202-35P core average *1	72.0	27.2	0.8	121.9	1.98
Average of Rise Pac crestal sediments *2	73.5	24.5	2.0	48.0	1.38
Average of Bauer Deep samples *3	63.8	23.1	13.1	4.9	0.91
Average pelagic clay *4	34.0	3.2	62.8	0.6	0.20

*1 Core 7202-35P located at 14°47.9′S and 113°30.1′W. From Boström et al. (1974a).
*2 Data from Boström and Peterson (1969).
*3 Data from Heath and Dymond (1977).
*4 From Cronan (1976b).

however, have advised caution when using the Al/(Fe + Mn + Al) ratio with regard to the origin of the sediment. The Fe/(Fe + Mn + Al) ratio here is surprisingly constant compared to other ratios and in all but one case has higher values than those tabulated for other metalliferous sediments. Iron enrichment is therefore a significant feature of these sediments, which together with the other major and trace element enrichments relative to pelagic clay, indicate they are indeed truly metalliferous sediments.

^{57}Fe Mössbauer spectroscopy results and discussion. ^{57}Fe Mössbauer spectroscopy is a very useful method for studying the chemistry of iron in solids and may readily be used to study metalliferous sediments. It provides information on valence state, magnetic states, coordination number, site symmetry and distortion and, in addition, does not suffer from interference by non-iron-containing components.

Mössbauer spectra of the four sediments investigated all show a doublet (Fig.5) typical of ferric iron in a paramagnetic or superparamagnetic state. In practice, the latter can be identified from low-temperature studies by the appearance of hyperfine structure.

The chemical shift (δ) and quadrupole splitting (interaction) (Δ) for the two peak fits to each experimental envelope are listed in Table V. All fits were generally acceptable using the criteria previously discussed. The δ values (0.11–0.14 mm s^{-1}) are consistent with iron in the ferric state and Δ values (0.59–0.69 mm s^{-1}) are characteristic of octahedrally coordinated Fe^{3+} in a nearest neighbour oxygen environment (Bancroft, 1973). The linewidths (Γ) which range from 0.49 to 0.60 mm s^{-1} are broad in comparison to those reported for ferric iron in well-crystallized silicate minerals by Bancroft et al. (1967), but similar to those quoted for ferric iron in clay minerals and poorly crystalline (X-ray amorphous) iron hydroxides and oxide hydroxides (see Table V). Coey and Readman (1973a) attributed the broad line width obtained for spectra of such an amorphous ferric gel [$Fe(OH)_3 \cdot 0.9\ H_2O$] to the variety of possible neighbourhoods for the iron arising from the absence of a crystalline lattice, which resulted in the superposition of many doublets with slightly different Δ values yet similar δ values. Kauffman and Hazel (1975) have further interpreted this broadness to be a mixed phase of α-FeOOH-like and β-FeOOH-like regions within each particle. The Mössbauer parameters for our four sediments fall within the range for these amorphous iron hydroxides which are variously named: ferric gel (Coey and Readman, 1973a, b), ferric sol (Kauffman and Hazel, 1975) and more specifically, ferrihydrite (Chukhrov et al., 1973).

Mössbauer parameters for the leached samples are significantly different from those for the corresponding parent material, indicating that the original character of the iron has been modified by this relatively mild chemical treatment. The precise nature of the modification is uncertain and under investigation. Increases in both Δ and δ parameters suggest solution and/or recrystallization to similar material but with smaller particle size. Alternatively, the difference between these spectra may be caused by solution of one of the components. Subsequent discussion therefore will be mainly concerned with the unleached sediment.

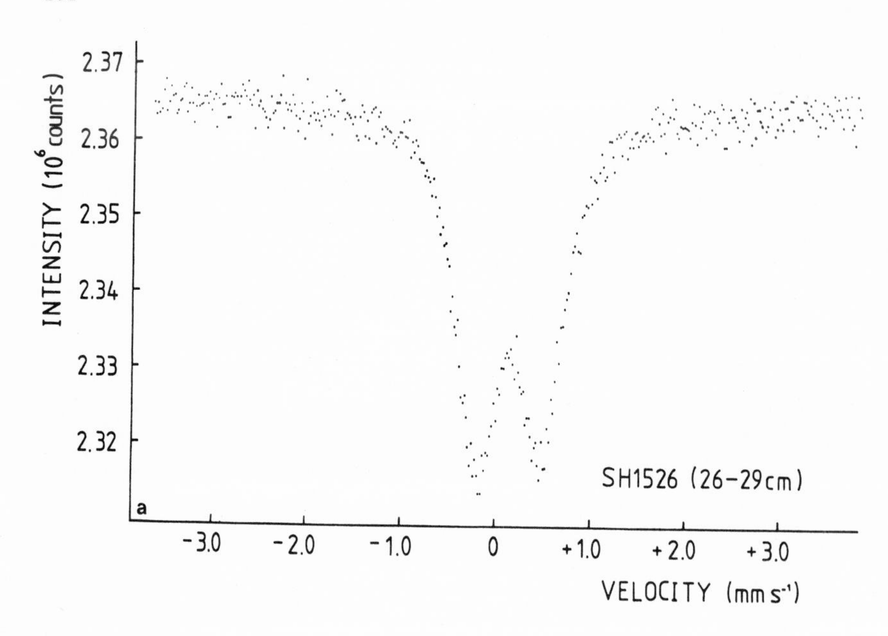

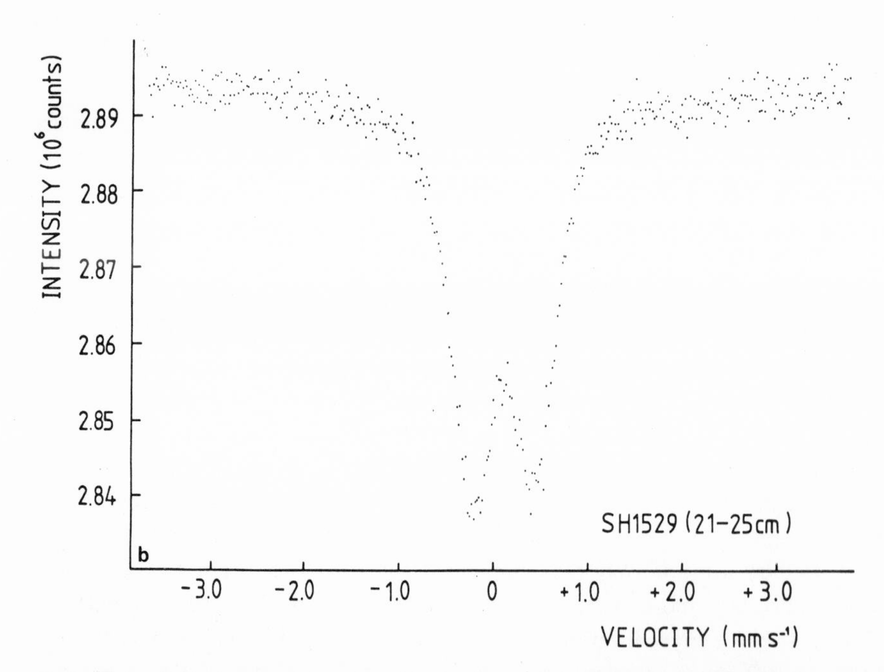

Fig.5. ^{57}Fe Mössbauer spectra of East Pacific Rise sediments (*a–f*) and Southwestern Pacific Basin sediments (*g–h*). G980 located at 32°12.4′S, 172°07.2′W and 4613 m depth. G1002 located at 22°32.6′S, 160°07.0′W and 4817 m depth.

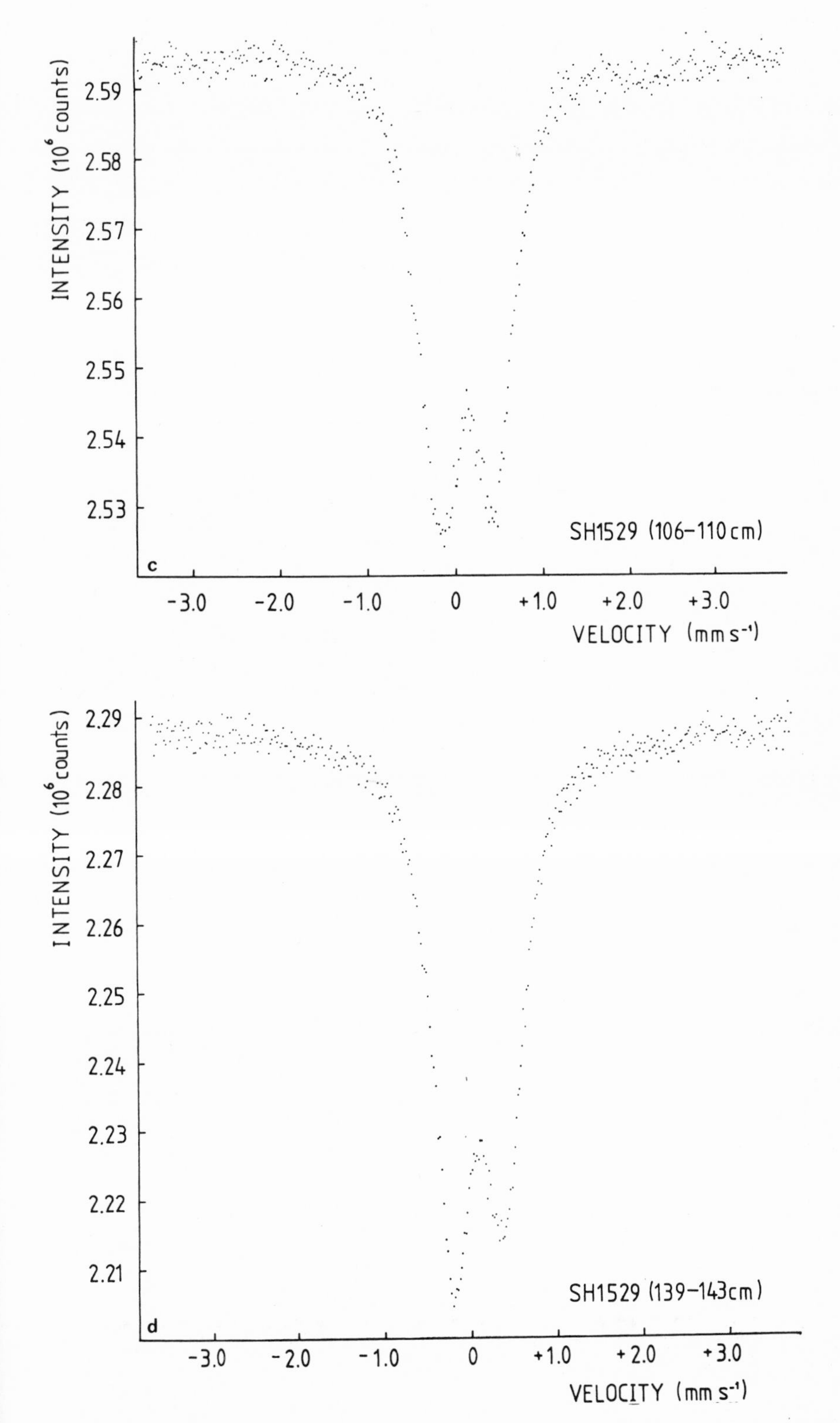

Fig. 5 (continued).

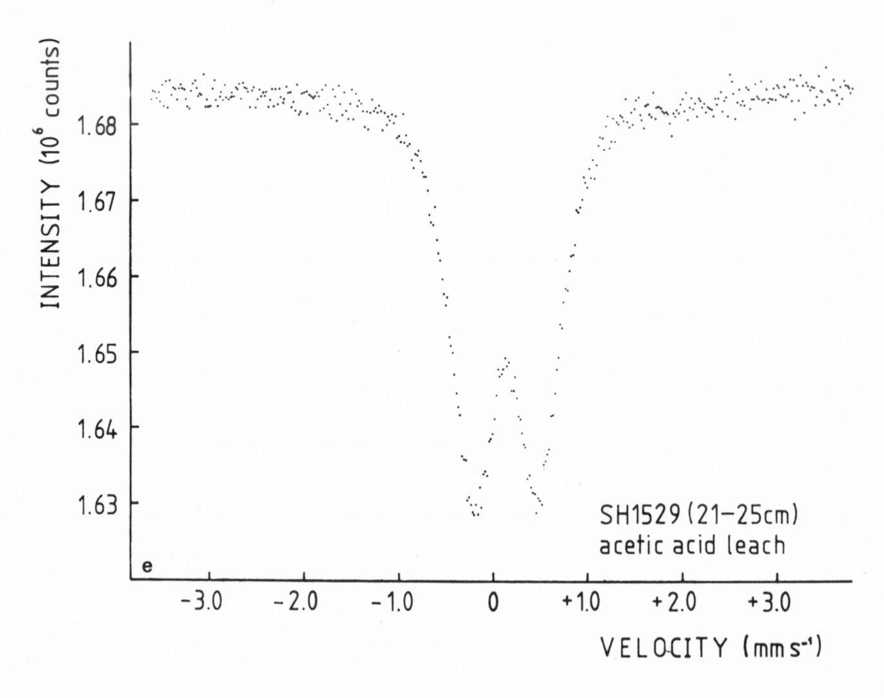

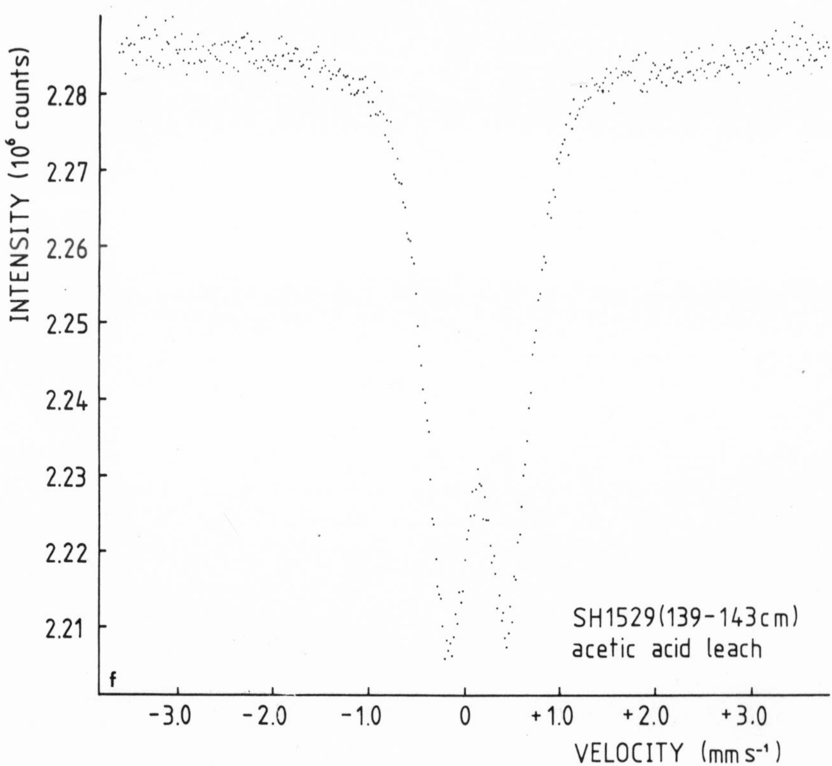

Fig. 5 (continued).

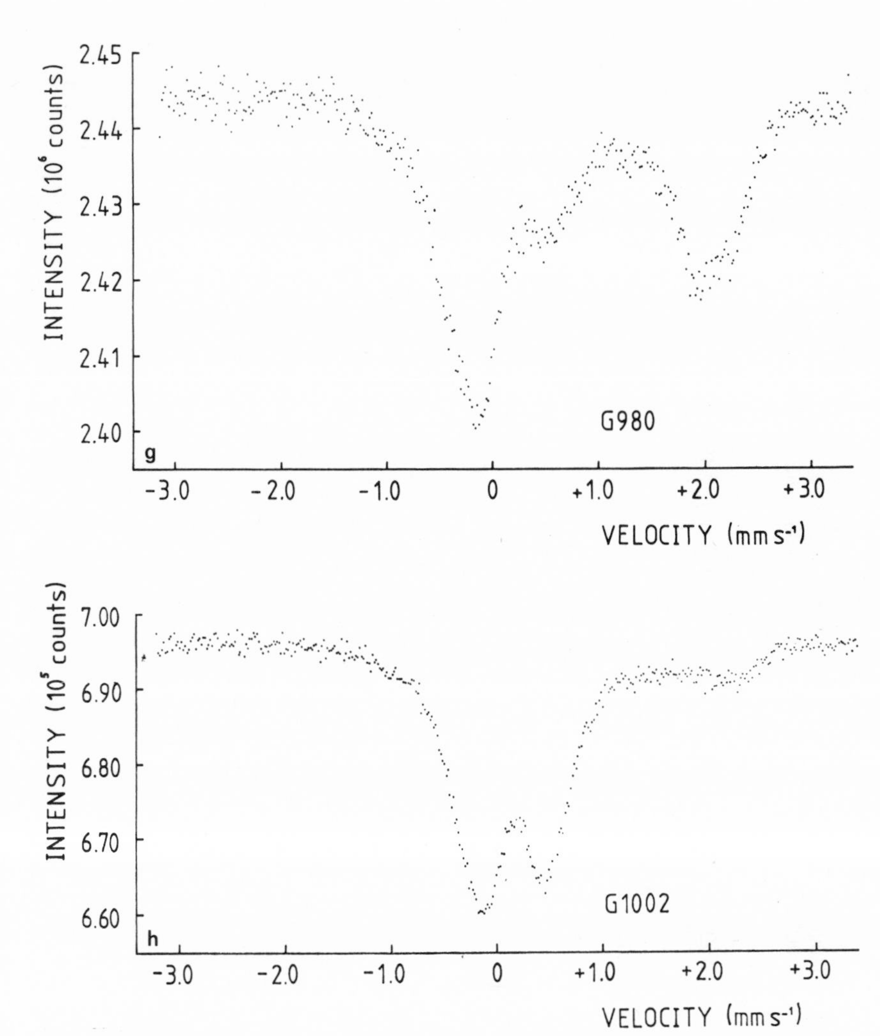

Fig. 5 (continued).

Some differences were observed between spectra of sediments from different depths. Sediments SH1526 (26–29 cm) and SH1529 (21–25 cm) are from similar depths below the sediment–seawater interface and the computer-fitted Mössbauer spectra have similar parameters (Δ: 0.674 ± 0.006, 0.649 ± 0.006 mm s^{-1}; δ: 0.139 ± 0.005, 0.125 ± 0.005 mm s^{-1}, respectively), whereas for sediments from core SH1529 there is a significant decrease in Δ from 0.649 ± 0.006 mm s^{-1} to 0.589 ± 0.005 mm s^{-1} with increasing depth, yet δ values show no systematic change (Table V). This decrease in Δ is interpreted as an increase in crystallinity attributed to a diagenetic ageing process whereby the hydrothermal ferric gel precipitate has in part recrystallized, probably to superparamagnetic goethite with parameters that compare with those (Δ = 0.55 ± 0.05 mm. s^{-1}, δ =

TABLE V

Computer-fitted Mössbauer parameters for two peak fits and selected minerals

	$\Delta (mm\ s^{-1})$		$\delta (mm\ s^{-1})$ *1	$\Gamma_1 (mm\ s^{-1})$ *2	$\Gamma_2 (mm\ s^{-1})$
SH1526 (26–29 cm)	0.674 ± 0.006		0.139 ± 0.005	0.544 ± 0.014	0.550 ± 0.014
SH1529 (21–25 cm)	0.649 ± 0.006		0.125 ± 0.005	0.544 ± 0.014	0.533 ± 0.014
SH1529 (21–25 cm) acetic acid leach	0.691 ± 0.007		0.144 ± 0.006	0.563 ± 0.012	0.545 ± 0.012
SH1529 (106–110 cm)	0.618 ± 0.007		0.141 ± 0.006	0.584 ± 0.013	0.538 ± 0.013
SH1529 (139–143 cm)	0.589 ± 0.005		0.110 ± 0.004	0.493 ± 0.009	0.597 ± 0.011
SH1529 (139–143 cm) acetic acid leach	0.637 ± 0.007		0.141 ± 0.006	0.569 ± 0.010	0.541 ± 0.010
Natural ferric gel *3	0.72 ± 0.03		0.09 ± 0.01	0.48 ± 0.03	
Iron gel *4	0.65 ± 0.02		0.10 ± 0.02		
Iron sol *5	0.63 ± 0.05		0.15 ± 0.02	0.65 ± 0.05	
Superparamagnetic goethite *6	0.546 ± 0.007		0.153 ± 0.007		
Superparamagnetic goethite *7	0.55 ± 0.05		0.155 ± 0.05		
Akaganéite *8	0.72 ± 0.07		0.15 ± 0.07		
Lepidocrocite *9	0.50 ± 0.06		0.16 ± 0.05		
Fine particles, α-Fe_2O_3 *10	0.55–0.68 ± 0.01		0.07–0.10 ± 0.01		
Montmorillonite *11	1.07 ± 0.02	[Fe^{3+} in M(1)]	0.12 ± 0.01	0.44 ± 0.02	
	0.57 ± 0.02	[Fe^{3+} in M(2)]	0.07 ± 0.01	0.44 ± 0.02	
Nontronite *11	0.70 ± 0.02	[Fe^{3+} in M(1)]	0.05 ± 0.00	0.39 ± 0.01	
	0.28 ± 0.02	[Fe^{3+} in M(2)]	0.04 ± 0.00	0.39 ± 0.01	

*1 All values are referred to the copper source matrix used in this work.
*2 Respective widths, Γ_1 and Γ_2, of the left and right hand peaks comprising the ferric doublet.
*3 Parameters for natural ferric gel [$Fe(OH)_3 \cdot 0.9H_2O$] from Coey and Readman (1973a).
*4 Parameters for synthetic iron gel with no excess of structural water. From Mathalone et al. (1980).
*5 From Kauffman and Hazel (1975).
*6 From Childs and Johnston (1980).
*7 Fine particles of α-FeOOH after Shinjo (1966).
*8 From Dezsi et al. (1967).
*9 From Greenblatt and King (1969).
*10 From Kündig et al. (1966).
*11 From Rozenson and Heller-Kallai (1977). Montmorillonite (No. 4) and nontronite (No. 13) have $Fe^{3+}_{M(1)}/Fe^{3+}_{M(2)}$ ratios of 0.18 and 0.50 respectively.

0.155 ± 0.05 mm s^{-1}) reported by Shinjo (1966). The precise nature of this transformation is not known. It could be a single-step process or conceivably through a β-FeOOH (akaganéite) intermediate, as this phase is stabilized by chloride ions and has been reported in marine sediments, formed by hydrolysis of Fe(III) in seawater (Murray, 1978). Subsequent transformation to the more stable α-FeOOH would be expected. Further such speculation would require a detailed knowledge of the interstitial fluids from these sediments.

Assuming an accumulation rate of greater than 5 mm non-carbonate matter per 1000 years in this region (Boström, 1973) and an estimated 80% calcium carbonate content, then 118 cm of sediment represents less than 50,000 years and any diagenetic changes must have been effected within this time.

The Mössbauer spectrum of SH1529 (139–143 cm) is different from the rest and shows a marked asymmetry reflected in both the intensity of absorption and the width of each peak (Fig.5, Table V). To explain this, two possibilities are considered. First, this could be a superparamagnetic doublet, probably goethite, exhibiting magnetic relaxation associated with intermediate particle sizes (see Johnston and Glasby, 1978). Such a spectrum would be expected to have a less intense, broader right-hand peak than that observed. In such cases, the ferric doublet exhibits non-Lorentzian line shapes and therefore cannot be computer-fitted by the method used here.

The second possibility arises from further consideration of the X-ray diffraction data. Montmorillonite was the only crystalline, iron-containing mineral detected, and depending on the relative proportion of this mineral to amorphous ferric gel, the ferric and any ferrous iron in the octahedral sites of montmorillonite (Rozenson and Heller-Kallai, 1977) would contribute to the spectrum. This contribution would be detectable if the mineral concentration is greater than about 5% (Bancroft, 1973). The second possibility is not believed to be the cause of the asymmetry because initial attempts to fit more than two peaks have failed and there is no evidence to suggest that the mineralogical composition of SH1529 (139–143 cm) is markedly different from that of the younger sediments within the same core. In addition, two peak fits were statistically acceptable for the other three sediments and therefore it appears that the asymmetry is due to goethite of larger particle size which is exhibiting magnetic relaxation.

Strictly speaking, the Δ and δ parameters obtained for SH1529 (139–143 cm) by the computer-fitting procedure used here will only be approximate and greater uncertainty is expected in the δ parameter than obtained by statistical considerations of error. It is proposed to study this effect further using low-temperature Mössbauer spectroscopy coupled with selective leaching procedures to remove particular forms of iron from the respective sediments.

Marked asymmetry of Mössbauer spectra has also been reported for other metalliferous sediments from the East Pacific Rise and the Bauer Deep by Bagin et al. (1975). These authors similarly attributed the asymmetry to magnetic relaxation effects of superparamagnetic goethite particles. The results of their studies give δ (relative to the copper matrix) and Δ values of 0.22 mm s^{-1} and 0.55 mm s^{-1}, respectively, with half widths ranging from 0.43 to 0.81 mm s^{-1} for the right hand peak and attendant particle sizes from 140–300 Å. Unfortunately, no attempt was made by these authors to investigate possible causes for such differences in particle size, nor was mention made of any differences between the mode of occurrence of iron in sediments within long cores or between localities (EPR versus Bauer Deep).

As previously mentioned, the partitioning procedure of Heath and Dymond (1977) indicates that smectite and an iron hydroxide phase are the principal components of Bauer Deep sediments. Sayles and Bischoff (1973) found that smectite comprised over 90% of the non-carbonate mineral fraction of the Bauer Deep sediments which they investigated (see also Sayles et al., 1975). In contrast, Bagin et al. (1975) concluded from their study that "mobile" iron (amorphous hydroxide) accounts for about 90% of the total iron in the metalliferous sediments of the Bauer Deep. This suggests that Mössbauer spectra of sediments from this region need careful re-examination with, perhaps, computer-fitting to determine more precisely the parameters and hence yield a greater understanding of the iron mineralogy of these Bauer Deep sediments.

Dymond et al. (1973) observed that metalliferous sediments display surprisingly little tendency to recrystallize with time, and that the crystalline size remains constant so far as could be determined by their methods (X-ray diffraction). Mössbauer studies are, however, much more sensitive to diagenetic changes which are the result of recrystallization of superparamagnetic substances. Diagenetic changes of the type proposed for the EPR sediments investigated here have been inferred by Dymond et al. (1977) to account for the mineralogical variations in D.S.D.P. Site 319 sediments. There, poorly crystallized ferric hydroxide deposited when the site was at the Galapagos Rise crest has subsequently recrystallized to goethite. Traces of goethite have also been reported in metalliferous sediments from the East Pacific Rise (Dasch et al., 1971; Heath and Dymond, 1977). There are, however, significantly more numerous reports of amorphous iron oxyhydroxides or dark brown colloidal material in metalliferous sediments (see, for example, Boström and Peterson, 1966, 1969; Bischoff, 1969a; Von der Borch and Rex, 1970; Cronan, 1972b; Cronan and Garrett, 1973; Boström et al., 1974a; Sayles et al., 1975; Bischoff and Rosenbauer, 1977; Bloch, 1978; Corliss et al., 1978).

Generally, the presence of goethite is noted in older metalliferous sediments such as the basal metalliferous sediment (Late Cretaceous) in the northeastern Indian Ocean (Pimm, 1974a), supporting the idea that diagenetic processes take place slowly, transforming the initial iron(III) hydroxide precipitate to the more stable goethite modification. Haematite, usually the final product of such ageing processes provided the conditions are suitable, has also been reported as a minor constituent of the largely goethite-containing basal iron oxide-rich facies of the northeastern Indian Ocean (Pimm, 1974b).

Interestingly, a possible ageing process of ferric gel to particles of larger size has been proposed by Hrynkiewicz et al. (1972), based upon the reduction in Δ parameter from exterior to centre in an iron–manganese nodule, probably as a result of the agglomeration of particles inside the nodule.

Finally, a comparison of Mössbauer spectra of metalliferous sediments and ocean basin sediments illustrates significant differences in the mode of occurrence of iron and in iron mineralogy. Sediments from Stns G980 and G1002 from the Southwestern Pacific Basin (Meylan et al., 1975) contain ferrous iron in addition to ferric iron, the ferrous iron occurring in lesser amount in the station furthest from New Zealand (G1002) (see Fig.5). Meylan et al. (in prep.) attribute this to decreased detrital sedimentation of clay minerals (illite and chlorite) relative to probable *in situ* formation of goethite.

Summarizing the ^{57}Fe Mössbauer spectroscopy of metalliferous sediments from the East Pacific Rise, ferric iron is the predominant species. It occurs mainly in amorphous iron gels or iron oxide-hydroxides, while iron in X-ray crystalline minerals is of minor importance. The nature and mode of occurrence of iron in metalliferous sediments is different from that in pelagic sediments from the deep ocean basins, reflecting the different origin and deposition of the iron. Preliminary work suggests that an ageing process takes place, transforming the ferric gel to an oxide-hydroxide mineral of larger particle size, probably goethite. The precise pathway is, however, not clear.

Bauer Deep

The Bauer Deep is a basin on the Nazca Plate, located between the actively spreading East Pacific Rise and the Galapagos fossil spreading centre to the east (Fig.4). The metalliferous sediments of the basin have received attention as part of the International Decade of Ocean Exploration (I.D.O.E.) Nazca Plate Project (see, for example, Dymond et al., 1973, 1977; Rosato et al., 1975; Heath and Dymond, 1977), and D.S.D.P. Site 319 (Leg 34) was drilled in the Bauer Deep (Yeats and Hart et al., 1976; Dymond et al., 1977).

The effect of the $CaCO_3$ compensation depth is evident in Bauer Deep sediment cores (Sayles and Bischoff, 1973). At depths above 4000 m, the cores are entirely calcareous. Cores from depths of about 4100 m have a surficial calcareous sediment which grades downward into carbonate-free ferromanganoan sediment. At depths below 4200 m, the surficial sediment is metalliferous, with little or no $CaCO_3$. The surface layer is of variable thickness (from about 1 m to at least 8 or 9 m).

Non-detrital smectite (montmorillonite) makes up the bulk of the metalliferous sediment (70–90%) and is the most important Fe-bearing phase. It occurs as yellowish microcrystalline aggregates up to 50 μm in diameter, the larger particles commonly rounded or nearly spherical in shape (Sayles and Bischoff, 1973; Dymond and Eklund, 1978). Iron and manganese oxides, including δ-MnO_2 and todorokite, occurring primarily as micronodules, comprise 10–20% of the sediment. The reducing acid-soluble fraction contains, in addition to the Fe–Mn hydroxides, goethite which is amorphous or poorly crystalline according to Sayles and Bischoff (1973), and moderately well crystallized according to Dymond et al. (1973). The coarse fraction of the sediment contains, in addition to micronodules and smectite aggregates, variable quantities of phillipsite and phosphatic fish debris. Volcanic glass (both acidic and basic), biotite flakes, palagonite and sponge spicules are common trace components. Ferromagnesian and terrigenous grains (other than biotite) are very rare (Dymond et al., 1973). At D.S.D.P. Site 319, metalliferous components visually estimated as representing from a few to 20% of the total sediment occur throughout 114 m of Early Miocene to Quaternary nannofossil oozes (Yeats and Hart et al., 1976).

Galapagos Spreading Centre

The Galapagos Spreading Centre forms part of the oceanic ridge system of the eastern Pacific, separating the Cocos Plate from the Nazca Plate (Fig.6), and has been well described by Klitgord and Mudie (1974), Williams et al. (1974), Sclater et al. (1974), Crane (1978), and Anderson et al. (1978). According to Schilling et al. (1976), the REE, Fe and Ti concentrations in basalts dredged along the Galapagos Spreading Centre reveal that a juvenile mantle plume is welling up beneath the Galapagos Islands. This is expressed at the sea floor by hydrothermal plumes in the general area of the spreading centre. The nature and chemical characteristics of these plumes have been well documented by Edmond et al. (1977), Klinkhammer et al. (1977), Lupton et al. (1977a), Weiss (1977), Weiss et al. (1977), Corliss et al. (1978), Jenkins et al. (1978), and Corliss et al. (1979).

Two types of hydrothermal encrustations have been found in the area encompassed by the suspected plume. Moore and Vogt (1976) found rapidly accumulated manganese crusts 20–60 mm thick at two sites near the Galapagos spreading axis. The uppermost 1 millimetre of one crust contained 49.1% Mn and less than 0.1% Fe; these values are representative of the entire crust. Burnett and Morgenstein (1976) and Burnett and Piper (1977) examined a 15 mm thick manganese crust from the Hess Deep (axis of Galapagos Spreading Centre), and found it to have a contrasting composition. The Hess Deep crust contained 21.8% Fe, 16.7% Mn, and greater concentrations of nickel, cobalt and copper than the crusts studied by Moore and Vogt. A minimum accretion rate of 800 mm/10^6 years was determined for it, more than 100 times faster than the average deep-sea nodule accretion rate.

Corliss et al. (1976) conducted a bottom photography and dredging programme about 20–25 km south of the Galapagos Rift near 86.5°W. Sediment mounds were found resting on about 30 m of sediment overlying basaltic basement in a zone of high heat flow. Acoustic reflecting layers disappeared beneath the mounds. Bottom photos showed ledges and massive outcrops of brown to yellow-orange material. Dredging recovered fragments of manganese crust containing exceptionally well-crystallized todorokite–birnessite associated with orange amorphous Fe oxides and yellow well-crystallized nontronite. Corliss et al. suggested that the sediment mounds were deposited around hydrothermal vents. This has been confirmed by further deep-tow surveying (Londsdale, 1977a). Additional studies of the hydrothermal mound area have resulted from Pleiades Leg II (Corliss et al., 1978), D.S.D.P. Leg 54 (Scientific Party, D.S.D.P. Leg 54, 1977; Hekinian et al.,

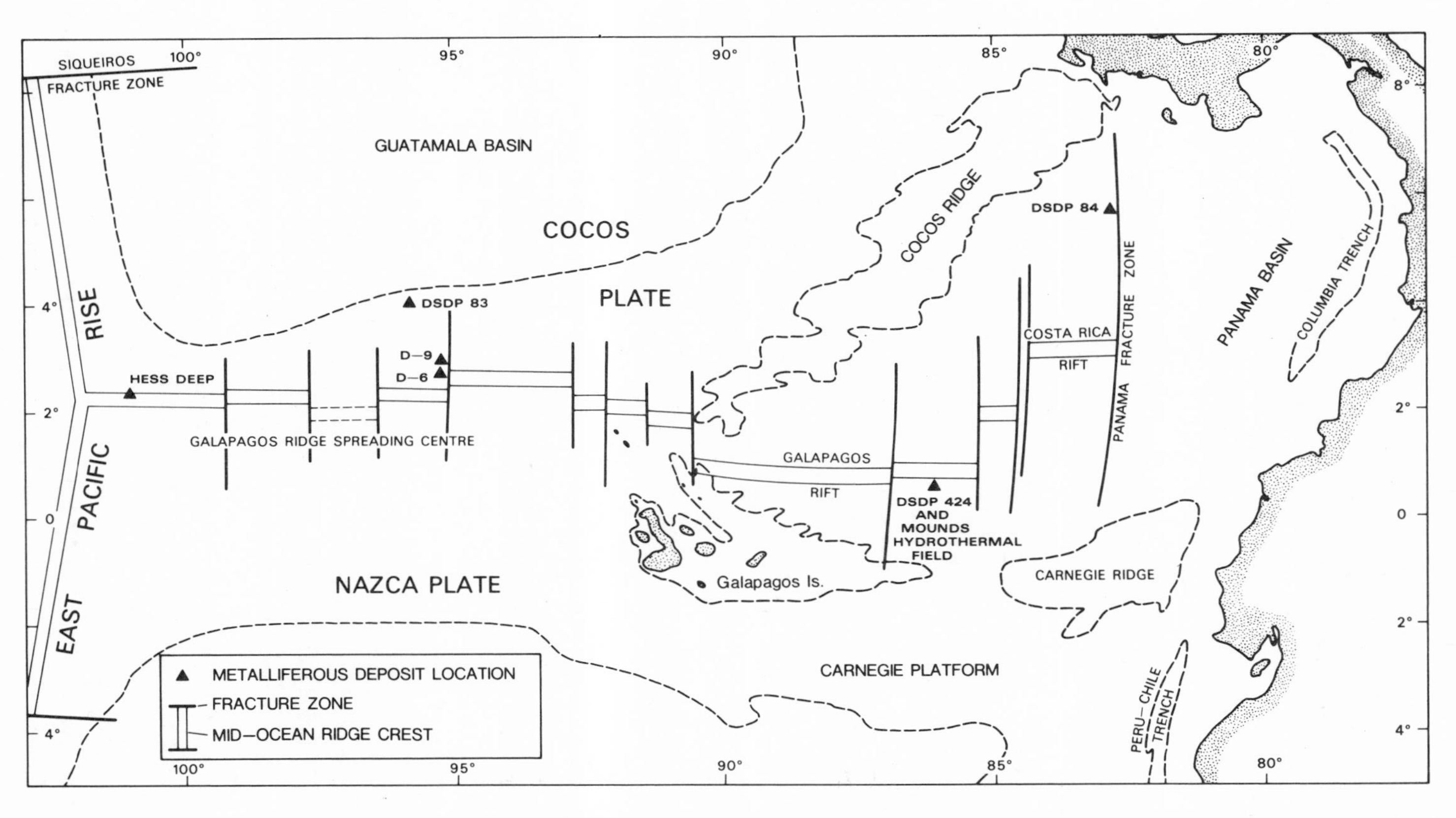

Fig. 6. Map of metalliferous deposits in area of Galapagos Spreading Centre. Bathymetry schematic; sea-floor tectonic pattern modified after Klitgord and Mudie (1974) and Dymond and Veeh (1975), among others. Analytical data for samples from specific deposit locations are given in Tables X–XII.

1978; and Natland et al., 1979), and dives by the research submersible *Alvin* (Corliss et al., 1979, among others).

The release of the hydrothermal fluids into seawater has permitted the establishment of unique benthic communities around vents, with consequent local high biological productivity (Ballard, 1977; Corliss and Ballard, 1977; Lonsdale, 1977b; Corliss et al., 1979). The primary producers are sulphur-oxidizing and heterotrophic bacteria, with thermal springs providing the basic energy source.

Northeastern Pacific

Bischoff and Rosenbauer (1977) have observed that metalliferous globules are a ubiquitous minor component of the Clipperton Oceanic Formation, which is found above basalt basement and beneath much of the Ni- and Cu-rich manganese nodule belt of the eastern tropical Pacific (Tracey et al., 1971). This may indicate that the deposition of hydrothermal precipitates is not confined to spreading centres.

A metalliferous fraction is also said to be present in the Deep Ocean Mining Environmental Study (D.O.M.E.S.) Site C Quaternary sediments (Fig.7) (Bischoff and Rosenbauer, 1977). Correction for the contribution of normal pelagic clay led Bischoff and Rosenbauer to believe that the hydrothermal fraction amounted to about 40% in one sample. Silica and magnesium are major components of the corrected composition, which is analogous to sediments designated as metalliferous. Marchig (1978), Glasby et al. (1979), and Hein et al. (1979) have, however, contested this conclusion and suggested that these are normal pelagic sediments formed under low sedimentation conditions. Hein et al. state that "the isotopic composition of D.O.M.E.S. Stn 18A does not support a hydrothermal origin for the smectite in the metalliferous sediment", while noting that Fe hydroxide precursor material for the smectite may have been derived from East Pacific Rise volcanism.

Southwestern Pacific Basin

The locations and mode of occurrence of metalliferous sediments in the South Pacific have been well documented by Glasby and Lawrence (1974) and Glasby (1976) (Fig.8). Nayudu (1971) analyzed two metalliferous sediment samples from the crest of the Pacific–Antarctic Ridge (PAR), but failed to recognize them as such (Glasby et al., in press).

Zemmels (1977) has further elucidated the geochemistry of metalliferous sediments on the crest and flanks of the Pacific–Antarctic Ridge adjacent to the Southwestern Pacific Basin. He found that sediments remote from the PAR crest have compositions which can be explained as a mixture of lithogenous materials and ferromanganese precipitates. Zemmels also noted widespread but discontinuous distribution of Fe enrichment and Cu–Zn–Pb anomalies in surface sediments of the Ridge and adjacent areas. He stated

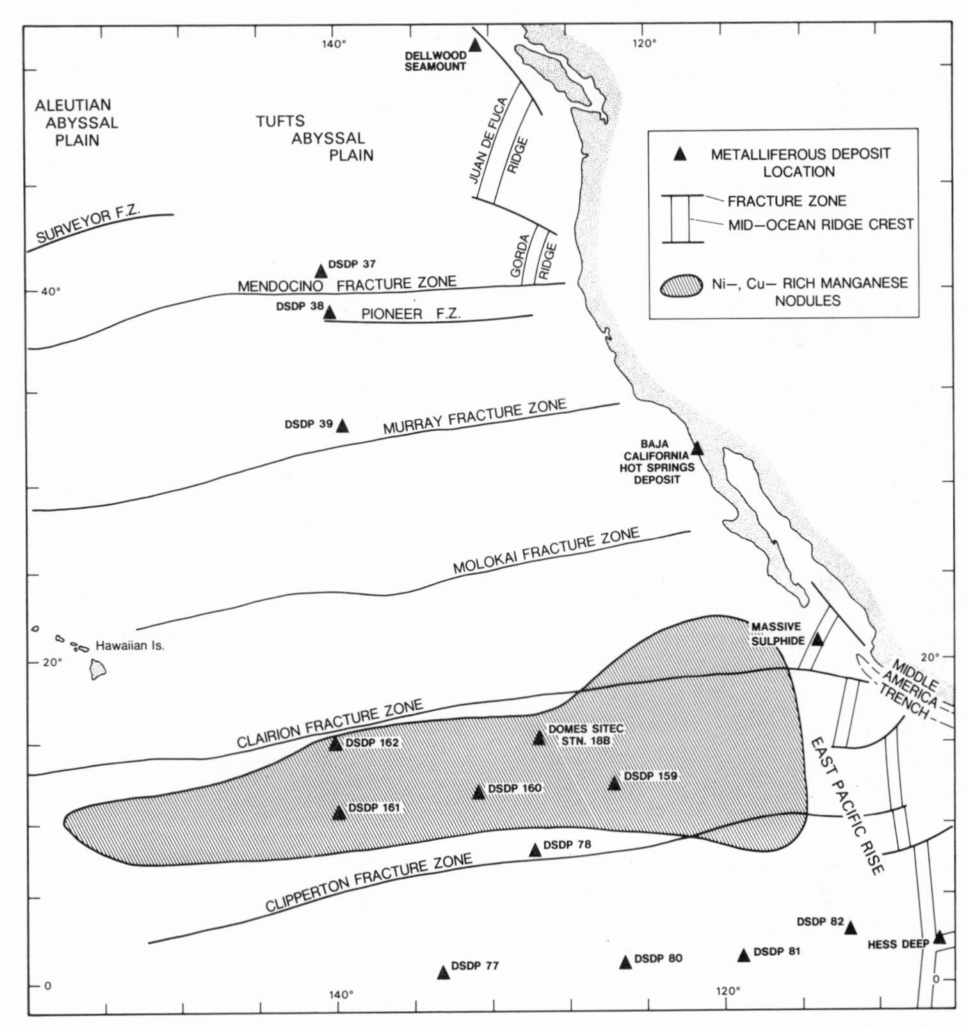

Fig.7. Map of metalliferous deposits in northeastern Pacific Ocean. Sea-floor tectonic pattern from Horn et al. (1972). Distribution of Ni–Cu-rich manganese nodules modified from Emery and Skinner (1977), among others. Sample descriptions and analytical data are given in Tables VI, X–XII.

that such a pattern is best explained by isolated, open hydrothermal convection cells rather than an axial hydrothermal system.

Analysis of 230 marine surface sediments from the southwestern Pacific by Cronan and Thompson (1978) revealed only a few sites of possible metal enrichment, most notably northwest of Fiji. Sillitoe (1972) has mentioned the possibility that diffuse spreading centres in marginal basins, such as those of the southwestern Pacific, may represent a third type of environment for the generation of massive sulphide deposits, in addition to mid-ocean ridges and island arcs.

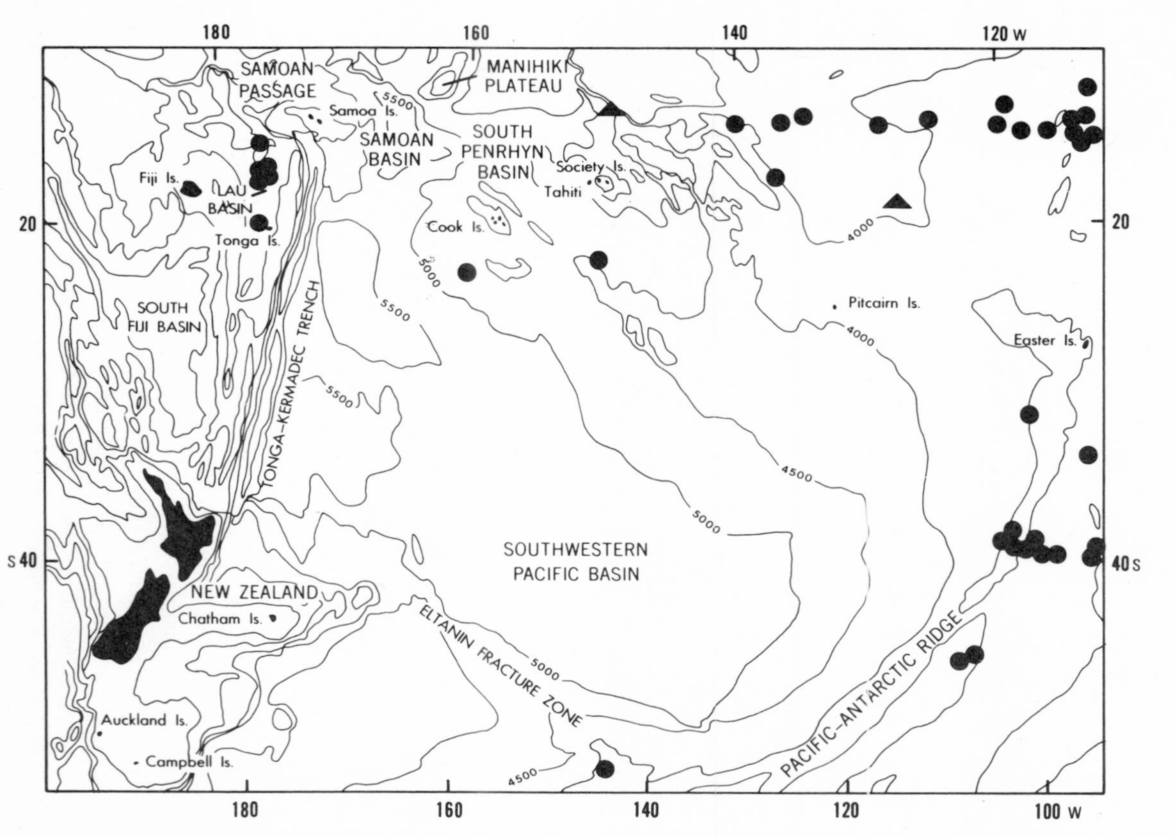

Fig. 8. Map of metalliferous deposits in southwestern Pacific Ocean. (Modified from Glasby and Lawrence, 1974.) Circles represent metalliferous sediment locations; triangles indicate Fe–Mn crust locations. Analytical data for Lau Basin samples are given in Tables X–XII.

Andrushchenko et al. (1975) have reported an unusual assemblage of volcanic lithologies and metamorphosed manganese nodules at 22°41′S, 160°50′W, south of Rarotonga in the Cook Islands. Abundant blocky and platy masses of pillow lavas overlie altered brick-red clays with a breccia-like structure. Mn-encrusted blocks of tuff breccias and manganese nodules are found adjacent to pillow lava areas. The nodules occur in a variety of forms (spherical, oval, platy and intergrown), and are most densely distributed on marly pelagic clays, with 10–50% $CaCO_3$. Some of the nodules have been altered by post-volcanic hydrothermal processes. These samples must represent a localized occurrence of such processes, because regionally there is no evidence of anything other than normal pelagic sedimentation in which red clay forms the dominant sediment type and manganese nodules occur in abundance (Glasby et al., 1979).

Mid-Atlantic Ridge

A variety of ferromanganoan deposits have been reported at the ocean bottom along the crest of the Mid-Atlantic Ridge (Fig.9). In the median valley near 45°N, Cronan 1972a) located highly ferruginous sediments, while less ferruginous deposits were sampled from the Eastern and Western Crest Mountains. M.R. Scott et al. (1974) and R.B. Scott et al. (1976) have described a very pure manganese oxide crust (42 mm thick) collected during the Trans-Atlantic Geotraverse (TAG) from the Mid-Atlantic Ridge median valley at 26°N, and suggested that it was formed at a submarine hot spring. The material has a grey, submetallic luster and conspicuous laminations 5–10 mm thick; some layers are very porous and exhibit bladed growth structures. Additional work on the TAG field has concentrated on the structural setting of the hydrothermal system that produced the crusts (M.R. Scott et al., 1974; McGregor and Rona, 1975; Rona, 1976a, b; Rona et al., 1976; Temple et al., 1976).

The manganese crusts are found on basalt talus and as a breccia matrix in and adjacent to the MAR rift valley. They have accumulated at a rate of about 200 mm/10^6 years, and consist of 39% Mn and 0.01–0.07% Fe. A 0.1°C temperature anomaly was found in near-bottom water over the inferred site of present discharge, and relatively high metal concentrations (especially Mn and Fe) were detected in suspended particulates in the area (Rona et al., 1975; Rona, 1976a).

The TAG hydrothermal field conditions have persisted for at least $1.4 \cdot 10^6$ years (Rona, 1976a, b). Manganese oxide crusts actively accumulate on the rift valley wall, and then become relict as they are moved away from the discharge zone by sea-floor spreading and are partially buried by off-axis extrusive volcanism. The sealing of a given volume of hydrothermally impregnated talus/breccia by mineralization and burial is inferred to occur over a period of about 100,000 years. A linear zone of hydrothermal mineral deposition extends away from the rift valley "hot spot" in the direction of sea-floor spreading. In such a zone at a point 18 km from the rift axis, Mn-crusts having two distinct layers were observed by R.B. Scott et al. (1976). The basal layer is of hydrothermal origin (40%

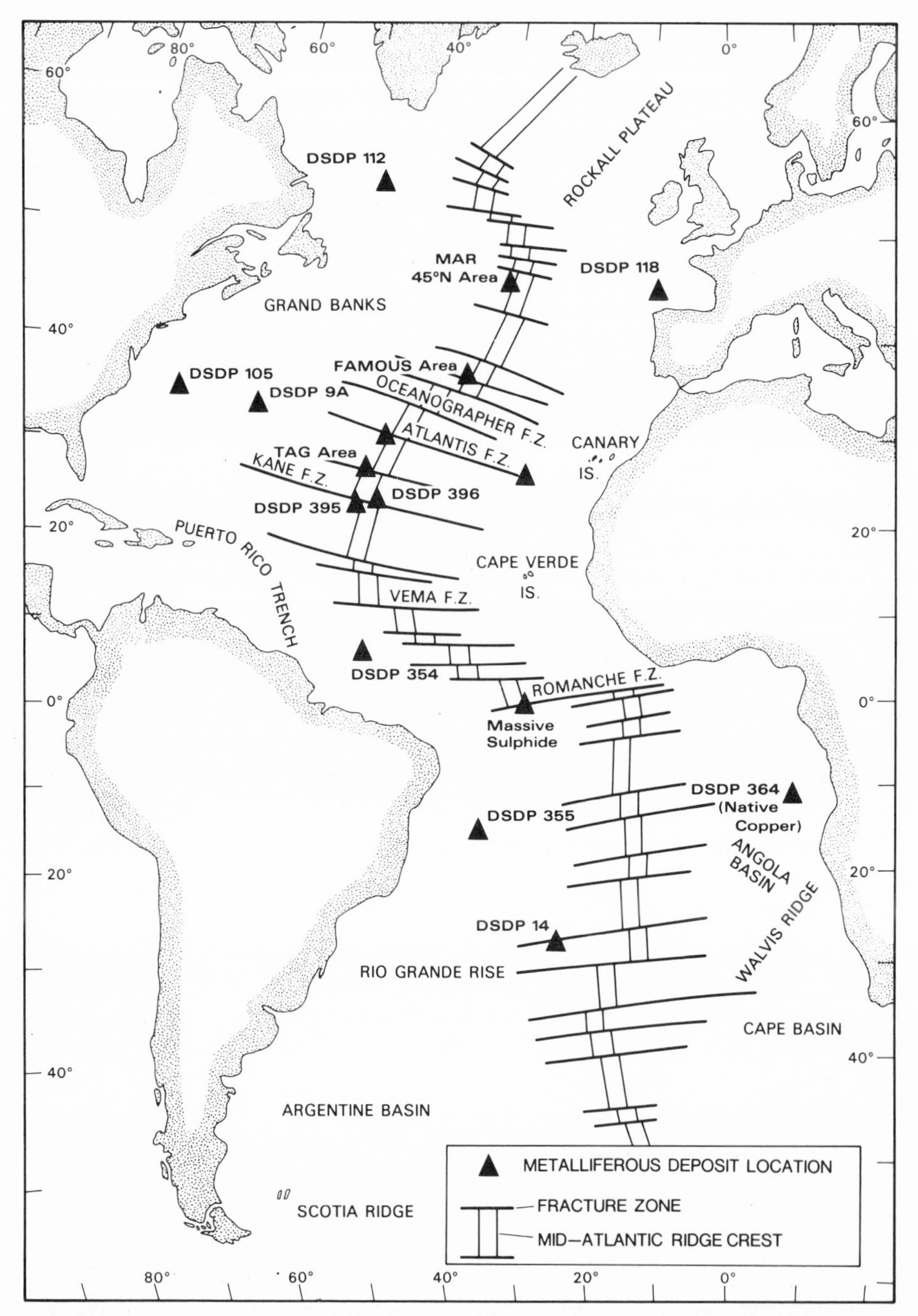

Fig.9. Map of Atlantic Ocean metalliferous deposits. Geographic base map and sea-floor tectonic pattern modified from "The Physical World", map insert, *National Geographic Magazine,* November 1975. Sample descriptions and analytical data are given in Tables VIII, X–XII.

Mn, $\leqslant$0.1% Fe, 0.15% Ni + Cu + Co), while the upper layer probably has a hydrogenous origin (>16% Fe, 0.4% Cu + Ni + Co). Apparently, hydrothermal activity is replaced by hydrogenous ferromanganese precipitation on ocean crust older than $0.7 \cdot 10^6$ years on the ridge-crest highlands.

Another area of the Mid-Atlantic Ridge which has been extensively studied, as part of the French–American Mid-Ocean Undersea Study (FAMOUS), is the region at about 36°30′ to 37°N (see especially Heirtzler and Van Andel, 1977, and accompanying papers in the *Geological Society of America Bulletin,* volume 88, numbers 4 and 5). A detailed review of the relief and faulting is presented in the Structural Setting section.

The principal constituents of the deposits recovered from Transform Fault "A" in the northern part of the FAMOUS area are a green clay-rich material consisting of hydromica, smectite and an amorphous Fe–Si compound, and concretions of ferromanganese oxides (Hoffert et al., 1978). Close to the small fissure-like vents, the green clay-rich material is abundant, but farther away the ferromanganese concretions are the dominant components. The hydrothermal deposits from Transform Fault "A" have consistently low Ni, Cu, Co and Zn contents, which sets them apart from manganese nodules and hydrogenous ferromanganese coatings on sea-floor basalts.

M.R. Scott and Salter (1977) studied sediment cores collected from the median valley of the Mid-Atlantic Ridge in the FAMOUS area. The sediment consists of 80–90% $CaCO_3$. On a carbonate-free basis, the sediment is enriched in Fe, Mn, Cu, Ni, Co, Cr and Zn compared to average Atlantic pelagic clay. The metal enrichments are most pronounced in the reddish, upper 150–250 mm of the cores, but even the bottom metre of the cores displays thin, dark, metal-rich layers.

In the Romanche and Vema Fracture Zones of the equatorial Atlantic, Bonatti et al. (1976b, c) and Bonatti and Honnorez (1976) have found evidence to support the hypothesis that "massive" sulphide deposits are present in the oceanic crust and that "massive" sulphide ores of ophiolitic complexes were formed in ancient spreading centres. Sulphide-bearing metabasalts probably originally emplaced at the ridge axis, and now covered with manganese crust up to 3 mm thick, were recovered by dredging. The copper–iron sulphides (mainly chalcopyrite; pyrite and pyrrhotite also being present) are assumed to have been derived from the oceanic crust by hydrothermal activity. The Cu–Fe sulphides form both "disseminated"- and "stockwork"-type deposits. Iron hydroxides, containing as much as 13% Cu, are a common alteration product of the chalcopyrite.

Siesser (1976) has reported the presence of native copper of presumed hydrothermal origin in D.S.D.P. cores from the Angola Basin (Site 364). He noted that native copper had also been reported at Site 105 in the northwestern Atlantic, but there it occurred in sediments immediately overlying basement rock, rather than far above basement as at Site 364.

Indian Ocean

The occurrence of active-ridge metalliferous sediments in the Indian Ocean was established by Boström and D.E. Fisher (1971). Subsequently, McArthur and Elderfield (1977) studied samples from the Mid-Indian Ocean Ridge and the Marie Celeste Fracture Zone, which offsets the Ridge north of its triple junction (Fig.10). Ferromanganese encrustations were collected from the valley wall of the Ridge, and what are probably sulphides in decorated vesicles of the basalts were recognized. Nevertheless, the overall composition of the Mn crust indicates a hydrogenous rather than hydrothermal origin: Fe 20.0%, Mn 12.5%, Cu 0.11%, Co 0.08%, Ni 0.14%, Zn 0.060%. One would also suspect a strong hydrothermal influence in the siliceous to calcareous clays of the fracture zone and indeed the Al/(Al + Fe + Mn) ratios (0.24–0.35) are in the range used to classify sediments as metal rich and of volcanic origin (Fig.1a).

Further examples of hydrothermal ore deposition in the Indian Ocean are given by Rozanova and Baturin (1971) and Cann et al. (1977). The latter workers reported two

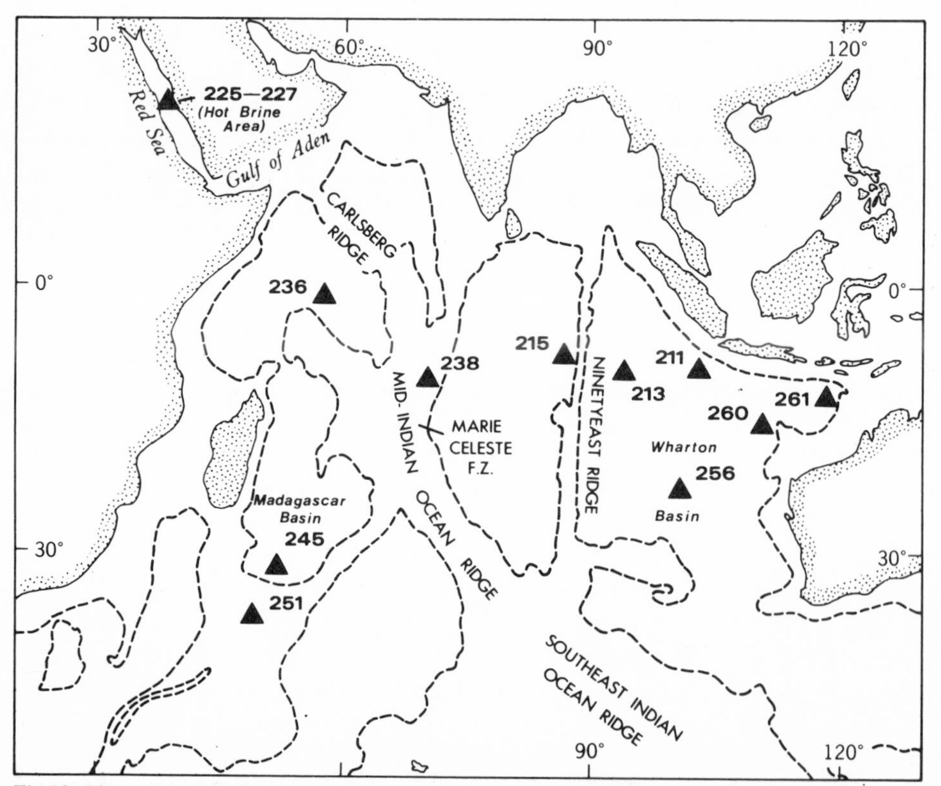

Fig.10. Map of Indian Ocean metalliferous deposits. (Modified from Veevers and Heirtzler et al., 1974.) Sample descriptions and analytical data are given in Tables IX–XII.

principal types of material encrusting basalt lava from near the axial region of the Gulf of Aden: (1) spongy brown to hard black lumps and coatings of manganese oxide overlying, (2) friable yellow to green massive smectite which apparently is a direct precipitate from hydrothermal emanations. Lesser amounts of orange powdery iron oxides were also noted.

Red Sea

The metalliferous sediments that have generated the most interest, from both scientific and economic viewpoints, are those found in the axial trough of the Red Sea (Degens and Ross, 1969; Ross et al., 1973; Bäcker, 1976). These deposits have been discussed in several chapters of this series; see particularly Degens and Ross, Chapter 4, Vol. 4.

Apparently, the mineralized sediments are restricted to the area underlying the present hot brines (Whitmarsh, 1974) or to intermediate areas between the known brine pools (Ross et al., 1969), i.e., the metalliferous sediments are confined to a relatively few deeps and their connecting sills. The deeps represent bathymetric lows within the axial trough that have been shaped by faulting and extrusion of basalt, and possibly also by salt flowage (Ross et al., 1969; Coleman, 1974; Bäcker et al., 1975; Bignell, 1975; Gass, 1977). The axial trough itself was formed by rifting that began at the beginning of the Pliocene (Coleman, 1974).

Red Sea metal-rich sediments are surficial deposits that extend, at least in some deeps, to basalt basement. For instance, drilling at D.S.D.P. Site 226 in the southwestern part of Atlantis II Deep encountered basalt beneath about 6 m of metalliferous deposits (Whitmarsh and Weser et al., 1974). Bäcker and Schoell (1972) reported the occurrence of several lava flows intercalated in hydrothermal sediments of the Atlantis II Deep. Piston coring in the Atlantis II Deep indicated a minimum thickness of 8 m of iron-rich sediment at three sites several kilometres apart, and seismic profiling and acoustic pinger returns suggested that a large portion of the Deep contains 20–30 m of heavy metal sediment (Ross et al., 1969). The same survey indicated that both the Chain and Discovery Deeps have relatively small amounts (probably less than 10 m) of heavy metal deposits. Areas outside the known deeps have only a thin veneer of metal-rich sediment, which would probably be underlain by carbonate ooze or fine-grained clastics.

The sediments of the Red Sea geothermal areas have been divided into a number of facies. Bischoff (1969a) recognized seven laterally correlative facies, from top to bottom: (1) detrital, (2) iron-montmorillonite, (3) goethite-amorphous, (4) sulphide, (5) manganosiderite, (6) anhydrite, and (7) manganite. Hackett and Bischoff (1973) added another, the magnetite facies. Amann et al. (1973) identified five units: (1) amorphous, (2) upper sulphide, (3) central oxic, (4) lower sulphide, and (5) detrital oxidic-pyritic. The sediments are well bedded on a scale ranging from a few millimetres to several metres, those recovered from below the brine–seawater interface being characterized by striking colours of black, buff, ochre, orange, blue, green and brown, while sediments from upper flank or sill areas have more subdued colours of grey, buff or orange.

The age of onset of metalliferous sediment formation is not well known, but Hackett and Bischoff present evidence to show that metalliferous sediments of the Atlantis II Deep are approximately 12,900 years old. The metalliferous sediments accumulate at a much higher rate (700 mm/10^3 years) than normal Red Sea sediments (100 mm/10^3 years), and this explains the masking of normal Red Sea sediments by metalliferous sediments in the brine deeps (Manheim, 1975).

Metalliferous sediment formation probably has been episodic. Bignell et al. (1974) and Bignell and Ali (1976), for example, have shown that there have been at least four separate periods of metalliferous sedimentation. Metalliferous sediment occurrence is much more pronounced in the eastern basin of the Nereus Deep compared to the western basin, and metal contents of the sediments suggest that the brines discharge from the faulted sides of the Deep and then flow into the deeper parts of the basin. Bignell (1975) has related the episodic nature of the brine discharge to changes in sea level in the Red Sea.

Submarine volcanic vents

Hydrothermal springs issuing forth from the underwater flanks of certain volcanoes represent another mode of emplacement of metalliferous deposits (Fig.3). Hot jets emerge from the submarine slopes of Banu Wuhu (Indonesia), coating extrusive fragments with a bright orange, soft, greasy iron hydroxide under which there is sometimes a denser, thin, bluish black slightly broken crust of Fe and Mn hydroxide (Zelenov, 1964). The extrusives on which the coatings are found show no signs of hydrothermal alteration, only a red stain penetrating 10–20 mm into the rock. Analogous deposits have been reported from Santorini Island (eastern Mediterranean) (Butuzova, 1966; Puchelt et al., 1973; and Smith and Cronan, 1975a, b); at Matupi Harbor (New Britain) (Ferguson and Lambert, 1972); and near the Pacific coast of northern Baja California (Vidal et al., 1978), among other places. Differential precipitation of Mn and Fe has produced dispersion haloes around hydrothermal sources, with metal sulfides, then iron silicates and/or oxides, and finally manganese oxides being deposited at increasing distances from the vents (Cronan, 1976a). Volcanic effluent processes also contribute to the formation of the previously discussed varieties of metalliferous sediments associated with mid-ocean ridge systems, but the sites of their vents are more difficult to locate because of greater water depths. Nevertheless, deep-tow surveys and submersibles have been used to locate hydrothermal vents in the Galapagos Spreading Centre and FAMOUS area of the Mid-Atlantic Ridge, as previously discussed.

MARINE DISTRIBUTION AND NATURE OF BASAL DEPOSITS

Pacific Ocean

Metalliferous sediments immediately overlying volcanic basement have been recovered during D.S.D.P. drilling at widely separated sites, especially in the eastern Pacific (Von der Borch and Rex, 1970; Von der Borch et al., 1971; H.E. Cook, 1972; and Cronan, 1973, 1976b) (Fig.11, Table VI). As noted by Cronan (1974), the basal sediments over much of the western Pacific do not show iron enrichment. Except for several marginal sites, only at D.S.D.P. Sites 66 (between the Line and Phoenix Islands) and 307 (southeast of the Shatsky Rise) do basal metal-rich sediments occur (Table VII). At Site 66, basement basalt is overlain by about 5 m of multi-coloured altered hyaloclastite and chemical sediments, including ferromanganiferous clay and evidence of iron oxide replacement (Winterer et al., 1971). A similar deposit occurs at Site 307. The apparent

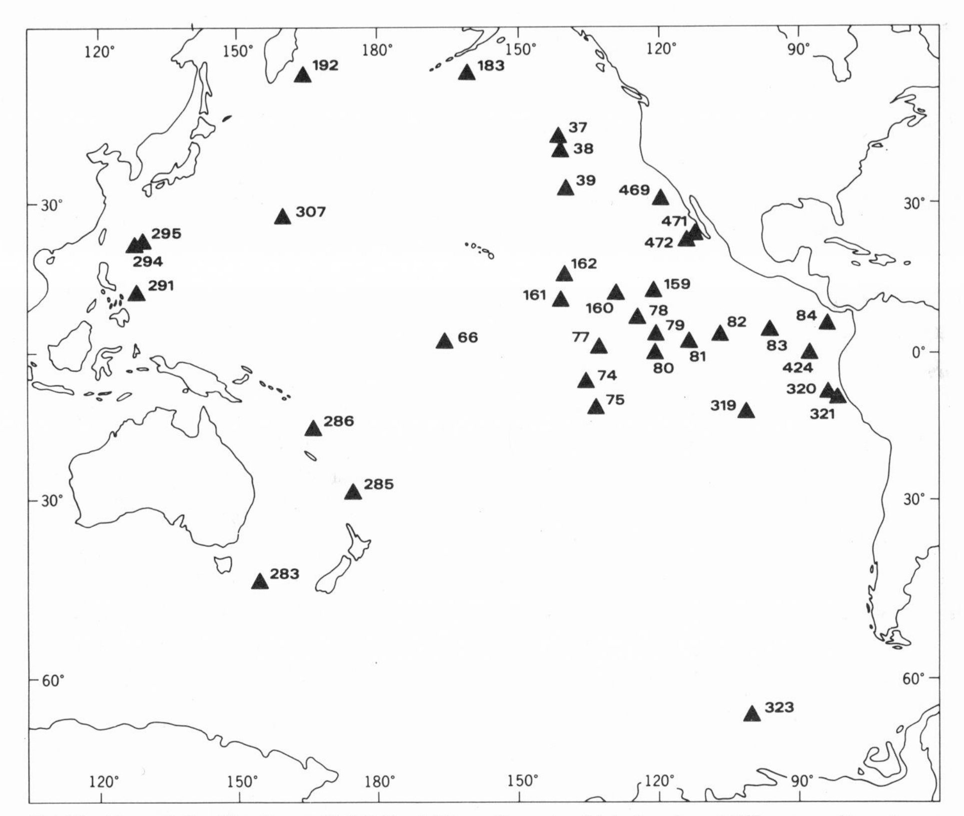

Fig.11. Map of Pacific Ocean D.S.D.P. drilling sites at which basal metalliferous sediments were recovered. Sample descriptions and analytical data are given in Tables VI, VII, X–XII.

TABLE VI

Characteristics of central and eastern Pacific basal metalliferous and hydrothermally altered sediment units

D.S.D.P. Site	Unit thickness (m)	Sub-bottom depth (m)	Age	Lithology
37	4	25–29	Early Eocene (?)	Very dark brown amorphous iron oxide sediment with traces of zeolites. Includes white layers and nodules of pure zeolites and two altered ash beds.
38	15	33–48	Early Eocene	Brown amorphous iron oxide sediment overlying 9 m of dusky yellow-brown foraminiferal–nannofossil ooze containing subequal amount of amorphous iron oxides.
39	7	10–17	Early Eocene	Dusky brown amorphous iron oxide sediment with thin interbeds of calcareous–nannofossil ooze near base.
66	5	187–192	Late Cretaceous	Three brownish-black layers 45, 33 and 18 cm thick in interval 190.6–192.0 m, consisting of iron–manganese oxide clays, interbedded with semi-indurated montmorillonite sandstone composed of altered hyaloclastic fragments, and with variegated clays rich in palagonite, volcanic glass and iron–manganese micronodules, and minor fish debris.
74	2	100–102	Middle to Late Eocene	Line Islands Oceanic Formation. Five dark yellowish-brown zones of iron oxides plus nannofossils, 50–1600 mm thick, interbedded at sharp contacts with very pale brown to white nannofossil ooze beds, 50–400 mm thick.
75	8.1	74.0–82.1	Early Oligocene	Base of Marquesas Oceanic Formation. Eight dark iron-rich zones of nannofossil ooze, 300–2400 mm thick, interbedded with light-coloured nannofossil ooze beds, 100–300 mm thick.
77	10.2	470.8–481.0	Late Eocene	Line Islands Oceanic Formation. Dusky brown nannofossil–clay mudstone, firmly indurated, overlying yellowish-brown baked limestone, well indurated and brecciated.

TABLE VI (continued)

D.S.D.P. Site	Unit thickness (m)	Sub-bottom depth (m)	Age	Lithology
78	9.8	310.5–320.3	Early Oligocene to Late Oligocene	Line Islands Oceanic Formation. Greyish-orange clay–radiolarian-nannofossil chalk overlying white baked foraminiferal packstone–grainstone. Massively bedded, well indurated, and containing reddish-brown amorphous iron–manganese oxides.
79	0.1+	above 413.6–413.7	Early Miocene	Base of Marquesas Oceanic Formation. Very pale orange to white foraminiferal–calcareous nannofossil chalk. Hydrothermal alteration in this baked zone exhibited by tridymite euhedra lining interior of foram tests, and yellowish-green clay and brown iron oxides replacing nannofossil chalk.
80	17.6+	above 185.0–202.6	Early Miocene	Line Islands Oceanic Formation. Greyish-brown and moderate brown nannofossil–clay–chalk mudstone, gradationally bedded; beds are 50–150 mm thick, with occasional 1–2 mm laminae. Contains amorphous iron–manganese oxides.
81	27	about 382–409	Early Miocene	Line Islands Oceanic Formation. Pale orange foraminiferal–radiolarian–nannofossil chalk with manganese dendrites overlying pale orange foraminiferal–nannofossil chalk. Basal metre partly brecciated; exhibits baking and deuteric mineralization.
82	20.4	202.6–223.0	Late Miocene	Line Islands Oceanic Formation. Very pale orange foraminiferal–nannofossil–radiolarian–ooze chalk overlying white foraminiferal–nannofossil–radiolarian–ooze chalk. Intensely mottled and indistinctly bedded, and containing abundant manganese (?) dendrites and scattered patches of olive green hydrothermal clay and deuteric iron oxides.

(continued)

TABLE VI (continued)

D.S.D.P. Site	Unit thickness (m)	Sub-bottom depth (m)	Age	Lithology
83	0.2+	above 232.9–233.1	Middle Miocene	Line Islands Oceanic Formation. Very pale orange, greyish-orange and moderate brown clay–foraminiferal–nannofossil chalk, containing yellow hydrothermal (?) clay and amorphous iron–manganese oxides.
84	3	251–254	Late Miocene	San Blas Oceanic Formation. Pale green montmorillonitic–calcareous–nannofossil chalk replaced by chert. Thin layer of glauconite at base of this baked zone.
159	37	71–108	Late Oligocene/ Early Miocene to Early Miocene	Nannofossil marl ooze, moderate yellow-brown and greyish orange grading down to dark moderate brown. Ferruginous, particularly below 100 m sub-bottom.
160	1+	above 108–109	Lower Oligocene	Dark brown calcareous clay rich in ferruginous aggregates.
161	44+	above 200–244	Middle to Late Eocene	Dark brown clayey–ferruginous–radiolarian ooze, indurated, fairly intensely mottled and burrowed.
162	6+	above 144–150	Middle Eocene	Dark to dusky brown ferruginous, locally ashy, locally zeolitic brown claystone grading down to foraminiferal–nannofossil marl and nannofossil chalk.
319	34	76–110	Early to Middle Miocene	Light to dark yellowish-brown, brown and dark brown, foram-trace to foram-rich, iron-bearing to iron-rich nannofossil ooze.
320	81.5+	above 73.5–155	Late Oligocene to Early Miocene	Pale brown, yellowish-brown and dark greyish-brown, foram-trace to foram-rich, clay-bearing to clay-rich, iron-bearing to iron-rich nannofossil ooze.
321	66	58–124	Late Eocene to Early Miocene	Interbedded pale brown, yellowish-brown and dark brown nannofossil ooze, ferruginous nannofossil ooze, foram-rich nannofossil ooze, and iron/foram/zeolite-bearing nannofossil ooze.

TABLE VI (continued)

D.S.D.P. Site	Unit thickness (m)	Sub-bottom depth (m)	Age	Lithology
424	~15–20	Surface or near surface to about 20 m	Holocene or Pleistocene	Green hydrothermal mud or mixture of iron–manganese chips and green hydrothermal mud. Underlain by 15–20 m of foraminiferal–nannofossil ooze showing no evidence of hydrothermal mineralization except for thin interbeds of green hydrothermal mud. Foram-nanno ooze unit underlain by basalt.

References: Sites 37–39 – Von der Borch and Rex (1970), McManus et al. (1970); Site 66 – Winterer et al. (1971), Heath and Moberly (1971); Sites 74 and 75 – Tracey et al. (1971), Von der Borch et al. (1971); Sites 77–84 – Hays et al. (1972); Sites 159–162 – Van Andel and Heath et al. (1973); Sites 319–321 – Yeats and Hart et al. (1976); Site 424 – Scientific Party, D.S.D.P. Leg 54 (1977).

absence of the basal ferruginous sediment over much of the western Pacific may be due to a high degree of dilution by volcanoclastics and nannofossil ooze, although at many sites basement was not reached.

The Pacific basal ferruginous sediments have been designated as part of the Line Islands Oceanic Formation (Tracey et al., 1971). (At some sites the ferruginous unit comprises the entire formation.) It is a relatively thin, diachronous unit that is interpreted to represent the initial deposits on or stratigraphically near newly formed oceanic crust (H.E. Cook, 1972). As such, it is laterally contiguous with metalliferous sediments on the East Pacific Rise that have not yet been blanketed by biogenic oozes and transported by sea-floor spreading into the deeper waters of the Pacific (Fig.12).

It is implied in Fig.12 that metalliferous components, which characterize the lower part of the Line Islands Oceanic Formation, gradually become less abundant upward in the sediment column. It must be pointed out, however, that there is not always a gradual transition to more normal pelagic sediments, as is the case at D.S.D.P. Sites 37, 38, 39, 159 and 162. In some holes (D.S.D.P. Sites 74, 75, 160 and 161), there is a sharp contact with the overlying calcareous ooze, possibly representing a hiatus of indeterminant length. At D.S.D.P. Site 66, the basal metal-bearing zeolitic pelagic clay and volcanoclastic sand is overlain with little transition by radiolarian ooze, but the lack of calcareous organisms throughout the section suggests that Site 66 did not originate on a topographic high comparable to the modern East Pacific Rise (Heath and Moberly, 1971).

D.S.D.P. Sites 320 and 321 have metalliferous components concentrated near the base of the sediment column, but also contain metal-rich units higher in the section. At D.S.D.P. Site 319 in the Bauer Basin, ferruginous material is distributed throughout the

TABLE VII

Characteristics of northern and western Pacific basal metalliferous and hydrothermally altered sediment units

D.S.D.P. Site	Unit thickness (m)	Sub-bottom depth (m)	Age	Lithology
183	24	476.5–501.5	Early to Middle Eocene	Unit consists of five sediment types, from top to bottom: (1) dusky yellow clay, (2) bluish-white nannofossil limestone, (3) dark yellow-brown ferruginous clay, (4) orange calcareous ironstone, and (5) indurated pyrite-bearing greyish-green aragonitic limestone.
192A	~5	about 1039–1044	Late Cretaceous	Interbedded light brown to dark reddish-brown ferruginous–calcareous claystone and ferruginous–nannofossil chalk.
283	12	576–588	Paleocene	Dark greenish-grey silty claystone with pale red feeding trails and fecal pellets; also burrow-mottled, organic-rich, and pyrite-, glauconite-, and manganese micronodule-bearing.
285	~0.2	about 564.6–564.8	early Middle Eocene	Very dark grey and dark reddish-brown sandy siltstone. Contains devitrified glass shards and secondary haematite as a result of heating by intrusive diabase.
286	~1	about 648–649	Middle Eocene	Moderate orange-pink and dark grey feldspar-bearing, ferric-oxide altered, glass shard clayey siltstone. Manganese–iron oxides distributed as lenses or nodules, or in disseminated grains.
291	17	101–118	Late Eocene	Dark brown ferruginous zeolite clay and silty clay with nannofossil- and radiolarian-rich interbeds. Abundance of iron oxide (haematite and goethite) increases downwards, particularly near base of unit.
294	14.5	97.5–112.0	Eocene (?)	Blackish-red to dark red ferruginous silt-rich clay and silty clay, with 5–30% goethite plus haematite. Unit contains thin dark yellowish-orange layers and patches of palagonite, clays and zeolites near base.

TABLE VII (continued)

D.S.D.P. Site	Unit thickness (m)	Sub-bottom depth (m)	Age	Lithology
295	57 (?)	101–158 (?)	Eocene (?)	Same as Site 294.
307	~0.5	about 297.5–298	Late Cretaceous	Lithified moderate red-brown ferruginous siliceous pelagic claystone. Contains chert fragments and manganese blebs scattered throughout. Unit overlies haematite- and calcite-veined hyaloclastite.

References: Sites 183 and 192A – Creager and Scholl et al. (1973), Natland (1973); Site 283 – Kennett and Houtz et al. (1974); Sites 285 and 286 – Andrews and Packham et al. (1975); Sites 291, 294, 295 – Karig and Ingle et al. (1975); and Site 307 – Larson and Moberly et al. (1975).

sediment column. Dymond et al. (1977) found that hydrothermal metal accumulation decreases upward in the section, whereas hydrogenous metal precipitation becomes more important. This indicates, as might be expected, that hydrothermal influence wanes as a section of crust moves away from a spreading centre.

A number of D.S.D.P. Leg 9 sites (77, 78, and 80–83) terminated in intrusive basalt which has baked the overlying sediments, but it is felt that the basal metalliferous

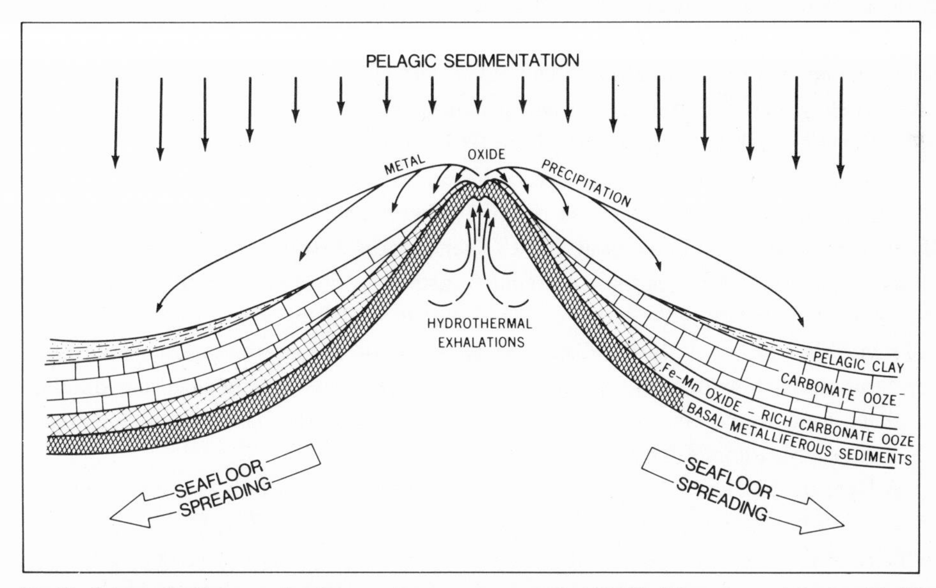

Fig.12. Schematic diagram of sedimentation processes on East Pacific Rise crest and flanks. (Modified from Boström and Peterson, 1969, Von der Borch and Rex, 1970, and Broecker, 1974.)

material is representative of hydrothermal precipitates formed at a rise crest (Hays et al., 1972). The Line Islands Oceanic Formation at these Leg 9 sites, where the upper contact was cored, shows either a moderately abrupt transition (through a metre or less) to, or a very sharp contact with, superjacent formations, according to lithologic descriptions (H.E. Cook, 1972). In disagreement with this observation is the report by Hays et al. (1972) that the amorphous iron and manganese, which are responsible for colouration used to distinguish formations, are concentrated in the lower part of the formation, decreasing upward (transitionally).

The principal metal-bearing phase in Pacific basal sediments is a mixture of colloidal material and microscopic sub-spherical grains, apparently similar to the East Pacific Rise metalliferous components (Von der Borch and Rex, 1970). Metal oxides also occur as micronodules and as distinct in-fillings within foraminiferal tests and diatom frustules. Although the ferruginous spherules are commonly reported to be X-ray amorphous (e.g., by Von der Borch and Rex, 1970), Dymond et al. (1973) have shown that similar metal-bearing particles in Bauer Deep sediments are microcrystalline iron-montmorillonite.

The basal ferruginous unit of the eastern Pacific has a relatively uniform thickness of 10–25 m over a 4000 km west–east distance, according to D.S.D.P. Leg 9 drilling (H.E. Cook, 1972), but thinner sections were found on other legs (Table VI). At a few locations, several metres of essentially pure metal-bearing facies have been found, e.g., at a depth of 33–39 m at D.S.D.P. Site 38 (Von der Borch and Rex, 1970), but more commonly the ferruginous matter is admixed with variable proportions of calcareous nannofossils and phillipsite. At Sites 74 and 75, the dark yellowish-brown beds (50–2400 mm thick) that contain up to 50% metalliferous components alternate with very pale brown to white nannofossil ooze beds, 50–400 m thick (Von der Borch et al., 1971). The basal sediments at Leg 9 sites are also laminated, and are moderately- to well-indurated. At several sites, partly flattened burrows or manganese dendrites were observed, and at Site 81, the metal-rich chalks were brecciated (H.E. Cook, 1972).

D.S.D.P. drilling has also recovered basal ferromanganoan sediments related to seamount volcanism rather than to ocean rise hydrothermal activity (Natland, 1973). Basal sediments cored above alkali basalt at Site 183 beneath the Aleutian Abyssal Plain include iron-rich clays, a goethite-bearing calcareous ironstone, and a pyrite-bearing nonfossiliferous aragonitic limestone, the latter probably precipitated from a near-bottom, Sr- and metal-rich "hot" brine. At Site 192A, atop Meiji Guyot, the northernmost of the Emperor Seamounts, 5 m of iron- and manganese-enriched clays interbedded and diluted with chalk were found on extrusive basalt pillow lavas at the base of a sedimentary section almost one kilometre thick.

A layer of metal-rich sediment above basaltic basement has been cored at D.S.D.P. Sites 294/295 in the West Philippine Basin (Bonatti et al., 1976d, 1979) (Table VII). This demonstrates that metalliferous sediment can be generated at minor spreading centres in marginal basins. The deposits contain up to 38% Fe and 9% Mn, with most of the iron occurring in hydroxides, haematite and smectite. The trace element chemistry and REE patterns (e.g., negative cerium anomaly) are similar to those of mid-ocean ridge hydrothermal deposits.

Atlantic Ocean

Basal metalliferous sediments have been encountered more commonly in the western than eastern Atlantic (Fig.13). Evidence from D.S.D.P. Leg 3 drilling along 30°S in

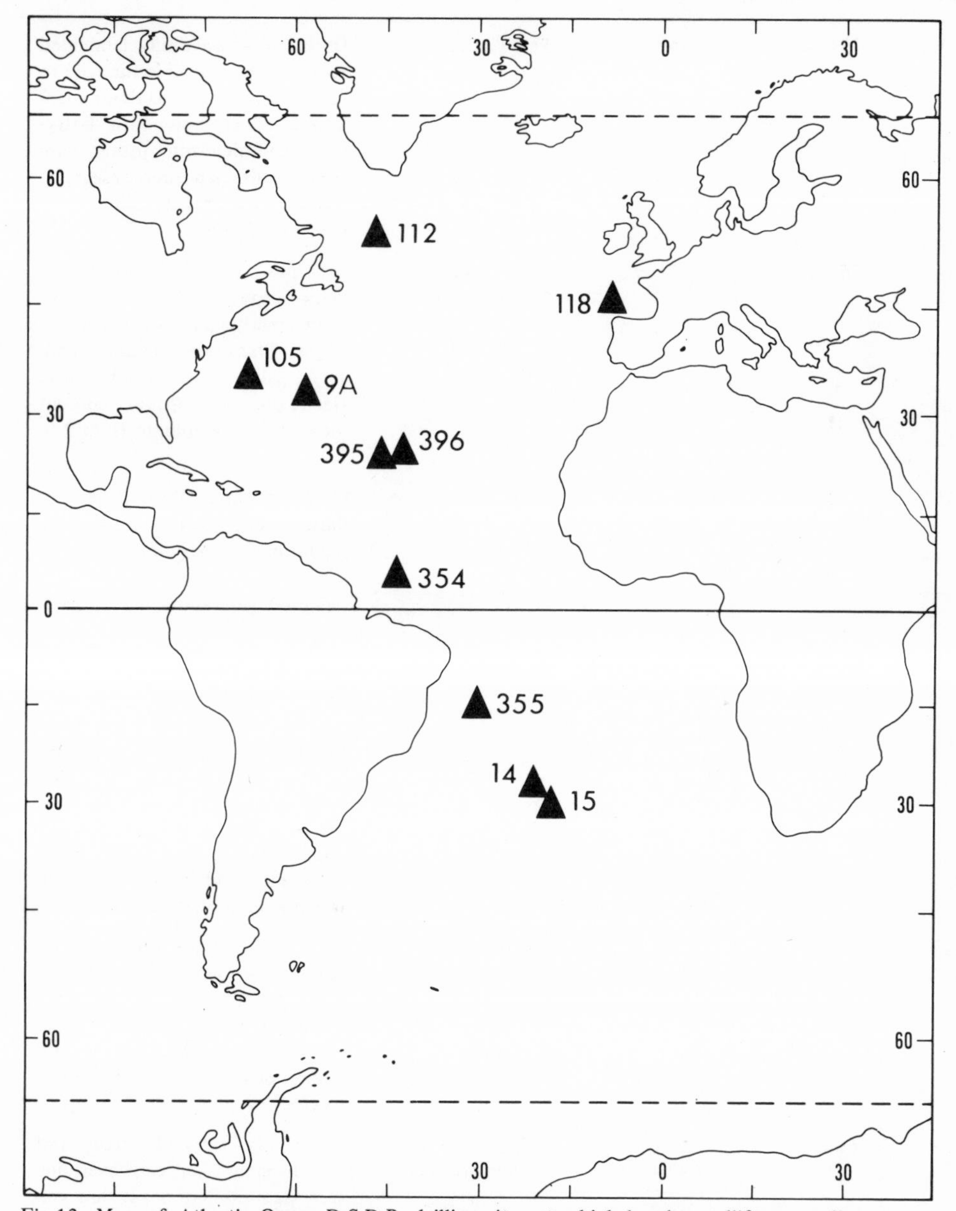

Fig.13. Map of Atlantic Ocean D.S.D.P. drilling sites at which basal metalliferous sediments were recovered. Sample descriptions and analytical data are given in Tables VIII, X–XII.

TABLE VIII

Characteristics of Atlantic basal metalliferous and hydrothermally altered sediment units

D.S.D.P. Site	Unit thickness (m)	Sub-bottom depth (m)	Age	Lithology
9A	3+	above 831–834	Late Cretaceous	Dark red-brown thinly laminated zeolitic clay, semi-indurated to fissile shale. Contains dense manganese concretions with botryoidal and dendritic forms; pure clinoptilolite in pockets or segregations; and haematite, rhodochrosite and siderite.
14	36	71–107	Late Eocene to Late Oligocene	Grampus Ooze. Interbeds of yellow-brown and dark yellow-brown marly-nannofossil-chalk oozes. Transitional contact with very pale brown chalk ooze. Haematite and opaque content increases down column from 2 to 15%.
15	9+	123–below 132	Early Miocene	Grampus Ooze. Dark brown to dark reddish-grey foram-bearing nannofossil–marly–chalk ooze. Opaques and haematite 10–25%. Uncored gap above basalt basement at –140.6 m sub-bottom.
105	45.9+	568.5–below 615.4	Late Jurassic	Laminated, indurated, reddish-brown clayey limestone with thin greyish and greenish lenses and interbeds. Displays flow and burrow structures. Contains recrystallized nannofossil ooze clasts, macrofossils; barite and native copper veinlet bounded by green palagonite; and haematite, which is common in reddish zones.
112	9+	above 652–661	Paleocene (?)	Indurated dark reddish-brown nannofossil-trace clay, with 20 mm palagonite sill. Colouring matter is predominantly haematite; also contains goethite and amorphous iron–manganese oxides.
118	66+	above 687–753	Paleocene to Middle Miocene	Chemically altered grey and brown nannofossil clay overlying recrystallized red clays; thin yellow-grey baked zone immediately above basalt sill.

TABLE VIII (continued)

D.S.D.P. Site	Unit thickness (m)	Sub-bottom depth (m)	Age	Lithology
354	36	850–886	Late Cretaceous	Well-indurated, pale red ferruginous marly calcareous chalk; some burrows filled with black material. Iron–manganese oxides 5%; occur as discrete subhedral grains and aggregates. Abrupt upper contact is sedimentary hiatus.
355	44	405–449	Late Cretaceous	Light grey-brown nannofossil ooze; altered (baked?) at lower contact to micritic limestone. Contains interbeds of reddish-brown ferruginous nannofossil ooze, which are associated with veins of sparry calcite. Overlain by units which are also ferruginous.
395	~4	about 80–84	Pliocene	Dark yellowish-brown and dark brown manganese micro-nodule-bearing calcareous clay. Separated from basalt basement by 4–12 m of talus (?): brownish-yellow nannofossil ooze with basalt cobbles.
396	8	117–125	Middle Miocene to Pliocene (?)	Dark yellowish brown to dark brown calcareous clay interbedded with pale yellowish brown marly nannofossil ooze. Calcareous sediment at basalt contact recrystallized (baked?) to limestone.

References: Site 9A – Peterson and Edgar et al. (1970); Sites 14 and 15 – Maxwell et al. (1970); and Boström et al. (1972); Site 105 – Hollister and Ewing et al. (1972) and Lancelot et al. (1972); Sites 112 and 118 – Laughton and Berggren et al. (1972); Sites 354 and 355 – Supko and Perch–Nielsen et al. (1977); Sites 395 and 396 – Melson and Rabinowitz et al. (1978).

the South Atlantic indicates that iron-rich crestal sediments have been emplaced along the Mid-Atlantic Ridge more or less uninterruptedly since the middle Eocene (Boström, 1970b). However, basal sediments on the inactive Rio Grande Ridge show no iron enrichment (Boström et al., 1972). Core samples from Sites 14 and 15 included basal sediments analogous to those found in the Pacific (Table VIII). These sediments contain at least 50% $CaCO_3$. Basal units from other Leg 3 sites (16–19) on the flanks of the Mid-Atlantic

Ridge consist predominantly of calcareous ooze, with a very minor metalliferous fraction.

Over much of the Atlantic, metalliferous components may be dispersed in basal anaerobic (carbonaceous) muds or evaporite deposits, rather than in typical pelagic sediments.

Indian Ocean

Basal metalliferous sediments were recovered in the Indian Ocean during D.S.D.P. Legs 22 and 24–27 (Fig.10, Table IX). They range in character from units that are almost

TABLE IX

Characteristics of Indian Ocean basal metalliferous and hydrothermally altered sediment units

D.S.D.P. Site	Unit thickness (m)	Sub-bottom depth (m)	Age	Lithology
211	~66.5	about 335–401.5	Tertiary	Includes: (1) moderate brown homogeneous amorphous iron oxide, (2) moderately indurated dark grey, pyrite-rich ash almost totally replaced by iron oxides, and (3) moderately indurated red, ash-rich iron oxide.
213	~5	about 147–152	Late Paleocene	Greyish-brown clay-bearing to clay-rich iron oxide sediment. Contains oolitic-like crystals, (goethite?) associated with red scales of haematitic and less crysstalline masses of limonitic material. Unit overlain and partly interbedded with calcareous nannofossil ooze.
215	2	149–151	Paleocene	Brown to dark brown iron oxide-rich nannofossil ooze grading down to dark greyish-brown iron oxide-rich clay nannofossil ooze.
236	4.1	301.0–305.1	Late Paleocene	Pale green, moderate yellowish-brown and dusky brown clayey nannofossil chalk. Contains a thin layer of yellowish-green ferruginous clay with about 23% iron, 400 mm above basalt basement.

(continued)

TABLE IX (continued)

D.S.D.P. Site	Unit Thickness (m)	Sub-bottom depth (m)	Age	Lithology
238	34.5	471.5–506.0	Early to Late Oligocene	Semilithified variegated (mostly shades of orange or brown, also pinks, greens and yellow) nannofossil chalk with intercalated horizons of volcanic ash and zeolite sands. Iron-oxide stained throughout; includes some layers of concentrated amorphous iron oxide globules mixed with calcareous material.
245	21	368–389	Early Paleocene	Moderate yellow-brown to dusky yellow-brown grading downward to brownish-black or olive black ferromanganoan clayey nannofossil chalk.
215A	18.3	468.2–486.5	Early Miocene	Very pale yellowish-brown to white garnet-rich calcite (micarb) chalk, with faint limonite mottling. Contains about 20% authigenic garnet and traces of iron oxides; garnet grains 2–3 μm in size.
256	~0.05	about 251	Early Cretaceous	Brown ferruginous coccolith detrital clay. Contains 25–30% translucent and opaque ferruginous material.
260	2	321–323	Early Cretaceous	Semilithified moderate brown to dark red-brown calcareous radiolarian clay.
261	5	527.5–532.5	Late Jurassic	Semilithified dark moderate brown to dark reddish-brown nannofossil claystone.

References: Sites 211, 213, 215 – Von der Borch and Sclater et al. (1974); Sites 236 and 238 – R.L. Fisher and Bunce et al. (1974); Site 245 – Simpson and Schlich et al. (1974); Site 256 – Davies and Luyendyk et al. (1974); Sites 260 and 261 – Veevers and Heirtzler et al. (1974).

completely composed of iron oxides (Sites 211 and 213), to the more common ferruginous nannofossil oozes and chalks (Sites 215, 236, 238 and 245), to ferruginous clays (Sites 256, 260 and 261), to a garnet-rich chalk (Kempe and Easton, 1974) (Site 251A).

Warner and Gieskes (1974) analyzed the ferromanganoan nannofossil chalk overlying a diabase sill, and found it to be chemically similar to basal sediments from the Pacific and Atlantic, as well as to East Pacific Rise crestal surface sediments, but it has a noticeably lower metal content. Leclaire (1974) observed that the ferromanganiferous component appears not as coatings, but as amorphous agglomerates of very fine grains, which to him suggested rapid formation.

SEDIMENT CHEMISTRY

Compositional variation

Metalliferous sediments are notably enriched compared to other deep-sea sediments in Fe, Mn, other transition metals, and some additional elements as well. Phosphorus, for example, is associated with the iron oxyhydroxides of metalliferous sediments, and this is a major sink for the removal of this element from seawater (Froelich et al., 1977; Bloch, 1978). Gurvich et al. (1976) noted that East Pacific Rise sediments are enriched in antimony.

Compositionally, metalliferous sediments show variations which cannot be entirely explained by detrital or biogenic dilution (Cronan, 1974). Possible chemical variations in source material and the variable influence of several geochemical processes must be called upon to rationalize the compositional differences of metal-rich sediments from different areas and depositional environments.

Data on chemical composition on various types of metalliferous deposits are presented in Tables X–XII. At the top of Tables XI and XII, data on the composition of average pelagic clay are presented for comparison (deposit numbers 1–3; note that these numbers refer to the same samples on all three tables). Note also that these tables on metalliferous deposits include data on several types of material, including ferromanganese crusts, and that some analyses are reported on a carbonate-free basis, while others are not.

Boström et al. (1969) observed that the carbonate-free fraction of sediments from active oceanic ridges is characterized by very low Al and Ti contents and by high Fe and Mn contents compared with sediments from other volcanic regions and from inactive ridges. These unconsolidated aluminium-poor ferromanganoan sediments are not restricted to mid-ocean ridge crests, but occur on a regional scale over large areas of the ocean floor [Boström and Peterson, 1969; see Boström et al., 1969, for a world-ocean map of the Al/(Al + Fe + Mn) ratio in sediments]. Compared to flank sediments, East Pacific Rise crestal sediments show high enrichments of Cd and B, moderate enrichments of Fe, Mn, V, Cr, and As, a slight enrichment of Zn, slight depletions of Ni, Co, Cu, and more marked depletions of Al, Si, Ti and Mo (deposit numbers 22–24; Boström and Peterson, 1969).

For the Mid-Atlantic Ridge, a crestal Zn enrichment is not apparent compared to the flanks, but Pb and Tl are slightly enriched, and Sn is somewhat depleted (deposit numbers 4–6; Horowitz, 1970). Cronan (1972a) noted that local variations in sediment composition occur at the crest of the Mid-Atlantic Ridge near 45°N; sediments from the median valley are more ferruginous than those of the Eastern and Western Crest Mountains. The valley sediments are also enriched in As and Hg, while Mn is only slightly enriched, and Al is not markedly depleted.

Considering absolute abundances of various metals, differences have been noted between mid-ocean ridge areas. Mid-Atlantic Ridge sediments near 45°N contain less Fe and Mn than similar deposits from the East Pacific Rise, but their content of As is somewhat higher (deposit numbers 7, 22–24, 28; Cronan, 1972a). Horowitz (1970) determined element concentrations of Pb, Tl and Zn, and concluded that the EPR system is most active (high concentrations), the Indian Ocean Ridge intermediate, and the northern MAR least active (low concentrations) (see deposit numbers 4–6, 87, 22–24). Lau Basin Rise sediments contain less Mn, Ni, Cu and Pb than do East Pacific Rise sediments (deposit numbers 80–84 and 22–24; Bertine, 1974), but more Mn than Mid-Atlantic Ridge sediments.

Basal sediments cored during D.S.D.P. drilling resemble mid-ocean ridge sediments in chemical composition, although, of course, variations do exist (deposit numbers 14–18, 45–76, 79, 86 and 88–90). The most apparent difference is in Al content, which is greater in basal sediments from the eastern Pacific than in EPR crestal sediments, but Lau Basin Rise sediments contain even more aluminium. This may reflect increasingly greater proportions of incorporated detrital material. A comparison of basal sediments of modern ocean basins and umbers overlying basalt in an ancient ophiolite complex (deposit numbers 104, 105) shows the umbers to be much more enriched in manganese and deficient in iron.

Horowitz and Cronan (1976) have analyzed a large number of sediment samples from D.S.D.P. cores in the North Atlantic (including deposit numbers 14–16), and compared their results with chemical compositions determined for other sediments which have been categorized as metalliferous. They found that North Atlantic basal sediments, when compared to overlying sediments, are enriched in Fe, Mn, Mg, Ni, Cr and Pb and depleted in Al, Ti, Cu, Zn and Li. However, the absolute amounts of the enriched metals are generally not as high as those reported for Mid-Atlantic Ridge surface sediments now being deposited. Observed bulk values, along with metal partitioning patterns, indicate that there is a much larger detrital contribution in basal North Atlantic than in basal Pacific sediments.

Chemical analyses of sediments from the TAG hydrothermal area indicate somewhat geographically variable metal enrichment, and that the degree of enrichment is much lower than observed in East Pacific Rise sediments (M.R. Scott et al., 1978). However, calculated non-detrital metal accumulation rates showed that Fe, Mn and trace elements are being added to sediment in the TAG area at rates well above background values for the North Atlantic Ocean and at rates higher than those reported for the FAMOUS area.

TABLE X

Chemical composition of metalliferous marine deposits: sample identification and description

Deposit No.	Locality	Latitude	Longitude	Depth (m)
1	World Ocean	–	–	–
2	Pacific Ocean	–	–	–
3	Atlantic Ocean	–	–	–
4	MAR, east flank, N. Atl.	19–36°N	13–37°W	3093–5970
5	MAR, crest, N. Atl.	24–30°N	37–53°W	2826–5384
6	MAR, west flank, N. Atl.	20–27°N	53–77°W	2777–6177
7	MAR, 45 N.	45–46°N	27–30°W	–
8	MAR, Atlantis F.Z., N. Atl.	30°08′N	42°29′W	–
9	Atlantis F.Z., N. Atl.	26°07′N	25°21′W	–
10	MAR near TAG area, N. Atl.	26°34′N	44°30′W	–
11	MAR near TAG area, N. Atl.	26°17′N	45°06′W	–
12	MAR, TAG area, N. Atl.	26°08′N	44°45′W	~3400
13	MAR, FAMOUS area, N. Atl.	36°57′N	33°04′W	2690
14	N. Atl., D.S.D.P. Site 9A	32°46′N	59°12′W	4981
15	N. Atl., D.S.D.P. Site 112	54°01′N	46°36′W	3657
16	N. Atl., D.S.D.P. Site 118	45°03′N	9°01′W	4901
17	W. Equat. Atl., D.S.D.P. Site 354	5°54′N	44°12′W	4052
18	W. Equat. Atl., D.S.D.P. Site 355	15°43′N	30°36′W	4896
19	EPR, seamount, S. Pac.	10°38′S	109°36′W	1790–2130
20	EPR, seamount, S. Pac.	10°38′S	109°36′W	1790–2130
21	EPR, seamount, S. Pac.	10°38′S	109°36′W	1790–2130
22	EPR, east flank, S. Pac.	12–14°S	97–108°W	3560–4210
23	EPR, crest, S. Pac.	12–15°S	110–120°W	2990–3685
24	EPR, west flank, S. Pac.	14°S	122–140°W	3790–4050
25	EPR north of 10°S	7°S	106–107°W	–
26	EPR, E. Pac	10–25°S	111–115°W	–
27	EPR, Pac.-Antarctic	44–63°S	76–160°W	2584–4927
28	EPR, Pac.-Antarctic	53–64°S	135–166°W	2932–3766
29	Galapagos Rise, E. Equat. Pac.	9°02′S	97°36′W	4080
30	Galapagos Spreading Centre, Hess Deep	2°18′N	101°02′W	3150
31	Galapagos Ridge, E. Equat. Pac.	2°42′N	95°14′W	2930
32	Galapagos Ridge, E. Equat. Pac.	2°58′N	95°10′W	2560
33	Galapagos Rift, D.S.D.P. Sites 424, 424A, B	0°36′N	86°08′W	2685–2708
34	Galapagos Rift hydrothermal mounds	0°34–37′N	86°05–09′W	–
35	Galapagos Rift hydrothermal mounds	0°25′N	86°06′W	–
36	Galapagos Rift hydrothermal mounds	0°34–37′N	86°05–09′W	–
37	Galapagos Rift hydrothermal mounds	0°34–37′N	86°05–09′W	–
38	Bauer Deep, E. Equat. Pac.	8°22′S	102°14′W	4124

Original sample number	Type of material	No. of samples	Sample abbreviation	Reference No. *
–	Average pelagic clay	?	PelagClayAvg	35
–	Average pelagic clay	22	PacClayAvg	22
–	Average pelagic clay	?	AtlClayAvg	15
Various	Surface, near-surface sediment	16–20	MAR/NAtl	27
Various	Surface, near-surface sediment	6–12	MAR/NAtl	27
Various	Surface, near-surface sediment	5–11	MAR/NAtl	27
Deep Drill 22-71/343-498	Surface sediment	16	MAR/45N	15
T3-71D 160-10G	Fe–Mn crust	4	MAR/ATLFZ	34
T3-71D 148-2B	Fe–Mn crust	4	ATLFZ	34
T3-71D 254-15-2	Fe–Mn crust	1	MAR/TAG	34
T3-71D 255-19-3	Fe–Mn crust	1	MAR/TAG	34
T3-72D 253-13-21	Mn crust, 42 mm thick	7	MAR/TAG	34
74-26-15b, A-C, -16	Clay-rich hydrothermal material	5	MAR/FAM	26
9A-5-1 to -5-2	Basal sediment	6	DSDP 9A	28
112-16-1	Basal sediment	3	DSDP 112	28
118-18-2 to -19-1	Basal sediment	5	DSDP 118	28
354-17-2, 117 to -18-4, 128	Basal sediment	3	DSDP 354	19
355-17-3, 140 to -20-2, 115	Basal sediment	7	DSDP 355	19
Amph D2	Red ferruginous rock	3	EPRsmtSPac	6
Amph D2	Black porous Mn-oxide rock	1	EPRsmtSPac	6
Amph D2	Mn crust on basalt	1	EPRsmtSPac	6
Risepac 46G to 60G	Near-surface sediment	4–6	EPR/SPac	9, 27
Risepac 63G to 74G	Near-surface sediment	9	EPR/SPac	9, 27
Risepac 75G to 82G	Near-surface sediment	5–6	EPR/SPac	9, 27
1-2	Near-surface sediment	2	EPR/EqPac	24
3-9	Near-surface sediment	7	EPR/EqPac	24
ELTANIN cores	Surface sediment	16	EPR/PacAntc	27
20, 65, 79, 81	Surface low-carbonate sediment	4	EPR/PacAntc	30
ATLANTIS II 54-11	Near-surface, subsurface sediment	2–4	GalapR	33
MENDELEEV 1052, 0-3 mm	Fe–Mn crust	1	GalapRHess	11
D-6, 0-3 mm	Mn crust	1	GalapR	29
D-9, 0-3 mm	Mn crust	1	GalapR	29
Various	Green hydrothermal clay	3–11	GalapHTM	25
TB1-TB7	Mn oxyhydroxide	7	GalapHTM	14
DM1, DM2	Fe–Mn crust on basalt	2	GalapHTM	14
A1, A2	Amorphous Fe oxide	2	GalapHTM	14
N1-N5	Fe-rich nontronite	5	GalapHTM	14
ATLANTIS II 54-14	Near-surface sediment	3–4	BauerD	33

(continued)

TABLE X (continued)

Deposit No.	Locality	Latitude	Longitude	Depth (m)
39	Bauer Deep, E. Equat. Pac.	8° 22′S	102° 14′W	4124
40	Bauer Deep, E. Equat. Pac.	8° 22′S	102° 14′W	4124
41	Bauer Deep, E. Equat. Pac.	8° 22′S	102° 14′W	4124
42	Bauer Deep, E. Equat. Pac.	–	–	–
43	Bauer Deep, E. Equat. Pac.	10–13°S	97–104°W	–
44	Central Basin, Nazca Plate	18–24°S	95–106°W	–
45	Bauer Deep, D.S.D.P. Site 319	13°01′S	101°32′W	4286
46	E. Equat. Pac., D.S.D.P. Site 320	9°00′S	83°32′W	4477
47	E. Equat. Pac., D.S.D.P. Site 320	9°00′S	83°32′W	4477
48	E. Equat. Pac., D.S.D.P. Site 321	12°01′S	81°54′W	4817
49	E. Equat. Pac., D.S.D.P. Site 321	12°01′S	81°54′W	4817
50	E. Equat. Pac., D.S.D.P. Site 321	12°01′S	81°54′W	4817
51	Bauer Deep, D.S.D.P. Site 319	13°01′S	101°32′W	4286
52	E. Equat. Pac., D.S.D.P. Site 320	9°00′S	83°32′W	4477
53	E. Equat. Pac., D.S.D.P. Site 321	12°01′S	81°54′W	4817
54	N.E. Pac., D.S.D.P. Site 37	40°59′N	140°43′W	4882
55	N.E. Pac., D.S.D.P. Site 37	40°59′N	140°43′W	4882
56	N.E. Pac., D.S.D.P. Site 37	40°59′N	140°43′W	4882
57	N.E. Pac., D.S.D.P. Site 37	40°59′N	140°43′W	4882
58	N.E. Pac., D.S.D.P. Site 38	38°42′N	140°21′W	5137
59	N.E. Pac., D.S.D.P. Site 38	38°42′N	140°21′W	5137
60	N.E. Pac., D.S.D.P. Site 38	38°42′N	140°21′W	5137
61	N.E. Pac., D.S.D.P. Site 38	38°42′N	140°21′W	5137
62	N.E. Pac., D.S.D.P. Site 38	38°42′N	140°21′W	5137
63	N.E. Pac., D.S.D.P. Site 38	38°42′N	140°21′W	5137
64	N.E. Pac., D.S.D.P. Site 39	32°48′N	139°34′W	4929
65	N.E. Pac., D.S.D.P. Site 39	32°48′N	139°34′W	4929
66	N.E. Pac., D.S.D.P. Site 39	32°48′N	139°34′W	4929
67	N.E. Pac., D.S.D.P. Site 39	32°48′N	139°34′W	4929
68	E. Equat. Pac., D.S.D.P. Site 77B	00°29′N	133°14′W	4291
69	E. Equat. Pac., D.S.D.P. Site 80	00°58′S	121°33′W	4411
70	E. Equat. Pac., D.S.D.P. Site 81	1°27′N	113°49′W	3865
71	E. Equat. Pac., D.S.D.P. Site 82	2°36′N	106°57′W	3707
72	E. Equat. Pac., D.S.D.P. Site 83	4°03′N	95°44′W	3646
73	E. Equat. Pac., D.S.D.P. Site 159	12°20′N	122°17′W	4484
74	E. Equat. Pac., D.S.D.P. Site 160	11°42′N	130°53′W	4940
75	E. Equat. Pac., D.S.D.P. Site 161A	10°40′N	139°57′W	4939
76	E. Equat. Pac., D.S.D.P. Site 162	14°52′N	140°03′W	4854
77	Dellwood Smt., N.E. Pac.	50°46′N	130°53′W	600–800
78	N.E. Pac., Domes Site C	15°12′N	126°59′W	4295
79	Cent. Equat. Pac., D.S.D.P. Site 66	2°24′N	166°07′W	5293
80	Lau Basin Rise, S.W. Pac.	20°24′S	176°46′W	2730
81	Lau Basin Rise, S.W. Pac.	17°36′S	177°45′W	2406
82	Lau Basin Rise, S.W. Pac.	16°48′S	176°40′W	2698
83	Lau Basin Rise, S.W. Pac.	16°29′S	177°33′W	2328
84	Lau Basin Rise, S.W. Pac.	15°16′S	176°50′W	2163

Original sample number	Type of material	No. of samples	Sample abbreviation	Reference No. *
AII 54-14, 467-478 cm	Subsurface sediment	1	BauerD	17
ATLANTIS II 54-14 (<2 μ)	Subsurface sediment	1	BauerD	17
ATLANTIS II 54-14 (2-20 μ)	Subsurface sediment	1	BauerD	17
Various	Metalliferous sediment	1–8+	BauerD	17
10-16	Near-surface sediment	7	BauerD	24
17-22	Near-surface sediment	6	CentDNaz	24
319-D	Basal sediment	21	DSDP 319	10
320-F	Basal sediment	5	DSDP 320	10
320-G	Basal sediment	8	DSDP 320	10
321-L	Basal sediment	1	DSDP 321	10
321-M	Basal sediment	2	DSDP 321	10
321-N	Basal sediment	6	DSDP 321	10
319-14 to -19	Basal sediment	6	DSDP 319	18
320-1, -2	Basal sediment	2	DSDP 320	18
321-1 to -6	Basal sediment	6	DSDP 321	18
D.S.D.P. 37-4-3	Basal sediment	1	DSDP 37	17
D.S.D.P. 37-4-5	Basal sediment	1	DSDP 37	17
37-4-3,5 (<2 μm)	Basal sediment	2	DSDP 37	17
37-4-3,5 (2–20 μm)	Basal sediment	2	DSDP 37	17
D.S.D.P. 38-5-3	Basal sediment	1	DSDP 38	17
38-5-3 (<2 μm)	Basal sediment	1	DSDP 38	17
38-5-3 (2–20 μm)	Basal sediment	1	DSDP 38	17
D.S.D.P. 38-6-4	Basal sediment	1	DSDP 38	17
38-6-4 (<2 μm)	Basal sediment	1	DSDP 38	17
38-6-4 (2–20 μm)	Basal sediment	1	DSDP 38	17
D.S.D.P. 39-2-3	Basal sediment	1	DSDP 39	17
D.S.D.P. 39-2-6	Basal sediment	1	DSDP 39	17
39-2-3, 6 (<2 μm)	Basal sediment	2	DSDP 39	17
39-2-3, 6 (2–20 μm)	Basal sediment	2	DSDP 39	17
77B-52-1, 52-2, 53-1	Basal sediment	6	DSDP 77	23
80-5-3, 80–82 cm	Basal sediment	1	DSDP 80	23
81-6-3, 7-1	Basal sediment	4	DSDP 81	23
82-6-6	Basal sediment	2	DSDP 82	23
83-8-1	Basal sediment	1	DSDP 83	23
159-12-1, 12-6, 13-2, 13-4	Basal sediment	4 **	DSDP 159	16
160-13-1	Basal sediment	1	DSDP 160	16
161A-12-1, 13-1, 14-1, 14-2	Basal sediment	5 **	DSDP 161A	16
162-17-1, 17-4	Basal sediment	2 **	DSDP 162	16
Dredge IOUBC 70-16-12D	Red, layered Fe-oxide	1	Dell/Pac	32
Stn. 18B, 34–36 cm	RSO-rich near-surface sediment	1	DOMES/C	5
66.0-9-38, 101-103 cm	Basal sediment	1	DSDP 66	17
7Tow 123-72G	Surface sediment	2–10	LauBR	1
7Tow 123-84G	Surface sediment	1	LauBR	1
7Tow 123-88G	Surface sediment	1	LauBR	1
7Tow 123-94G	Surface sediment	2–7	LauBR	1
7Tow 123-105G	Surface sediment	2–8	LauBR	1

(continued)

TABLE X (continued)

Deposit No.	Locality	Latitude	Longitude	Depth (m)
85	Lau Basin, Peggy Ridge, S.W. Pac.	16°55′S	176°50′W	1664–1990
86	W. Philippine Basin, D.S.D.P. Site 291	12°49′N	127°50′E	5217
87	MOR, Indian Ocean	4–34°S	57°–114°E	2550–5986
88	W. Indian Ocean, D.S.D.P. Site 212	19°12′S	99°18′E	6230
89	W. Indian Ocean, D.S.D.P. Site 213	10°13′S	93°54′E	5600
90	E. Indian Ocean, D.S.D.P. Site 261	12°57′S	117°54′E	5667
91	Gulf of Aden	12°34′N	47°39′E	2260–2550
92	Gulf of Aden	12°34′N	47°39′E	2260–2550
93	Red Sea, Atlantis II and Discovery Deeps	21°14′N–21°25′N	38°01′E–38°04′E	1949–2175
94	Red Sea, Atlantis II and Discovery Deeps	21°14′N–21°25′N	38°01′E–38°04′E	1949–2175
95	Red Sea, Atlantis II and Discovery Deeps	21°14′N–21°25′N	38°01′E–38°04′E	1949–2175
96	Red Sea, Atlantis II and Discovery Deeps	21°14′N–21°25′N	38°01′E–38°04′E	1949–2175
97	Red Sea, Nereus Deep	23°10′N	37°17′E	~2350
98	Red Sea, Nereus Deep	23°08′N	37°17′E	~2350
99	Red Sea, Nereus Deep	23°12′N	37°15′E	~2450
100	Red Sea, Nereus Deep	23°13′N	37°13′E	~2600
101	Red Sea, Nereus Deep	23°11′N	37°12′E	~2400
102	Matupi Harbor, New Britain, S.W. Pac.	4°S	152°E	18–36+
103	Cyprus, ophiolite complex, near sulfide ore	–	–	on land
104	Molinello, Northern Appenines ophiolite complex, Italy	–	–	on land
105	Gambatesa, Northern Appenines ophiolite complex, Italy	–	–	on land

* See at bottom of Table XII
** Compositional data presented as weighted average.

The Bauer Deep contains metalliferous sediments which show general patterns of element enrichment and depletion similar to those reported by Boström and Peterson (1969) for the non-carbonate fraction of East Pacific Rise sediments (Sayles and Bischoff, 1973). Bauer Deep sediments, however, contain more Si and Ni and less Fe and Mn in the non-carbonate fraction (Dymond et al., 1973; Sayles and Bischoff, 1973; Heath and Dymond, 1977; deposit numbers 23, 38–44). The higher Si content is due to the greater content in the Bauer Deep of poorly crystalline authigenic iron-rich montmorillonite, while the higher Ni values may be the result of a greater proportion of iron–manganese micronodules (Dymond et al., 1973). These authors also analyzed the $<2\ \mu m$ and 2–20 μm size fractions of Bauer Deep and D.S.D.P. basal sediments. They found that the $<2\ \mu m$ fraction was slightly enriched in Fe (as stated above), and also Si, Al, Cu, Zn, and

Original sample number	Type of material	No. of samples	Sample abbreviation	Reference No. *
Tow-86D, nos. A, B	Barite in hyaloclastite	2	LauBbar	2
-3, 5, 8, 11, 31, 36	Basal sediment	6	DSDP 291	8
Various	Surface, near-surface sediment	18	IndOcMOR	27
12-38-2, 98	Basal sediment	1	DSDP 212	31
13-16-4, 138 and 145	Basal sediment	1	DSDP 213	31
1	Basal sediment	1	DSDP 261	13
Stn. 6243-2, 3, 5	Fe–Mn oxide lumps, crusts	6	Aden	12
Stn. 6243-10, 12, 24, 31	Green massive smectite	4	Aden	12
Various	Fe-montmorillonite facies	43	RedSea	4
Various	Goethite-amorphous facies	43	RedSea	4
Various	Sulfide facies	43	RedSea	4
Various	Manganite facies	2	RedSea	4
0-14	Metalliferous sediment	4 **	RedSeaNerD	3
96	Metalliferous sediment	1–2 **	RedSeaNerD	3
86	Metalliferous sediment	4 **	RedSeaNerD	3
87	Metalliferous sediment	4 **	RedSeaNerD	3
90	Metalliferous sediment	1	RedSeaNerD	3
C1-C8	Near-surface sediments	8	MatHarb	20
Not identified	Fe–Mn oxide sediments	13–14	Cyprus	21
Z-17	Massive Mn mineralization	1	MolApp	7
Z-51	Massive Mn mineralization	1	GamApp	7

Mg relative to the 2–20 μm fraction (deposit numbers 40, 41, 56, 57, 59, 60, 62, 63, 66, 67). Mn and Ni are both greatly enriched in the 2–20 μm fraction, presumably co-precipitated in a hydrated manganese oxide phase.

Several distinct hydrothermal and hydrogenous phases have been recognized in an area of mounds near the Galapagos Rift (Hekinian et al., 1978; Corliss et al., 1978; deposit numbers 33–37). The smectite phase or clay-rich hydrothermal material is characterized by high silicon and iron contents, and low values for manganese and trace metals. This is true of similar material in other deposits; see deposit numbers 13, 33, 37 and 92.

The Red Sea metalliferous sediments attain greater concentrations of elements, such as Zn and Cu, than any other marine sediments (Bischoff, 1969a; deposit numbers

Table XI

Chemical composition of metalliferous marine deposits: major and minor elements
Major elements

Deposit No.	Sample abbreviation	Percent								
		Si	Al	Ti	Ca	Mg	Na	K	Fe	Mn
1	PelagClayAvg	25.0	8.4	0.46	2.9	2.1	4.0	2.5	6.5	0.67
2	PacClayAvg	23.0	9.2	0.73	2.9	2.1	4.0	2.5	6.5	1.25
3	AtlClayAvg	–	9.04	–	–	–	–	–	5.02	0.40
4	MAR/NAtl	–	–	–	–	–	–	–	–	–
5	MAR/NAtl	–	–	–	–	–	–	–	–	–
6	MAR/NAtl	–	–	–	–	–	–	–	–	–
7	MAR/45N	–	5.79	–	–	–	–	–	7.96	0.41
8	MAR/ATLFZ	–	–	–	–	–	–	–	18.1	9.8
9	ATLFZ	–	–	–	–	–	–	–	16.1	14.1
10	MAR/TAG	–	–	–	–	–	–	–	16.4	11.2
11	MAR/TAG	–	–	–	–	–	–	–	18.6	11.0
12	MAR/TAG	–	–	–	–	–	–	–	0.06	39.1
13	MAR/FAM	18.2	0.4	0.09	1.7	1.86	1.41	2.36	24.0	3.42
14	DSDP9A	–	4.10	0.19	–	0.75	–	–	6.00	0.17
15	DSDP112	–	5.26	0.29	–	1.95	–	–	6.95	0.12
16	DSDP118	–	8.52	0.48	–	1.30	–	–	4.70	0.41
17	DSDP354	–	–	0.23	–	–	0.44	0.83	2.25	0.06
18	DSDP355	–	–	0.12	–	–	0.45	0.82	2.18	0.40
19	EPRsmtSPac	6.9	<0.5	–	1.8	0.63	–	0.34	30.8	1.65
20	EPRsmtSPac	3.8	0.2	–	1.8	3.5	–	1.07	5.5	38.72
21	EPRsmtSPac	5.8	1.5	–	3.7	0.24	–	3.2	17.8	19.67
22	EPR/SPac	13.2	4.7	0.17	–	–	–	–	9.5	2.4
23	EPR/SPac	0.6	0.5	0.02	–	–	–	–	18.0	6.0
24	EPR/SPac	15.0	4.6	0.31	–	–	–	–	11.5	3.6
25	EPR/EqPac	25.72	0.73	–	–	–	–	–	6.05	1.37
26	EPR/EqPac	6.12	0.51	–	–	–	–	–	30.20	9.92
27	EPR/PacAntc	–	–	–	–	–	–	–	–	–
28	EPR/PacAntc	18.8	2.56	0.50	4.74	1.43	2.64	0.60	12.59	7.76
29	GalapR	22.0	3.37	0.14	1.28	2.11	0.87	0.86	12.2	3.88
30	GalapRHess	–	–	–	–	–	–	–	21.8	16.7
31	GalapR	–	–	–	–	–	–	–	0.02	52.7

32	GalapR	–	–	–	–	–	–	–	16.6	25.5
33	GalapHTM	23.8	0.12	0.02	0.22	2.54	1.52	2.69	21.6	0.10
34	GalapHTM	0.79	0.19	–	1.52	1.36	3.0	0.60	0.26	50.0
35	GalapHTM	11.3	1.32	–	2.3	0.88	1.11	0.46	13.8	18.8
36	GalapHTM	13.2	0.32	–	1.4	–	–	–	27.4	7.85
37	GalapHTM	20.0	0.45	–	5.1	1.29	0.53	1.11	19.3	0.44
38	BauerD	21.0	1.31	0.12	1.56	2.54	0.61	0.75	14.8	3.54
39	BauerD	14.5	1.20	–	–	–	–	–	14.4	4.61
40	BauerD	18.5	1.38	–	–	2.65	–	–	18.0	2.64
41	BauerD	17.3	1.52	–	–	2.53	–	–	16.8	4.34
42	BauerD	13.5	2.31	–	–	–	–	–	14.1	4.60
43	BauerD	17.43	3.24	–	–	–	–	–	15.83	5.74
44	CentDNaz	20.59	6.74	–	–	–	–	–	12.11	3.96
45	DSDP319	–	1.5	0.10	–	–	–	–	23.3	6.8
46	DSDP320	–	2.6	0.14	–	–	–	–	21.2	6.2
47	DSDP320	–	1.4	0.11	–	–	–	–	23.6	8.3
48	DSDP321	–	3.3	0.14	–	–	–	–	19.5	4.6
49	DSDP321	–	4.0	0.17	–	–	–	–	16.3	4.4
50	DSDP321	–	2.1	0.36	–	–	–	–	20.7	9.4
51	DSDP319	6.24	1.04	–	33.6	–	–	–	26.5	7.8
52	DSDP320	10.15	1.00	–	35.0	–	–	–	20.4	9.4
53	DSDP321	13.83	3.87	–	29.0	–	–	–	22.6	7.9
54	DSDP37	8.91	2.59	–	–	–	–	–	22.2	7.66
55	DSDP37	8.09	1.95	–	–	–	–	–	21.6	5.65
56	DSDP37	9.64	3.32	–	–	1.20	–	–	26.1	3.72
57	DSDP37	–	2.44	–	–	1.24	–	–	21.6	14.4
58	DSDP38	6.11	2.38	–	–	–	–	–	27.8	6.46
59	DSDP38	6.92	3.21	–	–	1.04	–	–	33.3	4.76
60	DSDP38	6.29	2.90	–	–	0.93	–	–	34.3	15.1
61	DSDP38	4.25	1.54	–	–	–	–	–	21.1	7.56
62	DSDP38	3.17	1.31	–	–	0.74	–	–	18.6	4.30
63	DSDP38	0.40	0.14	–	–	0.20	–	–	2.83	1.43
64	DSDP39	8.76	3.41	–	–	–	–	–	23.9	5.63
65	DSDP39	6.52	2.48	–	–	–	–	–	25.0	6.97
66	DSDP39	9.48	3.94	–	–	1.17	–	–	29.0	5.57
67	DSDP39	7.92	3.05	–	–	1.08	–	–	28.4	9.26
68	DSDP77	–	0.98	–	15.44	1.85	0.16	0.45	7.50	1.98
69	DSDP80	–	0.25	–	33.51	0.88	0.04	0.19	3.34	1.29
70	DSDP81	–	0.87	–	26.11	2.54	0.20	0.23	4.11	1.82

(continued)

Table XI (Major elements continued)

Deposit No.	Sample abbreviation	Percent								
		Si	Al	Ti	Ca	Mg	Na	K	Fe	Mn
71	DSDP82	–	0.51	–	17.54	1.46	0.06	0.28	3.89	1.36
72	DSDP83	–	0.67	–	23.69	2.38	0.04	0.13	3.42	0.87
73	DSDP160	–	–	–	–	–	–	–	17.00	4.14
74	DSDP161A	–	–	–	–	–	–	–	8.52	1.86
75	DSDP161A	–	–	–	–	–	–	–	5.27	1.00
76	DSDP162	–	–	–	–	–	–	–	24.99	8.32
77	Dell/Pac	8.6	1.0	–	1.30	0.70	–	0.30	28.5	2.05
78	DOMES/C	23.36	4.64	0.30	1.66	2.56	0.55	1.93	9.09	4.32
79	DSDP66	8.05	1.37	–	–	–	–	–	28.5	6.97
80	LauBR	22.6	5.2	0.29	5.3	2.4	4.8	0.8	11.0	1.3
81	LauBR	22.8	7.8	0.28	2.0	2.8	6.3	1.1	14.2	1.8
82	LauBR	20.3	5.9	0.34	5.1	2.0	6.8	0.9	11.0	1.4
83	LauBR	18.6	6.1	0.26	12.8	2.0	5.7	0.9	10.6	1.1
84	LauBR	18.6	4.0	0.50	10.2	2.7	6.0	0.8	11.9	2.2
85	LauBbar	–	–	–	–	–	–	–	0.06	–
86	DSDP291	27	2.0	0.08	1.1	2.0	2.4	1.4	10.0	1.47
87	IndOcMOR	–	–	–	–	–	–	–	–	–

88	DSDP212	21.29	6.27	0.52	0.67	1.97	1.05	2.88	10.28	–
89	DSDP213	17.97	4.69	0.34	3.00	2.10	–	2.07	16.2	3.2
90	DSDP261	7.4	1.5	0.1	28.6	0.5	0.49	0.6	2.2	0.42
91	Aden	3.65	0.69	0.11	1.54	1.76	2.98	1.39	2.67	37.92
92	Aden	21.99	0.16	0.02	0.50	1.83	1.25	3.19	22.94	<0.1
93	RedSea	11.5	0.9	–	3.4	–	–	–	24.9	1.0
94	RedSea	4.1	0.6	–	2.4	–	–	–	43.0	0.6
95	RedSea	11.6	0.8	–	1.8	–	–	–	16.3	0.6
96	RedSea	3.5	0.4	–	2.0	–	–	–	20.4	17.8
97	RedSeaNerD	–	–	–	4.80	–	–	–	28.30	16.90
98	RedSeaNerD	–	–	–	2.03	–	–	–	27.00	8.25
99	RedSeaNerD	–	–	–	12.10	–	–	–	6.84	2.58
100	RedSeaNerD	–	–	–	12.40	–	–	–	7.45	1.29
101	RedSeaNerD	–	–	–	17.00	–	–	–	5.83	1.53
102	MatHarb	–	–	0.38	3.3	1.6	–	–	9.15	0.40
103	Cyprus	–	–	–	–	–	–	–	36.4	2.56
104	MolApp	14.5	0.06	<0.006	2.14	0.04	<0.07	0.08	0.06	46.13
105	GamApp	38.7	0.34	0.012	0.23	0.09	0.13	0.12	0.30	9.45

*1 See at bottom of Table XII.
*2 Carbonate- and silica-free basis.
*3 Carbonate- and salt-free basis.
*4 All elements except strontium on carbonate-free basis.
*5 Yes ± indicates that material contains negligible $CaCO_3$.

TABLE XI (continued)

Minor elements

Deposit No.	Sample abbreviation	ppm								$CaCO_3$-free basis *5	% $CaCO_3$	Reference No. *1
		Ni	Cu	Co	Zn	Pb	Cr	Ba	Sr			
1	PelagClayAvg	225	250	74	165	80	90	2,300	180	No	–	35
2	PacClayAvg	320	740	160	–	150	93	–	–	No	–	22
3	AtlClayAvg	–	–	–	–	–	–	–	–	Yes?	–	15
4	MAR/NAtl	–	–	–	50.5	43.6	–	–	–	Yes	48	27
5	MAR/NAtl	–	–	–	86.5	57.1	–	–	–	Yes	47	27
6	MAR/NAtl	–	–	–	96.0	38.8	–	–	–	Yes	10	27
7	MAR/45N	–	–	–	–	–	–	–	–	Yes	81.0	15
8	MAR/ATLFZ	1,280	880	2,720	–	–	–	–	–	No	–	34
9	ATLFZ	2,200	750	7,200	–	–	–	–	–	No	–	34
10	MAR/TAG	900	300	9.480	–	–	–	–	–	No	–	34
11	MAR/TAG	1,100	405	9,450	–	–	–	–	–	No	–	34
12	MAR/TAG	340	43	19	–	–	–	–	–	No	–	34
13	MAR/FAM	53	68	6	16	–	20	79	230	Yes ±	–	26
14	DSDP9A	24	28	–	62	48	55	–	–	Yes	–	28
15	DSDP112	56	39	–	118	63	76	–	–	Yes	–	28
16	DSDP118	70	62	–	103	81	76	–	–	Yes	–	28
17	DSDP354	10	35	10	40	–	47	–	–	No	47.9	19
18	DSDP355	31	47	15	36	–	21	–	–	No	73.8	19
19	EPRsmtSPac	317	85	62	–	–	22	105	583	No	–	6
20	EPRsmtSPac	4,500	>500	290	–	–	210	1,700	420	No	–	6
21	EPRsmtSPac	3,200	220	6,800	–	–	87	670	940	No	–	6
22	EPR/SPac	690	990	215	285	144	36	–	–	Yes	58	9, 27
23	EPR/SPac	430	730	105	380	152	55	–	–	Yes	80	9, 27
24	EPR/SPac	660	930	245	295	182	28	–	–	Yes	47	9, 27
25	EPR/EqPac	212	558	–	301	–	–	13,200	–	Yes	82.8	24
26	EPR/EqPac	642	1,450	–	594	–	–	6,000	–	Yes	68.2	24
27	EPR/PacAntc	–	–	–	>169	79.1	–	–	–	Yes	34	27
28	EPR/PacAntc	5,214	2,173	1,302	–	–	180	–	–	No	–	30
29	GalapR	1,232	1,042	215	145	–	–	23,000	–	Yes	14	33
30	GalapRHess	1,950	735	508	543	–	–	–	–	Yes ±	–	11
31	GalapR	100	43	<23	105	–	–	–	–	Yes ±	–	29
32	GalapR	6,300	936	909	1,765	–	–	–	–	Yes ±	–	29

33	GalapHTM	16	14	22	35	60	11	–	–	Yes ±	–	25
34	GalapHTM	469	103	4.7	378	–	3.8	1,706	–	Yes ±	–	14
35	GalapHTM	6,155	390	375	649	–	28.6	1,100	–	Yes ±	–	14
36	GalapHTM	62	48	3.2	150	–	8.3	705	–	Yes ±	–	14
37	GalapHTM	41	50	4.3	146	–	9.0	953	–	No	–	14
38	BauerD	775	915	180	498	–	–	11,000	–	Yes	23	33
39	BauerD	615	770	67	370	178	13	9,400	–	Yes	2.2	17
40	BauerD	470	830	–	610	–	–	–	–	Yes	–	17
41	BauerD	760	945	–	465	–	–	–	–	Yes	–	17
42	BauerD	820	910	67	330	–	13	–	–	Yes	–	17
43	BauerD	1,066	1,171	–	413	–	–	18,600	–	Yes	17.8	24
44	CentDNaz	1,307	985	–	311	–	–	17,400	–	Yes	20.7	24
45	DSDP319	575	1,240	160	780	–	–	8,800	–	Yes *[2]	–	10
46	DSDP320	360	1,200	88	705	–	–	9,200	–	Yes *[2]	–	10
47	DSDP320	675	1,300	97	700	–	–	8,000	–	Yes *[2]	–	10
48	DSDP321	330	750	115	750	–	–	1,800	–	Yes *[2]	–	10
49	DSDP321	270	650	77	765	–	–	5,700	–	Yes *[2]	–	10
50	DSDP321	805	1,100	80	685	–	~37	1,700	–	Yes *[2]	–	10
51	DSDP319	665	1,390	114	590	–	23	6,200	–	Yes *[3]	83.55	18
52	DSDP320	615	1,160	73	640	–	23	5,600	–	Yes *[3]	87.20	18
53	DSDP321	870	1,160	153	660	–	27	2,700	–	Yes *[3]	71.96	18
54	DSDP37	630	830	87	530	–	19	4,100	350	Yes *[4]	–	17
55	DSDP37	600	800	55	480	–	16	2,300	364	Yes *[4]	–	17
56	DSDP37	338	915	–	630	–	–	–	–	Yes	–	17
57	DSDP37	1,440	832	–	595	–	–	–	–	Yes	–	17
58	DSDP38	710	1,490	92	660	–	12	1,500	341	Yes *[4]	–	17
59	DSDP38	650	1,700	–	940	–	–	–	–	Yes	–	17
60	DSDP38	1,040	1,630	–	760	–	–	–	–	Yes	–	17
61	DSDP38	800	920	70	570	–	9.2	15,000	1,050	Yes *[4]	76.9	17
62	DSDP38	810	780	–	545	–	–	–	–	No	–	17
63	DSDP38	315	125	–	80	–	–	–	–	No	–	17
64	DSDP39	570	1,220	100	680	–	23	2,000	269	Yes *[4]	–	17
65	DSDP39	590	1,160	110	690	–	16	1,600	315	Yes *[4]	–	17
66	DSDP39	505	1,315	–	868	–	–	–	–	Yes	–	17
67	DSDP39	900	1,372	–	800	–	–	–	–	Yes	–	17
68	DSDP77	200	430	240	270	<20	–	1,530	610	No	–	23
69	DSDP80	0	330	0	130	<20	–	0	750	No	–	23
70	DSDP81	210	280	120	180	<20	–	150	560	No	–	23
71	DSDP82	90	320	240	160	<20	–	1,050	490	No	–	23

(continued)

TABLE XI (Minor elements continued)

Deposit No.	Sample abbreviation	ppm								$CaCO_3$-free basis? *5	% $CaCO_3$	Reference No. *1
		Ni	Cu	Co	Zn	Pb	Cr	Ba	Sr			
72	DSDP83	160	0	0	280	<20	–	0	520	No	–	23
73	DSDP160	429	739	82	461	122	–	–	–	Yes	–	16
74	DSDP161A	994	994	56	262	56	–	–	–	Yes	–	16
75	DSDP161A	103	454	24	158	37	–	–	–	Yes	–	16
76	DSDP162	1,084	1,656	135	272	189	–	–	–	Yes	–	16
77	Dell/Pac	62	15	<10	450	–	12	267	–	Yes ±	–	32
78	DOMES/C	600	1,300	–	–	–	–	–	–	Yes ±	–	5
79	DSDP66	680	930	–	590	–	–	–	315	Yes	–	17
80	LauBR	69	230	72	145	30	60	–	–	Yes	–	1
81	LauBR	215	330	125	380	109	60	–	–	Yes	–	1
82	LauBR	90	240	–	150	–	40	–	–	Yes	–	1
83	LauBR	151	348	212	217	76	77	–	–	Yes	–	1
84	LauBR	196	327	130	199	47	115	–	–	Yes	–	1
85	LauBbar	90	115	130	110	–	–	–	17,000	Yes ±?	–	2
86	DSDP291	119	295	46	195	<44	24	730	–	No	4.0	8
87	IndOcMOR	–	–	–	154	79.2	–	–	–	Yes	35	13
88	DSDP212	–	–	–	–	–	–	–	–	Yes ±	–	31
89	DSDP213	280	523	–	350	168	39	–	–	No	–	31
90	DSDP261	18	210	1	40	10	<1,000	40	1,000	No	70 ±	13
91	Aden	395	82	30	311	–	–	1,160	325	Yes ±	–	12
92	Aden	4	0	4	2	–	–	75	44	Yes ±	–	12
93	RedSea	–	6,000	–	26,000	–	–	–	–	No	–	4
94	RedSea	–	2,000	–	6,000	–	–	–	–	No	–	4
95	RedSea	–	36,000	–	98,000	–	–	–	–	No	–	4
96	RedSea	–	1,000	–	11,000	–	–	–	–	No	–	4
97	RedSeaNerD	9	259	12	11,030	450	15	–	–	No	–	3
98	RedSeaNerD	15	239	12	7,940	386	39	–	–	No	–	3
99	RedSeaNerD	24	101	22	1,564	40	15	–	–	No	–	3
100	RedSeaNerD	14	171	19	726	73	18	–	–	No	–	3
101	RedSeaNerD	33	87	39	784	60	44	–	–	No	–	3
102	MatHarb	14	55	17	469	<100	–	380	480	No	–	20
103	Cyprus	250	1,749	141	1,324	202	30	–	–	No	–	21
104	MolApp	39	310	80	50	–	<10	3,500	–	Yes ±	–	7
105	GamApp	20	97	45	21	–	18	1,600	–	Yes ±	–	7

93–96). It is interesting to note, however, that not all Red Sea deeps are so metal-rich. Nickel and cobalt contents in Nereus Deep sediments are less than in average pelagic clay, and copper content is about the same as in such clay (Bignell et al., 1974; deposit numbers 1, 97–101).

The brines overlying metalliferous sediments in the Atlantic II and Discovery Deeps are themselves metal-enriched compared to ordinary seawater (Brewer and Spencer, 1969). Lead and manganese are particularly enriched in the brines, and Fe, Zn, Cu, Co and Ni also occur in greater concentrations than in seawater. Complexation of elements with chloride ion derived from associated evaporites plays a role in this enrichment (Dunham, 1970; Nriagu and Anderson, 1970; Tooms, 1970; Shanks and Bischoff, 1977; Mossman and Heffernan, 1978). In the brines, metal concentrations are much higher in solution than in the particulate form (Hartmann, 1973).

Not all marine metalliferous deposits occur in the form of unconsolidated sediments. Manganese nodules are a concretionary type of metal-rich material. Ferromanganese crusts draping submarine volcanic rocks often resemble in structure the ferromanganese oxides of nodules, but may differ somewhat in composition, possibly reflecting a lack of metal input from underlying sediments, a diagenetic process that can enrich nodules with Mn, Ni and Cu (Lynn and Bonatti, 1965; Raab, 1972). The most apparent differences in composition are the much lower Cu contents and somewhat higher Co contents of crusts (Bonatti and Joensuu, 1966; M.R. Scott et al., 1974; Piper et al., 1975; Frank et al., 1976; Lyle et al., 1977; Burnett and Piper, 1977; see deposit numbers 8–11, 19, 21, 30, 77). Certain anomalous crusts have extremely high Mn and low Fe contents (Bonatti and Joensuu, 1966; M.R. Scott et al., 1974; Moore and Vogt, 1976; Cann et al., 1977; see deposit numbers 12, 20, 31, 32, 91), possibly because the iron has been separated out in a sulphide phase under reducing conditions.

Metalliferous sediments deposited near known hydrothermal sources have been sampled from Matupi Harbour, New Britain; Santorini Volcano in the Mediterranean Sea; and off northern Baja California. At Matupi Harbour, hot, acid springs carry Fe, Mn and Zn in concentrations comparable to those of Red Sea hot brines (up to 100 ppm Fe, 100 ppm Mn, and 2.5 ppm Zn), but only about 10% as much Cu and Pb (slightly less than 0.1 ppm of both). Green et al. (1978) have concluded that the marked enrichment of Mn, Fe and Zn in the thermal waters is not the result of high-temperature hydrothermal leaching of the subsurface loosely consolidated Quaternary basalt–andesite pyroclastic rocks, but rather is the result of relatively shallow, but extensive, interaction of warm acidic waters with the pyroclastics. The Matupi Harbour sediments are enriched in Fe, Mn and Zn, but very little of the iron occurs in a sulphide form (Ferguson and Lambert, 1972; see deposit number 102).

At Santorini Volcano, sediment compositional differences over distances no greater than a kilometre have been observed in samples taken progressively farther from the shallow water, fumarolic sources (Smith and Cronan, 1975b; Cronan et al., 1977). Iron content decreases from 40% in the inner exhalative zone to 7% in the outer exhalative

TABLE XII

Chemical composition of metalliferous marine deposits: trace elements

Deposit No.	Sample abbreviation	ppm								
		Ag	As	B	Cd	Ce	Hg	La	Mo	Rb
1	PelagClayAvg	0.11	13	230	0.42	345	–	115	27	110
2	PacClayAvg	–	–	300	–	–	–	140	45	–
3	AtlClayAvg	–	18	–	–	–	–	–	–	–
4	MAR/NAtl	0.115	–	–	–	–	–	–	–	–
5	MAR/NAtl	0.047	–	–	–	–	–	–	–	–
6	MAR/NAtl	0.038	–	–	–	–	–	–	–	–
7	MAR/45N	–	174	–	–	–	0.41	–	–	–
13	MAR/FAM	–	–	538	–	–	–	–	–	–
19	EPRsmtSPac	–	–	277	–	–	–	–	–	–
20	EPRsmtSPac	–	–	85	–	–	–	–	–	–
21	EPRsmtSPac	–	–	230	–	–	–	–	–	–
22	EPR/SPac	4.12	33	155	1.0	–	–	–	100	–
23	EPR/SPac	6.22	145	500	4.0	–	–	–	30	–
24	EPR/SPac	4.66	98	140	1.1	–	–	–	127	–
27	EPR/PacAntc	2.02	–	–	–	–	–	–	–	–
31	GalapR	–	–	–	–	–	–	–	–	–
32	GalapR	–	–	–	–	–	–	–	–	–
34	GalapHTM	–	14	–	–	2.3	–	2.7	–	–
35	GalapHTM	–	150	–	–	160	–	134	–	–
36	GalapHTM	–	42	–	–	4.2	–	6.7	–	–
37	GalapHTM	–	4.6	–	–	3.3	–	4.0	–	–
39	BauerD	–	–	–	–	34	–	83	–	–
45	DSDP319	–	–	490	–	–	–	–	–	–
46	DSDP320	–	–	355	–	–	–	–	–	–
47	DSDP320	–	–	655	–	–	–	–	–	–
48	DSDP321	–	–	1200	–	–	–	–	–	–
49	DSDP321	–	–	670	–	–	–	–	–	–
50	DSDP321	–	–	310	–	–	–	–	–	–
51	DSDP319	–	–	–	–	30	–	128	–	–
52	DSDP320	–	–	–	–	27	–	80	–	–
53	DSDP321	–	–	–	–	73	–	169	–	–
54	DSDP37	–	–	–	–	50	–	92	–	20
55	DSDP37	–	–	–	–	29	–	66	–	35
58	DSDP38	–	–	–	–	74	–	188	–	9.5
61	DSDP38	–	–	–	–	35	–	171	–	3.6
64	DSDP39	–	–	–	–	81	–	177	–	51
65	DSDP39	–	–	–	–	81	–	187	–	19
68	DSDP77	–	–	–	–	–	–	–	–	–
77	Dell/Pac	–	–	–	–	2.2	–	0.66	–	–
79	DSDP66	–	–	–	–	–	–	–	–	21
85	LauBbar	–	160	–	–	–	–	–	–	–
86	DSDP291	–	–	<77	–	–	–	<24	–	–
87	IndOcMOR	0.74	–	–	–	–	–	–	–	–
89	DSDP213	–	–	–	–	–	–	–	–	–
91	Aden	–	–	–	–	–	–	–	–	11
92	Aden	–	–	–	–	–	–	–	–	48

	Sb	Sc	Sn	Th	Tl	U	V	Y	$CaCO_3$-free Basis? *4	% $CaCO_3$	Reference No.
300	1.0	19	1.5	7	0.8	1.3	120	90	No	–	35
–	–	25	–	–	–	–	450	150	No	–	22
–	–	–	–	–	–	–	–	–	Yes?	–	15
–	–	–	2.81	–	2.27	–	–	–	Yes	48	27
–	–	–	2.17	–	2.53	–	–	–	Yes	47	27
–	–	–	3.33	–	2.25	–	–	–	Yes	10	27
–	–	–	–	–	–	–	–	–	Yes	81.0	15
–	–	–	–	–	–	–	95	–	Yes ±	–	26
–	–	–	–	–	–	–	–	–	No	–	6
–	–	–	–	–	–	–	–	–	No	–	6
–	–	–	–	–	–	–	–	–	No	–	6
–	–	–	–	–	–	–	180	–	Yes	58	9, 27
–	–	–	–	–	–	–	450	–	Yes	80	9, 27
–	–	–	–	–	–	–	300	–	Yes	47	9, 27
–	–	–	2.20	–	4.56	–	–	–	Yes	34	27
–	–	–	–	0.01	–	6.0	–	–	Yes ±	–	29
–	–	–	–	6.8	–	9.5	–	–	Yes ±	–	29
–	26	0.70	–	0.16	–	–	–	–	Yes ±	–	14
–	18	9.3	–	10.4	–	–	–	–	Yes ±	–	14
–	6.0	1.1	–	0.03	–	–	–	–	Yes ±	–	14
–	1.8	1.6	–	0.42	–	–	–	–	No	–	14
–	13	11	–	–	–	–	–	–	Yes	2.2	17
–	–	–	–	–	–	–	640	205	Yes *1	–	10
–	–	–	–	–	–	–	440	100	Yes *1	–	10
–	–	–	–	–	–	–	835	115	Yes *1	–	10
–	–	–	–	–	–	–	405	255	Yes *1	–	10
–	–	–	–	–	–	–	270	400	Yes *1	–	10
–	–	13	–	–	–	–	740	115	Yes *1	–	10
–	7.0	8.3	–	0.87	–	–	–	–	Yes *2	83.55	18
–	3.0	5.6	–	–	–	–	–	–	Yes *2	87.20	18
–	8.0	18.4	–	4.6	–	–	–	–	Yes *2	71.96	18
–	9	7	–	–	–	4.33	–	–	Yes *3	–	17
–	13	10	–	–	–	5.83	–	–	Yes *3	–	17
–	16	15	–	–	–	5.10	–	–	Yes *3	–	17
–	20	8	–	–	–	1.97	–	–	Yes *3	76.9	17
–	26	17	–	–	–	5.40	–	–	Yes *3	–	17
–	19	15	–	–	–	5.60	–	–	Yes *3	–	17
–	–	–	–	–	–	–	150	–	No	–	23
–	–	–	–	<0.01	–	0.25	–	–	Yes ±	–	32
–	–	–	–	–	–	–	–	–	Yes	–	17
–	–	–	–	–	–	–	–	–	Yes ± ?	–	2
–	–	6	–	–	–	–	145	55	No	4.0	8
–	–	–	0.62	–	1.62	–	–	–	Yes	35	27
–	–	–	–	–	–	–	713	–	No	–	31
325	–	–	–	–	–	–	–	6	Yes ±	–	12
560	–	–	–	–	–	–	–	2	Yes ±	–	12

(continued)

TABLE XII (continued)

Deposit No.	Sample abbreviation	ppm								
		Ag	As	B	Cd	Ce	Hg	La	Mo	Rb
93	RedSea	–	–	–	–	–	–	–	–	–
94	RedSea	–	–	–	–	–	–	–	–	–
95	RedSea	–	–	–	–	–	–	–	–	–
96	RedSea	–	–	–	–	–	–	–	–	–
97	RedSeaNerD	–	–	–	–	–	0.67	–	–	–
98	RedSeaNerD	–	–	–	–	–	0.34	–	–	–
99	RedSeaNerD	–	–	–	–	–	0.09	–	–	–
100	RedSeaNerD	–	–	–	–	–	0.09	–	–	–
101	RedSeaNerD	–	–	–	–	–	0.04	–	–	–
102	MatHarb	–	–	–	–	–	–	<125	–	–
103	Cyprus	–	–	–	–	–	–	–	32	–
104	MolApp	–	–	–	–	5.0	–	1.72	–	–
105	GamApp	–	–	–	–	–	–	–	–	–

*[1] Carbonate- and silica-free basis.
*[2] Carbonate- and salt-free basis.
*[3] All elements except rubidium on carbonate-free basis.
*[4] Yes ± indicates that material contains negligible $CaCO_3$.

References:
1 – Bertine (1974)
2 – Bertine and Keene (1975)
3 – Bignell et al. (1974)
4 – Bischoff (1969a)
5 – Bischoff and Rosenbauer (1977)
6 – Bonatti and Joensuu (1966)
7 – Bonatti et al. (1976a)
8 – Bonatti et al. (1979)
9 – Boström and Peterson (1969)
10 – Boström et al. (1976)
11 – Burnett and Piper (1977)
12 – Cann et al. (1977)

zone, while less than 200 ppm Mn occurs in the inner zone deposits and values as high as 5500 ppm are reached in the outer zone. Aluminium also increases with distance from fumaroles. Zn and Cu of hydrothermal origin occur in both exhalative zones, Zn having a maximum of 140 ppm in sediments of the outer zone. Similar results have been obtained off White Island, New Zealand (Giggenbach and Glasby, 1977). Vidal et al. (1978) have examined the deposits found near hydrothermal springs off Baja California, and noted sea-floor dispersion haloes of only a few metres in size.

Metal accumulation rates

Studies of metal accumulation rates have been used to determine the location and operation of various marine metallogenic processes (e.g., Boström et al., 1973, and Leinen and Stakes, 1979). Dymond and Veeh (1975), Sayles et al. (1975, 1976), and McMurtry et al. (in press) have considered metal accumulation rates in the southeastern Pacific. Despite the fact that Bauer Basin sediment is enriched in Fe, Mn, Cu, Co, Ni, Zn and Ba

								$CaCO_3$-free Basis? *4	% $CaCO_3$	Reference No.
Sb	Sc	Sn	Th	Tl	U	V	Y			
–	–	–	–	–	–	–	–	No	–	4
–	–	–	–	–	–	–	–	No	–	4
–	–	–	–	–	–	–	–	No	–	4
–	–	–	–	–	–	–	–	No	–	4
–	–	–	–	–	–	–	–	No	–	3
–	–	–	–	–	–	–	–	No	–	3
–	–	–	–	–	–	–	–	No	–	3
–	–	–	–	–	–	–	–	No	–	3
–	–	–	–	–	–	–	–	No	–	3
–	27	–	–	–	–	158	55	No	–	20
–	–	–	–	–	–	–	–	No	–	21
–	<3	–	0.12	–	2.9	–	<3	Yes ±	–	7
–	<3	–	–	–	–	–	4	Yes ±	–	7

– P.J. Cook (1974)
– Corliss et al. (1978)
– Cronan (1972a)
– Cronan et al. (1972)
– Dymond et al. (1973)
– Dymond et al. (1976)
– Emelyanov (1977)
– Ferguson and Lambert (1972)
– Fryer and Hutchinson (1976)
– Goldberg and Arrhenius (1958)
– Hays et al. (1972)
– Heath and Dymond (1977)

25 – Hekinian et al. (1978)
26 – Hoffert et al. (1978)
27 – Horowitz (1970)
28 – Horowitz and Cronan (1976)
29 – Moore and Vogt (1976)
30 – Nayudu (1971)
31 – Pimm (1974a)
32 – Piper et al. (1975)
33 – Sayles and Bischoff (1973)
34 – M.R. Scott et al. (1974)
35 – Turekian and Wedepohl (1961)

compared to normal pelagic clay (Sayles et al., 1975), the Bauer Basin accumulates metalliferous components about an order of magnitude slower than the East Pacific Rise (Dymond and Veeh, 1975) (Table XIII).

Even on the crest of the EPR, accumulation rates vary by more than an order of magnitude (McMurtry et al., in press). Values near the equator (6°S) are approximately the same as normal authigenic accumulation rates, whereas toward 20°S, values increase by an order of magnitude. McMurtry et al. also point out that metal accumulation rates on the East Pacific Rise near 20°S are unusually high and that the extrapolation of these values over the entire world rift system would lead to an overestimation of the hydrothermal flux to the oceans.

Metal accumulation rates in the FAMOUS area of the Mid-Atlantic Ridge are not particularly high compared to the East Pacific Rise (M.R. Scott and Salter, 1977). Rates in the Marie Celeste Fracture Zone of the Indian Ocean are also relatively low, indicating that truly metalliferous sediments in this area must be more localized than previously realized (McArthur and Elderfield, 1977). Values for Ti and Al, attributable to terrige-

TABLE XIII

Metal accumulation rates in surface sediments

Reference	Area	Sample	Average sedimentation rate ($mm/10^3$ yr)	Accumulation rate ($\mu g/cm^2/10^3$ yr)			
				Fe	Mn	Al	Ti
(1)	North Pacific	Range mid-point	–	5600	620	11050	535
(2)	East Pacific Rise	Average	–	63000	24000	720	74
(3)	West flank, East Pacific Rise	V19-61	7.0	10000	2030	410	–
		V19-64	4.5	3800	620	410	–
(3)	East Pacific Rise	V19-54	15.0	82000	28000	610	–
(3)		KK71-109	10.8	710	860	4000	–
(3)		Y71-7-45	9.3	11000	5800	780	–
(3)	East flank, East Pacific Rise	KK71-106	4.3	1600	500	330	–
(3)	Bauer Basin	Y71-7-36	1.4	4400	1100	620	–
(3)		KK71-115	1.9	5900	2100	1300	–
(4)		Core 138	–	9000	3300	–	–
(3)	Between Bauer Basin,	DSDP 319	1.1	3500	1000	640	–
(3)	Galapagos Rise	KK71-132	2.0	1700	950	1900	–
(5)	FAMOUS area, Mid-Atlantic Ridge	Maximum	2.7–9.6	10400	400	9100	–
(6)	Marie Celeste Fracture Zone, Indian Ridge	Range mid-point	–	4300	550	2050	140

(1) Boström et al. (1973)
(2) Boström (1973)
(3) Dymond and Veeh (1975)
(4) Sayles et al. (1976)
(5) M.R. Scott and Salter (1977)
(6) McArthur and Elderfield (1977)

nous sources, are higher than EPR values, while Fe and Mn, mostly attributable to hydrothermal input, are lower.

STRUCTURAL SETTING AND ASSOCIATED DEPOSITS

Discounting metal-rich heavy mineral sands found off coasts and on shelves, and the organic-rich metalliferous muds of continental shelf basins, all marine metalliferous sediments are formed within the framework of an evolving volcanic ocean crust. The most extensive deposits are derived from hydrothermal exhalations at mid-ocean ridge crests (divergent plate boundaries), with contemporaneously deposited pelagic sediments on ridge flanks and in adjacent deep ocean basins containing a progressively lower content of hydrothermal metalliferous components as distance from the ridge crest increases. Iron- and manganese-rich sediments formed at ridge crests are transported away from their site of origin by sea-floor spreading, and are subsequently buried by normal pelagic sedimentation.

Si	Ba	Cu	Ni	Zn	U	Th
–	–	33	24.5	–	–	–
55000	–	–	95(3)	–	–	–
–	–	–	27	–	0.27	<0.1
–	–	–	17	–	0.11	0.08
–	–	–	160	–	4.0	<0.2
–	–	–	70	–	–	–
–	–	–	56	–	0.31	0.1
–	–	–	5	–	0.06	0.1
–	–	–	16	–	0.05	0.07
–	–	–	46	–	–	–
–	800	80	60	40	–	–
–	–	–	19	–	–	–
–	–	–	46	–	–	–
–	–	30.7	15.7	21.6	–	–
6900	–	–	–	–	–	–

Other structural features that have served as sites for the accumulation of precipitates from hydrothermal solutions include:

(1) small, restricted embayments around volcanic islands such as Santorini;

(2) basins associated with volcanic arcs or ridges, e.g., the North Fiji Basin and the Bauer Deep; and

(3) deep fracture zones such as the Romanche, which has yielded pyrite concretions, and the Atlantis, where iron and manganese oxides have been found (Cronan, 1976a).

The detailed bathymetry and structure of possible sites of generation of metalliferous sediments have been determined at a number of locations, including the TAG area of the Mid-Atlantic Ridge (26°N) (McGregor and Rona, 1975; Rona, 1976b; Rona et al., 1976; Temple et al., 1976; Udintsev et al., 1977), the Project FAMOUS area of the Mid-Atlantic Ridge (36°30′–37°N) (Needham and Francheteau, 1974; ARCYANA, 1975, 1977; Ballard, 1975; Ballard et al., 1975; Heirtzler, 1975; Macdonald et al., 1975; Ballard and Van Andel, 1977; Ramberg and Van Andel, 1977; Ramberg et al., 1977; Luyendyk and Macdonald, 1977; Macdonald and Luyendyk, 1977; Hoffert et al., 1978), the Galapagos inner rift (Crane, 1978), the Galapagos Hydrothermal Mounds area (Lonsdale, 1977a), the East Pacific Rise at 21°N (Normark, 1976; Crane and Normark, 1977), and the axial trough of the Red Sea (Ross et al., 1969; Bäcker and Schoell, 1972; Bäcker et al., 1975; Garson and Krs, 1976).

Transform Fault Valley "A", FAMOUS Area of Mid-Atlantic Ridge

A description of two hydrothermal vents and their structural setting in the valley of Transform Fault "A" between offset segments of the Mid-Atlantic Ridge has been presented by ARCYANA (1975). The deepest axial region of the valley is V-shaped and characterized by relatively steep slopes, giving way upward to slopes with smaller average gradients. Delimiting the sides of the valley are breaks in slope and large escarpments more than 100 m high; the valley has an average width of about 3 km and a total vertical relief of 600 m.

The sea floor of the valley has been intensely fractured and is marked by a series of steps and scarps trending parallel to the valley walls, with the height of the scarps varying from about one-half metre to several tens of metres, and the spacing between them varying from several metres to several tens of metres. Two hydrothermal deposits, about 100 m apart, have been emplaced on hills toward the top of the scarp which forms the southern limit of the deep portion of the transform valley. Both deposits cover an area of about 40 by 15 m and occur as red, yellow or green pisolitic layers up to 1 m thick near emissive vents or fissures, which downhill from the vents become black travertine-like crusts 100–500 mm thick lying on sediment.

Because no hydrothermal deposits were sighted in the Mid-Atlantic Ridge rift valley itself, ARCYANA (1975) speculated that the transform fault zone, which is the seat of presumably deep vertical fractures that have been active for a long time and involve a high level of shearing and microfracturing, is a particularly suitable setting for fluid circulation and efficient leaching of crustal rocks by hot seawater.

TAG geothermal area, North Atlantic

In the TAG geothermal area of the Mid-Atlantic Ridge, closely spaced fracture valleys and proximity to intrusive heat sources have promoted a vigorous hydrothermal circulation. Photographic transects of the southeast wall of the rift valley indicated that a high frequency of occurrence of faulting appears most conducive to fluid emission, whereas the amplitude of an individual fault does not seem to be related to discharge magnitude (Temple et al., 1976). Ten major fault blocks having an average strike of N26°E were crossed; the rift valley strikes at N25°E. The average width of the fault blocks is 320 m (range 160–620 m), and throw ranges from 10 to 220 m. Average fault block slope is 15°, and some blocks are backtilted 1–2°. Within the major blocks occur numerous small steps, 5–160 m wide (width decreases upslope), and with an average throw of 15 m.

The phototraverse crossed a lithologic sequence consisting of 47% sediment, 39% talus and 11% breccia. Pillow lavas (3%) were seen only near the top of the traverse. On only 5% of the traverse was hydrothermally Mn-coated breccia and talus observed, but these occurrences were spaced over one-third of the minor fault steps, which have 1–20 m throws, average slopes of 18%, and average spacing of 65 m.

Red Sea

The Red Sea is unique in that marginal evaporite beds provide a source for brines that circulate through fractured country rock, dissolving and transporting metals to the sea floor. Brine pools and their associated metalliferous muds are localized at cross-cutting faults in the axial rough zone (Bignell, 1975; Schoell and Bignell, 1976), but only in the central Red Sea is the surrounding structure conducive to the transport of formation waters to this zone. The northern Red Sea lacks an axial rough zone, and the southern Red Sea is characterized by horst-and-graben structures which form long, flat-bottomed terraces and deeps. However, the graben structures flanking the axial rough zone are only a few hundred metres high. In contrast, the sea floor of the central Red Sea drops directly from the coastal reefs to a depth of 500–600 m, from where the bottom slopes only slightly over a distance of about 40 km to the edge of the main trough (at about 1100 m), from which it reaches the final depth after a steep slope (Bäcker and Schoell, 1972).

Ophiolite sequences

Metalliferous sediments occur at times as the uppermost member of ophiolite sequences. These are cross-sections of oceanic crust formed at divergent plate boundaries, typified by an exposure in the Troodos Massif of Cyprus. Oceanic sediments there overlie pillow lavas, which are superimposed on basalt sheets or dikes, which in turn lie above ultramafic rocks. In the Troodos deposits, Cu, Fe and sometimes Zn sulphides are embedded in the pillow lavas; pods of Cr ore are found near the top of the ultramafics; and asbestos occurs deeper in the ultramafics (Hammond, 1975a). Lateritic nickel deposits are sometimes found in sections of oceanic crust where the ultramafics have been exposed and weathered, such as in New Caledonia.

Massive sulphide ores

Massive sulphides are high-grade ore bodies which often occur as large, nearly pure lenses with a polymetallic mineral assemblage containing Cu, Zn, Pb, Au and Ag. They are found in modern island arcs, and therefore represent metalliferous deposits associated with plate convergence (some massive sulphides are found in older arc material incorporated into continental margins).

Several types of massive sulphides are recognized (Hammond, 1975a). The Kuroko type is thought to have formed by submarine volcanic processes late in the evolution of volcanic island chains (Ohmoto, 1978); the metals are laid down in shallow, nearshore environments along with marine sediments and explosive fragments. Obvious modern analogues should be the Santorini and Matupi Harbour deposits, but these deposits are not particularly rich in sulphides at and near the sediment surface. The Besshi type of massive sulphide, containing predominantly Fe and Cu, is also found in island arcs. They are thought to be submarine volcanic emissions deposited on underwater slopes of volcanoes early in their evolution.

At the Troodos Massif, massive sulphide deposits are closely associated, both spatially and genetically, with fine-grained, siliceous, tuffaceous and iron-rich sediments which lie directly on pillow basalt (Fryer and Hutchinson, 1976). Manganese content increases with increasing distance, both stratigraphically upward and laterally away from associated pyritic deposits, presumably reflecting a change from reducing to oxidizing conditions away from submarine fumarolic sources. Similar conditions are thought to be responsible for differential dispersion of iron and manganese around Santorini Volcano (Smith and Cronan, 1975b), but massive sulphide ore bodies beneath the metalliferous sediments there are not yet known.

ORIGIN

Metalliferous sediments certainly form as the result of more than one process, but the most widely accepted model is based on deposition from submarine hydrothermal fluids generated during mid-ocean rift volcanism (Corliss, 1971, 1976). As summarized by Hammond (1975b), this model proposes that the interaction of seawater with cooling volcanic rock is the principal means by which many metals are extracted and concentrated into ore bodies, so that metalliferous sediments are of a hydrothermal rather than magmatic origin. Corliss (1976) lists the evidence for the hydrothermal model as:

(1) observations of the effects of hydrothermal circulation in both dredged and cored oceanic crustal rocks, e.g., metal leaching and serpentinization;

(2) the nature of the nearly ubiquitous basal metalliferous sediment layer in the oceans;

(3) the pattern of heat flow over spreading centres (Boström and Peterson, 1969, initially related East Pacific Rise sediments to high heat flow; Anderson and Halunen, 1974, reported many high values among their heat flow data for the East Pacific Rise crest, but the data also showed a significant scatter);

(4) anomalies in vertical profiles of suspended matter and dissolved gases in deep oceans, e.g., the "plume" of excess ^{3}He and Fe carried away from the crest of the East Pacific Rise (Broecker, 1974);

(5) observations of submarine hot springs in Indonesia, the Mediterranean, and on

the crest of the Galapagos Rift;

(6) experimental investigations of seawater–basalt interaction; and

(7) observations of ore deposits which can be interpreted as resulting from submarine hydrothermal systems.

Excellent accounts of the hydrothermal flux at mid-ocean ridges have been presented by Lister (1972, 1974), Williams et al. (1974), Wolery and Sleep (1976), Williams (1976), Davis and Lister (1977), Parmentier and Spooner (1978), and Sleep and Wolery (1978). Direct evidence of hydrothermal circulation can, however, be ambiguous. For example, in the FAMOUS area of the Mid-Atlantic Ridge, heat flow data supports the concept of hydrothermal circulation in basement rocks (Williams et al., 1977), whereas deep-water temperature measurements can fail to detect a hydrothermal influence (Fehn et al., 1977) (cf. Garner and Ford, 1969).

Other hypotheses that may partially account for the origin of metalliferous sediments have also been proposed. These include:

(1) deposition under reducing conditions in restricted ridge valleys (Turekian and Bertine, 1971);

(2) ponding of very fine terrigenous sediments with high transition metal/aluminium ratios (Turekian and Imbrie, 1966);

(3) authigenic precipitation of Fe and Mn from seawater (Bender, et al., 1971);

(4) the submarine weathering of tholeiitic basalts (Thompson, 1973; Bertine, 1974);

(5) hydrothermal alteration of oceanic basalts (Humphris and Thompson, 1976, 1978a, b; Elderfield et al., 1977; Seyfried and Bischoff, 1977; Spooner et al., 1977a, b; Arnórsson, 1978); and

(6) bacterial action, for example operative in production of the structures that Von der Borch et al. (1971) noted in Pacific basal sediments. None of the hypotheses, including the hydrothermal model, can unequivocally explain all the features of metalliferous sediments, as pointed out by Cronan (1974).

As with manganese nodules, the two principal concerns regarding origin of metalliferous sediments are the sources of the constituent elements and the mode of introduction of these elements into the deposits. Boström and Peterson (1969) decided that the most likely ultimate origin of mineralizing emanations is the upper mantle, and that only a minor fraction has been leached out of buried sediments (cf. Boström, 1967). While the upper mantle may be the ultimate source of metals, submarine basalts and the magmatic processes attending their emplacement on and near the sea floor are a more immediate source of metals. Corliss (1971) found that the slowly cooled interior portions of fine-grained holocrystalline Mid-Atlantic Ridge basalts are depleted, relative to quenched flow margins (pillow basalts), in elements (Mn, Fe, Co, REE, Cu and Pb) that are enriched in pelagic sediments (particularly metalliferous varieties) and manganese nodules. Many of these elements are excluded from the solid phases that crystallize from a melt, and are thus concentrated in residual liquids.

Additional elements are mobilized by deuteric alteration of early-formed olivine and the formation of immiscible sulphide liquids. Corliss suggested that these components of a melt occupy accessible sites, e.g., intergranular boundaries, in the hot solid rock mass, and are mobilized by dissolution as chloride complexes in seawater introduced along contraction cracks that form during cooling and solidification. These solutions may then represent metal-bearing "hydrothermal exhalations".

Ascending hydrothermal solutions presumably are acidic and reducing (Boström and Peterson, 1966). They would contain not only chloride supplied by seawater, but also sulphur (as sulphates which help to precipitate metals on the sea floor) and carbon dioxide, which is in the bicarbonate form in water expelled by submarine hot jets of Banu Wuhu, where the main ion composition is identical to seawater (Zelenov, 1964).

Several lines of evidence point to seawater as the fluid agent for metal transport at other locations as well. The waters from crater lakes of Deception Island (South Shetland Islands) are only slightly modified seawater when the major saline components are considered (Elderfield, 1972); and sediment pore fluids on the East Pacific Rise at 5–10°S display remarkably few variations from bottom water concentrations of dissolved Cl, SO_4, Mg, Ca and K in sediment types varying from carbonate ooze to metalliferous deposits (Bischoff and Sayles, 1972). Dymond et al. (1973) found that oxygen isotope data indicate isotopic equilibrium between seawater and North Pacific basal and Bauer Deep metalliferous sediments. They also found that the Sr-isotope ratio in these metal-rich sediments is the same as present-day seawater, and that sulphur isotopic compositions suggest that seawater sulphur dominates metalliferous sediments, with little or no contribution of magmatic or bacteriologically reduced sulphur (cf. Kaplan et al., 1969).

Dymond et al. (1973) also stated that uranium and rare-earth elements in the sediments appear to be derived from seawater (cf. Bender et al., 1971; Courtois and Trueil, 1977; Graf, 1977), and Berner (1973) and Froelich et al. (1977) have suggested that the phosphorus in East Pacific Rise surface sediments may be derived dominantly from seawater. Although other evidence supports the contention that the rare earths in metalliferous sediments are derived from seawater (Piper, 1974; Robertson and Fleet, 1976), the data are not entirely unambiguous and Piper et al. (1975) have concluded that the rare earths in an iron-rich deposit from the northeastern Pacific have been derived from the leaching of underlying basalt by seawater (cf. Ludden and Thompson, 1978). Finally, Chapman and Spooner (1977) have shown that seawater is the source of hydrothermal fluids in the Troodos Massif, and Muehlenbachs and Clayton (1976) have used oxygen isotope analysis of D.S.D.P. basalts to document the circulation of seawater in oceanic crust. It should be emphasized, however, that hydrothermal activity may remove some elements from seawater, such as magnesium (Bloch and Hofmann, 1978).

Whereas solutions that have circulated through rocks and sediments of hydrothermal areas show major ionic compositions similar to seawater, some minor components are clearly of different origin. Bender et al. (1971), Dasch et al. (1971) and Dymond et al. (1973) determined Pb-isotope ratios of metalliferous sediments from the East Pacific

Rise, and discovered that the values resemble those of oceanic tholeiite basalt, implying a magmatic source for the lead. This conclusion is supported by O'Nions et al. (1978). Banu Wuhu submarine hot springs, Deception Island crater lakes, and East Pacific Rise pore fluids all show very high dissolved silica and manganese, suggesting a volcanic source for these elements (Zelenov, 1964; Elderfield, 1972; and Bischoff and Sayles, 1972, respectively).

Hydrothermal circulation in the crustal rock of mid-ocean ridges may occur to depths of 10 km (Hammond, 1975b). Mineralizing solutions may reach the sediment–water interface through contraction cracks, tectonically weakened zones and fissures, volcanic vents, or by migration through sediments (Boström and Peterson, 1969; Corliss, 1971). Because disseminated sulphides occur in rocks of the oceanic crust, a reducing environment at elevated temperatures must be maintained to delay sulphide precipitation until hydrothermal solutions ascend to the sea floor (Cronan, 1976a), where the metal-rich fluids normally encounter cool, oxidizing, alkaline conditions. A rapidly changing Eh may be a major factor controlling variations in metal abundances and ratios in sea-floor deposits (Fryer and Hutchinson, 1976). Studying sediments of the Red Sea and those associated with the Troodos ore bodies of Cyprus, these authors found that Zn, Cu and Au are concentrated in high Fe/Mn ratio sediments proximal to fumaroles; that Ni is concentrated in low Fe/Mn distal sediments; and that Pb, Ag, Sn and Mo are relatively unaffected by the oxidation of fumarolic brine solutions by normal seawater.

The Bauer Deep and other parts of the Nazca Plate are apparently sites of polygenetic metalliferous sediments. McMurtry and Burnett (1975) suggest a direct volcanic origin from iron and manganese in the Bauer Basin, emanated from local sources. Elemental distributions suggest that the parent hydrothermal solution was an iron–silica-rich brine containing Mn, minor Ni and Cu, and traces of other transition metals (Heath et al., 1975). A smectite phase crystallized as this brine cooled at the sea floor. [Sayles and Bischoff (1973) pointed out that this diagenetic reaction must occur near the sediment–water interface, because below about 200 mm extensive reactions would perturb pore fluid composition.] Oxidation of dissolved iron in the brine formed an "amorphous" oxyhydroxide which co-precipitated the remaining transition metals from the brine and admixed seawater. Manganese micronodules that continue to grow in the sediment add Cu, Ni and associated elements in a manner comparable to the hydrogenous development of ferromanganese nodules.

Basal sediments at D.S.D.P. Site 319 in the Bauer Deep appear to have a dominantly hydrothermal source of metals, whereas hydrogenous precipitation is important in the near-surface sediments (Dymond et al., 1975) which have been deposited at some distance from the rise crest.

The deposition of metalliferous sediments in the Red Sea involves factors additional to those of other volcanic areas. The heavy-metal-rich brines are most likely derived from the leaching of evaporites and are not directly attributable to normal sea-floor spreading processes (Whitmarsh, 1974). According to Lupton et al. (1977b), ^{3}He and ^{4}He are the only Red Sea brine components which can be shown unequivocally to be derived from hydrothermal circulation in basalts rather than evaporites.

Oxygen isotopic evidence indicates that all brines in existing deeps are not present-day Red Sea deep or surface water, but are derived instead from ancient Red Sea deep water (Schoell and Bignell, 1976; Schoell and Faber, 1978). Atlantis II Deep brines are palaeo-Red Sea waters that were trapped during the climatic optimum 8000–5000 years B.P., when Red Sea deep water was slightly depleted in heavy isotopes compared to present day waters. After metal-rich brines were discharged into local deeps, minerals precipitated by several processes (Bischoff, 1969a), after which compaction and diagenesis increased the metal content of the sediments.

SIGNIFICANCE TO GEOCHEMICAL EXPLORATION FOR ORES

Except for the Red Sea deposits, which may be rich enough to mine (Bischoff and Manheim, 1969; Heckett and Bischoff, 1973; Bäcker and Post, 1978, but cf. Walthier and Schatz, 1969), the principal economic significance of metalliferous sediments may be as potential indicators of the existence of associated sulphide ore bodies, such as those of the Troodos Massif, Cyprus. It is doubtful that in the foreseeable future this relationship would even be used to locate sulphide ores in the basalts beneath deep-sea metalliferous basal sediments. However, it may prove to have exploration value around volcanic islands or terrestrially emplaced ophiolite sequences. Assuming a known spatial relationship between the hydrothermal source vents for metalliferous sediments and the subsurface location of sulphide ore bodies, the initial exploration task would be to find the hydrothermal vents. Cronan (1976a) has noted differential precipitation of manganese and iron in dispersion haloes around hydrothermal sources at Santorini Volcano, in the Red Sea, and at Matupi Harbour (New Britain). Metal sulphides, iron silicates and/or oxides, and manganese oxides successively form the principal metal-bearing facies with increasing distance from the source. Cronan (1976a) reports that the dispersion halo of manganese around Atlantis II Deep deposits in the Red Sea is up to 20 km in diameter, while the sulphide deposit is much smaller. Manganese haloes would therefore be the targets during initial grid sampling. On a larger scale, the Fe/Al ratio appears to be the most useful ratio in geochemical exploration for submarine exhalative deposits (Smith and Cronan, 1975b), because it seems to reflect the varying contribution of hydrothermal and detrital components. Mn/Al, Zn/Al and Cu/Al ratios also increase toward exhalative zones. Before these exploration concepts are applied to new areas, however, it must be demonstrated, probably by drilling, that promising sulphide ore bodies do indeed exist in close proximity to known metalliferous sediment deposits.

SUMMARY

Metalliferous sediments are a variety of deep-sea sediment that accumulate transition metals at a faster rate than normal pelagic sediments. Certain strata-bound ores, such as the umbers of the Troodos Massif, or potential ores, such as the sedimentary accumulation in the Atlantis II Deep of the Red Sea, represent end members of the process of hydrothermal metal enrichment in sediments. In ophiolite sequences generated at ancient spreading centres, metalliferous sediments are found immediately above volcanic pillow basalts that may contain intercalated massive sulphide ores.

Metal-enriched sediments in the deep sea have been deposited in two general forms – black shales which are the result of accumulation of lithogenous detritus and abundant organic matter in basins with restricted circulation or high productivity, and iron-rich deposits which are here referred to as metalliferous sediments, and are formed primarily by hydrothermal activity at the crest of mid-ocean ridges or on the flanks of some isolated volcanoes. The components of both types of metal-enriched sediment can be derived from detrital, hydrogenous, hydrothermal or biogenic sources. Reactions involving organic matter, particularly adsorption and reduction, lead to metal enrichment in black shales, whereas the hydrothermal leaching of metals from underlying basalt, with subsequent precipitation from seawater or the interstitial fluids of sediments, results in formation of metalliferous sediments.

Hydrothermal metal enrichment of deep-sea sediments may be syngenetic, involving settling of exhalative precipitates, or epigenetic, involving infiltration by hydrothermal fluids, possibly accompanied by baking. Syngenetic activity may be focused on hydrothermal vents, such as at the Galapagos Spreading Centre, and epigenetic activity may occur primarily at intrusive sills.

Transition metal enrichment in metalliferous sediments usually takes the form of X-ray amorphous ferric hydroxides, commonly as RSO's (red semi-opaque oxides), which are minute iron globules disseminated in sediment. Iron smectites and ferromanganese concretions also contribute to transition metal enrichment. Goethite and haematite may occur as ageing products of the more poorly crystalline ferric hydroxide phase. Metalliferous components are often diluted by calcareous material on the flanks of mid-ocean ridges.

ACKNOWLEDGMENTS

Dr. D.S. Cronan (Imperial College, London) is kindly thanked for providing the East Pacific Rise metalliferous sediment samples and Dr. L. Aldridge (Chemistry Division, D.S.I.R., Lower Hutt, New Zealand) for making available the automatic X-ray diffraction scanning facility. The Mining Ventures Division of Shell Oil Company, Houston, Texas, provided partial support for preparation of the manuscript. Mr. Donald Temple (University of Wisconsin-Milwaukee) drafted the figures, and Ms. Lora Hingston (Department of Oceanography, University of Hawaii) typed the manuscript.

REFERENCES

Amann, H., Bäcker, H. and Blissenbach, E., 1973. Metalliferous muds of the marine environment. *5th Annu. Offshore Technol. Conf.*, 1: 345–353.

Anderson, R.N. and Halunen, A.J., 1974. Implications of heat flow for metallogenesis in the Bauer Deep. *Nature (London)*, 251: 473–475.

Anderson, R.N., Hobart, M.A., Von Herzen, R.P. and Fornari, D.J., 1978. Geophysical surveys on the East Pacific Rise – Galapagos Rise system. *Geophys. J.R. Astron. Soc.*, 54: 141–166.

Andrews, J.E. and Packham, G. et al., 1975. *Initial Reports of the Deep Sea Drilling Project, 30.* U.S. Govt. Printing Office, Washington, D.C., 753 pp.

Andrushchenko, P.F., Gradusov, B.P., Yeroshchev-shak, V.A., Yanshima, R.S. and Barisovskiy, S.Ye., 1975. Composition and structure of metamorphosed ferromanganese nodules, new vein formations of manganese hydroxides, and the surrounding pelagic sediments in the southern basin of the Pacific Ocean floor. *Int. Geol. Rev.*, 17: 1375–1392.

Anonymous, 1977. Volcanic processes in ore genesis. Proceedings of a joint meeting of the volcanic studies group of the Geological Society of London and the Institution of Mining and Metallurgy held in London on 21 and 22 January, 1976. *Spec. Publ. Geol. Soc. London*, 7: 188 pp.

ARCYANA, 1975. Transform fault and rift valley from bathyscaph and diving saucer. *Science*, 190: 108–116.

ARCYANA, 1977. Rocks collected by bathyscaph and diving saucer in the FAMOUS area of the Mid-Atlantic Rift Valley: petrological diversity and structural setting. *Deep-Sea Res.*, 24: 565–589.

Arnórsson, S., 1978. Major element chemistry of the geothermal sea-water at Reykjanes and Svartsengi, Iceland. *Mineral. Mag.*, 42: 209–220.

Arrhenius, G. and Bonatti, E., 1965. Neptunism and vulcanism in the ocean. In: M. Sears (Editor), *Progress in Oceanography, 3.* Pergamon Press, New York, N.Y., pp.7–22.

Atkinson, R.J., Posner, A.M. and Quirk, J.P., 1968. Crystal nucleation in Fe(III) solutions and hydroxide gels. *J. Inorg. Nucl. Chem.*, 30: 2371–2381.

Atkinson, R.J., Posner, A.M. and Quirk, J.P., 1977. Crystal nucleation and growth in hydrolysing iron(III) chloride solutions. *Clays Clay Miner.*, 25: 49–56.

Bäcker, H.,, 1973. Rezente hydrothermal-sedimentäre Lagerstättenbildung. *Erzmetall*, 26: 544–555.

Bäcker, H., 1976. Fazies und chemische Zusammensetzung rezenter Ausfällungen aus Mineralquellen im Roten Meer. *Geol. Jahrb., Reihe D*, 17: 151–172.

Bäcker, H. and Post, J., 1978. Untersuchung von Erzschlämmen in Roten Meer. *Meerestechnik*, 9: 109–114.

Bäcker, H. and Schoell, M., 1972. New deeps with brines and metalliferous sediments in Red Sea. *Nature (London), Phys. Sci.*, 240: 153–158.

Bäcker, H., Lange, K. and Richter, H., 1975. Morphology of the Red Sea central graben between Subair Islands and Abul Kizaan. *Geol. Jahrb., Reihe D*, 13: 79–123.

Bagin, V.I., Bagina, O.A., Bogdanov, Yu, A., Gendler, T.S., Lebedev, A.I., Lisitsyn, A.P. and Pecherskiy, D.M., 1975. Iron in metalliferous sediments in the Bauer Deep and East Pacific Rise. *Geochem. Int.*, 12(2): 105–125.

Ballard, R.D., 1975. Photography from a submersible during Project FAMOUS. *Oceanus*, 18(3): 31–39.

Ballard, R.D., 1977. Notes on a major oceanographic find. *Oceanus*, 20(3): 35–44.

Ballard, R.D. and Van Andel, Tj.H., 1977. Morphology and tectonics of the inner rift valley at 36°50′N on the Mid-Atlantic Ridge. *Geol. Soc. Am. Bull.*, 88: 507–530.

Ballard, R.D., Bryan, W.B., Heirtzler, J.R., Keller, G., Moore, J.G. and Van Andel, Tj., 1975. Manned submersible observations in the FAMOUS area: Mid-Atlantic Ridge. *Science*, 190: 103–108.

Bancroft, G.M., 1973. *Mössbauer Spectroscopy: An Introduction for Inorganic Chemists and Geochemists.* McGraw–Hill, London, 252 + xii pp.

Bancroft, G.M., Maddock, A.G. and Burns, R.G., 1967. Applications of the Mössbauer effect to silicate mineralogy, I. Iron silicates of known crystal structure. *Geochim. Cosmochim. Acta,* 31: 2219–2246.

Bass, M.N., 1976. Rare and unusual minerals and fossils(?) in sediments of Leg 34. In: *Initial Reports of the Deep Sea Drilling Project, 34.* U.S. Govt. Printing Office, Washington, D.C., pp.611–626.

Baturin, G.N., Kochenov, A.V. and Trimonis, Ye.S., 1969. Composition and origin of iron-ore sediments and hot brines in the Red Sea. *Oceanology,* 9: 360–368.

Bender, M.L., Broecker, W., Gornitz, V., Middel, U., Kay, R., Sun, S.-S. and Biscaye, P., 1971. Geochemistry of three cores from the East Pacific Rise. *Earth Planet. Sci. Lett.,* 12: 425–433.

Berner, R.A., 1973. Phosphate removal from sea water by adsorption on volcanogenic ferric oxides. *Earth Planet. Sci. Lett.,* 18: 77–86.

Bertine, K.K., 1974. Origin of Lau Basin Rise sediment. *Geochim. Cosmochim. Acta,* 38: 629–640.

Bertine, K.K. and Keene, J.B., 1975. Submarine barite-opal rocks of hydrothermal origin. *Science,* 188: 150–152.

Bignell, R.D., 1975. Timing, distribution and origin of submarine mineralization in the Red Sea. *Trans. Inst. Min. Metall.,* 84B: 1–6.

Bignell, R.D. and Ali, S.S., 1976. Geochemistry and stratigraphy of Nereus Deep, Red Sea. *Geol. Jahrb., Reihe D,* 17: 173–186.

Bignell, R.D., Tooms, J.S., Cronan, D.S. and Horowitz, A., 1974. An additional location of metalliferous sediments in the Red Sea. *Nature (London),* 248: 127–128.

Bischoff, J.L., 1969a. Red Sea geothermal brine deposits: their mineralogy, chemistry and genesis. In: E.T. Degens and D.A. Ross (Editors), *Hot Brines and Recent Heavy Metal Deposits in the Red Sea.* Springer, New York, N.Y., pp.368–401.

Bischoff, J.L., 1969b. Goethite–hematite stability relations with relevance to seawater and the Red Sea brine system. In: E.T. Degens and D.A. Ross (Editors), *Hot Brines and Recent Heavy Metal Deposits in the Red Sea.* Springer, New York, N.Y., pp.402–406.

Bischoff, J.L., 1972. A ferroan nontronite from the Red Sea geothermal area. *Clays Clay Miner.,* 20: 217–223.

Bischoff, J.L. and Manheim, F.T., 1969. Economic potential of Red Sea heavy metal deposits. In: E.T. Degens and D.A. Ross (Editors), *Hot Brines and Recent Heavy Metal Deposits in the Red Sea.* Springer, New York, N.Y., pp.535–541.

Bischoff, J.L. and Rosenbauer, R.J., 1977. Recent metalliferous sediments in the North Pacific manganese nodule area. *Earth Planet. Sci. Lett.,* 33: 379–388.

Bischoff, J.L. and Sayles, F.L., 1972. Pore fluid and mineralogical studies of Recent marine sediments: Bauer Depression region of East Pacific Rise. *J. Sediment. Petrol.,* 42: 711–724.

Blissenbach, E., 1972. Continental drift and metalliferous sediments. *Oceanol. Int.,* 72: 412–416.

Blissenbach, E. and Fellerer, R., 1973. Continental drift and the origin of certain mineral deposits. *Geol. Rundsch.,* 62: 812–840.

Bloch, S., 1978. Phosphorus distribution in smectite-bearing basal metalliferous sediments. *Chem. Geol.,* 22: 353–359.

Bloch, S. and Hofmann, A.W., 1978. Magnesium metasomatism during hydrothermal alteration of new oceanic crust. *Geology,* 6: 275–277.

Bonatti, E., 1967. Mechanisms of deep-sea volcanism in the South Pacific. In: P.H. Abelson (Editor), *Researches in Geochemistry, 2.* Wiley, New York, N.Y., pp. 453–491.

Bonatti, E., 1975. Metallogenesis at oceanic spreading centers. *Annu. Rev. Earth Planet. Sci.,* 3: 401–431.

Bonatti, E., 1978. The origin of metal deposits in the oceanic lithosphere. *Sci. Am.,* 238(2): 54–61.

Bonatti, E. and Honnorez, J., 1976. Sections of the earth's crust in the equatorial Atlantic. *J. Geophys. Res.,* 81: 4104–4116.

Bonatti, E. and Joensuu, O., 1966. Deep-sea iron deposit from the South Pacific. *Science,* 154: 643–645.

Bonatti, E., Kraemer, J. and Rydell, H., 1972. Classification and genesis of submarine iron–manganese deposits. In: D.R. Horn (Editor), *Ferromanganese Deposits on the Ocean Floor.* National Science Foundation, Washington, D.C., pp.149–166.

Bonatti, E., Zerbi, M., Kay, R. and Rydell, H., 1976a. Metalliferous deposits from the Apennine ophiolites: Mesozoic equivalents of modern deposits from oceanic spreading centers. *Geol. Soc. Am. Bull.*, 87: 83–94.

Bonatti, E., Guerstein-Honnorez, M.B. and Honnorez, J., 1976b. Copper–iron sulfide mineralizations from the equatorial Mid-Atlantic Ridge. *Econ. Geol.*, 71: 1515–1525.

Bonatti, E., Honnorez-Guerstein, M.B., Honnorez, J. and Stern, C., 1976C. Hydrothermal pyrite concretions from the Romanche Trench (Equatorial Atlantic): Metallogenesis in oceanic fracture zones. *Earth Planet. Sci. Lett.*, 32: 1–10.

Bonatti, E., Kolla, V. and Stern, C., 1976d. Metallogenesis in marginal basins: Fe-rich basal sediments from the Philippine Sea. *EOS Trans. Am. Geophys. Union*, 57 (12): 1014 (Abstr.).

Bonatti, E., Kolla, V., Moore, W.S. and Stern, C., 1979. Metallogenesis in marginal basins: Fe-rich basal deposits from the Philippine Sea. *Mar. Geol.*, 32: 21–37.

Boström, K., 1967. The problem of excess manganese in pelagic sediments. In: P.H. Abelson (Editor), *Researches in Geochemistry, 2.* Wiley, New York, N.Y., pp.421–452.

Boström, K., 1970a. Submarine volcanism as a source for iron. *Earth Planet. Sci. Lett.*, 9: 348–354.

Boström, K., 1970b. Geochemical evidence for ocean floor spreading in the South Atlantic Ocean. *Nature (London)*, 227: 1041.

Boström, K., 1973. The origin and fate of ferromanganoan active ridge sediments. *Stockholm Contrib. Geol.*, 27: 149–243.

Boström, K., 1976. Particulate and dissolved matter as sources for pelagic sediments. *Stockholm Contrib. Geol.*, 30: 15–79.

Boström, K. and Fisher, D.E., 1969. Distribution of mercury in East Pacific sediments. *Geochim. Cosmochim. Acta*, 33: 743–745.

Boström, K. and Fisher, D.E., 1971. Volcanogenic uranium, vanadium and iron in Indian Ocean sediments. *Earth Planet. Sci. Lett.*, 11: 95–98.

Boström, K. and Peterson, M.N.A., 1966. Precipitates from hydrothermal exhalations on the East Pacific Rise. *Econ. Geol.*, 61: 1258–1265.

Boström, K. and Peterson, M.N.A., 1969. The origin of aluminum-poor ferromanganoan sediments in areas of high heat flow on the East Pacific Rise. *Mar. Geol.*, 7: 427–447.

Boström, K. and Valdés, S., 1969. Arsenic in ocean floors. *Lithos*, 2: 357–360.

Boström, K., Peterson, M.N.A., Joensuu, O. and Fisher, D.E., 1969. Aluminum-poor ferromanganoan sediments on active oceanic ridges. *J. Geophys. Res.*, 74: 3261–3270.

Boström, K., Farquharson, B. and Eyl, W., 1971. Submarine hot springs as a source of active ridge sediments. *Chem. Geol.*, 10: 189–203.

Boström, K., Joensuu, O., Valdés, S. and Riera, M., 1972. Geochemical history of South Atlantic Ocean sediments since late Cretaceous. *Mar. Geol.*, 12: 85–121.

Boström, K., Kraemer, T. and Gartner, S., 1973. Provenance and accumulation rates of opaline silica, Al, Ti, Fe, Mn, Cu, Ni and Co in Pacific pelagic sediments. *Chem. Geol.*, 11: 123–148.

Boström, K., Joensuu, O., Kraemer, T., Rydell, H., Valdés, S., Gartner, S. and Taylor, G., 1974a. New finds of exhalative deposits on the East Pacific Rise. *Geol. Fören. Stockholm. Förh.*, 96: 53–60.

Boström, K., Joensuu, O., Moore, C., Boström, B., Dalziel, M. and Horowitz, A., 1974b. Geochemistry of barium in pelagic sediments. *Lithos*, 6: 159–174.

Boström, K., Joensuu, O., Valdés, S., Charm, W. and Glaccum, R., 1976. Geochemistry and origin of East Pacific sediments sampled during DSDP Leg 34. In: *Initial Reports of the Deep Sea Drilling Project, 34.* U.S. Govt. Printing Office, Washington, D.C., pp.559–574.

Brewer, P.G. and Spencer, D.W., 1969. A note on the chemical composition of the Red Sea brines. In: E.T. Degens and D.A. Ross (Editors), *Hot Brines and Recent Heavy Metal Deposits in the Red Sea.* Springer, New York, N.Y., pp.174–179.

Broecker, W.S., 1974. *Chemical Oceanography.* Harcourt Brace Jovanovich, New York, N.Y., 214 pp.

Burnett, W.C. and Morgenstein, M., 1976. Growth rates of Pacific manganese nodules as deduced by uranium-series and hydration-rind dating techniques. *Earth Planet. Sci. Lett.*, 33: 208–218.

Burnett, W.C. and Piper, D.Z., 1977. Rapidly-formed ferromanganese deposit from the eastern Pacific Hess Deep. *Nature (London)*, 265: 596–600.
Butuzova, G.Y., 1966. Iron ore sediments of the fumarole field of Santorini Volcano, their composition and origin. *Proc. Acad. Sci. U.S.S.R., Earth Sci. Sect.*, 168: 215–217.
Butuzova, G.Yu., Lisitsyna, N.A. and Gradusov, B.P., 1976. Authigenic montmorillonite in bottom sediments of Station No. 655 south of the Gulf of California. *Dokl. Akad. Nauk, S.S.S.R.*, 231: 243–245.
Cann, J.R., Winter, C.K. and Pritchard, R.G., 1977. A hydrothermal deposit from the floor of the Gulf of Aden. *Mineral. Mag.*, 41: 193–199.
Chapman, H.J. and Spooner, E.T.C., 1977. ^{87}Sr enrichment of ophiolitic sulphide deposits in Cyprus confirms ore formation by circulating seawater. *Earth Planet. Sci. Lett.*, 35: 71–78.
Chester, R. and Aston, S.R., 1976. The geochemistry of deep-sea sediments. In: J.P. Riley and R. Chester (Editors), *Chemical Oceanography, 6.* Academic Press, London, 2nd ed., pp.281–390.
Chester, R. and Hughes, M.J., 1967. A chemical technique for the separation of ferromanganese minerals, carbonate minerals and adsorbed trace elements from pelagic sediments. *Chem. Geol.*, 2: 249–262.
Childs, C.W. and Johnston, J.H., 1980. Mössbauer spectra of proto-ferrihydrite at 77K and 295K, and a reappraisal of the possible presence of akaganeite in New Zealand soils. *Aust. J. Soil Res.*, 18: 245–250.
Chukhrov, F.V., 1973. On the genesis problem of thermal sedimentary iron ore deposits. *Miner. Deposita*, 8: 138–147.
Chukhrov, F.V., Zvyagin, B.B., Ermilova, L.P. and Gorshkova, A.I., 1973. New data on iron oxides in the weathering zone. In: *Proceedings of the International Clay Conference, 1972.* Consejo Superior de Investigaciones Cientificas, Madrid, pp.333–341.
Chukhrov, F.V., Zvyagin, B.B., Gorshkov, A.I., Yermilova, L.P. and Balashova, V.V., 1974. Ferrihydrite. *Int. Geol. Rev.*, 16: 1131–1143.
Chukhrov, F.V., Ermilova, L.P., Zvyagin, B.B. and Gorshkov, A.I., 1975. Genetic system of hypergene iron oxides. In: *Proceedings of the International Clay Conference, 1975.* Applied Publishing Ltd., Wilmette, Illinois, pp.275–286.
Chukhrov, F.V., Gorshkov, A.I., Zvyagin, B.B., Yermilova, L.P. and Sapolnova, L.P., 1977a. Transformation of ferric compounds into iron oxides. *Int. Geol. Rev.*, 19: 766–774.
Chukhrov, F.V., Zvyagin, B.B., Gorshkov, A.I., Yermilova, L.P., Korovushkin, V.V., Rudnitskaya, Ye.S. and Yakubovskaya, N.Yu., 1977b. Feroxyhyte, a new modification of FeOOH. *Int. Geol. Rev.*, 19: 873–890.
Coey, J.M.D. and Readman, P.W., 1973a. Characterisation and magnetic properties of natural ferric gel. *Earth Planet. Sci. Lett.*, 21: 45–51.
Coey, J.M.D. and Readman, P.W., 1973b. New spin structure in an amorphous ferric gel. *Nature (London)*, 246: 476–478.
Coleman, R.G., 1974. Geologic background of the Red Sea. In: *Initial Reports of the Deep Sea Drilling Project, 23.* U.S. Govt. Printing Office, Washington, D.C., pp.813–819.
Cook, H.E., 1972. Stratigraphy and sedimentation. In: *Initial Reports of the Deep Sea Drilling Project, 9.* U.S. Govt. Printing Office, Washington, D.C., pp.933–943.
Cook, P.J., 1974. Major and trace element geochemistry of sediments from Deep Sea Drilling Project, Leg 27, sites 259–263, eastern Indian Ocean. In: *Initial Reports of the Deep Sea Drilling Project, 27.* U.S. Govt. Printing Office, Washington, D.C., pp.481–497.
Corliss, J.B., 1971. The origin of metal-bearing submarine hydrothermal solutions. *J. Geophys. Res.*, 76: 8128–8138.
Corliss, J.B., 1976. Sea water, sea-floor spreading, subduction, and ore deposits. *Am. Geophys. Union Monogr.*, 19: 297 (Abstr.).
Corliss, J.B. and Ballard, R.D., 1977. Oases of life in the cold abyss. *Natl. Geogr.*, 152: 440–453.
Corliss, J.B., Dymond, J., Lyle, M., Doerge, T., Crane, K., Lonsdale, P., Von Herzen, R.P. and Williams, D., 1976. Sediment mound ridges of hydrothermal (?) origin along the Galapagos Rift. *EOS Trans. Am. Geophys. Union*, 57(12): 935–936 (Abstr.).
Corliss, J.B., Lyle, M., Dymond, J. and Crane, K., 1978. The chemistry of hydrothermal mounds near the Galapagos Rift. *Earth Planet. Sci. Lett.*, 40: 12–24.

Corliss, J.B., Dymond, J., Gordon, L.I., Edmond, J.M., Von Herzen, R.P., Ballard, R.D., Green, K., Williams, D., Bainbridge, A., Crane, K. and Van Andel, Tj.H., 1979. Submarine thermal springs on the Galapagos Rift. *Science,* 203: 1073–1083.

Courtois, C. and Treuil, M., 1977. Distribution des terres rares et de quelques éléments en trace dans les sediments récents des fosses de la Mer Rouge. *Chem. Geol.,* 20: 57–72.

Crane, K., 1978. Structure and tectonics of the Galapagos inner rift, 86°10′W. *J. Geol.,* 86: 715–730.

Crane, K. and Normark, W.R., 1977. Hydrothermal activity and crestal structure of the East Pacific Rise at 21°N. *J. Geophys. Res.,* 82: 5336–5348.

Creager, J.S. and Scholl, D.W. et al., 1973. *Initial Reports of the Deep Sea Drilling Project, 19.* U.S. Govt. Printing Office, Washington, D.C., 913 pp.

Cronan, D.S., 1972a. The Mid-Atlantic Ridge near 45°N, XVII: Al, As, Hg, and Mn in ferruginous sediments from the median valley. *Can. J. Earth Sci.,* 9: 319–323.

Cronan, D.S., 1972b. Iron-rich basal sediments from the eastern equatorial Pacific. *Science,* 175: 61–63.

Cronan, D.S., 1973. Basal ferruginous sediments cored during Leg 16, Deep Sea Drilling Project. In: *Initial Reports of the Deep Sea Drilling Project, 16.* U.S. Govt. Printing Office, Washington, D.C., pp. 601–604.

Cronan, D.S., 1974. Authigenic minerals in deep-sea sediments. In: E.D. Goldberg (Editor), *The Sea, 5.* Wiley–Interscience, N.Y., pp.491–525.

Cronan, D.S., 1976a. Implications of metal dispersion from submarine hydrothermal systems for mineral exploration on mid-ocean ridges and in island arcs. *Nature (London),* 262: 567–569.

Cronan, D.S., 1976b. Basal metalliferous sediments from the eastern Pacific. *Geol. Soc. Am. Bull.,* 87: 928–934.

Cronan, D.S. and Garrett, D.E., 1973. Distribution of elements in metalliferous Pacific sediments collected during the Deep Sea Drilling Project. *Nature (London), Phys. Sci.,* 242: 88–89.

Cronan, D.S. and Thompson, B., 1978. Regional geochemical reconnaissance survey for submarine metalliferous sediments in the southwestern Pacific Ocean – a preliminary note. *Trans. Inst. Min. Metall.,* 87B: 87–89.

Cronan, D.S., Van Andel, T.H., Heath, G.R., Dinkelman, M.G., Bennett, R.H., Bukry, D., Charleston, G., Kaneps, A., Rodolfo, K.S. and Yeats, R.S., 1972. Iron rich basal sediments from the eastern equatorial Pacific: Leg 16, Deep Sea Drilling Project. *Science,* 175: 61–63.

Cronan, D.S., Damiani, V.V., Kinsman, D.J.J. and Thiede, J., 1974. Sediments from the Gulf of Aden and western Indian Ocean. In: *Initial Reports of the Deep Sea Drilling Project, 24.* U.S. Govt. Printing Office, Washington, D.C., pp.1047–1110.

Cronan, D.S., Smith, P.A. and Bignell, R.D., 1977. Modern hydrothermal mineralization examples from Santorini and the Red Sea. In: *Volcanic Processes in Ore Genesis. Spec. Publ. Geol. Soc. London,* 7: p.80.

Cruickshank, M.J., 1974. Mineral resources potential of continental margins. In: C.A. Burke and C.L. Drake (Editors), *The Geology of Continental Margins.* Springer, New York, N.Y., pp.965–1000.

Dasch, E.J., Dymond, J.R. and Heath, G.R., 1971. Isotopic analysis of metalliferous sediment from the East Pacific Rise. *Earth Planet. Sci. Lett.,* 13: 175–180.

Davies, T.A. and Luyendyk, B.P. et al., 1974. *Initial Reports of the Deep Sea Drilling Project, 26.* U.S. Govt. Printing Office, Washington, D.C., 1129 pp.

Davis, E.E. and Lister, C.R.B., 1977. Heat flow measured over the Juan de Fuca Ridge: evidence for widespread hydrothermal circulation in a highly heat transportive crust. *J. Geophys. Res.,* 82: 4845–4860.

Degens, E.T. and Kulbicki, G., 1973. Hydrothermal origin of metals in some East African Rift lakes. *Miner. Deposita,* 8: 368–404.

Degens, E.T. and Ross, D.A. (Editors), 1969. *Hot Brines and Recent Heavy Metal Deposits in the Red Sea.* Springer, New York, N.Y., 600 pp.

Dézsi, I., Keszthely, L., Kulgawczuk, D., Moinár, B. and Eissa, N.A., 1967. Mössbauer study of β- and δ-FeOOH and their disintegration products. *Phys. Status Solidi,* 22: 617–629.

Duke, N.A. and Hutchinson, R.W., 1974. Geological relationships between massive sulfide bodies and ophiolitic volcanic rocks near York Harbour, Newfoundland. *Can. J. Earth Sci.,* 11: 53–69.

Dunham, K.C., 1970. Mineralization by deep formation waters: a review. *Trans. Inst. Min. Metall.,* 79B: 127–136.

Dymond, J. and Eklund, W., 1978. A microprobe study of metalliferous sediment components. *Earth Planet. Sci. Lett.,* 40: 243–251.

Dymond, J. and Veeh, H.H., 1975. Metal accumulation rates in the Southeast Pacific and the origin of metalliferous sediments. *Earth Planet. Sci. Lett.,* 28: 13–22.

Dymond, J., Corliss, J.B., Heath, G.R., Field, C.W., Dasch, E.J. and Veeh, H.H., 1973. Origin of metalliferous sediments from the Pacific Ocean. *Geol. Soc. Am. Bull.,* 84: 3355–3372.

Dymond, J., Corliss, J.B. and Heath, G.R., 1975. Nazca Plate metalliferous sediments, II. Chemical composition and metal accumulation rates of metalliferous sediments from DSDP Site 319. *EOS Trans. Am. Geophys. Union,* 56(5): 446 (Abstr.).

Dymond, J., Corliss, J.B. and Stillinger, R., 1976. Chemical composition and metal accumulation rates of metalliferous sediments from Sites 319, 320 and 321. In: *Initial Reports of the Deep Sea Drilling Project, 34.* U.S. Govt. Printing Office, Washington, D.C., pp.575–588.

Dymond, J., Corliss, J.B. and Heath, G.R., 1977. History of metalliferous sedimentation at Deep Sea Drilling Site 319, in the South Eastern Pacific. *Geochim. Cosmochim. Acta,* 41: 741–753.

Edmond, J.M., Gordon, L.I. and Corliss, J.B., 1977. Chemistry of the hot springs on the Galapagos Ridge axis. *EOS Trans. Am. Geophys. Union,* 58(12): 1176 (Abstr.).

Elderfield, H., 1972. Effects of volcanism in water chemistry, Deception Island, Antarctica. *Mar. Geol.,* 13: M1–M6.

Elderfield, H., Gunnlaugsson, E., Wakefield, S.J. and Williams, P.T., 1977. The geochemistry of basalt–sea water interactions: evidence from Deception Island, Antarctica, and Reykjanes, Iceland, *Mineral. Mag.,* 41: 217–226.

El Wakeel, S.K. and Riley, J.P., 1961. Chemical and mineralogical studies of deep-sea sediments. *Geochim. Cosmochim. Acta,* 25: 110–147.

Emelyanov, E.M., 1977. Geochemistry of sediments in the western central Atlantic, DSDP Leg 39. In: *Initial Reports of the Deep Sea Drilling Project, 39.* U.S. Govt. Printing Office, Washington, D.C., pp.477–492.

Emery, K.O. and Skinner, B.J., 1977. Mineral deposits of the deep-ocean floor. *Mar. Min.,* 1: 1–71.

Fehn, U., Siegel, M.D., Robinson, G.R., Holland, H.D., Williams, D.L., Erickson, A.J. and Green, K.E., 1977. Deep-water temperatures in the FAMOUS area. *Geol. Soc. Am. Bull.,* 88: 488–494.

Ferguson, J. and Lambert, I.B., 1972. Volcanic exhalations and metal enrichments at Matupi Harbor, New Britain, T.P.N.G. *Econ. Geol.,* 67: 25–37.

Field, C.W., Dymond, J.R., Corliss, J.B., Dasch, E.J., Heath, G.R., Senechal, R.G. and Veeh, H.H., 1976. Metallogenesis in the southeast Pacific Ocean: Nazca Plate project. *Am. Assoc. Pet. Geol., Mem.* No. 25, 539–550.

Fischer, A.G., Arthur, M.A., Herb, R. and Silva, I.P., 1977. Middle Cretaceous events. *Geotimes,* 22(4): 18–19.

Fisher, D.E. and Boström, K., 1969. Uranium rich sediments on the East Pacific Rise. *Nature (London),* 224: 64–65.

Fisher, R.L. and Bunce, E.T. et al., 1974. *Initial Reports of the Deep Sea Drilling Project, 24.* U.S. Govt. Printing Office, Washington, D.C., 1183 pp.

Francheteau, J., Needham, H.D., Choukroune, P., Juteau, T., Séguret, M., Ballard, R.D., Fox, P.J., Normark, W., Carranza, A., Cordoba, D., Guerroro, J., Rargin, C., Bougault, H., Cambon, P. and Hekinian, R., 1979. Massive deep-sea sulphide ore deposits discovered on the East Pacific Rise. *Nature (London),* 277: 523–528.

Frank, D.J., Meylan, M.A., Craig, J.D. and Glasby, G.P., 1976. Ferromanganese deposits of the Hawaiian Archipelago. *Hawaii Inst. Geophys. Rep.,* HIG-76-14: 71 pp.

Froelich, P.N., Bender, M.L. and Heath, G.R., 1977. Phosphorus accumulation rates in metalliferous sediments on the East Pacific Rise. *Earth Planet. Sci. Lett.,* 34: 351–359.

Fryer, B.J. and Hutchinson, R.W., 1976. Generation of metal deposits on the sea floor. *Can. J. Earth Sci.,* 13: 126–135.

Garner, D.M. and Ford, W.L., 1969. Mid-Atlantic Ridge near 45°N, IV. Water properties in the median valley. *Can. J. Earth Sci.*, 6: 1359–1363.

Garson, M.S. and Krs, M., 1976. Geophysical and geological evidence of the relationship of Red Sea transverse tectonics to ancient fractures. *Geol. Soc. Am. Bull.*, 87: 169–181.

Gass, I.G., 1977. The age and extent of the Red Sea oceanic crust. *Nature (London)*, 265: 722–723.

Giggenbach, W.F. and Glasby, G.P., 1977. The influence of thermal activity on the trace metal distribution in marine sediments around White Island, New Zealand. *N.Z. Dep. Sci. Ind. Res. Bull.*, 218: 121–126.

Glasby, G.P., 1976. Manganese nodules in the South Pacific: a review. *N.Z. J. Geol. Geophys.*, 19: 707–736.

Glasby, G.P. and Lawrence, P., 1974. Metalliferous sediments, submarine volcanism and submarine geothermal activity in the South Pacific Ocean. *N.Z. Oceanogr. Inst. Chart, Misc. Ser.*, 39.

Glasby, G.P., Hunt, J.L., Rankin, P.C. and Darwin, J.H., 1979. Major element analyses of marine sediments from the Southwest Pacific. *N.Z. Soil Bur. Rep.*, 36: 127 pp.

Glasby, G.P., Meylan, M.A., Margolis, S.V. and Bäcker, H., in press. Manganese deposits of the Southwestern Pacific Basin. *Proc. 25th Int. Geol. Congr.*

Goldberg, E.D. and Arrhenius, G.O.S., 1958. Chemistry of Pacific pelagic sediments. *Geochim. Cosmochim. Acta*, 13: 153–212.

Graf, J.L., 1977. Rare earth elements as hydrothermal tracers during the formation of massive sulfide deposits in volcanic rocks. *Econ. Geol.*, 72: 527–548.

Green, D.C., Hulston, J.R. and Crick, I.H., 1978. Stable isotope and chemical studies of volcanic exhalations and thermal waters, Rabaul caldera, New Britain, Papua New Guinea. *BMR J. Aust. Geol. Geophys.*, 3: 233–239.

Greenblatt, S. and King, F.T., 1969. Mössbauer spectra of some magnetic iron hydroxides precipitated by porous silica. *J. Appl. Phys.*, 40: 4498–4500.

Gurvich, Ye.C., Bogdanov, Yu.A., Kurinov, A.D. and Katargin, N.V., 1976. Antimony in the metalliferous sediments of the Pacific Ocean. *Oceanology*, 16: 279–283.

Hackett, J.P. and Bischoff, J.L., 1973. New data on the stratigraphy, extent, and geologic history of the Red Sea geothermal deposits. *Econ. Geol.*, 68: 553–564.

Halbach, P., Rehm, E. and Marchig, V., 1979. Distribution of Si, Mn, Fe, Ni, Cu, Co, Zn, Pb, Mg, and Ca in grain-size fractions of sediment samples from a manganese nodule field in the central Pacific Ocean. *Mar. Geol.*, 29: 237–252.

Hammond, A.L., 1975a. Minerals and plate tectonics: a conceptual revolution. *Science*, 189: 779–781.

Hammond, A.L., 1975b. Minerals and plate tectonics (II): seawater and ore formation. *Science*, 189: 868–869, 915, 917.

Hartmann, M., 1973. Untersuchungen von suspensiertem Material in den Hydrothermallaugen des Atlantis-II-Tiefs. *Geol. Rundsch.*, 62: 742–754.

Hays, J.D. et al., 1972. An interpretation of the geologic history of the eastern equatorial Pacific from the drilling results of *Glomar Challenger*, Leg 9. In: *Initial Reports of the Deep Sea Drilling Project, 9.* U.S. Govt. Printing Office, Washington, D.C., pp.909–931.

Heath, G.R., 1969. Mineralogy of Cenozoic deep sea sediments from the equatorial Pacific Ocean. *Geol. Soc. Am. Bull.*, 80: 1997–2018.

Heath, G.R. and Dymond, J., 1977. Genesis and transformation of metalliferous sediments from the East Pacific Rise, Bauer Deep, and Central Basin, northwest Nazca Plate. *Geol. Soc. Am. Bull.*, 88: 723–733.

Heath, G.R. and Moberly, R., 1971. Noncalcareous pelagic sediments from the western Pacific, Leg 7, Deep Sea Drilling Project. In: *Initial Reports of the Deep Sea Drilling Project, 7.* U.S. Govt. Printing Office, Washington, D.C., pp.987–990.

Heath, G.R., Dymond, J. and Eklund, W.A., 1975. Nazca Plate metalliferous sediments, IV. Partitioning of transition metals amongst mineral phases in metalliferous sediments from the Bauer Deep and adjacent East Pacific Rise. *EOS Trans. Am. Geophys. Union*, 56(6): 446 (Abstr.).

Hein, J.R., Yeh, H.-W. and Alexander, E.R., 1979. Distribution, mineralogy, chemistry and oxygen isotopes of clay minerals from the north equatorial Pacific manganese nodule belt. *Proc. 6th Int. Clay Conf.*, Oxford.

Heirtzler, J.R., 1975. Where the earth turns inside out. *Natl. Geogr.*, 147: 586–603.

Heirtzler, J.R. and Van Andel, Tj.H., 1977. Project FAMOUS: Its origin, programs, and setting. *Geol. Soc. Am. Bull.*, 88: 481–487.

Hekinian, R., Rosendahl, B.R., Cronan, D.S., Dmitriev, Y., Fodor, R.V., Goll, R.M., Hoffert, M., Humphris, S.E., Mattey, D.P., Natland, J., Petersen, N., Roggenthen, W., Schrader, E.L., Srivastava, R.K. and Warren, N., 1978. Hydrothermal deposits and associated basement rocks from the Galapagos spreading center. *Oceanol. Acta,* 1: 473–482.

Herman, Y. and Rosenberg, P.E., 1969. Mineralogy and micropaleontology of a goethite-bearing Red Sea core. In: E.T. Degens and D.A. Ross (Editors), *Hot Brines and Recent Heavy Metal Deposits in the Red Sea.* Springer, New York, N.Y., pp.448–459.

Hoffert, M., Perseil, A., Hékinian, R., Choukroune, P., Needham, H.D., Francheteau, J. and Le Pichon, X., 1978. Hydrothermal deposits sampled by diving saucer in transform fault "A" near 37°N in the Mid-Atlantic Ridge, FAMOUS area. *Oceanol. Acta,* 1: 73–86.

Hollister, C.D. and Ewing, J.I. et al., 1972. *Initial Reports of the Deep Sea Drilling Project, 11.* U.S. Govt. Printing Office, Washington, D.C., 1077 pp.

Horn, D.R., Horn, B.M. and Delach, M.N., 1972. Ferromanganese deposits of the North Pacific. *Tech. Rept. No. 1, NSF GX-33616.* IDOE-NSF, Washington, D.C., 78 pp.

Horowitz, A., 1970. The distribution of Pb, Ag, Sn, Ti and Zn in sediments on active oceanic ridges. *Mar. Geol.,* 9: 241–259.

Horowitz, A. and Cronan, D.S., 1976. The geochemistry of basal sediments from the North Atlantic Ocean. *Mar. Geol.,* 20: 205–228.

Hrynkiewicz, A.Z., Pustówka, A.J., Sawicka, B.D. and Sawicki, J.A., 1972. Mössbauer effect analysis of Fe–Mn nodules from various Pacific Ocean locations. *Phys. Status Solidi,* 10: 281–287.

Humphris, S.E. and Thompson, G., 1976. Seawater and the formation of ores. *Oceanus,* 19(4): 40–44.

Humphris, S.E. and Thompson, G., 1978a. Hydrothermal alteration of oceanic basalts by seawater. *Geochim. Cosmochim. Acta,* 42: 107–125.

Humphris, S.E. and Thompson, G., 1978b. Trace element mobility during hydrothermal alteration of oceanic basalts. *Geochim. Cosmochim. Acta,* 42: 127–136.

Jenkins, W.J., Edmond, J.M. and Corliss, J.B., 1978. Excess ^{3}He and ^{4}He in Galapagos submarine hydrothermal waters. *Nature (London),* 272: 156–158.

Johnston, J.H. and Glasby, G.P., 1978. The secondary iron oxidehydroxide mineralogy of some deep sea and fossil manganese nodules: a Mössbauer and X-ray study. *Geochem. J.,* 12: 153–164.

Kaplan, I.R., Sweeney, R.E. and Nissenbaum, A., 1969. Sulfur isotope studies on Red Sea geothermal brines and sediments. In: E.T. Degens and D.A. Ross (Editors), *Hot Brines and Recent Heavy Metal Deposits in the Red Sea.* Springer, New York, N.Y., pp. 474–498.

Karig, D.E. and Ingle, J.C. Jr. et al., 1975. *Initial Reports of the Deep Sea Drilling Project, 31.* U.S. Govt. Printing Office, Washington, D.C., 927 pp.

Kauffman, K. and Hazel, F., 1975. Infrared and Mössbauer spectroscopy, electron microscopy and chemical reactivity of ferric chloride hydrolysis products. *J. Inorg. Nucl. Chem.,* 37: 1139–1148.

Kempe, D.R.C. and Easton, A.J., 1974. Metasomatic garnets in calcite (micarb) chalk at Site 251, southwest Indian Ocean. In: *Initial Reports of the Deep Sea Drilling Project, 26.* U.S. Govt. Printing Office, Washington, D.C., pp.593–601.

Kennedy, P. and Roser, B., in preparation. X-ray fluorescence analysis of trace elements in rocks and soils.

Kennett, J.P. and Houtz, R.E. et al., 1974. *Initial Reports of the Deep Sea Drilling Project, 29.* U.S. Govt. Printing Office, Washington, D.C., 1197 pp.

Klinkhammer, G., Bender, M. and Weiss, R.F., 1977. Hydrothermal manganese in the Galapagos Rift. *Nature (London),* 269: 319–320.

Klitgord, K.D. and Mudie, J.D., 1974. The Galapagos spreading centre: a near-bottom geophysical survey. *Geophys. J.R. Astron. Soc.*, 38: 563–586.

Knedler, K.E., Cronan, D.S.,Glasby, G.P., Collen, J.D., Halunen A.J., Johnston, J.H., Rankin, P.C. and Wingfield, R.T.R., in preparation. Mineralogy and geochemistry of sediments from the Yasawa Trough, N.W. of Viti Levu, Fiji. *South Pac. Mar. Geol. Notes.*

Kündig, W., Bömmell, H., Constabaris, G. and Lindquist, R.H., 1966. Some properties of supported small α-Fe_2O_3 particles determined with the Mössbauer effect. *Phys. Rev.*, 142: 327–333.

Lancelot, Y., Hathaway, J.C. and Hollister, C.D., 1972. Lithology of sediments from the western North Atlantic; Leg 11, Deep Sea Drilling Project. In: *Initial Reports of the Deep Sea Drilling Project, 11.* U.S. Govt. Printing Office, Washington, D.C., pp.901–949.

Larson, R.L. and Moberly, R. et al., 1975. *Initial Reports of the Deep Sea Drilling Project, 32.* U.S. Govt. Printing Office, Washington, D.C., 980 pp.

Laughton, A.S. and Berggren, W.A. et al., 1972. *Initial Reports of the Deep Sea Drilling Project, 12.* U.S. Govt. Printing Office, Washington, D.C., 1243 pp.

Leclaire, L., 1974. Late Cretaceous and Cenozoic pelagic deposits – paleoenvironment and paleo-oceanography of the central western Indian Ocean. In: *Initial Reports of the Deep Sea Drilling Project, 25.* U.S. Govt. Printing Office, Washington, D.C., pp.481–513.

Leinen, M. and Stakes, D., 1979. Metal accumulation rates in the central equatorial Pacific during Cenozoic time. *Geol. Soc. Am. Bull.*, 90: 357–375.

Lister, C.R.B., 1972. On the thermal balance of a mid-ocean ridge. *Geophys. J.R. Astron. Soc.*, 26: 515–535.

Lister, C.R.B., 1974. Water percolation in the oceanic crest. *EOS Trans. Am. Geophys. Union*, 55: 740–742.

Lonsdale, P., 1977a. Deep-tow observations at the Mounds Abyssal Hydrothermal Field, Galapagos Rift. *Earth Planet. Sci. Lett.*, 36: 92–110.

Lonsdale, P., 1977b. Clustering of suspension – feeding macrobenthos near abyssal hydrothermal vents at oceanic spreading centers. *Deep-Sea Res.*, 24: 857–863.

Lopez, C. and Corliss, J., 1976. On the distribution of major and trace elements in the individual phases constituting distal metalliferous sediments of the Nazca Plate. *EOS Trans. Am. Geophys. Union*, 57(4): 269 (Abstr.).

Ludden, J.N. and Thompson, G., 1978. Behavior of rare earth elements during submarine weathering of tholeiitic basalt. *Nature (London)*, 274: 147–149.

Lupton, J.E., Weiss, R.F. and Craig, H., 1977a. Mantle helium in hydrothermal plumes in the Galapagos Rift. *Nature (London)*, 267: 603–604.

Lupton, J.E., Weiss, R.F. and Craig, H., 1977b. Mantle helium in the Red Sea brines. *Nature (London)*, 266: 244–246.

Luyendyk, B.P. and Macdonald, K.C., 1977. Physiography and structure of the inner floor of the FAMOUS rift valley: observations with a deep-towed instrument package. *Geol. Soc. Am. Bull.*, 88: 648–663.

Lyle, M., Dymond, J. and Heath, G.R., 1977. Copper–nickel-enriched ferromanganese nodules and associated crusts from the Bauer Basin, northwest Nazca Plate. *Earth Planet. Sci. Lett.*, 35: 55–64.

Lynn, D.C. and Bonatti, E., 1965. Mobility of manganese in diagenesis of deep-sea sediment. *Mar. Geol.*, 3: 457–474.

Macdonald, K.C. and Luyendyk, B.P., 1977. Deep-tow studies of the structure of the Mid-Atlantic Ridge crest near lat 37°N. *Geol. Soc. Am. Bull.*, 88: 621–636.

Macdonald, K., Luyendyk, B.P., Mudie, J.D. and Spiess, F.N., 1975. Near bottom geophysical study of the Mid-Atlantic Ridge median valley near lat 37°N: preliminary observations. *Geology*, 3: 211–215.

MacEwan, D.M.C., 1961. Montmorillonite minerals. In: G. Brown (Editor), *The X-ray Identification and Crystal Structure of Clay Minerals.* Mineralogical Society, London, pp.143–207.

Manheim, F.T., 1975. Geological and geochemical significance of Red Sea evaporites. *Proc. Offshore Technol. Conf.*, Houston, Pap. OTC 2391: 545–548.

Marchig, V., 1978. Brown clays from the central Pacific – metalliferous sediments or not? *Geol. Jahrb., Reihe D,* 30: 3–25.

Mathalone, Z., Ron, M. and Biran, A., 1970. Magnetic ordering in iron gel. *Solid State Commun.,* 8: 333–336.

Maxwell, A.E. et al., 1970. *Initial Reports of the Deep Sea Drilling Project, 3.* U.S. Govt. Printing Office, Washington, D.C., 806 pp.

McArthur, J.M. and Elderfield, H., 1977. Metal accumulation rates in sediments from Mid-Indian Ocean Ridge and Marie Celeste Fracture Zone. *Nature (London),* 266: 437–439.

McGregor, B.A. and Rona, P.A., 1975. Crest of Mid-Atlantic Ridge at 26 N. *J. Geophys. Res.,* 80: 3307–3314.

McManus, D.A. et al., 1970. *Initial Reports of the Deep Sea Drilling Project, 5.* U.S. Govt. Printing Office, Washington, D.C., 827 pp.

McMurtry, G.M., in press. Metallogenesis on oceanic plates: the East Pacific Rise and Bauer Basin. *CPEMRC Mem.* 2. Tulsa, Okla., AAPG.

McMurtry, G.M. and Burnett, W.C., 1975. Hydrothermal metallogenesis in the Bauer Deep of the southeastern Pacific. *Nature (London),* 254: 42–44.

McMurtry, G.M., Veeh, H.H. and Moser, C., in press. Sediment accumulation rate patterns on the Nazca Plate. *Nazca Plate Mem., Geol. Soc. Am.*

Melson, W.G. and Rabinowitz, P.D. et al., 1978. *Initial Reports of the Deep Sea Drilling Project, 45.* U.S. Govt. Printing Office, Washington, D.C., 717 pp.

Meylan, M.A., Bäcker, H. and Glasby, G.P., 1975. Manganese nodule investigations in the Southwestern Pacific Basin, 1974. *N.Z.O.I. Oceanogr. Field. Rep.* 4, 24 pp.

Meylan, M.A., Glasby, G.P., McDougall, J.C. and Kumbalek, S.C., in preparation. Distribution, mineralogy and geochemistry of marine sediments from the Southwestern Pacific and Samoan Basins.

Miller, A.R., Densmore, C.D., Degens, E.T., Hathaway, J.C., Manheim, F.T., McFarlin, P.F., Pocklington, R. and Jockela, A., 1966. Hot brines and recent iron deposits in deeps of the Red Sea. *Geochim. Cosmochim. Acta,* 30: 341–359.

Moore, W.S. and Vogt, P.R., 1976. Hydrothermal manganese crusts from two sites near the Galapagos spreading axis. *Earth Planet. Sci. Lett.,* 29: 349–356.

Mossman, D.J. and Heffernan, K.J., 1978. On the possible primary precipitation of atacamite and other metal chlorides in certain stratabound deposits. *Chem. Geol.,* 21: 151–159.

Mroczek, E., 1977. *Computer Fitting of Mössbauer Spectra.* Thesis, Victoria University of Wellington, Wellington, N.Z. (unpublished).

Muehlenbachs, K. and Clayton, R.N., 1976. Oxygen isotope composition of the oceanic crust and its bearing on seawater. *J. Geophys. Res.,* 81: 4365–4369.

Müller, G. and Förstner, U., 1973. Recent iron ore formation in Lake Malawi, Africa. *Miner. Deposita,* 8: 278–290.

Murray, J.W., 1978. β-FeOOH in marine sediments. *Trans. Am. Geophys. Union,* 59: 411–412 (Abstr.).

Murray, J. and Renard, A., 1891. *Deep-Sea Deposits. Report of the Scientific Results of the Voyage of the HMS Challenger, 5.* Eyre and Spottiswoode, London, 525 pp.

Natland, J.H., 1973. Basal ferromanganoan sediments at D.S.D.P. Site 183, Aleutian Abyssal Plain, and Site 192, Meiji Guyot, northwest Pacific, Leg 19. In: *Initial Reports of the Deep Sea Drilling Project, 19.* U.S. Govt. Printing Office, Washington, D.C., pp.629–641.

Natland, J.H., Rosendahl, B., Hekinian, R., Dmitriev, Y., Fodor, R.V., Goll, R.M., Hoffert, M., Humphris, S.E., Mattey, D.P., Peterson, N., Roggenthen, W., Schrader, E.L., Srivastava, R.K. and Warren, N., 1979. Galápagos hydrothermal mounds: stratigraphy and chemistry revealed by deep-sea drilling. *Science,* 204: 613–616.

Nayudu, Y.R., 1971. Lithology and chemistry of surface sediments in subantarctic regions of the Pacific Ocean. *Antarct. Res. Ser.,* 15: 247–282.

Needham, H.D. and Francheteau, J., 1974. Some characteristics of the rift valley in the Atlantic Ocean near 36°48′N. *Earth Planet. Sci. Lett.,* 22: 29–43.

Normark, W.R., 1976. Delineation of the main extrusion zone of East Pacific Rise at lat 21°N. *Geology, 4*: 681–685.

Norrish, K. and Hutton, J.T., 1969. An accurate X-ray spectrographic method for the analysis of a wide range of geological samples. *Geochim. Cosmochim. Acta,* 33: 431–453.

Nriagu, J.O. and Anderson, G.M., 1970. Calculated solubilities of some base-metal sulphides in brine solutions. *Trans. Inst. Min. Metall.,* 79B: 208–212.

Ohmoto, H., 1978. Submarine calderas: a key to the formation of massive sulfide deposits. *Econ. Geol.,* 73: 312–313 (Abstr.).

O'Nions, R.K., Carter, S.R., Cohen, R.S., Evensen, N.M. and Hamilton, P.J., 1978. Pb, Nd and Sr isotopes in oceanic ferromanganese deposits and ocean floor basalts. *Nature (London),* 273: 435–438.

Parmentier, E.M. and Spooner, E.T.C., 1978. A theoretical study of hydrothermal convection and the origin of the ophiolitic sulphide ore deposits of Cyprus. *Earth Planet. Sci. Lett.,* 40: 33–44.

Peterson, M.N.A. and Goldberg, E.D., 1962. Feldspar distributions in South Pacific pelagic sediments. *J. Geophys. Res.,* 67: 3477–3492.

Peterson, M.N.A., Edgar, N.T., Von der Borch, C.C. and Rex, R.W., 1970. Cruise leg summary and discussion. In: *Initial Reports of the Deep Sea Drilling Project, 2.* U.S. Govt. Printing Office, Washington, D.C., pp.413–427.

Peterson, M.N.A. et al., 1970. *Initial Reports of the Deep Sea Drilling Project, 2.* U.S. Govt. Printing Office, Washington, D.C., 499 pp.

Pimm, A.C., 1974a. Mineralization and trace element variation in deep-sea pelagic sediments of the Wharton Basin, Indian Ocean. In: *Initial Reports of the Deep Sea Drilling Project, 22.* U.S. Govt. Printing Office, Washington, D.C., pp. 469–476.

Pimm, A.C., 1974b. Sedimentology and history of the Northeastern Indian Ocean from Late Cretaceous to Recent. In: *Initial Reports of the Deep Sea Drilling Project, 22.* U.S. Govt. Printing Office, Washington, D.C., pp.717–803.

Piper, D.Z., 1973. Origin of metalliferous sediments from the East Pacific Rise. *Earth Planet. Sci. Lett.,* 19: 75–82.

Piper, D.Z., 1974. Rare earth elements in the sedimentary cycle: a summary. *Earth Planet. Sci. Lett.,* 14: 285–304.

Piper, D.Z., Veeh, H.H., Bertrand, W.G. and Chase, R.L., 1975. An iron-rich deposit from the Northeast Pacific. *Earth Planet. Sci. Lett.,* 26: 114–120.

Puchelt, H., Schock, H.H., Schroll, E. and Hanert, H., 1973. Recent marine iron ores off Thera, Greece. I. Geochemistry, genesis, mineralogy. II. Bacterial genesis of iron hydroxide sediments. *Geol. Rundsch.,* 62: 786–812.

Raab, W., 1972. Physical and chemical features of the Pacific deep sea manganese nodules and their implications to the genesis of nodules. In: *Ferromanganese Deposits on the Ocean Floor.* Lamont-Doherty Geological Observatory and IDOE-NSF, pp.31–50.

Ramberg, I.B. and Van Andel, Tj.H., 1977. Morphology and tectonic evolution of the rift valley at lat 36°30′N, Mid-Atlantic Ridge. *Geol. Soc. Am. Bull.,* 88: 577–586.

Ramberg, I.B., Gray, D.F. and Raynolds, R.G.H., 1977. Tectonic evolution of the FAMOUS area of the Mid-Atlantic Ridge, lat 35°50′ to 37°20′N. *Geol. Soc. Am. Bull.,* 88: 609–620.

Revelle, R.R., 1944. Scientific results of cruise VII of the CARNEGIE during 1928–1929 under command of Captain J.P. Ault. Oceanography-II. 1. Marine bottom samples collected in the Pacific Ocean by the Carnegie on its seventh cruise. *Carnegie Inst. Washington Publ.,* 556: 1–180.

Rex, R.W., 1970. X-ray mineralogy studies, Leg 3. In: *Initial Reports of the Deep Sea Drilling Project, 3.* U.S. Govt. Printing Office, Washington, D.C., pp.509–581.

Riley, J.P. and Chester, R., 1971. *Introduction to Marine Chemistry.* Academic Press, London, 465 pp.

Robertson, A.H.F., 1975. Cyprus umbers: basalt–sediment relationships on a Mesozoic ocean ridge. *J. Geol. Soc. London,* 131: 511–531.

Robertson, A.H.F. and Fleet, A.J., 1976. The origins of rare earths in metalliferous sediments of the Troodos Massif, Cyprus. *Earth Planet. Sci. Lett.,* 28: 385–394.

Robinson, B.W., 1977. Isotopic equilibrium in hydrothermal barites. *N.Z. Dep. Sci. Ind. Res. Bull.,* 218: 51–56.

Rona, P.A., 1976a. Hydrothermal manganese deposits of Mid-Atlantic Ridge crest (latitude 26°N). *Abstr. 25th Intl. Geol. Congr.,* 2: 355 (Abstr.).
Rona, P.A., 1976b. Pattern of hydrothermal mineral deposition: Mid-Atlantic Ridge crest at latitude 26°N. *Mar. Geol.,* 21: M59–M66.
Rona, P.A., 1977. Plate tectonics, energy and mineral resources: basic research leading to payoff. *EOS Trans. Am. Geophys. Union,* 58: 629–639.
Rona, P.A., 1978. Criteria for recognition of hydrothermal mineral deposits in oceanic crust. *Econ. Geol.,* 73: 135–160.
Rona, P.A., McGregor, B.A., Betzer, P.R., Bolger, G.W. and Krause, D.C., 1975. Anomalous water temperatures over Mid-Atlantic Ridge crest at 26° North latitude. *Deep-Sea Res.,* 22: 611–618.
Rona, P.A., Harbison, R.N., Bassinger, B.G., Scott, R.B. and Nalwalk, A.J., 1976. Tectonic fabric and hydrothermal activity of Mid-Atlantic Ridge crest (Lat. 26°N). *Geol. Soc. Am. Bull.,* 87: 661–674.
Rosato, V.J., Kulm, L.D. and Derks, P.S., 1975. Surface sediments of the Nazca Plate. *Pacific Sci.,* 29: 117–130.
Ross, D.A., Hays, E.E. and Allstrom, F.C., 1969. Bathymetry and continuous seismic profiles of the hot brine region of the Red Sea. In: E.T. Degens and D.A. Ross (Editors), *Hot Brines and Recent Heavy Metal Deposits in the Red Sea.* Springer, New York, N.Y., pp. 82–97.
Ross, D.A., Whitmarsh, R.B., Ali, S.A., Bordreaux, J.E., Coleman, R., Fleisher, R.L., Girdler, R., Manheim, F., Matter, A., Nigrini, C., Stoffers, P. and Supko, P.R., 1973. Red Sea drillings. *Science,* 179: 377–380.
Rozanova, T.V. and Baturin, G.N., 1971. Hydrothermal ore shows on the floor of the Indian Ocean. *Oceanology,* 11: 874–879.
Rozenson, I. and Heller-Kallai, L., 1977. Mössbauer spectra of dioctahedral smectites. *Clays Clay Miner.,* 25: 94–101.
Rydell, H., Kraemer, T., Boström, K. and Joensuu, O., 1974. Postdepositional injection of uranium-rich solutions into East Pacific Rise sediments. *Mar. Geol.,* 17: 151–164.
Sayles, F.L. and Bischoff, J.L., 1973. Ferromanganoan sediments in the equatorial East Pacific. *Earth Planet. Sci. Lett.,* 19: 330–336.
Sayles, F.L., Ku, T.-L. and Bowker, P.C., 1975. Chemistry of ferromanganoan sediment of the Bauer Deep. *Geol. Soc. Am. Bull.,* 86: 1423–1431.
Sayles, F.L., Ku, T.-L. and Bowker, P.C., 1976. Elemental accumulation rates in the Bauer Deep: a correction. *Geol. Soc. Am. Bull.,* 87: 1396.
Schilling, J.G., Anderson, R.N. and Vogt, P., 1976. Rare earth, Fe, and Ti variations along the Galapagos spreading center, and their relationship to the Galapagos mantle plume. *Nature (London),* 261: 108–113.
Schoell, M. and Bignell, R.D., 1976. The Red Sea brines and metalliferous sediments; their genesis in time and space. Paper presented to *Int. Geol. Congr.,* Sydney, Section 108.1.
Schoell, M. and Faber, E., 1978. New isotopic evidence for the origin of Red Sea brines. *Nature (London),* 275: 436–438.
Schwertmann, U., 1966. Die Bilding von Goethit und Hämatit in Böden und Sedimenten. In: *Proceedings of the International Clay Conference, 1966, 1.* pp.159–165.
Schwertmann, U. and Fischer, W.R., 1973. Natural "amorphous" ferric hydroxide. *Geoderma,* 10: 237–247.
Schwertmann, U. and Taylor, R.M., 1972a. The transformation of lepidocrocite to goethite. *Clays Clay Miner.,* 20: 151–158.
Schwertmann, U. and Taylor, R.M., 1972b. The influence of silicate on the transformation of lepidocrocite to goethite. *Clays Clay Miner.,* 20: 159–164.
Scientific Party, D.S.D.P. Leg 48, 1976. Glomar Challenger sails on Leg 48. *Geotimes,* 21(12): 19–23.
Scientific Party, D.S.D.P. Leg 54, 1977. Glomar Challenger completes 54th cruise. *Geotimes,* 22(11): 19–23.
Sclater, J.G., Von Herzen, R.P., Williams, D.L., Anderson, R.N. and Klitgord, K., 1974. The Galapagos spreading centre: heat-flow low on the north flank. *Geophys. J.R. Astron. Soc.,* 38: 609–637.

Scott, M.R. and Salter, P.F., 1977. Metal accumulation rates in sediments from the FAMOUS area on the Mid-Atlantic Ridge. *EOS Trans. Am. Geophys. Union,* 58(6): 420 (Abstr.).

Scott, M.R., Scott, R.B., Rona, P.A., Butler, L.W. and Nalwalk, A.J., 1974. Rapidly accumulating manganese deposit from the median valley of the Mid-Atlantic Ridge. *Geophys. Res. Lett.,* 1: 355–358.

Scott, M.R., Scott, R.B., Betzer, P.R., Butler, L.W. and Rona, P.A., 1978. Metal-enriched sediments from the TAG hydrothermal field. *Nature (London),* 276: 811–813.

Scott, R.B., Rona, P.A., McGregor, B.A. and Scott, M.R., 1974. The TAG hydrothermal field. *Nature (London),* 251: 301–302.

Scott, R.B., Malpas, J., Rona, P.A. and Udintsev, G., 1976. Duration of hydrothermal activity at an oceanic spreading center, Mid-Atlantic Ridge (lat 26°N). *Geology,* 4: 233–236.

Seyfried, W. and Bischoff, J.L., 1977. Hydrothermal transport of heavy metals by seawater: the role of seawater/basalt ratio. *Earth Planet. Sci. Lett.,* 34: 71–77.

Shanks, W.C. and Bischoff, J.L., 1977. Ore transport and deposition in the Red Sea geothermal system: a geochemical model. *Geochim. Cosmochim. Acta,* 41: 1507–1519.

Shinjo, T., 1966. Mössbauer effect in antiferromagnetic fine particles. *J. Phys. Soc. Jpn.,* 21: 917–922.

Siesser, W.G., 1976. Native copper in DSDP sediment cores from the Angola Basin. *Nature (London),* 263: 308–309.

Sillitoe, R.H., 1972. Formation of certain massive sulphide deposits at sites of seafloor spreading. *Trans. Inst. Min. Metall.,* 81: B141–148.

Simpson, E.S.W. and Schlich, R. et al., 1974. *Initial Reports of the Deep Sea Drilling Project, 25.* U.S. Govt. Printing Office, Washington, D.C., 1184 pp.

Skornyakova, I.S., 1965. Dispersed iron and manganese in Pacific Ocean sediments. *Int. Geol. Rev.,* 7: 2161–2174.

Sleep, N.H. and Wolery, T.J., 1978. Egress of hot water from midocean ridge hydrothermal systems: some thermal constraints. *J. Geophys. Res.,* 83: 5913–5922.

Smith, P.A. and Cronan, D.S., 1975a. Chemical composition of Aegean Sea sediments. *Mar. Geol.,* 18: M7–M11.

Smith, P.A. and Cronan, D.S., 1975b. The dispersion of metals associated with an active submarine exhalative deposit. *Proc. Third Oceanology Int.,* pp.111–114.

Spooner, E.T.C., Beckinsale, R.D., England, P.C. and Senior, A., 1977a. Hydration, ^{18}O enrichment and oxidation during ocean floor hydrothermal metamorphism of ophiolitic metabasic rocks from E. Liguria, Italy. *Geochim. Cosmochim. Acta,* 41: 857–871.

Spooner, E.T.C., Chapman, H.J. and Smewing, J.D., 1977b. Strontium isotopic contamination and oxidation during ocean floor hydrothermal metamorphism of the ophiolitic rocks of the Troodos Massif, Cyprus. *Geochim. Cosmochim. Acta,* 41: 873–890.

Stanaway, K.J., Kobe, H.W. and Sekula, J., 1978. Manganese deposits and the associated rocks of Northland and Auckland, New Zealand. *N.Z. J. Geol. Geophys.,* 21: 21–32.

Stephens, J.D. and Wittkop, R.W., 1969. Microscopic and electron beam microscopic study of sulfide minerals in Red Sea mud samples. In: E.T. Degens and D.A. Ross (Editors), *Hot Brines and Recent Heavy Metal Deposits of the Red Sea.* Springer, New York, N.Y., pp.441–447.

Stoffers, P. and Ross, D.A., 1974. Sedimentary history of the Ross Sea. In: *Initial Reports of the Deep Sea Drilling Project, 23.* U.S. Govt. Printing Office, Washington, D.C., pp.849–865.

Stone, A.J., 1967. Appendix to Bancroft, G.M., Maddock, A.G., Ong, W.K., Prince, R.H. and Stone, A.J., 1971. Mössbauer spectra of iron(III) diketone complexes. *J. Chem. Soc. (Abstr.),* 1967: 1971.

Strangway, D.W., McMahon, B.E. and Bischoff, J.L., 1969. Magnetic properties of minerals from the Red Sea thermal brines. In: E.T. Degens and D.A. Ross (Editors), *Hot Brines and Recent Heavy Metal Deposits in the Red Sea.* Springer, New York, N.Y., pp.460–473.

Supko, P.R. and Perch-Nielsen K. et al., 1977. *Initial Reports of the Deep Sea Drilling Project, 39.* U.S. Govt. Printing Office, Washington, D.C., 1139 pp.

Temple, D.G., Scott, R.B. and Rona, P.A., 1976. Geology of the hydrothermal field at 26°N, Mid-Atlantic Ridge: bottom photograph interpretations of the TAG area. *EOS Trans. Am. Geophys. Union,* 57(12): 932 (Abstr.).

Thompson, G., 1973. A geochemical study of the low-temperature interaction of sea-water and oceanic igneous rocks. *EOS Trans. Am. Geophys. Union*, 54: 1015–1019.

Tooms, J.S., 1970. Review of knowledge of metalliferous brines and related deposits. *Trans. Inst. Min. Metall.*, 78B: 116–126.

Towe, K.M. and Bradley, W.F., 1967. Mineralogical constitution of colloidal "hydrous ferric oxides". *J. Colloid Interface Sci.*, 24: 384–392.

Tracey, J.L. et al., 1971. *Initial Reports of the Deep Sea Drilling Project, 8.* U.S. Govt. Printing Office, Washington, D.C., 1037 pp.

Turekian, K.K. and Bertine, K.K., 1971. Deposition of molybdenum and uranium along the major ocean ridge systems. *Nature (London)*, 229: 250.

Turekian, K.K. and Imbrie, J., 1966. The distribution of trace elements in deep-sea sediments of the Atlantic Ocean. *Earth Planet. Sci. Lett.*, 1: 161–168.

Turekian, K.K. and Wedepohl, K.H. 1961. Distribution of elements in some major units of the earth's crust. *Geol. Soc. Am. Bull.*, 72: 175–192.

Udintsev, G.B., Litvin, V.M., Marova, N.A., Rudenko, M.V., Budanova, L.Ya. and Rona, P.A., 1977. New data on the morphostructure of the central part of the Mid-Atlantic Ridge. *Oceanology*, 17: 544–551.

Upadhyay, H.D. and Strong, D.F., 1973. Geological setting of the Betts Cove copper deposits, Newfoundland: an example of ophiolite sulfide mineralization. *Econ. Geol.*, 68: 161–167.

Van Andel, Tj.H. and G.R. Heath et al., 1973. *Initial Reports of the Deep Sea Drilling Project, 16.* U.S. Govt. Printing Office, Washington, D.C., 949 pp.

Veeh, H.H. and Boström, K., 1971. Anomalous $^{234}U/^{238}U$ on the East Pacific Rise. *Earth Planet. Sci. Lett.*, 10: 372–374.

Veevers, J.J. and Heirtzler, J.R. et al., 1974. *Initial Reports of the Deep Sea Drilling Project, 27.* U.S. Govt. Printing Office, Washington, D.C., 1060 pp.

Vidal, V.M.V., Vidal, F.V. and Isaacs, J.D., 1978. Coastal submarine hydrothermal activity off northern Baja California. *J. Geophys. Res.*, 83: 1757–1774.

Vine, J.D. and Tourtelet, E.B., 1970. Geochemistry of black shale deposits – a summary report. *Econ. Geol.*, 65: 253–272.

Von der Borch, C.C. and Rex, R.W., 1970. Amorphous iron oxide precipitates in sediments cored during Leg 5, Deep Sea Drilling Project. In: *Initial Reports of the Deep Sea Drilling Project, 5.* U.S. Govt. Printing Office. Washington, D.C., pp.541–544.

Von der Borch, C.C., Nesteroff, W.D. and Galehouse, J.S., 1971. Iron-rich sediments cored during Leg 8 of the Deep Sea Drilling Project. In: *Initial Reports of the Deep Sea Drilling Project, 8.* U.S. Govt. Printing Office, Washington, D.C., pp.829–833.

Von der Borch, C.C. and Sclater, J.G. et al., 1974. *Initial Reports of the Deep Sea Drilling Project, 22.* U.S. Govt. Printing Office, Washington, D.C., 890 pp.

Walthier, T.N. and Schatz, C.E., 1969. Economic significance of minerals deposited in the Red Sea deeps. In: E.T. Degens and D.A. Ross (Editors), *Hot Brines and Recent Heavy Metal Deposits in the Red Sea.* Springer, New York, N.Y., pp.542–549.

Warner, I.B. and Gieskes, J.M., 1974. Iron-rich basal sediments from the Indian Ocean: Site 245, Deep Sea Drilling Project. In: *Initial Reports of the Deep Sea Drilling Project, 25.* U.S. Govt. Printing Office, Washington, D.C., pp.395–403.

Weiss, R.F., 1977. Hydrothermal manganese in the deep sea: scavenging residence time and $Mn/^3He$ relationships. *Earth Plant. Sci. Lett.*, 37: 257–262.

Weiss, R.F., Lonsdale, P., Lupton, J.E., Bainbridge, A.E. and Craig, H., 1977. Hydrothermal plumes in Galapagos Rift. *Nature (London)*, 267: 600–603.

Whitmarsh, R.B., 1974. Summary of general features of Arabian Sea and Red Sea Cenozoic history based on Leg 23 cores. In: *Initial Reports of the Deep Sea Drilling Project, 23.* U.S. Govt. Printing Office, Washington, D.C., pp.1115–1123.

Whitmarsh, R.B., Weser, O.E. et al., 1974. *Initial Reports of the Deep Sea Drilling Project, 23.* U.S. Govt. Printing Office, Washington, D.C., 1180 pp.

Williams, D.L., 1976. Submarine geothermal resources. *J. Volcanol. Geothermal Res.*, 1: 85–100.

Williams, D.L., von Herzen, R.P., Sclater, J.G. and Anderson, R.N., 1974. The Galapagos spreading centre: lithospheric cooling and hydrothermal circulation. *Geophys. J.R. Astron. Soc.*, 38: 507–608.

Williams, D.L., Lee, T.-C., Von Herzen, R.P., Green, K.E. and Hobart, M.A., 1977. A geothermal study of the Mid-Atlantic Ridge near 37°N. *Geol. Soc. Am. Bull.*, 88: 531–540.

Winterer, E.L. et al., 1971. *Initial Reports of the Deep Sea Drilling Project, 7.* U.S. Govt. Printing Office, Washington, D.C., 1757 pp.

Wolery, T.J. and Sleep, N.H., 1976. Hydrothermal circulation and geochemical flux at mid-ocean ridges. *J. Geol.*, 84: 249–275.

Yeats, R.S. and Hart, S.R. et al., 1976. *Initial Reports of the Deep Sea Drilling Project, 34.* U.S. Govt. Printing Office, Washington, D.C., 814 pp.

Zelenov, K.K., 1964. Iron and manganese in exhalations of the submarine Banu Wuhu Volcano (Indonesia). *Proc. Acad. Sci. U.S.S.R., Earth Sci. Sect.*, 155: 94–96.

Zemmels, I., 1977. Sedimentary geochemical processes near the Pacific-Antarctic Ridge. *Antarct. J. U.S.*, 12(4): 78–79.

Zemmels, I., Harrold, P.J. and Cook, H.E., 1977. X-ray mineralogy data from the FAMOUS area of the mid-Atlantic Ridge – Leg 37. In: *Initial Reports of the Deep Sea Drilling Project, 37.* U.S. Govt. Printing Office, Washington, D.C., pp.895–905.

Chapter 4

RECENT HEAVY-METAL ACCUMULATION IN LIMNIC SEDIMENTS

ULRICH FÖRSTNER

INTRODUCTION

The number of investigations on heavy-metal accumulations in recent limnic sediments has increased greatly during the last decade because:

(1) The physico-chemical factors involved in the formation of *unconsolidated sediments* are more easily studied than those of *comparable deposits in the geological past.*

(2) *Limnic sediment geochemistry,* both from recent fluviatile and lacustrine environments, *can be used as a guide to the economic exploitation of mineralizations.*

(3) *Heavy metals* are among the most toxic forms of *environmental pollution,* constituting a threat both to aquatic life and to the quality of drinking water. By analyzing sediments it is possible to determine the *provenance, distribution, extent, and also the possible hazard of metal contamination.*

In areas where mineral exploration is followed by large-scale mining and processing activities – invariably accompanied by the environmental problems of mine wastes and atmospheric emissions – aspects (2) and (3) may coincide: "Both the exploration and environmental geochemist can be looking for the same type of areas, those with high metal concentrations, but obviously from a different motivation" (Allan, 1974).

Recent sedimentary environments of metal accumulation

Natural sources of metal enrichment in recent aquatic sediments are: (1) detrital minerals from magmatic, metamorphic, and sedimentary source rocks; (2) concentrations occurring within the water column, e.g., by precipitation, adsorption, and organo-metallic interactions; and (3) diagenetic transformations taking place after deposition of sediment particles originating from sources (1) and (2).

Influences from *allochthonous processes* (source 1) may affect *economical metal accumulations* both in marine and limnic environments. *Placers,* for example, occur in the marine (offshore) milieu as well as in the form of alluvial (stream) deposits (see Hails, 1976). Pelagic sediments, on the other hand, constitute typical areas of *autochthonous metal enrichment,* mainly because of the relatively small input of detrital components. The exploitable metal accumulations in the *pelagic sea* are thought to originate mainly from the *water column* (source 2), although diagenetic effects are also present (see Glasby and Read, 1976). Two major advantages in studying *limnic environments,* in particular for *diagenetic processes* (source 3), are the wider range of hydrological parameters and, as is often the case, their greater accessibility as compared with recent marine deposits.

Lake sediments

With respect to *physical processes* in the lacustrine environment, Sly (1978) has emphasized that, since lakes are essentially closed (or nearly closed) systems, and since the ratios of land drainage and lake area is often high, sediment loadings and sedimentation rates are substantially higher than in marine environments. Therefore, lake sediments are ideal records of *man's history,* in particular of more recent events, such as industrialization. In this context, it is of note that freshwater systems have been centers of important human developments ever since the earliest days of civilization and that man's impact on these systems is still growing. This is especially true with respect to artificial lakes and dams, whose global surface area has been estimated in the vicinity of 100,000 km^2 (for the year 1970; Lerman, 1978).

More recently, the *dynamic aspects of lake sedimentation* have received particular attention. Recycling of mineralized organic matter and pore-fluid transfer processes are now recognized as essential components of models devised to describe the nutrient dynamics of lakes and reservoir systems. Similarly, the study of sediment mineralogy reveals the characteristic processes taking place within such systems (B.E. Jones and Bowser, 1978). This can be exemplified both by major mineral phases, such as various carbonate species (Müller et al., 1972), and by minor components, e.g., phosphates, sulfides and Fe and Mn oxyhydrates. The latter compounds are regarded as essential *accumulative phases* for heavy metals in aquatic systems (Jenne, 1976).

The spatial and temporal occurrence of metal accumulation on a geological scale has recently been discussed by Degens and Stoffers (1976, 1977). Their concept is based on the consideration that density stratification in lakes and oceans, which generates a feedback mechanism between oxidizing and reducing conditions, is the main prerequisite for the formation of many strata-bound ore deposits in the past and the present.

In this Handbook, lacustrine examples have been described from Lake Kivu (Degens and Ross, 1976) and from freshwater ferromanganese deposits (Callender and Bowser, 1976). The present chapter gives a general review of the factors and processes affecting metal accumulations in recent limnic sediments.

MAJOR FACTORS OF METAL ACCUMULATION IN RECENT LIMNIC SEDIMENTS

The interpretation of sedimentological data as functions of the aquatic medium must take into account a number of geological, mineralogical, hydrological, and biological processes controlled both by internal and external factors: (1) *allochthonous influences,* which can be subdivided into "natural" (or "geochemical") and "civilizational" effects, and (2) *autochthonous influences* comprising the mechanisms of precipitation, sorption, enrichment in organisms and organo-metallic complexing during sedimentation, as well as the post-depositional effects of diagenesis.

These effects will be treated in detail in two major sections of the present compilation. As a general introduction to the study of metal accumulation, the "major factors of enrichment" and "geochemical background" are discussed here with respect to examples of recent and ancient lake deposits.

Metal distribution in recent lacustrine sediments

In order to evaluate the major influences controlling the distribution of trace metals in aquatic sediment samples, approximately 100 lakes situated in southern and western Australia, South America, East and South Africa, central and eastern Europe and western Asia were investigated. The following presentation is based on the data from 87 examples, which have previously been described by the author of this study and his colleagues (compilations, see Förstner, 1977a, b and 1978).

Sediment samples were collected during the period 1970–1975 [1]. Samples from the deeper lakes were taken from the center by a sediment grabber; sediment from the playa lakes and pans was usually collected about 300–500 m from the lake shores.

The pelitic fraction (<2 μm) was separated in settling tubes and dissolved by HF/ HNO_3. Fe, Mn, Zn, Cr, Ni, Cu, Li, and Sr were measured by conventional atomic absorption spectroscopy; Pb, Co, Cd, and Hg by flameless AAS techniques.

The studied lakes originated in a variety of ways (classification after Hutchinson, 1957) and are situated in areas with climates ranging from tropical humid (Amazon Basin) to arid (western Australia). They vary in area from 1 km^2 (Aci Crater Göl) to more than 30,000 km^2 (Lake Malawi), and in depth from less than 1 m (Dasht-i-Nawar) to 700 m (Lake Malawi). Some, for example the salt and clay pans of Australia, contain water only after heavy rainfalls and may be described as "dry lakes" (Langbein, 1961). The salinity of the lake waters ranges from less than 0.02 g/l (Lago Tupé, Brazil) to 450 g/l (Tuz Gölü during periods of maximum evaporation). The chemical composition of the

[1] Sediment material from Amazon lakes was provided by Dr. G. Irion and Dr. F. Reiss (Plön), from Turkish lakes by Dr. G. Irion, and sediments and analytical data from lakes in central Europe were placed at our disposal by Dr. A. Abdul-Razzak, Dr. M. Blohm, Dr. F. Wagner (former students of our institute) and by Dr. A. Hamm (Munich). The Geološki Zavod Skopje (Yugoslavia) kindly mediated the sampling on Lakes Ochrid and Presper. Sediments from Lakes Bled, Bohinj, and Moste Dam were sampled in cooperation with the Geološki Zavod Ljubljana (Slovenia). Colleagues from the Department of Natural Sciences of the University of Kabul assisted in the investigation of material from lakes in Afghanistan. The samples from lakes in Persia were collected by Mr. D. Bogumil, samples from lakes in Libya by Prof. Dr. H.J. Pachur (Berlin), material from reservoirs in South Africa by Dr. G.T.W. Wittmann (University of Pretoria) and researchers of the CSIR (Pretoria).

water also varies widely. Water with a very low salinity has a composition similar to rain water, with a predominance of sodium, chlorine, and bicarbonate ions (Gibbs, 1970); pH-values range from 4.7 to 10.0. Freshwater lakes of the moderately humid zones are characterized by the predominance of calcium and bicarbonate. With increasing salinity, magnesium and sodium contents exceed that of calcium. Further, hydrochemical differentiation has two distinct directions: the "normal" case of evaporation of river water leads to the formation of soda lakes, with a prevalence of sodium and carbonate/bicarbonate ions (Mackenzie and Garrels, 1966); examples of this are the African rift lakes (Kilham, 1971) as well as the deeper lakes in Turkey. A different evolution is found in the semi-arid playa lakes, which are subject to seasonal flooding and drying up; high concentrations of magnesium seem typical for this type of lake.

The mineralogical composition of the lake sediments reflects the lithological and hydrochemical influences. Clastic sedimentation prevails both in extremely arid and in humid zones. Clay mineralogy and the contents of Fe/Al-hydroxides and -oxides vary widely. Lake sediments of the humid zones are characterized by higher concentrations of organic carbon, whereas in the arid regions organic material is scarce or absent.

Authigenesis of minerals occurs mainly in moderately humid and in semi-arid climatic zones. Following the work of Müller et al. (1972), formation and diagenesis of carbonate minerals are regarded as proper criteria for distinguishing the different types of lacustrine geochemical evolution. Precipitation of sodium-carbonate minerals is restricted to the hyper-saline alkaline lakes, as in the East African Rift Zone (Eugster and Hardie, 1978). Formation of primary Ca–Mg minerals (calcite, high-Mg calcite, aragonite, hydrous Mg carbonates) and of secondary carbonates (dolomite, huntite and magnesite) is dependent on the Mg/Ca ratio of the lake or pore water, respectively.

Thus, according to these criteria, seven types of lacustrine sediments can be distinguished, which to a certain extent indicate geographical, climatic, and genetic associations (Table I).

I. Arid lakes in Australia (26 examples) are representative of the dominance of clastic sedimentation which reflects the composition of the Precambrian magmatic rocks of the lake floors and the surrounding areas (a detailed description will be given below). Precambrian rocks poor in carbonates also characterize the lithology of the freshwater reservoirs studied in *South Africa* (5 examples; Wittmann and Förstner, 1975, 1976 and unpublished data).

II. Lakes of the Amazon Basin (9 examples) at the junction of the Rio Negro and Rio Solimões near Manaus also exhibit clastic sedimentation, which is influenced by various kinds of weathering in the wet tropical climate (see below).

III. Lakes of the East African Rift Zone (15 examples) can be characterized by the dominance of sodium and bicarbonate/carbonate ions in their water chemistry. Some examples exhibit precipitates of sodium carbonate (Lake Natron); other lakes contain water with low salt concentrations (Lake Malawi). Sediments from Lake Amboseli (Kenya/Tanzania) situated outside the "Rift-Valley" proper contain Mg calcite and dolomite and should therefore also be related to the playa carbonate type (IV-3). Additional details will be given in a later section (p.218 ff.)

IV. Lakes in Europe and western Asia can be classified by the mode of Ca–Mg carbonate sedimentation: (1) Sixteen examples of freshwater lakes contain variable percentages of low-Mg calcite. (2) Hydrochemistry of Balaton (Hungary) and Hamun-i-Puzak (Afghanistan) is slightly brakish with variable Mg/Ca ratios between one and approximately seven; high-Mg calcite is found in both lakes. (3) Carbonate sedimentation in playa lakes with characteristic Ca-rich dolomite occurs in four of the lakes studied. (4) Five alkaline lakes in western Asia are characterized by relatively constant water levels and by high Mg/Ca ratios, leading to the formation of aragonite. Examples of this group were found in crater lakes of the Serir Tibesti in Libya (North Africa); according to the chemical analysis amorphous Mg silicates are present.

Analytical data of Pb, Cu, Ni, Cr, Zn, Sr and Mn in the $<2\ \mu m$-fraction of lake sediments are summarized in Fig.1. The outer solid lines represent the frequency distribution for all 87 samples of each element under consideration. Separate histograms are given for the lakes in Europe (shaded areas) and in western Asia (dotted areas).

A general inspection of the frequency curves indicates largely log-normal distribution of the elements Pb, Cu, Ni, Cr, and Zn, which are most likely to be present when the element concentrations in the various rock sources are very similar *and* when the subsequent change in the metal content due to internal factors (e.g., precipitation and diagenesis) is relatively small. However, there are characteristic deviations from log-normal values for the examples of *strontium* and *manganese.* With regard to Mn, the distribution curve exhibits a distinct asymmetry to low metal values. The behavior from diagenetic effects, where reducing conditions remobilize manganese leading to a depletion of Mn in the sediments, will be explained later. In the case of Sr, the higher concentrations (up to 50% in some samples) result from itş incorporation into carbonate minerals, particularly in aragonite.

A more detailed review of the data in Fig.1 indicates some anomalous high Ni and, in particular, Cr values. These deviations typically originate from the influence of gabbroid rocks. In the European lakes concentrations of Pb, Zn, and Cu exhibit on the average higher values than is the case for the overall distribution of these elements in the 87 samples. This apparently points to a civilizational influence, in particular for the enrichment of Pb and Zn.

Further evidence of the above-mentioned element associations can be gathered from the statistical evaluations of the analytical data. In the upper section of Table II, the correlation coefficients from linear regression analysis of possible element pairs is indi-

TABLE I

Important data relating to lakes (examples) in this study

Lake (Country) [a]	Lake type [b]	Surface (km^2)	Depth (m)	Salt (g/l)	Water
					major cations
Ia "Arid" lakes in Australia					
Lake Torrens (S.A.)	9, 63	14,000	<1	280	Na
Curtain Springs (N.T.)	3, 63	10	<1	380	Na, Mg
Lake Moore (W.A.)	63	800	<1	280	Na
II "Amazon" lakes					
Lago Tupe (BR)	57	0.7	<14	0.02	SiO_2, Na
L. Cabaliana (BR)	56	100	<15	0.05	Ca, SiO_2
III African "[illegible]" lakes					
Lake [illegible] (ETH)	9	500	250	16	Na
Lake Natron (EAK, EAT)	9	600	<10	75	Na
Lake Malawi (MW, MO)	9	30,800	706	0.2	Na, Ca
IV Lakes in Europe and Asia					
(1) Freshwater lakes					
Plöner See (D)	40	30	60	0.32	Ca
L. Constance (D, CH, A)	28c	539	252	0.45	Ca, Mg
Blejsko Jezero (YU)	28c	1.7	28	0.75	Ca
Band-e-Amir (AFG)	45	10	60	0.33	Ca
(2) Subbrackish lakes					
Balaton (H)	8	600	12	0.5	Mg, Ca
Hamun-i-Puzak (AFG)	63	390	3	0.8	Na, Mg
(3) Playa lakes					
Neusiedler See (A, H)	3	280	<2	1.5	Na, Mg
Dasht-i-Nawar (AFG)	16	40	<1	6	Na, Mg
Tuz Gölü (TR)	8	1,100	<2	450	Na, Mg
(4) Alkaline lakes					
Van Gölü (TR)	16	3,600	550	19	Na, K
Ob-i-Istada (AFG)	8	130	3	27	Na, Mg

Additional lake examples:
Group Ia. Pernatty Lagoon, Lakes MacFarlane, Finniss, Windabout, Hart, Harris, Gairdner (South Australia), Wallambin, Moore, Bulga Downs, Barlee, Ballard, Goongarrie, Carey, Lefroy, Cowan, Gilmore, Yindarlgooda, Seabrook, Deborah, Brown (Western Australia).
Group Ib. (Artificial reservoirs in South Africa) Hartbeespoort, Rietvlei, Marais, Buffelspoort, Linleyspoort Dam.
Group II. Lago Buiucu, Muratu, Jacaretinga, Lago dos Passarinhos, Calado, Manacapura, Aiapua.
Group III. Lakes Hannington, Nakuru, Magadi, Naivasha, Baringo (Kenya), Ngorongoro, Manyara, Eyasi, Amboseli (Tanzania), Zoutpan (South Africa).
Group IV. (1) Thuner See, Starnbergersee, Kochelsee, Walchensee, Tegernsee, Schliersee, Chiemsee, Waginger See, Prespanski Jezero, Ohridsko Jezero, Moste Dam, Zbilje Dam, Bohinjsko Jezero (Yugoslavia); (2) Hatam Abad, Chalus, Ali-Abad (Persia); (3) Tuzla Crater Göl; (4) Burdur Gölü, Aci Crater Göl, Ercek Gölü (Turkey), Serir Tibesti (Libya).
Abbreviations:
[a] Countries: S.A. = South Australia, N.T. = North Territory, W.A. = Western Australia, BR = Brazil,

hemistry [c]		Sediment fraction <2μm [d]			Ref. [e]
najor nions	pH	authigenous material	detrital clay min.	org. C %	
l	6.8	–	Ill, Kaol	<0.2	(1)
l, SO_4	6.4	–	Kaol, Ill	<0.2	(1)
l	6.6	–	Kaol, Ill	<0.2	(1)
l, HCO_3	4.7	–	Kaol	3.9	(2– 3)
ICO_3, Cl	6.4	–	Ill, Sm	2.5	(2– 3)
ICO_3, CO_3	10.0	(Cc)	Sm, Ill	1.0	(4)
O_3, HCO_3	10.0	(Cc)	Ill, Sm	0.4	–
ICO_3	8.1	(Sm, Diat)	Ill, Sm	3.7	(5)
ICO_3	7.7	Cc, Diat	Ill, Chl	4.0	(7)
ICO_3, SO_4	7.6	(Cc)	Ill, Chl	1.3	(7– 9)
ICO_3	7.7	(Cc), Diat	Ill, Chl	2.8	(6)
ICO_3	7.6	Cc	Ill, Chl	2.7	(10–11)
ICO_3, SO_4	8.6	Cc, MgC	Sm, Ill	1.6	(12–13)
ICO_3, SO_4	7.6	MgC	Ill, Chl	0.4	(11)
ICO_3, SO_4	9.0	MgC, Dol	Ill, Chl	3.5	(14, 15)
ICO_3, Cl	8.7	MgC, Dol, Sm	Sm, Ill	0.6	(11)
l, SO_4	7.5	MgC, Dol, Hn	Ill, Chl	<0.3	(16, 17)
O_3, Cl	9.9	Aragonite	Chl, Ill	0.8	(18)
l, SO_4	8.9	Aragonite	Ill, Chl	0.7	(11)

ETH = Ethiopia, EAK = Kenya, EAT = Tanzania, MW = Malawi, MO = Mozambique, ZA = Republic of South Africa, D = Germany, CH = Switzerland, A = Austria, YU = Yugoslavia, H = Hungary, TR = Turkey, AFG = Afghanistan.

[b] Lake Type (Hutchinson, 1957): 3–9 = tectonic basins, 10–19 lakes associated with volcanic activity, 23–42 = lakes formed by glacial activity, 43–47 = solution lakes, 48–59 = lakes due to fluviatile action, 60–63 = lake basins formed by wind.

[c] Water Chemistry: part of data after Livingstone (1963).

[d] Sediment Type, Carbonate minerals (after Müller and others, 1972): Cc = calcite, MgC = high-Mg calcite, Dol = dolomite, Hn = huntite, Ar = aragonite, Clay minerals: Ill = illite, Kaol = kaolinite, Chl = chlorite, Sm = smectite, Diat = silica of diatoms.

[e] References: (1) Förstner (1977c), (2) Reiss (1975), (3) Irion and Förstner (1975), (4) Baumann et al. (1975), (5) Müller and Förstner (1973), (6) Molnar et al. (1978), (7) Abdul-Razzak (1974), (8) Müller (1966), (9) Müller (1971), (10) Jux and Kempf (1971), (11) Förstner and Bartsch (1970), (12) Müller (1969), (13) F. Wagner (1974), (14) Schroll and Wieden (1960), (15) Blohm (1974), (16) Irion (1970), (17) Irion (1972), (18) Irion (1973).

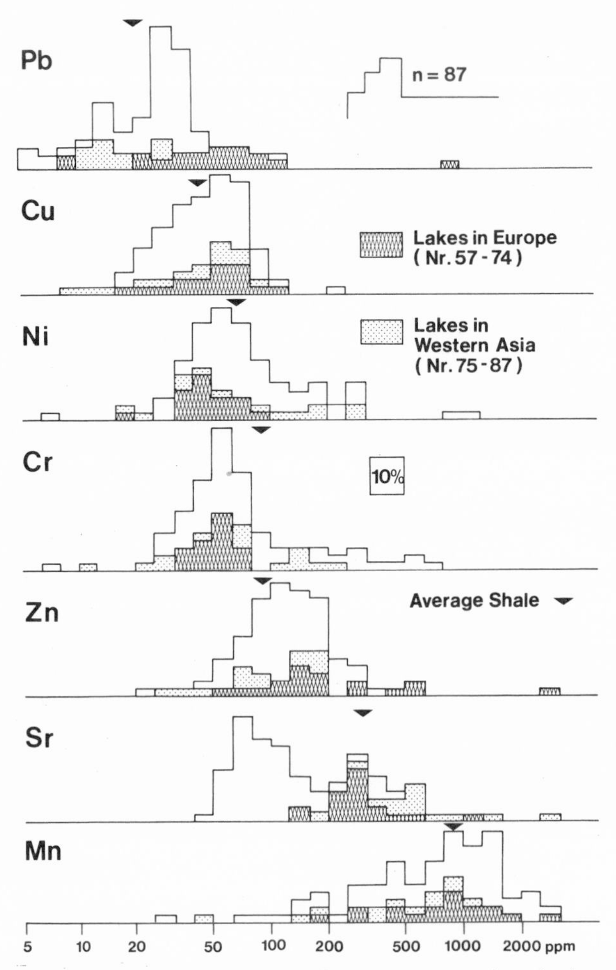

Fig. 1. Frequency distribution of metal concentrations in the pelitic fractions (<2 μm) of sediment samples from 87 lakes in different climatic zones. (After Förstner, 1978.)

cated. In the lower section of the table only values of more than 95% significance are given, whereby (a) there is a high degree of positive correlation between the elements Fe, Cr, Ni, Co, and Mn, and to a lesser degree Cu, indicating the influences from the lithology of the lake's catchment area, (b) the strong correlation between the Sr values and total carbonate contents can partly be associated with allochthonous influences; authigenic formations of carbonate minerals should additionally be considered as a possible mechanism of strontium enrichments in some lake sediments, (c) elevated concentrations of carbonate, Sr, and organic carbon obviously effect the decrease of the Fe concentrations in lake sediments, as is indicated by the negative correlations between these components, (d) the carbon content also appears to be a diluting factor for the Cr concentrations.

The metal associations, which are designated by (e), have been suggested to be partially enriched by man-made influences. A high degree of positive correlation exists particularly between the metals Zn, Pb, Cd and Hg; organic carbon, and to a lesser extent Cu, also indicates association with cultural activities.

Geochemical background values for freshwater sediments

The average composition of 87 samples of pelitic lake sediments was calculated for arithmetic and statistical mean values (Table III). Deviations from the mean are given either as "variation coefficients" or as "confidence limits" (range for 90% of the data for each element studied). Variation coefficients are relatively low for Fe (90% of the data fall between 11,5000 and 67,300 ppm) and Cu (20–90 ppm), whereas higher variations occur for Pb, Zn, and Cd. These higher variations are an additional indication of the presence of man-made effects.

For the determination of characteristic metal concentrations of a sediment – for example, when prospecting for ore deposits – it is sometimes possible to dispense with the differentiation between lithogenic and anthropogenic components. Nonetheless, when attempting to determine the "extent of pollution" in a lake or river by means of the heavy-metal load in sediments, it is of primary importance to establish the natural level of these substances, i.e., the "precivilizational" level (Shimp et al., 1971) and then subtract it from existing values for metal concentrations in order to derive the amount ascribable to enrichment caused by anthropogenic influences. In practice, however, there are several difficulties with this procedure: in the first place, it is often impossible to make a clear distinction between the "civilizational" and "geochemical" amounts of heavy metals in a given sediment sample, e.g., either because of their characteristic isotope compositions or due to specific bonding conditions. Furthermore, the carrier substances of the metals, i.e., the sediments in the broadest sense, can differ substantially in their granulometric and material composition, such that the conditions may vary widely from place to place. Finally, it has often been observed that the ultimate deposit on the bed of a river, lake, or ocean has quite a different composition compared with the suspended material in any one of these aquatic systems, particularly owing to the decomposition of the organic substances.

In order to obtain an ideal comparative basis for environmental studies, the following criteria should be fulfilled to achieve representative values for metal concentrations: (a) a large number of sediment samples must be analyzed, which correspond to Recent deposits in their grain size distribution, material composition and conditions of origin, and (b) they must be free of contamination by civilizational influences. In practice, these criteria cannot be fulfilled simultaneously. Several attempts have been made to solve this problem, which in short is called the "background question".

Average shale. The rock standard is a global value in general use and satisfies the basic requirement of being uncontaminated and based, for most elements, on a large number

TABLE II

Correlation matrix of metal concentration, organic carbon (OC), and carbonate contents (CM) from 87 samples (pelitic fractions)

	Fe	Mn	Zn	Cr	Ni	Cu	Pb
Fe	x	0.397	0.022	0.543	0.346	0.345	−0.065
Mn	0.397 [a]	x	0.097	0.258	0.111	0.261	0.018
Zn			x	−0.005	−0.072	0.334	0.957
Cr	0.543 [a]	0.258 [a]		x	0.719	0.486	−0.038
Ni	0.346 [a]			0.719 [a]	x	0.193	−0.053
Cu	0.346 [a]	0.261 [a]	0.334 [e]	0.486 [a]		x	0.299
Pb			0.957 [e]			0.299 [e]	x
Co	0.453 [a]	0.309 [a]		0.498 [a]	0.520 [a]	0.358 [a]	
Hg			0.838 [e]			0.289 [e]	0.796 [e]
Cd			0.940 [e]			0.254 [e]	0.900 [e]
Li							
Sr	−0.399 [c]						
OC	−0.218 [c]		0.525 [e]				0.419 [e]
CM	−0.634 [c]			−0.234 [d]			

In the lower half of the matrix only *r*-values significant at 0.05 (single underlined) and 0.01 (double underlined) levels are indicated. For footnotes *a–e*, see text.

of individual analyses. Extending this standard by incorporating the grain size (see below), one obtains for example, the "argillaceous sediment" standard. The samples used for setting such a standard, however, stem from different environments which, in some cases, such as under reducing conditions, are not comparable. In addition, the composition of the material in the study area can also vary as in the case of higher concentrations of carbonate minerals. Nevertheless, a comparison with a shale standard – the examples from the Turekian and Wedepohl (1961) compilation are given in Table III – is a quick and practical means of tracing significant metal enrichments which may constitute a source of dangerous environmental pollution or an economically interesting mineral deposit. Once the sources responsible for these accumulations in sediments have been traced, they can subsequently be evaluated more precisely by more refined – yet time-consuming – procedures, which will be described below.

Ancient lake sediments. In order to take those environmental factors into consideration which influence sediment formation, the global average should be supplemented with additional information about the regions of formation. This naturally increases the number of individual data to be dealt with, not to mention the difficulty of classifying the investigated rock samples into a specific clearly defined formation zone. The number of lacustrine examples, in particular those from an area acutely affected by pollutant influences, will be relatively small as compared to examples from the marine environment. A comparison of "shales" with deep-sea clays (e.g., Turekian and Wedepohl, 1961) shows

o	Hg	Cd	Li	Sr	OC	CM
.453	–0.094	–0.117	0.170	–0.399	–0.218	–0.634
.309	0.176	0.063	0.203	–0.188	0.044	0.189
.011	0.829	0.940	–0.025	–0.038	0.525	0.133
.498	–0.129	–0.129	–0.082	–0.164	–0.044	–0.234
.520	–0.143	–0.128	–0.145	–0.142	–0.172	–0.201
.358	0.278	0.254	–0.109	–0.221	0.209	–0.150
.034	0.796	0.900	–0.089	–0.021	0.419	0.133
	–0.011	–0.089	0.063	–0.115	0.060	–0.105
	x	0.848	–0.002	–0.034	0.443	0.209
	0.848 [e]	x	–0.025	–0.012	0.526	0.196
			x	–0.019	–0.104	0.011
				x	–0.016	0.682
	0.443 [e]	0.526 [e]			x	0.303
				0.682 [b]	0.303	x

that great variations can be expected within the marine deposition area. It is questionable whether the coastal region, which is strongly affected by metal pollution, is adequately represented by a marine average including deep-sea deposits. An improvement of the results over the use of a global shale standard can only be obtained, therefore, if the values of rock samples from defined formation environments are used for comparison with the actual data.

Fig.2 shows the metal data of a fossil lake (Förstner, 1977c), the Ries crater in southern Germany, which was formed by a meteorite about 14 million years ago. It was filled at first with predominantly clastic and later by increasingly chemical or biogenic lake deposits, which have since reached a thickness of 320 m. From the deposits of this crater, 25 samples were selected from a core, whereby the pelitic fraction was separated and ten trace metals were measured. With the exception of a sample from the clastic sedimentation sequence, which can be interpreted as a weathering product of gabbroic rock (high percentages of Cr, Ni, Co, V, Cu), the metal content of all other samples varied only very slightly either within the total pelitic fraction (shaded area) or within the carbonate-free substance (external contour line). If the average values of the individual profile curves of the elements are then compared (Table III), it is noticeable that there is a very close correlation between the data represented here and in the global shale standard of Turekian and Wedepohl (1961).

Short, dated sediment cores. The expense involved in drilling operations to obtain sedi-

TABLE III

Average values of metal contents (ppm), carbonate contents (%) and the percentage of organic carbon in the pelitic fraction (<2 μm) of 87 sediment samples from lakes compared with data of shales [1] and of sediments from the Tertiary Ries Lake [2]

	Mean values of 87 lake sediment samples				Average shale [1]	Ries Lake [2]
	arithmetic mean	variation coefficient	median	range for 90% of the values		
Iron	40,930	(47)	43,400	(11,500–67,300)	46,700	18,200
Manganese	912	(80)	760	(100–1800)	850	406
Zinc	172	(177)	118	(50–250)	85	105
Chromium	100	(120)	62	(20–190)	90	59
Nickel	107	(143)	66	(30–250)	68	51
Copper	51	(57)	45	(20–90)	45	25
Lead	43	(231)	34	(10–100)	20	16
Cobalt	21	(102)	16	(4–40)	19	15
Mercury	0.55	(116)	0.35	(0.15–1.50)	0.4	0.5
Cadmium	0.60	(171)	0.40	(0.10–1.50)	0.3	0.2
Lithium	57	(90)	45	(15–200)	66	203
Strontium	304	(191)	151	(60–750)	300	252
Org. C (%)	1.6	(107)	–	(<0.2–3.7)	–	3.5
Carbonate (%)	16.4	(140)	–	(0–70)	–	36

[1] Turekian and Wedepohl (1961).
[2] Mean of 25 samples taken from a depth range of 0–265 m: Förstner (1977d).

ment cores from great depth can hardly be justified merely to establish background values. Likewise, the evaluation of local enrichments in comparison to a global standard, e.g., the shale composition, is often not satisfactory due to the presence of lithological anomalies. In such a situation, the investigation of short sedimentary cores proved to be very useful, since they provide an historical record of the various influences on the aquatic system by indicating both the natural background levels *and* the man-induced accumulation of certain elements over an extended period of time.

The significance of sedimentary cores in the assessment of the quality of aquatic systems is well illustrated by examples from the lacustrine environment. First observations of cultural effects on lakes, as indicated in the variations in the sediment stratigraphic records, were made on Lake Zürich and described by Nipkow (1920). The influences of human activities on the rate of eutrophication were studied by Hutchinson and Wollack (1940) from Linsley Pond, Connecticut; Mortimer (1941) from Lake Windermere, England; Murray (1956) from Wisconsin lakes; and by Ohle (1956) from lakes in northern Germany. These reports suggest that lake sediments "may be regarded as a response of the conditions in an aquatic system" (Züllig, 1956). Man-made effects have been evaluated from the distribution of phosphorus and other nutrients (Livingstone

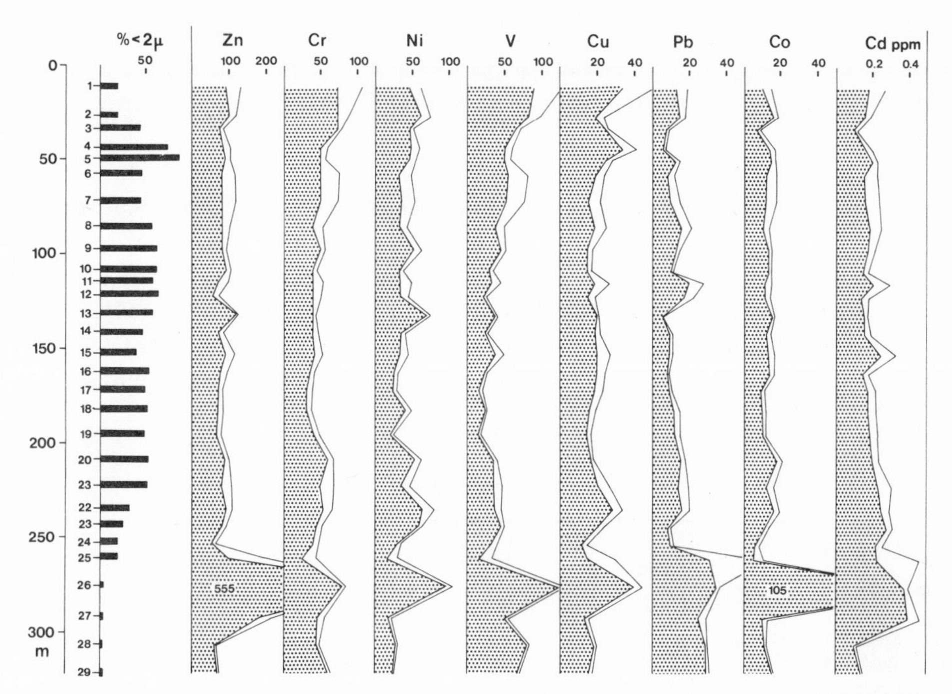

Fig. 2. Metal concentrations in the pelitic sediment fractions from the Tertiary Ries Lake in southern Germany. (Dotted areas: bulk analysis; outer solid line: corrected for carbonate-free sediment fractions; after Förstner, 1977d.)

and Boykin, 1962; Whiteside, 1965; Frink, 1967), from the distribution of iron monosulfide (Müller, 1967) and in changes of diatom assemblages (Stockner and Benson, 1967; Duthie and Sreenivasa, 1971); pollen variations in Recent lake sediments reflect historical changes in land use (Vuorela, 1970; Solomon and Kroener, 1971; Anderson, 1973; Kemp et al., 1974). During the last decade, lake-sediment analyses have increasingly been employed as a tool to trace sources of less degradable pollutants, such as halogenated hydrocarbons (Leshniowsky et al., 1970) and heavy metals. A compilation "Man-made chemical perturbations of lakes" was prepared by Stumm and Baccini (1978) for Lerman's study on "Lakes – Chemistry, Geology, Physics".

In lakes and marine coastal basins, where such investigations are for the most part carried out, the average annual rate of deposition of relatively fine-grained material (clay and fine to medium silt fractions) is from 1 to 5 mm. A core profile of approximately 1 m in length, which is relatively inexpensive to obtain by a piston corer of the Kullenberg type (Kullenberg, 1947), covers an historical period of at least 200 years, corresponding to the phase of strongest industrial development, and therefore to the highest inputs of man-derived, partly xenobiotic materials. Sedimentary cores are particularly useful when the individual phases of development, which are recorded in the different layers of sediment material, can be dated by means of certain granulometric characteristics, by pollen assemblages or by isotope measurements. Among the latter methods (Krishnaswami and Lal, 1978), the determination of the ^{210}Pb distribution has proved to be advantageous (Krishnaswami et al., 1971; Ritchie et al., 1973; Robbins and Edgington, 1975; Dominik et al., 1978). An example is given in Fig.3 (after Aston et al., 1973) indicating the distribution of mercury in a 1 m sediment core from Lake Windermere (England). According to the ^{210}Pb data, the bottom section of the core represents the sedimentation period from 520 to 1250 A.D., which appears to reflect the background value of about 120 ppb Hg. Since 1400 there has been a gradual increase of the Hg concentrations in sediments presumed to correlate with man's activities. Such activities include: denudation of land surface, industrialization, mining and quarrying, burning of

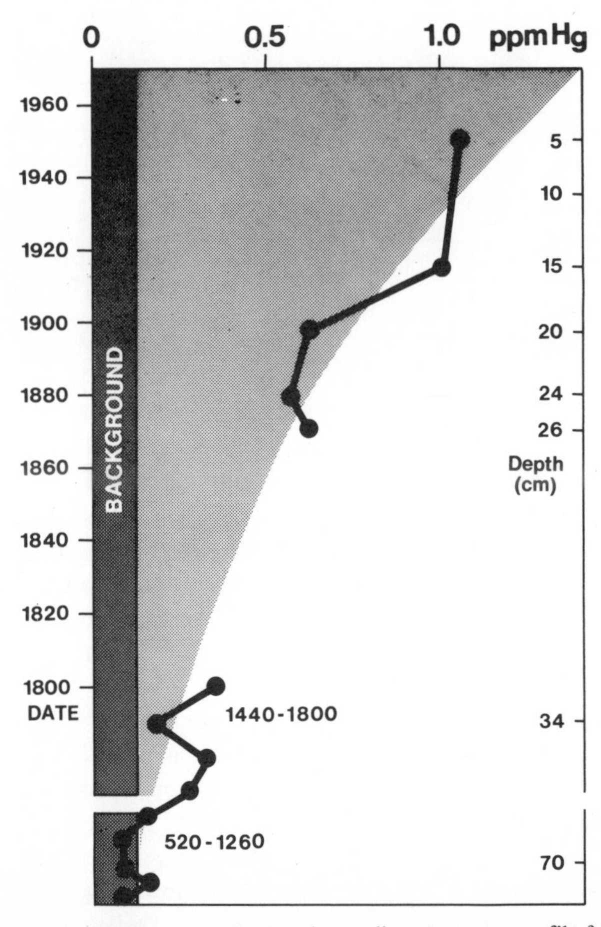

Fig. 3. Mercury concentrations in a sedimentary core profile from Lake Windermere, England. (Data from Aston et al., 1973.)

fossil fuels, and sewage disposal. Much of the anthropogenic Hg initially released in the atmosphere is adsorbed by soils and plant material after precipitation and later discharged via surface erosion into the aquatic environment. Of equal importance, some Hg is introduced directly into the water systems from sewage effluents (Aston et al., 1973).

Natural and anthropogenic metal fluxes into lake sediments. Evaluation of both geochemical and civilizational inputs of heavy metals into lacustrine sediments is increasingly established by means of flux calculations. Examples were given from Lake Erie by Kemp et al. (1976) and from Lake Windermere by Hamilton-Taylor (1979). The sedimentary flux of any metal may be calculated as:

$$F = R(1 - \phi)\rho C \qquad [\mu g\ cm^{-2}\ yr^{-1}]$$

where R is the sedimentation rate (cm/yr), ϕ is the porosity, ρ is the dry density of the sediment (g/cm^3) and C is the metal concentration (ppm). *Natural fluxes* in sediments from Lake Windermere (values in parentheses are for Lake Erie) were determined as follows: *Zn* 1.9–7.5 (6.5–52.6), *Pb* 0.7–2.6 (1.6–16.4), *Cu* 0.3–1.2 (1.9–16.3) and *Hg* 0.001–0.005 (0.004–0.07) $\mu g\ cm^{-2}\ yr^{-1}$. The data from Lake Windermere show the present-day natural fluxes (second value) being 4–5 times higher than the pre-cultural fluxes from the same catchment area; it is assumed that this is simply a function of the increased overall rate of deposition (anthropogenic effect?). *Anthropogenic metal fluxes* into Windermere sediments (values in parentheses are for Erie sediments) were measured for *Zn* as 16.9 – maximum 47.9 (10.9–113.1), for *Pb* as 9.9, max. 25.1 (5.4–44.3), for *Cu* as 1.8, max. 2.8 (2.6–13.0), and for *Hg* as 0.049 (0.06–0.49) $\mu g\ cm^{-2}\ yr^{-1}$, indicating that for both examples the anthropogenic metal input is highest for Hg, followed by Pb, whereas the cultural effects on fluxes of Zn and Cu are significantly lower (smaller than the differences between precultural and present-day natural fluxes of these metals into Lake Windermere).

Further analyses from Lake Windermere deposits (Hamilton-Taylor, 1979) suggest that incorporation of the metals into the sediments results from the deposition of metal-rich particles rather than by sorption processes at the sediment–water interface. The metal–particle association may occur before or after entry into the lake.

ALLOCHTHONOUS INFLUENCES ON THE HEAVY-METAL COMPOSITION IN RECENT LACUSTRINE SEDIMENTS

Metal accumulation in freshwater sediments induced by outside influences into the depositional areas, i.e., by discharges of particles already enriched in heavy metals, are described in this section. Apart from the predominant detrital inputs from mineralization zones and cultural waste materials, a group of influencing factors exists affecting metal accumulation by both clastic and authigenic processes. The effects of geothermal activities will be discussed here using the example of Hg enrichments in river and lake sediments of New Zealand. (The precipitation of sulfide minerals from volcanic emanations will be described in the section on "Autochthonous factors").

Grain-size effects

Grain-size distribution of clastic sediments is primarily influenced by physical processes of transportation and deposition. Thus, very strong variations of grain sizes within a river's sedimentary system occur; however, even in lakes and coastal basins significant changes of the grain sizes may be present within a sedimentary sequence, so that distinct effects on the metal concentrations of the sediment samples taken from different depositional environments are apparent. Examples from lake sediments are given in Fig.4 (from Deurer, 1978), indicating that the distribution of trace elements is particularly increased in the fine-grained fractions both for geochemical accumulations (Ni in Burdur Gölü, Turkey) and enrichments from man's activities (Zn in Lake Constance). Effects of this type have induced many researchers to opine that without a correction with respect to the grain-size influences, at least the analysis of river sediments will have little or no value

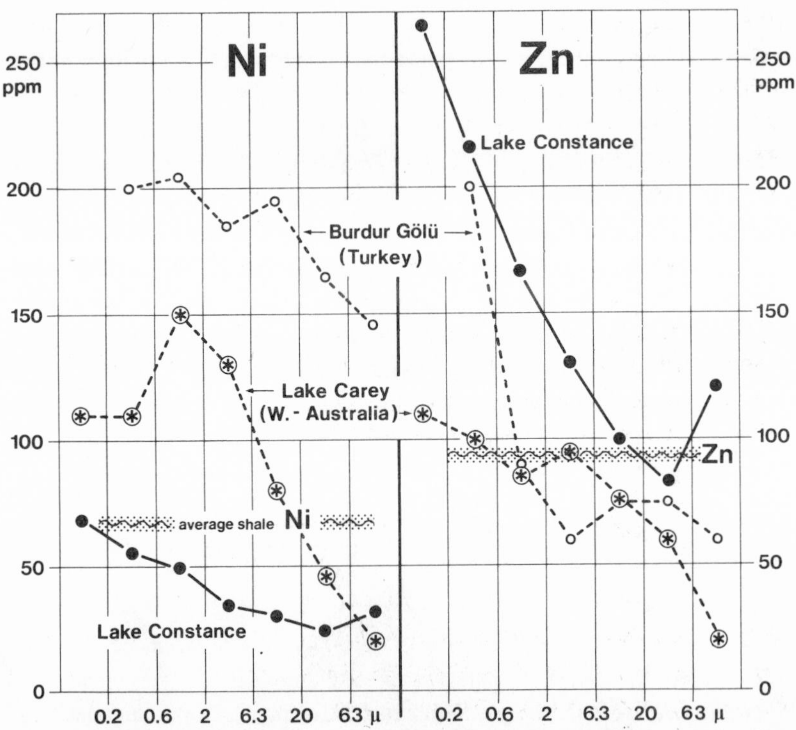

Fig. 4. Grain-size variations of the concentrations of nickel and zinc in sediments from Lake Carey, Burdur Gölü and Lake Constance. (Data from Deurer, 1978.)

TABLE IV

Methods of reducing grain-size effects

Grain-size distribution	
Extrapolation (e.g. <16 μm)	De Groot et al. (1971)
	Renzoni et al. (1973)
	Lichtfuss and Brümmer (1977)
Separation (<2 μm)	Banat et al. (1972)
	Helmke et al. (1977)
(<63 μm)	Allan et al. (1972)
	Davaud et al. (1977)
(<200 μm)	Thornton et al. (1975)
Specific surface area	Oliver (1973)
Acid-soluble fraction	
0.1 *N* hydrochloric acid	Piper (1971); Duinker et al. (1974)
0.3 *N* HCl	Malo (1977)
1 *N* HCl	Gross et al. (1971)
Mineral separation e.g., quartz correction	Thomas (1972, 1973)
Comparison to conservative elements, e.g. aluminum	Bruland et al. (1974)
Relative atomic variation	Allan and Brunskill (1977)

in tracing economic mineralization zones or in assessing the sources of pollution. In practice, there are several methods to reduce grain-size effects (Table IV); a detailed discussion is given by Förstner and Wittmann (1979).

Relative atomic variation (RAV) has been introduced by Allan and Brunskill (1977) from studies on sediments from Lake Winnipeg. Correlation coefficients were calculated for all possible pairs of elements, e.g., for the heavy metals Ni, Cu, Co, Cr, Zn, Fe and Mn. Where significant correlations existed, the slopes of the linear regression were determined, which represent the index of the RAV. It has been suggested that the analogy in RAV values for different sites implies a similarity or a homogeneity in the large number of processes occurring in the geochemical cycle, including weathering, transport, deposition, and diagenesis. At the same time, differences in the RAV values can be used to evaluate the influences of both cultural contamination and natural metal enrichments.

Natural allochthonous influences

Several forms of natural influences, especially the lithogenic, hydrogenic and atmospheric influences affecting the lacustrine environment, are dependent upon local conditions, such as relief, weathering, erosion, and chemical conditions. From theoretical considerations based on "normal" metal content in both river water and suspended material, one might expect some elements, for example, Na and Ca, to be supplied to the lacustrine sediments by aqueous solutions and by solid particles in similar quantities. Other metals, such as Fe and Cr, are accumulated primarily from suspended solids. Deposition of all metals from a river inflow, for example, by total evaporation within the lake basin would result in the ratios of "lithogenic" to "hydrogenic" contributions of approximately 0.5 : 1 for Na and Ca, 1 : 1 to approx. 5 : 1 for Cd, Li, Mg, Sr, K, Hg, Zn, and Cu, and 20 : 1 to approx. 50 : 1 for Pb, Ni, Co, Cr, Fe, and Mn, respectively (see next chapter of the present volume of the Handbook series.). In the case of stable water bodies, the more soluble metal compounds will be maintained in solution over long periods and may be discharged by groundwater or surface outflow. As a result, the metal content of these lake sediments are predominantly determined by the composition of the suspended solids.

Simplified conditions for evaluating the natural influences on lake sediments can be found in arid regions. According to Grim (1968), "the clay mineral composition of desert saline sediments is controlled almost entirely by the composition of the source area." Metal concentrations in lacustrine sediments from these regions, therefore, are more expected to reflect the chemical anomalies of the particular watershed than the secondary processes occurring within the lake area. In the following section we shall first review the conditions in some Australian dry lakes. The second part is concerned with the "dilution" of trace metals as a result of weathering processes in extremely humid climates; examples are given of Amazon lakes. The third section deals with natural allochthonous influences on lacustrine sedimentation resulting from hydrothermal solutions; examples will be given of Hg accumulations in river and lake deposits from New Zealand.

Trace metals in Australian lake sediments. Seventy percent of the Australian landmass is covered by desert regions, which lie mainly between latitudes 20° and 30°S. After heavy rains a large number of local depressions within this area are filled with saline water, thus forming temporary lakes, which are known as "salinas" or "claypans" (Hutchinson, 1957; Reeves, 1968). Several lake assemblages can be subdivided on the basis of their geographical and geological locality (Fig.5).

(a) The Lake Province of the Eyre Peninsula with Lake Torrens, Lake MacFarlane, Lake Gairdner, Lake Harris, Lake Everard and a large number of smaller dry lakes predominantly filled by groundwater seepage during winter. The crystalline basement rocks are covered in the northeast by clastic sediments.

(b) The Great Artesian Basin with the remains of the Pleistocene Lake Dieri, Lake Eyre, Lake Gregory, Lake Blanche, Lake Callabonna, and Lake Frome. This depression (at its deepest part about 20 m below sea level) originates from epeirogenetic warping occurring during the beginning of Mesozoic times and contains a thick sequence of post-Paleozoic sediments.

(c) The "Salt Lake Division" or "Salinaland" covers the Western Australian Shield desert from Lake Disappointment on the Tropic of Cancer in the north to Lake Dundas at 33°S. The bedrock of the extremely flat basins consists of Precambrian granites, gneisses, migmatites, and basic volcanics (or greenstones), covered by thin layers of silt or sand.

Sediment and water samples were taken from nine lakes of the Eyre Peninsula and from sixteen lakes in the southern part of the Salt Lake Division. One specimen stems from the Lake Province of the northwestern margin of the Great Artesian Basin (Förstner, 1977d).

Fig.5b shows the trace-metal distribution of 26 samples from Australian lakes. Most of the elements studied here are present in concentrations slightly below the standard shale composition. Only the median concentrations of Pb and Ni in the pelitic fractions of the Australian lakes sediments are higher than the average shale values.

In Table V, the mineralogical and chemical composition of two examples from the southern Australian lake region (Lake MacFarlane, Lake Gairdner) is compared with two samples from lakes of the western Australian Greenstone Province (Lake Goongarrie and Lake Yindarlgooda). The concentrations of Cd and Hg are somewhat higher in the sedi-

TABLE V

Mineral content and element distribution in the pelitic fractions of sediments of lakes in Australia (columns 1 and 2) and in the Central Amazon (columns 3–6) (After Förstner, 1977d and Irion and Förstner, 1975)
Metal data in ppm; n.a. = no analysis.

	Southern Australia [1]	Western Australia [2]	Amazon River [3]	Varzea lakes [4]	Shields soils [5]	Ria lake [6]
Clay minerals						
Smectite	x	xx	xxxx	xxx	–	–
Illite	xxx	x	xx	xxx	x	x
Kaolinite	xxx	xxx	xx	xxx	xxxxxx	xxxxxx
Chlorite	–	x	x	x	–	–
Talc	–	x	–	–	–	–
Metals						
Iron	36,650	62,950	64,400	56,000	61,000	12,000
Magnesium	31,610	55,340	11,800	9,800	400	1,200
Potassium	26,400	9,550	18,100	18,800	300	3,000
Sodium	5,040	965	3,900	6,200	600	3,300
Calcium	4,180	6,570	11,300	6,100	700	2,800
Manganese	705	1,100	732	664	54	86
Zinc	92	85	204	193	41	80
Nickel	59	1,000	–	–	–	–
Chromium	39	490	90	75	36	54
Vanadium	n.a.	n.a.	69	79	82	42
Lithium	39	19	56	63	7	20
Lead	33	27	34	34	39	32
Copper	25	44	60	66	30	32
Cobalt	6	68	20	30	1	5

[1] Lake MacFarlane, Lake Gairdner, [2] Lake Goongarrie, Lake Yindarlgooda, [3] suspended matter of the Amazon above the mouth of the Rio Negro, [4] lake sediments (ϕ of five lakes; L. Muratu, L. Cabaliana, L. dos Passarinhos, L. Jacaretinga, L. Buiucu), [5] soil samples (ϕ of three soils of the Precambrian Shield and Tertiary), [6] sediment from Lago Tupé (Ria Lake).
x to xxxxxx relative content of clay minerals (estimated from X-ray diagram).

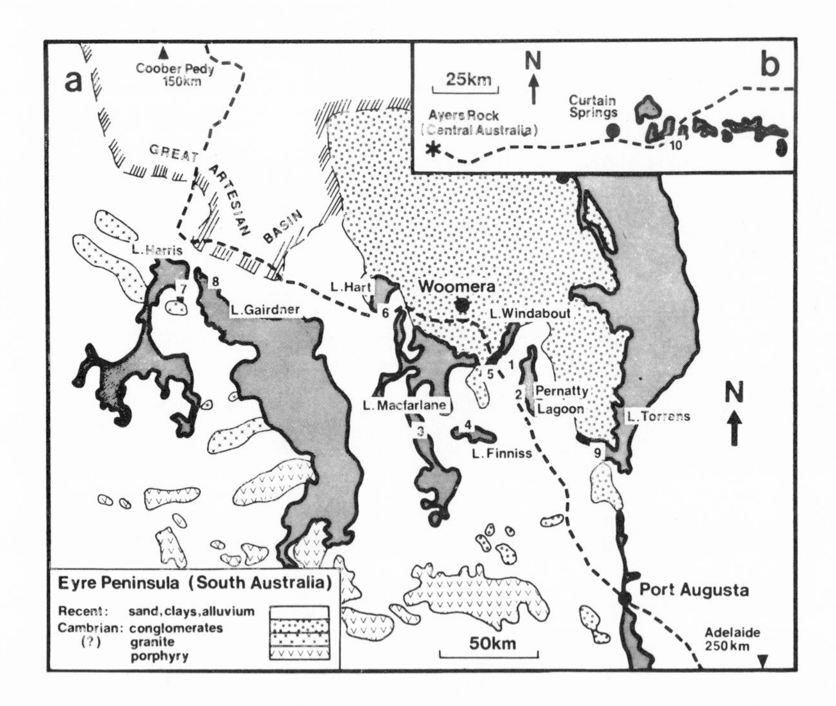

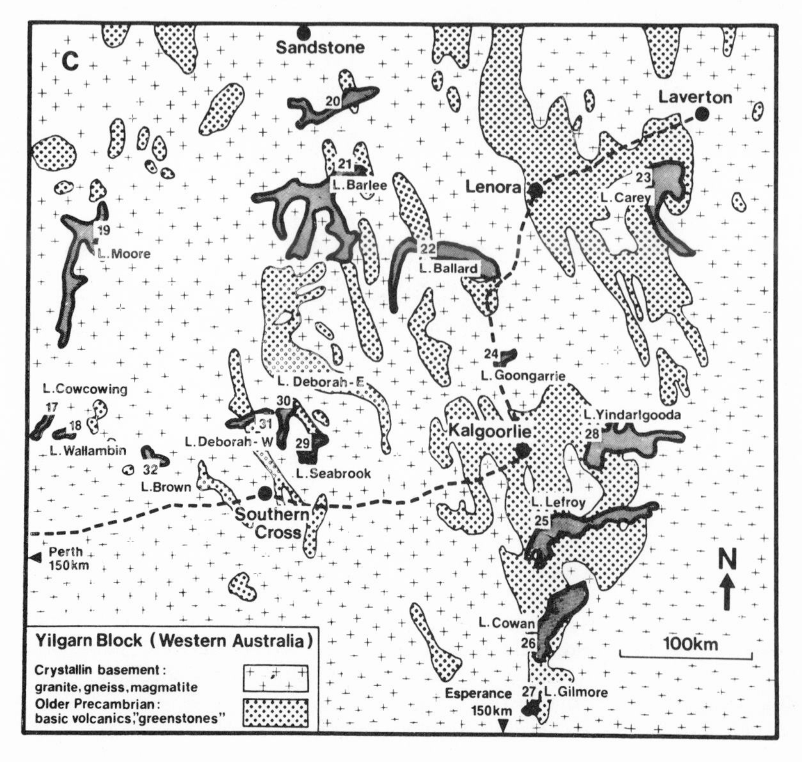

Fig. 5a, b, and c (explanation on next page).

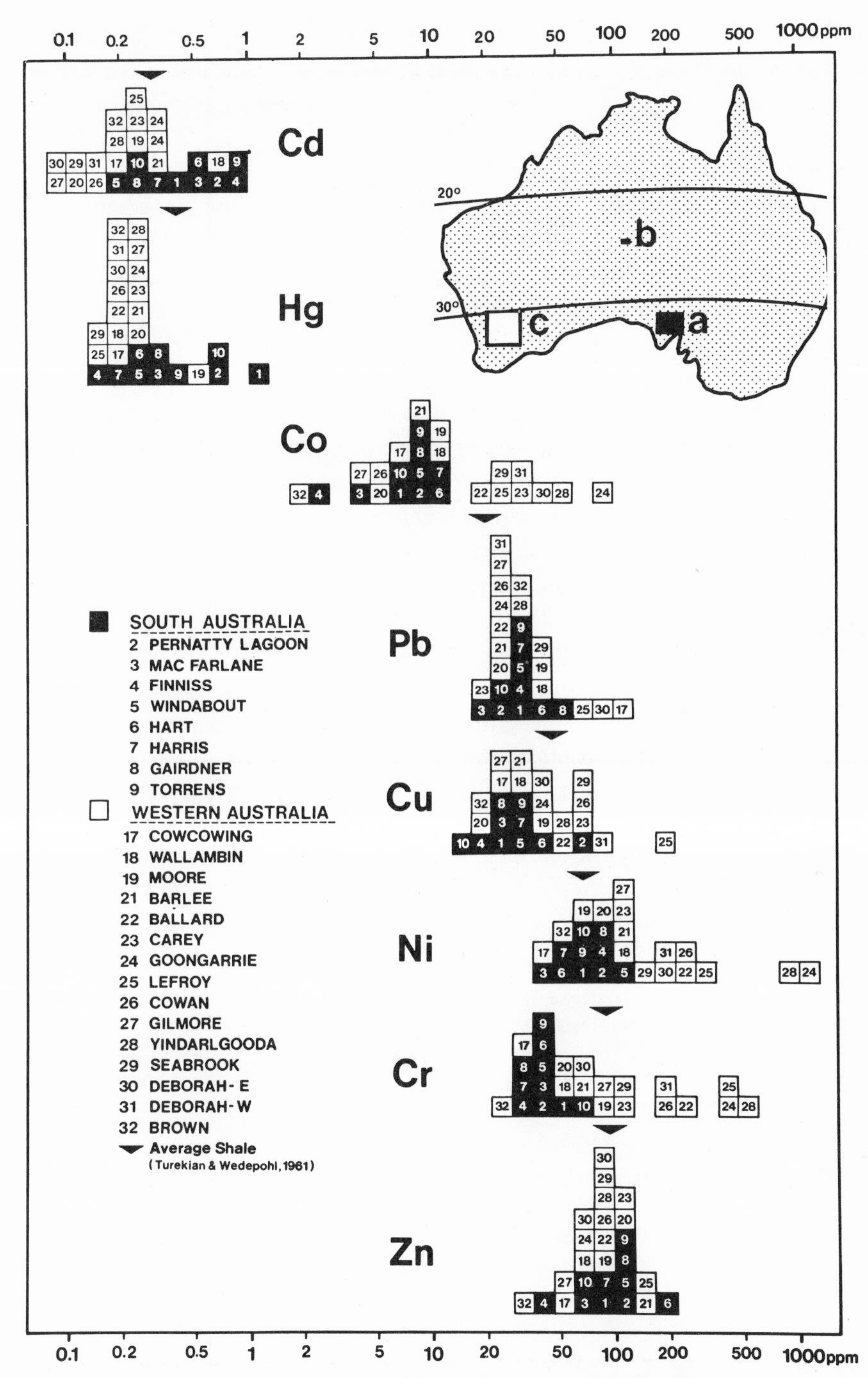

Fig. 5. Geological maps (Fig. 5a, b, and c) of the study areas in Australia and histograms showing the distribution of trace metal contents of the pelitic lake sediments. (After Förstner, 1977d.)

ments from the south Australian Lake Province compared with data from the western Salt Lake Division; the enrichment of K and Na is probably due to the higher content of illitic clay minerals. A similar effect is found for Li. The sediments from some lakes in western Australia contain a characteristic mineral, *talc* – $Mg_3Si_4O_{10}(OH)_2$ – which is probably formed during weathering of pyroxenites and peridotites of the Younger Greenstone series (David and Browne, 1950). Co, Ni, and Cr are enriched on an average by a factor of 10 or more in the lake deposits of the ultrabasic belt around Kalgoorlie. Economic nickel mineralizations have also been reported from that area; more detailed descriptions for the use of lake sediments in the exploration of economic mineralization will be given below.

Trace metals in Amazon lake sediments. Dilution of allochthonous trace metal content due to weathering effects is shown by examples from "river" lakes at the confluence of the Rio Negro and Rio Solimões in the upper Amazon lowlands near Manaus, Brazil (Fig.6). Two basic types of lakes can be distinguished in the Amazon Basin: the *Varzea Lakes* situated in the flood plains of the "white water" rivers relatively rich in particulates and electrolytes, e.g., the Amazon River, and the *Ria Lakes* (i.e., river lakes) in flooded Pleistocene river valleys with clear and "black water" rivers, poor in suspended material and dissolved ions (Irion and Förstner, 1975).

The deposits in the "white water" rivers originate from the Andes, the highland, and the southwestern Amazon Basin. In addition to characteristic smectite and illite components, river and lake sediments contain kaolinite; a high degree of correlation is displayed by values of the major elements – Na, K, Mg, Ca, and Fe (see Table V, columns 3 and 4).

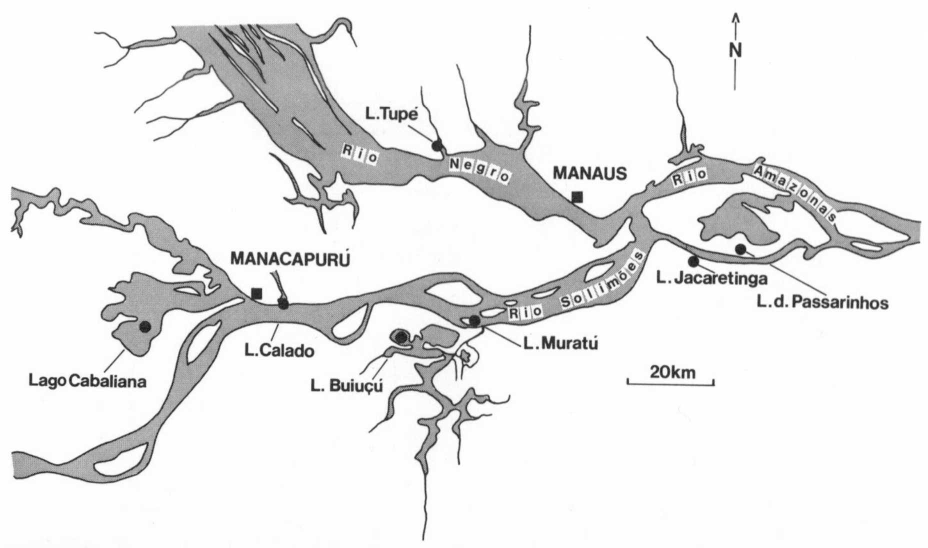

Fig. 6. Locations of lake examples in the Amazon region.

The catchment area of the "clear" and "black" water rivers includes the crystalline rocks of the Guayana and Brazilian Shield and their Tertiary lacustrine-fluvial erosion products. These rocks are strongly weathered (column 5). The recent lake and river deposits consist chiefly of kaolinite clay with an exceptionally low metal content (column 6).

The trace elements Mn, Co, and Li follow the pattern of distribution displayed by the amount of major ions. The levels in Zn, Cr, and Cu are also determined by changes in mineral composition. However, no differentiation was found for Pb and V. Leaching of trace metals, in this case under the influence of solutions poor in ions with pH values as low as 4, may also take place within the aquatic environment. (See following chapter "Trace Metals in Fresh Waters" in the present volume.)

Geothermal mercury enrichment in New Zealand rivers and lakes. Natural allochthonous influences on lacustrine sedimentation include the discharge of metals by hydrothermal solutions. Undisputed examples of this phenomenon, resulting in recent sulphide ore formation in the lacustrine environment, were reported from late-stage volcanic lakes in Hokkaido, Japan. At the same time, however, there is still a great difference of opinion

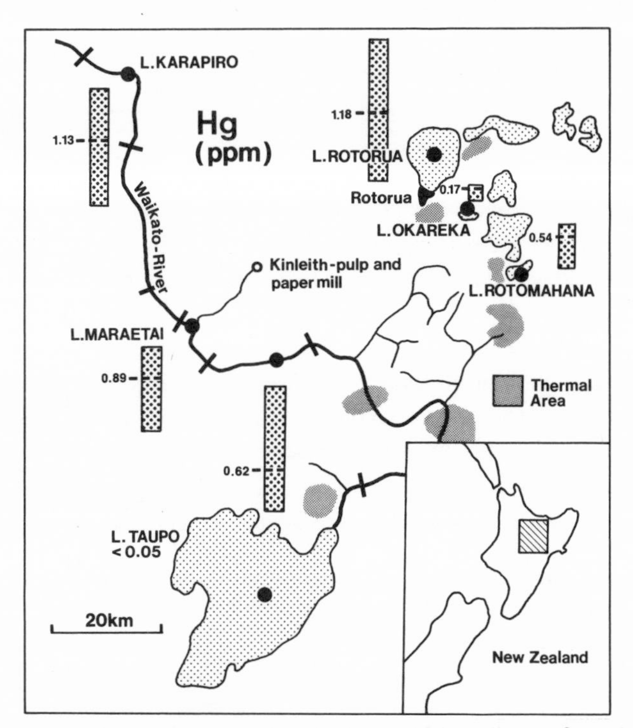

Fig. 7. Mercury concentrations in river and lake sediments from the geothermal area of New Zealand's North Island. (After Weissberg and Zobel, 1973.)

concerning the enrichment of metals in East African "Rift" lakes (see below).

Based on samples from New Zealand's North Island, Weissberg and Zobel (1973) show that discharge from natural hot springs, or from drill holes producing hot water or steam for geothermal power, may constitute an important source of mercury pollution. Their research into the effects of a pulp and paper mill, which discharges effluents into Lake Maraetai, led to the discovery of the magnitude and extent of natural Hg "pollution" originating from such geothermal sources (Fig.7). A detailed investigation of this particular area showed that the Hg accumulation in the sediments due to the pulp and paper mill does not exceed the values obtained for the industrially unpolluted Lake Rotorua. The increase of Hg, therefore, can only be attributed to natural enrichment. This explanation is also substantiated by the fact that the Hg concentrations in sediments showed no apparent variation with increasing depth (i.e., of age) of sediments. According to Weissberg and Zobel (1973), the magnitude of the problem is made clear by the following points: (1) Hg, although of extremely low concentrations in hot spring waters, has been accumulating in some sedimentary basins from natural geothermal discharges over the centuries, whereas industrial sources of Hg are of recent origin; (2) natural geothermal discharges are less localized and more difficult to control than industrial sources of Hg; (3) in affected areas, Hg concentrations in trout may rise to 3 mg/kg (0.5 mg/kg of fish is commonly considered to be the maximum concentration acceptable for human consumption).

Man's impact on metal distribution in limnic sediments

The effects of metals released into the aquatic environment by human activities on the composition of aquatic sediments has been studied during the last ten years in numerous marine and freshwater environments [1]. Areas of major concern are found in Japan, North America, and Europe (see next chapter "Trace Metals in Fresh Waters" of the present volume in the Handbook series).

The total extent of the heavy-metal pollution in sediment, however, is still unknown. A first attempt to tackle this question necessarily involves a comparison of the consumption of heavy metals with the natural concentration of these elements in unpolluted sediments, which is then recorded in the "Index of Relative Pollution Potential"

[1] Reviews and compilations are given, for example, in (1) D.D. Hemphill (Editor), "Trace Substances in Environmental Health." Symposium Volumes (annual), Univ. of Columbia, Missouri, 1967–1979, (2) the Proceedings of the International Conference in Heavy Metals in the Environment, Toronto, Oct. 27–31, 1975 (three volumes, Inst. for Environmental Studies, Univ. of Toronto, Ontario, Canada 1976/1977), (3) the chapter "Heavy Metals" of the Annual Literature Review for the June edition of the Journal Water Pollution Control Federation, Washington D.C., (4) P.A. Krenkel (Editor), "Heavy Metals in the Quatic Environment", Pergamon Press, Oxford, (1975), (5) U. Förstner and G. Wittmann, "Metal Pollution in Aquatic Environments", Springer-Verlag: Berlin-Heidelberg-New York, 1979, and (6) Proceedings of the International Conference "Management and Control of Heavy Metals in the Environment, 18–21 September 1979, Imperial College, London.

(Förstner and Müller, 1973a). It has been found that this index proves to be particularly high for Zn, Pb, Hg, and Cd. A similar approach was used by Nikiforova and Smirnova (1975) by calculating the technogenic migration of a metal and the degree of its utilization in the noosphere through its "Technophility Index", that is, the ratio of the annual output of a metal to its "Clarke" (mean concentration in the earth's crust). The higher the TP of a metal, the more intensively it is involved in technogenic migration (all values $\times 10^7$):

Mn	= Fe	< Ni	< Cr	< Zn	< Cu	< Hg	= Pb	< Cd
5	5	10	20	50	100	150	150	700

The TP-index varies with time, and thus each metal can be characterized by its own rate of TP growth. The TP of Pb for example, has grown two and a half times from the beginning of the century up to the present time and will have further increased four and a half times by the year 2000.

Sediment enrichment factors for elements. A fundamental investigation into the question of element enrichment in Recent sediments and the possibility of internal comparison was performed by Kemp et al. (1976) on sediment cores from Lake Erie. Kemp et al.'s (1974) first observation held true for most of the examples from the Great Lakes, according to which the concentration of a number of trace elements is much greater near the sediment/water interface than at the *Ambrosia* pollen horizon or below this chronological marker of approximately 120 BP (corresponding to early agricultural development following forest clearance). To quantify the results, Kemp et al. (1976) then introduced a ratio that is designated as the Sediment Enrichment Factor (SEF):

$$SEF = \frac{E_s/Al_s - E_a/Al_a}{E_a/Al_a}$$

E_s = the observed elemental concentration in the surface cm of the sediment,
E_a = the observed elemental concentration below the *Ambrosia* horizon,
Al_s = the aluminum concentration in the surface cm of the sediment,
Al_a = the aluminum concentration below the *Ambrosia* horizon.

Table VI depicts the Sediment Enrichment Factors and the respective interpretation from Kemp et al. (1976) for three profiles from the western, central, and eastern basin of Lake Erie. Those elements labelled "conservative" are of particular importance when determining comparative values of metals in deposits, e.g., in order to diminish grain size effects (Table IV).

In order to estimate the relative enrichment of trace metals by cultural effects, Table VII depicts selected examples of core sediment studies on both lacustrine and coastal marine environments. It is evident that Zn, Pb, Cd, and Hg are on the average par-

TABLE VI

Sediment enrichment factors in Lake Erie deposits (values between −0.2 and +0.2 are listed as zero) (After Kemp et al., 1976; with permission of Environment Canada, Fisheries and Marine Service)

Element	Western basin	Central basin	Eastern basin	Element behavior
Silicon	0	0	0	
Potassium	0	0	0	
Titanium	0	0	0	Conservative elements
Sodium	0	−0.3	0	
Magnesium	0	0	0	
Mercury	13.2	9.7	7.3	
Lead	3.4	2.6	3.7	
Zinc	1.7	1.4	2.2	Enriched elements
Cadmium	4.0	0.8	2.7	
Copper	1.4	1.1	0.8	
Organic carbon	1.1	1.3	1.7	
Nitrogen	1.2	2.1	2.0	Nutrient elements
Phosphorus	0.5	0.3	0.8	
Carbonate-C	0.4	−0.4	0	Carbonate elements
Calcium	−0.3	−0.4	−0.3	
Iron	0	0.3	0	
Manganese	0.4	0.4	1.1	Diagenetically mobile elements
Sulfur	0.3	0	1.0	

TABLE VII

Sediment enrichment factors (SEF) of heavy metals in selected examples from lacustrine and marine environments

	Cd	Co	Cu	Cr	Fe	Hg	Ni	Pb	Zn	Ref.
Lake Vänern	2.6	n.a.	1.2	n.a.	n.a.	(8) [a]	0.8	2.3	3.7	(1)
Lake Constance	2.9	0.5	1.1	3.0	1.0	2.0	0.9	2.7	3.1	(2)
Lake Geneva	4.6	1.5	2.7	1.3	n.a.	15	1.0	2.3	3.6	(3)
Lake Erie	7.3	1.6	3.7	2.9	1.5	8.3	2.1	4.7 [b]	3.6	(4)
Kieler Bight (Baltic Sea)	7.5	1.4	2.0	n.a.	0.9	n.a.	1.6	4.1	2.7	(5)
German Bight (North Sea)	7.5	2.2	1.8	1.5	1.4	9.4	1.3	8.2	4.0	(6)
Mean SEF	5.3	1.4	2.1	2.2	1.2	8.7	1.3	4.1	3.5	

References: (1) Håkanson (1977), mean values from 8 surficial samples (0–1 cm) vs. natural background ($\bar{x}$ from 68 preindustrial samples + 1 standard deviation), Cd background is estimated; (a) Hg probably local enrichment; (2) Förstner and Müller (1974a), sediment core from central part (250 m water depth); (3) Vernet (1976), mean SEF from 13 coastal stations; (4) Walters et al. (1974), sediment core from central basin (No. 13-2), (b) Kemp et al. (1976), mean of SEF from two cores of central basin; (5) Erlenkeuser et al. (1974), core A-GC from 28 m water depth; (6) Förstner and Reineck (1974), core 1/315 from 22 m water depth.

ticularly strongly enriched (the latter two elements and their accumulation in lacustrine sediments will be discussed in more detail below). Suess and Erlenkeuser (1975) have pointed out that the order of enrichment in anthropogenic-influenced sediment, Cd > Pb > Zn (Hg is not considered), corresponds to the accumulation of metals in fossil fuel residues. The latter are also reflected in the metal enrichment of air-borne particulates. An example of anthropogenic metal enrichment in fluviatile sediments is shown in Fig.8, which compares the metal concentrations in actual Rhine sediment with those found in ancient deposits from the Cologne area (26 m core in fossil Rhine sediments). The greatest increases in concentration caused by anthropogenic influences are shown to be the most dangerous heavy metals. The concentration levels for Cu, Zn, Pb, Hg, and Cd in the sediments of the lower Rhine originate to more than 90% from man-made sources. Only 2% of the Cd in the fine-grained sediments are supplied from natural sources; the other 98% results from anthropogenic influences, in particular from the highly Cd-contaminated tributaries Neckar and Main (Förstner and Müller, 1974a).

Lake sediments as indicators of metal pollution

The significance of sediments in assessing the conditions in aquatic systems is best illustrated by examples from the lacustrine environment since freshwater lakes have been the center of important cultural developments ever since the earliest days of civilization. As a consequence of increased population and industrialization densities, the threat of pollution has become most acute in such areas. Table VIII (Förstner, 1976) summarizes the major influencing factors on metal enrichment in lake sediments from characteristic examples (United States) by listing (1) the background levels of minor elements in the deeper part of sediment, (2) the maximum values in the upper layers of the core profiles, and (3) the factors of enrichment as the quotient (2): (1).

Mixed sewage inputs: Lake Michigan. Low concentrations in heavy metals are exemplified by the concentrations of Fe, Co, and Ni in most of the lake sediment sequences. These findings are in accordance with the results from highly polluted river sediments, e.g., form the lower Rhine section where the contents of Fe, Co, and Ni are influenced particularly by geochemical factors (Fig.8).

Moderate enhancement of the Zn, Pb, Hg, and Cd values in Lake Michigan sedimentary cores by factors of 2 to about 7 seems to originate mainly from mixed effluent inputs from industrial, domestic and agricultural sources.

Specification of effluents into domestic and industrial discharges can be made by the use of additional indicators, e.g., the contents of N, P, and organic carbon within the same sedimentary profiles. Förstner et al. (1974) compared the mode of accumulation of N and P with the concentrations of heavy metals in a dated core from central Lake Constance. Similar trends of both Zn/Pb and N/P concentrations indicate that these substances originate simultaneously in the public sewage system, whereas the increase of Cd

TABLE VIII

Distribution of minor elements in sedimentary profiles from lakes in the United States (After Förstner, 1976; with permission of Springer-Verlag)

	Lake Michigan [1]			Wisconsin lakes [2]			Lake Washington [3]			Lake Erie [4]		
	background	max. value	*F*	background	max. value	*F*	background	max. value	*F*	back-ground	max. value	*F*
Zinc	120	317	2.5	15	92	6	60	230	4	7	42	6
Chromium	77	85	1	7	49	7	n.d.			13	42	4.5
Nickel	54	44	1	34	50	1.5	(iron:		1)	40	95	2.5
Copper	44	75	1.5	22	268	12	15	50	3	18	59	4
Lead	40	145	3.5	14	124	9	20	400	20	n.d.		
Arsenic	11	22	2	(2	51	25)	10	200	20	0.6	3.2	5.5
Mercury	0.04	0.2	5	0.24	1.12	5	0.1	1.0	10	0.04	4.48	12
Cadmium	n.d.			2.5	4.6	2	n.d.			0.14	2.4	17

[1] Lake Michigan (Ruch et al., 1970; Shimp et al., 1971; Kennedy et al., 1971; Frye and Shimp, 1973), [2] Lake Monoma/Wisconsin (Lake Minocqua) (Syers et al., 1973; Shukla et al., 1972; Iskandar and Keeney, 1974), [3] Lake Washington (Barnes and Schell, 1973; Crecelius and Piper, 1973; Schell, 1974; Crecelius, 1975), [4] Lake Erie (Walters et al., 1974).
Data of background and maximum value in parts per million (ppm). *F* = factor of enrichment.

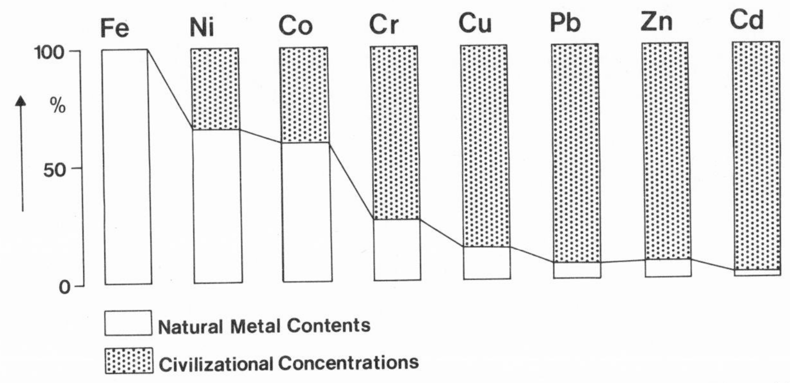

Fig. 8. Sources of heavy metals in pelitic sediments from the lower Rhine River. (After Förstner and Müller, 1973a.)

and Cr within the upper layers of the sediment sequences can be explained by industrial emissions from the electroplating and tanning industries, respectively. Mixed inputs from atmospheric sources and direct effluents are probably responsible for moderate (2–10-fold) increases of metal concentrations in surface sediments from large freshwater lakes in densely populated areas of northern and central Europe (Vänern, Bodensee, Léman, Lago Maggiore) since there is a similar accumulation of pollutants such as N- and P-compounds, DDT, PCB, polycyclic aromatic hydrocarbons and radioactive elements (Ravera, 1964; Ravera and Premazzi, 1971; Förstner et al., 1974; Vernet, 1976; Grimmer and Böhnke, 1977; Håkanson, 1977).

Algicides and herbicides: Wisconsin lakes. Metal pollution resulting from the use of pesticides is demonstrated by examples from lakes in Wisconsin: Cu concentrations in the sediments from Lake Monona increase with depth to a maximum (534 ppm Cu at 20–25 cm sediment depth) and then decline sharply (Fig.9a). These effects are a result of the intermittent treatment of the lakes with copper sulfate to control algal growth during 1918–1944. Since the rates of increase of Cu concentrations in sediments from other lakes, e.g., Lake Michigan are generally low, local Cu accumulations may, as a rule, be taken as indicators of previous application of copper for algal control. Arsenic values were not available from Lake Monona and were therefore taken from a study of Lake Minocqua in northeastern Wisconsin (Fig.9b). The characteristic increase of As in these sediments was explained by the fact that several Wisconsin lakes had been treated with sodium arsenite to reduce the population of noxious weeds. It is assumed that fertilizers and, above all, household detergents are responsible for the general enrichment of As concentrations observed, for example, in the sediments from Lake Michigan.

Airborne particulate fallout: Lake Washington. Characteristic sources of atmospheric metal pollution are ore smelters and coal-fired power plants, as well as heavy city traffic.

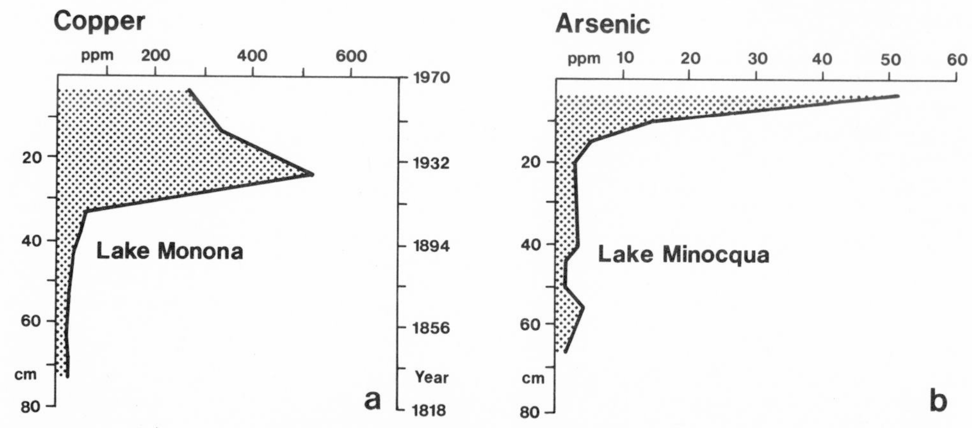

Fig. 9. Variations of copper (left) and arsenic (right) with sediment depth in lakes of Wisconsin. (After Iskandar and Keeney, 1974 and Shukla et al., 1972.)

An instructive example of atmospheric influences on metal concentrations in lacustrine deposits is given by Crecelius and Piper (1973) from sedimentary core studies of Lake Washington (Fig.10). Enhancement of *lead* concentrations from a background level of 20 ppm Pb during the 1880's corresponds to the development of land along the western slope of Lake Washington. The population of Seattle rose from 7600 in 1880 to over 100,000 in 1900. During this same period the Tacoma smelter, 50 km south of

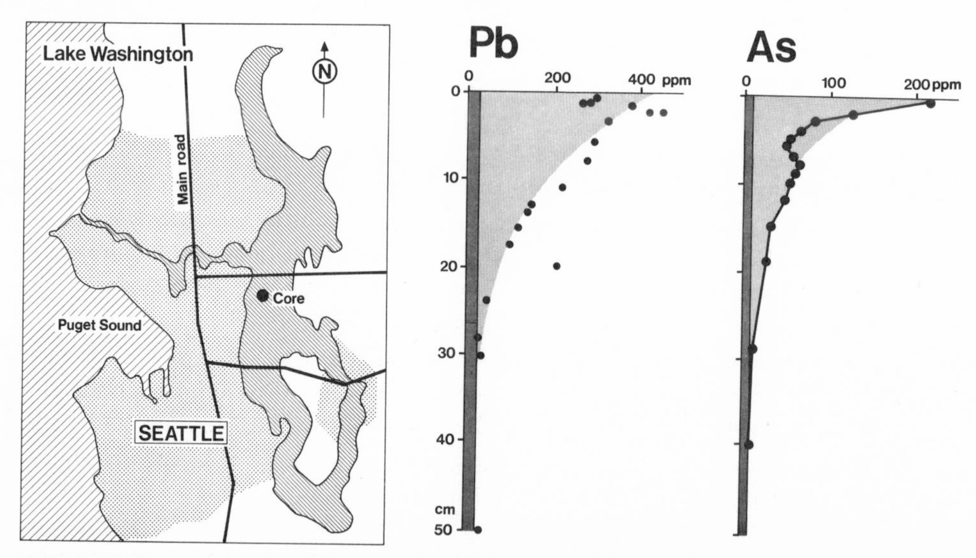

Fig. 10. Vertical distribution of Pb and As in sediments from Lake Washington. (After Crecelius and Piper, 1973.)

Seattle, began its operations. Atmospheric discharges from the smelter are held responsible for a 5- to 10-fold increase in the Pb concentrations observed in the sediments up until 1916. Conversion of the smelters from Pb to Cu production is reflected by a sharp decrease in the Pb values in sediments deposited between 1916 and 1920. Above this layer, Pb increases strongly to its present value of 400 ppm, probably as a result of the use of gasoline additives, which began in the 1920's. Similar enrichment of *arsenic* in present deposits from background concentrations of about 10 ppm to more than 200 ppm is attributed to atmospheric fallout from the Tacoma smelter.

Industrial effluents: Lake Erie. Trace and minor element concentrations in sedimentary cores from eight stations in Lake Erie were determined by Walters et al. (1974). Table VIII lists the data of a core from the central basin of Lake Erie near Cleveland at the mouth of the Cuyahoga River. The graphic example (Fig.11) seems to be representative of a moderate pollution in that area, whereby, according to the distribution of *mercury* in the sedimentary record, early cultural activity might have taken place beginning in about 1835. Major increases in the values of *zinc, arsenic* and *copper* during 1939–1955 reflect the general growth of industry during World War II and the Korean conflict. Strong enrichment of *chromium* and metal occurred during the late 1940's, which corresponds for the most part to the growth of the Cleveland electroplating industry. It is assumed that the establishment of chemical plants in the Cleveland area in 1949 caused a characteristic increase in the Hg concentrations. The major break in 1955 would then correspond to the opening of the Detrex chlorine-alkali plant near Ashtabula, Ohio.

Among the toxic metal contaminants, Hg and Cd have received the most attention – *cadmium* because of its relative high mobility and its possible release during chemical and physical changes in polluted sediments (redox changes, dredging activities; Förstner, 1979), and *mercury* in respect to its distinct tendency to form residues in aquatic ecosystems (Walters et al., 1974).

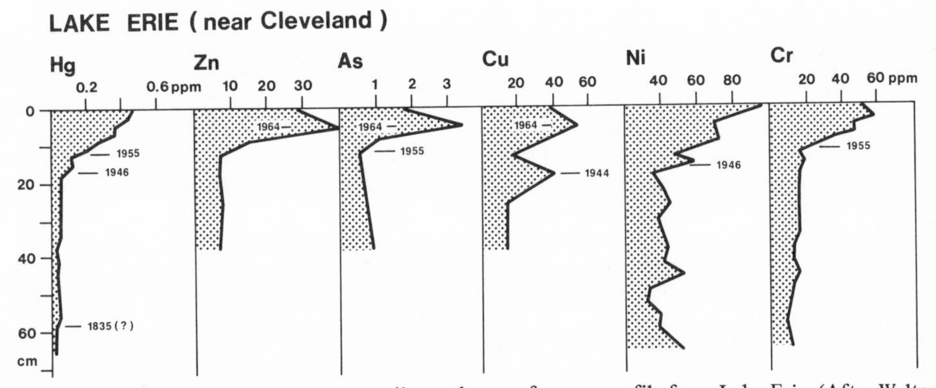

Fig. 11. Metal enrichment in the upper sediment layers of a core profile from Lake Erie. (After Walters et al., 1974.)

Ionic mercury is quickly removed from solution and immobilized at a level above its very low background concentrations (Jonasson, 1970). Microbial processes, however, are capable of converting originally inorganic Hg into more mobile and more toxic methylated forms. Methylation rates are particularly affected by a good supply of nutrients and elevated water temperatures. Bioturbation of the upper sediment layers by macroorganisms, e.g., tubificids and mussles may prolong the release of methylated mercury from polluted bottom mud for decades (Jensen and Jernelöv, 1969; Armstrong and Hamilton, 1973).

Several methods for the restoration of Hg-contaminated water bodies have been suggested (Jernelöv and Lann, 1973): (1) dredging of polluted sediments, (2) converting Hg to Hg sulfide with a low methylation rate (anaerobic conditions), (3) binding of Hg to silica or coprecipitation with hydrous Fe- and Mn-oxides (aerobic conditions), (4) increasing the pH to form volatile dimethyl mercury rather than monomethyl mercury, (5) isolating the polluted sediments from the water body by means of physical barriers, such as polymer film overlays, blankets or plugs of waste wool, sand and gravel overlays. As to the latter treatments, which seem to be particularly promising, Thomas (1974) reports the case of an unintentional, but very effective covering of mercury-polluted sediments by waste products from the mining industry.

Fig. 12 shows the distribution of tailings from a taconite processing plant in the Silver Bay of Lake Superior. From the analytical data obtained for a sediment core taken in the close proximity of the area in question (shaded area), it is evident that Hg enrichment, which has occurred at sediment depths of 20 to 8 cm, has been interrupted in the upper part of the profile and that the Hg content has decreased again to its original value before pollution occurred. This opens up the possibility of restoring large expanses of polluted areas with lasting effect.

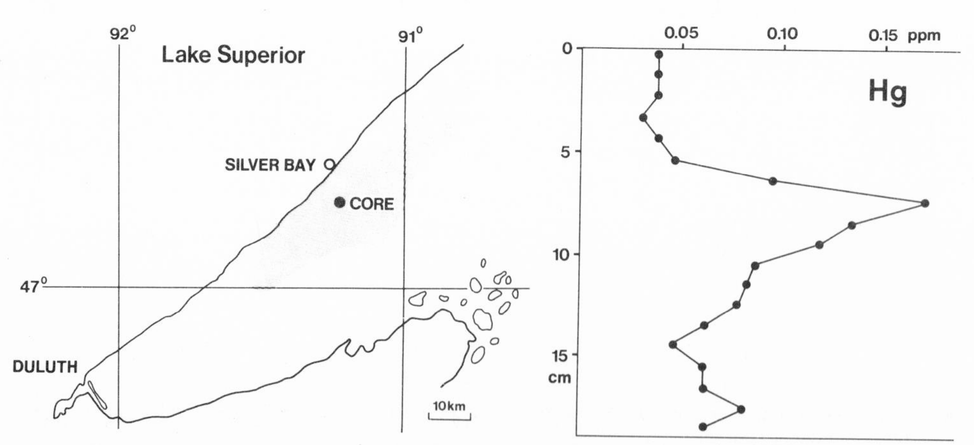

Fig. 12. Dispersion of mine tailings near Silver Bay of Lake Superior and vertical distribution of mercury in lake sediments. (After Thomas, 1974.)

Recently T.A. Jackson (1978) has suggested that small lakes or settling ponds with attendant H_2S production serving as disposal sites for toxic metal wastes may be an effective and economical method for preventing metal pollution in natural waters. As in Schist Lake in Canada, the metals would be largely trapped and immobilized as sulfides or as a sulfide-organic complex in the bottom sediments, and the waters draining the disposal site would be relatively free of heavy-metal contamination. Sewage from the nearest community would be a cheap, convenient, and virtually limitless source of available P and N for the algal bloom necessary for the process – in addition, sulfate could be used to stimulate SO_4-reducing bacteria.

Geochemical reconnaissance of limnic sediments

Heavy-metal analyses of limnic sediments have been standard practice in mineral exploration. The *stream sediment sampling* method is preferentially employed in remote areas in order to obtain a preliminary idea of the possible mineralization zones (Boyle et al., 1955; Hawkes et al., 1957). The variation in heavy-metal content of stream sediments can be characterized by the following model as a function of potential controlling factors (Dahlberg, 1968):

$$T = f(L, H, G, C, V, M, e)$$

T represents the resulting trace-metal concentration, L the influence of lithological units, H hydrological effects, G geological features, C cultural (man-made) influences, V the type of vegetational cover, M effects of mineralized zones and e refers to the error plus effects of additional factors not explicitly defined in the model. In mineral exploration the main problem is to maximize the factor M, i.e., to eliminate other effects as far as possible. However, the effects of the cultural influences and the mineralized zones interfere strongly and, as has been shown by a large number of investigations in the field of mineral exploration, traditional mining regions are particularly susceptible to contamination by metals. Airborne dust and particulate material from smelter stacks, wind-blown particles and leachate from tailings all contribute to an increase in the metal concentrations in the river sediments (see below).

At the same time, seasonal variations in stream sediments seem to be relatively slight and insufficient to cause serious problems in routine application of geochemical sediment surveys, as has been shown by the work of Govett (1960), Barr and Hawkes (1963), Garrett (1969), Howarth and Lowenstein (1971), Bølviken and Sinding-Larsen (1973), and Chork (1977). It was demonstrated in the latter study that measurement errors for Cu, Pb, Zn, Co, Ni, Mn, and Fe resulting from sampling, sample preparation, and analysis are small compared to the regional geochemical variation. The sampling error, which constitutes the main source of procedural errors (Miesch, 1967), is greatly in excess of the laboratory error for these elements.

Three categories are used in mineral exploration to characterize the distribution of metals: background, threshhold and anomaly. If the "log-normal distribution of the elements as a fundamental law of geochemistry" (Ahrens, 1957) is taken into account, it is possible to determine by statistical analysis the "threshhold" and "anomaly" levels as deviations from this distribution. An example of this is given in Fig.13 (after Hilmer, 1972) from measurements of the *cobalt* content of 670 individual samples from river sediments taken from the area around the Meggen deposit (for geology see Chapter 9, this volume, in this Handbook series) in northern Germany. From the breaks in the straight line, which result from the overlapping of two logarithmic distributions with the frequency distribution, it is possible to distinguish the three categories "background", "threshold" and "anomaly". For the area around Meggen, the following data shown in Table IX were statistically determined by Hilmer (1972).

It is proposed that during initial prospecting of an area where the contrast between background and anomaly is found to be weak, more refined sampling or analyzing techniques, or both, may have to be employed to reduce the procedural error to a level that allows the geochemical variation to be detected. However, at the location of a well-defined anomaly, in which the anomalous values contrast strongly with the background,

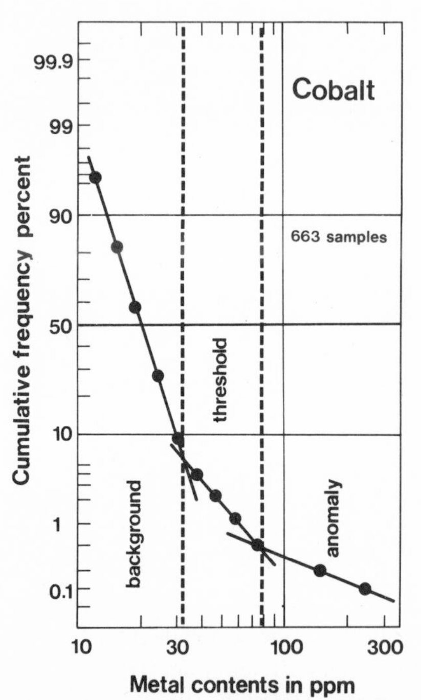

Fig. 13. Distribution of cobalt in 663 samples of fluviatile sediments from the lead–zinc district of Meggen, F.R.G. (After Hilmer, 1972.)

TABLE IX

Background, threshold, and anomaly levels (in ppm) for trace metals in river sediments in the area around Meggen deposit (After Hilmer, 1972)

	Background	Threshold	Anomaly
Zinc	<320	320–1500	>1500
Lead	<150	150– 700	>700
Nickel	<110	110– 300	>300
Copper	<55	55– 110	>110
Cobalt	<35	35– 75	>75

existing conventional techniques of sampling and analysis alone may be adequate (Chork, 1977).

For 20 years *lake sediment geochemistry* has also been used as a guide to mineralization. A center of recent activities in this respect is Canada, which has a bigger area covered by freshwater lakes than any other country. As early as 1956, Schmidt established anomalous metal distribution patterns in lake sediments bordering on areas of mineralization in New Brunswick and Quebec. From studies in Saskatchewan, Arnold (1970) and Dyck (1971) inferred that at low sampling density, lake sediments reveal the outline of metalliferous areas as accurately as stream sediments. Lake sediment geochemistry was first described by Allan (1971) in a survey covering an area of 3800 km^2 in the Coppermine region. Since then, thousands of mountain lakes (Appalachia, Cordillera), arctic lakes, prairie lakes, and, in particular, shield lakes have been analyzed, both for geochemical exploration and environmental management. Fig.14 and Table X give a compilation of lake sediment heavy-metal geochemistry study sites in Canada up to the middle of the 70's (Allan, 1977; after Nichol et al., 1975).

The general sequence of events leading to the ultimate *dispersion of metals into lake sediments* was depicted by Allan et al. (1974). According to the authors, the fate of metal ions originally derived from chemical weathering and mechanical disintegration of host rock is controlled by many factors involving atmospheric precipitation, water movement, soil movement, changes in redox and pH conditions, absorption–desorption processes, chemical complexation, precipitation and hydrolysis, uptake by and decay of vegetation and biochemical–bacterial interactions. Whether or not a specific flush load of freshly leached metal ions eventually reaches a lake system intact or widely dispersed depends on the relative interplay of these factors. A comparison of the predominance of metal in host rocks and in lake deposits is limited to a much narrower area than the corresponding levels for bedrock of the catchment area (Fig.15). This means that a relatively small number of samples will generally suffice in the exploration of lake sediments as compared to the number needed for soil or rocks. It can be assumed that by careful selection of lake sediment samples, the minimum number could be even further reduced. It may

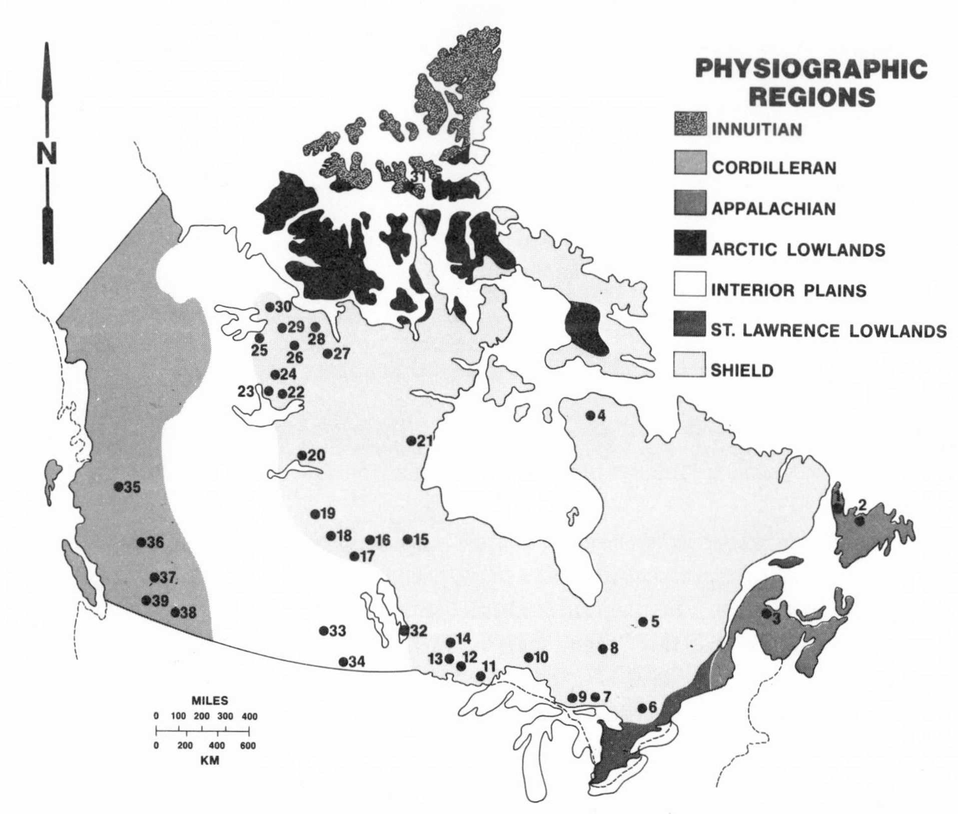

Fig. 14. Lake sediment geochemical studies in relation to physiographic regions of Canada. (After Allan 1977; reproduced with permission of the author.)

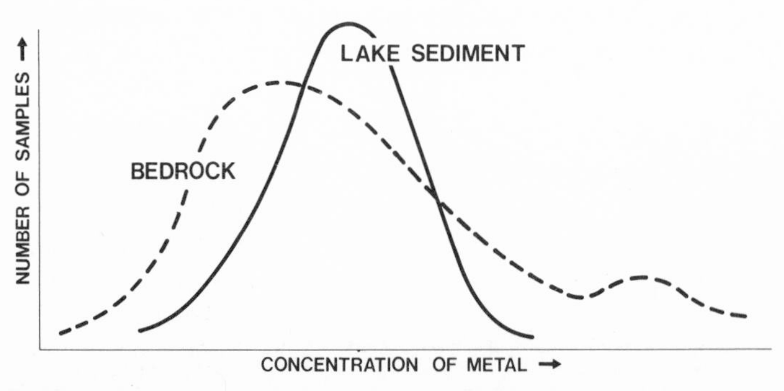

Fig. 15. Schematic presentation of the dispersion process of metals into lake sediment: the effect of dispersion on relative concentrations. (After Allan et al., 1974.)

TABLE X

Lake sediment heavy metal geochemistry study sites and investigators [1] (After Allan, 1977)

Site [2] number	Site name	Investigators [3]
1	Northern Newfoundland	Davenport et al. (1975)
2	Northern Newfoundland	Hornbrook et al. (1975)
3	Bathurst	Schmidt (1956)
4	Raglan	Allan (1973)
5	Chibougamau	Schmidt (1956) Allan and Timperley (1975)
6	Clyde Forks	Allan et al. (1974)
7	Sudbury	Allan and Timperley (1975)
8	Timmins-Val d'or	Hornbrook and Gleeson (1973)
9	Elliot Lake	Closs (1975)
10	Manitouwadge	Coker and Nichol (1975)
11	Shebandowan	Coker and Nichol (1975)
12	Sturgeon Lake	Coker and Nichol (1975)
13	Northwestern Ontario	Brunskill (1971)
14	Red Lake	Allan and Timperley (1975)
15	Wintering Lake	Bradshaw (1975)
16	Fox Lake	Bradshaw et al. (1973)
17	Flin Flon	Arnold (1970)
18	Northern Saskatchewan	Hornbrook et al. (1975)
19	Mudjatic	Haughton et al. (1973)
20	Beaverlodge	Dyck (1971)
21	Kaminak Lake	Klassen (1975), Klassen et al. (1974)
22	Harding Lake	Allan et al. (1972)
23	Yellowknife	Nickerson (1972), Jackson and Nichol (1974)
24	Indian Lake	Allan et al. (1972), Nickerson (1972)
25	Terra Mine	Allan et al. (1972)
26	Bear-Slave	Allan and Cameron (1973)
27	Hackett River	Allan et al. (1973), Cameron and Durham (1974)
28	High Lake	Allan et al. (1972)
29	Muskox Intrusion	Allan et al. (1972)
30	Coppermine River	Allan (1971)
31	Little Cornwallis	Allan (1974)
32	Lake Winnipeg	Allan and Brunskill (1977)
33	Qu'Appelle Lakes	Allan (unpubl.)
34	Boundary [4]	Allan (unpubl.)
35	Babine Lake	St. John (unpubl.)
36	Central B.C.	Mehrtens et al. (1972)
37	Rayfield River	Hoffman and Fletcher (1972)
38	Kamloops Lake	St. John (unpubl.)
39	Okanagan Lakes	St. John (unpubl.)

[1] This is a simplified version, but with additional references, of a table in Nichol et al. (1975).
[2] See Fig. 14 for location.
[3] This list includes to the best of the author's knowledge, all published data up to October 1975. Nearly all of these studies involved more than one lake. There may be other references where only one lake was sampled or where the heavy-metal investigations were only a minor aspect of the study. Some reference to unpublished data is made for sites in the Prairies and in Southern B.C. because these are locations where data are still relatively scarce. Reference is not made to anticipated results from the Shield, because there is already a reasonable volume of published data for this area.
[4] The boundary site is a reservoir used to cool thermal effluent from an adjacent lignite burning power station. All of the other sites referred to here are natural lakes. (Allan, 1977.)

also be possible to break down the wide range of the measurement data by separating the lake sediment samples according to grain sizes, whereby satisfactory conclusions can be drawn as to the movement of metal in the catchment area of the lake with relatively few samples.

During their studies of streams and snow-melt waters in the Arctic, Timperley and Allan (1974) found a number of peculiarities unique to *permafrost terrain.* For example, once the spring thaw has completely taken its course, there is little sign of base metals such as Zn, Cu, or Hg in the stream bed materials. In the presence of salt accumulations from the winter snow cover, these metals are rapidly carried away in soluble form. At the same time, however, elements such as As, Pb and Ag which are physically trapped in iron oxide precipitates survive in the stream sediments in much greater proportion and are therefore more useful in geochemical prospecting.

While little organic matter is involved in the transport processes in these lakes and rivers (except perhaps in the case of uranium dispersion; Timperley and Allan, 1974), there is a distinct correlation between almost all trace-metal contents (in particular for copper) and the organic carbon content of the sediments in the *lakes of northwest Ontario.* Hence, it can be concluded that sampling and analysis of lake sediments rich in organics may be a viable exploration procedure in the search for Ni–Cu and massive base metal sulfide deposits in the southern areas of the Canadian Shield (Coker and Nichol, 1975).

Mining activities are often associated with higher metal levels in the environment. Difficulties arise, therefore, in distinguishing natural metal anomalies, which are the prerequisite for the economic interest in ore exploitation, from the subsequent effects of ore extraction procedures. This is especially valid for areas where mining operations have been carried out for some time and where environmental contamination by mining operations occurs in one or more of the following ways: (1) contamination by tailings introduced as solids into drainage systems, (2) by leachates from on-shore tailing piles, (3) by effluents from mines or mining plants, and (4) by airborne particulate from crushers and smelters.

Examples of such an *exploration-environmental* geochemistry "overlap" was given by Allan (1974) from the large *nickel mining area at Sudbury* (Ontario, Canada). Fig.16 shows the distribution of Ni in sediment cores in and around the Sudbury Basin. Approximately 150 cores of lake sediments were collected from sites of mining activities of the Superior Province of the Southern Canadian Shield. The mining region of Sudbury consists of a Proterozoic norite intrusion, surrounded by Archaean granites. The intrusion, referred to as Sudbury Nickel Eruptive, contains many Ni–Cu sulfide ore deposits. The oldest mines at Sudbury have been in operation for about 80 years (apart from the Ni contamination around Sudbury, the effects of As at Red Lake and of Cu at Chibougamau were also studied in detail by Allan and Timperley, 1975).

The cores were taken from the center of the lakes (as that point is most likely to reflect the overall variation within the lake in relation to man's activities). Most of the

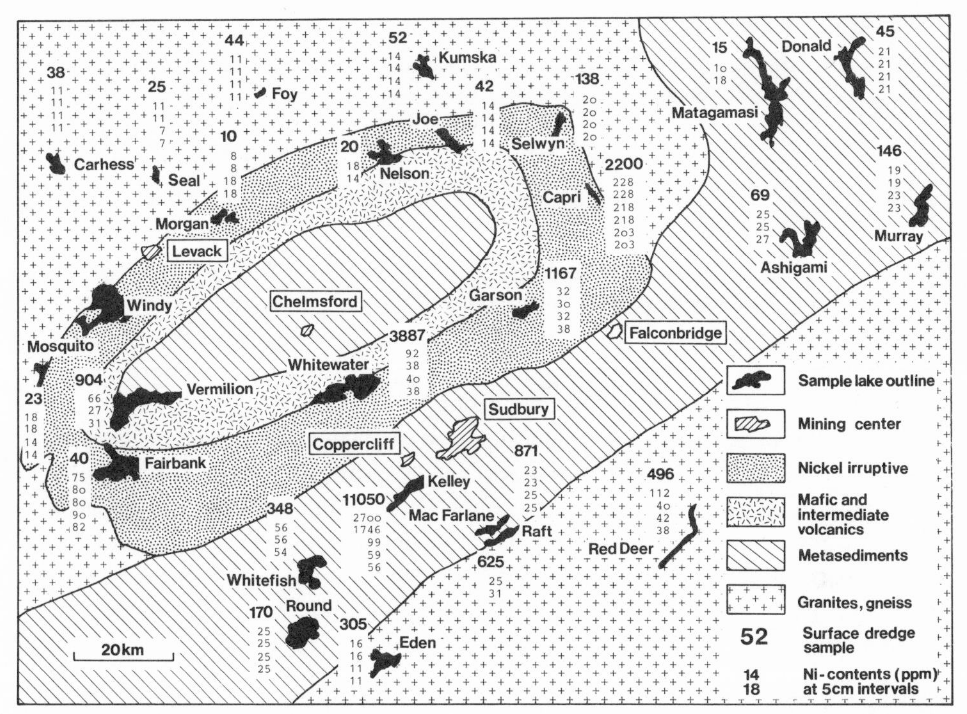

Fig. 16. Distribution of nickel in sediment cores from lakes in the Sudbury mining area of Canada. (After Allan, 1974.)

lakes had accumulated less than 10 cm of sediment within the last hundred years. Thus, samples of sediment cores taken from below this depth reflect the natural levels which occurred prior to ore exploitation, whereas samples within the upper 10 cm level can be expected to indicate the extent of subsequent mining activities. This is substantiated by the fact that only surface samples, collected with the aid of a grab, contain abnormally high nickel levels. In general, the Ni levels in the different sections of core samples do not decrease significantly with increase in depth, neither are they consistently higher in lakes within or without the Sudbury Basin. Most Ni concentrations in the 10–15 cm section of core samples vary between 10 and 50 ppm; exceptions are the Lakes Capri (north of Falconbridge) with an average of 200 ppm Ni in the deeper part of the core, and Lake Fairbank, which contains elevated Ni concentrations of 80 ppm in that section. In Kelley Lake, adjacent to the Coppercliff smelters, the Ni content, even at levels down to 15 cm of the sediment was elevated to such an extent that "it may prove economically feasible to reclaim this or other lakes for their metal content" (Allan, 1974).

AUTOCHTHONOUS PROCESSES AFFECTING METAL CONCENTRATIONS IN RECENT LIMNIC SEDIMENTS

Of the various processes taking place within a freshwater system that are potentially able to enrich metals in the sediment, two groups of enrichment mechanisms can be distinguished according to their spatial and temporal occurrence (Förstner and Müller, 1975): (1) the formation of minerals and organic phases directly from the water by means of such effects as precipitation, sorption, enrichment in organisms, organo-metallic interactions, and (2) diagenetic effects involving the redistribution of metals after the deposition of sediment particles.

A similar distinction has recently been made by B.F. Jones and Bowser (1978) in their contribution to Lerman's (1978) compilation:

(1) *Endogenic fractions,* referring to minerals originating from processes occurring within the water column. Metal accumulation mainly results from chemical precipitation or absorptive uptake of metals from aquatic solution. In addition, settling of particulates, filtering organisms, and flocculation represent other important mechanisms for the enrichment of chemical elements. Endogenic processes exhibit a distinct temporal character, predominantly as a result of the variations of the organic productivity, e.g., for carbonate precipitation.

(2) *Authigenic fractions* are those minerals which result from processes occurring within the sediments once they are deposited. Both the allogenic (or allochthonous particles) and endogenic fractions are subjected to changes in the physico-chemical environment, which are mainly the result of the decomposition of buried organisms. There is quite often a physical or chemical mixing by bottom-dwelling organisms, which promote the transfer of solutes from sediments into the overlying water via pore fluids. Lakes show particularly high fluxes and significantly higher pore-fluid concentrations than most of the other aquatic environments.

A compilation of *mineral phases* resulting from endogenic and authigenic sources in *freshwater lake sediments* is given by B.F. Jones and Bowser (1978). In *saline lake sediments* additional mineral phases occur: Na carbonates, Na–K–Mg–Ca–Sr sulfates, chlorides, nitrates, borates, and silicates (such as kenyaite, magadiite and analcime). Reviews on the types, occurrence, and genetic implications are presented by Reeves (1968), Müller (in Füchtbauer and Müller, 1970), and Hardie et al. (1978). The significance of *organic matter* in autochthonous metal accumulation has been reviewed by Saxby (1976).

A compilation of examples of authochthonous mineral formations and their effects on the enrichment or dilution of heavy metals in Recent lake deposits is presented in Table XI; these data, however, should not be generalized in all cases, since the selective separation of phases is often incomplete (see below).

TABLE XI

Effects of autochthonous mineral phases on the enrichment (+, ++) or reduction (–, ––) of heavy-metal concentration in lake sediments

Mineral phase	Type of source		Lake example(s)	Effect (examples)
	endogenic	authigenic		
Carbonates				
Aragonite	x		Aci Crater Göl [2] (Turkey)	–– all heavy metals
Calcite	x	x	Bande Amir [2] (Afghanistan)	– all heavy metals
Mg calcite	x	x	Balaton [2] (Hungary)	– Fe, ± heavy metals
Dolomite		x	L. Neusiedl [2] (Hungary, Austria)	± Mn, Zn
			Dasht-i-Nawar [2] (Afghanistan)	– Ni, Cr, etc.
Rhodochrosite		x	–	?
Siderite		x	Black Sea [3]	+ Mn, ± heavy metals
			Birket Ram [4] (Israel)	no effect
Sulfates				
Gypsum	x		Australian [5] lakes	–– all heavy metals
Glauberite	x		Curtain Springs [5]	–– all heavy metals
Alunite	x		Lake Brown [2]	± Pb, – heavy metals
Opaline silica	x		Lake Malawi [2]	–– heavy metals, – Cu
Sulfides				
Hydrotroilite	x	x	L. Constance [5]	+ Ni, Cu, Cr
Pyrite		x	Black Sea [6]	± Pb, Zn, Mn; ++ Cu, Ni, Co; ± Cr, Zn
			Lake Kivu [7]	++ Ni
Nontronite		x	L. Malawi [5]	± heavy metals, + Co, Mn
Hydrous Fe/Mn oxides	x	x	Australian lakes [8]	++ Cr, Ni, Co
Phosphates		x	–	?

[1] After Jones and Bowser (1978); [2] Förstner (1978); [3] Stoffers and Müller (1978); [4] Singer and Navrot (1978); [5] Förstner (unpubl. data); [6] Volkov and Fomina (1974); [7] Degens et al. (1972); [8] Förstner (1977d).

Dilution of heavy metals by endogenic minerals in lake sediments

In this section examples are described where a significant reduction in heavy-metal concentration is caused by the formation of minerals in both freshwater and saline lake environments, in respect to carbonates, sulfates, and silica. The data refer to examples given in Table I of this study (see pp.184–185).

The effects of intra-lacustrine dilution of metals by *Ca–Mg carbonates* are shown by comparing carbonate-free lake sediments (groups I–III) with deposits of the sub-brackish and playa lakes (group IV/2–3) which contain on an average of about 50% Mg–Ca carbonates in the pelitic fractions. Data from Table XII indicate that dilution by calcite, dolomite (and huntite in Tuz Gölü) is strongest for Fe (a reduction of 75%), less strong for Ni, Cr, Zn, and Mn (50%); for the contents of Cu, Pb, Co and Cd there is little dilution.

Dilution by aragonite sedimentation is shown by a comparison of pelitic sediments from Burdur Gölü ($<$20% $CaCO_3$, mainly aragonite) with those of the Aci Crater Göl (90% aragonite), both lakes situated in Anatolia/Turkey. The data indicate that in Lake Aci Crater, there is a strong reduction of Fe, Cr, Ni, Hg and Co, as well as a somewhat diminished Zn and Pb content. Unexpected, however, is the strong dilution of Mn in the Aci Göl sediments. The latter might partly be explained by the low incorporation of Mg into the structure of aragonite.

Precipitation of (low-Mg) *calcite* in the lakes of Bande-Amir in the Hindukush Mountains of central Afghanistan effects a significant reduction of all heavy-metal examples studied (Fe, Mn, Zn, Cr, Ni, Cu, Pb, Co); the dilution is, however, not as strong as with aragonite in the above-mentioned example from the Aci Crater Göl. The effects of *Mg-calcite* formations which are considered to be mostly diagenetic, have been studied on samples from Lake Balaton in Hungary. There is no characteristic reduction of most of the heavy-metal concentrations, except for Fe (similar to the findings from the example of Hamun-Puzak in southwestern Afghanistan). Selective chemical extraction of carbonate phases from the bulk sample (see below) of Lake Balaton, however, indicates a significant dilution effect on the contents of Cr, Ni, Co and – to a lesser degree – Zn and Cu (Schmoll and Förstner, 1979).

The effects of *sulfate formations* such as gypsum and glauberite, were investigated by analysis of large crystals separated from fine-grained sediments from dry lakes in western and central Australia. There is a strong dilution of all heavy metals studied. *Alunite,* $KAl_3(SO_4)_2(OH)_6$, was observed as a minor constituent in the pelitic fractions of six lakes studied in Australia. In Lake Brown, 60–70% of the total sediment consists of this mineral which is believed to have formed from feldspars during laterization of gneisses via an intermediate kaolinite stage (King, 1953). Compared with alunite-free lake deposits (e.g. Lake Moore), the contents of Mn, Cr, Co, and Fe in Lake Brown are particularly reduced. Pb, however, shows no decrease, and this is probably due to the crystal-chemical similarity of K and Pb.

TABLE XII

Autochthonous factors affecting the dilution of metals (ppm) in lacustrine sediments [1]

	Ca–Mg carbonates [1]		Aragonite		Alunite		Diatoms	
	I–III (50)	IV-2/3 (6)	Burdur Gölü	Aci Crater Göl	Lake Moore	Lake Brown	Lake Malawi Core No. 5 10–23 cm	35–50 cm
Iron	47,100	16,900	46,000	4,400	34,500	9,500	77,800	29,000
Manganese	902	475	480	46	640	25	700	190
Zinc	124	63	95	25	90	30	155	87
Chromium	91	42	130	12	92	21	92	36
Nickel	119	46	318	20	75	35	90	26
Copper	46	34	35	11	35	20	60	40
Lead	29	21	15	8	35	32	17	10
Cobalt	19	16	23	2	11	3	22	10
Mercury	0.41	0.46	0.19	0.03	0.54	0.19	0.32	0.16
Cadmium	0.42	0.39	0.20	0.20	0.24	0.18	0.20	0.15

[1] See classification in Table I.

In the sediments of highly productive lakes, *opaline silica* often constitutes one of the major endogenic formations. Table XII lists the data from Lake Malawi (described in more detail below), where allochthonous sediments are locally mixed by silica from *diatoms* (forming almost pure "diatomite" in the southern part of the lake, overlain by a few centimeters of diatom-free sediment). The dilution of heavy metals by silica in the Lake Malawi sediments differs from that noted for aragonite and alunite, in that there is relatively weaker reduction for the concentration of Fe, Mn, Zn, Cu and Co, probably due to the essential function of these elements within the biologic life cycles.

Heavy-metal accumulation in endogenic phases of lake sediments

Metal enrichment in sulfides (and related organic phases). Accumulation of trace metals in sulfidic phases is found to result both from direct precipitation (endogenic processes) and from diagenetic effects (post-depositional transformation). This is apparent from the major sulfide phases in aquatic sediments — the associations of iron and sulfur. The mechanisms of these formations and transformations have been outlined by Berner (1971) and Sweeney and Kaplan (1973) (for details see Chapters 6 and 10 of Vol. 2 in this Handbook series).

Unstable iron sulfide of the hydrotroilite type ($FeS \cdot H_2O$) are for the most part finely dispersed (black mud) and are instantly oxidized under aerobic conditions, which usually makes the separate analysis of these components and their sorbed trace metals impossible. Following the description of FeS concretions from Lake Constance surface sediments by G. Wagner (1971), the separation of coarse-grained sulfide particles from the fine-grained matrix was attempted using sieving. Initial investigations of a concentrate of such aggregates (Förstner, unpubl. data) indicate an approximate six-fold enrichment of iron compared with the composition of the surrounding sediment. When compared to the composition of the fine-grained sediment matrix, Co, Cu, and Cr were two times greater, Ni 1.5 times, whereas Pb, Zn, and Mn did not undergo a significant increase. It may, therefore, be concluded that coprecipitation with iron sulfides is less effective in concentrating trace metals than is the incorporation into hydrous iron oxides (see below).

Metal concentrations in both endogenic and diagenetic sulfidic phases have been studied intensively from the *Black Sea,* presently the largest existing anoxic water body (500,000 km^3).

Vertical profiles of water samples were analyzed by Brewer and Spencer (1974) during cruise 49 of the R/V "Atlantis II" of the Woods Hole Oceanographic Institution in early 1969. Their data show that the distribution of elements such as Mn, Fe, Cu, and Zn is considerably affected by redox reactions at the boundary between oxygenated surface waters and sulfide-containing deep water (Fig.17). Cu, Zn, (and Pb) are immediately precipitated in the anoxic water layer (up to 2000 m thickness) when introduced by detritus and tripton through the pycnocline. Mn remains partly in solution as the redox values of −100 to −200 mV hinder the precipitation in the form of manganese sulfide. At

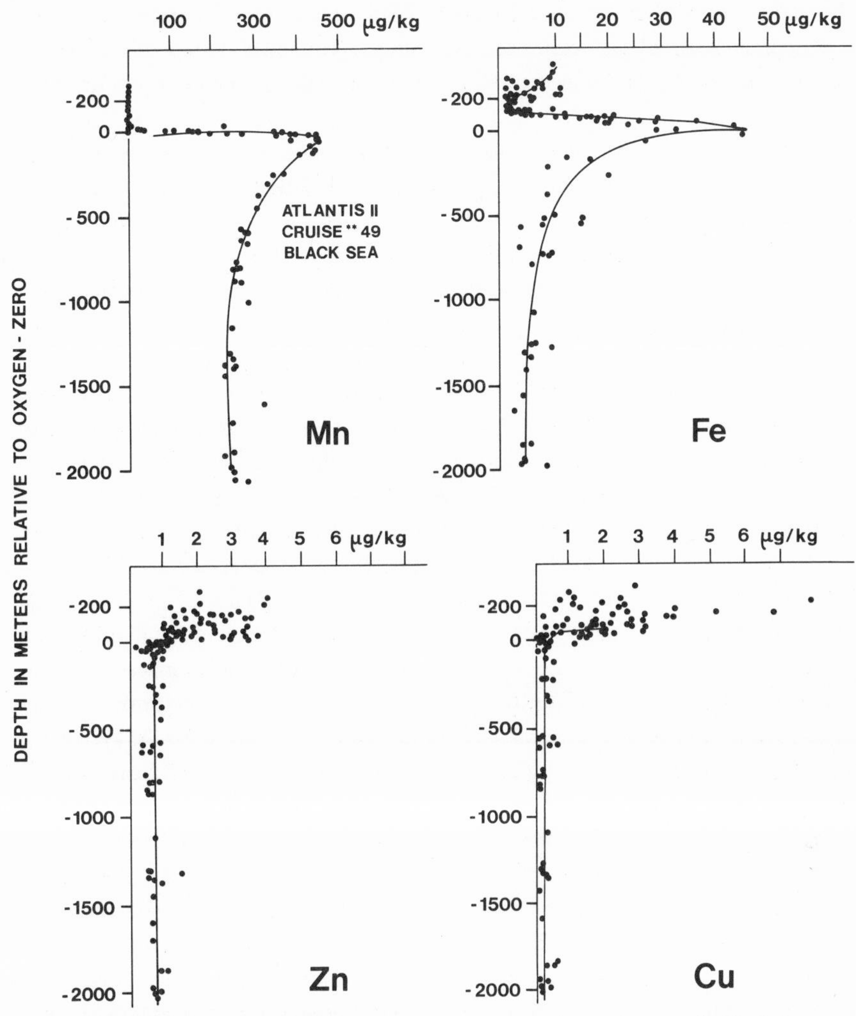

Fig. 17. Vertical distribution of dissolved manganese, iron, zinc and copper in Black Sea waters. (After Brewer and Spencer, 1974.)

the O_2–H_2S interface, a manganese oxide phase is formed by in situ precipitation; when the particles sink through the interface, they are redissolved and again move across the phase boundary into the oxygenated environment. At the same time the sulfide components of Fe, Zn, Cu settle to the sea floor. Degens and Stoffers (1977) have suggested that by these processes "metal enrichment of economic value may occur only if the supply of detritus is greatly reduced."

It seems that the simple model outlined on the vertical distribution of trace metals in Black Sea water must be modified or further developed in some respects. Generally, most of the trace metals seem to occur in substantially lower concentrations compared to ferrous iron in interstitial and other anoxic waters. Jenne (1976) has therefore suggested that these metals may be expected to coprecipitate with iron sulfides, rather than as discrete sulfide crystals. From investigations on the distribution of the microelements V, As, Cr, Cu, Pb, Co, Ni, and Zn in the pyrite and magnetic sulfides in Black Sea deposits, Butuzova (1969) found that all these elements – with the exception of V, Cr, and Zn – had accumulated in the iron sulfides. Volkov and Fomina (1974) and Philipchuk and Volkov (1974) found characteristic enrichment of Mo (36×), Cu (20.5×), Ni (17.6×), and Co (9.3×) in the pyrites from the modern deep-water deposits of the Black Sea when compared with the enclosing sediments. A comparison with the enrichment of trace metals in the pyrite minerals of the pre-Holocene Neoeuxinian deposits on the continental slope along the Anatolian shore of this water body, where the enrichment is significantly lower, indicates that the behavior of the elements in such enrichment processes depends on the sediment composition, which is mainly influenced by the concentrations of sulfur and organic carbon. Volkov and Fomina (1974) have suggested that some proportion of the Cu, Co, Ni, and V in Black Sea sediments is organically bound and that the metals are extracted from the water by settling action of organic substances.

Further information concerning the effects of organic and sulfidic phases for the enrichment of trace metals in Black Sea sediments has been presented from studies on core sediments from Leg 42B of the Deep Sea Drilling Project. A detailed investigation of the material from hole 379A in the central part of the Black Sea was performed by Calvert and Batchelor and published in the DSDP Initial Reports of April 1978. Fig.18 is a compilation of data for organic carbon (%), sulfur (%), and for selected heavy metals from 59 bulk sediment analyses: the fluctuations of organic carbon and total S contents correlate very well in the upper 250 m of the core (penetration 624.5 m; 2171 m water depth) where several horizons rich in pyrite and organic matter occur. Extreme enrichment of C (14.75%) and S (1.81%) was found in a sapropelic sample at 99 m depth; this sample also has the highest concentrations of Cu (97 ppm) and Mo (175 ppm), and relatively high concentrations of Ni and Zn. Factor analyses indicate that organic carbon, S, Cu, and Mo are closely interrelated, whereas Zn and Pb seem more associated with feldspars and Ni with chloritic alumosilicates. Whether the enrichment of Cu and Mo is due to organic or to sulfidic phases is not clear, since the distinction of these associations by selective extraction is still impossible. It has, however, been suggested that Mo is coprecipitated by FeS (Bertine, 1972); Sugawara et al. (1961) came to similar conclusions in studies on a Japanese lake.

At the same time, however, the vertical distribution of copper and nickel in the core profile from the Black Sea (Fig.18) does not indicate a significant enrichment of these metals in sulfidic phases – at least the data do not suggest that this mechanism could result in large-scale metal accumulations, or perhaps even deposits of economic interest. Degens and Ross (1972) have indicated that the distinctive sapropel in the modern sediments of the Black Sea was probably deposited during a period of increasing

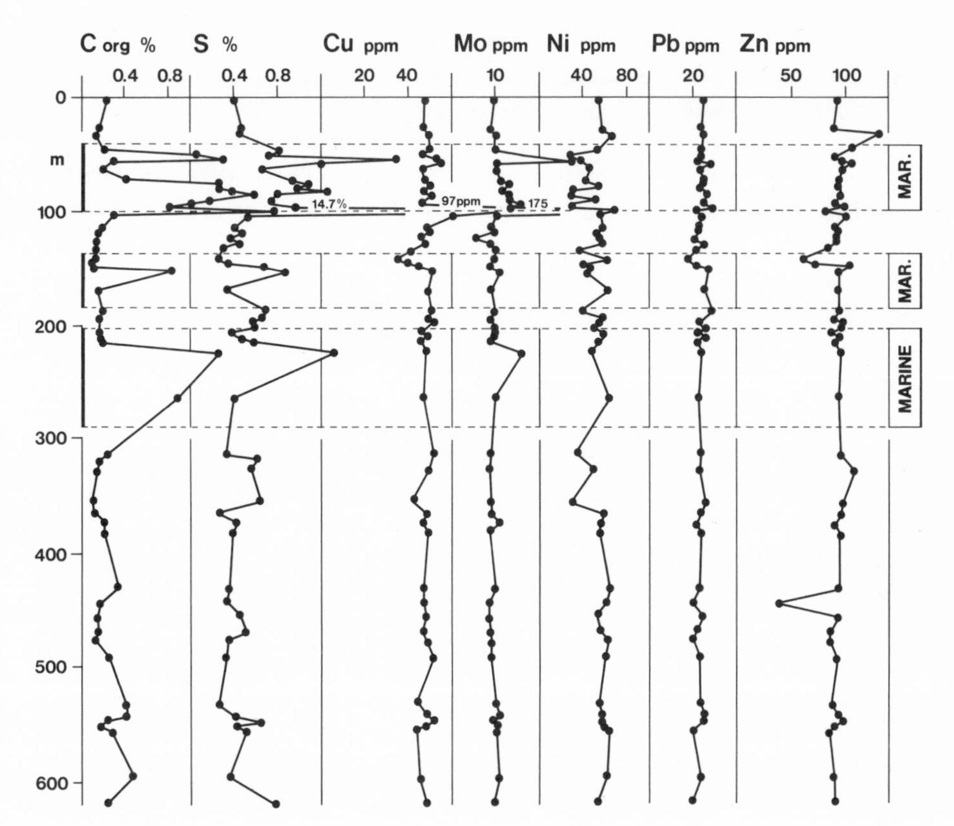

Fig. 18. Concentrations of organic carbon, sulfur and distribution of selected heavy metals in a sediment core profile from the Black Sea. (After Calvert and Batchelor, 1978.)

saline water influx into the late Pleistocene lake from the Mediterranean. Calvert and Batchelor (1978) suggest that earlier sapropels and organic- and pyrite-rich muds most probably represent similar periods when the water in the lake changed from fresh to saline (these variations are substantiated by the chlorinity data from pore waters). Thus, major enrichment of trace metals are associated with marine sapropelic sediments, whereas the lacustrine stages during the history of the Black Sea do not indicate any characteristic increase of the metal concentrations above normal levels.

The seemingly contradictory findings of a considerable precipitation of sulfide minerals from the water column, on the one hand, and the failure to detect significant enrichments of sulfidic heavy-metal phases in the corresponding sediments of the Black Sea, on the other, also become evident from other stratified water bodies, such as *Lake Kivu,* East Africa. This example has been thoroughly described by Degens and Ross (1976) in Volume 4, Chapter 4 of this Handbook. Microcrystalline sphalerite on resin globules are analyzed as suspended particles of up to more than 7 mg Zn/l in the anoxic water column below 60 m. This 200-fold enrichment of *zinc* in the deeper water has been suggested to originate from leaching of rock formations by hydrothermal solutions (Degens et al., 1972). Evidence of hydrothermal activities is mirrored, among others, in temperature and salinity; concentrations of methane in the lake water result from bacterial processes, partly promoted by thermocatalytic effects (Tietze, 1978). In the sediments of Lake Kivu, *nickel* is enriched in some layers in the form of Ni pyrite, and higher concentrations of Pb and Mo are associated with organic matter. The absence of significant accumulations of Zn in the sediments is explained by Degens et al. (1972) as due to their associations with organic chelates, which escape from rapid sedimentation since they are buoyant, finely dispersed globules. Suspended sulfide spherules have been carried to Lake Tanganyika via the Ruzizi River outflow from Lake Kivu, particularly at times of more intense hydrothermal activity (Degens et al., 1973). This could explain the layers of framboid pyrites in the upper two stratigraphic units of Lake Tanganyika sediments (Degens et al., 1971). It should also be considered that dissolved organic and inorganic material can be released from the compacting strata at the sediment/water interface via the pore water (Brewer and Spencer, 1974; Wolf and Chilingar, 1976; see also the section on diagenetic effects, p.230).

An important area of present-day formation of massive and disseminated sulfur and sulfide deposits are the *post-magmatic hot zones* around active volcanic belts. Metal accumulations were first described from the marine region near the island Vulcano in the Tyrrhenian Sea, Italy by Bernauer (1933, 1935, and 1939; recent studies have been performed by Honnorez, 1969; Honnorez et al., 1973; Wauschkuhn and Gröpper, 1975). During the last decade similar effects have been found from examples in the lacustrine environment, e.g., from the volcanic zone of Taupo, New Zealand (Browne, 1971), from the volcanic lakes in Kamchatka and the Curiles (Ozerova et al., 1971) and from *Hokkaido,* the northern island of Japan (Saito et al., 1967; Yagi and Hunahashi, 1970; Wauschkuhn, 1973; Wauschkuhn et al., 1977). From the latter studies the following findings are notable with regard to sulfidic metal accumulations (both endogenic and diagenetic) in recent lake sediments:

(1) The sediments usually consist of allogenic quartz, authigenic native S and a clay fraction from the weathering of the country rocks. The main *authigenic sulfide ore minerals* in four Japanese lakes studied are *pyrite* and *marcasite,* but pyrrhotite, arsenopyrite, chalcopyrite, galena, and sphalerite are also present.

(2) A comparison of lower Yumoto Lake, which is less affected by *chemical changes,* with Ojunuma Lake, indicating strong alterations in the sediment chemistry in the deeper deposits (16–25 m sediment depth), is given in Table XIII, and intimates that there is a 7- to 10-fold increase of *Cu, Pb, Ni,* and *As* in the latter, more contaminated lake; the enrichment of Fe, Co, and Zn is 2 to 4 times greater.

(3) With regard to the *mechanisms of metal enrichment,* it is found that fumarolic gases from post-volcanic processes carry H_2S into the lake water, where it is then oxidized to sulfuric acid with the aid of microorganisms. This acid in turn attacks the country rock

TABLE XIII

Temperature, pH and Eh values of the water and chemical composition of the sediments (mean from 6 samples each) of two lake examples from Hokkaido, Japan (After Wauschkuhn et al., 1977)

	T (°C)	pH	Eh (mV)	S (%)	Fe (%)	Co (ppm)	Cu (ppm)	Zn (ppm)	Pb (ppm)	Ni (ppm)	As (ppm)
Yumoto	70	4.6	+	65.9	6.1	45	30	65	75	30	<0.1
Ojunuma	120	2.4	−150	71.9	12.0	114	213	242	575	348	1.8

and produces deep *weathering*. Characteristic depletion in the heavy metals of Co, Cr, Fe, Mn, Ni, Pb, and Zn takes place in the dacite rocks, through which the fumaroles pass. However, the altered rocks are at the same time enriched in S and As. No evidence has been given for a direct transport of heavy-metal concentrations from the volcanic centers via the fumaroles into the lake water.

(4) As a result of physico-chemical changes (*T, p*, pH, Eh, pS), the elements are partly concentrated in the lake sediments mainly as *metal sulfides*. The processes are mediated and accelerated by microorganisms.

Wauschkuhn et al. (1977) found that the metal concentrations increase in the (sulfur-free) sediment fraction with increasing water depth. *Zinc* (490 ppm) and especially *lead* (3390 ppm) show significantly greater accumulation in the sulfur foam when compared with the metal-rich sediment layers of Ojunuma Lake (see Table XIII). The latter effect possibly results from similar processes as those depicted by Degens et al. (1972) from Lake Kivu (see above), where the interaction of zinc sulfide with organic globules partly reduced the incorporation of this metal into the lake sediments. It seems that a distinct connection of inorganic processes involving sulfidic (and partly oxidic) phases and organic mechanisms also affects the accumulation of heavy metals. However, as is the case in the former example, the separation of both effects in the resulting deposits is not possible.

Metal enrichment in organic substances. The significance of organic matter in ore genesis has been thoroughly discussed by Saxby (1976) (Vol. 2, Chapter 5 of this Handbook). With regard to the limnic environment, it has become clear that solubilization of metals is achieved by the combined processes of complexation and reduction (Ong et al., 1970; Theis and Singer, 1974; Ridley et al., 1977), whereas the incorporation of the dissolved metals involves the mechanisms of adsorption onto organic compounds and their subsequent flocculation, polymerization, and precipitation (Jonasson, 1977). The latter processes are particularly significant in the estuarine mixing zone (Sholkovitz et al., 1978).

A high positive correlation between the contents of organic matter and heavy-metal concentrations has often been observed in aquatic sediments. This, however, may not always be interpretated as being due to direct bonding of heavy metals to organic substances. In more strongly polluted areas, for instance, contamination by heavy metals, such as Zn, Pb, Cd, and Hg, simply coincides with the accumulation of organic substances, both of which are derived from urban, industrial or agricultural sewage effluent. Extraction of metals from fulvic and humic acids in sediments from the less polluted Lake Malawi (see below) indicates that only Zn, Cu, and V are significantly accumulated in the (more reactive) organic matter, whereas Fe, Mn, Cr, Pb, and Co are more or less diluted by organic substances.

Jonasson (1976) found a significant enrichment of heavy metals in the organic fraction of deep-water sediments from Perch Lake in Lanark County/Ontario – it is assumed that gelatinous colloidal substances, which are formed from dissolved organic acids, spores, pollen and decayed leaves, take up the metal ion from water. Similar effects have been observed by Tobschall et al. (1978) from gels and sediments from Gjøkvatn, Fiskevatn, and Sortbrysttjer Lakes in subarctic northern Norway. The gels occurring in the deeper parts of the lakes consist of flocculated organic colloids, possibly humates and fulvates, and residual organic matter derived from detritus of land plants, algae, plankton, and pollen. Compared to the organic sediments, the heavy metals Ni, Cu, Zn, Pb, Hg, Th, and U are enriched in the organic gels by factors of 2 to 500.

Significant accumulation of *uranium in organic matter* was reported by Degens et al. (1977, 1979) from sediments of the Black Sea and a Holocene lake in central Ontario. In the *Black Sea* the highest content of uranium was determined in the brackish–marine coccolith ooze at the bottom of the euxenic abyssal plain. The bulk of this element seems to be bound to planktonic matter and coccoliths seem to be the prime host for uranium in modern Black Sea sediments (other planktonic organisms share this affinity). Metal-ion coordination to specific proteins and polysaccharides is suggested as the major mechanism of metal transport from the central cell to the outer membranes, where biomineralization takes place. Values of more than 100 ppm U_3O_8 are found in the coccolith ooze; the total concentration of U_3O_8 in the sediment ash of the top 1 m strata of the Black Sea basin is $6.7 \cdot 10^6$ metric tons [1].

Bow Lake is an elongated lake, approximately 1 km^2 in area, in the vicinity of Bancroft, Ontario. Until only a few years ago the lake was used by the Faraday uranium mine as a waste dump and large amounts of uranium still reach the lake by leaching (Degens et

[1] The economic exploitation remains under discussion. Degens et al. (1977) pointed out that "although U_3O_8 values of 100 g $tonne^{-1}$ are not economically significant now, the situation may change in the years to come should global demand for uranium expand at predicted rates". Although heat combustion of Recent Black Sea mud will release more energy than is required for drying and thermal degradation of mineral matter, there are considerable environmental hazards to be expected from the high sulfur content of the mud. Borchert (Letter to the Editor in "Öffentliche Anzeigen für den Harz", August 3, 1978), for example, believes that even in 1000 years, the geochemical uranium enrichment in the mud of the 2000 m deep Black Sea will not be sufficient to render exploitation economical. He points out that many continental oil shales and phosphate deposits are more concentrated and more economical uranium sources.

al., 1979). Particulate organic substances collected with a plankton net have a uranium content of 200–400 ppm. Organic-rich sediments (10–30% organic carbon) from water depths of 10–35 m are principally diatomite of reducing character and attain between 150 and 440 ppm U at the surface and between 50 and 100 ppm U at 1 m sediment depth.

According to Degens et al. (1977): "It seems that *reducing conditions* in the depositional environment are only essential for the *preservation* of uranium-enriched detritus, and not for the fixation. Following sedimentation, a series of *organic molecules may pick up* additional increments of uranium as well as other heavy metals from sediment and water."

Metal sorption onto hydrous Fe and Mn oxides in rivers. Accumulation of heavy metals in hydrous Fe and Mn oxides is considered to be mainly the result of diagenetic (or "authigenic", according to B.E. Jones and Bowser, 1978) transformation processes. Compared with the large number of investigations performed on the metal concentrations in ferro-manganese deposits for marine (Glasby and Read, 1976, Vol. 7, this Handbook) and lacustrine (Callender and Bowser, 1976, Vol. 7, this Handbook) environments, there are only limited data available as to the significance of *endogenic* processes of sorption and coprecipitation with hydrous Fe and Mn oxides occurring, for example, in rivers.

A practical example of the important role of *Fe and Mn oxides* on the distribution of trace metals in fluviatile systems is described by Cameron and co-workers (Cameron, 1974; Allan et al., 1974; Cameron and Ballantyne, 1975; Jonasson, 1976) from arctic areas of Canada. These regions represent somewhat simplified conditions for the study of the interactions between hydrous Fe and Mn oxides and dissolved metal ions because of the scarcity of organic materials.

Two watersheds supplied with metal ions by similar volcanogenic Cu–Zn-sulfide mineralization areas were compared. The first location, Hackett River in the Northwest Territories, exhibits a pH in excess of 7 due to the presence of carbonate-enriched volcanic rocks. The second area, at Agricola Lake, N.W.T., is virtually identical to the Hackett River except that the underlying volcanic rocks are devoid of carbonate minerals; here, pH is commonly between 3 and 4. In the alkaline waters in the area near the source of the Hackett River, Fe and Mn were initially present at very low levels, whereas in the Agricola Lake area, Fe and Mn concentrations reached 800 μg/l and 90 μg/l, respectively. At *Agricola Lake,* the levels of dissolved Zn, Cu, and As were found to fall off very rapidly, i.e., the content of Zn from about 1000 μg/l at the source to 1 μg/l 5 miles downstream. In the *Hackett River* area, in contrast, the dispersion of soluble Zn and Cu was significantly more extensive. According to Jonasson (1977) this is attributable to the absence of an oxide scavenger coprecipitating trace metals below the source in the Hackett River area, whereas rapid hydrolysis of the high initial content of Fe and Mn in the Agricola Lake waters resulted in the coprecipitation of Cu, Zn, and As as the pH increases with dilution. Recent studies on the selective extraction of metals from different sedimentary phases, some of which will later be described with respect to the limnic environments, have indicated that apart from the diagenetic effects (see below), freshwater precipitates of hydrous Fe/Mn-oxides may act as significant sinks of heavy metals. Cutshall (1967) for example, found iron oxides to be the most important single sediment component in the retention of ^{51}Cr in sediments from the Columbia River. Gibbs (1973) demonstrated that up to 50% of Cu, Cr, Co, and Ni is present in the form of hydrous Fe- and Mn-oxide coatings in (carbonate-free) suspended sediments of the Amazon and Yukon Rivers. Perhac (1974) determined that 5–50% of Co, Cu, and Zn is associated with the hydrous iron and manganese oxide fraction in fluviatile sediments of the Tennessee River. Lastly, a discussion of endogenic processes of metal accumulation on hydrous Fe and Mn-oxides in freshwater environments has been given by Jenne (1976).

Diagenetic effects on metal accumulation in limnic sediments

Chemical interactions between limnic sediments and the overlying or flowing water is largely regulated through solute transfer associated with pore-fluid processes. A schematic listing of these mechanisms has been compiled by Lerman (1977) and is given in a modified version by B.E. Jones and Bowser (1978) in Fig.19. Influxes *to* the sediment are by means of allogenic and endogenic particulate transport and burial of pore fluids, whereas fluxes *from* the sediment are largely dissolved solutes, which arise from molecular diffusion, compaction, bioturbation, and groundwater discharge in coarser-grained porous sediment.

These effects are especially important with regard to the dynamics of nutrient elements (see, for example, various papers in the symposium volume "Interactions between Sediments and Fresh Water", edited by H. Golterman, 1977), but should also be considered in connection with the accumulation of heavy metals in recent limnic deposits. Price (1973), Elderfield and Hepworth (1975), and Krom (1976) in investigations on recent marine deposits suggest that besides the well-known diagenetic redistribution of manganese, there might be an influence on the concentrations of other metals such as Cu, Ni, and Zn to some extent; enrichment of these elements at the sediment surface must not necessarily be interpreted as a civilizational pollution effect. Obviously, these processes must be studied in areas little subjected to man-made influences.

We have investigated the example of *Lake Malawi* situated at the southern end of the East African Rift Valley (Müller and Förstner, 1973). Lake Malawi (formerly known as Lake Nyasa) is about 560 km long and at its greatest width 75 km wide. It consists of a single basin with a greatest depth of 704 m (sketch map in Fig.20). The lake is surrounded by rocks mainly from the Precambrian "Basement Complex", which consists of various gneisses, schists, metamorphosed sediments with intrusions of syenites and associated rocks. At present the lake is stably stratified below 250 m with an anoxic hypolymnion. The pH value decreases from about 8.5 in the oxic epilimnion to about 7.7 in the hypolymnion (Eccles, 1975); the pore solutions (which were first studied several days

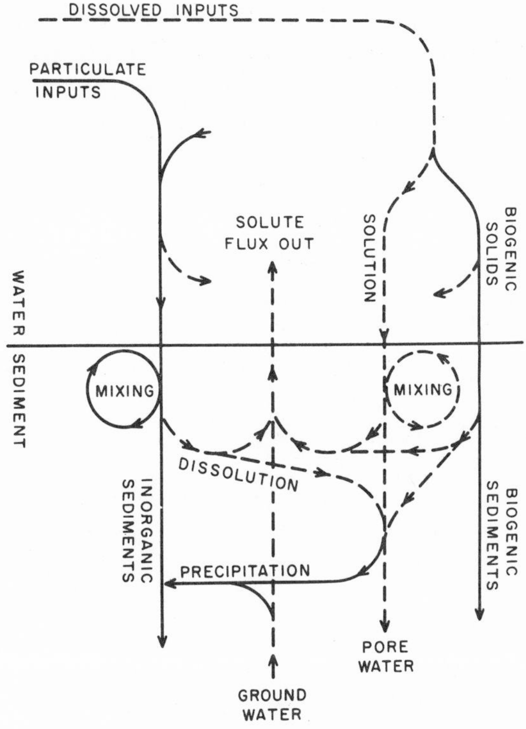

Fig. 19. Transport and reaction paths between sediments and lake water. (From Jones and Bowser, 1978; modified after Lerman, 1977; with permission of Springer-Verlag.) Solid phase routes are indicated by solid lines; solute phase routes are shown by dashes.

after extraction) had a very low carbonate content and thus a reduced buffer capacity, with pH values down to less than 3.

In water exhibiting low pH values the concentrations of dissolved Fe in the pore solutions normally are below 50 μg/l, but may increase to more than 1000 μg/l. Contrary to the strong dependencies of the Fe concentrations on the acidity of the pore water, there are largely irregular fluctuations determined for the distribution of Mn; for this metal, low values center around 100 μg Mn/l, but the majority of the data is above 1000 μg Mn/l (maximum concentrations 15 mg/l). Cu and Zn values are below 20 μg/l and cannot be correlated to the pH values.

Hydrous Fe and Mn oxides and nontronite. For these sediment investigations 320 samples were taken from 23 cores. Fig.21 lists the distribution of organic carbon, iron, and of various trace metals in the <2 μm fractions of eight sediment cores (1–2.60 m in length) from different parts of the lake. On the average, most of the trace metal data correspond to standard shale values. In the cores from the southern part of the lake – F_1, F_2, and G – sediments below approximately 20 cm depth are depleted in most metals,

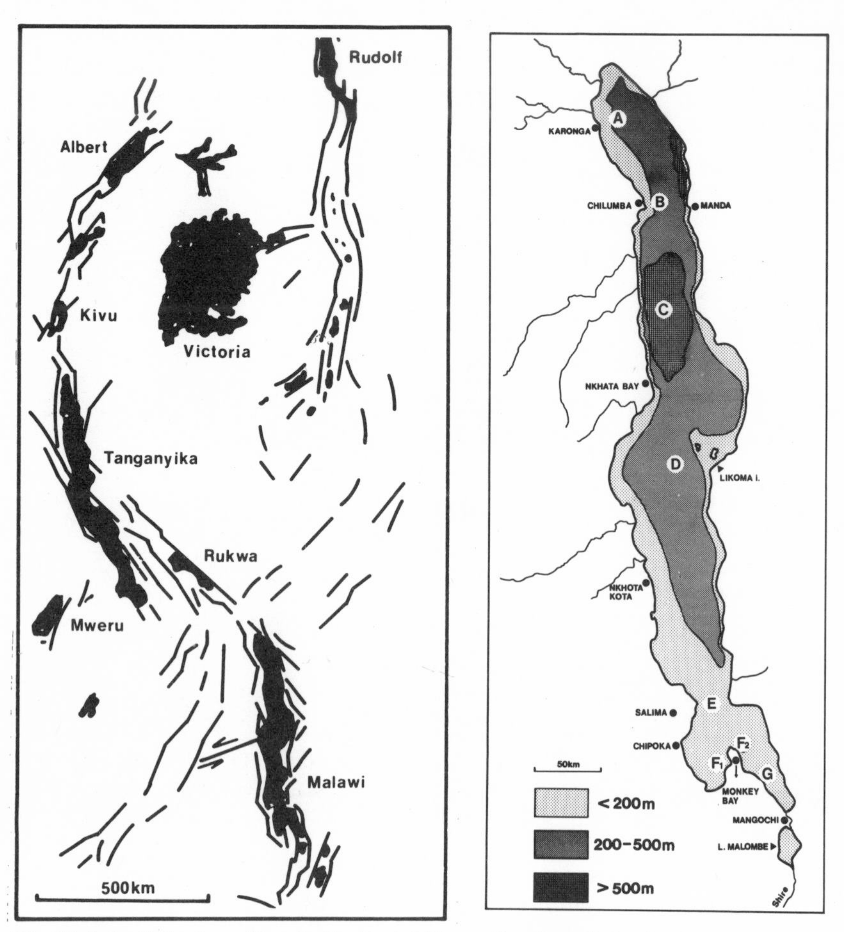

Fig. 20. Lake Malawi, position in the East African rift valleys (left, after Illies, 1970), depth contours and sampling stations for data in Fig.21 (right, after Müller and Förstner, 1973).

due to the higher content of silica from diatoms (see above). At the same time, there is a characteristic increase of Mn concentrations and to a lesser extent of Co, Fe, and V in the sedimentary cores from station C (704 m water depth) and D (384 m). In fact, the calculation of the correlation coefficients between pair groups of elements indicates a particular strong correspondence between the sediment concentrations of Fe, Co, Mn, and V for the data from core D (upper section of the matrix in Table XIV). The strongly positive correlation of the metal values to the content of organic carbon and the negative correlation to Li (probably from clay minerals) suggests the occurrence of diagenetic effects, which partly result from organic processes.

For comparison purposes, the lower part of Table XIV shows the correlations in core B (392 m depth) from the northern part of Lake Malawi. Here, detrital influences

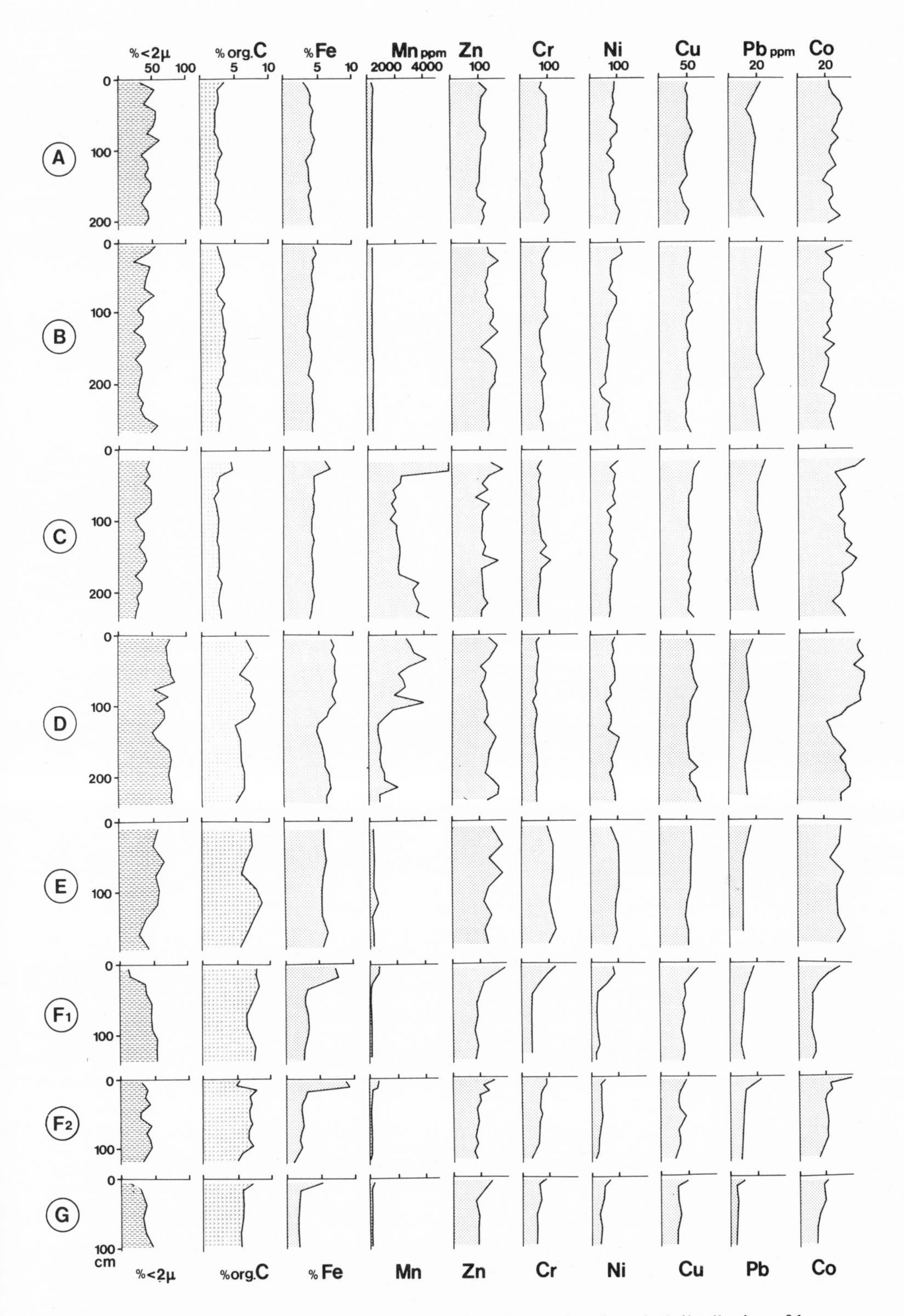

Fig. 21. Contents of <2 μm fraction and organic carbon (percent) and vertical distribution of heavy metals in the pelitic fraction of selected sediment core examples from Lake Malawi.

TABLE XIV

Correlation matrix of analytical data from pelitic sediments in core D (upper part of the diagram, 27 samples) and in core B (lower part, n = 23)

	Core D									
	Org C	Fe	Co	Mn	Zn	V	Cu	Ni	Cr	Li
C_{org}	x	0.683	0.699	0.734	0.069	0.140	0.133	–0.300	–0.057	–0.508
Fe	–0.561	x	0.925	0.819	–0.146	0.683	0.396	0.098	0.225	–0.557
Co	–0.003	0.046	x	0.773	–0.056	0.613	0.369	0.158	0.318	–0.556
Mn	–0.218	–0.230	0.162	x	–0.021	0.371	0.171	–0.119	0.001	–0.821
Zn	0.283	–0.038	–0.246	0.057	x	–0.056	–0.006	0.078	–0.047	0.091
V	0.088	0.351	–0.186	–0.525	0.151	x	0.534	0.433	0.421	–0.037
Cu	–0.079	0.250	–0.080	–0.488	0.302	0.507	x	0.235	0.298	–0.063
Ni	–0.065	0.455	0.355	–0.663	0.128	0.360	0.541	x	0.518	0.106
Cr	–0.269	0.251	0.441	–0.420	0.022	0.250	0.499	0.607	x	0.186
Li	0.032	–0.269	0.185	0.332	0.115	–0.210	–0.236	–0.249	0.316	x
	Core B									

Once underlined = significant at 0.05 level, doubly underlined = significant at 0.01 level.

dominate. There are positive correlations observed between the mainly lithogenic Cr, Ni, Cu, and V, which originate from basaltic source rocks. Fe seems to be diluted by organic substances; the significant negative correlations of the lithogenic elements to Mn could possibly be explained as arising through a process of diagenetic *remobilization* of the latter element.

In an earlier study on Lake Malawi (Müller and Förstner, 1973), it was suggested that the enrichment of Fe in the shallower aerated parts (<250 m water depth) is associated with the formation of limonite (with opal), nontronite and, to a minor extent, vivianite, all of which occur at the sediment–oxic-water interface from solutions enriched in metals by leaching of sediments from the deeper, anoxic parts of the lake. Whether these processes are affected by geothermal solutions present in that area or simply by changes in the redox conditions in deeper parts of the lake, cannot be conclusively decided. However, the analysis from Fe-rich (up to 19.3% Fe) sediment (bulk samples from core No. 8 off Nkhota Kota, 238 m depth) did not reveal a charateristic enrichment of other heavy metals, such as Ni, Cr, Cu, and Zn; only Co (two-fold increase compared to the average composition of Lake Malawi sediments) and Mn (2–4-fold increase) indicate the same trend of diagenetic accumulation as shown above from the data obtained in core D.

Phosphates. According to Williams et al. (1971) the associations of phosphorus in limnic sediments fall into four categories: (1) sorbed phosphorus on surfaces of minerals and organic substances, (2) coprecipitate or minor component of an amorphous phase, (3) constituent of an organic ester, and (4) component of a discrete mineral, such as apatite or vivianite. Since as yet no selective extraction procedure of phosphorus compounds from sediments is known, no quantitative data are available on the contribution of phosphates to the trace-metal concentration in limnic sediments. From theoretic phase relationships together with several field observations, Nriagu (1974) concluded that Pb phosphates such as pyromorphite and plumbogummite (very low solubility constants compared to lead carbonate, hydroxide, sulfide and sulfate) are important in the fixation of lead in the sediments. For Mn, Zn, and Cu, complex formation reactions of inorganic phosphates might significantly affect the distribution of the metal ion, the phosphates, or both (Stumm and Morgan, 1970). Significant enrichment of trace elements from marine phosphates, e.g., of U (McKelvey, 1956), As and heavy metals (Altschuler et al., 1978) can be transferred into recent aquatic sediments by the use of phosphate fertilizers and household detergents (Angino et al., 1970; Langmyhr et al., 1977).

Diagenetic carbonates. Effects of dolomite formations, the major diagenetic carbonate phase in limnic sediments, on the distribution of heavy metals have been described above in addition to the allogenic and endogenic carbonate minerals. Two minor phases are considered here: $MnCO_3$ (rhodochrosite) and $FeCO_3$ (siderite).

The formation of manganese carbonate is considered the limiting factor for the Mn concentrations in interstitial solution under anoxic marine conditions (Holdren et al.,

1975). Rhodochrosite has been obtained from sediment cores from Green Bay, Lake Michigan (Callender et al., 1974). The effects of coprecipitation of other heavy metals in these phases, however, have not yet been studied.

Under the Eh conditions of approximately +100 mV to 300 mV *and* higher concentrations of bicarbonate (>60 mg/l HCO_3^-), $FeCO_3$ may form (see Eh–pH diagram from Hem, 1970), and may coexist with ferromanganese oxy-hydroxides. In this respect, Callender and Bowser (1976, Vol. 7, this Handbook) have observed the occurrence of amorphous $Fe(OH)_3$ and siderite in the muds of Lake Michigan and Ontario. In addition, Stoffers and Müller (1978) have analyzed siderite-rich sediments from sapropelic deposits of the Black Sea from the DSDP-Leg 42B, where no enrichment in trace metals could be detected. Similar findings have been described by Singer and Navrot (1978) from siderite-containing sediments of Lake Birket Ram in Israel.

PHASES OF HEAVY-METAL ACCUMULATION IN LIMNIC SEDIMENTS

There are four major mechanisms of metal accumulation in sedimentary particles: (1) adsorptive bonding in fine-grained substances, (2) coprecipitation by hydrous Fe and Mn oxides and by carbonates, (3) associations with organic molecules, and (4) incorporation in crystalline minerals. This categorization – which had first been undertaken by Gibbs (1973) following his observations on particulate substances from the Amazon and Yukon Rivers – can be expanded to include all main types of metal associations occur-

TABLE XV

Metal associations in various phases of limnic sediments (After Förstner and Patchineelam, 1976)

Phase		Metal association
Mineral detritus (mainly silicates)		Metal bonding mostly in inert positions
Heavy metal –hydroxides –carbonates –sulfides		Precipitation as a result of exceeding solubility product in the water course
Clay minerals (sorption)	pH	Physico-sorption (electr. attraction)
		Chemical sorption (exchange H^+ in SiOH, AlOH and $Al(OH)_2$
Bitumen, lipids Humic substances Residual organics	pH	Physico-sorption, Chemical sorption ($COOH^-$, OH-groups)
		Complexes
Hydrous Fe/Mn oxides	pH	Physico-sorption, Chemical sorption
		Coprecipitation by exceeding the solubility product
Calcium carbonate	pH	Physico-sorption, Pseudomorphosis (supply and time)
		Coprecipitation

ring in both natural and man-affected water systems (Table XV):

(1) Heavy metals are transported and deposited as major, minor, or trace constituents in the detrital minerals derived from rocks and soils, in organic residues, and in solid waste material. The composition of the clay minerals is of prime importance for the natural background levels of fine-grained sediments (Hirst, 1962).

(2) Sorption and cation exchange substances with large surface areas. A generalized sequence of the capacity of solids to sorb heavy metals was found to occur in the order: manganese oxides > humic substances > hydrous iron oxides > clay minerals (Guy and Chakrabarti, 1975).

(3) A rise in pH and the oxygen content – e.g., in the estuarine mixing zone – promotes the formation of metal hydroxides, carbonate, and other heavy metal precipitates. Hydrous Fe and Mn oxides constitute significant "sinks" of heavy metals by the effect of coprecipitation (Jenne, 1968; G.F. Lee, 1975).

(4) In waters rich in organic matter, minerals may be solubilized by the combined processes of complexation and reduction; incorporation of metals into the sediment involves the mechanisms of adsorption, flocculation, polymerization, and precipitation (see above).

The various organic and inorganic phases are interrelated: clay minerals, carbonates, and other suspended materials form the nucleation centers for the deposition of Fe and Mn hydroxides (Aston and Chester, 1973; Jenne, 1976). Humic acids are increasingly adsorbed to clay minerals at higher salinities (Rashid et al., 1972). Characteristic effects between Fe, Mn, and P with organic substances in both river water (dissolved) and sea water (partly flocculated) were found in estuaries (Sholkovitz, 1976).

Mainly because of these interactions, there are still many problems involved in the chemical separation of different metal associations in sediments. However, the developments in the last decade have shown that with standardized extraction procedures it should be possible to get more information on both the formation and the technological properties (e.g., for ore processing) of mineralization in sediments and sedimentary rocks than is available from bulk analyses.

Extraction of different chemical phases of heavy metals [1]

Chemical extraction procedures were first used in pedology to ascertain the availability of certain trace substances important for the fertility of soil. In sedimentary geochemistry, similar methods were employed for the differentiation of the metal contents in detrital and authigenic mineral fractions from carbonate rocks and shales (Hirst and Nicholls, 1958; Chester, 1965; Gad and Le Riche, 1966). Presently, the most advanced methods include the successive extraction of metal contents in interstitial water, in ion-exchangeable, easily reducible, moderately reducible, various organic, carbonate and residual sediment fractions, which are compiled in Table XVI.

---

[1] See also note added in proof, pp. 269–270.

TABLE XVI

Extraction sequence for the termination of particle-associated heavy metals

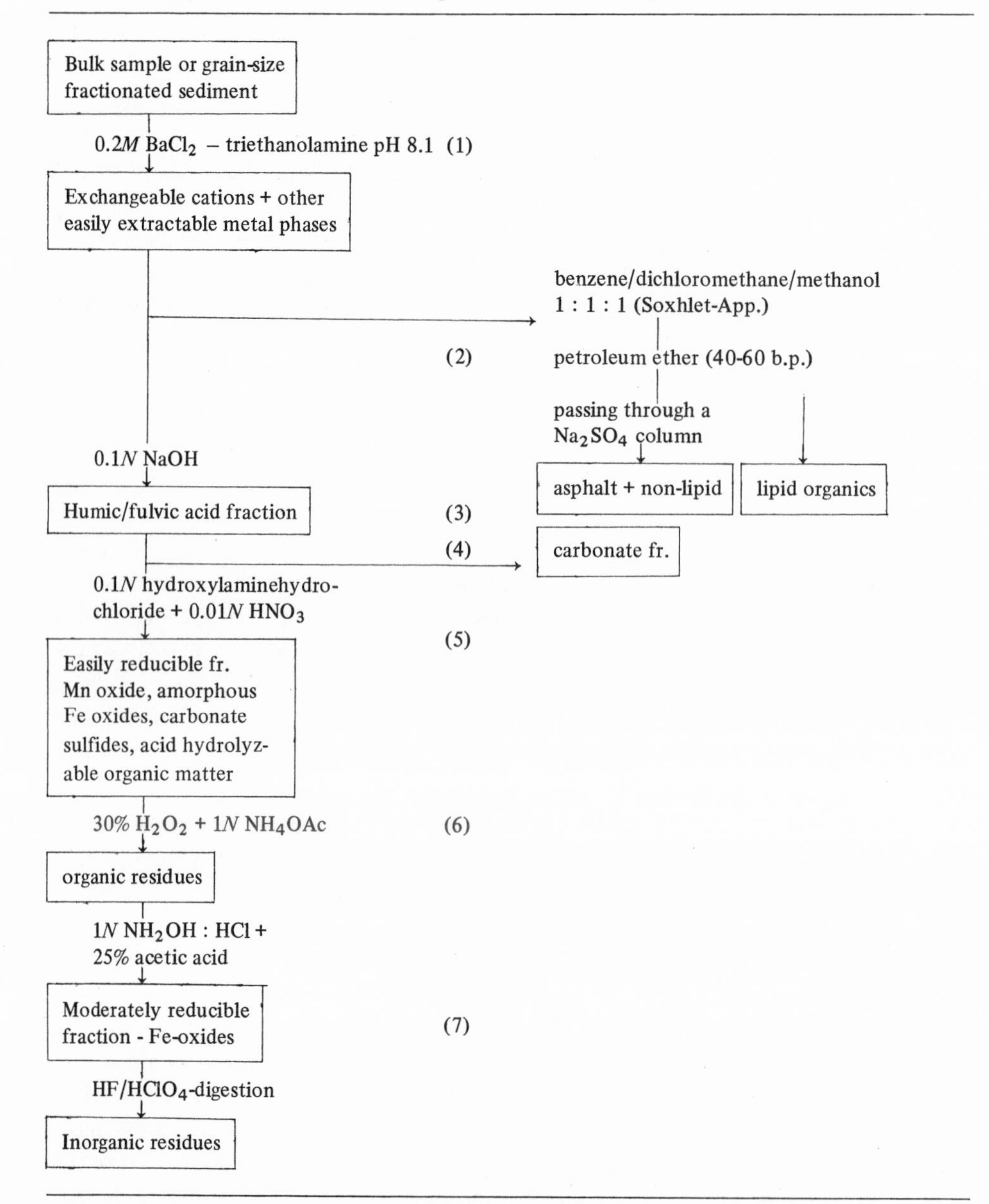

[1] M.L. Jackson (1958); [2] Cooper and Harris (1974); [3] Volkov and Fomina (1974); [4] Deurer et al. (1978); [5] Chao (1972); [6] Engler et al. (1974); [7] Chester and Hughes (1967).

Separation of heavy-metal associations with Fe and Mn hydroxides

The determination of the hydrogenous metal content in recent sediments (including authigenic and diagenetic formations), in particular those components which are bound to carbonate minerals and hydrous Fe and Mn oxides, and those which are adsorbed on clay minerals and other substances, has long been a source of interest in the detection of trace elements incorporated in ferro-manganese nodules. The historical development of the procedures has been described, among others, by Chester and Hughes (1967). Goldberg and Arrhenius (1958) used *EDTA* to establish the partition of different elements between detrital igneous minerals and authigenic phases in pelagic sediments. The soluble minerals present in the samples investigated were mainly the oxide minerals and microcrystalline apatite. Other minerals were less soluble: arranged in the approximate order of increasing dissolution, the minerals investigated appear as follows: biotite, titanite, goethite, magnetite, rutile, hematite, lepidocrocite, nontonite, and augite. The use of this chelating agent was, however, rejected by the present authors because (1) the stripping of adsorbed trace elements from clay mineral surfaces is very slow and (2) the EDTA was difficult to destroy by some of the trace-element analysis techniques subsequently applied to the solution.

Arrhenius and Korkish (1959) consequently tried to separate the Fe fraction of the ferro-manganese nodules with *1M hydrochloric acid.* This, however, also caused an alteration of the lattice structures of the clay minerals, and its chemical action on marine sediments designed to liberate only authigenic phases into solution was not satisfactory. As an alternative, Arrhenius and Korkish (1959) used dilute *acetic acid* in a specific chemical attack on pelagic sediments.

Experiments carried out by Chester and Hughes (1967) indicated, however, that treatment with 25% acetic acid will not completely dissolve the iron oxide material of the ferro-manganese nodule, even at a temperature of 100°C. In spite of these difficulties (and mainly to save time) the acid-soluble metal content in the sediments is still often determined with 0.1*M* hydrochloric acid (Piper, 1971), 0.3*M* HCl (Malo, 1977), or with 0.1*M* nitric acid (A.S.G. Jones, 1973).

Arrhenius and Korkish (1959) have shown that ferro-manganese nodules contain a reducible fraction consisting of the oxides of manganese, which can be separated using 1*M* hydroxylamine hydrochloride as reducing agent. Chester and Hughes (1967) confirmed these findings, but found that under the conditions used, the reducing agent will also dissolve up to approximately 50% of the iron oxide minerals present. The same authors investigated the effects of a combined *acid-reducing agent of 25% acetic acid and 1M hydrochloric acid*: At room temperature, approximately 90% of the total iron and 96% of the total Mn in the nodule was dissolved. Celestite, apatite, barite, feldspars, amphiboles, pyroxenes, micas, zircon, glauconite, and zeolites are not soluble to any extent in the combined acid-reducing agent solution, but there is a leaching of a large fraction of the biologically derived forms present in pelagic sediments. It must be noted, however, that this method does not cover the hydrogenous metal content in the opal and authigenic clay mineral phases. The amount of metals bound to hydrous Fe and Mn oxides can be determined by subtracting the carbonate-bound and cation-exchangeable metal content from the data obtained by the acid-reducing agent. Another method of extracting Fe and Mn oxides was developed by Aguilera and Jackson (1953) and was employed by Holmgren (1967), whereby reduction is achieved with *sodium dithionite and complexing* with *sodium citrate.* This method, which helps to determine the "moderately reducible phases" in the more recent fraction sequences (see Gupta and Chen, 1975), consists of a solution of 100 ml distilled water, 8 g sodium citrate, and 2 g sodium dithionite. The mixture is shaken for 14 hours on a mechanical shaking device.

The *oxalic acid extraction* is favoured by Schwertmann (1964): 100 mg of sample is shaken in a pH 3 mixture of 0.2*M* ammonium oxalate and 0.2*M* oxalic acid for 2 hours in the dark. This leaching solution removes amorphous ferric hydroxides and poorly crystalline ferromanganese oxyhydrates but does not attack crystalline goethite (Landa and Gast, 1973) or iron-rich smectite. A comparative study of the various extraction methods given here for the leaching of moderately reducible metal compounds by Heath and Dymond (1977) brought about the best results for this technique, particularly with respect to the analysis after replicate leaching. The first oxalic treatment removed about one-half the total Fe in pelagic sediments (Bauer Deep, see below); subsequent treatments remove little additional Fe. In contrast, the other leaches (0.1*M* hydrochloric acid, hydroxylamine hydrochloride–acetic acid, and sodium dithionite–sodium citrate–sodium bicarbonate) removed significant additional Fe after the first leach by attacking iron-rich smectite.

Hydrogenous metal phases in marine sediments. Since the greater portion of these analyses have been performed as yet on marine – mainly pelagic – sediments, such examples are given in Table XVII. In the North Atlantic, the hydrogenous character of trace metals decreases in the order Mn > Co > Cu > Ni > Cr > V > Fe. It is evident that the supply of V, Cr, and Ni is governed largely by the input of detrital material from the continents. Chester and Messiha-Hanna (1970) have shown from the geographical distribution of *nickel* that (1) in areas close to the continent >50% of the Ni has a lithogenous origin, (2) in intermediate areas, between 26 and 60%, and (3) in mid-ocean areas, <30% of the Ni is held in lithogenous fractions of the sediment. A similar order can be found for the hydrogenous components in the North Pacific region (Chester and Hughes, 1969), although here it is Mn, Co, and Ni which are enriched in the hydrogenic phases. The reason for this may be that less detrital material has been deposited in these areas, thus leading to a general rise in the levels of most metals.

Strongly enriched amounts of Mn, Co, and Ni occur in the hydrogenous phases of the sediments from the Bauer Deep in the East Pacific (Sayles et al., 1975). The adsorption and incorporation of metals from sea water on the micronodules or on the Fe and Mn oxide colloids best explain the element relations observed for Fe, Mn, Cu, Zn, and Ni

TABLE XVII

Metals in non-lithogenous (soluble in acid-reducing agent) sediment fractions (percent of total sediment)

	Fe	Mn	Co	Ni	Cu	Zn
North Atlantic						
Surface sediments [1]	18	68	58	45	56	–
Basal sediments [2]	6	57	–	41	18	20
Reykjanes Ridge [3] Crest	12	46	–	54	39	16
North Pacific [4]	4	83	82	75	32	–
East Pacific [5]	24	98	90	84	51	49
Pacific [6]						
Red-brown clay	40	99	96	95	70	60
Siliceous mud	22	92	79	68	47	43
Calcareous ooze	50	99	96	88	85	83
Micronodules	82	99	98	97	96	94

[1] Chester and Messiha-Hanna (1970); [2] Horowitz and Cronan (1976); [3] Horowitz (1974); [4] Chester and Hughes (1969); [5] Sayles et al. (1975); [6] Förstner and Stoffers (in prep.).

in the oxide fraction of the sediment. The correlation between the acid-reducing solubilized elements are significantly positive for Mn and Fe. The minor metals Cu, Zn, and Ni likewise correlate well with Mn; Co also shows a positive correlation with Mn at the sediment—water interface and under oxidizing conditions (Sayles et al., 1975). Our data from the northern and southern Pacific (Förstner and Stoffers, in prep.) generally confirm these findings, indicating a higher percentage of hydrogenous metal forms in the red-brown clay, calcareous ooze, and in particular, in the micro-nodules (Table XVII).

Evaluation of diagenetic effects on metal phases in limnic sediments

In a previous section it has been demonstrated from examples of Lake Malawi sediments that the processes of the redox conditions mainly affect the concentrations of Fe, Mn, and Co. The distribution of "crystalline" (mainly detrital) and "hydrogenous" phases in selected cores was additionally studied in that lake. Table XVIII compares the total metal concentrations with contents extracted by 1*M* hydroxylamine hydrochloride +25% acetic acid (Chester and Hughes solution) in cores B and D. It is obvious that the increase of Mn, Fe, and Co in core D, from the central part of the lake, in relation to core B is mainly associated with an increase in the percentage of hydrogenous phases, whereas the other metals do not significantly follow this development (except perhaps Zn, which does not seem to be affected by either one). Compared with the examples from the marine environments (Table XVII), there is a remarkable lower percentage of non-lithogenous Cu, Ni and Cr in the sediments of Lake Malawi.

Two additional extraction steps have been applied to the pelitic sediments from Lake Malawi: determination of cation-exchangeable elements by treatment with 1*M*

TABLE XVIII

Comparison of mean concentrations and 1*M* hydroxylamine-hydrochloride + 25% acetic-acid-extractable metal contents in cores B and D of Lake Malawi

	Average concentration		Hydrous Fe/Mn oxide extract	
	core B	core D	core B	core D
C_{org}	2.97 ± 0.4%	6.46 ± 0.8%	–	–
Fe	4.01 ± 0.3%	6.43 ± 0.8%	0.75% (19%) [1]	2.76% (43%) [1]
Co	23 ± 4 ppm	38 ± 8 ppm	4.8 ppm (21%)	15 ppm (40%)
Mn	351 ± 37 ppm	2990 ± 1079 ppm	153 ppm (43%)	1793 ppm (62%)
Zn	141 ± 23 ppm	147 ± 33 ppm	68 ppm (49%)	79 ppm (54%)
V	162 ± 50 ppm	193 ± 22 ppm	27 ppm (17%)	30 ppm (16%)
Cu	50 ± 5 ppm	59 ± 7 ppm	6.2 ppm (13%)	7.7 ppm (13%)
Ni	71 ± 17 ppm	83 ± 10 ppm	4.3 ppm (6%)	4.8 ppm (6%)
Cr	83 ± 9 ppm	59 ± 5 ppm	2.5 ppm (3%)	2.0 ppm (3%)
Li	43 ± 4 ppm	36 ± 6 ppm	0 (0%)	0 (0%)

[1] (in brackets): percentages of metal in hydrous Fe/Mn oxide extract from total metal concentration in bulk sample.

ammoniumacetate (M.L. Jackson, 1958) and the separation of humic acids with 0.1*N* NaOH (Volkov and Fomina, 1974). The data (examples of cores B and D in Fig.22) show that only Cu and V are characteristically enriched in the organic fraction of the sediment. No certain behavior could be discerned for Zn and Cr in organic substances, whereas Fe, Co, Mn, and Li are found to be *not* associated with humic or fulvic acids to any significant extent in the deposits of Lake Malawi. Among the exchangeable cations, only Zn and Mn occur in detectable amounts. It has been suggested by Stumm and Morgan (1970) that higher quantities of manganese in the ion exchange phase could be sorbed under aerobic conditions in the form of Mn(II) by the ferric oxides and oxidized fractions of manganese; the distinct correlation between the concentrations of manganese in 1*M* hydroxylamine extractable and exchangeable fractions of the sediments in core D would indicate a similar mechanism in the deposits of Lake Malawi.

Extraction of trace metals from carbonates

The associations of trace metals with carbonate has been given only little or no attention in contrast to that for the hydrous Fe and Mn oxide phases, although it has been shown in laboratory experiments that coprecipitation with calcium carbonate is an effective elimination mechanism for heavy metals from aqueous solutions (Popova, 1961). The most recent investigations of Salomons and Mook (1978) on the transition region from the Rhine to the IJsselmeer (lake) in Holland show that a decrease of the dissolved loads of Zn, Cd, and Ni is directly related to the precipitation of carbonate minerals, which is chiefly a result of an increase in pH.

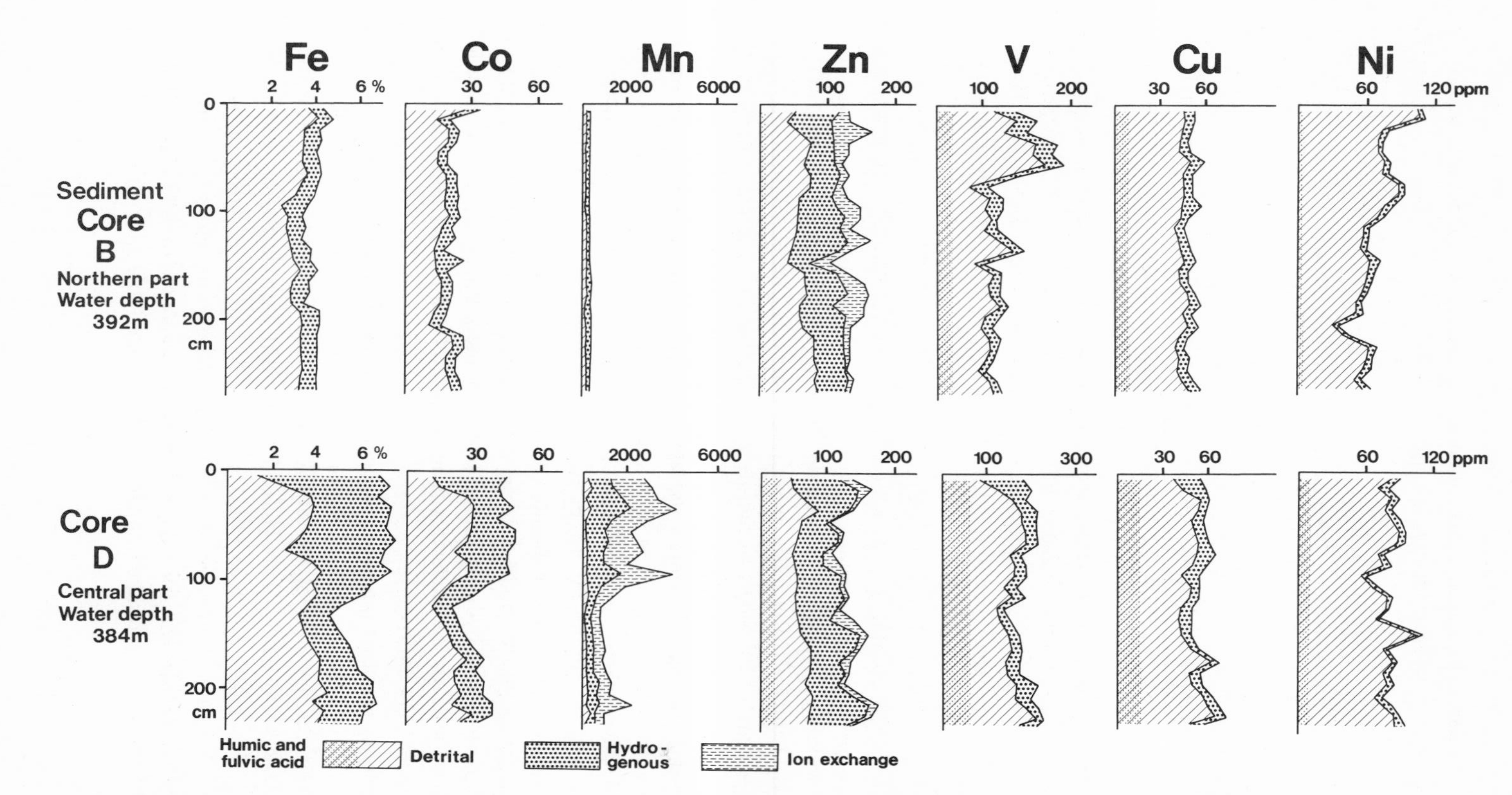

Fig. 22. Phase partitioning of iron, cobalt, manganese, zinc, vanadium, copper and nickel in sedimentary cores from Lake Malawi.

Treatment by acids of varying strengths influences not only the carbonates but also other solid phases, e.g., the iron oxide hydrates and clay minerals. Ray et al. (1957) have shown that dilute hydrochloric acid will attack the lattice structures of certain clay minerals (for example, Fe chlorites). Although this is not the case with dilute acetic acid, Hogson (1960) was able to show that 2.5% acetic acid will strip the exchangeable fraction of adsorbed (cobalt) ions from the surface of montmorillonite minerals. Attempts have also been made to free the carbonate sediment content by means of complexing agents (Hill and Runnels, 1960; Glover, 1961), but the exchangeable cation content was here likewise affected.

Two methods recently developed at the Institute of Sediment Research, Heidelberg University, seem to be useful for the selective extraction of carbonate-associated trace elements:

(1) CO_2(aq) treatment. Dissolution of carbonates was achieved according to the method suggested by Patchineelam (1975). CO_2 from a gas cylinder is passed through a suspension containing 0.2 g sample material in 100 ml distilled H_2O for 12 hours during constant agitation. The procedure is then repeated (usually 3–5 times) until no further remobilization of calcium occurs.

(2) Ion exchange. Following the method of Lloyd (1954) for the removal of carbonates from clay substances, extraction of carbonate-associated metals was conducted using a cation exchanger (Deurer et al., 1978). The strong acid exchanger (Merck I, exchange capacity 4.5 meq/g) is saturated with H^+-ions in 10% HCl, and washed with distilled H_2O until Cl^--free. The pulverized sample is pretreated with 0.2*N* $BaCl_2$-triethanolamine and 0.1*N* NaOH to remove exchangeable heavy metals and cations bonded to humic material. To a suspension of 0.5 g sample material and approximately 20 ml H_2O, a 4–5 g exchanger is added and agitated in an open reaction vessel for 3 to 12 hours. The procedure is ended as soon as a control vessel containing 0.25 g dolomite – less soluble than calcite – reveals complete solution. The exchanger particles are separated from the suspension by sieving, washed and twice regenerated with 50 mg 10% HCl suprapur reagent. Metal contents are determined in the total regenerated solutions. The advantage of this method is that no free H^+-ions can form which could in turn react with clay minerals and hydrous oxides. However, the affinity of the carbonate ions and the hydrogen ions for the exchanger is sufficient to form H_2CO_3, which in turn then disintegrates into CO_2 and H_2O.

Examples studied from lake sediments of different climatic zones have been listed in Table XIX in accordance with increasing percentages of Ca and Mg carbonates. The samples from the dry lakes of Western Australia are carbonate-free (example from Lake Carey); those from East Africa (Lake Hannington) as well as from moderate humid European areas (Lake Constance, Ochridsko and Blejsko Jezero) consist chiefly of detrital Ca and Mg carbonates. In contrast, the sediments from the semi-arid lakes Balaton and Neusiedl contain authigenic formations of high-Mg calcite and/or dolomite.

By comparing the total carbonate percentage in Table XIX with the corresponding

TABLE XIX

Carbonate-associated heavy metals in lake sediments (After Deurer et al., 1978; with permission of Pergamon Press)

	Carbonate (%)	Carbonate species (%)			Percent of total metal in carbonate association		
		Cc	MgC	Dol	Fe	Mn	Zn
(1) Lake Carey (W. Australia)	0	0	0	0	0	0	0
(2) Lake Hannington (Kenya)	5	5	–	–	1	26	18
(3) Lake Constance (Federal Republic of Germany, Austria, Switzerland)	28	20	–	8	7	62	43
(4) Neusiedler See (Austria, Hungary)	37	–	12	25	10	34	15
(5) Ochridsko Jez. (Yugoslavia)	44	44	–	–	10	55	43
(6) Balaton (Hungary)	55	–	41	14	20	68	13
(7) Blejsko Jezero (Yugoslavia)	72	70	–	2	26	46	67

Cc = calcite; MgC = high-magnesium calcite; Dol = dolomite.

carbonate-associated metal content (given as percent of the total metal concentration), it is possible to deduce the effects of either enrichment or depletion caused by the carbonate component. If the metal content associated with carbonate is lower than the total carbonate, a dilution effect results, even when – as in the case of the results for Fe – the metal content increases along with the carbonate percentage of the sample. It appears that the carbonate fraction is generally capable of bonding only up to $\frac{1}{3}$ to $\frac{1}{5}$ of the Fe associated with the other sediment components. In the case of Mn concentrations, it was found that most of the investigated samples, with the exception of the one from Lake Bled, revealed an enrichment by carbonate. With this knowledge in mind, it was suggested that the decrease of Mn in carbonate-rich sediments could be explained by certain diagenetic effects. The Zn values associated with carbonate seem to be widely independent of the total percentage. Relatively low Zn levels in lakes with authigenic Ca and Mg carbonates, accompanied by high salinity (Nos. 4 and 6, Table XIX) can possibly be accounted for by the formation of soluble zinc-chloro-complexes, which influence the distribution coefficient of Zn during coprecipitation with calcite. Exceptionally high Zn concentrations are associated with carbonate in sediment samples from Lakes Ochrid, Bled, and Constance. The Lake Constance samples even indicate a significant enrichment attributable to the carbonate component.

Metal bonding on organic material

In the study on sediments of the Saanich Inlet, Presley et al. (1972) carried out a *hydrogen peroxide treatment* after separating the acid-reducing soluble metal components. The metal content of *sulfides* and *organic matter* is predominantly found in the "*oxidizable fraction*". Within the fjord investigation area, only 10% or less of the total Fe, Ni, Cr, and Co was removed from the sill-region sediment by oxidation; this percentage increased progressively away from the sill to the deeper water of the fjord. Iron sulfides were therefore probably the major contributor to the oxidizable fractions. A higher percentage of the total Cu was leached by peroxide than that of any other metal. This is probably an indication of the often-proposed close association of *copper and organic matter* (Presley et al., 1972).

In order to determine stable metal–organic complexes from soils, Schnitzer and co-workers developed extraction methods which have since found widespread application. Griffith and Schnitzer (1975) cited the following working method for the isolation of heavy metal–organic complexes: 300 g of air-dried soil was weighed into a 4.5-l polypropylene flask, 2 l of 0.5*N* NaOH was added; the air in the flask and in the solution was displaced by N_2 and the system was shaken intermittenly for 24 hours at room temperature. The dark colored supernatant solution, after separation from the residual soil by centrifugation was acidified to pH 2 with 6*N* HCl and allowed to stand for 24 hours at room temperature to allow for the coagulation of the humic acid fraction. The soluble material (fulvic acid) was removed from the humic acid by centrifugation, lyophilized and dried in a vacuum desiccator over P_2O_5 at room temperature. The dried fulvic acid preparation contained large amounts of NaCl, which was eliminated by a further purification process.

In a detailed investigation on "the influence of organic material and processes of sulfide formation on distribution of some trace elements in deep water sediment of the Black Sea", Volkov and Fomina (1974) described the extraction of the bitumens. This was done initially with ethyl alcohol to remove the water present in the sample, after which it was mixed for a period of 40 hours with an alcohol/benzene mixture (1 : 1). The quantity of material extracted by the alcohol and alcohol/benzene mixture was determined after drying at 80°C. Free *humic substances* were extracted by treatment with 300 ml of 0.1*N* NaOH; the solution was filtered and humic acids were precipitated from the clear filtrate by addition of H_2SO_4 until a pH of 3 was obtained. After coagulation, the humic acid precipitates were dissolved in a hot solution of 0.05*N* NaOH and the solution was transferred to a measuring flask. The solution containing the fulvic acids, after removal of the humic acids, was also transferred to a measuring flask. The organic carbon and the trace element content was then determined.

Volkov and Fomina (1974) found that the relation between copper and organic material is manifested more or less uniformly. The largest amounts of Cu (and of Co, Ni, and V) were found in the free fulvic acid fractions; the latter three elements were not observed in the free humic acids. The amount of Cu ranged from 11% to 21% of the total content of the organic material.

The organic fraction of sediments from the River Blyth in Northumberland was analysed by Cooper and Harris (1974) according to the following extraction procedure: A subsample (300 g) was refluxed with 0.1*N* NaOH solution resulting in a brown solution, to which concentrated HCl was added after filtration to reach a pH of 1. After the solution was allowed to stand, a *humic acid* precipitate was filtered off. Extraction of another subsample (100 g) was carried out in a Soxhlet apparatus using a benzene/dichloromethane/methanol mixture (1 : 1 : 1). After washing with water, the sample was dried and the concentrate extracted with petroleum ether in an ultrasonic bath. The soluble material was separated from the dispersed insoluble material by passing the sample through a 50 mm column of anhydrous sodium sulfate. The material remaining in the column was eluted with dichloromethane; this constitutes the *asphalt fraction.* The petroleum-ether-soluble material was further separated by column chromatography over silica gel resulting in a petroleum ether eluate, a benzene eluate (= *less-polar lipids*), and a methanol eluate (= *more-polar lipids*)." In addition to these organic fractions there is *acid-hydrolysable* and *acid-resistant organic* matter; the latter is extracted by hydrogen peroxide treatment (see above), the former is included in the acid-reducing extraction step (together with carbonates, oxides, and sulfides).

A comparison of the metal contents in the organic fractions of two examples is given in Table XX (Cooper and Harris, 1974) from (1) a river sand above an effluent, containing 0.2% organic substances and from (2) river sand at tidal head, which reflects with approximately 2% organic substances a stronger influence of pollution. The data in Table XX show that the asphaltic fraction is particularly important for Zn, Cu, Ni, Pb, and Cd (the latter three elements show a distinct enrichment in the more polluted sample), in addition to the acid-resistant organics (Mn, Cu, Pb, Cr) and humic acids (Cu, Zn). Zinc seems to be closely associated with acid-hydrolyzable fractions; the more polluted sample indicates a strong enrichment of Ni in the more polar lipids. It should be noted,

TABLE XX

Metal concentrations in organic fractions (in ppm of each fraction) of sandy sediment samples from River Blyth, Northumberland (After Cooper and Harris, 1974; with permission of Pergamon Press)

	Acid-resistant organics		Humic acid		Asphalt		Lipids			
							less polar		more polar	
	(1)	(2)	(1)	(2)	(1)	(2)	(1)	(2)	(1)	(2)
Mn	370	18	93	1	150	130	22	1	7	2
Cu	190	60	270	120	150	310	1	24	11	1
Zn	(0.2)	240	200	35	730	490	9	95	34	20
Pb	45	38	18	19	38	110	1	2	0.5	1
Ni	65	85	62	35	100	230	1	160	0.5	110
Cr	11	9	1	4	29	5	1	2	0.5	1
Cd	0.7	0.3	1	0.5	2	15	2	4	0.7	1

(1) = above effluent, (2) = below effluent.

however, that only in a clayey sample (from River Blyth) containing 13.6% organic matter, the contribution of these substances to the metal in the sediment is of some significance (Zn = 8%, Pb = 15%, Ni = 22% and Co = 60%; see below).

Metal associations in less-polluted lake sediments

The samples selected for a study on the heavy-metal phases in lake sediments (Schmoll and Förstner, 1979) stem from different carbonate environments, which are still relatively less affected by environmental pollution. An exception is Lake Bled in Slovenia/Yugoslavia where sewage input is now known to cause distinct changes in the water chemistry (e.g., nutrients and zinc). The examples of Fig.23 (see also Table I of this study) that are from humid climates generally contain higher concentrations of organic matter than those situated in semi-arid or arid regions (from right to left). The concentrations of organic carbon varies from 0.16% in Lake Amboseli to 2.77% in Lake Bled. More expressive, however, is the concentration of "active" organic material such as humic substances. After extraction with 0.1*N* NaOH and a subsequent comparison with standard humic materials (e.g., Eloff, 1965) percentages between 0.014% for Dasht-i-Nawar and 0.70% for Lake Bled sediments are arrived at.

The examples of Fe, Zn, and Ni are given in Fig.23, whereby each represents a characteristic type of chemical metal association in lake sediments. Thus, *iron* occurs chiefly residually associated. The second most important forms are hydroxidic and carbonate associations – all other phases are of only minor importance. *Zinc* is principally associated either residually or with carbonates, and an association with humic substances or hydroxides seems not to be of importance (see also the example from Lake Malawi in Fig.22). The Zn concentration in the water-soluble fraction is surprisingly high for some samples, mainly from the more arid regions. *Nickel* behaves relatively uniformly – the residual forms predominate. Otherwise, only the hydroxidic association is to a certain extent of note (the same is found for Cr and Co).

General differences in the investigated samples seem to originate chiefly through climatic conditions that are, for their part, responsible for the other characteristic parameters such as "hydroxide content", "carbonate contents and carbonate species", and "percentage of humic substances". For example, arid samples clearly contain higher amounts of iron oxidhydrates and manganese oxide. Water-soluble metals (Fe, Zn, Cu, Cr, and Co) occur chiefly in the arid samples, whereas the sorptive bonding form (Fe, Mn, Cu, Co) occurs more often in the more humid examples. The carbonate-bound Fe clearly depends on the carbonate contents in the samples; this general relation seems to be true for Mn as well as for Zn, although there are greater deviations for the latter metal (see Table XIX).

A definite selectivity of a certain carbonate species for heavy metals could not be determined. It appears that Mn and Zn are adsorbed relatively well in the aragonite phase of the sediment from Van Gölü, whereas Cu is not found in these minerals. At the same time, Cu seems to be relatively well represented in calcite and Mg calcite.

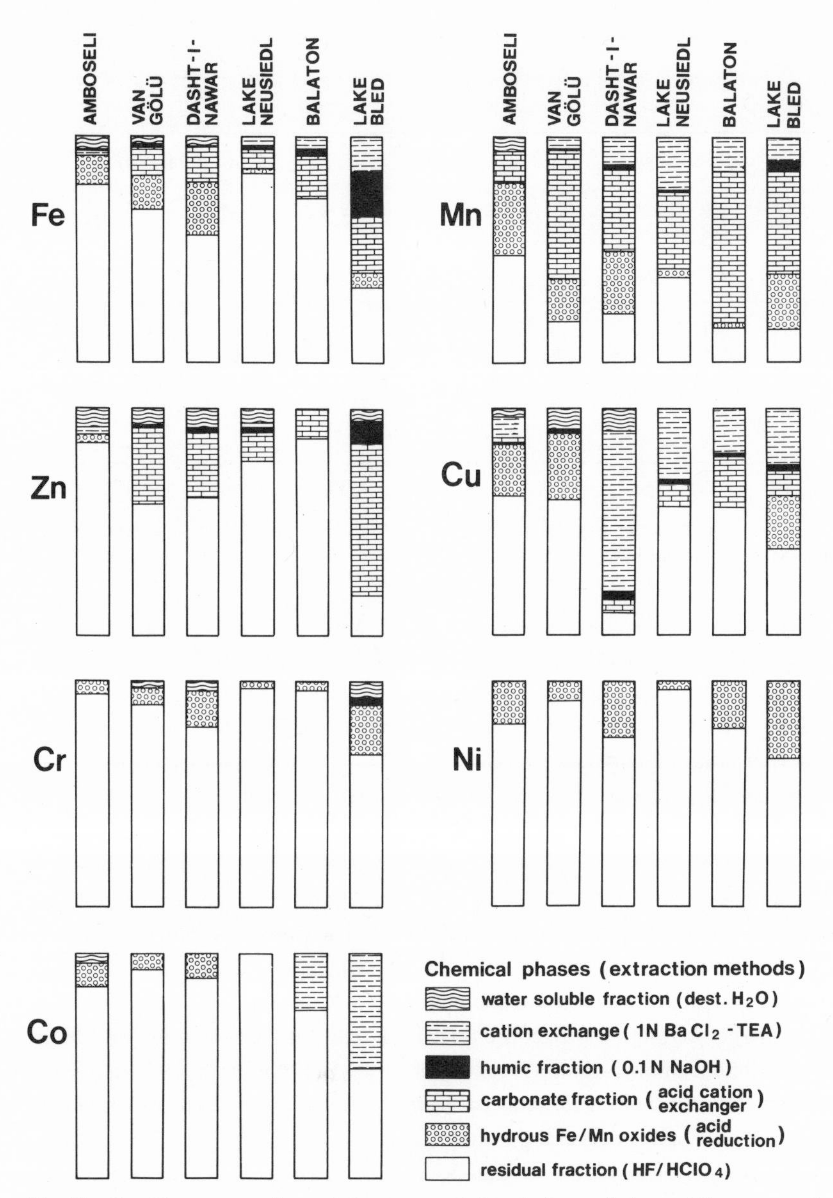

Fig. 23. Graphic presentation of the types of chemical associations of heavy metals in calcareous lake sediments. (After Schmoll and Förstner, 1979.)

Metal phases in contaminated dredged sediments

Progress in the determination of sediment-associated trace metals has been registered since 1974 in studies undertaken by the Environmental Effects Laboratory of the

TABLE XXI

Sediment associations of Ni, Zn, and Cd in harbor sediments (After Brannon et al., 1976; with permission of Pergamon Press)

	Mobile Bay			Ashtabula			Bridgeport		
Organic carbon (%)	2.0			2.4			2.7		
Inorganic carbon (%)	0.1			0.6			2.2		
Sulfur (ppm)	903			240			2680		
Fraction in percent	Ni	Zn	Cd	Ni	Zn	Cd	Ni	Zn	Cd
Exchangeable cations	1	–	–	<1	<1	–	<1	<1	–
Organics and sulfides	7	58	17	10	38	76	26	53	96
Easily reducible	2	4	3	2	5	2	2	13	4
Moderately reducible	–	–	–	23	27	–	3	17	–
Residual fraction	90	38	80	65	29	22	69	17	–
Metal concentration	(127)	243	3.6	(185)	444	4.1	(182)	1027	(15.2)

Metal concentrations in brackets refer to total particulate content; other values include interstitial water concentrations (all values in mg/kg).

U.S. Army Engineer Waterways Experiment Station in Vicksburg, Miss. (Engler et al., 1974; Brannon et al., 1976), with the participation of the Environmental Engineering Programs of the University of Southern California, Los Angeles (Chen et al., 1975). These studies centered on the question of the land application of dredged materials after it had been established that contaminants such as Cu, Zn, and Cd occupy the least stable of the sediment fractions and that the sediment physico-chemistry dominates the mobility and availability of the contaminants as well as the indigenous metals (C.R. Lee et al., 1976).

Engler et al. (1974) devised a selective extraction procedure to functionally separate interstitial water, exchangeable, easily reducible, moderately reducible, organic, and residual fractions from marine and freshwater sediments. The procedure precludes atmospheric oxidation at sensitive steps (during sampling and during extraction of interstitial water, exchangeable phases and easily reducible phases).

The results of this extraction procedure for Ni, Zn, and Cd are shown in Table XXI as performed on sediment samples from three harbor areas (Brannot et al., 1976): Mobile Bay, Alabama; Ashtabula on Lake Erie; and Bridgeport, Connecticut. Each of these test locations exceeds the one before in metal pollution.

From the table it is clearly evident that as the metal levels in the sediments rise (usually pointing to a rise in metal pollution), the metal concentrations in the residual fraction decrease. It can, therefore, be assumed that the additional metals that are carried into the water systems are in relatively unstable bonding forms.

Grain-size variations of chemical phases

According to the findings on the grain-size distributions determined above (Fig.4), one should expect a characteristic influence of the particle size of the sediment sample on

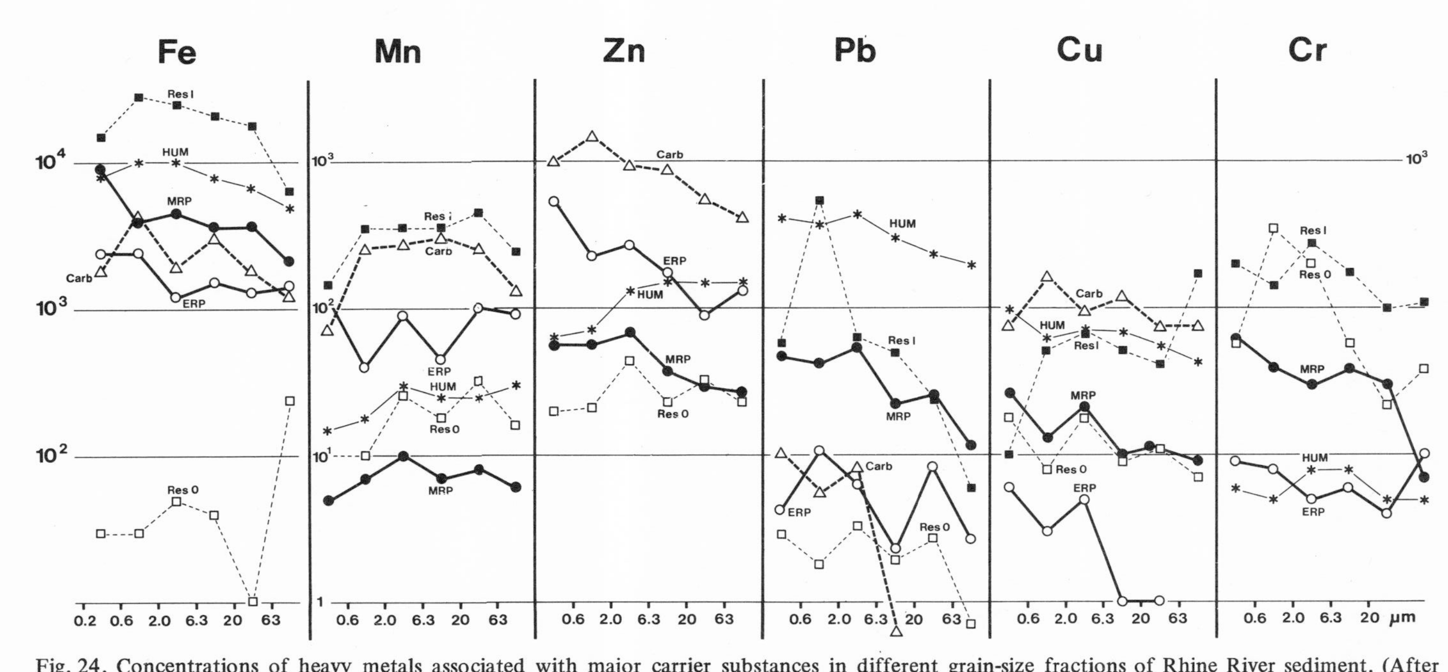

Fig. 24. Concentrations of heavy metals associated with major carrier substances in different grain-size fractions of Rhine River sediment. (After Förstner and Patchineelam, 1979; with permission of the American Chemical Society.)

ResI = residual inorganic, *ResO* = residual organic, *HUM* = humic acids, *Carb* = carbonate associations, *MRP* = moderately reducible phases, *ERP* = easily reducible phases. All values in mg/kg.

the chemical associations of trace metals. An example of chemical extractions of trace metals from different grain-size fractions is given in Fig.24 for a sample from the strongly polluted lower Rhine River (Förstner and Patchineelam, 1979). A distinct reduction of the metal content as grain size increases is observed for most of the chemical associations studied. These effects are particularly evident for the Zn and Pb associations with carbonates, easily and moderately reducible fractions, but can also be observed in the residual organic and inorganic fractions, such as for Cr, or in the humate fractions as for Fe, Pb, and Cu. It should be noted that these effects would be even more pronounced if the mechanical fractionization would more accurately separate individual particles according to their grain size. This is, however, not the case and "coatings", e.g., of Fe and Mn oxides, carbonates, organic substances on relatively inert materials, such as quartz grains, still play a characteristic role for the bonding of heavy metals in coarser grain-size fractions.

In a recent study of the suspended sediment from the Yukon and Amazon Rivers, Gibbs (1977) suggested that hydroxidic coatings, which can consist of X-ray amorphous, microcrystalline and of more "aged" crystalline forms, along with material in the particles (residual phases) are the major transporter of transition metals. Considering only the available forms and disregarding the residual fraction, metallic coatings become the major transporter of Mn, Fe, Co, and Ni and vastly increase in importance as a transporter for Cr and Cu. From analyses of grain-size-fractionated samples Gibbs (1977) found a distinct increase in the trace-metal/Fe-oxide ratios in the medium-to-coarse silt fractions. It was suggested that these effects are related to differences in the thickness of the iron-hydroxide layers. Thick layers on coarse material are considered among other factors to be a result of the weathering environment; the coarser material in the soil would have a higher permeability, bringing about a greater supply of the precipitating (and coprecipitated) ions. This would mean that a considerable portion of the metal accumulations in Fe- and Mn-hydroxide phases found in limnic sediments are of allochthonous origin.

Phase concentration factors of heavy metals in limnic sediments

In order to appraise the enrichment and dilution effects of each of the extractable metal associations, it is necessary to correlate the analyzed concentration to the amount of the respective carrier material found in the sediment. Förstner and Patchineelam (1979) have called this the "phase concentration factor" (PCF), which is calculated as the percentage of a certain metal association against the percentual contents of the reference phases ("phase" is defined here as one of the major sediment fractions, e.g., Fe/Mn-hydroxides, humic acids, etc.). Table XXII shows the PCF values from four of the lake examples of Fig.23 for the following reference phases: (1) the carbonate bonding of the total carbonate contents, (2) the residual association, which is the substance remaining after extraction of the carbonate portion, the humic component and the oxyhydrate phases, (3) humic associations, the contents of 0.1*N* NaOH extractable organic matter,

TABLE XXII

Phase concentration factors for heavy metals in lake sediments (examples from Fig. 23)

	Lake Amboseli				Van Gölü				L. Neusiedl				Lake Bled			
	(1)	(2)	(3)	(4)	(1)	(2)	(3)	(4)	(1)	(2)	(3)	(4)	(1)	(2)	(3)	(4)
Fe	0.1	1.0	29	–	0.3	1.2	18	–	0.3	1.2	27	–	0.4	1.4	28	–
Mn	0.8	0.6	14	49	1.4	0.3	9	49	1.1	0.6	11	58	1.4	0.7	5	na
Cr	<0.1	1.2	<5	7	<0.1	1.6	4	20	<0.1	1.5	4	26	<0.1	3.0	5	na
Ni	<0.1	1.0	<5	20	<0.1	1.5	<5	20	<0.1	1.5	<5	45	<0.1	2.9	<5	na
Co	<0.1	1.1	3	14	<0.1	1.6	3	15	<0.1	1.5	<3	20	<0.1	2.2	24	na
Zn	0.2	1.1	<10	5	0.9	1.0	9	<10	0.5	1.2	24	<10	0.9	0.8	16	na
Cu	0.1	0.8	<10	34	<0.1	1.1	26	77	0.3	0.9	25	<10	0.2	1.7	3	na

(1) = carbonates, (2) = residues, (3) humic acid, (4) FeOOH, na = not analyzed.

and (4) the hydroxidic bonding of the amount of FeOOH (calculated from $NH_2OH \cdot HCl$ + acetic acid extractable iron content). If the computed value lies above 1, a metal enrichment has occurred in the respective phase; if the value is less than 1, the metal concentration in the sample has been reduced. The PCF value is not influenced by the total content of the element and is therefore well suited for comparison of samples that occassionally differ greatly in the total metal concentration from the lake sediments investigated here.

The PCF values of Fe in *carbonate* lie between 0.1 and 0.4; in the case of Mn, however, it is found that most of the investigated samples reveal an enrichment by carbonate. For Cr, Ni, and Co, no association with carbonate was found. Carbonate phases of Zn and Cu are highly variable. One of the major consequences of these results is that the correction for "carbonate-free matter", which has in the past often been used for comparison of sediments containing different percentages of carbonates, cannot be applied for all heavy metals in the same manner.

With the exception of Mn in all samples, Zn in Lake Bled and Cu in Lakes Amboseli and Balaton, the *residual phase* – i.e., the most resistant mineral associations after application of extraction procedures (1) to (4) – causes a more or less clear increase of heavy metals in the sediments investigated. A large portion of these metals is apparently bound in inert lattice structures of clay minerals and especially in heavy minerals, whereas the contribution of quartz and feldspar minerals seems to be relatively low. Compared with the amounts of *humus substances,* the Fe contents which are associated with these substances are enriched 20–30 times. These values lie somewhat lower for Mn, Cr, and Ni (<5–15) and Co, Zn, and Cu (<3–25). Similar values are reported from samples of Southern Californian waters by Chen et al. (1976), who found 2–15 times higher concentrations of trace metals within the organic fraction than the total sediments on a weight basis.

Not considering the possibility that discrete trace metal-oxide/hydroxide phases are formed in the sediments, the enrichment factors for Mn in the *hydroxide phase* is 50–70, for Cu <10–80, Ni 20–60, Cr 10–40 and for Co 15–25. Thus, the enrichment in the hydrous oxide phases of these metals seems to be greater than that in the humic substances. At the same time, however, Zn is relatively poorly enriched in the hydrous oxide phases in the sediments (<10 times). In Lake Malawi, where no carbonate phase is present, the PCF values of Zn (compared to the 1*M* hydroxylamine-extractable Fe) are found as high as 30.

Release of heavy metals from sediments

Trace metals temporarily immobilized in the bottom sediments and suspended matter of limnic sediments may be released into open water as a result of physico-chemical changes, such as (1) low pH, (2) increased salinity, (3) increased input of organic chelators, (4) microbial activity, and (5) change in the redox conditions.

The effects of a *low pH* on the metal concentrations in aquatic systems has been discussed in detail in the next chapter of this Handbook from examples of acidic mine effluents.

Increased *salinity* in a water body leads to competition between dissolved cations and adsorbed trace metal ions and can result in partial replacement of the latter. Such effects should particularly be expected in the estuarine environment. It has been shown that part of the decrease of sediment-associated heavy metals along the river/sea mixing zone in polluted estuaries is due to a dilution of contaminated river particulates by relatively clean marine sediments (Müller and Förstner, 1975).

The interaction with *organic substances* plays an increasingly important role in the transport of heavy metals in both surface and ground water since the amount of organic complexing material is still increasing from secondary sewage treatment effluents. An additional impact on heavy metal remobilization from sediments may occur from the growing use of synthetic complexing agents (e.g., NTA in detergents to replace polyphosphates).

Microbial processes in the transport and enrichment of metals has long been recognized as an essential mechanism in mineralization, particularly in conjunction with the formation of stratiform sulfide deposits (Bastin, 1926; Temple and Le Roux, 1964; see for compilation Chapter 6 in Vol. 2 of this Handbook). Remobilization of heavy metals from sediments by microbial activity – which is increasingly being used for economical extraction of low-grade ores and wastes, e.g., of Cu and U – takes place (1) by the destruction of organic matter to lower molecular weight compounds, which are more capable of complexing metal ions, (2) by changes in the physical properties of the environment, e.g., the oxidation–reduction potential and pH conditions and (3) by the conversion of inorganic compounds into metal complexates. Such processes as noted in (2) are particularly responsible for the occurrence of acidic mine waters.

With regard to the remobilization of heavy metals from lacustrine sediments, the *changes of redox conditions* are of major interest. Oxidative conditions must be present on the lake bottom, as reducing conditions would cause the dissolution of the hydrous Fe/Mn-oxides. Sly and Thomas (1974) have examples from Lake Erie: Iron-oxide layers were rare there, suggesting that the low redox conditions release all of the available Fe and Mn from the sediments to the hypolimnion. The ferrous iron released is in excess of that required to combine with available sulphide present in the sediment as a product of the microbial reduction of SO_4^{2-}. Dissolution of Fe/Mn nodules has been observed in the Kingston Basin of Lake Erie, indicating recent changes in lake conditions in the area.

The potential health hazard arising from advanced eutrophication and subsequent dissolution of oxides has been emphasized by Edgington and Callender (1970) in a study of the minor element geochemistry of the Lake Michigan ferro-manganese nodules. These nodules contain unexpectedly large concentrations of As, up to 345 ppm with an average of 180 ppm As. Since ferro-manganese concretions and nodules are stable only under aerobic conditions, as was previously stated, further eutrophication would significantly raise the level of arsenic concentrations in the water. According to the calculations of Edgington and Callender (1970), the As-release from the total dissolution of the Green Bay nodules in Lake Michigan would increase the content of As, which is highly toxic to mammals and also carcinogenic, up to 90 μg/l, about twice the permissible value in drinking water.

Generally, however, the processes of *immobilization* of trace metals by the formation of *hydrous Fe/Mn oxides* seem to prevail. Among the various boundaries where these mechanisms may be effective, one should note the following: (a) the discharge of groundwaters to the surface through *springs,* (b) the *thermocline regions* in eutrophic lakes, (c) the surface of carbonate minerals such as calcite, where there is a *microzone of higher pH* (G.F. Lee, 1975), and (d) areas where *neutralization of acidic waters* takes place, such as the junction of rivers exhibiting different pH values (e.g., Deer Creek/Snake River in Colorado; Theobold et al., 1963), and from mixing zones of acid river water with seawater (e.g., Mullica River/Great Bay in New Jersey; Coonley et al., 1971). These processes effect the greatest part of the trace metals discharged from the continents to be deposited near the mouth of the river (see next chapter), so that the exchange of heavy metals within the estuarine zone is very restricted (Turekian, 1977).

CONCLUSIONS

This compilation presents several examples of heavy-metal enrichment in recent lacustrine sediments that are influenced by allochthonous and autochthonous processes. Apart from the few cases where anthropogenic influence has predominated, even the strongest heavy-metal enrichments in lacustrine sediments lie well below economically exploitable levels. In the case of natural allochthonous influences, dispersion effects are so great that the metal content of lake deposits is usually much lower compared to local metal enrichments in characteristic source rocks (e.g. samples from lakes in the vicinity of ultrabasic rock complexes such as those near the Sudbury mining area). In contrast with findings on marine sediments such as those from the Red Sea, no exploitable enrichment of any great extent have been found in limnic sediments from hydrothermally influenced regions (examples from Japan and New Zealand). In Lake Kivu, considerable amounts of zinc occur in the aqueous phase as suspended globules, whereas no comparable increase in zinc concentrations is detected in the sediments. In Black Sea sediments the highest metal (as well as the highest sulphur and organic carbon) enrichment rates are found in the marine sedimentation phases, whereas only moderate metal levels are present in the limnic deposits.

From the critical viewpoint of the uniformity of process principle (i.e. "the present is the key to the past"), however, these data have quite well indicated characteristic enrichment processes, which, in theory, could be mechanisms leading to the formation of economically attractive strata-bound deposits. In this context, the diagenetic processes taking place within the range of oxic and anoxic conditions are of particular interest (Degens and Stoffers, 1977). Hydrous Fe/Mn-oxides, on the one hand, and organic substances and sulfides, on the other, are seen to be the dominant enrichment phases in each of these environments (the latter two phases have not as yet been successfully separated with presently available selective extraction procedures). In the transfer of metals between the different accumulative phases, organo-metallic complexes perform a major function. Thus, it is this kind of interaction that must be the major subject in future detailed study.

ACKNOWLEDGEMENTS

The own research described in this chapter was performed within the framework of a programme on the "Geochemie umweltrelevanter Spurenstoffe" of the "Deutsche Forschungsgemeinschaft" (German Research Society). I thank R.J. Allan (Winnipeg), G. Müller and P. Stoffers for data and suggestions. I am grateful to D. Godfrey for his assistance in the preparation of the English version of this study.

REFERENCES

Abdul-Razzak, A.K., 1974. Geochemisch-sedimentpetrographischer Vergleich lakustrischer Sedimente aus verschiedenen Klimabereichen. *Chem. Erde,* 33: 154–184.

Aguilera, N.H. and Jackson, M.L., 1953. Iron oxide removal from soils and clays. *Soil Sci. Soc. Am. Proc.,* 17: 359–384.

Ahrens, L.H., 1957. The lognormal distribution of the elements – a fundamental law of geochemistry. *Geochim. Cosmochim. Acta,* 11: 205–212.

Allan, R.J., 1971. Lake sediments: a medium for regional geochemical exploration of the Canadian Shield. *Can. Inst. Min. Metall. Bull.,* 64: 43–59.

Allan, R.J., 1973. Surficial dispersion of trace metals in Arctic Canada: a nickel deposit, Raglan area, Cape Smith–Wakeham Bay belt, Ungawa. *Geol. Surv. Can. Pap.,* 73-1B: 9–19.

Allan, R.J., 1974. Metal contents of lake sediment cores from established mining areas: an interface of exploration and environmental geochemistry. *Geol. Surv. Can. Pap.,* 74-1B: 43–59.

Allan, R.J., 1977. Natural versus unnatural heavy metal concentrations in lake sediments in Canada. Proc. Int. Conf. *Heavy Metals in the Environment,* Toronto, 1975, Vol. II/2: 785–808.

Allan, R.J., 1978. Regional geochemical anomalies related to plate tectonic model for the northwestern Canadian Shield. *J. Geochem. Explor.,* 10: 218–228.

Allan, R.J. and Brunskill, G.J., 1977. Relative atomic variation (RAV) of elements in lake sediments. In: H.L. Goltermann (Editor), Proc. Int. Symp. *Interactions between Sediments and Fresh Water,* Amsterdam, Sept. 6–10, 1976. Junk, Den Haag, pp. 108–120.

Allan, R.J. and Cameron, E.M., 1973. U, Zn, Pb, Mn, Fe, Cu, Ni, and K contents in sediments: Bear-Slave operation. *Geol. Surv. Can. Maps,* 1972: 9–15.

Allan, R.J. and Jackson, T., 1978. Heavy metals in bottom sediment of the mainstem Athabasca River system in the AOSERP (Alberta Oil Sands Environmental Research Program) study area. *Prep. for the Alberta Oil Sand Environment Research Program by Fisheries and Environment Canada, Freshwater Institute. AOSERP Report* 34, 72 pp.

Allan, R.J. and Timperley, M.H., 1975. Prospecting using lake sediments in areas of industrial heavy metal contamination. *Prospecting in Areas of Glaciated Terrain – 1975*, pp.87–111.

Allan, R.J., Cameron, E.M. and Durham, C.C., 1972. Lake geochemistry – a low sample density technique for reconnaissance, geochemical exploration and mapping of the Canadian Shield. Proc. *4th Int. Geochemical Exploration Symp.*, pp.131–160.

Allan, R.J., Cameron, E.M. and Durham, C.C., 1973. Reconnaissance geochemistry using lake sediments of a 36,000 square mile area of the northwestern Canadian Shield. *Geol. Surv. Can. Pap.*, 72-50: 70 pp.

Allan, R.J., Cameron, E.M. and Jonasson, I.R., 1974. Mercury and arsenic levels in lake sediments from the Canadian Shield. Proc. *1st Int. Mercury Congress*, Barcelona, Spain, 43 pp.

Altschuler, Z.S., 1978. Trace elements as discriminants of origin in marine phosphorites. Abstr. *10th Int. Congress of Sedimentology*, Jerusalem, July 9–14, 1978, pp.16–17.

Anderson, T.W., 1973. Historical evidence of land use in a pollen profile from Osoyoos Lake, British Columbia. *Geol. Surv. Can. Pap.*, 73-2 (A): 178–180.

Angino, E.E., Magnuson, L.M., Waugh, T.C., Galle, O.K. and Bredfeldt, J. 1970. Arsenic in detergents: possible danger and pollution hazard. *Science*, 168: 389–390.

Armstrong, F.A.J. and Hamilton, A.L., 1973. Pathways of mercury in a polluted north-western Ontario lake. In: P.C. Singer (Editor), *Trace Metals and Metal-Organic Interactions in Natural Waters.* Ann Arbor Science Publishers, pp.131–156.

Arnold, R.G., 1970. The concentrations of metals in lake waters and sediments of some Precambrian lakes in the Flin Flon and La Ronge areas. *Sask. Res. Counc. Geol. Div. Circ.*, 4: 30 pp.

Arrhenius, G.O.S. and Korkish, J., 1959. Uranium and thorium in marine minerals. *Int. Ocean. Congress Am. Assoc. Adv. Sci. preprints*, 497 pp.

Aston, S.R. and Chester, R., 1973. The influence of suspended particles on the precipitation of iron in natural waters. *Estuarine Coastal Mar. Sci.*, 1: 225–231.

Aston, S.R., Bruty, D., Chester, R. and Padgham, R., 1973. Mercury in lake sediments: a possible indicator of technological growth. *Nature*, 241: 450–451.

Banat, K., Förstner, U. and Müller, G., 1972. Schwermetalle in Sedimenten von Donau, Rhein, Ems, Weser und Elbe im Bereich der Bundesrepublik Deutschland. *Naturwissenschaften*, 59: 525–528.

Barnes, R.S. and Schell, W.R., 1973. Physical transport of trace metals in the Lake Washington watershed. In: M.G. Curry and G.M. Gigliotti (Compilers) Cycling and Control of Metals, Proc. of an *Environmental Resources Conference, National Environment Research Center, U.S. Environmental Protection Agency*, Cincinnati/Ohio, pp.45–53.

Barr, D.A. and Hawkes, H.E., 1963. Seasonal variations in copper content of stream sediments in British Columbia. *Trans. Am. Inst. Min. Eng.*, 226: 342–346.

Bastin, E.S., 1926. A hypothesis of bacterial influence in the genesis of certain sulfide ores. *J. Geol.*, 34: 773–792.

Baumann, A., Förstner, U. and Rohde, R., 1975. Lake Shala: water chemistry, mineralogy and geochemistry of sediments in an Ethiopian Rift lake. *Geol. Rundsch.*, 64: 593–609.

Bernauer, F., 1933. Rezente Erzbildung auf der Insel Vulcano, *Fortschr. Mineral. Kristallogr. Petrogr.*, 17: p.28.

Bernauer, F., 1935. Rezente Erzbildung auf der Insel Vulcano, I. *Neues Jahrb. Mineral. Geol., Abt. A*, 69: 60–92.

Bernauer, F., 1939. Rezente Erzbildung auf der Insel Vulcano, Teil II. *Neues Jahrb. Mineral. Geol., Abt. A*, 76: 54–71.

Berner, R.A., 1971. *Principles of Chemical Sedimentology.* McGraw, New York, N.Y. 240 pp.

Bertine, K.K., 1972. The deposition of molybdenum in anoxic waters. *Mar. Chem.*, 1: 43–53.

Blohm, M., 1974. *Sedimentpetrographische Untersuchungen am Neusiedler See/Österreich.* Thesis, Univ. of Heidelberg, 85 pp. (unpublished).

Bølviken, B. and Sinding-Larsen, R., 1973. Total error and other criteria in the interpretation of stream sediment data. In: M.J. Jones (Editor), *Geochemical Exploration 1972.* Institution of Mining and Metallurgy, London, pp.285–295.

Boyle, R.W., Illsley, C.I. and Green, R.N., 1955. Geochemical investigation of heavy metal content of stream and spring waters in the Keno-Hill–Galena-Hill area, Yukon Territory. *Bull. Geol. Surv. Can.*, 32: 25 pp.

Bradshaw, P., 1975. Wintering Lake Cu–Ni prospect, Manitoba. *J. Geochem. Explor.*, 4: 188–189.

Bradshaw, P., Clews, D.R. and Walker, J.L., 1973. *Exploration Geochemistry (Barringer Research, Rexdale, Ont.)*, 50 pp.

Brannon, J.M., Engler, R.M., Rose, J.R., Hunt, P.G. and Smith, I., 1976. *Distribution of Manganese, Nickel, Zinc, Cadmium, and Arsenic in Sediments and the Standard Elutriate.* Environmental Effects Laboratory, U.S. Army Eng. Waterways Exp. Sta., Vicksburg, Miss., Misc. Pap., D-76-18, 38 pp.

Brewer, P.G. and Spencer, D.W., 1974. Distribution of some trace elements in the Black Sea and their flux between dissolved and particulate phases. In: *The Black Sea – Geology, Chemistry, and Biology. Am. Assoc. Pet. Geol. Mem.*, 20: 137–143.

Browne, P.R.L., 1971. Mineralisation in the broadlands geothermal field, Taupo volcanic zone, New Zealand. *Soc. Min., Geol. Japan*, Spec. Issue, 2: 64–75.

Bruland, K.W., Bertine, K., Koide, M. and Goldberg, E.D., 1974. History of metal pollution in southern California coastal zone. *Environ. Sci. Technol.*, 8: 425–432.

Brunskill, G.J., 1971. Chemistry of surface sediments of sixteen lakes in the Experimental Lakes area, northwestern Ontario. *J. Fish. Res. Board Can.*, 28: 277–294.

Butuzova, G.Y., 1969. K mineralogii i geochimii sul'fidov zhelza v osadkakh Chernogo morya (Mineralogy and geochemistry of iron sulfides in Black Sea sediments). *Litol. Polezn. Iskop.*, 4: 3–16.

Callender, E. and Bowser, C.J., 1976. Freshwater ferromanganese deposits. In: K.H. Wolf (Editor), *Handbook of Strata-Bound and Stratiform Ore Deposits, 7.* Elsevier, Amsterdam, pp.341–394.

Callender, E., Bowser, C.J. and Rossman, R., 1974. Geochemistry of ferromanganese and manganese carbonate crusts from Green Bay, Lake Michigan. *Trans. Am. Geophys. Union*, 54: 340.

Calvert, S.E. and Batchelor, C.E., 1978. Major and minor element geochemistry of sediments from Hole 379A, Leg 42B, Deep Sea Drilling Project. *Initial Reports of the Deep Sea Drilling Project, 42.* U.S. Govt. Printing Office, Washington, D.C., pp.527–539.

Cameron, E.M., 1974. Geochemical methods of exploration for massive sulphide mineralization in the Canadian Shield. Proc. *5th Int. Geochemical Exploration Symp.*, Vancouver, pp.21–49.

Cameron, E.M. and Ballantyne, S.B., 1975. Experimental hydrogeochemical surveys on the High Lake and Hackett River areas, Northwest Territories. *Geol. Surv. Can. Pap.*, 75-29: 19 pp.

Cameron, E.M. and Durham, C.C., 1974. Geochemical studies in the eastern part of the Slave Province, 1973. *Geol. Surv. Can. Pap.*, 74-27: 20 pp.

Chao, L.L., 1972. Selective dissolution of manganese oxides from soils and sediments with acidified hydroxylamine hydrochloride. *Soil Sci. Soc. Am. Proc.*, 36: 764–768.

Chen, K.Y., Gupta, S.K., Sycip, A.Z., Lu, J.C.S., Knezevic, M. and Choi, W.W., 1976. *The Effect of Dispersion, Settling and Resedimentation on Migration of Chemical Constituents During Open Water Disposal of Dredged Material.* Contract Rept., U.S. Army Eng. Waterways Exp. Sta., Vicksburg, Miss., 221 pp.

Chester, R., 1965. Geochemical criteria for differentiating reef from non-reef facies in carbonate rocks. *Bull. Am. Assoc. Pet. Geol.*, 49: 258–276.

Chester, R. and Hughes, M.J., 1967. A chemical technique for the separation of ferromanganese minerals, carbonate minerals and adsorbed trace elements from pelagic sediments. *Chem. Geol.*, 2: 249–262.

Chester, R. and Hughes, M.J., 1969. The trace element geochemistry of a North Pacific pelagic clay core. *Deep-Sea Res.*, 16: 639–654.

Chester, R. and Messiha-Hanna, R., 1970. Trace-element partition patterns in North Atlantic deep sea sediments. *Geochim. Cosmochim. Acta*, 34: 1121–1128.

Chork, C.Y., 1977. Seasonal, sampling and analytical variations in stream sediment surveys. *J. Geochem. Explor.*, 7: 31–47.

Closs, L.G., 1975. *Geochemistry of Lake Sediments in the Elliot Lake Region, District of Algoma.* Internal Rep. Ont. Division Mines on Proj. 74-21, 55 pp.

Coker, W.B. and Nichol, I., 1975. The relation of lake sediment geochemistry to mineralization in the northwest Ontario region of the Canadian Shield. *Econ. Geol.*, 70: 202–218.

Coonley, L.S., Baker, E.B. and Holland, H.D., 1971. Iron in the Mullica River in Great Bay, New Jersey. *Chem. Geol.*, 7: 51–63.

Cooper, B.S. and Harris, R.C., 1974. Heavy metals in organic phases of river and estuarine sediment. *Mar. Pollut. Bull.*, 5: 24–26.

Crecelius, E.A., 1975. The geochemical cycle of arsenic in Lake Washington and its relation to other elements. *Limnol. Oceanogr.*, 20: 441–451.

Crecelius, E.A. and Piper, D.Z., 1973. Particulate lead contamination recorded in sedimentary cores from Lake Washington, Seattle. *Environ. Sci. Technol.*, 7: 1053–1055.

Cutshall, N.H., 1967. *Chromium-51 in the Columbia River and Adjacent Pacific Ocean.* Thesis, Oregon State Univ., Corvallis (cited in E.N. Jenne, 1976).

Dahlberg, E.C., 1968. Application of a selective simulation and sampling technique to the interpretation of stream sediment copper anomalies near South Mountain, Pa. *Econ. Geol.*, 63: 409–417.

Davaud, E., Rapin, F. and Vernet, J.-P., 1977. Contamination des sédiments côtiers par les métaux lourds. In: Rapports 1977 Comm. Internationale pour la *Protection des Eaux du Léman contre la Pollution,* Lausanne, 16 pp.

Davenport, P.H., Hornbrook, E.H.W. and Butler, A.J., 1975. Regional lake sediment geochemical survey for zinc mineralization in western Newfoundland. *Geochemical Exploration – 1974,* pp.556–578.

David, T.W. and Browne, W.R., 1950. *The Geology of the Commonwealth of Australia.* Arnold, London, Vol. I, 852 pp, Vol. II, 618 pp.

Degens, E.T. and Ross, D.A., 1976. Strata-bound metalliferous deposits found in or near active spreading centers. In: K.H. Wolf (Editor), *Handbook of Strata-Bound and Stratiform Ore Deposits, 4.* Elsevier, Amsterdam, pp.165–202.

Degens, E.T. and Stoffers, P., 1976. Stratified waters as a key to the past. *Nature,* 263: 22–27.

Degens, E.T. and Stoffers, P., 1977. Phase boundaries as an instrument for metal concentration in geological systems. In: D.D. Klemm and H.-J. Schneider (Editors), *Time- and Strata-Bound Ore Deposits.* Springer, Heidelberg, pp.25–45.

Degens, E.T., von Herzen, R.P. and Wong, H.K., 1971. Lake Tanganyika: water chemistry, sediments, geological structure. *Naturwissenschaften,* 58: 229–241.

Degens, E.T., Okada, H., Honjo, S. and Hathaway, J.C., 1972. Microcrystalline sphalerite in resin globules suspended in Lake Kivu, East Africa. *Miner. Deposita,* 7: 1–12.

Degens, E.T., von Herzen, R.P., Wong, H.K., Deuser, W.G. and Jannasch, W., 1973. Lake Kivu: Structure, chemistry and biology of an East African Rift lake. *Geol. Rundsch.*, 62: 245–277.

Degens, E.T., Khoo, F. and Michaelis, W., 1977. Uranium anomaly in Black Sea sediments. *Nature,* 269, 566–569.

Degens, E.T., v. Bronsart, G., Wong. H.K. and Khoo, F., 1979. Bow Lake, Ontario/Canada: Uranlagerstätte in *statu nascendi.* Abstr. *69. Jahrestagung Geologische Vereinigung,* Heidelberg, 22–24 Febr. 1979, p.7.

De Groot, A.J., Goeij, J.J.M. and Zengers, C., 1971. Contents and behaviour of mercury as compared with other heavy metals in sediments from the Rivers Rhine and Ems. *Geol. Mijnbouw,* 50: 393–398.

Deurer, R., 1978. *Bindungsarten von Schwermetallen in Seesedimenten verschiedener Klimazonen unter besonderer Berücksichtigung des Bodensees.* Thesis, Univ. Heidelberg, 147 pp.

Deurer, R., Förstner, U. and Schmoll, G., 1978. Selective chemical extraction of carbonate-associated metals from recent lacustrine sediments. *Geochim. Cosmochim. Acta,* 42: 425–427.

Dominik, J., Förstner, U., Mangini, A. and Reineck, H.-E., 1978. Pb-210 and Cs-137 chronology of heavy metal pollution in a sediment core from the German Bight (North Sea). *Senckenbergiana Marit.,* 10: 213–227.

Duinker, J.C., Van Eck, G.T.M. and Nolting, R.F., 1974. On the behavior of copper, zinc, iron and manganese in the Dutch Wadden Sea. *Neth. J. Sea Res.,* 8: 214–239.

Duthie, H.C. and Sreenivasa, M.R., 1971. Evidence for the eutrophication of Lake Ontario from the sedimentary diatom succession. Proc. *14th Conf. Great Lakes Res.*, pp.1–13.

Dyck, W., 1971. Lake sampling re. stream sampling for regional geochemical surveys. *Geol. Surv. Can. Pap.*, 71-1(B): 70–71.

Eccles, D.H., 1975. An outline of the physical limnology of Lake Malawi (Lake Nyasa). *Limnol. Oceanogr.*, 19: 730–742.

Edgington, D.H. and Callender, E., 1970. Minor element geochemistry of Lake Michigan ferromanganese nodules. *Earth Planet. Sci. Lett.*, 8, 97–100.

Elderfield, H. and Hepworth, A., 1975. Diagenesis, metals and pollution in estuaries. *Mar. Pollut. Bull.*, 6: 85–87.

Eloff, J.N., 1965. Extraction of soil humic compounds with sodium hydroxide plus stannous chloride. *S. Afr. J. Agric. Sci.*, 8: 673–680.

Engler, R.M., Brannon, J.M. and Rose, J., 1974. *A Practical Selective Extraction Procedure for Sediment Characterization.* 168th Meeting ACS, Atlantic City, 17 pp.

Erlenkeuser, H., Suess, E. and Willkomm, H., 1974. Industrialization affects heavy metal and carbon isotope concentration in recent Baltic sea sediments. *Geochim. Cosmochim. Acta*, 38: 823–842.

Eugster, H.P. and Hardie, L.A., 1978. Saline Lakes. In: A. Lerman (Editor), *Lakes – Chemistry, Geology, Physics.* Springer, New York, N.Y. pp.237–293.

Förstner, U., 1973. Petrographische und geochemische Untersuchungen an afghanischen Endseen. *Neues Jahrb. Mineral. Abh.*, 118: 268–312.

Förstner, U., 1976. Lake sediments as indicators of heavy-metal pollution. *Naturwissenschaften*, 63: 465–470.

Förstner, U., 1977a. Metal concentrations in Recent lacustrine sediments. *Arch. Hydrobiol.*, 80: 172–191.

Förstner, U., 1977b. Metal concentrations in freshwater sediments – natural background and cultural effects. In: H.L. Goltermann (Editor), Proc. Int. Symp. *Interactions between Sediments and Fresh Water*, Amsterdam, Sept. 6–10, 1976. Junk, Den Haag, pp. 94–103.

Förstner, U., 1977c. Geochemische Untersuchungen an den Sedimenten des Ries-Sees (Forschungsbohrung Nördlingen 1973). *Geol. Bavarica*, 75: 37–48.

Förstner, U., 1977d. Mineralogy and geochemistry of sediments in arid lakes of Australia. *Geol. Rundsch.*, 66: 146–156.

Förstner, U., 1978. *Metallanreicherungen in rezenten See-Sedimenten – geochemischer background und zivilisatorische Einflüsse.* Mitt. Nationalkommitee der B.R. Deutschland für das Internationale Hydrologische Programm der UNESCO, Vol. 2, Koblenz, 66 pp.

Förstner, U., 1980. Cadmium in polluted sediments. In: J.O. Nriagu (Editor), *Biogeochemistry of Cadmium.* Wiley, New York, N.Y., in press.

Förstner, U. and Bartsch, G., 1970. Die Seen von Bande-Amir, Dasht-i-Nawar, Ob-i-Istada und Hamun-i-Puzak (Zentral- u. Südwestafghanistan). Messungen der Wassertiefen und der Salzgehalte. *Kabul Sci., Kabul Univ.*, 6: 19–23.

Förstner, U. and Müller, G., 1973a. Heavy metal accumulation in river sediments, a response to environmental pollution. *Geoforum*, 14: 53–62.

Förstner, U. and Müller, G., 1973b. Anorganische Schadstoffe im Neckar. *Ruperto Carola*, 51: 67–71.

Förstner, U. and Müller, G., 1974a. Schwermetallanreicherungen in datierten Sedimentkernen aus dem Bodensee und aus dem Tegernsee. *TMPM Tschermaks Min. Petr. Mitt.*, 21: 145–163.

Förstner, U. and Müller, G., 1974b. *Schwermetalle in Flüssen und Seen.* Springer, Berlin, 225 pp.

Förstner, U. and Müller, G., 1975. *Factors controlling the distribution of minor and trace metals (heavy metals, V, Li, Sr) in recent lacustrine sediments.* IX Int. Sedimentol. Congr., Nice, pp.57–63.

Förstner, U. and Patchineelam, S.R., 1976. Bindung und Mobilisation von Schwermetallen in fluviatilen Sedimenten. *Chem. Ztg.*, 100: 49–57.

Förstner, U. and Patchineelam, S.R., 1979. Chemical associations of heavy metals in polluted sediments. In: M. Kavanaugh and J.O. Leckie (Editors), *Particulates in Water: Characterization, Fate, Effects and Removal.* ACS Advances in Chemistry Series, in press.

Förstner, U. and Reineck, H.-E., 1974. Die Anreicherung von Spurenelementen in den rezenten Sedimenten eines Profilkernes aus der Deutschen Bucht. *Senckenbergiana Marit.*, 6: 175–184.

Förstner, U. and Wittmann, G.T.W., 1979. *Metal Pollution in the Aquatic Environment.* Springer, Berlin, 486 pp.

Förstner, U., Müller, G. and Wagner, G., 1974. Schwermetalle in Sedimenten des Bodensees – natürliche und zivilisatorische Anteile. *Naturwissenschaften,* 61: p.270.

Frink, C.R., 1967. Nutrient budgets and rational analysis of eutrophication in a Connecticut lake. *Environ. Sci. Technol.*, 1: 425–427.

Frye, J.C. and Shimp, N.F., 1973. Major, minor and trace elements in sediments of Late Pleistocene Lake Salina compared with those in Lake Michigan sediments. *Ill. State Geol. Surv., Environ. Geol. Note,* 60: 14 pp.

Füchtbauer, H. and Müller, G., 1970. *Sedimente und Sedimentgesteine. Sedimentpetrologie, Teil II.* Schweizerbart, Stuttgart, 726 pp.

Gad, M.A. and Le Riche, H.H., 1966. A method for separating the detrital and non-detrital fractions of trace elements in reduced sediments. *Geochim. Cosmochim. Acta,* 30: 841–846.

Garrett, R.G., 1969. The determination of sampling and analytical errors in exploration geochemistry. *Econ. Geol.*, 64, 568–574.

Gibbs, R.J., 1970. Mechanisms controlling world water chemistry. *Science,* 170: 1088–1090.

Gibbs, R.J., 1973. Mechanisms of trace metal transport in rivers. *Science,* 180: 71–73.

Gibbs, R.J., 1977. Transport phases of transition metals in the Amazon and Yukon Rivers. *Geol. Soc. Am. Bull.*, 88: 829–843.

Glasby, G.P. and Read, A.J., 1976. Deep-sea manganese nodules. In: K.H. Wolf (Editor), *Handbook of Strata-Bound and Stratiform Ore Deposits, 7.* Elsevier, Amsterdam, pp. 295–340.

Glover, E.D., 1961. Method of solution of calcareous materials using the complexing agent EDTA. *J. Sediment. Petrol.*, 31: 622–626.

Goldberg, E.D. and Arrhenius, G.O.S., 1958. Chemistry of Pacific pelagic sediments. *Geochim. Cosmochim. Acta,* 13: 153–212.

Golterman, H.L. (Editor), 1977. *Interactions Between Sediments and Fresh Water,* Junk, The Hague, 570 pp.

Govett, G.J.S., 1960. Seasonal variation in the copper concentration in drainage systems in Northern Rhodesia. *Trans. Inst. Min. Metall.*, 70: 177–189.

Griffith, S.M. and Schnitzer, M., 1975. The isolation and characterization of stable metal-organic complexes from tropical volcanic soils. *Soil Sci.*, 120: 126–131.

Grim, R.E., 1968. *Clay Mineralogy.* McGraw-Hill, New York, N.Y., 2nd ed., 596 pp.

Grimmer, G. and Böhnke, H., 1977. Schadstoffuntersuchungen an datierten Sedimentkernen aus dem Bodensee. I. Profile der polyzyklischen Kohlenwasserstoffe. *Z. Naturforsch.*, 32(c): 703–707.

Gross, M.G., Black, J.A., Kalin, R.J., Schramel, J.R. and Smith, R.N., 1971. Survey of marine waste deposits, New York metropolitan region. *Mar. Sci. Res. Center, State Univ. N.Y., Tech. Rep.*, 8: 72 pp.

Gupta, S.K. and Chen, K.Y., 1975. Partioning of trace metals in selective chemical fractions of nearshore sediments. *Environ. Lett.*, 10: 129–158.

Guy, R.D. and Chakrabarti, C.L., 1975. Distribution of metal ions between soluble and particulate forms. Abstr. Int. Conf. *Heavy Metals in the Environment,* Toronto, D-29.

Hails, J.R., 1976. Placer deposits. In: K.H. Wolf (Editor), *Handbook of Strata-Bound and Stratiform Ore Deposits, 3,* Elsevier, Amsterdam, pp.213–244.

Håkanson, L., 1977. Sediments as indicators of contamination – investigations in the four largest Swedish lakes, *SNV PM 839/NLU* 92, 159 pp.

Hamilton-Taylor, J., 1979. Enrichments of zinc, lead, and copper in Recent sediments of Windermere, England. *Environ. Sci. Technol.*, 13: 693–697.

Hardie, L.A., Smoot, J.P. and Eugster, H.P., 1978. Saline lakes and their deposits: a sedimentological approach. In: A. Matter and M.E. Tucker (Editors), *Modern and Ancient Lake Sediments, 2.* Spec. Publ. Int. Assoc. Sedimentol., pp.7–41.

Haughton, D.R., Smith, J.W. and Arnold, R.G., 1973. *Geochemistry of bedrock, overburden, lake and stream sediment in the Mudjatic area, Saskatchewan.* Sas. Res. Council Geol. Div. Rep. No. 12.

Hawkes, H.E., Bloom, H. and Riddell, J.E., 1957. Stream sediment analysis discovers two mineral deposits. Proc. *6th Commonwealth Min. Metall. Congr.*, pp.259–268.

Heath, G.R. and Dymond, J., 1977. Genesis and transformation of metalliferous sediments from the East Pacific Rise, Bauer Deep, and Central Basin, northwest Naszca plate. *Geol. Soc. Am. Bull.*, 88: 723–733.

Helmke, P.A., Koons, R.D., Schomberg, P.J. and Iskandar, I.K., 1977. Determination of trace element contamination of sediments by multi-element analysis of clay size fraction. *Environ. Sci. Technol.*, 11: 984–989.

Hill Jr., W.R. and Runnels, R.T., 1960. Versene, new tool for study of carbonate rocks. *Bull. Am. Assoc. Pet. Geol.*, 44: 631–632.

Hilmer, E., 1972. *Geochemische Untersuchungen im Bereich der Lagerstätte Meggen, Rheinisches Schiefergebirge.* Thesis, Technische Hochschule, Aachen.

Hirst, D.M., 1962. The geochemistry of modern sediments from the Gulf of Paria. II. The location and distribution of trace elements. *Geochim. Cosmochim. Acta,* 26: 1147–1187.

Hirst, D.M. and Nicholls, G.D., 1958. Techniques in sedimentary geochemistry. 1. Separation of the detrital and non-detrital fractions of limestones. *J. Sediment. Petrol.*, 28: 461–468.

Hoffman, S.J. and Fletcher, K., 1972. Distribution of copper at the Dansey-Rayfield river property, south central British Columbia, Canada. *J. Geochem. Explor.*, 1: 163–180.

Hogson, G.F., 1960. Cobalt reactions with montmorillonite. *Soil Sci. Soc. Am. Proc.*, 24: 165–168.

Holdren, G.R., Bricker, O.P. and Matisoff, G., 1975. A model for the control of dissolved manganese in the interstitial waters of the Chesapeake Bay. In: T.M. Church (Editor), *Marine Chemistry in the Coastal Environment. Am. Chem. Soc. Symp. Ser.*, 18: 364–381.

Holmgren, G.S., 1967. A rapid citrate-dithionite extractable iron procedure. *Soil Sci. Soc. Am. Proc.*, 31: 210–211.

Honnorez, J., 1969. La formation actuelle d'un gisement sousmarin de sulfures fumerolliens à vulcano (mer tyrrhénienne). Parti I. *Mineral. Deposita,* 4: 114–131.

Honnorez, J., Honnorez-Guerstein, M., Valette, J. and Waschkuhn, A., 1973. Present-day formation of an exhalative sulfide deposit at Vulcano (Tyrrhenian Sea). Part II: Active crystallization of fumarolic sulfides in the volcanic sediments of the Baia di Levante. In: G.C. Amstutz and A.J. Bernard (Editors), *Ores in Sediments.* IUGS, Series A, No. 3. Springer, Heidelberg, pp. 139–166.

Hornbrook, E.H.W. and Gleeson, C.F., 1973. Regional lake bottom sediment moving average-residual anomaly maps, Noranda–Val d'Or area. *Geol. Surv. Can. Open File*: 127.

Hornbrook, E.H.W., Garrett, R.G., Lynch, J.J. and Beck, L.S., 1975. Regional sediment geochemical reconnaissance data – east central Saskatchewan. *Geol. Surv. Can. Open File*: 266.

Horowitz, A., 1974. The geochemistry of sediments from the northern Reykjanes Ridge and the Iceland-Faeroes Ridge. *Mar. Geol.*, 17: 103–122.

Horowitz, A. and Cronan, D.S., 1976. The geochemistry of basal sediments from the north Atlantic Ocean. *Mar. Geol.*, 20: 205–228.

Howarth, R.J. and Lowenstein, P.L., 1971. Sampling variability of stream sediments in broad-scale regional geochemical reconnaissance. *Trans. Inst. Min. Metall.*, 80: B363–B372.

Hutchinson, G.E., 1957. *A Treatise on Limnology.* Wiley, New York, N.Y., 1015 pp.

Hutchinson, G.W. and Wollack, A.C., 1940. Studies on Connecticut lake sediments. II. Chemical analyses of a core from Linsley Pond. *Am. J. Sci.*, 238: 493–517.

Irion, G., 1970. Mineralogisch-sedimentpetrographische und geochemische Untersuchungen am Tuz Gölü (Salzsee) Türkei. *Chem. Erde,* 29: 167–196.

Irion, G., 1972. Lithium als Anreicherungsprodukt in zwei türkischen Salzseen. *Naturwissenschaften,* 59: 167.

Irion, G., 1973. Die anatolischen Salzseen, ihr Chemismus und die Entstehung ihrer chemischen Sedimente. *Arch. Hydrobiol.,* 71: 517–557.

Irion, G. and Förstner, U., 1975. Chemismus und Mineralbestand amazonischer See-Tone. *Naturwissenschaften,* 62: 179.

Iskandar, K. and Keeney, D.R., 1974. Concentration of heavy metals in sediment cores from selected Wisconsin lakes. *Environ. Sci. Technol.,* 8: 165–170.

Jackson, M.L., 1958. *Soil Chemical Analysis.* Prentice Hall, Englewood Cliffs, N.J., 498 pp.

Jackson, R.G. and Nichol, I., 1974. Factors affecting trace element dispersion in lake sediments in the Yellowknife area, N.W.T. Abstr. *19th Congr. Int. Assoc. Limnol.,* p.93.

Jackson, T.A., 1978. The biogeochemistry of heavy metals in polluted lakes and streams at Flin Flon, Canada, and a proposed method for limiting heavy-metal pollution of natural waters. *Environ. Geol.,* 2: 173–189.

Jenne, E.A., 1968. Controls on Mn, Fe, Co, Ni, Cu, and Zn concentrations in soils and water: the significant role of hydrous Mn and Fe oxides. Advances in Chemistry Ser. 73, *Trace Inorganics in Water. Am. Chem. Soc.,* pp.337–387.

Jenne, E.A., 1976. Trace element sorption by sediments and soils–sites and processes. In: W. Chappel and K. Petersen (Editors), *Symposium on Molybdenum, 2.* Marcel Dekker, New York, N.Y., pp.425–553.

Jensen, S. and Jernelöv, A., 1969. Biological methylation of mercury in aquatic organisms. *Nature,* 233: 753–754.

Jernelöv, A. and Lahn, H., 1973. Studies in Sweden on feasibility of some methods for restoration of mercury-contaminated bodies of water. *Environ. Sci. Technol.,* 7: 712–718.

Jonasson, I.R., 1970. *Mercury in the Natural Environment: A Review of Recent Work.* Geological Survey of Canada, Department of Energy, Mines and Resources/Queen's Printer for Canada, Ottawa, 45 pp.

Jonasson, I.R., 1976. Detailed hydrogeochemistry of two small lakes in the Grenville Geological Province. *Geol. Surv. Can. Pap.,* 76-13: 37 pp.

Jonasson, I.R., 1977. Geochemistry of sediment/water interactions of metals, including observations on availability. In: H. Shear and A.E.P. Watson (Editors), *The Fluvial Transport of Sediment-Associated Nutrients and Contaminants.* IJC/PLUARG: Windsor/Ont., pp.255–271.

Jones, A.S.G., 1973. The concentration of copper, lead, zinc and cadmium in shallow marine sediments, Cardigan Bay, Wales. *Mar. Geol.,* 14: M1–M9.

Jones, B.F. and Bowser, C.J., 1978. The mineralogy and related chemistry of lake sediments. In: A. Lerman (Editor), *Lakes – Chemistry, Geology, Physics.* Springer, New York, N.Y. pp.179–235.

Jux, U. and Kempf, E.K., 1971. Stauseen durch Travertinabsatz im zentralafghanischen Hochgebirge. *Z. Geomorphol. Suppl.,* 12: 107–137.

Kemp, A.L.W., Anderson, T.W., Thomas, R.L. and Mudrochova, A., 1974. Sedimentation rates and recent sediment history of lakes Ontario, Erie and Huron. *J. Sediment. Petrol.,* 44: 207–218.

Kemp, A.L.W., Thomas, R.L., Dell, C.I. and Jaquet, J.-M., 1976. Cultural impact on the geochemistry of sediments in Lake Erie. *J. Fish. Res. Board Can.,* 33: 440–462.

Kennedy, E.J., Ruch, R.R., Shimp, N.F., 1971. Distribution of mercury in unconsolidated sediments from southern Lake Michigan. *Ill. State Geol. Surv., Environ. Geol. Note,* 44: 18 pp.

Kilham, P., 1971. Geochemical evolution of closed basin water. *Abstr. Geol. Soc. Am.,* 3: 770–772.

King, D., 1953. Origin of alunite deposit at Pidinga, South Australia. *Econ. Geol.,* 48: 689–703.

Klassen, R.A., 1975. *Lake Sediment Geochemistry in the Kaminak Lake Area, District of Keewatin, N.W.T.* Thesis, Queen's Univ., Kingston, Ontario.

Klassen, R.A., Nichol, I., Shilts, W.W., 1974. Lake geochemistry in the Kaminak Lake area, District of Keewatin, N.W.T. Abst. *19th Congr. Int. Assoc. Limnol.,* p. 106.

Krishnaswami, S. and Lal, D., 1978. Radionuclide limnochronology. In: A. Lerman (Editor), *Lakes – Chemistry, Geology, Physics.* Springer, New York, N.Y., pp.153–177.

Krishnaswami, S., Lal, D., Martin, J.M. and Meybeck, M., 1971. Geochronology of lake sediments. *Earth Planet. Sci. Lett.,* 11: 407–414.

Krom, M.D., 1976. *Chemical speciation and diagenesis reactions at the sediment–water interface in a Scottish Fjord.* Abstr. Joint Oceanogr. Assembly, Edinburgh, p.89.

Kullenberg, B., 1947. The piston core sampler. *Sven. Hydrogr.–Biol. Komm. Skr. Ser. 3, Hydrogr.*, 1: 2.

Landa, E.R. and Gast, R.G., 1973. Evaluation of crystallinity in hydrated ferric oxides. *Clays Clay Miner.*, 21: 121–130.

Langbein, W.B., 1961. Salinity and hydrology of closed lakes. *U.S. Geol. Surv. Prof. Pap.*, 412: 20 pp.

Langmyhr, F.J., Solberg, R. and Thomassen, Y., 1977. Atom absorption spectrometric determination of thirteen minor and trace metals in phosphate rock concentrates. *Anal. Chim. Acta*, 92: 105–109.

Lee, C.R., Engler, R.M. and Mahloch, J.L., 1976. *Land application of waste materials from dredging, construction, and demolition processes.* Office, Chief of Engineers, U.S. Army, Misc. Paper D-76-5, 42 p.

Lee, G.F., 1975. Role of hydrous metal oxides in the transport of heavy metals in the environment. In: P.A. Krenkel (Editor), *Heavy Metals in the Aquatic Environment*, Pergamon, Oxford, pp.137–147.

Lerman, A., 1977. Migrational processes and chemical reactions in interstitial waters. In: E.D. Goldberg, I.N. McCave, J.J. O'Brien and J.H. Steel (Editors), *The Sea, 6.* Wiley, New York, N.Y., pp.695–738.

Lerman, A. (Editor), 1978. *Lakes – Chemistry, Geology, Physics.* Springer, New York, N.Y., 363 pp.

Leshniowsky, W.O., Dugan, P.R., Pfister, R.M., Frea, J.I. and Randies, C.I., 1970. Adsorption of chlorinated hydrocarbon pesticides by microbial floc and lake sediment and its ecological implications. Proc. *13th Conf. Great Lakes Research*, pp.611–618.

Lichtfuss, R. and Brümmer, G., 1977. Schwermetallbelastung von Elbe-Sedimenten. *Naturwissenschaften*, 64: 122–125.

Livingstone, D.A., 1963. Chemical composition of rivers and lakes. In: M. Fleischer (Editor), *Data of Geochemistry, 6th ed. U.S. Geol. Surv. Prof. Pap.*, 440-G: 64 pp.

Livingstone, D.A. and Boykin, J.C., 1962. Vertical distribution of phosphorous in Linsley Pond mud. *Limnol. Oceanogr.*, 7: 57–63.

Lloyd, R.M., 1954. A technique for separating clay minerals from limestones. *J. Sediment. Petrol.*, 24: 218–220.

Mackenzie, F.T. and Garrels, R.M., 1966. Chemical mass balance between rivers and oceans. *Am. J. Sci.*, 264: 507–525.

Malo, B.A., 1977. Partial extraction of metals from aquatic sediments. *Environ. Sci. Technol.*, 11: 277–282.

McKelvey, V.E., 1956. Uranium in phosphate rocks. *U.S. Geol. Surv. Prof. Pap.*, 300: 477.

Mehrtens, M.B., Tooms, J.S. and Troup, A.G., 1972. Some aspects of geochemical dispersion from base metal mineralization within glaciated terrain in Norway, North Wales, and British Columbia, Canada. *Geochem. Explor.*, 1972: 105–115.

Miesch, A.T., 1967. Theory of error in geochemical data. *U.S. Geol. Surv. Prof. Pap.*, 574-A; 17 pp.

Molnar, F.M., Rothe, P., Förstner, U., Stern, J., Ogorelec, B., Sercelj, A. and Culiberg, M., 1978. Sedimentology and geochemistry of recent deposits from Lakes Bled and Bohinj in Slovenia, Yugoslavia. *Geologija (Ljubljana)*, 21: 93–164.

Mortimer, C.H., 1941. The exchange of dissolved substances between mud and water in lakes. *J. Ecol.*, 29: 280–329.

Müller, G., 1966. Die Sedimentbildung im Bodensee. *Naturwissenschaften*, 53: 237–248.

Müller, G., 1967. Beziehungen zwischen Wasserkörper, Bodensediment und Organismen im Bodensee. *Naturwissenschaften*, 54: 454–466.

Müller, G., 1969. Sedimentbildung im Plattensee/Ungarn. *Naturwissenschaften*, 56: 606–615.

Müller, G., 1970. In: H. Füchtbauer and G. Müller. *Sedimente und Sedimentgesteine.* Schweizerbart, Stuttgart, 726 pp.

Müller, G., 1971. Sediments of Lake Constance. *Guidebook VIII. Int. Sedimentol. Congr.*, Heidelberg, pp. 237–252.

Müller, G. and Förstner, U., 1973. Recent iron ore formation in Lake Malawi, Africa. *Mineral. Deposita,* 8: 278–290.

Müller, G. and Förstner, U., 1975. Heavy metals in sediments of the Rhine and Elbe Estuaries: mobilization or mixing effect? *Environ. Geol.,* 1: 33–39.

Müller, G., Irion, G. and Förstner, U., 1972. Formation and diagenesis of inorganic Ca–Mg carbonates in the lacustrine environment. *Naturwissenschaften,* 59: 158–164.

Murray, R.C., 1956. Recent sediments of three Wisconsin Lakes. *Geol. Soc. Am. Bull.,* 67: 833.

Nichol, I., Coker, W.B., Jackson, R.G. and Klassen, R.A., 1975. Relation of lake sediment composition to mineralization in different limnological environments in Canada. *Prospecting in Areas of Glaciated Terrain – 1975,* pp.112–125.

Nickerson, D., 1972. An account of a lake sediment geochemical survey conducted over certain volcanic belts within the Slave structural province of the Northwest Territories, during 1972. *Geol. Surv. Can. Open File,* 129, 22 pp. and maps.

Nikiforova, E.M. and Smirnova, R.S., 1975. Metal technophility and lead technogenic anomalies. Abstr. *Int. Conf. Heavy Metals in the Environment,* Toronto, pp.C-94–96.

Nipkow, H.F., 1920. Vorläufige Mitteilungen über Untersuchungen des Schlammabsatzes im Zürichsee. *Z. Hydrol. (Aarau)*: 1–23.

Nriagu, J.O., 1974. Lead orthophosphates – IV: Formation and stability in the environment. *Geochim. Cosmochim. Acta,* 38: 887–898.

Ohle, W., 1956. Die Seen als Opfer der Abwasserkalamität. *Ber. Abwassertechn. Ver.* 7: 268–276.

Oliver, B.G., 1973. Heavy metal levels of Ottawa and Rideau River sediments. *Environ. Sci. Technol.,* 7: 135–137.

Ong, H.L., Swanson, V.E. and Bisque, R.E., 1970. *Natural organic acids as agents of chemical weathering.* Geol. Surv. Research Paper C130-C137.

Ozerova, N.A., Nabokov, S.I. and Vinogradov, V.J., 1971. Sulphides of mercury, antimony, and arsenic, forming from the active thermal spring of Kamchatka and Kuril Islands. *Soc. Min. Geol. Jpn. Spec. Issue,* 2: 164–170.

Patchineelam, S.R., 1975. *Untersuchungen über die Hauptbindungsarten und die Mobilisierbarkeit von Schwermetallen in fluviatilen Sedimenten.* Thesis, Univ. Heidelberg, 137 pp.

Perhac, R.M., 1974. Heavy metal distribution in bottom sediment and water in the Tennessee River – Loudon Lake reservoir system. *Water Resources Research Center, Univ. Tenn., Knoxville, Res. Rep.* 40, 9 pp.

Philipchuk, M.F. and Volkov, I.I., 1974. Behaviour of molybdenum in processes of sediment formation and diagenesis in Black Sea. In: E.T. Degens and D.A. Rose (Editors), *The Black Sea, Geology, Chemistry and Biology, Am. Assoc. Pet. Geol. Mem.,* 20: 542–553.

Piper, D.Z., 1971. The distribution of Co, Cr, Cu, Fe, Mn, Ni and Zn in Framvaren, a Norwegian anoxic fjord. *Geochim. Cosmochim. Acta,* 35: 531–550.

Popova, T.P., 1961. Coprecipitation of some microconstituents from natural waters with calcium carbonate. *Geochemistry,* 12: 1256–1261.

Presley, B.J., Kolodny, Y., Nissenbaum, A. and Kaplan, I.R., 1972. Early diagenesis in a reducing fjord, Saanich Inlet, British Columbia. II. Trace element distribution in interstitial water and sediment. *Geochim. Cosmochim. Acta,* 36: 1073–1099.

Price, N.B., 1973. Chemical diagenesis in sediments. WHO I, Woods Hole, Mass., Tech. Rep. *WHOI-73-39,* 73 pp.

Rashid, M.A., Buckley, D.E. and Robertson, K.R., 1972. Interactions of a marine humic acid with clay minerals and a natural sediment. *Geoderma,* 8: 11–27.

Ravera, O., 1964. The radioactivity of *Viviparus ater, christ.,* and *jan.* freshwater molluscs, in relation to that of the sediments: *Bull. FEPE* 10: 61–65.

Ravera, O. and Premazzi, G., 1971. A method to the study of the history of any persistent pollution in a lake by the concentration of Cs-137 from fallout. Proc. Int. Symp. *Radioecology Applied to the Protection of Man and his Environment,* Rome, pp. 703–719.

Ray, S., Gault, H.R. and Dodd, C.G., 1957. The separation of clay minerals from carbonate rocks. *Am. Mineral.,* 42: 681–686.

Reeves Jr., C.C., 1968. *Introduction to Paleolimnology. Developments in Sedimentology 11.* Elsevier, Amsterdam, 228 pp.

Reiss, F., 1976. Charakterisierung zentralamazonischer Seen auf Grund ihrer Macrobenthos-Fauna. *Amazoniana,* 1976: 123–134.

Renzoni, A., Bacci, E. and Falcia, L., 1973. Mercury concentration in the water, sediments and fauna of an area of the Tyrrhenian Coast. *Rev. Int. Oceanogr. Med.,* 31–32: 17–45.

Ridley, W.P., Dizikes, L.J. and Wood, J.M., 1977. Biomethylation of toxic elements in the environment. *Science,* 197: 329–332.

Ritchie, J.C., McHenry, R.J. and Gill, A.C., 1973. Dating Recent reservoir sediments. *Limnol. Oceanogr.,* 18: 254–263.

Robbins, J.A. and Edgington, D.N., 1975. Determination of recent sedimentation in Lake Michigan using lead-210 and cesium-137. *Geochim. Cosmochim. Acta,* 39: 285–304.

Ruch, R.R., Gluskoter, H.J. and Shimp, N.F., 1973. Occurrence and distribution of potentially volatile trace elements in coal. *Environ. Geol. Notes,* 61: 43 pp.

Ruch, R.R., Kennedy, E.J. and Shimp, N.F., 1973. Distribution of arsenic in unconsolidated sediments from southern Lake Michigan. *Environ. Geol. Note, Ill. Geol. Surv.,* 37: 16 pp.

Saito, M., Bamba, T., Sawa, T., Narita, E., Igarashi, T., Yamada, K. and Satoh, H., 1967. *Metallic and Non-Metallic Mineral Deposits of Hokkaido, Japan.* Geol. Surv. Japan, 11 pp. and 4 maps.

Salomons, W. and Mook, W.G., 1978. Processes affecting trace metals in Lake IJssel. Abstr. *10th Int. Congr. Sedimentology,* Jerusalem, pp.569–570.

Saxby, J.D., 1973. Diagenesis of metal-organic complexes in sediments: formation of metal sulphides from cysteine complexes. Chem. Geol., 12: 241–288.

Saxby, J.D., 1976. The significance of organic matter in ore genesis. In: K.H. Wolf (Editor), *Handbook of Strata-Bound and Stratiform Ore Deposits, 2.* Elsevier, Amsterdam, pp.111–133.

Sayles, F.L., Ku, T.-L. and Bowker, P.C., 1975. Chemistry of ferromanganoan sediment of the Bauer Deep. *Geol. Soc. Am. Bull.,* 86: 1423–1431.

Schell, W.R., 1974. *Sedimentation rates and mean residence times of Pb and 210-Pb in Lake Washington, Puget Sound estuaries and a coastal region.* Abstr. Meet. Amer. Soc. Limnol. Oceanogr., June 23–27.

Schmidt, R.C., 1956. *Adsorption of Cu, Pb, and Zn on Some Common Rock Forming Minerals and its Effect on Lake Sediments.* Thesis, McGill Univ., Montreal, Que., 181 pp.

Schmoll, G. and Förstner, U., 1979. Chemical associations of heavy metals in lacustrine sediments. I. Calcareous lake sediments from different climatic zones. *Neues Jahrb. Mineral. Abh.,* in press.

Schnitzer, M. and Wright, J.R., 1957. Extractions of organic matter from podsolic soils by means of dilute inorganic acids. *Can. J. Soil Sci.,* 37: 89–95.

Schroll, E. and Wieden, P., 1960. Eine rezente Bildung von Dolomit im Schlamm des Neusiedlersees. *Tschermaks Mineral. Petrogr. Mitt.,* 7: 286–289.

Schwertmann, U., 1964. Differenzierung der Eisenoxide des Bodens durch photochemische Extraktion mit saurer Ammoniumoxalat-Lösung. Z. Pflanzenernähr. Düng. Bodenkde., 105: 194–202.

Shimp, N.F., Schleicher, J.A., Ruch, R.R., Heck, D.B. and Leland, H.V., 1971. Trace element and organic carbon accumulation in the most recent sediments of southern Lake Michigan. *Environ. Geol. Notes,* 41: 25 pp.

Sholkovitz, E.R., 1976. Flocculation of dissolved organic and inorganic matter during the mixing of river water and seawater. *Geochim. Cosmochim. Acta,* 40: 831–845.

Sholkovitz, E.R., Boyle, E.A. and Price, N.B., 1978. The removal of dissolved humic acids and iron during estuarine mixing. *Earth Planet. Sci. Lett.,* 40: 130–136.

Shukla, S.S., Syers, J.K. and Armstrong, D.E., 1972. Arsenic interference in the determination of inorganic phosphate in lake sediments. *J. Environ. Qual.,* 1: 292–295.

Singer, A. and Navrot, J., 1978. Siderite in Birket Ram lake sediments. Abstr. *10th Int. Congr. on Sedimentology,* July 9–14, Jerusalem, pp.616–617.

Sly, P., 1978. Sedimentary processes in lakes. In: A. Lerman (Editor), *Lakes – Chemistry, Geology, Physics.* Springer, New York, N.Y., pp.65–89.

Sly, P. and Thomas, R.L., 1974. Review of geological research as it relates to an understanding of Great Lakes limnology. *J. Fish. Res. Board Can.*, 31: 795–825.

Solomon, A.M. and Kroener, D.F., 1971. Suburban replacement of rural land uses reflected in the pollen rain of northeastern New Jersey. *N.J. Acad. Sci. Bull.*, 16: 30–44.

Stockner, J.G. and Benson, W.W., 1967. The succession of diatom assemblages in the recent sediment of Lake Washington. *Limnol. Oceanogr.*, 12: 513–532.

Stoffers, P., 1975. *Sedimentpetrographische, geochemische und paläoklimatische Untersuchungen an ostafrikanischen Seen.* Habilitationsschrift, Univ. Heidelberg, 117 pp.

Stoffers, P. and Müller, G., 1978. Mineralogy and lithofacies of Black Sea sediments – Leg 42B Deep Sea Drilling Project. In: D.A. Ross, Y.P. Neprochnov *et al.*, *Initial Reports of the Deep-Sea Drilling Project,* 42(2). U.S. Govt. Printing Office, Washington, D.C., pp.373–411.

Stumm, W. and Baccini, P., 1978. Man-made chemical perturbation of lakes. In: A. Lerman (Editor), *Lakes – Chemistry, Geology, Physics.* Springer, New York, N.Y., pp.91–126.

Stumm, W. and Morgan, J.J., 1970. *Aquatic Chemistry.* Wiley, New York, N.Y.

Sugawara, K., Okabe, S. and Tanaka, M., 1961. Geochemistry of molybdenum in natural waters. *J. Earth Sci., Nagoya Univ.*, 9: 114–136.

Suess, E. and Erlenkeuser, H., 1975. History of metal pollution and carbon input in Baltic Sea sediments. *Meyniana,* 27: 63–75.

Sweeney, R.E. and Kaplan, I.R., 1973. Pyrite framboid formation. Laboratory synthesis and marine sediments. *Econ. Geol.*, 68: 618–634.

Syers, J.K., Iskandar, I.K. and Keeney, D.R., 1973. Distribution and background levels of mercury in sediment cores from selected Wisconsin lakes. *Water Air Soil Pollut.*, 2: 105–118.

Temple, K.L. and LeRoux, N.M., 1964. Syngenesis of sulfide ores: sulfate-reducing bacteria and copper toxicity. *Econ. Geol.*, 59: 271–278.

Theis, T.L. and Singer, P.C., 1974. Complexation of iron (III) by organic matter and its effect on iron (II) oxygenation. *Environ. Sci. Technol.*, 8: 569–572.

Theobold Jr., P.K., Lakin, H.W. and Hawkins, D.E., 1963. The precipitation of aluminium, iron and manganese at the junction of Deer Creek with the Snake River in Sumit County, Colo. *Geochim. Cosmochim. Acta,* 27: 121–132.

Thomas, R.L., 1972. The distribution of mercury in the sediment of Lake Ontario. *Can. J. Earth Sci.*, 9: 636–651.

Thomas, R.L., 1973. The distribution of mercury in the surficial sediments of Lake Huron. *Can. J. Earth Sci.*, 10: 194–204.

Thomas, R.L., 1974. The distribution and transport of mercury in the sediments of the Laurentian Great Lakes System. Proc. Int. Conf. *Persistent Chemicals in Aquatic Ecosystems,* I-1-16.

Thornton, I., Watling, H. and Darracott, A., 1975. Geochemical studies in several rivers and estuaries for oyster rearing. *Sci. Total Environ.*, 4: 325–245.

Tietze, K., 1978. *Geophysikalische Untersuchung des Kivu-Sees und seiner ungewöhnlichen Methangaslagerstätte – Schichtung, Dynamik und Gasgehalt des Seewassers.* Thesis, Kiel Univ., 148 pp.

Timperley, M.J. and Allan, R.J., 1974. The formation and detection of metal dispersion halos in organic lake sediments. *J. Geochem. Explor.*, 3: 167–190.

Tobschall, H.J., Göpel, C. and Rast, U., 1978. Geochemistry of organic gels and organic sediments from selected lakes of northern Norway. Abstr. *10th Int. Congress on Sedimentology,* Jerusalem, pp.682–683.

Trudinger, P.A., 1976. Microbiological processes in relation to ore genesis. In: K.H. Wolf (Editor), *Handbook of Strata-Bound and Stratiform Ore Deposits, 2.* Elsevier, Amsterdam, pp.135–190.

Turekian, K.K., 1977. The fate of metals in the oceans. *Geochim. Cosmochim. Acta,* 41: 1139–1144.

Turekian, K.K. and Wedepohl, K.H., 1961. Distribution of the elements in some major units of the earth's crust. *Bull. Geol. Soc. Am.*, 72: 175–192.

Vaughan, D.J., 1976. Sedimentary geochemistry and mineralogy of the sulfides of lead, zinc, copper and iron and their occurrence in sedimentary ore deposits. In: K.H. Wolf (Editor), *Handbook of Strata-Bound and Stratiform Ore Deposits, 2.* Elsevier, Amsterdam, pp.317–363.

Vernet, J.-P., 1976. Etude de la pollution des sediments du Léman et du bassin du Rhône. Rapport sur les études et recherches entreprises dans le bassin lémanique. *Comm. Int. pour la Protection des Eaux du Lac Léman contre la Pollution, campagne 1976, Geneva,* pp.247–321.

Vernet, J.-P., Rapin, F. and Scolari, G., 1977. Heavy metal content of lake and river sediments in Switzerland. In: H.L. Golterman (Editor), Proc. Int. Symp. *Interactions between Sediments and Fresh Water,* Amsterdam, Sept. 6–10, 1976. Junk, Den Haag, pp.390–397.

Volkov, I.I. and Fomina, L.S., 1974. Influence of organic material and processes of sulfide formation on distribution of some trace elements in deep-water sediments of the Black Sea. *Am. Assoc. Pet. Geol. Mem.,* 20: 456–476.

Vuorela, I., 1970. The indication of farming in pollen diagrams from southern Finland. *Acta Bot. Fenn.,* 87: 1–140.

Wagner, F., 1974. *Die holocänen Karbonatablagerungen im Plattensee (Balaton), Ungarn: Mineralogie, Geochemie, Sedimentologie.* Thesis, Univ. Heidelberg, 79 pp. (unpublished).

Wagner, G., 1971. FeS-Konkretionen im Bodensee. *Int. Rev. Ges. Hydrobiol.,* 56: 265–272.

Walters Jr., L.J., Herdendorf, C.E., Charlesworth, L.J., Anders, H.K., Jackson, W.B., Skoch, E.J., Webb, D.K., Kovacik, T.L. and Sikes, C.S., 1972. Mercury contamination and its relation to other physicochemical parameters in the western basin of Lake Erie. Proc. *15th Conf. Great Lakes Res.,* pp.306–316.

Walters Jr., L.J., Wolery, T.J. and Myser, R.D., 1974. Occurrence of As, Cd, Co, Cr, Cu, Fe, Hg, Ni, Sb and Zn in Lake Erie sediments. Proc. *17th Conf. Great Lakes Res.,* pp.219–234.

Wauschkuhn, A., 1973. Rezente Sulfidbildung in vulkanischen Seen auf Hokkaido. *Geol. Rundsch.,* 62: 774–785.

Wauschkuhn, A. and Gröpper, H., 1975. Rezente Sulfidbildung auf und bei Vulcano, Äolische Inseln, Italien. *Neues Jahrb. Mineral. Abh.,* 126: 87–111.

Wauschkuhn, A., Schwartz, W., Amstutz, G.C. and Yagi, K., 1977. Fumarolic hot lakes on Hokkaido: geochemical, mineralogical and biochemical investigations of their significance for the formation of massive sulfide-deposits. *Neues Jahrb. Mineral. Abh.,* 129: 171–200.

Weissberg, B.G. and Zobel, M.G., 1973. Geothermal mercury pollution in New Zealand. *Bull. Environ. Contam. Toxicol.,* 9: 148–155.

Whiteside, M.C., 1965. Paleoecological studies of Potato Lake and its environment. *Ecology,* 46: 807–815.

Williams, J.D.H., Syers, J.K., Shukla, S.S., Harris, R.F., 1971. Levels of inorganic and total phosphorous in lake sediments as related to other sediment parameters. *Environ. Sci. Technol.,* 5: 1113–1120.

Wittmann, G.T.W. and Förstner, U., 1975. Metal enrichment of sediments in inland waters – the Hartbeespoort Dam. *Water,* 1: 76–82.

Wittmann, G.T.W. and Förstner, U., 1976. Metal enrichment of sediments in inland waters – the Jukskei and Hennops River drainage systems. *Water,* 2: 67–72.

Wolf, K.H. and Chilingar, G.V., 1976. Diagenesis of sandstone and compaction. In: G.V. Chilingar and K.H. Wolf (Editors), *Compaction of Coarse-Grained Sediments, II.* Elsevier, Amsterdam, pp.69–444.

Yagi, K. and Hunahashi, M., 1970. *Volcanoes and mineral deposits of the neighbouring area of Sapporo, Hokkaido.* Guidebook, 7th General Meeting of IMA, 38 pp.

Züllig, H., 1956. Sedimente als Ausdruck des Zustandes eines Gewässers. *Schweiz. Z. Hydrol.,* 18: 7–143.

NOTE ADDED IN PROOF

The methods of phase partitioning of metals in recent sediments (pp. 237–254, Table XVI) are at present the object of intensive discussions. Despite their greater specificity for chemical phases as compared to many pedological procedures, it must be clearly pointed out that most of these extraction steps are *not selective* as sometimes stated. *Readsorption* of metals can occur, for example, after extractions with dilute acids and H_2O_2 (Rendell et al., 1980). Reactions are influenced by the *duration of treatment* and by the *ratio of solid matter to volume of extractant.*

There are major objections against the use of $BaCl_2$-triethanolamine (step 1 in Table XVI) or $MgCl_2$ for the determination of exchangeable trace metals, because of the possible chelating effects of the organic agent (added to raise the pH of $BaCl_2$-solutions) and the formation of dissolved metal-chloro-complexes. Basic metal oxides may be formed during initial high pH-conditions, e.g. for humic acid extraction, with sodium hydroxide (step 3 in Table XVI). Leaching by $1M$ sodium acetate solution adjusted to pH 5 with acetic acid has been used by Tessier et al. (1979) for carbonate extraction (step 2 in Table XVI).

As a result of these discussions the following sequence of leaching procedures for the speciation of particulate trace metals is proposed:

(1) *Exchangeable cations:* $1M$ Ammonium acetate, pH 7, for 2 hrs at 1 : 20 solids–solution ratio, separated by centrifugation.

(2) *Carbonate fraction:* Acidic cation exchanger (Deurer et al., 1978) or buffered $1M$ NaOAc-solution (Tessier et al., 1979).

(3) *Easily reducible phases* (Mn-oxides, amorphous Fe-oxyhydrates): $0.1M$ $NH_2OH \cdot HCl$ + $0.01M$ HNO_3, pH 2 (Chao, 1972).

(4) *Organic fraction:* 30% H_2O_2 (90°C) + $1M$ NH_4OAc, pH 2.5 (Engler et al., 1974).

(5) *Moderately reducible phases* (poorly crystallized Fe-oxyhydrates): $0.2M$ ammonium oxalate + $0.2M$ oxalic acid, pH 3 (Schwertmann, 1964).

(6) *Non-silicate iron phases:* Citrate-dithionite extraction (Holmgren, 1967).

(7) *Detrital silicates:* $HF/HClO_4$-digestion.

Prior to the determination of exchangeable cations (1) the *total water soluble* and *soluble complexed metals* can be extracted and analyzed (Hart and Davies, 1979; Khalid et al., 1979).

A simplified procedure using steps (3), (4) and (7) of the above-mentioned sequence can be applied for routine analysis of polluted sediments (W. Salomons, Inst. for Soil Fertility, Haren, The Netherlands): Step (3) would then include the extraction of the soluble, exchangeable and carbonate-associated metal fractions, which, in addition to the organic fraction (4), are expected to be predominantly accessible for organisms. The residual fraction (detrital silicates, moderately reducible and non-silicate iron phases) consists of metals, which are very strongly bound to the sediment and probably do not take part in short-term geochemical processes.

REFERENCES

Hart, B.T. and Davies, S.H.R., 1979. Speciation of Fe, Cd, Cu, Pb and Zn in the Yarra River and Estuary, Australia. Proc. Int. Conf. *Management and Control of Heavy Metals in the Environment,* London, Sept. 17–21, 1979, pp.466–471.

Khalid, R.A., Gambrell, R.P. and Patrick, W.H. jr., 1979. Chemical mobilization of cadmium in an estuarine sediment as affected by pH and redox potential. *Ibid.,* pp.320–324.

Rendell, P.S., Batley, G.E. and Cameron, A.J., 1980. Adsorption as a control of metal concentrations in sediment extracts. Manuscript submitted to *Environmental Science and Technology* (Aug. 1979).

Tessier, A., Campbell, P.G.C. and Bisson, M., 1979. Sequential extraction procedure for the speciation of particulate trace metals. *Anal. Chem.,* 51: 844–851.

Chapter 5

TRACE METALS IN FRESH WATERS (WITH PARTICULAR REFERENCE TO MINE EFFLUENTS)

ULRICH FÖRSTNER

INTRODUCTION

Analyses of trace metals (concentrations less than 1000 μg/l; Wilson, 1976) were first performed in marine waters and on surface and subsurface waters in traditional mining areas. From the beginning, these data have been related to questions of biological productivity and possible adverse effects on aquatic biota, as well as to methodological aspects.

Stock and Cucuel (1934) determined the Hg-content in samples from the proximity of the island of Helgoland (North Sea) using a procedure of electrochemical deposition of Hg on a copper wire. The results obtained by this method (0.05 μg Hg/l) are in accordance with values of Gardner (1975), who found average Hg-concentrations ranging from 0.0112 μg/l in the Southern Hemisphere to 0.0335 μg/l in the Northern (the increase of the latter concentrations probably results from the atmospheric Hg-discharges from the large industrial complexes of the U.S.A., Europe, and Japan). Noddack and Noddack (1939) performed trace-element analyses on waters from the Skagerrak; Wattenberg (1943) measured the contents of Cu, Zn, and other trace metals in waters from the Atlantic. In the early 50's, seawater from various areas was analyzed, e.g., by Chow and Thompson (1951; Cu – U.S. West Coast), Black and Mitchell (1952; Cu, Pb – Scottish coast), Griel and Robinson (1952; Ti), Atkins (1953; Cu – English Channel), Lewis and Goldberg (1954; Fe), Emery et al. (1955; P), Morita (1955; Cu, Zn – coasts of Japan), Mullin and Riley (1956; Cd), Rona et al. (1956; U), and Sugawara et al. (1956; V).

Trace-metal analyses on continental waters were commenced in the U.S.A., in central Europe, the U.S.S.R. and Japan, mainly in areas where local anomalies from ore mineralizations had been expected. Huff (1948) studied the concentrations of Cu, Pb, and Zn in five samples from the Colorado River; Turekian and Kleinkopf (1956) analyzed the abundance of Cu, Mn, Pb, Ti, Ni, and Cr in surface waters of Maine. Investigations and literature compilations have been performed during the 60's and early 70's by Merrill et al. (1960; Be – Delaware and Hudson rivers), Durum and Haffty (1961), Livingstone (1963; surface waters), White et al. (1963; subsurface waters), Silvey (1967; California), Turekian et al. (1967; Neuse River, North Carolina), Bradford et al. (1968; Sierra Nevada), Voegell and King (1968; Mo – Colorado), Angino et al. (1969; lower Kansas River basin), Linstedt and Kruger (1969; V – Colorado), Weiler and Chawla (1969; Great Lakes), Durum et al. (1971; surface waters), Bradford (1971; California), Mills and Oglesby (1971; Cayuga Basin), Robbins et al. (1972; tributaries to Lake Michigan), Mathis and Cummings (1973; Illinois River), and Proctor et al. (1973; Missouri River). In central Europe early investigations have been performed by Heide and Singer (1954) and Heide et al. (1957) on the concentrations of Cu, Zn, and Pb in the Saale River (German Democratic Republic), by Heyl (1954) in the Siegerland mining area (Federal Republic of Germany), and by Fricke and Werner (1957) in mineral waters of Nordrhein-Westfalen (Federal Republic of Germany). Both the European and Asian surface waters of the U.S.S.R. have been studied intensively; compilations and original data are given by Konovalov (1959), Udodov and Parilov (1961), Kontorovich et al. (1963), Konovalov et al. (1967a, b), Konovalov and Ivanova (1972), and Konovalov and Nazarova (1975). An early review of data on acidic mine drainage – which is the central aspect of the present compilation – was given by Smirnow (1954). Of the early Japanese investigations on trace metals in natural waters one should particularly mention the studies performed at Nagoya University (Morita, 1955; Sugawara et al., 1956; Sugawara and Okabe, 1960; Kanamori and Sugawara, 1965).

During the last few years the number of trace-metal analyses has greatly increased in both geoexploration and environmental management. This progress has been partly due to improvement in analytical techniques which have significantly lowered the cost and increased the effectiveness of these methods. However, a critical review of many investigations clearly shows that the difficulties in the analysis of trace elements have not been fully overcome and that great care is still necessary to obtain valid data. Significant problems are still being encountered in respect to adequate sampling and storage procedures. Especially when taking water samples, contamination is often considerable due to unprotected sampling devices made of metal. It is assumed, therefore, that many of the earlier data on dissolved heavy-metal contents are too high. This is particularly true for the analysis of seawater, where – due to decisive breakthroughs in analytical and sampling techniques – metal values of one to two orders of magnitude lower than the data from the 60's and early 70's have recently been determined for Cd (Boyle et al., 1976; Martin et al., 1976; Bender and Gagner, 1976), for Cr (Cranston and Murray, 1978), for Cu (Boyle and Edmond, 1975; Moore and Burton, 1976; Boyle et al., 1977), Pb (Schaule and Patterson, 1978), Ni (Sclater et al., 1976), Se (Measures and Burton, 1978), and Zn (Bruland et al., 1978). Similar developments, even if not so spectacular, have been found or may be expected for freshwater analysis, particularly in less polluted waters, e.g., for parts of the Amazon River system (Boyle, 1978).

The present review does not claim to have discussed these questions in full detail. Its main purpose is to give an overview on the major sources of metals in surface waters leading to significant local or regional anomalies and particular problems arising from acidic effluents. Major aspects of actual interest include furthermore: transport phases of metal, dependencies on water discharge and biological productivity, background data, anthropogenic effects and the changes of trace metal concentrations at the river/sea interface.

SOURCES AND EFFECTS OF METALS IN NATURAL WATERS

In general, there are five sources of heavy metals in inland waters:

(1) Geological weathering. This is the source of "background levels". It is to be expected that in areas characterized by metal-bearing formations, these metals will also occur in high levels in the water of the area.

(2) Industrial processing of ores and metals. During the processing of ores, metal-bearing dust particles are formed, which may be only partially filtered out by air-purification systems. Appreciable quantities of metals go to waste during chemical metal-refinement processes (e.g., galvanizing and pickling) by way of heavy-metal solutions, which are often discharged without any reclamation measures.

(3) The use of metals and metal compounds. Examples are the use of Cr-salts in tanneries, Cu-compounds as plant protection agents, and tetramethyl-Pb as an anti-knock agent. Examples of metal pollutants in industrial waste streams (sources "2" and "3") are listed in Table I (from Barnhart, 1978).

(4) Heavy metals in animal and human excretions. Heavy metals are present in human and animal food – Zn in particular at relatively high concentrations. These metals are concentrated in excretions and mainly find their way to the water environment. The adult human excretes between 7 and 20 mg of Zn per day.

(5) Leaching of metals from garbage and solid-waste dumps. The contribution of

TABLE I

Some hazardous materials in industrial waste effluents (After Barnhart, 1978; with permission of the American Chemical Society)

Industry	As	Cd	Cr	Cu	Pb	Hg	Se	Zn
Mining and metallurgy	x	x	x	x	x	x	x	x
Paints and dyes		x	x	x	x	x	x	
Pesticides	x				x	x		x
Electrical and electronic				x	x	x	x	
Cleaning and duplicating	x		x	x	x		x	
Chemical manufacturing			x	x		x		
Explosives	x			x	x	x		
Rubber and plastics						x		x
Batteries		x			x	x		x
Pharmaceuticals	x					x		
Textiles			x	x				
Petroleum	x				x			
Pulp and paper						x		
Leather			x					

this source to the heavy-metal pollution of inland waters merits close attention. Mine dumps especially can be a serious source of pollution in connection with acidic solutions (see pp. 286–295). Acidity of surface waters imposes problems in all aspects of metal enrichment, ranging from the toxification of drinking water to problems concerning the growth and reproduction of aquatic organisms (Beamish and Harvey, 1972), the increased leaching of nutrients from the soil and the ensuing reduction in soil fertility (Whitby et al., 1976; Tamm, 1976), the increased availability and toxicity of metals with regard to essential plants (Lucas and Davis, 1961; Linnman et al., 1973) and finally to the undesirable acceleration of Hg-methylation in sediments (Fagerström and Jernelöv, 1972).

Viewed from the standpoint of *environmental hygiene*, metals may be classified according to three criteria: (a) non-critical, (b) toxic but very insoluble or very rare, and (c) very toxic and relatively accessible. Such a classification has been offered by Wood (1974) as given in Table II. Here, special importance must be attached to the non-metals As and Se aside from the heavy metals Hg, Cd, and Pb.

Water-quality criteria

Although it has been well-established that many inorganic constituents enter inland waters from natural or man-made sources, their significance with regard to surface water quality depends on many interdependent factors. Not only is the abundance and widespread occurrence of a particular constituent of importance, but also its availability in the form of solubilized species. For example, the toxicity of Cu is strongly reduced with the formation of Cu-organic complexes (Davey et al., 1974). Organic ligands, such as fulvic acids, NTA and EDTA, can inhibit the uptake of metals and thus may raise the toxic threshold (Sibley and Morgan, 1975; Andrew, 1976; Zitko, 1976; V.M. Brown, 1976). Experiments of G.A. Jackson and Morgan (1978) have shown that free-ion activity is a good indicator of Cu-toxicity to phytoplankton [1].

When setting permissible limits or ultimate goals for drinking-water standards, cognizance must be taken of the bioaccumulation via the food chain. Moreover, it is imperative not only to impose limits which protect man's health on the basis of trace-metal quantities in surface water from which potable water is extracted, but also to consider the environmental impact of these waters discharged to the environment. Such considerations involve the ecosystem as a whole, self-purification of river systems, biological-treatment plants, the effects of trace-metal enrichment on biological-purification treatment, the effects on crustaceans, fish, and ultimately on man. With regard to the different types of drinking-water contaminants, the trace metals have received considerable attention in terms of their toxic effects during the past few years. Unfortunately, it has to be admitted that many basic questions regarding this group of elements still remain unan-

---

[1] These aspects are discussed, for example, in several papers of the symposium volume "Toxicity to Biota of Metal Forms in Natural Water", edited by R.W. Andrew, P.V. Hodson, and D.E. Konasewich, Int. Joint Commission Great Lakes Research, Windsor, Ontario (1976).

TABLE II

Classification of metals according to toxicity and availability (After Wood, 1974)

Non-critical			Toxic but very insoluble or very rare		Very toxic and relatively accessible		
Na	C	F	Ti	Ga	Be	As	Au
K	P	Li	Hf	La	Co	Se	Hg
Mg	Fe	Rb	Zr	Os	Ni	Te	Tl
Ca	S	Sr	W	Rh	Cu	Pd	Pb
H	Cl	Al	Nb	Ir	Zn	Ag	Sb
O	Br	Si	Ta	Ru	Sn	Cd	Bi
N			Re	Ba		Pt	

swered. As pointed out before, the question of chemical speciation poses one of the most difficult problems to be resolved by the chemist, pharmacologist, and toxicologist, especially regarding synergistic effects encountered in natural waters. Drinking-water standards have been proposed by various governmental bodies in accordance with toxicity data obtained from clinical investigations and various other studies, such as animal experiments. A brief summary is given in Table III compiled by Hattingh (1977), with additional data from the Federal Republic of Germany (Schöttler, 1977).

TRANSPORT PHASES OF HEAVY METALS IN RIVER WATER

In freshwater systems, the major difficulties in obtaining natural background data for trace metals arise from the great variability of rock formations, from the fluctuations in the water transport system (especially in rivers), and from the differentiation of particulate and dissolved metal species.

In particular two analytical techniques have been applied in *metal-ion speciation*: anodic stripping voltammetry (ASV) and ultrafiltration. The first procedure divides the metal species into two categories – electroactive (aqueous ions and "labile" complexes) and electroinactive (organic complexes and colloidal species). Ultrafiltration and dialysis is used to divide the metal species into different size fractions. The species that pass through the smallest pore size are generally taken to be free metal ions or small complexes.

A general schema of metal speciation – mainly based on the particle size fractions – has been given by Stumm and Bilinski (1972), and is reproduced in Table IV. In practice, the first step applied in separation of particulate from soluble metals involves filtration through a 0.45-μm pore-size membrane filter. The group of filtrable metal species can be subdivided into the following categories (Wilson, 1976):

(1) undissolved forms including colloidal and very finely divided materials;

(2) dissolved inorganic species, which, however, may exist in different oxidation

TABLE III

Drinking-water quality criteria (After Hattingh, 1977; Schöttler, 1977)

Parameter	USPHS (1962)	Japan (1968)	USSR (1970)	WHO European (1970)	WHO (1971)	SABS (1971)	NAS (1972)	Australia (1973)	EPA (1975)	FRG (1975)
As	10	50	50	50	50	50	100	50	50	40
Ba	1000	–	4000	1000	–	–	1000	1000	1000	–
Cd	10	–	10	10	10	50	10	10	10	6
Cr	50	50	100	50	–	50	50	50	50	50
Cu	1000	10 000	100	50	50	1000	1000	10 000	–	–
Pb	50	100	100	100	100	50	50	50	50	40
Hg	–	1	5	–	1	–	2	–	2	4
Se	10	–	1	10	10	–	10	10	10	8
Ag	50	–	–	–	–	–	–	50	50	
Zn	5000	100	1000	5000	5000	5000	5000	5000	–	2000

All values are in μg/l

TABLE IV

Metal species in aquatic systems (After Stumm and Bilinski, 1972; with permission of Pergamon Press)

Metal species	Range of diameters (μm)	Examples
Free aquated ions		$Fe(H_2O)_6^{3+}$; $Cu(H_2O)_6^{2+}$
Complex ionic entities		AsO_4^{3-}, UO_2^{2+}, VO_3^-
Inorganic ion-pairs and complexes		$CuOH^+$, $CuCO_3^0$, $Pb(CO_3)_2^{2-}$ $AgSH^0$, $CdCl^+$, $Zn(HO)^-$
Organic complexes, chelates and compounds	0.001	$Me - OOCR^{n+}$, HgR_2 $CH_2 - C = O$ O H_2N Cu O NH_2 $O = C - CH_2$ 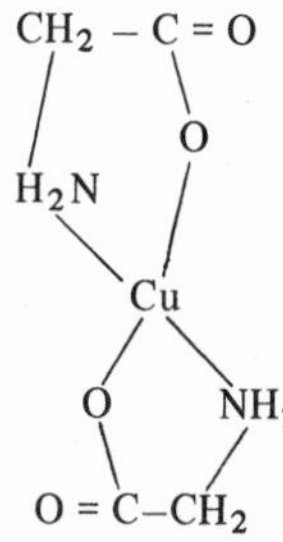
Metals bound to high-molecular-weight organic materials	0.01	Me-humic/fulvic acid Me-polymers
Highly dispersed colloids		FeOOH, Mn(IV) hydrous oxides
Metals sorbed on colloids	0.1	$Me \cdot aq^{n+}$, $Me_n(OH)_y$, $MeCO_3$, etc. on clays, FeOOH, organic
Precipitates, mineral particles, organic particles		$ZnSiO_3$, $CuCO_3$, CdS in FeS, PbS
Metals present in live biota	(Me = metal; R = alkyl)	Metals in algae

IN TRUE SOLUTION — DIALYSABLE — MEMBRANE FILTRABLE — FILTRABLE

(3) dissolved metal-organic species, which can be subdivided into (3a) truly dissolved metal complexes and chelates, e.g. those with compounds such as amino acids, EDTA, and (3b) colloidally dispersed metal-organic associations.

As a practical example the distribution of Mn in the grain size spectrum in a sample from the Amazon River is reproduced in Fig. 1 (from Gibbs, 1977). The size separation for the various fractions to be analyzed was accomplished by centrifugation of the fractions >0.1 μm. For the finer solids and dissolved materials, a series of membranes of 300 Å, 100 Å, 30 Å, 10 Å, and 5 Å were used. An additional phase differentiation of the

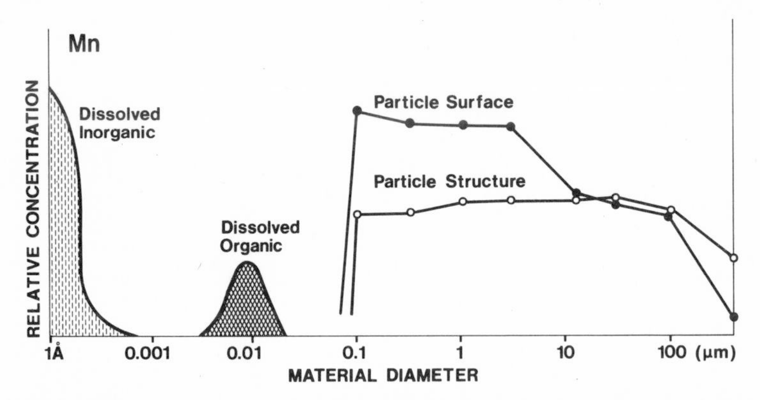

Fig. 1. Mn phases in the grain size spectrum of suspension samples from the Amazon River. (After Gibbs, 1977.)

solid particles – the procedures used are described below – indicates the decrease of the surface-bound metal content with increasing grain sizes, while the structurally bonded (more or less inertly incorporated) metal concentrations are fairly constant over the grain-size spectrum from 0.1 μm to 100 μm. The concentrations of both solid phases rapidly decrease to zero at approximately 0.1-μm particle diameter. The left-hand modes of dissolved organics (0.003–0.02 μm) and dissolved inorganic (<0.0008 μm) Mn-constituents are sufficiently small in particle size to assure physical transport of the material with the water mass. Conversely, the majority of right-hand Mn-modes (Fig. 1) are of sufficiently large particle size to be separated from the water mass.

The *source* of trace metals in aquatic systems significantly determines their distribution *ratio between the aqueous and solid phases.* For example, the bulk of the detrital trace-element particulates never leaves the solid phase from initial weathering to ultimate deposition. Similarly, metal dust particles (e.g., from smelters) and effluents containing heavy metals associated with inorganic and organic matter, undergo little or no change after being discharged into a river. This is attributable to the average residence times (in the order of days or weeks), which are too short for the establishment of stable, dynamic equilibrium between water and suspended material (Bowen, 1975).

Estimates on the world river transport of transition metals to oceans suggest that less than 3% is associated with dissolved species (Gibbs, 1977): Fe = 0.13%, Co = 0.3%, Ni = 0.5%, Cu = 1%, Cr = 2.5%, and Mn = 3%. It should be noted, however, that there are very large fluctuations even in the less-polluted systems. Gibbs (1977) gives the example of the Amazon River system, where the contributary Rio Negro carries very little sediment, so that almost the entire load of transition metals is transported in the dissolved-complexed phase. When the Rio Negro reaches the main channel of the Amazon River, the sediment-related metal transport surpasses the minor load carried in solution. Characteristic heavy-metal transport in less-polluted systems in a moderate climate (lower Mississippi River) is displayed in Table V by Trefry and Presley (1976). Only 11.1% (Cd)

TABLE V

Percentage particulate-associated metals of total metal discharge (solid and aqueous)

	Mississippi River [1]	Polluted U.S. rivers [2]	Polluted F.R.G. rivers [3]	Rhine River Netherlands [4]
Na	–	–	0.5	–
Ca	–	–	2.5	–
Sr	–	21	–	–
B	–	30	–	–
Cd	88.9	–	30	45
Zn	90.1	40	45	37
Cu	91.6	63	55	64
Hg	–	–	59	56
Cr	98.5	76	72	70
Pb	99.2	84	79	73
Al	–	98	98	–
Fe	99.9	98	98	–

[1] Trefry and Presley (1976); [2] Kopp and Kroner (1968); [3] Heinrichs (1975); [4] DeGroot et al. (1973).

to less than 0.02% (Fe) of the heavy-metal input to the Gulf of Mexico is transported in a dissolved-complexed state.

A similar sequence of the ratios between particulate and dissolved heavy-metal phases has also been found for polluted systems; typically, however, the dissolved discharges in polluted waters are significantly higher than in the less-polluted systems, particularly for metals such as Cd, Zn, and Cu.

In Table V, metal fractions in particulate form are indicated as percentages of the total metal discharges from U.S. rivers (Kopp and Kroner, 1968), rivers in West-Germany (Heinrichs, 1975) and from the Rhine River in the Netherlands (De Groot et al., 1973). The order of sequence of the mobility for a few selected metals is as follows: alkali and alkali-earth metals are predominantly present in a dissolved form; for trace metals such as B, Zn, and Cd the ratio of dissolved species to particulate species is between 2 : 1 and 1 : 1; Cu, Hg, Cr, and Pb exhibit ratios of the solid phases to the aqueous phases of between 2 : 1 and 4 : 1, whereas Fe and Al (and Mn under normal Eh conditions) are almost totally transported as solid particles.

DEPENDENCIES OF METAL TRANSPORT FROM WATER DISCHARGE AND ANNUAL CYCLES

The water discharge from river systems is one of the characteristic factors which can influence the ratio of metal concentrations in dissolved (filtrable) and solid (non-filtrable) species. Fig. 2 reproduces data given by Wilson (1976), which follows a graph by Hellmann (1970), for the development tendency of the non-filtrable and filtrable frac-

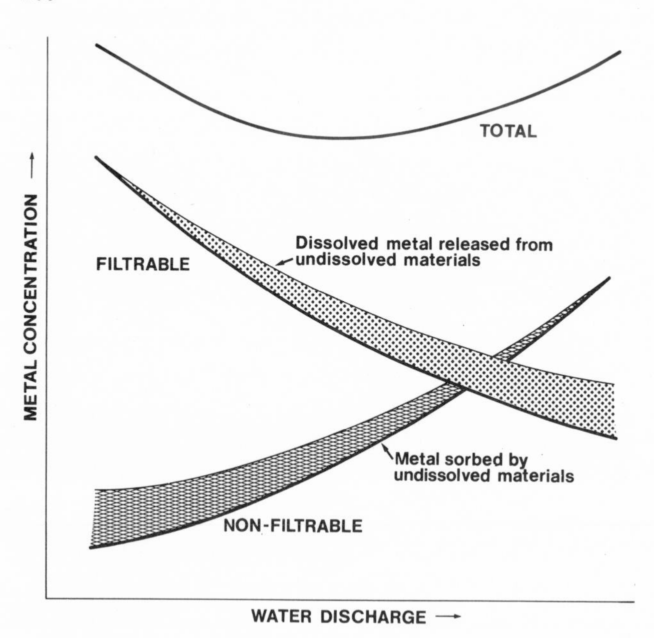

Fig. 2. Schematic presentation of the transport modes of trace metals in rivers. (After Hellmann, 1970, modified by Wilson, 1976.)

tions of the metal load with increasing water discharge. A decrease in the filtrable fraction is the result of dilution, whereas an increase of the non-filtrable fraction is mainly due to the resuspension ("flushing effect", see below) of particles from the river bed and its banks. The shaded areas of the upper section for both curves indicate minor effects of sorption, as in the case of the solid fraction, and remobilization from the particulates which increases the dissolved metal load to some extent. The decrease in the amount of sorbed cations with increasing discharge is due to (1) the higher percentage of relative coarse-grained material, which usually exhibits lower exchange capacities (see Chapter 4, pp.194–195); (2) a smaller amount of dissolved cations due to dilution; and (3) the shorter residence times of both solids and dissolved ions in the river channel, which in turn influence the attainment of equilibrium between both phases. However, metal cations are increasingly released from solid substances into the aqueous phase at higher water discharge rates owing to desorption and dissolution processes. There does not appear to be a significant variation in the total metal load with changing water discharges; the decrease of this curve for intermediate water flows in Fig. 2 should therefore be considered as being rather hypothetical.

Investigations into the dependency of trace-metal contents from water discharges was carried out, for example, on the Rhine River by Schleichert (1975). At the Koblenz sampling site on the middle section of the river, water samples were taken every working day between March 1973 and March 1974. In spite of considerable fluctuations, the following conclusions could be drawn concerning the metal contents of particulate matter of the highly polluted Rhine River system: (1) each discharge maximum can be ascribed to a concentration minimum, (2) the concentration changes of different trace elements occur more or less in the same form, and (3) extreme concentration peaks are rare, independent of discharge (these effects can be caused by short-term, local waste-water inputs as well as by remobilization of metals from deposits). Of the dissolved metal concentrations, only Cr showed this dependency, i.e., a decrease of metal content with increasing water discharge. Unfortunately, there is still no information regarding these phenomena from anthropogenically less-influenced systems. In this context, the findings of Aston and Thornton (1977) in their study of Cornish catchments are of interest, as they found a significantly smaller variation of heavy metals in both sediments and water of unmineralized areas than in the tributaries of the Carnon, Red and Gannel Rivers which are influenced by past and present mining and smelting industries.

Heavy-metal discharges in rivers may undergo characteristic developments in their *annual cycle*. From investigations on the temporal variability of metal transport by the Susquehanna River to the Chesapeake Bay, J.H. Carpenter et al. (1975) and Troup and Bricker (1975) found that the trace-metal concentrations correlate well with the amounts of solids discharged – the concentrations seemed to be highest in the spring and lowest in the summer and fall. This concurs with the findings of a major transport of heavy metals in associations with particulate matter mentioned above. Upon closer inspection, however, Mn, Ni, Zn, and Co exhibit large concentrations in January, and Cu, Cr, and Mn have concentration peaks in the late spring and early summer. When data are calculated for weight concentrations of metals in the solid fraction, it is found that all metals generally peak during December and January and secondary peaks occur for Co, Cr, Ni, Cu, and Mn in July. Since decaying organic matter is abundant in the Susquehanna River during these two periods, the high concentrations may be the result of metal bonding to such phases. Studies performed by Grimshaw et al. (1976) on the River Ystwyth in mid-Wales, where strong metal pollution from past mine operations is still obvious, indicate that metal concentrations in solution are highest during low flow periods, suggesting a dilution effect (which has been found by many other investigators, in particular from less polluted rivers, but also from polluted ones). For brief periods during the initial stages of storm runoff, there is a very significant increase of the metal concentrations in solution, apparently due to a flushing effect.

An important factor in controlling the trace metal content in natural waters is the ability of *planktonic material* to absorb some metals from solution. These effects have been thoroughly studied by Abdullah and Royle (1974a, b) in two surveys carried out on the Bristol Channel – where metal concentrations originate in runoff from mineralization zones and waste disposal – during April and June, 1971. The plot of the amount of acid extractable metal present in the suspended matter against the weight of solid in suspension (Fig. 3) shows a first-order relationship for Zn, Cd, and – approximately – for Cu during *April*. This uniformity suggests that the distribution of the particulates is con-

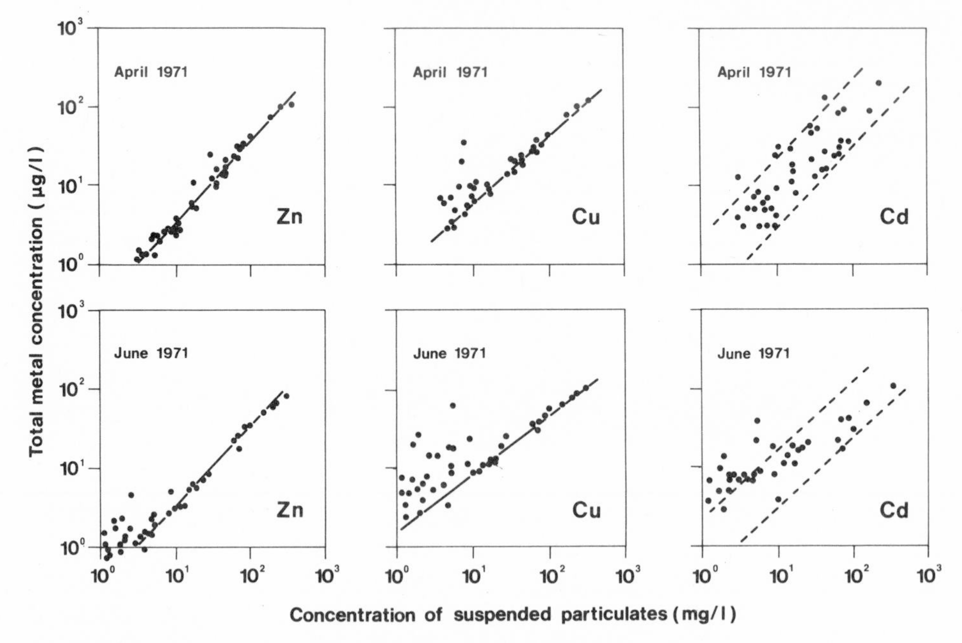

Fig. 3. Extractable metals in suspended matter from the Bristol Channel during April and June 1971. (After Abdullah and Royle, 1974a; with permission of Pergamon Press.)

trolled by mixing and turbulence and that little or no fractionation by settling takes place in the area studied. The positive anomalies found in the copper plot may be due to the contribution of particulate copper from north Devon and Cornwall runoff. For the *summer data* (June, 1971, below) Zn, Cu, and Cd plots show positive anomalies at stations situated in the outer part of the Channel, indicating that the additional metal is due to agencies other than runoff. The likelihood that plankton concentration of metal occurs, and not physical or chemical processes, may be deduced from the fact that the plot of the April data exhibits a linear relationship throughout the Channel (Abdullah and Royle, 1974a). Similar seasonal effects have been found by Knauer and Martin (1973) from studies on the uptake of Cu, Zn, Mn, Cd, and Pb by plankton in Monterey Bay, California.

Temporal variations in the metal distribution in three phases – biomass, allochthonous particulates and dissolved phase – were studied by Baccini (1976) from the Lake of the Four Cantons (Vierwaldstättersee). In the epilimnion the biogenic portion of the total particulate phase for Cu is approximately the same as the portion of the allochthonous phase and can even be higher during periods of high organic production. The distribution of Cu in the hypolimnion, however, indicates that the allochthonous portion dominates during periods of greatest sedimentation (May to September). These data show that Cu (and Zn) introduced in soluble form is transported into the particulate matter by plankton and that the Cu from decomposed sedimentary plankton is partially returned to solution, whereas the allochthonous particles are deposited on the lake bottom relatively unchanged.

DISSOLVED METAL CONCENTRATIONS IN LARGE FRESHWATER SYSTEMS

In order to determine the influence of both mineralization zones and civilizational effects, background values of metal concentrations are most desirable, which, although not "absolute", serve as *guidelines* in respect to major changes of the trace metal chemistry in inland waters (first column in Table VI). Most of these values stem from large inland waters, e.g., from the Amazon, Yukon, and Mississippi Rivers (2nd to 4th column). In these water bodies, civilizational effects are either still low or the inputs of contaminants are diluted by the large water mass.

The example of the *Amazon River* has shown that the various subsystems of large river systems may differ widely with respect to the concentrations of trace metals in both water and suspended matter. Irion and Förstner (1975) found in studies of Amazon "lake" sediments that deposits contributed by rivers from regions with heavily leached soils ("blackwater rivers", e.g., Rio Negro) are significantly depleted of Fe, Mn, Zn, Cu, and Co, when compared to the "whitewater" river deposits, e.g., from Rio Solimoes. Recent analyses of copper in the Amazon River system by Boyle (1978) indicate a similar distribution for dissolved metal species; the values of the black- or clear-water samples center around 0.5 μg/l, whereas the Cu-concentrations of the second group of rivers originating in the Andes, pre-Andes and southwestern Amazon lowlands, range between 1.5 and 2.0 μg/l.

For Gibbs (1977) it seems that in the Yukon River the content of metals such as Co, Cr, Cu, and Ni is slightly enriched in the dissolved phases, compared to the Amazon. This is probably due to lithogenic influences rather than civilization effects. The investigations on large rivers flowing through relatively unpolluted areas (Mississippi River) have shown that with improved sample extraction and storage and handling procedures, most of the data are significantly below those values which had been established as background data ten years ago (e.g., Turekian, 1969). However, there is still a good deal of controversy as to the origin of the sometimes strong divergences in the metal contents of inland waters, which are even greater than those for sea water. It seems that the amount of suspended matter, which can, on the one hand, partially pass through the filter (usually 0.45 μm pore size) and, on the other, can adsorb metals from solution, plays a considerably important role for the metal concentrations in water analyses. The Mississippi River, however, showed no significant seasonal changes in dissolved trace metals, although the suspended matter concentrations decrease from over 300 mg/l during periods of normal flow to 10 mg/l or less during the three month low flow period (Trefry and Presley, 1976).

The size of the water body also seems to be important in areas more strongly pol-

TABLE VI

Trace metals in freshwater systems (examples)
Values in $\mu g/l$

Metal	Background [1]	Amazon R. [2]	Yukon R. [2]	Mississippi [5]	L. Michigan [7]	Danube [9]	L. Constance [10]	Rhine [11]
As	1 [7]	–	–	3 [6]	1	1.5	3.7	13
Cd	0.07 [3]	0.07 [3]	–	0.1	0.3	<1	–	5.5
Co	0.05 [2]	0.06 [2]	0.1	–	0.2	<0.75	–	10
Cr	0.5 [5]	2 [2]	2.3	0.5	1.7	0.6	5.7	33
Cu	1 [4]	1 [4]	2	2	5	5	8.2	34
Hg	0.01	–	–	<0.1 [6]	0.03	0.5	0.18	0.65
Ni	0.3 [2]	0.27 [2]	0.43	1.5	3	3	5	20
Pb	0.2 [5]	–	–	0.2	1.5 [8]	3	–	57
Se	0.1 [7]	–	–	–	0.08	2.2	1.3	6.5
Zn	10	–	–	10 [6]	16	20	37	330

[1] Förstner and Wittmann (1979) from refs. 2–7 and other studies; [2] Gibbs (1977); [3] Boyle et al. (1976); [4] Boyle (1978); [5] Trefry and Presly (1976); [6] U.S. Geol. Survey, Baton Rouge (1972/76); [7] Copeland and Ayers (1972); [8] Edgington and Robbins (1975); [9] Schroll et al. (1975), low water at six stations of the Austrian section; [10] Quentin and Winkler (1974): mean values 1971–1973; [11] Inst. f. Wasser-, Boden- und Lufthygiene. Bundesgesundheitsamt Berlin F.R.G. (unpubl. data, Working Group "Metals", German Research Society 1971/73).

luted, such as in the Lake Michigan area. An increase for most of the metals by a factor of approximately three, compared to the background values, is determined to be mainly the result of the influence of atmospheric contaminants (Klein, 1975). An indication of this is the characteristic increase in dissolved Pb-values in the water samples from Lake Michigan. Although the Danube River dewaters an area which is both densely populated and industrialized, very few of the trace elements are found to exceed the geochemical averages (Schroll et al., 1975); this seems also partly the consequence of the relatively high water discharge during most of the year. Similar metal concentrations have been observed in water from Lake Constance, which, being the largest drinking water reservoir in Europe, provides water for large regions of southern Germany. However, when compared to the background values of Table VI, first column, a four- (As, Zn, and Cu) to more than ten-fold (Hg, Ni, and Se) increase of the concentrations of trace elements can be determined.

The last column in Table VI refers to the lower Rhine River, which can be considered as one of the most heavily polluted large river systems on earth. Metals such as Pb, Fe, and Mn commonly exceed the acceptable maximum values for use as drinking water. In some cases Cd, Cr, Hg, and Se are present in critical concentrations. Between Lake Constance and the Dutch/German border, the concentrations of Zn and Cd increase by factors of 45 and 35, respectively (Heinrichs, 1975).

River–sea interface

A characteristic decrease of the dissolved-metal concentrations has been observed in the estuaries and has been partly ascribed to the "non-conservative" behavior of some of these elements (see Burton and Liss, 1976). This is due to the effects of coprecipitation with hydrous Fe-oxides and flocculation of organic substances (Sholkovitz et al., 1978). Such an effect is particularly pronounced in areas where rivers of low pH reach the sea. Studies performed by Foster et al. (1978) in the estuary of Afon Goch and Dulas Bay, North Wales, which drain an area mined for copper until the end of the last century, exhibit significant reduction of concentrations of dissolved transition elements at higher salinities. Examples of non-conservative removal of iron have been given by Coonley et al. (1971) from the Mullica River in New Jersey, estuaries in British Columbia (Williams and Chan, 1966), in three estuaries in southeastern U.S.A. (Windom et al., 1971), in the Gulf of St. Lawrence (Bewers et al., 1974), and in the Merrimack Estuary (Boyle et al., 1974). Fig. 4 gives examples from investigations of Duinker and Nolting (1977) showing the concentrations of dissolved species of Cu, Zn, and Cd at salinity values with the range of 0.4–35‰ in the estuarine mixing zone of the Rhine River. The decrease of the concentrations is explained by the removal from the dissolved state – relative to conservative mixing of freshwater and seawater – in the early stages of mixing: roughly 40% for Cd, 50% for Cu and 30% for Zn. No evidence of non-conservative behavior has been found for lead in the Rhine Estuary. However, the influence of the relatively clean seawater on polluted river waters during estuarine mixing should not be neglected. Müller and Förstner

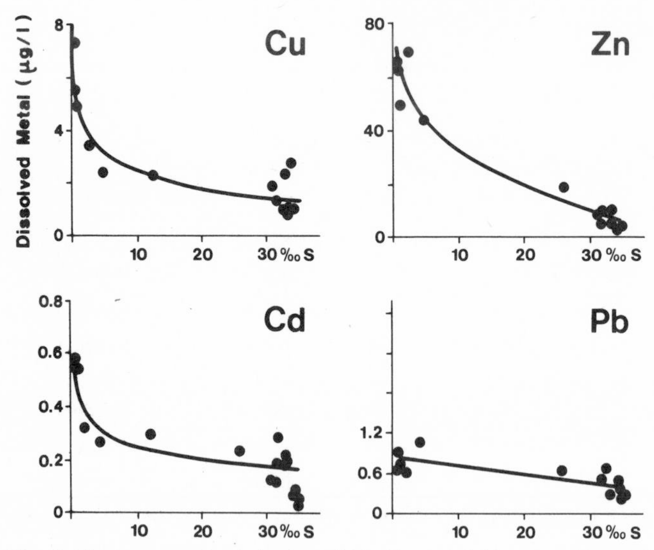

Fig. 4. Non-conservative behavior of Cu, Zn and Cd in estuarine mixing zones of the Rhine River in the Netherlands. (After Duinker and Nolting, 1977; with permission of Pergamon Press.)

(1975) found characteristic decreases of trace metal concentrations in the Elbe River approaching the North Sea (German Bight). Concentrations of Cd fell from 1.4 µg/l to 0.2 µg/l, Pb from 11.4 to 1.5 µg/l, Cu from 18 to 3.5 µg/l, Ni from 18.6 to 3.8 µg/l, Cr from 15.5 to 2.5 µg/l, and Zn from 194 to less than 10 µg/l. It seems that the decrease of dissolved (and solid) metal concentrations during estuarine mixing is strongest for those metals particularly enriched by civilizational effects, e.g., from domestic and industrial effluents.

METAL ENRICHMENT IN MINE EFFLUENTS

Mine effluents and tailings from mining and processing of ores represent a major source of heavy metals released to the environment and may have serious effects on the water quality of rivers and lakes, as well as on any biotopes, particularly on the fish populations (for further information on this subject, see Aplin and Argall, 1973 and Down and Stocks, 1977). One of the first descriptions of the problems arising from the strong enrichment of toxic trace elements both in dissolved and solid phases, is found in the Report of the 1868 River Pollution Commission in Britain, which studied the dispersal of metals from lead, zinc, and arsenic mines in mid-Wales (cit. Lewin et al., 1977):

> "All these streams are turbid, whitened by the waste of the lead mines in their course; and flood waters in the case of all of them bring down poisonous slime which, spreading over the adjoining flats, either befoul or destroy grass, and thus injure cattle and horses grazing on the dirtied herbage, or, by killing the plants whose roots have held the land together, render the shores more liable to abrasion and destruction on the next occasion of high water."

There are many examples of deleterious effects of mine effluents on freshwater ecosystems. In some streams of *Wales* enrichments of Pb, Cu, and Zn leached from the outcrops of mineralized zones and spoil heaps of disused mines still cause a high mortality rate in fish and other organisms (Abdullah and Royle, 1972). As early as 1924, Carpenter suspected that the complete lack of fish in several rivers in the Aberystwyth District of Cardiganshire could be the consequence of the pollution by the nearby lead mines. The unproductivity of certain fields in north Cardiganshire was explained by Griffith (1918) as a result of toxic levels of Pb and Zn in the soils. More recent investigations by Abdullah et al. (1972) on the distribution of transition metals in Welsh rivers clearly reflect the influence of mineralization zones. The rivers and lakes in regions where no mineral deposits are known, show Cd-levels ranging between 0.1 and 0.6 μg/l, whereas the annual average Cd-levels in rivers of the mineralized regions are found to range between 1.2 and 4.7 μg/l, with the highest recorded concentration being 20 μg/l. There seems to be a characteristic influence from this area on the waters of the adjacent Irish Sea (Fig. 5). The highest concentrations of Cd in the Bristol Channel are possibly derived from industrial effluents entering the area from the Avon and Severn Estuary. In Cardigan Bay, however, which is relatively free from industrial effluent and, because of a low population density, little domestic waste is present; runoff from the mineralized zones and sites of former mining activity is the main source of cadmium and of other trace metals (Abdullah et al., 1972). Laboratory tests and natural evidence indicate a distinct toxicity of waters from past and present mining areas of *southwest England* due to elevated concentrations of Cu, Pb, As, and Zn. Experiments on the Pacific oyster (*Crassostrea gigas*) with Zn-rich mine-adit water ranging between 100 and 500 μg/l reveal a decrease in growth, an increase in mortality rate (90% of the larvae died within two days at 500 μg/l Zn), and an increased incidence of abnormal larvae (Brereton et al., 1973). Oysters placed in the heavily contaminated Restronguet Creek contained initially 250 ppm Cu in dry matter, rising to 1500 ppm after one month, 2500 ppm after two months, and 6000 ppm after six months (Thornton, 1977). Data supplied by Anderson et al. (1976) from the Tamar Valley in the west of England suggest a significantly greater prevalence of dental caries in young residents of the Bere Alston area, where the soils are heavily contaminated by lead.

Studies related to the chemical composition of present mine drainage from the "Erzgebirge" in *Saxony* (East Germany), which has been exploited for centuries (ore processing dates back to the middle ages), have revealed that there is practically no plant growth in the vicinity of the major mine districts. The soils surrounding these areas have been markedly enriched with As and Pb, and mine drainage from percolation contains high levels of Pb, Zn, and in many cases, Ba (Leutwein and Weise, 1962). Flood-plain sediments of the Innerste River, originating in the Goslar–Oker area of the *Harz Mountains* in Germany, exhibit concentrations of Pb and Zn up to 20,000 ppm; poisoning of animals (e.g., cattle) by these deposits had been reported as early as the 18th and 19th centuries (Nowak and Preul, 1971). In *Poland*, Pasternak (1973, 1974) investigated the waste

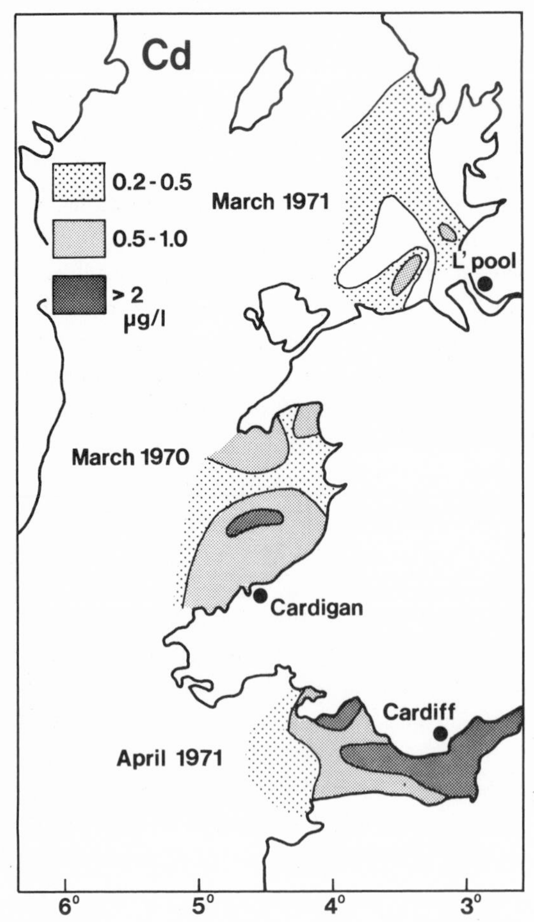

Fig. 5. Cd-enrichment in waters of the Irish Sea derived from industrial, domestic and mine waste effluents. (After Abdullah et al., 1972.)

waters from the regions of the Bolesław and Miasteczko Slaskie (Upper Silesia), where large deposits of Pb and Zn are mined and processed. The undergound water from the mine at Bolesław, which discharges directly into the Sztoła River, is enriched with Pb and Zn at maximum values of 300 mg Pb/l and 1800 mg Zn/l; the concentrations of Pb and Zn in flotation effluents reached 10.3 mg Pb/l and 1.7 mg Zn/l, respectively (Pasternak, 1973). The process of self-purification in some of the receiving waters is inhibited by the high concentrations of several metals as well as by the content of suspended material which prevents the penetration of light to the aquatic organisms. During the last few years no fish has appeared in the Sztoła River and the development of various algae is very poor (Pasternak, 1974).

Drainage water and stream sediments below the cinnabar mine at *Mount Avala near*

Belgrade (Serbia/Yugoslavia) have an abnormally high metal content, particularly of Fe, Ni, Hg, and As. The As-content in the Topciderska Reka stream was found to be up to 25 mg/l; the concentrations of Hg in the river sediments were as high as 6000 ppm. The very fine particles rich in As and Hg eventually reach the Danube and are deposited upstream of the Iron Gate Dam (Maksimović and Dangić, 1973).

In the *United States of America* the increasing awareness of the environment since the 60's has led to the recognition that metal accumulation in mine effluents is one of the main problems of water protection. In *Colorado* alone, 450 miles of surface streams are classified as affected by mill tailings and metal drainages (Wentz, 1974). Elevated contents of Ag, V, and other metals at Loma (Station of the Federal Water Pollution Control Administration; Kopp and Kroner, 1968), the station farthest upstream on the Colorado, stem partly from active mines and uranium plants at Rifle, Grand Junction and Gunnison. Part of these radioactive tailings have been utilized as landfill in construction projects and some 3000 homes in Grand Junction had to be abandoned since the highly penetrating γ-rays and Rn decay product could increase the risk of lung cancer to the inhabitants (Edsall, 1974; Anonymous, 1975). The high V-content in the Colorado River is probably due to the activities of an oil-shale extraction plant at Rifle. During the last few years particular attention has been devoted to the pollution effects of Mo in surface waters in the vicinity of the Mo-mining areas of Climax, Colorado (Runnels, 1973; Thurmann, 1974; Kaback, 1976). In Idaho, it is mainly the Kellog Smelterville area in Silver Valley (Miller et al., 1975) and the Cataldo Mission Flats in the catchment area of the Coeur d'Alene River (Galbraith et al., 1972) that are most affected by metal pollution. Airborne sources, consisting of particulates from smelter stack and dust blowing from smelter operations, as well as leachates from tailing ponds, contribute to an increase in the metal concentrations in the adjacent rivers, as has been repeatedly demonstrated in the Coeur d'Alene district (M. Ellis, 1940; Mink et al., 1972; Filby et al., 1974; Johnson et al., 1977; Rabe and Bauer, 1977).

Investigations in *Arizona* carried out by the U.S. Geological Survey (Durum et al., 1971) revealed extreme rates of increase for Cd (up to 130 μg/l), Co (4500 μg/l), and Zn (42,000 μg/l) in the Mineral Creek near Big Dome which could be attributed to the influence of acidic mine effluents (see below). Mining activities in the new lead belt of *Missouri*, the world's largest lead mining district, still pose some problems for water quality despite modern processing methods, as there is obviously a considerable metal transport by fine particulates (Jennett and Wixson, 1977). In the Mississippi drainage system, the *Tennessee River* gained particular attention due to the pollution influences from mining activities in the northern Tennessee zinc district (Derryberry, 1972; Perhac, 1972, 1974; Hildebrand et al., 1975). Acid mine drainage from abandoned pyrite mines in the North Anna River area of *Virginia* effected elevated concentrations of trace metals in water, sediments, and biota (Blood and Reed, 1975): the dissolved metal concentration in Contrary Creek were found to be up to 1580 μg/l for Cu, 190 μg/l for Pb, and 2670 μg/l for Zn. The concentrations in fish muscle tissue were as high as 22.7 mg Pb/kg and 16.9 mg Zn/

kg, which are among the highest reported values.

Comparable developments have also been reported from many mining areas of *Canada*, especially in those with sulfidic mineral occurrences, and particularly in places where, due to the lack of carbonates, the water has a low buffer capacity. Characteristic increases of heavy metals in freshwater systems result from atmospheric emissions containing both elevated contents of SO_2 and trace metals. An example is the Clearwater Lake in *Ontario* where the Ni-, Zn-, and Cu-concentrations are significantly higher than in the anthropogenically less-influenced Blue Chalk Lake (Dillon et al., 1977). From lakes near Sudbury, where very strong atmospheric inputs from smelter emissions occur, Wright and Gjessing (1976; data partly from Beamish, 1976) report dissolved concentrations of Ni and Cu of as much as 1850 μg/l and 1120 μg/l, respectively. Harvey (1976) has reviewed several examples of acid pollution problems from *Brunswick* mines: one of the worst cases where valuable fish resources were destroyed is the Brunswick No. 6 mine; the wastewater had the composition of 389 mg Zn/l, 31 mg Cu/l, and 131 Mg Fe/l, with a pH of 3.0 (Anonymous, 1972).

Deleterious metal inputs from mining activities include the discharge of tailings into *fjords,* such as those in *British Columbia.* D.V. Ellis (1977) described conditions in the Island copper mine near the north-end of Vancouver Island; Littlepage (1975) investigated the deposition of Mo-tailings in Alice Arm, on the northern mainland of the British Columbian Inlet; and Thompson (1977) cites the examples of two mining sites on Howe Sound and Rupert Island. In western *Greenland* deposits of mining waste contribute considerably to metal enrichments in fjord sediments (Bondam et al., 1976). In bottom waters of the Agfardlikavsa Fjord, Pb concentrations of up to 1000 μg/l have been recorded (Thomson, 1975). The pollution problems in the Sörfjord of western *Norway,* a 40 km long north–south extension of the Hardanger Fjord, were investigated by Skei et al. (1972, 1973). Three industrial factories, including a Zn-smelting plant, are reported to have released their metal-bearing wastes into a relatively shallow area of the fjord. During the period of investigation, approximately 100 tonnes (= metric-tons) of Cu, 1500 tonnes of Pb, and 2000 tonnes of Zn per year were discharged, strongly affecting the metal concentrations in the organisms (Stenner and Nickless, 1974). In green algae, the Pb concentration was 300 times greater, whereas Cd- and Zn-concentrations in mussels were 4–70 times greater than normal levels.

The most serious case of metal poisoning of humans by mine effluents is known from *Japan.* During 1947 an unusual and painful disease of a "rheumatic nature" was recorded in 44 patients from villages on the banks of the *Jintsu River,* Toyama Prefecture, Japan (Friberg et al., 1974). During subsequent years, it became known as the "itai-itai" disease (meaning "ouch-ouch") in accordance with the patient's shrieks resulting from painful skeletal deformations. It is estimated that approximately 100 deaths occurred as a result of the disease until the end of 1965. However, the cause of this disease was completely unknown until 1961 when sufficient evidence led to the postulation that Cd played a role in its development (Hagino and Yoshioka, 1961). It was found that the source of Cd-pollution of the Jintsu River was a zinc mine owned by the Makioko Co. situated some 50 km upstream from the afflicted villages. During World War II production of Zn and Pb from the mine was increased without sufficient accompanying treatment of the plant effluents and flotation sludge. The sludge from the plant became deposited downstream and caused considerable damage to the rice crops, which were irrigated or flooded by water from the Jintsu River. All patients were found to reside within 3 km of the river bank and in the low-lying rice-field areas which had been flooded by the polluted river water. After the mine constructed a retaining dam in 1955, pollution of the Jintsu River and the number of itai-itai cases rapidly declined. However, analytical data from the beginning of the 70's still indicated enhanced Cd-values in this area (Goto, 1973): 1 μg Cd/l in well water, 5–61 μg Cd/l (average 17 μg Cd/l) in mine waste water, and 1–9 μg Cd/l in river water.

The effects of the *Togane arsenic mine* north of Nagoya (Gifu-Prefecture) on the water of the Wada River was studied by Kato et al. (1973) from March 1972 to February 1973. The As-concentration in the river water was usually less than 30 μg/l. However, during the high-water period a maximum value of 1440 μg/l was determined near the closed mine (the mine ceased operations in 1957).

The earliest mining activity in the *Philippines* may be traced to ancient times when copper and gold had been traded and bartered between the Chinese mainland and other countries in Southeast Asia. At the turn of the century, after the American occupation, prospecting and mining activities were intensified, mainly in the mines of Mt. Province, Masbate, Surigao and in the Lepanto copper mines in Baguio (Lesaca, 1977). The latter region particularly suffers from the large quantities of mine tailings, which – despite provisions for settling ponds – still are partially disposed of into rivers, thereby causing excessive siltation (30,000 hectares of riceland have already been affected) and river quality degradation which affects agriculture and fishery resources.

Similar deleterious consequences on ecosystems have been reported from the disposal of mining wastes in Bougainville, *Papua New Guinea* (M.J.F. Brown, 1974). The influence of the Tui mine, Te Aroha and Maratoto silver mines in *New Zealand* on natural vegetation and inhabitants was investigated by Ward et al. (1976, 1977), who found some degree of pollution in the immediate vicinity of mining and processing plants.

A recent ecological survey conducted to evaluate the effects of a disused copper–lead–zinc mine at the Molonglo River in *Australia* (Weatherly and Dawson, 1973) revealed that the area is still disfigured by slime dumps, some 35 years after a flood that conveyed an enormous amount of tailing deposits from the mine area to the fairly productive adjacent flats. Today, 15 km downstream along the *Molonglo River,* the area is a virtual wasteland, whereby the release of zinc is regarded as the chief detrimental influence. Another example of large-scale destruction of organic life has recently occurred at *Rum Jungle* in northern Australia, 64 km south of Darwin, where U and Cu have been mined since the mid-1960's. An estimated 1300 tons of Cu have been released and dispersed onto the River Finiss floodplain. In addition, 90 curies of Ra, whose fate is still uncertain, has been leached from the tailing dump (Watson, 1975). A program to determine the immediate input of a new nickel refinery on metal levels in *Halifax Bay, North Queensland* is presently being performed by Knauer (1977). The refinery is located adjacent to the central part of the Great Barrier Reef and is continuously discharging liquid waste at the rate of some $15 \cdot 10^6$ l/day. The tailings contain significant quantities of Ni, Co, Fe, and Mn; Knauer (1977) found 130 μg/l Co and 1320 μg/l Ni in artifically prepared tailing supernatants.

Despite the relatively short history of Sn- and W-mining in *Tasmania,* the sparsely populated island has nevertheless suffered from mining activities. The coarse tailings and supernatants have caused Pb-, Zn-, Cu-, Mn-, Fe-, and H_2SO_4-enrichment in the *South Esk River* system, far beyond the borders of the mining areas (Tyler and Buckney, 1973). It was found that bordering farmers were unable to utilize the water from the creeks containing these mining effluents; fish and other biota were found to be even absent in some areas. Other parts of the river system less affected by the tailings still indicate a drop in both species diversity and abundance after winter floods, probably due to periodic movements of Cd and Zn down the river (Thorp and Lake, 1973). Analyses of the concentrations of Cd, Cr, Cu, Pb, Hg, Zn and other trace metals in filtered waters, suspended particulates, sediments, shellfish, airborne particulates, and sewage performed by Bloom and Ayling (1977) have shown that the *Derwent Estuary* of Tasmania is one of the most polluted areas of the world. Metallurgical liquid effluent discharges into the Derwent began several decades ago when an electrolytic Zn-refining plant went into operation. Dust falling in residential areas of Hobart contained up to 1450 ppm of Cd and 30,000 ppm of Pb. Approximately 42,000 ppm of Pb and 100,000 ppm of Zn were found in sediments near the wharf of the Zn-refining company. The highest ever recorded concentrations of Zn in oysters – 38,000 ppm dry weight – have been reported from the Derwent Estuary.

Finally, the investigations performed by Wittmann and Förstner (1976a, b; 1977a, b) indicate large-scale pollution by mine effluent from *Witwatersrand Goldfields in South Africa.* Here 247 slime dams are situated in a belt stretching approximately 120 km from Randfontein in the west to Nigel in the east (the total gold strike is over 380 km in length, containing seven goldfields). The levels of dissolved Mn, Co, and Ni exceed the normal surface water values by a factor of more than 10,000 for each individual metal; Fe, Cr, Zn, and sulfate concentrations are increased 1000-fold, whereas Pb and Cd in many instances are encountered at values exceeding a 100-fold enrichment. A comparison of these metal concentrations with drinking water standards reveals that the maximum values of all metals determined in gold/uranium mining effluents significantly surpass the permissible levels. The high Zn and Pb values are attributable to the cyanidation process for the recovery of Au, whereas high Mn values result from the oxidation of uraninite by pyrolusite (MnO_2) in sulphuric acid medium. However, tucholite, described as an "enigmatic hydrocarbon" (Feather and Koen, 1973) is a common cause of uraninite loss. It is therefore not surprising that the environmental impact from the uranium recov-

TABLE VII

Metal concentrations in inland waters affected by acidic mine effluents (examples)
All values in $\mu g/l$

	Cornwall (SW England)	Silesia (Poland)	Siberia (U.S.S.R.) [6]	Colorado (U.S.A.) [7]	Philippines [8]	Tasmania [9]	South Africa [10]
As	250 [1]	–	499	70	–	–	–
Cd	–	1,325 [4]	207	70	–	6,100	52
Co	–	13 [5]	368	–	–	–	3,300
Cr	–	17 [5]	–	–	120	–	4,000
Cu	1,160 [2]	62 [5]	20,710	3,900	953	1,350	5,400
Fe	23,000 [2]	3,185 [5]	–	213,000	176,100	20,500	550,000
Mn	2,400 [2]	315 [5]	1,624	8,000	–	22,500	206,000
Ni	–	14 [5]	900	460	80	–	6,400
Pb	530 [3]	23 [5]	2,071	300	443	–	290
Zn	10,000 [2]	43,100 [4]	5,770	17,000	1,280	105,000	26,000

[1] Tamar River (Aston et al., 1975); [2] Carnon R; [3] Gannel R. (Aston et al., 1974); [4] Granisczna Woda, inflow to [5] Mala Panew (Pasternak, 1974); [6] Maximum values from up to 4500 water samples (Udodov and Parilov, 1961); [7] Hill (1973); [8] Baguio Mining District, Agno and Bued Rivers (Lesaca, 1977); [9] Storys Creek in South Esk catchment (Tyler and Buckney, 1973); [10] West Driefontein Mine, West Wits Goldfield (Wittmann and Förstner, 1977a).

ery does not rest directly with heavy metal toxicity and pollution, but rather with the hazards associated with radioactivity (Wittmann, in Förstner and Wittmann, 1979).

A compilation of typical examples from the areas mentioned above is presented in Table VII. It is clear that these data of dissolved metal concentrations, which exceed by far the metal contents known from industrially or domestically influenced waters, are related to the occurrence of acid conditions in an aquatic system. The major process affecting the lower of pH-values (down to pH 2 to 3) is the exposure of pyrite (FeS_2) and of other sulfide minerals to atmospheric oxygen and moisture, whereby the sulfidic component is oxidized to sulfate (SO_4^{2-}) and the acidity (H^+-ions) is generated. Bacterial

TABLE VIII

Factors of environmental change from pollution by mines (After Jennett and Foil, 1979)

Mine waters
- Inorganic nutrients in the subterranean water
- High levels of CO_2 and carbonates
- Fuel spills
- Oil spills
- Hydraulic fluid spills
- Small mineral particles in mine effluent that produce turbidity
- Blasting agents – spills and partially oxidized compounds containing nutrients
- Highly variable mineral content of ore

Mill waters
- Chemical spills, both organic and inorganic
- Variable mineral content of ore may cause:
  - (a) Excessive use of reagents and loss of toxic chemicals to effluent
  - (b) Low recovery of heavy metals during pulses of very rich ore chemical reagents not adsorbed to concentrate and heavy metals are released in effluent
- Improperly placed concentrate piles allow dispersal of heavy metals either aerially or during runoff.
- High suspended and dissolved solids in effluent

Solid wastes
- Dams constructed of tailings wash directly into streams or blow onto soil and enter stream during runoff.
- Concentrated ore washes directly from storage piles into stream system or is blown onto soil and enters stream during runoff

Transport emission
- Concentrate-hauling vehicles are uncovered and high concentrations of heavy metals are blown onto soil and enter streams during runoff

Smelter emission
- Particulates build up into soil layer and enter stream during runoff

Tailings ponds
- Improper design relating to placement of ponds; insufficient size or number
- Insufficient retention time
- Release of toxic milling reagents to streams
- Release of organic and inorganic nutrients to streams
- Release of finely ground rock and mineral particles to streams

action (here *Thiobacillus ferroxidans*) can assist the oxidation of Fe^{2+}(aq) in the presence of dissolved oxygen. Water seeping from mine refuse has been passing increased metal concentrations into water for decades. The threat is especially great in waters with little buffer capacity, i.e., in carbonate-poor areas where dissolved-metal pollution can be spread over great distances. A compilation of factors which might effect environmental changes by mining activities is given in Table VIII from a study by Jennett and Foil (1979). The authors have studied the impact of the world's newest and largest Pb–Zn mining district, the "New Lead Belt" or "Viburnum Trend" of southeast Missouri on an unpolluted stream basin, which was until recently virgin woodland. It is shown that run-off transport is a major factor in moving heavy metals from one ecosystem to another. In this case, Cd was almost completely solubilized, Pb was generally particulate-associated, and Zn was approximately one-half dissolved.

There are many different *pathways* for the release of heavy metals from tailings (Andrews, 1977): (1) structural failures from improper operational techniques and design in regard to possible stimulus of catastrophic events, such as earthquakes and floods; (2) direct discharges of mill waste effluent or total tailing to surface waters; (3) dust from unstabilized, desiccated, wind-blown surfaces; (4) biological concentration in plants and ultimately in animals; (5) erosion of embankment surfaces; (6) leaching to the surface via capillary action due to a high groundwater level or leaching to subsurface waters by permeation.

CONCLUSIONS

Of the data on the heavy-metal contents in fresh water, the aspect of geochemical exploration has not been considered to any great depth, although it was this field which influenced to a great extent the development of water analysis for heavy metals, mainly in the search for useable mineral deposits. For some years now prospecting has been done with the aid of solid substances in water bodies for measurements on fluviatile sediments and, more recently, on lacustrine sediments. These methods will be discussed in another chapter of this Handbook (see Chapter 4 in this volume).

If the data presented here is evaluated under the aspect of water pollution, the following conclusions can be drawn:

Of all the sources of toxic heavy-metal pollution in inland waters, the waste water from sulfide ore processing plants presents a particular problem: as a consequence of the oxidation of these minerals and the consequent low pH, the concentration of a number of heavy metals can rise inordinately. The result of these changes have been known for over 100 years, and yet it is still very difficult to effectively protect water systems against these pollutants.

Methods of controlling the problem of acidic mine drainage include thermodynamic measures (elimination of oxygen and the maintenance of reducing conditions, e.g., by application of sewage sludge on the surface of the spoil heaps), kinetic effects (e.g., changes of the bacterial propagation cycle leading from Fe^{2+} to Fe^{3+} – in an abiotic system, the oxidation of ferrous iron is roughly a factor of 10^{-6} slower than in a system mediated by bacteria), and especially, the application of bactericides (Singer and Stumm, 1970). The environmental consequences and controls of tailings were reviewed by Andrews (1977), whose study includes proposals for seepage collection and handling, underwater disposal and alternative tailings disposal in order to reduce or prevent the release of toxic metals into surface waters. In the future, the possibility of applying physicochemical methods to water processing, as is presently being applied in other fields of metallurgy (e.g. electroplating), will have to be further investigated. Apart from the widely used neutralization and electrolysis methods, three others have shown promising results: Hill et al. (1971) have developed a system in collaboration with the U.S. Environmental Protection Agency whereby the waste stream is neutralized, the sludge removed and the neutralized water returned to the influence of a *reverse osmosis unit,* a procedure referred to as *neutrolysis* (Hill, 1973). Metals can also be effectively removed from mine drainage by *cementation,* i.e. by the electromotive force of other metals, e.g., by passing the Cu-bearing water through shredded iron. Further, favorable results can be expected from ion-exchange methods, particularly in diluted solutions.

REFERENCES

Abdullah, M.I. and Royle, L.G., 1972. Heavy metal content of some rivers and lakes in Wales. *Nature,* 238: 239–330.

Abdullah, M.I. and Royle, L.G., 1974a. A study of the dissolved and particulate trace elements in the Bristol Channel. *J. Mar. Biol. Assoc. U.K.,* 54: 581–597.

Abdullah, M.I. and Royle, L.G., 1974b. Cadmium in some British coastal and fresh water environments. Proc. Int. Symp. *Problems of the Contamination of Man and his Environment by Mercury and Cadmium,* Luxembourg, July 3–5, 1973, pp. 69–81.

Abdullah, M.I., Royle, L.G. and Morris, A.W., 1972. Heavy metal concentration in coastal waters. *Nature,* 235: 158–160.

Anderson, R.J., Davies, B.E. and James, P.M.C., 1976. Dental caries in a heavy metal contaminated area of the west of England. *Br. Dent. J.,* 141: 311–314.

Andrew, R.W., 1976. Toxicity relationships to copper forms in natural waters. In: R.W. Andrew, P.V. Hodson and D.E. Konasewich (Editors), *Toxicity to Biota of Metal Forms in Natural Waters.* Int. Joint Comm., Windsor, Ont., pp. 127–143.

Andrews, R.D., 1977. Tailings: Environmental consequences and a review of control strategies. Proc. Int. Symp. *Heavy Metals in the Environment,* Toronto, 1975, Vol. II/2: 645–675.

Angino, E.E., Galle, O.K. and Waugh, T.C., 1969. Fe, Mn, Ni, Co, Sr, Li, Zn, and SiO_2 in streams of the lower Kansas River basin. *Water Resour. Res.,* 5: 698–705.

Anonymous, 1972, 1973, 1974, 1975. *Water Resources Data for Louisiana.* U.S. Geol. Survey, Water Resources Division, Baton Rouge, La. (cited in Trefry and Presley, 1976).

Anonymous, 1972. *Northeastern New Brunswick Mine Water Quality Program.* Montreal Engineering Co., Fredericton, N.B. (cited in Harvey, 1976).

Anonymous, 1974. Australia Govt. Joint Technical Committee on *Mine Waste Pollution of the Molonglo River.* Final Report on Remedial Measures, June, 1974. Australian Govt. Publ. Serv., 46 pp.

Anonymous, 1975. *Controlling the Radiation Hazard from Uranium Mill Tailings.* NRC/ERDA Report to Congress, Comptroller General. U.S. Govt. Accounting Office, May 21, 1975.

Aplin, C.L. and Argall, G.O. (Editors), 1973. Tailing disposal today. *Proc. 1st Int. Tailing Symp.,* Tucson, Ariz., 1972. Miller/Freeman, San Francisco, Calif., 861 pp.

Aston, S.R. and Thornton, I., 1977. Regional geochemical data in relation to seasonal variations in water quality. *Sci. Total Environ.,* 7: 247–260.

Aston, S.R., Thornton, I., Webb, J.S., Purves, J.B. and Milford, B.L., 1974. Stream sediment composition, an aid to water quality assessment. *Water Air Soil Pollut.,* 3: 321–325.

Aston, S.R., Thornton, I., Webb, J.S., Milford, B.L. and Purves, J.B., 1975. Arsenic in stream sediments and waters of south-west England. *Sci. Total Environ.,* 4: 347–358.

Atkins, W.R.G., 1953. The seasonal variation in the copper content of sea water. *J. Mar. Biol. Assoc. U.K.,* 31: 493–502.

Baccini, P., 1976. Untersuchungen über den Schwermetallhaushalt der Seen. *Schweiz. Z. Hydrol.,* 38: 121–128.

Barnhart, B.J., 1978. The disposal of harzardous wastes. *Environ. Sci. Technol.,* 12: 1132–1136.

Beamish, R.J., 1976. Acidification of lakes in Canada by acid precipitation and the resulting effects on fishes. *Water Air Soil Pollut.,* 6: 501–514.

Beamish, R.J. and Harvey, H.H., 1972. Acidification of the La Cloche Mountain Lakes, Ontario, and resulting fish mortalities. *J. Fish. Res. Board Can.,* 29: 1131–1143.

Bender, M.L. and Gagner, C.L., 1976. Dissolved copper, nickel and cadmium in the Sargasso Sea. *J. Mar. Res.,* 34: 327–339.

Bewers, J.M., MacAulay, I.D. and Sundby, B., 1974. Trace metals in the waters of the Gulf of St. Lawrence. *Can. J. Earth Sci.,* 11: 939–950.

Black, W.A.P. and Mitchell, R.L., 1952. Trace elements in the common brown algae and in sea water. *J. Mar. Biol. Assoc. U.K.,* 30: 575–588.

Blood, E.R. and Reed, J.R., 1975. Heavy metals in a lake affected by acid mine drainage. Abstr. Int. Conf. on *Heavy Metals in the Environment,* Toronto, C-162/163.

Bloom, H. and Ayling, G.M., 1977. Heavy metals in the Derwent Estuary. *Environ. Geol.,* 2: 3–22.

Bondam, J., Asmund, G. and Schrøder, S., 1976. Miljøkontrol ved Marmorilik in Nordvest Grønland. In: *Cadmium Forskning i Danmark.* Rapport Danmarks Tekniske Højskole 1976, pp. 21–32.

Bowen, H.J.M., 1976. Residence times of heavy metals in the environment. Proc. Int. Conf. on *Heavy Metals in the Environment,* Toronto, Vol. 1, pp. 1–19.

Boyle, E.A., 1978. Trace element geochemistry of the Amazon and its tributaries. *EOS,* 59: 276.

Boyle, E.A. and Edmond, J.M., 1975. Copper in surface water south of New Zealand. *Nature,* 253: 107–109.

Boyle, E.A., Collier, R., Dengler, A.T., Edmond, J.M., Ng, A.C., and Stallard, R.F., 1974. On the chemical mass-balance in esturaries. *Geochim. Cosmochim. Acta,* 38: 1719–1738.

Boyle, E.A., Sclater, F. and Edmond, J.M., 1976. On the marine chemistry of cadmium. *Nature,* 263: 42–44.

Boyle, E.A., Sclater, F.R. and Edmond, J.M., 1977. The distribution of dissolved copper in the Pacific. *Earth Planet. Sci. Lett.,* 37: 38–54.

Bradford, G.R., 1971. Trace elements in the water resources of California. *Hilgardia,* 41: 45–53.

Bradford, G.R., Bair, F.L. and Hunsker, V., 1968. Trace and major element content of 170 High Sierra lakes in California. *Limnol. Oceanogr.,* 13: 526–529.

Brereton, A., Lord, H., Thornton, I. and Webb, J.S., 1973. Effects of zinc on growth and development of larvae of the Pacific oyster *Crassostrea gigas. Mar. Biol.,* 19: 96–101.

Brown, M.J.F., 1974. A development consequence, disposal of mining waste on Bougainville, Papua, New Guinea. *Geoforum,* 18: 19–27.

Brown, V.M., 1976. Aspects of heavy metal toxicity in freshwater. In: R.W. Andrew, P.V. Hodson and D.E. Konasewich (Editors), *Toxicity to Biota of Metal Forms in Natural Water.* Int. Joint Comm., Windsor, Ont., pp. 59–75.

Bruland, K.W., Knauer, G.A. and Martin, J.H., 1978. Zinc in northeast Pacific water. *Nature,* 271: 741–743.

Burton, J.D. and Liss, P.S., 1976. *Estuarine Chemistry*. Academic Press, London, 229 pp.

Carpenter, J.H., Bradford, W.L. and Grant, V., 1975. Processes affecting the composition of estuarine waters (H_2CO_3, Fe, Mn, Zn, Cu, Ni, Cr, Co, and Cd). In: L.E. Cronin (Editor) *Estuarine Research*, 1. Academic Press, New York, N.Y., pp. 137–152.

Carpenter, K.E., 1924. A study of the fauna of rivers polluted by lead mining in the Aberystwith District of Cardiganshire. *Ann. Appl. Biol.*, 11: 1–23.

Chow, T.J. and Thompson, T.G., 1951. The determination and distribution of copper in sea water. I. Spectrographic determination of copper in sea water. *J. Mar. Res.*, 11: 124–127.

Coonley, L.S., Baker, E.B. and Holland, H.D., 1971. Iron in the Mullica River in Great Bay, New Jersey. *Chem. Geol.*, 7: 51–63.

Copeland, R.A. and Ayers, J.C., 1972. *Trace Element Distributions in Water, Sediment, Phytoplankton, Zooplankton, and Benthos of Lake Michigan.* Environ. Research Group, Inc., Ann Arbor, Mich., (cited in Klein, 1975).

Cranston, R.E. and Murray, J.W., 1978. Dissolved chromium species in seawater. *EOS*, 59: p. 306 (abstract 0. 142).

Davey, E.W., Morgan, M.J. and Erickson, S.J., 1974. A biological measurement of the copper complexation capacity of seawater. *Limnol. Oceanogr.*, 19: 993–997.

De Groot, A.J., Allersma, E. and Van Driel, W., 1973. Zware metalen in fluviatile en marine ecosystemen. Symp. *Waterloopkunde in Dienst van Industrie en Milieu.* Publ. No. 110 N, Sect. 5.

Derryberry, O.M., 1972. Investigation of mercury contamination in the Tennessee Valley region. In: R. Hartung and B.D. Dinman (Editors), *Environmental Mercury Contamination.* Ann Arbor Science Publ., Ann Arbor, Mich., pp. 76–79.

Dillon, P.J., Yan, N.D., Scheider, W.A. and Conroy, N., 1977. *Acidic Lakes in Ontario, Canada, Their Extent and Responses to Base and Nutrient Additions.* Paper presented at Jubilee Symp. on Lake Metabolism and Lake Management, Uppsala Univ., Aug. 1977. 25 pp.

Down, C.G. and Stocks, J., 1977. *Environmental Impact of Mining.* Applied Science Publishers, London, 371 pp.

Duinker, J.C. and Nolting, R.F., 1977. Dissolved and particulate trace metals in the Rhine estuary and the Southern Bight. *Mar. Pollut. Bull.*, 8: 56–71.

Durum, W.H. and Haffty, J., 1961. Occurrence of minor elements in water. *U.S. Geol. Surv. Circ.*, 445: 11 pp.

Durum, W.H., Hem, J.D. and Heidel, S.C., 1971. Reconnaissance of selected minor elements in surface waters of the United States. *U.S. Geol. Surv. Circ.*, 643: 49 pp.

Edgington, D.N. and Robbins, J.A., 1976. Records of lead deposition on Lake Michigan sediments since 1800. *Environ. Sci. Technol.*, 10: 266–273.

Edsall, J.T., 1974. Hazards of nuclear fission power and the choice of alternatives. *Environ. Conserv.*, 1: 21–30.

Ellis, D.V., 1977. Pollution controls on mine discharges to the sea. Proc. Int. Conf. on *Heavy Metals in the Environment*, Toronto, 1975 Vol. II/2: 677–685.

Ellis, M., 1940. Pollution of the Coeur d'Alene River and adjacent waters by mine wastes. *Spec. Sci. Rep. 1, U.S. Bureau of Fisheries* (cited in Johnson et al., 1977).

Emery, K.O., Orr, W.L. and Rittenberg, S.C., 1955. Nutrient budgets in the ocean. In: *Essays in the Natural Sciences in Honor of Captain Allan Hancock.* University of Southern California Press, Los Angeles, Calif.

Fagerström, T. and Jernelöv, A., 1972. Aspects of the quatitative ecology of mercury. *Water Res.*, 6: 1193–1202.

Feather, C.E. and Koen, G.M., 1973. The significance of the mineralogical and surface characteristics of gold grains in the recovery process. *J. South Afr. Inst. Min. Met.* 73: 223–234.

Filby, R.H., Shah, K.R. and Funk, W.H., 1974. Role of neutron activation analysis in the study of heavy metal pollution of a lake–river system. Proc. 2nd Int. Conf. *Nuclear Methods in Environmental Research*, pp. 10–23.

Förstner, U. and Wittmann, G.T.W., 1979. *Metal Pollution in the Aquatic Environment.* Springer, New York, N.Y., 486 pp.

Foster, P., Hunt, T.E. and Morris, A.W., 1978. Metals in an acid mine stream and estuary. *Sci. Total Environ.*, 9: 75–86.

Friberg, L., Piscator, M., Nordberg, G.F. and Kjellström, T., 1974. *Cadmium in the Environment.* CRC, Cleveland, Ohio, 356 pp.

Fricke, K. and Werner, H., 1957. Geochemische Untersuchungen von Mineralwässern auf Kupfer, Blei und Zink in Nordrhein-Westfalen und angrenzenden Gebieten (vorl. Mitt.). *Heilbad Kurort,* pp. 45–46.

Gailbraith, J.H., Williams, R.E. and Siems, P.L., 1972. Migration and leaching of metals from old mine tailings deposits. *Ground Water,* 10: 33–44.

Gardner, D., 1975. Observations on the distribution of dissolved mercury in the ocean. *Mar. Pollut. Bull.*, 6: 43–46.

Gibbs, R.J., 1977. Transport phases of transition metals in the Amazon and Yukon Rivers. *Geol. Soc. Am. Bull.*, 88: 829–843.

Goto, M., 1973. Inorganic chemicals in the environment – with special reference to the pollution problems in Japan. *Environ. Qual. Saf.*, 2: 72–77.

Griel, V. and Robinson, R.J., 1952. Titanium in sea water. *J. Mar. Res.*, 11: 173–179.

Griffith, J.J., 1918. Influence of mines upon land and livestock in Cardiganshire. *J. Agric. Sci.*, 9: 365–395.

Grimshaw, D.L., Lewin, J. and Fuge, R., 1976. Seasonal and short-term variations in the concentration and supply of dissolved zinc to polluted aquatic environments. *Environ. Pollut.*, 11: 1–7.

Hagino, N. and Yoshioka, K., 1961. A study on the cause of Itai-Itai disease. *J. Jpn. Orthop. Assoc.*, 35: 815 (cited in Friberg et al., 1974).

Harvey, H.H., 1976. Aquatic environmental quality: problems and proposals. *J. Fish. Res. Board Can.*, 33: 2634–2670.

Hattingh, W.H.J., 1977. Reclaimed water: a health hazard? *Water S.A.*, 3: 104–112.

Heide, F. and Singer, E., 1954. Der Gehalt des Saalewassers an Kupfer und Zink. *Naturwissenschaften,* 37: 541.

Heide, F., Lerz, H. and Böhm, G., 1957. Der Gehalt des Saalewassers an Blei und Quecksilber. *Naturwissenschaften,* 44: 441.

Heinrichs, H., 1975. *Die Untersuchung von Gesteinen und Gewässern auf Cd, Sb, Hg, Tl, Pb und Bi mit der flammenlosen Atomabsorptions-Spektralphotometrie.* Thesis, Univ. Göttingen, 82 pp.

Hellmann, H., 1970. Die Charakterisierung von Sedimenten auf Grund ihres Gehaltes an Spurenmetallen. *Dtsch. Gewässerkd. Mitt.*, 14: 160–164.

Heyl, K.E., 1954. *Hydrochemische Untersuchungen im Gebiet des Siegerlander Erzbergbaus.* Thesis, Univ. Heidelberg, 72 pp.

Hildebrand, S.G., Andren, A.W. and Huckabee, J.W., 1975. Distribution and bioaccumulation of mercury in biotic and abiotic compartments of a contaminated river-reservoir system. *Oak Ridge Natl. Lab. Rep.*, CONF-751058-1, 28 pp.

Hill, R.D., 1973. Control and prevention of mine drainage. In: *Cycling and Controls of Metals.* Nat. Environ. Res. Center, U.S. Environ. Protection Agency, Cincinnati, Ohio, pp. 91–94.

Hill, R.D., Wilmoth, R.C. and Scott, R.B., 1971. Neutrolysis treatment of acid mine drainage. 26th Annu. Purdue Industrial Waste Conf., Lafayette, Indiana 1971.

Huff, L.C., 1948. A sensitive field test for heavy metals in water. *Econ. Geol.*, 43: 675–681.

Irion, G. and Förstner, U., 1975. Chemismus und Mineralbestand amazonischer See-Tone. *Naturwissenschaften,* 62: 476.

Jackson, G.A. and Morgan, J.J., 1978. Trace metal–chelator interactions and phytoplankton growth in seawater media: theoretical analysis and comparison with reported observations. *Limnol. Oceanogr.*, 23: 268–283.

Jackson, T.A., 1978. The biogeochemistry of heavy metals in polluted lakes and streams at Flin Flon, Canada, and a proposed method for limiting heavy-metal pollution of natural waters. *Environ. Geol.*, 2: 173–189.

Jennett, J.C. and Foil, J.L., 1979. Trace metal transport from mining, milling, and smelting watersheds. *J.W.P.C.F.*, 51: 378–404.

Jennett, J.C. and Wixson, B.G., 1977. The new lead belt: aquatic metal pathways control. Proc. Int. Conf. on *Heavy Metals in the Environment,* Toronto, 1975, Vol. II/1: 247–255.

Johnson, R.D., Miller, R.J., Williams, R.E., Wai, C.M., Wiese, A.C. and Mitchell, J.E., 1977. The heavy metal problem of Silver Valley, Northern Idaho. Proc. Int. Conf. on *Heavy Metals in the Environment,* Toronto, 1975, Vol. II/2: 465–485.

Kaback, D.S., 1976. Transport of molybdenum in mountainous streams, Colorado. *Geochim. Cosmochim. Acta,* 40: 581–582.

Kanamori, S. and Sugawara, K., 1965. Geochemical study of arsenic in natural waters. *J. Earth Sci. Nagoya Univ.,* 13: 36–45.

Kato, K., Takahashi, T., Morishita, Y., Mori, H., Umemura, M., Watanabe, N., Hayakawa, T. and Yamada, F., 1973. Influence of the Togane arsenic mine on the Wada River. *Gifu-ken Eisei Kenkyusho Ho,* 18: 31–38.

Klein, D.H., 1975. Fluxes, residence times and sources of some elements to Lake Michigan. *Water Air Soil Pollut.,* 41: 3–8.

Knauer, G.A., 1977. Immediate industrial effect on sediment metals in a clean coastal environment. *Mar. Pollut. Bull.,* 8: 249–254.

Knauer, G.A. and Martin, J.H., 1973. Seasonal variations of cadmium, copper, manganese, lead, and zinc in water and phytoplankton in Monterey Bay, California. *Limnol. Oceanogr.,* 18: 597–604.

Konovalov, G.S., 1959. Removal of microelements by the principal rivers of the USSR. *Dokl. Akad. Nauk. USSR,* 129: 1034–1038.

Konovalov, G.S. and Ivanova, A.A., 1972. Content and regime of trace elements in the water and suspended substances in the Volga River basin. *Sov. Hydrol.,* 1972: 506–514.

Konovalov, G.S. and Nazarova, L.N., 1975. Mapping trace elements in river waters. *Gidrokhim. Mater.,* 62: 37–42.

Konovalov, G.S., Ivanova, A.A. and Kolesnikova, T.Kh., 1967a. Rare and dispersed elements (microelements) in the water and suspended substances of rivers in the European USSR. *Sov. Hydrol., 1967:* 520–533.

Konovalov, G.S., Ivanova, A.A. and Kolesnikova, T.Kh., 1967b. Microelements in the water and suspended substances of rivers in the Asiatic USSR. *Sov. Hydrol., 1967:* 533–542.

Kontorovich, A.E., Sadikov, M.A. and Shvartsev, S.L., 1963: Abundances of certain elements in the surface and ground waters of the northwestern part of the Siberian Platform. *Dokl. Akad. Nauk USSR,* 149: 168–173.

Kopp, J.F. and Kroner, R.C., 1968. *Trace Metals in Waters of the United States.* Fed. Water Pollut. Control. Admin., Div. Pollut. Surveillance.

Lesaca, R.M., 1977. Monitoring of heavy metals in Philippine rivers, bay waters and lakes. Proc. Int. Conf. on *Heavy Metals in the Environment,* Toronto, 1975, Vol. II/1: 285–307.

Leutwein, F. and Weise, L., 1962. Hydrogeochemische Untersuchungen an erzgebirgischen Gruben- und Oberflächenwässern. *Geochim. Cosmochim. Acta,* 26: 1333–1348.

Lewin, J., Davies, B.E. and Wolfenden, P.J., 1977. Interactions between channel change and historic mining sediments. In: K.J. Gregory (Editor), *River Channel Changes.* Wiley, New York, N.Y., pp. 353–367.

Lewis, G.J. and Goldberg, E.D., 1954. Iron in marine waters. *J. Mar. Res.,* 13: 183–197.

Linnman, L., Andersson, A., Nilsson, K.O., Lind, B., Kjellström, T. and Friberg, L., 1973. Cadmium uptake by wheat from sewage sludge used as a plant nutrient source. *Arch. Environ. Health,* 27: 47–57.

Linstedt, K.D. and Kruger, P., 1969. Vanadium concentrations in Colorado River basin waters. *J.A.W. W.A.,* 61: 85–88.

Littlepage, J., 1975. Heavy metals in a northern inlet. Abstr. Int. Conf. on *Heavy Metals in the Environment,* Toronto, 1975, C-159–161.

Livingstone, D.A., 1963. Chemical composition of rivers and lakes. In: M. Fleischer (Editor), *Data of Geochemistry,* 6th ed. *Geol. Surv. Prof. Pap.,* 440-G: 64 pp.

Lucas, R.E. and Davis, J.F., 1961. Relationships between pH values of organic soils and availabilities of 12 plant nutrients. *Soil Sci.*, 92: 177–182.

Maksimovic, Z. and Dangic, A., 1973. Mercury mine at Mount Avala, a source of environmental pollution by mercury and arsenic. *Geol. Ann. Balk. Poluostrva*, 38: 349–358.

Martin, J.H., Bruland, K.W. and Broenkow, W.W., 1976. Cadmium transport in the California current. In: H.L. Windom and R.A. Duce (Editors), *Marine Pollutant Transfer.* D.C. Heath Co, Lexington, pp. 159–184.

Mathis, B.J. and Cummings, T.F., 1973. Selected metals in sediments, water, and biota in the Illinois River. *J.W.P.C.F.*, 45: 1573–1583.

Measures, C.I. and Burton, J.D., 1978. Determination of selenium(IV) and total selenium in oceanic waters. *EOS*, 59: 307 (0 148).

Merrill, J.R., Lyden, E.F.X., Honda, M. and Arnold, J.R., 1960. The sedimentary geochemistry of the beryllium isotopes. *Geochim. Cosmochim. Acta,* 18: 108–129.

Miller, R.J., Johnson, R.D., Williams, R.E., Wai, C.M., Wiese, A.C. and Mitchell, J.E., 1975. Heavy metal problem of Silver Valley, North Idaho. Abstr. Int. Conf. on *Heavy Metals in the Environment,* Toronto, 1975, C-64/65.

Mills, E.L. and Oglesby, R.T., 1971. Five trace elements and vitamin B_{12} in Cayuga Lake, New York. *Proc. 14th Conf. Great Lakes Res.*, pp. 256–267.

Mink, L.L., Williams, R.E. and Wallace, A.T., 1972. Effect of early day mining operations on present day water quality. *Ground Water,* 10: 17–26.

Moore, R.M., and Burton, J.D., 1976. Concentrations of dissolved copper in the eastern Atlantic Ocean, 23°N to 47°N. *Nature,* 264: 241–243.

Morita, Y., 1955. Distribution of copper and zinc in various phases of the earth materials. *J. Earth Sci. Nagoya Univ.*, 3: 33–45.

Müller, G. and Förstner, U., 1975. Heavy metals in sediments of the Rhine and Elbe estuaries: Mobilization or mixing effect? *Environ. Geol.*, 1: 33–39.

Mullin, J.B. and Riley, J.P., 1956. The occurrence of cadmium in seawater and in marine organisms and sediments. *J. Mar. Res.*, 15: 103–122.

Noddack, I. and Noddack, W., 1939. Die Häufigkeit der Schwermetalle in Meerestieren. *Ark. Zool.*, 32-A: 1–30.

Nowak, H. and Preul, F., 1971. Untersuchungen über Blei- und Zinkgehalte in Gewässern des Westharzes. *Geol. Jahrb., Beih.*, 105: 68 pp.

Pasternak, K., 1973. The spreading of heavy metals in flowing waters in the region of occurrence of natural deposits of the zinc and lead industry. *Acta Hydrobiol.*, 15: 145–166.

Pasternak, K., 1974. The influence of the pollution of a zinc plant at Miasteczki Slaskie on the content of micro-elements in the environment of surface waters. *Acta Hydrobiol.*, 16: 273–297.

Perhac, R.M., 1972. Distribution of Cd, Co, Cu, Fe, Mn, Ni, Pb, and Zn in dissolved and particulate solids from two streams in Tennessee. *J. Hydrol.*, 15: 177–186.

Perhac, R.M., 1974. *Heavy Metal Distribution in Bottom Sediments and Water on the Tennessee River–Loudon Lake Reservoir System.* Water Resources Res. Center, Univ. Tennessee, Knoxville. Res. Report No. 40.

Proctor, P.D., Kisvarsanyi, G., Garrison, E. and Williams, A., 1973. Heavy metal content of surface and ground waters of the Springfield–Joplin areas, Missouri. In: D.D. Hemphill (Editor), *Trace Substances in Environmental Health.* Univ. of Missouri, Columbia, Vol. VII: 63–73.

Quentin, K.-E. and Winkler, H.A., 1974. Vorkommen und Nachweis von anorganischen Schadstoffen im Oberflächen- und Grundwasser. *Zentralbl. Bakteriol., Parasitenkd., Infektionskr. Hyg., Abt. 1: Orig., Reihe B,* 158: 514–523.

Rabe, F.W. and Bauer, S.B., 1977. Heavy metals in lakes of the Coeur d'Alene River Valley, Idaho. *Northwest Sci.*, 51: 183–197.

Robbins, J.A., Landström, E. and Wahlgren, M., 1972. Tributary inputs of soluble trace metals to Lake Michigan. *Proc. 15th Conf. Great Lakes Res.*, pp. 270–290.

Rona, E., Gilpatrick, L.O. and Jefrey, L.M., 1956. Uranium determination in sea water. *Trans. Am. Geophys. Union,* 37: 697–701.

Runnels, D.D., 1973. Detection of molybdenum enrichment in the environment through comparative study of stream drainages, central Colorado. In: D.D. Hemphill (Editor), *Trace Substances in Environmental Health.* Univ. of Missouri, Columbia, Vol. 7: 99–104.

Schaule, B. and Patterson, C., 1979. The occurrence of lead in the northeast Pacific, and the effects of anthropogenic inputs. In: M. Branica (Editor), *Proc. Int. Experts Discussion on Lead: Occurrence, Fate, and Pollution in the Marine Environment.* Pergamon, Oxford, in press.

Schleichert, U., 1975. Schwermetallgehalte der Schwebstoffe des Rheins bei Koblenz im Jahresablauf – eine gewässerkundliche Interpretation. *Dtsch. Gewässerkd. Mitt.,* 19: 150–157.

Schöttler, U., 1977. *Ausbreitung und Eliminierung von Spurenmetallen bei Infiltration und Untergrundpassage.* Literaturstudie, Veröff. Inst. Wasserforschung, Dortmund, Nr. 27, 99 p.

Schroll, E., Krachsberger, H. and Dolezel, P., 1975. Hydrogeochemische Untersuchungen des Donauwassers in Österreich in den Jahren 1971 und 1972. *Arch. Hydrobiol. Suppl.,* 44: 492–514.

Sclater, F.R., Boyle, E.A. and Edmond, J.M., 1976. On the marine geochemistry of nickel. *Earth Planet. Sci. Lett.,* 31: 119–128.

Sholkovitz, E.R., 1976. Flocculation of dissolved organic and inorganic matter during the mixing of river water and seawater. *Geochim. Cosmochim. Acta,* 40: 831–845.

Sholkovitz, E.R., Boyle, E.A. and Price, N.B., 1978. The removal of dissolved humic acids and iron during estuarine mixing. *Earth Planet. Sci. Lett.,* 40: 130–136.

Sibley, T.H. and Morgan, J.J., 1975. Equilibrium speciation of trace metals in fresh water/sea water mixtures. Proc. Int. Conf. on *Heavy Metals in the Environment,* Toronto, 1975, Vol. I, pp. 319–338.

Silker, W.B., 1964. Variations in elemental concentrations in the Columbia River. *Limnol. Oceanogr.,* 9: 540–545.

Silvey, W.D., 1967. Occurrence of selected minor elements in the waters of California. *U.S. Geol. Surv. Water Supply Pap.,* 1535-L: 17 pp.

Singer, P.C. and Stumm, W., 1970. Acidic mine drainage: the rate-determining step. *Science,* 167: 1121–1123.

Skei, J.M., Price, N.B., Calvert, S.E. and Hogdahl, O., 1972. The distribution of heavy metals in sediments of Sörfjord, West Norway, *Water Air Soil Pollut.,* 1: 452–461.

Skei, J.M., Price, N.B. and Calvert, S.E., 1973. Particulate metals in waters of Sörfjord, West Norway. *Ambio,* 2: 122–124.

Smirnov, S.S., 1954. *Die Oxidationszone sulfidischer Lagerstätten.* Akademie-Verlag, Berlin, 312 pp.

Stenner, R.D. and Nickless. G., 1974. Distribution of some heavy metals in organisms in Hardangerfjord and Skjerstadfjord, Norway. *Water Air Soil Pollut.,* 3: 279–291.

Stock, A. and Cucuel, F., 1934. Die Verbreitung des Quecksilbers. *Naturwissenschaften,* 22: 390–393.

Stumm, W. and Bilinski, H., 1972. Trace metals in natural waters. Difficulties of interpretation arising from our ignorance of their speciation. In: S.H. Jenkins (Editor), *Advances in Water Pollution Research.* Pergamon, New York, N.Y., pp. 39–52.

Sugawara, K. and Okabe, S., 1960. Geochemistry of molybdenum in natural waters. *J. Earth Sci. Nagoya Univ.,* 8: 93–107.

Sugawara, K., Naito, H. and Yamada, S., 1956. Geochemistry of vanadium in natural waters. *J. Earth Sci. Nagoya Univ.,* 4: 44–61.

Tamm, C.O., 1976. Acid precipitation: biological effects in soil and forest vegetation. *Ambio* 5: 235–238.

Thompson, J.A.J., 1977. Copper in marine waters – effects of mining wastes. Proc. Int. Conf. on *Heavy Metals in the Environment.* Toronto, 1975, Vol. II/1: 273–284.

Thornton, I., 1977. Some aspects of environmental geochemistry in Britain. Proc. Int. Conf. on *Heavy Metals in the Environment,* Toronto, 1975, Vol. II/1: 17–38.

Thorp, V.J. and Lake P.S., 1973. Pollution of a Tasmanian river by mine effluents. II. Distribution of macro-invertebrates. *Int. Rev. Ges. Hydrobiol.,* 58: 885–892.

Thurman, M.E., 1974. *Statistical Study of Content of Molybdenum in Stream Sediment, Adjacent to a Molybdenum Mill.* Trace Contam. Proc., Univ. Calif., Berkeley, Aug. 29–31, 1974, NSF, pp. 196–204.

Trefry, J.H. and Presley, B.J., 1976. Heavy metal transport from the Mississippi River to the Gulf of Mexico. In: H.L. Windom and R.A. Duce (Editors), *Marine Pollution Transfer.* Heath and Co: Lexington, pp. 39–76.

Troup, B.N. and Bricker, O.P., 1975. Processes affecting the transport of materials from continents to oceans. In: T.M. Church (Editor), *Marine Chemistry in the Coastal Environment. ACS Symp. Ser.,* 18: 135–151.

Turekian, K.K., 1969. The oceans, streams and atmosphere. In: K.H. Wedepohl (Editor), *Handbook of Geochemistry, I.* Springer, New York, N.Y., pp. 297–323.

Turekian, K.K. and Kleinkopf, M.D., 1956. Estimates of the average abundances of Cu, Mn, Pb, Ti, Ni, and Cr in surface waters of Maine. *Bull. Geol. Soc. Am.,* 67: 1129–1132.

Turekian, K.K., Harriss, R.C. and Johnson, D.G., 1967. The variations of Si, Na, Ca, Sr, Ba, Co, and Ag in the Neuse River, North Carolina. *Limnol. Oceanogr.,* 12: 702–706.

Tyler, P.A. and Buckney, R.T., 1973. Pollution of a Tasmanian River by mine effluents. I. Chemical evidence. *Int. Rev. Ges. Hydrobiol.,* 58: 873–883.

Udodov, P.A. and Parilov, Y.U.S., 1961. Certain regularities of migration of metals in natural waters. *Geochemistry,* 8: 763–769.

Voegell, P.T. and King, R.U., 1968. Occurrence and distribution of molybdenum in the surface water of Colorado. *U.S. Geol. Surv. Water-Supply Pap.,* 1535–N.

Ward, N.I., Brooks, R.R. and Reeves, R.D., 1976. Copper, cadmium, lead, and zinc in soils, stream sediments, waters and natural vegetation around the Tui Mine, Te Aroha, New Zealand. *N.Z.J. Sci.,* 19: 81–89.

Ward, N.I., Brooks, R.R. and Roberts, E., 1977. Silver in soils, stream sediments, waters and vegetation near a silver mine and treatment plant at Maratoto, New Zealand. *Environ. Pollut.,* 13: 269–280.

Watson, G.M., 1975. *Rum Jungle Environmental Studies.* Summary Report, Aust. Atomic Energy Comm. E 366, 21 pp.

Wattenberg, H., 1943. Ergänzung zu der Mitteilung "Zur Chemie des Meerwassers; über die in Spuren vorkommenden Elemente". *Z. Anorg. Allg. Chem.,* 251-B: 86–92.

Weatherley, A.H. and Dawson, P., 1973. Zinc pollution in a freshwater system: analysis and proposed solutions. *Search,* 4: 471–476.

Weiler, R.R. and Chawla, V.K., 1969. Dissolved mineral quality of Great Lakes waters. *Proc. 12th Conf. Great Lakes Res.,* pp. 801–818.

Wentz, D.A., 1974. Effect of mine drainage on the quality of streams in Colorado, 1971–1972. *Colo. Water Cons. Board Circ.,* 21.

Whitby, L.M., Stokes, P.M., Hutchinson, T.C. and Myslik, G., 1976. Ecological consequence of acidic and heavy-metal discharges from the Sudbury smelters. *Can. Mineral.,* 14: 47–57.

White, D.-E., Hem, J.D. and Waring, G.A., 1963. Chemical composition of subsurface waters. In: M. Fleischer (Editor), *Data of Geochemistry,* 6th ed., *U.S. Geol. Surv. Prof. Pap.,* 440-F: 25 pp.

Williams, P.M. and Chan, K.S., 1966. Distribution and speciation of iron in natural waters – transition from river water to a marine environment. *J. Fish. Res. Board Can.,* 23: 575–593.

Wilson, A.L., 1976. *Concentrations of Trace Metals in River Waters: A Review.* Water Research Center. Tech. Rep., 16: 60 pp.

Windom, H.L., Beck, K.C. and Smith, R., 1971. Transport of trace elements to the Atlantic Ocean by three southeastern streams. *Southeast. Geol.,* 12: 169–181.

Wittmann, G.T.W. and Förstner, U., 1976a, b, 1977a, b. Heavy metal enrichment in mine drainage. *S.Afr.J.Sci.,* 72: 242–246, 365–370; 73: 53–57, 374–378.

Wood, J.M., 1974. Biological cycles for toxic elements in the environment. *Science,* 183: 1049–1053.

Wright, R.F. and Gjessing, E.T., 1976. Changes in the chemical composition of lakes. *Ambio,* 5: 220–223.

Zitko, V., 1976. Structure activity relationships and the toxicity of trace elements to aquatic biota. In: R.W. Andrew, P.V. Hodson and D.E. Konasewich (Editors), *Toxicity to Biota of Metal Forms in Natural Water.* Int. Joint Comm., Windsor, Ont., pp. 9–43.

Chapter 6

THE NATURE AND ORIGIN OF ARCHAEAN STRATA-BOUND VOLCANIC-ASSOCIATED NICKEL–IRON–COPPER SULPHIDE DEPOSITS

D.I. GROVES and D.R. HUDSON

INTRODUCTION

Genetic and economic importance

Deposits that are now considered to be volcanic-associated deposits were known in Canada (e.g. Naldrett, 1966) and Rhodesia (e.g. Le Roex, 1964) prior to the discovery of Kambalda in 1966. However, it was this discovery, the subsequent nickel rush in Western Australia in the late 1960's and early 1970's and consequent interest in the nature and origin of a number of similar deposits discovered in rapid succession, combined with the pioneering work of M.J. Viljoen and R.P. Viljoen (1969) on what are now known to be ultramafic komatiites (e.g. Arndt et al., 1977), that contributed to the rapid realization that this was an important new class of deposit (e.g. Naldrett and Gasparrini, 1971; Hudson, 1972; Imreh, 1973; Naldrett, 1973). The deposits thus represent one of the most recently recognized classes of strata-bound ore deposits, and a review of their major characteristics and suggestions as to their genesis is warranted.

The discovery of Kambalda also had a major impact on exploration philosophies for Ni-sulphide ores (Woodall and Travis, 1969; Hudson, 1972). Mining companies and prospectors were quick to recognize the gross geological and geochemical characteristics of the Kambalda mineralization, and successfully apply this information as a new strategy for base-metal exploration in the Archaean greenstone belts. In the simplest terms, the Kambalda exploration "model", as developed in Western Australia, was based on prospecting the basal contact of ultramafic complexes for gossanous expressions of sulphide mineralization. Not all explorationists followed this philosophy precisely, with the result that some of the nickel deposits found had little or no similarity to Kambalda (e.g. the Carr Boyd mineralization in a layered gabbroic complex, and the large tonnage, low grade, disseminated dunitic deposits typified by Mt. Keith). This "model" changed the emphasis of exploration strategy which, prior to the discovery of Kambalda, was largely based on sulphide ores associated with large layered noritic or gabbroic intrusions, such as Sudbury, Bushveld, Great Dyke, Noril'sk etc., many of which also occur within the older stable shield areas.

Available data suggest that reserves of nickel ore (>1% Ni) in all sulphide deposits of Western Australia represent in excess of 8% of the western world's sulphide nickel reserves, with the volcanic-associated deposits contributing about one half of these reserves. Estimation of reserves from *all* volcanic-associated ores is very difficult because in several cases assignment of deposits to the class is tenuous, but available data suggest they represent a minimum of 6% of western world's reserves. They are therefore an important class of deposits from an economic as well as a genetic and exploration viewpoint.

Definition

Lambert (1976) presents a classification scheme for stratiform base-metal sulphide deposits of volcano-sedimentary associations (Table I). He includes "Yilgarn-type" nickel deposits (equivalent to volcanic-associated Ni–Fe–Cu sulphide deposits) in this scheme, emphasizing their grouping with other previously reviewed stratiform to strata-bound base-metal deposits (Table I) and their close genetic association with Archaean ultramafic lavas and/or shallow intrusives.

Naldrett and Cabri (1976) present a classification, modified from that of Naldrett (1973), of ultramafic and related mafic bodies as a framework within which to consider Ni–Cu (and platinoid) ore bodies. This classification has recently been modified by committee members of IGCP Project 161 (Besson et al., 1979) and the modified version is presented in Table II.

As there is increasing evidence that greenstone belts may form in intracratonic basins (e.g. Barley et al., 1979), the ultramafic rocks are divided into three main classes rather than the two (orogenic vs. non-orogenic) of Naldrett and Cabri (1976). The modified classification distinguishes the two major types of komatiite host to Ni–Cu deposits, viz. peridotite flow and dunitic pods, and we have assigned as many deposits as possible to these two groups (Table II). Synvolcanic layered sills within greenstone belts may be of either tholeiitic or komatiitic affinity (e.g. Hallberg and D.A.C. Williams, 1972; D.A.C. Williams, 1972), and although some deposits occur within sills of determined affinity (e.g. the tholeiitic sill at Carr Boyd, Purvis et al., 1972), there are many others whose classification is unclear. Although realizing that there may be an important distinction, there are commonly insufficient data to assign deposits to these two groups (Table II).

Volcanic-associated Ni–Fe–Cu sulphide deposits are those that occur at or towards the base of gravity-differentiated peridotitic lenses of komatiitic affinity that are normally less than 50 m, but up to 150 m thick. These lenses rarely have well-developed spinifex-textured tops, but do have similar asymmetric mineralogical and geochemical profiles to spinifex-textured flows (cf. Pyke et al., 1973; Barnes et al., 1974) that normally occur in overlying thick sequences of komatiitic flows, implying emplacement as horizontal or near-horizontal flows or less likely subvolcanic sills. An important aspect of the deposits is their intimate association with mafic–ultramafic volcanic sequences consisting of komatiitic and tholeiitic lava flows with lesser fragmentals and volcaniclastic

TABLE I

Working classification scheme for stratiform base-metal sulphide deposits of volcano-sedimentary associations (From Lambert, 1976)

Ore types	Minor metals	Main rock associations	Examples	Ages	Recent reviews
Pb–Zn–Ag "McArthur-type"	Cu	tuffaceous mudstones and siltstones	McArthur, Mount Isa, Hilton Broken Hill (Australia), Sullivan (Canada), Rammelsberg (Germany)	Proterozoic Palaeozoic	Lambert (1976)
Zn–Cu–Pb–Ag–Au "Kuroko-type"		felsic tuffs and lavas of calc-alkaline suite, "volcanic" sediments, mudstones	Kuroko (Japan); New Brunswick, etc. (Canada, U.S.A.); Captains Flat, Woodlawn, Rosebery (Australia); Skellefte deposits (Sweden)	Proterozoic Phanerozoic	Ishihara (1974), Lambert and Sato (1974), Solomon (1976)
Zn–Cu–Ag–Au "Superior-type"	Pb	andesitic to rhyolitic tuffs and lavas of calc-alkaline (?) suite, "volcanic" sediments	Noranda, etc. (Canada), West Shasta (U.S.A.)	Archaean Phanerozoic	Sangster and Scott (1976)
Cu–Au "Cyprus-type"	Zn–Ag	tholeiitic basalts (pillowed), chalks and marls, mudstones	Skouriotissa, etc. (Cyprus); Ergani Maden, etc. (Turkey); Besshi, etc. (Japan); Atlantis II muds (?) (Red Sea)	Phanerozoic	Constantinou and Govett (1973)
Ni "Yilgarn-type" *	Cu	ultramafic lavas and/or shallow intrusives, "volcanic" sediments	Kambalda, etc. (Australia)	Archaean	

* Equivalent to volcanic-associated Ni–Fe–Cu sulphide deposits.

TABLE II

Classification of sulphide Ni/Cu deposits in terms of gross tectonic setting and association with ultramafic and mafic bodies with emphasis on Archaean deposits of komatiitic affinity (From Besson et al., 1979, adapted from Naldrett and Cabri, 1976)
Numerous examples are given for deposits of komatiitic affinity, but only typical deposits for other deposit types

Deposit type	Region	Examples
A. Synvolcanic deposits (largely restricted to Precambrian greenstone belts)		
1. Deposits associated with komatiitic suite		
(a) Deposits directly associated with volcanic rocks	*W. Australia:*	Kambalda, Windarra, South Windarra, Scotia, Nepean, Wannaway, Redross, Mt Edwards, Spargoville
	Canada:	Langmuir, Texmont, McWatters, Sothman, Hart, Dundonald, Marbridge, Alexo
	Rhodesia:	Shangani, Inyati-Damba, Trojan, Hunters Road, Matopos, Epoch, Selukwe, Lower Gwelo
(b) Deposits in dunitic lenses and pods	*W. Australia:*	Mt Keith, Perseverance (Agnew), Yakabindie, Forrestania, Black Swan
	Canada:	Dumont
(c) Deposits of uncertain type in tectonically reworked terrains	*Canada:*	Thompson belt, Shebandowan
2. Deposits associated with tholeiitic suite		
(a) Deposits in synvolcanic stratiform intrusions	*W. Australia:*	Carr Boyd
	Canada:	Lynn Lake, Dumbarton
	U.S.S.R.:	Kola Peninsula deposits
(b) Deposits in anorthositic bodies	No known mineralization	

3. Deposits for which komatiitic or tholeiitic percentage is uncertain		
(a) Deposits in synvolcanic stratiform intrusions	*W. Australia:*	Mt Sholl
	Canada:	Montcalm, Ontario; Ungava deposits
	Rhodesia:	Empress, Perseverance (?)
(b) Deposits in iron formations and related metasediments	*W. Australia:*	Sherlock Bay, some Windarra ores
(c) Deposits in tectonically reworked terrains	*Botswana:*	Pikwe-Selibe
B. Deposits associated with intrusive bodies emplaced in cratonic areas		
1. Deposits in large layered complexes unrelated to flood basalts		
(a) Sheetlike		
(i) With repetitive layering	Bushveld, S. Africa; Stillwater, U.S.A.	
(ii) Without repetitive layering	Sudbury, Canada	
(b) Dyke-like	Great Dyke, Rhodesia; Jimberlana, W. Australia	
2. Deposits in intrusions equivalent to flood basalts	Insizwa, S. Africa; Duluth, U.S.A., Noril'sk, U.S.S.R.	
3. Deposits in other medium and small-sized intrusions	Losberg, S. Africa	
4. Deposits in alkaline ultramafic rocks	No known mineralization	
C. Deposits associated with mafic and ultramafic bodies emplaced during orogenesis		
1. Deposits in synorogenic intrusions	Rona, Norway; La Perouse, Alaska; Hitura and Kohtalahti, Finland	
2. Deposits in tectonically emplaced bodies		
(a) Deposits in ophiolite complexes	Queen of Bronz, Oregon, U.S.A., Table Mtn, Newfoundland	
(b) Deposits in possible diapirs	No known mineralization	
(c) Deposits of unknown association		
3. Deposits in Alaskan-type complexes	Salt Chuck, Alaska	
4. Deposits in bodies of uncertain type		

TABLE III

Combined production plus reserve figures for some volcanic-associated Ni deposits

	Million tonnes (approx.)	Grade % Ni (approx.)
Western Australia		
Kambalda–St Ives (14 *)	30	3.2
Windarra (3)	6	1.4
Wannaway	4.5	1.2
Mt Edwards	2.5	2.2
South Windarra	2.5	1.1
Scotia	1.5	3.1
Spargoville (3)	1.2	2.4
Redross (2)	1.0	3.5
Nepean (2)	0.8	4.0
Canada		
Texmont	3.5	1.0
Langmuir (2)	1.5	1.8
McWatters	0.7	1.0
Sothman	0.6	1.0
Marbridge (3)	0.5	2.3
Alexo	0.1	4.1
Rhodesia		
Shangani (2)	20	0.9
Damba-Silwane (2)	7	0.9
Epoch	2.5	0.8

The figures are not intended to be authentic statements, but estimates based on most recent figures published by the respective companies. Due to government control, figures are unobtainable for most Rhodesian deposits. Deposits are listed in order of decreasing tonnage. * Number of main ore shoots.

and exhalative sediments. Individual ore shoots are generally less than 5 million tonnes with most less than 2 m tonnes (Table III), but are normally high grade (2–4% Ni). They generally show a consistent vertical zonation, with sulphides becoming more concentrated towards the base of the ore profile, and a relatively restricted range in bulk Ni/Cu ratio of 10 : 1 to 20 : 1 with most deposits between 10 and 15 : 1 [$Cu/(Cu + Ni) \leqslant 0.1$]. Well-defined volcanic-associated deposits appear to be restricted to Archaean greenstone belt terrains. Their distribution within the Yilgarn Block of Western Australia, the Rhodesian Craton of Southern Africa and the Superior Province of Canada is shown in Figs. 1–3, respectively.

Comparison with other deposits

The deposits can generally be distinguished from dunite-associated Ni–Cu deposits (see Table II), the other major type of Ni–Cu deposit of komatiite association (e.g. Binns

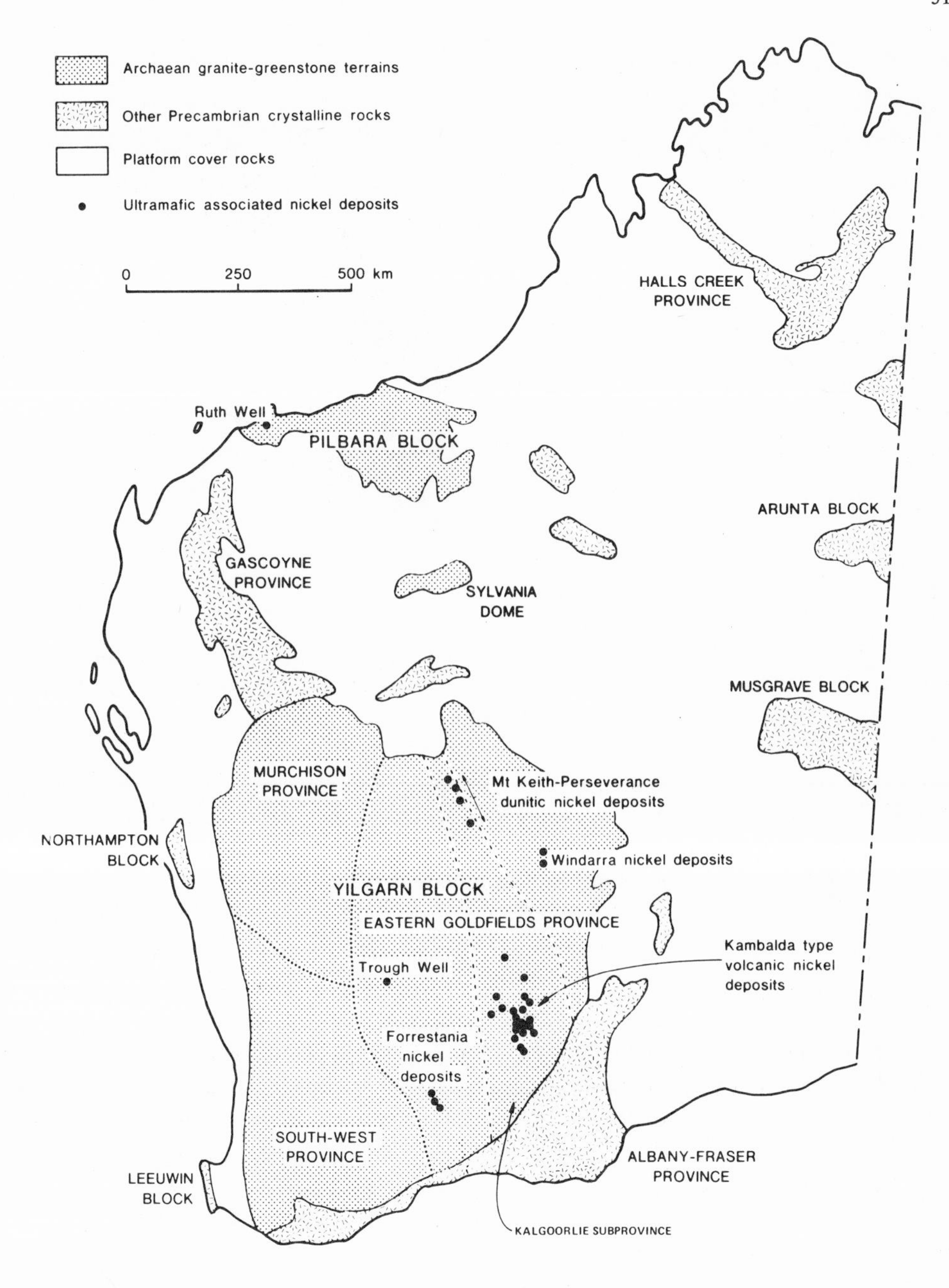

Fig. 1. Location of ultramafic-associated nickel deposits in Western Australia in relation to major tectonic units. The large majority of volcanic-associated Ni–Fe–Cu sulphide ores occur within the Kalgoorlie Subprovince of the Eastern Goldfields Province of the Yilgarn Block.

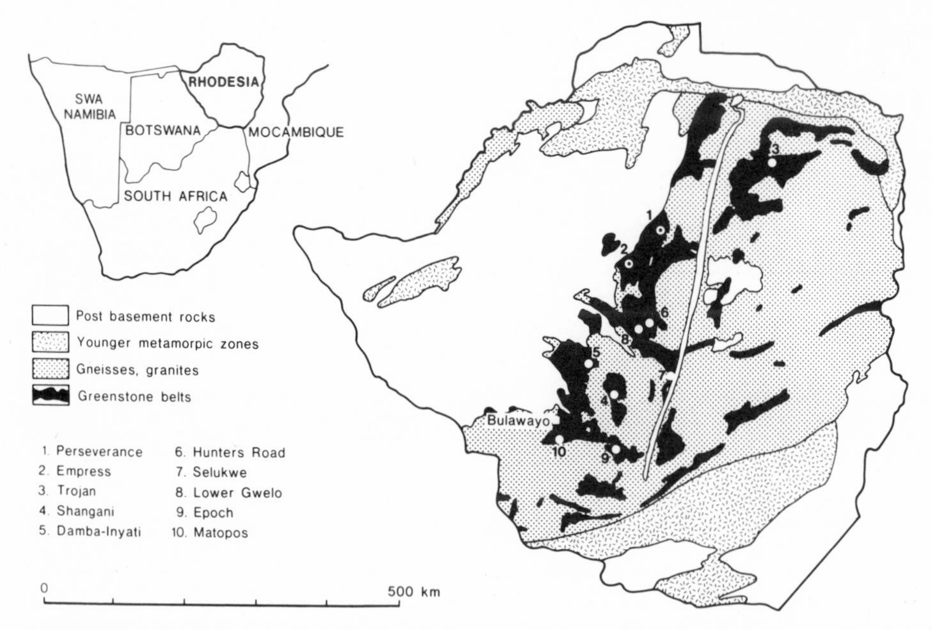

Fig. 2. Location of volcanic-associated Ni–Fe–Cu sulphide ores (localities 3–10) and other nickel sulphide ores in Rhodesia in relation to major tectonic units (after D.A.C. Williams, 1979). Localities 1 and 2 represent deposits in komatiitic (?) stratiform sills.

et al., 1977). The dunite-associated deposits occur within thicker, up to ca. 1000 m thick, lenses or pods of dunite that commonly show no asymmetric variation in mineralogy or geochemistry although marginal peridotites or pyroxenites may be present (e.g. Naldrett and Turner, 1977). Mineralization typically occurs as centrally disposed disseminated sulphides in low-grade metamorphic environments (e.g. Mt. Keith: Burt and Sheppy, 1975; Black Swan: Groves et al., 1974), although marginal more massive sulphides occur in higher-grade metamorphic environments such as Perseverance (e.g. Martin and Allchurch, 1975) and Forrestania (e.g. Binns et al., 1977). Individual ore lenses are normally much larger than the volcanic-associated ores, for example the Mt. Keith deposit contains ca. 290 million tonnes of 0.6% Ni and Perseverance (three main ore shoots) contains ca. 45 million tonnes of ca. 2% Ni (Gee et al., 1976). Bulk Ni/Cu ratios are also higher, generally in the range 25 : 1 to 60 : 1 [Cu/(Cu + Ni) $\leqslant$ 0.05]. The present configuration of the most important dunitic bodies is as discontinuous lenses or tectonic pods along major lineaments up to 200 km long, but whether the bodies were originally large dyke-like bodies (e.g. Binns et al., 1977) emplaced as crystal mushes, tectonic slices of sills or feeder chambers for overlying komatiitic flows (e.g. Naldrett and Turner, 1977) is still unclear. An important aspect in Western Australia is the marked spatial separation between the two well-defined groups of both volcanic-associated and dunite-associated deposits (Fig. 4).

Fig. 3. Location of ultramafic-associated nickel deposits in Canada in relation to major structural provinces of the Canadian Shield. (After Kilburn et al., 1969.)

The volcanic-associated deposits may be distinguished from most other deposits shown in Table II on the basis of bulk composition of the hosting body. All other deposits occur within ultramafic units that form part of layered bodies that are normally dominated by mafic rocks and are therefore less Mg-containing intrusions, whose MgO content (calculated volatile-free unless otherwise stated) is much less than the minimum 36% MgO weighted mean composition of hosts to volcanic-associated ores (e.g. Groves et al., 1979) or the mean composition of ca. 49–50% MgO for dunitic hosts (Binns et al., 1977). A major problem exists in classifying deposits that occur in high-grade metamorphic belts of

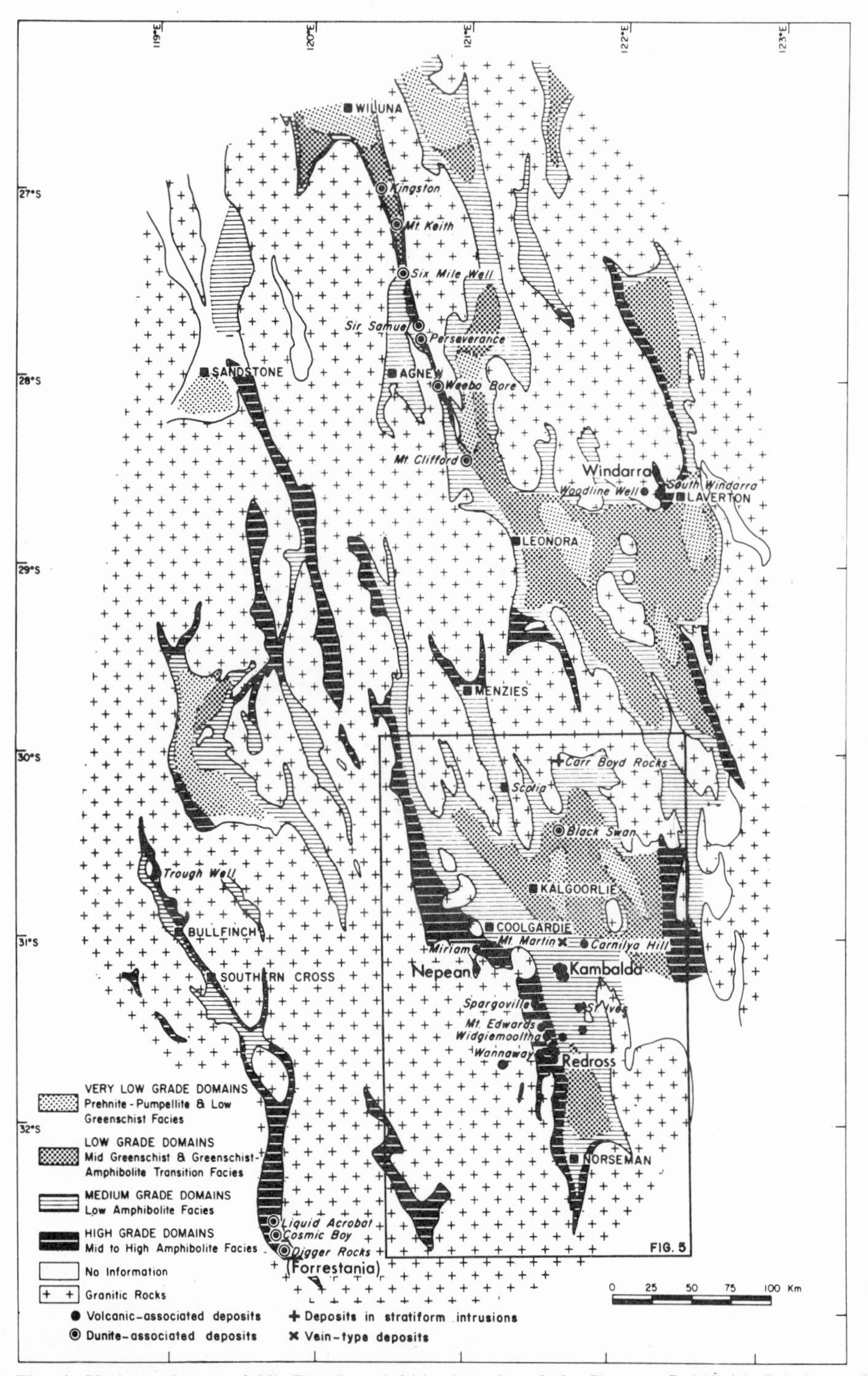

Fig. 4. Various classes of Ni–Fe–Cu sulphide deposits of the Eastern Goldfields Province of the Yilgarn Block, Western Australia, in relation to metamorphic grade. (After Barrett et al., 1977.) Note that additional deposits have been discovered in the Forrestania area.

high strain (e.g. Thompson belt deposits of Canada), where tectonic slicing of host rocks and separation of low-strength sulphide ores from host rocks is a common feature. In such cases, there is no substitute for detailed structural and metamorphic mapping in unravelling the ore associations, but the *bulk* Ni/Cu ratio of the ores may be useful as a guide to the composition of the hosting magma, if the ores were originally of magmatic origin (e.g. Rajamani and Naldrett, 1978) as bulk Ni/Cu ratio does not appear to be significantly affected by high-strain metamorphism (e.g. Barrett et al., 1977). On the basis of this ratio, the problematical Thompson belt deposits fall within Class A.1.b (Table II), if the Cu/(Cu + Ni) ratios given by Naldrett (1973) are considered. Naldrett and Cabri (1976) further emphasize the komatiitic affinity of the Thompson belt deposits, and they show strong similarities to the Perseverance (Agnew)–Mt. Keith and Forrestania belts of Western Australia.

Type examples

The type examples of volcanic-associated Ni–Fe–Cu deposits are the Kambalda deposits which alone constitute approximately 2% of the world's sulphide nickel reserves and to date have produced more sulphide nickel than all other Western Australian deposits combined. Thirteen large ore shoots occur around the Kambalda Dome (Figs. 5 and 6), if the various spatially related ore lenses within the Juan, Durkin and Fisher Shoots are considered as single shoots, and several more shoots are present to the south of Kambalda on an extension to the same structure (Fig. 5). A considerable literature exists on the Kambalda deposits including the specific studies of Woodall and Travis (1969), Ewers and Hudson (1972), Keele and Nickel (1974), Nickel et al. (1974), Ross (1974), Ross and Hopkins (1975), Bavinton and Keays (1978), Ostwald and Lusk (1978), Ross and Keays (1979), Marston and Kay (in press.), and the more general studies that embrace Kambalda by Hudson (1972), Lusk (1976), Keays and Davison (1976), Barrett et al. (1977), Binns et al. (1977), Groves et al. (1977), Donnelly et al. (1978), and Groves et al. (1979). Despite this abundance of information, Western Mining Corporation (Kambalda Nickel Operations) hold a wealth of unpublished data on individual ore shoots and traverses across the Kambalda Dome, which they hope to assemble into a major review article in the near future. As much of these data are not available to us, we prefer not to review Kambalda separately as the type example, but instead look at the general features shown by all deposits and discuss any exceptions to our generalizations. However, considerable emphasis is still placed on the Kambalda deposits.

Nature of review

This chapter is mainly a review of the better documented Western Australian deposits with information contributed from Canadian and Rhodesian deposits where available and appropriate. Our search for information on Canadian and Rhodesian deposits indi-

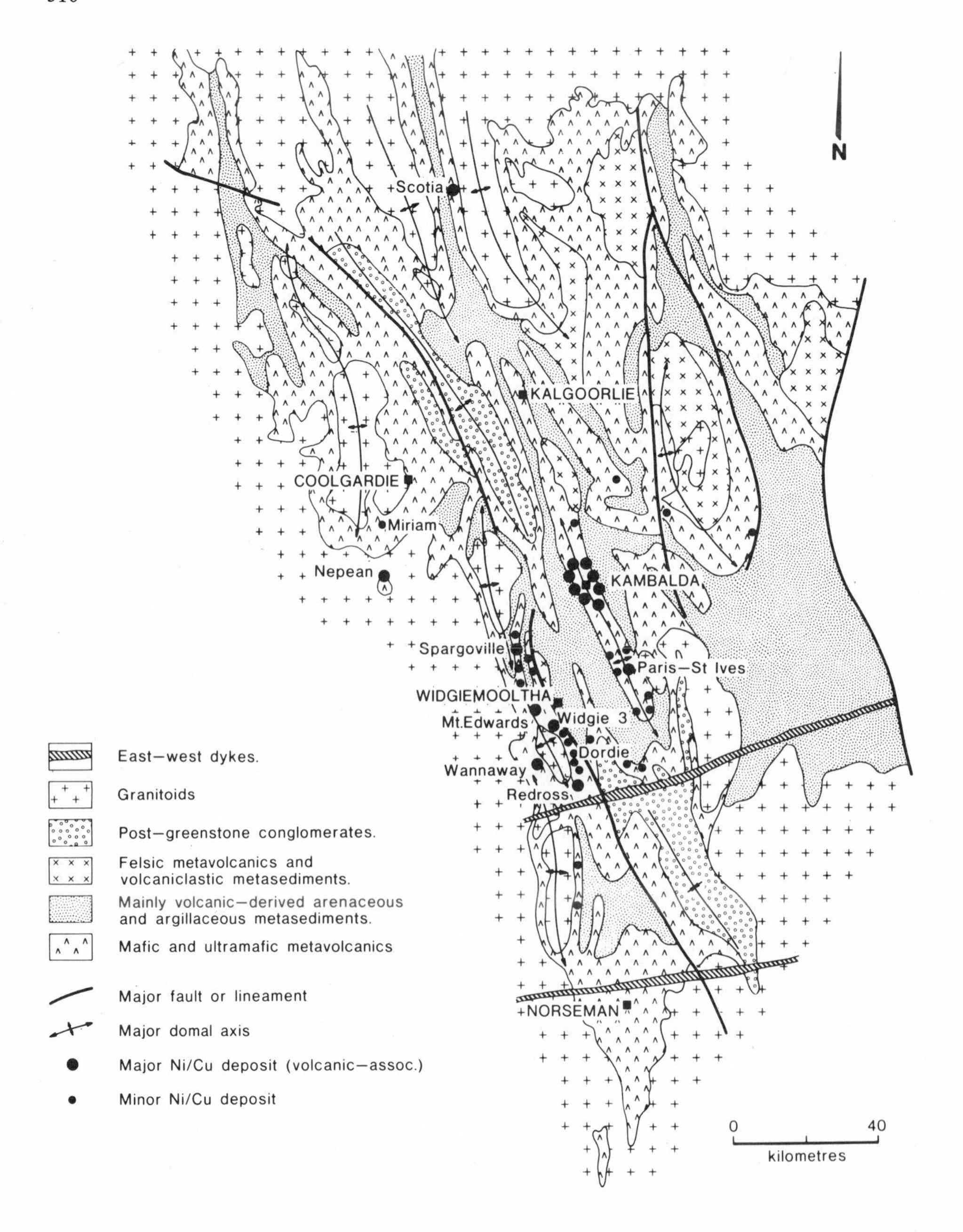

Fig. 5. Volcanic-associated Ni–Fe–Cu sulphide ores of the Kalgoorlie Subprovince in relation to major structures, granitoid domes and major "greenstone" lithological units.

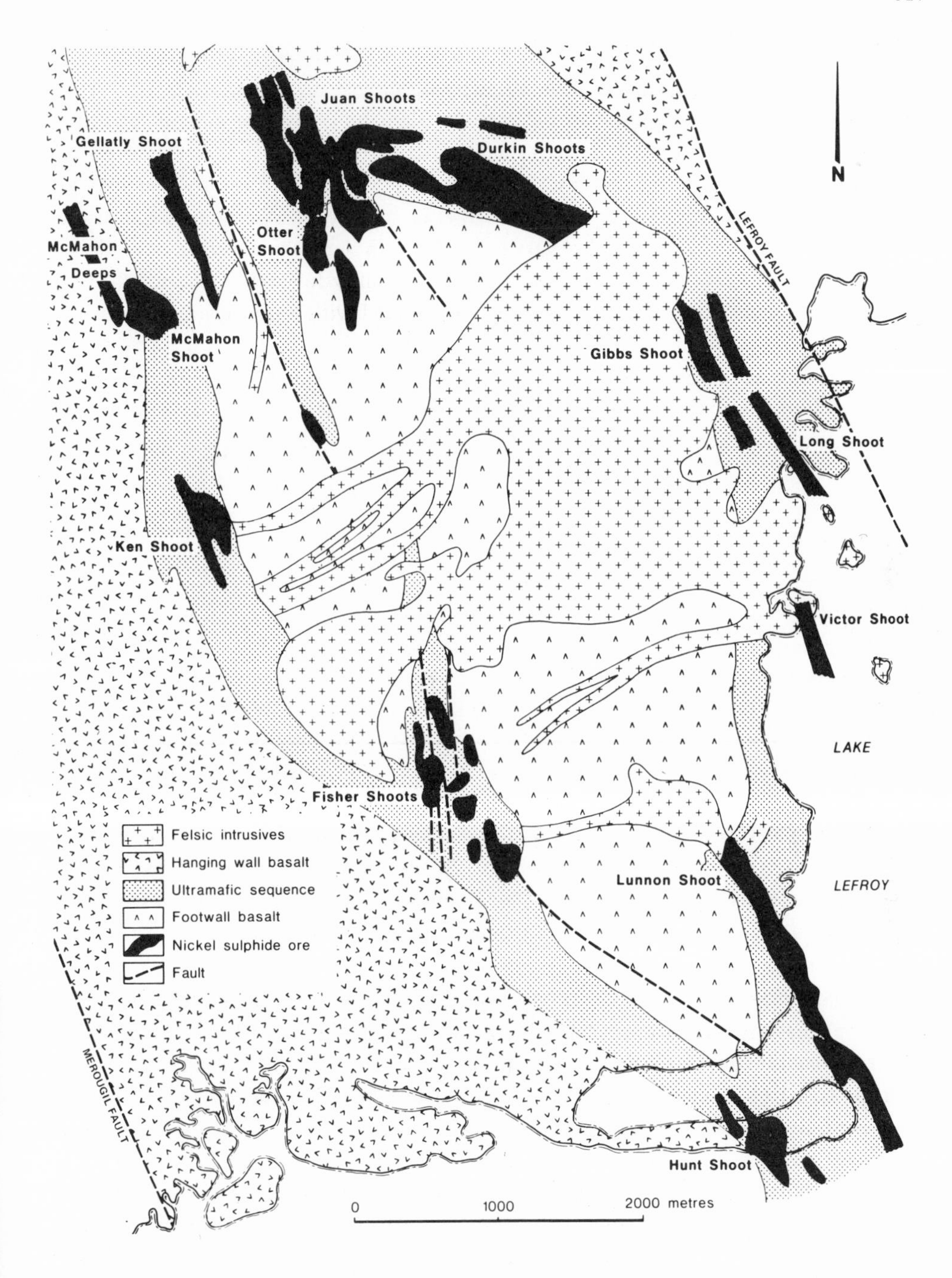

Fig. 6. Plan of the volcanic-associated Ni–Fe–Cu sulphide deposits and associated rock types of the Kambalda Dome. (After Ross and Hopkins, 1975.)

cates that several critical parameters are commonly not documented in published literature, so that confident classification of deposits cannot always be made. Important information that is commonly lacking includes data on structural and metamorphic setting, the geochemistry of host rocks and particularly detailed mineralogical and/or geochemical profiles through ore-bearing ultramafic bodies to indicate whether they are gravity-differentiated, peridotitic komatiite flows or subvolcanic sills. Many deposits that resemble the volcanic-associated Ni–Cu ores of Western Australia and that were classed by Naldrett (1973) as Class A1(ii) type, have significantly lower Ni/Cu ratios than the Western Australian examples, and probably equilibrated with less-magnesian melts than the latter.

Although all deposits have been modified to some extent by deformation and metamorphism, an attempt is made throughout to stress those aspects of the ores that are most significant to their initial formation.

TEMPORAL AND TECTONIC SETTING

Distribution and age

As with most groups of stratiform base-metal deposits (e.g. Sangster and Scott, 1976), the volcanic-associated Ni–Fe–Cu sulphide deposits are not uniformly distributed between and within Archaean greenstone terrains, but tend to occur in distinctive provinces. They are most widespread in the Eastern Goldfields Province of the Yilgarn Block, Western Australia, and are less well-developed in the Rhodesian Craton and the Canadian Shield (Figs. 1–3). Despite the extensive development of Ni–Cu ores in the Kola Peninsula (summarized in Smirnov, 1977), no deposits have been described that can be definitely ascribed to this class. There appear to be no descriptions of significant deposits of this class from granitoid–greenstone terrains of other cratons nor from Archaean high-grade gneiss belts.

Importantly, this type of deposit is extremely rare in two of the oldest (3.3–3.5 Gyr), well-preserved greenstone belts in the Barberton Mountain Land, South Africa and the Pilbara Block, Western Australia: one small, non-economic deposit occurs at Ruth Well in the latter (Tomich, 1974). Although geochronological data are equivocal, economic deposits appear to be restricted to greenstone belts that formed at post-3.0 Gyr and were clearly stabilized following granitoid emplacement at ca. 2.7–2.6 Gyr: supporting geochronology of the Eastern Goldfields Province is summarized by Archibald et al. (1978) and for the Superior Province by Goodwin (1977). This is particularly evident in the Rhodesian Craton where the deposits are virtually restricted to greenstone belts of younger age (D.A.C. Williams, 1979; Wilson, in prep.) despite the complex evolution of this craton involving several phases of greenstone belt formation (Wilson et al., 1978).

Tectonic setting

The importance of tectonic setting to the occurrence of volcanic-associated Ni–Cu deposits can best be demonstrated by reference to their distribution with respect to contrasting tectonic units in Western Australia (Fig. 1).

The greenstone belts of the eastern Pilbara Block, containing no significant Ni–Cu deposits, have a relatively simple lithological distribution on the broadest scale (e.g. Hickman, 1975; Lipple, 1975; Barley et al., 1979) with basal volcanic-dominated sequences (Warrawoona Group) overlain by thick clastic sedimentary sequences (Gorge Creek Group): in this regard they are similar to the Barberton Mountain Land (e.g. Anhaeusser, 1971). Within the Warrawoona Group, the basal units are commonly dominated by ultramafic–mafic associations of both komatiitic and tholeiitic affinity, whereas upper units consist of interfingering mafic–ultramafic and felsic volcanic (commonly calc-alkaline) associations. Barley et al. (1979) have studied metasedimentary units within both the mafic and felsic volcanic associations in the east Pilbara and conclude that they were deposited in extensive intracratonic shallow-water basins. The rapid change to widespread clastic sedimentation represents a change in tectonic style, possibly related to uprise of granitoids. The resultant pattern of synformal keels of greenstone belts between large, complex, roughly equidistant granitoid domes may result from diapiric uprise of the latter into a more dense greenstone layer of relatively constant thickness, but further study is required to verify this suggestion.

The younger, mineralized Eastern Goldfields Province contrasts markedly with this in consisting of essentially linear belts of greenstones with more complex stratigraphy (e.g. Gemuts and Theron, 1975) separated by "corridors" of granitoids. Archibald et al. (1978) consider that this pattern developed because the distribution of greenstone basins and resultant variable thickness of greenstone sequences was controlled by major pre-greenstone crustal fracture zones which later controlled diapiric rise of granitoids and dynamic metamorphism along linear zones. Their model implies a tectonically more unstable environment of initial greenstone formation than those envisaged for the Pilbara Block.

Within the Eastern Goldfields Province, all but the Windarra–South Windarra deposits and a few minor occurrences of volcanic-associated deposits are confined to a zone within the Norseman–Wiluna greenstone belt (e.g. I.R. Williams, 1974; see Figs. 1 and 4). This zone is commonly bounded by high-grade metamorphic belts of high strain (dynamic-style belts of Binns et al., 1976) and is best interpreted as an important rift zone during greenstone development (Archibald et al., 1978), although this is almost certainly an oversimplification of a complex situation. I.R. Williams (1974) pointed out that the mineralized area corresponded to a zone (the Kalgoorlie Subprovince) in which oxide-type BIF, common in flanking greenstone terrains, was absent, whereas sulphidic metasediments are abundant (e.g. Donnelly et al., 1978; Groves et al., 1978). Even in the Windarra area, sulphidic metasediments are abundant in the ore environment (e.g. Seccombe et al., 1977). The occurrence of plane-laminated shales and greywacke sequences and the general lack of shallow-water metasediments, such as described by Barley et al. (1979), within volcanic-dominated sequences suggests deeper-water conditions than for the Pilbara basin(s), but insufficient research has been carried out on the sedimentary sequences within and outside the mineralized zone in the Eastern Goldfields Province to allow meaningful comparison. The occurrence of sulphidic sediments alone *cannot* be used to indicate very deep-water conditions, and even a water-depth contrast with the flanking zones of oxide-type BIF cannot be substantiated, as nowhere has a lateral transition from oxide-type to sulphide-type iron formation been documented to support the facies concept used by Goodwin (1973) in Canada.

It appears from the Western Australian setting that the occurrence of volcanic-associated Ni–Fe–Cu sulphide ores is favoured within younger greenstone belts in more active, fracture-controlled greenstone basins in which interflow sulphidic sediments were formed as an integral part of volcanic activity, but oxide-facies iron formations were rare or absent. Mineralized greenstone basins of the Superior Province, Canada, are also interpreted to form in an overall rifting regime, although Goodwin (1973, 1977) suggested that ovoid basins, contrasting in style with those envisaged in the Eastern Goldfields Province, formed within this broader tectonic environment. A summary of the occurrence of Ni–Cu sulphide deposits in the Timmins area by Pyke and Middleton (1971) suggests the importance of sulphidic sediments in the ore environment, as in the Eastern Goldfields Province. Wilson (in prep.) concluded that the Younger Greenstones of Rhodesia, which host the Ni–Cu mineralization, originated in a rifting environment within a pre-existing basement of gneisses and older greenstone belts, and several authors (e.g. Anderson et al., in press; M.J. Viljoen et al., 1976; D.A.C. Williams, 1979) indicate the presence of sulphidic sediments in the mineralized environment.

Structural, metamorphic and stratigraphic setting

The structural and metamorphic setting of the Western Australian volcanic-associated deposits has been described by Barrett et al. (1977).

The deposits are generally clustered around the periphery of granitoid-cored domes or anticlines (Figs. 4, 5 and 6), some of which are demonstrably syntectonic diapirs (Archibald et al., 1978). The Kambalda Dome may represent the subvolcanic expression of felsic volcanics rather than a post-greenstone intrusive granitoid (Archibald et al., 1978). Some evidence exists, such as variation in thickness of the ultramafic sequence across the dome, that suggests its location on a pre-existing volcanic structure, but no such evidence for other domes has been recorded. The domes typically have a north–south elongation, normally with steeply dipping east and west flanks and more shallowly plunging north and south extremities. With the exception of Kambalda, where the granitoid rock is essentially massive, the greenstones and granitoids have compatible linear tectonite fabrics.

Within this regional framework, the ores have a marked structural control, being essentially tabular bodies that have their greatest elongation subparallel to penetrative linear fabrics in enclosing rocks and/or the plunge of regional and parasitic folds. In some deposits (e.g. Kambalda), the tabular ore bodies occur as a series of complex troughs bounded by faults (e.g. Ross and Hopkins, 1975 – Fig. 7). In other cases (e.g. Redross), the immediate controls of ore shoots are zones of ductile or brittle shear, but the parallelism with regional linear fabrics is maintained.

Most ore bodies are associated with basal ultramafic units within trough-like embayments in the footwall rocks (e.g. Kambalda – Ross and Hopkins, 1975; Scotia – Christie, 1975; Widgiemooltha occurrences – Dalgarno, 1975; Windarra – Seccombe et al., 1977), although their distribution with respect to embayments is more complex than this at

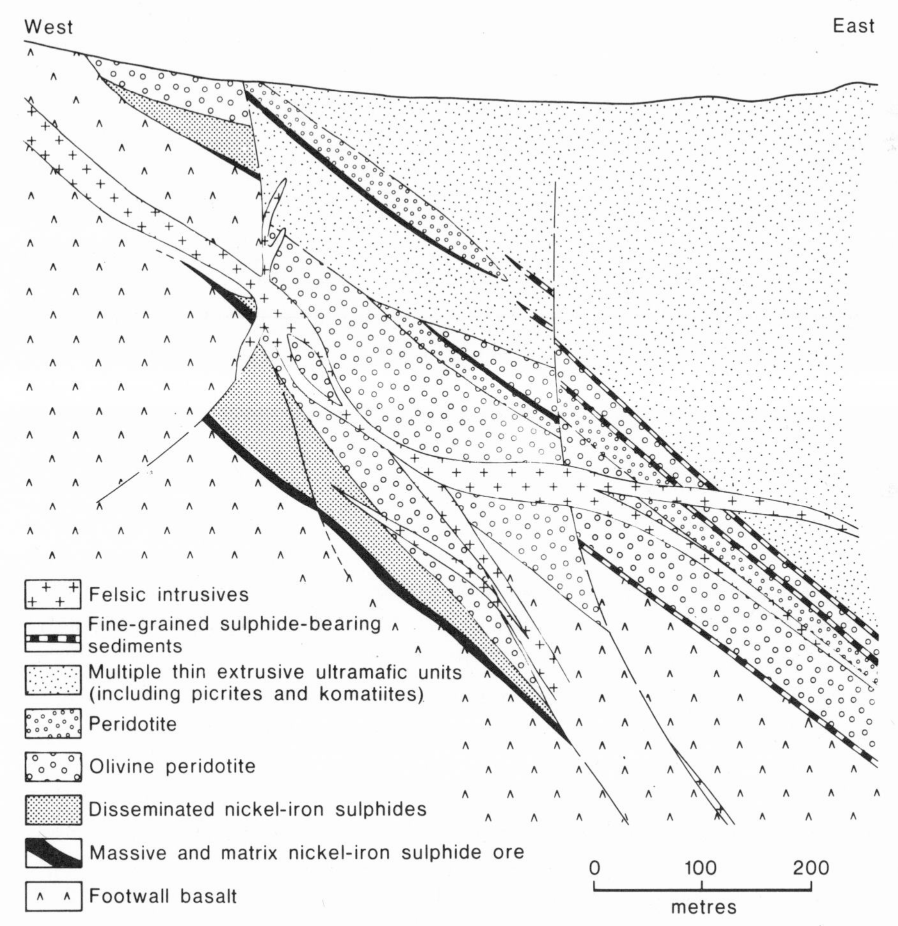

Fig. 7. Generalised east–west section of the ore environment at Lunnon Shoot, Kambalda. (After Ross and Hopkins, 1975.)

Redross and Nepean (Barrett et al., 1976, 1977). The alignment of these embayments parallel to regional structure, the presence of reverse faults in some cases (e.g. Kambalda), and bounding or associated shear zones in others, all point to the importance of deformation in controlling their present configuration. However, exhaustive studies around the Kambalda Dome by Western Mining Corporation have supported the evidence presented by Ross and Hopkins (1975) that these structural embayments represent zones in which there are discontinuities in stratigraphy, the most obvious being the common termination of sulphidic metasediments within the zone bounded by reverse faults, the basal ultramafic unit has its thickest development, and in which there are "stacked" ores with hanging-wall ores occurring stratigraphically above the basal or contact ores (Fig. 7: see also Bavinton and Keays, 1978, their fig. 2). This evidence suggests that the embayments existed at the time of formation of the ultramafic pile and subsequently represented important heterogeneities which acted as the loci for high strain during deformation.

Binns et al. (1976) produced a metamorphic framework of the Eastern Goldfields Province (Fig. 4) in which the occurrence of Ni–Cu deposits was considered by Barrett et al. (1977). Although the metamorphic grade varies from prehnite–pumpellyite facies to high-amphibolite facies, economic volcanic-associated Ni–Cu deposits are confined to amphibolite-facies domains (Fig. 4). Most individual deposits occur in zones of high strain (dynamic-style domains of Binns et al., 1976) of mid- to high-grade amphibolite facies, but the notable exception is the Kambalda group which occurs in a lowermost amphibolite grade, low-strain (static-style of Binns et al., 1976) environment where structures and textures of volcanic rocks are commonly well preserved.

Barrett et al. (1977) and Binns et al. (1977) argued that there was a direct relationship between metamorphism and the occurrence of at least some Ni–Cu *ores*, particularly of the dunite-associated type, but also considered that the metamorphic pattern reflected a more fundamental geotectonic and/or stratigraphic control of mineralization at the magmatic stage. For example, there is evidence that the high-strain belts may be coincident with important crustal fractures that controlled magma eruption (Archibald et al., 1978), and that the higher-grade, low-strain environments may be broadly coincident with lower greenstone sequences containing the most extensive sequences of ultramafic rocks (e.g. Binns et al., 1976). Stratigraphic studies of greenstone belts between Kambalda and Norseman (e.g. Gemuts and Theron, 1975), even when constrained by lack of continuity of outcrop, correlation around and across granitoid bodies and structural complication (e.g. Archibald et al., 1978), still suggest the strong probability that the volcanic-associated Ni–Fe–Cu deposits in this area are *broadly* stratigraphically controlled, although individual deposits may occur at different horizons within a regionally recognizable ultramafic-rich sequence (Sequence 3 of Table IV) that is low in the stratigraphic sequence. This correlation is difficult to extent to the more northern deposits (e.g. Windarra area).

When assessing the relative importance of stratigraphic, structural and metamorphic

TABLE IV

Generalized stratigraphy of the Norseman–Kalgoorlie area showing proposed stratigraphic position of volcanic-associated nickel deposits according to Gemuts and Theron (1975)

	TOP
Sequence 8 –	Polymictic conglomerate, cross bedded greywacke (e.g. Kurrawang Conglomerate).
Sequence 7 –	Acid tuffaceous rocks and acid volcanic breccia. Some acid extrusive rocks (equates with Black Flag metasediments).
Sequence 6 –	Komatiitic basalt, ultramafic rocks, minor chert, black slate and tholeiitic basalt (e.g. Yilmia mafic–ultramafic rocks).
Sequence 5 –	Conglomerate, arkosic greywacke and argillite with minor tholeiite basalt (e.g. Merougil Creek Conglomerate).
Sequence 4 –	Grades into sequence 5. Greywacke, chert and slate. Acid extrusive rocks and feldspar-porphyry intrusives. Minor tholeiitic basalt (e.g. Causeway Beds).
Sequence 3 –	Black chert. Tholeiitic basalt and komatiitic ultramafic rocks, mainly peridotitic komatiites with minor komatiitic basalts. Intercalated sulphide-bearing black shales and cherts. VOLCANIC-ASSOCIATED NICKEL DEPOSITS.
Sequence 1 – and 2	Sequence 1 (Penneshaw Beds) consists of minor tholeiitic basalt and lithic tuffs, greywacke and shale, overlain by Sequence 2 (Noganyer Group) of banded iron formation, conglomerate, sandstone and shale.
	BOTTOM

setting, it should be instructive to compare the Western Australian deposits with similar ones in Rhodesia and Canada. This is however not straightforward, because regional studies, such as those of Binns et al. (1976), have not generally been reported for these terrains, and the structural and metamorphic aspects of individual ore deposits have commonly received much less attention than petrological and geochemical aspects.

Available data on Canadian deposits suggest that some, at least, occur around the periphery of granitoid-cored domes and that in some areas there is a broad stratigraphic control of deposits (Pyke and Middleton, 1971), and embayments may be important in localizing ore: A.H. Green (1978) suggests that such features are palaeosurface features partly modified by folding and faulting (cf. Ross and Hopkins, 1975). Most authors dealing with the Abitibi Belt in general, and Ni–Cu deposits in particular, suggest greenschist- or subgreenschist-facies conditions of metamorphism. Jolly (1978) presents a metamorphic map of the Abitibi Belt (his fig. 4), showing considerable variation in metamorphic grade. From this map, most Ni–Cu deposits appear to be in greenschist or greenschist/amphibolite transition-facies environments.

Data from Rhodesia (e.g. M.J. Viljoen et al., 1976; D.A.C. Williams, 1979, Anderson et al., in press; M.J. Viljoen and Bernasconi, in press; Moubray et al., in press) support the occurrence of some deposits (e.g. Shangani, Inyati, Damba) around granitoid domes, the importance of embayments or troughs, and the structural control of the deposits parallel to linear fabrics in host rocks (e.g. Shangani) or in shear zones (e.g. Perseverance). Most ore environments are reported as greenschist facies without supporting evidence, with the exception of Perseverance where mid-amphibolite facies metamorphism is suggested (e.g. Anderson et al., in press): it is however not completely clear whether Perseverance is a volcanic-associated deposit or not. Wilson (in prep.), in an excellent review of the Rhodesian Craton, points out the temporal and stratigraphic control of the nickel deposits with the Shangani, Inyati and Hunters Road deposits occurring within ultramafic sequences that are time-equivalents of the Reliance Formation, and the Perseverance deposit occurring within a bimodal sequence, comprising both ultramafic and felsic lithologies, of equivalent stratigraphic position. D.A.C. Williams (1979) also stresses the strong stratigraphic control on the location of ore deposits.

Thus, apart from metamorphic grade, which may in part be due to lack of critical study, the volcanic-associated Ni–Fe–Cu deposits of the three cratons have rather similar settings. The occurrence of deposits in groups associated with a particular ultramafic-dominated sequence, or part of a sequence, appears to be particularly evident as a common feature.

LITHOLOGICAL ASSOCIATIONS

Introduction

In all mineralized environments, the host and country rocks have been metamorphosed and deformed. However, in some environments (e.g. Kambalda) there is excellent preservation of original textures and structures of the volcanic rocks and even in high-grade, high-strain environments deformation has commonly been heterogeneous with zones of good preservation of original features. Comparison of volatile-free compositions of metavolcanic rocks from varied metamorphic environments and studies using Pearce chemical-fractionation diagrams (e.g. Barrett et al., 1976) suggest that, although there has been addition of H_2O and CO_2 and loss of alkalies and CaO from some lithologies, most rocks may be classified into related groups on the basis of their geochemistry. Hence, the pre-metamorphic lithologies can generally be established for most deposits.

For the above reasons, igneous terminology is used in much of the following sections and the textures and structures of lithologies showing low internal strain are selectively described. The Kambalda lithologies are used as examples wherever possible, as these are best constrained by surface mapping, deep diamond drilling and careful description. However, it should always be understood that metamorphic mineralogy and fabrics vary considerably between and even within deposits. As an example, the reader is referred to the contrasting mineralogies and fabrics of geochemically similar tholeiitic metabasalts (amphibolites) and peridotitic komatiites from Nepean (Barrett et al., 1976) and Kambalda (Ross and Hopkins, 1975).

When allowance is made for structural disruption and repetition of sequences by folding and strike-slip faulting, generalized stratigraphic sections can be erected for several of the better documented deposits in Western Australia (Fig. 8): these are remarkably similar in a number of areas. There is an important association of tholeiitic metabasalts, komatiitic ultramafics, and metabasalts and sulphide-bearing cherty or shaley metasediments.

In most deposits, the main mineralized ultramafic unit represents the first of a thick sequence (240–600+ m at Kambalda) of progressively thinner and less-magnesian spinifex-textured komatiitic units: the second and third units may also be mineralized. A similar sequence is noted in Rhodesia and in Canada. In Western Australia, the footwall series is normally a thick, monotonous sequence (2000+ m at Kambalda) of tholeiitic metabasalts terminated by, and commonly containing, thin interflow sulphidic metasediment units. A major exception is Windarra (Fig. 8), where the mineralized ultramafic unit overlies an interflow metasediment above a barren ultramafic unit that is in turn underlain by an extensive, thick banded iron formation representing the footwall sequence to the South Windarra deposit (e.g. Roberts, 1975). Footwall rocks in Rhodesia and the

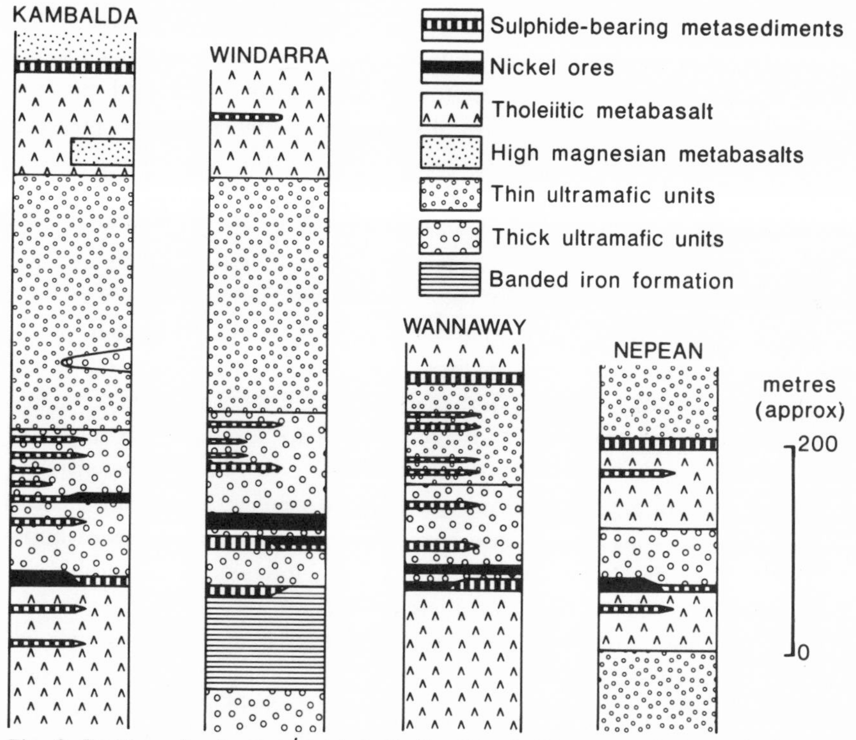

Fig. 8. Position of volcanic-associated Ni–Fe–Cu sulphide ores and ultramafic host rocks in the stratigraphic sequence at Kambalda, Windarra, Wannaway and Nepean. (After Groves et al., 1979.)

Abitibi Belt, Canada, differ markedly from those in Western Australia. They are normally felsic volcanoslastic sequences containing agglomerates, tuffs and volcaniclastic metasediments, and the mineralized ultramafic unit commonly appears to represent the start of a new cycle of ultramafic/mafic volcanism (e.g. D.A.C. Williams, 1979). Hanging-wall sequences (280–380+ m at Kambalda) are dominated by metabasalts or metadolerites, apparently of mixed tholeiitic and komatiitic parentage.

The various lithologies within the mineralized sequences are described below with particular emphasis on data from Western Australia. The country rocks and later intrusives are described first and the ultramafic sequence is treated immediately prior to the discussion of the ore environment.

Metabasalt sequences

Footwall metabasalts. Footwall metabasalts or amphibolites from most Western Australian volcanic-associated nickel deposits are compositionally and geochemically similar. The best-documented and best-preserved sequence occurs at Kambalda, where the footwall metabasalt succession has been drilled to a depth of almost 1700 m, and this is described below as a typical example: compositional data for other metabasalts (or amphibolites) are compared with the Kambalda metabasalts. The Kambalda footwall metabasalt comprises a compositionally monotonous sequence of fine-grained massive metabasalts with thin, discontinuous horizons of pillow metabasalt, metabasalt breccia, rare amygdaloidal metabasalt, and thin metasedimentary intercalations.

Massive metabasalts are fine- to medium-grained, dark green, and some retain their igneous texture with preservation of thin tabular crystals of saussuritized plagioclase. Interstitial glass and/or pyroxene has, however, recrystallized to fibrous and felted masses of green amphibole, with variable amounts of chlorite, epidote, carbonate and biotite.

Pillowed metabasalts have a crudely ellipsoidal outline accentuated by a dark green selvedge of chlorite or uralite after glass. Pillows commonly vary from 20 to 60 cm in diameter, and they are readily recognized both in diamond drill core and in underground workings (Fig. 9). Individual pillows vary in colour from dark greenish black interiors through a paler green to yellow green immediately adjoining the black pillow margin. Copper and Zn are depleted in epidotized pale green pillow margins, and Zn is enriched in pillow skins. Cores of some pillows have been altered to irregular patches of buff coloured carbonate. Veins of white carbonate cut the pillow interiors, but either thin or terminate when they reach the pillow skin. Interpillow spaces are filled with mixtures of fragmental pillow margins, metasediment, sulphide and carbonate. Some interpillow sulphides, which form rounded irregular masses or droplets in the tricuspate interstices (Plate I, C) have similar compositions to overlying Ni–Fe–Cu sulphide ores, suggesting that ore formation occurred while interpillow spaces were still void or before interpillow sediment had lithified.

Fig. 9. Pillowed tholeiitic metabasalts with interpillow sulphides in the footwall sequence of Lunnon Shoot, Kambalda.

Hanging-wall metabasalts. Ross and Hopkins (1975) recognize two formations of metabasalt that overlie the ultramafic sequence at Kambalda. The lower metabasalts consist of an alternating sequence (80–100 m thick) of variolitic and massive metabasalts, with development of ocelli of the former in the latter along lithological interfaces. Variolitic metabasalts are enriched in silica, alumina and alkalies relative to massive metabasalts. The upper hanging-wall metabasalts are massive in character with compositions ranging from normal tholeiite into magnesia-rich varieties (Table V).

Available data from elsewhere suggest that sequences are dominated by tholeiitic metabasalts similar to footwall metabasalts, but that magnesian metabasalts may also be present. For example, at Scotia some magnesian metabasalts with distinctive skeletal textures, that appear to typify the komatiite suite, are intercalated with komatiitic ultramafics (Christie, 1975).

At Windarra, strongly foliated and lineated amphibolites of tholeiitic parentage dominate the hanging-wall sequence, and magnesian amphibolite appears to be geographically restricted to an area north of Mt. Windarra rather than occurring towards the base of the sequence as at Kambalda and Scotia. Ferruginous biotite–garnet amphibolite layers of uncertain parentage also occur.

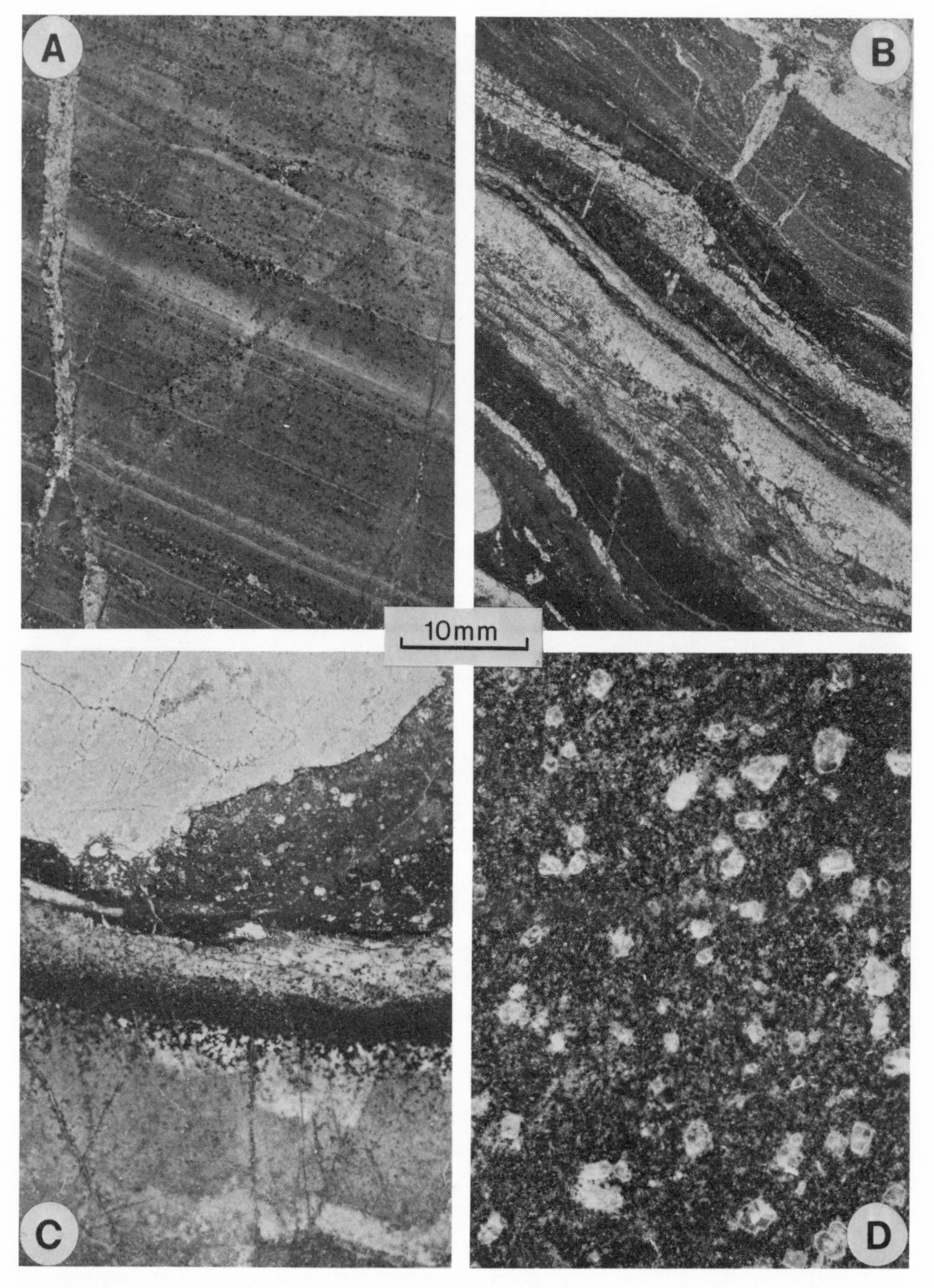

PLATE I

Rock types associated with Ni–Fe–Cu sulphide ores at Kambalda.
A. Finely laminated cherty metasediment.
B. Sulphide-bearing black shaley metasediment.
C. Irregular patches and subspherical droplets of sulphide in interpillow spaces in tholeiitic metabasalt.
D. Porphyritic felsic intrusive.

Geochemistry. Analyses of both the footwall and hanging-wall metabasalts at Kambalda have been given by Hallberg (1970), Roddick (1974), and Ross and Hopkins (1975). Average analyses are provided in Table V.

Footwall metabasalts are normal tholeiites and, as shown in Table V, are compositionally indistinguishable from the average mafic metavolcanic rock from the Kalgoorlie Subprovince (Hallberg and Glikson, in prep.) and the average Eastern Goldfields metabasalt (Hallberg and D.A.C. Williams, 1972). Footwall metabasalts or amphibolites from the Redross, Mt. Edwards, Nepean and Miriam nickel occurrences have similar average compositions, indicating that they also are normal tholeiites. Metabasalts from Mt. Monger and some metabasalts from Scotia (Table V) have compositions that are closer to the field of iron-rich tholeiites (Naldrett and Turner, 1977; Arndt et al., 1977), but these are rare relative to normal tholeiites in ore environments.

Unlike the footwall metabasalts, the Kambalda hanging-wall metabasalts do not represent a monotonous, nonfractionated magma. They range from variolitic metabasalts of intermediate composition through normal tholeiites to magnesia-rich metabasalts (Table V) that may have komatiitic affinities. Similar compositional variability has been noted in hanging-wall amphibolite sequences at Windarra, and further study may prove it to be a characteristic of the waning of komatiitic volcanism.

Variation diagrams have been used by many authors in order to separate spatially associated tholeiitic and komatiitic magma types which may show considerable overlap in MgO content (e.g. Sun and Nesbitt, 1977). Recent studies by Arndt et al. (1977) and Naldrett and Turner (1977) have suggested that a distinction may be made between "tholeiitic" and komatiitic lavas on the basis of $CaO{-}MgO{-}Al_2O_3$ and Al_2O_3 vs. FeO^* $(FeO^* + MgO)$ diagrams. Plots of metabasalts from Western Australian nickel ore environments (Fig. 10) do not, however, coincide with the tholeiite field of Arndt et al. (1977) or Naldrett and Turner (1977), except for the iron-rich tholeiite from Mt. Monger. Most metabasalts associated with nickel deposits are indistinguishable from the primitive tholeiite of Hallberg and D.A.C. Williams (1972) and overlap with the field of komatiitic basalt (Figs. 10 and 11). Further work is required before the true relationships between these two important rock suites can be clearly defined in terms of either texture, composition or genesis. Variation diagrams, such as those cited above, delineate the total field of mafic and ultramafic rocks associated with the nickel deposits, but arbitrary subdivisions do not appear to have significance, except perhaps on a very restricted and local scale.

Metasedimentary rocks

Interflow metasediments warrant description and discussion as they have been considered to be the source of S for nickel ores on a regional scale (Prider, 1970), on a local scale (Seccombe et al., 1977), and as facies variants of volcanogenic nickel ores (Lusk, 1976). They are not simply sulphide facies of iron formations (cf. Goodwin, 1973). They

TABLE V

Analyses of metabasalts from Western Australian volcanic-associated nickel deposits

	Kambalda			Redross 4	Mt Edwards 5	Nepean 6
	1	2	3			
SiO_2	51.5	53.7	53.0	51.6	51.8	50.9
TiO_2	0.7	0.7	0.6	0.7	0.7	0.7
Al_2O_3	14.8	15.8	12.8	14.1	14.5	15.6
Fe_2O_3			1.7	1.4	2.2	1.3
FeO	10.6 *	8.8 *	8.7	9.0	8.0	8.5
MnO	0.2	0.1	0.2	0.2	0.2	0.2
MgO	8.1	7.7	9.8	8.0	7.9	7.7
CaO	10.8	8.0	9.7	10.8	12.0	11.0
Na_2O	2.2	2.2	1.6	2.4	2.3	2.7
K_2O	0.4	0.7	0.2	0.3	0.3	0.2
P_2O_5		0.1	0.1	–	0.1	
$H_2O(t)$	0.4	3.2	2.3	1.5	1.6	1.0
CO_2		2.8	0.3	0.4	0.2	
Cr		218	195	421	338	
Cu	133	76	98	82	79	
Ni	175			166	75	
Rb	13	11	7		22	
Sr	141	228	131	93	132	
V				252	235	
Y	19				23	
Zn	98				89	
Zr	61	65	67		40	

* Total iron as FeO; ** Loss on ignition.

1 = Kambalda footwall basalt: Average of 16 major-element analyses of massive and pillowed tholeiitic metabasalt (Hallberg, 1970; Roddick, 1974; Ross and Hopkins, 1975). Minor elements are average of 195 metabasalts (Hudson, unpublished).
2 = Kambalda variolitic hanging-wall metabasalt: Average of 15 major-element analyses (Roddick, 1974; Ross and Hopkins, 1975). Minor elements are average of 3 analyses (Ross and Hopkins, 1975).
3 = Kambalda upper hanging-wall metabasalt: Average of 4 major-element anlayses (Roddick, 1974; Ross and Hopkins, 1975). Minor elements are average of 2 analyses (Ross and Hopkins, 1975).
4 = Redross footwall amphibolite: Average of 4 analyses (UWA, unpublished analyses).
5 = Mount Edwards tholeiitic amphibolites: Average of 6 analyses (Hallberg, unpublished data).
6 = Nepean amphibolite: Average of 3 analyses (Barrett et al., 1976).

are much less continuous, more aluminous and Fe-poor than most described iron formations and no lateral transitions to carbonate- or oxide-facies have been observed. They exhibit similarities to "exhalites" associated with volcanogenic massive Cu–Zn sulphide deposits in Canada (Sangster, 1978).

Occurrence. In Western Australia, thin interflow metasediment units, commonly sulphide-

iriam	Scotia	Mt Monger	Windarra			
	8	9	10	11	12	13
50.9	50.4	52.1	50.7	49.1	51.3	51.3
0.6	1.0	1.9	1.1	1.7	0.9	1.0
14.8	14.5	14.8	10.3	15.2	14.7	14.8
2.3		2.9	1.4	1.9	1.6	
8.5	10.8 *	9.6	11.4	12.4	9.0	10.5 *
0.2	0.2	0.3	0.1	0.1	0.2	
7.7	6.4	4.6	12.5	5.7	6.9	6.7
11.7	12.8	9.9	9.6	9.3	10.6	10.8
1.9	2.0	3.0	2.5	1.6	2.6	2.7
0.1	0.2	0.3	0.6	0.5	0.2	0.2
0.1	0.1	0.2	0.1	0.2	0.1	0.1
1.4	2.6 **	1.0	0.2 **	0.5 **	1.4	1.2
0.1					0.1	
12	242	370	905	155	398	367
70	134	31			105	107
66	160	151			163	170
3	11	3			10	9
04	101	126			121	105
84	253		272	345	301	320
21	10	41	23	36	20	
94	103	125			98	
43	67	116	71	117	61	61

7 = Miriam tholeiitic metabasalts: Average of 12 major-element and 24 minor-element analyses of pillowed and massive tholeiitic metabasalts (Hallberg, unpublished).
8 = Scotia amphibolites: Average of 8 major-element analyses (Hallberg, unpublished). Minor elements from 1 metabasalt (Hallberg, unpublished).
9 = Mt Monger metabasalt: Average of 4 analyses (D.A.C. Williams, 1972).
10 = Windarra olivine metabasalt: Average of 3 analyses (Roberts, 1975).
11 = Windarra quartz tholeiitic metabasalt (Roberts, 1975).
12 = Mafic volcanic rocks from the Kalgoorlie Subprovince: Average of 195 analyses of lower to middle greenschist facies mafic volcanic rocks (Hallberg and Glickson, in prep.).
13 = Eastern Goldfields average tholeiitic metabasalts: Average of 337 analyses (Hallberg and D.A.C. Williams, 1972).

bearing, occur within footwall metabasalts, the mineralized ultramafic sequence and overlying metabasalts, but are commonly best developed in the lower part of the ultramafic sequence (e.g. Kambalda, Windarra; Fig. 8). They vary in thickness from a few centimetres to ca. 10 m thick, with most units being between 0.5 and 3 m thick. They have variable lateral continuity, with some units being difficult to correlate over distances of a few hundred metres whereas others, particularly those on contacts representing major

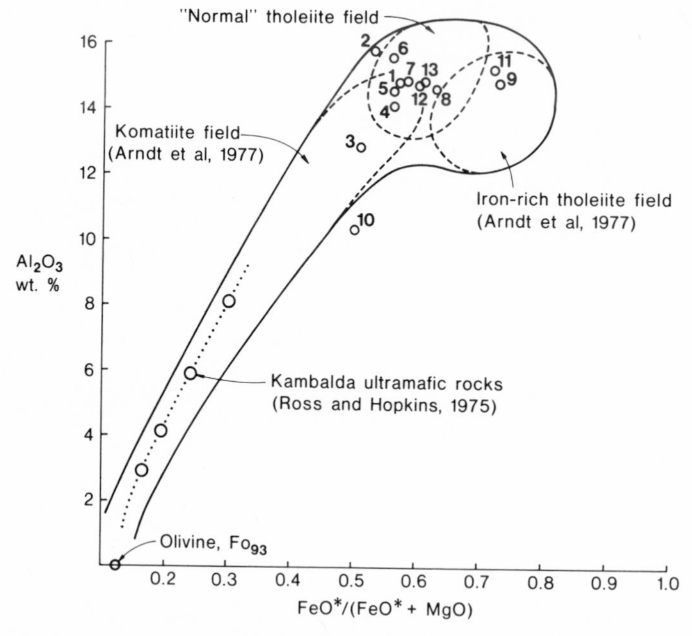

Fig. 10. Al_2O_3 vs. FeO*/(FeO* + MgO) plot of compositions of tholeiitic metabasalts, komatiitic metabasalts and komatiitic peridotites associated with Western Australian volcanic-associated Ni–Fe–Cu sulphide ores. Localities for numbered metabasalts are given in Table V.

changes in volcanism (e.g. basal contact of ultramafic sequence), are traceable over tens of kilometres along strike (e.g. Bavinton and Keays, 1978).

In several mineralized environments (e.g. Kambalda), the metasediment units are lateral to, and at the same stratigraphic horizon as, the nickel ores but are not present in the mineralized section (Fig. 7). There is some evidence (O.A. Bavinton, pers. comm., 1978) that the basal metasediment post-dates the mineralized ultramafic unit at Kambalda, but in other deposits there are commonly laterally discontinuous and thinner units of metasediment beneath the ore zone (e.g. Windarra).

In Rhodesia and Canada, poorly described sulphide-bearing "cherts" or "shales" commonly terminate the underlying felsic metasediment and metavolcanic sequence, and there are similar horizons within the ultramafic sequence.

Mesoscopic and microscopic features. The interflow metasediments from Western Australia may be roughly divided into three gradational types (e.g. Bavinton and Keays, 1978). Most common is a cherty variety, next in abundance is a carbonaceous, shaley or slatey variety, and least abundant is a chlorite- or amphibole-rich variety that is commonly schistose. Individual interflow horizons may be comprised solely of one type or may consist of intercalations of the various types. Metasomatic reaction zones rich in chlorite,

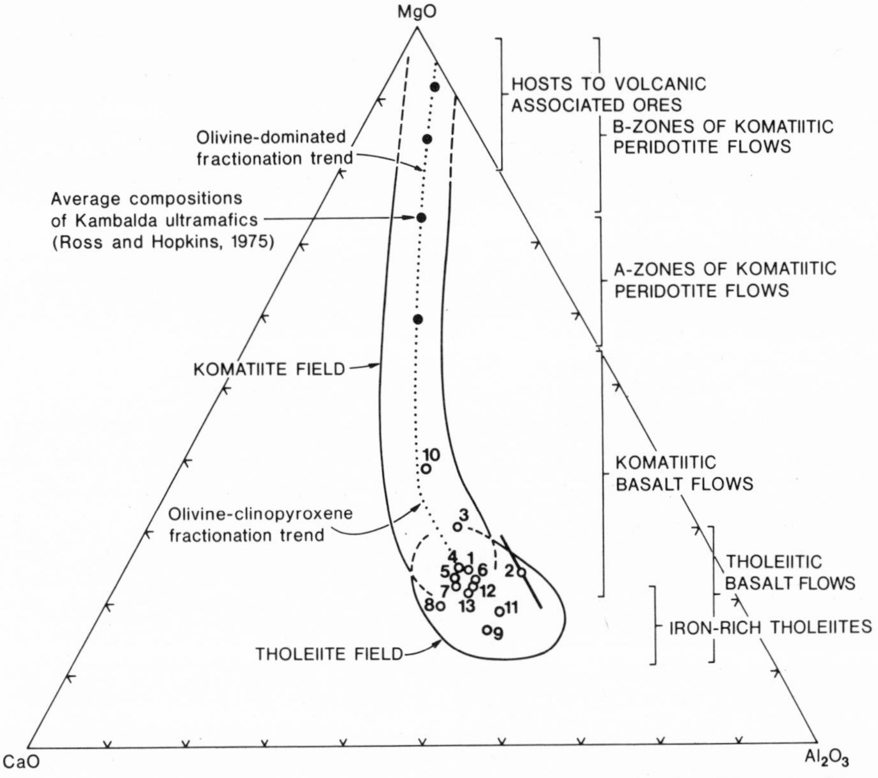

Fig. 11. $MgO-CaO-Al_2O_3$ plot of compositions of tholeiitic metabasalts, differentiated komatiitic metabasalts and komatiitic peridotites associated with Western Australian volcanic-associated Ni–Fe–Cu sulphide ores. Analyses of numbered metabasalts are given in Table V.

amphibole and phlogopite may be developed on contacts between metasediments and enclosing metavolcanic rocks.

The metasediments are normally well layered with siliceous layers, 5 mm to several centimetres thick, alternating with discontinuous layers rich in mafic minerals and/or sulphides (Plate I, B). Finer scale layering (0.1 mm scale) is developed in some specimens (Plate I, A). Siliceous layers are dominated by fine-grained ragged microcrystalline aggregates (low metamorphic grade) to coarsely recrystallized polygonal aggregates (high metamorphic grade) of quartz, plagioclase (An_2–An_{35}) and minor microcline. Alternate layers are dominated by amphibole (tremolite, cummingtonite, hornblende and/or actinolite), chlorite, biotite or phlogopite and white mica. Carbon-rich layers consist of poorly recrystallized carbonaceous material at low metamorphic grades and graphite at higher grades.

Most textures are metamorphic in origin, although the fine-grained microcrystalline aggregates are consistent with chemical sedimentation, and a coarse-grained fragmental (tuffaceous?) texture is developed in some Kambalda interflow rocks (Bavinton and Keays, 1978).

The sulphide content of the rocks is highly variable, with S contents in the range 1–20 wt.% S. Hexagonal and/or monoclinic pyrrhotite is normally dominant, but pyrite may be locally abundant. In some environments, pyrrhotite has replaced pyrite and more rarely magnetite has replaced pyrrhotite. There is also generally a higher pyrrhotite/pyrite ratio in higher-grade metamorphic environments, but this is not universal. Pentlandite occurs as small flames within pyrrhotite or discontinuous aggregates at its grain boundaries. Chalcopyrite, or more rarely cubanite, and sphalerite are locally abundant, and galena, gersdorffite, arsenopyrite, violarite and marcasite may also occur. Magnetite with lesser ilmenite and rutile are commonly present. Textures in pyrrhotite-dominated aggregates range from tectonite fabrics to annealed grain aggregates, but relict pyrite nodules with concentrically arranged inclusions also occur.

Geochemistry. The Western Australian interflow metasediments have a highly variable composition (Bavinton and Keays, 1978; Groves et al., 1978) that is in part due to a highly variable sulphide content (Table VI), but considerable variation is also apparent in the non-sulphide fraction (Fig. 12). Despite this, the rocks tend to have high total FeO, Al_2O_3 and TiO_2, variable but commonly high MgO, CaO, Na_2O, K_2O, S and C, and very low P_2O_5 and MnO over a wide range of SiO_2. Available data on Rhodesian and Canadian interflow rocks suggests lower alkali contents than those of Western Austra-

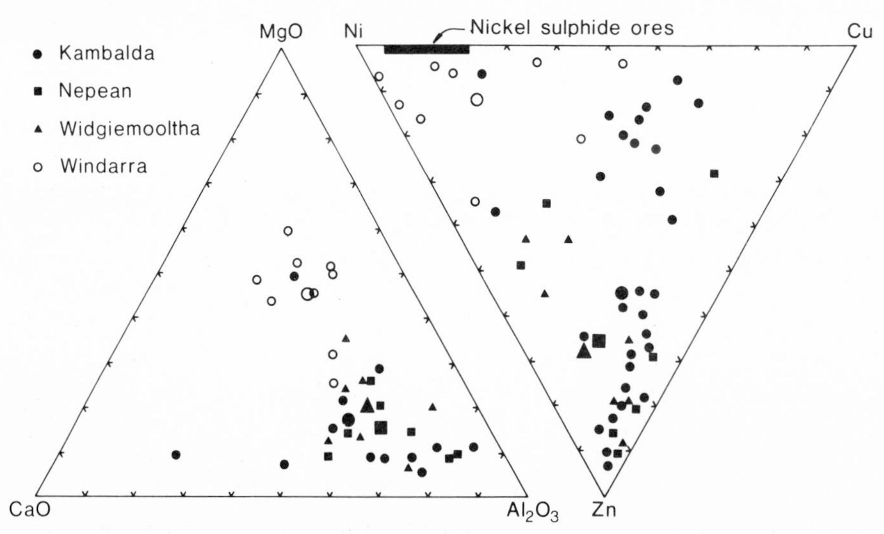

Fig. 12. Compositions of sulphide-bearing metasedimentary rocks from Western Australian Ni–Fe–Cu sulphide deposits. Bulk compositions are plotted in terms of MgO–CaO–Al_2O_3; metal composition of the sulphide fraction is plotted in terms of Ni–Cu–Zn. (Data from Groves et al., 1978 and Hudson, unpublished.) Large symbols represent mean analyses.

TABLE VI

Arithmetic mean and standard deviation for composition of sulphide-rich metasediments in volcanic-associated ore environments (Data from Groves et al., 1979, Bavinton and Keays, 1978 and Chapman and Groves, 1979)

	Kambalda		Nepean		Widgiemooltha		Windarra	
	Mean	S.D.	Mean	S.D.	Mean	S.D.	Mean	S.D.
%								
SiO_2	52.95	8.77	48.01	3.36	51.02	7.74	52.00	10.05
TiO_2	0.45	0.10	0.46	0.07	0.38	0.11	0.48	0.20
Al_2O_3	11.35	2.50	11.65	2.14	11.44	2.31	8.95	2.93
FeO	12.45	5.61	18.95	5.84	13.23	7.11	12.08	6.12
MnO	0.17	0.33	0.10	0.05	0.06	0.03	0.19	0.06
MgO	3.55	4.23	2.93	2.07	3.94	2.08	12.64	5.35
CaO	5.79	4.36	4.05	2.70	4.37	1.70	6.05	2.26
Na_2O	3.01	2.32	1.99	0.58	3.87	2.11	2.72	2.27
K_2O	1.50	1.15	1.50	1.10	0.66	0.63	1.82	1.89
P_2O_5	0.11	0.03	0.06	0.01	0.06	0.02	0.08	0.10
C	0.24	0.33	2.43	2.60	4.82	3.47	0.0	0.0
S	4.3	2.18	8.56	2.60	4.97	3.65	2.56	2.86
ppm								
Ni	632	1200	1021	1460	863	1310	1270	1001
Co	66	24	153	72	103	44	152	20
Cu	433	297	860	469	483	289	449	710
Zn	991	618	3680	3083	2773	2525	140	52
Cr	246	371	220	105	274	214	1340	1473
As	39	108	–		14	27	–	
No. of specimens	12		7		7		9	
ppb								
Au	80 (146) *	208	8.4	3.0	7.7	3.3	3.9	1.3
Pd	9.8 *							
Ir	0.3 *							

* From Bavinton and Keays (1978).
– Below detection limit.

lia (e.g. A.H. Green, 1978). Correlation between elements is generally poor (e.g. Bavinton and Keays, 1978 – Fig. 4). The rocks also contain significantly high concentrations of Ni, Co, Cu, Zn and Cr, commonly with Zn > Cu > Ni > Co, although absolute and relative abundances are variable (Table VI; Fig. 12). Gold contents are highly variable with most interflow metasediments containing 2–10 ppb Au (Chapman and Groves, 1979), whereas Kambalda metasediments have Au contents up to ca. 2000 ppb Au (log. av. 30 ppb Au: Bavinton and Keays, 1978). This is significant in view of the occurrence of stratiform gold deposits in Archaean sulphidic and carbonate-bearing cherts elsewhere (e.g. Fripp, 1976). Precious-metal studies of Kambalda metasediments indicate arithmetic mean values of 9.8 ppb Pd and 0.28 ppb Ir with Pd/Ir ratios in excess of 35.

Available geochemical data (Groves et al., 1978) suggests that interflow metasediments as a group from different areas are indistinguishable with the exception of Windarra (Table VI). In individual localities, there appears to be no significant difference between their composition within footwall and hanging-wall metabasalts and at the base of, or within, the ultramafic sequence (Bavinton and Keays, 1978; Groves et al., 1978; Chapman and Groves, 1979). In most cases, calculation of the mafic portion of the metasediments indicates that, with the exception of Windarra, there is a low ultramafic component (Groves et al., 1978). There appears to be no consistent relationship between metasediment composition and nickel ore, and in particular platinoid compositions do not overlap (Bavinton and Keays, 1978), although some metasediments may be Ni-enriched adjacent to massive ore.

Stable isotope and S/Se data. Available δ^{34}S values and S/Se ratios for sulphides from interflow metasediments are summarized in Figs. 13 and 14. The relatively low S/Se ratios (5000–20,000) for most metasediments are similar to those of nickel ores (Fig. 13) and are consistent with a magmatic source of sulphur. Slightly positive δ^{34}S values for Kambalda interflow metasediments are also consistent with direct exhalation or leaching of juvenile sulphur. The higher S/Se ratios combined with slightly negative δ^{34}S values for Windarra metasediments suggest that fractionation of magmatic sulphur, perhaps due to slight oxidation of reduced species (e.g. Rye and Ohmoto, 1974; Yamamoto, 1976) occurred in this case. There is no evidence for bacterial sulphate reduction in generation of sulphides in the metasediments (Donnelly et al., 1978).

Carbon-isotope data (Donnelly et al., 1978) do not give unequivocal evidence of the origin of non-carbonate carbon, but suggest a biogenic component.

Origin of metasediments. The available data suggest that the interflow metasediments represent a mixture of volcaniclastic detritus derived from basaltic and less commonly ultramafic volcanic rocks and chemical-sedimentary material derived largely from volcanic exhalations between extrusions; organic debris was also probably present. There are indications that S, Au and some base metals (Cu, Zn) were derived by direct exhalation rather than sea-floor leaching of submarine lavas, and that metamorphic mobiliza-

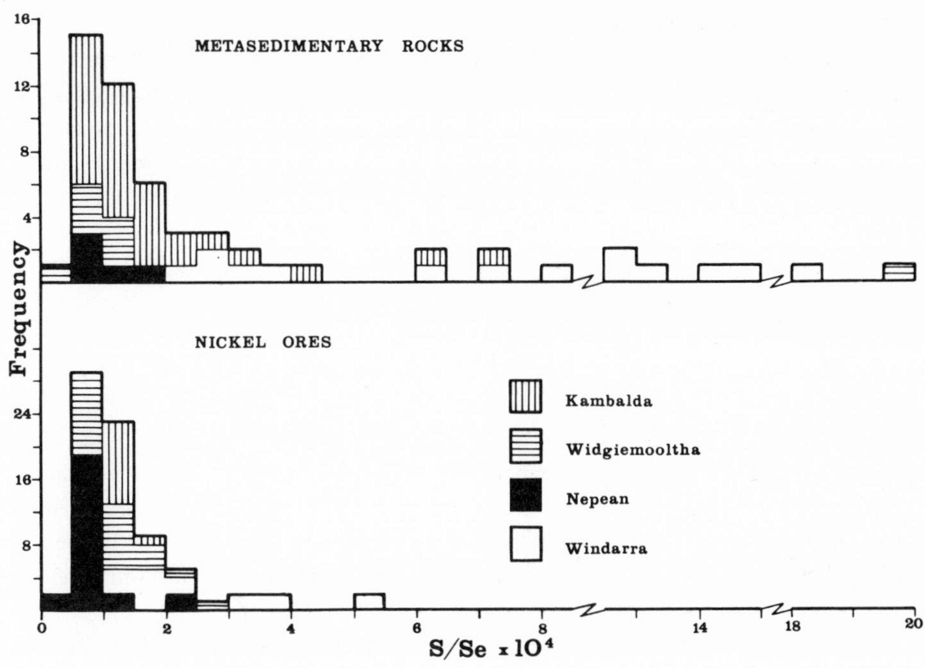

Fig. 13. Comparison of the sulphur/selenium ratio of Western Australian volcanic-associated Ni–Fe–Cu sulphide ores and associated sulphide-bearing metasedimentary rocks. (From Groves et al., 1979.)

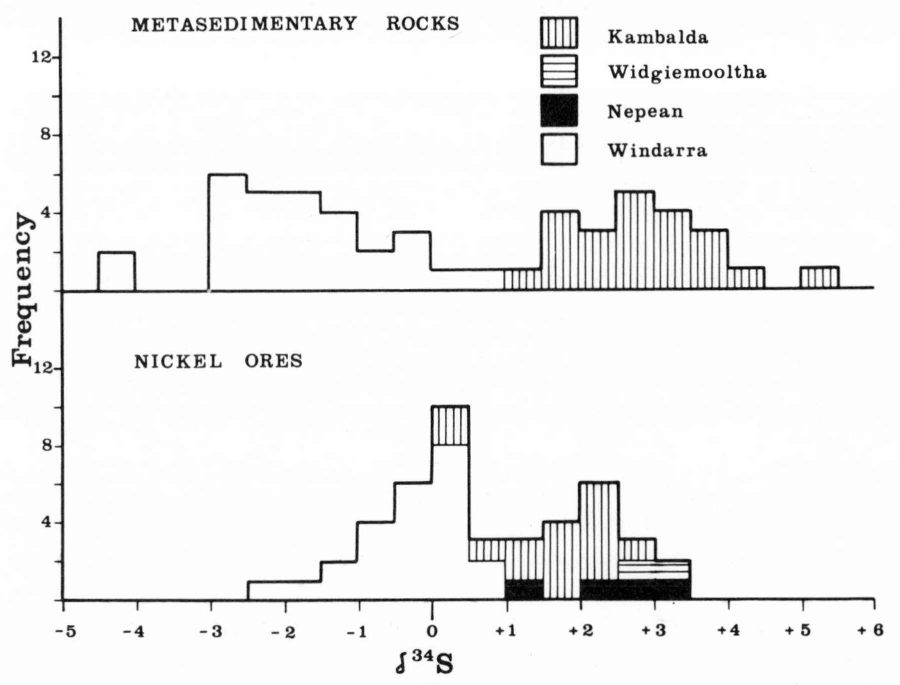

Fig. 14. Variation in sulphur isotope ratio, $\delta^{34}S$‰, in Western Australian volcanic-associated Ni–Fe–Cu sulphide ores and associated sulphide-bearing metasedimentary rocks. (From Groves et al., 1979.)

tion of metals was localized (e.g. Chapman and Groves, 1979).

The metasediments do not appear to be specifically associated with any particular volcanic suite, although they are most commonly developed in the waning stages of basaltic volcanism and initiation, rather than waning of ultrabasic volcanism. Most data (e.g. low positive $\delta^{34}S$, low S/Se, low MnO and high S) support deposition in anoxic environments, and the lack of current structures suggest deposition below wave base. The metamorphic imprint in mineralized environments hampers further interpretation.

The most similar modern counterparts (e.g. Bonatti, 1975) are metalliferous sediments forming around shallow-water volcanic vents (e.g. Thera; Vulcano – Honnorez et al., 1973; Matupi Harbour – Ferguson and Lambert, 1972).

Later intrusives

Later intrusives in Western Australian mineralized environments are briefly described below. They clearly post-date ore in a number of cases and hence have no direct significance to initial sulphide formation, but they are important both in modifying the ore and in establishing subsequent sequences of alteration and deformation events within the ore.

Granitic rocks. The granitoid-cored domes that control the present geographic distribution of ore deposits are normally the synkinematic granitoids of Archibald et al. (1978). They are typically biotite adamellites and/or granodiorites (Table VII) that contain enclaves of banded granodioritic gneiss in some cases. The cores of the domes are typified by massive or weakly deformed granitoids, whereas the marginal granitoids are typically strongly lineated and/or foliated. Some domes (e.g. Widgiemooltha, Pioneer; Fig. 5) are polydiapirs whose emplacement was penecontemporaneous with prograde metamorphism in surrounding greenstone belts; their age is generally 2.7–2.65 Gyr (Archibald et al., 1978).

The major exception is the Kambalda Dome which consists of a largely undeformed "sodic granite" (Ross and Hopkins, 1975), whose composition is similar to overlying dykes and sills of porphyritic felsic rocks. This "granite" is significantly more sodic and has much higher Na_2O/K_2O and MgO/CaO ratios than the other intrusive granitoids of the Eastern Goldfields Province (Table VII), and is considered to be related to felsic volcanism rather than granitoid emplacement (Archibald et al., 1978). It has been dated by the Pb–Pb method at 2.76 ± 0.07 Gyr (Oversby, 1975).

Felsic and mafic intrusives. A variety of metamorphically recrystallized felsic and mafic intrusives occur in the ore environments. They cut the sulphide ores, and are generally massive with foliated margins in places. The lower degree of strain exhibited by these intrusives compared to the country rocks combined with their compatible metamorphic mineral assemblages suggest that thermal events outlasted deformational events during the main metamorphic episode (e.g. Barrett et al., 1977).

TABLE VII

Geochemistry of felsic and mafic intrusive rocks in mineralised environments.

	1	2	3	4	5	6	7
SiO_2	72.90 (2.4)	73.17 (1.9)	72.53	72.03	60.86	70.24	51.22
TiO_2	0.26 (0.16)	0.25 (0.12)	0.27	0.31	0.66	0.38	0.94
Al_2O_3	14.47 (0.84)	14.37 (0.70)	15.70	15.71	15.22	15.36	13.63
Fe_2O_3	0.82 (0.46)	0.76 (0.33)	0.26	0.64	1.29	1.28	0.83
FeO	1.10 (0.66)	1.04 (0.52)	1.04	0.86	3.22	1.14	11.08
MnO	0.04 (0.03)	0.04 (0.02)	0.00	n.d.	0.07	0.15	0.08
MgO	0.52 (0.39)	0.47 (0.35)	1.54	0.62	4.44	1.68	8.36
CaO	1.80 (0.73)	1.70 (0.62)	0.48	1.13	2.98	2.74	9.61
Na_2O	4.30 (0.63)	4.34 (0.49)	5.43	5.56	5.44	6.35	3.22
K_2O	3.72 (1.1)	3.78 (0.77)	2.19	2.28	2.13	0.86	0.23
P_2O_5	0.10 (0.09)	0.09 (0.05)	n.d.	n.d.	n.d.	0.14	0.13
H_2O^+	–	–	0.44	0.37	0.88		
H_2O^-	–	–	0.14	0.09	0.11	0.07	0.05
CO_2	–	–	n.d.	0.05	2.32		
Total:			100.02	99.65	99.62	100.39	99.38

1 = Average (and standard deviation) of 188 Archaean granitoids (Archibald et al., 1978).
2 = Average (and standard deviation) of 26 synkinematic granitoids (Archibald et al., 1978).
3 = Sodic granite from Kambalda analysed by O'Beirne (1968): from Ross and Hopkins (1975).
4 = Average of 7 sodic rhyolite porphyries from Kambalda analysed by O'Beirne (1968): from Ross and Hopkins (1975).
5 = Average of 4 dacitic intrusives from Kambalda analysed by O'Beirne (1968): from Ross and Hopkins (1975).
6 = Felsic porphyry from Windarra analysed by Leahey (1973): from Roberts (1975).
7 = Sodic metadolerite from Windarra analysed by Leahey (1973): from Roberts (1975).

Felsic intrusives are commonly porphyritic (Plate I, D) and include sodic rhyolite porphyries and dacites at Kambalda (Ross and Hopkins, 1975) and various types of feldspar porphyry at Windarra (J.B. Roberts, 1975). Pegmatites and aplites are more common at Nepean (Sheppy and Rowe, 1975).

Mafic intrusives are rare at Kambalda, but metamorphosed sodic dolerite and ferrodolerite dykes are commonly disposed subparallel to the axial surface of the Mount Windarra fold in the Windarra area. Typical analyses of both felsic and mafic rocks are given in Table VII.

Metasomatism and alteration. Metasomatic zones, commonly mineralogically zoned, typically occur up to 1 m from the margins of felsic and less commonly mafic intrusives into the ultramafic sequence. From the contact of the intrusive outwards into the ultramafic rock they commonly have a succession of biotite-rich, amphibole-rich and chlorite-rich zones and may contain sulphides. They are similar to metasomatic reaction zones developed between interflow metasediments and ultramafic rocks. Ross and Hopkins (1975) consider their formation to be related to intrusion of felsic dykes whereas Barrett et al. (1977) suggest they, at least in part, formed during metamorphism.

Other alteration, except along discrete shear zones, is rare. In particular, chloritic alteration pipes similar to those beneath Archaean volcanogenic Cu–Zn deposits (Sangster and Scott, 1976) are not developed. Some localized carbonation and alteration does, however, occur rarely in footwall metabasalt below ore shoots at Kambalda and sulphide-rich pods may be associated with such zones. No detailed information is available on such alteration.

Proterozoic dykes. In the Widgiemooltha area (Fig. 5), several unmetamorphosed, east–west trending mafic dykes of Proterozoic age cut the mineralized sequences. These dykes, termed the Widgiemooltha Dyke Suite (Sofoulis, 1965) include the mineralized Jimberlana Intrusion (Travis, 1975) to the south near Norseman.

Ultramafic stratigraphy

In this section, the broader features of the ultramafic sequences are outlined, but the mineralized ultramafic unit is described in more detail in the later section on the ore environment. The data below are generalized and the features mentioned are not present in *all* deposits: major exceptions are noted, but in a review such as this not all exceptions can be discussed.

Ultramafic sequence. Ultramafic sequences are typically variable in thickness both within and between areas, but the maximum thickness of the sequence in several areas (e.g. Kambalda, Windarra, Scotia) is in the range 600–700 m. The sequence is traceable over tens of kilometres around granitoid-cored domes.

The sequences typically comprise a series of ultramafic units, which can be recognized on the basis of one or more of the following criteria (e.g. Ross and Hopkins, 1975): chilled contacts, distribution of distinctive textures, marked mineralogical or compositional changes at unit boundaries, compositional trends between units, the presence of flow-top or other ultramafic breccias at the tops of units, the presence of intercalated metasediments or more rarely metabasalt flows or metasomatic reaction zones marking their former presence. A schematic diagram emphasizing these features is presented in Fig. 15, and in accord with interpretation of similar units elsewhere (e.g. Pyke et al., 1973; Arndt et al., 1977) the units are considered to be ultramafic lava flows. They show an extreme range in thickness from ca. 0.5 m to 150 m. Thicker units (generally >10 m thick), that can commonly be traced laterally over hundreds of metres, typically dominate the lower part of the sequence while thinner (<10 m thick) commonly less continuous units dominate the upper part (Fig. 8). Exceptions to this generalization include the Nepean deposit where ultramafic units of markedly different mineralogy and compo-

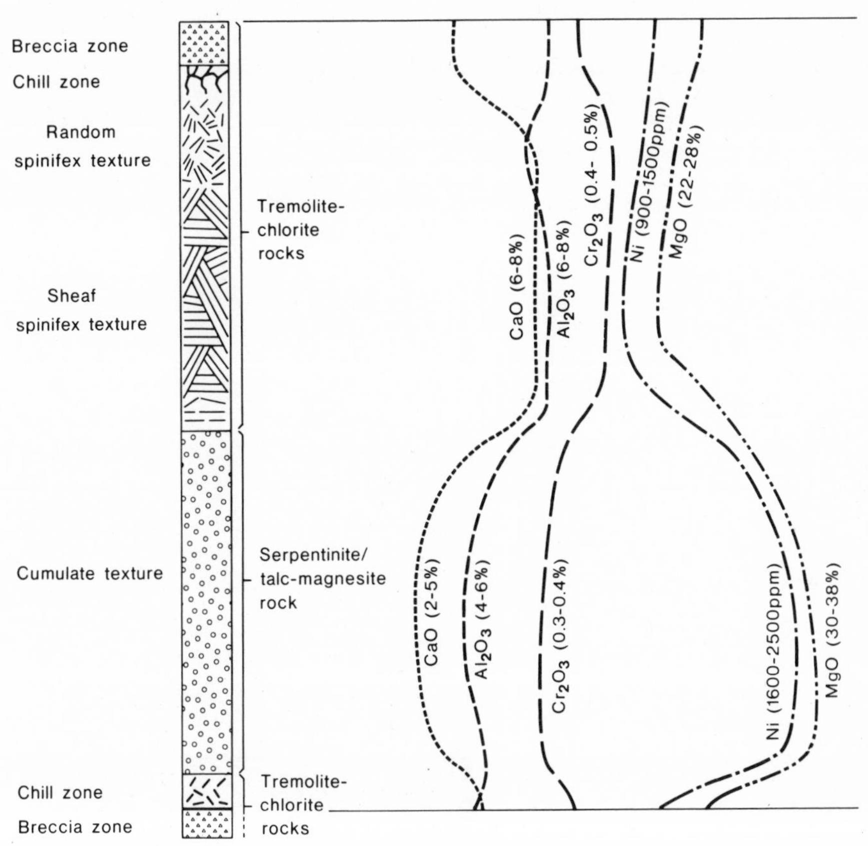

Fig. 15. Schematic section through a typical komatiitic peridotite flow that has differentiated into an upper spinifex-textured A zone and a lower olivine-rich cumulate B zone. See Plate II for examples of textures in these ultramafic rock types.

sition are separated by amphibolite lenses (e.g. Sheppy and Rowe, 1975; Barrett et al., 1976).

In areas of low strain (e.g. Kambalda, Wannaway), the lithological and textural variations within each flow unit are best displayed by the thinner units, which commonly have better developed and relatively thicker A zones and basal chill zones than the thicker units. In these well-preserved flows (Fig. 15), the upper interface is commonly marked by a breccia zone, generally less than 1 m thick, consisting of rounded to angular fragments of underlying ultramafic lithologies in a fine-grained chloritic groundmass. The underlying fine-grained ultramafic is weakly brecciated and grades downwards into a spinifex-textured zone, commonly with finer-grained, random spinifex texture (Plate II, A) overlying sheaf spinifex texture (e.g. Ross and Hopkins, 1975) which is normally coarse-grained with individual blades up to several centimetres long (Plate II, B). More rarely amygdaloidal rocks indicate initially vesicular flows. The mineralogy of the rocks is vari-

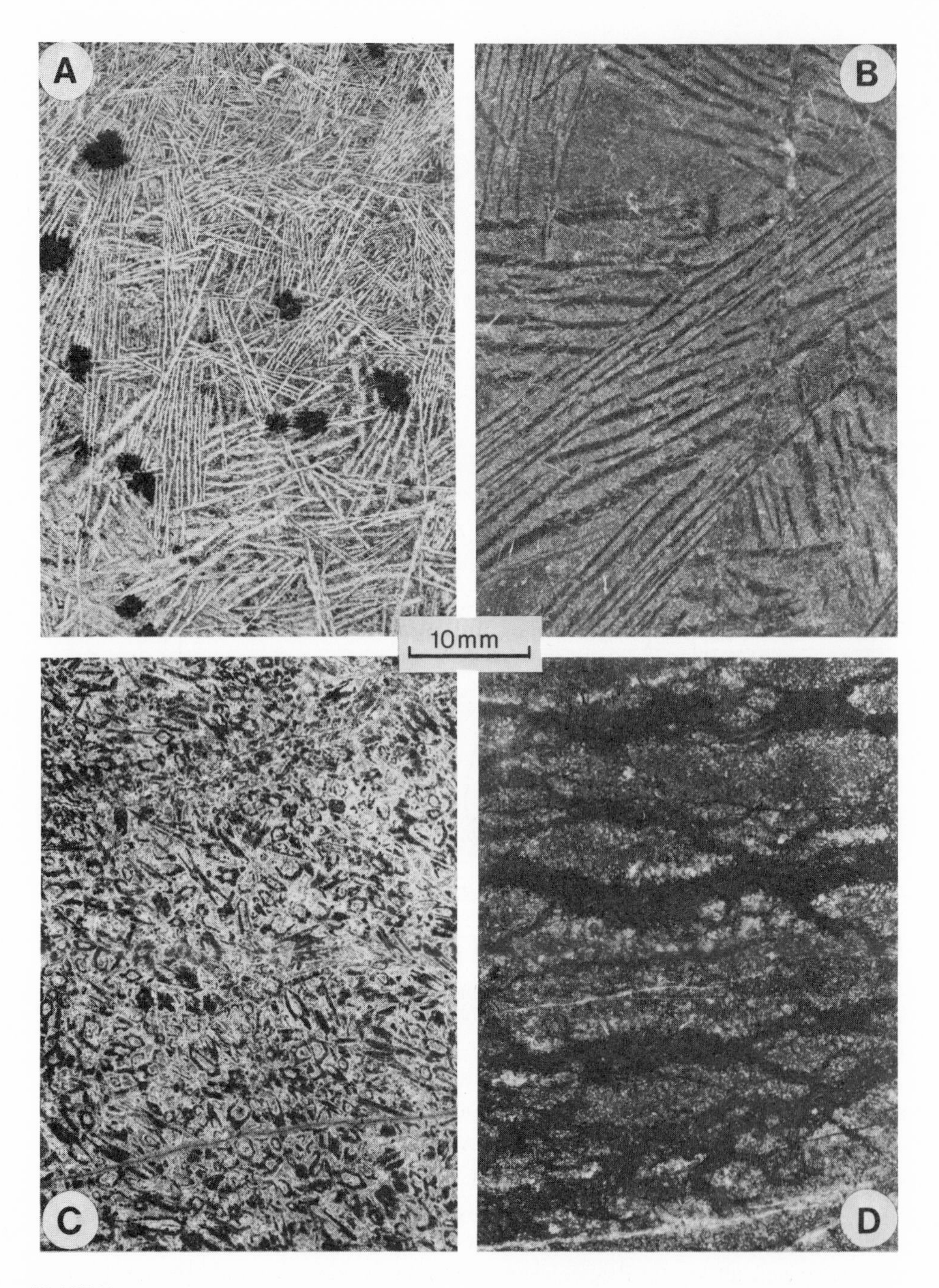

PLATE II

Ultramafic host rocks.

A. Fine-grained sheaf-spinifex texture preserved in a chlorite–tremolite rock with porphyroblastic metamorphic olivine (black), Miriam.

B. Coarse sheaf spinifex in chlorite–tremolite rock, Kambalda.

C. Serpentinised cumulus olivine in peridotitic B-zone, Kambalda.

D. Dusty magnetite veins in serpentinised peridotite, Kambalda.

able (see below), but the coarse blades are normally pseudomorphs after olivine in a groundmass interpreted to represent altered acicular clinopyroxene and recrystallized glass. These textures typify the ultramafic komatiite suite and have been attributed to quenching or supercooling (e.g. Donaldson, 1976). There is normally a sharp boundary between this upper A zone and a lower B zone, which has a close-packed cumulate or adcumulate texture commonly defined by pseudomorphs after equant olivine grains (Plate II, C), although textural preservation may be poor (Plate II, D). A thin basal chill zone may be present. It should be stressed that the proportion of A and B zones varies considerably between flows, and any of the subzones may be absent from particular flows.

In areas of high metamorphic grade, igneous textures are commonly poorly preserved, but mineralogical and geochemical variation (see below) suggests that the ultramafic sequences represent deformed equivalents of those discussed above.

Mineralogy of ultramafic rocks. There are few primary igneous minerals remaining in the ultramafic rocks from mineralized environments, and care must be taken to distinguish metamorphic and relict igneous phases (e.g. olivine, pyroxene and spinel), particularly in the higher-grade metamorphic environments. The discrimination between colourless, porphyroblastic olivine of metamorphic origin overprinting spinifex texture (Plate II, A) and brownish or pinkish recrystallized olivine of igneous origin occurring within spinifex blades is relatively simple, but distinction in close-packed B zones is more difficult (e.g. Barrett et al., 1976), although a distinctive bladed texture may be developed by metamorphic olivine in olivine–talc–(orthopyroxene) or olivine–sulphide rocks (Plate IV, B).

Where igneous olivines are preserved in B zones (e.g. Durkin and Victor Shoots, Kambalda; Wannaway) they are commonly equant, 0.5–1.5 mm diameter grains, and normal compositions are in the range Fo_{88} to Fo_{95} with 0.2 to 0.4% Ni (Nesbitt, 1971; McQueen, 1979; UWA unpublished data). Relict olivines from A zones have a range of compositions (commonly Fo_{76}–Fo_{90}), but are consistently less magnesian and less Ni-rich than cumulate olivines. Metamorphic olivines formed by dehydration or decarbonation of serpentinites and talc–magnesite rocks are generally more magnesian, but Ni-poor relative to igneous olivines in the same lithology. Chromites occur throughout the ultramafic units and show no tendency towards basal concentration: they may in fact be depleted in the cumulate, olivine-rich zone. Normally, they are small, less than 0.1 mm in diameter, strongly zoned grains with a chromite core and chromian magnetite rim (Groves et al., 1977). In terms of RCr_2O_4–RAl_2O_4–RFe_2O_4 (Fig. 16), they are similar to other igneous chromites from ultramafic rocks, but they normally have *mg* values [100 Mg/(Mg + Fe^{2+} + Zn + Mn + Ni)] of less than 20 due to metamorphic modification. However, in rocks with igneous olivine preserved, chromites have narrow rims and more normal igneous *mg* values of ca. 35. Chromites from mineralized environments have distinctive high Zn contents, normally >1 at.% Zn, when compared to those from unmineralized sequences (Groves et al., 1977). Relict clinopyroxene is extremely rare or

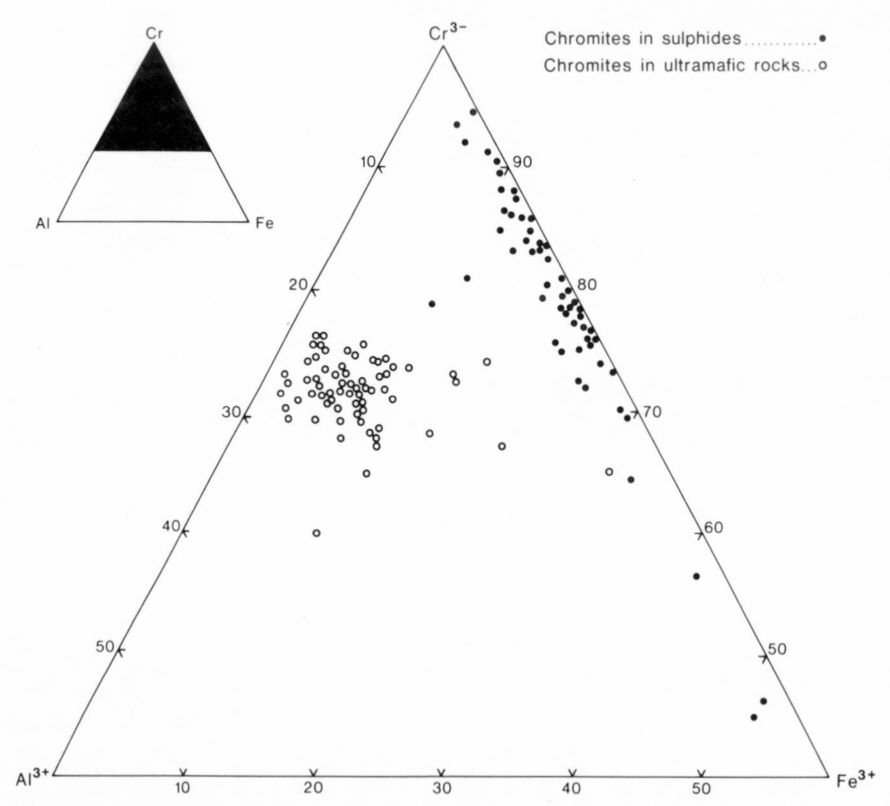

Fig. 16. Contrast in composition between chromites from Western Australian Ni–Fe–Cu sulphide ores (closed circles) and associated ultramafic host rocks (open circles).

absent, although it is considered to have been the third main component of the ultramafic rocks by analogy with well-preserved sequences from lower metamorphic grade areas (e.g. Arndt et al., 1977).

The rocks are normally thoroughly altered. This alteration probably occurred partly prior to metamorphism, continued during prograde metamorphism and at least in high-grade areas, there was subsequent retrogressive alteration. Assemblages vary from area to area and between deposits in the same locality dependent on maximum temperatures of prograde metamorphism and $a(CO_2)$ in alteration fluids. The stability of various minerals present in the ultramafic rocks is shown schematically in Table VIII. In low- to medium-grade areas, serpentinites and talc–magnesite rocks are typical of B zone lithologies in the thicker flows with the proportion of chlorite, tremolite–actinolite and dolomite–calcite increasing in A zone lithologies and chill zones: the proportion of these minerals also increases progressively up the sequence. In higher-grade environments (e.g. Nepean), olivine–talc ± anthophyllite ± enstatite rocks constitute cumulate zones with tremolite–chlorite–olivine rocks representing less-magnesian lithologies. Retrogressive

TABLE VIII

Schematic diagram showing the variation in relict and metamorphic mineral assemblages in peridotitic and pyroxenitic komatiites with metamorphic grade (Adapted from Binns et al., 1976)
The boundary between low and medium grades approximates to the greenschist–amphibolite facies transition.

Metamorphic domains	Very low grade	Low grade	Medium grade	High grade
(a) *Relict Mineralogy*				
Clinopyroxene	———	– – –		
Chromite	———	——— – – –	—modified— – –	– –
Olivine (forsterite)	– – – –		– – – –	
(b) *Metamorphic Mineralogy*				
Lizardite	———	– – – –		
Magnetite	———	——— – –		
Magnesite	———	———	———	– – – –
Dolomite	———	———	———	– – – –
Chlorite	———	———	———	– – –
Tremolite–actinolite	———	———	———	———
Antigorite		– – –	———	– –
Chromian magnetite		– – –	———	——— – –
Talc		– – – –	———	———
Olivine (forsterite)			– – – –	———
Chromite			– – – – –	———
Anthophyllite				– – – – –
Enstatite				– – – – –
Spinel				– – – – –

serpentinization is widespread in such rocks, and metamorphosed rodingites may occur at the basal contact (Barrett et al., 1976).

Geochemistry and differentiation of ultramafic sequence. Rocks within the ultramafic sequence have a wide range of composition with variation from ca. 20 to 45% MgO on a volatile-free basis. They are best described as komatiitic peridotites and pyroxenites. Their compositional variation is shown in Fig. 11. Several studies (e.g. Ross and Hopkins, 1975; Barrett et al., 1976) have demonstrated that there have been only minor variations in element ratios during alteration and that bulk compositional variation is largely controlled by olivine fractionation at the magmatic stage. Rocks show similar variation on a volatile-free basis (Table IX) to those documented by several authors including Barnes et al. (1974), Arndt et al. (1977) and Naldrett and Turner (1977).

Systematic geochemical variation occurs at two scales. There is a distinctive variation on the scale of each unit (Fig. 15) with a zone of high MgO and Ni, low CaO and Al_2O_3, and commonly lower Cr, and total Fe corresponding to the B zone with inverse relative concentration in the A zone and basal chill zone where developed. This combined with the mineralogical variation is consistent with fractional crystallization and gravity

TABLE IX

Geochemistry of some selected ultramafic rocks to demonstrate geochemical variation within units and between units within the ultramafic sequences. All analyses are calculated volatile-free. For further analyses see Ross and Hopkins (1975: Table 7).

	Lower thick units				Upper thick units				
	1	2	3	4	5	6	7	8	9
%									
SiO_2	43.7	45.9	44.0	51.0	47.8	46.9	45.5	47.4	45.6
TiO_2	0.15	0.33	0.09	0.26	0.44	0.19	0.28	0.27	0.40
Al_2O_3	3.0	6.6	2.1	4.1	9.2	4.8	6.2	5.9	7.9
Cr_2O_3	0.35	0.44	0.68	0.53	0.36	0.33	0.39	0.41	0.47
Fe_2O_3	4.4	1.9	2.4	4.0	1.2				
FeO	2.9	8.8	6.2	5.5	9.0	9.3	9.6	10.2	12.3
MnO	0.08	0.19	0.12	0.16	0.18	0.13	0.12	0.17	0.22
MgO	44.5	29.3	42.5	29.9	23.1	36.0	30.9	30.3	24.9
CaO	0.1	6.0	1.4	4.2	7.7	2.2	6.7	4.7	7.6
Na_2O	0.01	0.33	0.04	0.06	0.76	0.07	0.14	0.45	0.55
K_2O	0.03	0.04	0.02	0.03	0.12	<0.01	0.01	0.18	0.05
ppm									
Ni	2900	1650	2450	1750	950	1900	1700	1300	850

1 = Average of 4 serpentine (olivine)–chlorite–talc rocks from lower (B?) zones, Nepean (Barrett et al., 1976).
2 = Average of 4 tremolite–chlorite–olivine rocks from upper (A?) zones, Nepean (Barrett et al., 1976).
3 = Average of 3 serpentine–talc–carbonate rocks from lower (B?) zones, Windarra (UWA, unpublished analysis).
4 = Average of 5 tremolite–chlorite–carbonate–serpentine rocks from upper (A?) zones, Windarra (UWA, unpublished analyses).
5 = Average of 2 tremolite–chlorite rocks, Nepean (Barrett et al., 1976).
6 = Average of 2 talc–carbonate–serpentine rocks from B2 zone of ultramafic flow, Windarra (UWA, unpublished analyses).
7 = Average of 3 spinifex-textured talc–carbonate–chlorite–serpentine rocks from A zones of some ultramafic flows, Windarra (UWA, unpublished analyses).
8 = Average of 3 serpentinites from B2 zone, Mount Clifford (Barnes et al., 1974).
9 = Average of 4 spinifex-textured A zone serpentinites, Mount Clifford (Barnes et al., 1974).

settling of olivine to form the B zone beneath an A zone representing largely magnesium-rich liquid with some olivine phenocrysts; chromites were too small to settle at the same rate as olivines. Ross and Hopkins (1975) also present evidence for an increase in Zn and Cu adjacent to flow margins.

Larger-scale variation is shown by the general decrease in the MgO content of the flows upwards; this is most clearly seen in the composition of B zones where MgO contents decrease from ca. 45% to ca. 28% MgO up the sequence (Fig. 17). Probable silicate liquid compositions appear to be around 28% MgO, possibly as high as 32% MgO, for thick basal flows but commonly around 23% MgO for the thinner flows. Ross and Hopkins (1975) and Ross and Keays (1979) suggest that the silicate liquid composition at Kambalda was ca. 23% MgO for all ultramafic flows. Rocks of this composition have similar Pd and Ir contents irrespective of whether they occur in ore-bearing or barren environments, and the Pd/Ir ratio (6–10) is very similar to that of the sulphide ores (Ross and Keays, 1979).

Emplacement of ultramafic sequence. Although the evidence discussed above points strongly to an extrusive origin for the ultramafic sequence, no feeder zones or extrusive centres have yet been recognized for the mineralized sequences in Western Australia. Several authors (e.g. Archibald et al., 1978) have speculated that both basaltic and ultramafic volcanics were derived via fissure eruptions, but there is no direct evidence. Naldrett (1973) postulated a relationship between the flows and a subsurface magma reservoir, and Naldrett and Turner (1977) suggested that dunitic peridotite pods in the Yakabindie area may represent remnants of such feeder zones for largely unmineralized komatiitic flow sequences.

D.A.C. Williams (1979) presents evidence from several mineralized ultramafic sequences in Rhodesia (e.g. Damba, Epoch, Hunters Road) for the existence of feeder pipes for ultramafic flows, and points out the similarity with Naldrett's (1973) model. D.A.C. Williams (1979) also interprets the Shangani deposit as occurring within an ultramafic complex representing a volcanic neck that marks the initial stages of komatiitic volcanism in the Shangani area. The complex (Fig. 18) has a trench-like or fissure-like form with unlayered peridotite in the stem and in the lower part of each lobe and passes upwards into an alternation of komatiitic metabasalts with quench textures and peridotite lenses with some interflow metasediment units. It appears to be roughly contemporaneous with felsic volcanism, as shown by overlying felsic agglomerates and tuffs. The shape is reminiscent of the Kambalda troughs (Fig. 7), but these appear to be everywhere floored by metabasalt and have no stem.

The initial origin of suites of ultramafic komatiites of varying composition is still strongly debated. Hypotheses include equilibrium partial melting (e.g. Nisbet et al., 1977), in which discrete crystal–liquid systems evolve to varying degrees of partial melt prior to separation of the crystalline residue, complete polybaric assimilation of a varying proportion of a crystalline fraction (Bickle et al., 1977) or high-level fractionation

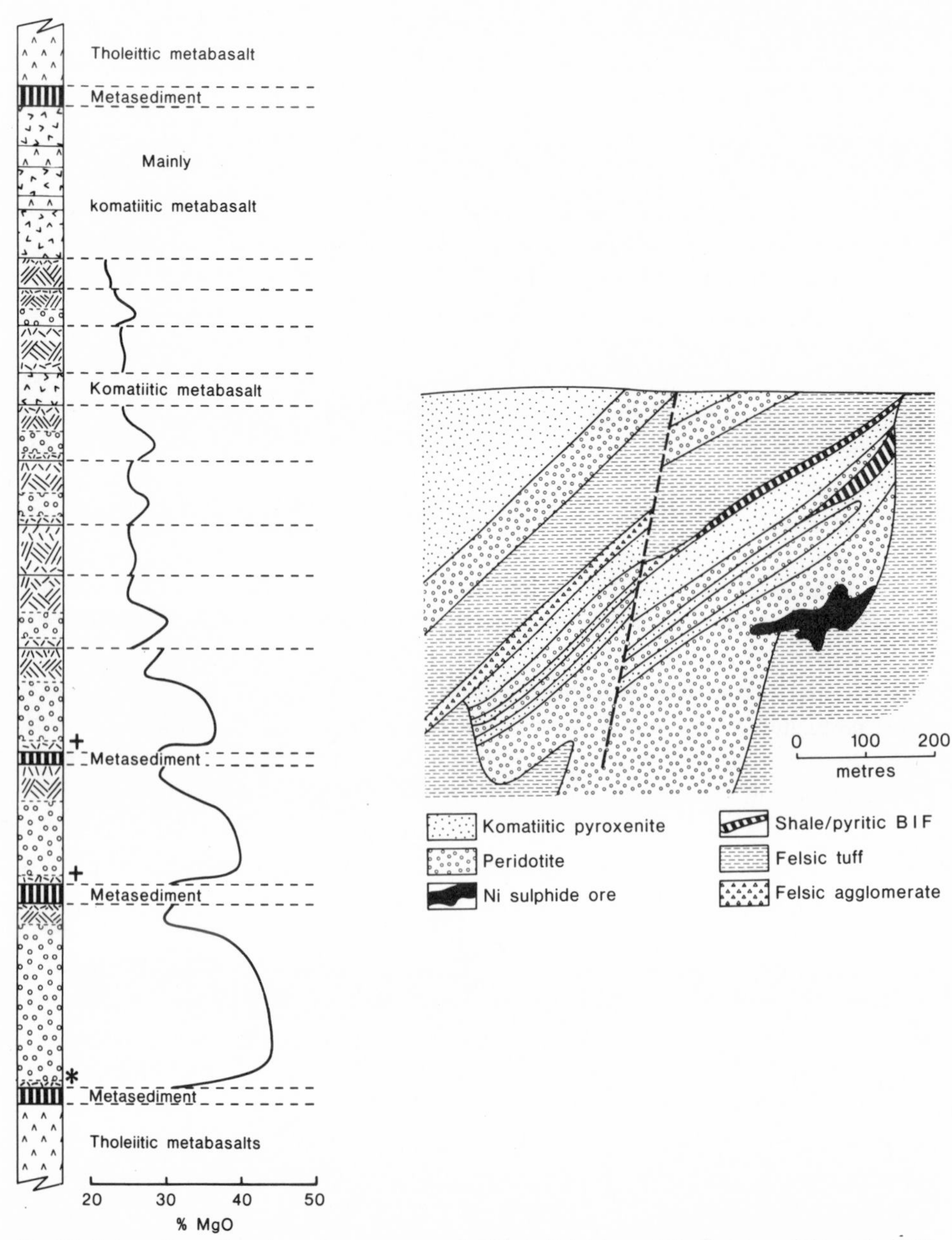

Fig. 17. Schematic section through a sequence of differentiated komatiitic flows overlying a tholeiitic metabasalt footwall. The irregular upward trend to thinner and less magnesian flows is shown, together with the position of sulphide-bearing metasediments and major (*) and minor (+) Ni–Fe–Cu sulphide ores.

Fig. 18. Generalized section through the Shangani volcanic-associated Ni–Fe–Cu sulphide deposit showing its neck-like form. (After D.A.C. Williams, 1979.)

(Arndt et al., 1977; Sun and Nesbitt, 1977) or accumulation of olivine (e.g. McIver and Lenthall, 1974). All models involve high degrees of partial melting of mantle source or the crystalline residue of a previous melting (e.g. Naldrett and Turner, 1977). Most authors suggest that the komatiitic magmas were generated in rising mantle diapirs that were initiated at depths of 200 to 250 km, but the depth of separation of the magmas is debated. A model presented by Naldrett and Turner (1977), in which komatiitic magmas form from the residuum of previous mantle melting and separate at depths equivalent to those of present continental crust, is discussed below with regard to formation of sulphide ores.

Whatever the ultimate origin of the ultrabasic komatiite magmas, most authors agree that they were extruded at temperatures in the vicinity of 1650 ± 50°C (e.g. D.H. Green et al., 1975). They appear to have been erupted as highly magnesian silicate liquids carrying varying proportions of olivine phenocrysts, which together with subsequently crystallized olivine crystals settled to form the distinctive mineralogical and chemical assymmetry of the flows. There has been no satisfactory explanation for the systematic decrease in MgO content of the successively erupted ultrabasic magmas.

THE ORE ENVIRONMENT

Introduction

Despite the many genetic models proposed to explain ore generation, a characteristic association exists of the volcanic-associated deposits with specific, thick, highly magnesian ultramafic units, and there is overwhelming evidence that at least the disseminated sulphides have a magmatic origin. It is also impossible to divorce discussion of magmatic models of ore concentration from those involving generation and emplacement of the ultramafic units. For this reason, the mineralization is described below as an integral part of the host ultramafic units.

Mineralized ultramafic units

Nature of units. The mineralized ultramafic units are the lowermost units in the ultramafic pile. They are typically the thickest in the ultramafic sequence, ranging from ca. 20 m (Nepean) to ca. 150 m (Juan Shoot; Marston and Kay, in press) thick. Ross and Hopkins (1975) show that the units are thickest over ore and thin towards the flanks of the ore zone at Lunnon Shoot, Kambalda. The thickness of the ultramafic units overlying the ore zone is typically ca. 50 m. Their lateral extent is poorly defined. At Windarra, the mineralized unit appears to be continuous for over 1500 m along strike and over 800 m down dip (e.g. J.B. Roberts, 1975), but the possibility of several discrete units developed at the same stratigraphic horizon cannot be discounted. At Lunnon Shoot, the mineralized ultramafic unit is at least 300 m wide and 1500 m long within the trough-like zone of mineralization (Ross and Hopkins, 1975), but its limits outside this structure have not yet been clearly defined. Christie (1975) suggests that the mineralized unit at Scotia is a lens up to 50 m thick extending for ca. 500 m along strike and more than 450 m down dip.

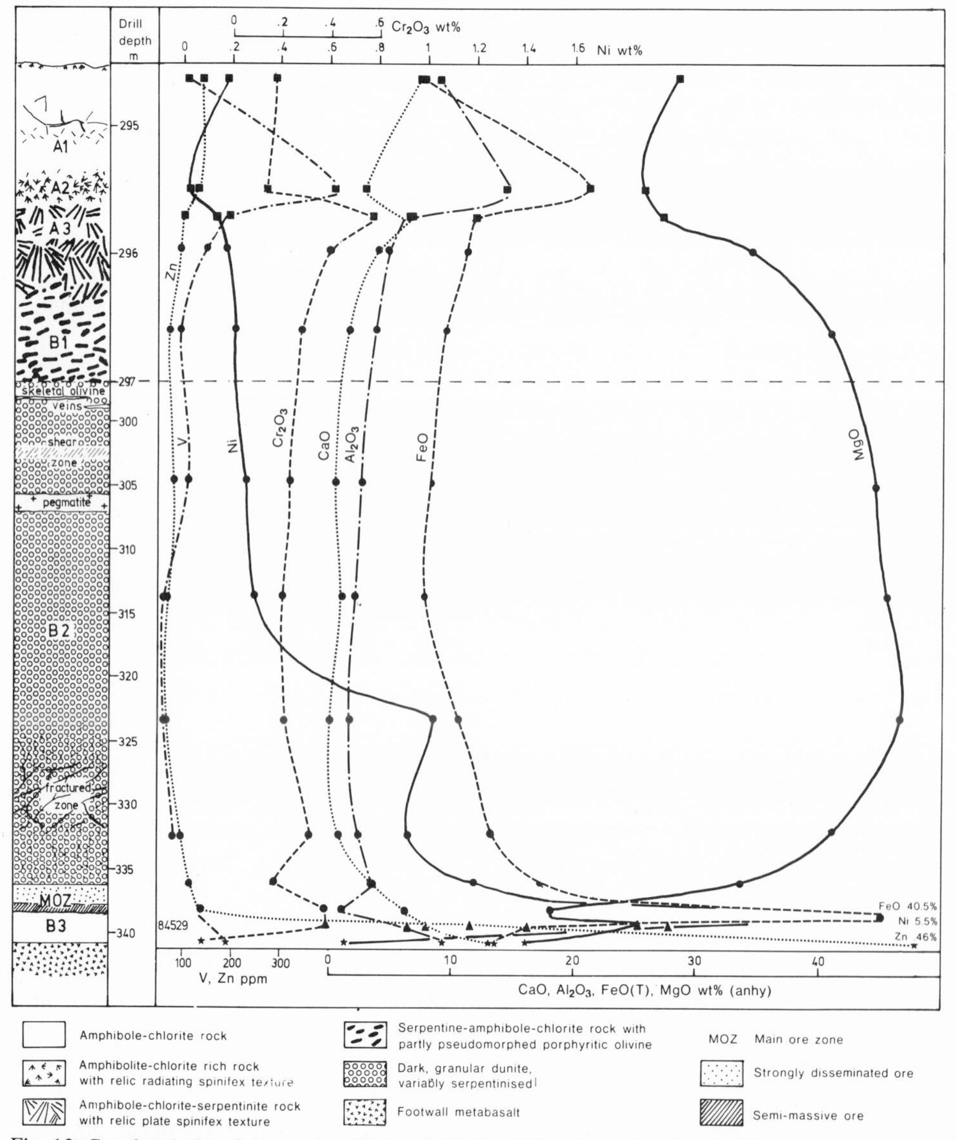

Fig. 19. Geochemical and textural variation through a differentiated ore-bearing komatiitic peridotite at Wannaway, Western Australia. Note the change in vertical scale from A zone to B zone in the diagram. (After McQueen, 1979.)

The units are similar in many respects to other units within the sequence (cf. Fig. 19 with Fig. 15), but are all typified by a thick cumulate (B) zone and a thin, telescoped upper (A) zone. Upper chill zones and spinifex-textured zones may be present (e.g. Kambalda, Wannaway), but in many localities they are absent or unrecognized. Where massive ores are developed, basal chill zones are commonly absent or poorly defined, although thin tremolite- and chlorite-rich zones, probably representing chill zones, occur beneath ore in some cases (e.g. Widgiemooltha area: McQueen, 1979). The basal zone is generally complicated by the occurrence of metasomatic reaction zones that make interpretation of original features difficult.

Mineralogy and geochemistry. The mineralized ultramafic units are the most magnesian in the ultramafic sequence and are therefore dominated in different environments by assemblages including antigorite, talc, magnesite, relict olivine (Fo_{92}–Fo_{94}; ca. 3000 ppm Ni), metamorphic olivine and retrograde lizardite (Table VIII). Olivines are more Fe-rich in contact with sulphides (e.g. Barrett et al., 1976). Tremolite, chlorite and dolomite are only abundant in the lower metre or upper few metres of the units. Zinc-rich chromite occurs throughout most units and, although total Cr_2O_3 is variable, the upper zones are commonly slightly enriched in Cr_2O_3 relative to the cumulate zones. An example of mineralogical variation within a single mineralized ultramafic unit from Nepean is shown in Fig. 20.

The cumulate zones of mineralized units characteristically contain 42–45% MgO and less than 5% combined CaO and Al_2O_3 on a volatile-free basis with overlying zones containing 28–30% MgO. Their bulk composition, estimated in Table X and shown in Fig. 11, is dominated by the chemistry of their thick cumulate zones (Fig. 19), and MgO contents are typically ca. 40%. Most elements show a linear correlation with MgO and SiO_2, except for CaO which may be redistributed during alteration (Barrett et al., 1976). Importantly, Ni shows a strong positive correlation with MgO and there is no significant difference in Ni content with respect to MgO between mineralized units and those from unmineralized environments (Fig. 21). Although data are limited, the mineralized ultramafic units appear to be S-saturated (0.2–0.4 wt.% S), whereas most overlying unmineralized units and the spinifex-textured zones of flows from unmineralized sequences appear to be S-undersaturated (Fig. 22). The host ultramafic unit at Juan Shoot, Kambalda, is also S-undersaturated (Marston and Kay, in press).

The mineralized ultramafic units in Western Australia are considered most likely to be flows as they are commonly bounded on both margins by metasediment units, they have spinifex-textured upper zones, thin basal chill zones and are an integral part of flow sequences. Where textural preservation is poor and metasediment units are poorly developed (e.g. Nepean), distinction between an extrusive and concordant intrusive origin is impossible, but an extrusive origin is preferred by analogy with similar, though less deformed and metamorphosed units elsewhere.

The composition of the silicate liquids forming the flows is not always clear. Most

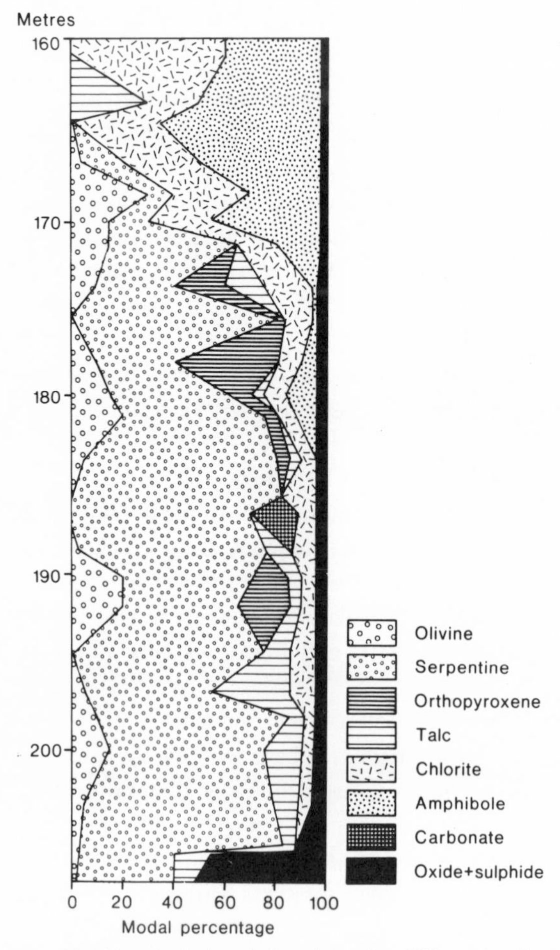

Fig. 20. Mineralogical variation through a differentiated, ore-bearing komatiitic peridotite at Nepean, Western Australia. (After Hudson, 1973.) Note increase in chlorite and amphibole up the section.

probable chill zones, or uppermost zones in areas of poor textural preservation, have compositions of 28 ± 2% MgO, but these zones may contain some intratelluric phenocrysts. Ross and Hopkins (1975) suggest that metapicrites with 23% MgO represent the silicate liquid composition at Kambalda, and the most magnesian natural liquids recognized have 32–33% MgO (e.g. D.H. Green et al., 1975). Assuming an olivine composition of $Fo_{92.5}$, the estimated percentage of intratelluric olivine phenocrysts in the mineralized units (Table X) varies from 45 to 65% (23% MgO silicate liquid), 35 to 55% (28% MgO silicate liquid), and 15 to 35% (32% MgO silicate liquid). The flows therefore appear to have been crystal mushes on extrusion, unless syn-extrusion olivine concentration occurred selectively in the ore zones from which all bulk compositions were calculated.

TABLE X

Estimated compositions (volatile-free) of some ultramafic hosts to volcanic-associated nickel deposits from Eastern Goldfields Province. Western Australia

	1	2	3	4	5	6	7
%							
SiO_2	45.3	47.3	44.6	45.8	40.8	49.1	45.7
TiO_2	0.16	0.13	0.18	0.49	0.13	0.15	0.21
Al_2O_3	3.6	2.7	3.6	3.4	1.65	2.7	3.1
Cr_2O_3	0.19	0.69	0.34	0.27	n.d.	n.d.	0.54
FeO(t)	8.6	8.6	8.5	8.3	9.5	8.7	8.8
MnO	0.13	0.12	0.10	0.13	0.14	n.d.	0.13
MgO	40.1 (40.7 *)	38.8 (39.5)	40.7 (36.0)	39.5 (40.0⁺)	45.0	39.9	40.0
CaO	1.8	1.5	1.9	2.0	0.28	1.15	1.5
Na_2O	0.05	0.05	0.12	n.d.	n.d.	n.d.	0.05
K_2O	0.02	0.04	0.03	0.05	n.d.	0.27	0.08
P_2O_5	0.01	0.02	0.02	n.d.	n.d.	n.d.	0.02
ppm							
Ni	2650	2500	2450	n.d.	>2000	2800	ca. 2500

There are no published calculations of complete rock compositions of the mineralized ultramafic rocks based on continuous sampling and the estimates above are based on limited analyses. MgO values shown in brackets are those calculated independently from a larger number of partial analyses representing continuous geochemical profiles (Groves et al., 1979) and are more accurate estimates of the composition of the complete mineralized unit. Nickel contents are also calculated for a larger number of samples and are rounded off to 50 ppm. Sulphur contents of included samples are less than 0.5%.

1 = Lunnon Shoot, Kambalda: average of 22 talc–carbonate rocks and serpentinites (Ross and Hopkins, 1975). * Estimates for Juan Shoot (Marston and Kay, in press).

2 = Windarra: average of 7 serpentinites and talc–carbonate rocks (UWA, unpublished analyses).

3 = Nepean: weighted average of 4 serpentine (olivine)–talc–chlorite rocks and 4 tremolite–chlorite–olivine rocks (Barrett et al., 1976).

4 = Wannaway: average of 12 serpentinites (Anaconda Australia Inc. unpublished analyses + estimate from McQueen, 1979).

5 = Scotia: estimate for serpentinised dunite/peridotite given by Christie (1975).

6 = Mount Edwards: serpentinised talc–carbonate–olivine rock (porphyritic peridotite) by I.N.A.L. staff (1975).

7 = Eastern Goldfields Province: average of 56 ultramafic hosts to volcanic-associated deposits from Binns et al. (1977). Analysed rocks from Lunnon Shoot, Kambalda, Windarra, Nepean, Wannaway, Scotia, Mount Edwards, Carnilya Hill, and Spargoville with equal weight given to each locality.

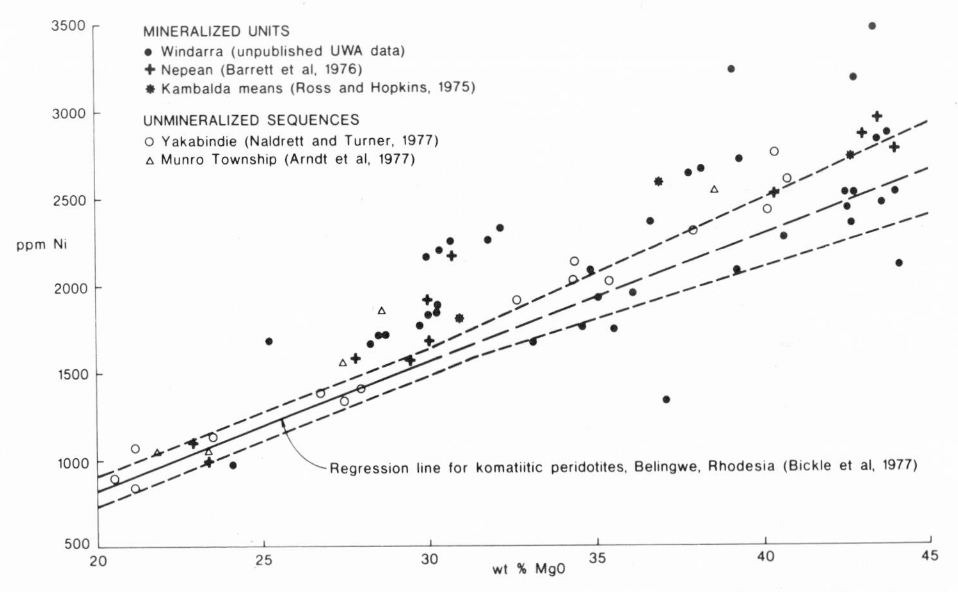

Fig. 21. Plot of MgO content vs. Ni for some selected unmineralized komatiitic ultramafic rocks and host rocks to volcanic-associated Ni–Fe–Cu sulphide ores.

Available geochemical data (e.g. Ross and Hopkins, 1975; Barrett et al., 1976) are consistent with olivine fractionation as the dominant control of geochemical variation (Fig. 19) within the units, as for other units in the sequence.

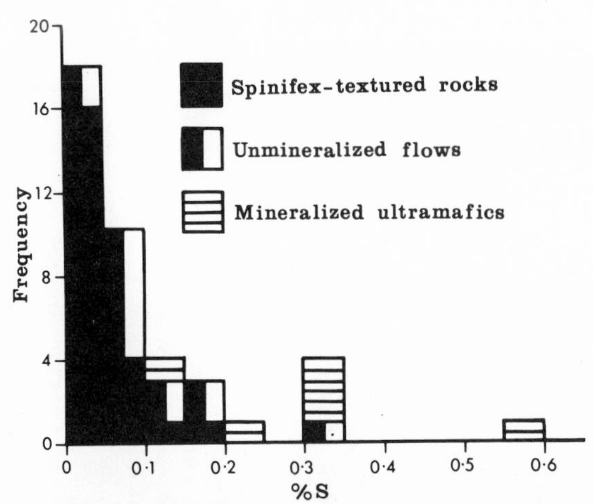

Fig. 22. Variation of sulphur content of mineralized and unmineralized ultramafic rocks from Western Australia. (From Groves et al., 1979.)

Primary sulphide ores

Position of ores in the ultramafic sequences. Significant concentrations of Ni–Fe–Cu sulphides are almost entirely restricted to the basal portion of individual ultramafic units. The sulphide ores can thus be considered as part of the cumulate (or B) zone of differentiated ultramafic flows. In any ultramafic sequence, however, sulphides are not found at the base of all such differentiated units. Almost without exception the major sulphide ores (probably ca. 80% of total ore production) lie on or close to the contact between the lowermost ultramafic unit and footwall rocks (e.g. Kambalda, Redross, Wannaway, Spargoville and Nepean in Western Australia and most Rhodesian and Canadian examples) with a thin metasediment horizon separating sulphide ore from other footwall rocks in some deposits (e.g. Scotia, Windarra). The second and third ultramafic units in the sequence may also be mineralized in many cases.

Ore zones along the contact tend to be elongate and in many cases are contained within embayments or linear depressions in the footwall rocks as discussed above. The embayments coincide with a general absence of sulphide-bearing metasediments even in those deposits in which contact metasediments are a persistent feature. At Lunnon Shoot, Kambalda, for example, the contact ores lie within a 250 m wide corridor in which sulphide-bearing shales and cherts are absent (Fig. 23). Contact between ore and metasediment is rarely observed in any of the Kambalda ore shoots (Ross and Hopkins, 1975).

At Kambalda, "hanging-wall" mineralization occurs at the base of ultramafic units that overlie the main ore-bearing ultramafic units. Most occurrences are within 100 m of the contact and have a close spatial relationship with contact mineralization (Fig. 7).

Definition of ore types. Many terms have been used to define the different types of sulphide mineralization. Most have some unsatisfactory features and generally do not allow for the infinite variety in ore types that has resulted from a complex interaction of mag-

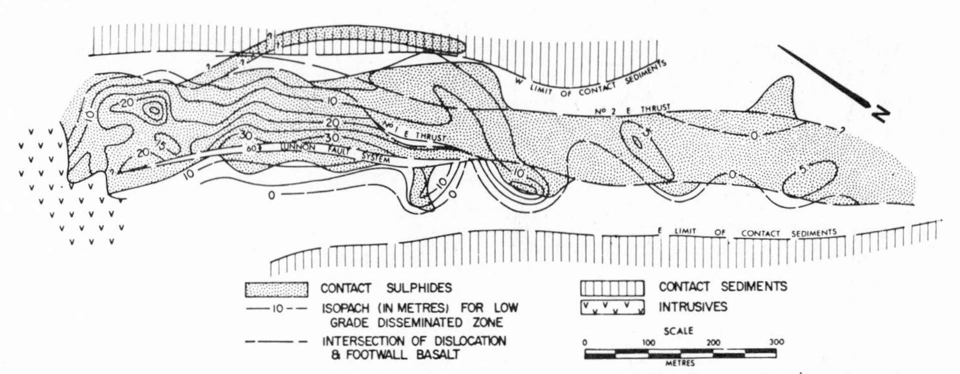

Fig. 23. Plan of the interpreted original distribution of contact sulphides, low-grade disseminated zone (isopachs) and basal sediment horizon at Lunnon Shoot, Kambalda. (From Ross and Hopkins, 1975.)

matic, tectonic, metamorphic and weathering processes.

Spatially based definitions, such as *contact* mineralization, for the ore zone at the ultramafic metabasalt contact, *hanging-wall* mineralization, for the ore zones at the base of overlying ultramafic units or within the upper section of the basal ultramafic unit, and *offset* mineralization, for ores located on fractures and faults within both the foot-wall and ultramafic sequence, have been used by Woodall and Travis (1969) and Ross and Hopkins (1975). The offset mineralization represents only a low percentage of the total ore reserve at Kambalda, but the ores are diverse in character and their brecciated and deformed nature reflects the major role that tectonism and metamorphism has played in the formation of their present physical features. In other deposits, such as Redross and Windarra, breccia ore zones make up a more significant percentage of the ore reserve. Sulphides, commonly chalcopyrite-rich, that fill veins and fractures in footwall meta-basalt beneath the contact mineralization at Kambalda, have been referred to as *stringer* sulphides (Woodall and Travis, 1969), and the zone of fracturing and mineralization described as the stringer zone.

Sulphide ores in both contact and hanging-wall positions consist of varying proportions of sulphide, oxide and silicate minerals. The Lunnon Shoot ore section described by Woodall and Travis (1969), Ewers and Hudson (1972) and Ross and Hopkins (1975) has become generally accepted as a type section through the volcanic-associated ores. It typically comprises a zone of concordant massive sulphide (Fig. 24) at the base, overlain by a zone in which sulphide is interstitial to olivine or more commonly olivine alteration products (matrix or net-textured ore), overlain in turn by weak disseminations of sulphide in an ultramafic host (disseminated ore), and finally barren ultramafic passing up-

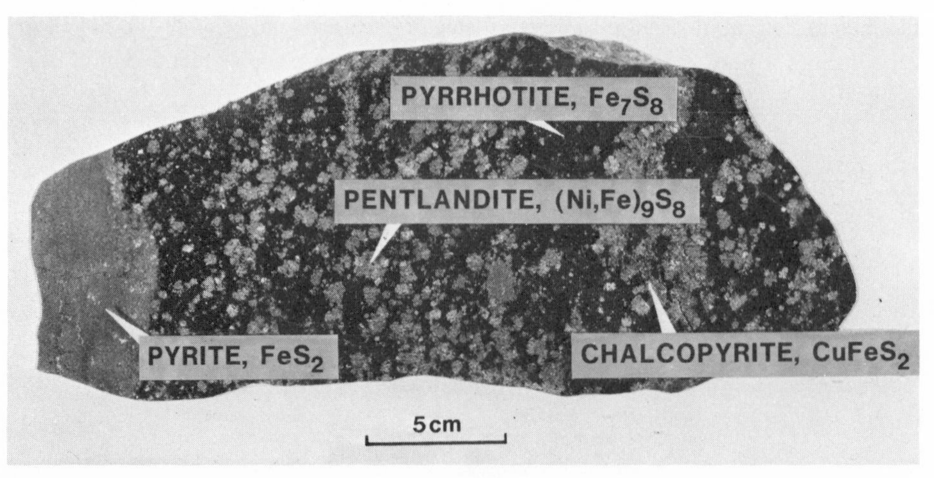

Fig. 24. Massive Ni–Fe–Cu sulphide ore from Kambalda, Western Australia showing the major sulphide minerals. A crude banding is developed by the pyrite concentration and by the grouping of subspherical aggregates of pentlandite in pyrrhotite.

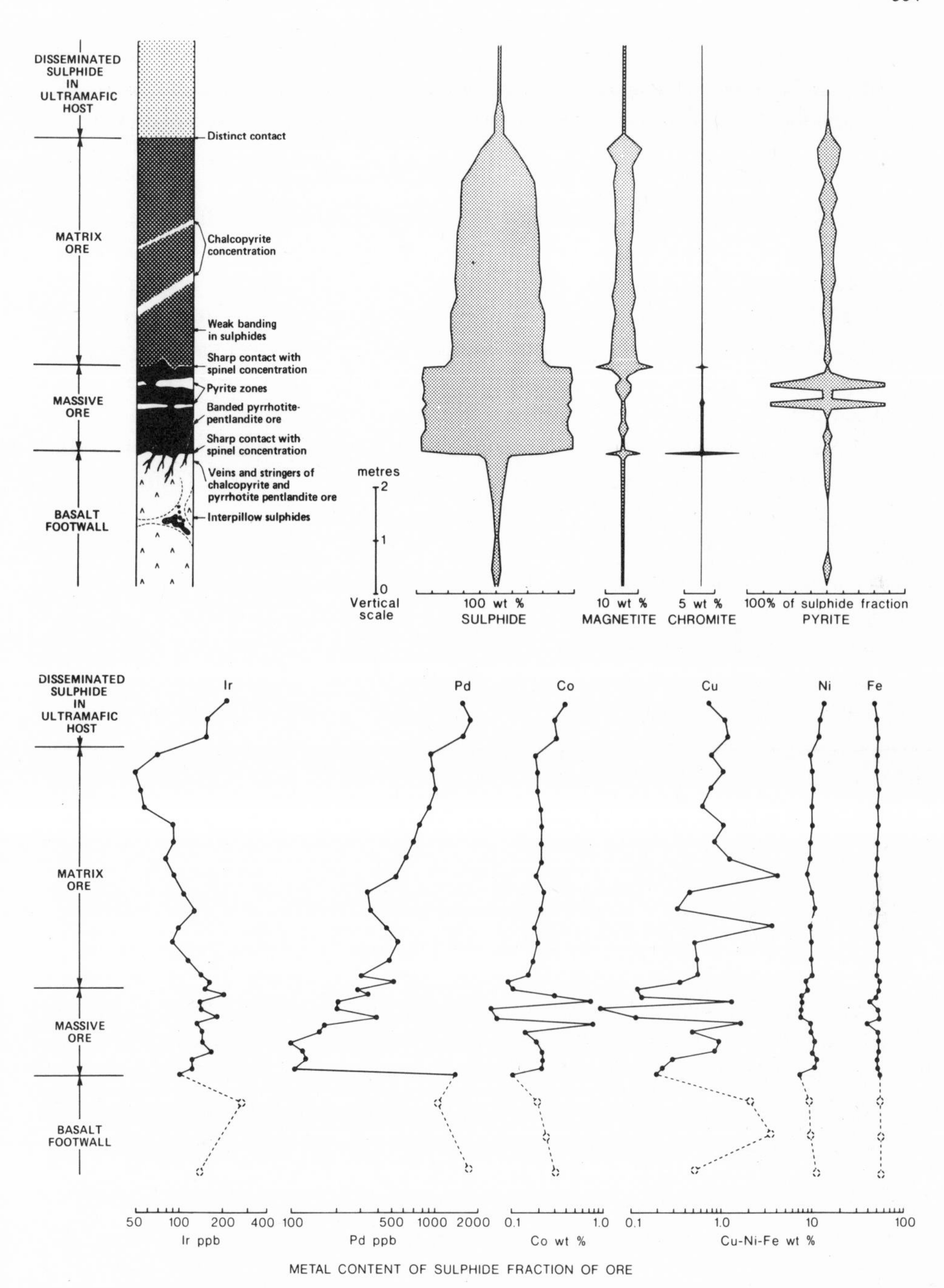

Fig. 25. Mineralogical and chemical variation through a specially selected ore section at Lunnon Shoot, Kambalda, described by Ewers and Hudson (1972). (Pd and Ir data from Keays and Davison, 1976; other data from Ewers and Hudson, 1972.)

wards into a spinifex-textured zone (Fig. 25). It should be noted, however, that while massive sulphides normally underlie more disseminated sulphides, there may be considerable variation in ore sequences both within and between deposits. Massive ores are virtually absent from many Rhodesian ores (e.g. D.A.C. Williams, 1979).

Massive sulphide ores. Ores containing 90% or more sulphide plus oxide minerals are found in most volcanic-associated deposits. They vary greatly in their physical and chemical characteristics, thickness, percentage of the total sulphide reserve, and spatial relationships with other ores. Some examples of massive ore are illustrated in Plate III. They normally range from essentially isotropic massive ores in which pentlandite occurs as coarse spherical to ellipsoidal granular aggregates in pyrrhotite (e.g. some Spargoville and Kambalda ores) to strongly layered massive ores (Plate III, A and C) in which banding is developed by grouping of pentlandite grains into subparallel, irregular, lenticular aggregates and laminae (e.g. Kambalda, Scotia) and to breccia ores (Plate III, B) which contain blocks and angular (commonly folded) fragments of ultramafic host rocks, footwall and metasomatic reaction zone lithologies and younger intrusive rocks (e.g. Redross, Windarra, Kambalda offset ores). Isotropic and layered massive ores are typically present in contact positions and breccia ores commonly represent offset mineralization, but contact breccia ores and offset massive ores also occur.

Layered massive sulphide ores were first described by Woodall and Travis (1969) from Lunnon Shoot, Kambalda. Since then, numerous mineralogical studies (e.g. Thornber, 1972; Bennett et al., 1972a, b; Ewers and Hudson, 1972; Ross and Hopkins, 1975) and textural studies (Bayer and Siemes, 1971; Barrett et al., 1977; Ostwald and Lusk, 1978) have been made on the ores. Most layering is crudely parallel to the major bounding surfaces of the massive ore. Thus, layering follows the upper and lower surfaces of contact ore, margins of tongues of offset ore, surfaces of faults along which ore has been emplaced and margins of post-ore intrusives (Ross and Hopkins, 1975). In some Kambalda ores, the layering is developed both by pentlandite laminae and by lenticular aggregates of pyrite. In many cases, the pyrite foliations are at a low angle to pentlandite foliations (Plate III, A), which in turn may be slightly oblique to the orientation of *ab* planes of pyrrhotite (Barrett et al., 1977; Ostwald and Lusk, 1978). Gangue minerals commonly are arranged in bands that accentuate the sulphide mineralogical layering.

Pyrite layers may locally comprise up to 60% of the massive ore at Lunnon Shoot, Kambalda. Descriptions by Woodall and Travis (1969) placed the main pyrite layer at the top of the massive contact ore. Subsequent studies have demonstrated that at both Lunnon Shoot and other Kambalda ore shoots the pyrite occurs throughout the massive ore section and may be represented by spherical or lenticular aggregates as well as layers. It is commonly associated with chalcopyrite in these concentrations. Pyrite is not as abundant in most volcanic-associated ores as at Kambalda, and layered massive ores in these deposits consist essentially of pentlandite laminae in either monoclinic or hexagonal pyrrhotite.

Spinel-rich bands may occur at the base (e.g. Ewers and Hudson, 1972) or through-

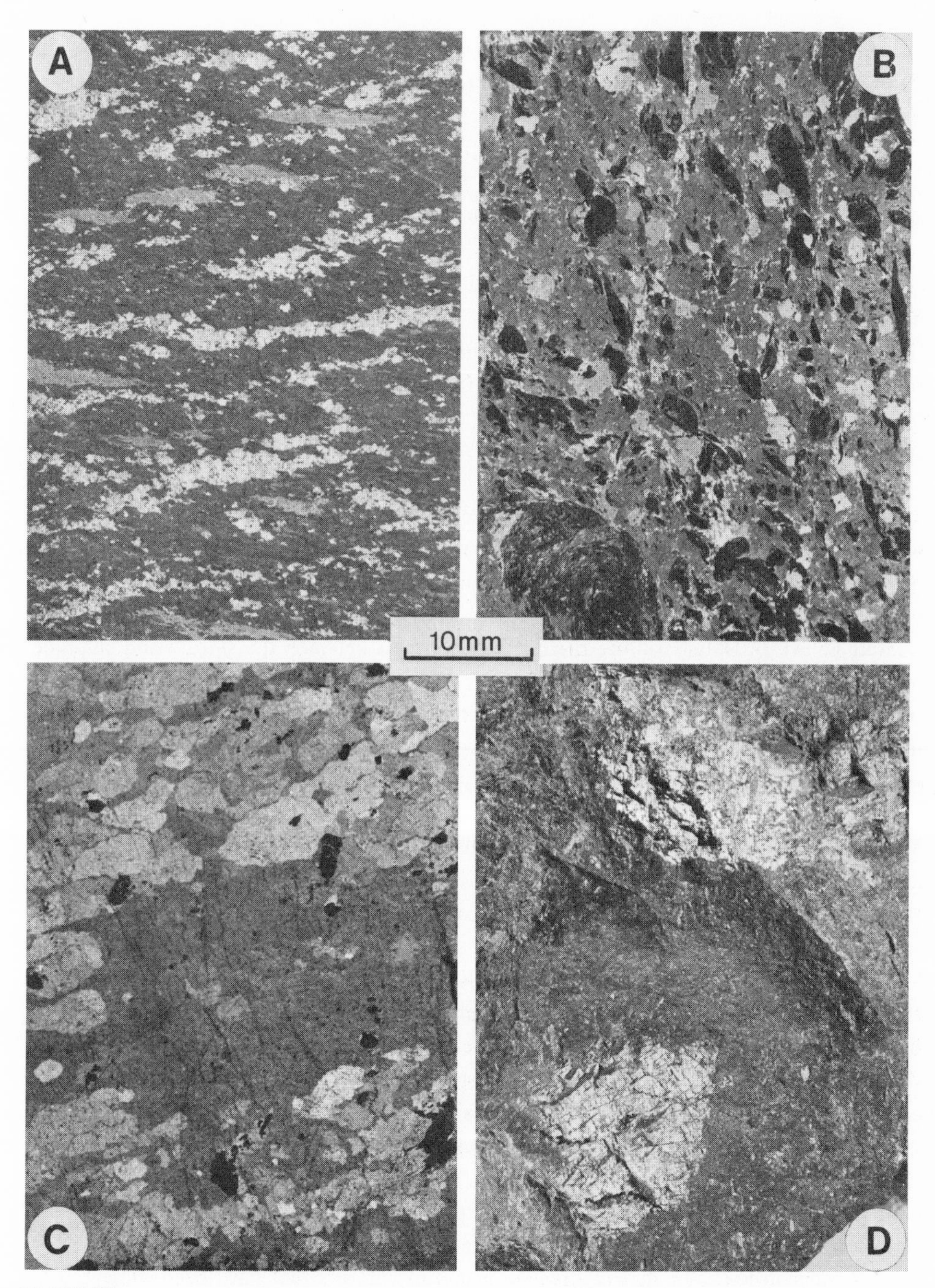

PLATE III

Massive sulphide ores from Western Australia.
A. Banded ore with pentlandite (white), pyrite (light grey) and pyrrhotite (dark grey), Kambalda.
B. Breccia ore with pentlandite (light grey), pyrrhotite (dark grey) and gangue (black), Windarra.
C. Coarse pentlandite (light grey) in pyrrhotite (dark grey), Nepean.
D. Millerite (light grey) in violarite (dark grey), Otter Shoot, Kambalda.

out the massive ore (e.g. Barrett et al., 1977) at Kambalda. They are generally less common in other deposits. They comprise zoned ferrochromites (Groves et al., 1977) and less commonly magnetite.

The massive millerite–violarite ores at Otter Shoot, Kambalda (Plate III, D) are unique and warrant special comment. These form part of a supergene-altered contact ore assemblage that has been described (Keele and Nickel, 1974) as consisting of two different primary sulphide assemblages – a normal pyrrhotite–pentlandite–pyrite assemblage and the Ni-rich assemblage comprising pentlandite–millerite–pyrite. Spherical blebs of millerite–pentlandite–pyrite–magnetite ore occur within the hanging-wall ultramafic (Plate IV, D). Whilst it is possible that the millerite ores are of primary magmatic origin, their compositions are sufficiently different from both the associated Kambalda ores and other similar deposits that the possibility of modifying processes (see below) must be considered.

Matrix ores. Matrix ores are found in a closer and more predictable spatial relationship with the ore-bearing ultramafic unit, and invariably occur near the base of the cumulate or B-zone. They have sharp, commonly tectonic, contacts with massive ore. They consist on average of about 50 wt.% sulphide (Plate IV, A), but show variation between about 20 and 65 wt.%. Details of the original relationships between the silicate and sulphide portions of the ore have been obscured in many cases by alteration and metamorphic recrystallization. Nevertheless, sufficient textural preservation remains to indicate that matrix ores consisted essentially of a framework of ovoid to subhedral cumulus olivines surrounded by sulphide. This is confirmed by rare occurrences of mineralization in relict peridotites from very low-grade metamorphic environments (e.g. Dundonald, Canada: Muir and Comba, 1979). Despite the alteration, the sulphide portion of the ore remains as a continuous, interconnecting and electrically conducting network that surrounds talc, serpentine, amphibole, chlorite, and carbonate pseudomorphs of original silicate minerals.

Both Woodall and Travis (1969) and Ross and Hopkins (1975) prefer to describe these ores as disseminated, pointing out that in the lower-grade, upper portions of the matrix ore zone gangue minerals predominate over sulphides. Whilst this is certainly true, the continuity of sulphides (which is the basis for the definition of matrix ores) persists even when the sulphide content is as low as 20%. The main portion of the ore zone is certainly better described as matrix ore rather than disseminated ore. The term "matrix" was first introduced by Hancock et al. (1971) to describe the Spargoville ores and was adopted by Ewers and Hudson (1972). Similar textures in Canadian Ni–Fe–Cu sulphide ores have been described as "net-textured" (Naldrett, 1973).

Triangular-textured matrix ores are found in deposits that have been subjected to a sufficiently high grade of metamorphism to develop an interlocking three-dimensional framework of bladed metamorphic olivine in sulphide (e.g. Nepean, Plate IV, B). In these ores, there has been subsolidus equilibration between sulphides and olivine, such that the

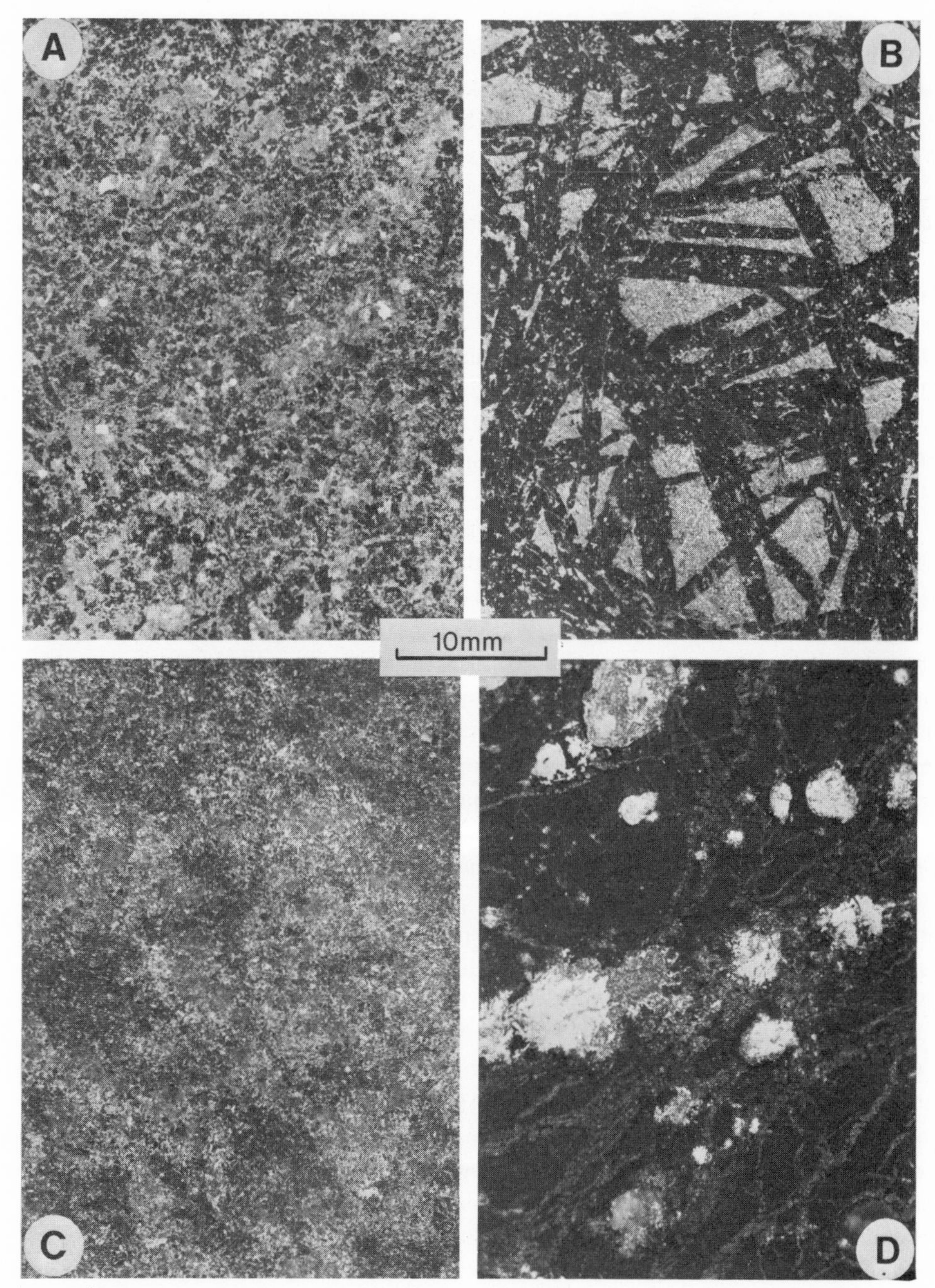

PLATE IV

Matrix and disseminated sulphide ores from Western Australia.

A. Matrix ore with 50–60 wt.% sulphide (grey) surrounding olivine and serpentine (black), Nepean.

B. Triangular textured matrix sulphides (grey) interstitial to partially serpentinised bladed metamorphic olivine (black), Nepean.

C. Disseminated sulphide (light grey) in serpentinised ultramafic, Windarra.

D. Millerite-bearing sulphide "blebs" (white) in hanging-wall ultramafic, Otter Shoot, Kambalda.

latter is significantly more Fe-rich than olivines within immediately adjacent ultramafic rocks (e.g. Hudson, 1973; Barrett et al., 1976).

Sulphides in matrix ores consist of pyrrhotite, pentlandite, pyrite and chalcopyrite. Whilst the ratio of one mineral to another varies greatly from one deposit to another, within any particular ore shoot the matrix ore tends to be more-or-less homogeneous and similar in composition to the associated massive ores. The significance of variations in magnetite to sulphide ratios and relative changes of Cu and Ni in matrix ores is discussed below.

Disseminated ore. Weak disseminations of sulphide, generally less than 5%, occur in the ultramafic overlying matrix ore. The boundary between matrix ore and disseminated ore is usually sharp, being marked by a rapid fall in the amount of sulphide in the rock, an increase in magnetite content, and by absence of electrical conductivity between sulphides.

Disseminated sulphides commonly consist of sulphide aggregates whose mineralogy may be quite different from those of the matrix and massive ore zones. Most disseminated sulphides are composed of assemblages that are more oxidized and Ni-enriched than the underlying ores. At Nepean, Hudson (1973) has described a sequence of mineralogical variations, from the pyrrhotite–pentlandite–pyrite assemblage of the matrix ore through pyrrhotite–pentlandite, pentlandite–millerite to heazlewoodite-bearing disseminated sulphides; and D.A.C. Williams (1979) records millerite-bearing assemblages from disseminated ores at Damba–Silwane in Rhodesia. The change in the mineralogy of disseminated sulphide generally accompanies a progressive fall in total sulphide abundance.

Blebby sulphides of similar composition to other disseminated sulphides may be present throughout mineralized ultramafic units. These blebs, up to 2 cm diameter, consist of sulphides that are intergrown with, or partially replaced by, carbonate, chlorite or antigorite. They have been interpreted to represent original immiscible sulphide droplets (e.g. Keele and Nickel, 1974) or amygdale fillings (e.g. Willett, 1975; D.A.C. Williams, 1979).

Lunnon Shoot ore section. The best-documented single ore section is a specially selected drill hole through Lunnon Shoot from which continuous sections have been analysed for total sulphide, pyrite, magnetite and chromite contents and the sulphide fraction (Fe, Ni, Cu, Co and S contents) by Ewers and Hudson (1972), and Pd and Ir concentrations by Keays and Davison (1976). Analyses of sulphide minerals are also given over the entire section (Ewers and Hudson, 1972). These data are summarized in Fig. 25, and represent important constraints on any ore genetic models.

The major structural and mineralogical features of the ore section from the base upwards are:

(1) interpillow sulphides, having a similar composition to massive ore, within metabasalt footwall;

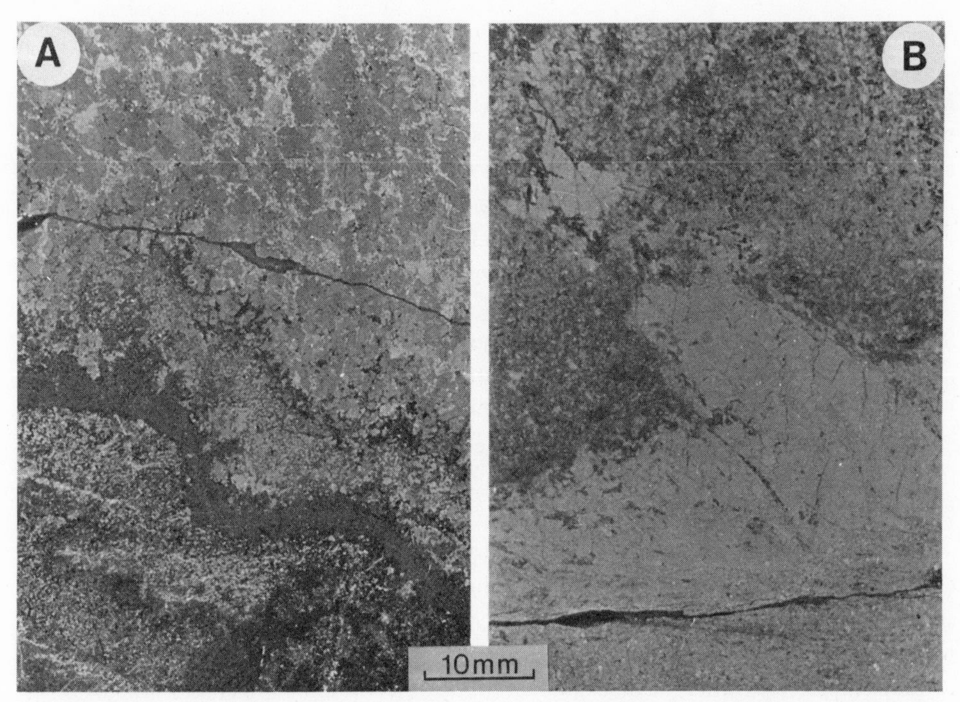

Fig. 26. Sharp contacts between ore types at Kambalda, Western Australia. A. Contact between footwall metabasalt and massive ore is marked by a spinel band (black). B. The sharp but irregular contact between massive ore and overlying matrix ore. Spinel is also concentrated at this interface (see Fig. 25). (After Ewers and Hudson, 1972.)

(2) chalcopyrite-rich stringers extending below massive ore into footwall metabasalt;

(3) a sharp contact between footwall metabasalt and massive ore marked by a thin spinel (ferrochromite and magnetite) layer (Fig. 26A): this may be absent in other sections of Lunnon Shoot ore (Barrett et al., 1977);

(4) layered massive ore passing upwards into pyrite-rich zones that are Co- and Cu-enriched;

(5) a sharp but irregular contact between massive and matrix ores (Fig. 26B), marked by a further ferrochromite ± magnetite concentration: this concentration is very persistent (Barrett et al., 1977);

(6) matrix ore of similar bulk composition to massive ore, containing zones of chalcopyrite enrichment and decreasing in sulphide content upwards with a rapid decrease over the upper metre or so. The ratio of matrix: massive ore is ca. 3 : 1;

(7) sharp contact with disseminated ore marked by an accumulation of magnetite and increase in concentration of Ni and Co in the sulphide fraction; and

(8) barren ultramafic with very weakly disseminated sulphides.

Mineralogy of sulphide ores

Introduction. The opaque minerals present in Western Australian volcanic-associated nickel deposits are listed in Table XI. There are a large number of rare phases in the ores and the oxidized zone above the ores (not shown in Table XI) has been particularly fruitful for mineralogists, with a large number of rare and new minerals being discovered. However, only six common opaque minerals in most ores have been recognized, namely pyrrhotite, pentlandite, pyrite, chalcopyrite, magnetite and ferrochromite.

The ore system can thus be considered simply in terms of the system Fe–Ni–Cu–S–O with Co and Cr as important, but minor additional components. Relevant experimental work in this system includes the studies of Kullerud (1963), Naldrett et al. (1967), Naldrett (1969), Kullerud et al. (1969), Craig and Kullerud (1969), Shewman and Clark (1970), Barton (1973) and Ewers et al. (1976). The compositions of pentlandite in equilibrium with other Fe–Ni sulphides are summarized by Harris and Nickel (1972) and Misra and Fleet (1973).

Studies of the system Fe–Ni–S at high temperature indicate that sulphide liquids commence crystallization as Fe-rich solid solutions above 1100°C and that these monosulphide solid solutions (*mss*) progressively contain more Ni with declining temperature to ca. 990°C, where *mss* extends continuously from $Fe_{1-x}S$ to $Ni_{1-x}S$. This *mss* persists to low temperatures where it breaks down to the observable ore minerals. Pyrite or pentlandite may first appear at 600°C if the composition of the *mss* is adjacent to the S-rich or S-poor limits respectively at that temperature (Fig. 28), and mineral phases continually crystallize from *mss* to much lower temperatures; pyrite and pentlandite will not crystallize simultaneously until below 300°C. At low concentrations, as in the nickel ores, Co can be accommodated in the low-temperature Fe–Ni sulphides. Significant Cu concentrations may occur in high-temperature *mss*, but chalcopyrite will crystallize at low temperature; for example chalcopyrite starts crystallizing at 450°C from *mss* containing 1.5 wt.% Cu, approximately the percentage in volcanic-associated ores. Oxygen will dissolve in high-temperature Fe (+ Ni)-rich melts and depresses the solidification temperature. The first solid to crystallize from an oxy-sulphide melt is *mss* leaving an oxygen-enriched liquid until a magnetite-*mss* cotectic is reached.

Irrespective of the primary origin of the ore, the experimental work is highly relevant to crystallization of the ore minerals, as all Western Australian ores have been metamorphosed under conditions where temperatures are at or above 500°C. Metal diffusion rates are extremely rapid in Fe–Ni sulphides at such temperatures (e.g. Klotsman et al., 1963; Ewers, 1972; Condit et al., 1974; McQueen, 1979), so that during peak metamorphism the volcanic-associated ores would have consisted largely of *mss* with some pyrite or pentlandite in S-rich and S-poor ores, respectively, together with the spinel phases. If Cu-rich layers existed prior to metamorphic heating, localized areas of intermediate solid solution (*iss*), rather than Cu-bearing *mss,* may have been developed. As metamorphic cooling is likely to be more prolonged than simulated magmatic cooling in experiments,

TABLE XI

Opaque mineralogy of Western Australian volcanic-associated nickel deposits

Primary zone		Supergene zone
Common	Rare	
Native elements		
	Awaruite	Copper
	Bismuth	Gold
	Carbon	
	Copper	
	Gold	
	Nickel	
	Platinum	
Oxides		
Ferrochromite	Ilmenite	Cryptomelane
Magnetite	Pyrophanite	Goethite
	Rutile	Haematite
	Spinel	
	Trevorite	
Sulphides		
Chalcopyrite	Argentian Pentlandite	
Pentlandite	Bismuthinite	Covellite
Pyrite	Bornite	Greigite
Pyrrhotite	Bravoite	Marcasite
	Chalcocite	Millerite
	Cinnabar	Pyrite
	Cubanite	Smythite
	"Chalcopentlandite"	Violarite
	Galena	
	Godlevskite	
	Heazlewoodite	
	Joseite A	
	Mackinawite	
	Millerite	
	Molybdenite	
	Parkerite	
	Polydymite	
	Sphalerite	
	Troilite	
	Vaesite	
	Valleriite	
	Violarite	
Tellurides		
	Altaite	
	Calaverite	
	Hessite	
	Melonite	
	Michenerite	
	Rucklidgeite	
	Tellurobismuthite	
	Testibiopalladite	

(continued)

TABLE XI (continued)

Primary zone		Supergene zone
Common	Rare	
Arsenides and antimonides		
	Arsenopyrite Cobaltite Gersdorffite Maucherite Niccolite Skutterudite Sperrylite Stibnite Ullmanite	

sulphide minerals should have compositions typical of low-temperature equilibration (e.g. Misra and Fleet, 1973).

Pyrrhotite. Pyrrhotite is the dominant sulphide mineral of the ores, forming a matrix to the other opaque phases. This is in agreement with the experimental studies that predict pyrrhotite as the residual phase resulting from progressive exsolution of excess S (as (pyrite), Ni (as pentlandite) and Cu (as chalcopyrite) from cooling *mss.* Monoclinic pyrrhotite (Fe_7S_8) and hexagonal pyrrhotite (Fe_9S_{10}) are both dominant in different ore deposits, with monoclinic pyrrhotite being the most common in pyrite-rich ores (e.g. Kambalda). Intergrowths of these two types may occur, troilite may be exsolved in hexagonal pyrrhotite in pyrite-free examples, or hexagonal pyrrhotite may be replaced by monoclinic pyrrhotite during retrogressive alteration (e.g. Nepean). The (001) or (0001) planes of pyrrhotite have been important in controlling exsolution, and metamorphic and supergene alteration of pyrrhotite, and are themselves commonly strongly oriented (Fig. 27A). A lamellar magnetic domain structure may be defined in monoclinic pyrrhotite by application of magnetic colloid (Bennett et al., 1972a). This is thought to result from a lamellar intergrowth of pyrrhotite with exsolved submicroscopic magnetite or S-rich, Fe-sulphides such as smythite (Bennett et al., 1972b).

Pyrrhotites typically contain less than 1% (Ni + Co) with monoclinic pyrrhotite from Kambalda having 0.5–0.6% Ni and 0.06–0.08% Co (Ewers and Hudson, 1972; Nickel et al., 1974), monoclinic and hexagonal pyrrhotite from Spargoville having 0.24–0.65% Ni (Ramsden, 1975), and hexagonal pyrrhotite from Nepean having both Ni and Co contents at, or below 0.1% (Hudson, 1973; Barrett et al., 1976). These low Ni and Co contents and low Co/Ni ratios are typical of pyrrhotites that equilibrated at low temperature (cf. Misra and Fleet, 1973).

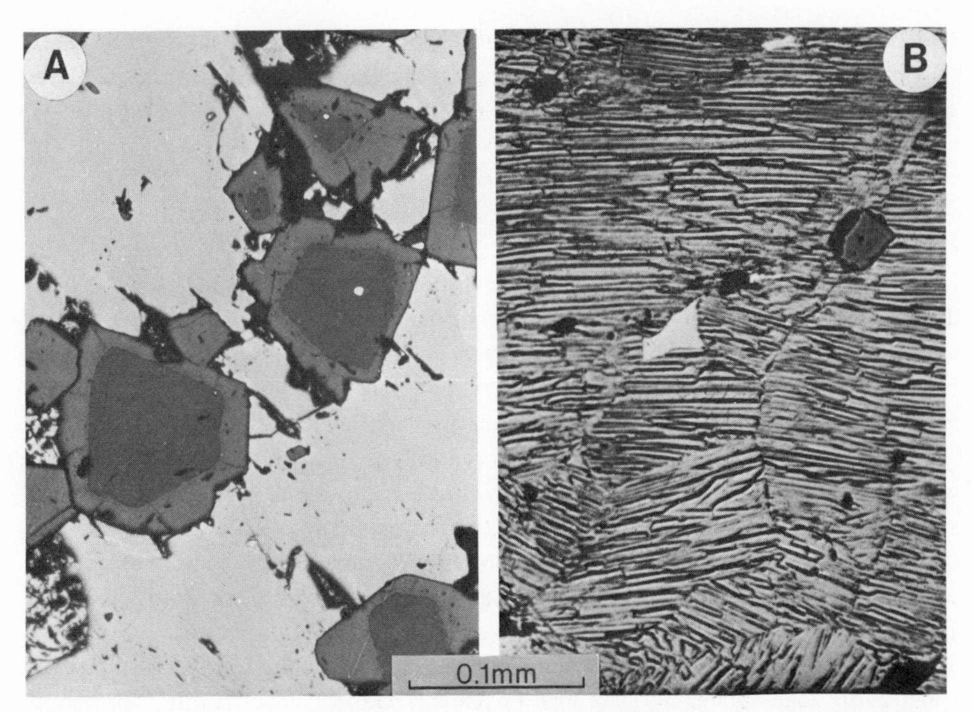

Fig. 27. Photomicrographs of polished sections of ore from Kambalda, Western Australia. A. Zoned spinel euhedra in sulphide from the spinel band at the base of massive ore zone, Lunnon Shoot. Magnetite rims occur on a ferrochromite that is believed to have crystallised from the sulphide melt. B. Magnetic domains parallel to (001), defined by magnetic colloid, indicating the strong orientation of monoclinic pyrrhotite in banded massive ores.

Pentlandite. Pentlandite typically occurs as relatively coarse-grained discrete grains (e.g. Fig. 31) or discontinuous, lens-like grain-aggregates in a pyrrhotite matrix, in agreement with relations in the experimental system. Distinct layers rich in coarse-grained pentlandite may be present in some ores (e.g. Nepean). Less commonly, pentlandite occurs as small flame-like exsolutions along (001) or (0001) planes of pyrrhotite.

The composition of pentlandite varies in accord with the nature of co-existing sulphide assemblages. The most Ni-poor pentlandites (19–27 at.% Ni) coexist with hexagonal pyrrhotite ± troilite or pyrite (Barrett et al., 1976), whereas pentlandite coexisting with dominantly monoclinic pyrrhotite + pyrite assemblages contains 28 ± 2 at.% Ni with Ni/Fe ratios close to unity (Ewers and Hudson, 1972; Ramsden, 1975). More Ni-rich pentlandite occurs in millerite- or heazlewoodite-rich disseminated ores (Ewers and Hudson, 1972; Hudson, 1973). The Co content of pentlandites is highly variable. Most pentlandites from pyrrhotite-dominated assemblages contain 0.3–0.8 wt.% Co, but pentlandite from zones rich in coarse-grained pyrite may be strongly depleted in Co (to 0.1%) relative to pentlandite in pyrrhotite-dominated layers (e.g. Ewers and Hudson, 1972). Pentlandite from weakly disseminated sulphide aggregates may contain in excess of

10 wt.% Co (Hudson, 1973). Despite the relatively low Co contents of most pentlandites, they contain the bulk of Co in the sulphide ores. The composition of pentlandite is generally in agreement with low-temperature equilibration (cf. Misra and Fleet, 1973), although a wide range in composition of pentlandites from ores that have undergone retrogressive alteration may indicate local disequilibrium (e.g. Barrett et al., 1976).

Pyrite. Pyrite occurs in a variety of forms. It may be scattered throughout pyrrhotite matrix as discrete, unfractured subhedal to euhedral grains; occur as coarse, zoned euhedra (several centimetres in side length) in pyrite-rich layers (e.g. Lunnon Shoot); as fine- to medium-grained, commonly fractured subhedra or euhedra in more typical pyrite-rich layers; as vein-like forms replacing pyrrhotite along grain boundaries; or as pyrite veins, some of which replace pyrrhotite-rich massive ore (Barrett et al., 1977).

Unlike pyrrhotite and pentlandite, that generally show a small range in composition within any one ore type in each deposit, pyrite has highly variable Ni and Co contents and Co/Ni ratios. Concentrations range from less than 0.1% to 2.5% Co and from less than 0.1% to 3.5% Ni in the single ore section from Lunnon Shoot (Ewers and Hudson, 1972) where pyrite is abundant and several different textural types are present, but have a more restricted range in relatively pyrite-free ores (e.g. Nepean). The variation is partly due to compositional zoning: for example, the coarse-grained pyrites from the Lunnon Shoot pyrite zones commonly have high Co and low Ni cores, but rims with significantly lower Co/Ni ratios. The variation may also be due to crystallization of pyrite from metamorphically generated *mss* over a wide temperature range (e.g., McQueen, 1979), combined with the occurrence of pre-metamorphism pyrites that were preserved in a matrix of *mss* in relatively S-rich ores subjected to metamorphic temperatures at the lower end of the range (e.g. Kambalda). The occurrence of late-formed pyrite replacing pyrrhotite further complicates the picture.

Chalcopyrite. Chalcopyrite occurs as discrete grains within a pyrrhotite matrix, as an interstitial phase in pyrite-rich layers in which pyrite and chalcopyrite may also form a symplectite-like intergrowth (Barrett et al., 1977), as irregular veinlets in footwall rocks and irregular patches in matrix ores. It has low Co and Ni contents ($<0.1\%$). In places it contains, or is associated with, minor sphalerite whose composition (ca. 12 ± 2 mole % FeS) further suggests low-temperature equilibration of sulphide phases (Groves et al., 1975).

Ferrochromite. One of the most interesting and genetically important minerals occurring in massive and matrix ores is a distinctive chromite, termed ferrochromite by Groves et al. (1977), with very low Al and Mg and low, but significant, Ti, Mn and Zn contents. These ferrochromites, rimmed by variable thickness of magnetite, occur in most Western Australian volcanic-associated ores (Groves et al., 1977), some Rhodesian ores (e.g. Trojan, Perseverance: UWA unpublished data) and importantly similar, more V-rich chromites occur in association with sulphides from *unmetamorphosed* deposits in the La Perouse (Alaska) gabbro (Czamanske et al., 1976) and from Insizwa, South Africa (UWA unpublished data). In the volcanic-associated ores, the rimmed ferrochromites most commonly occur as discrete grains or small clusters of grains (e.g. Fig. 27B), but ferrochromite-rich layers, commonly with fractured ferrochromite grains veined by sulphides, may be developed, particularly at Kambalda. At Lunnon Shoot, the most persistent layer occurs at the base of matrix ores, but layers may also be developed at the base of and within massive ore.

Ferrochromites contrast markedly with chromites in the overlying ultramafic unit that are considered to have crystallized from the silicate magma. The ferrochromites may be up to ten times larger than the largest "lithophile" chromites and have non-overlapping, sharply contrasting compositions (Fig. 16). The experimental demonstration by Ewers et al. (1976) of the solubility of Cr in sulphide melts and the crystallization of "chalcophile" chromite euhedra with mantles of magnetite similar to those of natural ores, combined with the restriction of ferrochromite to the sulphide fraction of a number of nickel ores, is strong evidence for their crystallization from oxy-sulphide liquids during the formation of the nickel ores (Ewers et al., 1976; Groves et al., 1977).

Some chromites occurring within the ore zone in metasomatic reaction zones have compositions that are constrained by coexisting metamorphic silicate phases and are almost certainly metamorphic chromites (Groves et al., 1977). Similarly, aluminous chromites within the high metamorphic grade Nepean ores are probably metamorphic in origin.

Magnetite. The variety of magnetite types that occur in the nickel ores has been documented by Groves et al. (1977). The most abundant type is ragged magnetite occurring most commonly at boundaries between sulphide and silicate grains in matrix and disseminated ores: it is responsible for the high magnetite content of these ore types (Fig. 25). Concentrations of this type of magnetite commonly occur at the upper boundary of matrix ores (Fig. 25). Its composition is similar to magnetite in overlying altered ultramafic rocks with both having increasing Cr contents at higher metamorphic grades.

It is difficult to determine whether this magnetite formed with the sulphides (e.g. Ewers and Hudson, 1972) or formed during later serpentinization of the overlying ultramafic unit (e.g. Barrett et al., 1976; Groves et al., 1977). In the latter case, migration and fixation of Fe at the upper contact of significant sulphide concentration must be invoked. Small subhedral, Cr-poor magnetites that are completely enclosed within sulphides in massive ores are best explained in terms of magmatic crystallization (Groves et al., 1977), presumably after available Cr was fixed in ferrochromite.

Other minor types of magnetite include small blebs exsolved (?) in pyrrhotite, secondary magnetite veinlets in pentlandite cleavages, retrograde magnetite veinlets in sulphides and fine-grained magnetite-pyrite intergrowths representing oxidation of pyrrhotite during retrograde serpentinization (e.g. Nepean).

TABLE XII

Bulk compositions of ore types from the Eastern Goldfields Province

Deposit		Fe	Co	Ni	Cu	S	Ni/Fe	Ni/Co	Ni/Cu
A: Composite samples of mined ore [1] *and concentrates* [2] *(from Ross and Keays, 1979)*									
Lunnon Shoot, Kambalda [1]		47.8	0.30	12.1	1.02	38.8	0.25	40	11.7
McMahon Shoot, Kambalda [1]		45.9	0.31	12.0	1.05	40.7	0.26	38	11.5
Fisher Shoot, Kambalda [1]		46.6	0.34	13.7	1.14	38.2	0.29	40	12.1
Juan Shoot, Kambalda [1]		46.7	0.29	14.1	0.95	38.0	0.30	48	14.8
Ken Shoot, Kambalda [1]		39.1	0.33	15.0	1.19	44.4	0.38	46	12.6
Durkin Shoot, Kambalda [1]		43.9	0.32	17.2	1.24	37.3	0.39	53	13.9
Nepean [1]		41.8	0.34	20.1	1.28	36.5	0.48	60	15.7
Windarra [2]		39.5	0.32	14.9	1.36	43.9	0.38	47	10.9
Scotia [2]		39.4	0.36	23.1	1.45	35.7	0.59	63	15.9
B: Average composition of ore types calculated from individual core or specimen analyses: from Hudson (1972, 1973) [1,2]*, Ewers and Hudson (1972)* [3]*, Barrett et al. (1976)* [4]*, Seccombe et al. (1977)* [5]*, Ross and Keays (1979)* [6]*, McQueen (1979)* [7] *and Marston and Kay (in press)* [8]*. * (Calculated from ore reserve data.)*									
Lunnon, Kambalda [3]	massive	50.6	0.22	9.2	0.48	39.5	0.18	42	19 (13)
Lunnon, Kambalda [3]	matrix	50.5	0.19	9.6	1.21	38.6	0.19	51	8 (13)
Lunnon, Kambalda [3]	average	50.6	0.20	9.4	0.87	39.0	0.19	47	11 (13)
Juan, Kambalda [8]	massive layered	44.9	0.17	13.6	0.65	40.5	0.30	80	21 (13)
Nepean [2]	matrix	39.9	0.3	22.4	0.7	36.6	0.56	75	32 (15)
Nepean [4]	matrix/triangular	41.3	0.30	21.1	0.93	37.2	0.51	70	23 (15)
Nepean [4]	massive	42.3	0.30	20.0	0.21	36.7	0.47	67	95 (15)
Windarra [5]	disseminated	45.3	0.25	14.7	1.0	38.7	0.32	55	15 (11)
Windarra [5]	massive	49.6	0.14	13.0	0.2	37.2	0.26	94	65 (11)
Windarra [5]	breccia	53.0	0.06	5.5	0.4	41.9	0.10	98	14 (11)
Wannaway [7]	average	47.7	0.31	14.3	1.12	36.6	0.30	46	13 (12)
Wannaway [7]	average (serp.)	50.7	0.25	11.0	1.19	36.8	0.22	44	9 (12)
Wannaway [7]	average (unserp.)	43.3	0.39	19.0	0.96	36.3	0.44	48	20 (12)
Mount Edwards [1]	average	51.0	0.3	11.0	1.0	37.0	0.22	37	11 (11)
Scotia [1]	average	40.0	0.3	20.0	1.0	39.0	0.50	67	20 (12)
Redross [6]	average	45.8	0.31	14.4	1.29	38.2	0.31	46	11 (13)

All compositions as percentages in terms of 100% sulphides. Details of analyses, etc., given in references.

Ore compositions. Bulk-ore compositions are presented in Table XII and Fig. 28, and are compared to those of interflow metasediments in Fig. 12. Composite samples of mined ore have a relatively restricted range of Ni/Cu ratios from 12 to 16 and Ni/Co ratios from 40 to 60 (Table XIIA). Compositional estimates based on a number of smaller samples (Table XIIB) provide a much wider spread and consistently higher Ni/Cu, and to a lesser extent Ni/Co, relative to compositions based on composite samples and ore reserves. This is due to selective concentration of Cu (and Co) on a relatively small scale, particularly in footwall stringers which are not normally included in the sampling.

Nickel/Fe ratios calculated for small-scale sample sets agree well with those determined on composite samples. The total range of Ni/Fe ratios is proportionally greater than that of reliable Ni/Cu and Ni/Co ratios, showing a total variation from 0.10 to 0.56, or 0.18 to 0.56, if Windarra breccia ore is excluded. The wide range is well illustrated by Fig. 28, which also demonstrates the restriction of bulk-ore compositions to the experimentally determined field of monosulphide solid solution (*mss*) at high temperatures. There is no consistent difference of Ni/Fe or Ni/Co ratios between massive and matrix ores, but the latter have consistently lower Ni/Cu ratios than the massive ores. At Wannaway (McQueen, 1979), disseminated to matrix ores in serpentinized peridotite have significantly lower Ni/Fe ratios (mean 0.30) than those in serpentinized peridotite (mean Ni/Fe = 0.44). Zinc contents of analysed ores are less than 0.14% Zn and normally less

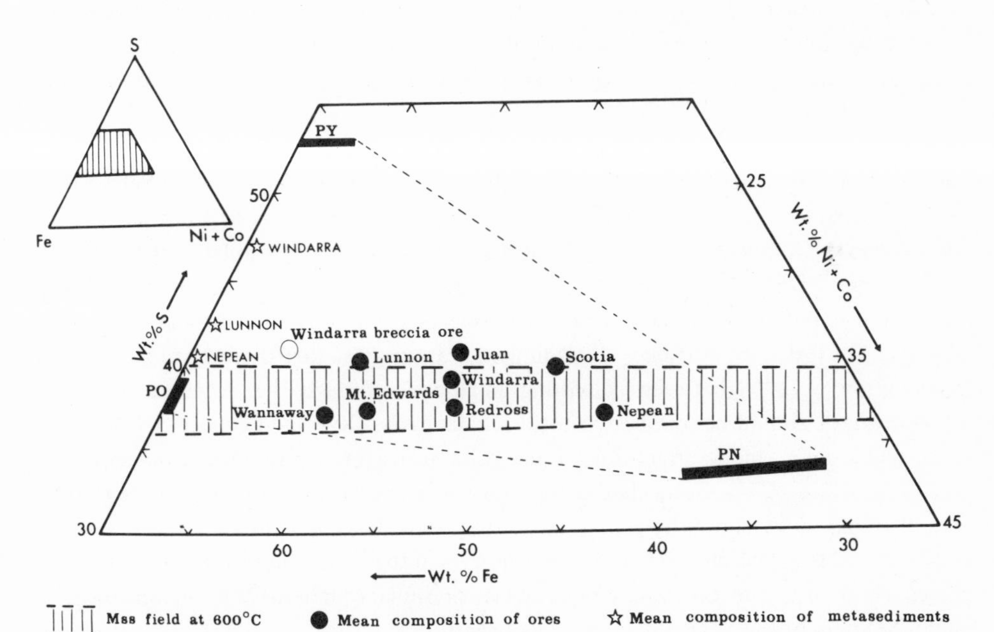

Fig. 28. Variation in the composition of some Western Australian volcanic-associated Ni–Fe–Cu sulphide ores and their contained sulphide minerals. (From Groves et al., 1979.) Values for Wannaway are of total ore including the contact vein. More precise compositions for Redross ore are given by McQueen (1979).

than 0.02% Zn in 100% sulphides (e.g. Barrett et al., 1976; Seccombe et al., 1977).

The ores contain significant concentrations of platinoids and Au (Keays and Davison, 1976; Ross and Keays, 1979). In bulk ores (100% sulphides), they exhibit ranges of 1000–2500 ppb Pt, 500–3500 ppb Pd and 200–450 ppb Ir. Each element, particularly Pd and Ir, shows a strong positive correlation with Ni. The Kambalda ores have arithmetic mean Pd and Ir contens of ca. 360 ppb and 60 ppb respectively (ca. 1560 and 255 ppb in 100% sulphides), and a mean Pd/Ir ratio of ca. 6.0 which is similar to that of sulphide-free metapicrites within the Kambalda sequence but contrasts with the Pd/Ir ratio of ca. 35 for Kambalda interflow metasediments (Bavinton and Keays, 1978). Palladium/Ir ratios for Redross and Nepean ores are ca. 6.5 and 8.5 respectively (Ross and Keays, 1979). Gold contents in 100% sulphides range from 150 to 2500 ppb (Ross and Keays, 1979) and Au/(Pd + Ir) ratios of 0.26 contrast with ratios of ca. 14.5 for Kambalda interflow metasediments (Bavinton and Keays, 1978). Naldrett et al. (1979) show that sulphides from volcanic-associated ores have characteristic levels of platinoid elements and Au close to or slightly lower than average values in chondrites, whereas mafic-associated nickel deposits (e.g. Class B, Table II) tend to have higher (Pt + Pd + Rh)/(Ru + Ir + Os) ratios.

Sulphur/Se ratios of ores are normally in the range 5000–15,000 which is typical of magmatic sulphides (e.g. Hatten, 1977). The major exception in Western Australia is Windarra, which shows a wide range of S/Se to ca. 50,000, and the Langmuir deposit, Canada has a mean S/Se ratio of ca. 30,000 (A.H. Green, 1978). Sulphur/Se ratios for ores and metasediments from the same localities are remarkably similar (Fig. 13). Available data suggest that there is no significant fractionation of Se between major sulphide minerals in the ores.

$\delta^{34}S$ values for most ore minerals are in the range 0 to +3.5‰ (Fig. 14), being typical values of magmatic sulphides from elsewhere (e.g. M. Shima et al., 1963; Schwartsz, 1973). The Windarra sulphides are, however, significantly more ^{32}S enriched with $\delta^{34}S$ values ranging from –2.5 to +1‰. Langmuir sulphides show a similar range of $\delta^{34}S$ (A.H. Green, 1978). Once again the similarity between $\delta^{34}S$ values of sulphides from ores and metasediments in the same environments from Western Australia is notable (Fig. 14). As for S/Se, available, albeit limited, data suggest that there is no systematic fractionation of $\delta^{34}S$ between coexisting sulphides, although the expected order of ^{34}S enrichment should be pyrite, pyrrhotite, chalcopyrite (Rye and Ohmoto, 1974). This may result from a kinetic effect due to unmixing of discrete ore minerals from *mss* (e.g. Seccombe et al., 1977). Solid-state diffusion of S is several orders of magnitude slower than diffusion of Fe and Ni in sulphides, so it appears likely that the diffusion of metal ions is the most important rate-controlling process in the appearance of discrete sulphide phases within *mss*. The exsolving mineral phase presumably inherits the sulphur isotopic (and S/Se) ratio of the originally homogeneous *mss*.

The silicate fraction of matrix and disseminated ores is dominated by the normal mineralogy of the hosting ultramafic unit and has a similar geochemical composition (Table XIII). The massive ore normally contains less than 10% silicate fraction which has

TABLE XIII

Geochemistry of silicate fraction of some sulphide ores compared to mean composition of ultramafic hosts to volcanic-associated ores (From Binns et al., 1977)

	1	2	3	4	5	6	7
SiO_2	35.5	39.0	46.5	48.8	44.1	48.8	45.7
TiO_2	0.07	0.13	n.d.	n.d.	0.19	0.13	0.21
Al_2O_3	1.8	2.2	0.3	0.4	3.9	2.8	3.1
Fe_2O_3	10.7	4.0	15.0(t)	3.0(t)	1.9	3.0(t)	
FeO	5.9	5.4			9.6		8.8(t)
MnO	0.39	0.42	n.d.	n.d.	0.12	n.d.	0.13
MgO	21.9	35.9	15.0	44.6	36.1	42.5	40.0
CaO	16.5	12.1	22.6	3.0	3.8	1.8	1.5
Na_2O	0.04	0.02	n.d.	n.d.	0.12	0.3	0.05
K_2O	0.04	0.05	n.d.	n.d.	0.04	n.d.	0.08
P_2O_5	0.04	0.02	n.d.	n.d.	0.01	n.d.	0.02

All analyses quoted volatile-free.
1 = Massive ore, Wannaway (McQueen, 1979).
2 = Massive ore, Wannaway (McQueen, 1979).
3 = Massive ore, Lunnon Shoot (Ewers and Hudson, 1972).
4 = Matrix ore, Lunnon Shoot (Ewers and Hudson, 1972).
5 = Disseminated ore, Wannaway (McQueen, 1979).
6 = Very disseminated ore, Lunnon Shoot (Ross and Hopkins, 1975).
7 = Average of 58 mineralized ultramafic units (Binns et al., 1977).

an unusual composition, dominated by SiO_2, MgO, CaO and FeO and typically with very low Al_2O_3 (Table XIII). It contrasts with all lithologies in the sequence including metabasalts (Table V), metasediments (Table VI) and ultramafic rocks (Tables IX and X); it is most similar to the composition of amphibole-dominated metasomatic reaction zones or quartz–carbonate veinlets.

Geochemical profiles through the ore zone are rare, although available data suggest that maximum Cu and Ni concentrations are commonly offset (e.g. Barrett et al., 1976). The most intensively studied section is the Lunnon Shoot core, described by Ewers and Hudson (1972) and shown in Fig. 25. Nickel and Fe show a relatively constant concentration in the sulphide fraction of massive and matrix ores with minor depletion in the S-rich pyrite zones. Cobalt is essentially constant over most of the section with notable enrichment in the pyrite zone and depletion in adjacent zones, and enrichment, together with Ni, in disseminated sulphides. Copper shows a highly erratic distribution with notable features being erratic high Cu values in matrix ore, high Cu concentration in pyrite zones with adjacent zones depleted in Cu, Cu depletion at the base of massive ore, and enrichment in footwall metabasalt.

Palladium and Ir vary systematically through the ore profile (Keays and Davison, 1976; see Fig. 25). Iridium remains essentially constant in massive ore, but decreases in concentration systematically through the matrix ore, whereas Pd increases upwards throughout the ore section, showing minor relative depletion in zones adjacent to anomalous Cu concentrations. The base of the massive ore also has low Pd contents, whereas the adjacent sulphides in footwall metabasalts are strongly enriched in Pd. Remobilized sulphides from elsewhere (e.g. Nepean) also appear to be Pd-rich (Keays and Davison, 1976). The profile thus reveals a marked increase in Pd/Ir ratio up the ore section and a significantly higher Pd/Ir ratio for matrix relative to massive ore (see also Ross and Keays, 1979), even when allowance is made for remobilization. Keays and Davison (1976) have interpreted this variation in terms of fractional crystallization of magmatic sulphide liquid from the base upwards, with Ir and Pd enrichment in early- and late-crystallized sulphides, respectively.

STRUCTURAL AND METAMORPHIC MODIFICATION OF ORE

Introduction

Evidence from Western Australia indicates that the Ni–Fe–Cu sulphide ores shared the deformational and metamorphic history of enclosing rocks, which in many cases involved three major deformational events (Archibald et al., 1978), amphibolite facies metamorphism, and some minor retrogressive alteration and associated deformation. These events were important influences on the present configuration, textures and, in some cases, composition of the ores (Barrett et al., 1977). Available data suggest that metamorphic and deformational events were similarly important in modifying Rhodesian deposits (e.g. D.A.C. Williams, 1979; Anderson et al., in press) and Canadian deposits (e.g. A.H. Green, 1978), although peak metamorphic temperatures were probably lower.

The structural setting and orientation of the ore bodies, together with the metamorphic imprint on ore environments, have been emphasized above, and detailed metamorphic and structural histories are presented for a number of representative deposits by Barrett et al. (1977). These are not repeated here, but a schematic history of events at the Windarra deposit is presented in Fig. 29; other deposits have essentially similar histories. The major effects of ore modification discussed by Barrett et al. (1977) are briefly reviewed below.

Mesoscopic features

The sulphide ores, and in particular the massive ores, have represented low-strength layers throughout deformation. Remobilization of massive ores relative to more disseminated ores, with relative movement of at least tens of metres, is common to all deposits and massive ore may be thickened in fold closures and footwall and hanging-wall irregularities. Remobilization has occurred prior to and during metamorphism in several deposits, and massive ore veins cutting metamorphosed but largely undeformed porphyry dykes

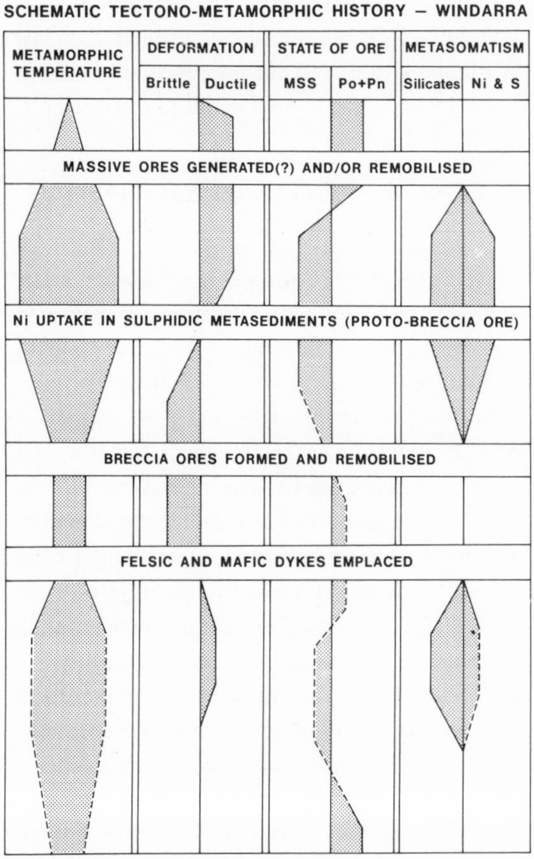

SCHEMATIC TECTONO-METAMORPHIC HISTORY – WINDARRA

METAMORPHIC TEMPERATURE | DEFORMATION: Brittle, Ductile | STATE OF ORE: MSS, Po+Pn | METASOMATISM: Silicates, Ni & S

MASSIVE ORES GENERATED(?) AND/OR REMOBILISED

Ni UPTAKE IN SULPHIDIC METASEDIMENTS (PROTO-BRECCIA ORE)

BRECCIA ORES FORMED AND REMOBILISED

FELSIC AND MAFIC DYKES EMPLACED

Fig. 29. Diagrammatic representation of the complex tectonometamorphic history of the Windarra deposit which is typical of many Archaean volcanic-associated Ni–Fe–Cu sulphide ores.

attest to late-stage mobilization of massive sulphides in a number of cases (see Barrett et al., 1977 – Fig. 11). The common development of intrafolial folds in and adjacent to massive ores probably reflects locally accelerated strain rates within this ore type.

Patchy development of massive sulphide stringers in matrix ore and matrix concentrations in disseminated ores suggest metamorphic upgrading of sulphides, but direct evidence for metamorphic formation of major massive sulphide concentrations is rare. A major obstacle to complete understanding of the genesis of massive ores is that once formed they moved independently of more disseminated ores during deformation, thus destroying much of the evidence for their mode of formation. However, Barrett et al. (1976) present evidence for formation of metamorphic massive sulphides around some flexures and in a major shear zone in which sulphides are remobilized from the basal contact of one unit to the upper contact of an adjacent unit at Nepean. They also describe the formation of contact massive sulphides in thin slivers of matrix ore faulted into footwall amphibolites at the same mine. Diffusion of S and metals along high chemical potential gradients provide an explanation for the latter case, but the precise mechanism of upgrading is not clear in the other cases. In high-grade metamorphic environments, dehydration reactions in silicate phases such as:

$$5\ Mg_3Si_2O_5(OH)_4(\text{serpentine}) \rightarrow 6\ Mg_2SiO_4(\text{forsterite}) + Mg_3Si_4O_{10}(OH)_2(\text{talc}) + 9\ H_2O$$

may cause a small relative increase in sulphide volume.

Breccia ores appear to form late in the deformational history (Fig. 29), when metamorphic temperatures had declined sufficiently to allow brittle fracture of deformed country rocks and metasomatic reaction zones and incorporation into the lower strength sulphides. Whether breccia ores represent mobilized massive, more disseminated or even Ni-enriched metasediments, is not always clear.

Mineralogical layering of ores is almost certainly the result of deformational and/or metamorphic segregation. The layering is normally subparallel to the boundaries of massive ore layers or lenses even in remobilized ores that cut or are remobilized along the margins of late porphyry dykes: pre-metamorphic layering would anyway have been destroyed during metamorphic heating in which the ores largely reverted to *mss* (McQueen, 1979). In structurally complicated areas, there may be complex relationships between layered and unlayered massive ores and in S-rich ores (e.g. Kambalda), fine-grained pyritic ore and pyrite lenses may be superimposed on layered pyrrhotite-rich ore (e.g. Barrett et al., 1977 – Fig. 10). The latter has been interpreted to result from stress-induced S diffusion (Barrett et al., 1977). Solid-state diffusion of S is extremely slow, and experiments show that S diffuses much faster along fractures and in zones of high porosity than in more massive parts of the sample, probably due to vapour transport along grain boundaries (Condit et al., 1974). Thus S diffusion is thought to have occurred at a late stage in the metamorphic history, when there was significant brittle rather than ductile deformation and/or when massive ores underwent thermal contraction.

All massive and many matrix ores show extreme segregation of Cu as chalcopyrite, which is enriched in footwall veinlets, in pyrite-rich bands and at the extremities of and around silicate inclusions in breccia ores. Stress-induced diffusion of Cu rather than mechanical segregation of chalcopyrite is thought responsible for the observed segregation (Barrett et al., 1977). Ores that suffered significant pre-metamorphic deformation show a wide spread in Ni/Fe ratios on the large sample scale (e.g. Nepean: Barrett et al., 1976), presumably due to segregation of more ductile pyrrhotite from pentlandite during deformation of *mss*-poor ore. In contrast, massive ores that suffered predominantly synmetamorphic deformation have a very limited range of Ni/Fe ratios and show little evidence of segregation of pyrrhotite from pentlandite at this scale (e.g. Lunnon Shoot, Barrett et al., 1977). This is consistent with their deformation as *mss* (Fig. 29). Some S-rich ores (e.g. Lunnon Shoot) show a trend towards S-rich compositions at constant Ni/Fe ratio, consistent with late S diffusion.

Microscopic features

The least altered, but minor example of the volcanic-associated ores at Dundonald, Canada (Muir and Comba, 1979), contains some basal massive ore with colloform-like pentlandite in a devitrified originally glassy matrix. However, the textures from all other described deposits appear to be secondary and solely of metamorphic origin (e.g. Bayer and Siemes, 1971; Barrett et al., 1977; Ostwald and Lusk, 1978); typical fabrics are figured by these authors.

The most common fabric in pyrrhotite-rich, layered massive ore is a foliation defined by flattened and optically oriented pyrrhotite grains, whose *c*-axes are at a high angle to planes of dimensional preferred orientation. This foliation is commonly slightly oblique to foliations defined by segregations of pentlandite aggregates and pyrite aggregates (Plate III, A). In some cases, these fabrics are demonstrably superimposed on the gross ore layering (e.g. Ostwald and Lusk, 1978). In unlayered massive ore, matrix ore and even some layered ores dimensional preferred orientation is lacking in pyrrhotite and more equant, polygonal grain aggregates with 120° triple junctions are dominant. Transition from one fabric to the other is visible on the polished section-scale in some sections, with the equant grains resulting from recrystallization of oriented aggregates, particularly at the margins of coarse pyrite or pentlandite grains. In many ores, kink bands are developed within pyrrhotite grains at a high angle to the foliation, and more rarely one or more sets of deformation twins may be present.

Pyrite and spinel grains are commonly unfractured in pyrrhotite-dominated layers, although spinels may be enclosed within rotated chloritic inclusions. In pyrite and/or spinel-rich layers, these minerals are commonly fractured and veined by pyrrhotite and/or chalcopyrite, presumably due to their strength contrast (e.g. Graf and Skinner, 1970). Less commonly, pyrite occurs as strongly flattened aggregates, in places with pentlandite, defining a strong foliation. Veinlike pyrite may also occur along the grain boundaries of oriented or annealed aggregates of pyrrhotite and within fractures in ferrochromite, supporting the mesoscopic evidence for late-stage S diffusion.

The distribution of chalcopyrite reveals the mobility of Cu on a small scale. It occurs as discrete concentrations, commonly at the interface between pentlandite- and pyrrhotite-rich aggregates, on the margins of silicate inclusions, or in the interstices of pyrite grains in pyrite-rich layers. It appears to have migrated into dilatant zones. Chalcopyrite-rich segregations may show preferred orientation with individual grains having several generations of deformation twins, some kinked, superimposed on lensatic twins.

The ores clearly display partly annealed tectonite fabrics. These almost certainly developed late in the deformational-metamorphic history following at least partial unmixing of sulphide minerals from *mss*. Earlier fabrics, if developed, are not normally preserved, although the rare occurrence of pyrrhotite foliation markedly oblique to axial surfaces of folds defined by pentlandite layers support a complex deformational history for the ore. Comparison of kinking and deformation twinning within sulphide grains with experimental deformation studies (e.g. B.R. Clark and Kelly, 1973; Atkinson, 1974; Kelly and B.R. Clark, 1975; Roscoe, 1975) suggest that they formed at low temperatures ($<300°C$) and probably reflect late and relatively insignificant deformational events in the ore zone; this is supported by their restriction to late shear zones in some ores.

Mineralogical changes

The present sulphide mineralogy of massive and matrix ores is generally consistent with its exsolution from metamorphically reconstituted *mss*, so that there is little direct evidence for mineralogical changes occurring during the early stages of prograde metamorphism. However, metamorphic chromites were generated in some high-grade metamorphic environments, particularly in metasomatic reaction zones (Groves et al., 1977), recrystallized magnetites in matrix ore became more Cr-rich and excess magnetite may have been generated at the upper boundary of significant sulphide concentration. Introduction of As probably accompanied carbonation of ultramafic rocks, as ores in talc-carbonate hosts (e.g. Redross, some Spargoville ores) tend to be enriched in arsenic with significant concentrations of gersdorffite, niccolite and ullmanite in ore assemblages.

The composition of disseminated and blebby sulphides commonly does not lie within the field of *mss* at peak metamorphic temperatures, and significant changes in mineralogy are inferred to have occurred prior to, or during prograde metamorphism. The occurrence of millerite- and heazlewoodite-bearing assemblages attests to Ni uptake by sulphides during metamorphism (cf. Ewers, 1972), and the common occurrence of millerite-, pyrite-, and magnetite-rich aggregates is indicative of recrystallization under higher fO_2 and/or fS_2 conditions than those of initial pyrrhotite–pentlandite formation (e.g. Eckstrand, 1975). Sulphide mineralogy is controlled by Fe-related redox mechanisms related to alteration of the dominant silicate phases of the disseminated ores.

In contrast to the limited detectable effects of prograde metamorphism on the mineralogy of massive and matrix ores, retrogressive alteration in areas of high metamorphic grade (e.g. Nepean, Wannaway) has resulted in a number of important mineralogical changes. Most changes in sulphide mineralogy result from oxidation of pre-existing sulphides during retrogressive serpentinization. Effects include the replacement of pentlandite along cleavages, cracks and grain boundaries by anastomosing vermiform to dendritic aggregates of nickeliferous mackinawite (approx. $Fe_{0.95}Ni_{0.08}S$) or bladed magnetite, and the replacement of hexagonal pyrrhotite by monoclinic pyrrhotite or microcrystalline aggregates of pyrite and magnetite. Secondary ragged magnetite may be concentrated at the upper boundary of matrix ore (Barrett et al., 1976).

A chrome-bearing variety of valleriite:

$Fe_{1.21}Cu_{0.78}Ni_{0.01}S_{2.00}1.46[Mg_{0.43}Cr_{0.18}Mn_{0.05}Fe_{0.30}(OH)_2]$ to
$Fe_{1.58}Cu_{0.28}Ni_{0.15}S_{2.00}1.79[Mg_{0.51}Cr_{0.30}Fe_{0.16}Ti_{0.03}(OH)_2]$

commonly replaces chromite grains within the ore zone (Nickel and Hudson, 1976) and a non-chromian valleriite may occur around sulphide grain boundaries and along retrogressive shear zones (Barrett et al., 1976; McQueen, 1979). Nickel and Hudson (1976) suggest that the formation of valleriite at the expense of chromite results from internal redistribution of Cu, Fe, Ni, Mg and S, but there is some evidence from Wannaway to suggest that S may be redistributed or introduced via alteration fluids (cf. McQueen, 1979; Groves and Keays, 1979; Keays et al., in press). At this deposit (McQueen, 1979), the bulk composition of the sulphide fraction of ore within retrogressively serpentinized peridotite is significantly more Fe-rich than that in unserpentinized peridotite (Table XII), suggesting the formation of additional Fe sulphides during alteration and implying addition of S if Ni is fixed in the sulphide fraction.

Empirical observations indicate that most phases produced during retrogressive alteration are not stable at the temperatures of peak prograde metamorphism. It is therefore *possible* that mineralogical changes equivalent to those accompanying retrogressive alteration formed prior to peak prograde metamorphism, but were subsequently destroyed due to the progressive formation of metamorphic *mss*. This raises the possibility that bulk compositions of ores, particularly matrix/disseminated ores were modified prior to peak metamorphism, although normally no direct evidence exists for such changes: evidence for such changes are recorded from Wannaway by McQueen (1979).

Associated gold mineralization

Volcanic-associated Ni–Fe–Cu sulphide deposits in Archaean greenstone belts have a close spatial association with gold deposits. For example, in Canada the Timmins gold camp is close to a number of important volcanic-associated Ni–Fe–Cu sulphide deposits (see Fig. 3); and in Western Australia, the Ni–Fe–Cu sulphide ores of Lunnon Shoot, Kambalda, are within 100 m of the workings of the Red Hill gold mine. Many other nickel deposits are close to old gold mines.

A broad stratigraphic correlation between known gold and nickel mineralization was pointed out by Gemuts and Theron (1975), with most gold production coming from Sequence 3 rocks (see Table IV) with only minor gold occurrences in Sequences 4 and 6. The occurrence of the more important gold deposits, in lower-grade metamorphic environments than the nickel deposits in the Eastern Goldfields Province, was emphasized by Binns et al. (1976) and Gee et al. (1976). A possible genetic relationship between the gold deposits and carbonation of sulphide-bearing ultramafic rocks has been suggested by Keays et al. (in press). They envisage release of Au during talc-carbonation and migration via carbonated S-bearing fluids down the metamorphic gradient to be precipitated as the fluids encountered cooler rocks.

At Kambalda, Au occurs in association with a suite of heavy-metal tellurides and sulphides in quartz-carbonate veins. Altaite, melonite, rucklidgeite, volynskite, hessite, michenerite and testibiopalladite have been noted (Hudson et al., 1978), together with

TABLE XIV

Gold and precious-metal production * from nickel mines in the Coolgardie Goldfield (mainly Kambalda) Western Australia, 1975

Gold	Silver	Palladium	Platinum	Ruthenium
35	140	147	62	9

All values in kg.

* Production from 324,272 tonnes of concentrates containing 39,367 tonnes Ni, 315 tonnes Cu, 79 tonnes Co.

galena, sphalerite, molybdenite, chalcopyrite, hawleyite and gersdorffite. Reaction between telluride veins and Ni–Fe sulphide ore gives rise to melonite, which corrodes and embays pentlandite and pyrrhotite. Gold concentrations in the veins can be quite spectacular and some excellent specimen pieces have been recovered. Veins such as these make an important contribution to the total Au and precious metal recovered from volcanic-associated deposits (Table XIV). It is unclear whether the Au is locally remobilized from the Fe–Ni sulphide ores or sulphidic metasediments, both of which contain significant Au concentrations (Bavinton and Keays, 1978; Ross and Keays, 1979), or is introduced from outside the immediate ore environment.

OXIDATION AND SUPERGENE ALTERATION OF ORES [1]

Introduction

Ni- and Fe-sulphide minerals are unstable in the presence of oxygen or oxygenated groundwaters, and rapidly oxidize to secondary or supergene sulphide minerals and finally to iron oxides and hydroxides. As a consequence, they are not generally exposed at the Earth's surface except in areas, such as Canada and Scandinavia, where recent glacial activity has stripped the weathering mantle.

In Australia and southern Africa deep chemical weathering has acted on ancient land surfaces for many millions of years. The profiles of alteration extend to several hundreds of metres in depth with sulphides being generally absent from above the water table. Oxidation reactions include the simple replacement of metal sulphides by metal oxides, but also encompass the more general chemical oxidation reactions, such as the conversion of pyrrhotite to pyrite, which release metal ions into solution and generate free electrons.

Except where recent dissection has cut through the deep-weathering profiles of these ancient terrains, direct evidence of sulphide mineralization is not observed. In Western Australia, despite intensive, if intermittent, exploration for Au and base metals, the

[1] See Chapter 3 by Lelong et al., Vol. 3, in this Handbook series.

presence of volcanic-associated Ni–Fe–Cu sulphide deposits was not recognized for some 70 years. The fact that deposits like Kambalda were not recognized earlier is ascribable to a combination of factors, perhaps the most important of which were the lack of understanding of the volcanic-associated type of nickel deposit and the difficulties of exploration in a terrain characterized by deep weathering, poor outcrop and transported overburden.

Virtually all of the discoveries of nickel deposits in Western Australia resulted from the correct identification of gossans. These compositionally variable ferruginous and siliceous remnants of sulphide mineralization represent the surface expression of a weathering profile that comprises an oxide zone, a zone of supergene sulphides, a transition zone and primary ore (Fig. 30). Whilst the nature of the profile is variable both on a local and regional scale, there are some unifying features that can be recognized in virtually all deposits. Woodall and Travis (1969) gave the first account of supergene weathering based on Durkin and Lunnon Shoots, Kambalda. Their description has proved to be applicable to many occurrences in which there is continuity within the sulphide portion of the ore (i.e. either massive or matrix ore). An understanding of the detailed mineralogical changes that accompany supergene alteration, and a chemical model for the oxidation process has been given by Nickel et al. (1974).

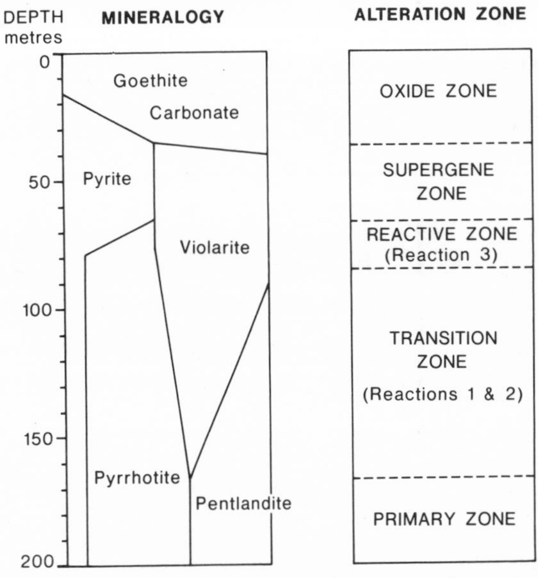

Fig. 30. Schematic diagram showing zones of supergene alteration indicated by different mineral assemblages and mineral reactions as developed above primary nickel ores of Western Australia.

Transition zone

Supergene alteration commences at the base of the transition zone by the development of minute, dispersed flecks of violarite (Vi_{pn}) in pentlandite (Fig. 31). With increased alteration, the flecks of violarite increase in number, eventually coalescing to give complete replacement of pentlandite. Although the development of violarite tends to pervade the entire pentlandite grains, some preferred orientation of violarite along cleavage traces

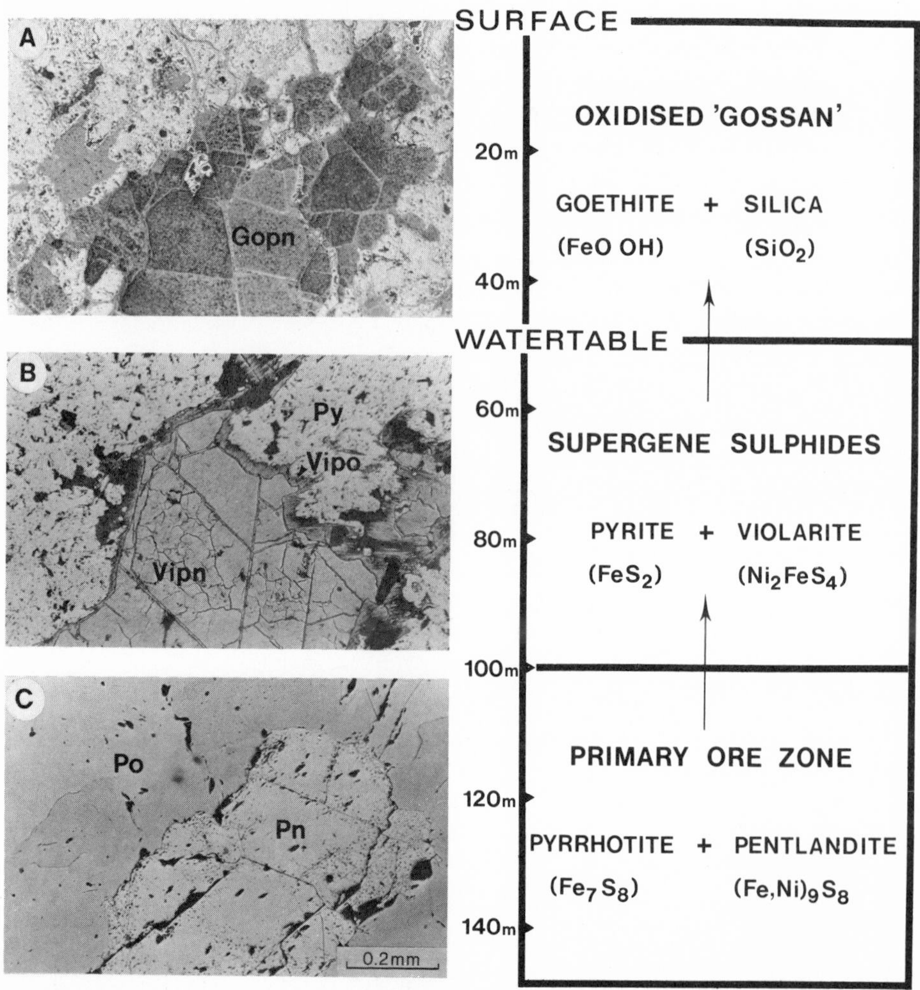

Fig. 31. Textural and mineralogical changes during supergene alteration of Ni–Fe–Cu sulphide ores. A. Relict octahedral cleavage traces of pentlandite preserved in goethite. B. Pentlandite replaced by blocky violarite (Vi_{pn}; note cleavage traces and shrinkage cracks) and pyrrhotite replaced by lamellar violarite (Vi_{po}) and porous pyrite (Py). C. Specks of violarite in pentlandite (Pn) indicate the start of supergene alteration. Pyrrhotite (Po) is unaltered.

of pentlandite is apparent. A system of shrinkage cracks in violarite is invariably present and suggests a loss of volume (perhaps as high as 15%) during replacement. This shrinkage tends to enhance the octahedral cleavage traces of the original pentlandite and gives the violarite a distinctly "blocky" appearance (Fig. 31).

The composition of violarite replacing pentlandite is $Ni_{1.9}Fe_{1.3}S_{4.0}$ (Nickel et al., 1974). It thus has a higher metal/S ratio and lower Ni/Fe ratio than the empirical violarite formula, Ni_2FeS_4. A generalized equation for the replacement of pentlandite by violarite, eq. 1 below, indicates that both Ni and Fe are released from pentlandite during oxidation. Iron is fixed within the supergene profile as siderite, which at Kambalda is in contrast with the magnesite in primary ores and the Ni-rich carbonates in the oxide zone. In other deposits which are low in carbonate, Fe is fixed in supergene magnetite (Nickel et al., 1977).

Oxidation of pyrrhotite leads to an increase in S/metal ratio. Thus, hexagonal pyrrhotite alters progressively to monoclinic pyrrhotite and then to smythite; monoclinic pyrrhotite alters directly to a smythite-like phase. This alteration proceeds inward from grain boundaries, and is influenced by the strong *ab* structural direction in pyrrhotite to give a lamellar, comb-like or feathery margin to the oxidizing pyrrhotite grains (Fig. 31).

In the presence of Ni, released during the replacement of pentlandite by violarite, monoclinic pyrrhotite alters to lamellar violarite (Vi_{po}), either as a direct oxidation replacement or via the intermediate smythite-like phase. The lamellar violarite is consistently lower in Ni/Fe ratio than violarite derived from pentlandite, with an average composition, given by Nickel et al. (1974), of $Ni_{1.7}Fe_{1.5}S_4$. Cobalt contents similarly reflect the difference in composition between the violarite precursors, with values up to 1 wt.% Co in pentlandite and Vi_{pn} and generally less than 0.1% Co in pyrrhotite and Vi_{po}.

For constant numbers of S atoms per unit volume of replaced pentlandite and pyrrhotite, the coupled reactions for formation of blocky and lamellar violarite have been given by Nickel et al. (1974) as:

$$\underset{\text{pn}}{Ni_{4.9}Fe_{4.1}S_{8.0}} \rightarrow \underset{\text{Vi}_{\text{pn}}}{Fe_{2.6}Ni_{3.8}S_{8.0}} + 1.1Ni^{2+} + 1.5Fe^{2+} + 5.2e^{-} \quad (1)$$

$$\underset{\text{po}}{0.33Fe_7S_8} + 1.1Ni^{2+} \rightarrow \underset{\text{Vi}_{\text{po}}}{0.33Ni_{3.3}Fe_{3.0}S_8} + 1.32Fe^{2+} + 0.44e^{-} \quad (2)$$

The ratio of Vi_{pn} to Vi_{po} suggested in the eqs. 1 and 2 is of the order of 3 to 1, which is in general agreement with the observed ratio.

Oxidation of hexagonal pyrrhotite in the presence of Ni in solution gives rise to comb-like alteration fringes of nickeliferous monoclinic pyrrhotite (Nickel et al., 1977). The cores of hexagonal pyrrhotite are unstable and alter to secondary pyrite, even before pentlandite has been completely replaced by violarite. Conversion of nickeliferous monoclinic pyrrhotite to lamellar violarite follows the replacement of pyrrhotite cores by pyrite.

Reactive zone

Replacement of pentlandite by blocky violarite is complete at the top of the transition zone and, as a consequence, no further Ni is available for formation of lamellar violarite. Residual cores of pyrrhotite grains are then progressively dissolved and partially replaced by mixtures of secondary pyrite, marcasite and siderite according to the equation:

$$\underset{\text{Po}}{Fe_7S_8} \rightarrow \underset{\text{Py}}{4FeS_2} + 3Fe^{2+} + 6e^- \tag{3}$$

Colloform textures in pyrite and open solution cavities indicate that replacement of pyrrhotite is not pseudomorphous.

The secondary pyrite and marcasite have generally higher Ni contents than the replaced pyrrhotite, and are higher in Ni and lower in Co than pyrite from the primary ore zone. A feature of the replacement of pyrrhotite is that it takes place over a restricted depth interval of about 10 m (Woodall and Travis, 1969), which has been called the "reactive zone" by Thornber (1972).

Supergene zone

Within the supergene zone, pentlandite is completely replaced by blocky violarite and pyrrhotite by lamellar violarite, secondary pyrite and marcasite (Fig. 31). Nickel/Fe ratios in both types of violarite show an increase upwards through the zone. Despite the overall increase, the relative difference between Vi_{pn} and Vi_{po} is maintained by virtue of the greater porosity of Vi_{pn} (Nickel et al., 1974), which allows easier ingress of Ni-sulphate containing groundwaters. Further evidence of the Ni-rich groundwaters can be seen in the upward increase in Ni contents of siderite, and in the development of fibres, encrustations and stalactites of the Ni sulphate minerals retgersite, morenosite and nickelhexahydrite in drill holes and mine openings. High Ni contents in secondary pyrite, and the development of rare Ni minerals are also attributable to these Ni-rich solutions that are derived from the overlying oxide zone.

Oxide zone

Above the present water table, sulphides break down to give Ni-rich carbonates including gaspeite and reevesite, and goethite. Violarite is the most readily decomposed sulphide with pyrite persisting into the lower portion of the oxide zone.

Textural preservation within the zone is very variable and is complicated by the irregular and complex precipitation modes of silica and colloform or crystalline goethite.

Nevertheless, goethite and silica pseudomorphs of the pentlandite octahedral cleavage traces and the comb-like alteration of pyrrhotite to lamellar violarite are generally observable by careful study (Fig. 31). The textures provide a valuable guide for exploration and faithfully record the remarkable sequence of mineralogical changes that have occurred during the supergene alteration process.

Supergene alteration of disseminated sulphides

Supergene alteration of disseminated pyrrhotite–pentlandite–pyrite assemblages results in sulphide assemblages that are mineralogically similar to those that result from alteration of massive or matrix ores. However, the degree of supergene alteration, at any given depth, is generally less for disseminated sulphides than for massive or matrix sulphides, and alteration is extremely variable from grain to grain. As a consequence, it is not uncommon to find disseminated grains of pyrrhotite and pentlandite at depths that correspond to the supergene (violarite–pyrite) zone of massive or matrix ores.

Electrochemical model for supergene alteration

The concept that the supergene alteration of Kambalda massive or matrix ores is dependent on the ore body behaving like a giant electrochemical redox cell, was suggested by Thornber (1972). In this model, anodic reactions at depth liberate metal ions and electrons. Electrons are conducted through the sulphide ore to a zone near the water table, where they take part in reactions of the type:

$$O_2 + 2H_2O + 4e^- \rightarrow 4(OH)^- \qquad (4)$$

The consumption of free electrons in this zone of high oxygen activity creates a potential gradient, and encourages the electron-liberating anodic reactions at depth. Both the pervasive nature of violarite alteration of pentlandite, and the difference in depth of alteration between massive/matrix compared with disseminated ores, is well explained by this electrochemical model. Oxidation reactions, which extend well below the water table, are thus an expected phenomenon in conducting ore bodies that intersect the water table, but might not be expected in "blind" ore bodies (Thornber, 1975a).

Gossans

Gossans, the ferruginous and siliceous surface remnants of sulphide mineralization, have long been a major exploration target for mining companies and prospectors. A review of sulphide weathering and gossan development has recently been given by Blain and Andrew (1977).

In Western Australia, the discovery of new nickel sulphide deposits has been almost exclusively determined by the ability to find and correctly identify nickel gossans. Discovery of buried or blind ore bodies has proved to be largely unsuccessful or impractical except in established mining camps, such as Kambalda, where statistically based grid drilling has been used extensively.

Discrimination of gossans from other ironstones is made difficult in ancient lateritized landscapes by the intense chemical leaching of ore metals and by the variable nature of the precipitation and dissolution of Fe and SiO_2. Most gossans have developed over a long period of time and the effects on textural preservation and geochemistry of fluctuating water tables, changing geomorphological position and burial by alluvium are only just beginning to be understood (Butt and Sheppy, 1975). However, the gross structures of the ores are commonly preserved (compare Fig. 32 with Fig. 24 and Plate III).

Study of microtextures of gossans (Roberts and Travis, 1973) indicates that even the most altered and silicified gossans, with nickel contents as low as 0.1% Ni, contain small relict areas with textures diagnostic of both the primary sulphide minerals and those developed during supergene weathering. Goethite pseudomorphs of the octahedral cleavage traces of pentlandite, and the comb-like texture of lamellar violarite after pyrrhotite

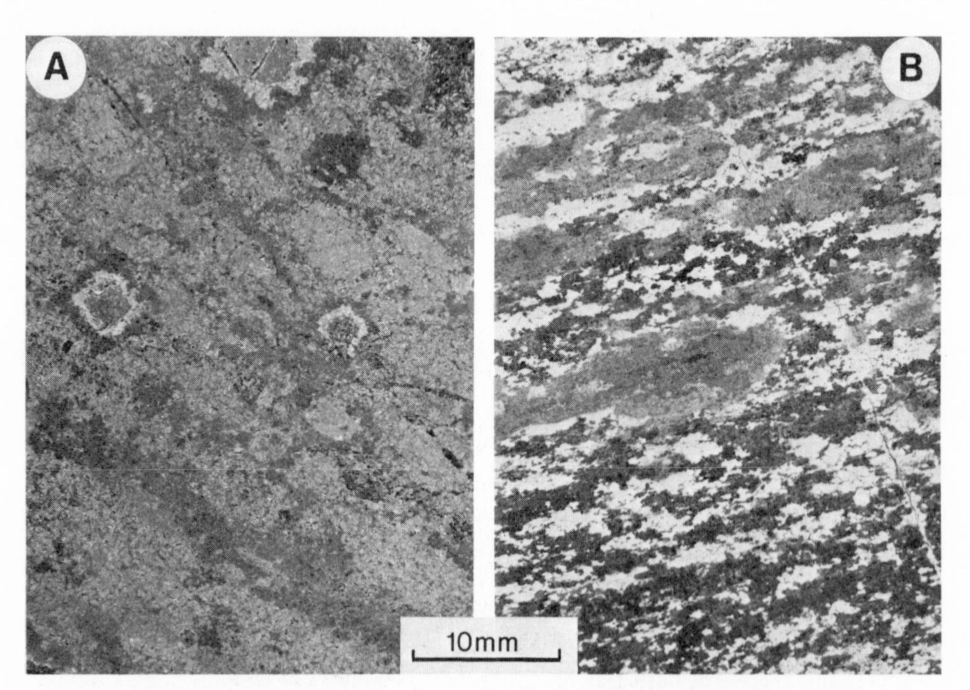

Fig. 32. Oxidised ore and gossan developed from banded massive ore. A. Chalcopyrite rims (white) around euhedral pyrite (grey, centre top) in banded violarite after pentlandite (dark grey) and pyrrhotite in "transition zone" ore, Lunnon Shoot, Kambalda. B. Relict-texture in gossan after banded pentlandite–pyrrhotite ore (cf. Plate III, A), Redross.

(Fig. 35), have proved particularly useful in discriminating Ni gossans from other Ni-bearing ironstones. Nevertheless, the task of microscope study of gossans requires patience and experience, and is generally undertaken in conjunction with a geochemical evaluation.

Chemical analysis has been widely used as a technique for distinguishing true gossans from other accumulations of iron oxides. Most analytical techniques consist of acid digestion of the sample followed by an atomic absorption determination of elements including Ni, Cu, Zn, Co, Pb, Mn and Cr.

Moeskops (1976) and Blain and Andrew (1977) have summarized the main findings of a number of studies based on Western Australian nickel gossans. High values of both Cu (500–5000 ppm) and Ni (1000–5000 ppm), associated with generally low values of Cr (<500 ppm), Mn (<500 ppm) and Zn (<100 ppm), were found by Cochrane (1973) to be characteristic of nickel gossans. Scatter diagrams have been used by Clema and Stevens-Hoare (1973) to identify Ni gossans on the basis of their high (Ni + Cu) and low (Mn + Cr) values relative to the total (Ni + Cu + Zn + Pb + Mn + Cr). A log/log plot of Cr versus (Cu × Ni) enables a similar distinction to be made between Ni gossans and other ironstones (Joyce and Clema, 1974). Wilmshurst (1975) suggests that during gossan formation Ni suffers a relatively greater depletion than Cu. Thus, Ni/Cu ratios of less than the average primary ore value of about 15 are favourable indicators of sulphide ores, whilst higher Ni/Cu ratios are indicative of lateritic nickel deposits. However, as Blain and Andrew (1977) point out, the extreme variability of gossan geochemistry necessitates the collection of large numbers of samples before assessment can be made with confidence. Moreover, they suggest that plots that distinguish gossans from other ironstones in Western Australia may have no relevance for gossans developed in other terrains.

Precious metals are partitioned into the sulphide phase of Ni deposits and, as they are relatively inert during weathering, have been used by Wilmshurst (1975) and Travis et al. (1976) to identify Ni gossans. Values of above 15–30 ppb Pd and 5–10 ppb Ir are believed to be positive indicators.

Geochemical exploration

The chemical environment that surrounds an oxidizing Ni–Fe sulphide ore body is highly anomalous (Thornber, 1975b). Metal ions, notably Fe and Ni, are liberated into the groundwater together with sulphate from the breakdown of sulphides. Measured and calculated changes in Eh and pH have been given by Thornber (1975b).

Adsorption of Ni and Cu on to soil minerals surrounding the ore zone, together with a mechanical surface dispersion of gossan fragments establishes a favourable situation for surface geochemical exploration. Butt and Sheppy (1975), Dalgarno (1972), Hall et al. (1973), Mazzucchelli (1972), and Smith (1976) give examples of its successful application. Fig. 33 shows the distribution of Ni in soils around the Kambalda Dome (cf. Fig. 6) as an example of metal dispersion that may be picked up during geochemical exploration.

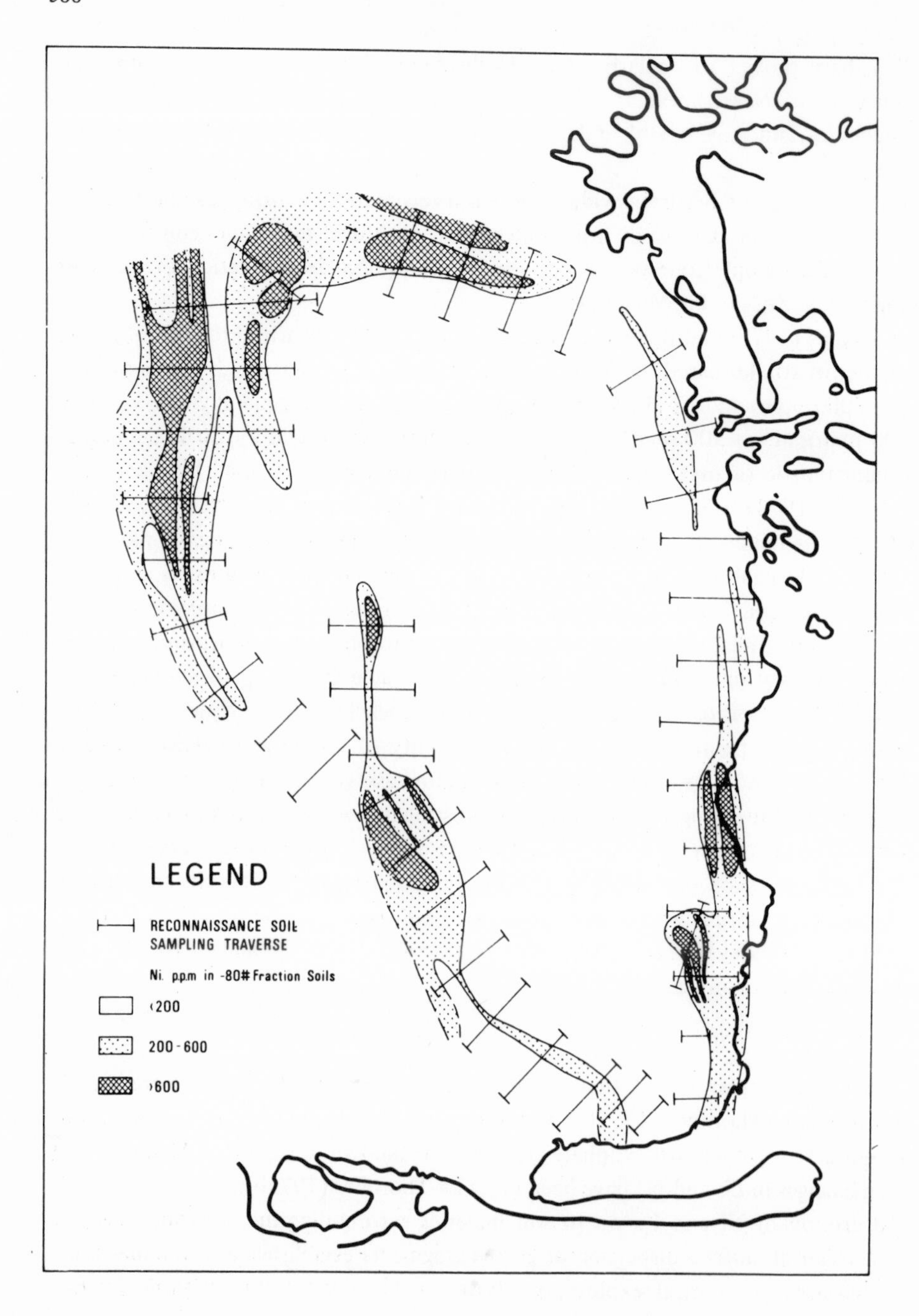

Fig. 33. Plan of nickel values found in a reconnaissance geochemical soil survey, Kambalda. (After Mazzucchelli, 1972.) Compare the anomalous soil values with the position of ore shoots in Fig. 6.

GENETIC MODELS

Introduction

Despite weathering, the strong metamorphic imprint on the volcanic-associated ores and their restriction to amphibolite-facies metamorphic environments in Western Australia, there is strong evidence for the existence of pre-metamorphic syngenetic sulphides. Significant ores appear to be restricted to the younger Archaean greenstone belts where they occur in groups in specific tectonic settings; many appear to be stratigraphically controlled on a broad scale and they are commonly closely stratigraphically controlled on a local scale. Importantly, ores are associated with specific, highly magnesian ultramafic units in thick komatiitic sequences, and there is convincing evidence for their presence prior to the metamorphic climax in a number of cases. Minor examples (e.g. Dundonald) occur in essentially unmetamorphosed sequences. In many of these respects they resemble the stratiform volcanogenic pyritic Cu–Zn–Pb deposits of Archaean terrains, although the latter are most common in sequences dominated by felsic metavolcanic rocks (Sangster and Scott, 1976).

Nature of syngenetic sulphides

Previous hypotheses relating to the genesis of volcanic-associated ores are summarized in Table XV. There have been only two origins seriously considered for syngenetic sulphides.

Most authors favour a magmatic origin for at least the disseminated and matrix sulphides, and envisage that the sulphide ores were derived from immiscible oxy-sulphide liquids that separated from ultrabasic magma at some stage (Table XV). Most models involve settling of immiscible oxy-sulphide liquid droplets that are included within an erupted, phenocryst-rich ultrabasic magma, with subsequent cooling and exsolution of ore minerals from *mss*.

Lusk (1976), however, emphasized the similarities of the Ni–Fe–Cu sulphide ores to stratiform pyritic Cu–Zn deposits that are generally accepted to form from submarine volcanic exhalations (e.g. Sangster and Scott, 1976). In particular, Lusk (1976) pointed out the similarity of the relationships between massive Ni–Fe–Cu ores and sulphidic metasediments in Western Australia with that between Archaean volcanogenic Cu–Zn ores and "exhalites" in Canada. Lusk (1976) explained the occurrence of disseminated and matrix ores intimately associated with ultramafic flows by invoking melting of part of volcanogenic massive ores by subsequent high-temperature ultrabasic flows with formation and settling of immiscible oxy-sulphide liquids within these overlying flows (Table XV).

Available data appear overwhelmingly in favour of the existence of immiscible oxy-sulphide liquids in the generation of the initial sulphide, including those that now com-

TABLE XV

Summary of genetic models of formation of volcanic-associated Ni–Fe–Cu sulphide deposits
Where models have been applied to specific deposits only, this is indicated after the reference.

Syngenetic models

(A) Ores formed by flow and gravity separation of immiscible oxy-sulphide liquids from crystal-rich ultramafic flows fed from magma reservoirs in which partial crystallization of olivine and separation of oxy-sulphide liquid has occurred (e.g. Ewers and Hudson, 1972; Hudson, 1972, 1973). Sulphur derived directly from the mantle (Naldrett, 1973; Naldrett and Cabri, 1976), or derived via crustal contamination (A.H. Green 1978 – Langmuir; Groves et al., 1978 – Windarra).

(1) Gravity separation of magmatic sulphide liquid in situ ("billiard-ball" model: Naldrett, 1973; Usselman et al., 1979).

(2) Eruption of magmatic oxy-sulphide lavas (Ross and Hopkins, 1975 – Lunnon Shoot) or syn-eruption separation of oxy-sulphide liquid (Groves et al., 1979; Muir and Comba, 1979 – Dundonald; McQueen, 1979 – Widgiemooltha ores; Marston and Kay, in press – Juan Shoot, Kambalda.

(B) Ores formed from submarine volcanic exhalations (cf. volcanogenic Cu–Zn–Pb ores).

(1) Massive ores formed as proximal equivalents of the sulphidic metasediments via volcanic exhalations and more disseminated ores formed by their incomplete melting as ultramafic flows were extruded over them (Lusk, 1976).

Metamorphic models

(A) Ores formed by sulphurization or metal diffusion.

(1) Sulphur released during alteration or metamorphism reacted with Fe, Ni and Cu in ultramafic units to produce ores (Naldrett, 1966 – Porcupine District – subsequently retracted by Naldrett, 1973; Prider, 1970).

(2) Nickel and Cu migrated down high chemical potential gradients from Ni-rich ultramafic rocks to S-rich metasediments during metamorphism (Seccombe et al., 1977 – Windarra; McQueen, 1979 – ? Wannaway).

(B) Ores formed by metamorphic modification of pre-existing, less-concentrated magmatic sulphides.

(1) *Some* massive ores formed in specific dilatant zones, shear zones and metasomatic reaction zones (Woodall and Travis, 1969 – Kambalda; Hudson, 1973 – Nepean; Barrett et al., 1976 – Nepean; Barrett et al., 1977; Binns et al., 1977; Groves et al., 1979; McQueen, 1979 – Redross).

---

prise massive ore. Those data which strongly support a magmatic origin, despite metamorphic modification, are listed below (sources of data are given above):

(1) The intimate association of the ores and their specific position within ultramafic units that have a very consistent composition and persistent position within the komatiite sequence.

(2) The geochemical similarity of massive and matrix ores from within individual ore shoots and their restriction to, or occurrence just on the S-rich side of, the field of *mss* at magmatic temperatures.

(3) The very consistent mean Ni/Cu ratios of bulk ores of 10–15 [$Cu/(Cu + Ni) \leqslant 0.1$], which are in agreement with a fundamental control by partitioning between immiscible oxy-sulphide liquids and peridotitic magmas (Rajamani and Naldrett, 1978; Duke and Naldrett, 1978). Ni/Co ratios also have a limited range. These consistent ratios con-

trast markedly with the variable Cu/Zn and Zn/Pb ratios of volcanogenic ores, and other ores formed by hydrothermal processes.

(4) The high precious metal tenor of the ores, their relatively low Pd/Ir ratios, the good correlation between Ni and platinoids, and their distinctive chondritic precious metal abundance pattern. Mean Pd/Ir ratios for Kambalda ores of ca. 6.0 are similar to Pd/Ir ratios of metapicrites, but are significantly lower than mean ratios of Kambalda metasediments (Pd/Ir = 35). Au/(Pd + Ir) ratios of Kambalda ores of 0.26 contrast with ratios of 14.5 for metasediments.

(5) The existence of distinctive "chalcophile" ferrochromites in massive and matrix ores. These ferrochromites appear to have a widespread occurrence in both metamorphosed and unmetamorphosed Ni–Fe–Cu sulphide ores, but appear restricted to such ores and have only been crystallized experimentally from oxy-sulphide melts.

(6) The variation of Cu, Pd and Ir and the magnetite distribution within the selected Lunnon Shoot ore section, while modified by deformation and metamorphism, is broadly consistent with expected fractional crystallization trends within a cooling oxy-sulphide liquid.

Features that specifically argue *against* the volcanic-exhalative model include:

(1) The lack of widespread wall-rock alteration beneath ore zones. Even where interpillow sulphides occur, there is normally no significant alteration other than that in unmineralized interpillow spaces, suggesting that the interpillow fillings also represent magmatic rather than hydrothermal sulphides.

(2) The formation of the ores at the beginning of the komatiitic volcanic cycle rather than during the waning stages of komatiitic volcanism.

(3) The lack of a specific association of the sulphidic metasediments (the proposed distal equivalents of volcanogenic massive ores) with ultrabasic volcanism, and their later formation in some cases (O.A. Bavinton, pers. comm., 1978).

(4) Sharp contacts between massive ores and metasediments in most instances.

(5) The lack of overlapping compositions between ores and metasediments in terms of Fe–Ni–Cu–Zn and precious metals: the common high Zn contents of metasediments is particularly evident.

(6) The marked contrast between compositions of the non-sulphide fractions of massive ores and metasediments.

While it is argued that most ores do not represent volcanic-exhalative deposits, *possible* exceptions include the basal breccia ores and hanging-wall F Shoot at Windarra (e.g. Seccombe et al., 1977; Groves et al., 1979), some basal ores at Wannaway (McQueen, 1979) and possibly some Kambalda hanging-wall ores. The Windarra ores, in particular, have geochemical characteristics intermediate between normal massive ores and sulphidic metasediments and a silicate fraction essentially equivalent to that of the metasediments. It may be significant that the metasediments from this deposit have a more ultramafic character and a sulphide composition that is closer to the massive ores than those from other localities.

The Sherlock Bay deposit of the west Pilbara Block (Miller and Smith, 1975) may be an example of a volcanic-exhalative deposit. It does not, however, show such an intimate relationship with komatiitic volcanism as the volcanic-associated massive ores. Country rocks include sequences of calc-alkaline metavolcanics, and the metasediments within which Ni- and Cu-sulphide concentrations occur are non-aluminous carbonate-, sulphide- and oxide-type iron formations that contrast with the metasediments in massive ore environments of the Eastern Goldfields Province (Groves et al., 1978). Ore grades of 0.5% Ni and 0.1% Cu, and a Ni/Cu ratio of 5.0, contrast markedly with most of the volcanic-associated deposits.

Source of magmatic sulphides

There is considerable evidence that immiscible oxy-sulphide liquid droplets were not exsolved from an initially S-undersaturated magma, but that basal ultrabasic magmas were erupted carrying such droplets. Important evidence includes:

(1) Neither the mineralized ultramafic units nor their contained olivines are depleted in Ni (relative to MgO, Fig. 22) or platinoid elements relative to unmineralized ultramafic units. Duke and Naldrett (1978) and Ross and Keays (1979) show that ultrabasic liquids from which Ni–Fe–Cu sulphides exsolve are rapidly depleted in Ni and platinoids due to partitioning relations between sulphide and silicate melts. These data also argue strongly against a sulphur source for ore formation from in-situ contamination of ultrabasic flows by assimilation of sulphidic metasediments, and against sulphurization hypotheses (Table XV).

(2) Nickel/Fe ratios of bulk ores are variable, even when allowance is made for metamorphic alteration and variable magnetite content, despite the geochemical similarity of mineralized ultramafic units. Early separated sulphide liquids forming in situ from similar highly magnesian ultrabasic liquids should have similar compositions (Rajamani and Naldrett, 1978; Duke and Naldrett, 1978).

(3) Most ultramafic units have S contents considerably below saturation levels determined by Shima and Naldrett (1975), suggesting that ultrabasic magmas were not normally close to saturation with respect to S; a condition necessary for in-situ formation of significant magmatic sulphides.

The variable Ni/Fe ratios of ores, together with small variations in Ni/Cu and Ni/Co ratios and variation in precious-metal abundances, are best explained if the present ores represent tapping of oxy-sulphide liquid reservoirs at various stages of their development, i.e., when the volume of such liquid was variable relative to the total volume of ultrabasic magma, as later-formed oxy-sulphide liquids may be relatively Ni-depleted (e.g. Duke and Naldrett, 1978).

There are severe problems with upward movement of phenocryst-rich and sulphide-rich magmas from mantle depths. Superheating is likely to cause dissolution of olivine phenocrysts in the silicate liquid and high density (ca. 4.0 g cm^{-3}) oxy-sulphide droplets are unlikely to be entrained in low viscosity, magnesian liquids unless ascent was extremely rapid. It thus appears likely that the highly magnesian crystal mushes and sulphides formed at crustal levels at an intermediate stage. Although evidence is equivocal, the most likely model to explain the formation of ultrabasic komatiite magmas is the two-stage melting model of a rising mantle diapir envisaged by Naldrett and Turner (1977) and shown diagrammatically in Fig. 34. Magmas generated at crustal levels, may undergo fractional crystallization at even higher levels with the formation of basal zones enriched in olivine crystals and oxy-sulphide liquids from which the ore-bearing ultrabasic magmas were tapped. This would leave silicate liquids depleted in sulphur, as observed in overlying flows.

Naldrett (1973) and Naldrett and Cabri (1976) suggest that sulphide-rich magmas were generated because the source region of the Archaean mantle at ca. 200 km depth was enriched in sulphur due to melting of sulphides and their downward percolation below 100 km depth (Fig. 34). Certainly, the widespread occurrence of S-saturated Archaean metabasalts (Naldrett et al., 1978), that represent the early-formed melts from the mantle diapir in the proposed model, support the suggestion of a S-enriched mantle. The similarity of S/Se ratios and $\delta^{34}S$ values for metasediments and ores

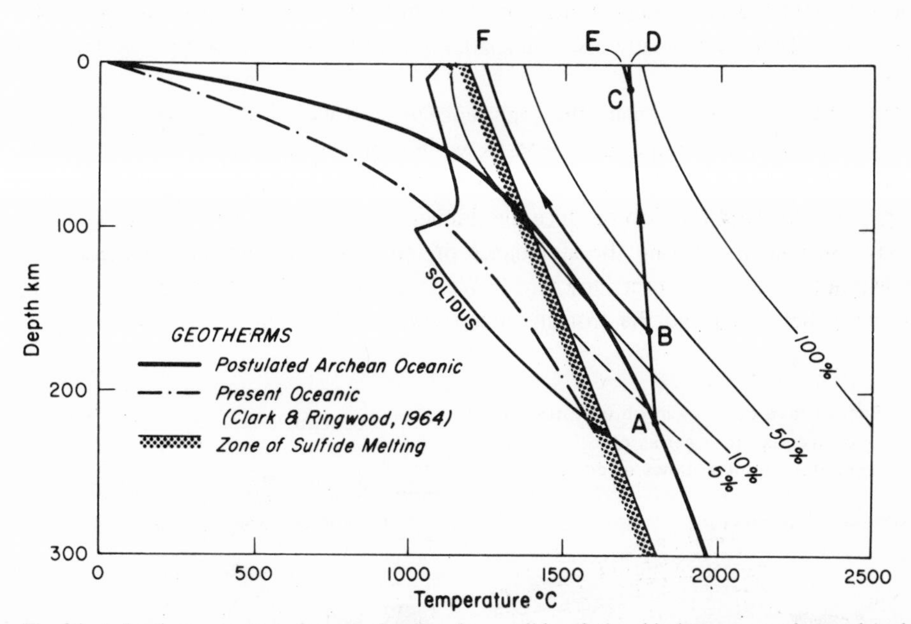

Fig. 34. A depth-temperature diagram showing the possible relationship between modern and Archaean oceanic geotherms, and melting of komatiitic magmas and mantle sulphides. (After Naldrett and Cabri, 1976.) A mantle diapir (S-rich) generated at *A* rises to *B* at which stage ca. 25% melting occurs, and basaltic liquid separates and rises along a non-adiabatic path *BF*. The residuum of partial melting rises from *B* to *C*, resulting in ca. 30% further melting and separation of komatiitic magma that rises along path *CE*.

from a number of deposits, and in particular the negative $\delta^{34}S$ values and high S/Se ratios of ores from Windarra and Langmuir (Seccombe et al., 1977; A.H. Green, 1978), raise the possibility that sulphur was at least in part derived via contamination from a crustal source (Groves et al., 1979). The proposed high-level generation and/or fractional crystallization of the ultrabasic magmas allows the possibility of such contamination. Although there is no *direct* evidence to support the hypothesis, it is interesting to note the common role of assimilation in the generation of other types of Ni–Fe–Cu sulphide ores (e.g. Haapala, 1969; Papunen, 1971; Kovalenker et al., 1975; Mainwaring and Naldrett, 1977; Naldrett et al., 1979).

Emplacement of sulphide-rich magmas

The occurrences in Rhodesia of feeder zones and possible volcanic necks or vents (e.g. Fig. 18) both containing and beneath mineralization, raises the possibility of subvolcanic magma chambers from which fractional crystallates were erupted. Magmas containing abundant phenocrysts, may have had sufficiently high viscosities to carry oxy-sulphide droplets to the surface. There is, however, normally no evidence that dunitic hosts (Mt. Keith-type) to mineralization represent feeders for the ultramafic hosts to volcanic-associated deposits, although the Mt. Clifford deposit is a possible example (Travis, 1975). There is normally a strong spatial separation of the two types of deposit (Fig. 4).

It is also tempting to explain the occurrence of most Western Australian and Canadian deposits and some Rhodesian deposits within trough-like embayments (see for example Fig. 23) as resulting from fractional crystallization within a surface lava pool, which periodically overflowed to produce the thinner, more liquid-rich flows (see Fig. 35). This may conveniently explain the abundance of mineralization and lack of metasediments within the zone. However, nowhere in Western Australia has the feeder to such a "pool" been discovered and the troughs, unless severely flattened during deformation,

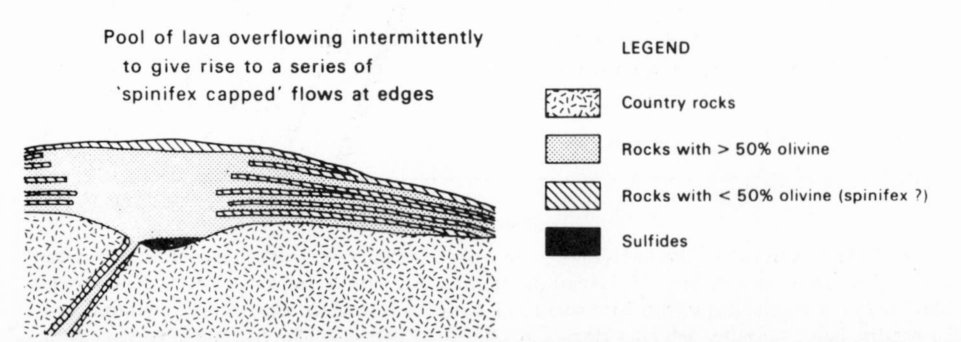

Fig. 35. One possible reconstruction of the volcanic environment of Ni–Fe–Cu sulphide deposits. (After Naldrett, 1973.) Compare with the Shangani section, Fig. 18.

are too shallow to represent the source of the overlying flows.

It seems most likely in the Western Australian case that the mineralized units represent differentiated flows that were extruded across the sea floor, with the troughs perhaps acting as riffles for phenocryst and sulphide liquid accumulation. Nevertheless, the high-viscosity phenocryst-flows could not have flowed far from their source.

While there is general agreement concerning the eruption of sulphide- and olivine-rich ultramafic flows, there has been considerable discussion of the *precise* mechanism by which discrete layers of disseminated, matrix and massive ore may form at the magmatic stage; the origin of massive ores is particularly problematical with metamorphic upgrading suggested as an alternative to magmatic formation for at least some ores (Barrett et al., 1977).

Ewers and Hudson (1972) first suggested a model involving formation of a basal pool of oxy-sulphide liquid (massive ore), which was partly displaced upwards by the weight of overlying olivines to replace less dense silicate liquid (matrix ore) and was overlain by small sulphide liquid droplets trapped between crystallized olivines (disseminated ore). They suggested separation of components by a combination of flow and gravity, but did not specify a precise mechanism. Naldrett (1973) proposed a simple static model termed the "billiard-ball model", in which in-situ gravity separation was envisaged as the major process. Theoretical calculations by Usselman et al. (1979) indicate that a pool of sulphide liquid (massive ore) would only be preserved beneath a ca. 60 m thick gravity-separated flow, if the concentration of olivine phenocrysts was below ca. 10% at the time of emplacement. For a higher percentage of phenocrysts, all sulphide liquid should be displaced upwards into cumulate olivine aggregates by the weight of the overlying column of olivine (cf. Barrett et al., 1977). As the phenocryst content of the mineralized units appears to have been always in excess of 10% and their thickness normally approximates that of the calculated case, it appears unlikely that massive sulphides can form by such a mechanism.

Ross and Hopkins (1975) emphasized a number of features from Lunnon Shoot that were inconsistent with a single-stage settling model. These included the direct contact of massive ore and metabasalt without an intervening silicate quench zone, the persistent high S.G. ferrochromite layer at the base of matrix ore, the slightly different spatial distribution of massive and more disseminated sulphides, the differing composition of the silicate fraction of massive and matrix ores, and the distinct contact between matrix and disseminated sulphides. They suggested a model involving segregation of magmatic components during vertical flow due to viscosity contrasts between sulphide liquid, silicate liquid and olivine crystals. The massive ore, matrix ore and disseminated ore plus overlying ultramafic unit were considered to be emplaced as separate flow units. A problem with the model is the spatial overlap of the ore types, which is unlikely to occur for completely separate flows.

A third, and more likely, explanation for the formation of magmatic massive sulphides is essentially a modification of the Ross and Hopkins (1975) model (see A3, Table XV). It involves gravity separation of olivine phenocrysts and oxy-sulphide liquid during horizontal flow of the erupted magma with differential flow between silicate liquid, olivine-rich magma (cf. Lajoie and Gélinas, 1978) and oxy-sulphide liquid due to density and viscosity contrast. The lower-viscosity basal sulphide liquid could then become separated and slightly precede the olivine mush, becoming more rapidly cooled in sea water so that displacement upwards into overlying olivine was not possible. Some sulphides possibly filled open interpillow spaces in underlying basalts. If massive sulphides were concentrated in sea-floor troughs, it is possible that preferential olivine accumulation also occurred in these depressions, so that calculations of profiles above ore zones indicate higher intratelluric phenocryst contents than existed initially in the extruded magma.

This model overcomes most of the objections to magmatic formation of massive ore raised by Barrett et al. (1977). The occurrence of ferrochromites and silicate inclusions throughout the massive ores remains a problem, as their marked density contrasts should lead to rapid separation of these phases to the base and top of massive ore respectively (cf. Goryainov and Sukhov, 1975). These features could, however, be interpreted to be the products of metamorphic modification (see below).

Crystallization of sulphide ores

Provided there has been no extensive modification of the bulk composition of ores during metamorphism, the magmatic crystallization history of the oxy-sulphide liquid can be determined by reference to the relevant data in the experimental Fe–Ni–Cu–S–O system.

It appears that most oxy-sulphide liquids would have begun crystallizing from the base upwards at $1125 \pm 25^{\circ}C$ and been completely solid as *mss* at $1050 \pm 10^{\circ}C$ (e.g. Ewers and Hudson, 1972; Hudson, 1973). Experimental data suggests that early crystallization of *mss* should lead to enrichment of the residual liquid in oxygen (Naldrett, 1969), and Ewers and Hudson (1972) and Hudson (1973) have interpreted the concentration of magnetite within, and particularly towards the upper part of matrix ore as a result of such fractional crystallization; this is complicated, however, by the generation of metamorphic magnetite (Groves et al., 1977). Ferrochromites clearly crystallized before magnetite: experiments by Ewers et al. (1976) suggest that they crystallized prior to and during solidification of *mss*. The concentration of Cu in matrix ores relative to massive ores may also result from fractional crystallization, but interpretation must be equivocal because of the obvious mobility of Cu during metamorphism. The distribution pattern of Ir and Pd within the ore section at Lunnon Shoot (Fig. 25), particularly the steady increase in Pd/Ir ratio from the base to the top of the matrix ore is also consistent with fractional crystallization (Keays and Davison, 1976). The spatial variation in Pd within Lunnon massive ore, however, appears to be strongly related to metamorphic processes with strong depletion in Pd at the base of massive ore and corresponding enrichment in adjacent footwall stringers (Fig. 25). The lack of fractionation of Ir in massive ore supports the rapid cooling of this layer predicted by the dynamic separation model.

Once solid, the *mss* would have remained essentially unchanged to 500 ± 100°C, although subsolidus equilibration with olivine would probably have occurred (e.g. T. Clark and Naldrett, 1972). At these temperatures, either pyrite or pentlandite would have exsolved dependent on whether the bulk composition lay on the S-rich or S-poor side of the *mss* field. Exsolution would have continued to low temperatures (<200°C) until the Ni content of residual pyrrhotite was below 0.6%. Cobalt distribution was probably influenced by nucleation of some pyrite in layers, although many pyrite-rich layers appear metamorphic in origin. The solubility of oxygen in *mss* is shown by the occurrence of submicroscopic lamellae of magnetite in pyrrhotite.

During metamorphism, sulphide ores reverted to *mss* with residual chromite plus magnetite and probably some residual pyrite *or* pentlandite and intermediate solid solution (*iss*), dependent on ore composition and layering. The majority of present sulphides crystallized from this metamorphic *mss* during slow cooling with the crystallization sequence following the magmatic sequence, except for late formation of pyrite via sulphur diffusion.

Metamorphic modification

The role of metamorphism and deformation in modifying the ores has been emphasized in previous sections. These processes appear responsible for:

(1) the present orientation of the ore bodies;

(2) the occurrence of offset mineralization, thickened ore sections, stringer mineralization, breccia ores and sulphides in metasomatic reaction zones;

(3) the mineralogical layering, fabrics and microtextures of ores;

(4) the generation of pyrite-rich zones;

(5) changes in bulk Ni/Fe ratios in some cases (e.g. Wannaway);

(6) redistribution of Cu (as chalcopyrite) throughout the ore section, and redistribution of Pd at the base of massive ores;

(7) severe mineralogical and chemical modification of low-grade disseminated sulphide ores, particularly Ni and Co enrichment;

(8) generation of additional magnetite (more rarely chromite) in ore sections;

(9) subsolidus re-equilibration between silicate and sulphide phases (cf. Binns and Groves, 1976);

(10) retrogressive alteration of ores in high-grade metamorphic environments (e.g. Nepean); and

(11) possible release of Au to metamorphic fluids.

A critical question is, however, whether deformational/metamorphic processes significantly upgrade sulphide concentrations, and in particular whether they can generate massive ores (e.g. Groves et al., 1976). There is evidence, reviewed previously, for generation of massive ores from matrix ores in specific structural situations at Nepean (e.g. Barrett et al., 1976), and for the possible formation of breccia ores at Windarra by Ni diffusion into sulphidic metasediments during metamorphism (Seccombe et al., 1977). There are strong suggestions of similar processes operating at Redross and Wannaway (McQueen, 1979). However, normally no definitive evidence exists for the transformation of the majority of matrix ores into massive ores except on a small scale, despite the preferential occurrence of the latter in dilatant zones. A major problem in making an unequivocal interpretation is that movement of the very low-strength massive ores occurs relative to all other lithologies and masks original relationships. A largely unresolved problem with a magmatic origin for massive ore is the non-sulphide fraction (Table XIII), which is geochemically unlike any of the ultramafic lithologies and is dispersed throughout the ore instead of concentrated at the top. It is possible that this represents metasomatic reaction zone fragments incorporated in massive ore during deformation and/or hydrothermal carbonate-quartz veinlets filling (?) cooling joints in massive ore (cf. carbonate-filled, late-stage fractures in present massive ores; Barrett et al., 1977), which were subsequently reoriented into the sulphide foliation during penetrative deformation: significant contraction of massive sulphides is likely during cooling (e.g. Morimoto and Kullerud, 1965; Rau, 1976).

An explanation for the occurrence of 50–65 wt.% sulphides in matrix ores is also equivocal. This is higher than the amount which could be accommodated interstitially to closely packed, discrete olivine crystals at the magmatic stage (cf. Visscher and Bolsterli, 1972), and an origin by metamorphic upgrading has been suggested by Barrett et al. (1976) and Barrett et al. (1977). Usselman et al. (1979) have suggested that chains of olivine crystals may form rather than discrete crystals and that volumes of sulphide liquid equivalent to those of matrix ores could occur interstitially to such crystal chains. However, no detailed textural analysis has been carried out on well-preserved cumulate zones of mineralized ultramafic units to test for the existence of chains of crystals.

With present information, it seems most likely that many Western Australian massive ores formed at the magmatic stage. This is supported by the occurrence of massive ores, albeit of minor significance, at the base of an essentially unmetamorphosed and undeformed komatiitic peridotite at Dundonald, Canada. Most deposits can thus be considered as metamorphosed ores, although metamorphic ores occur in specific environments.

Summary genetic model

A schematic genetic model for the generation of volcanic-associated Ni–Fe–Cu sulphide ores is presented in Fig. 36, with particular reference to the Western Australian examples. This model draws heavily on hypotheses generated by Woodall and Travis (1969), Hudson (1972), Naldrett (1973), Ross and Hopkins (1975), Naldrett and Cabri (1976), Naldrett and Turner (1977), Arndt et al. (1977), Barrett et al. (1977), Groves et

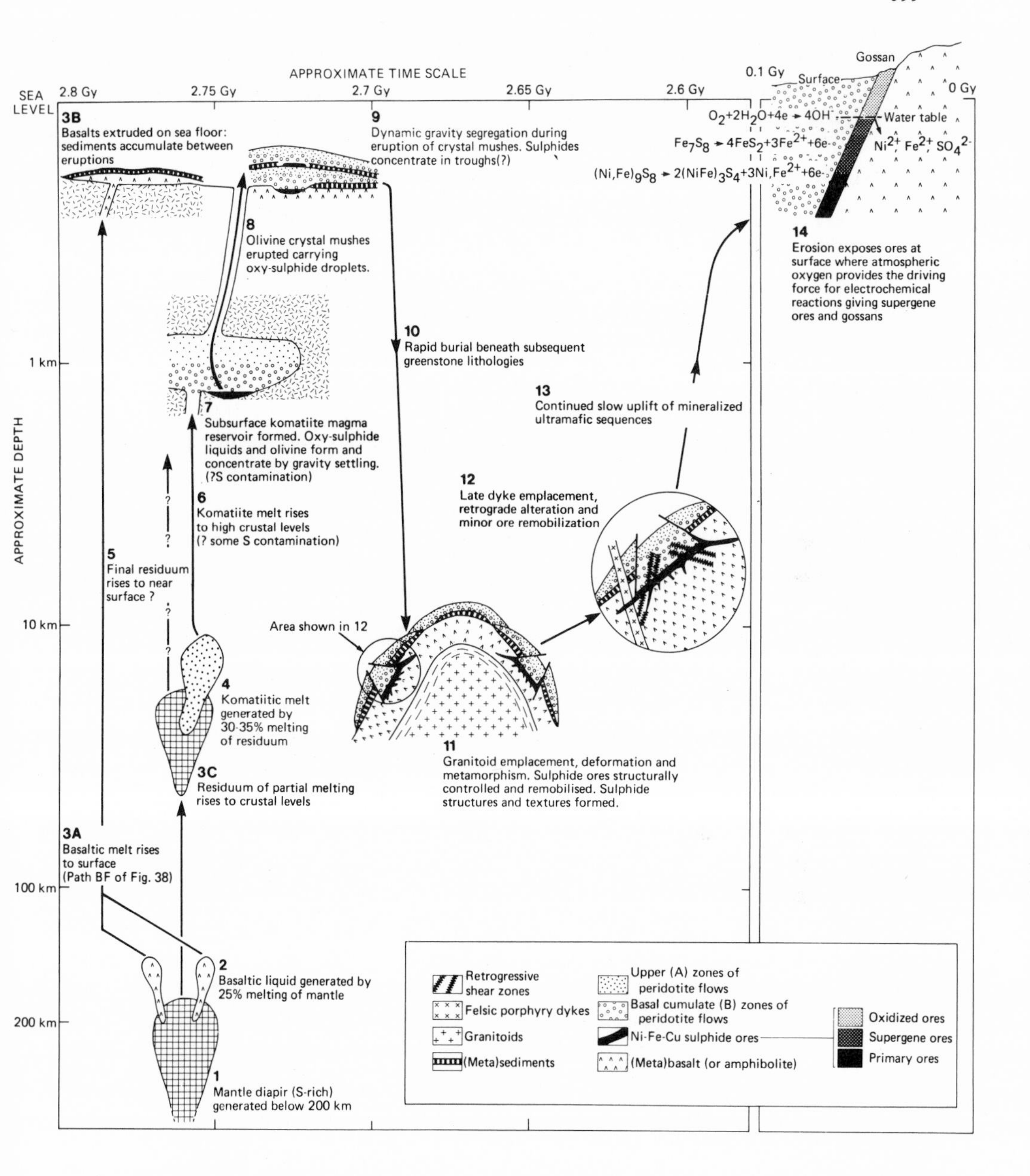

Fig. 36. Schematic diagram showing the possible evolution in depth and time of volcanic-associated Ni–Fe–Cu sulphide ores with particular relevance to Western Australian deposits (see acknowledgments in text).

al. (1979), D.A.C. Williams (1979), and Marston and Kay (in press). It is briefly described in the following "Summary" section.

SUMMARY

Volcanic-associated Ni–Fe–Cu sulphide ores of komatiitic affinity are a recently recognized class of strata-bound ore deposit, representing a genetically and economically important part of the metallogenesis of Archaean cratons in Western Australia, Rhodesia and Canada. The Kambalda deposits can be regarded as the type examples. They occur in groups (or camps) within younger Archaean (<3.0 Gyr) greenstone belts that formed in rifting environments in which discontinuous sulphide-rich tuffaceous and exhalative sediments were an integral part of volcanic activity; significant ores are not recorded from older Archaean terrains. Despite tectono-metamorphism, many primary features of the deposits can be recognized.

The deposits show an intimate relationship with sequences of komatiitic volcanics that form part of a mixed tholeiite–komatiite sequence (Western Australia) or a new cycle of komatiitic volcanism (Rhodesia, ?Canada). There is commonly a broad regional-scale and local-scale stratigraphic control of mineralization. The sulphide ores occur at or towards the base of thick (15–150 m), highly magnesian (>36% MgO) komatiitic peridotite units that represent ultramafic flows in most cases. These units represent the basal, less commonly the second or third, units in thick sequences of progressively thinner and less magnesian komatiitic peridotite and pyroxenite flows that exhibit distinctive spinifex-textured upper zones and cumulate lower zones. Although fissure eruptions are suspected, no feeder zones for mineralized ultramafic sequences have been recorded from Western Australia, but feeders are recorded from Rhodesia where mineralization may also occur within a volcanic neck in one instance. Sulphidic metasediments occupy some interflow positions, and the ultramafic komatiite sequence is followed by a mixed tholeiitic and komatiitic basalt suite.

The ores, consisting largely of pyrrhotite, pentlandite, chalcopyrite, pyrite, magnetite and ferrochromite, normally occupy trough-like embayments in footwall rocks from which interflow sediments are commonly absent or attenuated. Ore types include layered massive ore, massive ore, breccia ore, matrix ore and disseminated (or droplet) ore. Where several varieties occur, massive sulphides are commonly overlain by matrix and disseminated ores, but there is considerable variation in individual ore sections. Ores have consistent Ni/Cu ratios of 10–15 $[Cu/(Cu + Ni) \leqslant 0.1]$, significant precious-metal tenors and distinctive chondritic precious-metal abundance patterns. Sulphur/Se ratios and $\delta^{34}S$ values are commonly in the normal magmatic range, but some deposits have significantly higher S/Se ratios and more negative $\delta^{34}S$ values. There may be systematic metal variations through the ore zone with Cu and Pd enriched, and Ir depleted, in matrix ore relative to massive ore.

The ore bodies are typically situated around granitoid-cored domes or anticlines and their orientation is structurally controlled. They and the enclosing ultramafic rocks have suffered varying degrees of deformation and metamorphism. In Western Australia, metamorphic environments range from lowermost amphibolite-facies domains of low strain to upper amphibolite-facies domains of high strain, but in Rhodesia and Canada greenschist- or rarely sub-greenschist-facies environments are recorded. Most presently observed structures, and textures of the ores are metamorphic in origin and most sulphides have exsolved from a metamorphic *mss*. Tectonometamorphism has also resulted in remobilization of ores, redistribution of chalcopyrite and pyrite, changes in composition of disseminated, and more rarely more massive, ores and some retrogressive alteration of ores. Some massive and breccia ores appear to have been generated and matrix ores possibly upgraded during metamorphism, but there is no definitive evidence that most massive ores formed by metamorphic processes, although massive ores are not generally well developed in greenschist-facies environments.

Despite the metamorphic imprint, there is strong evidence that the ores formed from magmatic oxy-sulphide liquid droplets carried by phenocryst-rich ultrabasic melts that were extruded at temperatures above 1600°C. Gravity separation of olivine phenocrysts and sulphide liquid droplets during eruption, with differential flow between silicate liquid, olivine-rich ultrabasic mush and sulphide liquid due to density and viscosity contrasts, best explains the formation of magmatic massive ores, interpillow sulphides and observed ore profiles. The lower-viscosity basal sulphide layer is thought to have separated and slightly preceded the olivine mush, with both being concentrated in sea-floor depressions. Fractional crystallization of ore occurred during cooling over the range ca. 1150–1040°C. Ferrochromites and magnetites would have commenced crystallization during this temperature interval and continued to crystallize to lower temperatures. Magmatic sulphide minerals exsolved from *mss* below 600°C.

The mechanism for formation of the sulphide-rich ultrabasic magmas is even more speculative. A possible model, largely based on hypotheses by A.J. Naldrett and co-workers, is presented in Fig. 36. This involves uprise of a mantle diapir (?S-enriched) from below 200 km, separation of basaltic liquid at 25% partial melting above 200 km, continued adiabatic rise of the residuum until 30–35% further melting allows separation of komatiitic melt at crustal levels. The formation of a subvolcanic magma reservoir from this komatiitic melt is postulated to explain the generation and emplacement of phenocryst-rich magmas carrying oxy-sulphide liquid droplets with variable Ni/Fe ratios. The latter may be explained by extraction of previously separated oxy-sulphide liquid from reservoirs containing variable proportions of such liquid relative to olivine crystals and silicate liquid. Contamination of some reservoirs by exhalative sulphides in crustal sequences may account for anomalous S/Se and $\delta^{34}S$ values of some ores.

After formation, the volcanic-associated ores were buried in rapidly accumulated greenstone basins, deformed, metamorphosed and intruded by late-Archaean felsic dykes and pegmatites (Fig. 36). They commonly record the latest deformation and alteration events observed in the ore environment; massive ores in particular behaved as very low strength layers throughout deformation.

In the recently glaciated terrains of Canada, sulphides may outcrop at surface but in the deeper-weathering environments of southern Africa and particularly Western Australia, oxidation has played a further important role in modification of the deposits. The system approximates a corrosion cell in which aeration near the water table results in oxidation of primary sulphides below the water table to transitional sulphur-rich violarite–pyrite assemblages, and subsequent oxidation of these assemblages to sulphates and gossanous oxides nearer the surface. Gossans may faithfully record distinctive textures produced by replacement of pentlandite and pyrrhotite by violarite, and may also have definitive base-metal and precious-metal geochemistry. Dispersion of base metals also commonly occurs in soils adjacent to weathering ore bodies. Hence, the recognition of indigenous gossans and identification of dispersion haloes in soils have played a major role in exploration for volcanic-associated Ni–Fe–Cu sulphide deposits in Western Australia. Exploration has recently entered a new phase with statistically controlled grid drilling of the favourable ultramafic horizon by Western Mining Corporation Ltd. over the entire Kambalda dome. Nevertheless, correct identification of potential mineralized environments by geologic deduction must play an ever-increasing role in the search for further "blind" ore bodies not detected by gossan discovery and/or surface geochemistry.

ACKNOWLEDGMENTS

For one of us (D.I.G.) this review represents the culmination of a project "Metamorphism and Nickel Mineralization" funded by the Australian Mineral Industries Research Association, the Australian Research Grants Committee and the University of Western Australia. Colleagues involved in this project who contributed considerably to data included in this review include F.M. Barrett, R.A. Binns, R.J. Gunthorpe, R.J. Marston and K.G. McQueen. D.R.H.'s contribution is part of a continuing programme of research on nickel mineralization conducted by CSIRO, Division of Mineralogy. Other colleagues whose dicussion was invaluable or who allowed access to unpublished data or draft manuscripts were O.A. Bavinton, M.J. Bickle, M.J. Donaldson, J.M. Duke, O.A. Eckstrand, W.E. Ewers, A.H. Green, J.A. Hallberg, R.E.T. Hill, B.D. Kay, R.R. Keays, A.J. Naldrett, E.H. Nickel, J.R. Ross, M.R. Thornber, M.J. Viljoen and D.A.C. Williams. This review would not have been possible without the active co-operation of a number of mining companies who provided both access and company data on a number of deposits. These include Anaconda Australia Inc., Great Boulder Mines Ltd., Metals Exploration N.L., Poseidon Ltd., and particularly Western Mining Corporation Ltd. F.M. Barrett and R.J. Marston made invaluable comments on a draft manuscript. C. Steel drafted the figures.

This represents a contribution to the International Geological Correlation Programme Projects No. 91 – Metallogeny of the Precambrian and No. 161 – Magmatic Sulfide Ores Associated with Mafic and Ultramafic Rocks.

REFERENCES

Anderson, I.G., Varndell, B.J. and Westner, G.J., in press. Some geological aspects of the Perseverance nickel deposit, Rhodesia. In: M.J. Viljoen (Editor), *Metallogenesis '76. Trans. Geol. Soc. S.Afr.*

Anhaeusser, C.R., 1971. The Barberton Mountain Land, South Africa – a guide to the understanding of the Archaean geology of Western Australia. In: J.E. Glover (Editor), *Symposium on Archean Rocks. Geol. Soc. Aust. Spec. Publ.*, 3: 103–119.

Archibald, N.J., Bettenay, L.F., Binns, R.A., Groves, D.I. and Gunthorpe, R.J., 1978. The evolution of Archaean greenstone terrains, Eastern Goldfields Province, Western Australia. *Precambrian Res.*, 6: 103–131.

Arndt, N.T., 1977. Ultrabasic magmas and high-degree melting of the mantle. *Contrib. Mineral. Petrol.*, 64: 205–221.

Arndt, N.T., Naldrett, A.J. and Pyke, D.R., 1977. Komatiitic and iron-rich tholeiite lavas of Munro Township, northeast Ontario. *J. Petrol.*, 18: 319–369.

Atkinson, B.K., 1974. Experimental deformation of polycrystalline galena, chalcopyrite and pyrrhotite. *Trans. Inst. Min. Metall.*, 83B: 19–28.

Barley, M.E., Dunlop, J.S.R., Glover, J.E. and Groves, D.I., 1979. Sedimentary evidence for a distinctive Archaean shallow-water volcanic–sedimentary facies, eastern Pilbara Block, Western Australia. *Earth Planet. Sci. Lett.*, 43: 74–84.

Barnes, R.G., Lewis, J.D. and Gee, R.D., 1974. Archaean ultramafic lavas from Mount Clifford. *Annu. Rep. Geol. Surv. West. Aust.*, 1973: 59–70.

Barrett, F.M., Groves, D.I. and Binns, R.A., 1976. Importance of metamorphic processes at the Nepean nickel deposit, Western Australia. *Trans. Inst., Min. Metall.*, 85B: 252–273.

Barrett, F.M., Binns, R.A., Groves, D.I., Marston, R.J. and McQueen, K.G., 1977. Structural history and metamorphic modification of Archean volcanic-type nickel deposits, Yilgarn Block, Western Australia. *Econ. Geol.*, 72: 1195–1223.

Barton Jr., P.B., 1973. Solid solutions in the system Cu–Fe–S, Part I. The Cu–S and CuFe–S joins. *Econ. Geol.*, 68: 455–465.

Bavinton, O.A. and Keays, R.R., 1978. Precious metal values from interflow sedimentary rocks from the komatiite sequence at Kambalda, Western Australia. *Geochim. Cosmochim. Acta*, 42: 1151–1163.

Bayer, H. and Siemes, H., 1971. Zur Interpretation von Pyrrotin-Gefügen. *Miner. Deposita*, 6: 225–244.

Bennett, C.E.G., Graham, J. and Thornber, M.R., 1972a. New observations on natural pyrrhotites, Pt. I. Mineragraphic techniques. *Am. Mineral.*, 57: 445–462.

Bennett, C.E.G., Graham, J., Parks, T.C. and Thornber, M.R., 1972b. New observations on natural pyrrhotites, Pt. II. Lamellar magnetite in monoclinic pyrrhotite. *Am. Mineral.*, 57: 1876–1880.

Besson, M., Boyd, R., Czamanske, G.K., Foose, M.P., Groves, D.I., Von Gruenewalt, G., Naldrett, A.J., Nilsson, G., Page, N.J., Papunen, H. and Peredery, W.V., 1979. IGCP Project No. 161 and a proposed classification of Ni–Cu–PGE sulfide deposits. *Can. Mineral.*, 17: 143–144.

Bickle, M.J., Ford, C.E. and Nisbet, E.G., 1977. The petrogenesis of peridotitic komatiites: evidence from high-pressure melting experiments. *Earth Planet Sci. Lett.*, 37: 97–106.

Binns, R.A. and Groves, D.I., 1976. Iron–nickel partition in metamorphosed olivine–sulfide assemblages from Perseverance, Western Australia. *Am. Mineral.*, 61: 782–787.

Binns, R.A., Gunthorpe, R.J. and Groves, D.I., 1976. Metamorphic patterns and development of greenstone belts in the Eastern Yilgarn Block, Western Australia. In: B.F. Windley (Editor), *The Early History of the Earth.* Wiley London, pp. 303–313.

Binns, R.A., Groves, D.I. and Gunthorpe, R.J., 1977. Nickel sulfides in Archaean ultramafic rocks of Western Australia. In: A.V. Sidorenko (Editor), *Correlation of the Precambrian, 2.* Nauka, Moscow, pp. 349–380.

Blain, C.F. and Andrew, R.L., 1977. Sulphide weathering and the evaluation of gossans in mineral exploration. *Miner. Sci. Eng.*, 9(3): 119–150.

Bonatti, E., 1975. Metallogenesis at oceanic spreading centers. *Annu. Rev. Earth Planet. Sci.*, 3: 401–431.

Bull, A.J. and Mazzucchelli, R.H., 1975. Application of discriminant analysis to the geochemical evaluation of gossans. In: I.L. Elliott and W.K. Fletcher (Editors), *Geochemical Exploration 1974.* Elsevier, Amsterdam, pp. 219–226.

Burt, D.R.L. and Sheppy, N.R., 1975. Mount Keith nickel sulphide deposit. In: C.L. Knight (Editor), *Economic Geology of Australia and Papua-New Guinea, 1. Metals.* Australas. Inst. Min. Metall., Melbourne, pp. 159–168.

Butt, C.R.M. and Sheppy, N.R., 1975. Geochemical exploration problems in Western Australia exemplified by the Mt. Keith area. In: I.L. Elliott and W.K. Fletcher (Editors), *Geochemical Exploration 1974.* Elsevier, Amsterdam, pp. 391–415.

Chapman, D.G. and Groves, D.I., 1979. A preliminary study of the distribution of gold in Archaean interflow sulphidic metasediments, Yilgarn Block, Western Australia. In: J.E. Glover and D.I. Groves (Editors), *Gold Mineralization.* Publs. Geol. Dep. and Extension Service, Univ. West. Aust., 3: 76–88.

Christie, D., 1975. Scotia nickel sulphide deposit. In: C.L. Knight (Editor), *Economic Geology of Australia and Papua-New Guinea, 1. Metals.* Australas. Inst. Min. Metall., Melbourne, pp. 121–125.

Clark, B.R. and Kelly, W.C., 1973. Sulfide deformation studies, I. Experimental deformation of pyrrhotite and sphalerite to 2000 bars and 500°C. *Econ. Geol.*, 68: 332–352.

Clark, T. and Naldrett, A.J., 1972. The distribution of Fe and Ni between synthetic olivine and sulfide at 900°C. *Econ. Geol.*, 67: 939–952.

Clema, J.M. and Stevens-Hoare, N.P., 1973. A method of distinguishing nickel gossans from other ironstones on the Yilgarn Shield, Western Australia. *J. Geochem. Explor.*, 2: 393–402.

Cochrane, R.H.A., 1973. A guide to the geochemistry of nickeliferous gossans and related rocks from the Eastern Goldfields. *Annu. Rep. Geol. Surv. West. Aust.*, 1972: 69–76.

Condit, R.H., Hobbins, R.R. and Birchenall, C.E., 1974. Self-diffusion of iron and sulfur in ferrous sulfide. *Oxid. Met.*, 8: 409–455.

Constantinou, G. and Govett, G.J.S., 1973. Geology, geochemistry and genesis of Cyprus sulphide deposits. *Econ. Geol.*, 68: 843–858.

Craig, J.R. and Kullerud, G., 1969. Phase relations in the Cu–Fe–Ni–S system and their application to magmatic ore deposits. In: H.D.B. Wilson (Editor), *Magmatic Ore Deposits. Econ. Geol. Monogr.*, 4: 344–358.

Czamanske, G.K., Himmelberg, G.R. and Goff, F.E., 1976. Zoned Cr, Fe-spinel from the La Perouse layered gabbro, Fairweather Range, Alaska. *Earth Planet. Sci. Lett.*, 33: 111–118.

Dalgarno, R., 1972. Geochemistry of the Redross nickel prospect, Widgiemooltha area, Western Australia. *Jt Spec. Group Meet., Geol. Soc. Aust., Canberra. Abstr.*, 99: B12–B14.

Dalgarno, C.R., 1975. Nickel deposits of the Widgiemooltha dome – Redross, Wannaway, Widgiemooltha, Dordie. In: C.L. Knight (Editor), *Economic Geology of Australia and Papua-New Guinea, 1. Metals.* Australas. Inst. Min. Metall., Melbourne, pp. 82–86.

Donaldson, C.H., 1976. An experimental investigation of olivine morphology. *Contrib. Miner. Petrol.*, 57: 187–213.

Donnelly, T.H., Lambert, I.B., Oehler, D.Z., Hallberg, J.A., Hudson, D.R., Smith, J.W., Bavinton, O.A. and Golding, L.Y., 1978. A reconnaissance study of stable isotope ratios in Archaean rocks from the Yilgarn Block, Western Australia. *J. Geol. Soc. Aust.*, 24: 409–420.

Duke, J.M. and Naldrett, A.J., 1978. A numerical model of the fractionation of olivine and molten sulfide from komatiite magma. *Earth Planet. Sci. Lett.*, 39: 255–266.

Eckstrand, O.R., 1975. The Dumont serpentinite: A model for control of nickeliferous opaque mineral assemblages by alteration reactions in ultramafic rocks. *Econ. Geol.*, 70: 183–201.

Ewers, W.E., 1972. Nickel–iron exchange in pyrrhotite. *Proc. Australas. Inst. Min. Metall.*, 241: 19–25.

Ewers, W.E. and Hudson, D.R., 1972. An interpretative study of a nickel–iron sulfide ore intersection, Lunnon Shoot, Kambalda, Western Australia. *Econ. Geol.*, 67: 1075–1092.

Ewers, W.E., Graham, J., Hudson, D.R. and Rolls, J.M., 1976. Crystallization of chromite from nickel–iron sulphide melts. *Contrib. Miner. Petrol.*, 54: 61–64.

Ferguson, J. and Lambert, I.B., 1972. Volcanic exhalations and metal enrichments at Matupi Harbor, New Britain, T.P.N.G. *Econ. Geol.*, 67: 25–37.

Fripp, R.E.P., 1976. Gold metallogeny in the Archaean of Rhodesia. In: B.F. Windley (Editor), *The Early History of the Earth.* Wiley, London, pp. 455–466.

Gee, R.D., Groves, D.I. and Fletcher, C.I., 1976. Archaean geology and mineral deposits of the Eastern Goldfields. *25th Int. Geol. Congr. Excursion Guide,* 42A: 56 pp.

Gemuts, I. and Theron, A., 1975. The Archaean between Coolgardie and Norseman – stratigraphy and mineralization. In: C.L. Knight (Editor), *Economic Geology of Australia and Papua-New Guinea, 1. Metals.* Australas. Inst. Min. Metall., Melbourne, pp. 66–74.

Goodwin, A.M., 1973. Archean iron-formation and tectonic basins of the Canadian Shield. *Econ. Geol.*, 68: 915–933.

Goodwin, A.M., 1977. Archean basin–craton complexes and the growth of Precambrian shields. *Can. J. Earth Sci.*, 14: 2737–2759.

Goryainov, I.N. and Sukhov, L.G., 1975. Phase state of material that formed copper–nickel sulfide shoots. *Dokl. Akad. Nauk S.S.S.R.*, 221: 173–176.

Graf, J.L. and Skinner, B.J., 1970. Strength and deformation of pyrite and pyrrhotite. *Econ. Geol.*, 65: 206–215.

Green, A.H., 1978. *Evolution of Fe–Ni Sulphide Ores Associated With Archean Ultramafic Komatiites, Langmuir Township, Ontario.* Thesis, Univ. Ontario (unpublished).

Green, D.H., Nicholls, I.A., Viljoen, M.J. and Viljoen, R.P., 1975. Experimental demonstration of the existence of peridotitic liquids in earliest Archaean magmatism. *Geology,* 3: 11–14.

Groves, D.I. and Keays, R.R., 1979. Mobilization of ore-forming elements during alteration of intrusive dunites, Mt. Keith–Betheno, Western Australia. *Can. Mineral.*, 17(2): 373–390.

Groves, D.I., Hudson, D.R. and Hack, T.B.C., 1974. Modification of iron–nickel sulfides during serpentinization and talc–carbonate alteration at Black Swan, Western Australia. *Econ. Geol.*, 69: 1265–1281.

Groves, D.I., Binns, R.A., Barrett, F.M. and McQueen, K.G., 1975. Sphalerite compositions from Western Australian nickel deposits, a guide to equilibria below 300°C. *Econ. Geol.*, 70: 391–396.

Groves, D.I., Barrett, F.M., Binns, R.A., Marston, R.J. and McQueen, K.G., 1976. A possible volcanic-exhalative origin for lenticular nickel sulphide deposits of volcanic association with special reference to those in Western Australia: discussion. *Can. J. Earth Sci.*, 13: 1646–1650.

Groves, D.I., Barrett, F.M., Binns, R.A. and McQueen, K.G., 1977. Spinel phases associated with metamorphosed volcanic-type iron–nickel sulfide ores from Western Australia. *Econ. Geol.*, 72: 1224–1244.

Groves, D.I. Barrett, F.M. and McQueen, K.G., 1978. Geochemistry and origin of cherty metasediments within ultramafic flow sequences and their relationship to nickel mineralization. In: J.E. Glover and D.I. Groves (Editors), *Archaean Cherty Metasediments: Their Sedimentology, Micropalaeontology, Biogeochemistry and Significance to Mineralization.* Publ. Geol. Dep. and Extension Service, Univ. West. Aust., 2: 57–69.

Groves, D.I., Barrett, F.M. and McQueen, K.G., 1979. The relative roles of magmatic segregation, volcanic exhalation and regional metamorphism in the generation of volcanic-associated nickel ores of Western Australia. *Can. Mineral.*, 17(2).

Haapala, P.S., 1969. Fennoscandian nickel deposits. In: H.D.B. Wilson (Editor), *Magmatic Ore Deposits. Econ. Geol. Monogr.*, 4: 262–275.

Hall, J.S., Both, R.A. and Smith, F.A., 1973. A comparative study of rock, soil and plant chemistry in relation to nickel mineralization in the Pioneer area, Western Australia. *Proc. Australas. Inst. Min. Metall.*, 247: 11–22.

Hallberg, J.A., 1970. *The Petrology and Geochemistry of Metamorphosed Archaean Basic Volcanic and Related Rocks Between Coolgardie and Noreseman, Western Australia.* Thesis, Univ., West. Aust. (unpublished).

Hallberg, J.A., 1972. Geochemistry of Archaean volcanic belts in the Eastern Goldfields Region of Western Australia. *J. Petrol.,* 13: 45–56.

Hallberg, J.A. and Williams, D.A.C., 1972. Archaean mafic and ultramafic rock associations in the Eastern Goldfields Region, Western Australia. *Earth Planet. Sci. Lett.,* 15: 191–200.

Hallberg, J.A. and Glikson, A.Y., in prep. Archaean granite–greenstone terrains of Australia.

Hancock, W., Ramsden, A.R., Taylor, G.F. and Wilmshurst, J.R., 1971. Some ultramafic rocks of the Spargoville area, Western Australia. In: J.E. Glover (Editor), *Symposium on Archaean Rocks. Geol. Soc. Aust. Spec. Publ.,* 3: 269–280.

Harris, D.C. and Nickel, E.H., 1972. Pentlandite compositions and associations in some mineral deposits. *Can. Mineral.,* 11: 861–878.

Hatten, H.J., 1977. Geochemistry of selenium. *Geochim. Cosmochim. Acta,* 41: 1665–1678.

Hickman, A.H., 1975. Precambrian structural geology of part of the Pilbara region. *Geol. Surv. West. Aust. Annu. Rep.,* 1974: 68–73.

Honnorez, J., Honnorez-Guerstein, B., Vallette, J. and Wauschkuhn, A., 1973. Present day formation of an exhalative sulfide deposit at Vulcano (Tyrrhenian Sea), Part II. Active crystallization of fumarolic sulfides in the volcanic sediments of the Baia de Levante. In: G.C. Amstutz and A.J. Bernard (Editors), *Ores in Sediments.* Springer, New York, N.Y., pp. 139–166.

Howard, H.J., 1977. Geochemistry of selenium: formation of ferroselite and selenium behaviour in the vicinity of oxidizing sulfide and uranium deposits. *Geochim. Cosmochim. Acta,* 41: 1665–1678.

Hudson, D.R., 1972. Evaluation of genetic models for Australian sulphide nickel deposits. *Australas. Inst. Min. Metall. Newcastle Conf.,* pp. 59–68.

Hudson, D.R., 1973. Genesis of Archaean ultramafic associated nickel–iron sulphides at Nepean, Western Australia. *Australas. Inst. Min. Metall. West. Aust. Conf.,* pp. 99–109.

Hudson, D.R., Robinson, B.W., Vigers, R.B.W. and Travis, G.A., 1978. Zoned michenerite–testibiopalladite from Kambalda, Western Australia. *Can. Mineral.,* 16(2): 211–126.

Imreh, L., 1973. L'utilisation des coulées ultrabasiques dars la recherche minière: equisse structurale et lithostraphique de la Motte (Abitibi-Est, Québec, Canada). *Symposium Volcanism and Associated Metallogenesis,* Bucharest, pp. 291–314.

INAL Staff, 1975. BHP/INAL nickel sulphide occurrences of the Widgiemooltha area. In: C.L. Knight (Editor), *Economic Geology of Australia and Papua-New Guinea, 1. Metals.* Australas. Inst. Min. Metall., Melbourne, pp. 86–89.

Ishihara, S. (Editor), 1974. *Geology of Kuroko Deposits.* Min. Geol. Spec. Issue, 6. Tokyo, 435 pp.

Jolly, W.T., 1978. Metamorphic history of the Archaean Abitibi Belt. *Geol. Surv. Can., Pap.,* 78-10: 63–78.

Joyce, A.S. and Clema, J.M., 1974. An application of statistics to the chemical recognition of nickel gossans in the Yilgarn Block, Western Australia. *Proc. Australas. Inst. Min. Metall.,* 252: 21–24.

Keays, R.R. and Davison, R.M., 1976. Palladium, iridium, and gold in the ores and host rocks of nickel sulfide deposits in Western Australia. *Econ. Geol.,* 71: 1214–1228.

Keays, R.R., Groves, D.I. and Davison, R.M., in press. Ore element remobilization during progressive alteration of sulphide-bearing ultramafic rocks at the Black Swan nickel deposit, Western Australia: results of a precious metal study. *Econ. Geol.*

Keele, R.A. and Nickel, E.H., 1974. The geology of a primary millerite-bearing sulfide assemblage and supergene alteration at the Otter Shoot, Kambalda, Western Australia. *Econ. Geol.,* 69(7): 1102–1117.

Kelly, W.C. and Clark, B.R., 1975. Sulfide deformation studies, III. Experimental deformation of chalcopyrite to 2000 bars and 500°C. *Econ. Geol.,* 70: 431–453.

Kilburn, L.C., Wilson, H.D.B., Graham, A.R., Ogura, Y., Coats, C.J.A. and Scoates. R.F.J., 1969. Nickel sulfide ores related to ultrabasic intrusions in Canada. In: H.D.B. Wilson (Editor), *Magmatic Ore Deposits. Econ. Geol. Monogr.,* 4: 276–293.

Klotsman, S.M., Timofeyev, A.N. and Trakhtenberg, I.Sh., 1963. Investigation of the diffusion properties of the chalcogenides of transition metals, IV. Temperature dependence of the anisotropy of nickel and sulphur self-diffusion in nickel monosulphide. *Phys. Met. Metallogr.*, 16(5): 92–98.

Kovalenker, V.A., Gladyshev, G.D. and Nosik, L.P., 1975. Isotopic composition of sulfide sulfur from deposits of Talnakh ore node in relation to their selenium content. *Int. Geol. Rev.*, 17: 725–734.

Kullerud, G., 1963. The Fe–Ni–S system. *Carnegie Inst. Washington Yearb.*, 62: 175–189.

Kullerud, G., Yund, R.A. and Moh, G.H., 1969. Phase relations in the Cu–Fe–S, Cu–Ni–S and Fe–Ni–S systems. In: H.D.B. Wilson (Editor), *Magmatic Ore Deposits. Econ. Geol. Monogr.*, 4: 323–343.

Lajoie, J. and Gélinas, L., 1978. Emplacement of Archean peridotitic komatiites in La Motte Township, Quebec. *Can. J. Earth Sci.*, 15: 672–677.

Lambert, I.B., 1976. The McArthur zinc–lead–silver deposit: Features, metallogenesis and comparisons with some other stratiform ores. In: K.H. Wolf (Editor), *Handbook of Strata-bound and Stratiform Ore Deposits, II. Regional Studies and Specific Deposits, 6.* Elsevier, Amsterdam, pp. 535–585.

Lambert, I.B. and Sato, T., 1974. The Kuroko and associated ore deposits of Japan: a review of their features and metallogenesis. *Econ. Geol.*, 69: 1215–1236.

Leahey, J.E., 1973. *The Geology and Nickel–Copper Mineralogy of the D Shoot, Mount Windarra, Western Australia.* Thesis, Univ. Tasmania (unpublished).

Le Roex, H.D., 1964. Nickel deposit on the Trojan claims, Bindura District, Southern Rhodesia. In: S.H. Haughton (Editor), *The Geology of Some Ore Deposits in Southern Africa, 2.* Geol. Soc. S. Afr., Johannesburg, pp. 509–520.

Lipple, S.L., 1975. Definitions of new and revised stratigraphic units of the eastern Pilbara region. *Geol. Surv. West Aust. Annu. Rep.*, 1974: 58–63.

Lusk, J., 1976. A possible volcanic-exhalative origin for lenticular nickel sulfide deposits of volcanic association with special reference to those in Western Australia. *Can. J. Earth Sci.*, 13: 451–469.

Mainwaring, P.R. and Naldrett, A.J., 1977. Country-rock assimilation and the genesis of Cu–Ni sulfides in the Water Hen Intrusion, Duluth Complex, Minnesota. *Econ. Geol.*, 72: 1269–1284.

Marston, R.J. and Kay, B.D., in press. The distribution and petrology of nickel ores at Otter–Juan Shoot complex in relation to ore genesis theories at Kambalda, Western Australia. *Econ. Geol.*

Martin, J.E. and Allchurch, P.D., 1975. Perseverance nickel deposit, Agnew. In: C.L. Knight (Editor), *Economic Geology of Australia and Papua-New Guinea, 1. Metals.* Australas. Inst. Min. Metall., Melbourne, pp. 149–155.

Mazzucchelli, R.H., 1972. Secondary geochemical dispersion patterns associated with the nickel sulphide deposits at Kambalda, Western Australia. *J. Geochem. Explor.*, 1: 103–116.

McIver, J.R. and Lenthall, D.H., 1974. Mafic and ultramafic extrusives of the Barberton Mountain Land in terms of the CMAS System. *Precambrian Res.*, 1: 327–343.

McQueen, K.G., 1979. Experimental heating and diffusion effects in Fe–Ni sulfide ore from Redross, Western Australia. *Econ. Geol.*, 74: 140–148.

Miller, L.J. and Smith, M.E., 1975. Sherlock Bay nickel–copper. In: C.L. Knight (Editor), *Economic Geology of Australia and Papua-New Guinea, 1. Metals.* Australas. Inst. Min. Metall., Melbourne, pp. 168–174.

Misra, K.C. and Fleet, M.E., 1973. The chemical composition of synthetic and natural pentlandite assemblages. *Econ. Geol.*, 68: 518–539.

Moeskops, P.G., 1976. Yilgarn nickel gossan geochemistry – a review including new data and considerations. *25th Int. Geol. Congr., Sydney, Abstr.*, 2: 449–450.

Morimoto, N. and Kullerud, G., 1965. Pentlandite: thermal expansion. *Carnegie Inst. Washington Yearb.*, 64: 204–205.

Moubray, R.J., Brand, E.L., Hofmeyr, P.K. and Potter, M., in press. The Hunters Road nickel prospect. In: M.J. Viljoen (Editor), *Metallogenesis '76. Trans. Geol. Soc. S. Afr.*

Muir, J.E. and Comba, C.D.A., 1979. The Dundonald deposit: an example of volcanogenic nickel–sulphide mineralization. *Can. Mineral.*, 17(2): 351–360.

Naldrett, A.J., 1966. The role of sulphurization in the genesis of iron–nickel sulphide deposits of the Porcupine district, Ontario. *Trans. Can. Inst. Min. Metall.*, 69: 147–155.

Naldrett, A.J., 1969. A portion of the system Fe–S–O between 900 and 1080°C and its application to sulfide ore magmas. *J. Petrol.*, 10: 171–201.

Naldrett, A.J., 1973. Nickel sulfide deposits – their classification and genesis, with special emphasis on deposits of volcanic association. *Trans. Can. Inst. Min. Metall.*, 76: 183–201.

Naldrett, A.J. and Cabri, L.J., 1976. Ultramafic and related rocks: their classification and genesis with special reference to the concentration of nickel sulphides and platinum-group elements. *Econ. Geol.*, 71: 1131–1158.

Naldrett, A.J. and Gasparrini, E.L., 1971. Archaean nickel sulphide deposits in Canada: their classification, geological setting and genesis with some suggestions as to exploration. In: J.E. Glover (Editor), *Symposium on Archaean Rocks. Geol. Soc. Aust. Spec. Publ.*, 3: 201–226.

Naldrett, A.J. and Turner, A.R., 1977. The geology and petrogenesis of a greenstone belt and related nickel sulfide mineralization at Yakabindie, Western Australia. *Precambrian Res.*, 5: 43–103.

Naldrett, A.J., Craig, J.R. and Kullerud, G., 1967. The central portion of the Fe–Ni–S system and its bearing on pentlandite exsolution in iron–nickel sulfide ores. *Econ. Geol.*, 62: 826–847.

Naldrett, A.J., Goodwin, A.M., Fisher, T.L. and Ridler, R.H., 1978. The sulfur content of Archean volcanic rocks and a comparison with ocean floor basalts. *Can. J. Earth Sci.*, 15: 715–728.

Naldrett, A.J., Hoffman, E.L., Green, A.H., Chen-Lin Chou, Naldrett, S.R. and Alcock, R.A., 1979. The composition of Ni-sulfide ores with particular reference to their content of PGE and Au. *Can. Mineral.*, 17: 403–416.

Nesbitt, R.W., 1971. Skeletal crystal forms in the ultramafic rocks of the Yilgarn Block, Western Australia: evidence for an Archaean ultramafic liquid. In: J.E. Glover (Editor), *Symposium on Archaean Rocks. Geol. Soc. Aust. Spec. Publ.*, 3: 331–347.

Nesbitt, R.W. and Sun, S.S., 1976. Geochemistry of Archaean spinifex-textured peridotites and magnesian and low-magnesian tholeiites. *Earth Planet. Sci. Lett.*, 31: 433–453.

Nickel, E.H. and Hudson, D.R., 1976. The replacement of chrome spinel by chromian valleriite in sulphide-bearing ultramafic rocks in Western Australia. *Contrib. Miner. Petrol.*, 55: 265–277.

Nickel, E.H., Ross, J.R. and Thornber, M.R., 1974. The supergene alteration of pentlandite–pyrrhotite ore at Kambalda, Western Australia. *Econ. Geol.*, 69: 93–107.

Nickel, E.H., Allchurch, P.D., Mason, M.G. and Wilmhurst, J.R., 1977. Supergene alteration at the Perseverance nickel deposit, Agnew, Western Australia. *Econ. Geol.*, 72: 184–203.

Nisbet, E.G., Bickle, M.J. and Martin, A., 1977. Mafic and ultramafic lavas of the Belingwe Greenstone Belt, Rhodesia. *J. Petrol.*, 18(4): 521–566.

O'Beirne, W.R., 1968. *The Acid Porphyries and Porphyroid Rocks of the Kalgoorlie Area.* Thesis, Univ. Western Australia (unpublished).

Ostwald, J. and Lusk, J., 1978. Sulfide fabrics in some nickel sulfide ores from Kambalda. *Can. J. Earth Sci.*, 15(4): 501–515.

Oversby, V.M., 1975. Lead isotopic systematics and ages of Archaean acid intrusives in the Kalgoorlie–Norseman area, Western Australia. *Geochim. Cosmochim. Acta*, 39: 1107–1125.

Papunen, H., 1971. Sulfide mineralogy of the Kotalahti and Hitura nickel–copper ores, Finland. *Ann. Acad. Sci. Fenn., Ser. A.*, 109: 1–74.

Prider, R.T., 1970. Nickel in Western Australia. *Nature*, 226: 691–693.

Purvis, A.C., Nesbitt, R.W. and Hallberg, J.A., 1972. The geology of part of the Carr Boyd Rocks Complex and its associated nickel mineralization, Western Australia. *Econ. Geol.*, 67(8): 1093–1113.

Pyke, D.R. and Middleton, R.S., 1971. Distribution and characteristics of the sulphide ores of the Timmins area. *Can. Inst. Min. Trans.*, 74: 157–168.

Pyke, D.R., Naldrett, A.J. and Eckstrand, O.R., 1973. Archaean ultramafic flows in Munro Township, Ontario. *Geol. Soc. Am. Bull.*, 84: 955–978.

Rajamani, V. and Naldrett, A.J., 1978. Partitioning of Fe, Co, Ni and Cu between sulfide liquid and basaltic melts and the composition of Ni–Cu sulfide deposits. *Econ. Geol.*, 73: 82–93.

Ramsden, A.R., 1975. Compositions of coexisting pyrrhotites, pentlandites and pyrites at Spargoville, Western Australia. *Can. Mineral.*, 13: 133–137.

Rau, H., 1976. Energetics of defect formation and interaction in pyrrhotite $Fe_{1-x}S$ and its homogeneity range. *J. Phys. Chem. Solids,* 37: 425–429.

Roberts, D.E. and Travis, G.A., 1973. Textural evaluation of nickel sulphide gossans. *Australas. Inst. Min. Metall., Western Australian Conference, Perth, 1973. Abstract:* p. 97.

Roberts, J.B., 1975. Windarra nickel deposits. In: C.L. Knight (Editor), *Economic Geology of Australia and Papua-New Guinea, 1. Metals.* Australas. Inst. Min. Metall., Melbourne, pp. 129–143.

Roddick, J.C.M., 1974. *Responses of Strontium Isotopes to Some Crustal Processes.* Thesis, Aust. National Univ. (unpublished).

Roscoe, W.E., 1975. Experimental deformation of natural chalcopyrite at temperatures up to 300°C over the strain rate range 10^{-2} to 10^{-6} sec^{-1}. *Econ. Geol.,* 70: 454–472.

Ross, J.R., 1974. *Archaean Nickel Sulphide Mineralization, Lunnon Shoot, Kambalda.* Thesis, Univ. Calif., Berkeley (unpublished).

Ross, J.R. and Hopkins, G.M., 1975. Kambalda nickel sulphide deposits. In: C.L. Knight (Editor), *Economic Geology of Australia and Papua-New Guinea, 1. Metals.* Australas. Inst. Min. Metall., Melbourne, pp. 100–121.

Ross, J.R. and Keays, R.R., 1979. Precious metals in volcanic-type nickel sulfide deposits in Western Australia, Part I. Relationship with the composition of the ores and their host rocks. *Can. Mineral.,* 17: 417–436.

Rye, R.O. and Ohmoto, H., 1974. Sulfur and carbon isotopes and ore genesis: a review. *Econ. Geol.,* 69: 826–842.

Sangster, D.F., 1978. Exhalites associated with Archaean volcanogenic massive sulphide deposits. In: J.E. Glover and D.I. Groves (Editors), *Archaean Cherty Metasediments: their Sedimentology, Micropalaeontology, Biogeochemistry and Significance to Mineralization.* Publ. Geol. Dep. and Extension Service, Univ. West Aust., 2: 70–81.

Sangster, D.F. and Scott, S.D., 1976. Precambrian stratabound massive Cu–Zn–Pb sulfide ores of North America. In: K.H. Wolf (Editor), *Handbook of Strata-bound and Stratiform Ore Deposits, II. Regional Studies and Specific Deposits, 6.* Elsevier, Amsterdam, pp. 129–222.

Schwartsz, H.P., 1973. Sulfur isotope analyses of some Sudbury, Ontario, ores. *Can. J. Earth Sci.,* 10: 1444–1459.

Seccombe, P.K., Groves, D.I., Binns, R.A. and Smith, J.W., 1977. A sulphur isotope study to test a genetic model for Fe–Ni sulphide mineralization at Mt. Windarra, Western Australia. In: B.W. Robinson (Editor), *Stable Isotopes in the Earth Sciences.* DSIR Bull., 220: 187–200.

Sharpe, J.W.N., 1964. The Empress nickel-copper deposit, Southern Rhodesia. In: S.H. Haughton (Editor), *The Geology of Some Ore Deposits in Southern Africa, 2.* Geol. Soc. S. Afr., Johannesburg, pp. 497–508.

Sheppy, N.R. and Rowe, J., 1975. Nepean nickel deposit. In: C.L. Knight (Editor), *Economic Geology of Australia and Papua-New Guinea, 1. Metals.* Australas. Inst. Min. Metall., Melbourne, pp. 91–99.

Shewman, R.W. and Clark, L.A., 1970. Pentlandite phase relations in the Fe–Ni–S system and notes on the monosulfide solid solution. *Can. J. Earth Sci.,* 7: 67–85.

Shima, M., Gross, W.H. and Thode, H.G., 1963. Sulfur isotope abundances in basic sills, differentiated granites, and meteorites. *J. Geophys. Res.,* 68(9): 2835–2847.

Shima, H. and Naldrett, A.J., 1975. Solubility of sulfur in an ultramafic melt and the relevance of the system Fe–S–O. *Econ. Geol.,* 70: 960–967.

Smirnov, V.I., 1977. Deposits of nickel. In: *Ore Deposits of the USSR, II.* Pitman, London, pp. 3–79.

Smith, B.H., 1976. Some aspects of the use of geochemistry in the search for base-metal sulphides in lateritic terrain in Western Australia. *25th Int. Geol. Congr. Abstr.,* 2: 458–459.

Solomon, M., 1976. "Volcanic" massive sulphide deposits and their host rocks – a review and an explanation. In: K.H. Wolf (Editor), *Handbook of Strata-bound and Stratiform Ore Deposits, II. Regional Studies and Specific Deposits, 6.* Elsevier, Amsterdam, pp. 21–54.

Sofoulis, J., 1965. Explanatory notes on the Widgiemooltha 1 : 250,000 geological sheet, Western Australia. *Rec. Geol. Surv. West. Aust.,* 1965/10.

Sun, S.S. and Nesbitt, R.W., 1977. Chemical heterogeneity of the Archaean mantle, composition of the earth and mantle evolution. *Earth Planet. Sci. Lett.,* 35: 429–448.

Thornber, M.R., 1972. Pyrrhotite – The matrix of nickel sulphide mineralization. *Australas. Inst. Min. Metall., Newcastle Conf.*, pp. 51–58.

Thornber, M.R., 1975a. Supergene alteration of sulphides, I. A chemical model based on massive nickel sulphide deposits at Kambalda, Western Australia. *Chem. Geol.*, 15: 1–14.

Thornber, M.R., 1975b. Supergene alteration of sulphides, II. A chemical study of the Kambalda nickel deposits. *Chem. Geol.*, 15(2): 117–144.

Tomich, B.N.V., 1974. *The Geology and Nickel Mineralization of the Ruth Well Area, Western Australia.* Thesis, Univ. West. Aust. (unpublished).

Travis, G.A., 1975. Nickel–copper sulphide mineralization in the Jimberlana Intrusion. In: C.L. Knight (Editor), *Economic Geology of Australia and Papua-New Guinea, 1. Metals.* Australas. Inst. Min. Metall., Melbourne, pp. 75–78.

Travis, G.A., Keays, R.R. and Davison, R.M., 1976. Palladium and iridium in the evaluation of nickel gossans in Western Australia. *Econ. Geol.*, 71: 1229–1243.

Usselman, T.M., Hodge, D.S., Naldrett, A.J. and Campbell, I.H., 1979. Physical constraints on the localization of nickel sulfide ore in ultramafic lavas. *Can. Mineral.*, 17(2): 361–372.

Viljoen, M.J. and Bernasconi, A., in press. The geochemistry, regional setting and genesis of the Shangani–Damba nickel deposits, Rhodesia. In: M.J. Viljoen (Editor), *Metallogenesis '76. Trans. Geol. Soc. S. Afr.*

Viljoen, M.J. and Viljoen, R.P., 1969. Evidence for the existence of a mobile extrusive peridotitic magma from the Komati Formation of the Onverwacht Group. *Geol. Soc. S. Afr. Spec. Publ., Upper Mantle Project*, 2: 87–112.

Viljoen, M.J., Bernasconi, A., van Coller, N., Kinloch, E. and Viljoen R.P., 1976. The geology of the Shangani nickel deposit, Rhodesia. *Econ. Geol.*, 71: 76–95.

Visscher, W.M. and Bolsterli, M., 1972. Random packing of equal and unequal spheres in two and three dimensions. *Nature*, 239: 504–507.

Watmuff, I.G., 1974. Supergene alteration of the Mount Windarra nickel sulphide ore deposit, Western Australia. *Miner. Deposita*, 9: 199–211.

Willett, G.C., 1975. Possible amygdaloidal sulfides in the Otter Shoot, Kambalda, Western Australia. *Econ. Geol.*, 70: 1127.

Williams, D.A.C., 1972. Archaean ultramafic, mafic and associated rocks, Mt. Monger, Western Australia. *J. Geol. Soc. Aust.*, 19: 163–188.

Williams, D.A.C., 1979. The association of some nickel sulfide deposits with komatiitic volcanism in Rhodesia. *Can. Mineral.*, 17(2): 337–350.

Williams, I.R., 1974. Structural subdivision of the Eastern Goldfields Province, Yilgarn Block. *West. Aust. Geol. Surv. Annu. Rep.*, 1973: 53–59.

Wilmshurst, J.R., 1975. The weathering products of nickeliferous sulphides and their associated rocks in Western Australia. In: I.L. Elliott and W.K. Fletcher (Editors), *Geochemical Exploration 1974.* Elsevier, Amsterdam, pp. 417–436.

Wilmshurst, J.R., 1976. The recognition of gossans and related rocks. *25th Int. Geol. Congr., Sydney, Abstr.*, 2: 464–465.

Wilson, J.F., in prep. A preliminary reappraisal of the Rhodesian basement complex.

Wilson, J.F., Bickle, M.J., Hawkesworth, C.J., Martin, A., Nisbet, E.G. and Orpen, J.L., 1978. Granite–greenstone terrains of the Rhodesian Archaean craton. *Nature*, 271: 23–27.

Woodall, R. and Travis, G.A., 1969. The Kambalda nickel deposits, Western Australia. *Proc. 9th Commonw. Min. Metall. Congr., London*, 2: 517–533.

Yamamoto, M., 1976. Relationship between Se/S and sulfur isotope ratios of hydrothermal sulfide minerals. *Miner. Deposita*, 11: 197–209.

Chapter 7

THE SIGNIFICANCE OF PYRITIC BLACK SHALES IN THE GENESIS OF ARCHEAN NICKEL SULPHIDE DEPOSITS

TIM HOPWOOD

INTRODUCTION

Although almost all Archean massive Ni-sulphide deposits are hosted in complexly deformed and metamorphosed country rocks [1] (in some cases with notable structural elongation of orebodies), most work on these deposits has emphasized igneous-geochemical aspects rather than the structural, stratigraphic and metamorphic-textural relationships (see Barrett et al., 1977). As a consequence of this, geochemical processes of magmatic differentiation tend to dominate genetic concepts for the deposits (Lambert, 1976, Handbook Vol. 6, p. 581; Wolf, 1976, Handbook Vol. 1, pp. 73–74; cf. Groves and Hudson, 1981). Although such models may be applicable in part, the present author believes that a simple magmatic model (Hudson, 1973; Naldrett, 1973) does not adequately account for many geological observations in these deposits. This chapter suggests an alternative mechanism involving the reaction and contamination of the ultramafic melts with pyritic black shales which provide a sulphur source for mineralization and a transport mechanism for Ni^{2+}, etc., in the form of metal carbonyls.

Contrasted massive Ni-sulphide host environments, and definition of environment under discussion

As summarized in Fig. 1, ultramafic host-rocks as well as their associated pentlandite–pyrrhotite orebodies appear to fall naturally into four classes (cf. Naldrett, 1973) depending upon the tectonic setting, the degree of metamorphic alteration of the ultramafic host, and relationship of the sulphide deposit to the host-rock. The host-rocks of types 1 and 2 below are characterized by gravity-differentiated rock varieties with obvious igneous textures which are essentially unaltered. Types 3 and 4 host-rocks are subject to extensive metamorphic alteration forming serpentine–tremolite–chlorite–dolomite–siderite–talc–magnetite assemblages (Barret et al., 1977). Such ultramafic

[1] The term host-rocks refers to rocks which contain the sulphide deposits; the term country rocks refers to rocks enclosing intrusive ultramafic bodies.

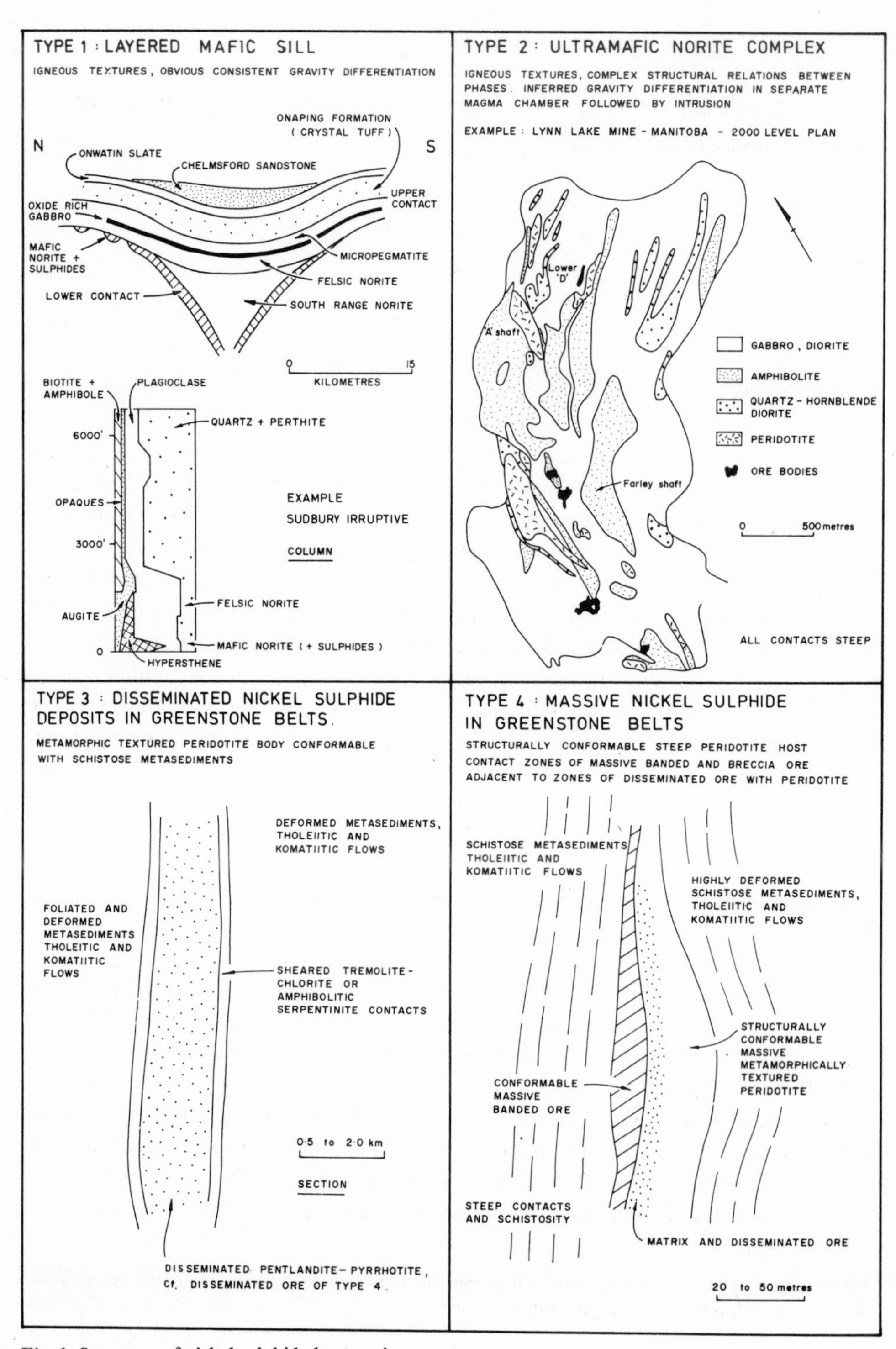

Fig. 1. Summary of nickel sulphide host environments.

bodies tend to be smaller than those of types 1 and 2 above, and, in contrast to types 1 and 2, are structurally conformable.

On the basis of textural characteristics, tectonic setting, nature of the host ultramafic body and the relationship of the sulphide mineralization to the host-rock, ultramafic-nickel deposits can be broadly divided into four types as follows:

Type 1: *layered mafic sill.* Deposits of this type occur in tectonically stable areas with concentrations of massive and/or disseminated pentlandite–pyrrhotite present at the base of large sub-horizontal stratiform gravity-layered mafic sills. Although the mineralization occurs within complex breccias at the base of such intrusions, texturally the host-rocks are undeformed and unmetamorphosed and *obviously* igneous. Gravity-differentiated sequences from gabbro through to diorite or granophyre are recognizable. Examples of this type are the deposits associated with Sudbury, Ontario (Souch et al., 1969; Card et al., 1972) and the Stillwater Complex, Montana (Jackson, 1961; Barker, 1975).

Type 2: *ultramafic norite complex.* In this type, the mafic host-rocks for these deposits are steep intrusive plug-like ultramafic norite–gabbro complexes. These intrusions are composed of a series of igneous phases which compositionally are interpreted to represent a gravity-differentiated sequence, although such phases are arranged in a complex structural pattern and the relative distribution of phases is not systematically arranged in an obvious gravity-differentiated sequence as in type 1. Concentrations of disseminated Ni–Fe sulphides form multiple steep ore-lenses often associated with the more silica-rich amphibolitic rock types. Tectonically, the intrusions are located within metamorphosed greenstone belts and, although partly altered, the host textures are recognizably igneous. Gravity-differentiation is inferred to take place in a separate magma chamber followed by multiple intrusion of each of the phases. Examples are the Lynn Lake Deposit, Manitoba (Sherrit Gordon Mines, 1972), Carr Boyd, Western Australia (Purvis et al., 1972; Groves and Hudson, 1981) or the Empress Mine, Rhodesia (Sharpe, 1964).

Type 3: *low-grade disseminated.* The host-rocks for this type are large steeply oriented lens-shaped intrusions of serpentinite or serpentinitized pyroxenite, which form conformable sheets within deformed and metamorphosed greenstone belts (with associated sediments). These ultramafic bodies are not simply or systematically differentiated, in the sense of type 1 above, but are characterized by metamorphic textures superposed upon igneous textures. These serpentinite bodies contain irregular pod-like concentrations of disseminated pyrrhotite–pentlandite, referred to as "matrix-" or "lace-" sulphide. Such deposits are generally large-tonnage/low-grade. Examples of this style of mineralization are Mystery Lake, Manitoba (Coats et al., 1972) and Mt. Keith, Western Australia (Burt and Sheppy, 1975; Groves and Hudson, 1981).

Type 4: *massive sulphide/altered ultramafic host.* These deposits occur within the same regional tectonic environment as that of type 3 and are usually associated with the contact of the altered lenticular ultramafic host body with greenstone-sediment stratigraphy. The deposits are structurally composite, commonly consisting of two adjacent texturally contrasted sulphide zones. These comprise: (a) a zone of massive pyrrhotite–pentlandite, sometimes banded; and (b) an adjacent zone of disseminated pentlandite–pyrrhotite. The disseminated zone of "matrix-" or "lace-" sulphide (resembling the textural style of the sulphide of type 3 above) occurs within the ultramafic, and the massive sulphide zone is generally close to or on a contact of the ultramafic with metasedimentary rocks. The ultramafic host-rocks are altered, comprising serpentine–tremolite–chlorite–dolomite–siderite–talc–magnetite assemblages (Barrett et al., 1976, 1977). Although remnant igneous olivines (Fo_{90-92}) are recognizable, the textures of these rocks are not obviously igneous in the sense of type 1 above; they are metamorphic textures. Individual host bodies to the mineralization are chemically differentiated, often (but not always) with more magnesian compositions at the base; this is interpreted by petrologists as indicating the altered equivalent of a gravity-differentiated body (Ewers and Hudson, 1972; Ross and Hopkins, 1975). There is a repeated association of pyritic black shales (or the oxidized equivalent banded iron formation) either close to or adjacent to the host ultramafic, or within the associated basalt-sediment stratigraphy (Naldrett, 1966; Prider, 1970; Gilmour, 1976, Handbook, Vol. 1, p. 114; King, 1976, Handbook, Vol. 1, pp. 168 and 177). The genetic significance of this association has not been fully investigated or explained in the literature, and this is the central aim of this paper.

GEOLOGICAL RELATIONSHIPS OF ARCHEAN MASSIVE NICKEL-SULPHIDE DEPOSITS

Regional setting

Regionally, the Archean greenstone belts (Coats et al., 1972; Gemuts and Theron, 1975), containing these nickel deposits, are characterized by the following associations:

(1) Stratigraphic sequences with large volumes of basic and spinifex-textured komatiitic flow rocks (Nesbitt, 1971a; Pyke et al., 1973; Barrett et al., 1977; cf. also Groves and Hudson, 1981).

(2) Pyritic black shales or sulphide-bearing banded iron formations. These rocks commonly consist of schistose carbonate–quartz–sericite–pyrite–graphite assemblages at lower metamorphic grades (greenschist facies), or in areas unaffected by alteration associated with mineralization. They characteristically have zinc contents between 400 and 1500 ppm. In some cases, these pyritic black shales are oxidized in surface outcrop or during metamorphism to result in haematite- or magnetite-bearing siliceous banded iron formations (Drake, 1972; Hopwood, 1974).

(3) A common deformation style, often observed within these sedimentary volcanic greenstone sequences, comprises tight, often complex folds parallel to a steep metamorphic mineral streaking style of lineation developed within the schistosity. This lineation is often consistent in orientation over large areas (except where refolded by later deformations). In some areas of greenstone belts, fold styles associated with the steep mineral

streaking and schistosity are confined to schist zones, between areas which are relatively less strained, and with little development of schistosity; ore body elongation is commonly parallel to this mineral streaking lineation (Hopwood, 1974, 1978; Binns et al., 1976).

(4) Zones of intrusion, or extrusion, of many magnesian-rich ultramafic bodies occur regionally associated with zones of active deformation. Many of these bodies appear to have been peridotitic in composition, and several intrusive bodies are characterized by a central dunitic core, with a thin metapyroxenitic border zone (Wilson et al., 1969; Nesbitt, 1971a; Hallberg and Williams, 1972). Although there are many known magnesian ultramafic bodies in both the Canadian and Australian Archean terranes, very few of these are ore-bearing, and most of the barren bodies have relatively minor sulphide content.

(5) Regionally, the distribution of these ultramafics appears generally associated with elongate lineaments defined by gravity boundaries. These lineaments are conformable with regional schistosity and/or schist zones and possibly represent deep crustal faults (Thompson Belt, Wilson and Brisbin, 1961; Superior Province, Goodwin and Ridler, 1970, J. Parry, pers. comm.; Yilgarn Block, W. Gailbraith, pers. comm., I. Gemuts, pers. comm., Binns et al., 1976).

Comparison of deposits

By comparing a series of related deposits of similar tectonic settings, rock-type associations, and structural characteristics, it is possible to distinguish those observations which are recurrent and therefore likely to be genetically significant, from those observations that are merely ancillary. These repeated observations and rock-type associations ultimately suggest a common geological environment for the deposits, with implicit similar or related ore-genesis mechanisms.

Table I presents a condensed comparative summary of data currently available from the literature, along with the author's observations on ten Australian and Canadian deposits (Figs. 2 and 3) considered to be representative of the Archean massive Ni-sulphide deposits (type 4 above). It must be pointed out that this tabulation is intended primarily for comparison, and is dependent on the work of many authors. For some detailed descriptions of specific deposits the reader is referred to the original works, listed with each deposit in Table I.

Ore zones and host ultramafic

These deposits are characterized by three adjacent zones (Fig. 1) of sulphide mineralization passing from *massive* sulphide at the base or wall-rock contact of the ore-zone, through *matrix* sulphide into disseminated sulphide within a serpentinite or talc–carbonate altered ultramafic host-rock:

(1) The *massive sulphide* is composed predominantly of iron and nickel sulphides, texturally massive or banded, with relatively minor interlensed gangue material. The

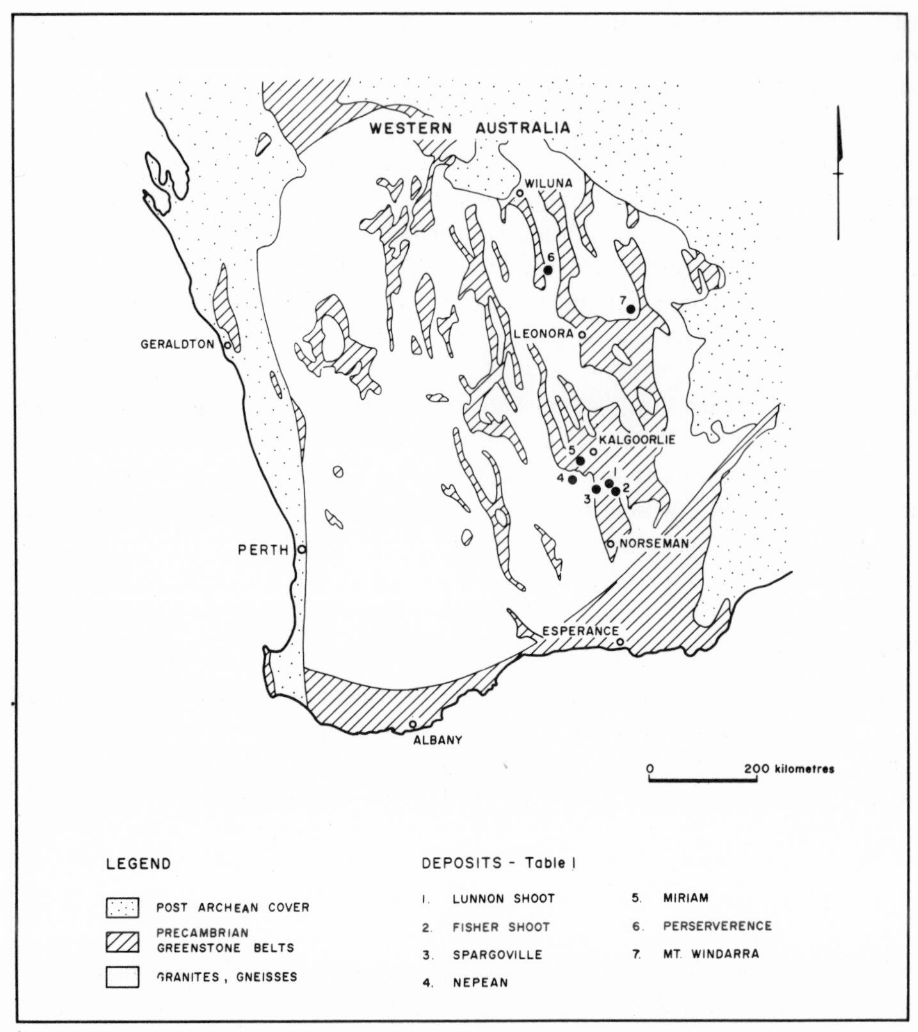

Fig. 2. Localities of Australian deposits compared on Table I.

sulphides usually comprise more than 75% of the ore zone. In many cases, there are breccia ores containing numerous graphitic, chloritic and quartzitic metasedimentary fragments (Table I).

(2) *Matrix sulphide* is a term introduced by Hancock et al. (1971) and Ewers and Hudson (1972). It is equivalent to the net-textured ore illustrated by Naldrett (1966, p. 492, and 1973, p. 59). The ore is texturally random and non-schistose and is characterized by an interconnected array of sulphides within silicate gangue. This style of texture is presumably dependent on the relative sulphide/silicate-gangue concentrations. Round grains of olivine, 1–10 mm in diameter interstitial to the sulphides and now pseudo-

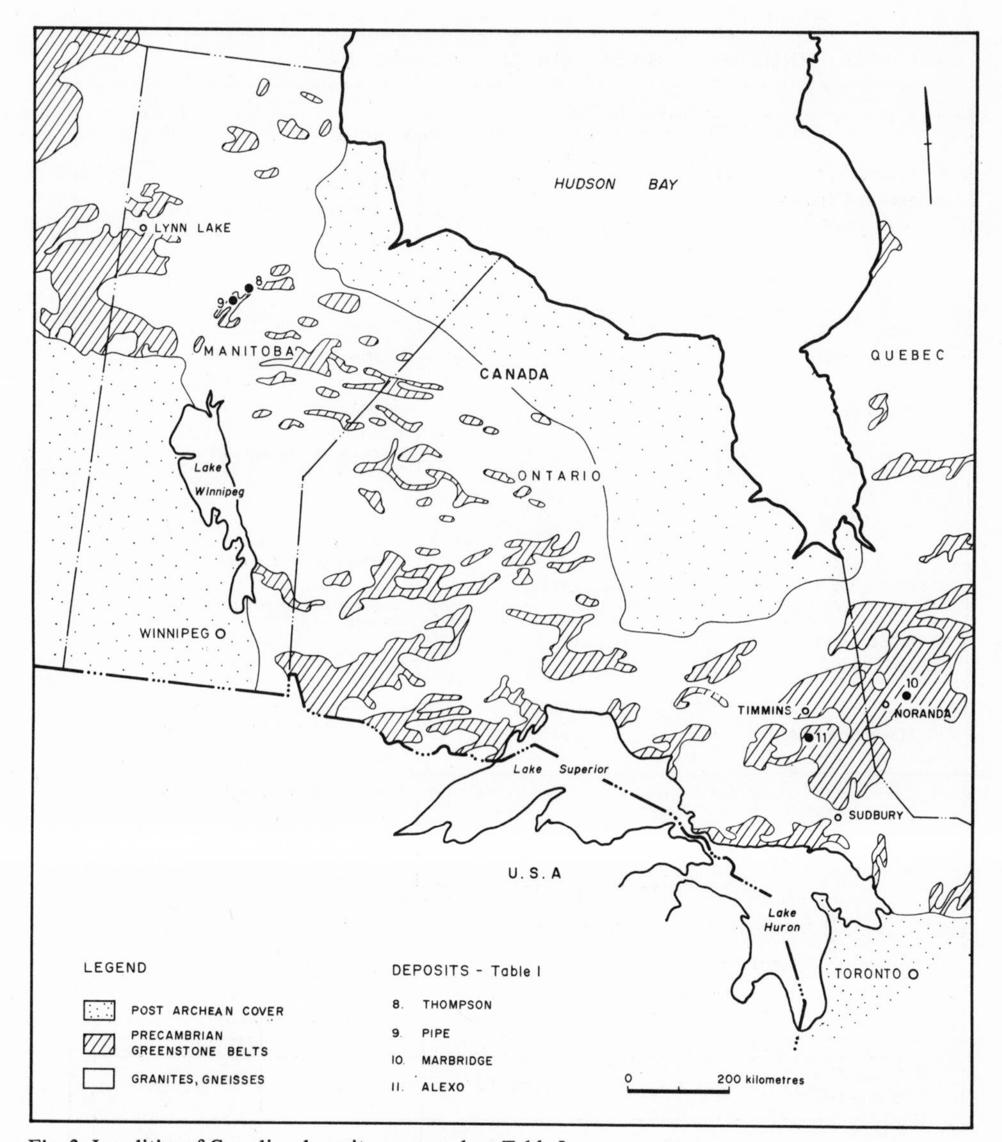

Fig. 3. Localities of Canadian deposits compared on Table I.

morphed by serpentine–carbonate, can occasionally be recognized. Sulphide contents in this style of ore range from approximately 20% to 65%.

(3) *Disseminated sulphide.* With decrease in sulphide content (to less than 10%), a lack of interconnection of the sulphides (and electrical conductivity) exists, so that truly disseminated textural styles result.

As described by Ewers and Hudson (1972), Naldrett (1973), and Ross and Hopkins

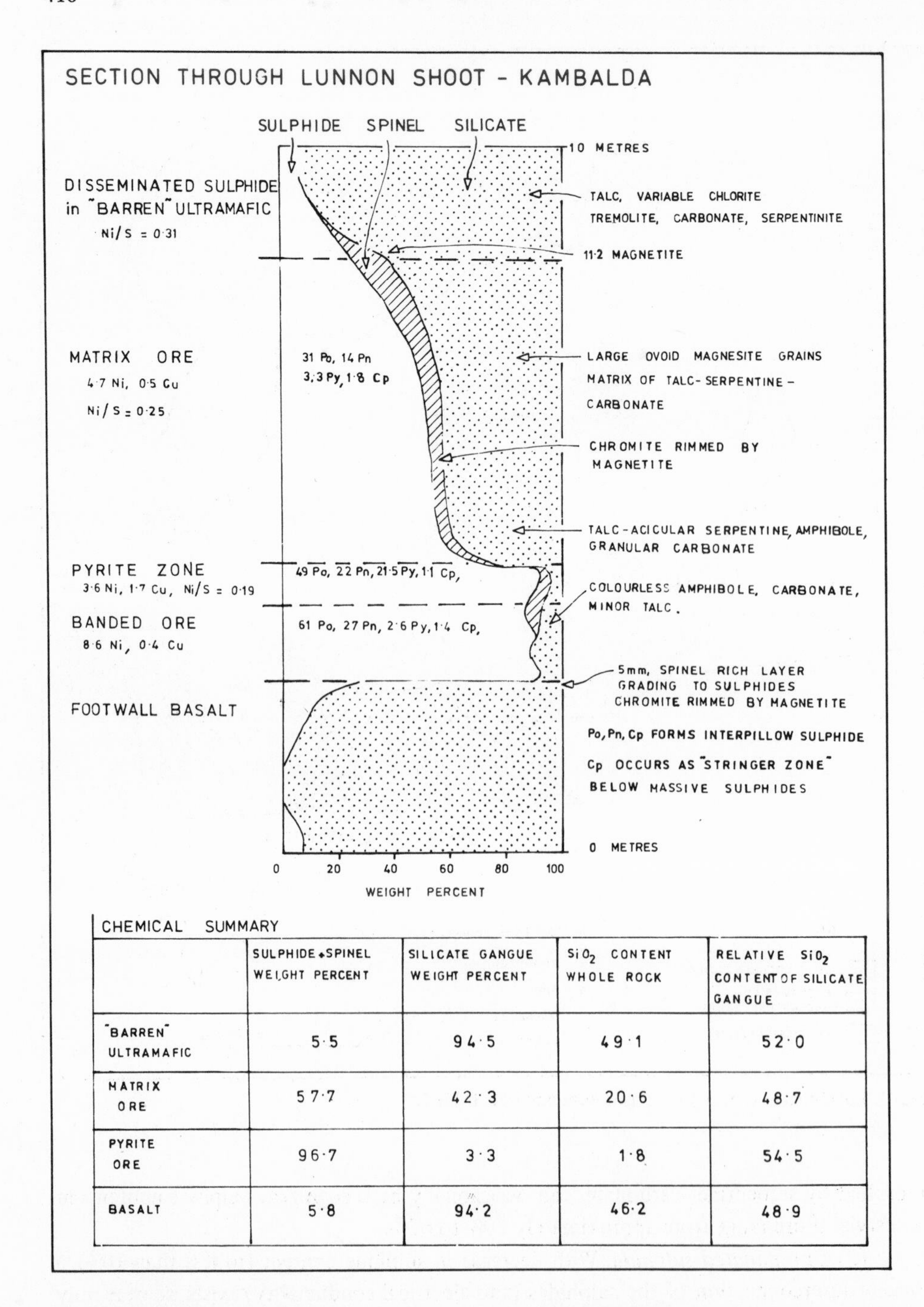

CHEMICAL SUMMARY

	SULPHIDE+SPINEL WEIGHT PERCENT	SILICATE GANGUE WEIGHT PERCENT	SiO_2 CONTENT WHOLE ROCK	RELATIVE SiO_2 CONTENT OF SILICATE GANGUE
"BARREN" ULTRAMAFIC	5·5	94·5	49·1	52·0
MATRIX ORE	57·7	42·3	20·6	48·7
PYRITE ORE	96·7	3·3	1·8	54·5
BASALT	5·8	94·2	46·2	48·9

Fig. 4. Relationships of ore zones to ultramafic host-rocks, Lunnon Shoot Kambalda, Western Australia, Deposit 1 on Table I, Fig. 2.

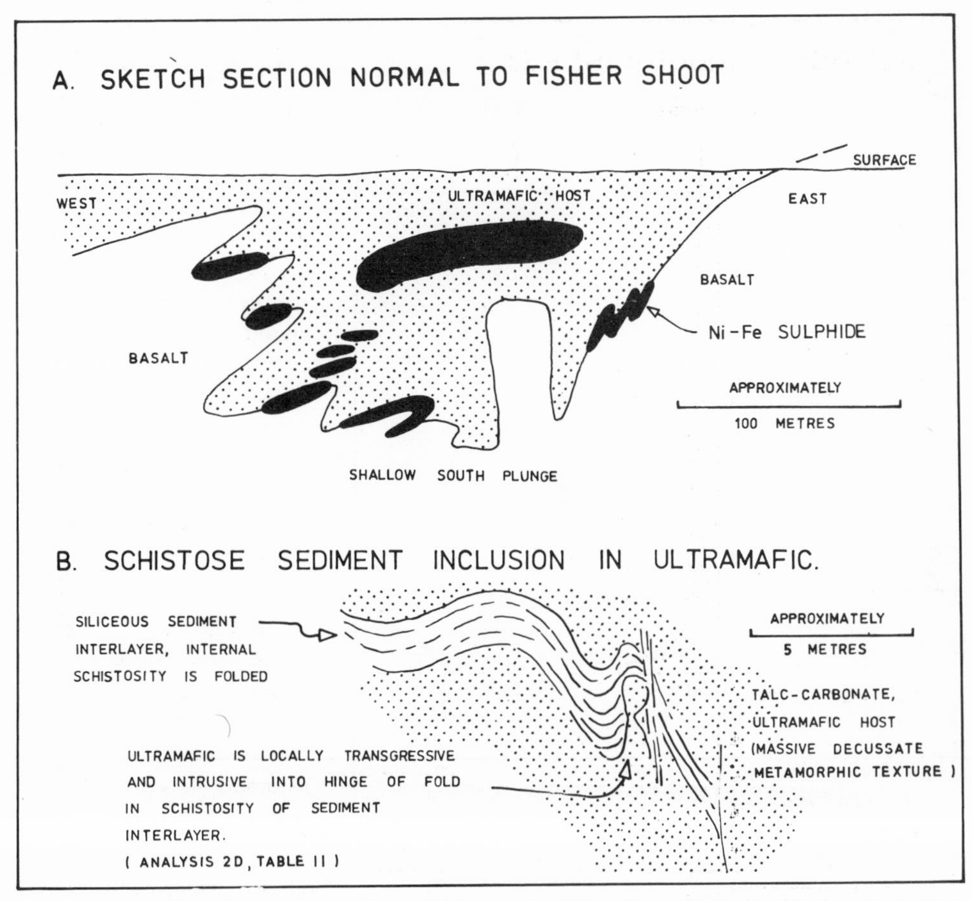

Fig. 5. Relationships of ore zones to ultramafic host-rocks, Fisher Shoot Kambalda, Western Australia, deposit 2 on Table I. A. Hanging-wall ore bodies B. Schistose sediment inclusion in ultramafic; photomicrograph, Fig. 12C.

(1975), the massive–matrix and matrix–disseminated boundaries are relatively sharp (for igneous textural boundaries) in Lunnon Shoot and Alexo, but this does not consistently appear to be the case, even in sulphide deposits which are thought to be immiscible-magmatic (e.g. the Katiniq Sill, Wilson et al., 1969, p. 299).

Of the deposits listed in Table I, seven comprise composite massive–matrix ore zones, usually with the massive sulphide zone close to a metasediment contact, and the matrix ore zone within the ultramafic body. These are Lunnon (Fig. 4), Perserverence, Mt. Windarra, Thompson, Pipe, Marbridge No. 1 and No. 3, and Alexo. Spargoville and Nepean consist of matrix-sulphide concentrations along one contact of the ultramafic host (Fig. 6). In contrast, Fisher Shoot and Miriam are conformable sheet-like concentrations of matrix ore situated wholly within the host ultramafic body at some distance

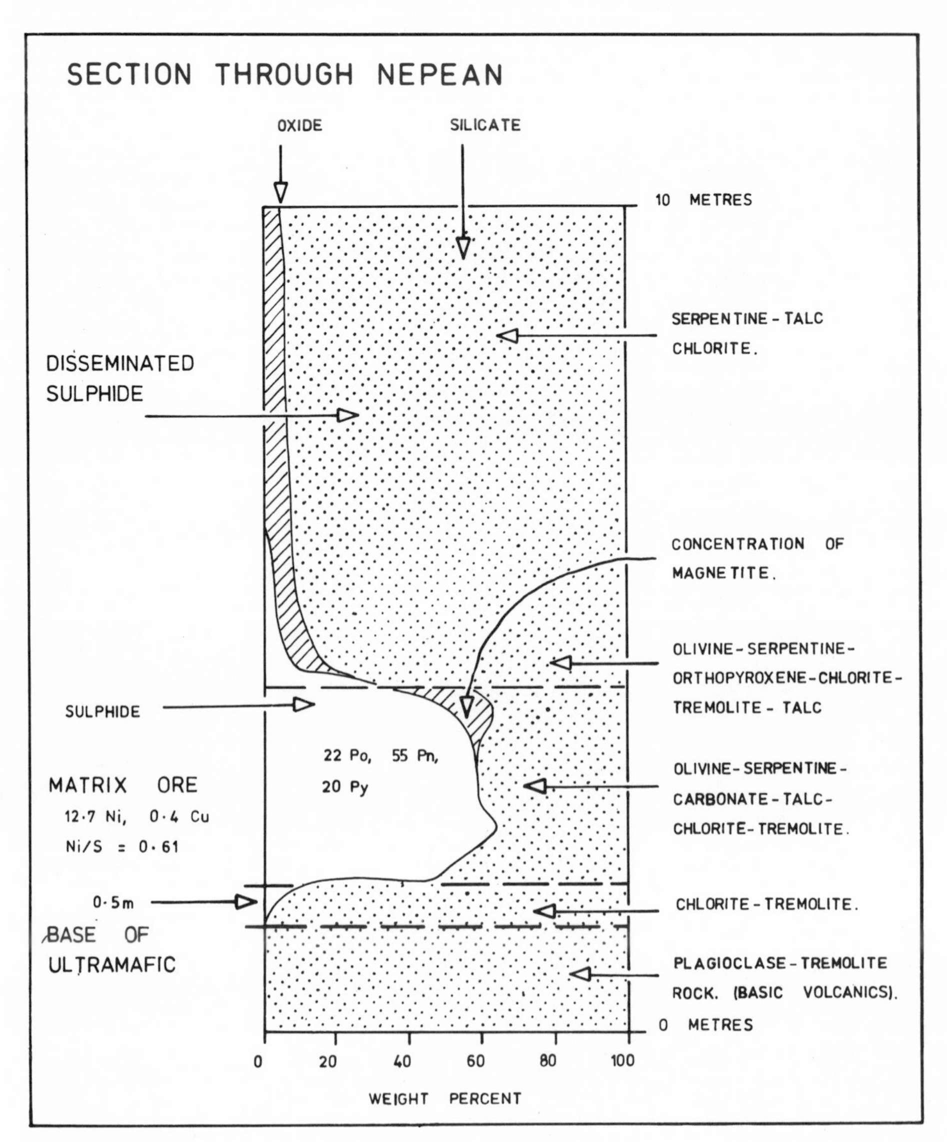

Fig. 6. Relationships of ore zone to ultramafic host, Nepean, Western Australia, deposit 4 on Table I, Fig. 2.

from a metasediment contact of the ultramafic, termed hanging-wall mineralization (Figs. 5 and 7).

Notably, the Thompson deposit is associated with a particularly small ultramafic body as seen in plan (Fig. 9), and although the ore zone is in contact with this ultramafic body for part of the length of the deposit, the massive sulphide ore zone extends for

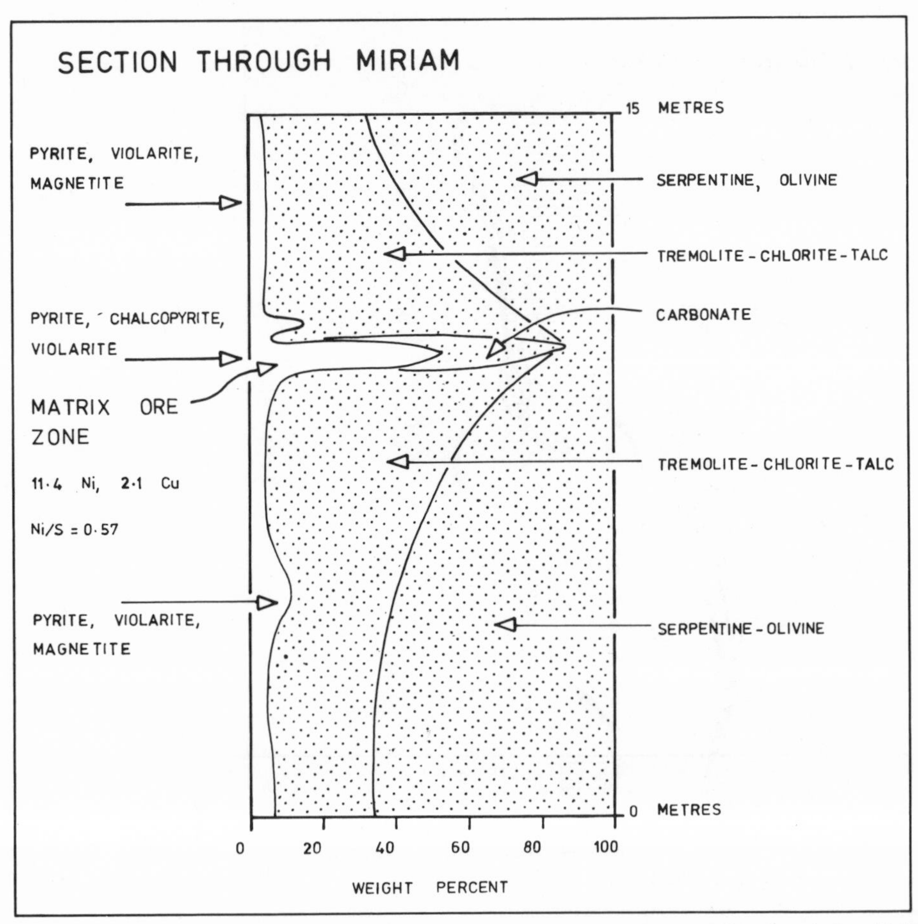

Fig. 7. Relationships of ore zone to ultramafic host, Miriam, Western Australia, deposit 5 on Table I, Fig. 2.

some considerable distance into what appear to be metasediments. Another relationship is that of Marbridge No. 1, where two ultramafic lenses, each containing a matrix ore zone, occur on opposite sides of a sigmoidally curved massive sulphide zone (Fig. 11).

Figs. 4–11 show examples of these contrasted relationships between massive, matrix or disseminated sulphide ore zones and host ultramafics.

The ultramafic rock adjacent to each of these ore zones contains random blades of tremolite–chlorite texturally over-printed on the original igneous texture (on both sides of the ore zone in the case of Miriam). Where data are available, a decrease in sulphur content from the massive ore zone is noticable, through the matrix ore zone to very low values in the disseminated sulphide zone within the host ultramafic. In some deposits, there are relatively sharp decreases in sulphur content at the massive–matrix and matrix–

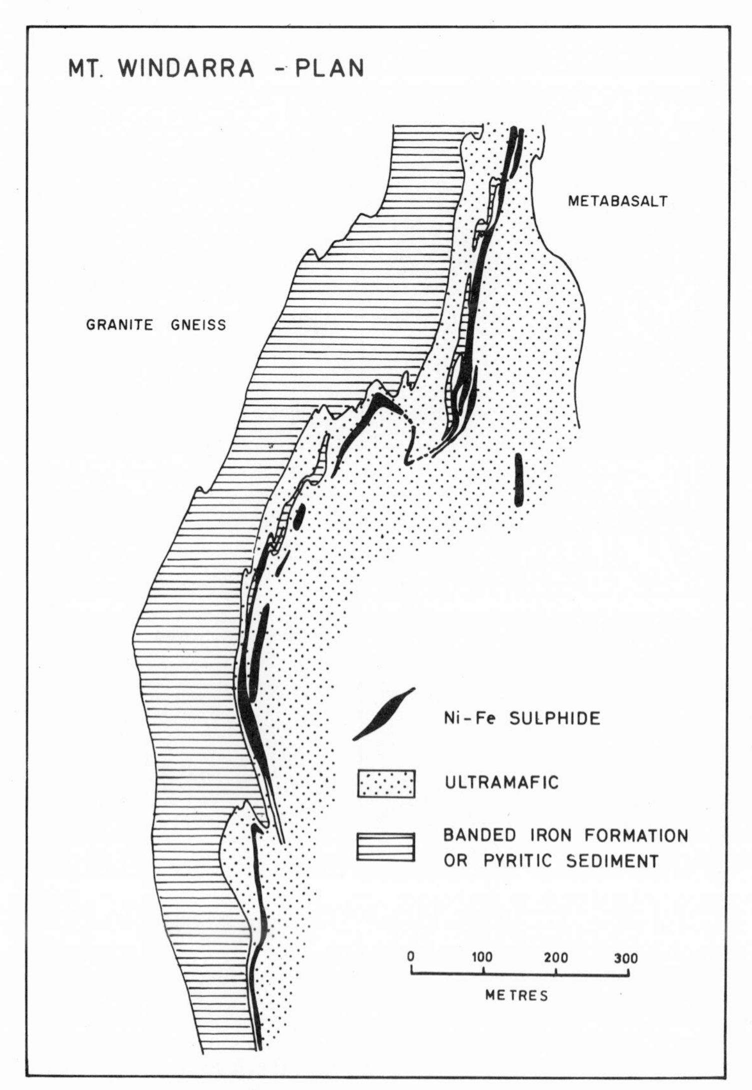

Fig. 8. Relationship between pyritic metasediment, host ultramafic and mineralization, Mt. Windarra, Western Australia, deposit 7 on Table I, Fig. 2.

disseminated boundaries (Ross and Hopkins, 1975). Associated with the decrease in sulphur content from massive ore to matrix ore to disseminated sulphide, an increase in the Ni/S ratio of the sulphide fraction is present, i.e. the sulphides are progressively more Ni-rich with dispersion in the ultramafic. This is reflected in changes in sulphide assemblages from pyrrhotite–pentlandite–pyrite of the massive or matrix zones, to assemblages containing millerite and violarite in the disseminated sulphide zones. This continuous variation is consistent with the observations of Graterol and Naldrett (1971) for

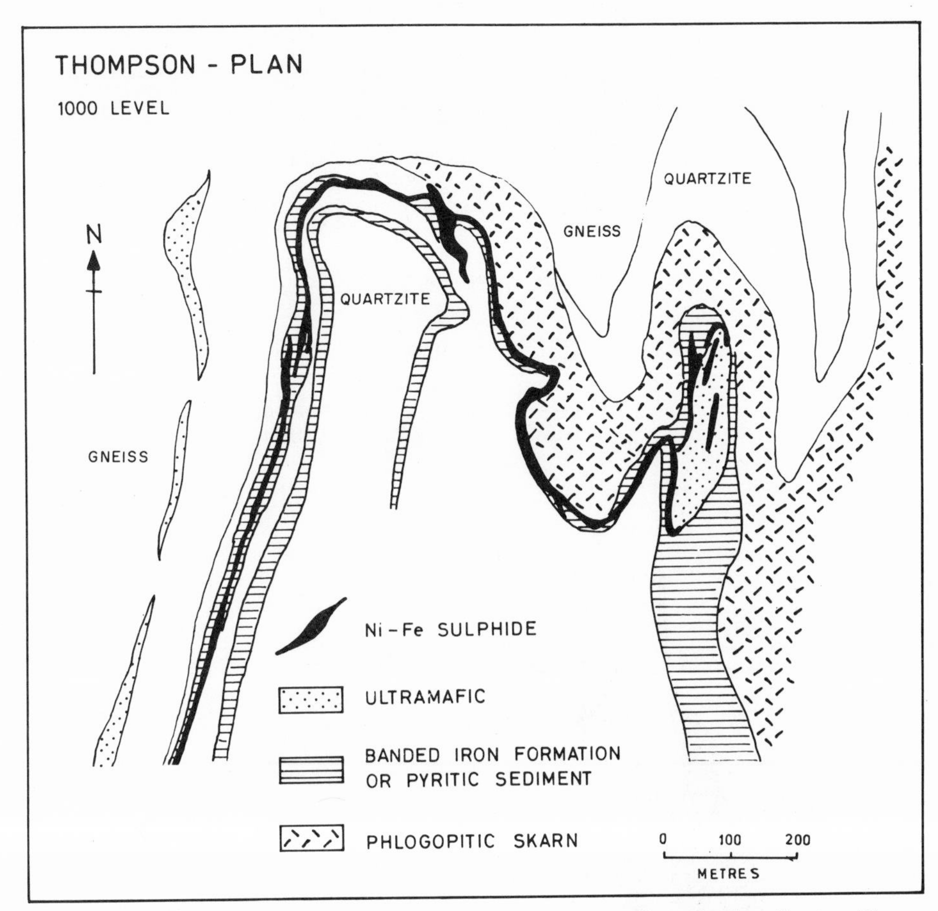

Fig. 9. Relationship between pyritic metasediment, host ultramafic and mineralization, Thompson Mine, Canada, deposit 8 on Table I, Fig. 3.

Marbridge, by Hudson (1973) for Nepean, and phase equilibrium work by Harris and Nickel (1972) and Misra and Fleet (1973) that the composition of pentlandite is variable, depending upon the associated mineralogy. In some cases, there is a zone of relative Ni-depletion (0.3% reduced to 0.03% Ni) in the silicates of the matrix ore of the disseminated zone. Common textural relationships observed between pyrrhotite, pentlandite and pyrite are:

(1) Pyrite occurs in the core of pyrrhotite grains or has pyrrhotite rims.

(2) Pentlandite penetrates the basal parting of pyrrhotite as flame structures extending from fractures or grain boundaries.

(3) Violarite is present in fine cracks or veins in pentlandite, closely associated with fine magnetite. The distribution of chalcopyrite and cobalt does not seem to be particularly systematic, except that they tend to occur with more pyritic or sulphur-rich assem-

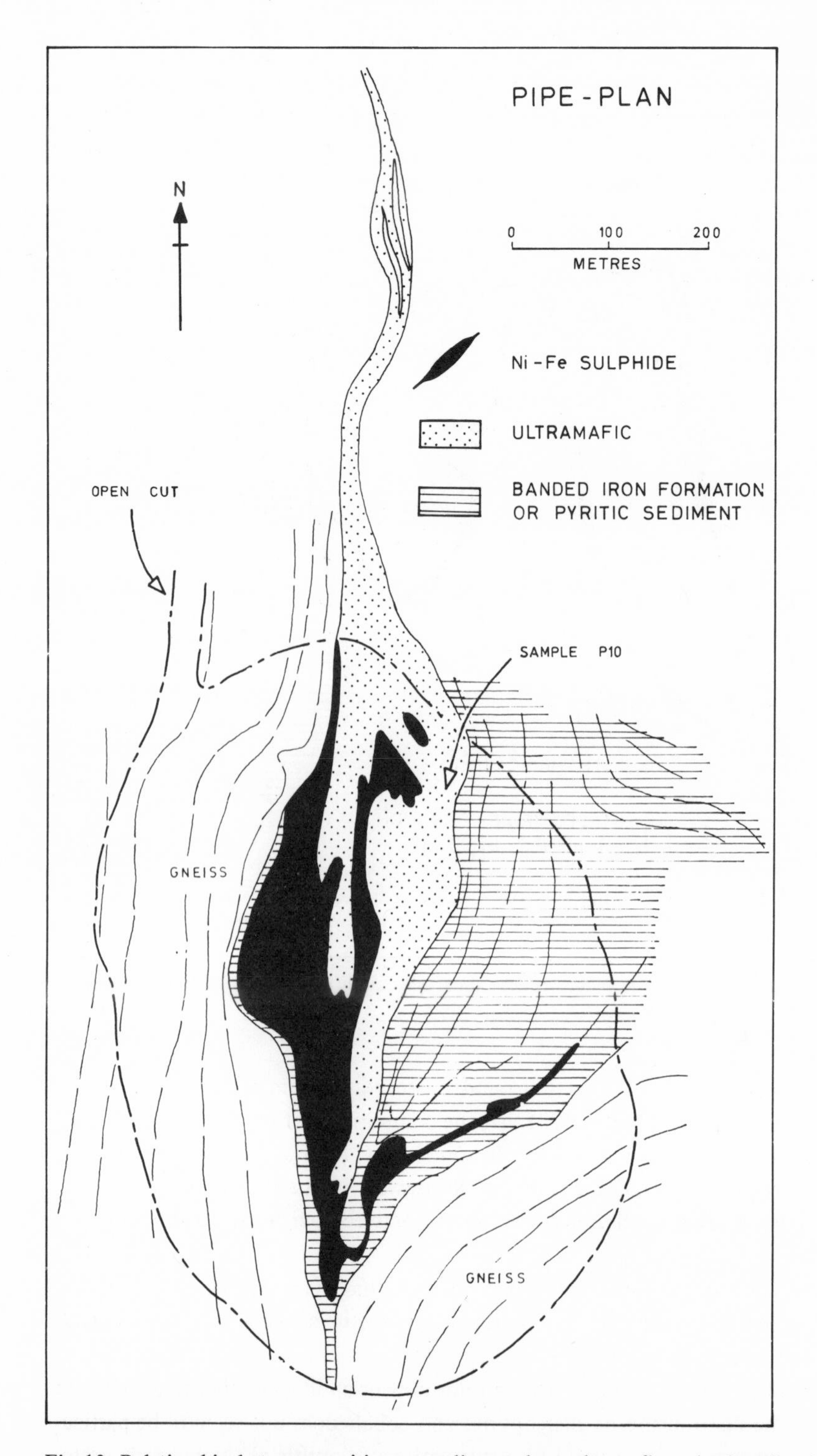

Fig. 10. Relationship between pyritic metasediment, host ultramafic and mineralization, Pipe Mine, Canada, deposit 9 on Table I, Fig. 3.

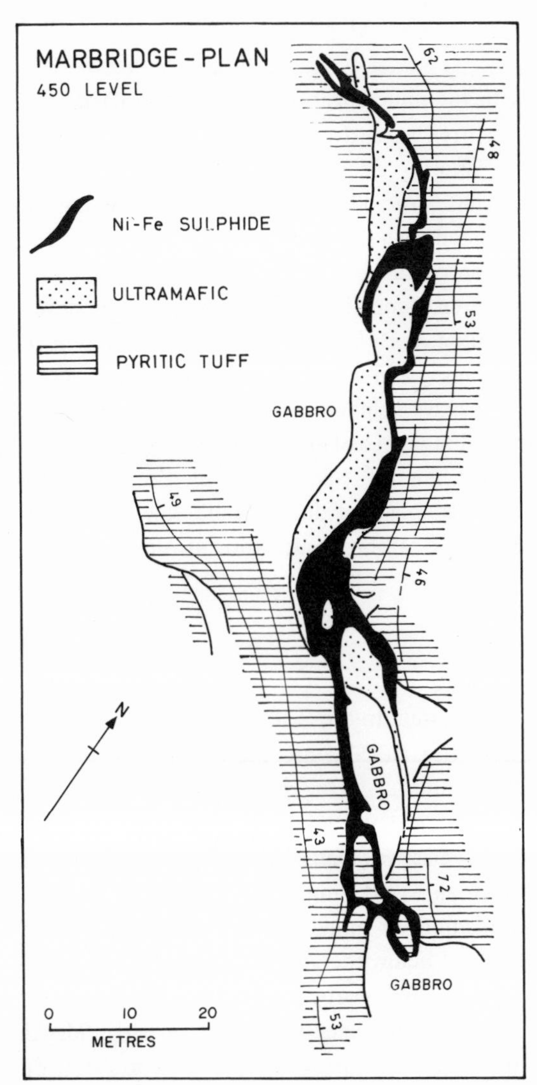

Fig. 11. Relationship between pyritic metasediment, host ultramafic and mineralization, Marbridge, Canada, deposit 10 on Table I, Fig. 3.

blages. Chalcopyrite also is concentrated in stringer zones associated with carbonate-rich alteration of host metavolcanics of the contact zone of Kambalda and Nepean. The ore zone at Miriam, which is carbonate-rich, has an anomalously high chalcopyrite content.

Carbonate alteration

Most ultramafic bodies associated with the mineralization appear to be affected in some way by carbonate alteration, which is often extensive, although patchily or irregularly distributed. These alteration products comprise calcite, magnesite and/or dolomite. Sometimes the carbonates form pseudomorphs after olivines (Fig. 12D). Disseminated sulphides often appear to be texturally associated with small carbonate-rich regions in the host ultramafic. Extensive carbonate alteration occurs in the host ultramafics of Lunnon Shoot and Fisher Shoot, Spargoville, Perserverence and Mt. Windarra; it is patchily developed in ultramafics at Marbridge, and Pipe Mine. Carbonate alteration is also present in metavolcanics adjacent to the ore zones in Lunnon and Nepean. Skarns occur adjacent to the ultramafic at Pipe, and overlapping the massive sulphide zone at Thompson (Fig. 9). The "skarn" zone at Thompson has an unusual mineral assemblage comprising carbonate-phlogopite-diopside-clinozoizite-tremolite-microcline (cf. Nepean, and see Zone C, Table III). The skarn zone at Pipe consists of an assemblage of dolomite-diopside-serpentine-olivine-quartz-feldspar. No detailed description of carbonate alteration at Alexo has been made available.

Chromite and magnetite

Spinels in these deposits are variably associated. In the Pipe, Thompson, Marbridge, Alexo and Mt. Windarra occurrences, magnetite euhedra are present in the massive sulphide zone, and sometimes magnetite veinlets occur in the matrix ore zone. The most common observation is that magnetite tends to concentrate in the matrix ore zone, and usually at the top of an ore zone near the massive–matrix or matrix–disseminated boundaries (Lunnon, Nepean, Miriam, Perserverence). At Lunnon and Nepean (Ewers and Hudson, 1972; Barrett et al., 1977), concentrations of zoned spinels with chromite cores and magnetite rims are observed both above and below the massive sulphide ore zone on the contacts; ferrochromites appear to be concentrated nearby pyrite-rich zones in the ore stratigraphy. Chromites from the least-altered olivine-bearing host-rocks have high Mg-Al contents compared with ore-zone chromites; they are surrounded by a narrow rim of chrome-rich magnetite. The chromites of ore-bearing peridotites are notably Zn-rich compared with barren ultramafics (Groves et al., 1977, Papunen et al., 1979). Chromite-bearing metasomatic chlorite–amphibole–biotite rocks develop along contacts of host ultramafic rocks; these chromites are considered metamorphic by Groves et al. (1977; cf. also Groves and Hudson, 1981). Chromite–magnetite is recorded at Marbridge by Naldrett and Gasparinni (1971), but there is no indication of zoning. Discrete euhedra of chromite is also recorded from Miriam, as well as from Mt. Windarra, where it is associated with sulphides. Magnetite is notably absent from Spargoville.

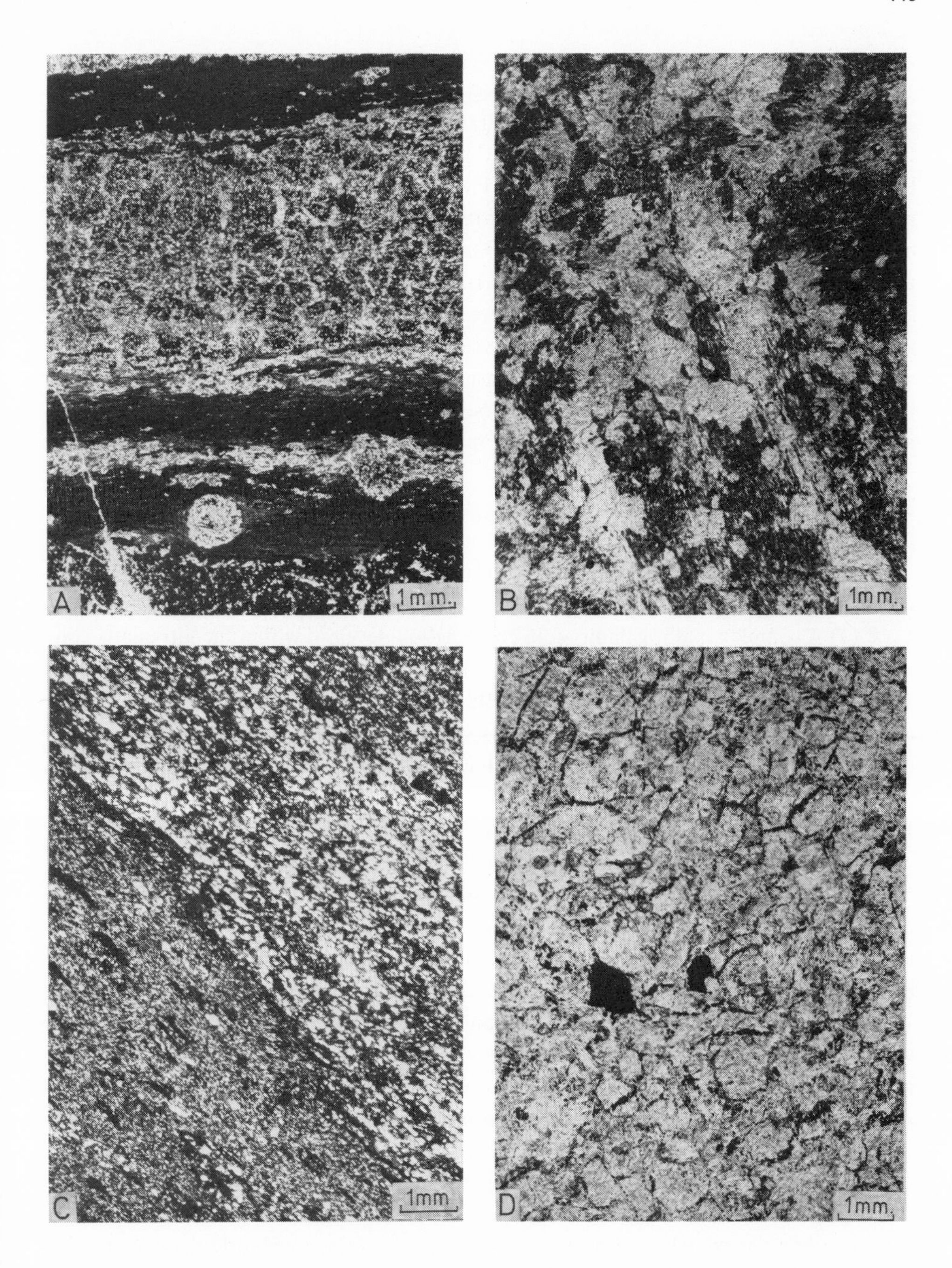

Fig. 12. A. Carbonate-bearing pyritic black argillite. Kambalda, Western Australia. B. Ultramafic host, Pipe Mine, Canada. C. Silicified metasediment interlayers within ultramafic host, Fisher Shoot, Western Australia, (Fig. 5B). D. Carbonate-altered ultramafic, Fisher Shoot.

Association of pyritic sediments

Pyritic sediments occur in the host stratigraphy adjacent to Ni-sulphide mineralization of the Spargoville, Nepean, Miriam, Marbridge and Alexo deposits. Units of sulphide-bearing iron formation are associated with the Thompson, Pipe and Mt. Windarra occurrences. In the Kambalda Dome, siliceous pyritic units are interlensed in the lower portion of the ultramafic stratigraphy, and pyritic black shales occur along the basal contact of the ultramafic with underlying tholeitic basalts (Fig. 12A). However, the pyritic black shale units are notably missing in a linear zone immediately either side of Lunnon Shoot at Kambalda (Woodall and Travis, 1969; Ross and Hopkins, 1975).

Figs. 8–11 illustrate the close relationship between pyritic stratigraphy and the distribution of nickel sulphide mineralization on the contacts of the ultramafic with pyritic metasediment at Mt. Windarra, Thompson, Pipe and Marbridge.

Structural relationships

The structural settings of these deposits are varied, but appear to consist of two predominant types, which occur individually or in combination:

(1) A complex multiple fault–fold combination develops within the ultramafic contact, resulting in a persistent embayment steeply elongate parallel to the massive ore zone. The faults are transgressive to the ultramafic contact, confining the ore zone; the folds commonly form en echelon or parallel warps, bound by the faults. In several deposits, the ore zones are structurally controlled forming elongate bodies parallel to a pronounced mineral streaking lineation or intrafolial folds in the adjacent metasediments (Drake, 1972; Barrett et al., 1977; Hopwood, 1978).

(2) A major fault zone appears to develop within the massive sulphide zone, on the margin of the ultramafic body. This zone is commonly indicated by extensive shearing, or by numerous fragments of pyritic black shale, quartzite, and other country rock material occurring within banded or massive (recrystallized) sulphides (ore breccias). In many cases, graphitic or chlorite schist fragments consist of intrafolial fold remnants (Fig. 13).

Examples of the first type, are Lunnon Shoot or Fisher Shoot at Kambalda (Ross and Hopkins, 1975, fig. 22). Siting of the ore zone in a fold-warp and sheared ultramafic contact, occurs at Perserverence, Nepean, Alexo, and Marbridge No. 1. The Miriam ore zone is present in a bicuspate warp in the ultramafic contact, combined with a carbonate-filled schist zone within the ultramafic body. The Mt. Windarra deposit forms a series of oreshoots parallel to intrafolial folds and mineral streaking lineations in the host metasediments (Figs. 8 and 13).

Ore breccias containing fragments of intrafolially folded graphitic shale or chlorite schist are developed at Thompson (illustrated by Zurbrigg, 1963, p. 458, plate 10E); at Pipe, Marbridge No. 1, Mt. Windarra (illustrated by Barrett et al., 1977, p. 1205, plates A and E); and at Spargoville and Perserverence. Deposits which appear to be related to

Fig. 13. Intrafolial fold hinge within pyrite-graphite schist fragment from Breccia Ore, Mt. Windarra, Western Australia.

major fault zones are Marbridge No. 1, Thompson, Perserverence and possibly Pipe. The composite sulphide–biotite schist zone, possibly representing such a major fault zone in sulphide metasediments at Thompson, has been refolded. The schistosity is folded about a major plunging combined anticline–syncline; thus these folds are presumably second-generation structures, refolding the earlier formed sulphide schist zone (Fig. 9). At the Pipe Mine, there is an increase in cataclasis of the metasediments towards the ultramafic contact, suggesting that this may also represent an active schist zone. Almost all the massive Ni-sulphide deposits of Table I are associated with major gravity boundaries or major tectonic lineaments (Zurbrigg, 1963; Binns et al., 1976).

Significance of structural data

The host metasediments, and commonly the mafic volcanics, usually have a well-developed schistosity (Figs. 12A, 12C) bearing a persistent mineral streaking lineation. Texturally, these metasedimentary host-rocks, especially in areas away from ore zones, are highly schistose. Yet, the ultramafic bodies containing the mineralization are massively textured even though they are composed of extremely deformable assemblages of serpentine, carbonate, talc, chlorite and tremolite (Figs. 12B, 12D). If these rocks had been altered prior to or during strain, they would now be schistose. Matrix ore zone textures are also massive, and are often characterized by radiating tremolite–chlorite–serpentine textures overprinting sulphides and former igneous olivines (Clark, 1965, fig. 10; Woodall and Travis, 1969, fig. 6C). The textures of many massive sulphide ore zones are recrystallized, even those that contain metasediment breccia fragments. Lunnon Shoot contact ore characteristically shows excellent deformation banding (Barrett et al., 1977).

Schistose metasediments are enclosed in massively textured ultramafic rock in Fisher Shoot, as illustrated in Figs. 5B and 12C.

An implication of these structural and textural relationships, combined with the ore breccias (containing folded graphitic and chloritic schist fragments), is that the metasedimentary rocks must have been deformed prior to the intrusion of the ultramafic bodies. Also, the ultramafics must have been intruded in their present steep orientations, and, except for minor second-generation shear zones (crenulate strain, Hopwood, 1974, p. 219) have not been substantially deformed or rotated. Vertical intrusion of the ultramafic bodies is required because the schistosity of the metasediments has been formed in its presently observed steep state, and such schistosity must have been formed prior to the intrusion of the ultramafic bodies. Static metasomatic alteration to form massive textures in the host ultramafic has continued after intrafolial folding and schistosity development in the metasediments.

Geochemical relationships between host ultramafic and ore zone

Table II gives the compositions of four zones A, B, C and D, where data are available for host-rock and ore zones of eight massive Ni-sulphide deposits. Fig. 14 and Table III summarize geochemical relationships between these zones, which are recognized as follows:

(1) *Zone A*: An upper zone of chlorite–tremolite assemblages, chemically of pyroxenitic composition (Ross and Hopkins, 1975), this zone generally has 22–35% MgO, 4–7% Al_2O_3, and 3–6% CaO. The average composition is summarized in Table III. The thickness of zone A plus zone B is generally between 20–50 m.

(2) *Zone B*: This is a central and/or lower MgO-rich, geochemically peridotitic zone (34–44% MgO), which together with zone A above constitutes the bulk of the ultramafic host to mineralization. This rock type is dunite, serpentinized peridotite, or talc–carbonate altered peridotite (Fig. 12D). In some cases, an original granular olivine (Fo_{90-95}) texture is recognizable, overprinted by the metamorphic alteration assemblage.

(3) *Zone C*: In the area approaching the ore zone the ultramafic host-rock becomes increasingly altered. This is a zone containing disseminated or matrix sulphides within a chlorite–tremolite gangue. It is generally 2–5 m thick. With increase in sulphides, it passes to the massive sulphide zones. This zone has been variously described as a contact zone or a reaction zone. The chemistry of this zone is summarized in Table III, and Fig. 14.

(4) *Zone D*: This is a metasedimentary unit which occurs at the basal contact of the ore zone. It is discontinuously preserved. These rocks, when available, are often finely banded and consist of pyritic black argillites, containing calcite, graphite, sericite and silica (Fig. 12A).

The contact zone C above, consists of anomalously high K_2O, Na_2O, and high Al_2O_3, and CaO. It is difficult to conceive of a magmatic differentiation process in ultramafic rock producing basal enrichment of K_2O and Na_2O. It is thought that the chemical variation of zone C, illustrated in Table III and Fig. 14, has been caused by contamination of the ultramafic melt by metasediment. The geochemical pyroxenitic–peridotitic variation of zone A to zone B is thought to be due to magmatic flow differentiation (Bhattacharji and Smith, 1964). "TM" is a theoretical melt composition intermediate between zones A and B. "TCM" is the theoretical contaminated melt composition calculated by adding zone D metasediment composition to the theoretical melt composition "TM". The theoretical contaminated melt composition is comparable for all major oxides within zone C, except SiO_2 which is too high, and FeO and MgO which are too low. Silicification of metasediments is observed at both Kambalda and Thompson, and this may indicate metasomatic mobility of SiO_2 during mineralisation. The increased FeO and MgO in the contaminated zone may be due to magmatic differentiation. The high K_2O, Na_2O, CaO and Al_2O_3 values of zone C would appear to be consistent with contamination of the melt by carbonate-bearing pyritic black argillites.

DISCUSSION

Ore genesis mechanisms in the literature

Mechanisms for the formation and concentration of nickel sulphides in these type 4 deposits (above) as described in publications, can be summarized as follows:

(1) *Immiscible sulphide melt.* The host ultramafic rock is thought to represent an intrusion of a melt of ultramafic composition, which contains an immiscible sulphide-melt phase. This mechanism requires gravity-concentration of the immiscible pentlandite–pyrrhotite phase in topographic depressions or troughs at the base of a subhorizontal sill (Hawley, 1962; Naldrett and Kullerud, 1967; Souch et al., 1969; Wilson et al., 1969). Variations of this concept are implied: (a) separate intrusion of magmatic sulphide (to explain massive Ni-sulphides extending into metasediments) advocated by Clark (1965) for Marbridge; or (b) extrusion of an olivine-rich differentiate containing a large immiscible sulphide fraction, as advocated more recently by Naldrett (1973). This concept, although readily applicable to large gravity-differentiated bodies such as Sudbury (type 2, page 413), has also been applied to the type 4 greenstone belt massive Ni-sulphide deposits by Hancock et al. (1971, p. 278), Nesbitt (1971b, p. 253), Fardon (1971, p. 256), Hudson (1972), Naldrett (1973), and Ross and Hopkins (1975). Although applicable to type 1 deposits, the extension of this model to include type 4 is not so obviously supported by the geological data.

The main support for this conclusion is the zone A to zone B pyroxenite to peridotite geochemical variation (Table III, Fig. 14) which is interpreted as a product of gravity-differentiation. In developing the immiscibility concept for these deposits, it is proposed that the massive nickel–iron sulphides are concentrated in depressions at the base of an ultramafic flow or sill, which is subsequently folded (with accompanying regional metamorphism), causing the sulphide body to be rotated into its now observed steep orientation.

TABLE II

Comparison of zone A, B, C, D compositions from individual Archean massive nickel sulphide deposits

Zone	Lunnon (1)				Fisher (2)			Spargoville (3)		
	1A	1B	1C	1D	2A	2B	2D	3B	3C(1)	3C(2)
SiO_2	48.4	44.0	49.1	46.2	44.6	44.1	70.9	50.4	28.3	42.8
Al_2O_3	8.2	3.0	10.3	7.0	5.4	4.6	9.8	1.5	2.1	12.4
Fe_2O_3	2.3	4.1	4.7	Total	Total		Total	8.9	15.5	22.4
FeO	8.2	4.6	2.3	26.5	11.6	10.5	6.1	1.8	4.7	4.8
MgO	23.6	42.7	28.8	2.1	31.4	36.0	3.4	31.0	40.0	6.3
CaO	8.1	0.7	1.4	13.2	5.3	2.9	1.4	4.5	7.9	6.1
Na_2O	0.1	0.03		3.9	0.01	<0.01	0.7	0.2	0.5	1.8
K_2O	0.06	0.01		0.3	0.01	0.01	6.6	0.2	0.24	2.2
H_2O^+	(6.4)	(4.1)		(6.6)	–	–	(2.9)		–	–
TiO_2	0.32	0.14		0.34	0.24	0.22	0.36	0.36	0.5	0.8
P_2O_5	0.05	0.01		0.11	0.03	0.02	0.05	0.01	0.05	0.05
MnO	0.20	0.15		0.16	0.17	0.13	0.04	0.29	0.28	0.20
Cr_2O_3	0.44	0.23	0.09	0.23	1.1	1.1	0.57			
NiO	0.11	0.27	0.11	0.01	0.24	0.36	0.03	0.8	(5.2)	(3.6)
CO_2	(0.22)	(17.0)		(2.6)	(7.9)	(4.5)	(0.60)	(1.0)	(10.8)	(2.4)
S	(0.11)	(0.15)	(1.79)	(11.7)			(1.04)	(1.5)	(5.5)	(4.6)
C				(3.1)			–			
Total	100.08	99.94	Partial	100.05	100.1	99.95	99.95	99.96	100.07	99.9
Al_2O_3 + CaO + Alkalis	16.5	3.7	11.7	24.4	10.7	7.5	18.5	6.4	10.7	22.5
$\frac{MgO}{Al_2O_3 + CaO + Alkalis}$	1.4	11.4	2.5	0.09	2.9	4.8	0.18	4.8	3.7	0.28

1A. Tremolite–chlorite ultramafic (Lunnon Shoot, av. of 14 samples from Ross and Hopkins, 1975, (1), Table 7).
1B. Talc–magnesite–chlorite. MgO-rich ultramafic (Lunnon Shoot, av. 10 samples from Ross and Hopkins, 1975, (4), Table 7).
1C. Ultramafic containing disseminated sulphides, Lunnon Shoot, Ewers and Hudson, 1972 (FeO from sulphides).
1D. Carbonate-bearing pyritic black argillite, Kambalda. Analyst R. Beevers, Adelaide.
2A. Tremolite–chlorite ultramafic, Fisher Shoot. Analyst R. Beevers, Adelaide.
2B. Talc–magnesite–tremolite–chlorite ultramafic, Fisher Shoot. Analyst R. Beevers, Adelaide.
2D. Siliceous metasedimentary interlayer at base of ultramafic host, Fisher Shoot. Analyst R. Beevers, Adelaide.
3B. Least-altered peridotite host, Spargoville. Analyst North Broken Hill Ltd.
3C(1). Talc–carbonate host containing matrix-ore, Spargoville. Analyst North Broken Hill Ltd.
3C(2). Sulphide-bearing tremolite–chlorite schist on ore contact, Spargoville. Analyst North Broken Hill Ltd.

Zone	Nepean (4)					Mt. Windarra (5)						
	4A	4B	4C	4D(1)	4D(2)	5B	5C(1)	5C(2)	5D	6B	7C	8B
SiO_2	44.0	41.6	43.6	53.3	46.8	50.4	47.4	51.9	53.4	41.2	50.6	41.8
Al_2O_3	2.7	1.2	6.4	0.8	22.3	1.4	3.3	6.4	12.0	0.09	4.4	2.3
Fe_2O_3	4.3	0.5	4.6	0.2	2.8	Total	10.7	Total	Total	Total	Total	Total
FeO	2.5	9.3	4.2	2.9	11.7	11.4	1.5	11.5	12.5	7.8	11.0	5.6
MgO	45.4	46.1	32.3	16.5	3.6	34.8	33.5	19.8	7.6	51.0	25.7	37.9
CaO	0.03	0.2	8.0	25.3	4.3	1.2	1.8	6.5	7.0	0.13	2.5	1.0
Na_2O	0.01	0.0	0.02	0.2	3.3	0.2	0.5	0.4	3.8	0.05	3.2	–
K_2O	0.03	0.1	0.02	0.01	2.5	0.2	0.3	1.7	1.3	0.01	0.1	<0.01
H_2O^+	(12.0)	(2.5)	(8.4)	(0.88)	(1.8)							
TiO_2	0.12	0.05	0.3	0.04	2.0	0.09	0.09	1.3	1.8	0.01	0.3	0.13
P_2O_5	0.01	0.00	0.03	0.03	0.23	0.1	0.09	0.03	0.13	0.02	0.06	
MnO	0.05	0.07	0.15	0.59	0.32	0.07	0.21	0.36	0.49	0.11	0.15	
Cr_2O_3	0.3	0.15	0.02	–						0.2	1.7	
NiO	0.38	0.6	0.28			0.20	0.53				0.3	0.22
CO_2				(2.8)		(2.2)	(15.09)	(0.3)	(3.9)		(1.3)	
S	<0.01	0.15				(0.54)	(1.03)	(0.26)	(0.26)		(0.45)	
C												
Total	99.8	100.02	99.92	99.9	99.9	100.1	99.9	99.9	100.0	100.6	100.0	Partial
Al_2O_3 + CaO + Alkalis	2.8	1.5	14.4	26.3	32.4	3.0	5.9	15.0	24.1	0.28	10.2	3.3
$\frac{MgO}{Al_2O_3 + CaO + Alkalis}$	16.3	30.7	2.2	0.6	0.1	11.6	5.7	1.32	0.3	182	2.5	11.5

4A. Serpentine–chlorite–talc ultramafic host, Nepean, from Barrett et al. (1976).
4B. Peridotite, olivine (71%)–chlorite–talc rock, Nepean, from Barrett et al. (1976).
4C. Chlorite–serpentine–diopside contact ore, ultramafic host rock, Nepean, from Barrett et al. (1976).
4D(1). Diopside–tremolite–calcite–chlorite tactite, Nepean, from Barrett et al. (1976).
4D(2). Garnet–staurolite–sillimanite schist, Nepean, from Barrett et al. (1976).
5B. Serpentinite host above disseminated ore, Mt. Windarra. Analyst North Broken Hill Ltd.
5C(1). Talc–carbonate host with disseminated sulphides, Mt. Windarra. Analyst, North Broken Hill Ltd.
5C(2). Tremolite–chlorite schist on contact with metasediments, Mt. Windarra. Analyst North Broken Hill Ltd.
6B. Dunite host, Perserverence, av. of two unmetamorphosed dunite; Dave Groves, pers. comm. 1979.
7C. Altered peridotite, contact with massive ore, Pipe Mine. Analyst R. Beevers, Adelaide.
8B. Metaperidotite, Marbridge, mean. Partial analysis; Clark (1965).

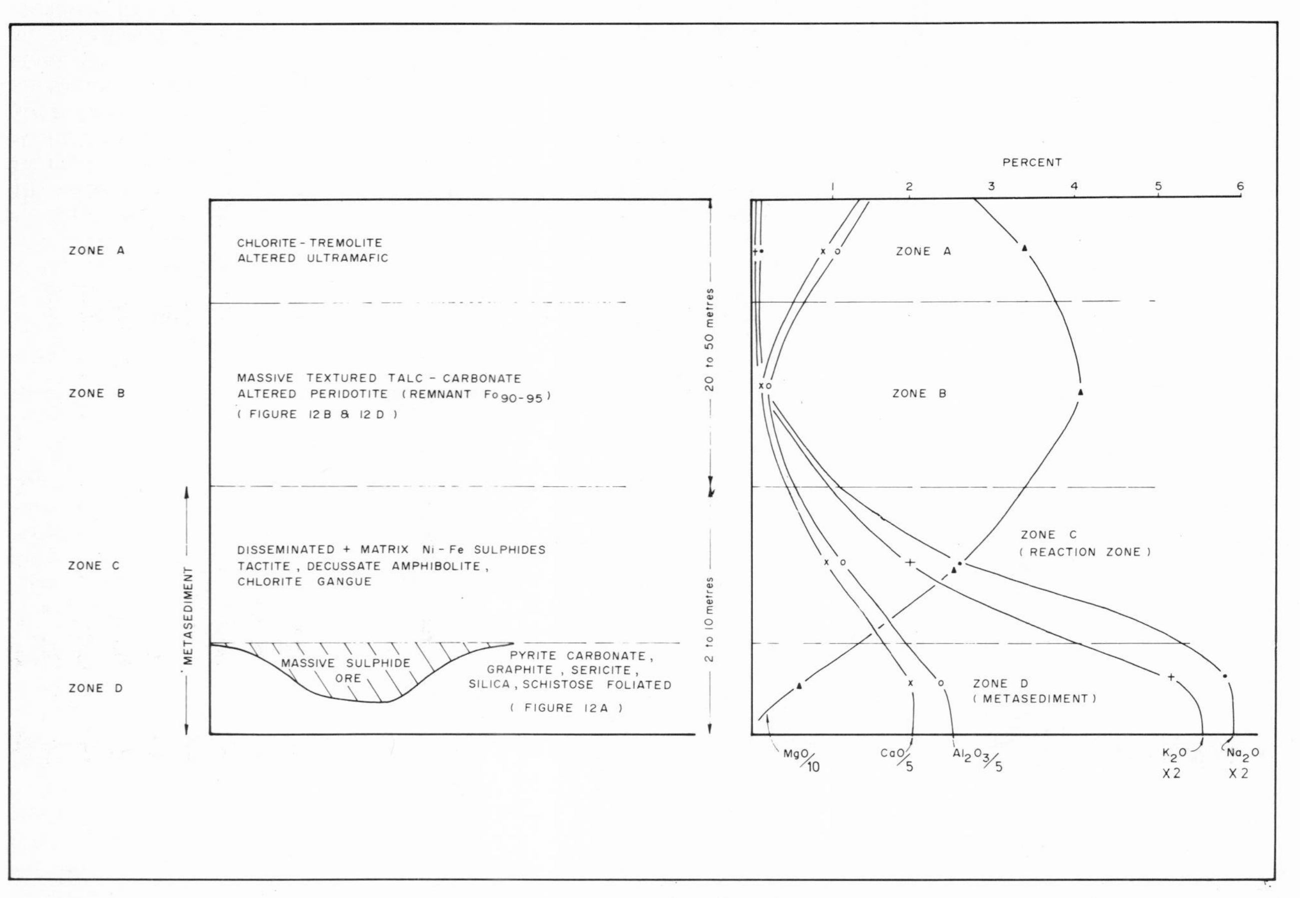

Fig. 14. Summary of geochemical relationships between ultramafic host metasediments and contact zone. For analyses see Table III.

TABLE III

Summary of geochemical relationships between ultramafic host metasediments and ore zone

	A	*B*	*TM*	*TMC*	*C*	*D*
SiO_2	45.7	45.3	45.5	49.8	44.8	54.1
Al_2O_3	5.4	2.0	3.7	7.05	6.5	10.4
Fe_2O_3	Total	Total	Total	Total	Total	Total
FeO	9.6	9.3	9.5	11.0	14.0	12.5
MgO	33.5	40.5	37.0	21.9	26.6	6.7
CaO	4.5	1.5	3.0	6.6	4.9	10.2
Na_2O	0.01	0.07	0.04	1.2	1.1	2.4
K_2O	0.03	0.08	0.06	1.08	0.8	2.1
H_2O^+	(9.2)	(3.3)				(3.0)
TiO_2	0.23	0.14	0.19	0.55	0.55	0.9
P_2O_5	0.03	0.03	0.03	0.07	0.05	0.11
MnO	0.14	0.13	0.14	0.23	0.23	0.32
Cr_2O_3	0.61	0.43	0.52	0.5	0.6	0.4
NiO	0.24	0.42			(1.7)	0.02
CO_2	(4.1)	(6.3)			(6.0)	(2.5)
S	(0.06)	(0.6)			(2.3)	(4.3)
C						(3.1)
Total	99.99	99.9	99.7	99.98	100.1	100.15
Al_2O_3 + CaO + Alkalis	10.00	3.7	6.8	16.0	13.3	25.1
$\frac{MgO}{Al_2O_3 + CaO + Alkalis}$	4.4	11.0	5.4	1.4	2.0	0.3

A. Average A-Zone, tremolite–chlorite host ultramafics from Table II.
B. Average B-Zone talc–carbonate altered peridotitites or serpentinites or dunites, from Table II.
TM. Theoretical melt composition $(A + B)/2$.
TCM. Theoretical contaminated melt composition $(TM + D)/2$ compared with reaction zone composition.
C. Average reaction or contact zone consisting of matrix or stringer sulphides in an amphibolite–chlorite gangue.
D. Metasedimentary contact rocks, pyrite, carbonate, graphite, sericite, silica.

(2) *Sulphurization.* Sulphur from pyritic sediments is thought to diffuse into an ultramafic melt (which becomes the ultramafic host body), supplying locally higher concentrations of sulphur than would normally be available in the melt, as indicated by the associated melt composition. This concept was introduced by Naldrett (1966) for Alexo, and Naldrett (1969), and developed further in Naldrett and Gasparrini (1971) on Marbridge, and Prider (1970) for Western Australian nickel deposits. Once incorporated into the melt, the sulphur reacts with pyroxene in the peridotite forming nickel–iron sulphides, which migrate and concentrate in a "structural trap". Naldrett (1973) subsequently rejected his concept of sulphurization as unable to explain net- or matrix-texture (by the

replacement of pyroxene) and to account for the upper sharp boundary of matrix ore with the ultramafic host-rock.

(3) *Hydrothermal replacement of host sediments.* This concept invokes replacement of sediments or of former pyritic units by migration of nickel ions, from an associated intrusive ultramafic body. This is thought to be induced by the breakdown of nickel-bearing olivines by sulphur, or during serpentinization, causing release of nickel ions for replacement of contact sediments (Boldt and Queneau, 1967, p. 15). Although they concluded that the major concentration of ore is magmatic, Barrett et al. (1976, 1977) suggested that metamorphic processes or serpentinizations may up-grade the nickel content of massive Ni-sulphide ores. It is difficult to account for a large proportion of nickel mineralization in this fashion because of the relatively small size of the host ultramafic body, and the present relatively high nickel content of the ultramafic rocks above the ore zone (Table III, zone B: average 0.42% NiO).

Observations inadequately explained by current mechanisms in the literature

The main deficiencies of the magmatic immiscibility concept are:

(1) Lack of melt sulphur-carrying capacity (discussed below).

(2) The ultramafic host-rock bodies associated with ore zones are too small to carry sufficient sulphur or nickel. This is also a problem for the metamorphic nickel-diffusion mechanism.

(3) No adequate explanation for the repeated association of pyritic metasediments with ore zones (Figs. 8–11, Table I).

(4) No adequate explanation of fragments of fold hinges of graphitic and chlorite schists in massive-sulphide ore breccias (Fig. 13).

(5) No adequate explanation for the high K_2O, Na_2O, CaO and Al_2O_3 contents of reaction zone (zone C) compositions.

(6) No adequate explanation for the tectonic structural control or elongation of ore zones, or association with footwall shear zones, or massive-sulphide breccia-ore zones.

(7) No adequate explanation for the massive undeformed textures of the host ultramafic, compared with the schistose metasediments of the wall-rock and as inclusions.

As a consequence of the extremely low viscosity (Bottinga and Weill, 1972) of MgO-rich melts (namely, 0.1 to 1 poise at 1650°C, cf. water at 0.01 poise at 20°C), such a melt is unlikely to carry immiscible sulphides from depths of 300 km as suggested by Naldrett (1973). The solubility of sulphur in ultramafic melts is low (Shima and Naldrett, 1975), with maximum values of 0.5%, i.e. of the order of 1 : 200. This would imply that if all the sulphur were to be derived from the ultramafic host, for ore bodies between 2 and 5 m thick, then a host ultramafic body would need to be 400–1000 m thick (observed host ultramafic thicknesses are between 20 and 100 m thick). The magmatic-sulphide immiscibility concept requires the ultramafic sill or flow to be horizontal to concentrate metal sulphides and to "riffle out" sulphides with flow of the melt over a basal topographic depression. As discussed above, structural data indicate the ultramafic bodies were intruded steeply (in their present orientations). For the breccia-ore, the source of sulphur needs to be close to the metasediments in order to explain the included fragments of folded metasedimentary graphitic and chloritic schist, in these ores.

Proposed ore genesis mechanism

Sulphur sources, metal sources. As a result of the repeated association of massive nickel sulphide mineralization with the *combined* association of pyritic black argillites and magnesian-rich ultramafic intrusive bodies, it would appear logical that the ultramafic body represents a nickel source and that pyrite-rich zones of the stratigraphy represent a possible sulphur source for the mineralization (see Lovering, 1961; Robinson and Strens, 1968; Hopwood, 1977). Because of the small size of the ultramafic host-bodies and the relative lack of depletion of NiO from the host ultramafic, it is necessary to propose some sort of nickel-scavenging processes to adequately explain the size and grade of the ore zones. This is inherent in Naldrett and Cabri's (1976) explanation of sulphides being "riffled out" of the melt by topographic irregularities.

Ultramafic melt. Fig. 15 illustrates variation between Al_2O_3, NiO, and Cr_2O_3 from Dundonald Sill (Naldrett and Mason, 1968). Although it may be argued that NiO is removed with cumulus olivine, this diagram illustrates the apparent incompatibility of high Al_2O_3 and high NiO in melts. Chemically, unless sulphur is provided, there are few other minerals other than olivine which will accept substantial NiO contents. The preferential partitioning of Ni in Mg-olivines observed in melts is caused by the presence of higher concentrations of alkali or SiO_2 and possibly Al_2O_3 in the melt (Irvine and Kushiro, 1976). According to crystal field theory and subsequently confirmed experimentally if nickel (and other transition metals) are provided with a more satisfactory site than available in a melt due to high alkali, SiO_2 or Al_2O_3 contents in that melt, then nickel will accept that site preferentially (Burns, 1970, p. 160; Mysen, 1976). In addition, if the Al_2O_3 + CaO + alkali contents are too high, i.e. >2.0%, it is difficult to get high NiO concentrations in the melt in the first place in order to precipitate the NiO-bearing olivines from that melt (Kellerud, 1963, and pers. commun., 1979; Irvine and Kushiro, 1976).

It is proposed that highly magnesian melts with very low Al_2O_3 + CaO + alkali contents are intruded into steep, actively deforming tectonic zones adjacent to major lineaments (Figs. 16–19). Eggler (1976, 1978) and Wyllie (1978, p. 708) believe that even small amounts of CO_2 can reduce solidus temperatures in peridotitic mantle by several hundred degrees, which suggests that large proportions of partial melting of peridotite mantle are possible, resulting in MgO-rich and Al_2O_3 + CaO-poor melts. Further, differential strain effects adjacent to major tectonic boundaries may also favor the formation of such melts by localizing melting. The presence of CO_2 in the melt will also aid in the

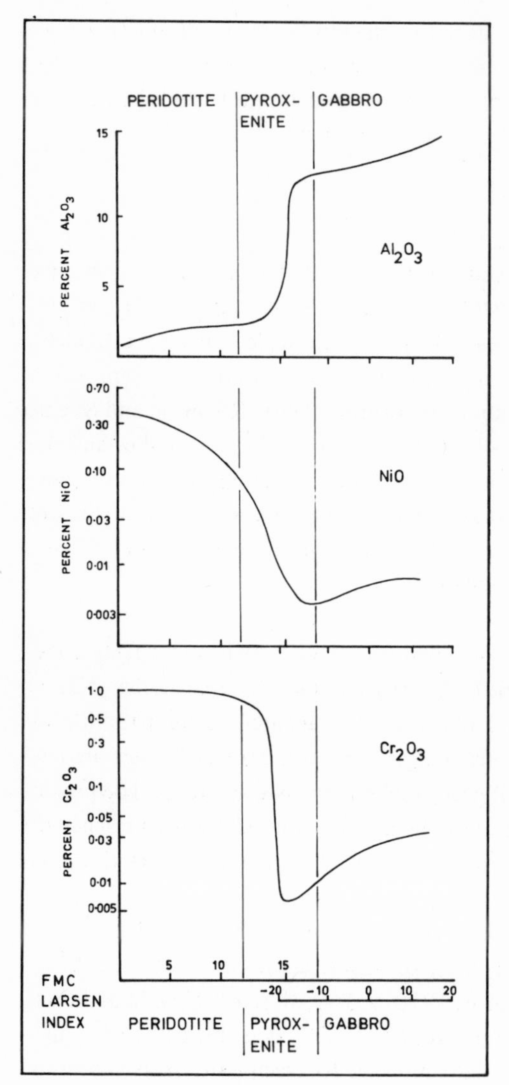

Fig. 15. Geochemistry from Dundonald Sill (Naldrett and Mason, 1968), illustrating antithetic variation of NiO and Cr_2O_3, with Al_2O_3.

generation of carbonyls (see below). As discussed above, such magnesian-rich melts will have an extremely low viscosity, resulting in a rapid adiabatic rise and intrusion adjacent to major fault or lineament zones. Such intrusive bodies, because of their low viscosity, are likely to be extremely structurally conformable with the regional schistosity. At the surface, fault-bound basins often develop adjacent to major lineaments in greenstone belts, forming sedimentary sequences including carbonate-bearing pyrite-rich graphitic

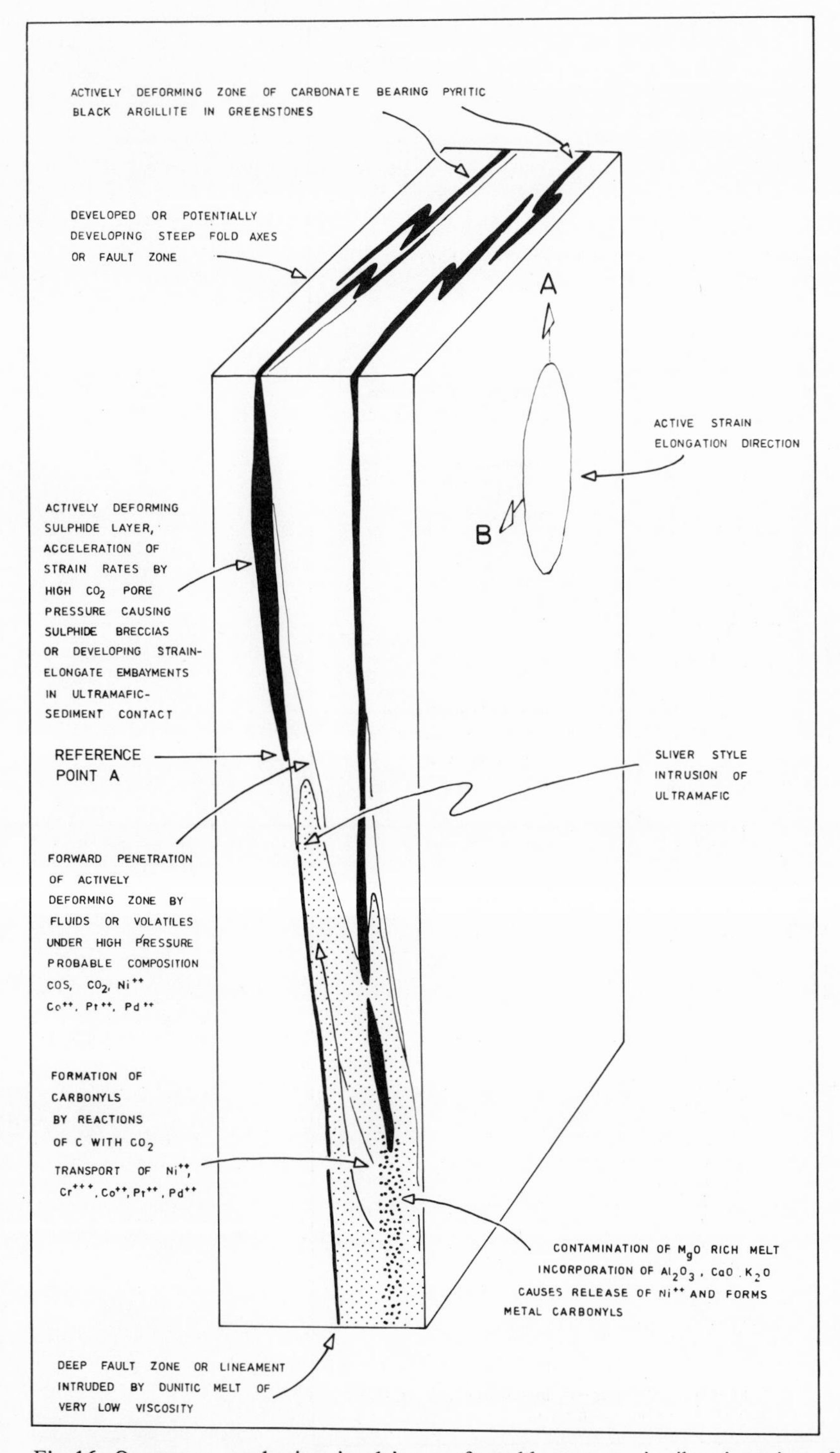

Fig. 16. Ore-genesis mechanism involving conformable syntectonic silver intrusion of low-viscosity ultramafic melt into actively deforming pyritic black shales. Contamination at depth releases Ni^{2+}, Co^{2+}, Cr^{3+}, Pt^{2+}, Pd^{2+}, reaction of C with CO_2 produces metal carbonyls, resulting in a continuous supply of reactive metal carbonyls in fluid or gaseous form. Note level of intrusion with respect to Reference Point A in Figs. 16, 17 and 18.

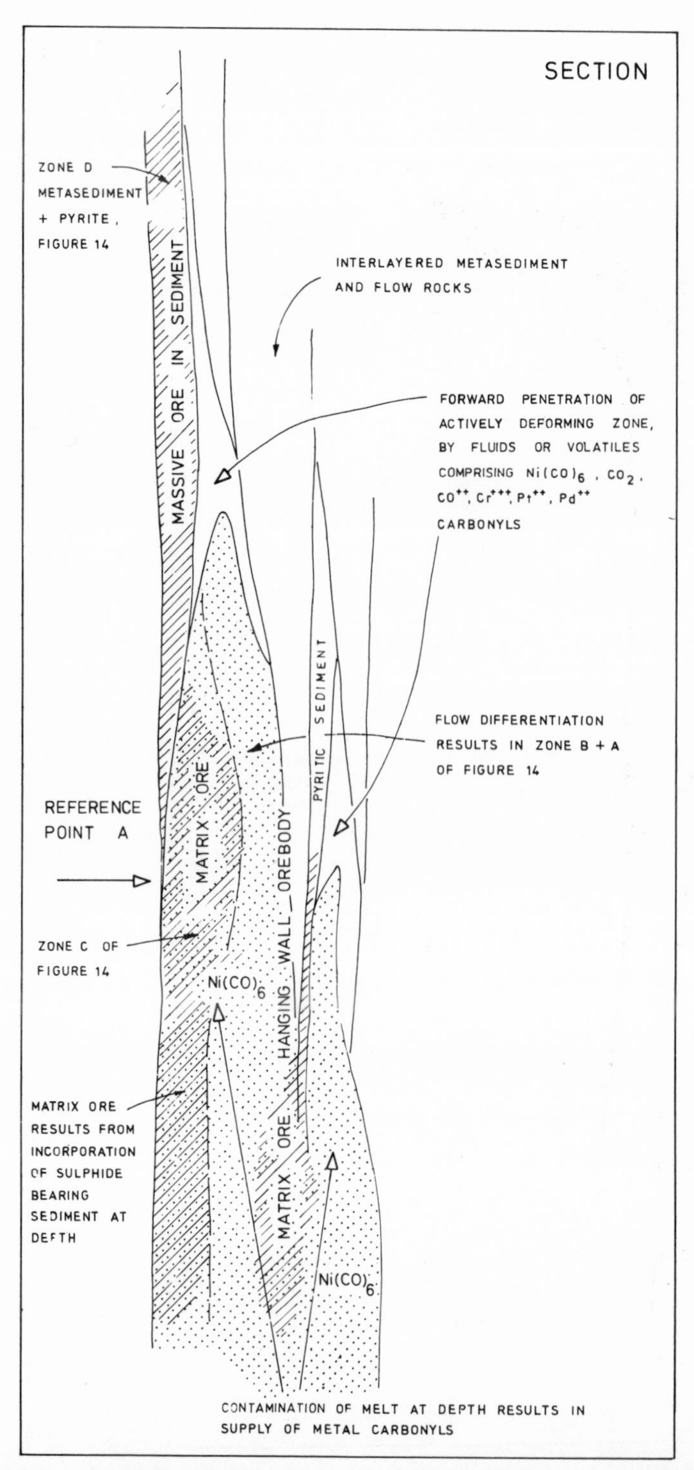

Fig. 17. Reaction of Ni, Co, Cr (Pt, Ir, Os) carbonyl volatiles with pyrrhotite (metasedimentary pyrite) results in massive pentlandite-pyrrhotite ore. Sulphides incorporated from original contaminating sediment incorporated at depth into the melt result in matrix ore.

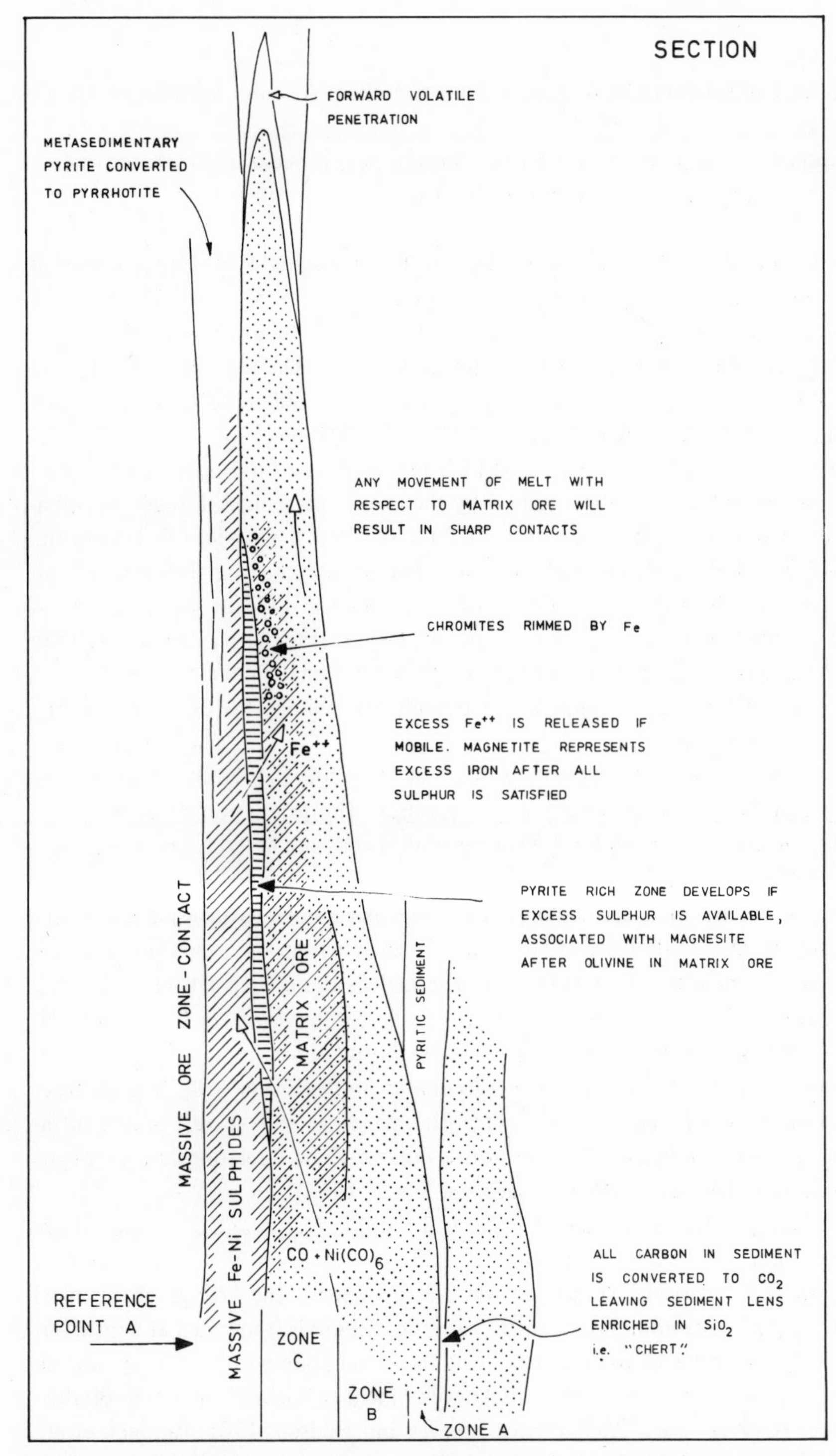

Fig. 18. Release of iron from reaction of Ni^{2+} with pyrrhotite results in Fe-rimmed chromites. Interpenetration of ultramafic melts and other pyritic metasedimentary units in the sequence results in hanging-wall orebodies.

argillites and/or banded iron formations, which (of course) are subsequently buried and metamorphosed.

The proposed structural relationships of this intrusive process are illustrated in Figs. 16, 17 and 18, and the sequence of events is summarized in Fig. 19.

Nickel release and transport mechanism. The melt is thought to be contaminated at depth by the incorporation of carbonate-bearing pyritic graphitic argillites. The temperatures of intrusion of the melt are estimated at 1650°C (see Naldrett and Cabri, 1976, p. 1134, who advocate *extrusive* melts at this temperature). The reaction is thought to take place under metamorphic conditions of approximately 600–750°C and 3–4 kbar (Barrett et al., 1976, p. B258; Barrett et al., 1977, p. 1210). Incorporation of the metasediment will be achieved by: (a) the initial high temperature of the melt; (b) the high-temperature metamorphic conditions; and (c) by the latent heat gained through precipitation of olivine from the melt. This contamination reaction has three effects: (1) the increase of Al_2O_3 + CaO + alkalis results in decreased affinity of the melt for nickel and other transition metals; (2) the combination of graphite + CO_2 in the presence of pyrrhotite is thought to generate carbonyls of Ni, Cr, Co (Pt, Os, Ir, Ru); and (3) the incorporation of substantial amounts of sulphur into the ultramafic melt.

Carbonyls are gaseous or liquid compounds of metals, $Ni(CO)_4$, $Co_2(CO)_8$, $Cr(CO)_6$, etc ..., generated by the reaction of graphite + CO_2 with siderophile metals. Although they decompose at atmospheric pressure at 150–300°C, their high-temperature stability is promoted preferentially by high pressures. The generation of carbonyls is accelerated by the presence of sulphur (Queneau et al., 1969; and M. Willis, personal communication, 1974).

The metasediment–melt contamination reaction is assumed to take place at depth and the ultramafic melt is thought to be intruded higher within the active deformation zone within pyritic metasediments (Fig. 16). The melt is thought to intrude as a steep flow-differentiated olivine-crystal mush, rich in volatile carbonyls of Ni, Co, Cr (Pt, Pd, Os, Ir, Ru), which will then react with massive pyrrhotite zones (metamorphosed sedimentary pyrite) adjacent to the intrusive ultramafic body (Ewers, 1972). Such reactions will produce massive pendlandite–pyrrhotite zones at the "base" of the ultramafic body (Fig. 17). The pyroxenite–peridotite Zone A/Zone B relationship is thought to be due to flow differentiation (Bhattacharji and Smith, 1964).

Sulphur incorporated in the melt at depth will result in matrix and disseminated ore (Fig. 18). Expulsion of Fe^{2+} from Ni-pyrrhotite reactions may result in chromites rimmed by magnetite, characteristic of spinels above the ore zones. The change of chemical environment caused by reaction of pyrrhotite with the carbonyls (generated at depth, but reacted with sulphur nearer to the surface), may also precipitate chromite in pyrite-rich areas (source of excess Fe^{2+}). The preferential development of zinc-rich spinels in productive ultramafics (Groves et al., 1977; cf. also Groves and Hudson, 1981; Papunen et al., 1979) is consistent with early contamination of the melt by pyritic black shales, which

characteristically contain high zinc contents (400–1500 ppm).

It is also noteworthy that carbonyls also transport Pd and Ir, which are notably concentrated in the ores of type 4 nickel deposits (Keays and Davison, 1976).

The contamination of the melt and production of metal carbonyls is believed to continue to take place at depth providing a steady stream of Ni-rich metasomatic fluids for reaction with pyrrhotite-rich zones higher in the deformation and reaction zone. This continuous supply of CO_2 and CO, combined with continued reactions in the ore zones, will induce steep structural elongation of ore zones, due to accelerated strain within the ore zones (as suggested in Hopwood, 1974, 1979) and induce the formation of breccia ores.

Following decrease in active strain, the ultramafic melt will be a heat source for some time, and with continued supply of CO and CO_2 at depth, there is likely to be continued post-strain growth of metamorphic assemblages, initially at high temperatures (600–700°C, 3–4 kbar) and subsequently undergoing serpentinization with decrease in temperature and with increased availability of water. This process will result in the observed decussate textures of the ultramafic hosts, contrasted with the schistose textures of included slivers of metasediment (Fig. 5A) or adjacent metasedimentary host-rocks to the intrusion. The sequence of events is summarised in Fig. 19.

Concluding note

The suggested mechanism provides answers to the following major problems:

(a) It provides an abundant source of sulphur for both massive sulphides and matrix-textured ores.

(b) It supplies a continuous metasomatic supply of Ni^{2+}, Co^{2+}, and Cr^{2+} (independent of the size of the host ultramafic) for reaction with the metasedimentary sulphur (pyrrhotite).

It also explains the following observations:

(1) The small size of ultramafic hosts with respect to the ore body size and grade.

(2) The repeated association of pyritic metasediments with the massive nickel sulphide deposits (Table I).

(3) The development of breccia ores, structural elongation of ore bodies and association with tectonic lineament zones.

(4) The spinel compositions and textures, particularly Zn-rich spinels of productive ultramafics.

(5) The massive ultramafic metamorphic textures and adjacent schistose metasediment textures, and extensive carbonate alteration of the ultramafic host.

(6) The high Pd and Ir contents of nickel sulphide ores.

(7) The high K_2O and Na_2O contents of Zone C reaction zone.

(8) It is also consistent with the geochemical requirements of pyroxenite/peridotite (Zone A/Zone B) observed compositional variation.

ACKNOWLEDGEMENTS

The development of the concepts in this chapter has taken place over a considerable period of time (since 1970). I would like to thank the many people in mining companies, whose deposits are listed in Table I, for providing access to data, numerous discussions, and constructive criticism of earlier drafts of this paper. I would also like to thank Abminco N.L. and North Broken Hill Ltd, for providing the chemical analysis of some samples. The carbonyl concept for nickel transport was first suggested to me by my good friend and colleague Owen P. Singleton. Melbourne University provided support for the writing of the first draft; the University of Illinois at Urbana, Ill., provided the opportunity to complete the final draft. Constructive criticism of the paper has been made by many people too numerous to list here but their help nevertheless is fully appreciated. I would like to thank Jim Krikpatrick, Ross Kennedy and Dr. K.H. Wolf for their critical review of the final manuscript.

REFERENCES

Andrews, P.B., 1975. Spargoville nickel deposits. In: C.L. Knight (Editor), *Economic Geology of Australia and Papua New Guinea, 1. Metals.* Australas. Inst. Min. Metall. Monogr., 5: 89–90.

Barker, R.W., 1975. Metamorphic mass transfer and sulphide genesis, Stillwater Intrusion, Montana. *Econ. Geol.,* 70: 275–298.

Barrett, F.M., Groves, D.I. and Binns, R.A., 1976. Importance of metamorphic processes at the Nepean nickel deposit, Western Australia. *Inst. Min. Metall. Trans. Sect. B*: 252–273.

Barrett, F.M., Binns, R.A., Groves, D.I., Marston, R.J. and McQueen, K.G., 1977. Structural history and metamorphic modification of Archean volcanic-type nickel deposits, Yilgarn Block, Western Australia. *Econ. Geol.,* 72: 1195–1223.

Bhattacharji, S. and Smith, C.H., 1964. Flowage differentiation. *Science*, 145: 150–153.

Binns, R.A., Gunthorpe, R.J. and Groves, D.I., 1976. Metamorphic patterns and development of greenstone belts in the Eastern Yilgarn Block, Western Australia. In: B.F. Windley, (Editor), *The Early History of the Earth.* Wiley, London, pp. 303–313.

Boldt, Jr., J.R. and Queneau, P., 1967. *The Winning of Nickel, Its Geology, Mining and Extractive Metallurgy,* Methuen, London, 487 pp.

Bottinga, Y. and Weill, D.P., 1972. The viscosity of magmatic silicate liquids: A model for calculation. *Am. J. Sci.,* 272: 438–475.

Burns, R.G., 1970. *Mineralogical Applications of Crystal Field Theory.* Cambridge Univ. Press, 224 pp.

Burt, D.R.L. and Sheppy, N.R., 1975. Mount Keith nickel sulphide deposits. In: C.L. Knight (Editor), *Economic Geology of Australia and Papua New Guinea, 1. Metals.* Australas. Inst. Min. Metall. Monogr., 5: 159–167.

Card, K.D., Naldrett, A.J. et al., 1972. The regional setting of Sudbury. In: D.J. Glass (Editor), *General Geology of the Sudbury-Elliot Lake Region. 24th Int. Geol. Congr. Montreal.*

Clark, L.A., 1965. Geology and geothermometry of the Marbridge nickel deposit, Malartic, Quebec. *Econ. Geol.,* 60: 792–811.

Coats, C.J.A., Quirke, T.T., Jr., Bell, C.K., Cranstone, D.A. and Campbell, F.H.A., 1972. Geology and mineral deposits of the Flin Flon, Lynn Lake and Thompson areas Manitoba, and the Churchill Superior Front of the Western Precambrian Shield. *24th Int. Geol. Congr. Excursion Guideb. A31/C31, Montreal,* pp. 64–71.

Cuthill, J., 1972. Geological notes on the Pipe Mine. *24th Int. Geol. Congr. Excursion Guideb. A31/C31 Montreal.*

Drake, J.R., 1972. *The Structure and Petrology of Banded Iron Formations at Mount Windarra, Western Australia.* Thesis, Univ. West. Australia (unpublished).

Eggler, D.H., 1976. Does CO_2 cause partial melting in the low velocity layer of the mantle? *Geology,* 4: 69–72.

Eggler, D.H., 1978. The effect of CO_2 upon partial melting of peridotite in the system Na_2O–CaO–Al_2O_3–MgO–SiO_2–CO_2 to 35 kb, with an analysis of melting in a peridotite–H_2P–CO_2 system. *Am. J. Sci.,* 278: 305–343.

Ewers, W.E., 1972. Nickel–iron exchange in pyrrhotite. Proc. Australas. Inst. Min. Metall., 241: 19–25.

Ewers, W.E. and Hudson, D.R., 1972. An interpretive study of a nickel–iron sulfide ore intersection, Lunnon Shoot, Kambalda, W.A. *Econ. Geol.,* 67: 1075–1092.

Fardon, R.S.H., 1971. The Western Australian nickel deposits. *Symposium on Archean Rocks. Geol. Soc. Aust. Spec. Publ.,* 3: 256.

Gemuts, I., 1975. Miriam nickel prospect, Coolgardie area. In: C.L. Knight (Editor), *Economic Geology of Australia and Papua New Guinea, 1. Metals.* Australas. Inst. Min. Metall. Monogr., 5: 98–99.

Gemuts, I. and Theron, A., 1975. The Archaean between Coolgardie and Norseman – stratigraphy and mineralization. In: C.L. Knight (Editor), *Economic Geology of Australia and Papua New Guinea. 1. Metals.* Australas. Inst. Min. Metall. Monogr., 5: 66–74.

Gilmour, P., 1976. Some transitional types of mineral deposits in volcanic and sedimentary rocks. In: K.H. Wolf (Editor), *Handbook of Strata-Bound and Stratiform Ore Deposits, 1.* Elsevier, Amsterdam, pp. 111–160.

Goodwin, A.M. and Ridler, R.H., 1970. The Abitibi Orogenic Belt. In: A.J. Baer (Editor), *Symposium on Basins and Geosynclines of the Canadian Shield. Geol. Surv. Can. Pap.* 70-40: 1–24.

Graterol, M. and Naldrett, A.J., 1971. Mineralogy of the Marbridge No. 3 and No. 4 deposits with some comments on low temperature equilibration in the Fe–NiS system. *Econ. Geol.,* 66: 886–900.

Groves, D.I. and Hudson, D.R., 1981. The nature and origin of Archaean stratabound volcanic-associated nickel-iron-copper sulphide deposits. In: K.H. Wolf (Editor), *Handbook of Strata-bound and Stratiform Ore Deposits. Vol.* 9 Elsevier, Amsterdam. pp. 305–410.

Groves, D.I., Barrett, F.M., Binns, R.A. and McQueen, K.G., 1977. Spinel phases associated with metamorphosed volcanic-type iron–nickel sulphide ores from Western Australia. *Econ. Geol.,* 72: 1224–1244.

Hallberg, J.A. and Williams, D.A.C., 1972. Archean mafic and ultramafic rock associations in the eastern Goldfields region, Western Australia. *Earth Planet. Sci. Lett.,* 15: 191–200.

Hallberg, J.A., Hudson, D.R. and Gemuts, I., 1973. An Archaean nickel sulphide occurrence at Miriam, Western Australia. *Proc. Australas. Inst. Min. Metall. Western Australia Conference,* pp. 121–127.

Hancock, W., Ramsden, A.R., Taylor, G.F., Wilmshurst, J.R., 1971. Some Ultramafic Rocks of the Spargoville Area, Western Australia. In: J.D. Glover (Editor), *Symposium on Archean Rocks. Geol. Soc. Aust. Spec. Publ.,* 3: 269–280.

Harris, D.C. and Nickel, E.H., 1972. Pentlandite compositions and associations in some mineral deposits. *Can. Mineral.,* 11: 861–878.

Hawley, J.E., 1962. The Sudbury Ores: their mineralogy and origin. *Can. Mineral.,* 7: 207.

Hopwood, T.P., 1974. *Structural Geology and Mineral Exploration, Vols. 1 and 2.* Aust. Mineral. Foundation, Adelaide, S.A.

Hopwood, T.P., 1977. *Geological Environments of Ore Deposits.* Workshop Course, Aust. Mineral. Foundation, Adelaide, S.A., 207 pp.

Hopwood, T.P., 1978. Conformable elongate orebodies and intrafolial folds parallel to the mineral streaking lineation. In: W.J. Verwoerd (Editor), *Mineralization in Metamorphic Terranes.* Geol. Soc. S. Afr., Spec. Publ., 4: 41–51.

Hudson, D.R., 1973. Evaluation of genetic models for Australian sulphide nickel deposits. *Proc. Australas. Inst. Min. Metall. Western Australia Conference*, pp. 59–68.

Hudson, D.R., 1973. Genesis of Archean ultramafic associated nickel iron sulphides at Nepean, Western Australia. *Proc. Australas. Inst. Min. Metall. Western Australia Conference*, pp. 99–109.

Irvine, T.N. and Kushiro, I., 1976. Partitioning of Ni and Mg between olivine and silicate liquids. *Carnegie Inst. Washington, Yearb.*, 75: 668–675.

Jackson, E.D., 1961. Primary textures and mineral associations in the ultramafic zone of the Stillwater Complex Montana. *U.S. Geol. Surv. Prof. Pap.*, 358: 106 pp.

King, H.F., 1976. Development of syngenetic ideas in Australia. In: K.H. Wolf (Editor), *Handbook of Strata-Bound and Stratiform Ore Deposits, 1.* Elsevier, Amsterdam, pp. 161–182.

Keays, R.A. and Davison, R.M., 1976. Palladium, iridium and gold in the ores and host rocks of nickel sulphide deposits in Western Australia. *Econ. Geol.*, 71: 1214–1228.

Kullerud, G., 1963. The Fe–Ni–S System. *Annu. Rep. Dir. Geophys. Lab., Carnegie Inst. Washington, Yearb.*, 62: 175–189.

Lambert, I.B., 1976. The McArthur zinc-lead-silver deposit: features, metallogenesis and comparisons with some other stratiform ores. In: K.H. Wolf (Editor), *Handbook of Strata-Bound and Stratiform Ore Deposits, 6.* Elsevier, Amsterdam, pp. 535–585.

Lovering, T.S., 1961. Sulphide ores formed from sulphide deficient solutions. In: *Korzhinsky Fetschrift Moscow, Izv. Akad. Nauk, SSSR*, Moscow, 1: 107–137 (Russian with English summary). Abstract in Econ. Geol. annotated bibliography 1964, 36: 38.

Martin, J.E., 1975. The Perserverence nickel deposit, geological summary. *Notes Australian Selection (Pty.) Ltd.*, 10 pp. (unpublished).

Martin, J.E. and Allchurch, P.D., 1973. Geology of the Perserverence nickel deposit, Western Australia. *Proc. Australas. Inst. Min. Metall. Western Australia Conference*, p. 93.

Martin, J.E. and Allchurch, P.D., 1975. The Perserverence nickel deposit, Agnew. In: C.L. Knight (Editor), *Economic Geology of Australia and Papua New Guinea, 1.* Metals. Australas. Inst. Min. Metall. Monogr., 5: 149–155.

Misra, K.C. and Fleet, M.E., 1973. The chemical compositions of synthetic and natural pentlandite assemblages. *Econ. Geol.*, 68: 518–539.

Mysen, B.O., 1976. Nickel partitioning between upper mantle crystals and partial melts as a function of pressure temperature, and nickel concentration. *Carnegie Inst. Washington, Yearb.*, 75: 662–668.

Naldrett, A.J., 1966. The role of sulphurization in the genesis of iron-nickel sulphide deposits of the Porcupine district, Ontario. *Trans. Can. Inst. Min. Metall.*, 69 (648): 147–155 (Bull., pp. 489–497).

Naldrett, A.J., 1969. Discussion of Papers Concerned with Sulfide Deposits. *Econ. Geol. Monogr.*, 4.

Naldrett, A.J., 1973. Nickel sulphide deposits – their classification and genesis, with special emphasis on deposits of volcanic association. *Bull. Can. Inst. Min. Metall.*, 66 (739): 45–63.

Naldrett, A.J. and Cabri, L.J., 1976. Ultramafic and related mafic rocks: their classification and genesis with special reference to the concentration of nickel sulphides and platinum-group elements. *Econ. Geol.*, 71: 1131–1158.

Naldrett, A.J. and Gasparinni, E.L., 1971. Archaean nickel sulphide deposits in Canada: their classification, geological setting and genesis with some suggestions as to exploration. In: J.D. Glover (Editor), *Symposium on Archaean Rocks.* Geol. Soc. Aust. Spec. Publ., 3: 201–226.

Naldrett, A.J. and Kullerud, G., 1967. A study of the Strathcona Mine and its bearing on the origin of the nickel-copper ores of the Sudbury district, Ontario. *J. Petrol.*, 8: 453–531.

Naldrett, A.J. and Mason, G.D., 1968. Contrasting Archean ultramafic igneous bodies in Dundonald and Clerque Townships, Ontario. *Can. J. Earth Sci.*, 5: 111–143.

Nesbitt, R.W., 1971a. Skeletal crystal forms in the ultramafic rocks of the Yilgarn block, Western Australia. Evidence for an Archaean ultramafic liquid. In: J.D. Glover (Editor), *Symposium on Archaean Rocks.* Geol. Soc. Aust. Spec. Publ., 3: 331.

Nesbitt, R.W., 1971b. The case for liquid immiscibility as a mechanism for nickel sulphide mineralization in the Eastern Gold Fields, Western Australia. In: J.D. Glover (Editor), *Symposium on Archaean Rocks.* Geol. Soc. Aust. Spec. Publ., 3: 253.

Papunen, H., Häkli, T. and Idman, H., 1979. Geological, geochemical and mineralogical features of sulphide-bearing ultramafic rocks in Finland. *Can. Mineral.*, 17: 217–232.

Prider, R.T., 1970. Nickel in Western Australia. *Nature*, 226: 691–693.

Purvis, A.C., Nesbitt, R.W. and Hallberg, J.A., 1972. The geology of part of the Carr Boyd rocks complex and its associated nickel mineralization, W.A. *Econ. Geol.*, 67: 1093–1113.

Pyke, D.R., Naldrett, A.J. and Eckstrand, O.R., 1973. Archaean ultramafic flows in Munro Township, Ontario. *Geol. Soc. Am. Bull.*, 84: 955–977.

Queneau, P., O'Neill, C.E., Illis, A. and Warner, J.S., 1969. Some novel aspects of the pyrometallurgy and vapometallurgy of nickel. *J. Met.*, p. 35–45.

Roberts, J.B., 1975. Windarra nickel deposits, In: C.L. Knight (Editor), *Economic Geology of Australia and Papua New Guinea. 1. Metals.* Australas. Inst. Min. Metall. Monogr., 5: 129–143.

Robinson, B.W. and Strens, R.G.J., 1968. Genesis of concordant deposits of bare metal sulphides: an experimental approach. *Nature*, 217: 535–536.

Robinson, W.B., Stock, E.C. and Wright, R., 1973. The discovery of evaluation of the Windarra nickel deposits, Western Australia. *Proc. Australas. Inst. Min. Metall. Western Australia Conference*, pp. 69–90.

Ross, J.R. and Hopkins, G.M.F., 1973. The Nickel Sulphide Deposits of Kambalda, Western Australia. *Proc. Australas. Inst. Min. Metall. Western Australia Conference*, pp. 119–120.

Ross, J.R. and Hopkins, G.M.F., 1975. Kambalda Nickel sulphide deposits. In: C.L. Knight (Editor), *Economic Geology of Australia and Papua New Guinea. 1. Metal.* Australas. Inst. Min. Metall. Monogr., 5: 100–120.

Sharpe, J.W.N., 1964. The Empress nickel-copper deposit, Southern Rhodesia. In: S.H. Houghton (Editor), *The Geology of some Ore Deposits in Southern Africa, II.* pp. 497–508.

Sheppy, N.R. and Rowe, J., 1975. Nepean Nickel Deposit. In: C.L. Knight (Editor), *Economic Geology of Australia and Papua New Guinea. 1. Metals.* Australas. Inst. Min. Metall, Monogr., 5: 91–98.

Sherritt Gordon Mines, 1972. Lynn Lake Mine. In: Geology and Mineral Deposits of the Flin Flon, Lynn Lake and Thompson Areas, Manitoba, and the Churchill–Superior Front of the Western Precambrian Shield. *24th Int. Geol. Congr. Excursion Guideb. A31/C31, Montreal,* pp. 49–54.

Shima, H. and Naldrett, A.J., 1975. Solubility of sulfur in an ultramafic melt and the relevance of the system Fe–S–O. *Econ. Geol.*, 70: 960–967.

Souch, B.E., Podolski, T. and Geological Staff of The International Nickel Company of Canada Ltd., 1969. The sulfide ores of Sudbury; their particular relation to a distinctive inclusion-bearing facies of the nickel irruptive. *Econ. Geol. Monogr.*, 4: 252–261.

Wilson, H.D.B. and Brisbin, W.C., 1961. Regional structure of the Thompson-Moak Lake nickel belt. *Trans. Can. Inst. Min. Metall.*, 64: 815–822.

Wilson, H.D.B., Kilburn, L.C., Graham, A.R. and Ramlar, K., 1969. Geochemistry of some ultrabasic intrusions. *Econ. Geol. Monogr.*, 4: 294–309.

Wolf, K.H., 1976. Conceptual models in geology. In: K.H. Wolf (Editor), *Handbook of Strata-Bound and Stratiform Ore Deposits, 1.* Elsevier, Amsterdam, pp. 11–78.

Woodall, R. and Travis, G.A., 1969. The Kambalda nickel deposits, Western Australia. *Proc. 9th Commonw. Min. Met. Congr. London, Pap. 26,* pp. 517–533.

Wyllie, P.J., 1978. Mantle fluid compositions buffered in peridotite-CO_2-H_2O by carbonates, amphibole and phlogopite. *J. Geol.*, 86: 687–713.

Zurbrigg, H.F., 1963. Thompson Mine geology. *Bull. Can. Inst. Min. Metall.*, 56 (614): 451–460.

Chapter 8

SEDIMENT-HOSTED SUBMARINE EXHALATIVE LEAD–ZINC DEPOSITS – A REVIEW OF THEIR GEOLOGICAL CHARACTERISTICS AND GENESIS

DUNCAN E. LARGE

INTRODUCTION

The problems of mineral-deposit classification, and the plethora of classification schemes, have been outlined and discussed by Gabelmann (1976), Wolf (1981) and Laznicka (1981). There are two alternatives: a morphological classification upon empirical features, or a genetic one based on a particular interpretation of these features. Gilmour (1971; see also his Chapter 4, Vol. 1 in this Handbook series) noted that the main problem of the morphological classification of mineral deposits is the impossibility of characterising a class of ore deposits by one or two features.

It will be shown that the class of deposit under discussion is characterised by a variety of host lithologies, several different styles of mineralisation, and contrasting geotectonic settings. The nearest morphological definition of the class is "stratabound, sediment-hosted, massive sulphide lead–zinc deposits". This definition suffers from being too long and, more importantly, the use of words that can have several meanings. For example the word "massive" could describe deposits that consist of 50% or more by volume of sulphides (e.g. Sangster and Scott, 1976), the texture of the sulphide minerals as well as the structure of the ore deposit itself. Each of the other descriptions in the morphological definition is also open to tedious semantic arguments. For reasons outlined below a genetic classification has been selected, and the class of deposits discussed here is defined as sediment-hosted submarine exhalative Pb–Zn deposits.

It is the purpose of this paper to illustrate that the various styles of mineralisation, as well as other geological features of this class, can be explained by the "submarine exhalative" model. "Submarine exhalative" deposits are those formed on the sea floor from hydrothermal solutions discharged into the sea. Schneiderhöhn (1944) first described the Devonian Lahn–Dill (volcanic-hosted) and Meggen (sediment-hosted) deposits of Germany as being submarine exhalative in origin (see Ouade's Chapter 6, Vol. 7, and Krebs' Chapter 9, this volume, respectively). Present examples include the Atlantis II deep in the Red Sea (Degens and Ross, 1976).

There are numerous fossil submarine exhalative deposits, and it is obviously necessary to classify them into smaller groups. This is most commonly accomplished on the basis of the host rocks to the mineralisation. The deposits in question are hosted by non-volcanic sediments, and they are thus distinguished from those that are hosted by volcanics and volcaniclastics. It will be shown that the sediment-hosted deposits are contained within thick sedimentary sequences in an Atlantic-type margin or intracontinental basin setting, whereas the volcanic-hosted deposits are located in crustal accretionary (e.g. Cyprus-type) or crustal subduction (e.g. Kuroko-type) tectonic settings (Mitchell and Garson, 1976). It is thus considered misleading to visualise a continuum between the volcanic- and sediment-hosted types of massive sulphide accumulation, although the volcanic-hosted types commonly have minor sedimentary intercalations, and minor tuffite horizons are often present in the sequence containing sediment-hosted deposits. The use of a "transitional" or "mixed volcanic–sedimentary" type is avoided.

Apart from submarine exhalative, numerous other genetic descriptors have been used in the classification of ore deposits (e.g. volcanogenic, epigenetic, sedimentogenic, volcanic–sedimentary, syngenetic, diagenetic) and they all suffer from what Gilmour (1962, p.452) referred to as the "wind of change blowing over the field of economic geology", i.e. the constant development of genetic theories. A good example is afforded by the history of geological thought on the Rammelsberg deposit (reviewed by Kraume, 1955, pp.265–286). Within 60 years hypotheses on the genesis of this deposit have swung from a relatively sophisticated synsedimentary, submarine exhalative theory at the turn of the century (e.g. Wiechelt, 1904) to the "bedded vein" epigenetic theory of Lindgren (1933, p.628), and recently back to a submarine exhalative model (e.g. Ramdohr, 1953; Gunzert, 1969). There is still controversy concerning the origin of various features of Rammelsberg (e.g. the "kniest"). All the above theories were interpretations of the observed geology in terms of the contemporary genetic theory (see Hannak, 1981, for a comprehensive review of the geology of the deposit). The submarine exhalative theory, therefore, must be critically examined, and accepted or rejected depending on how successfully it explains the various empirical features of the deposits under review.

GENERAL CHARACTERISTICS OF SEDIMENT-HOSTED, SUBMARINE EXHALATIVE MINERALISATION

The sediment-hosted, submarine exhalative Pb–Zn deposits include some of the world's most important ore bodies (Table I). The geology of these deposits has been reviewed in the literature and other volumes of this Handbook series (see indexes in Vols. 4 and 7), and it is not intended to repeat the data of each individual occurrence in this review. Certain common geological features, however, are summarised in the tables.

Table I is not complete, as it is only based on those deposits that are fully described in the literature; and no doubt more will be recognised that belong to this class. Sangster (1979) has shown that it is possible to reinterpret the geology of the Cobar deposits, Australia, which had previously been considered as epigenetic, by the submarine exhalative

TABLE I

Tonnage (prior to mining) and grade of selected sediment-hosted, submarine exhalative Pb–Zn deposits

Ore deposits	Sulphide tonnage (Mt)	Barite tonnage (Mt)	Zn(%)	Pb(%)	Cu(%)	Ag(g/t)
Rammelsberg	22	0.2	19	10	2	120
Meggen	50	10	10	1.3	0.02	3
Silvermines	13.9 *1	2.5 *3	8.3	2.4		23
	2.1 *2		3.4	4.5		32
Tynagh	9.4		3.2	3	0.3	28
Mount Isa	88.6 *4		6.3	6.9		149
	181.6 *5				3.0	
Hilton	35.6		9.6	7.7		180
Lady Loretta	8.6	?	18.1	6.7		109
McArthur River	190		9.5	4		45
Sullivan	155 *6		5.7	6.6		68
Tom	9	?	8.4	8.6		84

*1 Upper G and B stratiform ore zones, Silvermines.
*1 Lower G cross-cutting ore zone, Silvermines.
*3 Macgobar deposit, Silvermines.
*4 Stratiform lead–zinc–silver ore, Mount Isa (32.6 Mt already mined out).
*5 Copper ore deposit, Mount Isa (41.6 Mt already mined out).
*6 Estimated original size of Sullivan deposit (98 Mt already mined out).
References: Rammelsberg (Hannak, 1981), Meggen (Krebs, 1981), Silvermines (Morrissey et al., 1971; Taylor and Andrew, 1978), Tynagh (Morrissey et al., 1971), Mount Isa (Mathias and Clarke, 1975), Hilton (Mathias et al., 1973), Lady Loretta (Loudon et al., 1975), McArthur River (Lambert, 1976), Sullivan (Ethier et al., 1976), Tom (Freberg, 1976).

model in a sediment-hosted setting. In addition, only those deposits that have not been subjected to high-grade metamorphism have been included; original lithologic relationships tend to be obscured in metamorphic terrains, as well as the textural relationships of the sulphides themselves. Deposits that might belong to this category, but are hosted by high-grade metamorphic rocks, are Faro, Canada (Tempelman-Kluit, 1972), Ducktown, U.S.A. (Addy and Ypma, 1977), Broken Hill, Australia (Johnson and Klingner, 1975; see also Both and Rutland, 1976) and Gamsberg, South Africa (Rozendaal, 1978).

The submarine exhalative deposits under consideration here have many similarities that together characterise the class as a whole and which are outlined in the following.

(1) They contain one or more lens-like, tabular bodies of stratiform sulphides up to a few tens of metres in thickness. The lateral dimensions of these bodies are several hundred metres to one or two kilometres. The individual bodies, which together comprise a stratabound ore deposit, may be distributed through a stratigraphic interval of up to 1000 m (e.g. Mount Isa).

(2) Stockwork or vein-type mineralisation (here generally referred to as cross-

TABLE II

Stratiform and cross-cutting mineralisation in selected sediment-hosted, submarine exhalative Pb–Zn deposits

Deposit	Stratiform mineralisation	Cross-cutting mineralisation
Rammelsberg	Cu, Pb, Zn, Fe sulphides and $BaSO_4$ as massive and laminated ore.	"Kniest", cupriferous and silicified footwall sediments beneath stratiform deposit.
Meggen	Laminated Zn, Pb, Fe sulphides. $BaSO_4$ deposit peripheral to the sulphides along same horizon.	Minor, discordant Cu mineralisation beneath the stratiform ore.
Silvermines	Laminated Pb, Zn, Fe sulphides in the Upper G zone. Laterally equivalent barite (Macgobar).	Vein and cavity filling Pb, Zn and minor Cu sulphides in the Lower G Zone.
Tynagh	Fe-oxides and chert facies with minor sulphides.	Most Pb, Zn, Fe, and Cu sulphides are cavity-filling veins.
Mount Isa	Pb, Zn, and Fe sulphides as finely laminated and bedded ore.	"Silica-dolomite", cupriferous, silicified and brecciated zone adjacent to stratiform sulphides.
McArthur River	Pb, Zn, and Fe sulphides as finely laminated and bedded ore.	"Cooley" vein-type Cu–Pb sulphide deposits adjacent to the stratiform ore.
Sullivan	Pb, Zn, and Fe sulphides as finely laminated and massive ore.	Not reported, but significant brecciation of footwall sediments.
Tom	Laminated Pb, Zn, and minor Fe sulphides. Barite is interlaminated with the sulphides.	Disseminated Cu sulphides and silicification underlying stratiform ore.

References: Rammelsberg (Gunzert, 1969; Hannak, 1981), Meggen (Krebs, 1981), Silvermines (Taylor and Andrew, 1978), Tynagh (Derry et al., 1965; Russel, 1975), Mount Isa (Mathias and Clarke, 1975), McArthur (Lambert, 1976; Williams, 1978a, b), Sullivan (Ethier et al., 1976), Tom (Carne, 1976).

cutting mineralisation) is commonly found subjacent or adjacent to the stratiform mineralisation. The contact between the cross-cutting and stratiform mineralisations is usually distinct (Table II).

(3) The deposits are of variable size and metal content (Table I), but those of Proterozoic age tend to be very much larger than volcanic-hosted, submarine exhalative ore deposits. The Phanerozoic examples are, however, comparable in size to the volcanic-hosted deposits. Sediment-hosted deposits tend not to be clustered together into mining camps, as is often the case for the volcanic-hosted class. However, numerous epigenetic deposits (vein-type, carbonate-hosted-type), which are economically relatively unimportant, are often present in the immediate neighbourhood of the submarine exhalative mineralisation (e.g. the situation of Rammelsberg within the Harz mining district, Sperling, in press; epigenetic carbonate-hosted mineralisation in the vicinity of the McArthur River deposit, Walker et al., 1978; the presence of numerous vein-type deposits within the vicinity of the Sullivan mine, Thompson and Pantaleyev, 1976).

(4) The deposits are usually characterised by a simple sulphide mineralogy: pyrite and/or pyrrhotite, sphalerite, galena, and minor chalcopyrite; marcasite and arsenopyrite are occasionally also present (e.g. Rammelsberg). Trace quantities of other sulphides and sulphosalts have been reported from many deposits but do not comprise an important proportion of the total. The proportion of the different sulphides varies, although iron sulphides are normally the most abundant. Silver may be present as freibergite or contained within the crystal lattice of the galena.

(5) The stratiform sulphides may in part be massive (i.e. with no visible internal texture), imperfectly banded with a wispy or streaky texture (such as the melierterz at Rammelsberg) or, most commonly and characteristically, as laterally persistent beds from 1 mm to 1 m thick. These beds are often composed of monomineralic, fine-grained laminae (2–200 μm), especially in those deposits that have not been significantly metamorphosed or deformed (e.g. McArthur River). Pyrite framboids have been described from many localities (e.g. McArthur River, Mount Isa, Silvermines, Meggen and Rammelsberg). Minor coarse-grained pyrite and galena is reported from several deposits and is interpreted to be the result of remobilisation during deformation. The cross-cutting mineralisation consists of coarser-grained sulphides in cavity-fillings, veinlets and breccia-fillings.

(6) Barite often overlies or occurs as a lateral stratigraphic extension to the stratiform sulphide mineralisation (e.g. Meggen, Rammelsberg, Lady Loretta and Silvermines). The barite is fine grained, usually massive with an indistinct internal lamination. In the upper part of the Rammelsberg massive sulphide deposit the barite is intermixed with the sulphides and in the Tom the barite is finely interlaminated in the sulphides, but these are exceptions to the usual rule that associated barite is spatially separated from the sulphides with only a minor transitional zone (e.g. Meggen, see Weisser, 1972; Krebs, 1981).

(7) The footwall sediments to the stratiform sulphides are often hydrothermally altered, especially in the vicinity of cross-cutting mineralisation [see (2) above]. This alteration usually takes the form of silicification (e.g. the kniest at Rammelsberg).

(8) A distinct lateral and/or vertical zonation of the sulphides is commonly recognised in many of the deposits (Table III). This zonation is centred on a copper-rich core, with zinc and lead being more widely dispersed, and with barite on the periphery. In lateral zonation sequences the zinc is more widely dispersed than lead (e.g. Mount Isa, McArthur River, Tom) to a give a typical Cu → Pb → Zn → (Ba) sequence. In vertically zoned deposits the usual sequence of upwards zonation is Cu → Zn → Pb → (Ba) (e.g. (Rammelsberg). Iron is dispersed throughout the deposits as pyrite and/or pyrrhotite, sometimes forming an iron-rich core (e.g. Sullivan), or as hematite on the periphery of the stratiform facies (e.g. Tynagh).

(9) The stratiform sulphides are concordantly interbedded in marine sediments of different lithologies (black shales, conglomerates, breccias, turbidite sandstones, siltstones, cherts, dolomites and micritic limestones), which are subdivided into fine-grained lithologies (so-called autochthonous) and coarse-grained lithologies (so-called

TABLE III

Examples of metal zonation in the stratiform mineralisation of selected sediment-hosted, submarine exhalative Pb–Zn deposits

Deposit	Metal Zonation in the stratiform mineralisation
Rammelsberg	Vertical zonation Cu → Zn → Pb → Ba upwards through the stratiform ore.
Meggen	Mn-halo in the ore horizon sediments around the ore body. Barite ore body laterally equivalent to the stratiform sulphides.
Silvermines	In the stratiform Upper G zone the Pb/Zn ratio decreases towards the periphery of the zone away from the Silvermines fault.
Tynagh	Mn-halo in the stratiform iron-oxide facies and laterally equivalent sediments. Lateral increase of Zn towards the cross-cutting mineralisation.
Mount Isa	Lateral zonation Pb → Zn within the stratiform ore, away from the contact with the "silica-dolomite".
McArthur River	Lateral zonation within the stratiform mineralisation of Pb → Zn → Fe, away from the cross-cutting vein-type mineralisation.
Sullivan	Within the stratiform mineralisation there is a decreasing Pb/Zn ratio towards the periphery of the deposit. Central core zone is Fe-rich.
Tom	Within the stratiform mineralisation the Pb/Zn ratio decreases laterally away from the massive ore and the underlying cross-cutting mineralisation.

References: Rammelsberg (Gunzert, 1969), Meggen (Krebs, 1981), Silvermines (Taylor and Andrew, 1978), Tynagh (Russel, 1975), Mount Isa (Mathias and Clarke, 1975), McArthur River (Williams, 1978a, b), Sullivan (Ethier et al., 1976), Tom (Carne, 1976).

allochthonous) (Table IV). The former are indicative of the quiet, low-energy, euxinic environment that was present at the site of mineralisation, and the latter represent the detritus that was rapidly introduced into this environment by turbidity currents or slumping. The stratiform mineralisation is often interbedded or underlain by slump breccias (poorly rounded blocks of varying size), the constituents of which were derived from a local provenance or are intraformational.

(10) Thin tuffite horizons (1–10 cm) are found within the host sediments of many deposits. These tuffites contain distinctive glass shards, which are sometimes replaced by K-feldspar, clay minerals or sulphides, and are usually very fine grained. Without exception they are enriched in potassium that was probably introduced during their diagenetic alteration (see Table VI, in Krebs, 1981). No coarse-grained pyroclastic debris is described from these tuffites and their eruptive source is generally not known. Other evidence for penecontemporaneous igneous activity may include minor intrusions (e.g. the Hellroaring Creek stock near the Sullivan, Canada; Ethier et al., 1976).

(11) Many of the deposits in this class are spatially associated with a growth fault that was thought to have been active at the time of mineralisation. This fault activity is probably responsible for the sedimentary breccias mentioned in (9) above. Zonation within the stratiform and associated cross-cutting sulphides is often centred on these faults (e.g. McArthur River, Silvermines, Tynagh), which suggests that they may have

TABLE IV

Autochthonous and allochthonous lithologies of selected sediment-hosted, submarine exhalative Pb–Zn deposits

	Autochthonous	Allochthonous
Rammelsberg	Pyritic black shales.	Siltstones and sandstones, possibly turbidites.
Meggen	Black and gray shales, bituminous limestones.	Siltstones, sandstones and local limestone turbidites.
Silvermines	Gray shales, dolomitic shale, biomicrite, Waulsortian limestone and chert.	"Reef" limestone breccias and dolomite breccias.
Tynagh	Waulsortian "reef" limestone, black shaly mudstone, calcareous shale and chert.	Bioclastic limestone with crinoidal debris, calcarenites and volcanic debris.
Mount Isa	Dolomitic siltstone, slightly pyritic.	Tuffaceous horizons, no allochthonous clastic input recognised.
Lady Loretta	Carbonaceous pyritic shale.	Rhythmic dolomitic siltstones and sandstones, possible turbidites.
McArthur River	Carbonaceous, slightly pyritic silty dololutite, black cherts.	Graded tuffaceous dolomites, talus slope breccia and conglomerate.
Sullivan	Shale with pyrrhotite laminations.	Intraformational basal conglomerate, quartzwacke and quartzite of possible turbidite origin.
Tom	Gray siltstone, carbonaceous shales.	Argillite-clast breccia, and chert-pebble conglomerate.

References: Rammelsberg (Hannak, 1981), Meggen (Krebs, 1981), Silvermines (Taylor and Andrew, 1978), Tynagh (Morrissey et al., 1971), Mount Isa (Mathias and Clarke, 1975), Lady Loretta (Loudon et al., 1975), McArthur River (Walker et al., 1978a, b), Sullivan (Ethier et al., 1976), Tom (Carne, 1976).

been the conduits of the metal-bearing solutions. Subsequent reactivation of these faults may result in the formation of a significant crustal lineament.

(12) The deposits are hosted by sediments of various ages (Table V), most commonly Middle Proterozoic (1700–1400 Ma) and Lower–Middle Palaeozoic (500–300 Ma). Although the mineralisation within one geotectonic province is often of approximately the same age, they are not necessarily exactly contemporaneous as is exemplified by Meggen (Givetian) and Rammelsberg (Eifelian) in the Variscan trough.

(13) The deposits are often located in miogeosynclinal (passive continental–oceanic crustal margin) or basinal (intra-continental) preorogenic regimes. As a result of later tectonism many of the sediment-hosted, submarine exhalative sulphides, especially those in a palaeo-continental margin situation, are located in folded and cleaved host rocks (e.g. Meggen, Rammelsberg, Mount Isa, Sullivan, Tom). In contrast, those that are located in an intracontinental setting are not as severely deformed (e.g. McArthur River, Silvermines, Tynagh).

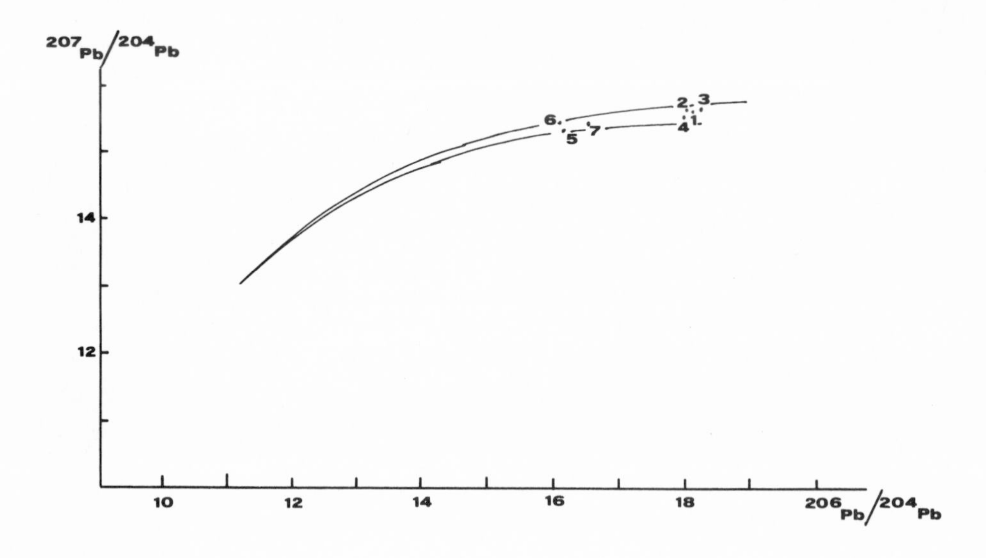

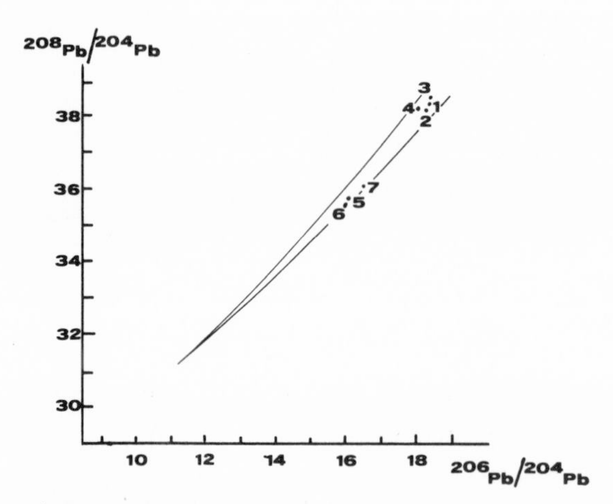

Fig.1. Two-stage lead growth curves based on a major differentiation of the U/Pb and Th/Pb systems 3750 Ma ago (from Köppel and Saager, 1976). The two curves in each diagram are drawn on the basis of using the values 9 and 10 for μ ($^{238}U/^{204}Pb$) since 3750 Ma. *1* = Rammelsberg; *2* = Meggen; *3* = Silvermines; *4* = Tynagh; *5* = McArthur River; *6* = Mount Isa; *7* = Sullivan. *References;* Meggen and Rammelsberg (Wedepohl et al., 1978); Tynagh and Silvermines (Greig et al., 1971); Mount Isa and McArthur River (Richards, 1975).

(14) The lead-isotope ratios for the galena in these deposits are homogeneous and non-radiogenic (Fig.1), and lie close to the "growth curve" as defined by Stacey and Kramers (1975). Galena-Pb from neighbouring vein-type mineralisation (e.g. post-Variscan vein mineralisation in the Harz, Wedepohl et al., 1978; Cretaceous vein-type

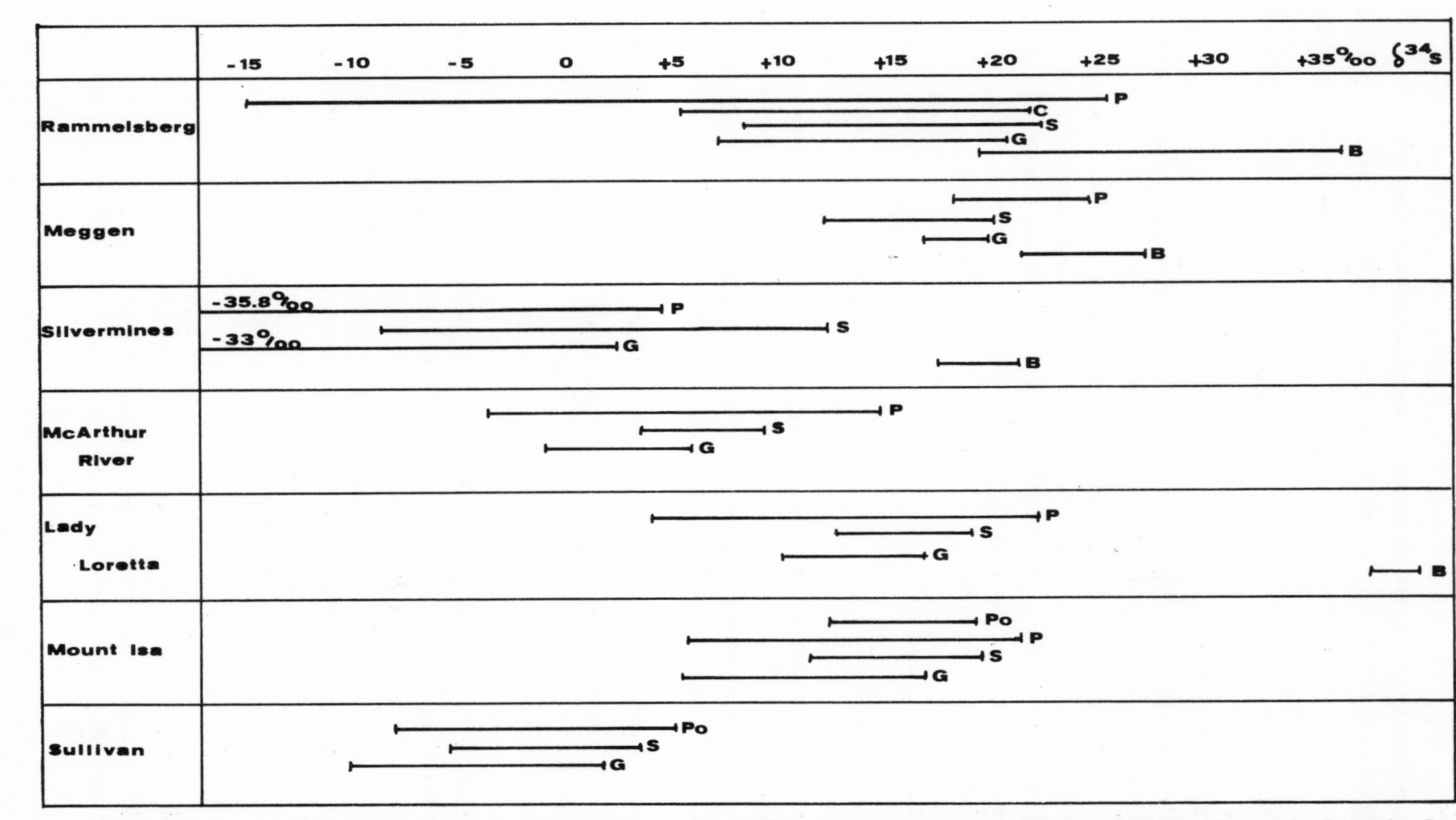

Fig. 2. The range of $\delta^{34}S$ from the various sulphides and barite in the sediment-hosted, submarine exhalative sulphide-barite deposits. *P* = pyrite; *Po* = pyrrhotite; *C* = chalcopyrite; *S* = sphalerite; *G* = galene; *B* = barite. *References*: Rammelsberg (Anger et al., 1966): Meggen (Buschendorf et al., 1963); Silvermines (Coomer and Robinson, 1976); McArthur River (Smith and Croxford, 1973); Lady Loretta (Carr and Smith, 1977); Mount Isa (Smith et al., 1978); Sullivan (Campbell et al., 1978).

mineralisation in the Purcell-Belt basin, Zartman and Stacey, 1971) is invariably more radiogenic and less conformable with respect to the growth curves.

(15) Sulphur-isotope studies from several deposits indicate that there are, isotopically, two types of sulphur within one deposit. The galena-S, sphalerite-S, pyrrhotite-S and some of the pyrite-S are in isotopic equilibrium. The remainder of the pyrite-S is isotopically very variable, and this variation may be consistent with the stratigraphic horizon within the ore deposit (e.g. at McArthur River, Smith and Croxford, 1973; and Lady Loretta, Carr and Smith, 1977; the pyrite-S becomes isotopically heavier vertically upwards). Pyrite-S with these isotopic characteristics is considered to be biogenically reduced from sea-water sulphate at the site of mineralisation, whereas the galena-S, sphalerite-S, pyrrhotite-S and the remainder of the pyrite-S is considered to have a hydrothermal source. Associated barite-S tends to (a) not be in isotopic equilibrium with the base-metal sulphide sulphur, (b) be isotopically heavier than the base-metal sulphide sulphur, and (c) be isotopically similar to, or slightly heavier than, the sulphur in the contemporaneous sea-water sulphate.

REGIONAL GEOLOGICAL SETTING

The sediment-hosted, submarine exhalative lead–zinc deposits are located in either epicontinental reentrants developed on an Atlantic-type palaeocontinental margin or in intracontinental basins (Table V and Fig.3). These basins, which have dimensions of several hundred kilometres, are usually characterised by markedly thick (several thousand metres), practically continuous sedimentary sequences and are termed *first-order* basins. Basic or bimodal (e.g. spilite–keratophyre) volcanics may also be present within the sequence, but the mineral deposits in question are not hosted by them.

Within the epicontinental reentrants, local vertical tectonic movements often lead to the formation of smaller basins and rises – commonly referred to as "Becken and Schwellen" or "Graben and Horst" structures. The basins so formed are generally discontinuous features, with dimensions of several tens of kilometres. They are termed *second-order* basins and are separated by rises of approximately equal dimensions. This palaeogeographic pattern is often found in epi- and intracontinental basins and is related to penecontemporaneous faulting and different rates of subsidence within the first-order basin. The development of such features has been described by Krebs (1979) in the Devonian of the Variscan Trough, by Glikson et al. (1976) in the Leichhardt River Fault Trough, as well as by Murray (1975) and Plumb and Derrick (1975) in the McArthur Basin.

Third-order basins are relatively small depressions several hundred metres to a few kilometres in lateral dimensions, and they are the depositories of the stratiform mineralisation. The depositional environment in the third-order basins is discussed on p.483.

The prevailing water depth at the site of sulphide precipitation varies from the

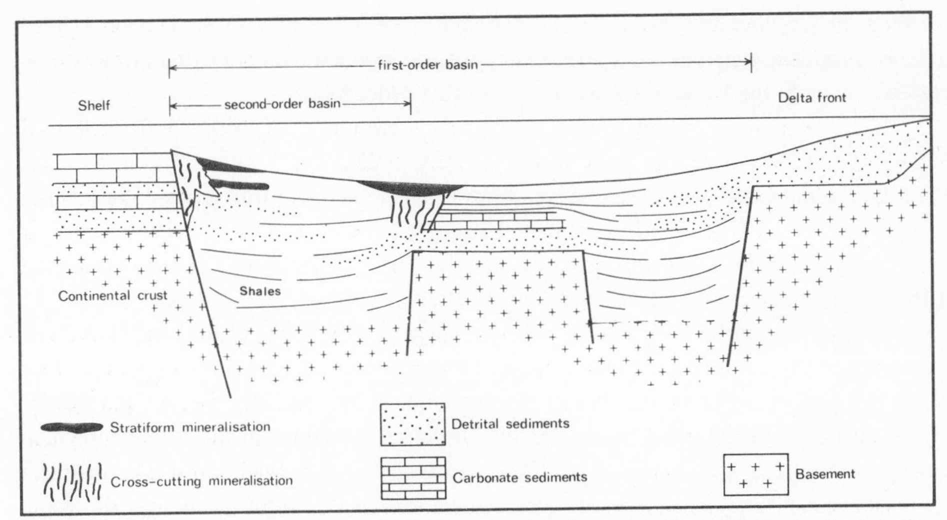

Fig.3. The geotectonic situation of sediment-hosted submarine exhalative sulphide–barite deposits – sketch of a hypothetical situation. The deposits are contained within first-order intra- or epicontinental basins in which there are smaller second-order basins and rises.

shallow marine (±100 m in the Irish deposits) to depths of 500–1000 m (Rammelsberg). Sedimentological depth indications in the outer shelf are few and can only be estimated from broad palaeontological and lithological features. It will be shown that the hydrostatic pressure at the site of sulphide deposition has a strong influence on the style of mineralisation by determining whether or not second-boiling occurred prior to the exhalation of the hydrothermal solutions (see also Finlow-Bates and Large, 1978).

Burke and Dewey (1973), Dunnet (1976), and Sawkins (1976) have noted that Mount Isa and Sullivan formed in rift environments on continental margins. The rifting may be related to "hot spot" activity. The regional geology in the first-order basins containing these and other sediment-hosted deposits is characterised by features indicative of "hot spot" related rifting (e.g. basalts that may be accompanied by acid volcanics to give a characteristic bimodal association, very thick clastic sedimentary sequences, and evidence for penecontemporaneous vertical tectonism; Burke and Dewey, 1973). It is suggested that the mineralised first-order basins may, in fact, be intracratonic rifts or epicratonic aulacogens that were formed above "hot spots" or anomalously high geothermal gradients.

Glikson et al. (1976) have proposed that the Mount Isa and Hilton deposits formed in an ensialic rifted trough (Leichhardt River Fault Trough) close to the continental margin. Rifting was followed by a compressive phase during which a micro-cratonic block (the Kalkadoon–Leichhardt Block) was welded onto the cratonic margin. This east–west compression resulted in the deformation of the host sediments and sulphides at Mount Isa (e.g. McClay, 1979). Glikson et al. (1976) noted the possible similarities between the Leichhardt River Fault Trough and Proterozoic aulacogens. Dunnet (1976) has also proposed that the Lady Loretta deposit is contained within a rifted intracratonic graben – namely the Paradise Greek Graben (a first-order basin).

In western North America, the depositional environment of the Proterozoic Purcell Supergroup, which hosts the Sullivan deposit (see Indexes in Volumes 4 and 7 of this Handbook series), has been described in terms of an embayment into a rifted continental margin (Stewart, 1972). The Purcell-Belt first-order basin may thus be a rifted trough or aulacogen that formed above a zone of high heat flow ("hot spot") at the continental margin (Kanasewich, 1968; Burke and Dewey, 1973).

An extensional tectonic regime was also suggested by Krebs (1960) and Burke and Sawkins (1978) for the Devonian of central Europe during the formation of various massive sulphide deposits (including Meggen and Rammelsberg). Werner and Rösler (1979) investigated the geochemistry of the Devonian volcanism in the Rhenohercynian zone of central Europe (including the Variscan trough), and showed that it is comparable to alkali volcanism (spilite–keratophyre association) in an intracontinental rift zone. They concluded that this rift zone may have formed above a hot spot. There is, however, no geological or geochemical evidence for the presence of oceanic tholeiites in the Variscan trough. The duration of this extension tectonism in the Devonian must have been quite short-lived as it was followed in the Lower Carboniferous by the Hercynian orogeny, which was marked by north–south compressional folding.

Russell (1968) has proposed that the Irish deposits lie close to N–S trending lineaments, which may be incipient extensional fractures related to the initial phases of opening of the Atlantic ocean. Within the Variscan palaeogeographic framework, the Irish deposits are situated in a basin (Central Irish Basin; fig.1 in Evans, 1976) within the northern continental foreland and thus escaped the intense Hercynian deformation of the geosynclinal sediments.

As such the Irish deposits can be compared to those at Lady Loretta and McArthur River, both of which are hosted by relatively undeformed platform sediments. This contrasts with the strongly deformed (folded and thrusted) "geosynclinal" sequences that host the Mount Isa, Meggen and Rammelsberg deposits.

In conclusion the submarine exhalative, sediment-hosted Pb–Zn deposits were formed within epicontinental and intracontinental rift zones (Fig.3). The rifting was probably related to hot-spot activity, and was closely followed by a compressive phrase.

Lineaments and hinge zones

Penecontemporaneous block faulting was the main control on the distribution of the sedimentary facies within the first-order basins, and lead to the development of second-order basins. Block faulting is a common feature within continental rift zones (Dickinson, 1974). The submarine exhalative, sediment-hosted Pb–Zn deposits all tend to be located close to the margins of first- and second-order basins, and are all close to fault zones that were considered to have been active during deposition of the host sediments Subsequent reactivation of these faults may lead to the formation of readily identifiable lineaments, such as those related to the north Australian Proterozoic Pb–Zn deposits (e.g. the Emu Fault at McArthur River, figs. 1 and 2 in Lambert, 1976; and the Mount Isa Fault at Mount Isa and Hilton, fig. 14 in Lambert, 1976).

Typical of a hinge zone that is not readily identifiable as a lineament is that which separates the Goslar Trough from the West Harz "Schwelle" in the vicinity of the Rammelsberg deposits. This hinge zone is marked by a facies and thickness change from more than 1000 m of clastic sediments in the trough to several tens of metres of carbonate sediments on the rise (Hannak, 1981; fig. 4).

Meggen lies near the SW–NE trending Middle Devonian shelf–basin boundary which was active as a mobile hinge line separating areas of different subsidence, facies and bathymetry over a long period (Krebs, 1981; fig. 1). Over 1300 m of middle and late Givetian shallow-marine sediments to the northwest of the shelf margin are correlated with 40–60 m of pelagic sediments to the southwest. Here the shelf appears to have subsided faster and accumulated a greater thickness of shallow-water sediments than the adjacent trough, although the low sedimentation rate in the trough could be in part due to a lack of clastic input and "starved" environment, possibly protected by shelf margin carbonates.

There are numerous lineaments on the earth's surface and the significance of many remains a mystery, but those that represent hinge zones may be regarded as prospective, especially at their intersection with other features. It is also important to note that the ore deposits rarely lie directly on the lineament, but within a zone about 10 km on either side of it, often along minor splay faults subparallel to the main structure.

Age of mineralisation

The ages of selected sediment-hosted, submarine exhalative deposits are presented in Table V. The ages of the Phanerozoic deposits are often well defined by biostratigraphical dating, and those for the Proterozoic deposits are the age ranges of the host sediments as deduced from isotopic dating.

On the basis of the deposits listed in Table V, there are two main periods of sediment-hosted submarine exhalative mineralisation – the Middle Proterozoic (1600–1400 Ma) and the Lower Middle Palaeozoic (500–320 Ma).

It is noteworthy that the Middle Proterozoic was a period of great crustal stability when thick, conformable sequences were deposited over wide areas of the earth's continental crust. It is thought that the continental cratons were welded together as a "Proto-Pangaea" in the Lower and Middle Proterozoic (Piper, 1976; Windley, 1977) during which time sedimentation was contained within very large intra- and epicontinental basins. The early part of this period (2200–2000 Ma) is marked by the huge banded-iron formation deposits of the Hammersley Basin (Australia) and Superior Province (North America) (see

TABLE V

First- and second-order basins containing sediment-hosted, submarine exhalative lead-zinc deposits

First-order basin	Dominant lithologies	Age range of sediments in basin	Second-order basins	Mineral deposits
Variscan Trough	Pelagic, dark shales, turbidite siltstones and sandstones, cephalopod limestones	Devonian–Lower Carboniferous	"Becken", Goslar Trough	Rammelsberg, Meggen
Central Irish Basin	Bioclastic limestones, micrites and argillaceous limestone	Lower–Middle Carboniferous		Tynagh, Silvermines
Leichhardt River Fault Trough	Dolomitic shales and siltstones	Middle Proterozoic	"Basins"	Mount Isa
Batten Trough (McArthur Basin)	Shallow marine dolomites, evaporites, siltstones and sandstones	Middle Proterozoic	Bulburra Depression	McArthur River
Purcell-Belt Basin	Turbidite sandstones, siltstones and minor carbonates	Middle–Upper Proterozoic		Sullivan
Selwyn Basin	Cherts, shales, siltstones, and calcarenites	Upper Ordovician–Carboniferous	Macmillan Pass Basin	Tom

References: Variscan Trough (Krebs and Gwosdz, in press), Central Irish Basin (Macdermot and Sevastopulo, 1972); Leichhardt River Fault Trough (Glikson et al., 1976; Smith, 1969), Batten Trough (Murray, 1975), Purcell-Belt Basin (Harrison, 1972), Selwyn Basin (Blusson, 1976; Dawson, 1977).

also Lambert and Groves, 1981, for a review).

In addition the Lower–Middle Proterozoic is interpreted as being of crucial importance as it marks the beginning of the accumulation of free oxygen in the hydrosphere and, subsequently, in the atmosphere (2000 Ma). It is striking that, in spite of this radical change in the chemistry of the earth's surface layers, the formation of massive stratabound and stratiform sulphide ore deposits continued from the Archaean (e.g. the Abitibi belt in Canada), through the Lower Proterozoic (e.g. Jerome, U.S.A. and Broken Hill, Australia) and into the Middle Proterozoic (e.g. sediment-hosted deposits in Table V). Massive sulphides are found throughout most of the Phanerozoic. All the above-mentioned deposits are described as being syngenetic or submarine exhalative. If this genetic interpretation is correct, then the overall chemistry of the earth's surface cannot play an important role in determining the formation of stratabound massive sulphide deposits. It is probably only significant that the palaeotopography and chemistry of a restricted environment is conducive to the precipitation and preservation of metal sulphides, and that this characteristic chemical requirement has probably remained constant through time. The nature of this chemical environment will be discussed later.

The deposits in question were formed during relatively brief periods of extensional tectonism (possibly failed rifting) that preceded the break-up of the Precambrian Proto-Pangaea and of the Phanerozoic Pangaea.

Sedimentary environment at the site of mineralisation

A euxinic environment is required for the precipitation and preservation of sulphides (see Vaughan, 1976) and reducing conditions may be represented by the accumulation of organic-rich (cf. Saxby's Chapter 5, Vol. 2) sediments in third-order basins. In shallow-marine environments, carbonaceous lithologies tend to contrast sharply with the surrounding sediments (e.g. McArthur River), whereas in deeper-marine milieus it is often difficult to distinguish the carbonaceous shale facies from the normal dark-gray to gray shale deposits (e.g. Rammelsberg).

The actual composition of the sedimentary host rocks can be very variable, and ranges from dolomites through dolomitic siltstones, to silty argillites and shales (Table IV). The fine-grained lithologies are commonly thinly interbedded with the sulphides, and are termed here the autochthonous host rocks as they are considered to represent the prevailing sedimentary environment at the site of mineralisation. These lithologies show no evidence of high-energy conditions (e.g. ripple marks, cross-beddings, scours, erosive channel margins). This is in accordance with the concept of a quiet euxinic environment at the site of ore deposition being essential for the precipitation and preservation of the stratiform sulphides.

Breccias, conglomerates and coarse clastic deposits are also often interbedded within the autochthonous sediments and the sulphides themselves (e.g. McArthur River, Lambert, 1976; and Meggen, Krebs, 1981). Graded beds, erosive channel margins and scouring features are associated with these lithologies, which are considered to have been very rapidly introduced into the third-order basin. They are termed the allochthonous host rocks (Table IV). The stimulus that triggers the inflow of the breccias is probably associated with synsedimentary fault activity or seismic shock. Their composition is related to the basin margin facies (e.g. reef carbonates, delta sands) where that can be identified, or they may be intraformational. Only rarely are clasts of stratiform mineralised sulphides to be found in the allochthonous breccias (e.g. McArthur River; Croxford, 1968).

One consequence of this variation in host-rock lithology is that the commonly used "shale-hosted" classification for the sediment-hosted, submarine exhalative deposits is misleading.

SOME COMPARISONS WITH THE GEOLOGICAL SETTING OF OTHER SEDIMENT-HOSTED Pb–Zn DEPOSITS

It has been noted that the submarine exhalative, sediment-hosted deposits are located in preorogenic tectonic environments that are characterised by penecontemporaneous block faulting. Subsequent post-sedimentary tectonism has strongly deformed (folded and thrusted) the sediments and their contained mineral deposits in the Variscan Trough, Selwyn Basin, Leichhardt River Fault Trough and the Belt-Purcell Basin.

In central Europe the Kupferschiefer (sediment-hosted, stratabound, polymetallic sulphide mineralisation; see Jung and Knitschke, 1976) is hosted by Permian post-orogenic sediments that unconformably overly the folded Variscan sequences containing Meggen and Rammelsberg. The footwall to the Kupferschiefer deposits, the Rotliegender sandstones, are molasse-type sediments derived from the Variscan orogen. As such the regional geologic setting of the Kupferschiefer sharply contrasts with that of Meggen and Rammelsberg. The Kupferschiefer is contained within a post-orogenic, molasse-type basin, and Meggen and Rammelsberg are located in a pre-orogenic, pre-flysch basin.

The general consensus on the genesis of the Kupferschiefer is that it was probably formed during the diagenesis of the host transgressive black shales, which obviously contrasts with the rapid syngenetic precipitation of the sediment-hosted massive sulphides from an exhaled hydrothermal solution. Apart from a different genesis, the Kupferschiefer has many geological features that distinguish it from the submarine exhalative deposits; it has a very wide lateral distribution within a distinct stratigraphic horizon, the mineralisation is generally not massive, and the metals are zoned about an oxidised facies of the footwall sandstones (see Jung and Knitschke, 1976, for further details).

The Maubach and Mechernich Pb–Zn deposits in northwest Germany are hosted by post-orogenic Triassic sandstones (Buntsandstein), which also unconformably overly the folded Variscan shales. The mineralisation, which consists of disseminated galena and very minor sphalerite in porous sandstones, is possibly similar to the sandstone-hosted deposits at Largentière, France, that have been interpreted by Samama (1976) as also being formed during the early diagenesis of post-orogenic sandstones. The Laisvall deposit, Sweden, may also have an origin comparable to that of Largentière (Rickard et al., 1979).

The geology of the carbonate-hosted lead–zinc deposits has been both generally discussed in this Handbook by Sangster (1976b), who distinguished between the Mississippi Valley-type and the Alpine-type, as well as in detail for particular areas by Hagni (1976), Hoagland (1976), Kyle (1981), and especially Laznicka (1981).

The Mississippi Valley-type deposits are hosted by reef or platform carbonates, and are uniformly considered to be epigenetic; the sulphides commonly fill cavities in brecciated carbonates, which may have been formed by a variety of means – solution collapse, karsting, dolomitisation as well as tectonic brecciation. These deposits generally lack many of the internal characteristics of the sediment-hosted, submarine exhalative sulphides – no stratiform mineralisation, little or no copper mineralisation, no distinctive metal zonation sequences, no hydrothermal alteration, as well as non-conformable (anomalous) and generally inhomogeneous Pb-isotope characteristics. In addition they are often associated with fluorite, a mineral which is rare in submarine exhalative mineralisation. The Irish deposits (Tynagh, Silvermines), in spite of their dominant cross-cutting style of mineralisation, are characterised by submarine exhalative features and are considered to be carbonate-hosted members of the class under discussion.

Carbonate-hosted, submarine exhalative deposits were most probably formed in relatively shallow-water depths (±100 m), and the low resultant hydrostatic pressure at the site of discharge will result in the second-boiling of the hydrothermal solution, prior to its exhalation (Finlow-Bates and Large, 1978; Fig.5 this paper). The consequent precipitation of sulphides in the feeder zone will thus be as cross-cutting veinlets and stockwork as well as cavity fillings. Subsequent to discharge of the hydrothermal fluid, any remaining metals in solution will be precipitated as stratiform sulphides (e.g. Upper G zone at Silvermines) or oxides (e.g. iron formation at Tynagh). The incorporation of the sulphides into cavities and fissures in the carbonates during subsequent lithification could lead to some of the diagenetic features described by Taylor and Andrew (1978) for the Silvermines deposit.

The Alpine-type deposits (Bleiberg, Raibl, Mezica, Salafossa; see Brigo et al., 1977) do not share many of the features that are considered to be characteristic of submarine exhalative mineralisation. Many of the so-called syngenetic features can be interpreted as deposition of sulphides in karst and large solution collapse cavities (Bechstädt, 1975). While it is acknowledged that the Alpine-type deposits may form a distinct class from the Mississippi Valley-type deposits (e.g. Sangster, 1976b), it is not considered that they are necessarily comparable to submarine exhalative mineralisation.

As discussed in the introduction, the definition of the class of deposits in question is based on the submarine exhalative theory. Other genetic theories, however, have been proposed for this class of deposits, and are briefly described below. Should the reader have any misgivings about the application of the submarine exhalative theory to any or all of the deposits in question, he is urged to follow up the alternative proposals in depth. Ancient or modern, the alternative theories have been, and are being, discussed at length in the literature.

Lindgren (1933, p.628) proposed an epigenetic model for the Rammelsberg deposit, which he described as an epigenetic vein. The mineralisation is considered to have occurred after lithification and folding of the host rocks. Taken in its historical context, this theory reflects the "hydrothermalist" views of the contemporary geologists and would not be seriously considered today for the ore types considered in this chapter. The present trend is a reinterpretation of epigenetic deposits in terms of a syngenetic origin, as is exemplified by Sangster's (1979) discussion of the Cobar deposits.

Williams (1978, 1979) has suggested an epigenetic model for the origin of the McArthur River deposit. In this model the base-metal mineralisation postdates the formation of diagenetic pyrite. The metals were introduced in sulphate-rich hydrothermal solutions and, after reduction of the sulphate to sulphide by organic carbon, were precipitated as a second generation of sulphides. It is important to realise that the mineralisation is considered to have occurred within about 100 m of the sediment–sea-water interface, prior to the lithification of the host sediments (Williams, 1979). This is essentially mineralisation during very early diagenesis of the sediments and, although Williams prefers to describe the model as being epigenetic, it is markedly different from the classical epigenetic model of Lindgren (1933). Halls et al. (1979) prefer the word "syndiagenetic" to describe the textures of some of the cavity-filling sulphides at Silvermines, Ireland, and they point out (p.B129) that "in contiguous parts of the same orebody there can be seen ore textures that range from those that are obviously epigenetic to those that are unequivocally syngenetic" [1].

Williams' model is based on very detailed mineralogical, geochemical and geological investigations of stratiform mineralisation at McArthur River. Before his conclusions which have been questioned (Scott and Lambert, 1979; Finlow-Bates, 1979c), can be generally applied to other deposits, further detailed investigations are obviously needed. Sulphide mineral paragenesis, determined on the basis of textural studies, does not necessarily indicate the sequence of mineralisation, or the time of mineralisation as compared to the time of deposition of the host sediments. In syngenetic ores the observed textural

[1] See Wolf's (1981) Chapter 1, Vol. 8 on terminologies and classifications, as well as Wolf (1976).

relationships of the sulphides reflect the diagenetic modification of the original sulphide gels and muds. R. Edmonds (pers. comm.) noted that the semantics of the argument bear many resemblances to those that raged during the "granitisation" controversy, about which Turner and Verhoogen commented (1960, p.365) "the validity of the assumption that a sequence of metasomatic effects observed in space actually reflects a sequence of stages of alteration in time may be questioned". It is clear, however, regardless of whether a submarine exhalative or an epigenetic theory (*sensu* Williams, 1978, 1979) is accepted, that many of the principles regarding the source and transport of the metals and sulphur are applicable to both models.

Origin of the metals and sulphur

Two of the possible sources of the metals in sediment-hosted, submarine exhalative deposits are:

(1) leached from underlying sediments and volcanics, and

(2) juvenile derivation from magmatic processes (volcanic or intrusive).

Bischoff and Seyfried (1978) have experimentally shown that sea water trapped in basalt tends to become increasingly acid at higher temperatures (300°C), and as such is able to leach metals from the basalt. Wedepohl (1974) concluded, on the basis of the average abundances of metals in various rock types, that Zn, Pb and Cu occur almost everywhere in the continental crust in sufficient quantities such that ore-forming solutions could be derived as a result of persistent leaching of these metals by circulating acid solutions.

The submarine exhalative sulphides are characterised by relatively homogeneous and uniform isotopic compositions of lead in galena and sphalerite. Isotopic homogenisation of lead would be favoured by the convective circulation of leaching solutions through a variety of rock types (Doe and Stacey, 1974). In those instances where the trace lead in the pyrite is relatively more radiogenic than that in the galena and sphalerite (e.g. McArthur River; Gulson, 1975), the pyrite is considered to have a different origin to that of the galena and sphalerite. However, the possibility that at least some of the lead and zinc is derived from an isotopically homogeneous magmatic source should not be ignored. At Sullivan evidence exists for at least a partial contribution of material derived from crustal melting and anatexis. The presence of the cassiterite–tourmaline association is reminiscent of the greisens found in many tin belts. Sullivan lies within the peri-Pacific tin province in which tin mineralisation is found associated with rhyolites, granites and pegmatites of various ages. The cassiterite at Sullivan is found in cross-cutting veins and is also contained with the stratiform sulphide ore (Freeze, 1966; Mulligan, 1975). Ethier and Campbell (1977) have proposed that the high concentrations of boron contained within the tourmaline-rich horizons of the Aldridge Formation sediments could have been introduced into the sedimentary environment by submarine exhalations. The generation of tin-bearing hydrothermal solutions may be related to the partial melting of continental crust (Garson and Mitchell, 1977). In the case of the Sullivan, the tin mineralisation is evidence for at least a partial contribution of magmatically derived hydrothermal solutions and metals to the ore body. (See Lehmann and Schneider, 1981.)

The potential source of the sulphur can also be discussed with reference to sulphur isotope studies on the deposits under review (see p.478, and Fig.2). Anger et al. (1966) first proposed a dual sulphur source to explain why the isotopic behaviour of pyrite differed from that of galena and sphalerite at Rammelsberg. Subsequently, Smith and Croxford (1973) and Carr and Smith (1977) have used this model to explain the isotopic features of the sulphides at McArthur River and Lady Loretta, respectively.

The source of the galena-S, sphalerite-S and pyrrhotite-S is interpreted to be deep-seated, which implies that it was introduced to the site of mineralisation by the same hydrothermal solution that also transported the metals. The ultimate origin of the sulphur is probably sea-water sulphate that was inorganically reduced at temperatures higher than 200°C during convective circulation through the underlying sedimentary prism (Hajash, 1975).

It is difficult to conceive of a magmatic (juvenile) origin of the sulphur in many of the deposits. Rye and Ohmoto (1974) have shown that it is virtually impossible for sulphide with $\delta^{34}S$ greater than +5‰ to have been precipitated from hydrothermal solutions containing magmatic (juvenile) sulphur, as the lower temperatures at the site of mineralisation will result in the sulphides being isotopically lighter.

The source of the pyrite-S is the biogenic reduction of sea-water sulphate in a reducing environment at the site of mineralisation. Bacteriological reduction of the sulphate will cause the resultant sulphide to be relatively enriched in the lighter ^{32}S isotope as a result of kinetic isotopic fractionation (Trudinger, 1976). Numerous factors can affect the degree of kinetic fractionation during reduction of sulphate to sulphide; notably sulphate concentration, temperature, pH, fO_2, concentration of bacteria, and nutrients available (Saxby, 1976). Sangster (1976a and Fig.4) has shown how the morphology and the presence or lack of sea-water circulation into the third-order basin (open and restricted basins respectively), in which the stratiform sulphides accumulate, affects the isotopic composition of the diagenetic pyrite-S. In a restricted basin the pyrite-S will isotopically become progressively heavier stratigraphically upwards. This is the model proposed for the pyrite at McArthur River (Smith and Croxford, 1973) and Lady Loretta (Carr and Smith, 1977). At Rammelsberg there is no evidence on the basis of the pyrite-S isotopes for a restricted basin, and the wide range and variable isotopic composition of the pyrite-S is probably related to different microenvironments at the site of biogenic pyrite formation.

The isotopic composition of the barite-S sediment-hosted deposits is generally the same as, or heavier than, that of the contemporaneous sea-water sulphate. The biogenic formation of sulphides in a restricted basin will result in the isotopic composition of the sea-water sulphate in the basin becoming increasingly isotopically heavier. If the barite sulphate is derived from sea water in a restricted basin, then it ought to reflect the

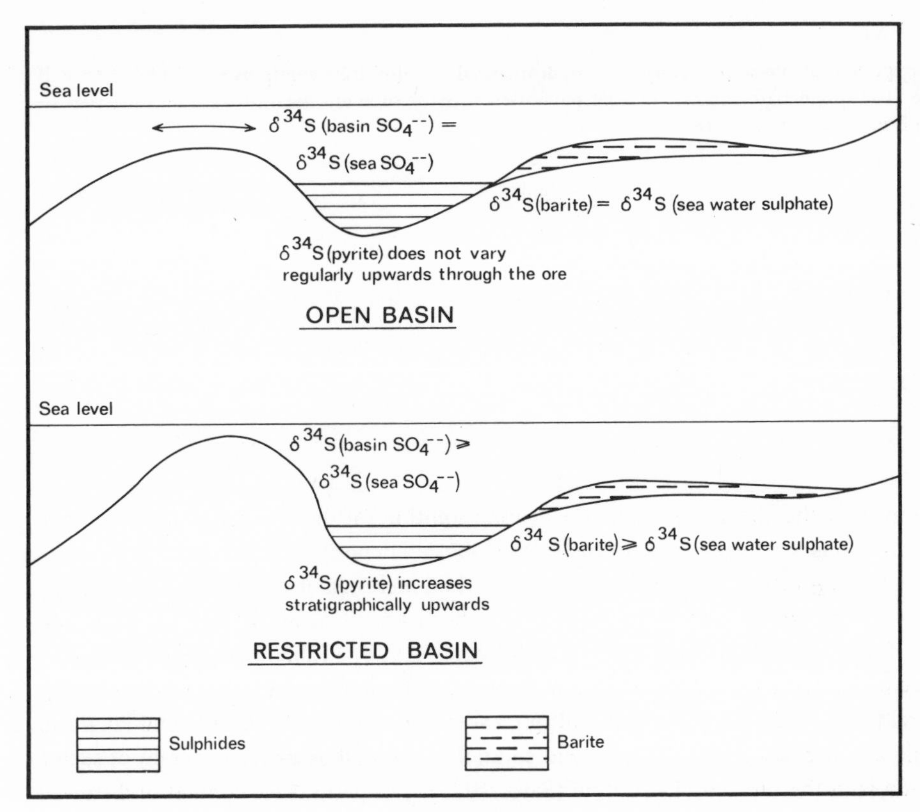

Fig.4. Sketch of an open basin and restricted basin to illustrate the effect of the basin morphology on the sulphur isotope composition of pyrite and barite relative to sea-water sulphate. (Modified from Sangster, 1976.)

isotopic variation of the sea-water sulphate. Such a situation may have occurred during the precipitation of the "Grauerz über Tage" barite deposit at Rammelsberg, which stratigraphically overlies the main sulphide bodies (see Hannak, 1981). This barite is isotopically heavier than both the contemporaneous sea-water sulphate and the barite associated with the sulphides. The minor amount of pyrite in this deposit is interpreted by Anger et al. (1966) to be biogenic in origin.

The ore-bearing solution

Due to the microscopic size of the sulphides in many sediment-hosted, submarine exhalative deposits, there are few published fluid-inclusion data and, consequently, it is not possible to directly identify the original nature of the ore solution as is the case for the Kuroko- and Mississippi Valley-type deposits (Roedder, 1976). It is presumed here, on the basis of the similar mineralogy, metal ratios and vertical and horizontal zonation

TABLE VI

The hypothetical physical chemistry of hydrothermal metal-bearing solutions as calculated by R.R. Large (1977) and Sato (1977) for volcanic-hosted massive sulphide deposits (The chemistry of sea water is included for comparison)

	Kuroko Ore Fluids		Sea water
	R.R. Large (1977)	Sato (1977)	
Temperature	150°–300°C	240°–250°C	20°–30°C
pH	3.5–5.0	3.7–4.7	8.0
fO_2	10^{-38}–10^{-50} atm	10^{-20}–10^{-70} atm	10^{-34} atm.
ΣS (reduced)	10^{-3}–10^{-2} *M*		10^{-2} *M*
NaCl concentration	0.5–3.0 *M*	1–3 *M*	0.5 *M*

patterns, that the chemistry of the ore-bearing solution for the sediment-hosted sulphide mineralisations is similar to that of the volcanic-hosted deposits.

On the basis of theoretical studies and fluid-inclusion investigations, the limiting chemical conditions for such an ore fluid have been described by R. Large (1977), Sato (1977), Shanks and Bischoff (1977) and Lydon (1978) (Table VI). Helgeson (1969) pioneered the study of thermodynamic equilibria in relation to the stability of various mineral phases and a hydrothermal solution, and he showed that Fe, Cu, Zn and Pb are all soluble as chloride complexes in an acid solution. The total base-metal content of such a solution is in the range of 1–10 ppm (Sato, 1977). The permissible content of dissolved reduced sulphur under the conditions outlined above is very low, and quite insufficient to provide enough sulphur for a massive sulphide deposit. The alternative proposal is that the metals are transported as soluble bisulphide complexes (Barnes, 1975). However, the solubility of metals complexed with bisulphides is less than that of the metal-chloride complexes, and the evidence from the fluid inclusion studies and recent geothermal brines suggests a chloride-rich solution (see Vaughan, 1976).

Shanks and Bischoff (1977), in a study of the Red Sea brines, agree that the concentration of metals greatly exceeds the concentration of the reduced sulphur in solution. However, the major soluble sulphur-bearing species, $NaSO_4^-$ and $MgSO_4^0$, are not affected by the chloride content. The sulphur is present as the soluble $MgSO_4^0$ and $NaSO_4^-$ phases which are in equilibrium with a very low concentration of H_2S and HS^-.

The possible origins of the solution are magmatic (juvenile), formation (connate) waters [1], and circulating sea water. Sato (1977) argues for a magmatic origin for the ore solution; Lydon (1978) prefers formation waters, although these are ultimately of sea-water origin modified by interaction with the host rocks; and Shanks and Bischoff (1977), Spooner (1977) and Sheppard (1977) conclude that circulating sea water is the

[1] For compaction fluids, see summary by Wolf (1976)

main leaching agent for the volcanic-hosted submarine exhalative deposits.

The proponents for the sea-water origin use the stable isotope (S, H and O) data and the geochemistry of the hydrothermal alteration associated with some volcanic-hosted deposits to support their theory. Sato (1977) opposes the sea-water origin of the ore-bearing solution for the Kuroko deposits on the basis of the lead isotopes, the lack of geochemical evidence for significant leaching in the vicinity of the mineralisations, and the inability of this model to provide an explanation for the association of the Kuroko deposits with a specific phase of Miocene felsic volcanism. He suggests that the stable isotope data could be explained by isotopic exchange with sea water during and/or after mineralisation.

With regard to the sediment-hosted stratabound ores, the lack of any significant volcanic centres in the near vicinity of, and contemporaneous with, the mineralisation argues against the magmatic origin for a major proportion of the metal-bearing solutions. Sea water, modified by interaction with the rocks through which it passes, is preferred as the transporting medium. Evaporite beds in the vicinity of some sediment-hosted deposits (e.g. McArthur River; Walker et al., 1978) have been cited as a source for the chloride-rich brines being derived from dissolution of evaporites. The presence of evaporites, however, has not been established in every case (e.g. Meggen, Rammelsberg, Sullivan). Although saline formation waters derived from evaporites might contribute to the leaching solution (e.g. the origin of the Red Sea brines), it is not considered that the presence of evaporites is essential to the origin of the metal-bearing hydrothermal solutions (see White, 1967).

Convective circulation of the metal-bearing solution

The model of sea water circulating within a convective cell beneath the sea floor has been described by Solomon (1976) for the volcanic-hosted deposits. The cell is calculated to have a diameter of about 10 km and to extend to a depth of 4–5 km. He observed that the volcanic deposits are often grouped in clusters spaced at 5–15 km intervals, and that this spacing may be related to the size of the convection cell.

As compared to the volcanic-hosted ores, the sediment-hosted deposits generally contain more metal and are individually more isolated from each other. These features may be explained as follows by a greater size of the convective cell.

Russell (1978) proposed that convective cells extended to depths of up to 14 km through the Lower Palaeozoic geosynclinal prism beneath the central Irish Carboniferous basin. At these depths the permeability of the prism is enhanced by fracturing and jointing. Cell spacing is calculated by Russell to have been 35–50 km. Only a fraction of 1% of the total zinc available in the source rocks (average content of 100 ppm Zn), in which the cell was active, would be required to provide sufficient zinc for the formation of the biggest Irish deposit (Navan). The approximate volume of this cell is an order of magnitude greater than the cell size calculated by Solomon (1976) for the volcanic-hosted deposits.

A heat source is required for the convective circulation of a hydrothermal solution, and on the surface this will be manifested as an area of anomalously high heat flow. It has been shown that most of the sediment-hosted deposits are associated with minor contemporaneous igneous activity, which was often regionally persistent both before and after the mineralising event. It is considered that this igneous activity is the manifestation of a regionally anomalous geothermal gradient, possibly related to a mantle hot spot, that would provide sufficient heat energy over a long period of time to convectively circulate very large volumes of connate water.

Sato (1977) has argued against the convective-cell hypothesis for the volcanic-hosted Kuroko deposits on the grounds that the intrusive rhyolitic domes, with which these deposits are associated, can not supply sufficient heat to convectively circulate enough of the metal-bearing solution to the site of precipitation. Therefore, he argues for a derivation of juvenile hydrothermal solutions direct from the magma. It is proposed here that the heat source to which the sediment-hosted deposits are related is a more persistent and regionally anomalous feature as compared to a relatively short-lived, near surface volcanic centre. The actual site of discharge of the metal-bearing solution is probably controlled by the presence of deep-seated fractures that penetrate these convective cells and were active during mineralisation. Movement on these fractures may seismically pump the solutions to the sea floor (Sibson et al., 1975). Fundamental fault zones and lineaments commonly separate crustal blocks that are subsiding at different rates, and this results in the sediment-hosted, submarine exhalative deposits being located at the margins of first- and second-order basins.

Elder (1976, fig.15.6) and Henley and Thornley (1979) have demonstrated that a circulatory hydrothermal system will be formed within permeable rock above a heat source. If fluid from this convective system is released at the surface, then this point of discharge becomes the focus of energy output from the whole convective cell, which could explain why sediment-hosted, submarine exhalative deposits tend to be individually isolated. Furthermore, the discharged fluid is derived from close to the heat source and thus could, according to Henley and Thornley (1979), contain a small magmatic component (Fig.5).

Precipitation of the metal sulphides

Major changes in the chemistry of the metal-bearing hydrothermal solution occur as it mixes with the sea water contained in the poorly lithified sediments at the exhalative centre, and after discharge into the sea water itself. Anderson (1973) and R. Large (1977) have shown that at solution temperatures of 100–300°C (a reasonable temperature range in the immediate vicinity of the exhalative centre) lead and zinc will be precipitated from acid chloride-rich hydrothermal solutions as a result of the following: a decrease in temperature, dilution of the solution, an increase in pH towards neutrality, and an increase in concentration of reduced sulphur. These four variables are discussed below.

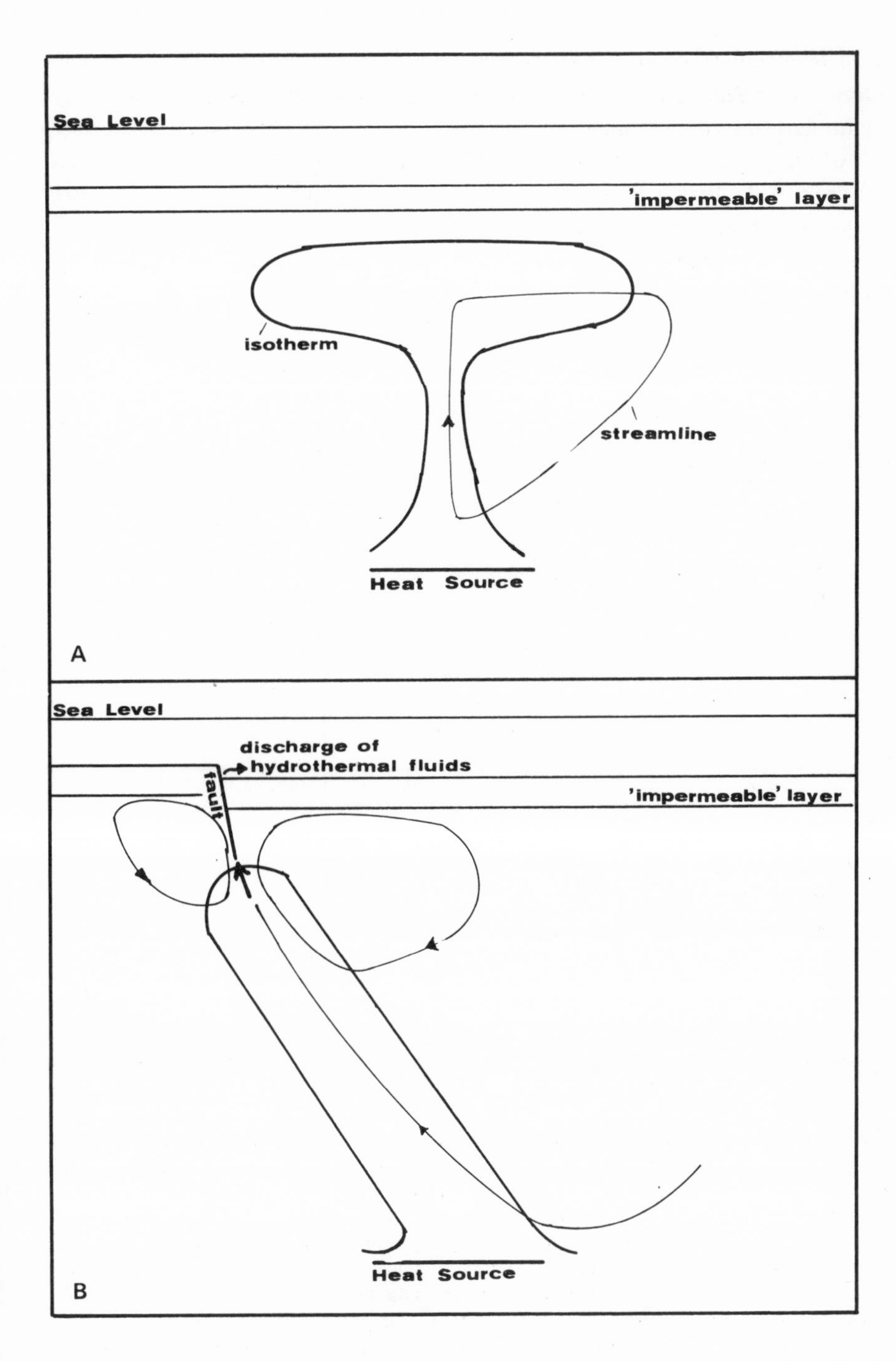

Fig.5. Convective flow in a sedimentary prism as generated by heat source at depth. A. No discharge. B. During discharge. (After Henley and Thornley, 1979 and Elder, 1976.)

Decrease in temperature as a result of boiling prior to the discharge of the hydrothermal solution (water depth and its control on the style of mineralisation). The possibility of the hydrothermal solution boiling prior to reaching the sediment–sea-water interface as a result of the progressive decrease in hydrostatic pressure has been discussed by Haas (1971), Ridge (1973), Finlow-Bates and Large (1978) and Henley and Thornley (1979). When the vapour pressure of the solution exceeds the hydrostatic pressure, the solution will boil and its temperature will concurrently decrease along the boiling point curve (Fig. 6) due to the loss of heat by vapourisation. The boiling may be explosive if the drop in pres-

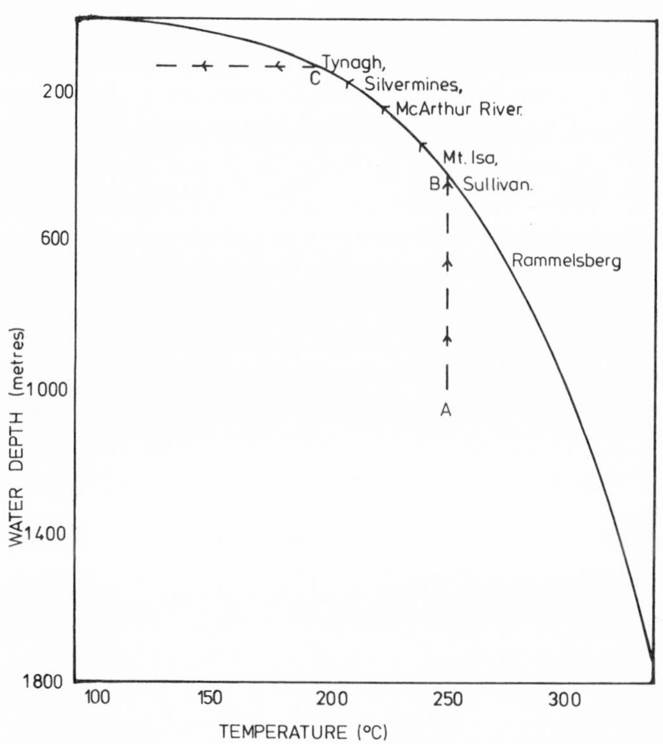

Fig. 6. The boiling point curve for a brine solution containing 5 wt.% NaCl at various depths of sea water (hydrostatic pressure). (From Haas, 1971.) A hypothetical solution at point A is under hydrostatic pressure within a fracture zone in the upper crust, and at a temperature of 250°C. As the solution rises to point B, its temperature remains constant (very low rate of heat loss to surrounding rocks). At point B, the solution commences to boil (vapour pressure is equal to hydrostatic pressure). The solution continues to rise to C, adiabatically boiling and decreasing in temperature (loss of latent heat of vapourisation). At point C (sea floor), the solution is discharged and rapidly decreases in temperature as a result of mixing with sea water. The solutions from which Tynagh, Silvermines and McArthur River were formed would probably have boiled in the feeder system prior to discharge; the solutions from which Mount Isa and Sullivan were formed would have boiled very close to the sea floor; solutions from which Rammelsberg was formed would not have boiled before discharge.

sure is sudden as a result of fracturing of the acquifer rocks, in which case there may be an associated brecciation of the host rocks at the site of mineralisation. Some silica and the most insoluble of the sulphides will be precipitated as cross-cutting mineralisation from this solution because of the temperature decrease. The remaining solution will be discharged onto the sea floor, and the subsequent sudden drop in temperature of the solution as a result of interaction with the sea water, will cause an immediate precipitation of the metal sulphides and silica on the sea floor.

The hydrostatic pressure (water depth) at the discharge site will determine whether or not boiling, and the consequent precipitation of sulphides, occurs in the exhalative vent prior to discharge. The relationships between water depth and the style of mineralisation are illustrated by various sediment-hosted deposits in Figs.6 and 7. Chalcopyrite, being the least soluble sulphide, is often precipitated during boiling in the feeder vents. Lead and zinc may remain dissolved in the solution until it has been discharged onto the sea floor, then galena and sphalerite are precipitated to constitute a part of the stratiform mineralisation.

Deposits in a deeper-water environment tend to have the base-metal sulphides represented in the stratiform facies and only minor chalcopyrite in the underlying cross-cutting

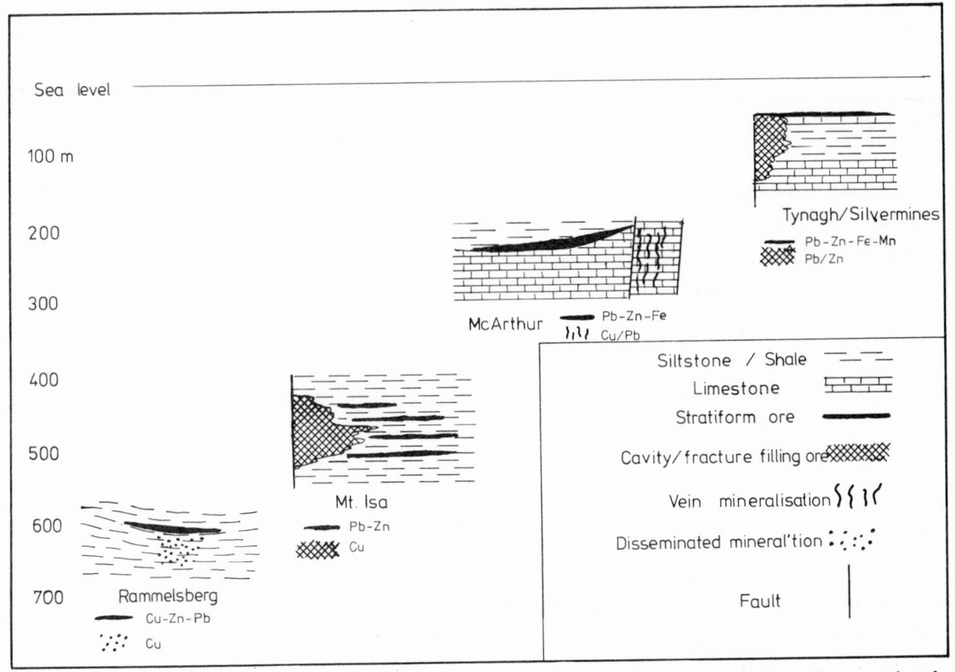

Fig.7. Relationship between style of mineralisation (stratiform and cross-cutting) to the postulated depth of water which was prevailing at the site of mineralisation during exhalation of the ore-bearing solutions. (Modified from Finlow-Bates and Large, 1978.)

facies (e.g. Rammelsberg), mineralisations in a shallower-water milieu have a large proportion of epigenetic, cross-cutting metalliferous parts with a minor stratiform facies, which may only be comprised of iron and manganese oxides (e.g. Tynagh). Thus, deposits that have grossly different proportions of stratiform to cross-cutting metalliferous facies but are in other respects similar, may be formed by submarine-exhalative events at different hydrostatic pressures.

It should be emphasised that the boiling is adiabatic, there being no external heat source, and that the hydrothermal solution is continually moving upwards to the discharge centre. For solutions with a chloride concentration considered to be most representative from the fluid inclusion data (e.g. 5% NaCl solution), it is most unlikely that the solution will boil to such an extent that the solubility product of halite will be exceeded. The absence of halite in submarine exhalative deposits, however, has been incorrectly cited (e.g. Ridge, 1973) as being indicative of the lack of boiling during sulphide precipitation.

As a result of the decrease in temperature of the hydrothermal solution, silica will also be precipitated together with the base-metal sulphides, and with the peripheral iron sulphide and oxide facies. Silicification is a common alteration component in the feeder zones (e.g. Kniest) and bedded cherts are sometimes found as lateral equivalent to the stratiform sulphide facies (e.g. Lady Loretta, Tynagh). The silicification and brecciation in the 'silica-dolomite' at Mount Isa may also be the result of submarine exhalative activity (Finlow-Bates, 1979a).

Dilution of the hydrothermal solution. The dilution of the hydrothermal solution by sea water will result in a decrease in temperature and an increase in pH. Sato (1972) and Turner and Gustafson (1978) have discussed the effects of mixing a chloride-rich hydrothermal solution with sea water, and have shown that the resulting range of fluid densities dictate the behaviour of the solution after exhalation. The main variables that control the density (i.e. prior to sea-water mixing) of such metal-bearing solutions are the temperature and salinity.

If the discharged hydrothermal solution is less dense than sea water (e.g. very hot, moderately salty, 2*M* NaCl), then it will be immediately dispersed and diluted by the sea water (type 3 of Sato, 1972, and fig.38 in Sangster and Scott, 1976), and the resultant precipitate will accumulate around the discharge centre. If the solution is denser than sea water (type 1 of Sato, 1972), it could flow as a density current away from the discharge centre along the sediment–sea-water interface, and thus retain its original chemical composition for a longer time, and be spread further, than the type 3 solutions. A solution with characteristics between the two extremes (i.e. approximately the same density as sea water, type 2 of Sato, 1972) may at first be less dense than sea water, but subsequently becoming, as a result of decreasing temperatures, relatively denser. Turner and Gustafson (1978) have shown that such a solution initially forms a buoyant plume above the point of discharge which, as the temperature of the solution falls and its density increases, subsequently collapses and releases the solution into a sidewards spreading flow. A new plume is then formed and the cycle repeats itself, so that there is an oscillatory flow of the solution away from the exhalative centre. On dilution with sea water, sulphides will be precipitated in the immediate vicinity of the exhalative centre and the cyclic nature of the plume collapse could thus result in the stratification of the ore deposit.

Proximal and distal deposits. Turner and Gustafson (1978), on the basis of experimental evidence, claim that hot, dense and saline brines (type 1 of Sato, 1972) could flow under gravity for long distances (tens to hundreds of kilometres) from the discharge centre with only minimal mixing and dilution by sea water. When trapped in a submarine depression (third-order basin) the solution would accumulate in stratified layers with the denser solution always flowing to the bottom of the pool. Sulphides and chert would thus be precipitated as a stratiform deposit with no footwall mineralisation or alteration, which are some of the features recognised by R. Large (1977) and Plimer (1978) as being indicative of a distal deposit. However, for the class of deposit in question, these indications that the stratiform sulphides were deposited in the immediate vicinity of the hydrothermal discharge centre are usually present. In addition, the metals are commonly zoned around a point that corresponds to the discharge centre. For those deposits where this evidence is lacking their geological relationships to the surrounding host rocks may be structurally complicated (e.g. Meggen).

It should be noted here that, in this review, the terms "distal" and its antonym "proximal" are used to describe the relative distance of the stratiform mineralisation from the discharge centre. This usage is prefered to that by which these terms describe the relative distance of the stratiform mineralisation from the volcanic centre that was the source of the tuffites commonly found in the host sequence (Plimer, 1978). The tuffites are undoubtedly distal, and no volcanic centres have been identified in the vicinity of sediment-hosted, submarine exhalative deposits. The reasons for the apparent lack of distal sediment-hosted deposits, indeed all submarine exhalative deposits, may be:

(a) The hydrothermal solutions were discharged into third-order depressions from which it was impossible for the dense, chloride-rich fluids (type I of Sato, 1972) to escape by gravity flow. Should such depressions be too small then the ore solution may overflow onto the surrounding sea floor (K.H. Wolf, written comm., 1979), in which case the marginal banded (interbedded sulphides and sediment) ore found at Rammelsberg may reflect this overflow activity.

(b) The hydrothermal solutions were of Sato's (1972) types 2 and/or 3 (i.e. those that are lighter or have approximately the same density as sea water); in which case the dilution, temperature decrease, and consequent precipitation of the metal sulphides will occur close to the exhalative centre. This is the scenario preferred by Solomon and Walshe (1979) for ore-forming hydrothermal solutions, and is described by them as the buoyant-plume model.

Increase of the pH towards neutrality. An increase of the hydrothermal solution's pH is to be expected on dilution by sea water. The pH of the hydrothermal solution prior to dilution is thought to be in the range 3.5–5.5 and that of sea water at 25°C is 8.0. At temperatures up to 250°C, which could be expected in the fumarolic centre, the neutral pH of sea water will be about 6.0.

Increase in reduced sulphur. Shanks and Bischoff (1977) have shown that as the temperature of the chloride-rich hydrothermal solution decreases to about 60°C, the concentration of H_2S and HS^- in that solution increases as that of $MgSO_4$ decreases. Additionally, in a euxinic environment such as in a reducing basin, sulphide biogenically reduced from sea-water sulphate is also present.

All of the changes discussed above can affect the hydrothermal solution during its discharge into the sea and while spreading out from the exhalative centre.

Helgeson (1969) and R. Large (1977), among others, demonstrated that the sulphides will precipitate out in order of their solubility constants. As the hydrothermal solution becomes oversaturated with sulphides for any of the above reasons, chalcopyrite is the first to precipitate and, as the solution cools, is followed by sphalerite and galena. This agrees with the earlier observation that copper is characteristically restricted to the core of a zoned metal sequence with peripheral lead–zinc and iron zones. Only Hilton, Australia (Mathias et al., 1973), where a copper-rich zone overlies the lead–zinc ore, appears to contradict this trend. The problem of zoning is undoubtedly very complex, and the possibility of variations in the hydrothermal solution chemistry as a major cause of zoning must not be ignored; this could well be the explanation of apparent contradictions such as at Hilton.

Solomon and Walshe (1979) suggest that chalcopyrite is rapidly precipitated at the discharge centre. Sphalerite and galena, which are subsequently precipitated, are entrained in the buoyant hydrothermal plume and are swept upwards to be deposited on top of the chalcopyrite. It is not, however, possible to explain the differences between lead and zinc in laterally and vertically zoned deposits (Table III).

Laterally zoned deposits are usually well banded, and each bed is visualised as representing a pulse of discharged hydrothermal fluid, and the lateral zonation within each bed as reflecting the temperature gradient away from the exhalative centre.

The repeated discharge of hydrothermal solutions in a series of pulses in a submarine environment could lead to the characteristic banding seen in so many of the sediment-hosted, submarine exhalative deposits. Finlow-Bates (1979b) has shown that each band at Mount Isa is comprised of a galena-rich core, which is bounded laterally and vertically by sphalerite and then pyrrhotite. Each zoned band represents the sulphide deposition from one such pulse of hydrothermal activity. Similar cyclicity has been reported by Zimmermann (1976, fig.6) from the banded ore at Sullivan.

Relationship of barite to stratabound, sediment-hosted sulphides

Stratabound barite deposits are associated with many stratabound sulphide deposits and are commonly found in zones peripheral to, or overlying, the sulphides. It has already been noted that the sulphur isotopes of these barite deposits are indicative of sea-water sulphate, and the barite is formed as the result of mixing of acidic barium-bearing hydrothermal solutions with sea water. The barium is probably transported together with the metals to the site of mineralisation in the hydrothermal solution.

Palaeogeographic reconstructions (e.g. Silvermines, Lady Loretta, Meggen) demonstrate that the barite was deposited on topographically higher, and in chemically more oxidising, environments as compared to the sulphide depository milieus. At Silvermines (Macgobar), the barite is associated with hematite and jasper (ferruginous chert) indicative of an oxidising sedimentary environment in contrast to that prevailing in the sulphide depositories (Mogul) where massive pyrite and dolomitic argillite host the base-metal sulphides. In the Tom deposit, interlaminated beds of sulphides and barite indicate that there is an overlap or transition of chemical environments in which both sulphides and barite are precipitated. The lack of pyrite at the Tom deposit may be indicative of a more oxidising environment of deposition in which there was no biogenic reduction of sulphate.

It should be pointed out, however, that just as several massive sulphide deposits are not associated with any barite (e.g. Mount Isa, Sullivan), there are many stratabound, sediment-hosted barite deposits with no associated base-metal sulphides. Notable among these are the Devonian barites in Nevada and Arkansas, U.S.A. (Shawe et al., 1967; Zimmermann, 1976). Similar accumulations of barite are also found in the Devonian of Germany (Krebs and Gwosdz, in press).

CONCLUSIONS

Sediment-hosted, submarine exhalative Pb-Zn deposits comprise a particular class of mineralisation that can be readily distinguished from other types of deposit. Mineralisation of this type is regionally restricted to embayments into passive continental margins or intracontinental basins that were formed during a "hot spot" induced rifting event. The host sediments reflect a low-energy, locally euxinic, pre-flysch environment. Penecontemporaneous faulting in the basin resulted in the occasional development of conglomerates and breccias. Evidence for volcanic activity is restricted to the presence of minor tuff horizons in the host sequence.

Fig.8 is an attempt to summarise some of the genetic concepts that have been mentioned in the text. A submarine exhalative origin is suggested as the best theory to explain the various characteristics of the mineralisation. The stratiform sulphide mineralisation is considered to have formed on the sea floor subsequent to the discharge of the metal-bearing hydrothermal solution. The alteration zone in the footwall sediments is often

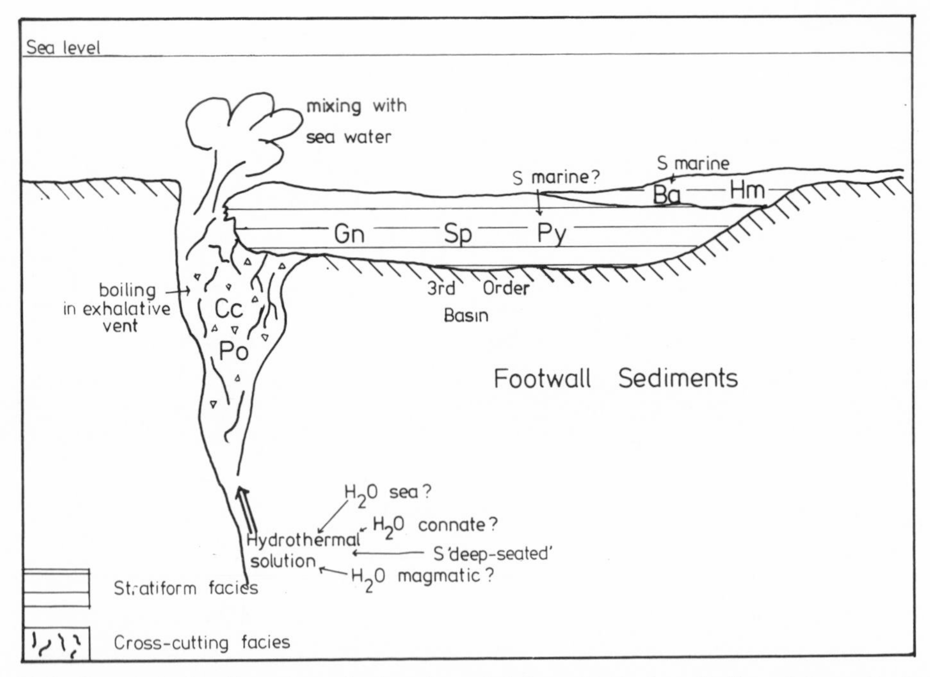

Fig.8. Hypothetical sediment-hosted submarine exhalative sulphide deposit. *Cc* – chalcopyrite; *Po* – pyrrhotite; *Gn* – galena; *Sp* – sphalerite; *Py* – pyrite; *Ba* – barite; *Hm* – haematite. In the figure, the location of the mineral symbols corresponds to the general sequence of mineral zonation.

associated with brecciation and minor sulphide mineralisation, and probably represents the exhalative vent through which the submarine exhalative solutions flowed to the sea floor. Adiabatic boiling in the exhalative vent will result in a temperature drop, and the consequent precipitation of sulphides and silica as cross-cutting mineralisation and alteration in the footwall. Explosive boiling, due to a sudden lowering of the confining pressure of the hydrothermal solution, can lead to brecciation of the footwall. Whether or not boiling of the solution occurs will, to a great extent, be determined by the prevailing water depth (hydrostatic pressure) at the site of mineralisation.

ACKNOWLEDGEMENTS

This paper summarises some of the results of a project on stratabound ore deposits being undertaken at the Federal Institute for Geosciences and Natural Resources, Hannover, F.R.G., and at the Institute of Geology and Palaeontology, Technical University of Braunschweig, F.R.G. The writer is indebted to all the members of the project team for stimulating discussions and exchange of ideas, especially W. Klau (project manager), W. Krebs, and T. Finlow-Bates, and to the numerous company geologists who patiently clarified the geology of their mines. The suggestions of Dr. K.H. Wolf, who critically read the draught manuscript, are gratefully acknowledged.

REFERENCES

Addy, S.K. and Ypma, P.J.M., 1977. Origin of the massive sulfide deposits at Ducktown, Tennessee: an oxygen, carbon and hydrogen isotope study. *Econ Geol.*, 72: 1245–1268.

Anderson, G.M., 1973. The hydrothermal transport and deposition of galena and sphalerite near 100°C. *Econ. Geol.*, 68: 480–492.

Anger, G., Nielsen, H., Puchelt, H. and Ricke, W., 1966. Sulphur isotopes in the Rammelsberg ore deposit (Germany). *Econ. Geol.*, 61: 511–536.

Barnes, H.L., 1975. Zoning of ore deposits: types and causes. *Trans. R. Soc. Edinburgh*, 69: 295–311.

Bechstädt, Th., 1975. Lead–zinc ores dependent on cyclic sedimentation (Wetterstein-limestone of Bleiberg–Kreuth, Carinthia, Austria). *Miner. Deposita*, 10: 234–248.

Bischoff, J.L. and Seyfried, W.E., 1978. Hydrothermal chemistry of sea water from 25 to 350°C. *Am. J. Sci.*, 278: 838–860.

Blusson, S.L., 1976. Selwyn Basin, Yukon and District of Mackenzie. In: *Report of Activities, Part A, Geol. Surv. Can. Pap.*, 76-19: 131–132.

Both, R.A. and Rutland, R.W.R., 1976. The problem of identifying and interpreting stratiform ore bodies in highly metamorphosed terrains: the Broken Hill example. In: K.H. Wolf (Editor), *Handbook of Strata-Bound and Stratiform Ore Deposits, 4.* Elsevier, Amsterdam, pp.261–326.

Brigo, L., Kostelka, L., Omenetto, P., Schneider, H.-J., Schroll, E., Schulz, O. and Strucl, I., 1977. Comparative reflections on four Alpine Pb–Zn deposits. In: Klemm, D.D. and Schneider, H.-J. (Editors), *Time- and Strata-Bound Ore Deposits.* Springer, Berlin, pp.274–293.

Burke, K. and Dewey, J.F., 1973. Plume-generated triple junctions: key indicators in applying plate tectonics to old rocks. *J. Geol.*, 81: 406–433.

Burke, K. and Sawkins, F.J., 1978. Were the Rammelsberg, Meggen, Rio Tinto and related ore deposits formed in a Devonian rifting event? *Econ. Geol.*, 73: 916–917 (Abstract).

Buschendorf, F., Nielsen, H., Puchelt, H. and Ricke, W., 1963. Schwefel-Isotopenuntersuchungen am Pyrit–Sphalerit–Baryt-Lager Meggen/Lenne (Deutschland) und an verschiedenen Devon-Evaporiten. *Geochim. Cosmochim. Acta*, 27: 501–523.

Campbell, F.A., Ethier, V.G., Krouse, H.R. and Both, R.A., 1978. Isotopic composition of sulphur in the Sullivan orebody, British Columbia. *Econ. Geol.*, 73: 248–268.

Carne, R.C., 1976. Stratabound barite and lead-zinc-barite deposits in the eastern Selwyn Basin, Yukon Territory. *Dep. Indian and Northern Affairs, Open File Rep.*, EGS 1976-16: 41 pp.

Carr, G.R. and Smith, J.W., 1977. A comparative isotopic study of the Lady Loretta zinc–lead–silver deposit. *Miner. Deposita*, 12: 105–110.

Coomer, P.G. and Robinson, B.W., 1976. Sulphur and sulphate-oxygen isotopes and the origin of the Silvermines deposits, Ireland. *Miner. Deposita*, 11: 155–169.

Croxford, N.J.W., 1964. Origin and significance of volcanic potash rich rocks from Mount Isa. *Trans. Inst. Min. Metall.*, 74: 33–43.

Croxford, N.J.W., 1968. A mineralogical examination of the McArthur River lead–zinc–silver deposit. *Australas. Inst. Min. Metall. Proc.*, 226: 97–108.

Croxford, N.J.W. and Jephcott, S., 1972. The McArthur lead –zinc deposit, N.T. *Australas. Inst. Min. Metall. Proc.*, 243: 1–26.

Dawson, K.M., 1977. Regional metallogeny of the northern Cordillera. In: *Report of Activities, Part A, Geol. Surv. Can. Pap.*, 77-1A: 1–4.

Degens, E.T. and Ross, D.A., 1976. Strata-bound metalliferous deposits found in or near active rifts. In: K.H. Wolf (Editor), *Handbook of Strata-Bound and Stratiform Ore Deposits, 4.* Elsevier, Amsterdam, pp.165–202.

Derry, D.R., Clark, G.R. and Gillatt, N., 1965. The Northgate basemetal deposit at Tynagh, County Galway, Ireland. *Econ. Geol.*, 60: 1218–1237.

Dickinson, W.M., 1974. Plate tectonics and sedimentation. *Sec. Econ. Paleontol. Mineral. Spec. Publ.*, 22: 1–27.

Doe, B.R. and Stacey, J.S., 1974. The application of lead isotopes to the problems of ore genesis and ore prospect evaluation: A review. *Econ. Geol.*, 69: 757–776.

Dornsiepen, U., 1977. *Zum geochemischen und petrofaziellen Rahmen der hangenden Schichten des Schwefelkies–Zinkblende–Schwerspatlagers von Meggen (Westf.)*, Thesis, T.U. Braunschweig, 77 pp. (unpublished).

Dunnet, D., 1976. Some aspects of the panantartic cratonic margin in Australia. *Philos. Trans. R. Soc. London*, Ser. A, 280: 641–654.

Elder, J., 1976. *The Bowels of the Earth.* Oxford Univ. Press, 222 pp.

Ethier, V.G. and Campbell, F.A., 1977. Tourmaline concentration in Proterozoic sediments of the southern Cordillera of Canada and their economic significance. *Can. J. Earth Sci.*, 14: 2348–2363.

Ethier, V.G., Campbell, F.A., Both, R.A. and Krouse, H.S., 1976. Geological setting of the Sullivan orebody and estimates of temperatures and pressure of metamorphism. *Econ. Geol.*, 71: 1570–1588.

Evans, A.M., 1976. Genesis of Irish base-metal deposits. In: K.H. Wolf (Editor), *Handbook of Strata-Bound and Stratiform Ore Deposits, 5.* Elsevier, Amsterdam, pp.231–256.

Finlow-Bates, T., 1979a. Chemical mobilities in a submarine exhalative hydrothermal system. *Chem. Geol.*, 27: 65–83.

Finlow-Bates, T., 1979b. Cyclicity in the lead–zinc–silver-bearing sediments at Mount Isa mine, Queensland, Australia, and rates of sulfide accumulation. *Econ. Geol.*, 74: 1408–1419.

Finlow-Bates, T., 1979c. Studies of the base metal sulfide deposits at McArthur River, Northern Territory, Australia: II. The sulfide-S and organic-C relationships of the concordant deposits and their significance – a discussion. *Econ. Geol.*, 74: 1697–1699.

Finlow-Bates, T. and Large, D.E., 1978. Water depth as a major control on the formation of submarine exhalative ore deposits. *Geol. Jahrb. Reihe D*, 30: 27–39.

Freberg, R.A., 1976. Tom property, Macmillan Pass, Yukon Territory (abstract). *Can Inst. Min. Metall. Meet.*, Vancouver, October 1976.

Freeze, A.C., 1966. On the origin of the Sullivan orebody, Kimberley, British Columbia. *Can. Inst. Min. Metall. Spec. Vol.*, 8: 263–294.

Gabelmann, J.W., 1976. Classifications of strata-bound ore deposits. In: K.H. Wolf (Editor), *Handbook of Strata-Bound and Stratiform Ore Deposits, 1.* Elsevier, Amsterdam, pp.79–110.

Garson, M.S. and Mitchell, A.H.G., 1977. Mineralisation at destructive plate boundaries: a brief review. In: *Volcanic processes in ore genesis. Geol. Soc. London Spec. Publ.*, 7: 81–97.

Gilmour, P., 1962. Notes on a non-genetic classification of copper deposits. *Econ. Geol.*, 57: 450–455.

Gilmour, P., 1971. Strata-bound massive pyritic sulfide deposits – review. *Econ. Geol.*, 66: 1239–1244.

Gilmour, P., 1976. Some transitional types of mineral deposits in volcanic and sedimentary rocks. In: K.H. Wolf (Editor), *Handbook of Strata-Bound and Stratiform Ore Deposits, 1.* Elsevier, Amsterdam, pp.111–160.

Glikson, A.Y., Derrick, G.M., Wilson, I.H. and Hill, R.M., 1976. Tectonic evolution and crustal setting of the middle Proterozoic Leichhardt River Fault Trough, Mount Isa region, northwestern Queensland. *B.M.R. J. Australas. Geol. Geophys.*, 1: 115–130.

Greig, J.A., Baardsgaard, H., Cumming, G.L., Folinsbee, R.E., Krouse, H.R., Ohmoto, H., Sasaki, A. and Smejkal, V., 1971. Lead and sulphur isotopes of the Irish base-metal mines in Carboniferous carbonate host rocks. *Soc. Min. Geol. Jpn. Spec. Issue*, 2: 84–92.

Gulson, B.L., 1975. Differnces in lead isotopic composition in the stratiform McArthur zinc–lead–silver deposit. *Miner. Deposita*, 10: 277–286.

Gunzert, G., 1969. Altes und neues Lager am Rammelsberg bei Goslar. *Erzmetall*, 22: 1–10.

Gwosdz, W. and Krebs, W., 1977. Manganese halo surrounding Meggen ore deposit, Germany. *Trans-Inst. Min. Metall. Sect. B*, 86: 73–77

Gwosdz, W., Krüger, H., Paul, D. and Baumann, A., 1976. Die Liegend-Schichten der devonischen Pyrit- und Schwerspat-Lager von Eisen (Saarland), Meggen und Rammelsberg. *Geol. Rundsch.*, 63: 74–93.

Haas, J.L., 1971. The effect of salinity on the maximum thermal gradient of a hydrothermal system at hydrostatic pressure. *Econ. Geol.*, 66: 551–556.

Hagni, R.D., 1976. Tri-state ore deposits: the character of their host rocks and their genesis. In: K.H. Wolf (Editor), *Handbook of Strata-Bound and Stratiform Ore Deposits, 6.* Elsevier, Amsterdam, pp.457–494.

Hajash, A., 1975. Hydrothermal processes along mid-ocean ridges: an experimental investigation. *Contrib. Mineral. Petrol.*, 53: 205–226.

Halls, C., Boast, A.M., Coleman, M.L. and Swainbank, I.G., 1979. Discussion of Silvermines orebodies, Ireland. *Trans. Inst. Min. Metall. Sect. B*, 88: 129–131.

Hannak, W., 1981. Genesis of the Rammelsberg ore deposit near Goslar, Upper Harz, Federal Republic of Germany. In: K.H. Wolf (Editor), *Handbook of Strata-Bound and Stratiform Ore Deposits*, 9. Elsevier, Amsterdam, pp. 551–642.

Harrison, J.E., 1972. Precambrian Belt basin of northwestern United States: its geometry, sedimentation and copper occurrences. *Geol. Soc. Am. Bull.*, 83: 1215–1240.

Helgeson, H.C., 1969. Thermodynamics of hydrothermal systems at elevated temperatures and pressures. *Am. J. Sci.*, 267: 729–804.

Henley, R.W. and Thornley, P., 1979. Some geothermal aspects of polymetallic massive sulfide formation. *Econ. Geol.*, 74: 1600–1612.

Hoagland, A.D., 1976. Appalachian zinc–lead deposits. In: K.H. Wolf (Editor), *Handbook of Strata-Bound and Stratiform Ore Deposits, 6.* Elsevier, Amsterdam, pp. 495–534.

Johnson, I.R. and Klingner, G.D., 1975. Broken Hill ore deposit and its environment. In: Knight, C.L. (Editor), *Economic geology of Australia and Papua New Guinea. Australas. Inst. Min. Metall. Monogr.*, 5: 476–491.

Jung, W. and Knitschke, G., 1976. Kuperfschiefer in the German Democratic Republic (G.D.R.) with special reference to the Kupferschiefer deposit in the southwestern Harz foreland. In: K.H. Wolf (Editor), *Handbook of Strata-Bound and Stratiform Ore Deposits, 6.* Elsevier, Amsterdam: pp. 353–405.

Kanasewich, E.R., 1968. Precambrian rift: genesis of strata-bound ore deposits. *Science*, 161: 1002–1005.

Köppel, V. and Saager, R., 1976. Uranium, thorium- and lead-isotope studies of strata-bound ores. In: K.H. Wolf (Editor), *Handbook of Strata-Bound and Stratiform Ore Deposits, 2.* Elsevier, Amsterdam, pp.267–316.

Kraume, E., 1955. Die Erzlager des Rammelsberges bei Goslar. *Beih. Geol. Jahrb.*, 18: 394 pp.

Kraume, E. and Jasmund, K., 1951. Die Tufflagen des Rammelsberges. *Beitr. Mineral Petrogr.*, 1: 443–454.

Krebs, W., 1960. Zur Schwellen- und Beckenfazies in der südwestlichen Dillmulde. *Z. Dtsch. Geol. Ges.*, 111: 773–774.

Krebs, W., 1979. Devonian basinal facies. *Spec. Pap. Palaeontol.*, 23: 125–139.

Krebs, W., 1981. The geology of the Meggen ore deposit. In: K.H. Wolf (Editor), *Handbook of Strata-Bound and Stratiform Ore Deposits*, 9, Elsevier, Amsterdam, pp. 509–549.

Krebs, W. and Gwosdz, W., in press. Ore controlling parameters of Devonian stratiform lead–zinc–barite ores in central Europe. *Geol. Jahrb.*

Kyle, J.R., 1981. Geology of the Pine Point lead–zinc district. In: K.H. Wolf (Editor), *Handbook of Strata-Bound and Stratiform Ore Deposits*, 9. Elsevier, Amsterdam, pp. 643–741.

Lambert, I.B., 1976. The McArthur lead–zinc–silver deposit: features, metallogenesis and comparisons with some other stratiform ores. In: K.H. Wolf (Editor), *Handbook of Strata-Bound and Stratiform Ore Deposits, 6.* Elsevier, Amsterdam, pp.535–585.

Lambert, I.B. and Groves, D.L., 1981. Early earth evolution and metallogeny. In: K.H. Wolf (Editor), *Handbook of Strata-Bound and Stratiform Ore Deposits,* 8, Elsevier, Amsterdam, in press.

Large, R.R., 1977. Chemical evolution and zonation of massive sulphide deposits in volcanic terrains. *Econ. Geol.,* 72: 549–572.

Laznicka, P., 1981. The concept of ore types – summary, suggestions and a practical test. In: K.H. Wolf (Editor), *Handbook of Strata-Bound and Stratiform Ore deposists,* 8, Elsevier, Amsterdam, in press.

Lehmann, B. and Schneider, H.-J., 1981. Strata-Bound tin deposits. In: K.H. Wolf (Editor), *Handbook of Strata-Bound and Stratiform Ore Deposits,* 9. Elsevier, Amsterdam, pp. 743–771.

Lindgren, W., 1933. *Mineral Deposits.* McGraw-Hill, New York, N.Y., 4th ed., 930 pp.

Loudon, A.G., Lee, M.K., Dowling, J.F. and Bourn, R., 1975. Lady Loretta silver–lead–zinc deposit. In: C.L. Knight (Editor), *Economic Geology of Australia and Papua New Guinea. Australas. Inst. Min. Metall. Monogr.,* 5: 377–382.

Lydon, J.W., 1978. Some criteria for categorizing hydrothermal basemetal deposits. In: *Report of Activities, Part A, Geol. Surv. Can. Pap.,* 78-1A: 299–302.

Macdermot, C.V. and Sevastopulo, G.D., 1972. Upper Devonian and Lower Carboniferous stratigraphical setting of the Irish mineralisation. *Geol. Surv. Ireland Bull.,* 1: 267–280.

Mathias, B. and Clarke, G.J., 1975. Mount Isa copper and silver–lead–zinc orebodies – Isa and Hilton mines. In: C.L. Knight (Editor), *Economic Geology of Australia and Papua New Guinea. Australas. Inst. Min. Metall. Monogr.,* 5: 351–376.

Mathias, B.V., Clark, G.J., Morris, D. and Russell, R.E., 1973. The Hilton deposit – stratiform silver–lead–zinc mineralisation of the Mount Isa type. In: N.H. Fisher (Editor), *Metallogenic Provinces and Mineral Deposits in the Southwest Pacific. Bur. Miner. Resour. Australas. Bull.,* 141: 33–58.

McClay, K.R., 1979. Folding in silver-lead-zinc orebodies, Mount Isa, Australia. *Trans. Inst. Min. Metall. Sect. B,* 88: 5–14.

McClay, K.R. and Carlile, D.G., 1978. Midproterozoic sulphate evaporites at Mount Isa mine, Queensland, Australia. *Nature,* 274: 240–241.

Mitchell, A.H.G. and Garson, M.S., 1976. Mineralisation at plate boundaries. *Miner. Sci. Eng.,* 2: 129–169.

Morrisssey, C.J., 1977. Reflection on ores and the apparent scarcity of oil in Irish Carboniferous sediments. In: P. Garrard (Editor), *Proceedings of the Forum on Oil and Ores in Sediments.* Imperial College, London: 147–160.

Morrissey, C.J., Davis, G.R. and Steed, G.M., 1971. Mineralisation in the Lower Carboniferous of central Ireland. *Trans. Inst. Min. Metall. Sect. B,* 80: 174–185.

Mulligan, R., 1975. Geology of Canadian tin occurrences. *Geol. Surv. Can. Econ. Geol. Rep.,* 28: 155 pp.

Murray, W.J., 1975. McArthur River, HYC lead–zinc and related deposits Northern Territory. In: C.L. Knight (Editor), *Economic Geology of Australia and Papua New Guinea. Australas. Inst. Min. Metall. Monogr.,* 5: 329–339.

Oehler, J.H. and Logan, R.G., 1977. Microfossils, cherts and associated mineralisation in the Proterozoic McArthur (H.Y.C.) lead–zinc–silver deposits. *Econ. Geol.,* 72: 1393–1409.

Piper, J.D.A., 1976. Palaeomagnetic evidence for a Proterozoic supercontinent. *Philos. Trans. R. Soc. London Ser A,* 280: 469–490.

Plimer, I.R., 1978. Proximal and distal stratabound ore deposits. *Miner. Deposita,* 13: 345–353.

Plimer, I.R. and Finlow-Bates, T., 1978. Relationship between primary iron sulphide species, sulphur source, depth of formation and age of submarine exhalative sulphide deposits. *Miner. Deposita,* 13: 399–410.

Plumb, K.A. and Derrick, G.M., 1975. Geology of the Proterozoic rocks of the Kimberley to Mount Isa region. In: C.L. Knight (Editor), *Economic Geology of Australia and Papua New Guinea. Australas. Inst. Min. Metall. Monogr.,* 5: 217–252.

Quade, H., 1976. Genetic problems and environmental features of volcano-sedimentary iron-ore deposits of the Lahn-Dill type. In: K.H. Wolf (Editor), *Handbook of Strata-Bound and Stratiform Ore Deposits,* 7. Elsevier, Amsterdam, pp.255–294.

Ramdohr, P., 1953. Mineralbestand, Strukturen und Genese der Rammelsberglagerstätte. *Geol. Jahrb.*, 67: 367–494.
Richards, J.R., 1975. Lead isotope data on three north Australian galena localities. *Miner. Deposita,* 10: 287–301.
Rickard, D.T., Willdén, M.Y., Marinder, N.-E. and Donnelly, T.H., 1979. Studies on the genesis of the Laisvall sandstone lead–zinc deposit, Sweden. *Econ. Geol.,* 74: 1255–1285.
Ridge, J.D., 1973. Volcanic exhalations and ore deposition in the vicinity of the sea floor. *Miner. Deposita,* 8: 332–348.
Roedder, E., 1976. Fluid inclusion evidence on the genesis of ores in sedimentary and volcanic rocks. In: K.H. Wolf (Editor), *Handbook of Strata-Bound and Stratiform Ore Deposits, 2.* Elsevier, Amsterdam, pp.67–110.
Rozendaal, A., 1978. The Gamsberg zinc deposit, Namaqualand. In: W.J. Verwoerd (Editor), *Mineralisation in Metamorphic Terrains.* J.L. van Scheik Ltd., Pretoria, pp.235–265.
Russell, M.J., 1968. Structural control of base-metal mineralisation in Ireland in relation to continental drift. *Trans. Inst. Min. Metall.,* 77: 117–128.
Russell, M.J., 1973. Base-metal mineralisation in Ireland and Scotland and the formation of the Rockall trough. In: D.H. Tarling and S.K. Runcorn (Editors), *Implications of Continental Drift to the Earth Sciences, 1.* Academic Press, London, pp.581–597.
Russell, M.J., 1975. Lithogeochemical environment of the Tynagh base-metal deposit, Ireland, and its bearing on ore deposition. *Trans. Inst. Min. Metall. Sec. B,* 84: 128–133.
Russell, M.J., 1978. Downward-excavating hydrothermal cells and Irish-type ore deposits: importance of the underlying thick Caledonian prism. *Trans. Inst. Min. Metall. Sect. B,* 87: 128–133.
Rye, R.O. and Ohmoto, H., 1974. Sulphur and carbon isotopes and ore genesis: a review. *Econ. Geol.,* 69: 826–842.
Samama, J.C., 1976. Comparative review of the genesis of the copper–lead–sandstone type of deposits. In: K.H. Wolf (Editor), *Handbook of Strata-Bound and Stratiform Ore Deposits, 6.* Elsevier, Amsterdam, pp.1–20.
Sangster, D.F., 1976a. Sulphur and lead isotopes in strata-bound ore deposits. In: K.H. Wolf (Editor), *Handbook of Strata-Bound and Stratiform Ore Deposits, 2.* Elsevier, Amsterdam, pp.219–266.
Sangster, D.F., 1976b. Carbonate-hosted lead–zinc deposits. In: K.H. Wolf (Editor), *Handbook of Strata-Bound and Stratiform Ore Deposits, 6.* Elsevier, Amsterdam, pp.447–456.
Sangster, D.F., 1979. Evidence for an exhalative origin for the deposits of the Cobar district, New South Wales. *BMR J. Aust. Geol. Geophys.,* 4: 15–24.
Sangster, D.F. and Scott, S.D., 1976. Precambrian stratabound massive Cu–Zn–Pb sulfide ores of North America. In: K.H. Wolf (Editor), *Handbook of Strata-Bound and Stratiform Ore Deposits, 6.* Elsevier, Amsterdam, pp.129–222.
Sato, T., 1972. Behaviour of ore-forming solutions in sea water. *Min. Geol.,* 22: 31–42.
Sato, T., 1977. Kuroko deposits: their geology, geochemistry and origin. In: *Volcanic Processes in Ore Genesis. Geol. Soc. London Spec. Publ.,* 7: 153–161.
Sawkins, F.J., 1976. Widespread continental rifting: some considerations of timing and mechanism. *Geology,* 4: 427–430.
Saxby, J.D., 1976. The significance of organic matter in ore genesis. In: K.H. Wolf (Editor), *Handbook of Strata-Bound and Stratiform Ore Deposits, 2.* Elsevier, Amsterdam, pp.111–134.
Schneiderhöhn, H., 1944. *Erzlagerstätten.* Fischer, Jena, 371 pp.
Schultz, R.W., 1971. Mineral exploration practice in Ireland. *Trans. Inst. Min. Metall. Sect. B,* 80: 238–258.
Scott, K.M. and Lambert, I.B., 1979. Studies of the base metal sulfide deposits at McArthur River, Northern Territory, Australia: II. The sulfide-S and organic-C relationships of the concordant deposits and their significance – a discussion. *Econ. Geol.,* 74: 1693–1694.
Shanks, W.C. and Bischoff, J.L., 1977. Ore transport and deposition in the Red Sea geothermal system: a geochemical model. *Geochim. Cosmochim. Acta,* 41: 1507–1519.
Shawe, D.R., Poole, F.G. and Brobst, D.A., 1967. Newly discovered bedded barite deposits in East Northumberland Canyon, Nye County, Nevada. *Econ. Geol.,* 64: 245–254.

Sheppard, S.M.F., 1977. Identification of the origin of ore-forming solutions by the use of stable isotopes. In: *Volcanic Processes in Ore Genesis. Geol. Soc. London Spec. Publ.*, 7: 25–41.

Sibson, R.H., Moore, J.M. and Rankin, A.H., 1975. Seismic pumping – a hydrothermal fluid transport mechanism. *J. Geol. Soc. London,* 131: 653–659.

Sinclair, A.J., 1966. Anomalous lead from the Kootenay arc, British Columbia. In: *Tectonic History and Mineral Deposits of the Western Cordillera in British Columbia and Neighbouring Parts of the United States – A Symposium. Can. Inst. Min. Metall. Spec. Vol.,* 8: 249–262.

Smith, W.D., 1969. Penecontemporaneous faulting and its likely significance in relation to Mount Isa ore deposition. *Geol. Soc. Australas. Spec. Publ.,* 2: 225–235.

Smith, J.W. and Croxford, N.J.W., 1973. Sulphur isotope ratios in the McArthur lead–zinc–silver deposit. *Nature,* 245: 10–12.

Solomon, M., 1976. "Volcanic" massive sulphide deposits and their host rocks – A review and explanation. In: K.H. Wolf (Editor), *Handbook of Strata-Bound and Stratiform Ore Deposits, 6.* Elsevier, Amsterdam, pp.21–54.

Solomon, M. and Walshe, J.L., 1979. The formation of massive sulfide deposits on the sea floor. *Econ. Geol.,* 74: 797–813.

Sperling, H., in press. On the metallogenetic setting of the ore deposits in the Harz. *Erzmetall.*

Spooner, E.T.C., 1977. Hydrodynamic model for the origin of the ophiolitic cupriferous pyrite ore deposits of Cyprus. In: *Volcanic Processes in Ore Genesis. Geol. Soc. London Spec. Publ.,* 7: 58–71.

Stacey, J.S. and Kramers, J.D., 1975. Approximation of terrestrial lead isotope evolution by a two-stage model. *Earth Planet. Sci. Lett.,* 26: 207–221.

Stewart, J.H., 1972. Initial deposits in the Cordilleran geosyncline: evidence of a late Precambrian (850 m.y.) continental separation. *Geol. Soc. Am. Bull.,* 83: 1345–1360.

Taylor, S. and Andrew, C.J., 1978. Silvermines orebodies, County Tipperary, Ireland. *Trans. Inst. Min. Metall. Sect. B,* 87: 111–124.

Tempelman-Kluit, D.J., 1972. The geology and origin of the Faro, Vangorda, and Swim concordant zinc–lead deposits, central Yukon Territory. *Geol. Surv. Can. Bull.,* 208: 73 pp.

Thompson, R.I. and Pantaleyev, A., 1976. Stratabound mineral deposits of the Canadian Cordillera. In: K.H. Wolf (Editor), *Handbook of Strata-Bound and Stratiform Ore Deposits, 5.* Elsevier, Amsterdam, pp.37–108.

Trudinger, P.A., 1976. Microbiological processes in relation to ore genesis. In: K.H. Wolf (Editor), *Handbook of Strata-Bound and Stratiform Ore Deposits, 2.* Elsevier, Amsterdam, pp.135–190.

Turner, F.J. and Verhoogen, J., 1960. *Igneous and Metamorphic Petrology.* McGraw-Hill, New York, N.Y., 694 pp.

Turner, J.S. and Gustafson, L.B., 1978. The flow of hot saline solutions from vents in the sea floor – some implications for exhalative massive sulphide and other ore deposits. *Econ. Geol.,* 73: 1082–1100.

Vaughan, D.J., 1976. Sedimentary geochemistry and the mineralogy of the sulfides of lead, zinc, copper and iron and their occurrence in sedimentary ore deposits. In: K.H. Wolf (Editor), *Handbook of Strata-Bound and Stratiform Ore Deposits, 2.* Elsevier, Amsterdam, pp.317–363.

Walker, R.N., Muir, M.D., Diver, W.L., Williams, N. and Wilkins, N., 1977. Evidence for major sulphate evaporite deposits in the Proterozoic McArthur group, Nothern Territory, Australia. *Nature,* 265: 526–529.

Walker, R.N., Logan, R.G. and Binnekamp, J.G., 1978. Recent geologic advances concerning the H.Y.C. and associated deposits, McArthur River, N.T. *Geol. Soc. Aust. J.,* 25: 365–380.

Wedepohl, K.H., 1974. Basic geochemical data of Zn, Pb, and Cu, and hydrothermal ore genesis. *Schriftenr. Erdwiss. Komm. Österr. Akad. Wiss.,* 1: 160–173.

Wedepohl, K.H., Delevaux, M.H. and Doe, B.R., 1978. The potential source of lead in the Permian Kupferschiefer bed of Europe and some selected Palaeozoic mineral deposits in the Federal Republic of Germany. *Contrib. Mineral. Petrol.,* 65: 273–281.

Weisser, J.D., 1972. Zur Methodik der Exploration Meggen. *Schr. Ges. Dtsch. Metallhütten Bergleute,* 24: 167–186.

Werner, C.-D. and Rösler, H.J., 1979. Aussagemöglichkeiten der initialen Magmatite für die Klärung struktureller Verhältnisse am Beispiel des mitteleuropäischen Variszikums. *Z. Geol. Wiss.,* 7: 305–313.

White, D.E., 1967. Mercury and base-metal deposits with associated thermal and mineral waters. In: H.L. Barnes (Editor), *Geochemistry of Hydrothermal Ore Deposits.* Holt, Reinhart and Winston, New York, N.Y., pp.575–631.

Wiechelt, W., 1904. Die Beziehungen des Rammelsberger Erzlagers zu seinen Nebengesteinen. *Berg-Hüttenm. Z.,* 63: 285–361.

Williams, N., 1978. Studies of the base metal sulfide deposits at McArthur River, Northern Territory, Australia: I. The Cooley and Ridge deposits. *Econ. Geol.*, 73:1005–1035.

Williams, N., 1979. Studies of the base metal sulfide deposits at McArthur River, Northern Territory, Australia: II. The sulfide-S and the organic-C relationships of the concordant deposits and their significance – a reply. *Econ. Geol.,* 74: 1695–1697.

Windley, B.F., 1977. *The Evolving Continents.* Wiley, Chichester; 385 pp.

Wolf, K.H., 1976. Ore genesis influenced by compaction. In: G.V. Chilingar and K.H. Wolf (Editors), *Compaction of Coarse Grained Sediments, 2.* Elsevier, Amsterdam, pp.475–676.

Wolf, K.H., 1981. Terminologies, structuring and classifications in ore and host-rock petrology. In: K.H. Wolf (Editor), *Handbook of Strata-Bound and Stratiform Ore Deposits,* 8. Elsevier, Amsterdam, in press.

Zartman, R.E. and Stacey, J.S., 1971. Lead isotopes and mineralisation ages in Belt supergroup rocks, northwestern Montana and northern Idaho. *Econ. Geol.,* 66: 849–860.

Zimmermann, R.A., 1976. Rhythmicity of barite-shale and of Sr in strata-bound deposits of Arkansas. In: K.H. Wolf (Editor), *Handbook of Strata-Bound and Stratiform Ore Deposits, 3.* Elsevier, Amsterdam, pp.339–353.

Chapter 9

THE GEOLOGY OF THE MEGGEN ORE DEPOSIT [1]

WOLFGANG KREBS

INTRODUCTION

Apart from the Rammelsberg ore body in the Harz Mountains (Hannak, this volume), the Meggen ore body is the largest sediment-hosted, stratiform lead–zinc–barite deposit in Central Europe. The predominantly banded sulphides of Meggen and Rammelsberg are concordantly interbedded in marine Devonian sediments, such as black shales, siltstones, sandstones and micritic limestones. Both ores are considered to be of a synsedimentary, submarine hydrothermal origin (syngenetic hydrothermal deposits of Barnes, 1974/75). The lens-like ore deposits show a distinct vertical and/or horizontal metal zonation as a result of varying temperatures and changing Eh and pH conditions. Epigenetically mineralized zones, known at Rammelsberg as "Kniest", represent channelways of ascending thermal brines to the discharge area on the sea floor. Table II shows the ore types, tonnage and average grade of the Meggen and Rammelsberg ore bodies.

The late Middle Devonian pyrite–sphalerite–barite ore of Meggen is located at the southeast flank of the Elspe syncline, Sauerland area, Rhenish Schiefergebirge (Fig. 1), which belongs to the Variscan (Hercynian) deformed Meso-Europe.

The sulphides of the Meggen ore have been mined since 1852. The Meggen mine belongs to the Sachtleben Bergbau GmbH, a subsidary company of the Metallgesellschaft AG. The total tonnage is about 50 million metric tons of sulphides and 10 million metric tons of barite. From 1852 to 1974, the Meggen mine produced approximately 35 million metric tons of sulphide ores. The present annual production is about 950,000 tonnes (Fuchs, 1978).

The geology of the Meggen ore body was monographically described by Ehrenberg et al. (1954) in the series *Monographien der Deutschen Blei–Zink-Lagerstätten.* Since 1954 many new contributions to the Meggen ore and to the geology of the Meggen area have been published as a result of new diamond drilling, new underground development and newly applied methods. Table I shows the specific topics on which new results have been obtained since 1954.

[1] Brief references have been made to these deposits in several volumes of this Handbook series; see indexes in Vols. 4 and 7.

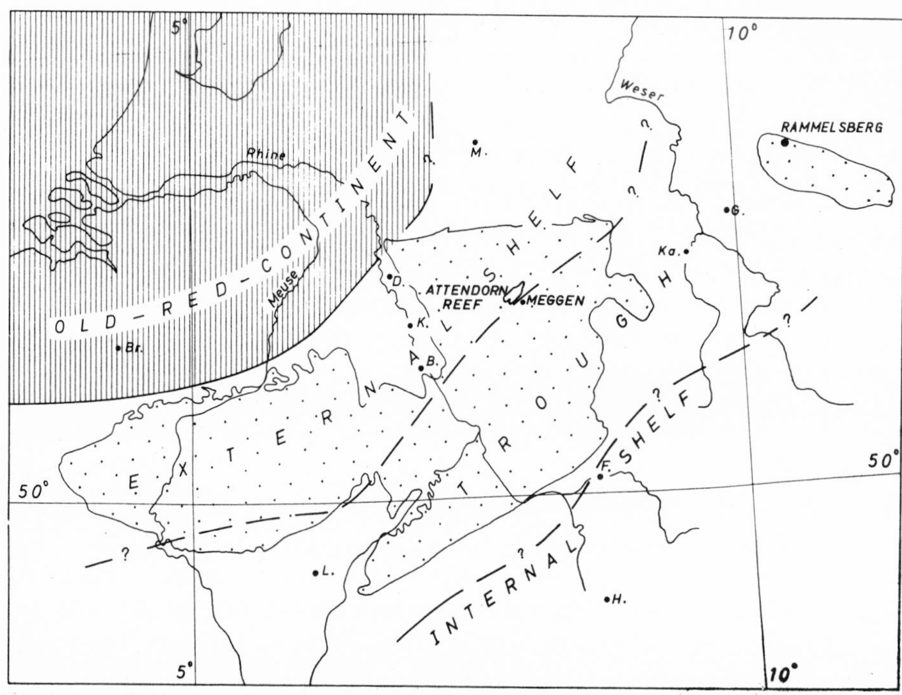

Fig. 1. Late Middle Devonian paleogeography in Central Europe and the location of the Meggen ore body very close to the external shelf margin. Stippled areas: Paleozoic outcrops. *B.* = Bonn, *Br.* = Brussels, *D.* = Düsseldorf, *F.* = Frankfurt, *G.* = Göttingen, *H.* = Heidelberg, *K.* = Köln, *Ka.* = Kassel, *L.* = Luxemburg, *M.* = Münster. (From Krebs, 1972b.)

TABLE I

Summary of topics on the geology of Meggen which have been published since Ehrenberg et al. (1954)

Topics	Authors
Footwall stratigraphy	Wellmer (1970), Gwosdz and Krüger (1972), Krüger (1973), Gwosdz et al. (1974)
Hanging-wall stratigraphy	Noa (1958), Wellmer (1966), Dornsiepen (1977 and in press)
Paleogeography	Krebs (1972a, b), Pilger (1972), Gwosdz (1972), Clausen (1978)
Tectonics	Fuchs and Ommer (1970), McMahon Moore (1971), Weisser (1972), Wellmer (1973), Fuchs (1978)
Ore (sulphide and barite)	Puchelt (1961), Buschendorf and Puchelt (1965), Zimmermann (1970), Gasser (1974), Gasser and Thein (1977)
Sulphur isotopes	Buschendorf et al. (1963)
Lead isotopes	Brown (1965), Wedepohl et al. (1978)
Tuffs	Dornsiepen (in press), Scherp (1978)
Geochemistry	Hilmer (1972), Friedrich et al. (1972), Gwosdz et al. (1974), Dornsiepen (1977), Gwosdz and Krebs (1977)
Origin	Pilger (1972), Scherp (1974), Dornsiepen (1977)

GEOLOGIC SETTING

The late Middle Devonian pyrite—sphalerite—barite ore deposit of Meggen is located in the Paleozoic fold belt of the Rhenish Schiefergebirge, which consists of a folded and thrusted Devonian and Carboniferous geosynclinal sequence. The Rhenish Schiefergebirge (Fig. 1) forms one of the Paleozoic massifs in Central Europe, which resulted from the folding of the Variscan (Hercynian) geosyncline. During Devonian and Carboniferous times the geosynclinal evolution is marked by a gradual northward migration of zones of maximum subsidence, subsequent sedimentation and deformation. The Rhenish Schiefergebirge belongs to the Rhenohercynian zone of the Variscan fold belt and is characterized by northwestward thrusting and folding towards the Caledonian fore-land.

During the Devonian, the Rhenohercynian zone of the Variscan geosyncline was subdivided into the following two facies belts:

(1) A northwestern external shelf sequence bordering the erosional area of the Caledonian-deformed Old Red Continent, consisting of thick shallow-water sediments, such as near-shore conglomerates, sandstones, siltstones, red beds, marls, shallow-water carbonates and shelf-margin reefs. The external shelf was characterized during Middle Devonian time by a continuously subsiding, predominantly shallow-marine environment which resulted in a sequence of more than 5000 m in thickness.

(2) A southeastern trough sequence comprising pelagic dark shales, cephalopod limestones, sandstone turbidites and acid as well as basic volcanics. The trough of the Variscan geosyncline represented a deeper-water environment (>200 m) with variable, but generally minor subsidence, so that about 1500 m of pelagic sediments correspond to the thick time-equivalent shelf deposits.

Paleogeographically, the late Middle Devonian Meggen ore deposit belongs to the trough, but is situated very close to the external shelf margin (Figs. 1 and 6).

In the vicinity of Meggen, the shelf margin is characterized by significant facies and thickness changes during Middle and Late Devonian time (Krebs, 1972a; Gwosdz, 1972; Buggisch and Gwosdz, 1973). South of Meggen there existed a centre of repeated eruptions of Lower Devonian acid submarine volcanics (Rippel, 1954), but more than 1000 m of Middle Devonian sediments were deposited between the youngest volcanic horizon and the shale-hosted Meggen ore.

STRATIGRAPHY

Fig. 2 shows the stratigraphic sequence of Middle Devonian and early Upper Devonian beds in the Meggen area. From a paleogeographic point of view the Meggen area can be subdivided during that time into two different environments: the Meggen Basin and the Meggen Reef.

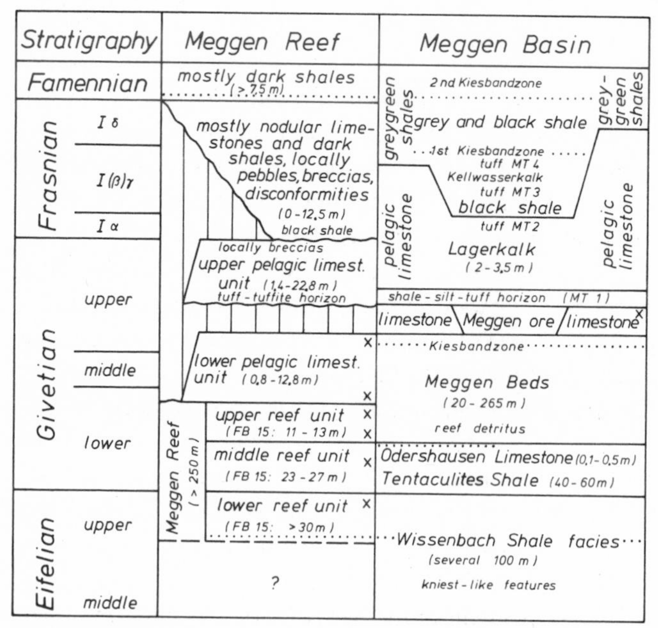

Fig. 2. Middle and Upper Devonian stratigraphy in the area of the Meggen Reef and the Meggen Basin (locations see Fig. 14A). x = transected by sedimentary dikes.

The Meggen Basin

The Wissenbach Shale, Tentaculites Shale and Meggen Beds are black to dark grey shales with variable amounts of silt and sand, which occur in thin bands or thicker beds. The silt and sand intercalations can be interpreted as turbidites, with components derived from the adjacent external shelf. The black shales of the Tentaculites Shale represent a regional transgression of the trough toward the shelf and are devoid of clastic intercalations. Tentaculitids, styliolinids, ostracods, goniatids, orthoceratids and conodonts demonstrate a pelagic environment and a water depth greater than that found in the shelf. Limestone turbidites, intercalated with the autochthonous shale sequence, indicate the contemporaneous growth and erosion of the adjacent Meggen Reef (Krebs, 1972a; Krüger, 1973).

The early Givetian Odershausen Limestone, up to a few tens of cm thick, forms a thin but widespread lithological and biostratigraphical marker horizon within the Meggen Basin.

The shale–silt sequence between the Odershausen Limestone and the Meggen pyrite–sphalerite–barite ore was formerly known as "Lenneschiefer" by the miners of Meggen, but has been newly defined as the Meggen Beds by Gwosdz and Krüger (1972). Krüger (1973) and Gwosdz et al. (1974) have shown that a shallow depression on the sea floor had already developed beneath the Meggen ore body during the deposition of the Givetian Meggen Beds. This submarine depression was characterized by a decreasing autochthonous carbonate content, a greater proportion of allochthonous reef debris, a higher proportion of pyrite nodules and sulphide dissemination, and a higher content of some base metals in the shales. In the uppermost Meggen Beds, 6–8 m below the Meggen ore horizon, pyrite nodules and lenses, locally with barite cores (Amstutz et al., 1971), represent a pre-phase of the subsequent pyrite–sphalerite–barite ore.

A thin silt intercalation within the Meggen ore (Ehrenberg et al., 1954; Gasser, 1974; Gasser and Thein, 1977) and a very thin silty and tuffitic horizon immediately above the Meggen ore body (Dornsiepen, 1977) suggest that the clastic influx from the external shelf continued during the time of mineralization.

The wavy banded, mm-thin black shale seams and wisps within the Meggen ore were definitely deposited at a slower rate than the interbedded sulphide layers.

The late Givetian to early Frasnian Lagerkalk, which overlies the pyrite–sphalerite–barite ore, is a lithologically uniform, micritic pelagic limestone up to 3.5 m thick. The Lagerkalk lithologically resembles the pelagic cephalopod limestones, which mostly are found on submarine elevations or ridges ("Schwelle").

According to Dornsiepen (1977), it is possible to subdivide the early Frasnian shale sequence above the Lagerkalk in the Meggen area into the following time-equivalent units (Fig. 3) : (1) black shales, partly with bedded black bituminous limestones ("Kellwasserkalk"), deposited in a small depression of the sea floor, which was situated more or less above the underlying sulphide ore body; and (2) green shales, with limestone nodules and layers, which mark the underlying barite margin and areas outside the ore body.

Dornsiepen (1977) has identified four thin and fine-grained tuff–tuffitic horizons in

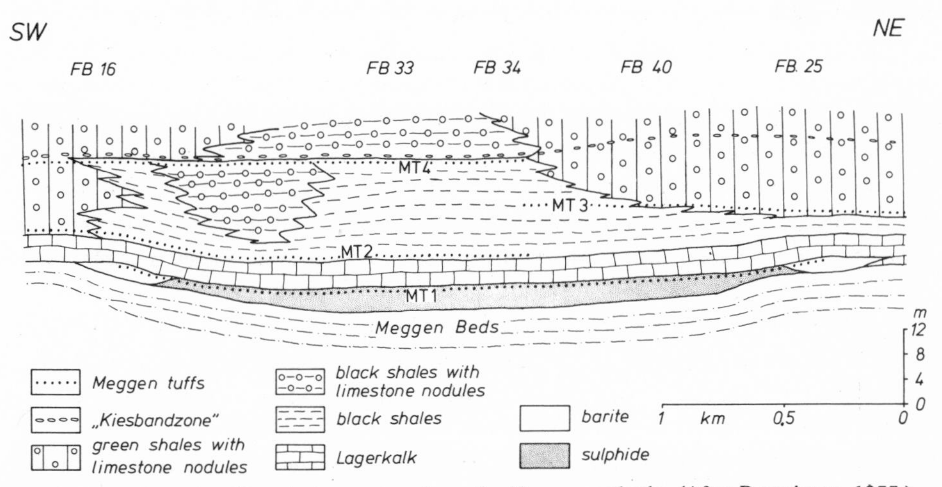

Fig. 3. Upper Devonian (Frasnian) sequence above the Meggen ore body. (After Dornsiepen, 1977.)

the hanging wall of the pyrite–sphalerite–barite deposit: one immediately above the Meggen ore mixed with silty material (MT 1) and three horizons (MT 2–MT 4) within the Frasnian shale–limestone sequence (Fig. 3). In addition to the tuff–tuffitic horizons, two pyrite nodule marker beds ("Kiesbandzones") are present (Noa, 1958; Wellmer, 1966; Weisser, 1972), indicating (1) the presence of volcanic activity after the Meggen ore formation and (2) post-phase sulphide mineralization (although a sedimentary origin for the pyritic nodules cannot be ruled out completely).

The lower pyrite nodule marker horizon, 20–25 m above the ore, is closely associated with a tuff band (MT 4). The tuff bands are commonly very thin (up to 20 cm) and can be used as stratigraphic marker horizons. They contain angular quartz, feldspar, biotite, chlorite, sericite, zircon and monazite in a sericite–chlorite matrix. According to Dornsiepen (1977 and in press), the zircons clearly demonstrate an igneous origin of the horizons MT 2–MT 4. The tuff horizons are indicated by high values of K_2O and Al_2O_3, and low values of Na_2O and SiO_2 (Table VI).

The Meggen Reef

The Meggen Reef, which is only known in the subsurface, forms a bioherm-like table reef, completely surrounded by basinal sediments. Neither the position of the reef base, nor the original thickness of the limestones (>250 m) is known because the base of the Meggen Reef has not been reached by drilling.

According to conodonts and corals, the reef limestones belong to the upper Eifelian and lower Givetian (Krebs, 1972b; 1978b). The Meggen Reef was built by stromatoporoids, corals and crinoids. There are no indications of a lagoonal facies. Grey pelagic limestones are increasingly abundant as intercalations in the upper part of the reef.

In the reef-core facies, the most important frame-builders are massive, partly irregularly growing, and tabular stromatoporoids. The matrix of the reef-core facies consists predominantly of crinoid debris. In contrast to the reef-flank facies, the reef-core facies is characterized by a low primary porosity. The types of the reef-flank facies are composed of broken and transported skeletal fragments (stromatoporoids, massive to dendroid tabulate corals, colonial and solitary rugose corals and brachiopods). In some types, the primary voids between skeletal fragments comprise up to 45% of the rock. Near the reef core the matrix consists of crinoid debris, whereas towards the basin there is a gradual transition from fine-grained lime muds with pelagic fossils into dark grey and black calcareous shales.

The reef limestones can be subdivided by a black shale tongue into an older upper Eifelian reef unit and a younger early Givetian reef unit. The black shale tongue has been correlated, using conodonts, with the widespread transgression of the early Givetian Tentaculites Shale (Fig. 2).

Reef growth terminated as a result of the decreasing rate of subsidence of the sea floor, so that the subsequent early Givetian to early Frasnian time was characterized by

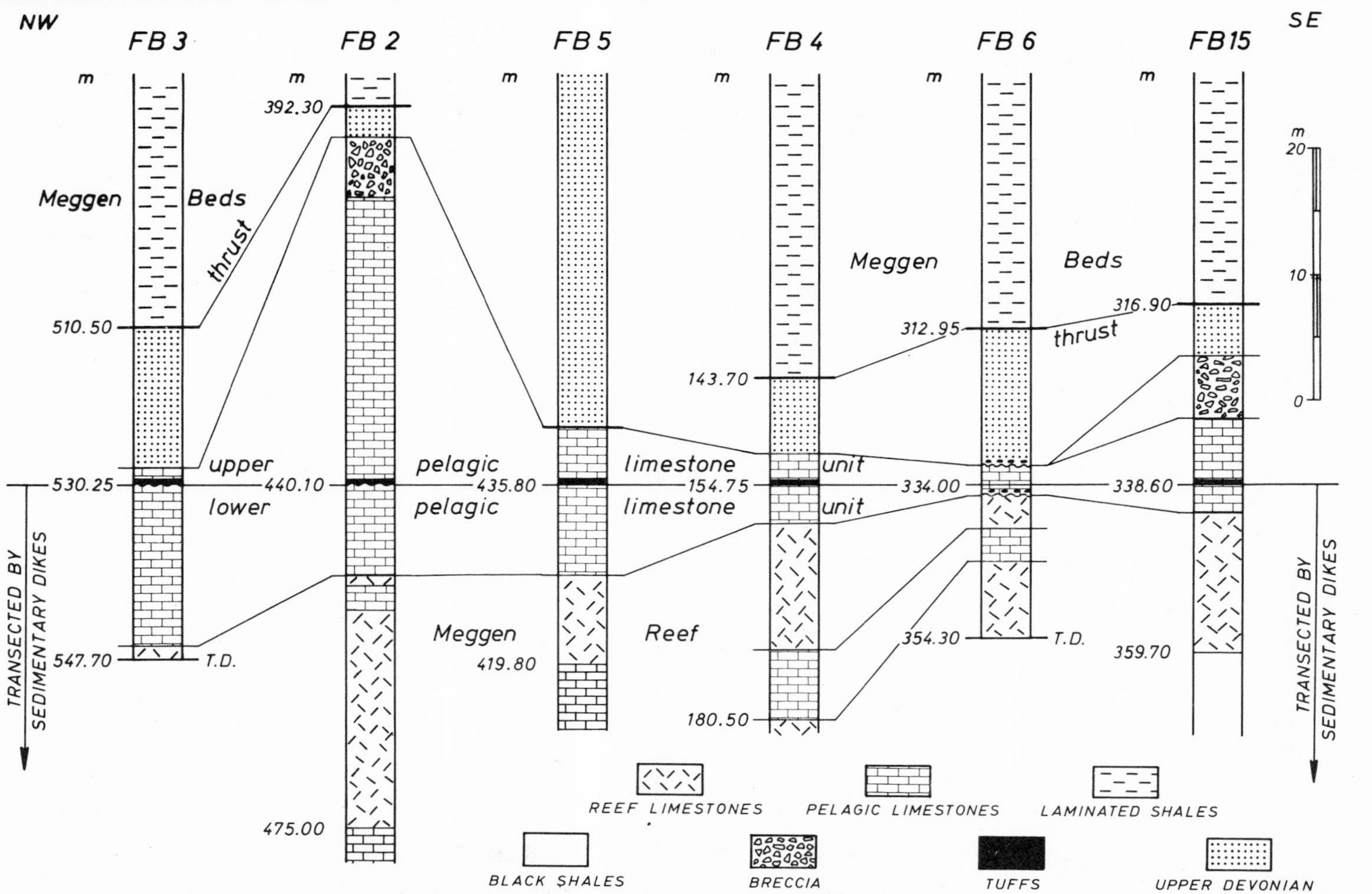

Fig. 4. The late Middle and early Upper Devonian sequence in the area of the Meggen Reef. Horizontal line = time-equivalent of the Meggen ore. For location of bore holes see Krebs (1972b, fig. 2).

the deposition of pelagic cephalopod limestones of minor thickness. The pelagic limestones are grey to dark grey, subordinately also reddish grey to red. Abundant styliolinids, ostracods, conodonts and cephalopods indicate their pelagic nature.

The cephalopod limestones on top of the Meggen Reef can be stratigraphically subdivided into two units (Fig. 4):

(1) A lower pelagic limestone unit (0.85–12.7 m) which comprises lower to upper Givetian. The cephalopod limestones are characterized by crinoid accumulations, calcite-filled cavities, pebbles, reworked conodonts and sedimentary dikes.

(2) An upper pelagic limestone unit (1.4–22.7 m) which has an upper Givetian age (*varcus*-zone). The base of the upper unit is locally marked by a disconformity and/or a thin tuff–tuffite horizon. The thin tuff–tuffitic horizon is apparently equivalent to the shale–silt–tuff marker horizon (MT 1) immediately above the Meggen ore. Only in a very few cases can sedimentary dikes be observed. In the FB 2 and FB 15 bore holes, the upper unit is overlain by limestone breccias showing a pre-fracture pyrite mineralization.

A significant feature within the reef limestones and the lower part of the overlying pelagic limestones are numerous neptunian dikes filled with pelagic limestones, crinoid limestones, black shales, greenish tuffitic marls and several generations of fibrous calcite (Krebs, 1972b). The widths of the neptunian dikes range from a few mm up to 20 cm and more.

On the basis of conodonts, the fissure fillings can be dated as predominantly lowermost Frasnian, and subordinately as uppermost Givetian and lower to middle Frasnian.

The formation of sedimentary dikes, the tuff eruptions and the ascent of metal-bearing solutions occurred after the deposition of the lower pelagic limestone unit in uppermost Givetian time. These events were probably represented by a hiatus in the pelagic limestone deposition (Fig. 2). Except for few scattered pyrite nodules, there are no traces of sulphide mineralization in the cephalopod limestones.

The Givetian pelagic limestones are stratigraphically followed by Upper Devonian (Frasnian to early Famennian) greenish, grey to dark grey and black shales with bands, layers and nodules of pelagic limestones. Locally, a hiatus exists between the underlying reef and/or pelagic limestones and the Upper Devonian sequence. The youngest Upper Devonian sediments are virtually eliminated by the large thrust fault between the Meggen Basin and Meggen Reef sequence (Fig. 4).

PALEOGEOGRAPHY

During late Middle to early Upper Devonian time, environments can be recognized in the Elspe syncline near Meggen, which are described next.

The Attendorn Reef

The Givetian to Frasnian reef limestones, which follow on clastic shelf sediments, form a shelf-margin reef (Fig. 5). The Attendorn carbonate complex can be subdivided into

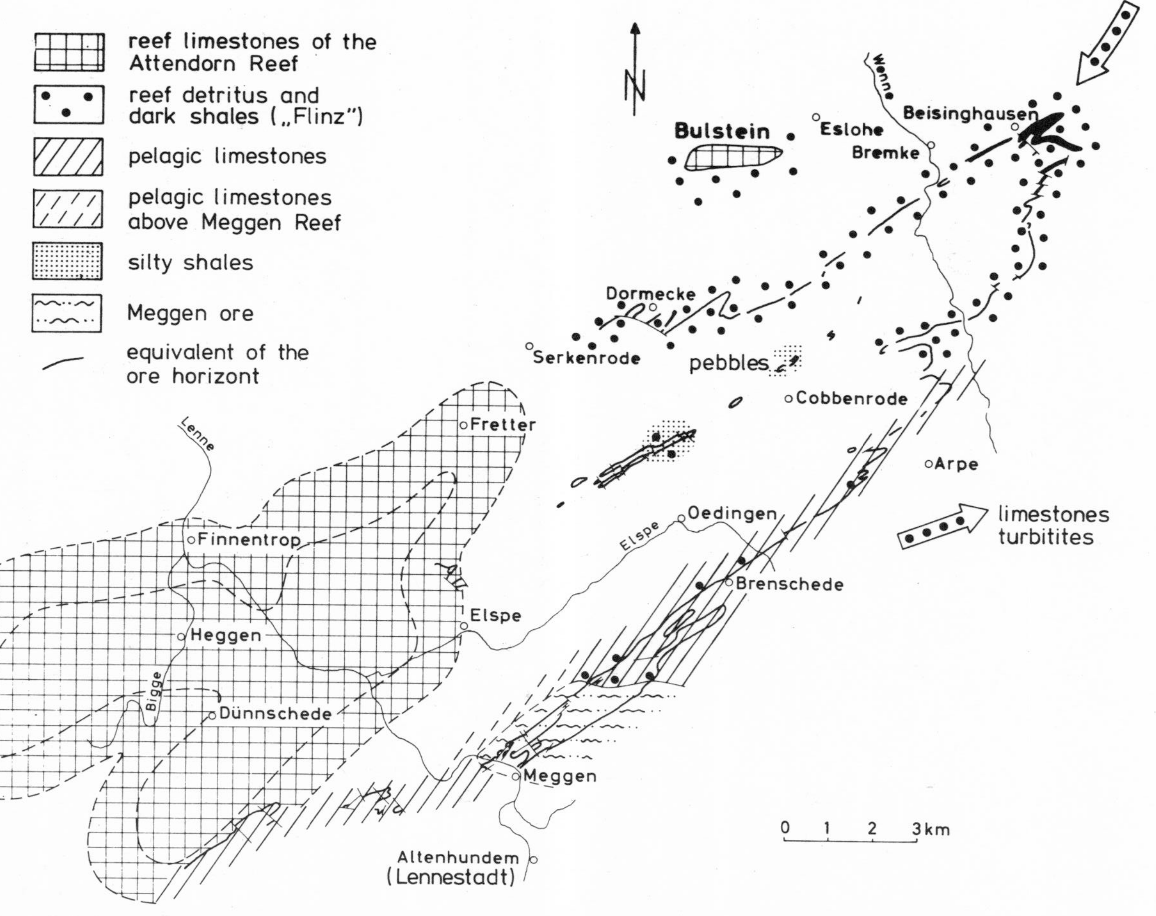

Fig. 5. Paleogeography during late Givetian time in the Attendorn–Elspe syncline. (From Krebs, 1972a.)

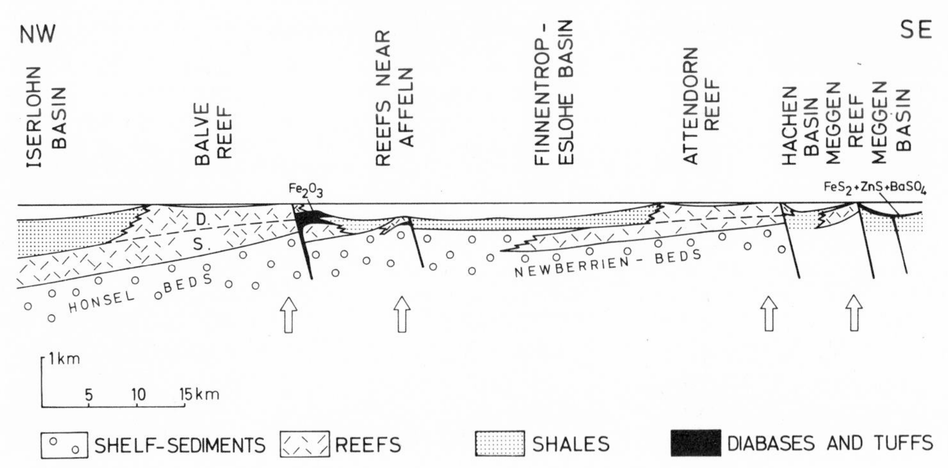

Fig. 6. Vertically exaggerated cross-section from the Iserlohn Basin to the Meggen Basin during late Middle Devonian time showing synsedimentary block faulting. (From Krebs, 1971.)

two units: (1) a widespread biostromal carbonate bank at the base (Schwelm facies) and (2) an atoll-like reef built on the carbonate platform (Dorp facies) consisting of fore-reef, reef-core and back-reef facies (Krebs, 1971; Gwosdz, 1972). According to Gwosdz (1972), the carbonate sedimentation starts at the boundary between middle and upper Givetian and ends in different levels of middle to upper Frasnian time. The thickness of the platform and reed limestones varies from 200–300 m in the southeast to 950 m in the northwest as a result of a tilted basement block (Fig. 6).

A pyritic black shale and limestone horizon of 4 m thickness, embedded about 60 m below the top of the reef limestones, does not represent an equivalent of the Meggen ore (Henke and Schmidt, 1922) but, as conodont investigations have shown, is biostratigraphically younger (Gwosdz, 1972).

The Hachen Basin

Between the southeastern margin of the Attendorn Reef at Grevenbrück and the Meggen Reef in Meggen, a thick Upper Devonian to early Upper Carboniferous basinal sequence of the Elspe syncline is developed. Due to the synclinal character there are no outcrops of the ore-equivalent horizon. Bore hole FB 37 between Grevenbrück and Meggen ended at 1012 m in Lower Carboniferous strata, so that in the centre of the Elspe syncline the ore-equivalent horizon near the Middle to Upper Devonian boundary can be expected at a depth of 1500 m and more (Weisser, 1972). Bore hole FB 15, however, also revealed late Givetian basinal sediments adjacent to the northwestern flank of the Meggen Reef. Of particular interest is the discovery from 1024.15 to 1024.35 m in bore hole FB 15 of fine-grained pyrite layers, bands and nodules which can be dated as late

Givetian. This observation suggests that, at least along the southeast flank of the Hachen Basin, a thin mineralized horizon is developed at the same stratigraphic level as the Meggen ore.

The Meggen Reef

In the paleogeographic development of the Meggen Reef, four stages can be recognized (Krebs, 1972b):

(1) During the late Eifelian and early Givetian an isolated table reef without a central lagoon was built on an as yet unknown foundation (probably a tilted fault block off the external shelf margin), which was completely surrounded by basinal sediments (Hachen and Meggen Basin).

(2) After reef growth stopped in the early Givetian, the morphological contrast between the dead reef and the time-equivalent basins persisted. The sunken reef, forming a seamount, was covered with thin, partly condensed pelagic limestones. The crinoid accumulations within the pelagic limestones suggest a benthonic colonization of a submarine rise on the Givetian sea floor. At the same time, basinal shales continued to be deposited in the adjacent depressions (Fig. 7).

(3) In uppermost Givetian time, the lower pelagic limestone unit and the underlying Meggen Reef were transected and brecciated by numerous sedimentary (neptunian) dikes and fissures. After a locally observed disconformity, a thin tuff–tuffitic horizon was

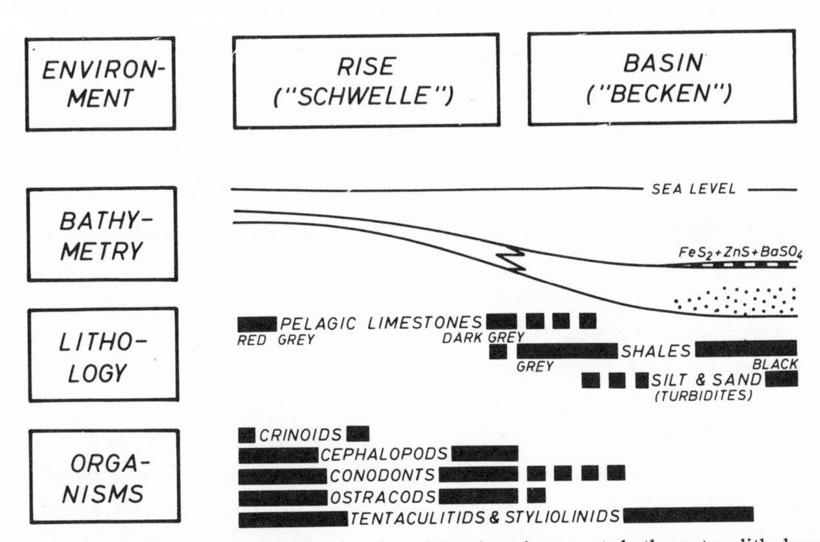

Fig. 7. Schematical diagram showing depositional environment, bathymetry, lithology and organisms of the pelagic limestones overlying the Meggen Reef and time-equivalent basinal sediments in the Meggen Basin. (From Krebs, 1972b.)

deposited at the base of the upper pelagic limestone unit. The same tuff horizon occurs immediately above the Meggen ore in the Meggen Basin (MT 1).

These observations lead to the conclusion that the submarine brecciation of the Meggen Reef and the lower pelagic limestone unit, acid tuff eruptions and ore formation were closely related with respect to this genesis. The brecciation and sedimentary dike formation probably occurred at the same time as the ascent of hydrothermal solutions along a deep-seated fault near the margin of the Meggen Reef.

(4) Upper Devonian pelagic limestones and shales cover, partly after a generally extensive sedimentary hiatus, the upper pelagic limestone unit and, locally, the older reef limestones. The Frasnian and early Famennian sequence is more or less similar to that of the Meggen Basin.

The Meggen Basin

The Middle Devonian dark grey to black shales of the Meggen Basin belong to the basinal facies of the Variscan geosyncline (Fig. 7). During Eifelian time, the shallow hemipelagic–pelagic environment was strongly influenced by the adjacent shelf (e.g. sandstone turbidites with abundant benthonic fossils). In Givetian time, however, a deeper bathyal environment was formed, in which distal turbidites predominate (Clausen, 1978).

Along the southeastern flank of the Elspe syncline, the Givetian Meggen Beds show very conspicuous changes in thickness from 15 up to 265 m (Gwosdz and Krüger, 1972; Gwosdz, 1972; Krüger, 1973; Clausen, 1978). Whereas the average thickness of the Meggen Beds below the Meggen ore is of the order of 20-30 m, there is a significant increase in thickness up to 265 m toward the southwest of the Meggen ore body. In particular the sandy intercalations within the Meggen Beds ("Untere sandige Zone" and "Obere sandige Zone") increase in thickness from 3 to 70 m and from 4 to 120 m, respectively. The increase in thickness of the sandy horizons toward the southwest can be explained by a submarine clastic fan derived from the nearby shelf. This interpretation is in agreement with the fact that the pelagic Meggen Beds pass rapidly into time-equivalent neritic shelf sediments (Wiedenest, Grevenstein and Newberrian Beds) toward the northwest (Gwosdz, 1972).

The Meggen ore was deposited within the Meggen Basin in a third-order depression (Krebs, 1978a). According to Krüger (1973), a gentle submarine depression was already present during the deposition of the Givetian Meggen Beds.

The third-order depression (Meggen depression) was bounded to the west by the submarine clastic fan of the Meggen Beds, to the northwest by the seamount-like elevation of the dead Meggen Reef and to the northeast by a slightly raised submarine platform which was characterized by a higher content of autochthonous micritic limestones and a lower content of clastic turbiditic material within the Meggen Beds (Krüger, 1973). Pelagic limestones (2-3 m thick) in the footwall of the eastern barite margin (Ehrenberg et al., 1954; Weisser, 1972), called "Liegender Lagerkalk", also indicate the more stable conditions in the northeast. Sedimentary dikes, breccias and reworked older conodonts in the Meggen Beds of the bore holes BI/3, BI/1 and BI/20 along the southeastern flank of the Elspe syncline probably reflect synsedimentary movements along the northeastern edge of the Meggen depression. Due to post-tectonic erosion, the former southeastern margin of the Meggen depression is unknown.

The barite, which surrounds the central sulphide ores like a halo, was precipitated in a slightly oxidizing environment on the flanks of the third-order depression. The central pyrite–sphalerite ore, however, was restricted to the inner and deeper part of the Meggen depression below the O_2/H_2S interface. According to Ehrenberg et al. (1954), the thin silt intercalations within the Meggen ore are only restricted to the central sulphide accumulations.

After the deposition of the pyrite–sphalerite–barite ore, a shallow depression was still present. The pelagic micritic limestones (the so-called "lageräquivalente Kalke") are diachronously distributed (Clausen, 1978). On either side, outside of the Meggen depression, the pelagic limestones continue into middle to early upper Frasnian time, whereas within the depression the Lagerkalk ends within the lowermost Frasnian. According to Dornsiepen (1977), the black Frasnian shales, which overly the Meggen sulphides, pass toward the flanks of the Meggen depression into grey-green shales with limestone nodules (Fig. 3). The black Frasnian shales in the third-order Meggen depression show a significantly high content of organic carbon (up to 5%). The dark shales in the second-order basins, on the contrary, generally contain up to 1% organic carbon (Baumann, in press).

In summary, the third-order Meggen depression was characterized before and after the ore precipitation by the following parameters (Krüger, 1973; Gwosdz et al., 1974; Dornsiepen, 1977; Gwosdz and Krebs, 1977):

- decrease of autochthonous calcium carbonate content;
- local accumulation of allochthonous shallow-water carbonate detritus;
- increase of organic carbon in the black shales;
- local silicification in the footwall;
- absence of trace fossils;
- local accumulations of pyrite nodules and layers (pre- and post-phases);
- higher content of Zn, Fe and Mn in the shales;
- presence of a significant manganese halo in the pelagic limestones.

TECTONICS

The Meggen ore and the associated Devonian shelf and basinal deposits were folded and thrust-faulted during the Variscan orogeny in Upper Carboniferous time. In the Sauerland area of the Rhenish Schiefergebirge, one of the major structures formed by the Variscan orogeny is the southwest–northeast trending Attendorn–Elspe syncline, which lies between the Ebbe anticline to the northwest and the Siegen anticline to the

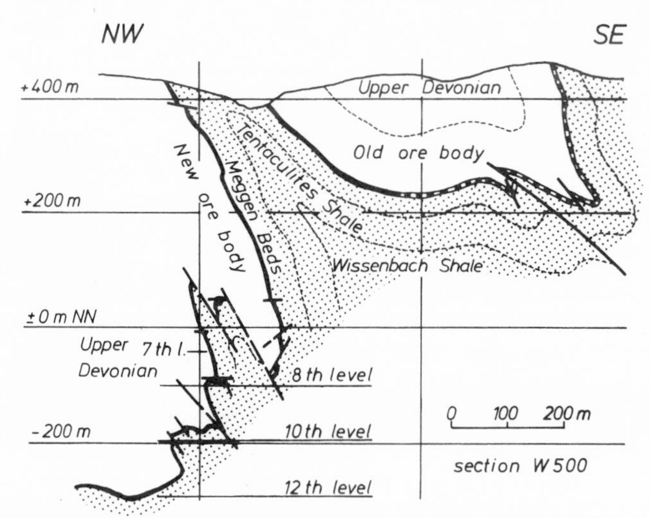

Fig. 8. Cross-section through the New and Old ore bodies along line W 500. (After Weisser, 1972.)

southeast. The Attendorn–Elspe syncline contains an Upper Devonian to early Upper Carboniferous series in its central part and is completely surrounded by Middle Devonian shelf and basin deposits. The syncline can be subdivided into the northwestern Attendorn syncline and the southeastern Elspe syncline (Fig. 17).

The axial planes and cleavage planes of the Variscan folds dip generally to the southeast; in addition the 50–60°-striking thrust-faults dip in the same direction (Fig. 8). The cleavage is well developed as a slaty cleavage in the argillaceous rocks and a fracture cleavage in the sandstone and thicker limestone beds. Due to the northwestward vergency of the folds, the southeastern limbs of the synclines and the northwestern limbs of the anticlines are overturned (Fig. 8). In a few cases, isoclinal folds are also developed. The axial planes plunge 10–15° towards the southwest or the northeast.

The Meggen ore crops out along the southeastern flank of the Elspe syncline over a length of 3.8 km from the southwest to the northeast (Fig. 9). Due to folding and subsequent erosion the originally continuous Meggen ore has been separated into two segments:

(1) The Old ore body ("Altes Lager") is confined to the Meggen syncline, a small oblong structure with an overturned southeastern limb, which contains early Upper Devonian strata in its inner part (Fig. 8). A narrow eroded anticline ("Luftsattel"), which is composed of a Middle Devonian sequence, separates the Old ore body from the New ore body ("Neues Lager").

(2) The New ore body occurs along the overturned and thrusted southeastern flank of the Elspe syncline and is known in the Meggen mine to a depth of about minus 400 m below sea level. Above the underlying Meggen Reef, the steeply dipping and overturned limbs of the Elspe syncline suddenly pass into extremely complicated fold structures with

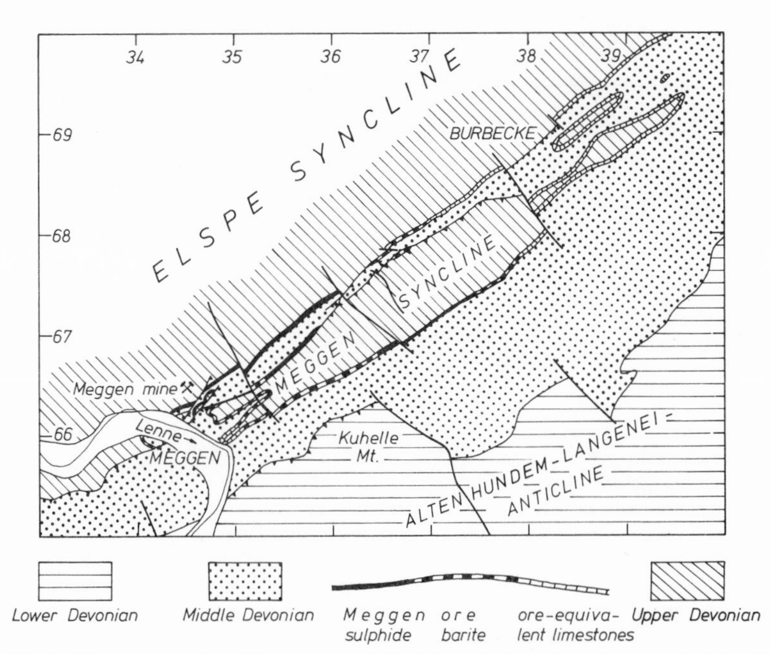

Fig. 9. The outcropping Meggen ore (sulphide and barite) and ore-equivalent limestones along the southeast flank of the Elspe syncline. (After Ehrenberg et al., 1954 and Clausen, 1978.)

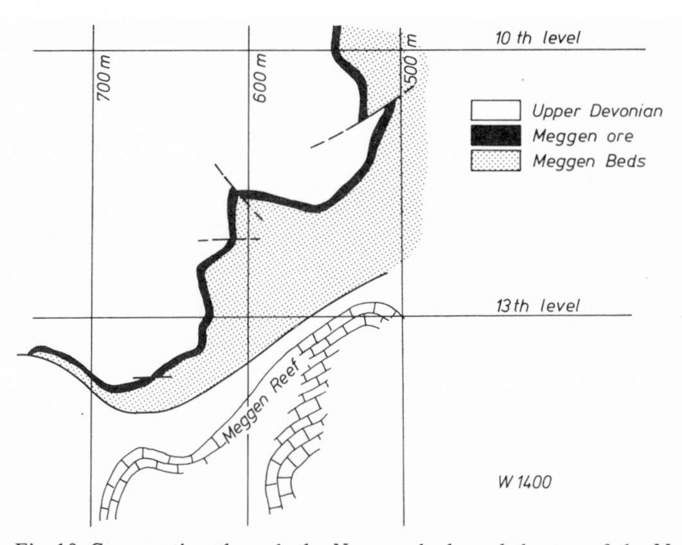

Fig. 10. Cross-section through the New ore body and the top of the Meggen Reef along line W 1400. (From W. Fuchs, Sachtleben Bergbau GmbH, unpublished.)

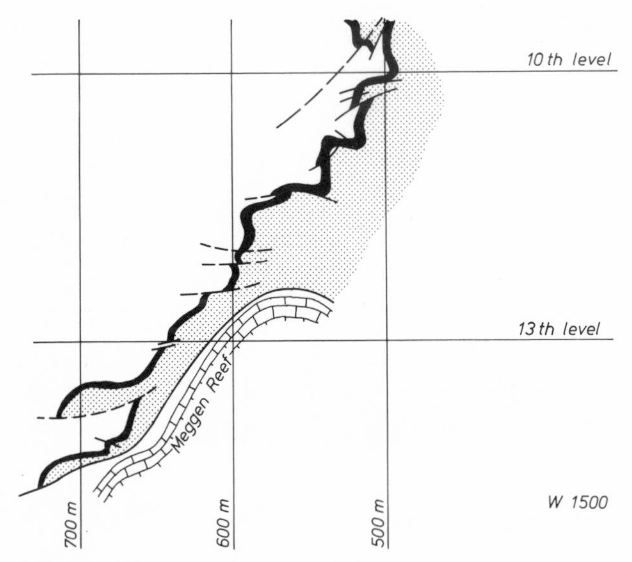

Fig. 11. Cross-section through the New ore body and the top of the Meggen Reef along line W 1500. For legend see Fig. 10. (From W. Fuchs, Sachtleben Bergbau GmbH, unpublished.)

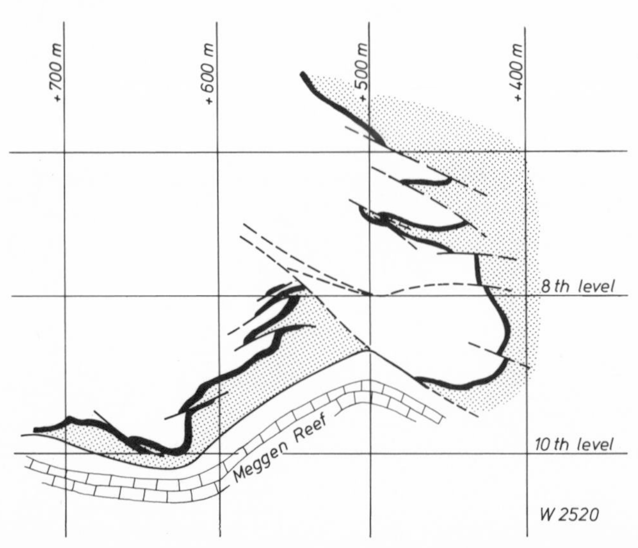

Fig. 12. Cross-section through the New ore body and the top of the Meggen Reef along line W 2520. For legend see Fig. 10. For position of line W 2520 see Fig. 15. (From W. Fuchs, Sachtleben Bergbau GmbH, unpublished.)

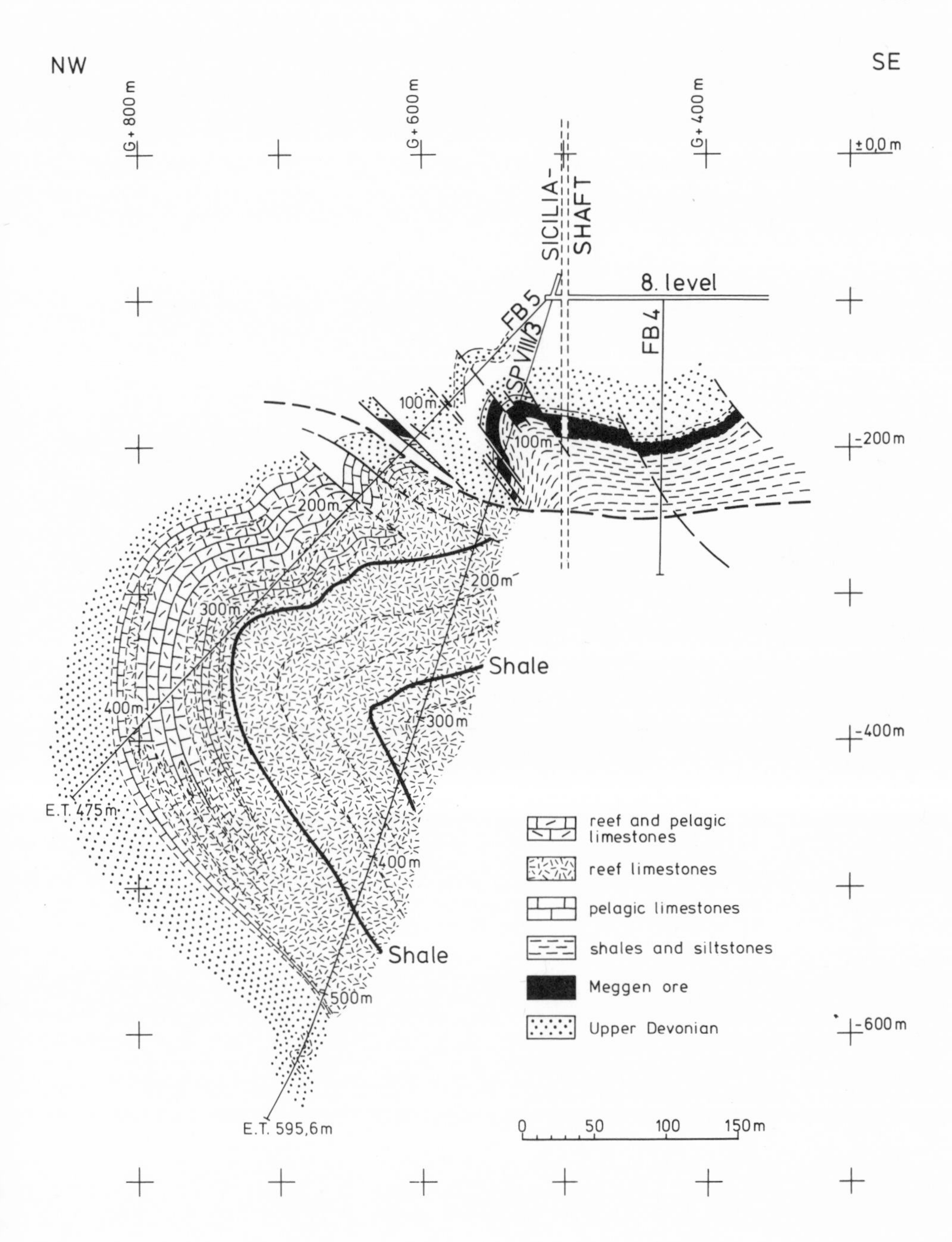

Fig. 13. Cross-section through the Meggen Reef (bore holes FB 5, Sp VIII/3 and FB 4). (From Krebs, 1978b.)

recumbent folds, flat-dipping thrust planes, tectonic repetitions of the ore body and northwest-dipping faults (Figs. 10–12). Figs. 10–12, reconstructed (by Dr. Fuchs, Sachtleben Bergbau GmbH Meggen) from underground workings and diamond drillings, illustrate impressively the strong influence of the topography of the folded competent carbonates on the strike and dip of the incompetent shaly series, including the interbedded Meggen ore horizon, which again acts as competent member within the deformed succession.

In the area of Meggen, the Meggen syncline (Old ore body) and the southeastern flank of the Elspe syncline (New ore body) belong tectonically to a Middle to Upper Devonian shale unit ("Schieferstockwerk"; Weisser, 1972). The ore and the associated predomi-

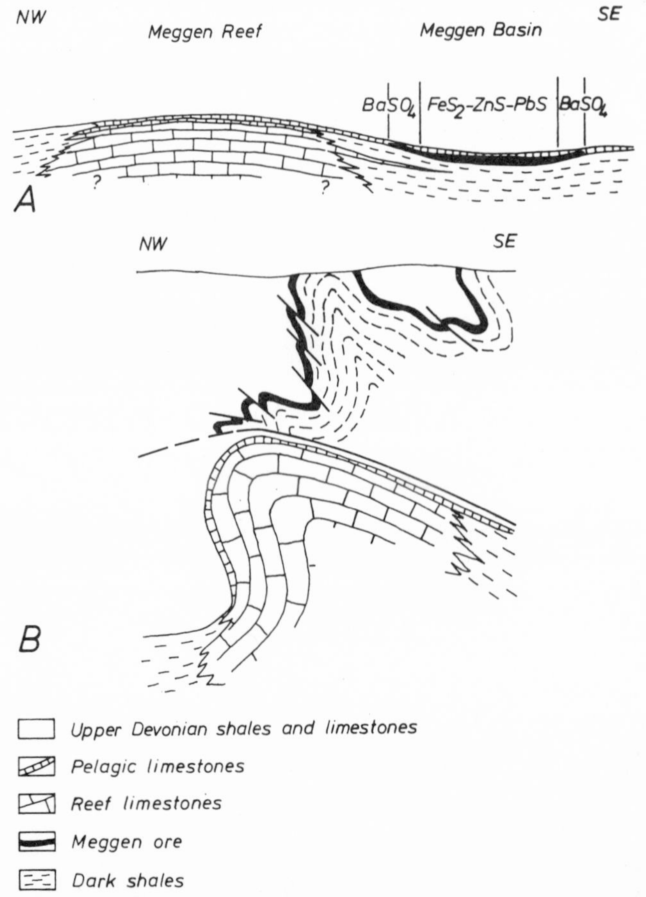

Fig. 14. A. Pre-orogenic position of the Meggen Reef and Meggen Basin. B. Post-orogenic situation of the Meggen Reef ("Kalkstockwerk") and the Meggen Basin ("Schieferstockwerk") separated by a large thrust plane.

nantly shaly sediments were originally deposited in a basin (Meggen Basin), which was part of the geosynclinal trough. The folded and thrusted shale unit, with a maximum vertical extent of 700 m, is separated from an underlying Middle Devonian limestone unit ("Kalkstockwerk"; Weisser, 1972) by a large thrust fault (Fig. 14). The limestone unit (Meggen Reef) is only known from underground workings in the Meggen mine and diamond drilling to a depth of 1000 m. The limestone unit is deformed to an overturned anticline with southeasterly dipping axial planes (Fig. 13), locally modified by thrust planes (e.g. FB 15). Only bore holes FB 5, Sp VIII and FB 15, which were drilled through the overturned anticline, again intersected late Middle Devonian and early Upper Devonian sediments which follow stratigraphically above the reef (Krebs, 1972b).

During Middle Devonian time, the Meggen Basin was originally situated toward the southeast of the Meggen Reef (Fig. 14A). Due to the compression of the Variscan orogeny, the shaly sediments of the Meggen Basin (shale unit) were folded and faulted along a flat-dipping overthrust ("Schubbahn") above the overturned anticlinal structure of the Meggen Reef (limestone unit; Fig. 14B). The dip of the large thrust, separating the two contrasting environments, obviously follows the convex-shaped crest of the competent Meggen Reef (Fig. 12). The continuation of that thrust-plane towards the northwest is unknown to date.

THE ORE

The Meggen pyrite–sphalerite–barite ore consists of a central sulphide body (50 million metric tons) with an average of 7% Zn and 1.0% Pb and a surrounding dark grey, fine-grained barite body (10 million metric tons) with an average grade of 96% $BaSO_4$ (Table II). The average thickness of the sulphide ore is about 4 m, that of the barite

TABLE II

Ore types, tonnage and average grade of the Meggen and Rammelsberg ore bodies (From Kraume, 1955, and Ehrenberg et al., 1954)

Ore body	Ore types	Ore tonnage (10^6 metric tons)	Average grade (%)			
			Zn	Pb	Cu	$BaSO_4$
Meggen	sulphide	50	7	1.0	0.02	<1
	barite	10	–	–	–	96
Rammelsberg	rich ore	22	19	9	1	22
	banded ore	2.5	8	4	0.5	5
	kniest	2.5	3	1.4	1.3	–
	gray ore	0.2	3	3	<0.1	85

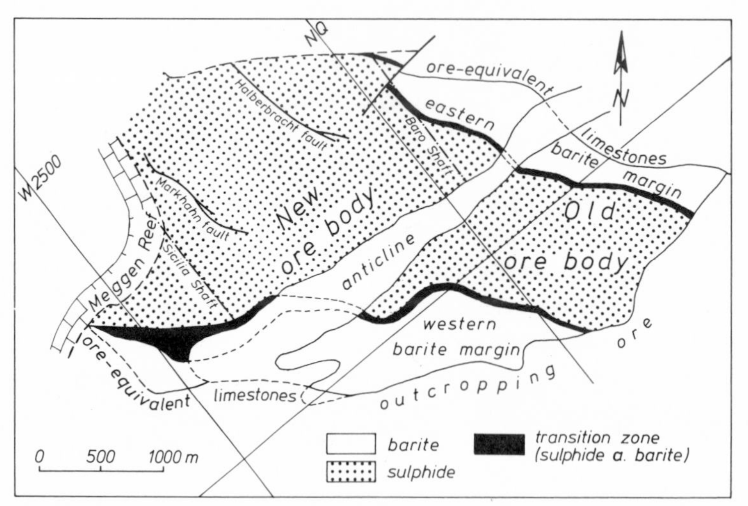

Fig. 15. Palinspastic map of the Meggen ore. (After Weisser, 1972, and Fuchs, 1978.)

between 1.5 and 3.5 m (Ehrenberg et al., 1954; Weisser, 1972). The maximum thickness of the sulphide locally attains 6–8 m.

In the sections where the marginal barite overlaps the peripheral sulphide ore, a narrow "transition zone" exists (Figs. 15 and 16). The barite margin, which is between 300 and 1500 m wide, wedges out into time-equivalent pelagic limestones (so-called "lager-äquivalente Kalke"). Fig. 16 shows the transition zone between the sulphide ore and barite on the 8th level of the Meggen mine (Dornsiepen, 1977).

The original oval-shaped pyrite–sphalerite–barite ore (Fig. 15), which covered an area of 10 km^2, was folded, partly eroded and separated into an "Old ore body" (Altes Lager) and a "New ore body" (Neues Lager). The names are based on the time of their discovery and are, therefore, unrelated to their geologic age. Both ore bodies are separated by an

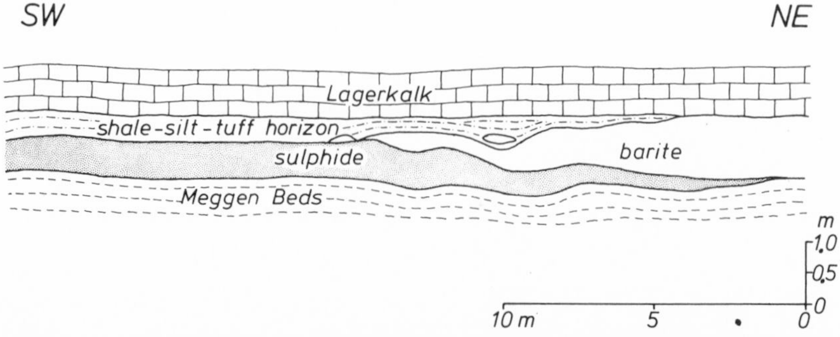

Fig. 16. Transition zone between the sulphide ore and the barite margin. New ore body, 8th level, cross-cut 69. (After Dornsiepen, 1977.)

anticlinal structure in which the footwall rocks are exposed (Fig. 8).

The mineralogy of the Meggen ore is quite simple. The sulphide ore consists of iron sulphides (72%), sphalerite (11%), galena (1%) and gangue minerals such as quartz, clay minerals, carbonates and barite (16%). The iron sulphides are mainly pyrite with minor amounts of marcasite and melnicovite. Only traces of chalcopyrite and boulangerite have been observed. Table III lists the major- and trace-element distribution in the Meggen and Rammelsberg ores. As, Ni, Co, Tl and traces of Au are related to pyrite, Cd to sphalerite and Ag to galena (Gasser, 1974; Gasser and Thein, 1977).

Gasser (1974) and Gasser and Thein (1977) have distinguished between a central, an intermediate and a marginal ore-facies. The pre-orogenic reconstruction of the Meggen ore shows a concentric zonation from pyrite–sphalerite–galena in the centre → pyrite → pyrite and interbedded gangue material → barite toward the margin. As, Co, Cu, Ni, Pb, Sb and Zn contents generally decrease towards the sulphide margin; whereas Ba, Fe, Mn, S, Tl and gangue material increase towards the margin (Table IV). Generally, the content of Zn decreases as the sulphides thin towards the periphery of the ore body. In contrast

TABLE III

Major- and trace-element composition of Meggen and Rammelsberg ore bodies

Major (%) and trace (ppm) elements		Rammelsberg (Kraume, 1955)	Meggen (Hilmer, 1972)
Zn		19	10
Pb		9	1.3
Fe		9	31
Cu		1	0.025
S	(%)	21	40
$BaSO_4$		22	0.29 [1]
SiO_2		10	9.4
CaO		4	1.2
Al_2O_3		3	2.4
Mn		1.3	0.2 [2]
Sb		800	110
As		500	830
Cd		500	310
Ag		160	3
Co		150	traces
Bi	(ppm)	70 [3]	n.d.
Sn		50 [4]	60
Hg		40	n.d.
Ni		20	20
Tl		10	190
Au		1.2	traces

[1] BaO; [2] Mn_3O_4; [3] locally up to 2100 ppm; [4] locally up to 730 ppm.

TABLE IV

Mean values of the groups of the Meggen deposit (From Gasser, 1974)

	Sand and silt (%)	S (%)	Fe (%)	Zn (%)	Pb (%)	Sb (ppm)	Ba (ppm)	Cu (ppm)	Mn (ppm)	As (ppm)	Ni (ppm)	Co (ppm)
Central group	15.70	41.85	33.48	7.92	1.05	217	169	40	1052	644	209	29
Intermediate group	18.57	41.42	34.07	5.45	0.49	144	382	18	1741	419	209	17
Marginal group	20.73	40.42	33.49	4.81	0.56	110	874	26	1227	270	130	16

to the Rammelsberg ore bodies, the elements studied do not show any stratigraphical vertical zonation.

Seiffert et al. (1952), Ehrenberg et al. (1954) and Gasser and Thein (1977) described a few fine-grained, often graded, sandy–silty bands within the sulphide ore, which were transported by turbidity currents into the Meggen depression. The same authors have also classified the sulphide deposit according to their textures and structures. Gasser (1974) and Gasser and Thein (1977) have described the following ore types:

(1) Type A_1: Grossly laminated to nodular pyrite, alternating with mm-thick black shale bands and sphalerite seams which contain up to 10% galena.

(2) Type A_2: As A_1, but distinguished by synsedimentary disturbance and folding, so that the bedding is completely obliterated. Sphalerite and galena often fill the interstices between the deformed pyrite layers.

(3) Type B: Mixed ore comprised of rapid horizontal and/or lateral transitions and repetitions between types A_1, A_2 and C.

(4) Type C: Finely laminated pyrite, alternating with paper-thin shale seams, with sparsely scattered sphalerite within pyrite and an absence of galena.

According to Gasser (1974), a standard section in the Meggen mine, 8th level, cross-cut 14, shows the following distribution of ore types: type A_1 and A_2 – 41%; type B – 34%; type C – 25%. Generally, the sulphide ores start with type A_1 and end with type C; but within the individual sections there are many irregularities in the vertical and horizontal distribution of ore types. The different ore types are very irregularly distributed within the central, intermediate and marginal ore facies. Gasser and Thein (1977), therefore, caution against reconstructions of the possible channelways of the ore-bearing solutions which are only based on ore textures (e.g. Ehrenberg et al., 1954).

The barite is dark grey, fine-grained, indistinctly developed finely laminated to homogeneous, which follows with a sharp contact above the thinning sulphide ore (Fig. 16). The barite, which surrounds the central sulphide body like a halo, has a lens-like shape in cross-section, thinning toward the sulphide ore as well as the time-equivalent pelagic limestones. Very thin black mudstone interbeds have the same composition and texture as those in the sulphide ore. The barite content of the ore varies between 94 and 96%. Only near the "transition zone", in which the barite overlies the marginal sulphide, are there nodules and thin bands of pyrite within the barite. According to Dornsiepen (1977), the base-metal content in these scattered sulphides is much lower than that in the "Kiesband-zonen" of the footwall and hanging-wall.

The average composition of the barite is according to Ehrenberg et al. (1954, pp. 176–177):

$BaSO_4$	95.58%
Fe_2O_3	0.49%
Al_2O_3	0.76%
MnO_2	0.07%
CaO	0.34%
MgO	0.08%
SiO_2	1.45%
FeS_2	0.28%
Loss on ignition	0.77%
Insoluble	0.10%
	99.92%

In the Old ore body the width of the barite ranges between 1 km and 1.5 km, whereas the barite margin of the New ore body is only 300–500 m wide. Immediately above the pyritic New ore body, lens-like barite concretions were locally reported which contained laminae of pyrite, galena and sphalerite (Ehrenberg et al., 1954).

In bore hole FB 16, fine-grained greenish tuff bands occur not only in the hanging wall (MT 1), but also directly beneath the barite horizon.

Macroscopically there are two types of barite: (1) lenses and layers of nodular, coarse-grained barite; and (2) lenses and layers of fine-grained barite. The dark colour of the barite is a result of finely distributed bitumen (Zimmermann, 1970). Microscopically, the Meggen barite is composed of (1) radially oriented barite fibres up to 1 mm in diameter which are particularly well-developed in the clay-rich types, and (2) coarser crystalline barite crystals which are formed due to tectonic deformation. In the latter type, the oblong to fibrous barite crystals are oriented parallel to the cleavage planes.

The average Sr content of the Old ore body attains 0.3%, that of the New ore body 0.79%. The barite of Meggen contains a total amount of 35,000 metric tons Sr (Buschendorf and Puchelt, 1965).

Footwall mineralization

In the Meggen Basin as well as in the Meggen Reef there are several horizons which show a conformable footwall sulphide mineralization:

(1) Ehrenberg et al. (1954, pp. 225 and 227) have described fine-grained pyrite bands and nodules in the uppermost Wissenbach Shale and in the Tentaculites Shale, both occurring in the area below the Meggen ore body.

(2) In the Meggen Reef, thin fine-grained pyrite layers and bands have probably a stratigraphic position similar to those in the adjacent Meggen Basin. In the reef unit S,

which corresponds to the uppermost Wissenbach Shale, bore hole FB 2 has intersected (878.80-879.30 m) a fine-grained laminated pyrite horizon. The sulphides do not show any enrichment of Zn and Pb. Finely laminated pyrite layers and lenses, alternating with reef-detritus beds, were also observed in the reef unit L of bore hole FB 2. From a stratigraphical point of view the unit L can probably be compared with the upper Tentaculites Shales or the Odershausen Limestone.

(3) In the uppermost Meggen Beds ("Obere tonige Zone") a few metres below the Meggen ore, a widespread conformable horizon with layers, lenses and nodules of fine-grained pyrite is developed ("Kiesbandzone"). According to Amstutz et al. (1971), the pyrite nodules locally contain baritic cores. Irregular lower surfaces and flat upper surfaces of the pyrite layers and lenses were explained as the result of settling and compaction. Sphalerite becomes more abundant towards the Meggen ore. In addition, other horizons of the Meggen Beds locally show pyrite nodules and/or disseminations of fine-grained iron sulphides.

According to Dornsiepen (1977), the average composition of the sulphide concentrate of pyrite nodules below the Meggen ore is: Zn – 520 ppm; Pb – 980 ppm; Cu – 240 ppm; Ni – 590 ppm.

(4) Discordant mineralization was reported beneath the Meggen ore body in brecciated and fractured, partly silicified Wissenbach Shales in the anticline between the Old and the New ore body. Fractures are filled with chalcopyrite, sphalerite and galena, which may be comparable to the Kniest of the Rammelsberg ore body (Ehrenberg et al., 1954, p. 225).

Hanging-wall mineralization

Noa (1958) and Wellmer (1966) discovered two iron sulphide horizons in the hanging wall of the Meggen ore which can be used as important marker horizons ("1st and 2nd Kiesbandzone").

The first pyrite-rich horizon ("1st Kiesbandzone") occurs 20–25 m above the Meggen ore and belongs stratigraphically to the boundary between the middle and upper Frasnian (Dornsiepen, 1977). The "1st Kiesbandzone" is immediately overlain by the tuff horizon MT 4 (Fig. 2). The horizon has a thickness of 0.5–1.0 m and consists of grey nodular limestones and green and black shales interbedded with mm-thick sulphide-rich laminations and disseminations of pyrite and marcasite.

According to Dornsiepen (1977), the "1st Kiesbandzone" is composed of: (1) fine-grained aggregates of pyrite and marcasite with a grain size up to 500 μm; (2) mostly idiomorphic pyritic crystals of 100–300 μm in diameter; and (3) framboidal pyrites, composed of accumulations of crystals 10–40 μm in diameter, are concentrated in conformable lenses and layers.

The average composition of the "1st Kiesbandzone" is in the sulphide concentrate

(Dornsiepen, 1977):

Zn 3300 ppm
Pb 1700 ppm
Cu 700 ppm
Ni 410 ppm .

In comparison to the Frasnian black shales above the Lagerkalk, the metals in the "1st Kiesbandzone" are enriched 20–25-fold for Zn, 12–17-fold for Pb and 13-fold for Cu.

A second pyrite-rich horizon ("2nd Kiesbandzone"), composed of pyrite and marcasite laminations, nodules and disseminations in shales and nodular limestones, occurs 45–50 m above the Meggen ore (Fig. 2). Stratigraphically the "2nd Kiesbandzone" belongs to the lower Famennian (Nehden stage).

ISOTOPE GEOCHEMISTRY

Sulphur isotopes

Sulphur isotopes of the sulphide ores and the surrounding barite of Meggen have been analysed by Buschendorf et al. (1963). The isotopic compositions of sulphur in the Meggen sulphides and sulphates differ only slightly from each other and show a markedly narrow spread of the $\delta^{34}S$ values:

Meggen ore	Range	Average
pyrites from the ore	+17.8–+24.1‰	+19.8‰
galena from the ore	+16.4–+19.4‰	+18.0‰
sphalerite from the ore	+11.9–+19.9‰	+16.4‰
barite	+20.8–+26.8‰	+23.4‰

In comparison the isotopic composition of sulphur in the Rammelsberg sulphide ores ranges in $\delta^{34}S$ from +7 to +20‰ and that of the Rammelsberg barites from +20 to +33‰ with a main value of +23‰ for barite (Anger et al., 1966).

The sulphur of the barites from Meggen and Rammelsberg was undoubtedly derived from the Devonian sea-water sulphate, because both mean $\delta^{34}S$ values correspond to the assumed Middle Devonian sea water composition of +23‰ (Buschendorf et al., 1963; Anger et al., 1966; Holser and Kaplan, 1966).

According to Barnes (1974/75) (see also Chapter 8, Vol. 2, in this Handbook), the average sulphur isotope ratios of sulphides in syngenetic hydrothermal deposits, such as the Meggen-Rammelsberg type, range from those for igneous sulphur, near 0‰ and seawater sulphur characteristic of each geological period. The isotopic composition of the Meggen sulphides was interpreted by Buschendorf et al. (1963) as being derived from metalliferous hydrothermal solutions. The enrichment of the heavier sulphur isotope ^{34}S in the Meggen sulphides was explained by the incorporation of sediments containing sulphur with higher $\delta^{34}S$ values during the rise of the hydrothermal solutions to the surface. Wedepohl et al. (1978), however, have discussed the possibility that metal-bearing thermal brines could have dissolved sulphides from the Lower Paleozoic sedimentary basement beneath the Meggen and Rammelsberg ore bodies. The isotopic composition of sulphur in older Paleozoic sandstones is reported by Wedepohl et al. (1978) to exceed +20‰, and that of older shales attains +13‰.

Lead isotopes

The lead-isotopic composition of Pb–Zn ores of Meggen and Rammelsberg was first reported by Brown (1965). Brown has already pointed out that the isotopic composition of lead in the Middle Devonian Meggen–Rammelsberg ore bodies is virtually identical with that of the epigenetic hydrothermal vein deposits in the Rhenish Schiefergebirge (Christian Levin, Lüderich, Ramsbeck, Siegerland) and Harz Mountains (Oberharz, St. Andreasberg). However, the hydrothermal vein deposits of Christian Levin in the Ruhr area, and other deposits in the Oberharz, have an Upper Carboniferous or younger age. In addition the galena-impregnated Triassic sandstones and conglomerates of Maubach and Mechernich, Eifel area, have about the same isotopic Pb composition as the above-mentioned deposits.

Wedepohl et al. (1978) have shown that the lead-isotopic composition of the hydrothermal syngenetic ore deposits (Meggen and Rammelsberg), hydrothermal epigenetic veins (Bad Grund/Harz Mountains and Ramsbeck/Sauerland), as well as the sedimentary syngenetic deposits (Permian Kupferscheifer; see Chapter 7, Vol. 6, in this Handbook) is almost identical within Central Europe. They conclude that the lead-isotope data suggest a common source from Precambrian detritus for all the genetically different Pb–Zn deposits. According to Wedepohl et al. (1978), the extraction of lead by thermal brines from country rocks of the Variscan geosyncline containing Precambrian detritus produces a comparable isotopic composition to diagenetically recycled lead extracted from the footwall of the Kupferschiefer which contains the same Precambrian detritus. The lead impregnations of the Triassic sandstones and conglomerates of Maubach and Mechernich were probably also formed by the extraction of subsurface lead from Precambrian detritus during the rifting of the Rheingraben system (Brown, 1965).

GEOCHEMISTRY OF ELEMENTS

The geochemical composition of the footwall and hanging-wall shales were described by Krüger (1973), Gwosdz et al. (1974) and Dornsiepen (1977). The Mn content of the

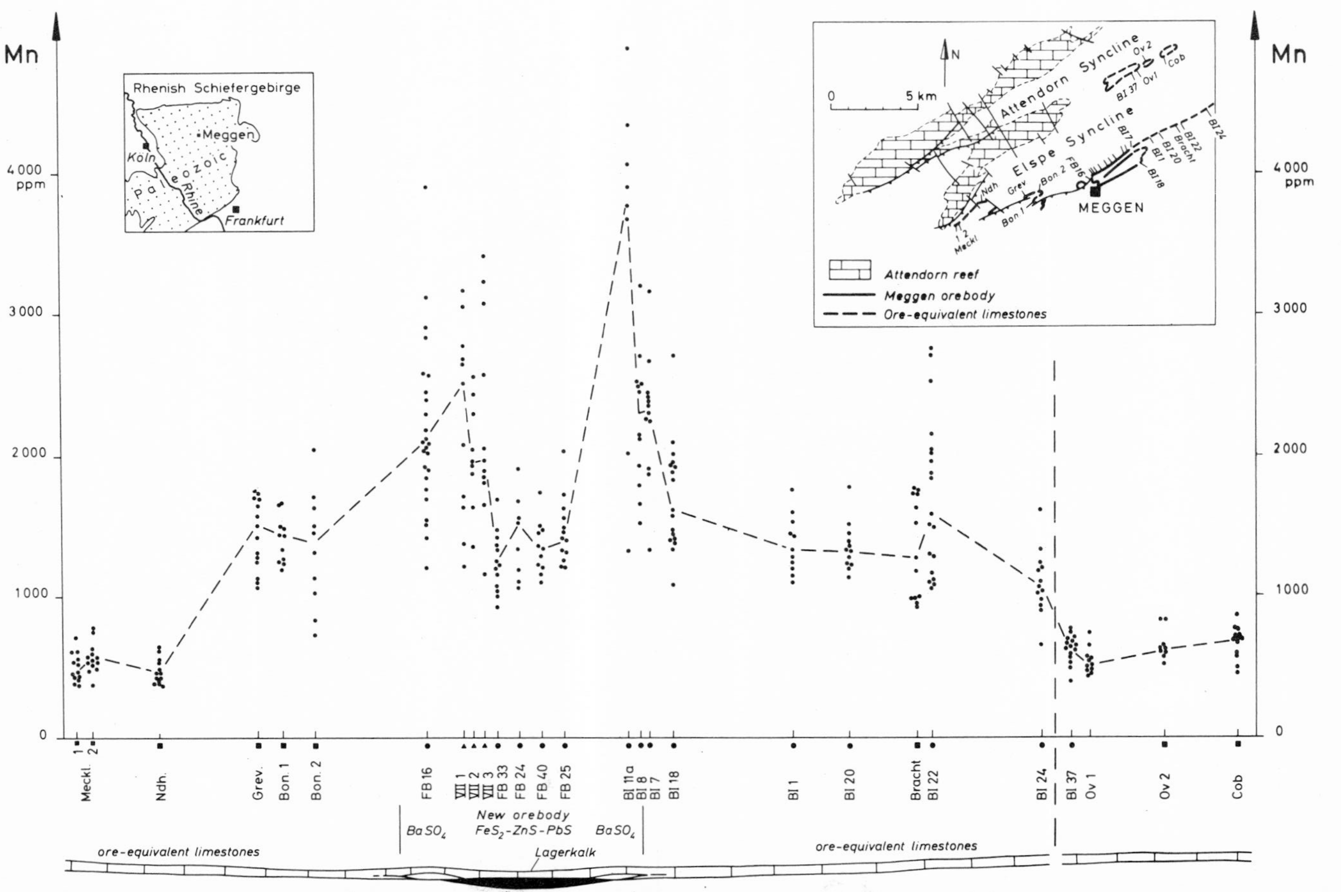

Fig. 17. Manganese distribution in the Lagerkalk and ore-equivalent limestones in the area of Meggen. The southeastern flank of the Elspe syncline is separated from an anticlinal structure between Elspe and Attendorn syncline by a broken vertical line; other line links median values. ■ quarry, ▲ underground working, ● drill hole. (From Gwosdz and Krebs, 1977.)

TABLE V

Geochemical distribution of some elements in rocks, ores, soils and stream sediments in the area of Meggen (After Friedrich et al., 1972)

	Pb	Zn	Cu	Cd	As	Ni	Co	Hg
Rocks (ppm)								
Meggen Beds	55	75	20				15	55
Limestones	30	50	15				15	15
Ore (%)								(ppb)
Massive sulphide	1.28	10.26	0.028	0.031	0.08	0.03	0.006	1100–8500
Pyrite lenses in shales (ppm)	130– 950	200–2100	70				40	100– 200
Soils (ppm)								(ppb)
Background								
A_0	155	70	55	1.08		40	20	
A_1	95	120	50	0.88	10	35	25	130
A_2	70	20	20	0.42		30	10	55
B_1	125	110	55	0.64	30	60	20	
B_2	95	100	45			45	15	
C	50	30	70			45	10	
Above ore								
A	265– 870	300–1600	60	1.62		80	50	440
B	1970	4700	115	1.06		105	60	360
C	2960	4960	150	0.72		105	40	150
Stream sediments (ppm)								
Background	55	90	25			60	20	
Anomaly	220–9860	280–4000	100–380			120–330	50–215	

Lagerkalk and the ore-equivalent limestones was analysed by Gwosdz and Krebs (1977). As Fig. 17 shows, outside of the Meggen ore deposit the background values of Mn range from 400 to 900 ppm. Within a distance of <2 km southwest and northwest of the ore body, however, the Mn content increases up to 1000–2000 ppm. In the vicinity of the barite margins the Mn content of the Lagerkalk distinctly rises to above 2000 ppm. The Mn is interpreted as being derived, together with the base metals, from a synsedimentary exhalative supply. Gwosdz and Krebs (1977) have pointed out that the lateral extension of the Mn halo to 5 km and more beyond the ore body signifies an important prospection aid for the location of hidden stratiform lead–zinc–copper–barite deposits of the Meggen–Rammelsberg type.

Hilmer (1972) and Friedrich et al. (1972) investigated the geochemistry of the soils above the outcropping Meggen ore as well as soils and stream sediments in the footwall and hanging-wall series in the Meggen area (Table V). In the soils above the Meggen ore, significant anomalies of Zn up to 2000 ppm have been found and above the gossan up to 5000 ppm Zn. The Zn anomalies are very narrow (15–20 m) and strictly limited to the outcropping and steeply dipping ore horizons. A rapid transition exists from the high, but narrow, anomalies to the background values of Zn(100–200 ppm) of the soils. Pb is particularly concentrated in the gossan, but no Pb anomalies are present where the overburden above the ore is thicker than 5 m. All other analysed elements (such as Cu, Cd, Co, Ni, As and Hg) do not show any significant anomalies in the soils above the Meggen ore. In the soils above the barite of the Old ore body only minor Zn values were found.

The Pb and Zn anomalies in the stream sediments in the southeast of the Meggen area are strongly influenced by mineralized hydrothermal veins cutting Middle and Lower Devonian sediments.

ORE-CONTROLLING PARAMETERS

Krebs (1978a) has listed the ore-controlling parameters for the shale-hosted stratiform lead–zinc–barite ores of the Meggen–Rammelsberg type in Western Germany:

(1) Relation to pelagic basinal sediments of the Variscan trough

The Meggen ore is confined to dark grey to black pelagic shales (Wissenbach Shale facies). Nektonic, planktonic and pseudoplanktonic organisms, as well as frequently interbedded sandstone or siltstone turbidites suggest a water depth of at least several 100 m during late Middle Devonian time. The basinal areas of the Variscan trough can be subdivided according to their size into first-, second- and third-order depressions (Krebs, 1978a).

(2) Relation to paleotopographical depressions on the sea floor which acted as a potential "ore trap"

These third-order depressions on the sea floor, with a diameter of several 100 m to a few km, form the preconditions for the accumulation of base metals because they act as morphological traps and prevent a mechanical dilution of metal-bearing brines. The euxinic environment, indicated by an unusually high content of organic carbon in the shales, acts as a geochemical trap preventing the oxidation and chemical dilution of base metals. (For further processes related to the importance of organic matter and bacteria, see Chapters 5 and 6, Vol. 2, in this Handbook.)

According to Krüger (1973) and Gwosdz et al. (1974), the third-order basin before and during the deposition of the Meggen ore was characterized by:

(a) a low calcium carbonate content obviously due to the influence of acid hydrothermal solutions (Bäcker, 1973; Shanks and Bischoff, 1977);

(b) a locally higher amount of allochthonous reef-detrital limestones which were accumulated in the third-order depression;

(c) the deposition of distal silty turbidites which continued during the deposition of the sulphides and ended immediately after the ore deposition (shale-silt-tuff horizon);

(d) higher contents of base metals, frequent pyrite layers and nodules and pyrite impregnations in the dark grey to black shales. The Zn values, for example, are significantly enriched in the Meggen Beds below the Meggen ore in contrast to the time-equivalent shales outside the third-order depression. These observations suggest that the minor quantities of fluids were repeatedly discharged into the sub-basin before the main mineralization took place.

(e) High concentrations of Mn which, in the Meggen Beds below the overlying barite margin, attain up to 1000 ppm (maximum 4000 ppm). Outside the overlying Meggen ore, however, the Mn content decreases to 400–600 ppm. Due to the decreasing primary calcium carbonate content of the Meggen Beds below the Meggen ore, no correlation exists between manganese and calcium carbonate.

(f) A higher content of organic carbon in contrast to the time-equivalent surrounding dark shales. The dense and saline brines, which filled the paleotopographically low areas in the third-order depressions, resulted in euxinic conditions. Here ocean currents and wave action were totally absent.

(3) Relation to paleogeographical turning points

Many of the shale-hosted stratiform lead–zinc–barite deposits in the Central European Variscan fold belt are stratigraphically related to widespread transgressions of the basinal facies upon shelf sediments. The rapid transgressions were obviously connected with a higher tectonic mobility, which resulted in subsidence and/or collapse of third-order basins, tilting of fault-bounded blocks, differential subsidence along lineaments, and the formation or reactivation of deep-seated faults.

The late Givetian Meggen ore body is more or less contemporaneous with the widespread transgression of late Givetian pelagic black shales (Flinz Shale) upon middle Givetian shelf sandstones and biostromal limestones (Newberrien Beds and *Sparganophyllum* limestone). The time of ore formation in the Meggen Basin was marked by the interruption of sedimentation on the adjacent Meggen Reef and intensive submarine fracturing of the reef and the overlying lower pelagic limestone unit (Fig. 2).

(4) Relation to culminations of volcanic activity or intrusive heat

The Meggen ore deposit is indeed contemporaneous with the eruption of the basic Hauptgrünstein volcanism in the Sauerland area (Balve, Meschede, Brilon, Adorf), but no traces of synchronous basic volcanics (Pilger, 1972) in the whole Attendorn–Elspe syncline have been found. On the other hand, the Meggen area is coincident with the northern flanks of the eruptive centres of the Lower Devonian keratophyre and keratophyre tuff horizons K 3–K 7 (Rippel, 1954). However, the vertical distance between the acid tuff horizon K 6 and the Meggen ore, is about 1000 m (Figs. 18 and 19).

At present the acid tuff horizons MT 1–MT 4 are only known in underground workings and diamond drill-holes above the Meggen ore (Dornsiepen, 1977). Chemical analyses of the MT 2–MT 4 horizons (Table VI) show some similarities to acid tuffs in the Rammelsberg area and other acid or bentonitic tuff horizons, although some alteration processes took place due to diagenetic processes and tectonic deformations (Dornsiepen, in press). The main alteration products are sericite, chlorite and calcite. Nevertheless, these thin tuff horizons MT 1–MT 4 indicate a reactivation of volcanic processes along the southeastern flank of the Elspe syncline which was spatially related to the Meggen ore. Dornsiepen (in press) assumes that the Meggen tuffs have been derived from an alkaline suite.

Ehrenberg et al. (1954) postulated a hypothetical subvolcanic pluton beneath the Meggen ore from which exhalative ore-bearing solutions ascended. Scherp (1974) supposed a long and persistent cooling time (8–10 m.y.) of the late Emsian plutonic intrusions which underly the vent centres K 3–K 7 south of Meggen. According to Hodgson and Lydon (1977), hydrothermal systems not associated with active volcanoes may still be related to deeper-seated intrusive heat. Recent systems of this type tend to show lower temperatures (98–245°C, averaging about 170°C). The low Cu and high Zn ores of Meggen would be expected from relatively low-temperature hydrothermal systems.

(5) Relation to hydrothermally altered and mineralized channelways

Many sediment-hosted stratiform lead–zinc–barite ores are associated with mineralized source channels and/or alteration and stringer zones. A striking example forms the "Kniest" below the Rammelsberg deposit with intensive silicifation of the footwall and

TABLE VI

Chemical composition (in %) of Meggen tuffs in comparison to other tuffs associated with stratiform lead-zinc ores (2,4 and 5) or to other Devonian tuff horizons (3 and 4)

	1 Meggen tuffs (10) * Dornsiepen (in press)	2 Rammelsberg tuffs (3) Kraume and Jasmund (1951)	3 Bentonites Eifel (7) Winter (1969)	4 Keratophyre tuff Lahn syncline (1) Winter (1969)	5 MacArthur River (2) Croxford and Jephcott (1972)	6 Mount Isa Croxford (1968)
SiO_2	47.66–48.49	48.5–50.6	43.50–51.90	49.50	56.36–68.50	60.06
TiO_2	0.34– 0.52	0.5– 0.9	0.21– 0.49	0.98	0.39– 0.45	0.20
Al_2O_3	24.06–26.14	27.7–30.5	22.40–24.80	13.80	12.14–13.43	15.92
Fe_2O_3	0.90– 1.44	1.9– 3.9				1.08
FeO	2.34– 6.99	0.5– 1.7	1.20– 3.90	6.70	4.19– 5.40	0.62
MnO	0.02– 0.04	0.01	–	–	0.01– 0.16	0.06
MgO	2.46– 3.27	1.1– 1.6	0.78– 2.12	1.32	0.17– 2.86	0.88
CaO	0.70– 4.91	0.4	0.32– 7.55	5.01	0.03– 4.16	2.82
Na_2O	0.22– 0.88	0.7– 1.6	0.09– 1.13	1.88	0.12– 0.15	0.15
K_2O	6.37– 6.73	4.8– 6.8	3.15– 6.91	5.80	8.38– 9.68	13.42
P_2O_5	0.08– 0.15	0.1	0.07– 0.5	0.34	0.02– 0.08	0.28
$H_2O^{\pm}$	5.10– 6.80	4.7– 6.3	13.70–20.40	8.60	1.19– 1.64	0.70
CO_2	0.16– 2.96	0.4– 0.8	0.01– 6.90	6.50	0.49– 6.38	2.97
S	0.6	0.3	–	–	2.70– 2.72	0.57

* Number of samples in parentheses.

fracture-filling mineralization of arsenopyrite, chalcopyrite, sphalerite and galena (Ramdohr, 1953; Kraume, 1955).

Ehrenberg et al. (1954) reported from the footwall shales between the Old and the New ore body hydrothermally altered, silicified and fractured portions filled with chalcopyrite, sphalerite and galena. Schmidt (1919) points out that, in the vicinity of Meggen veins of barite and sulphides can only be observed in the footwall series of the Meggen ore, which may be interpreted as possible synchronous channelways. The presence of Cu in the footwall shales of Meggen would indicate higher temperatures in contrast to the copper-poor stratiform sphalerite–pyrite mineralization.

Whereas at Rammelsberg the vertical metal sequence As–Fe–Cu–Zn–Pb–Ba indicates decreasing temperatures, and therefore there are many similarities to the volcanogenic sulphide deposits (Lambert and Sato, 1974; Brathwaite, 1974; Barnes, 1974/75; Lambert's Chapter 12, Vol. 6, in this Handbook), there is no clearly visible vertical metal zonation at Meggen.

(6) Relation to active faults and lineaments

All present-day hydrothermal systems occur in association with fissures and fault structures. "The most common structural association on the scale of an individual hydrothermal system in each case is with tilted fault blocks and half grabens, and within this structural setting hydrothermal systems are most commonly found where the block-bounding faults are intersected by transverse fractures or where they change their strike" (Hodgson and Lydon), 1977, p. 96).

The Meggen ore deposit lies near the southwest–northeast trending Middle Devonian shelf–basin boundary which was active as a mobile hinge line separating areas of different subsidence and bathymetry over a long period (Krebs, 1972a; Gwosdz, 1972; Fig. 6).

During middle and upper Givetian time, for example, shallow-water sediments (>1300 m thick) along the northwestern flank of the Attendorn syncline correspond stratigraphically to 40–60 m pelagic deposits along the southeastern flank of the Elspe syncline in the Meggen area (Gwosdz, 1972). The probably fault-controlled boundary between the Middle Devonian Meggen Reef and the time-equivalent shales in the Meggen Basin also runs subparallel to the shelf edge. Fig. 6 shows a highly exaggerated cross-section through the northern Rhenish Schiefergebrige from the Balve Reef to the Meggen Basin. The increase in thickness of Devonian reef limestones toward the northwest, the local occurrence of basic volcanics and rapid facies transitions can be interpreted as tilted fault-bounded blocks and abrupt thickness and facies changes along synsedimentary active faults.

According to Pilger (1957, 1972), the Meggen area is intersected by the NW–SE trending Unna–Giessen lineament. The Unna–Giessen lineament controls the eruption centres of the Middle Devonian diabases at Balve, as well as the vent centres of the Lower Devonian acid volcanics at the Ebbe and Müsen anticline (Rippel, 1954). The submarine destruction of the Meggen Reef by numerous neptunian dikes reflects on a minor scale the synsedimentary tensional movements during the formation of the Meggen ore (Krebs, 1972b, fig. 5).

In summary, the area of the Meggen ore is characterized by SW–NE trending synsedimentary fault systems and cross-cutting NW–SE striking lineaments which, during late Lower Devonian to early Upper Devonian time, controlled the subsidence, sedimentary thickness and facies, volcanic vent centres, heat source, and, finally, ascending ore-bearing hydrothermal solutions.

ORIGIN

Different explanations have been offered for the origin of the Meggen deposit, ranging from epigenetic theories through syngenetic sedimentary to the generally accepted syngenetic hydrothermal hypotheses. All these various hypotheses are discussed in detail by Ehrenberg et al. (1954, pp. 257–262).

Besides the thin tuff horizons MT 1–MT 4 above the Meggen ore, there are no indications of volcanic or intrusive activity in the Meggen area during the time of ore deposition. However, the position of the Meggen ore spatially coincides with a large region of repeated Lower Devonian volcanic activity (Rippel, 1954). Due to the uprising geothermal gradient in the upper mantle, a hot spot was formed during Lower Devonian time, which implies anatectic processes in the upper crust and the repeated ascent of acid magmas (volcanic horizons K 3–K 7). The late Emsian "Hauptkeratophyr" (K 4) attains a maximum thickness of 300 m and covers an area of about 2000 km^2. The sediments between the volcanic horizons K 4 and K 7 (quartz keratophyric and keratophyric effusions and/or pyroclastics) locally contain concretions of iron carbonate, iron sulphide, iron oxide and silicate forming the so-called "Sphärosideritscheifer". The distribution of this anomalous facies shows that the concretions undoubtedly coincide with the greatest thickness of the acid volcanics. They can be interpreted as distal exhalites (Solle, 1937), spatially and temporally related to the vent centres of the acid volcanism (Fig. 18).

According to McNitt (1970), hydrothermal systems associated with active Quaternary volcanic centres generally have higher temperatures (95–360°C, average 235°C) than those in areas of past volcanic activity (98–245°C, average 170°C). Most hydrothermal fields are associated with acid volcanic (dacite and rhyolite), rather than basaltic volcanics, because rising hot-water systems are "supplied by a plume of magmatic hydrothermal fluid given off from a subsurface felsic magma body" (Hodgson and Lydon, 1977, p. 96).

The anatectic processes during the late Lower Devonian, which persisted south of Meggen over a period of at least 5 m.y., produced an anomalous high geothermal gradient in the upper crust. According to Scherp (1974), subsurface heat intrusion continued during Middle Devonian time over a period of 8–10 m.y. due to the slow cooling of a large intru-

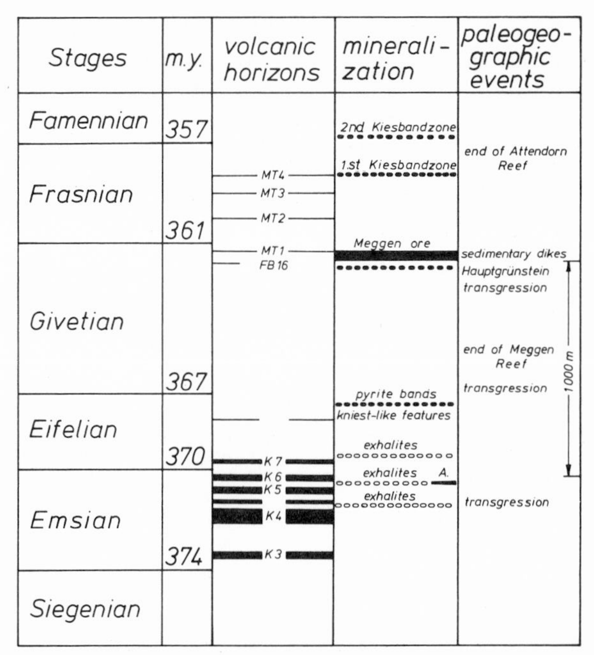

Fig. 18. Volcanic horizons, stratiform mineralization and palaeogeographic events during Lower to Upper Devonian time in the Meggen area and vicinity. *A.* = pyrite ore of Auerhahn/Müsen anticline. *FB 16* = tuff horizon below barite ore in the bore hole FB 16.

sive mass. Davis and Lister (1977) have pointed out that a sediment cover prevents the hydrothermal waters in the igneous lithosphere rising to the ocean. The same authors estimate that at the Juan de Fuca Ridge, at the top of the igneous crust, the temperatures, which generally range from 300 to 100°C, range from 200 to 25°C where there is no protecting sediment-cover above the volcanics.

The Cu-poor, Zn-rich ores of Meggen suggest a relatively low-temperature hydrothermal system (Hodgson and Lydon, 1977), which is in agreement with recent hydrothermal systems not associated with active volcanoes, but only with intrusive heat. From a theoretical point of view, Cu-rich ores should be expected in the older stratigraphic horizons below the stratiform Meggen ore along the channelways of uprising thermal brines. Many syngenetic-hydrothermal stratiform ore deposits are genetically associated with such an epigenetic Cu mineralization and hydrothermal alteration representing the channelways (Rammelsberg, Mt. Isa, McArthur River, some of the Irish lead–zinc deposits, Sullivan) (see indexes in Vols. 4 and 7 of this Handbook for chapters discussing these deposits).

The tectonic disturbances, as seen in the abrupt facies changes between shelf and basin in the Attendorn–Elspe syncline and in the submarine destruction of the Meggen Reef by neptunian dikes, reached their culmination in the late Givetian time (Fig. 18). At the same time, in the northeastern Sauerland the basic Hauptgrünstein was extruded and the hematite ores of the Lahn-Dill type were formed (Venzlaff, 1956). During the culmination of vertical tensional movements along the shelf margin and cross-cutting lineaments, deep-seated faults were formed or reactivated along which the rising column of convected fluids escaped onto the sea floor.

Fig. 19 shows the stratigraphic position of sediment-hosted, stratiform lead–zinc–barite deposits in Germany. The metal-bearing thermal brines were discharged into the third-order depression of the Meggen Basin. The sharp boundary between the sulphides and the barite can probably be explained by a distinct stratification in the third-order depression that coincided with the H_2S/O_2 interface. In the higher saline waters, which filled the paleotopographically low areas of the Meggen Basin, under reducing conditions and in the presence of dissolved sulphide, Fe, Zn, Pb and minor Cu-sulphides were precipitated, whereas Mn and Ba remained in solution. The reducing low-energy environment is represented by black shales rich in organic carbon. The decrease of autochthonous $CaCO_3$ in the Meggen Beds below the Meggen ore (Krüger, 1973; Gwosdz et al., 1974) and the dissolution of tourmaline in the footwall clastics of the Rammelsberg ore (Paul, 1975) indicate that the brines had an extremely low pH of the order of 3, which is comparable to recent brines at Matupi, New Guinea (see Chapter 12, Vol. 6 in this Handbook).

More or less normal, oxygenated sea water lay above the higher-density, reducing bottom waters with a sharp boundary. Here, Ba was rapidly precipitated as barite by the reaction with the seawater sulphate on the flanks above the deepest part of the Meggen Basin. The oxidation of sulphides was obviously only a minor process at Meggen.

According to Fuchs (oral communication, 1978), there is no barite margin along the northwestern flank of the Meggen Basin (Fig. 15), because in some boreholes the sulphide ore thins out into time-equivalent sediments without any intercalation of barite. The absence of barite near the edge of the Meggen Reef can be explained by the mechanism of tilted fault blocks, as shown in Fig. 6. Like the other fault blocks, the Meggen Basin was also tilted toward the northwest, so that in the deepest part euxinic conditions prevented the precipitation of barite. The decreasing depth of the Meggen Basin toward the southeast explains the increasing width of barite margin due to oxidizong conditions in the same direction (Fig. 15).

The irregularly distributed and highly distorted sulphides (type A_2) within the Meggen ore suggest that they may have been subject to seismic shocks rather than submarine slumping. The rhythmic sequence between the autochthonous black-shale bands and the sulphide layers (types A_1 and C) can probably be explained by the mechanism of "seismic pumping" at the discharge channels (Sibson et al., 1975).

Within the sulphide ores the Pb/Zn ratio is generally decreasing from the discharge centres. The highest Pb/Zn ratios within the sulphide ores were reported along the anticlinal area separating the Old and New ore bodies (Ehrenberg et al., 1954). The observations of a kniest-like Cu–Zn–Pb mineralization in the footwall shales of the same anticline is consistent with the assumption of a discharge area between the Old and New ore body (Fig. 15).

Extremely high Pb/Zn ratios as well as a relative high Cu content were observed from borehole 44 III near the tectonic boundary between the Meggen Basin and the Meggen Reef (Gasser and Thein, 1977). These data lead to the conclusion that additional channelways of submarine exhalative vents developed near the reef margin (Fig. 6).

The slight increase in Fe, Mn and Ba toward the sulphide margin shows some similarities to the Atlantis II deep in the Red Sea (Gasser and Thein, 1977; Chapter 4, Vol. 4, in this Handbook). In comparison with Fe, which precipitates under anoxic conditions as iron sulphide, Mn remains in solution and is transported more extensively laterally and/or vertically. This mechanism explains the fractionation between Fe and Mn (Degens and Stoffers, 1977) and the development of a manganese halo on top or beyond the distal portions of sulphide ores. Gwosdz and Krebs (1977) have shown that within the Lagerkalk and the ore-equivalent limestones a significant manganese halo is developed that extends >5 km beyond the ore body. Gasser and Thein (1977) discovered a thallium halo around the Meggen ore over several hundreds of meters, which is a feature also known from recent active geothermal systems and several stratiform sedimentary and volcanogenic sulphide deposits (Evers and Keays, 1977).

TABLE VII

Comparison of the Meggen with the Rammelsberg ore deposits

	Meggen	Rammelsberg
Basinal shales and siltstones as host rock	++	++
Distal turbidites in the foot wall	++	++
Deposition in third order depressions	++	+
Change in lithology below and above the ore horizon	++	–
Disconformities in adjacent environments	++	?
Sedimentary dikes in adjacent environments	++	–
Low carbonate content in the footwall sediments	++	++
Manganese halo in the surrounding sediments	++	+
Tuffs in the footwall sediments	–	++
Tuffs in the hanging-wall sediments	+	–
Relation to active faults and lineaments	++	+
Time-equivalent to volcanic activity or intrusive heat in distant regions	+	+
Pre-ore phases	+	+
Post-ore phases	+	–
Horizontal metal zonation of the ore	+	?
Vertical metal zonation of the ore	–	++
Stratiform copper sulphides	–	++
Banded ore textures	+	++
Association with fine-grained barite	++	++
Mineralized channelways	+	++
Strong deformation of the ore horizon	+	++
Weak metamorphism of the ore	–	+

+ developed; ++ well developed; – absent; ? doubtful.

The distinct vertical element zonation of the Rammelsberg ores in the sequence As–Fe–Cu–Zn–Pb–Ag–Ba is identical to that of sulphide ores of the Kuroko type or other volcanogenic sulphide deposits (Lambert and Sato, 1974; Brathwaite, 1974; Barnes, 1974/75; see indexes in Vols. 4 and 7 for chapters discussing Kuroko ores). The presence of "exotic" elements in the sulphides (such as As, Au, Sn and Bi) as well as the syngenetic stratiform copper ore near the base of the Rammelsberg sulphides indicate much higher temperatures than at Meggen. At Rammelsberg the water depth was sufficient for the brine to be exhalated on the sea floor prior to boiling, so that a copper-rich stratiform facies ("Kupfererz") was precipitated near the base of the Rammelsberg ore deposit (Finlow-Bates and Large, 1978).

There is still an open question as to whether the thermal brines extracted the metals from older sediments and igneous rocks (Scherp, 1974; Wedepohl et al., 1978) or whether the metals were derived from the magmatic source itself (Pilger, 1972; Dornsiepen, 1977). Scherp (1974) has suggested that barium, and possibly some metals, were derived from lower Paleozoic black shales. Wedepohl et al. (1978), on the other hand, on the basis of isotope data, favoured recycled Precambrian clastic detritus as the source for the metals. According to these authors, all important Central European Cu–Pb–Zn deposits (stratiform sulphides of the Meggen–Rammelsberg type – for certain differences and similarities between these two see Table VII – hydrothermal Pb–Zn veins in the Variscan basement, Permian Kupferschiefer and Triassic lead ores or Maubach and Mechernich) were ultimately derived from Precambrian rocks or repeatedly recycled Precambrian detritus.

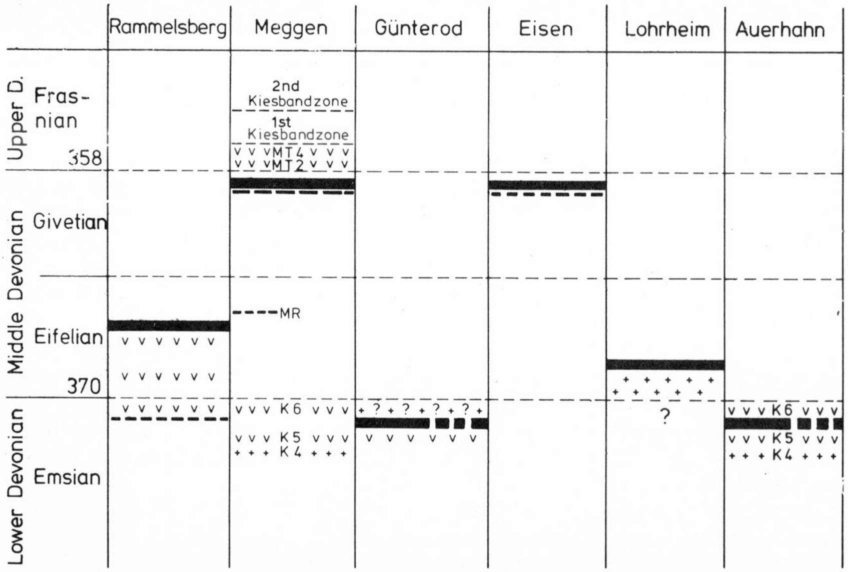

Fig. 19. Stratigraphic position of Devonian sediment-hosted stratiform lead–zinc–barite deposits in Germany. + = acid volcanics, v = tuffs, black = Pb–Zn–Ba ores. (From Krebs, 1978a.)

ACKNOWLEDGEMENTS

I thank the Metallgesellschaft AG for permission to publish this paper. The author also thanks D. Large (Braunschweig), W. Gwosdz (Braunschweig) and F.R. Edmunds (Brussels) whose discussions and comments have contributed to this paper. Finally, I thank D. Large for critical reading of the manuscript, Mrs. U. Busch for typing the text and Mr. G. Weber and Mr. H. Stosnach for drawing the figures.

REFERENCES

Amstutz, G.C., Zimmermann, R.A. and Schot, E.H., 1971. The Devonian mineral belt of Western Germany (the mines of Meggen, Ramsbeck and Rammelsberg). In: *Guidebook 8. Int. Sedimentol. Congr., 1971. Sedimentology of Parts of Central Europe* Kramer, Frankfurt, pp. 253–272.

Anger, G., Nielsen, H., Puchelt, H. and Ricke, W., 1966. Sulphur isotopes in the Rammelsberg ore deposit (Germany). *Econ. Geol.*, 61: 511–536.

Bäcker, H., 1973. Rezente hydrothermal-sedimentäre Lagerstättenbildung. *Erzmetall,* 26: 544–555.

Barnes, H.L., 1974/75. Zoning of ore deposits: types and causes. *Trans. R. Soc. Edinburgh,* 69: 295–311.

Baumann, A., in press. Anorganische Geochemie devonischer Schwarzschiefer im Rheinischen Schiefergebirge.

Brathwaite, R.L., 1974. The geology and origin of the Rosebery ore deposit, Tasmania. *Econ. Geol.,* 69: 1086–1101.

Brown, J.S., 1965. Oceanic lead isotopes and ore genesis. *Econ. Geol.*, 60: 47–68.

Buggisch, W. and Gwosdz, W., 1973. Beitrag zur Tektonik der Attendorn-Elsper Doppelmulde (Sauerland, Rheinisches Schiefergebirge). *Geol. Mitt.,* 12: 131–162.

Buschendorf, F. and Puchelt, H., 1965. Untersuchungen am Schwerspat des Meggener Lagers. Zur Geochemie des Barytes. I. *Geol. Jahrb.,* 82: 499–582.

Buschendorf, F., Nielsen, H., Puchelt, H. and Ricke, W., 1963. Schwefel-Isotopen-Untersuchungen am Pyrit–Sphalerit–Baryt-Lager Meggen/Lenne (Deutschland) und an verschiedenen Devon-Evaporiten. *Geochim. Cosmochim. Acta,* 27: 501–523.

Clausen, C.D., 1978. Erläuterungen zu Blatt 4814 Lennestadt. *Geologische Karte von Nordrhein-Westfalen 1 : 25.000,* 474 pp.

Croxford, N.J.W., 1968. A mineralogical examination of the McArthur lead-zinc-silver deposit. *Proc. Australas. Inst. Min. Metall.,* 226: 97–108.

Croxford, N.J.W. and Jephcott, S., 1972. The McArthur lead–zinc–silver deposit, N.T. *Proc. Australas. Inst. Min. Metall.,* 243: 1–26.

Davis, E.E. and Lister, C.R.B., 1977. Heat flow measured over the Juan de Fuca ridge: evidence for widespread hydrothermal circulation in a highly heat transportive crust. *J. Geophys. Res.,* 82: 4845–4860.

Degens, E.T. and Stoffers, P., 1977. Phase boundaries as an instrument for metal concentrations in geological systems. In: D.D. Klemm and H.-J. Schneider (Editors), *Time- and Stratabound Ore Deposits.* Springer, New York, N.Y., pp. 25–45.

Dornsiepen, U., 1977. *Zum geochemischen und petrofaziellen Rahmen der hangenden Schichten des Schwefelkies-Zinkblende-Schwerspat-Lagers von Meggen/Westfalen.,* Thesis, Technische Universität Braunschweig, 77 pp. (unpublished).

Dornsiepen, U., in press. Lithology and geochemistry of the rocks overlying the Meggen ore body. *Geol. Jahrb.*

Ehrenberg, H., Pilger, A. and Schröder, F., 1954. Das Schwefelkies–Zinkblende–Schwerspatlager von Meggen (Westfalen). *Beih. Geol. Jahrb.,* 12: 352 pp.

Evers, G.R. and Keays, R.R., 1977. Volatile and precious metal zoning in the Broadlands geothermal field, New Zealand. *Econ. Geol.,* 72: 1337–1354.

Finlow-Bates, T. and Large, D.E., 1978. Water depth as major control on the formation of submarine exhalative ore deposits. *Geol. Jahrb., Reihe D,* 30: 27–39.
Friedrich, G., Plüger, W. and Hilmer, E., 1972. Die Anwendung geochemischer Methoden bei der Prospektion im Lagerstättengebiet von Meggen. *Schr. Ges. Dtsch. Metallhütten Bergleute,* 24: 211–225.
Fuchs, W., 1978. Tektonik des Meggener Erzlagers. In: C.D. Clausen, Erläuterungen zu Blatt 4814 Lennestadt. *Geologische Karte von Nordrhein-Westfalen 1 : 25.000,* pp. 360–362.
Fuchs, W. and Ommer, P., 1970. Beispiele faziesgebundener Tektonik im Neuen Lager des Schwefelkies–Zinkblende–Schwerspatlagerstätte von Meggen. *Bergbauwissenschaften,* 17: 447–449.
Gaser, U., 1974. Zur Struktur und Geochemie der Stratiformen Sulfidlagerstätte Meggen (Mitteldevon, Rheinisches Schiefergebirge). *Geol. Rundsch.,* 63: 52–73.
Gasser, U. and Thein, J., 1977. Das syngenetische Sulfidlager Meggen im Sauerland (Struktur, Geochemie, Sekundärdispersion). *Forschungsber. Landes Nordhrein-Westfalen,* 2620: 171 pp.
Gwosdz, W., 1972. Stratigraphie, Fazies und Paläogeographie des Oberdevons und Unterkarbons im Bereich des Attendorn-Elsper Riffkomplexes (Sauerland, Rheinisches Schiefergebirge). *Geol. Jahrb., Reihe A,* 2: 71 pp.
Gwosdz, W. and Krebs, W., 1977. Manganese halo surrounding Meggen ore deposit, Germany. *Trans. Inst. Min. Metall., Sect. B,* 86: 73–77.
Gwosdz, W. and Krüger, H., 1972. Meggener Schichten (Devon, Sauerland, Rheinisches Schiefergebirge). *Neues Jahrb. Geol. Paläontol., Monatshefte,* 85–94.
Gwosdz, W., Krüger, H., Paul, D. and Baumann, A., 1974. Die Liegendschichten der devonischen Pyrit- und Schwerspat-Lager von Eisen (Saarland), Meggen und des Rammelsberges. *Geol. Rundsch.,* 63: 74–93.
Henke, W., and Schmidt, W.E., 1922. Erläuterung zu Blatt Attendorn. *Geol. Karte von Preussen und benachb. Bundesstaaten 1 : 25.000,* 58 pp.
Hilmer, E., 1972. *Geochemische Untersuchungen im Bereich der Lagerstätte Meggen, Rheinisches Schiefergebirge.* Thesis, Technische Hochschule Aachen, 162 pp. (unpublished).
Hodgson, C.J. and Lydon, J.W., 1977. Geological setting of volcanogenic massive sulphide deposits and active hydrothermal systems: some implications for exploration. *Can. Inst. Min. Metall. Bull.,* 95–106.
Holser, W.T. and Kaplan, I.R., 1966. Isotope geochemistry of sedimentary sulphates. *Chem. Geol.,* 1: 93–135.
Kraume, E., 1955. Die Erzlager des Rammelsberges bei Goslar. *Beih. Geol. Jahrb.,* 18: 394 pp.
Kraume, E. and Jasmund, K., 1951. Die Tufflagen des Rammelsberges. *Beitr. Mineral. Petrogr.,* 1: 443–454.
Krebs, W., 1971. Devonian reef limestones in the eastern Rhenisch Schiefergebirge. In: *Guidebook 8. Int. Sedimentol. Congr., 1971. Sedimentology of Parts of Central Europe.* Kramer, Frankfurt, pp. 45–81.
Krebs, W., 1972a. Die paläogeographisch-faziellen Aussagen zur Position des Meggener Lagers. *Schr. Ges. Dtsch. Metallhütten Bergleute,* 24: 187–196.
Krebs, W., 1972b. Facies and development of the Meggen Reef (Devonian, West Germany). *Geol. Rundsch.,* 61: 647–671.
Krebs, W., 1978a. Moderne Suchmethoden auf Buntmetalle im mitteleuropäischen Grundgebirge. *Erdöl Kohle,* 31: 128–133.
Krebs, W., 1978b. Meggener Riff. In: C.D. Clausen, Erläuterungen zu Blatt 4814 Lennestadt. *Geologische Karte von Nordrhein-Westfalen 1 : 25.000,* pp. 104–111.
Krüger, H., 1973. *Der sedimentologische und paläogeographische Rahmen der Liegendschichten des Schwefelkies-Zinkblende-Schwerspat-Lagers von Meggen/Westfalen.* Thesis, Technische Universität Braunschweig, 63 pp. (unpublished).
Lambert, J.B. and Sato, T., 1974. The Kuroko and associated ore deposits of Japan: a review of their feature and metallogenesis. *Econ. Geol.,* 69: 1215–1236.
McMahon Moore, J., 1971. Fold styles in the orebodies of Meggen and Rammelsberg, Germany. *Trans. Inst. Min. Metall. Sect. B.,* 80: 108–115.

McNitt, J.R., 1970. The geologic environment of geothermal fields as a guide to exploration. *Proc. U.N. Symp. on the Development and Utilization of Geothermal Resources, Pisa,* 1: 24–31.

Noa, W., 1958. *Feinstratigraphische Untergliederung des Oberdevons bei Meggen.* Thesis, Technische Universität Clausthal (unpublished).

Paul, D. 1975. Geologische Untersuchungen zur Rekonstruktion des Ablagerungsraumes vor und nach der Bildung des Rammelsberger Blei-Zink-Lagers im Oberharz. *Geol. Jahrb., Reihe* D, 12: 3–93.

Pilger, A., 1957. Über den Untergrund des Rheinischen Schiefergebirges und des Ruhrgebietes. *Geol. Rundsch.* 46: 197–212.

Pilger, A., 1972. Beziehungen des Meggener Lagers zum initialen Magmatismus. *Schr. Ges. Dtsch. Metallhütten Bergleute,* 24: 149–160.

Puchelt, H., 1961. *Das Barium–Strontium-Verhältnis in Schwerspäten der Lagerstätten Meggen/Lenne und Rammelsberg – zugleich ein Beitrag zur Klärung der genetischen Zusammenhänge exhalativ-sedimentärer Erzlager.* Thesis, Technische Universität Hannover, 242 pp. (unpublished).

Ramdohr, P., 1953. Mineralbestand, Strukturen und Genese der Rammelsberg-Lagerstätte. *Geol. Jahrb.*, 67: 367–494.

Rippel, G., 1954. Räumliche und zeitliche Gliederung des Keratophyrvulkanismus im Sauerland. *Geol. Jahrb.* 68: 401–456.

Scherp, A., 1974. Die Herkunft des Baryts in der Pyrit–Zinkblende–Baryt-Lagerstätte Meggen. *Neues Jahrb. Geol. Palaeontol. Monatshefte,* 1974: 38–53.

Scherp, A., 1978. Meggener Tuffite. In: C.D. Clausen, Erläuterungen zu Blatt 4814 Lennestadt. *Geologische Karte von Nordrhein-Westfalen 1 : 25.000,* pp. 310–312.

Schmidt, W.E., 1919. Über die Entstehung und die Tektonik des Lagers von Meggen nach neueren Aufschlüssen. *Jahrb. Preuss. Geol. Landesanst.*, 39: 23–72.

Seifert, H., Nickel, E. and Bruckmann, E., 1952. Studien am "Neuen Lager" der Kieslagerstätte von Meggen (Lenne). *Opusc. Mineral. Geol.*, 1: 70 pp.

Shanks, W.C. III and Bischoff, J.C., 1977. Ore transport and deposition in the Red Sea geothermal system: a geochemical model. *Geochim. Cosmochim. Acta,* 41: 1507–1519.

Sibson, R.H., McMahon Moore, J. and Rankin, A.H., 1975. Seismic pumping – a hydrothermal fluid transport mechanism. *J. Geol. Soc. London,* 131: 653–659.

Solle, G., 1937. Zur Entstehung der Kieselgallen. *Senckenbergiana,* 19: 385–391.

Venzlaff, H., 1956. Das geologische Bild des Hauptgrünsteinvulkanismus im nördlichen Sauerland. *Geol. Jahrb.*, 72: 241–294.

Wedepohl, K.H., Delevaux, M.H. and Doe, B.R., 1978. The potential source of lead in the Permian Kupferschiefer bed of Europe and some selected Paleozoic mineral deposits in the Federal Republic of Germany. *Contrib. Mineral. Petrol.*, 65: 273–281.

Weisser, J.D., 1972. Zur Methodik der Exploration Meggen. *Schr. Ges. Dtsch. Metallhütten Bergleute,* 24: 167–186.

Wellmer, F.-W., 1966. *Feinstratigraphische Gliederung der hangenden und liegenden Grenzschichten des Neuen Meggener Lagers in der Schwefelkiesgrube Meggen.* Thesis, Technische Universität Clausthal, 68 pp. (unpublished).

Wellmer, F.-W., 1970. Feinstratigraphische Gliederung in den sandigen liegenden Grenzschichten des Neuen Meggener Lagers – ein Beitrag zur Montangeologie des Schwefelkies–Zinkblende–Schwerspatlagers von Meggen/Westfalen. *Bergbauwissenschaften,* 17: 458–463.

Wellmer, F.-W., 1973. Störungsmechanische Analyse von Faltenabschiebungen im Gebiet von Meggen. *Glückauf-Forschungsh.*, 34: 109–112.

Winter, J., 1969. Stratigraphie und Genese der Bentonitlagen im Devon der Eifeler Kalkmulden. *Fortschr. Geol. Rheinld. Westfalen,* 16: 425–472.

Zimmermann, R.A., 1970. Sedimentary features in the Meggen barite–pyrite–sphalerite deposit and comparison with the Arkansas barite deposits. *Neues Jahrb. Mineral. Abh.*, 113: 179–214.

Chapter 10

GENESIS OF THE RAMMELSBERG ORE DEPOSIT NEAR GOSLAR/UPPER HARZ, FEDERAL REPUBLIC OF GERMANY

WOLFGANG W. HANNAK

THE HISTORICAL DEVELOPMENT OF IDEAS ON THE GENESIS OF THE RAMMELSBERG ORE DEPOSIT

Convincing criteria for a sedimentary origin of the Rammelsberg ore deposit have been mentioned repeatedly during more than 200 years of scientific study at Rammelsberg. Nevertheless, the concept of a sedimentary origin was still challenged, and sometimes strenuously opposed, as late as the early nineteen-fifties. One of the reasons for the dispute was the faulting in the boundary zone between the orebodies and the surrounding rocks, the importance of which was recognized only recently. Kraume (1955) made the following comments: The ideas on the genesis were so variable primarily because nothing definite was known about the external shape of the deposit, and about the tectonic relationship between the orebodies and the host rock. According to Kraume's (1955) investigations the differences between the structural deformation of the host rock and the orebodies were considered to be insignificant. As a result, in his second comprehensive treatise, Ramdohr (1953b) withdrew his previous and well-grounded opinion that there were strong structural differences between the orebodies and the host rock. He then followed the prevailing ideas of the time, viz. that there was a uniformly overturned sequence in which the orebodies were concordantly intercalated, and that any movements between the orebodies and the host rock were, on the whole, unimportant. With the "Lead–Zinc Ore Monograph" (Kraume, 1955), it was supposed that a comprehensive description of the deposit and interpretation of its genesis had finally been made. However, detailed structural mapping began in the same year and yielded surprising and fundamental data from which a completely new interpretation of the geologic–tectonic setting of the orebodies became possible (Hannak, 1956, 1963). These results led to a conclusive tectonic classification of the ore deposit and an understanding of its situation in the overall geologic–tectonic setting, both of which were previously lacking. From basic palaeogeographic concepts developed by Kraume (1955), a logical and complete series of arguments for the syngenetic origin of the ore could now be related in detail to the structural features. In addition, the palaeogeographic ideas were considerably reinforced by the new knowledge of the stratigraphy and facies of the surrounding sediments, and the special or characteristic setting of the ore deposits could be described in more detail than ever before.

Even if one wished, there would be insufficient space here to take account of all the errors in past investigations which, however, led to an increased knowledge of the genesis, although at the time they appeared to be as many steps behind. To describe the Rammelsberg deposit and to give a genetic interpretation without reviewing other studies is a problem which so far has had to be faced by every author. It is, therefore, impossible to provide a reasonably short description of the history, although it would be worthwhile to discuss the status of the Rammelsberg as the "father" of the stratiform polymetallic sulphide ore deposits.

An unusual number of authors have written about the Rammelsberg "problem". The present author regretfully had to refrain from citing all the available literature because of space limitations; moreover, there are valuable contributions in internal reports that are only obtainable with difficulty. In this respect we have to thank Kraume (1955), who collected much of the data and compared the various ideas and who also described in detail the epigenetic interpretations. (See the indexes in Volumes 4 and 7 of this Handbook series for brief references made to the Rammelsberg and Meggen ores, and refer to Chapter 9 in this volume for a comparison between these two deposits.)

Ever since the question of the origin of ores was first raised, the Rammelsberg deposit has been in the limelight. Both the ore and the setting of the orebodies are unusual. According to old documents, the Rammelsberg ore was already worked for silver in 968, and thus represented a kind of treasure chest for the German emperors. Even so, the mine had only been worked to shallow depths and at low production rates over many centuries. The second bigger sulphide orebody, the New Orebody ("Neues Lager"), was not discovered until 1859. The real extent of the deposit was not recognized before the turn of the century, and important structural features were not known until later. At least it was clear that the deposit did not match the usual picture of vein mineralization, as there were many being mined to considerable depths in the immediate vicinity of the "Oberharz" and thus allowed constant comparison.

In 1784 von Trebra inferred that the Rammelsberg ore deposit could be an ore seam, though he believed everybody would be astonished about this opinion.

In 1864, von Cotta still believed that the deposit consisted of innumerable small ore lenses conformably intercalated in the Wissenbach Shales. This concept was only abandoned in 1877, when Wimmer claimed that it consisted of large, compact orebodies.

Today it is well known that the mineralization is concentrated in two neighbouring, but isolated, orebodies containing more than $22 \cdot 10^6$ metric tons of ore within an area of less than 0.5 km^2; they consist of massive lead, zinc, copper and iron sulphides and barite, together comprising 85–95% of the orebodies. The problem of whether the ore was originally deposited as a single or as two separate main sulphide bodies of slightly different ages will be discussed later (Figs.1–3).

Strong arguments in favour of a sedimentary origin of the mineralization were

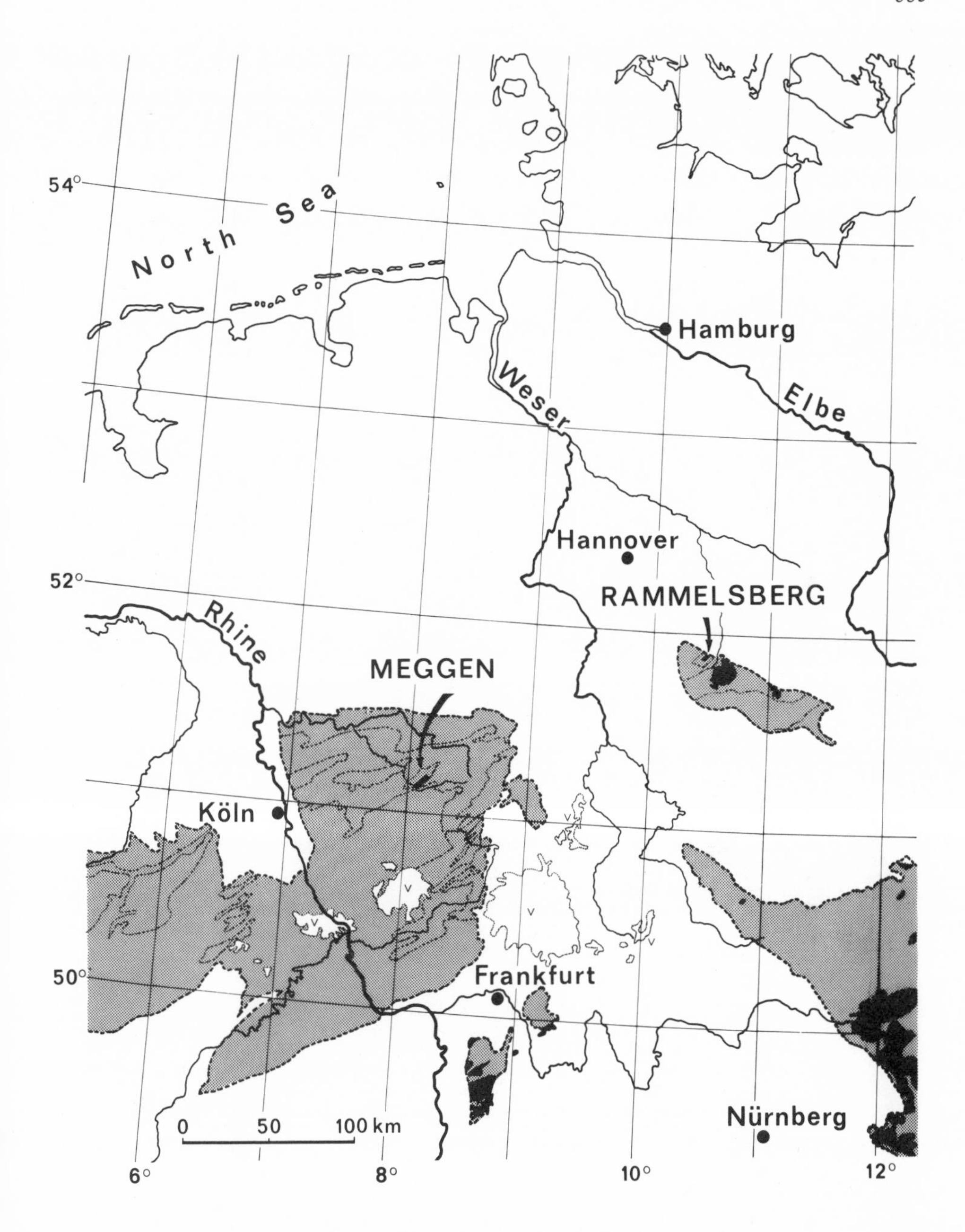

Fig. 1. General geological map showing the stratiform massive ore deposits of Rammelsberg and Meggen in the Variscan Palaeozoic massifs (stippled); Palaeozoic plutonites = black; Tertiary volcanites = v.

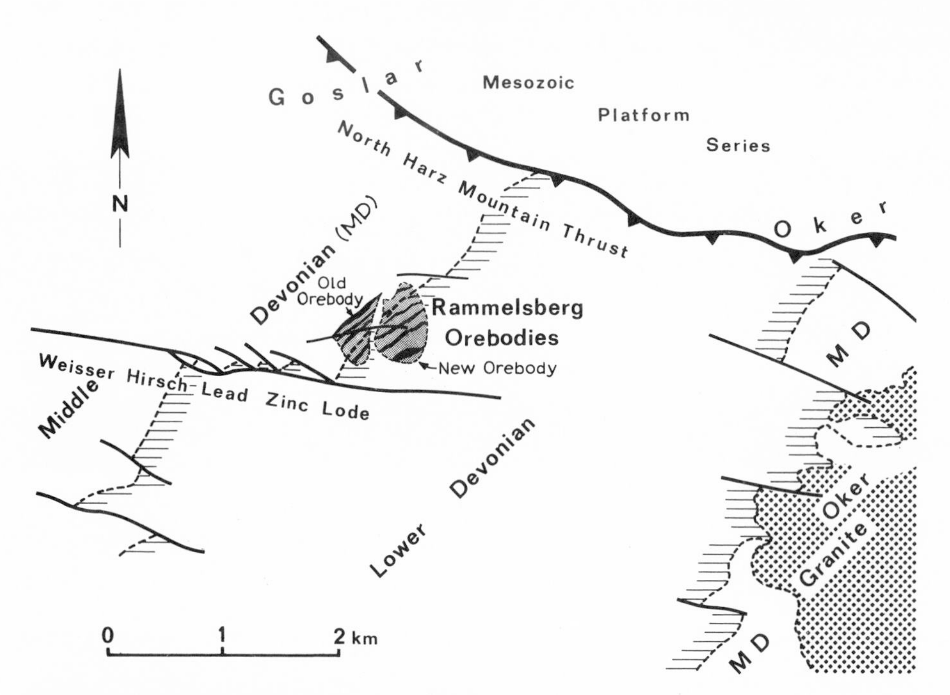

Fig. 2. Simplified geological map of the ore deposit area.

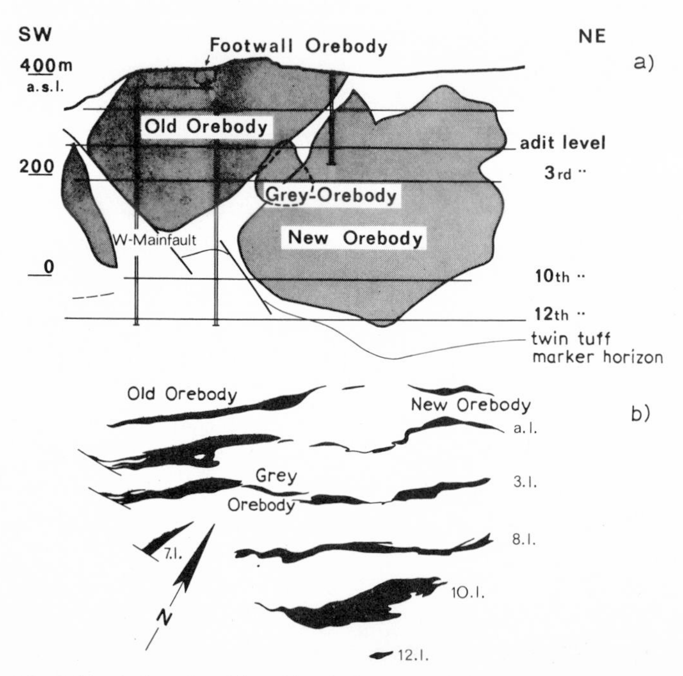

Fig. 3. Vertical section (a) and level planes (b) of the orebodies. (After Kraume, 1955.)

quoted before the introduction of ore microscopy. Together with many other sulphide occurrences [for example Spain (Huelva district), Ireland (Avoca) and Norway which today are virtually all unambiguously considered to be of a sedimentary origin], von Groddeck (1879) classified the Rammelsberg deposit as a sedimentary occurrence. He did so on the basis of unprejudiced observations of sedimentary ore textures and concordant contacts with the country rocks. In this connection he did not consider the origin of the metals. Later, particularly after the turn of the century, discussion of this question intensified. However, this led to many misinterpretations because most authors supposed an epigenetic origin of the orebodies.

Schuster (1867), Wimmer (1877), Babu (1893), Klockman (1893) and Söhle (1899) supported a sedimentary formation of the ores, which they thought were chemical precipitates of metal sulphides from sea water. To the discussion, Wiechelt (1904) contributed the concept of the participation of volcanically contributed hydrogen sulphide to the ore precipitation. Hornung (1905) supposed that chloride waters dissolved the metals from the footwall layers. At the contact of these solutions with sea water, the formation of the ores took place under the influence of hydrogen sulphide. Schot (1973) provides an excellent general review of these hypotheses.

Bergeat (1912) compared the Middle and Upper Devonian hematite ore occurrences (stratified red iron ore) of the "Oberharz" with the Rammelsberg. He saw a mutual origin in the initial geosynclinal volcanism which, somewhat later in time relative to the development of the Rammelsberg deposit, led to lava flows during the late Middle Devonian and Early Carboniferous. In his opinion, the origin of the ore is related to hydrothermal sulphide solutions, which were able to dissolve carbonate in the prevailing slightly acidic conditions. He interpreted the occurrence of sulphate (barite) as an oxidation product of sulphide. Much later Anger et al. (1966), on the basis of sulphur-isotope distribution proposed that, among other things, sea-water sulphate took part in the barite precipitation. Using the latest literature, Schot (1973) again discussed the isotopic distribution of sulphur. Bergeat (1902) did not regard the ore textures as primary and attached great importance to the influence of diagenesis.

Basic genetic studies with detailed, but not undisputed, ore-microscopic investigations, were carried out by Frebold (1927). It was finally left to Ramdohr (1928, 1953b) to give a detailed description of the deposit and its genesis. He strongly defined the concept of a sedimentary origin, with a subsequent metamorphic overprint during the Variscan orogenesis. Ramdohr used Bergeat's interpretation, as Frebold did before, and improved the knowledge on the genesis. Kraume (1955) gave a detailed account of the distribution of the ore minerals and trace elements. He also described the ore genesis in both general and detailed treatments.

The conduit, through which the metal-bearing solutions rose, was located by studying the non-ferrous metal mineralization of the underlying strate (Ramdohr, 1953b; Kraume, 1955; 1960a; Gundlach and Hannak, 1968).

Prior to folding, this path formed a steep, cylindrical zone (traceable into the

Lower Devonian beds) beneath the orebodies. This confirmed Bergeat's original concept.

Among the authors dealing with the tectonics, Köhler (quoted by Kraume, 1955) and W.E. Schmidt (1933a, b) should be mentioned, as well as Jahns (1955), who made the first modern structural analysis of the western extension of the deposit at about the same time as the present author started his own studies.

Investigations by Puchelt of the strontium distribution in barite followed (Hannak and Puchelt, 1965). Anger et al. (1966) investigated the sulphur-isotope distribution. In the meantime, Paul (1975) made a geochemical study of the distribution of base metals (Pb, Zn, Cu, Mn) Ca, and heavy minerals in the country rock. Detailed mapping of the deposit, and geophysical and geochemical studies within the immediate vicinity of the mineralizations, were carried out by the mining company. A doctorial thesis (600 pages) dealing with these aspects has been presented by Schot (1973), wherein the ore deposit is treated monographically, but unfortunately, the geological framework is not considered critically enough.

The concepts on the genesis that have so far been mentioned, represent almost the entire spectrum of all possible combinations of origin (hypogene–supergene), relationships with the country rock (syngenetic–epigenetic), and method of emplacement (ranging from external deposition through metasomatism to internal deposition). The possible relationships with orogenesis (pre-, syn- and post-orogenic) are also considered. The possibility of a truly magmatic origin related to "intruded" sulphide melt is also recorded (Vogt, 1894). In his textbook on the Rammelsberg ore deposits, Lindgren (1933) mentioned: "Vogt, following Freiesleben and Lossen, explains it as a deposit from magmatic solutions. Dahlgrün says that only pyrite is of syngenetic origin. Pb, Zn, Cu, are formed by plutonic hydrothermal agencies and altered by regional metamorphism". Lindgren (1933) himself says that "the deposit is a bedded vein . . . the distribution and structure of the ore itself are inconsistent with the theory of sedimentary deposition . . . their structure is that of a dynamometamorphic rock . . . could easily be duplicated from any area of fine-grained schist . . . While the surrounding slates are soft they evidently behaved quite differently from the sulphide mass". It is true that the tectonic behavior of the ore and its host rock is very different, but as far as the structural competence is concerned the opposite is true: the sulphide masses are the softer material.

Authors favouring epigenesis (e.g. Bornhardt, 1939, who is commonly cited by Ramdohr) have not been successful against the convincing arguments of the advocated syngenesis.

In the meantime, it has been ascertained that the deposit was formed in a special depression on the sea floor within the sedimentation regime of the Wissenbach Shales after a geosynclinal and magmatic preparation period (geological maturation process) during the Middle Devonian. From the end of the Early Devonian to the early Middle Devonian, this area formed a slope between the West Harz swell and the Goslar trough. The depression, as well as the deposit, were transformed during the Variscan orogeny into a major syncline. During this process a great number of *b*-axial structural elements, in particular fractured zones parallel to the deposit, were formed. They were used as proof in both syngenetic and epigenetic interpretations. Many proponents of epigenesis recognized that the ore is intensely and complexly deformed. Nevertheless, the specific tectonic interactions between ore and host rock, which depend on their very different geological natures, were not recognized for a long time. Because the tectonic structure of the orebody syncline was not clarified, it was not possible to reconcile the other data of individual observations. Meanwhile, the reasons which led to the formation of the orebody syncline and its specific structure are now understood. The causative tectonic factors are:

(1) the palaeogeographic flank position of the orebodies;

(2) the special depression during the formation of the orebodies; and

(3) the extremely high plasticity of the orebodies, in relation to the tectonic strain and their space limitation compared with the host rock.

It is the difference in tectonic competence between the orebodies and the host rocks which resulted in fault zones in between them. This shear-tectonism intensified the partial mobility of the host sediments. Thus, these sediments have an intense isoclinal folding which is otherwise generally unknown in the Upper Harz, and likewise the orebodies show an isoclinal folding, which is only known in highly metamorphosed and kinematically intensely deformed rocks.

This indicates that the genesis of the Rammelsberg deposit was not terminated subsequent to the cessation of sedimentation. Independent of the effects of diagenetic processes (which cannot be completely reconstructed and estimated), the ores received their final dominating structure from the Variscan orogeny during the Late Carboniferous. This has obliterated or destroyed many primary textures through recrystallization. Ramdohr (1928, 1953b) describes these textural changes and the formation of new minerals.

THE STRATIGRAPHIC AND PALAEOGEOGRAPHIC DEVELOPMENT OF THE ENVIRONMENT OF THE DEPOSIT AND THE INITIAL GEOSYNCLINAL TUFF VOLCANISM

The Lower Devonian sequences of quartzites and sandy pelite beds up to 1000 m thick are the earliest indicators of the palaeogeographic setting of the ore deposit. Channels, graded-bedding, cross-laminations, clay-fragment breccias and loadcasts, as well as fossil horizons, indicate that the area of sedimentation was a littoral–neritic environment until the end of the Early Devonian (see Fig.32a, c). The age of the sequence is determined from key fossils. After the end of the Early Devonian this sedimentary complex formed a NE–SW striking swell. The subsequent fault and fold tectonics resulted in a rotation of the region in an E–W direction. The mineralization is located on the northwestern flank of the swell (Fig.4). After Kraume (1955) had recognized it as the Okertal swell, Früh (1960) called it the West Harz uplifted region. As a result of a gravimetric survey, Horst (1967) discovered that the West Harz swell (the geosynclinal anticlinorium to

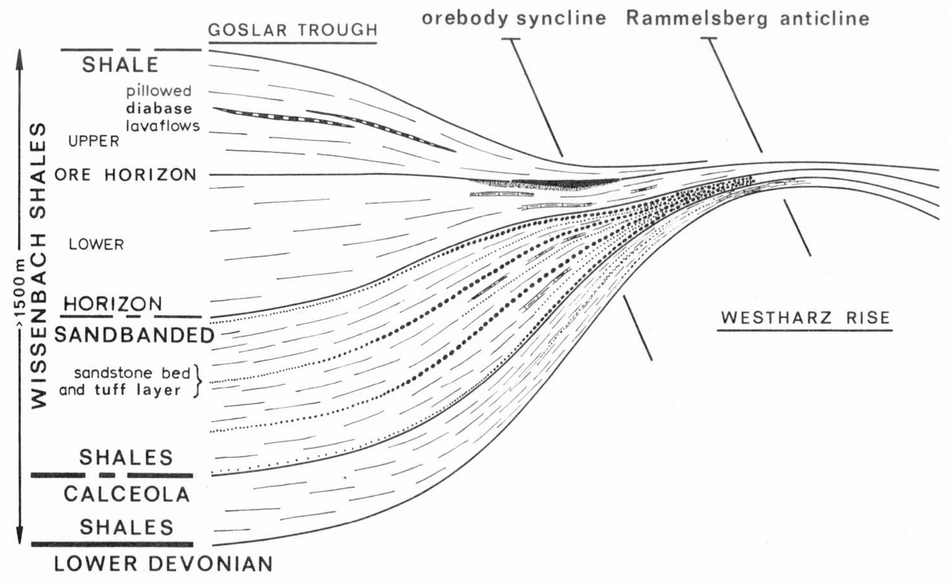

Fig. 4. Palaeogeographic cross-section.

the south of the ore deposit) corresponds to a gravimetric high. Similarly, the Goslar trough, north of the ore deposit, corresponds to a gravimetric low. The gravity high over the West Harz swell may be indicative of a magmatic intrusion. The significance of these possible intrusions in the development of the geosyncline is beyond the scope of this paper. Geosynclinal plutonic intrusions, which form the cores of folds, were discussed by Krebs and Wachendorf (1974) in connection with processes resulting in the formation of the ore deposit.

During the lower part of the Middle Devonian, the differentiation of the geosyncline resulted in a change to a main sedimentary sequence, which ended in fine-grained silty-pelitic sediments during the formation of the orebodies (Fig.4). The Sandbanded Shales represent a short break in this sedimentary pattern. This was followed by a tectonic stabilization of the geosyncline (Krebs, 1972). The first extrusions of diabases (spilitic rocks) occurred at this time, and are connected with the synsedimentary hematite (red iron ore) formation on the southeastern side of the West Harz uplifted region, in the so-called “diabase range” of the Upper Harz (Ramdohr, 1928). In the “Rheinisches Schiefergebirge”, the Meggen pyrite–barite–sphalerite orebody (see chapter by Krebs in this volume) was formed at the same time (Givetian) (Fig.1).

In the lower part of the Middle Devonian, a remarkable differentiation of the thicknesses and facies took place in the district of the ore deposit (Table I; Fig.4). During the Eifelian, the West Harz swell formed a region of minor subsidence compared to the region situated to the northwest. A fragmentary, at first sandy-pelitic, then marly-calcareous and

TABLE I

Stratigraphic section of the mine area

Middle Devonian Eifelian	Wissenbach Shales		
	Upper	Upper shale horizon with diabases	
	Lower		Ore horizon/orebodies
		Lower shale horizon	
			— — Twin tuff layer — —
		Sandbanded shales	
	Calceola Shales		
	Upper Speciosus Formation		
	370 m.y.		
Lower Devonian Emsian	Lower Speciosus Formation		
	Kahleberg Sandstone		

finally pelitic sedimentary sequence of about 150 m thick was deposited in the swell region prior to the deposition of the main mineralization. The primary (prefolding) distance between the top of the swell and the orebodies might have been 3–5 km.

At the slope between the swell and the trough, the facies of the *Calceola* Shales first changed from a limestone shale series to a marly shale sequence with a few sandstone layers, and then to a marly-bituminous shale sequence of about 100 m thick, northwest of what was to become the location of the orebody. Fossils indicate that the area of the deposit was a shallow-sea environment.

The formation of the Sandbanded Shales can be explained by turbidity currents flowing into the Goslar trough from the northeast along the slope of the uplifted area, as a result of geosynclinal movements which intensified the relief ratio. The consequence was the sandy–silty sediment sequences in the slope region, which corresponds to the present mining hanging wall. On top of the geosynclinal swell, these layers are missing or they are present in very reduced thicknesses. Towards the basin (northwest of the ore deposit) only fine-grained pelitic material, with a significant carbonate content and of varying thickness, was deposited. In this sedimentary sequence some sandstone beds of decimetre to metre thicknesses together with tuff layers form excellent key horizons in the sedimentary succession of the Rammelsberg mine (Fig.4).

The Middle Devonian sedimentary sequences in the Goslar trough, amount to a thickness of over 1000 m, which corresponds to the condensed incomplete sequence of less than 150 m on the swell. The prefolding distance between the top of the swell and the centre of the trough was less than 10 km.

The *Lower Wissenbach Shales* and the *Ore Horizon* do not show distinct variations of the sedimentary conditions in the vicinity of the ore deposit. Apart from the interbedded tuff layers, these horizons are composed of poorly banded, very uniform black shales (see Fig.33b). It seems certain that the accumulation of the ore took place in a local depression on the sea floor, which was situated directly on, or adjacent to, the feeder zone of the metal-bearing solutions. The formation of the ore trough may be genetically related to the intensified thermal activity. Synsedimentary tectonics could have promoted the shaping of the basin, although such a mechanism cannot be proved.

The channel way of the metal-bearing hydrothermal solutions is recognized by its mineralization in the basal layers. However, the conduit zone is not distinguished from the adjacent non-affected rocks by any particular tectonic features.

The geological evidence of the sedimentary environment just prior to ore accumulation indicates marine shelf conditions; at the time of ore precipitation, the sediments were mainly pelitic and slightly silty in nature. Frequently appearing syngenetic pyrite and black shales in the older units, which prior to tectonism were vertically beneath the ore deposits, indicate an euxinic, H_2S-rich and chemically reducing milieu. The sulphide was most likely formed through sulphate-reducing bacterial processes (see Trudinger, 1976) that may have been periodically active whenever ascending hypogene material was supplied. Dwarfed forms of brachiopods, goniatites and tentaculites, most of them totally replaced by pyrite, characterize both the banded and massive sulphide accumulations within the black-shale host rock. The "dwarfed", not fully developed, organisms are typical for a biological hostile environment which was the consequence of metal-bearing sulphide solutions of the hot-spring-type contaminating the sea water.

Subsequent changes, probably during the later Middle Devonian but certainly during the Late Devonian, resulted in a deeper shelf environment and more pelagic conditions. The rate of geosynclinal movements apparently diminished as indicated by the thinner units, their more lateral extent, and the weaker facies and lithologic differentiation. Over the topographic highs of the West Harz swell, the pelagic cephalopod limestones were comparatively thin, thus further demonstrating an area of minor subsidence.

The above information suggests that the ore originated near the end of an intense geosynclinal tectonic differential movement, which was accompanied by a distinct tuff-forming volcanism. Until about 1930, the only poorly and sporadically known thin beds of green and very fine-grained tuffs were merely described as metamorphic chloritic schists. Schmidt (1933a, b) was the first to recognize these typical beds as tuffs and he, therefore, presented evidence contrary to those researchers who did not accept the occurrence of volcanism prior to and during the accumulation of the Rammelsberg ore. Schmidt, however, assumed that the tuffs are confined to a relatively thin sedimentary or stratigraphic complex. On the other hand, Kraume (1955) was able to prove that the tuffs had a more extensive stratigraphic distribution, so that he concluded that they cannot be utilized as time or horizon markers. Nevertheless, the tuffs were found by Jahns (1955) and Hannak (1956) to be significant in lithostratigraphic subdivisions which formed the basis for the interpretation in more detail, especially of the fold structure. The presence of a single tuff bed on both sides of the New Orebody was believed by Schmidt (1933) to be evidence of a "plastic" tectonic injection from a deeper stratigraphic level, or tectonic stockwork. Consequently, he supported a parautochthonous localization of the orebodies. Abt (1958), on the other hand, maintained that a tuff bed stratigraphically separates the New and Old Orebodies from each other and assumed, therefore, that the main ore mineralization must have taken place during two episodes. Gunzert (1969), however, concluded that the origin of the same tuff horizon was closely associated with the main ore precipitation and did not refer to episodic processes. Today's detailed stratigraphic reconstruction of the Rammelsberg ores have reduced, or are confined, to studies of 1 m thick units, inasmuch as the major correlation problems are considered to be understood.

Tuffs have been known from the oldest stratigraphic section of the Lower Devonian, and these volcanic accumulations originated in remarkable uniformity during the whole period of tectonic subsidence until the late Middle Devonian. In the Early Devonian, thicker tuffs were formed only locally, whereas the younger ones are normally only a few centimetres or decimetres thick. During the subsequent volcanic episodes the tuffs increased in thickness to about 3 m in the Ore Horizon. Simultaneously, the time intervals between the tuffaceous accumulations (i.e. the hiatus from one tuff bed to the next) became shorter. One tuff unit within the Calceola Shale contains a chert bed, with a maximum thickness of several decimetres. This is one case where direct evidence has been observed that the volcanic-hydrothermal solutions also furnished silica (called "Hornstein"; also referred to as a possible sinter deposit by Gilmour, 1976, p.147, when compared with the Mt. Isa district).

During the geosynclinal cycle, the tuffs formed the first phase of the regional development of volcanics. The Rammelsberg orebodies belong to this first stage.

During the late Middle Devonian, as well as the Late Devonian, the second phase of volcanism is represented by diabase greenstones and associated sedimentary red iron-ore deposits in the "Oberharz" region. Diabase volcanism alone extended into the Lower Carboniferous. It is noteworthy that stratiform red iron ores and Pb–Zn–Cu sulphide–barite occurrences are separated from each other both spatially and temporally during the development of the Variscan geosynclinal of the "Harz" and "Rheinisches Schiefergebirge" districts. The iron ores were connected with the later basic volcanism, whereas the Pb–Zn–Cu sulphide ores were associated with the earlier acidic to intermediate volcanic (explosive) tuff activity.

A total of about twenty tuff beds has been identified, of which nearly ten can be used as horizon markers in underground mapping. The tuffs are unusually fine-grained (pelitic, ash-type) and their colour ranges from grey-green in centimetre-thick tuffs to vari-coloured dark green in thicker layers. Compositionally, the tuffs belong to acidic to weakly intermediate volcanics.

The main components (Kraume and Jasmund, 1951; Schumann, 1952) of the fine-grained tuff matrix consist of well-ordered light-coloured mica and quartz. The minor kaolinite present is not of clastic but of secondary origin. The quartz grains (>200 μm) contain rutile and fluid inclusions, which have not been examined further. The muscovite (sericite) occurs frequently as rod-like inclusions in the matrix up to 1 mm in diameter. They give the tuffs a speckled appearance.

Pyrite, biotite, chlorite, anatase, zircon, rutile, tourmaline and apatite are the accessory constituents. The pyrite is partly present in layered concretionary particles and associated with coarse-grained sparry calcite. Sphalerite is occasionally observable. Paul (1975) investigated in detail the zircon and there is no doubt that it was derived from an acidic magmatic source.

The source locality of the tuff volcanism is not known, and only little information is available on the distribution of the pyroclastic rocks beyond the mine. Their thin-bedded nature and easy weathering make it impossible to trace them in surface outcrops. It can be assumed that the centre(s) of eruption is (are) situated in the northeastern Harz foreland, where the Variscan orogen is overlain by Mesozoic platform sedimentary rocks several thousand metres thick.

The question of the source locality is, however, less significant, because during genetic interpretations it is only important to establish that a rhyolitic type of geosynclinal volcanism accompanied the geosynclinal epeirogenic movements. These processes are considered to be the main prerequisite for the origin of the Rammelsberg-type ore deposits. Bergeat (1902) was the first to recognize this interrelationship, which has been proved by the discovery of the conduits or channel ways of the hydrothermal fluids. Schot (1973) does not disagree with such an ore-forming mechanism, but prefers to rely on circulating connate or formation waters which received their thermal-fluid characteristics from magmatic intrusions (cf. White, 1968; White et al., 1971). The

TABLE II

Average chemical composition of tuffs based on sixteen analyses (after Abt, 1958 and Kraume, 1955)

	Arith. average (%)	Minimum content (%)	Maximum content (%)
SiO_2	52.4	39.68	65.91
Al_2O_3	22.4	12.91	30.04
K_2O	6.2	3.54	7.90
Na_2O	1.0	0.26	4.61
MgO	1.7	0.64	3.00
CaO	1.2	<0.1	(12.04) 6.06
CO_2	1.8	0.38	9.64
TiO_2	0.6	0.10	2.42
Fe (total)	2.9	1.32	5.23

magma, or hot rocks, at the same time also provided the heat for this dynamic hydrothermal cell. White et al. (1971) have discussed recent examples of a deep connate-water circulation system in relation to various types of ore deposits; however the applicability of this model to the submarine lead–zinc–copper–barite type of mineralization has to be tested.

Schot's (1973) possible mechanism explains the origin of the metal concentration, though plausible and convincing quantitative geochemical balance-calculations are very difficult to make. Considerations of the average metal contents in the sedimentary rocks proved that a complete removal of base metals through leaching of 1 km^3 rock would be required to account for 300,000 tonnes of ore (i.e., the average annual production rate of the Rammelsberg mine).

At present, similar tuffs are known from analogous ore districts, such as Mt. Isa (Stanton, 1972; see also Quade, 1976; Baumann and Tischendorf, 1976, p.59–62, 133, 232, 244, 287 and 289–294). They have also been recognized in the stratiform Pb–Zn carbonate-rock-hosted ore districts in the Alps. The Bawdwin Pb–Ag–Zn deposit, northern Shan State of Burma, is also considered to be a submarine-volcanic–sedimentary type (Hannak, 1972), which was for a long time thought to constitute a replacement-type ore within tuffs and rhyolites (Sommerlatte, 1959).

Acidic tuffs are associated with pyritic copper bodies in Avoca, Ireland. In the pyrite district of southern Spain (Strauss, 1970), the tuffs are the main constituents of the host rocks. Strauss and Madel (1974) proved that the pyritic orebodies of Tharsis belong to the eruption centers of acidic volcanics. Consequently, in addition to the black shales or slates, these acidic tuffs are typical indicators and horizon markers of the eugeosynclinal submarine-exhalative-sedimentary sulphides.

To put the Rammelsberg deposit into context with similar ore types, and to permit a comparison with other varieties of Pb–Zn mineralizations, the reader may wish to consult the works of Bogdanov and Kutyrev (1973) who presented five stratified Pb–Zn deposits. The Rammelsberg (and Meggen) ores belong to their "Atasu type". According to Lambert (1976, pp.578–582, his table II), the Rammelsberg ore is of the "McArthur-type"; thus contrasting them to and differentiating them from the Kuroko-, Superior-, Cyprus- and Yilgarn-types. Stanton (1972, p.502) described the Rammelsberg deposits in the section on "stratiform sulphides of marine and marine-volcanic association". Gilmour (1976, pp.123 and 124, his table I and fig.8) provided useful classificatory comparative models based on gradational variations of the ore host rocks, which includes the Rammelsberg deposits, thus contrasting them to similar as well as to dissimilar types of mineralizations.

TECTONISM OF THE HOST AND SURROUNDING ROCKS OF THE OREBODIES

The host rocks as well as the orebodies of the Rammelsberg are part of a complicated isoclinal fold zone overturned to the northwest. The mineralizations are situated in the tight slinglike isoclinal orebody syncline. Folding, the formation of schistosity, shearing and subsequent block faulting – the result of tranverse fault movements – took place during the Variscan orogeny.

The differential tectonization, which occurred on all scales down to microscopic dimensions, resulted in deformations generally related to the axial fold planes [i.e., due to lateral compression and vertical expansion (= dilation)]. The younger dislocations along E–W and N–S striking transverse faults were formed at the end of the orogenic deformation, as a consequence of lateral compression and horizontal dilation in the strike direction of the folds. During this tectonic stage, the Old Orebody ("Altes Lager") was separated from the Old Orebody West ("Altlager West") along the western main fault (Fig.3), whereas at the eastern margin the dislocations, along the eastern main fault, smaller sections of the New Orebody were cut and displaced. The fault-zone mylonites exhibit brecciated material of the banded and massive ores. For this reason, the transversal faults took on an important role in the past discussions on the Rammelsberg ore genesis. In particular the western main fault was believed to be an example of a conduit (feeder) for hydrothermal solutions, as judged by the presence of massive ore remnants.

The presently available data on the fabrics and the fabric symmetry indicate that both the massive [1] and banded [2] ores are pre-orogenic in origin. Contrary to this interpretation, there is no doubt that the transversal faults are late-orogenic in origin. They are even younger than the ore mineralization in the silicified lenses ("Kniest") and approximately of the same age as the Upper Harz Pb–Zn ore vein deposits.

The extreme differences in the mechanical properties of the massive sulphide ore versus the host rock has resulted in distinct differential movements and deformations in the vicinity of the orebodies as well as in the host rocks. Beyond the deposit, these fabrics are absent in the orebody syncline. The specific ore fabrics (described on pp.609–625) are the consequence of a very intense deformation confined to specific positions in the orebody syncline.

In order to better understand this differential or preferential tectonic pattern, it should be pointed out that no similarly intense deformational features have been recorded anywhere in the Upper Harz ("Oberharz") region. The orebodies are superimposed upon one another in an isoclinally folded fashion, with no younger sections recognizable in those parts of the orebody syncline exposed up to this date.

In cross-section (Fig.7), the massive and banded ores in both limbs of the syncline are superimposed and in contact with each other as observed over long distances. Only in some places are younger upper shales of the Ore Horizon present in the normally oriented limbs and also structurally underlying the ore in the overturned limbs of the syncline.

At present, the generally accepted interpretation suggests that the orebody syncline

---

[1] Massive ore ("Lagererz"): massive high-grade sulphide ore without any significant gangue components. Typical sedimentary structures are common.

[2] Banded ore ("Banderz"): interlaminated stratiform sulphides and shales.

passes through the Old and New Orebodies and that the major part of these orebodies is overturned (Kraume, 1960b). The Grey Orebody is believed to be fully overturned, which has been proved more recently by sedimentological and petrographic studies (e.g., presence of graded-bedding in the grey ore; Schot, 1973). The youngest ore mineralization is the Footwall Grey Orebody (= "Liegender Grauerzkörper"), of which the largest portion is in the normally oriented limb of the orebody syncline, recently confirmed by Gunzert (1979). Nevertheless, other interpretations have been offered regarding the position of the synclinal axial plane and, therefore, also of the tectonic plus stratigraphic relationships of the stratiform ore mineralization.

The Ore Horizon containing the orebodies reaches its maximum thickness at the lower nose in the core of the syncline (between the 11th and 12th mine levels); in contrast the Ore Horizon is of minor thickness in the outcrop region 800 m above the core measured along the orebodies. This is the consequence of tectonic thinning of the limbs, which at the same time resulted in the thickening at the core of the syncline (Fig.7).

The tectonically induced fabric of the orebodies has been formed by an internal kinematically high-grade deformation, which also has affected the immediately neighbouring host rocks. It can be demonstrated that the major and minor tectonic structures of the ore deposit and host rocks are related to the palaeogeographic position and the general tectonic fabric. For this reason, comparisons between the tectonic history of the host rock with that of the banded and massive ores are of great importance in documenting the genetic interpretation of the ore mineralization.

The distinctly restricted distribution of the orebodies, plus their intensely incompetent response to tectonism, resulted in major deformation on both the *ac*- and *ab*-planes ("Formungsebenen") due to the generally compressive strain, which was recorded in the fabrics down to micro-scale (i.e., intergranular movements), in contrast to the weaker host-rock deformation. This also led to the non-affine, highly plastic diastrophic deformations of the orebodies. The latter, therefore, are characterized by a complicated 3-axial deformation pattern, which when examined in smaller areas can be resolved to be of 2-axial deformations. This tectonic style recognized not properly for a long time, has in the past often resulted in wrong genetic reconstructions.

From the regional tectonic point of view, the Rammelsberg ore deposit lies in the Devonian Upper Harz anticline, which is more a stratigraphic than a regional tectonic unit (Fig. 2). Within this unit, the orebodies form the core of a major syncline and are confined to the deeper parts of a synclinorium (Figs.5 and 6). In the southeast of this orebody syncline, the level of flexure ("Faltenspiegel") rises over 1600 m nearly parallel to the steeply dipping overturned limbs in the direction of the Rammelsberg anticline; towards the northwest, this enveloping contour rises gradually for several kilometres with a small dip. The axial plane of the orebody syncline dips on the average of about 45°SE. The axial planes of the minor folds converge towards the top of this syncline. In the northwest of the orebody syncline, they dip 60°SE and in the southwest the dip changes to 30°SE (Figs.5 and 6), according to known structural deformation rules.

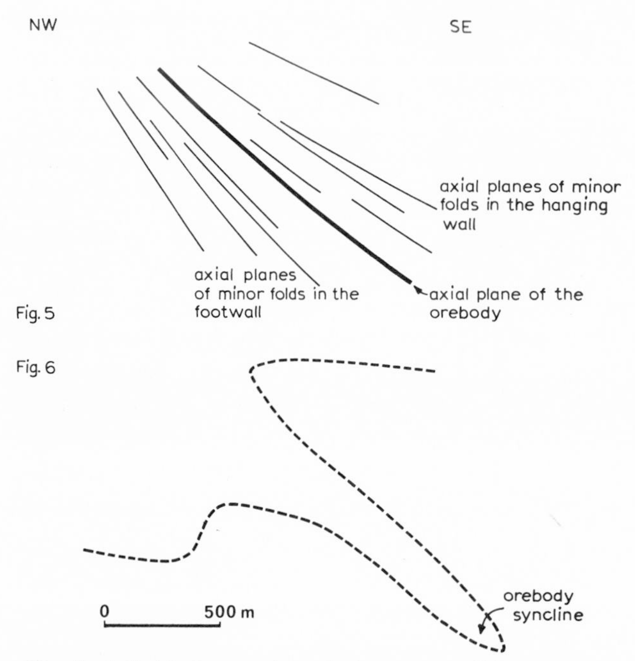

Figs. 5–6. Axial planes and enveloping contour ("Faltenspiegel") at the interface Sandbanded Shales to Lower Shale Horizon.

In general, the pelitic-silty sequences have been cleaved more or less parallel to the axial planes. Fracture cleavage is common in the more competent layers, whereas in coarse-grained clastic rocks the cleavage is normally absent. Sericite growth, as well as the remobilization of quartz and carbonate parallel to the cleavage joints, are indicative of the beginning of anchimetamorphism (= dynamo-metamorphism); the evidence of coalification of sedimentary carbon to anthracite supports this interpretation (Jacob, 1974, unpublished report of the BGR; cf. also Wolf and Chilingar, 1976, for a summary of the concepts involved).

The intensity of folding is considerable and is reflected in the high degree of isoclinal folding of the orebody syncline. A good measure of the extent of folding can be obtained from the angle of inclination of the central parts of the limbs where they are parallel to the axial plane, especially in the footwall of the orebodies. The thickness changes by folding of similar beds are more intense in the overturned than in the normal limbs (Fig.7), demonstrating that the overturned beds generally have undergone a higher degree of thinning.

In general, the tectonically produced thickness variations range between factors of 3 and 6, if the thickness of the same bed is measured at right angles and correlated from the axial plane ("Achsenebene") at the fold noses ("Faltenumbiegungen") to the central parts of limbs. A factor of 20 is quite frequently exceeded in the folded massive ores. The

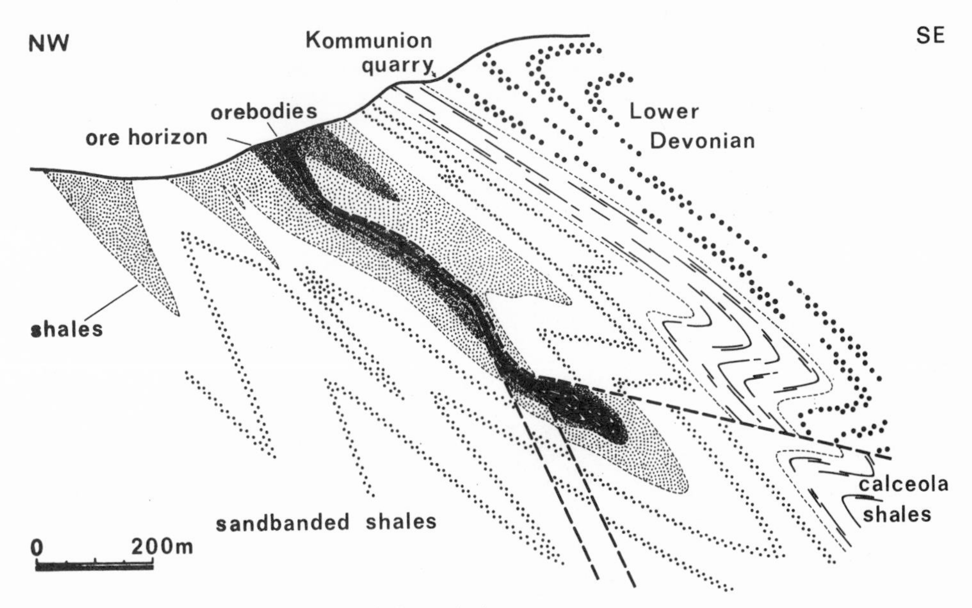

Fig. 7. Simplified cross-section through the orebody.

explanation of the generally greater tectonic mobility of the ore in contrast to the host rocks, lies solely in the greater ductility of the dominant ore minerals which are galena, sphalerite and chalcopyrite. Siemes (1967) experimentally proved the greater susceptibility to mobilization of galena versus the common silica host-rock minerals. The great competence of massive pyrite is well known. Consequently, during the folding the massive pyrite orebodies behaved very differently from the Rammelsberg ore deposit. Any number of examples can be provided, e.g. from the Southwest Iberian pyritic belt ("Kieserzgürtel") (Strauss, 1970). The Meggen pyrite orebody, therefore, is different from those of the Rammelsberg (cf. chapter by Krebs in this volume).

The shear strength of the host rocks was frequently exceeded in the fold hinges. Shearing parallel to the fold axes (i.e., *b*-parallel) and disharmonic folding introduced complications to the structural pattern described above, also affecting the rocks adjacent to the orebodies. The shearing between orebodies and the host rock – downthrown fault movements as well as overthrust displacements – increased the "degree of freedom" of the host rocks' folding. Several minor fold types, including drag folds, are widespread, and must also be considered as a general part of the host rocks' tectonized envelope in the vicinity of the ore deposit.

As mentioned earlier, the great differences in competence between the orebodies versus the host rocks (which can be numerically expressed) resulted in the development of the shear zones in the hanging wall and footwall of the orebodies. In principle, these movements (Fig.8) correspond to a "twin-shear system" (= "zweischarige Scherung").

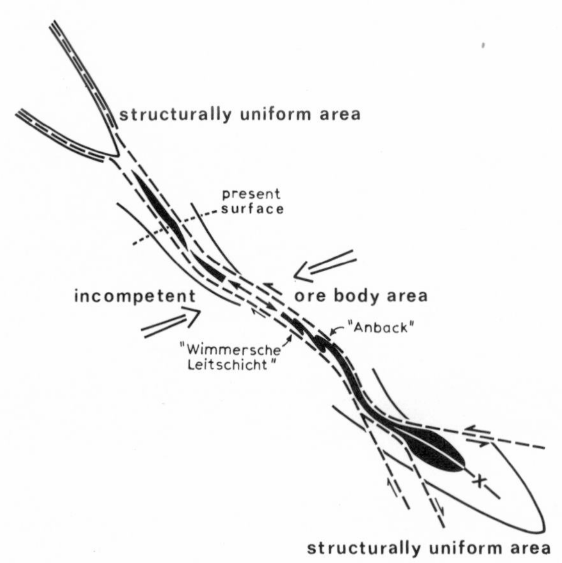

Fig. 8. Outline of tectonic movements in the orebody syncline (New Orebody).

Probably dependent on the stratigraphically different position, there were distinct differences in the development of the shear zones at the New and Old Orebodies. The following tectonic reconstruction suggests two main stages for the origin of the ore mineralization. As has been demonstrated already, even small differences in the stratigraphic position of the orebodies expressed by different stratigraphic levels were tectonically enhanced by the plastic mobilization of the limbs to the hinges of the isoclinal folds.

The style of displacement of the New Orebody is shown in the profile of Fig.8. From the upper portion of the orebody down to its thickened part, the folding is disharmonic. Both the hanging and footwall of the orebodies are marked by the dislocations of the so-called "Wimmersche Leitschicht" and "Anback". The hanging-wall shear zone directly follows the orebody and, therefore, constitutes a sharply defined discordant weak zone ("Anback"). In general, the footwall is within a few metres (up to 10 m) of the massive ore, being well marked by the mylonite zone of the so-called "Wimmersche Leitschicht" which can be several metres wide and in places splits into several minor fault branches. From the lower, thickened nose of the New Orebody, the faults diverge by changing in dip into the host rock with spreading angles ranging from parallel to the orebodies to about 50° in the deep-seated host rock.

The fault pair of the footwall, with its dip of 65°SE, underwent a total net slip of 100 m at the locality of greatest depth of the New Orebody. According to direct measurement of some subordinate displacements and the stratigraphic characteristics, the hanging-wall fault zone, dipping 15°SE, underwent 170 m overthrusting. The difference in the amount of displacement, as well as the varying of the fault zones, can be explained by an inclined compressive stress pattern, which also produced the overturned folds. This inclined stress field was also responsible for the differential thinning of the limbs. Genetic connections between the thrust movements, the "Kniest" feeder joint mineralization, as well as the Pb–Zn–Cu–pyrite–quartz vein-type mineralization in the overthrust region of the host rock, seem to be recognizable.

The development of that fault-pair system, its presence confined to the locality of ore occurrence, was controlled by the extreme difference in competence, i.e., variations in the rheological properties of the orebodies versus the host rock, as already mentioned above.

The dislocations are also the consequence of the intense maximum compressive narrowing of both the orebodies and the host rock. Because of the minimum amount of displacement and the maximum length of inclination of the New Orebody, an arithmetric "elongation ratio" of 1.25 : 1 of orebody to host rock is the result. In other words, the New Orebody was stretched plastically about 0.25 m for every 1.0 m of the host rock (i.e., about 25%). This discrepancy in length was compensated for by the above-described dislocations.

Earlier researchers were familiar with these dislocations which has led to varying interpretations. Schmidt (1933a, b) believed that the syngenetic orebody was injected from below – the opposite is actually true. The New Orebody appears to have been forced downwards into the locality of minimum normal stress – that is the core of the orebody syncline – to be intensely thickened to nearly 100 m of horizontal width. Vogt (1894) even visualized the massive ore to be intruded as a sulphide melt along the fault zones. Earlier publications described these dislocations as partial criteria to support an epigenetic origin. It is noteworthy here that for a long time the displacement zone of the orebody footwall served as a marker horizon to distinguish the ore-free layers from the ore-bearing units (termed "Wimmerian key bed" after Wimmer).

The tectonic deformation features described above are absent outside the orebodies along the strike.

At the Old Orebody, the fault tectonism appears to have been modified because of the shallower stratigraphic location and the consequently different tectonic position (Fig.14). There is no doubt that this specific tectonically produced structure excludes any involvement of epigenetic processes in the origin of the Rammelsberg ore deposit. Likewise, a metasomatic replacement of sedimentary rocks cannot be quoted as a possible process of ore formation, because the required lithologic and facies type of sediments, as well as the necessary mechanical competence differences, are absent in the host rock.

The fold patterns (folding style) of the banded and massive ores are comparable only to the fabrics of high-grade metamorphic, and also highly tectonized, sling-like folded rocks. In the whole region of the adjacent deposit area, there is no host rock that could have been tectonically deformed in such a way during the Variscan orogenesis, and it is even impossible to imagine. Additionally, no explanation can be offered for the totally unaltered carbonate rocks inside of the deposit. From this point of view a replacement origin can also be excluded.

It has been impossible to develop a palaeogeographic model that could incorporate a plausible epigenetic metasomatic ore-creating mechanism. One also has to consider that in the case of a metasomatic origin the generally expected accompanying, contemporaneously formed, replacement influences of the host rocks are absent. Tectonic-rheological factors also exclude a pretectonic pyritic formation together with a post-orogenic Pb–Zn–Cu mineralization of epigenetic origin (this two-stage genesis has been called "diplogenesis" by Lovering, 1963; cf. also Wolf, 1976, on multi-stage ore processes). An orebody composed solely of syngenetic pyrite would have had a different competence and the tectonic style would, therefore, also have been totally varying from those actually observed at the Rammelsberg. This argument alone is sufficient indication of the significance of comparative tectonic studies (i.e., to find differences and similarities) that culminate in the proposal of the most reasonable genetic hypothesis.

Below the contact of the fault zone with the New Orebody (Fig.8), the massive ore mineralization is harmonically folded with the host rock. Fault dislocations along the orebody contacts with the host rock are only locally present. The orebody became conspicuously thickened (up to 50 m) in the core of the syncline under the influence of a predominating tensional stress. In the massive ore isoclinal slinglike folds ("Schlingengefüge") are present, the characteristic features of which were deciphered by Kraume (1960b) who utilized the variation in recognizing ore types (Fig. 9). Within the east-striking unit interbedded with the banded ore, the adjacent host rock took part in this slinglike folding. At this part of the New Orebody, clearly recognizable discordances prove that primary slump structures were tectonically overprinted; this being supported also by the internal thickness changes in the Ore Horizon itself. Without question, the formation of *banded* ores at this location had already taken place prior to the sedimentation of the *massive* ore.

The effects of the shear tectonism on the flanks of the orebodies have been well recognized along the eastern side of the New Orebody. The angles of plunging fold axes

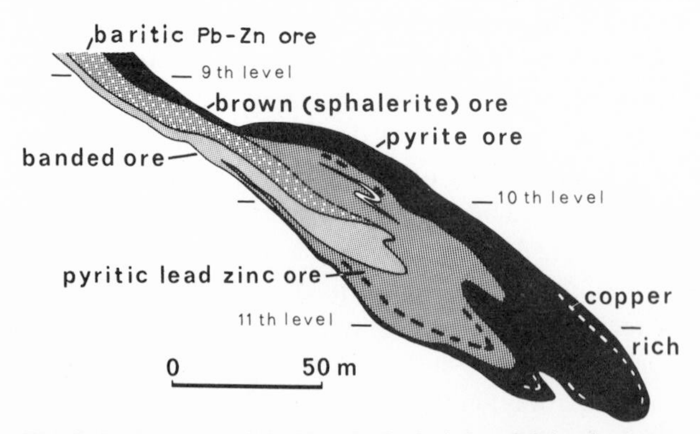

Fig. 9. Cross-section of the New Orebody below 9th level; ore types and internal folds. (After Kraume, 1960b.)

are 60°W or more, which is characteristic for the region above the fault junction (Figs.3, 7 and 8).

The *ac*-plane ("Formungsebene") of compressive strain is nearly horizontal in these parts of the orebody and, therefore, the ore is clearly horizontally elongated. Principally similar to the double shearing around the New Orebody, synclinal massive ore "injections" have in places resulted in forklike insertions into the host rock, e.g. banded ore sequences that vary widely in dimensions (Hannak, 1963). They may reach up to several tens of metres and can also be recognized in handspecimens as well as on a microscopic scale (Figs.10, 11 and 12). The shear planes are, in these instances, developed only around the massive ore. On decimetre to microscopic dimensions, the cleavage planes served as preferential "injection" planes.

Equivalent to the eastern part of the New Orebody, steeply to vertically plunging fold axes are to be expected along its western margin. It can be generalized that in the lower section of the New Orebody the boundary of the massive ore coincides approximately with the plunge of the local fold axes, whereas in its central sections horizontal axes predominate. In the strike section, the lower stratigraphic boundary plane of the Ore Horizon, fixed by the twin-tuff, exhibits analogous flexures of the fold axes in the adjacent host rock (Fig.3). Between the orebodies, the plunge of the axis varies widely and reaches a maximum of 40°E, which indicates a conspicuous deviation from the general 10°W-plunge beyond the immediate locality of the orebodies. The easterly plunge of the axis was increased even further by a northwest-striking fault that passes between the orebodies (Fig.3).

For the above reasons, it is not possible to "unfold" the distorted orebodies to

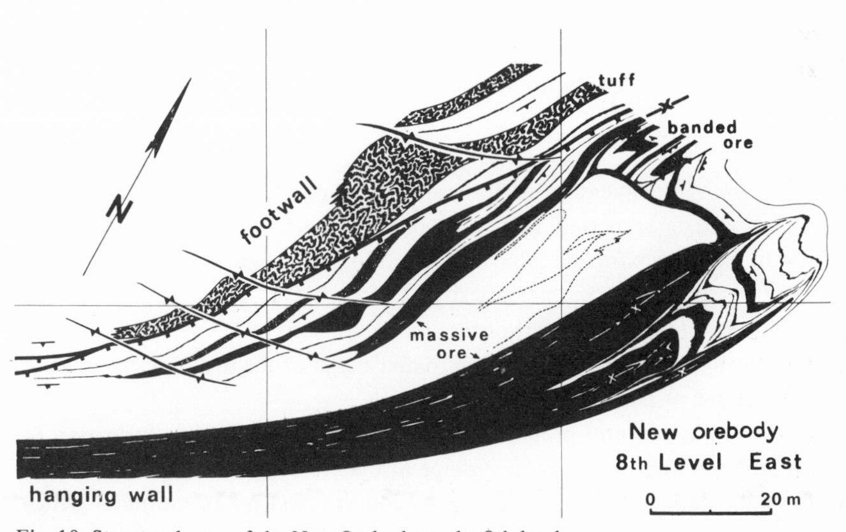

Fig. 10. Structural map of the New Orebody at the 8th level.

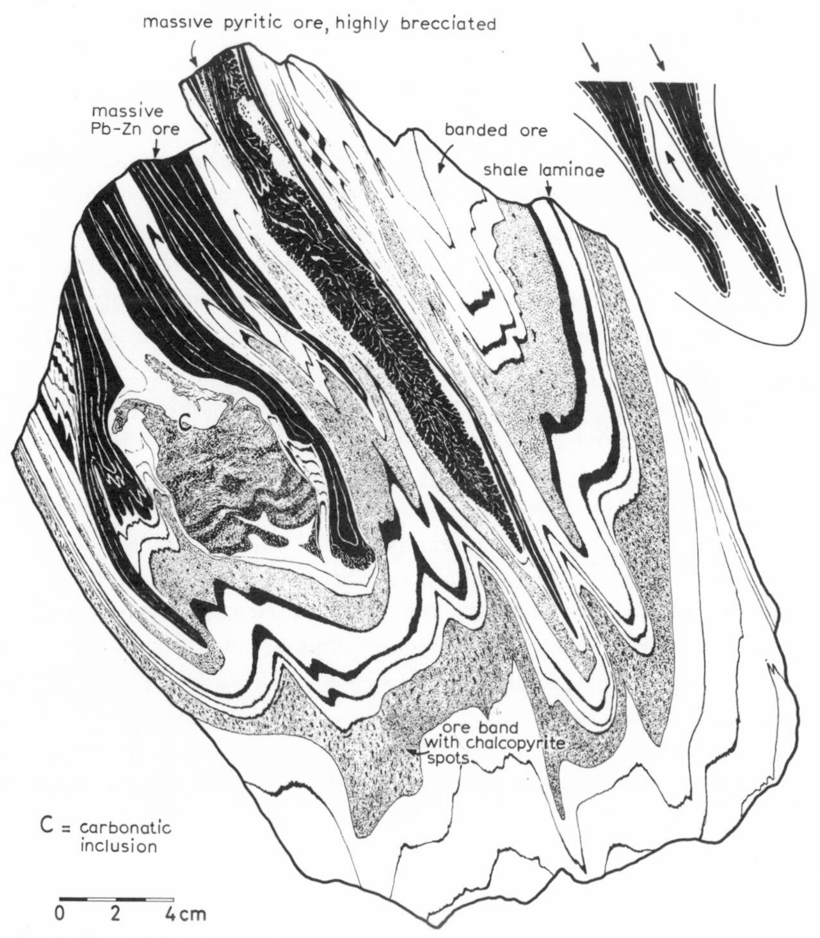

Fig. 11. Highly folded banded ore with pike-like synclines in the massive ore, New Orebody.

establish their primary synsedimentary position and form, because extremely non-affine plastic deformation does not permit such a theoretical "unfolding". Any attempt to unfold vertically to the approximated plunge of the fold axis (about 10°W) will be useless. However, it would be less problematical to unfold the orebodies vertically to a 30°E-plunging fold axis in the central part of the ore deposits (Fig.13). Nevertheless, even in this case, in order to obtain the primary pre-folded shape of the ore deposits, one has to consider a large degree of plastic non-affine unfolding.

Gunzert (1969) has recently proposed a model for the synchronous formation of both orebodies. In describing the pre-folding shape of the ore deposit, he assumed that at least two-thirds of the initial metalliferous concentration was removed by erosion – which in principle is unimportant for the problem under consideration. As the axis of unwinding,

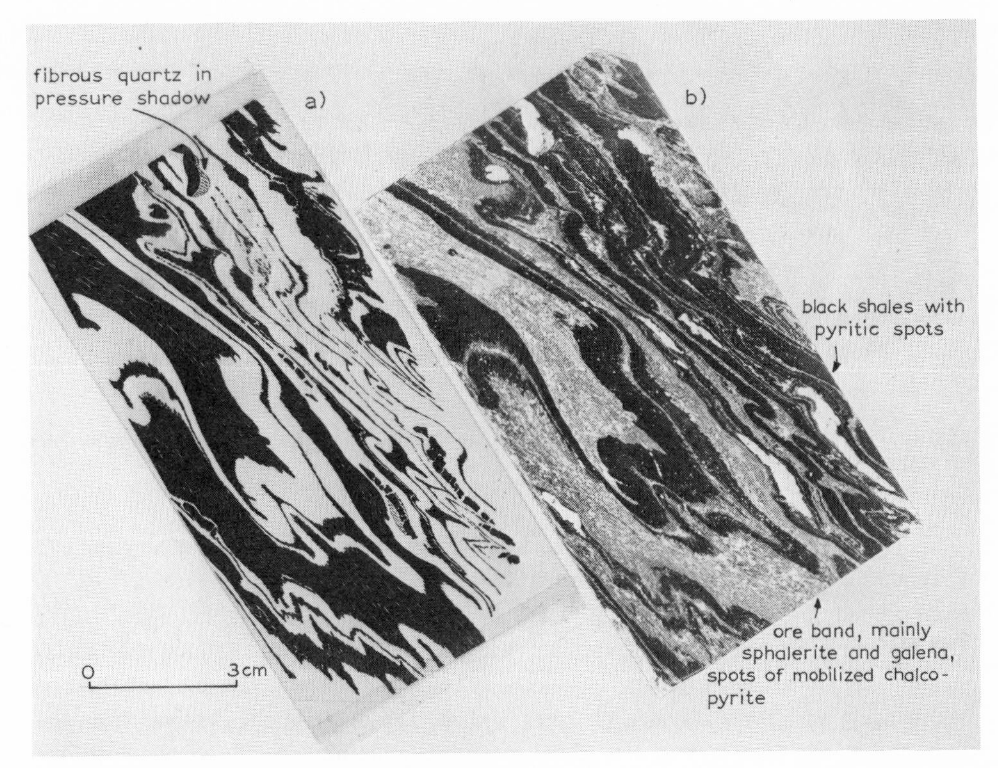

Fig. 12 (a, b). Highly tectonized banded ore, New Orebody.

he used an axis plunging 8°E. The procedure of unwinding parallel to this axis into a pre-folded position does not establish the likely original shape of the ore deposit, as a comparison with Fig.13 shows.

The variation in the tectonic deformation of the Old Orebody in contrast to the New Orebody can be explained by its younger stratigraphic position. The characteristic thickening of the synclinal core known from the New Orebody is absent at the Old Orebody. The latter's continuation below the surface is much smaller than is the case of the New Orebody and the vertical difference between the maximum depth below surface reached by the two orebodies is 180 m. Consequently, the maximum depth reached by the hinge of the orebody syncline in the Old Orebody is less than the point of the twin-shear-fault intersection associated with the New Orebody. Nevertheless, both the downthrow in the footwall and the upthrust of the hanging wall are similar in the Old Orebody, as already described for the New Orebody. Therefore, the elongation (= stretching) of the Old Orebody was developed in a zone with the highest degree of tectonic freedom, which was between the parallel-running shear zones in the footwall and hanging wall, respectively (Fig.14).

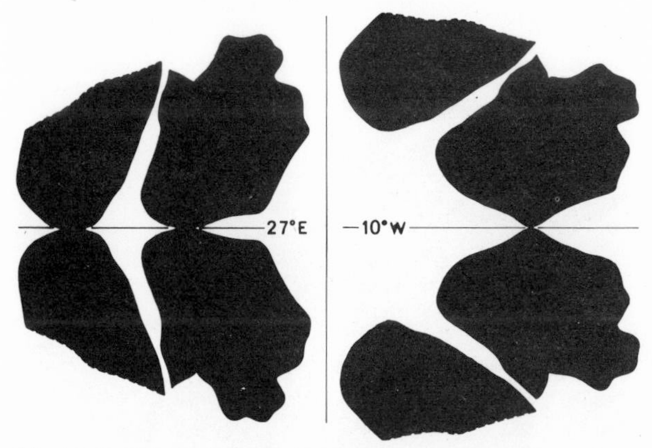

Fig. 13. Prefold position of the orebodies related to the plunge of fold axes (27°E) at the deposit (left) as well as to a general fold plunge (10°W).

The fundamental difference of the tectonic deformation between the New and Old Orebodies is recognizable by the external folds of the massive ore that extend from the main orebody into the hanging wall (known as the "Hangende Lagerabfaltungen" of the Old Orebody), especially in the so-called "Hangendes Trum" which follows one branch of the fault system (cf. Fig.14). These orebody folds can be traced into the host rocks of the hanging wall for a distance of up to 100 m. The observations available from the exposure in the mine suggest that these are disharmonic folds parallel to the host-rock bedding within the normal limb of a hanging-wall anticline, which has been thrusted to

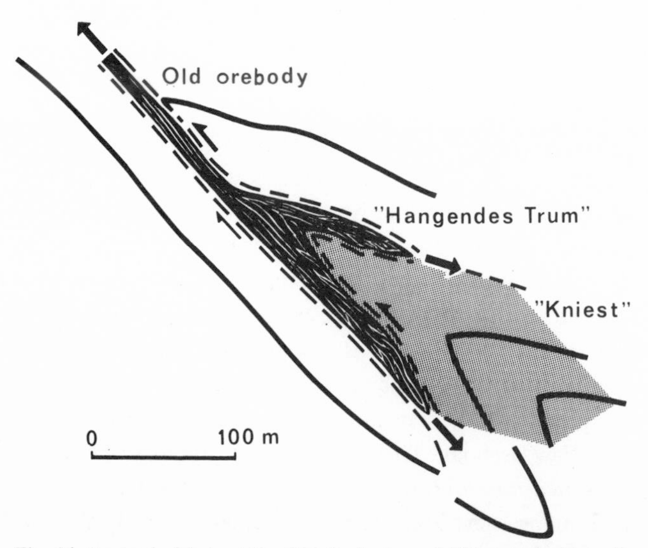

Fig. 14. Tectonic fabric at the Old Orebody and "Hangendes Trum".

take on an overlapping position above the orebody. This deformational picture can be explained by assuming that the transposed rock mass, under the given stress field, selected the "path of least resistance" ("Ausweichbewegung"), which was parallel to the bedding planes (Fig.14). These dislocated disharmonic folds of the Old Orebody are the consequence of the more intense congruent and plastic deformation of the massive ore in contrast to the host rocks, as has already been mentioned in the case of the New Orebody. Many researchers have proposed that the "Hangendes Trum" is a dislocated or transposed fold ("Abfaltung"), and they especially relied on the so-called "plough effect" ("Pflugwirkung") of the highly competent "Kniest" rock complex (cf. Fig.14). The "Kniest" is not the cause of the "plough" mechanism, but the result of the difference of the rock competences which, in turn, controlled the style of tectonism; however, the "Kniest" complex had some modifying effect. It can be assumed that the uninterrupted stretching of the host rock was insufficient to compensate for the "overstretching" of the more plastic massive ores along their contact zones. This differential behaviour promoted the plastic forklike injection of the massive ore along the bedding planes of the hanging-wall's transposed overlapping anticlinal limb. Moreover, one can deduce from these considerations that the bedding planes offered the least resistance to the plastically injected massive ore. Thus, in spite of their different tectonically produced shapes, the Old and New Orebodies behaved kinematically similarly. Consequently, the tectonism of the two orebodies cannot be examined separately because they interacted reciprocally during deformation.

Gunzert (1969) considered the above tectonic reconstruction as too complicated, and disregarded the downthrow in the footwall region because, in his opinion, it is insignificant. His arguments are convincing inasmuch as: (1) banded ore margins occur only along the western extremity of the Old Orebody as well as along the eastern border of the New Orebody; (2) in between the deposits this ore-bearing facies is absent; and (3) it seems likely, therefore, that a primary (prefold) single ore unit existed which was tectonically separated into two distinct sulphide bodies. Nevertheless, the relatively wide (20–50 m) unmineralized section between the orebodies remains to be explained (Fig.3). Inasmuch as Gunzert visualized the en-échelon arrangement of the different ore types in the Old Orebody (Kraume, 1955) to be the result of easterly translocation caused by the obliquely oriented, northeast-striking tectonic compression of the lens-shaped "Kniest" complex at the hanging wall, the just-mentioned 20 to 50 m-wide unmineralized section between the Old and New Orebodies is even more difficult to clarify.

Originally, the ore sequence of the Old Orebody must have been, following Gunzert's idea, a vertical succession of one above the other; however, this implies that the original horizontal distance between the Old and New Orebodies must have been even greater than it is now. Thus, in this case, the unmineralized gap between the two orebodies cannot be plausibly explained. An analogous effect is visualized by a general flattening of the plate-shaped orebodies by a 3-axial plastic deformation. In this instance, one has to presume that the original plate-shaped New Orebody (this would also apply to the Old

Orebody) had a smaller surface area, and that originally there was a greater distance between the two sulphide deposits than is the case in reality. Moreover, a stratigraphic repetition of a specific paragenetic metalliferous sequence, separated by a non-mineralized section, is not an unusual occurrence when a world-wide comparison is made.

PRIMARY ORE FABRICS, ORE-MINERAL PRAGENESES, DISTRIBUTION OF TRACE ELEMENTS AND SULPHUR ISOTOPES

The data on the primary ore fabrics in particular and their application to genetic interpretations of the Rammelsberg ore, has caused heated discussions among various reseachers. Ramdohr (1953b) especially argued against Schouten's (1937) treatise in which the latter unreservedly advocated a metasomatic origin. Ramdohr's comments are significant, so that his opinions and conclusions will be stressed here. He compared, among others, carbonaceous and bituminous sediments, citing their original above-average metal contents (cf. also Saxby, 1976, his table I; and Lelong et al., 1976, their table XII). With these data, he hoped to convince even the staunchest advocates of epigenesis that a syngenetic origin of the Rammelsberg ore would be the more logical interpretation.

Shadlum's (1971) questions of the metal source, as well as the localization of the ore of analogous Russian massive sulphide deposits, are still unanswered. He proposed a magmatic–metasomatic origin for the Pb–Zn–Cu mineralization, believing that only the pyrite is of sedimentary, that is to say, of diagenetic derivation. This is similar to Dahlgrün's (1929) opinion. For this reason Shadlum's interpretation is of interest, because he pointed to the fundamentally important correlation between the textures and the mineralogy (both composition and amounts) of the Rammelsberg ores. A diplogenetic origin (partly syngenetic and partly epigenetic) must be rejected with great certainty for the Rammelsberg ore deposit, and only a totally sedimentary (single-stage) mechanism in ore formation can be invoked. Therefore, a detailed analysis of textures, structures and parageneses is particularly necessary to find a plausible genetic theory.

Ramdohr's (1953b) investigations into the crystallization behaviour of pyrite in relation to other sulphide minerals are particularly noteworthy. He was able to prove that metasomatism (i.e., replacement) was not involved in the origin of the ore; and also that subsequent metamorphic modifications had played only a minor, insignificant role.

In addition to the five major minerals, namely sphalerite, galena, pyrite, barite and chalcopyrite, twenty-five accessories are known. Ramdohr (1953b) described gold, bismuth, arsenopyrite, linnaeite, gudmundite, valleriite, cubanite, tetrahedrite, bournonite, boulangerite, jamesonite, polybasite, proustite, wolframite, hematite, magnetite, rutile, anatase and three other unclassified sulpho-salts. Schot (1973) described Ramdohr's valleriite as mackinawite. The former also gave details of two previously unknown minerals of the Rammelsberg ores, namely kobellite and electrum.

The following gangue minerals have been identified: barite, dolomite, calcite, quartz, ankerite, biotite, muscovite, sericite, chlorite, albite and carbonaceous components up to the rank of graphite. These gangue constituents are partly of hypogene sedimentary and partly of clastic sedimentary derivation.

The supergene ore minerals of secondary weathering origin are covellite, chalcocite, bornite, tenorite and delafossite, but are not of importance.

The mineralogical and textural features of the unaltered protore are to be found in untectonized parts of the deposit (i.e., in "pressure shadows" = "Druckschatten"). This is, of course, based on the presupposition that dynamometamorphic processes during tectonism occurred and actually left an imprint on the rocks and ores. Under favourable conditions, the interior of massive pyrite could be "pressure-free", with primary inclusions remaining unaffected by dynamometamorphic overprints. On the other hand, additional protective envelopes are provided by "soft" (in the tectonic sense), highly incompetent ore minerals which surrounded the primary prefolded textures.

Ramdohr (1953b) described several different morphological types of pyrite from the Rammelsberg, which after precipitation remained unaffected by subsequent deformation stresses, except for brecciation.

Ramdohr's (1953b) observations are described in the following. Primary syngenetic forms of sphalerite and barite are relatively quite frequent. These two minerals, as well as pyrite, are characterized particularly by globular forms up to several millimetres in diameter (Plate I, d; Plate II, e–g; and Plate V, a and c).

Chalcopyrite and galena tend to be altered very early (already during diagenetic conditions), so that their primary textures are not normally preserved. In general, the primary textures of any of the ore minerals are a combination of colloform and crystallization features, several varieties of crusts and laminations (in particular of "cauliflower"-type), and globular aggregates of which some are either formed of concentrically arranged individual crystallites (= framboids), or are composed of radial-fibrous crystals, or have no particular internal configuration (Plate II). Another textural variety consists of a very finely crystalline (near the limit of microscopic resolution) mineral intergrowth. Among the primary textures are probably early-diagenetic pseudomorphs of barite after gypsum. Subsequently, also the barite was pseudomorphically replaced by other minerals.

The external form of the primary globular pyrite is vari-shaped (polymorphic); both concentric shell-like and skeletal forms have been recorded. The pyrite is intergrown with detrital clay fragments as well as with sphalerite, chalcopyrite, marcasite, galena and native bismuth (Schot, 1973). So-called "auto-intergrowths" ("Eigenverwachsungen") composed of mineralogically slightly different, but related, pyrite species (melnikovite?) were also identified. These variations are the result of rhythmical precipitation and/or coagulation. Mutual replacements of the individual pyrite microcrystals by the matrix minerals, or vice versa, have not been recognized. Sulphate-reducing bacteria may have acted as nuclei for the initial sulphide precipitation leading to the formation of framboids. Inorganic conditions of precipitation, however, must also be given consideration. Schot (1973) offered a detailed review of the numerous pyritic framboid types, discussing

a)
b)
c)
d)
e)
2 cm

their origin based on the most recently published information. The pyrite framboids often occur in such large concentrations as to constitute strata in the massive and banded sulphide orebodies, but are less widespread in the Wissenbach Shale. The framboids' diameter averages between 1 and 100 μm, whereas the rhythmically laminated and radially fibrous pyrite crusts and nodules reach up to several centimetres in thickness or diameter, respectively. They are frequently rich in marcasite. Inclusions of idiomorphic grains of sphalerite intergrown with chalcopyrite are occasionally present. One should also mention the presence of geodes several decimetres in diameter within the banded ore as well as in the black shales, with a diagenetic concretionary structure that incorporated the earlier-formed banded ore layers. Kraume (1955) has provided excellent photographs of them, illustrating among other features that these nodules (septarian type) nearly always have shrinkage cracks filled with carbonates or barite and often also concentric layers of pyrite. Fossils have frequently acted as nuclei for the crystallization or precipitation of the sulphide. Primary (syngenetic) hybrid or mixed sulphide gels upon crystallization gave rise to a complexly interwoven pattern of galena with pyrite, as well as of pyrite with sphalerite, and chalcopyrite with sphalerite (Plate IV). Pyrite crusts sometimes grade into those of marcasite. In these intergrowths, the marcasite differs from the pyrite only by the absence of other sulphide minerals. The crustal pyrite and knotty surfaces formed as geopetal (= top-and-bottom) structures in the banded ore, being the product of polar accretion or precipitation from a solution (Plate Xf). In addition to these distinctly colloform pyrite and marcasite, there are widespread intergrowths and gradations into idiomorphic pyrite. Idiomorphic pyrite crystals with a sieve (= diablastic) texture are thought to be early diagenetic (Plate III).

The pores or micro-cavities of the pyrite (= host mineral) are occupied in all instances by the surrounding (= matrix) minerals. The recognition of genuine idioblasts is not always possible. Two additional varieties of pre-tectonic sulphide constituents that formed in a predominantly colloid-inducing milieu should be mentioned, namely globular galena several hundred microns in diameter with shrinkage cracks and colloidal galena–sphalerite with distorted shells. These features are found especially in the massive ores. A considerable number of additional examples of intergrowth textures have been listed by Ramdohr (1953b) from whose work the above description has been quoted.

Most of these ore mineral textures indicate beyond doubt that the orebodies did not originate metasomatically, but support a sedimentary, including diagenetic, origin. A metasomatic process is also excluded by the presence of unaltered carbonate beds and

---

PLATE I

a. Highly cleaved sphalerite ore, Lower Devonian ore seam.
b. Pyritic banded ore, quartz stringer parallel to cleavage, small fault (Old Orebody).
c. Sphalerite ore with bands of thin black shales, Lower Devonian ore seam.
d. Sphalerite nodules, Lower Devonian ore seam.
e. Sphalerite ore, highly tectonized, Lower Devonian ore seam.

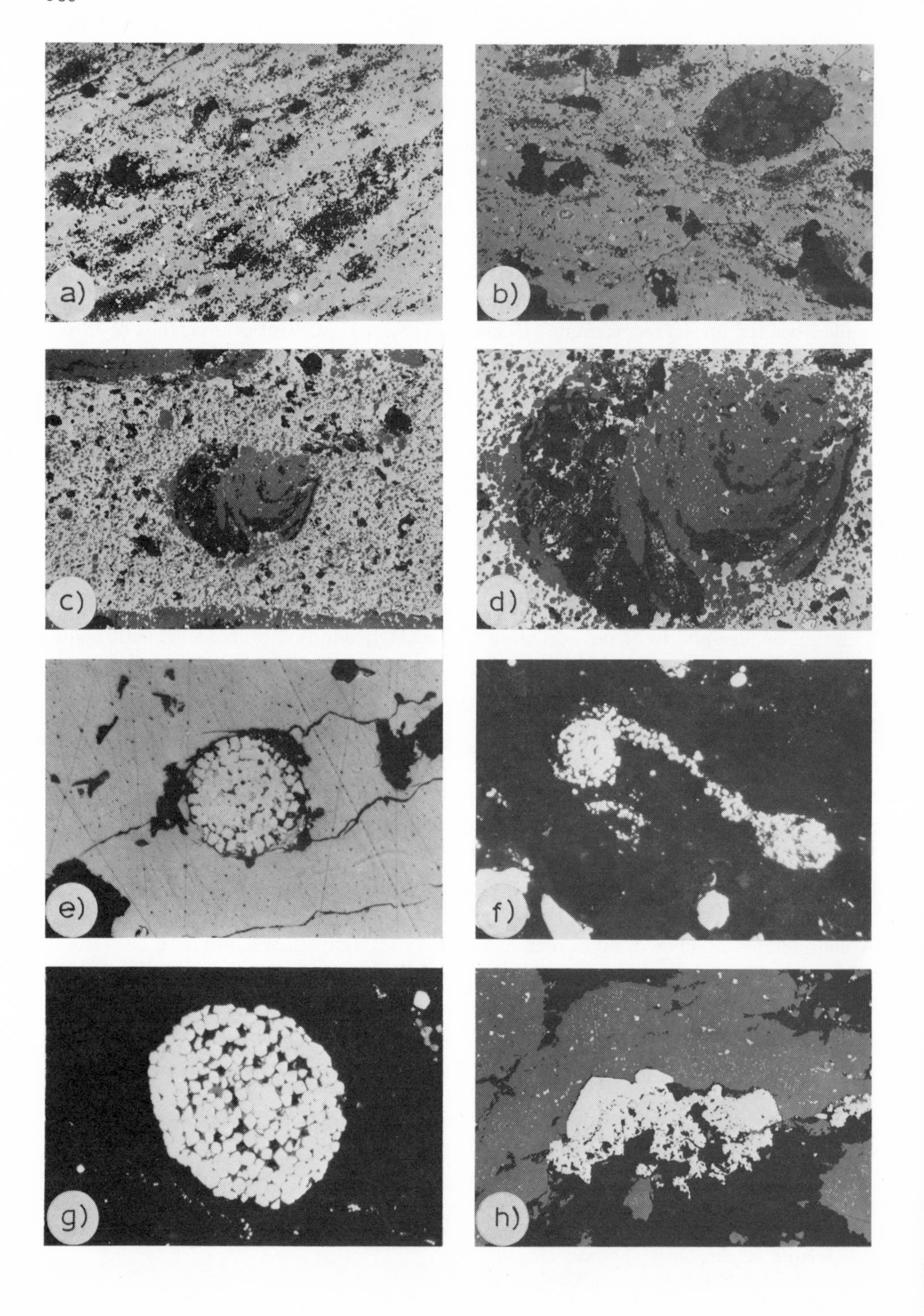
a)
b)
c)
d)
e)
f)
g)
h)

carbonate geodes closely associated with the Ore Horizon, the mineral constituents of which suggest a predominantly hypogene (= ascending) hydrothermal source and origin.

An even stronger and more conclusive criterion in support of a syngenetic sedimentary origin, with a colloidal and or ionic supply of the required chemical components, is furnished by the particularly characteristic conformable bedding and the interfingering of both ore and gangue minerals with shales. In general, however, the massive ores are unsuitable for sedimentological textural analyses because of their colloform fabric and the secondary high-grade tectonic fabric. In contrast, the banded ores and thin ore seams, in parts with a much lower degree of tectonic deformation features, are more amenable to such investigations. In such sections one can prove the existence of graded-bedding, cross-laminations, bedding due to grain-size variations, fractional sedimentation formed by changes during precipitation, horizontally continuous globular or knotty to nodular laminae, current-erosion channels, load casts, sedimentary mylonites and breccias of micro- to macroscopic size, as well as subaqueously formed slumps constituting larger sedimentary complexes (Plates I, II, VI and X).

Sphalerite globules, up to 5 mm in diameter, are particularly common in a Lower Devonian "pre-ore phase" seam ("Vorläuferbank"), and are the product of syngenetic sedimentary to early diagenetic processes (Gundlach and Hannak, 1968). Depending on the facies and the milieu of formation, these sphalerite globules are the host of predominantly carbonate and chalcopyrite inclusions. Barite spherulites are genetically comparable to the sphalerite globules. The barite nodules contain in their core primary clastic particles (mainly clay minerals and quartz) and some authigenic albite, which is characteristic of the Footwall Grey Orebody (Ramdohr, 1953b).

Oolitic carbonate concretions, with clearly discernible laminae, occur in strata in the so-called "Kniest" and the immediately adjacent rocks. Some of the concentric shells are composed of pyrite (Kraume, 1955). In part, a 2 to 3 cm-thick chalcopyrite seam (Lower Devonian massive sulphide ore seam) is unequivocally cross-laminated on a microscopic scale, as seen by the accompanying laminated sphalerite grain accumulations (Plate II). The "granoblastic" recrystallization texture of the chalcopyrite appears to have origi-

---

PLATE II

a. Cross-bedded copper ore; Lower Devonian ore seam. Light grey: chalcopyrite; dark grey: sphalerite; dark: gangue material. Magnification 30×.
b. Copper ore of the Lower Devonian ore seam with fractured concretionary sphalerite (mineral contents compare with a). Magnification 30×.
c. Load cast of sphalerite and gangue material in a band of galena. Lower Devonian ore seam. Magnification 20×.
d. Load cast (compare with c). Magnification 45×.
e. Framboidal pyrite in galena, Magnification 300×.
f. Framboidal pyrite in gangue material. Magnification 150×.
g. Framboidal pyrite, centre with chalcopyrite and sphalerite, dark: pelitic gangue material. Magnification 300×.
h. Two generations of pyrite, Lower Devonian ore seam. Magnification 180×.

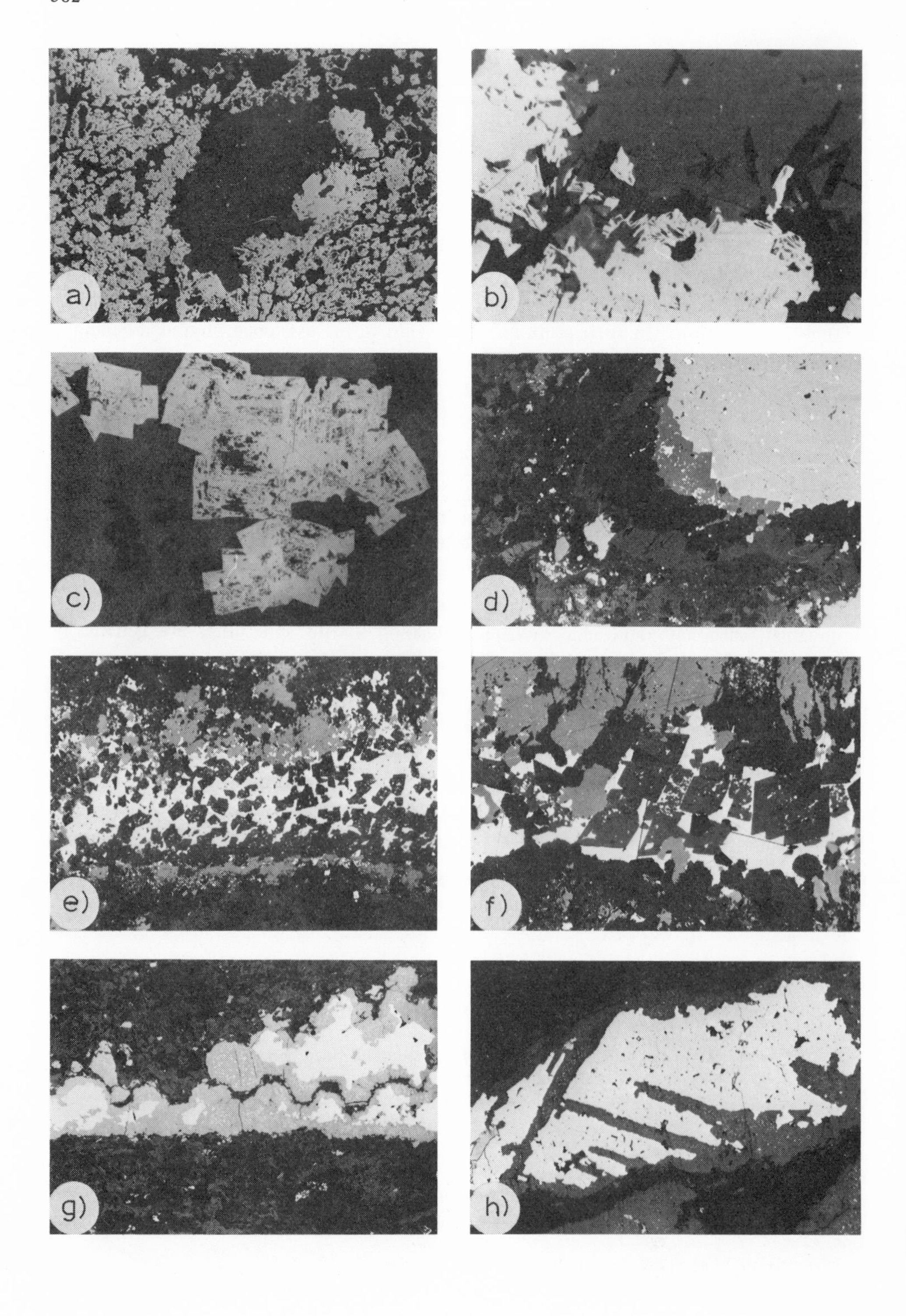
a)
b)
c)
d)
e)
f)
g)
h)

nated very early, probably during the stage of sedimentation followed by a process of reworking and redeposition. Analogous cross-laminations were observed in massive barite of the Footwall Grey Orebody (Fig.15), where pyrite globules, among others, delineate the cross-lamination. Significantly, a polar (= geopetal) grain-size decrease in specific barite layers suggest both graded-bedding and the formation of these barite grains prior to their accumulation (Plate VI). The precipitation of the barite, consequently, must have occurred from a chemically suitable supersaturated solution or suspension, which was succeeded by differential sedimentary deposition according to grain size.

Without any doubt, the above-listed sedimentary textures and structures are genuine top-and-bottom features. Moreover, they are suggesting a possible sedimentary reworking, retransportation and redeposition; the least they indicate is a settling of the particles from suspension. Impressive textbook examples of sedimentary ore textures are also well documented from the massive pyrite deposits of the Iberian belt, Tharsis, South Spain (Strauss and Madel, 1974).

The macro-sequences (i.e., the large-scale stratiform beds of banded sulphides) of the banded ore as well as of the Footwall Grey Orebody also have well-developed sedimentary characteristics. They support the hypothesis of a periodic or pulsating ascending supply of chemical components, a certain degree of turbulence in the depositional milieu in the vicinity of the thermal springs, and – extrapolating from this – an accumulation of the material under exogenic sedimentary conditions. Consequently, the micro- to macroscopic features prove without any doubt a primary sedimentary origin of the ore accumulations. Moreover, no criterion has been recorded that would even suggest a metasomatic process. The pyritized fossils (goniatites, brachiopods and tentaculites) present in the ore bands (Fig.24) (to be considered the product of diagenetic pyritization rather than of metasomatism), plus the mostly unaltered fossils adjacent to the sulphide bodies, are additional features offering information about the environment.

A more detailed description of the textures of the banded ore cannot be provided here as there are too many varieties – to do justice would require a lengthy treatment. Let it suffice to say that they are composed of a rhythmical alternation of shales and fine-grained ore, which constitute very thin laminae or beds up to several decimetres in thickness (Fig.33 and Plate I). In contrast to the massive ore, the layers of the banded ore

---

PLATE III

a. Recrystallized colloidal pyrite; sphalerite: grey. Magnification 60X.
b. Sericite partially replaced by pyrite; sphalerite: grey. Magnification 300X.
c. Zonal pyrite idioblasts with sieve texture in sphalerite. Magnification 180X.
d. Part of sphalerite nodule with "Druckschatten" mineralization. Magnification 60X.
e. Carbonate idioblasts in galena. Magnification 30X.
f. Carbonate idioblasts in galena. Magnification 180X.
g. Crinoid stem fragments replaced by pyrite. Magnification 30X.
h. Crinoid stem fragment replaced by pyrite, sphalerite and chalcopyrite. Magnification 30X.

(All figures Lower Devonian ore seam)

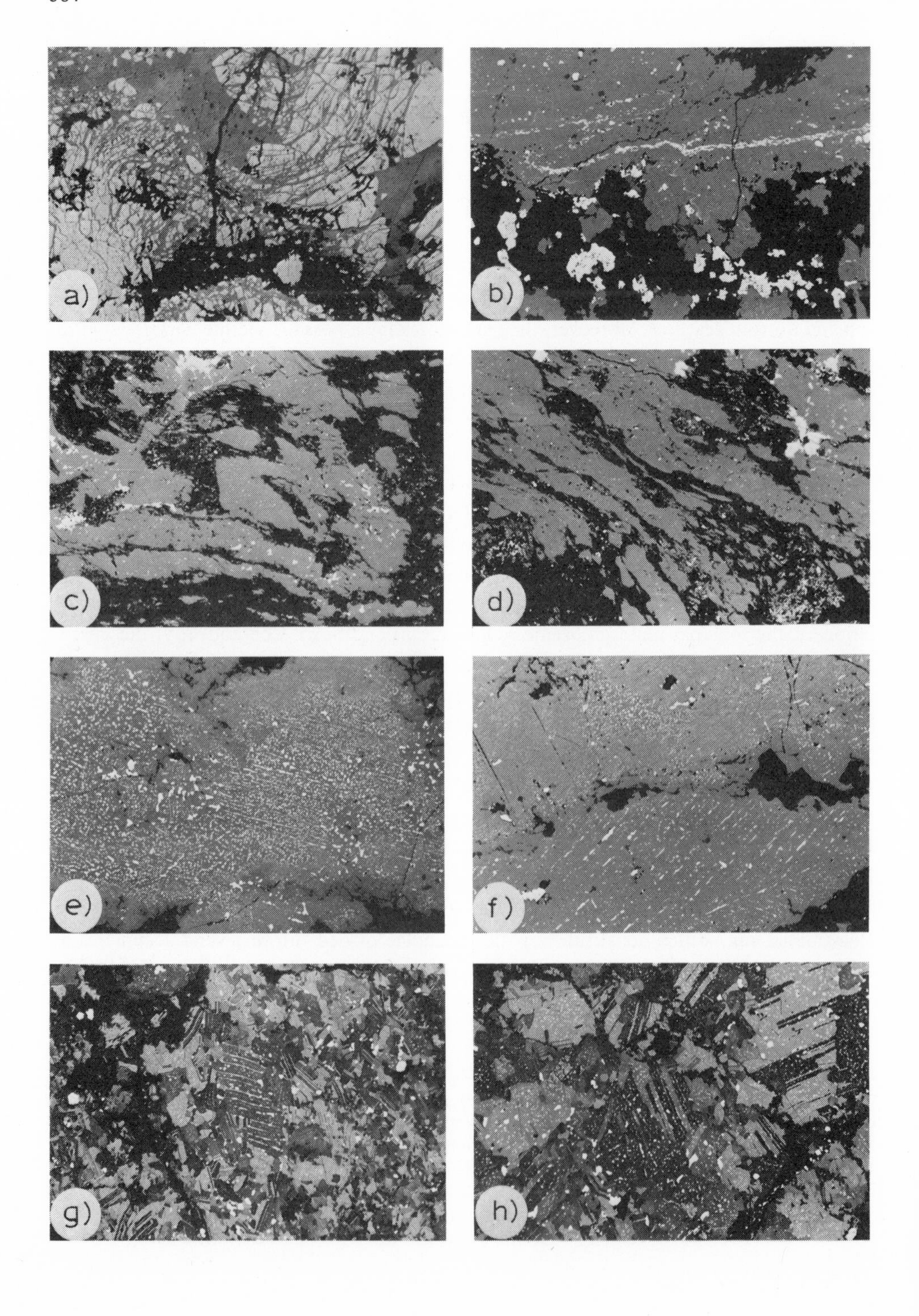
a)
b)
c)
d)
e)
f)
g)
h)

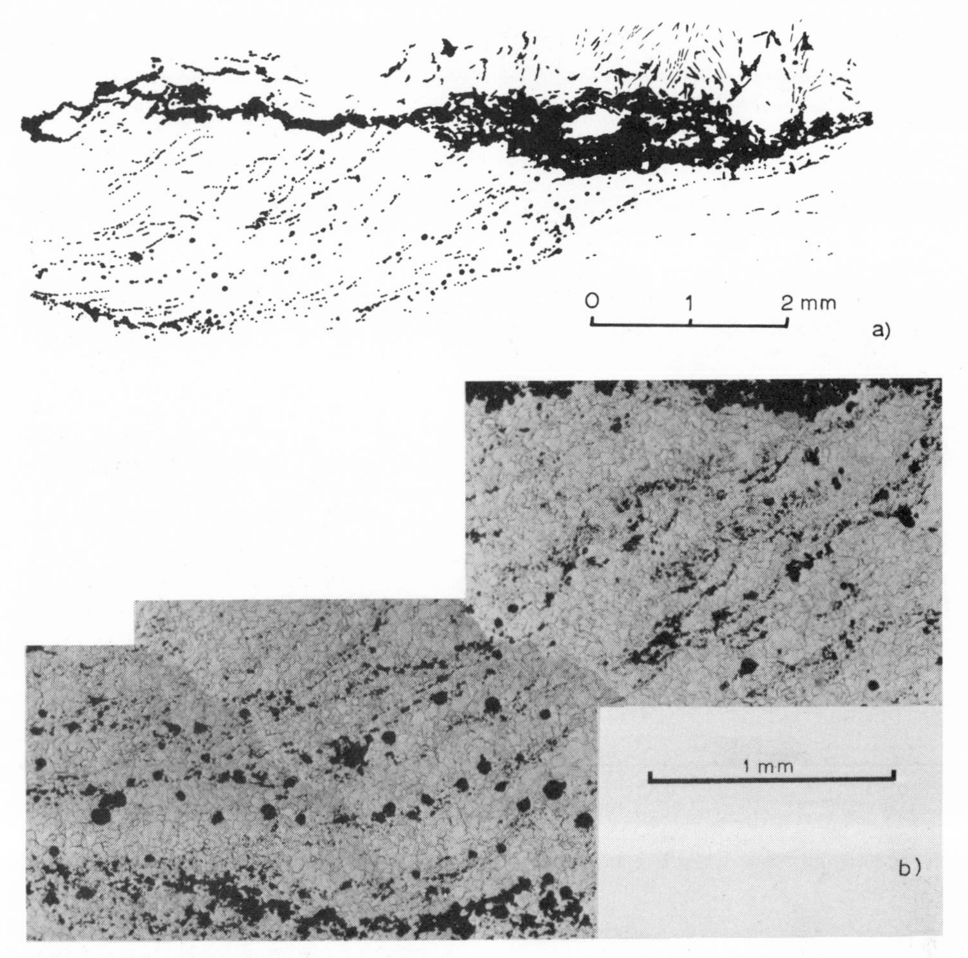

Fig. 15 (a, b). Cross bedding in massive barite (grey ore); black: pyrite, (sphalerite) and detrital material.

PLATE IV

a. Cataclastic pyrite in sphalerite. Magnification 60×.
b. Folded pyrite band in sphalerite. Magnification 60×.
c. Tectonical elongation and folding of sphalerite nodules. Magnification 10×.
d. Compare with c.
e. Sphalerite–chalcopyrite regular intergrowths. Magnification 45×.
f. Sphalerite–chalcopyrite regular intergrowths. Magnification 60×.
g. Sphalerite ore, etched with NaClO. Magnification 60×.
h. Sphalerite ore, etched with NaClO. Magnification 100×.

(All polished sections from ore of the Lower Devonian ore seam).

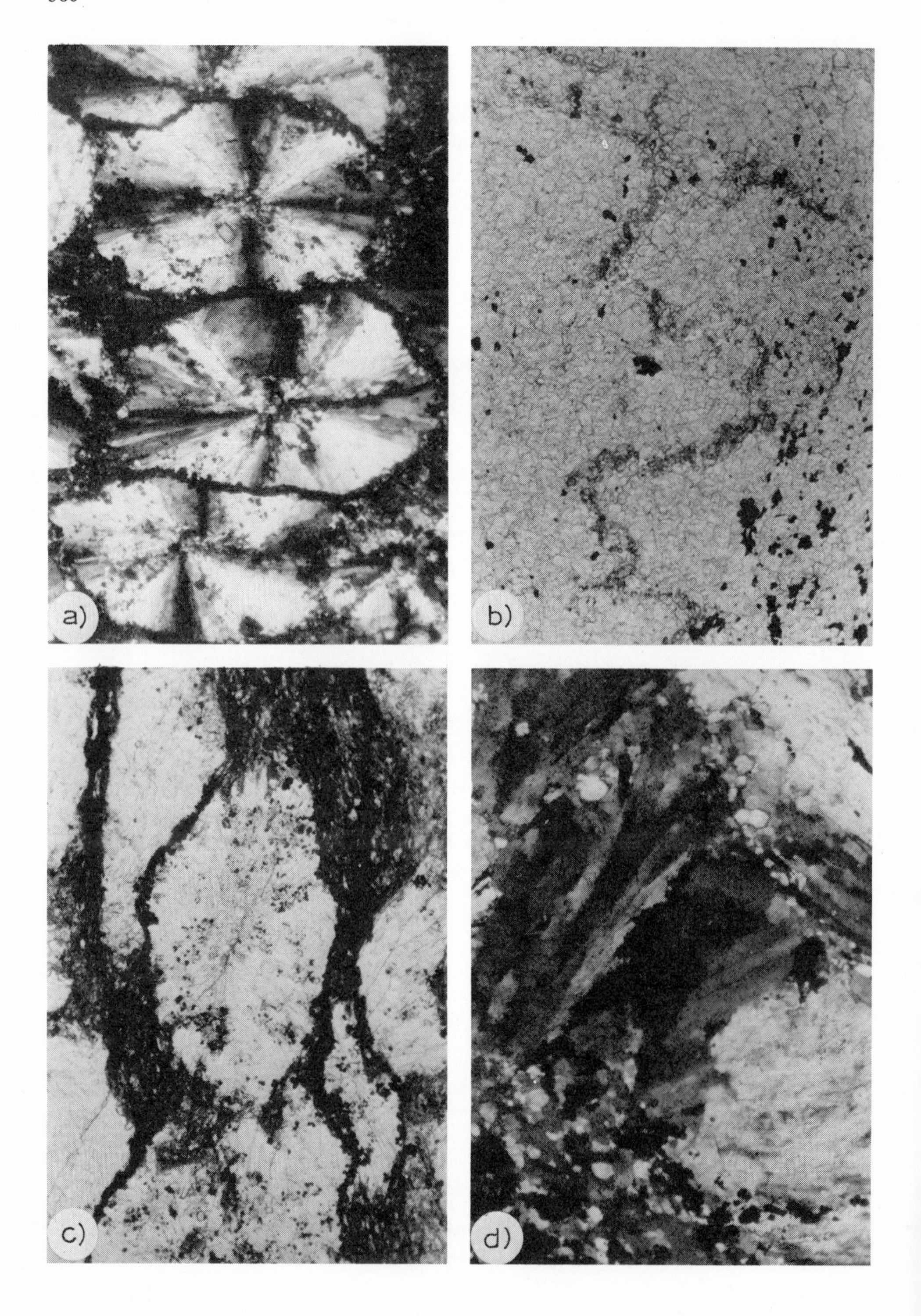
a)
b)
c)
d)

are composed of a mixture of ore mineral particles and "extra-environmentally" derived clastic silt and clay particles. The ore grains are predominantly sphalerite and galena, but there are also very Cu-rich beds, as well as nearly pure pyritic banded ore concentrations. The individual banded ore deposits partly consist of micro-layers (= laminae) which are first pyrite-rich, followed in profile by a Zn maximum lamina and finally are succeeded by a Pb-rich top lamina.

For further details on the sedimentary ore textures and structures, the reader is referred to the excellent photographs by Ramdohr (1953b), Kraume (1955) and Bornhardt (1939).

Summarizing the above, the pre-deformation (= primary) features in both massive and banded orebodies indicate that the mineralization took place in a unique sedimentary environment with a very narrow lateral transition zone into a more normal facies. The metal distribution at the margins of the main orebodies exponentially decreases to slightly above background. Beyond the margins of the banded ore, the metals diminish rapidly to very low values. The slightly anomalous metal contents continue into banded shales of the Ore Horizon over a distance of several thousand meters.

The *prephase ore mineralization,* as seen in the pre-deformational vertical cross-section of the orebody, consists of a cylindrical conduit, and the pre-deformational mineral parageneses of the impregnations and metasomatism are simple. Galena, sphalerite, chalcopyrite, pyrite and carbonates formed during this stage; and more recently Paul (1975) reported barite. Trace minerals are minimal. In general, the minerals are coarse-grained allotriomorphic (Table III; Fig. 17).

Ore impregnations of uneconomic value already are found in the Lower Devonian carbonate-cemented sandstone layers; the dominant ore minerals being sphalerite, with rare chalcopyrite and pyrite. The sphalerite constitutes interparticle fillings up to 3 mm in size and exhibits exsolution-like chalcopyrite intergrowths. In general, the carbonate matrix has been replaced and parts of the intergranular space are totally occupied by the ore minerals, but the quartz grains normally remained unaffected by the replacement process. However, some larger ore grains appear to have formed by the replacement of quartz also. Part of the impregnation-like sphalerite mineralizations certainly occurred after clastic-grain deposition, possibly under a sedimentary cover of 10–100 m overburden. The ore mineralization may have already taken place early-diagenetically within a loose unlithified sediment either after and/or during formation of the host rock. [Thus, this mineralization would be of the pre-cementation (or pre-lithification) to post-cemen-

PLATE V

a. Spherulitic barite with albite crystals in the core. Matrix: black shales, Footwall Grey Orebody. Magnification 10 × + Nicols.

b. Suture line in barite, Footwall Grey Orebody. Magnification 60×.

c. Compare with a. Structural elongation parallel to the cleavage. Magnification 60×.

d. Different types of syngenetic crystallization of barite, Footwall Grey Orebody. Magnification 90×.

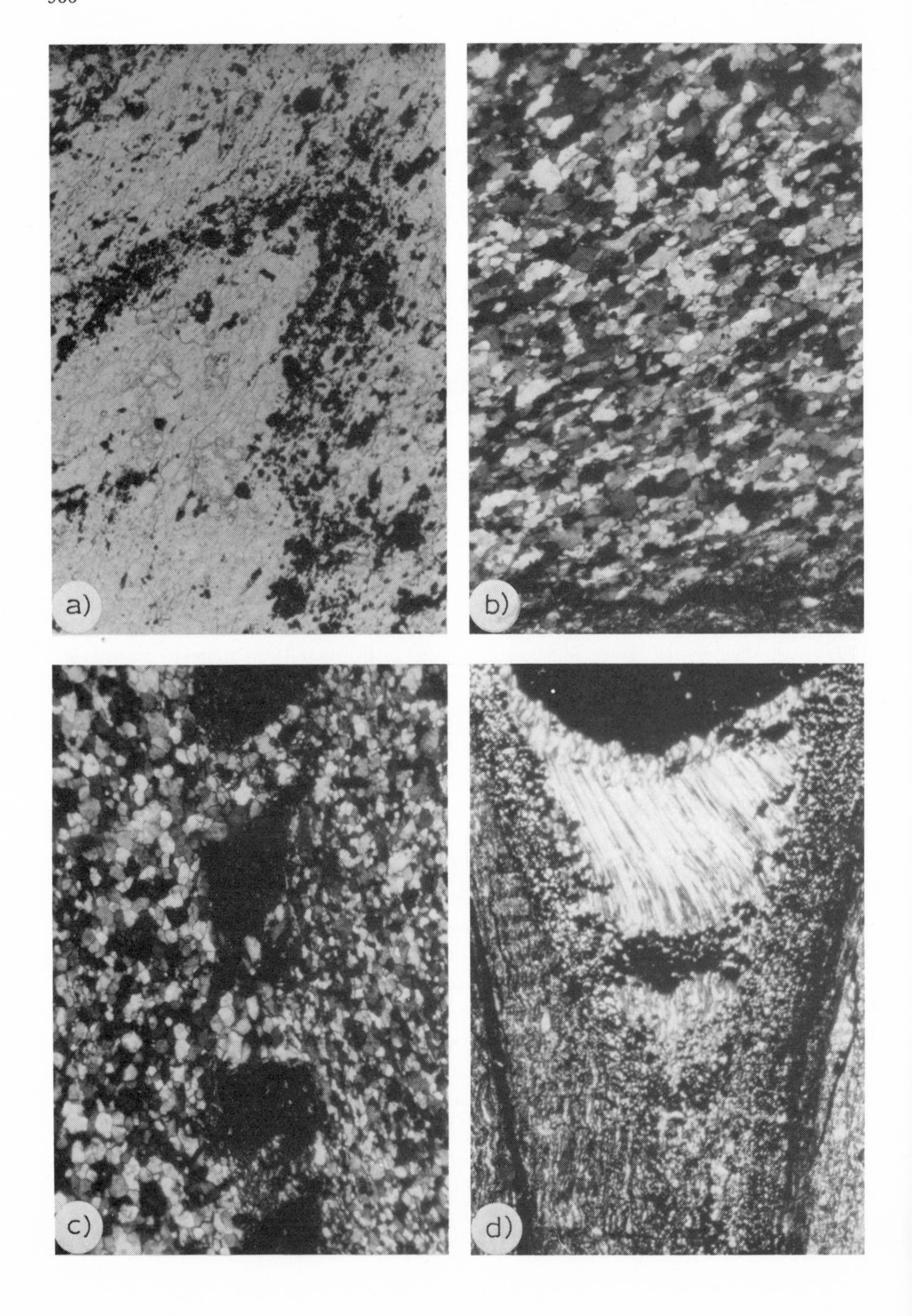
a)
b)
c)
d)

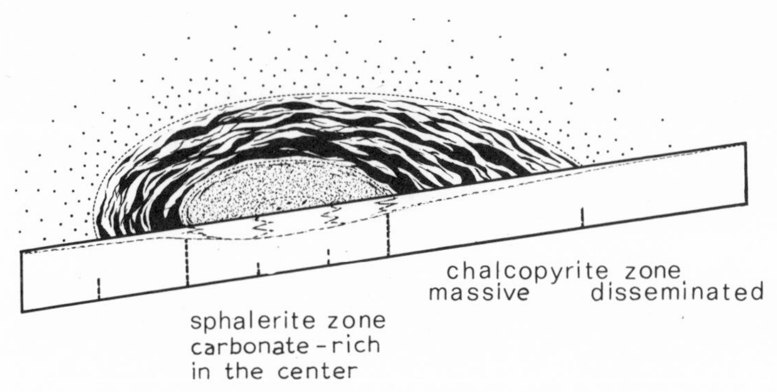

Fig. 16. Probable mineral zoning of the Lower Devonian sulphide ore seam.

tation diagenetic stages according to Wolf (1967), in Chilingar et al. (1967, table V, p.248); for similar approaches see also Wolf and Chilingarian (1976) and Wolf (1976).]

The exsolution-like chalcopyrite inclusions, as well as the minor occurrence of pyrite intergrowths in sphalerite, are explainable as a flocculation process of a polymetallic sulphide gel, because they are very similar to coagulation textures that originated in a genuine sedimentary milieu. Higher-temperature gradients in the upper sedimentary or stratigraphic regimen would, of course, result in more intense physico-chemical thermal conditions. In this respect, no answer can be found to the question as to whether either exsolution from a higher-temperature mixed sulphide phase or an accretionary crystallization from a gel has occurred – the reason being that the information about the original environment is incomplete. The sphalerite–chalcopyrite intergrowth (= exsolution feature) is no proof of a higher-temperature origin.

The sediments, which had not as yet undergone diagenesis or were only slightly altered at that stage, probably offered a sufficiently high permeability for hypogene metal-bearing solutions. One has to assume that these fluids, in contrast to the initial pore solution or connate water, were under higher pressure and also differed from them in their varying pH and Eh. According to Barnes (1974), these sulphide impregnations could have been precipitated by neutral bisulphide chloride solutions in an oxidizing environment. During this reaction an acidic milieu is created that, in turn, would explain the replacement of the carbonates.

PLATE VI

a. Folded massive barite ore. Foliation parallel to axial plane cleavage, Footwall Grey Orebody. Magnification 60X.

b. Foliation of massive barite ore parallel to the cleavage, graded bedding ("overturned"). Coarse grains at the top. Footwall Grey Orebody. Magnification 50X.

c. Boudinaged shale layer between different grained massive grey ore (barite) bands. Magnification 50X.

d. "Druckschatten" with fibrous quartz and "Knitterung". Magnification 30X.

TABLE III

Distribution of the mineralizations in the ore deposit area

	postkinematic fissure and *breccia mineralizations* in silicified lentils (*Kniest-mineralization*)	
UPPER CARBONIFEROUS		
∼∼∼ Variscan Orogeny ∼∼∼	dynamometamorphic overprint and folding of the orebodies	
MIDDLE DEVONIAN	*formation of the main orebodies*	Footwall Grey Orebody / *Old Orebody* / Grey Orebody / *New Orebody* — banded ore
—— Twin Tuff Layer ——		*prephase silicification* (= *Kniestformation*) under pulsated conditions, metasomatic, impregnative, sedimentary
	syngenetic prephase mineralizations	
LOWER DEVONIAN	fine disseminated pyritizations and lead, zinc, copper, pyrite seams	lead, zinc, copper, barite impregnations and replacements

In general, the same conditions apply to widely occurring replacements of the carbonate fossils in the sandy and pelitic sediments. They are often completely mineralized, but mostly monomineralogic in composition (Plate III). The replacement of the fossil carbonate occurred commonly pseudomorphically, and the ore minerals are coarser-grained with chalcopyrite predominating, but also sphalerite is quite frequent, whereas galena is less important. Completely pyritized fossils are widespread, but they are no proof of the hypogenetic supply of the ore-forming solutions. These pyrites are of a heterogeneous origin as demonstrated by the sulphur-isotope distribution, which has a wide range of $\delta^{34}S$. Especially the monomineralogic mineralizations should lend themselves to well-defined physico-chemical reconstruction of the precipitational milieu, but detailed studies are missing. The geological observations indicate that high proportions of the fossils were mineralized only after complete burial. However, it seems that some of the carbonate of the skeletons have already begun to be replaced shortly after accumulation without being appreciably embedded.

The probable age of these impregnative and metasomatic mineralizations in the channel way must range from the oldest proven sedimentary ore seams and the massive ore formation of the main ore deposits. The oldest possible age, based purely on a logical deduction, must be that of the host rock, the youngest the main-ore mineralization phase. A post-orogenic phase of origin in the Upper Carboniferous, and younger, is impossible, although well-defined post-kinematic mineralizations are not absent (cf. pp.625–631). These latter ore formations, however, are texturally and structurally quite different.

In the stratigraphic cross-secttion beneath the main orebodies, the prekinematic ore mineralization is more or less statistically uniformly distributed and, consequently, appear to suggest that they originated by a general hydrothermal impregnation of the conduit (Fig.17). The geochemical rock investigation by Paul (1975) confirms this concept. The locality and outline of the channel way has been clearly delineated by a general and distinct increase of Pb, Zn, Cu and Mn in contrast to the same stratigraphic units outside the vent. There is a concomitant decrease of Ca, as would be expected from the general replacement of carbonate from the sediments and the fossil skeletons. The black shales of the footwall's sedimentary sequence (i.e., Calceola- and Sandbanded Shales as well as the Shale Horizon) exhibit numerous stratified well-bedded pyrite of the impregnative disseminated type, which are most likely syngenetic. Framboidal pyrite is quite common. The source of the disseminated pyrite's constituents, like that of the pyritized fossils, is without question heterogeneous, as already pointed out.

The syngenetic prephase mineralizations (i.e., Pb–Zn–Cu–pyrite ore seam) were examined in detail during at least one study (Gundlach and Hannak, 1968; Kraume, 1960). The investigated, 1 to 13 cm-thick, ore seam is exposed in the "Kommunion" quarry over a distance of 180 m and belongs to the hydrothermal conduit region exposed on the Rammelsberg hill slope (Fig.32). The lithologic section dominantly consists of

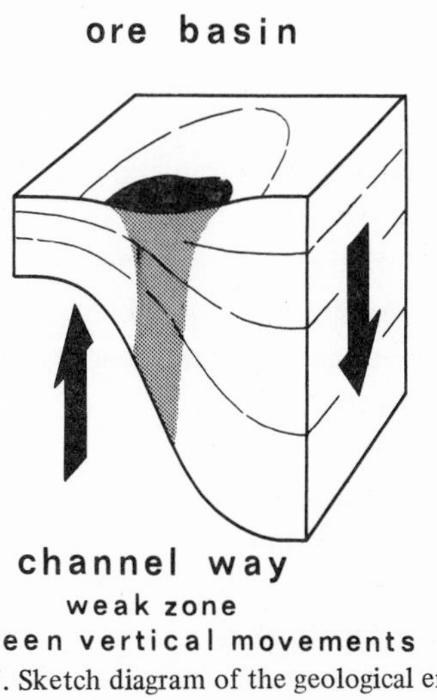

Fig. 17. Sketch diagram of the geological environment.

Lower Devonian quartzite-type facies and minor fine-grained, mainly argillaceous, beds of littoral origin. The ore seam is developed in a fine-grained shaly horizon 10–20 cm thick, laterally grading into a fossil-bearing unit and coarser-grained sediments. It is obvious, that the ore seam is better developed in the fine-grained facies than in the coarser equivalents. This indicates the influence of the general sedimentary conditions of the depositional environment on the submarine exhalative-sedimentary ore genesis. The variations of the depositional conditions under which the ore-bearing unit accumulated was caused by the interrelationships of the exogenic sedimentary conditions, on one hand, and the hypogene hydrothermal supply, on the other (Fig.32). A quartzitic sandstone bank abruptly overlying the ore seam indicates that the process of "ore sedimentation" was suddenly interrupted by the influx of the coarse clastics, i.e., pointing to an exogenous influence. Independent from the exogenic factors, the hydrothermal supply may have continued without visible sedimentary ore precipitation having occurred. The ore textures lead to the conclusion that the metals upon reaching the sea floor spontaneously flocculated under the changed pH and Eh conditions in a sulphidic–carbonatic surface environment. The relative proportions of the different sulphides were unimportant because they were precipitated as mixed gels. This interpretation is supported by the high trace-element content of the sulphides and the elements' considerable variation over short distances (Table IV).

A carbonate-rich inner zone of the ore seams should be mentioned. This carbonate could have been derived from the host rock of the conduit region of the footwall. Based on the information from the outcrops available at present, a reconstruction as shown in Fig.17 can be made. Assuming that the zoning is indicative of a central hydrothermal vent, it suggests that the carbonate- and sphalerite-concentrations only here were sufficient to be precipitated. According to this concept, the copper migrated further and accumulated as chalcopyrite along the outer margins of the ore seam. Galena occurs only as minor amounts and accompanies the sphalerite, whereas the pyrite is an ubiquitous mineral, i.e., is present throughout the ore seam. Relative-age relationships between the minerals, in the usual sense of petrographic paragenetic sequences, have not been observed, which substantiates the opinion that only a partial mineral-precipitation sequence was formed.

The sudden submarine supply of hydrothermal solutions seems to have coincided with suitable exogenic conditions and resulted in the precipitation of the sulphides. Silica was supplied in larger quantities only during the initial stage and diminished during the accumulation of the ore-bearing horizon. The ore sedimentation process was also suddenly terminated by an influx of coarse-grained clastics.

Northwest of the main orebodies, in the upper sections of the Sandbanded Shales ("Sandbandschiefer"), there are characteristic fine-grained, dark-grey to black, carbonate-rich beds, up to a few decimetres in thickness, the so-called carbonate banks (Fig.33a). Their silica content of approximately 15% accounts for the high degree of lithification and hardness, as well as for the conchoidal fracture. A fracture cleavage is often present.

These beds are mainly present in the Shale Horizon and in the Ore Horizon, particularly in close proximity of the ore deposit. Kraume (1955) supplied the compositional data in Table V. In some samples, Paul (1975) determined high siderite contents. As these SiO_2-rich beds are commonly absent in the sedimentary sequences within the wider

TABLE IV

Trace element distribution in the ore minerals of a Lower Devonian sulphide ore seam (After Gundlach and Hannak, 1968)

Chemical analysis of pyrite and chalcopyrite

	2.7 m	16.5 m	11 m	15 m
FeS_2	82.7%	54.1%	–	–
$CuFeS_2$	–	–	77.4%	89.1%
residue	5.0%	20.2%	6.0%	7.6%
As (ppm)	~10 000	~100	<10	~100
Sb (ppm)	~10 000	~1000	~1000	100–1000
Bi (ppm)	<3	<3	<3	~10
Ag (ppm)	~10	~10	~100	~100
Cu (ppm)	~10 000	~1000		
V (ppm)	<10	~100	<10	<10
Ga (ppm)	~10	~30	~10	~30
Ge (ppm)	~1	~1	~1	~1
In (ppm)	~100	~100	~1000	1000–10 000
Sn (ppm)	~1000	~10 000	~1000	~1000
Ni (ppm)	~100	~10	<10	10–100
Co (ppm)	10–100	10–100	<10	10–100

All samples carry galena, pyrite + 1% PbS.

Chemical analysis of galena

	12.0 m	20.6 m
PbS	35.5%	95.0%
residue	14.4%	2.0%
Bi (ppm)	21	≤3
Ag (ppm)	190	410
Sb (ppm)	100	460
Sn (ppm)	<10	<10
As (ppm)	<100	<100
Cu (ppm)	~100	~100
V (ppm)	<10	<10
Ge (ppm)	<1	<1
Ga (ppm)	<1	<1
Tl (ppm)	<10	<10

TABLE IV (continued)

Chemical analysis of sphalerite

	2.7 m	12.0 m	12.0 m	12.0 m	13.0 m
ZnS (%)	69.3	54.7	49.2	54.7	58.1
Fe (%) *	2.14	5.99	3.36	5.90	6.32
Cd (ppm) **	840	1030	1045	1030	1000
Ga (ppm) **	30	50	95	50	35
In (ppm) **	275	35	60	35	15
Ge (ppm)	<1	<1	<1	<1	<1
Sn (ppm)	~1000	~1000	~1000	~1000	~1000
Hg (ppm)	~100	~10	~10	10–100	10–100
Tl (ppm)	<10	<10	<10	<10	<10
Ga : In	0.11	1.43	1.58	1.43	2.33

	13.0 m	14.7 m	16.5 m	16.5 m	18.4 m	32.0 m
ZnS (%)	67.7	61.1	64.1	44.4	55.1	76.8
Fe (%) *	6.95	5.59	5.31	8.47	7.53	4.57
Cd (ppm) **	925	1000	1010	1500	815	1080
Ga (ppm) **	40	15	30	65	50	100
In (ppm) **	30	15	60	190	70	10
Ge (ppm)	<1	<1	<1	<1	<1	<1
Sn (ppm)	~1000	~1000	~1000	~1000	~1000	~1000
Hg (ppm)	~10	100–1000	10–100	~10	~10	10–100
Tl (ppm)	<10	<10	<10	<10	<10	<10
Ga : In	1.33	1.00	0.50	0.34	0.71	10.00

Other elements doubtfully to correlate with sphalerite.
Pb: >1% (galena)
Ag: 10–50 ppm (probably to correlate with galena)
Bi: <3 ppm
Cu: 0.5% (0.03–7.5% calculated as chalcopyrite)
V: <10–50 ppm
As: 100–~1000 ppm
Sb: <100 ppm
Co: ~10–~100 ppm
Ni: <10–~1000 ppm, on average 10–50 ppm
Insoluble residue 3.5–20%

* Without Fe in CuFeS.
** Calculated on the basis of ZnS with 64% Zn.

Rammelsberg region, and as they are restricted to the immediate vicinity of the ore-bearing locality, one can deduce a hypogene (hydrothermal) origin of the silica-rich solutions. Of similar chemical composition, and also indicative of the conditions of formation of the ore-containing unit, is the presence of diagenetic carbonate concretions, up to several decimetres in diameter.

TABLE V

Chemical composition of carbonate layers and nodules in the Ore Horizon shales (After Kraume, 1955)

		1	*2*	*3*	*4*	*5*	*6*	*7*	*8*	*9*	*10*	*11*
Zn	(%)	–	–	0.4	0.3	0.05	1.2	Sp.	–	0.2	0.1	0.1
Fe	(%)	1.4	1.8	6.8	5.9	6.7	5.8	4.6	5.5	6.1	6.1	4.0
Mn	(%)	1.4	1.2	10.1	1.7	0.8	–	–	–	1.6	1.7	1.8
BaO	(%)	–	–	–	1.8	–	–	–	–	–	–	–
CaO	(%)	44.5	41.2	18.0	22.3	22.5	24.6	30.6	24.5	20.2	19.1	17.7
MgO	(%)	1.4	1.2	1.3	5.9	7.5	11.2	10.0	9.4	10.2	10.3	0.2
CO_2	(%)	38.5	36.7	28.2	35.4	33.2	34.2	36.2	33.9	31.5	29.1	36.1
SiO_2	(%)	–	–	11.9	14.5	17.2	15.8	14.8	16.5	15.7	17.9	14.6
Al_2O_3	(%)	–	–	2.7	2.6	3.6	1.4	1.9	5.1	3.5	3.5	2.3
S	(%)	–	–	2.8	0.8	–	–	–	–	–	–	–

1 = Light-grey calcareous concretion.
2 = Light-grey limestone layer.
3 = Dark-grey calcareous concretion.
4–10 = Dark-grey carbonate banks.

SILICIFICATION PHASE (ORIGIN OF THE "KNIEST")

The stage of silicification and, therefore, the origin of the "Kniest", is complex. The prevailing concept maintains that the main stage of ore concentration was preceded by the formation of local lentoid silica concentrations in the area which is now the hanging-wall of the Rammelsberg deposit. These silica lenses constitute, in general, an irregularly shaped zone, and their maximum extent along the strike is about 500 m, with a similar down-dip extent. The extent of the silicified zone is less than one-quarter of the orebodies. The thickness is up to 100 m without considering deformation. In cross-section, the zone of silicification is wedge-shaped. Laterally, the zone passes discordantly into unaltered sedimentary rocks. Primary sedimentary interfingerings appear to be absent.

The silica impregnation has affected parts of the Sandbanded Shales, the Shale Horizon as well as the Ore Horizon. The silicification obviously succeeded pre-existing stratiform sulphide layers in the footwall region of the main orebodies consisting mostly of syngenetic impregnations of sphalerite and pyrite, as described above. Also massive ore lenses in places were already formed prior to silicification. These massive mineralizations extend over an area of several square metres and their thickness is up to several decimetres. It should be made clear, however, that the stratigraphic-tectonic relationships of these massive ore lenses have not been fully resolved (Kraume, 1955).

Sedimentary carbonate nodules, located in the so-called "granular Kniest", were derived from an above-average supply of carbonate in hypogene solutions on the sea floor under favourable conditions, in addition to the silica and sulphide. The oolitic carbonate grains of about 1 mm in diameter contain occasional concentric pyrite laminae. These nodules are anomalous for the particular sedimentary facies under consideration. The carbonates are manganosiderite in composition [i.e., containing 16% $FeCO_3$, 31% $MnCO_3$, 8% $(Ca,Mg)CO_3$]; thus, they show a chemical similarity to the so-called carbonate banks which were described above. In the unsilicified shales of the channel way, these oolitic carbonate nodules are also present. Generally, the carbonate content is low in the "Kniest" and its average composition, as compared to the unaltered shales, is given by Kraume (1955, p.70), where it is shown that the free quartz (about 60%) is twice the amount of the unaffected shales. Consequently, approximately 30% SiO_2 must be accounted for by silicification.

The bedding features of the "Kniest", i.e., lamination, grain-size variations, graded-bedding, correspond directly to the unsilicified sediments (namely Shale Horizon and Sandbanded Shale Horizon). The detrital quartz grains of the silicified sediments exhibit crystal growths, overgrowths and textural interfingering initiated by the hypogene supply of silica.

The cursory mineralogical description of the "Kniest" presented above indicates that a late-syngenetic to early-diagenetic silicification by hydrothermal solutions is the most plausible explanation. Kraume (1955), on the contrary, believes in a genuine sedimentary origin of the "Kniest". In any case, the question becomes important as to where the boundary lies between the sedimentary and diagenetic processes which were operative under the heterogene (= multiple-source) element supply. The absence of a schistosity in the "Kniest" requires a syngenetic, namely pre-tectonic, silica supply. Also, the spatial restriction of the "Kniest" to the orebody and the conduit zone confirmed this view. Not in agreement with this interpretation is, according to the information available as present, the conformable folding of the "Kniest" together with the unsilicified host rocks, as the silicified "Kniest" would have offered a much greater resistance to deformation. However, this problem can be overemphasized. Therefore, a post-tectonic silicification origin (in which the present writer temporarily believed) cannot be accepted.

The silicification appears to mark the introductory phase of the main ore formation approximately at the time of the accumulation of the twin-tuff layer, and may have lasted until the first appearance of the massive ores. During the formation of the main ore precipitation, no more silica of significant quantity was supplied.

THE MAIN ORE FORMATION

Opinions differ on the beginning of the phase of the main ore deposition. Kraume (1955) proposed that it was marked by the massive ore of the New Orebody, whereas underground-mapping data demonstrated that banded ore formations may be the initial stage of the main ore precipitation. These banded units consist predominantly of sphalerite, with some galena, pyrite and chalcopyrite. In the footwall of the eastern part of the New Orebody and the Old Orebody, banded ore formations commence immediately above the twin-tuff layer together with minor, local massive ore. The so-called Hanging Wall Ore Occurrence, which was formed during this very early phase of the main ore precipitation, has the only recorded significant presence of arsenopyrite.

According to new tectonic results, Kraume (1960b) offered a detailed diagrammatic scheme of the paragenesis of the New Orebody's massive mineralization. As shown in Fig.18, the average elemental distribution varies significantly from the bottom to the top, which in turn is reflected by the different ore types (Table VI a, b). The skepticism regarding the value of these ore variations as indicators of genetic conditions up to recent times was also based on the lack of understanding of both the tectonic structures and the stratigraphic/lithologic sequences.

The Red Sea model (cf. Degens and Ross, 1976) with its hot-brine subsurface cell below a normal sea-water environment, could be partially applied to explain the origin of the Rammelsberg massive ore – except that the regional and local tectonism is different. Therefore, an uninterrupted ore precipitation and accumulation requires subaqueous topographic depressions that act as depositional traps (at the same time stabilizing the sedimentary deposits by reducing, if not eliminating, reworking and possible removal) and cause a low influx of clastics that would "dilute" the ore mineralization, among others. The well-known physico-chemical principles based on modern experiments (cf. Vaughan, 1976), would apply here. These data also provide facts to support a syngenetic mechanism for the Rammelsberg ores.

The so-called *sulphur ore* ("Schwefelerz"/massive pure pyrite ore), which marks the initial stage of the New Orebody, is only seldom above 1 m thick. This massive ore, exhibiting only weak stratification, is not continuous in its distribution as is the case of the later-formed more widespread Pb–Zn ores. This individual bed of sulphur ore is

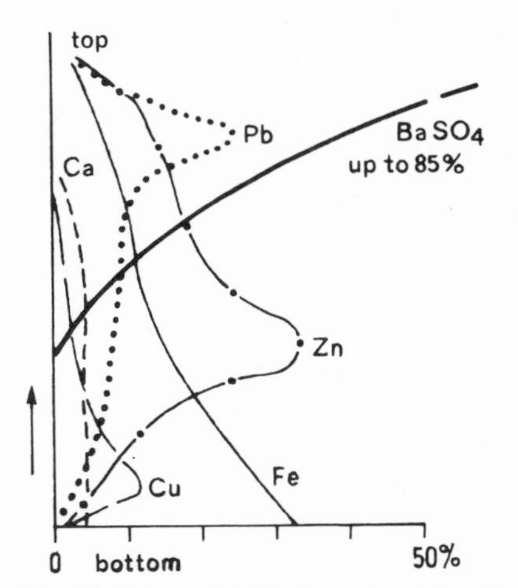

Fig. 18. Average distribution of metals and gangue material in the New Orebody (cf. Table VIa, b).

TABLE VIa

Stratification of ore types in the New Orebody (After Kraume, 1960b)

		Average metal content of characteristic ore types (%)					
		Pb	Zn	Cu	Fe	$BaSO_4$	CaO
Banded ore	Intercalation of shales and ore with quite different metal contents decreasing in metal contents upwards and laterally						
Grey ore	Mainly barite	3	3	<0.1	2	>80	<1
Lead–zinc ore	Baritic lead–zinc ore, partly with lead ore	10 25	18 10	<0.1 <0.5	5 4	>50 40	<1 <1
	Baritic lead–zinc ore	10	20	<1	8	28	<1
	Baritic carbonatic and pyritic lead–zinc ore, partly combined with mottled ore (Melierterz)	6 10	16 20	<1 3	16 11	18 20	6 3
Brown ore	Often with mottled ore (Melierterz); brown ore with increased contents of barite	8 15	22 33	4 <1	12 6	14 8	3 3
	Pyritic brown ore with low contents of barite, partly with copper ore	7 4	32 12	<2 10	17 21	<1 <1	4 4
Pyritic ore	Carbonatic, mainly pyritic ore free of barite with increased contents of zinc, partly with copper ore	5 4	12 8	1 12	22 28	<1 <1	4 4
Sulphur ore (pyrite ore)	Mainly carbonatic pyrite ore, free of barite	2	4	1	30	<1	1–4
Footwall	Wissenbach Shales with concretionary pyrite, seldom pyritic banded ore						

TABLE VIb

Paragenetic sequence of metals and gangue minerals in the massive orebodies

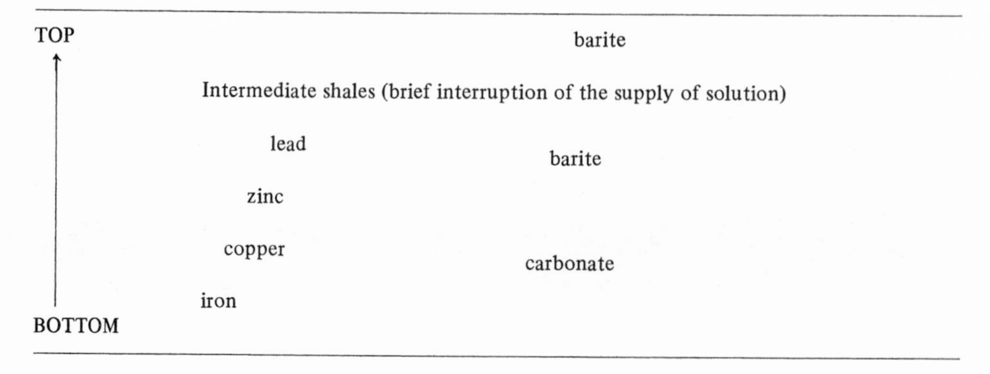

clearly separable from the younger types of mineralizations, the latter being characterized by very gradual transitions and imperceptible intercalations that make it difficult to establish boundary relationships. Distinct lenses of *copper ore* (= chalcopyrite/yellow ore) and pure *brown ore* (= sphalerite/black ore) have been recognized in pyrite-rich sphalerite ores ("kiesiges Erz") of the lower massive ore sequences. The *mottled ore* ("Melierterz") tends to have a wider distribution. The latter is interstratified with the more dominant younger *lead–zinc ore* generations, which are culminating in *barite-rich lead–zinc ore* ("barytisches Blei–Zinkerz") in parts of the grey ore ("Grauerz"/ massive barite) (Table VI a, b).

With the exception of the massive pyrite mineralization, all the other ore types are laminated in scale down to microscopic dimensions, caused by depositional-sedimentary processes and subsequently intensely overprinted by diagenetic and tectonic processes.

Predominantly, changes in the mineral composition, rarely variations in grain-size, are responsible for the layering. As already pointed out above, they reflect the original intermittent precipitation of colloidally dispersed material, as evidenced by the textures. The gradual compositional modifications of the thermal solutions is indicated by the change of the main constituents in the ore sequence. However, some physico-chemical variables must have remained constant over a longer period to allow for the development of a distinct or favoured ore mineralization, exemplified by the chalcopyrite in the "copper ore" and sphalerite in the "brown ore". Although subsequent reworking of the precipitates cannot be ruled out entirely, resedimentation on the whole was absent. Among the gangue minerals, carbonate is prevalent in the older ore sequences, whereas in the younger ores barite increases significantly (Table VIb).

It should be mentioned, that the Rammelsberg sedimentary ore sequence given in Tables VIa and VIb corresponds to the well-known paragenetic successions of plutonic-hydrothermal lead–zinc vein deposits. Their physico-chemical causes and effects, magma differentiation, lattice energy theories (cf. Vaughan, 1976), among others, have been discussed in many publications, which are also important when judging the genesis of other stratiform deposits of the Rammelsberg-type.

When the processes responsible for the origin of the massive orebodies were interrupted by an influx of exogenous clastic material, lenses of banded ore originated within the massive ore mineralization. Banded ores are developed along the outer margins of the orebodies as an intercalation zone between the massive ore and the host shales, except at the western rim of the New Orebody, i.e., the eastern rim of the Old Orebody. Consequently, corresponding to the main ore sequences pyrite, lead–zinc, copper-rich lead–zinc, and barite banded ores are present.

The paragenetic succession of the Old Orebody is similar to that of the New Orebody, but some differences in the distribution of the ore types do exist. Gunzert (1969) proposed a tectonically induced lateral shearing to the northeast predominantly parallel to the strike to explain the en échelon arrangement of the various ore types described by Kraume (1955). It should be made clear again, that no unequivocal explanation has been found to explain the banded ore-free interval between the New and Old Orebodies. Possible factors, among others, may have been the location of the feeder zones, variations in the subsurface thermal pattern of the hydrodynamic cell, and the distribution and stratification of the brines. Schot (1975) considered a penecontemporaneous origin of the ores, but within two separate topographic depressions.

According to our present information, the concept based on a second somewhat younger complete cycle of brine supply with a concomitant precipitation forming the Old Orebody, is more reasonable. As in the New Orebody sequence, the precipitation of pure barite in the Old Orebody succession is separated from the main sulphide deposit by a temporal interruption of shale interbeds; although first generations of pure barite lenses occur already in both the New and Old Orebodies. The Grey Orebody forms the largest quantity of pure barite as a separate concentration in the New Orebody mineral succession which is composed of massive barite. During the initial stage of barite precipitation it was accompanied by galena, with microscopically visible minor quantities of sphalerite, chalcopyrite and sulpho-salts (Schot, 1973). In contrast to the massive Grey Orebody, the Footwall Grey Orebody genetically tied to the Old Orebody comprises several sequences of alternating barite and shale beds (Fig. 22). Except for pyrite, other sulphides are rare or absent. Present information indicates that the Footwall Grey Orebody represents the final phase of the syngenetic mineralization of the Rammelsberg deposit.

The present writer opines that it serves no purpose to list here primary thicknesses of each ore type, because of the very high degree of tectonic overprint and the consequent large secondary structural modifications of the primary bed thicknesses – even relative thicknesses are irrelevant for the present genetic discussion.

In contrast to the major minerals, the accessories may have a variety of origins from the paragenetic point-of-view. Newly formed minerals, alterations and transformations, in general, took place isochemically and without noticeable compositional changes, so that the paragenesis of the ore minerals is analogous to the syngenetically formed trace-metal distribution. To complete this interpretation, and to avoid any misunderstanding, it has to be repeated that the mineralogy and its paragenesis consists of pre-, syn- and post-kinematic constituents, which partly resulted in gradational less-demarcated changes.

The tectonically produced textures will be described and discussed in detail later, whereas the possible connection between the properties of the main minerals and the accessories will be explored first, as this may throw additional light on the paragenesis. For example, sphalerite shows obvious iron variations (as evidenced by optical changes) in response to the original partial pressure of sulphur, which is controlled by the associated pyrrhotite and pyrite contents (Ramdohr, 1955, in Kraume, 1955). With an increase in pyrrhotite and pyrite, the sphalerite becomes lighter coloured until it is colourless as a result of the decreasing iron content in the sphalerite. In the Brown Ore layers, the pyrrhotite is the predominant constituent. Pyrite and marcasite are often alternating in the concentrically laminated intergrowths, and it is quite certain that the marcasite belongs to the primary precipitate as the degree of metamorphism was too low to invert it to pyrite. The marcasite's main occurrence is especially in the chalcopyrite-rich Mottled Ore ("Melierterz"). The magnetite appears to increase in the sphalerite-rich parts, but – except for the Grey Ore and the banded ore – is also represented in the other ore types. Some crystal forms of the magnetite (Ramdohr, 1953b) must have originated during the late-orogenic phase, because the deformations that would otherwise be expected are absent.

The northworthy and conspicuously high As-tenor of the Hanging Wall Ore Occurrence, represented by macroscopically visible arsenopyrite formed during the early stage, has been mentioned already. During the formation of the main ore deposits, the arsenopyrite content distinctly diminished. Ramdohr could find only small amounts grown on marcasite. Occasionally, the arsenopyrite contains inclusions of löllingite or gutmundite. Bismuthinite, frequently difficult to distinguish from emplektite and galanobismutite, often occurs with native bismuth and native gold, especially in chalcopyrite within the Mottled Ore. Schot (1973) recently described electrum. Linnaeite is also present in this paragenetic sequence and must be considered to be the host of the small cobalt content. Inasmuch as molybdenite and wolframite were rare discoveries made by Ramdohr, nothing can be said about their paragenetic position. The Pb–Sb sulpho-salts (e.g., jamesonite, boulangerite, and with a degree of doubt also zinkenite or plagionite) were preferentially concentrated in the Pb–Zn ores. Apart from the electrum (Schot, 1973), no specific Ag-bearing minerals have been recognized. As to the tetrahedrite, bournonite and pyrargyrite, they are late- to post-kinematic minerals as they do not show tectonic deformational features, and belong to the quantitatively unimportant paragenesis of remobilization origin.

The investigations of valleriite, more rarely of cubanite, led to the conclusion (Borchert, 1934; Ramdohr, 1953b) that the temperature reached in the orebody metamorphism was up to 225–250°C, but not much higher, because marcasite has been preserved in considerable proportions. Above this quoted temperature, marcasite becomes increasingly unstable. Schot (1973) re-identified the earlier-described valleriite as mackinawite (Ramdohr, 1953b). Also, Borchert's (1934) valleriite appears to be mackinawite – therefore, the overall genetic interpretation has fundamentally not altered. As numerous high-grade metamorphosed submarine-exhalative-sedimentary ore deposits demonstrate when, for example the host rocks have attained a gneiss facies, the primary pyrite grains also become unstable and recrystallize to very coarse idioblastic grains up to a centimeter or larger in diameter.

The data on the occurrence and distribution of trace elements is insufficient to attempt statistical analyses, a deficiency already mentioned by Kraume (1955). Using his information, some average values (Table VII) can be used for genetic interpretations, but caution has to be exercised as more research is required. One also has to consider the possibility that improved analytical methods may somewhat change the data.

The average compositions of ore concentrate bulk samples which were analysed at the same time as the trace elements were determined, indicate the following:

TABLE VII

Average chemical composition of run-of-mine ore (average of several years of analytical results) (After Kraume, 1955)

22% $BaSO_4$	800 ppm Sb	10 ppm Tl
19% Zn	500 ppm As	3 ppm Ge
9% Pb	500 ppm Cd	1.2 ppm Au
9% Fe	160 ppm Ag	0.04 ppm Pt
1% Cu	150 ppm Co	0.02 ppm Pd
21% S	70 ppm Bi	
10% SiO_2	50 ppm Sn	
4% CaO	40 ppm Hg	
3% Al_2O_3	20 ppm Ni	
1.3% Mn	20 ppm In	

Traces of Ga, Se, Te, Rh, Ru, Re and W were detected.

Note: Mo was not detected by the chemical analytical methods, although molybdenite was proved to be present by ore microscopic investigations (Ramdohr, 1953b).

(1) Ag, Au, Sb and Bi are associated with the Pb–Cu ore minerals;

(2) Cd, In and Hg are bound to sphalerite; and

(3) Co, Ni, Tl and As to pyrite (Table VIII).

The Mottled Ore ("Melierterz") is the main gold-bearing ore variety, and with an average of 3 ppm Au it is significantly richer than the average of the run-of-mine ore. In occasional cases, the Mottled Ore can have a gold tenor considerably above 30 ppm. Bismuth is linked to native gold and reaches, in the same ore type with 250 ppm, more than 3.5 times the total average amount.

Gold is predominantly present in native bismuth and bismuthite. Silver is distinctly bound to the galena mineralization and, consequently, reaches a maximum later in the paragenetic sequence, but more precise data are not available – also typical primary Ag-bearing minerals have not been recognized. High-silver tenors of about 140 ppm average in the barite of the Grey Ore have, as yet, not been properly explained, and due to the relatively low galena content it cannot be assumed that the silver is associated alone with this lead sulphide. Generally accepted geochemical rules explain the preferential association of Cd, In and Hg with sphalerite. More precise investigations of Co, Ni, Tl and As bound intimately to pyrite are also missing, so that no specific discussion can be provided.

Puchelt (pers. communication) studied the distribution of Sr in barite: the Sr tenor seems to be relatively high with an average of about 0.42% (Fig.19) of the New Orebody and with 0.8–1.3% (Fig.20) in the early barite precipitates of the Grey Orebody. Towards the outer extremity of the New Orebody (Fig.21), as well as in the rhythmic barite–shale sequences of the Footwall Grey Orebody (Fig.22), there is a change to lower values from 0.1 to 0.5% Sr.

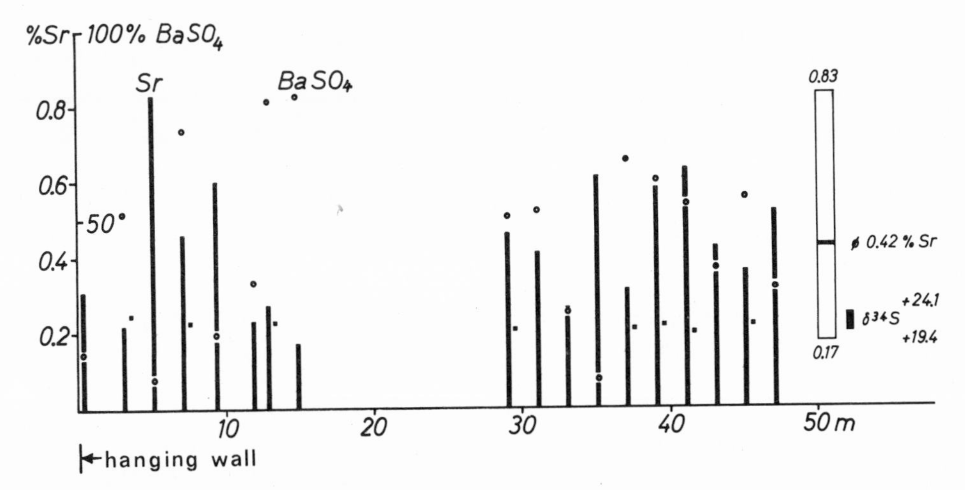

Fig. 19. Distribution of barite, strontium and sulphur isotopes in the New Orebody at the 10th level. (Sr-analyses from Puchelt, unpubl. data; sulphur-isotpe data from Anger et al., 1966.)

During the initial stage of the Grey Ore formation, only hypogene-hydrothermal sulphur was probably supplied [i.e., fed by magmatically derived solutions and/or magma-heat-driven cells of recirculated connate waters – White (1968) and White et al. (1971) – which is also believed to be operative in the Red Sea system as discussed by Schott (1976)

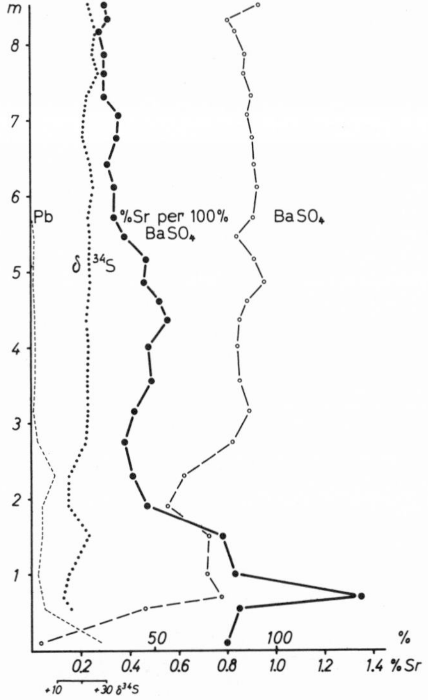

Fig. 20. Distribution of barite, lead, strontium and sulphur isotopes in the Grey Orebody from bottom to top. (Sr-analyses from H. Puchelt, unpubl. data; sulphur-isotope data from Anger et al., 1966.)

and Degens and Ross (1976)]. During further geological development of the Grey Ore, a hybrid origin of the sulphate from both hydrothermal solutions and from sea water can be assumed, as demonstrated by the increase in heavy sulphur in the Rammelsberg syngenetic barite.

The distribution of the S isotope ratio of sulphide-sulphur studies by Anger et al.

TABLE VIII

Analyses of the ore types in the Old and New Orebodies (After Kraume, 1955)

%											
Fe	Cu	Zn	Pb	S	Mn	MgO	CaO	CO_2	$BaSO_4$	SiO_2	Al_2O_3
Grey ore											
2.76	0.11	4.05	4.95	15.4	0.11	0.42	1.22	1.10	79.48	1.19	0.81
1.46	0.05	3.0	1.93	14.15	0.1	0.14	2.31	1.82	84.92	0.68	0.24
2.8	0.1	0.2	0.8	14.4	0.1	0.2	1.7	1.9	83.5	4.2	1.2
2.7	0.1	0.1	0.1	11.1	<0.1	0.2	0.5	0.5	67.7	18.6	6.1
Lead–zinc and lead ore											
1.46	0.05	13.3	28.0	18.55	0.18	0.15	<0.7	0.54	42.36	0.68	<0.1
3.74	0.08	10.40	16.70	13.65	1.32	1.65	1.39	2.60	51.16	1.60	0.41
2.7	0.49	15.4	15.6	17.83	<0.1	0.16	<0.1	0.78	53.68	1.24	0.31
3.7	0.7	14.3	59.9	20.0	0.1	0.2	0.1	0.3	0.5	0.2	0.2
4.7	0.1	17.9	7.9	16.5	0.2	0.6	0.8	1.4	54.1	1.3	0.6
Mottled ore											
7.8	2.56	30.9	14.8		0.24	0.42	0.70	1.08	18.28	0.60	<0.1
10.1	3.81	28.4	12.8		0.72	0.40	0.90	0.66	22.08	0.34	<0.1
14.85	5.19	21.2	9.3	26.62	1.38	0.40	0.45	0.80	10.34	4.24	4.93
14.2	3.65	22.95	10.40	28.2	1.23	0.50	2.45	2.13	14.94	0.96	0.26
Brown ore											
12.51	1.4	42.42	9.2		0.22	0.24	0.28	0.58	0.78	0.58	0.10
6.73	0.075	41.1	21.75		<0.1	0.36	0.11	0.90	0.92	0.16	<0.1
6.12	0.16	32.3	15.8	25.0	<0.1	0.26	<0.1	0.48	23.26	0.14	0.21
7.9	0.27	45.66	14.4	30.21	0.11	0.18	0.22	0.54	<0.1	0.48	<0.1
5.78	0.19	46.96	11.6		<0.1	0.62	1.04	1.00	1.66	2.3	<0.1
10.6	3.5	40.8	6.4	33.8	0.1	0.2	0.1	0.3	0.6	1.5	1.4
6.1	0.3	45.6	16.3	30.0	0.1	0.2	0.1	0.3	0.3	0.7	0.4
6.1	0.3	34.0	29.4	28.1	0.1	0.2	0.1	0.6	0.4	0.9	0.2
Copper ore											
29.52	24.12	2.1	0.5		0.25	0.36	0.78	0.76	<0.10	1.12	0.41
26.88	21.87	6.9	2.8		0.48	0.43	0.36	1.42	0.16	1.06	<0.1
34.4	7.3	5.6	1.3	43.0	0.1	0.3	0.1	0.9	0.5	3.2	2.2
34.5	9.6	1.2	2.9	40.4	1.5	0.4	0.1	3.0	0.1	2.1	0.4
30.55	15.6	3.6	2.3	36.97	0.62	0.12	<0.1	0.86	<0.1	4.3	3.38

(1966) supports a main hypogene-hydrothermal source of the ore-forming chemical elements (cf. also Sangster, 1976). The shift in the $\delta^{34}S$ ratio has been recognized by these authors as a kinetic effect of the ore-forming solutions supplied over a longer geologic period and believed to be the result of magmatic differentiation processes. Except for the pyrite, the initial or early sulphide precipitates begin with a $\delta^{34}S$ ratio of about

ppm												
Co	Ni	Cd	Hg	Jn	Tl	Ge	Sn	Ag	Au	As	Sb	Bi
20	21	180	30	13	<5		350	76	0.28	810	710	10
<10	<10	80	120	<5	<10		36	101	0.6	40	430	<10
<10	<10	<10	33	<10	27		24	60	0.2	160	120	<10
<10	31	<10	25	<10	<5		30	26	0.2	70	960	12
7	<10	580	61	<10	<10		57	187	1.06	80	330	12
50	<10	360	29	25	<5		53	203	0.35	390	870	<10
15	<10	650	53	<10	14		48	169	0.7	270	1230	<10
90	<10	600	58	28	<5		95	834	0.4	80	2800	260
120	210	460	71	<10	25		59	126	0.8	740	580	<10
380	Sp.	1100	93	36	<10	4	240	291	1.64	370	1500	110
350	Sp.	1000	54	35	<10	3	360	229	2.88	380	1500	77
410	19	610	22	34	9		510	232	9.52	740	1100	380
350	<10	640	24	50	<5		83	220	1.2	270	930	36
580	<10	1200	91	18	10	2.7	490	109	3.44	860	840	12
200	Sp.	1400	22	30	14	1.6	130			350	1300	<10
350	105	1040	127	<10	15		74	214	0.90	540	1160	10
64	<10	1100	67	<10	<5		240	88	0.20	240	1500	<10
66	Sp.	1400	180	10	14	4	60			560	930	<10
550	<10	920	58	65	135		310	173	0.6	1600	350	110
270	<10	1200	194	13	<5		47	315	1.0	310	870	<10
75	<10	1300	160	<10	22		15	381	1.1	460	1800	<10
850	Sp.	400	12	42	<10	3.8	280	88	0.14	2700	480	80
480	Sp.	170	16	120	<10	2.6	730	304	34	630	1400	2100
180	<10	150	28	<10	<10		74	88	0.8	1730	430	110
600	Sp.							86	0.2	1900		
26	<10	80	18	38	<10		200	249	0.8	870	770	420

(continued)

TABLE VIII (continued)

%											
Fe	Cu	Zn	Pb	S	Mn	MgO	CaO	CO_2	$BaSO_4$	SiO_2	Al_2O_3
Pyritic ore											
30.74	2.3	12.6	1.55	35.74	0.44	0.85	1.4	1.34	<0.1	7.94	1.14
27.48	0.24	8.9	2.65		0.64	0.48	12.0	9.68	0.52	1.56	<0.1
31.56	0.11	7.2	4.0	39.63	0.54	0.77	2.85	3.3	0.62	5.12	2.55
35.35	0.03	3.2	1.6	41.46	0.53	0.37	2.23	2.58	<0.1	10.96	0.66
24.2	0.8	9.5	5.8	28.5	6.4	2.7	6.9	14.3	0.2	0.4	0.2
34.2	0.9	11.0	2.9	45.4	0.3	0.6	0.1	1.3	0.1	1.7	1.2
19.7	0.5	16.5	8.3	29.8	2.9	0.8	5.0	5.0	4.4	1.3	0.5
17.7	0.7	27.0	7.9	32.9	1.3	0.6	0.1	2.4	7.9	1.5	0.4
Sulphur ore (massive pyrite)											
44.0	0.1			48.0							
36.4	1.2	4.8	1.9	40.5	3.1		1.2	6.5	0.1	0.5	0.7
41.1	1.4	4.9	1.5	49.7	0.1	0.2	0.1	0.2	0.1	0.4	0.1
34.7	0.1	0.7	0.5	38.0	0.2	0.2	1.2	1.2	7.3	13.2	1.9

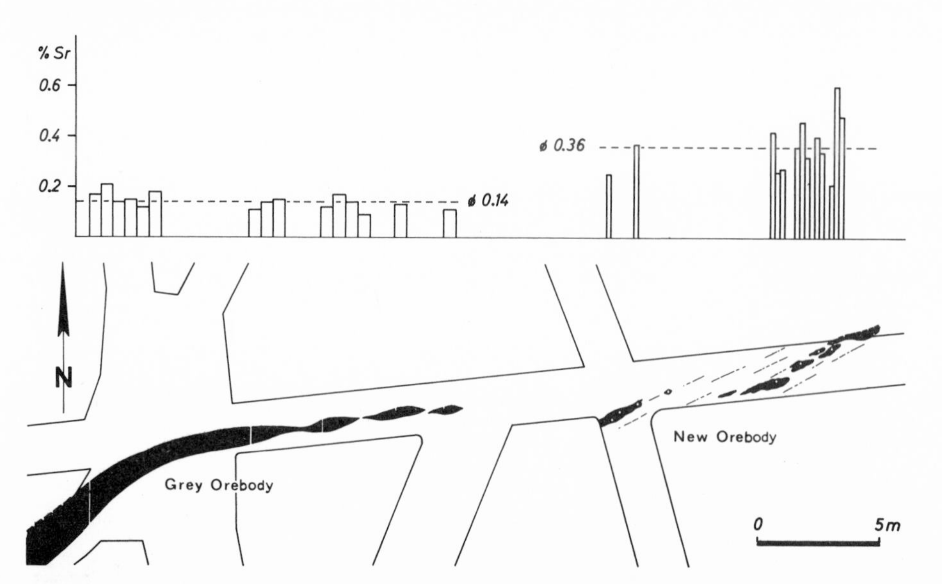

Fig. 21. Distribution of strontium in the margins of the Grey Orebody and the New Orebody. (Sr-analyses from H. Puchelt, unpubl. data.)

TABLE VIII (continued)

ppm												
Co	Ni	Cd	Hg	Jn	Tl	Ge	Sn	Ag	Au	As	Sb	Bi
4700	200	170	8	16	<10	1.6	120	197	1.06	17500	740	88
90	Sp.	380	Sp.	16	110	1.6	160			3200	450	<10
65	35	160	30	<10	8		24	91	0.6	2900	490	<10
<5	38	80	216	<10	<10		15	74	0.52	910	450	<10
		260						93	0.4	800	700	70
730	Sp.							56	0.1	1900		
340	<10	480	74	<10	42		176	102	0.2	1260	360	<10
225	75	690	46	33	26		208	104	0.4	3700	640	<10
	4000							70				
	1500									900		
450	<10	98	8	<10	<5		74	82	0.3	1800	340	56
<10	<10	40	20	<10	<5		24	24	0.2	150	340	<10

+7‰ and increases up to +20‰ in the youngest sulphide. A maximum $\delta^{34}S$ ratio of +36‰ was determined in the youngest barite (i.e., sulphate-sulphur) sequence. The following details of the $\delta^{34}S$ changes from bottom to top of the orebodies have been varified (Table IX).

It seems to be beyond doubt that bacterial activity in the sulphur cycle (e.g., sulphate-reducing organisms) is an important supplier of the sulphur required for pyrite. According to Anger et al. (1966), about 50% of this sulphur seems to be derived from the exogenic geochemical cycle (cf. Trudinger, 1976). Also, the original presence of organic matter should be significant (cf. Saxby, 1976). Four sources were considered for the barite (= sulphate)-sulphur, though they are genetically interrelated and overlapping or transitional:

(1) Hypogene-hydrothermal solutions of "magmatic" origin (including heat-driven circulating connate-water cells);

(2) oxidized sulphide-sulphur of hypogene-hydrothermal solutions;

(3) sea-water sulphate; and

(4) sea-water sulphate by bacterial processes enriched in heavy sulphur.

In principle, a hybrid sulphate supply from both the endogenic and exogenic geochemical cycles is believed to have been responsible in the formation of barite; nevertheless, it has to be assumed that barium is totally of hypogene-hydrothermal origin. Schot (1973) again discussed the data of Anger et al. (1966) utilizing the extensive litera-

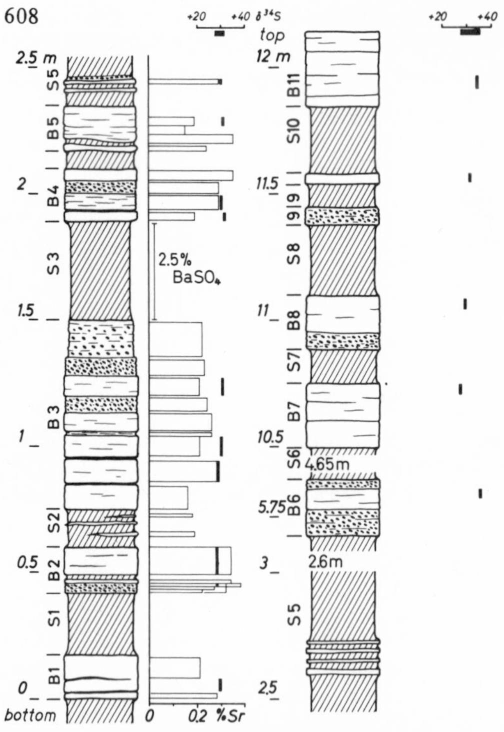

Fig. 22. Sequence of massive and sphaerolithic barite layers in the Footwall Grey Orebody from bottom to top. Surface outcrop in the normal limb of the orebody syncline (*B* = barite layer, *S* = shales). (Sr-analyses from H. Puchelt, unpubl. data, sulphur-isotope data from Anger et al., 1966.)

TABLE IX

Variation in the isotopic composition of sulphur (After Anger et al., 1966)

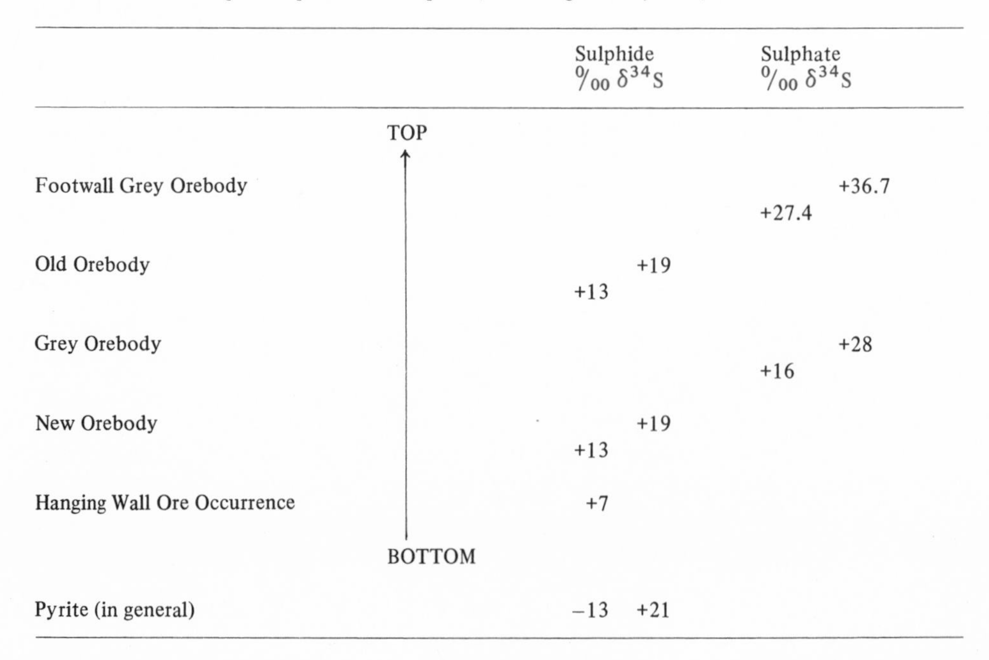

		Sulphide ‰ $\delta^{34}S$	Sulphate ‰ $\delta^{34}S$
	TOP ↑		
Footwall Grey Orebody			+36.7 / +27.4
Old Orebody		+19 / +13	
Grey Orebody			+28 / +16
New Orebody		+19 / +13	
Hanging Wall Ore Occurrence		+7	
	BOTTOM		
Pyrite (in general)		−13 +21	

ture on S-isotopes equilibrium. Analogous to Nielsen (1974), by using various thermodynamic models, he arrived at a "temperature of ore formation" averaging nearly 800°C – which of course is unrealistic and unacceptable. Therefore, the hypothesis of a *closed* geochemical system, on which such calculations are based, cannot be advocated for the Rammelsberg ore deposits.

TECTONIC ORE FABRIC, SYNKINEMATIC MINERAL FORMATION AND ALTERATION

The overall orogenic strain resulted in an intense internal plastic deformation of the orebodies. The subsequent influence of late-stage fault tectonism on the deposit boundaries, which was particularly well-developed in the hanging-wall as well as in the footwall of the orebodies (Fig.8), is a further consequence of the continuous internal plastic movements of the massive ores. The results are complex tight sling-like multiple isoclinal folds (e.g., refolded and ptygmatic types) as well as local partial fracture deformation of the massive ores, as is well known in intensely kinematically deformed metamorphic rocks. The non-affine deformation on all scales – especially folding – affected the massive ore down to microscopic (grain-size scale) dimensions, causing intense flattening and stretching of the primary sedimentary features. This strata-parallel shear transformation and folding produced flaser, schlieren and lensoid fabrics in all types of ores except in the massive pyrite ores. Many parts of the massive ore exhibit genuine augen structures (see Figs.26–28, Plate VII, c and d).

This tectonic development, and the consequent difficulties encountered in genetic reconstructions, is the reason for the absence until recently of a correct understanding of the laminated and banded structures. Ramdohr (1953b) has, in general, assumed that these fabrics are of primary sedimentary origin, but when each was examined in detail, other genetic interpretations could be offered. There is no doubt that intense shearing of a colloform multi-sulphide mineral fabric, with a statistical distribution of the various mineral constituents, can form a lamination that is very similar to sedimentary textures. Gundlach and Hannak (1968) demonstrated such a process by referring to a shear foliation developed in a Lower Devonian ore seam (Plates I and IV). To prove their case, it was necessary in each instance to continue the study in detail from the macro- through the mega- to the microscopic scales and correlate the data thus obtained to provide unequivocal, conclusive evidence for the origin of the textures and structures. Ramdohr (1953b and in Kraume, 1955) consistently followed such an approach in his reconstructions of the interrelationships of the fabrics from the macro- to the microscopic data.

Fig.23 demonstrates that such fabric convergences apply to intensely deformed vein ores. Therefore, Lindgreen's (1933) proposal that the Rammelsberg ore deposit is a metamorphosed bedded vein cannot be accepted, and can be refuted by the new observations. The doubtless Cu–Pb–Zn vein ore in Fig.23 had normal vein-type features, such as

Fig. 23. Highly tectonized vein ore, mainly sphalerite, Mühlenbach mine near Koblenz (Rheinisches Schiefergebirge, FRG).

coarse-grained mineral aggregations and typical mineral growth fabric, prior to undergoing lamination by shear transformation along an overthrust to form the schlieren structure. Therefore, the pre-deformed features of both the Rammelsberg and vein-type ores were of different origins. Both of these deformed ores microscopically exhibit quite clearly that one was of fine-grained sedimentary and the other of coarsely crystalline vein-type origin. Convergences have also been noticed in relation to the blastomylonitic dense ore of the Ramsbeck mine, Sauerland (Rheinisches Schiefergebirge), where vein-type ore was in parts intensely tectonized (Scherp, 1958). Incidentally, this ore deposit was incorrectly interpreted by Udabasa (1972) as of sedimentary origin – there is no question that it is a genuine epigenetic vein-type formation.

The schlieren fabrics of the tectonized vein ores are only individual local occurrences in the deposits described above and it should be mentioned that certain vein-type features are not totally overprinted. In contrast, the schlieren fabrics of the Rammelsberg massive ore are penetrative, i.e., developed everywhere without exception. In genetic interpretations, one should emphasize particularly the generally and widely occurring structural characteristics of a deposit, not any specific local features.

In summary, the following criteria support a primary synsedimentary origin of the Rammelsberg massive ores with its schlieren and laminated textures formed through strata-parallel plastic flattening, stretching and folding:

(1) Concordance of the orebody's margins with the internal lamination of the massive ore.

(2) Concordance of the sedimentary host-rock inclusions and the internal lamination of the massive ore.

(3) Concordance among the different ore and gangue mineral strata independent of their different rheological properties.

(4) Parallelism of the laminae with slab-shaped pyrite lenses, which have undergone little or no deformation in grain structure.

(5) Despite the high degree of immobility of the barite (similar to pyrite), there are no visible differences of the barite-rich beds from the other parallel structures. Grain-size modifications of the barite are absent or negligible. (Not to be considered in reconstructing the original depositional milieu are the several types of tectonic deformations merely reflecting extreme differences in competence to tectonic stress.)

(6) Concordance between the general ore lamination and the relict geopetal features. Lentoid marcasite crusts are parallel (conformable) to the laminae.

(7) Concordance between the distribution of ore types and the internal lamination of these ores.

(8) Harmonious sedimentary interfingering of banded ore, massive ore and shales.

(9) Sedimentary gradations between banded ore types with the massive ore caused by a higher proportion of clastic material and evidence of graded-bedding.

(10) In microscopic dimensions, the concordance of the unequivocal sedimentary beds of framboidal pyrite with the banded ore textures as well as with the shaly beds.

(11) The absence of larger quantities of the ore selections that could have originated by tectonic mobilization.

(12) Congruency of the internal folds in the massive ore, which has homogeneous mechanical or rheological properties.

(13) Lastly, but just as importantly, there is a systematic continuous gradation from the fold styles of the massive ores through the rich and poorly mineralized banded ores to the fold types of the host rocks, reflected especially by the congruency of symmetry.

Arguments against a synsedimentary origin, which are based on a replacement mechanism according to the proponents of the "epigenetic" hypothesis, are no longer acceptable. Their so-called "replacement" features are to be explained as tectonically deformed polymetallic-sulphide colloform structures, in which each of the sulphide components has a different susceptibility to subsequent deformations.

The complicated fold styles mentioned earlier (isoclinal sling-like folds) are not readily recognizable in normal outcrop and mine exposures. In polished handspecimen, the tight isoclinal folding is clearly visible, and is always related to intense plastic deformation. Factors controlling the fold styles are:

(1) The composition of the massive ore, whether with or without pyrite and/or host-rock inclusions such as carbonates;

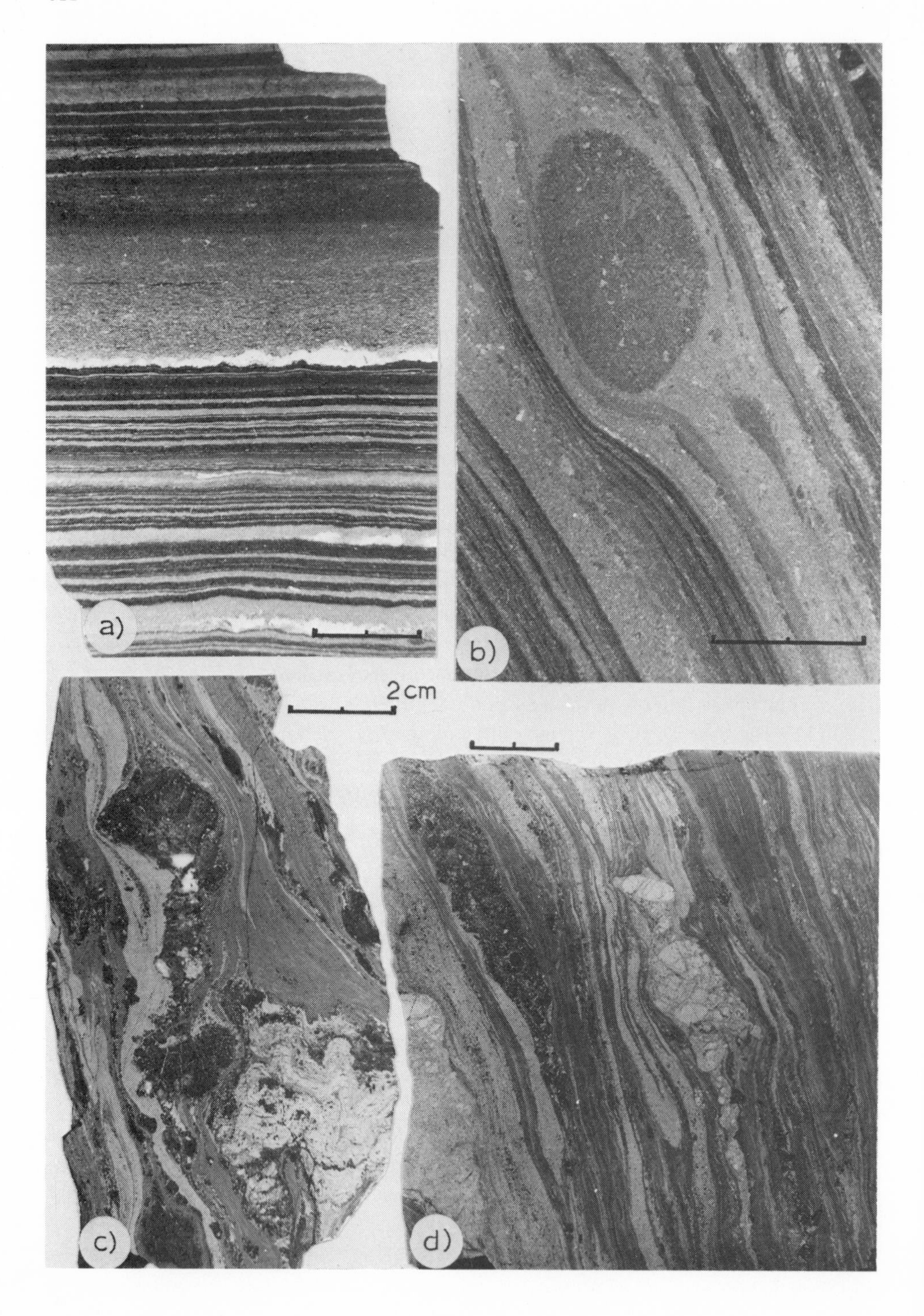
a)
b)
2cm
c)
d)

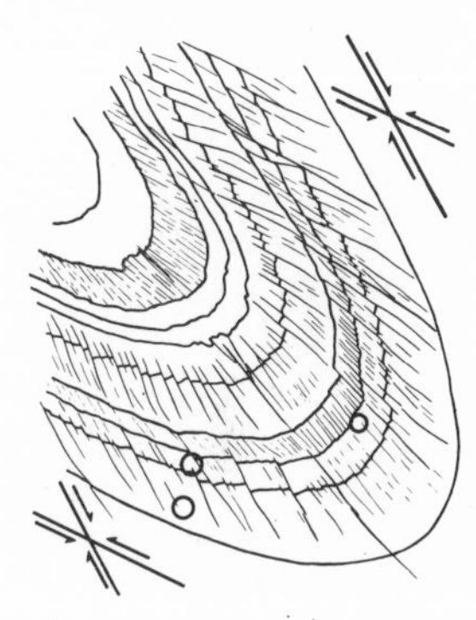

Fig. 24 (a, b). Banded Ore, syncline, flow and fracture cleavage. *O* = pyritized brachiopods and goniatides, main fracture lines filled with mobilized pyrite and sphalerite.

(2) the original extent of the individual ore layers and distribution of the host-rock inclusions; and

(3) the particular tectonic position.

These parameters explain the complicated multiplicity of the fold styles (Figs.11, 12, 24, 25, and Plates I, and VII–X). Ramdohr (1953b) has published a great number of illustrations of the typical structures, among which are the "tadpole" types, which are the result of very tight isoclinal folding of isolated short ore lenses. It is also possible that earlier-folded ore sections were subsequently plastically and intensely compressed and stretched, and then underwent isoclinal re-folding. This complex process of folding can be compared to the movements of an oil film on a turbulent, flowing water surface. An

---

PLATE VII

a. Banded ore with graded bedding.
b. Early diagenetic lithified mud pebble with disseminated chalcopyrite in banded ore host (sphalerite-rich) apparently with graded bedding.
c. Highly tectonized and isoclinal folded "Meliert" ore, carbonate augen and competent portions of pyrite.
d. Isoclinal sling-like folds in "Meliert" ore.

a)
b)
c)
2 cm
d)

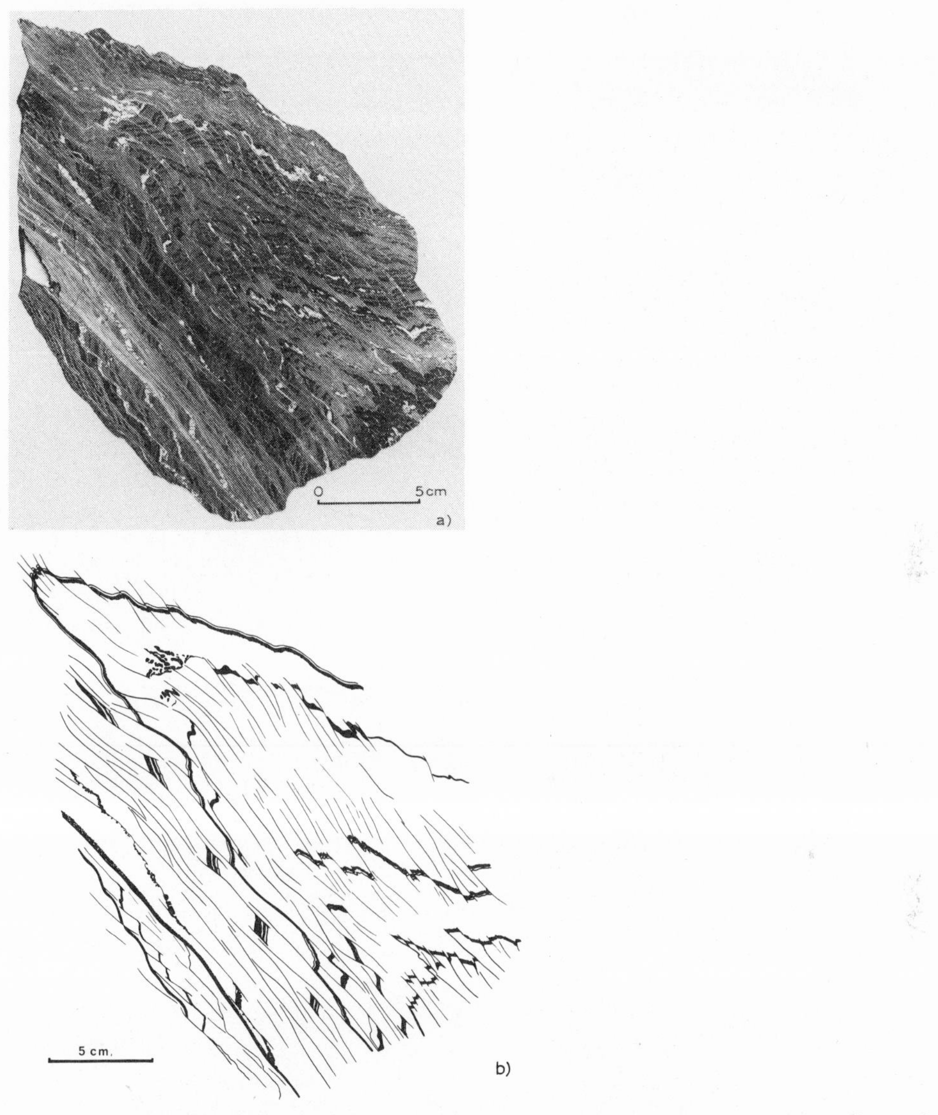

Fig. 25 (a, b). Banded ore, strong fracture cleavage, different thicknesses of the limbs according to the different tectonical movements.

PLATE VIII

Fold styles of different pyritic banded ore types. Very fine grained to massive pyrite layers are interbedded in black shales (a, b, c) and sphalerite-galena beds (d). The massive pyrite layer in b is boudined and the intersections are filled with fibrous quartz.

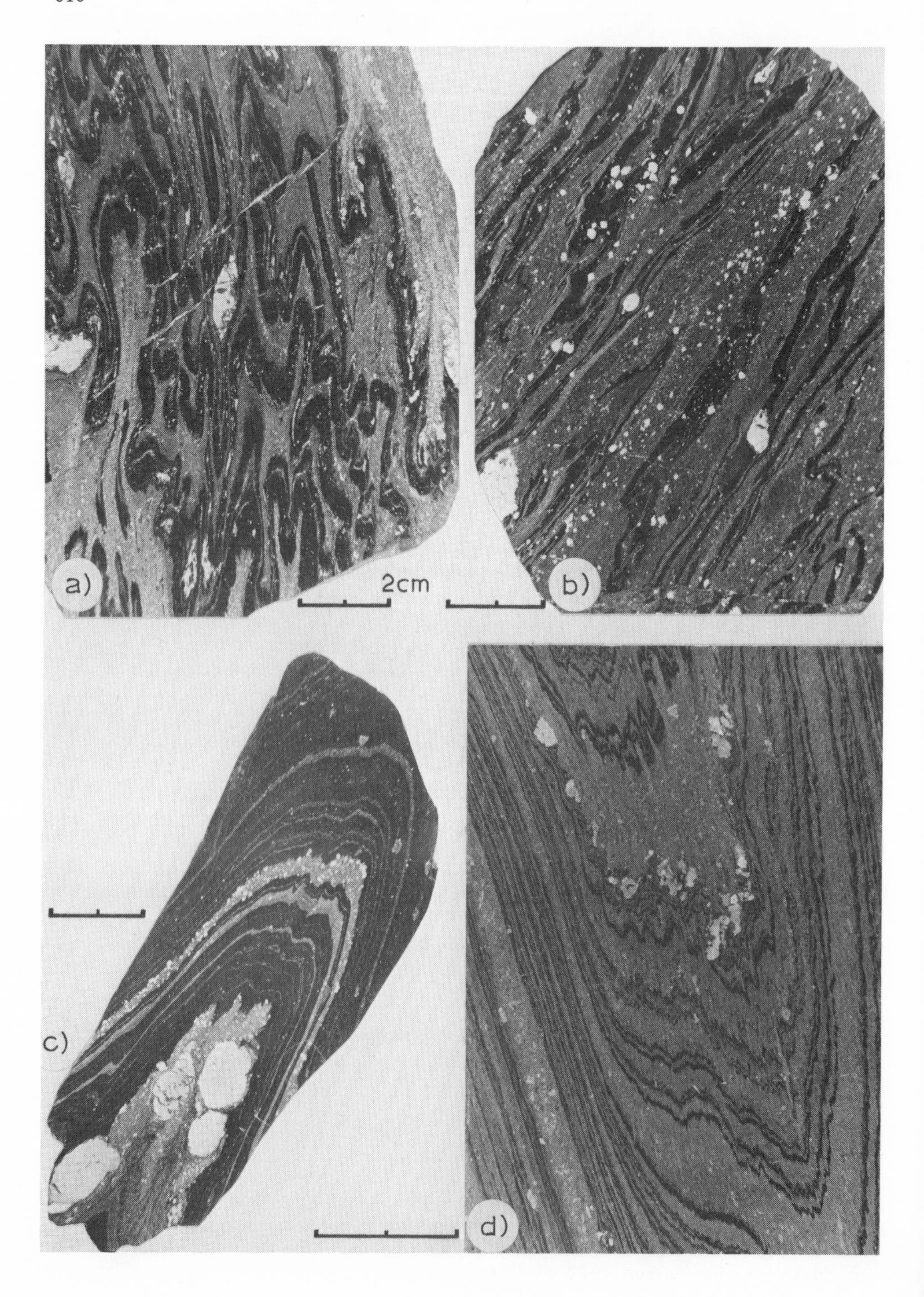
a)
2cm
b)
c)
d)

analysis of the possibly multiple refolded structures is nearly impossible because of the absence of top-and-bottom features. As a consequence of the high degree of partial mobility and differential movement, the deformation is non-affine and the structural elements cannot be "unfolded" or reconstructed to their original position. This is in full agreement with the preceding analysis of the tectonic behaviour of the orebodies, in contrast to the host rocks.

The deciphered structural data of the orebodies show that sections of intense compression alternate with those of weaker deformation, which can grade into those of a greater degree of tectonic stretching. Different structural types are present in close proximity to each other, and it is obvious that the massive ore – except for the competent pyrite and carbonate inclusions – was, in general, plastically stretched. The massive pyrite in particular, which has a very low ductility, exhibits multiple fracture deformation, depending on its structural location. In the massive ores, the pyrite was normally passively folded due to its great competence. The plastic movements of incompetent ore minerals into fracture zones of the more competent units within the massive ore, for instance fractured pyrite inclusions, demonstrate that diagenetic processes, such as sedifluction (slumping), were not responsible for this mobilization.

These observations make it quite clear that in the pyrite, independent of both its primary form and tectonic position, the internal stretching structures predominate even when they have been folded. The tectonic effect is illustrated by the total brecciation ("broken to bits" = 'kurz and klein zerbrochen', as Ramdohr, 1953b, has described it) (Plate IV). The destruction along an internal pair of shear planes is widespread, with an axis of intersection parallel to the fold axes. Boudinages are common in all dimensions from the micro- to macroscopic scale (Fig.33 and Plate VI). The internal stretching of the boundinaged layers took place without the influence of a shear stress, but as a result of the formation of joints vertical to the bedding and parallel to the fold axes. Depending on the size of these boundinages (from micro- over centimetre- to decimetre-dimensions), fibrous quartz has filled these joints, and/or surrounding material with higher susceptibility to plastic mobilization, moved into these open spaces by the process of "in-folding" from both sides. Boudinages on all scales are also present in the banded as well as in the grey ores. Under the influence of a preferentially strata-parallel shear stress, the pyrite developed a characteristic fracture cleavage with an angle steeply inclined towards the bedding planes and with rotations of the individual shear segments (Fig.26). In places, several segments have been rotated up to 90°. Analogous to this feature, tight crumpling folds ("Stauchfalten") have developed which are confined to more mobile ore layers between less deformed units (Fig.27). "Pressure shadows" are, therefore, quite common in the corresponding extension (= dilated) sections (Plate VI). As already mentioned, the

---

PLATE IX

Different fold styles of banded ore. a. Slinglike injection folds. b. Tight isoclinal folds. Ore bands consist mainly of sphalerite and galena. Nodules are of pyrite.

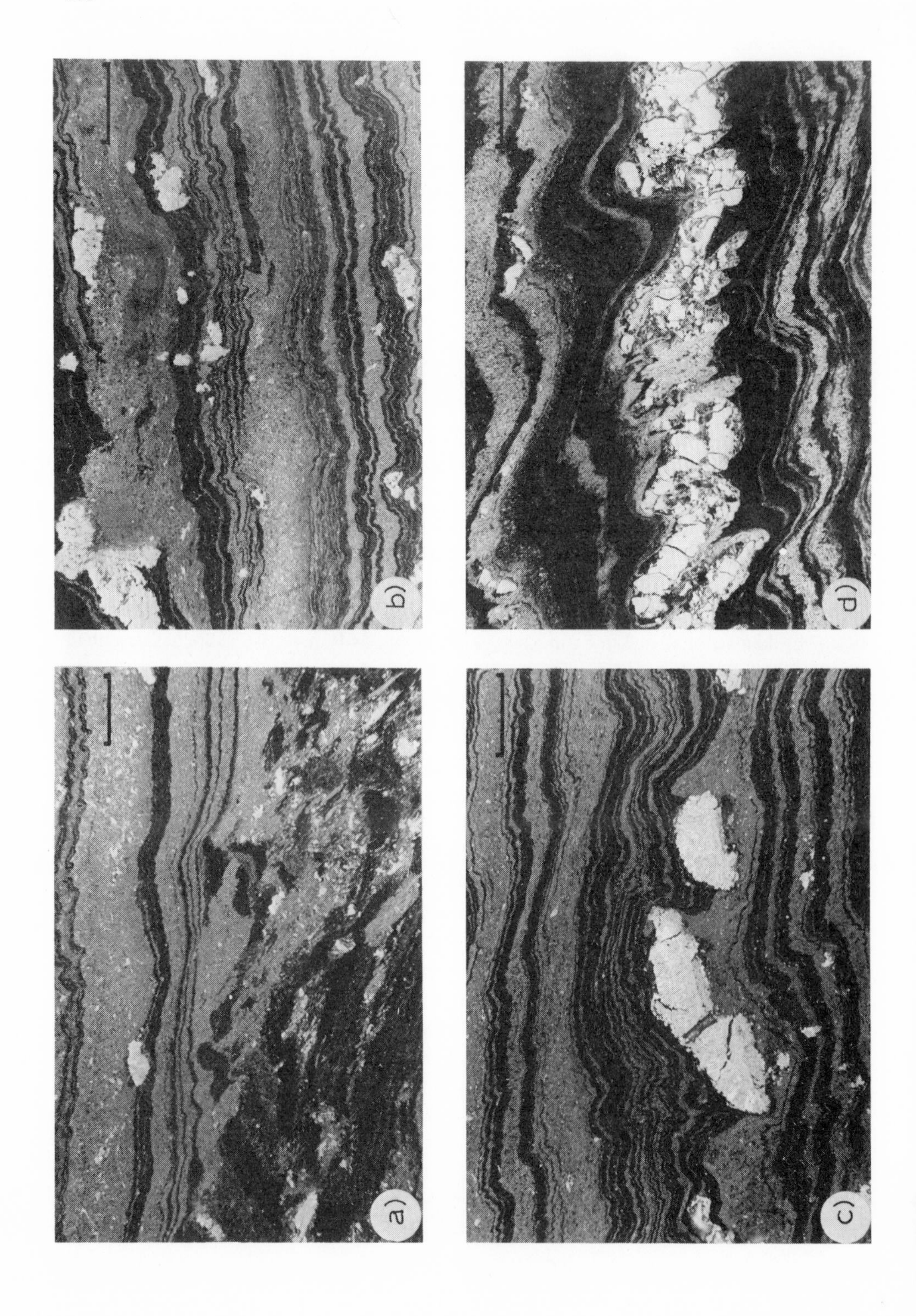
a)
b)
c)
d)

Fig. 26. "Meliert" ore, fracture cleaved pyrite layer.

foregoing structural examples are well-known from kinematically highly-deformed metamorphic complexes. Both internal and external rotations are characteristic of the general tectonic fabric of these rocks. Ramdohr (1953b) has provided a large number of impressive photographs of such features.

PLATE X

a. Disturbed sedimentation features produced by sedifluction processes in predominantly sphalerite-rich banded ore.
b. Sedifluction structures in top part leading to minor unconformities in predominantly sphalerite-rich banded ore.
c. Cauliflower pyrite, typically fractured and highgrade plastic deformation of the enclosing banded ore.
d. Pyritic banded ore, pyrite layers of different grain size and consequently of different grades of deformation. Massive pyrite layers strongly brecciated with interfragmental fissures filled by secondary quartz.

Fig. 27. Folded "Meliert" ore with pyrite nodules.

At the eastern extremity of the New Orebody, certain internal flaser and breccia fabrics suggest that the tensile strength was exceeded in specific parts of the ore deposits. In particular the copper ores have been affected. The explanation lies in the general tensional (= dilational) conditions during isoclinal folding of the ore. Analogous structural features are present also at the western rim of the New Orebody, as described by Kraume (1955) who advocated the same genetic interpretation (Fig.28). It is very unlikely that these structures have a primary sedimentary origin, e.g., submarine slumps.

In the transition zone from the massive ore to the banded ore mineralization (i.e., in the shales), as well as in the banded ore itself, the triaxial deformation – with its extreme "stretching" of the ore on the axial planes in contrast to the host rocks – resulted in the formation of well-developed disharmonic folding. The general tectonic behaviour of the orebodies, contrasted with their host rocks, has already been discussed (see pp.563–576); here the analogous deformation features, which range in scale down to metre dimensions, should be mentioned. Finger- or fork-like (resembling "injections") isoclinally in-folded parts of the massive ore into the host rock must be regarded as one of the principle characteristic features of the transition zones between the orebodies and the host rocks. Similar structural elements of the Mount Roseberry ore deposit have been described by Brathwaite (1972), and these fork-like massive ore "injections" correspond to those of the Rammelsberg deposit (Hannak, 1963). The already mentioned so-called "Hanging Wall Trum" (= "Hangendes Trum"), as well as the similar, smaller disharmoni-

Fig. 28. Brecciated and folded "Meliert" ore with flaser gneiss texture.

cally folded massive ores in the hanging-wall of the Old Orebody, also belong to this structural style. The so-called "bulges" (= Wülste) are to be considered as embryonic (= initial) features (Ramdohr, 1953b, and many others before him).

At the eastern margin of the New Orebody, minor synclines (= "Spezialmulden") of the massive ore were isoclinally and congruently (perfectly harmonious with the general tectonic symmetry) injected into the shales and banded ores over a distance of at least 10 m. The directions of maximum transposition are shown by the foliation planes in the core of these synclines, which under the given stress fields offered the lowest resistance. These tectonic relationships are definite proof of a longer period of continuous – or continual – plastic deformation of the ore, even after the flexural folding of the host rock was terminated. Figs.10, 11 and 12 provide typical examples. On the metre-scale, as well as in smaller dimensions, the dike-like massive ore injections clearly took place. Together with bulge-like thickness increases, brush-like forms originated in the cores of folds, and are associated with small pencil- to hair-shaped ore injections along the planes of foliations.

The tectonically produced structures in the banded ores are unusually varied, and the reasons for this are:

(1) varied ore-mineral associations (= parageneses) of the banded ores,

(2) variations in thicknesses; and

(3) changes in the corresponding tectonic position.

The intense folding down to the smallest dimensions is of particular interest. Combined with the shearing, cleavage and faulting, the folding is part of the high-grade structural complex. As a result of partial, incomplete and/or preferential shearing parallel to the bedding planes and other mechanical controls, the succession of strata that has been more intensely internally deformed alternate with those of apparent lesser deformation. Therefore, harmonic and disharmonic fold styles, combined with flexure-slip folds, are widespread.

Genuine tectonic disharmonic folds and drag folds were occasionally described as primary sedimentary features. Schot (1973), for example, interpreted drag folds as sedifluction products. Karl (1964) described a drag fold zone in pyritic banded ore (see his illustration) as follows: The folded and finely fractured pyrite layers in shales, resembling soft-rock deformation features, suggest a deformation of only partially lithified pyrite sand accumulation within pelite. As mentioned already, subaqueous sliding or slumping was a possible process in the origin of some ore structures, but the proof is usually only of an indirect nature. There is, for example, cleavage diagonally cutting the isoclinal folds (not symmetrical to the axial plane) which originated pre-kinematically. At the eastern extremity of the New Orebody (12th level in 1785 Ord.) a larger unconformity in the banded ore also suggests a synsedimentary origin tectonically overprinted.

Below this primary sedimentary-stratigraphic unconformity, the banded ore is isoclinally folded, whereas above it the bedding lies parallel to the gently undulating unconformity. It appeared to the author, therefore, that the isoclinal folding of the unit below the unconformity must have originated by synsedimentary slumping. On the contrary, the structures described by the above-cited authors are to be regarded as genuine synkinematic (orogenic) features. The folds described by Karl (1964) and Schot (1973), respectively, developed under conditions where quartz was remobilized and deposited into innumerable fractures – this is only known from tectonized rocks. The pelite (shale) was most likely already lithified by diagenesis at the time it underwent deformation, because there is an absence of pelitic encrustations or envelopes on the pyrite fracture fragments – which would be a characteristic criterion for a sedifluction origin as proposed by Karl. As a result, his claim that the fold structures originated under conditions of a very mobile wet pyritic sand layer cannot be accepted. Moreover, the framboidal pyrite-rich layers with their "ball-bearing" effect during structural deformation, in contrast to the greater competence of the massive pyrite ore, was not given its due consideration.

Ramdohr (in Kraume, 1955) discussed five conditions of formation required for the microscopic intergrowths of chalcopyrite and sphalerite. In addition to the primary formation by the precipitation of a gel or by alternating precipitation, he believed that migration of chalcopyrite into sphalerite is possible under synkinematic conditions. The partly oriented and partly irregularly distributed pyrrhotite in fracture fillings, which were formed during the development of boudinages, suggests that this mineral was predominantly formed under metamorphic (= synkinematic) conditions. The strong orientation of the pyrrhotite along two directions, making a 50–60° angle, also indicates a mechanically (deformational) formed twin shearing system. One of these directions is parallel to

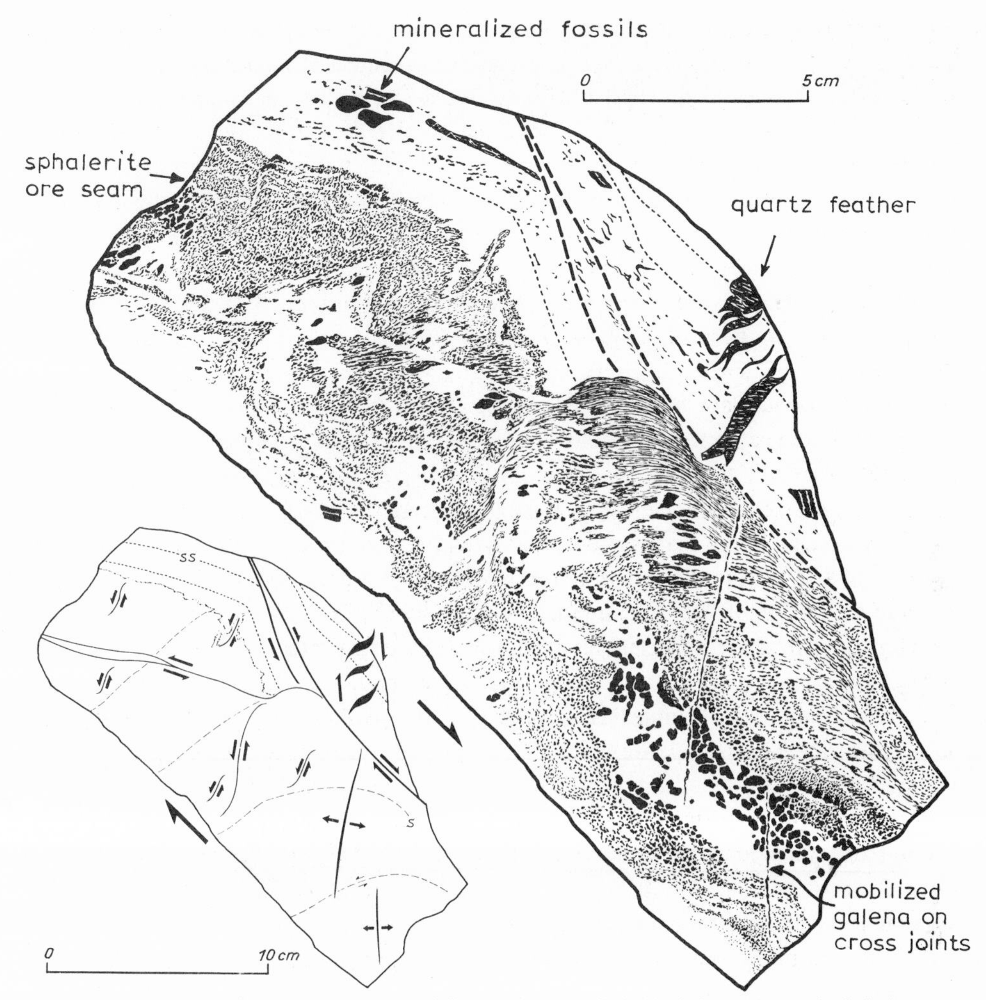

Fig. 29. Lower Devonian ore seam showing different degrees of plastic deformations of sphalerite; *ss* = bedding, *s* = cleavage.

the foliation, whereas the other is along the bedding planes. Analogously oriented galena, though less well-developed, has been observed. Excellently developed orientations of barite, which correspond to the foliation, are present in the grey ore beds of the Footwall Grey Orebody (Plate VI). Mica, as well as a chlorite of unknown specification, partly show a very distinct parallelism to the bedding planes. Judging from the various types of tectonic deformations of the phyllosilicates, Ramdohr (1953b) believed that the mica is an early-metamorphic product. Graphite occurs only rarely along shear planes, which was formed under extremely high tectonic stress. Slickensides are quite common in the massive ore. Polished shear planes (= slickensided) in adjacent ores are frequent, e.g., polished pyrite surfaces in contact with pyrite ore, etc. Galena, in contrast, has been "moved out" into a film over longer distances, and it is very common along shear planes of the banded ores. The formation of pressure shadows and mineral transformations are isochemical in origin and demonstrate, as already pointed out, that the ore and gangue minerals migrated only short distances (centimetres to decimetres, maximum) during mobilization. Every major mineral was involved in the kinematically produced remobilization. Among the ore minerals, galena and chalcopyrite were the most mobile, whereas the gangue minerals quartz and carbonates tended to be easily transposed. Sphalerite, pyrite and barite, in general, were more resistent. The degree of mineral mobility must have been directly dependent on local conditions, otherwise the observed deviations – even a reversal – from the listed sequence of relative susceptibility to remobilization could not be explained. At the present time, our knowledge is insufficient to determine to what extent solutions (with or without Cl/HS^- complexes) were involved, i.e., whether or not the mineral remobilization was entirely a solid-state phenomenon. Fibrous quartz, acicular mineral habit, and the presence of small amounts of cubanite, suggest an origin under higher pressure and temperature synkinematic conditions. Pressure twinning in bournonite and sigmoidal folding prove a progressive deformation accompanying the filling of the pressure-shadow open spaces. The strong orientation of the mobile galena in shear and cleavage planes in the massive ores has already been mentioned. The chalcopyrite in the chalcopyritic banded ores seems to have undergone accretive crystallization (= "Sammelkristallisation"), though there are reasonable doubts as to whether similar chalcopyrite spots did not already exist as a primary product (i.e., inherited sedimentary chalcopyrite). The same considerations apply to the occasionally present coarsely crystalline barite believed to have originated also by accretive crystallization (for terminology, see Folk, 1965). Ramdohr (1953b) has described well-developed orientations of boulangerite, jamesonite and bismuthinite, which were formed by intense shearing parallel to the primary stratification.

The multiple forms of folding, especially the preferential tight crumpling folding

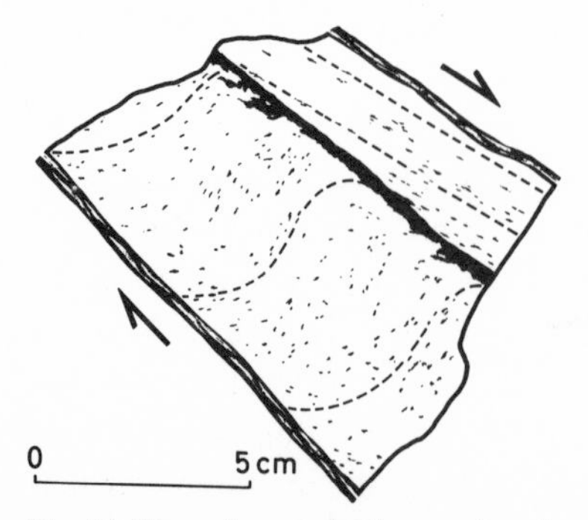

Fig. 30. Flow cleavage in the massive zinc ore, Lower Devonian ore seam. Black: galena.

(= "Stauchfalten") of less competent beds between more resistent units, are also observable on the microscopic scale.

Lastly, one should mention the numerous widespread idioblasts and xenoblasts characteristic of the microscopic metamorphic texture of the Rammelsberg ores. The poikilitic sieve textures are also typical. However, it should be made clear that the present writer doubts that all, especially the pyrite idioblasts, are of synkinematic origin as some could have already formed diagenetically. Ankerite idioblasts are locally present in great numbers (Plate III). Ramdohr (1953b) has described idioblasts of arsenopyrite and magnetite. As to the "replacement" question, one can only again quote Ramdohr (1953b) that "precise investigations have shown that actually nothing has been replaced".

The occurrence of very rare cubanite in association with mackinawite (Schot, 1973; according to Ramdohr, 1953b described as valleriite) and marcasite, induced Ramdohr (1953b) to follow the experimental results of Borchert (1934) to propose a maximum metamorphic temperature of 225–250°C. The physico-chemical conditions led to a fine-grained recrystallization of all the orebodies, but were insufficient to cause a "grain-growth"-type recrystallization in the macro-dimension (Folk, 1965). Ore-dressing difficulties are a result of the fine crystalline texture and the intimate intergrowths of the ore minerals in general. The intergrowths are often so small as to be near the limit of optical resolution. In spite of grinding 60% of the ore feed to $<40\ \mu m$, it is impossible to treat the ore concentrates as one would in the case of coarse-grained vein-type ore (Kraume, 1955).

ORE MINERALIZATION IN SILICIFIED LENSES (SO-CALLED KNIEST MINERALIZATION) AND OTHER POST-KINEMATIC MINERALIZATIONS

The ore mineralization in the silicified lenses of the hanging-wall, which has been called Kniest mineralization, is epigenetic in nature and comparable to genuine hydrothermal irregularly distributed stockwork ore deposits. This mineralization occupies curved overthrust planes, sets of fissures and their associated breccia zones, all of which are the result of local shear tectonism – i.e., the mineralization is controlled by late orogenic-tectonic elements as described below.

On average, the thicknesses of the ore stringers are several centimetres, rarely decimetres, in width. The main strike directions of the fissures are east-northeast and west-northwest. Barite veins occur more rarely, and are up to 1 m wide, having predominant north and northwest strikes with a steep dip to the east or northeast, respectively. The Kniest mineralization occupies only a part of the whole Kniest (silicified) zone.

The ore reserves of the Kniest mineralization were calculated as $2.5 \cdot 10^6$ metric tons, based on an average metal content of 3% Zn, 1.4% Pb, 1.3% Cu, 28 ppm Ag, and 0.2 ppm Au. Additionally, there are less intensely mineralized zones for which no ore reserve calculations have been made; their total metal-content lies between 1.5 and 3%.

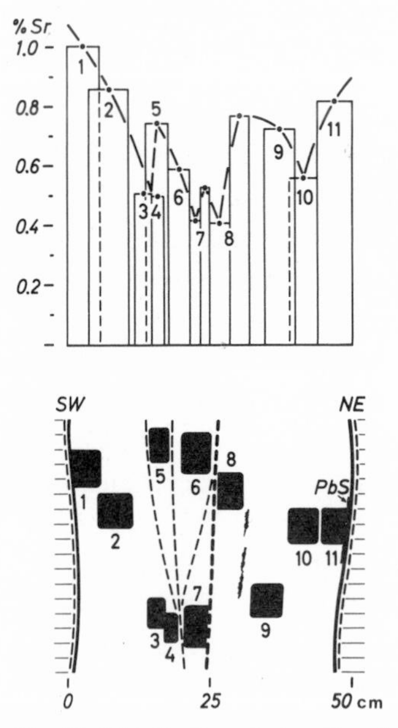

Fig. 31. Distribution of strontium in a barite vein. Black: traces of galena; "Kniest" mineralisation.

It is quite evident that these ores were not subjected to any high-grade internal textural or fabric deformations, as was the case of the massive ore, and there is not even the slightest resemblance between the fabrics and structures of the massive ore and Kniest ore. Generally, the minerals of the Kniest mineralization are coarsely crystalline and xenomorphic. Vein-type open-space fillings, such as superimposed mineral growth, and bilateral fissure fillings are particularly common. In the barite veins, this bilateral congruent symmetrical distribution is reflected by the Sr-contents (Fig.31). En toto, their characteristic features are distinctly different from both the syngenetic massive ore and the syn- to diagenetic mineralization in the conduit.

Kraume (1955) observed that the Kniest mineralization is best-developed where there is only a thin shale bed intercalated between the Kniest and the orebodies. Also, the metal contents are much higher in the immediate vicinity of the massive orebodies than in those further removed from the ore. Kraume also believed to have observed an inter-relationship between the abundance of the ore minerals in the orebodies and the adjacent

Fig. 32a. Sediment sequence of the Lower Devonian sandstone with ore seam outcrop (black line) in the Kommunion quarry.

b. Metal distribution of the ore seam section shown above.

c. Load casts in the overturned Lower Devonian sandstone, Kommunion quarry. (After Gundlach and Hannak, 1968.)

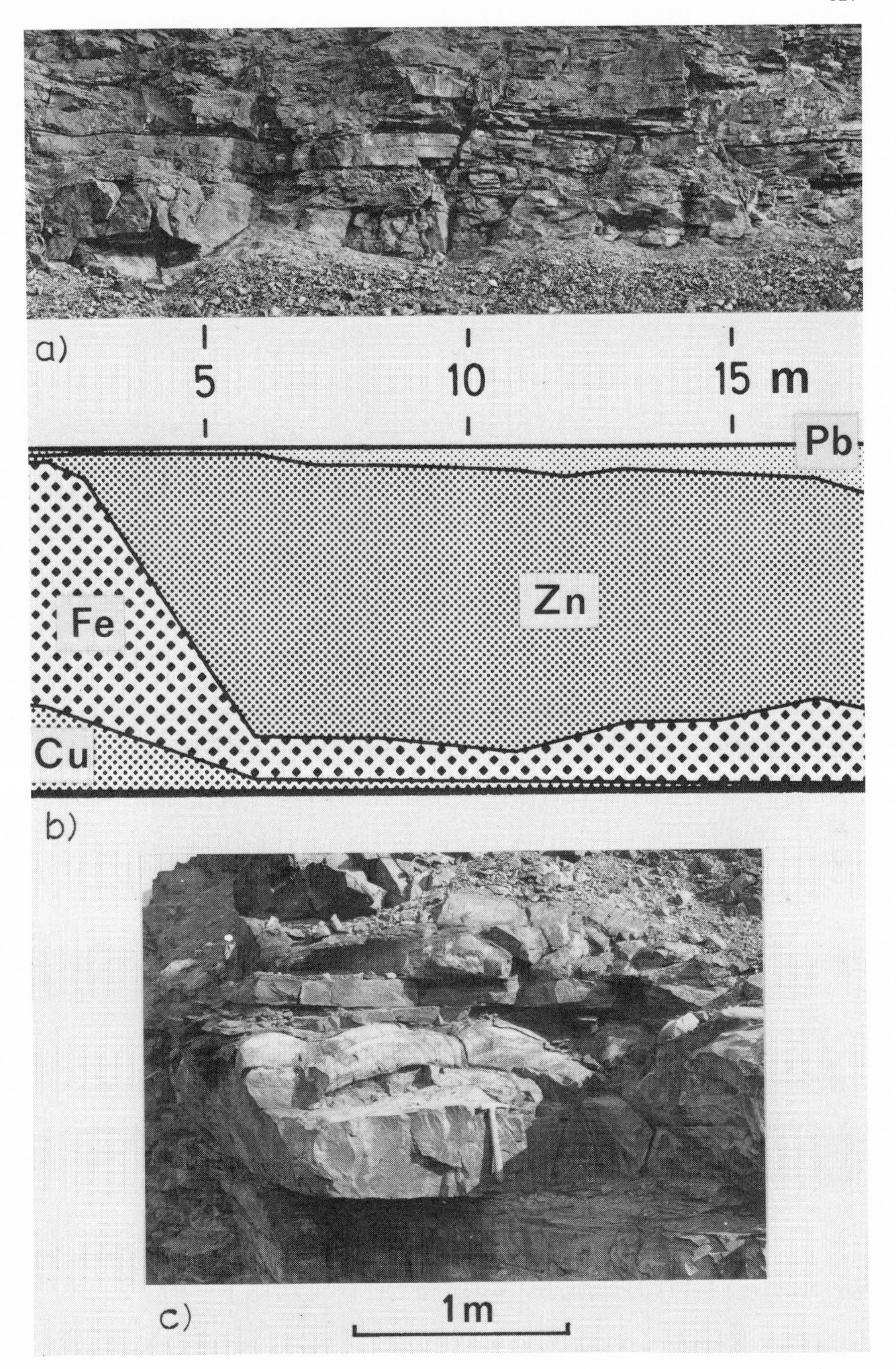
a)
5
10
15 m
Pb
Zn
Fe
Cu
b)
c)
1m

Fig. 33a. Banded ore, boudinage in the competent black "carbonate" layer (white to light grey: quartz and carbonates), banded ore slightly folded and flow cleaved.

b. Black shales (Ore Horizon) with fracture and flow cleaved carbonate layers.

Kniest mineralization: the pure Brown Ore (= sphalerite-rich ore) in the Old Orebody correlates with a Zn-rich mineralization in the adjacent Kniest, and the same relationship should exist for both copper and lead.

In addition to the main Kniest minerals (i.e., sphalerite, galena, chalcopyrite and pyrite) Wilke (in Kraume, 1955) listed marcasite, pyrrhotite, tetrahedrite, bournonite, galenobismutite, bismuthinite; and the gangue minerals as vein-type quartz, calcite, siderite, barite, chlorite and albite. Locally, a few breccias of the sedimentary massive ores are part of the Kniest mineralization, though no hydrothermal alteration or metasomatic replacements have been noticed (Table X).

TABLE X

Paragenesis of the Kniest mineralization (After Wilke, in Kraume, 1955)

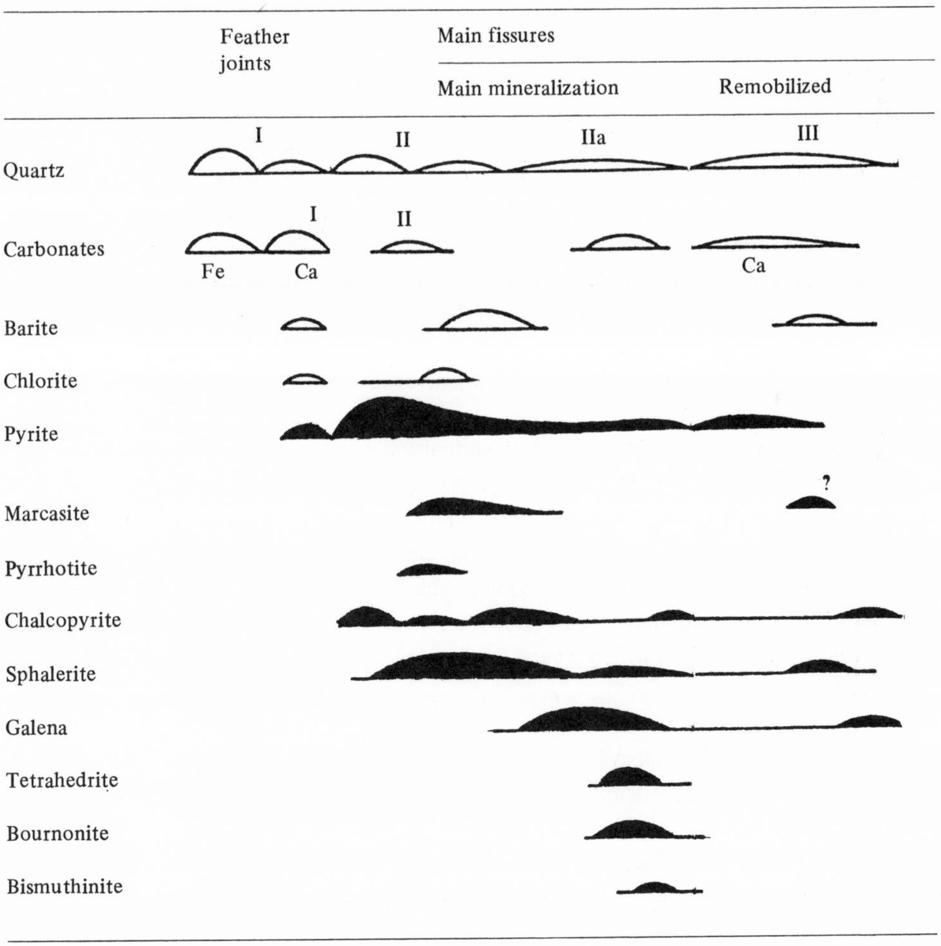

	Feather joints	Main fissures	
		Main mineralization	Remobilized
	I	II IIa	III
Quartz			
Carbonates	I (Fe, Ca)	II	Ca
Barite			
Chlorite			
Pyrite			
Marcasite			?
Pyrrhotite			
Chalcopyrite			
Sphalerite			
Galena			
Tetrahedrite			
Bournonite			
Bismuthinite			

The tectonic fabrics and their symmetries support a post-kinematic mineralization at the end of the folding process. The ore mineralization was introduced by a conspicuous vein quartz in the feather joints (= "Fiederspalten"). This quartz generation is noticeable far beyond the Kniest-lenses in the mine exposures, particularly in the whole hanging-wall region of the ore deposit. There is clearly a relationship between this vein-quartz mineralization and the overthrust tectonism. The Kniest mineralization was terminated by the barite vein formation coinciding with the tectonic phase of transverse faulting (Table X).

The spatial relationship of the Kniest mineralization to the orebodies and the correlative mineral composition between them, support Kraume's (1955) opinion that the ores were mobilized from the massive sulphide ore deposits and redeposited to form the Kniest mineralization – the latter is, therefore, the product of a synkinematic later-stage secretion of epigenetic origin. The vein quartz could, in general, be of lateral secretionary origin mobilized under orogenic conditions.

A syngenetic Kniest mineralization hypothesis, in connection with the massive ore formation as feeder-zone-type mineralization, is not acceptable. This opinion has been supported by the fact that the Kniest, when compared to the pre-tectonically formed orebodies, is totally restricted to the footwall. Ramdohr (1953) supported this interpretation. Also on analogous ore deposits it is the general opinion that this stockwork-type mineralization has to be compared with the feeder zone of the main ore precipitation as epigenetic synchronous fissure fillings. In the case of Rammelsberg, the writer cannot follow this interpretation because of the structural features present.

The hypothesis of a synkinematic ore mobilization is, consequently, the most plausible, though it is not fully convincing. The observed and known distances covered by the mobilized ore components are merely in the order of centimetres to decimetres, as already explained. It is evident that the unequivocally recognized remobilization of the ores was restricted to the massive ore and to their banded ore rims. A remobilization ore paragenesis probably formed by solution activity is present in cross joints, which are limited in the massive orebodies. The thickness of these fissure-fillings seldom exceeds several centimetres and they are often only millimetres wide. The fractures are occupied by generally coarse-grained, often idiomorphic galena, tetrahedrite, sphalerite, bournonite, chalcopyrite, calcite and barite, and contain the only well-developed crystals of these minerals in the Rammelsberg mine. Rare finds of several millimetre-thick smithsonite crusts, developed on other crystal surfaces in the 11th mine level, are considered merely to be a curiosity.

Numerous leaching rims (= dissolution fronts), extending several centimetres from the fissure faces into the massive ore, support the proposal that local mobilization of the massive ore has occurred. It should also be stressed here that, generally, the mineral composition of the fissure paragenesis depends on the adjacent massive ore, as mentioned above. Similar observations were made by Strauss (1970) in the southwest Iberian pyrite province, Spain. The relatively low trace-element contents of the fissure ores, and the absence of mineral intergrowths, are very characteristic. These features, together with the fluid-inclusion properties (cf. general treatment by Roedder, 1976) of sphalerite and gangue minerals led Ramdohr (in Kraume, 1955) to assume a maximum temperature of formation of 120°C. To complete our observations, it should be mentioned that in contrast to the Kniest mineralization, the ore produced through remobilization–reprecipitation are quantitatively insignificant.

The results are difficult to reconcile with the cited origin of the Kniest mineralization; even galenobismutite requires a higher temperature of formation (i.e., above 120°C). Moreover, the previous discussion has not included the epigenetic mineralizations in the overturned limb of the orebody syncline, which have been observed in the Lower Devonian wall-rock layers. These epigenetic mineralizations are confined to cleavage planes and to locally developed horizontal to low-dipping overthrusts. For example, one noteworthy mineralization is bound to the overthrust zone, which was discovered by a cross-cut at the 12th level in Ord. 1320 in the hanging-wall. This mineralization is of really epigenetic origin and it belongs symmetrically, as well as according to its age of formation, to the Kniest-type mineralization.

Quite frequently, researchers (including the present writer at one stage) were of the opinion that the Kniest mineralization coincided in time of formation with the Pb–Zn vein deposits of the Upper Harz area; but the special situation of the Kniest deposit was not considered enough. Unequivocally, the Kniest mineralization is older than the Upper Harz Pb–Zn veins, which only overlap in geologic time during the final respective the initial phases. Also overlooked was the fact that a symmetrically similar mineralization beyond the Rammelsberg district has not been found. On the contrary, there is generally a clear spatial coincidence of the post-orogenic mineralization with the syngenetic conduit-ore occurrences. Therefore, this locality is characterized by a distinct heterotectonic type of mineralization. Based on these considerations, one has to consider the possibility that the Kniest mineralization was the result of orogenic remobilization of the Middle Devonian channelway mineralization, probably of a deeper tectonically formed stockwork.

SUMMARY AND CONCLUSIONS

The Rammelsberg Pb–Zn–Fe–Cu sulphide–barite ore deposit [1] (i.e., the New and Old Orebodies) was formed along the sea floor during the early Middle Devonian in a particular sedimentary basin of the Variscan eugosyncline into which the magmatic-hydro-

[1] For comparisons of the Rammelsberg mineralization with those of the Sullivan, Balmat, Mt. Isa, Meggen, McArthur, and other districts, see in this Handbook: Chapter 4, Volume 1, pp.147, 149, 154; Chapter 6, Vol. 2, p.158; Chapter 6, Vol. 3 pp.277–278; Chapter 8, Vol. 3, pp.347–348; Chapter 6, Vol. 4, p.314; Chapter 1, Vol. 5, p.15; Chapter 4, Vol. 6, p.113; Chapter 5, Vol. 6, p.141; Chapter 12, Vol. 6, pp.578–582 (table II); Chapter 6, Vol. 7, pp.284–285.

thermal (exhalative) ore-forming solutions were discharged. The precipitation and accumulation of the ore took place in a chemical reducing low-energy environment, as suggested by the thermal sulphides and the organic carbon-rich black-shale facies (= "Wissenbacher Schiefer") of the host rocks [2]. The main sulphide precipitation is associated by a quartz pre-phase (Kniest formation), a sulphide "fore-runner" mineralization which had already begun to form during the Early Devonian. The final stage of the synsedimentary ore formation is represented by barite partly in separated occurrences. This indicates, at last, a change to chemically oxidized conditions. The time of the known sedimentary ore formation process of the Rammelsberg from the first Lower Devonian ore seam to the main orebodies could be in the order of 10^7 years, according to the thickness of sediments and average depositional conditions.

Since the end of the Early Devonian, the area of mineralization was situated along the slope of the Westharz ridge and the Goslar trough. In the proximity of the ridge, the sedimentary sequence is stratigraphically condensed reflecting stratigraphic gaps, being the result of reduced geosynclinal subsidence. In the basin itself, the fine-grained clastic sediments are of great thickness. As the local depression, in which the Rammelsberg mineralization occurred, coincides with the tectonic shear or hinge zone between the uplifted West Harz ridge and the subsided Goslar trough, the ore deposit is clearly situated along a geotectonically active zone.

The geosynclinal tectonism and the formation of the orebodies were associated with the origin of acidic to intermediate ash tuffs (pelitic pyroclastics) as a result of explosive, probably dacitic to rhyolitic, volcanism since the Early Devonian. Judging by the number and thickness of the tuff beds, an increase in the intensity of volcanism occurred up to the period of formation of the main orebodies. The volcanic centres, however, are unknown, but were probably situated in the north of the Harz Mountains, a region subsequently covered by Mesozoic platform sediments. Towards the end of the Middle Devonian, a predominantly effusive basic (spilitic) volcanism replaced the earlier, more acid tuff-forming explosive volcanism. This was accompanied by a decrease in the epeirogenic crustal movements. Coincident with the development of the West Harz ridge at the southeastern flank, the younger diabasic (basaltic) effusive volcanics in the Upper Harz Diabase Range are associated by sedimentary red iron ore (hematite) deposits.

The West Harz ridge must be regarded as a region of intense magmatic intrusive and probable anatectic activity, during which differentiation plus crustal assimilation processes formed keratophyric (dacitic–rhyolitic) melts. These magmas were probably the source of the ash tuffs as well as of the hydrothermal ore-bearing solutions. The metal-source problem related to heat cells, magma conditions, etc., is discussed in more detail by Krebs in this volume. A lateral-secretionary ore mobilization origin (for a

[2] See in this Handbook Vol. 1, Chapter 4 by Gilmour (table I and fig.8), who has conceptualized the host rocks of numerous ore deposits by using end-member lithologies, thus providing a comparative basis of the Rammelsberg (and Meggen) ores.

general summary of this process in the context of other mechanisms, see Wolf, 1976) by solutions, for instance from the geosynclinal sediments, cannot be accepted. Geochemical balance calculations to account for the source of the metals from the sedimentary pile did not furnish plausible data in the present author's opinion. It is also impossible that sea water could have been the source of the components required for the ore deposits, according to general element calculations of the different systems involved. On the other hand, it should be mentioned that, based on the very high metal content of the wall rocks, Scherp (1974) was able to account for the metals of the Meggen ore deposit (cf. chapter Krebs in this volume) by suggesting the sedimentary country rock as a possible source. For example, he calculated that all of the barium could have been derived by leaching from the footwall sediments.

A sedimentary metal sulphide formation is already known from the upper Lower Devonian stratigraphic section, which occurs as an ore seam up to 10 cm thick in a littoral clastic facies composed of quartzitic sandbank sequence with minor shale intercalation of several centimetres thickness in average. The high oxidation potential of the sedimentary environment in general present was only for short times and locally interrupted by an H_2S-rich reducing environment, which allowed for the formation of the sulphide ores. In general, the geochemical conditions, except for the rare exceptions described, were not conducive to the accumulation of sulphides, which is one factor explaining the relatively limited thickness of the ore seam.

Uneconomic Pb–Zn–Cu mineral concentrations of the replacement and impregnation–dissemination types are present from the older sedimentary rocks and extend over the whole stratigraphic sequence up to the Rammelsberg main orebodies. Including the known ore seam occurrences, the shape of the mineralized rock mass constitutes a vertical pre-deformational cylinder overlain by the Rammelsberg main orebodies (Fig.17). This cylinder represents the channel ways of the ascending hydrothermal fluids. The secondary calcium depletion and the dissolution of clastic tourmaline (Paul, 1975), removed by the hypogene ore-forming solutions, indicate that the fluids were acidic with a pH of about 3. Moreover, Paul proved the presence of secondarily introduced barite and a significant increase of manganese in the conduit wall rocks. Numerous blackshale facies and diffusely disseminated pyrite in the channelway wall rock repeatedly attest that they originated in an H_2S-rich, reducing sedimentary environment probably induced by hypogene sulphide-bearing solutions. It was the spatial coincidence of the channel way with this reducing environment that was the prerequisite for the accumulation of the Rammelsberg ore.

The formation of the main orebodies was preceded by a silica-rich prephase (cf. table III, in Gilmour, 1976) which resulted in impregnative silicifications under diagenetic conditions of parts of the Ore and Shale Horizon as well as the Sandbanded Shales; perhaps also forming sedimentary cherts within a very localized area, mainly restricted to the conduit zone.

The formation of the main orebodies commenced with a short difference in time at adjacent localities. At the top of the twin-tuff layer, the main ore mineralization was introduced by the sedimentary accumulation of the banded ores, which are probably located close to the submarine feeder zones. This banded ore mineralization constitutes a manifold and complexly intercalated sequence of ore and shale beds from a few millimetres up to a few decimetres thick. The ore bands are chiefly composed of pyrite, sphalerite, galena, minor chalcopyrite, in association with pelitic to silty clastic material of various amounts. The locally developed arsenic-rich Hanging Wall Ore Occurrence in the primary footwall region, respectively in the overturned limb of the orebody syncline, has to be correlated with the initial ore precipitations of the main ore phase. Therefore, arsenic is part of the prephase. Later, during the main ore phase it was followed by antimony.

The banded ore is believed to have originated by flocculation, followed by the sedimentary accumulation of crystallized particles, as well as by a more direct precipitation of colloidal and crystalline components on the sea floor, depending on the prevailing physico-chemical conditions. In places reworking and slumping resulted in resedimentation and the formation of breccias and slump structures, all of which subsequently experienced an intense tectonic overprint.

Accretionary processes, e.g., coagulation, of the precipitated mixed metal sulphides and barite formed spherulite-like concretions and nodules up to 5 mm in diameter, which constitute parts of the thin ore seam of the Lower Devonian, in the western rim of the Old Orebody and the Footwall Grey Orebody. Benthonic and nektonic fossils (goniatites, brachiopods and tentaculites) are quite numerous, as are pyritized embryonic (= dwarfed) forms in the banded ores. They were drifted into the chemically hostile marine environment, died, settled and became syndiagenetically pyritized. The origin of the framboidal pyrite was probably due in part to the activity of micro-organisms, as proposed for the Mt. Isa occurrences, for example (cf. various descriptions and discussions in Vols. 1, 2, 3, 6 and 7 of this Handbook). However, newly published data indicate that framboids can also originate by inorganic processes.

The subsequent two periods of massive ore formation, only slightly different in age, took place under conditions that reduced the influx of clastics in the vicinity of the hydrothermal feeder zones and were confined to the margins of the basin, where a consequent interfingering of the chemically precipitated ore with the clastic sediments occurred.

It should be pointed out, that the stratigraphic ore-mineral sequences of the main orebodies correspond to world-wide known normal hydrothermal epigenetic Pb–Zn vein parageneses, whereas the trace elements as well as the accessory minerals of the Rammelsberg exhibits their own individual distribution characteristics. Therefore, a genetic interrelationship on the question of source of both is evident. An answer could be the investigation of Wedepohl et al. (1978) into the lead-isotopic distribution in such distinct deposits as Rammelsberg, Bad Grund, Ramsbeck, and Kupferschiefer. It resulted in the establishment, that the lead source of all the deposits investigated is sedimentary material containing Precambrian detritus.

Pyrite, chalcopyrite, carbonates, sphalerite, galena and barite succeed each other in the vertical stratigraphic sequence and also have a distinct lateral overlap. Relict fabrics and structures of these major minerals indicate that the massive ores formed, depending on local conditions, by colloid flocculation, sedimentary accumulation of already crystallized material, chemical precipitation of colloids and mineral growing directly on the sea floor along the water–sediment interface. This explains the local increase in gold, bismuth and other rare trace elements, which are known to be characteristic of gel phases, but are usually not typical of veins. The gel phase, with its poly-metal sulphide composition, probably prevailed during the ore mineral precipitation. In addition to the sulphides formed from the hypogene sulphur exhalated on the sea floor, sea-water sulphate-reducing conditions probably induced by bacteria most likely also produced sulphide ore. The distribution of the S-isotopes of the Pb-, Zn-, and Cu-sulphides supports mainly a magmatic-hydrothermal source, and Anger et al. (1966) regard the shift of the $^{32}S/^{34}S$ ratio as a normal response to magmatic differentiation beginning with a ^{32}S-rich and ending with a ^{34}S-rich sequence. In contrast to all the other sulphide minerals, the sulphur of pyrite consists of about half of bacterial origin; and the barite sulphur appears to have had a three-fold origin, namely from sea-water sulphate, oxidized sulphur from bacterial processes, and hydrothermal sulphur.

Prior to compaction due to increasing overburden pressure, the sulphide coagulates with their high-water contents were approximately twice as thick when compared to the surrounding sediments, as can be observed from concretions. It seems logical to assume that the first (= early diagenetic) mobilization of the chemical components took place through reactions with the sea water as well as connate fluids, even though more extensive diagenetic alterations cannot be proved. During the subsequent geological history, as the stratigraphy demonstrates, the Rammelsberg orebodies were covered by Middle Devonian to Upper Carboniferous sedimentary sequences probably up to more than 1000 m in thickness.

Folding and metamorphism of the ore deposits occurred during the Late Carboniferous Variscan orogeny, when the ore-containing sedimentary basin was orthotectonically compressed to form the orebody syncline. The very high plasticity of the sulphide ores, plus the extreme degree of tectonic compression during folding, resulted in a partial severance of the orebodies from their surrounding host rocks. The final result: an isoclinal overturned fold with the ore deposit and the Ore Horizon superimposed (Figs.5–8). Younger layers of the Ore Horizon were squeezed out of the orebody syncline. The tectonic deformation was a triaxial type, and the intense plastic deformation of the orebodies formed a fabric which is only comparable to a high-grade metamorphic synkinematic sling like structure. Ramdohr (1953b) stated that the present features of the Rammelsberg ore are that of a highly complex metamorphic slate.

The original minerals recrystallized to a very fine-grained synkinematic product; also new minerals originated during this stage. The paragenesis marcasite (= palaeosome), magnetite, pyrrhotite, sericite, mackinawite, and very rare cubanite, suggests a temperature increase up to 250°C. However, it is also possible that magnetite, pyrrhotite and sericite originated during earlier diagenesis, according to favourable chemical conditions. The mobilization of the minerals over distances varying from microns to metres occurred synkinematically by solution activity along schistosity planes, to be precipitated as "pressure shadows" and joint-fillings that are ubiquitous in both massive and banded ores. The mineral movements, in general, took place only over relatively short distances, but differential susceptibility to mobilization has been noticed. Galena and chalcopyrite, followed by sphalerite, have the greatest mobility among the ore minerals, whereas pyrite has the lowest. Among the gangue minerals, quartz and carbonates are the easiest to mobilize tectonically. Coarsely crystalline, newly formed bournonite and tetrahedrite were precipitated from these solutions. Although the mineralization formed by this process of remobilization-reprecipitation is volumetrically very small, there are some local fissure-fillings up to several centimetres in width.

Postkinematic ore mineralizations, with economic grades in places, occur in silicified lenses (Kniest; quartz prephase of the main ore formation) as well as in the lower section of the primary footwall region. The mineralizations of the silicified lenses (= Kniest) are regarded as:

(1) synkinematically mobilized or remobilized material derived from the Old and New Orebody;

(2) epigenetic mineralization precipitated in the cylindrical channel-way region by solutions, which also supplied the metal for the main orebodies (feeder-zone mineralization);

(3) ore mineralization produced in connection with the Upper Harz vein deposits.

None of these three interpretations is fully convincing. A fourth alternative, contradicting the other hypotheses, is plausible, namely that the metals could have been mobilized during the Variscan orogeny from a deeper subsurface region of the channel way, and subsequently reprecipitated in the vicinity of the main orebodies.

Tectonism and erosion of the whole Harz Mountain block has exposed parts of the Rammelsberg ores, but supergene processes had only a minor influence, and therefore a well-developed gossan is absent.

Finally, it should be pointed out that the descriptions and discussions provided above are by necessity brief, i.e., in the form of a summary, and the utilized references are only a selection of a much more extensive list.

ACKNOWLEDGEMENTS

The author thanks Dr. K.H. Wolf for fruitful discussions and translation of most of the manuscript as well as Mr. D. Large who critically read the manuscript and translated several pages.

REFERENCES

Abt, W., 1958. Ein Beitrag zur Kenntnis der Lagerstätte des Rammelsberges auf Grund von Spezialuntersuchungen der Tuffe und der Tektonik. *Z. Dtsch. Geol. Ges.*, 110 (1): 152–204.

Amstutz, G.C., 1959. Syngenese und Epigenese in Petrographie und Lagerstättenkunde. *Schweiz. Mineral. Petrogr. Mitt.*, 39: 1–84.

Amstutz, G.C., Zimmermann, R.A. and Schot, E.H., 1971. The Devonian Mineral Belt of Western Germany (the mines of Meggen, Ramsbeck and Rammelsberg). In: *Sedimentology of parts of Central Europe. Int. Sedimentol. Congr.*, Guidebook, 8: 253–272.

Anger, G., Nielsen, H., Puchelt, H. and Ricke, W., 1966. Sulfur isotopes in the Rammelsberg Ore Deposit (Germany). *Econ. Geol.*, 61: 511–536.

Babu, L., 1888. Uber den Rammelsberg am Unterharz. *Berg- Hüttenmänn. Z.*, 47: 208–210.

Bäcker, H., 1973. Rezente hydrothermal-sedimentäre Lagerstättenbildung. *Erzmetall*, 26: 544–555.

Bäcker, H., and Richter, H., 1973. Die rezente hydrothermal-sedimentäre Lagerstätte Atlantis-II-Tief im Roten Meer. *Geol. Rundsch.*, 62: 697–741.

Bäcker, H. and Schoell, M., 1974. Anreicherung von Elementen zu Rohstoffen im marinen Bereich. *Chem. Ztg.*, 98: 299–305.

Barnes, H.L., 1967. *Geochemistry of Hydrothermal Ore Deposits.* Holt, Rinehart and Winston, New York, N.Y., 670 pp.

Barnes, H.L., 1974. *Prozesse der hydrothermalen Lagerstättenbildung.* Gemeinsame Diskussionstagung des GDMB-Lagerstättenausschusses und der GDMB-Sektion Geochemie, Karlsruhe.

Baumann, L. and Tischendorf, G., 1976. *Einführung in die Metallogenie/Minerogenie* VEB Deutscher Verlag für Grundstoffindustrie, Leipzig, 458 pp.

Berg, G., 1926. Zur Frage nach der Entstehung des Erzlagers des Rammelsberges. *Z. Prakt. Geol.*, 34: 23–27.

Bergeat, A., 1902. Über merkwürdige Einschlüsse im Kieslager des Rammelsbergs bei Goslar. *Z. Prakt. Geol.*, 10: 289–293.

Bergeat, A., 1914a. Untersuchungen über die Struktur des Schwefelkies–Schwerspatlagers zu Meggen a.d. Lenne als Unterlage für dessen geologische Deutung. *Neues Jahrb. Mineral.*, 39: 1–63.

Bergeat, A., 1914b. Das Meggener Kies–Schwerspatlager als Ausscheidung auf dem Grunde des mitteldevonischen Meeres. *Z. Prakt. Geol.*, 22: 237–249.

Bogdanov, Y.V. and Kutyrev, E.I., 1972. Die geologischen Verbreitungsbedingungen von stratiformen Cu- und Pb–Zn-Lagerstätten in der Sowjetunion. *Z. Angew. Geol.*, 18: 10 pp.

Bogdanov, Y.V. and Kutyrev, E.I., 1973. Classification of stratified Cu and Pb–Zn deposits and the regularities of their distribution. In: G.C. Amstutz and A.J. Bernard (Editors), *Ores in Sediments.* Springer, Heidelberg, pp.59–63.

Böhmer, K.F.v., 1974. Geognostische Beobachtungen über den östlichen Communion-Unterharz, vorzüglich zur Beantwortung der Frage: "Zu welcher Art von besonderen Lagerstätten gehört die Erzmasse im Rammelsberg?". *Bergm. J.*, 6 (1): 193–237.

Borchert, H., 1934. Über Entmischungen im System Cu–Fe–S und ihre Bedeutung als geologisches Thermometer. *Chem. Erde*, 9: 145–172.

Borchert, H., 1960. Geosynklinale Lagerstätten, was dazu gehört und was nicht dazugehört, sowie deren Beziehungen zu Geotektonik und Magmatismus. *Freiberg. Forschungsh.*, C, 79: 7–61.

Borchert, H., 1961. Zusammenhänge zwischen Lagerstättenbildung, Magmatismus und Geotektonik. *Geol. Rundsch.*, 50: 131–165.

Borchert, H., 1972. Zur Bildung marin-sedimentärer Eisen- und Manganerze in Verknüpfung mit spilitischen und keratophyrisch–weilburgitischen Gesteinassoziationen. *Miner. Deposita*, 71: 18–24.

Bornhardt, W., 1939. Die Entstehung des Rammelsberger Erzvorkommens. *Arch. Lagerstättenforsch.*, 68: 61pp.

Bornhardt, W., 1948. Zur Entstehung des Rammelsberger Erzvorkommens. *N. Arch. Landes-Volkskde. Nieders.*, 5: 123–181.

Brathwaite, R.L., 1972. The structure of the Rosebery ore deposit, Tasmania. *Proc. Aust. Inst. Min. Metall.*, 3/1972 (241): 1–13.
Buschendorf, F. and Puchelt, H., 1965. Untersuchungen am Schwerspat des Meggener Lagers. *Geol. Jahrb.*, 82: 499–582.
Buschendorf, F., Nielsen, H., Puchelt, H. and Ricke, W., 1963. Schwefel-Isotopen-Untersuchungen am Pyrit–Sphalerit–Baryt-Lager Meggen/Lenne (Deutschland) und an verschiedenen Devon-Evaporiten. *Geochim. Cosmochim. Acta*, 27: 501–523.
Chilingar, G.V., Bissell, H.J. and Wolf, K.H. 1967. Diagenesis of carbonate rock. In: G. Larsen and G.V. Chilingar (Editors), *Diagenesis in Sediments.* Elsevier, Amsterdam, pp.179–322.
Cissarz, A., 1957. Lagerstätten des Geosynklinalvulkanismus in den Dinariden und ihre Bedeutung für die Geosynklinale Lagerstättenbildung. *Neues Jahrb. Mineral. Abh.*, 91: 485–540.
Cotta, B.v., 1864. Uber die Kieslagerstätte am Rammelsberg. *Berg-Hüttenmän, Z.*, 23: 369–373.
Dahlgrün, F., 1929. Zur Klassifikation der jungpaläozoischen Erzgänge des Harzes. *Jahrb. Hallesch. Verb.*, 8: 163–171.
Dechen, H.v., 1845. Das Vorkommen des Schwerspates als Gebirgsschicht bei Meggen a.d. Lenne *Karstens Dechens Archiv Mineral.*, 19: 748–753.
Degens, E.T. and Ross, D.A., 1976. Strata-bound metalliferous deposits found in or near active rifts. In: K.H. Wolf (Editor), *Handbook of Strata-Bound and Stratiform Ore Deposits*, 4. Elsevier, Amsterdam, pp.165–202.
Dunnet, D. and Moore, J.M., 1969. Inhomogeneous strain and the remobilization of ores and minerals. In: *Remobilization of Ores and Minerals – Symposium.* Instituto di Giacimenti Minerai, Univ. Cagliari, pp.81–100.
Ehrenberg, H., Pilger, A., and Schröder, F. Das Schwefelkies–Zinkblende–Schwerspatlager von Meggen (West falen). *Beih. Geol. Jahrb.*, 12: 352 pp.
Eichmeyer, H., 1960. The Rammelsberg Mine. *Mine Quarry Eng.*, pp.424–431.
Eichmeyer, H. and Aly, 1969. 1000 Years of base metal mining at Rammelsberg in Harz Mountains. *World Min.*, 22 (4): 69–73.
Finlow-Bates, T. and Large, D.E., 1978. Water depth as a major control on the formation of submarine exhalative ore deposits. *Geol. Jahrb., Reihe D*, 30: 27–39.
Fleischer, M., 1955. Minor elements in some sulfide minerals. *Econ. Geol.*, 50 (2): 970–1024.
Folk, R.L., 1965. Some aspects of recrystallization in ancient limestones. In: *Dolomitization and Limestone Diagenesis. Soc. Econ. Paleontol. Mineral. Spec. Publ.*, 13: 14–48.
Förster, H. and Leonhardt, J., 1972. Mathematische Simulation ptygmatischer Strukturen. *Geol. Rundsch.*, 61 (3): 883–896.
Frebold, G., 1924. Probleme des Rammelsberges. *Z. Kristallogr.*, 59: 436–437.
Frebold, G., 1925. Über die Genesis kiesiger Erzlagerstätten vom Typus Meggen-Rammelsberg. *Jahresber. Nieders. Geol. Ver.*, 18: 23–58.
Frebold, G., 1927. Über die Bildung der Alaunschiefer und die Entstehung der Kieslagerstätten Meggen und Rammelsberg. *Abh. Prakt. Geol. Bergwirtsch.-Lehre*, 13: pp. VI and 119.
Frebold, G., 1928a. Wandlungen in den Anschauungen über die Entstehung des Rammelsberger Erzlagers. *Z. Dtsch. Geol. Ges.*, 79 (1927) B: 210–216.
Frebold, G., 1928b. Über verlegte Relikttexturen im Rammelsberger Melierterz. (Beiträge zur Kenntnis der Erzlagerstätten des Harzes, V.) *Zentralbl. Mineral., A:* 260–261.
Früh, W., 1960. Becken und Schwellen in Westharz-Abschnitt der Mittel- und Oberdevonmeeres. *Geol. Jahrb.*, 77: 205–240.
Fruth, I., 1966. Spurengehalte der Zinkblenden verschiedener Pb–Zn-Vorkommen in den Nördlichen Kalkalpen. *Chem. Erde*, 25: 105–125.
Fruth, I. and Maucher, A., 1966. Spurenelemente und Schwefelisotope in Zinkblenden der Blei–Zink-Lagerstätte von Gornò. *Miner. Deposita*, 1: 238–250.
Fuchs, W. and Ommer, Ph., 1970. Beispiele gebundener Tektonik im Neuen Lager der Schwefelkies–Zinkblende–Schwerspatlagerstätte von Meggen. *Berbauwissenschaften*, 17 (12).
Gasser, U., 1974. Zur Struktur der stratiformen Sulfidlagerstätte Meggen. *Geol. Rundsch.*, 63 (1): 52–73.

Gilmour, P., 1976. Some transitional types of mineral deposits in volcanic and sedimentary rocks. In: K.H. Wolf (Editor), *Handbook of Strata-Bound and Stratiform Ore Deposits, 1.* Elsevier, Amsterdam, pp.111–160.

Gräbe, R., 1972. Analyse der metallogenetischen Faktoren stratiformer sulfidischer Geosynklinallagerstätten. *Z. Angew. Geol.,* 18: 289–300.

Groddeck, A. v., 1879. *Die Lehre von den Lagerstätten der Erze. Ein Zweig der Geologie.* Veit, Leipzig, 351 pp.

Gundlach, H. and Hannak, W., 1968. Ein synsedimentäres submarin-exhalatives Buntmetallerz-Vorkommen im Unterdevon bei Goslar. *Geol. Jahrb.,* 85: 193–225.

Gunzert, G., 1969. Altes und Neues Lager am Rammelsberg bei Goslar. *Erzmetall,* 22 (1): 1–10.

Gunzert, G., 1979. Die Grauerzvorkommen und der tektonische Bau der Erzlagerstätte am Rammelsberg bei Goslar. *Erzmetall,* 32 (1): 1–7.

Gwosdz, W., Krüger, H., Paul, D. and Baumann, A., 1974. Die Liegendschichten der devonischen Pyrit-u. Schwerspatlager von Eisen (Saarland), Meggen und des Rammelsberges. *Geol. Rundsch.,* 63 (1): 74–93.

Hannak, W., 1956. Bericht zu den Strukturuntersuchungen im Erzbergwerk Rammelsberg. *Manuskript Unterharzer Berg-Hüttenwerke,* 48 pp.

Hannak, W., 1963. Zur tektonischen Stellung der Erzlager des Rammelsberges im Nebengestein. *Roemeriana,* 7: 91–108.

Hannak, W., 1972. Die Blei–Silber–Zink-lagerstätte von Bawdwin (Burma), ein Erzlager? *Nachr. Dtsch. Geol. Ges.,* 6: 93–94.

Hannak, W. and Kraume, E., 1966. Die Rammelsberger Zink–Kupfer-Erzlagerstätte. Fortschr. Mineral., 43: 104–107.

Hannak, W. and Puchelt, H., 1965. *Untersuchungen an den Baryten der Rammelsberger Lager.* Vortrag 43, Jahrestag der Deutschen Mineralogischen Gesellschaft. Hannover.

Hegemann, F., 1939. Die geochemischen und kristallchemischen Beziehungen von Mangan zur Pyrit. *Metallwirtschaft,* 18: p.705.

Hegemann, F., 1939. Die geochemischen und kristallchemischen Beziehungen von Mangan zu Pyrit. Entstehung der Kieslagerstätten. *Z. Angew. Mineral.,* 4: p.121.

Hegemann, F., 1948a. Über sedimentäre Lagerstätten mit submariner vulkanischer Stoffzufuhr. *Fortschr. Mineral.,* 27: p.54.

Hegemann, F., 1948b. Geochemische Untersuchungen über die Herkunft des Stoffbestandes sedimentärer Kieslager.- *Fortschr. Mineral.,* 27: p.45.

Hertel, L., 1966. Die Fremdelementführung der Bleiglanze als Hilfe zur Bestimmung der Bildungstemperatur. *Z. Erzbergbau. Metallhüttenwes.,* 19: 632–635.

Hilmer, E., 1972. *Geochemische Untersuchungen im Bereich der Lagerstätte Meggen, Rheinisches Schiefergebirge.* Thesis, Technische Hochschule Aachen, 162 pp.

Hornung, F., 1905. Ursprung und Alter des Schwerspates und der Erze im Harze. *Z. Dtsch. Geol. Ges.,* 57: 291–360.

Horst, M., 1967. Ergebnisse von geomagnetischen Messungen im Nordwestharz und seinem Harzvorland sowie Untersuchungen über die Konstruktion und Verwendbarkeit eines Kreiseldeklinatoriums. *Bergbauwissenschaften,* 14 (2): 66–80. Goslar.

Jahns, H., 1955. *Stratigraphische und tektonische Untersuchungen im Westfeld der Erzgrube Rammelsberg unter besonderer Berücksichtigung der 7. Sohle mit Grubenaufnahme.* Thesis, Technische Universität Clausthal.

Kalliokoski, J. and Cathles, L., 1969. Morphology, mode of formation and diagenetic changes in framboids. *Bull. Geol. Soc. Finland,* 41: 125–133.

Karl, F., 1964. Anwendung der Gefügekunde in der Petrotektonik. *Tektonische Hefte,* No. 5, 142 pp.

Kleinevoss, A., 1971. *Zur geochemischen Charakteristik des Quecksilbers unter besonderer Berücksichtigung der Hg-Verteilung in den Erzlagern des Rammelsberges und ihrer Umgebung.* Thesis, Technische Universität Clausthal, 190 pp.

Klockmann, F., 1904. Über den Einfluss der Metamorphose auf die mineralische Zusammensetzung der Kieslagerstätten. *Z. Prakt. Geol.,* 12: 153–160.

Klockmann, F., 1893. Die Erzlager des Rammelsberges. *Z. Prakt. Geol.*, 1: 475–476.
Koark, H.J., 1973. Zur Entstehung des tektonischen Stengelbaus an präkambrischen Eisen- und Sulfiderzkörpern der zentralschwedischen Leptitserie. *Mineral. Deposita*, 8 (1): 19–34.
Koark, H.J., 1974. Gesichtspunkte zu Hypothesen über metamorphogene Bildung von Sulfidgrosslagerstätten in den Svekofenniden Schwedens. *Geol. Rundsch.*, 63 (1): 165–180.
Köhler, G., 1882. Die Störungen im Rammelsberger Erzlager bei Goslar. *Z. Berg-, Hütten-Salinenwes.*, 30: 31–43.
Kraume, E., 1955. Die Erzlager des Rammelsberges bei Goslar. *Beih. Geol. Jahrb.*, 18: 394 pp.
Kraume, E., 1960a. Erzvorkommen in den tektonisch hangenden Schichten der Rammelsberger Erzlager bei Goslar. *Neues Jahrb. Mineral., Abh.*, 94: 479–494.
Kraume, E., 1960b. Stratigraphie und Tektonik der Rammelsberger Erzlager unter besonderer Berücksichtigung des Neuen Lagers unter der 10. Sohle. *Erzmetall*, 13 (1): 7–12.
Kraume, E. and Jasmund, K., 1951. Die Tufflagen des Rammelsberges bei Goslar. *Heidelb. Beitr. Mineral. Petrogr.*, 2: 443–454.
Krause, H.F. and Pilger, A., 1969. Möglichkeiten der Rejuvenation von Pb–Zn-Erzlagerstätten im Saxonikum. In: *Remobilization of Ores and Minerals–Symposium.* Instituto die Giacimenti Minerai, Univ. Cagliari, pp.101–127.
Krebs, W., 1972a. Die paläogeographisch-faziellen Aussagen zur Position des Meggener Lagers. In: *Geophysikalische Prospektionsmethoden, Heft 24.* Schriften der GDMB, Clausthal-Zellerfeld.
Krebs, W., 1972b. Facies and development of the Meggen Reef (Devonian, West Germany). *Geol. Rundsch.*, 61 (2): 647–671.
Krebs, W., 1973. *Die magmatischen und paläogeographischen Voraussetzungen für die Bildung der devonischen Sulfid–Baryt-Lagerstätten Rammelsberg und Meggen.* Vortrag, Geologische Vereinigung, 63. Jahrestag, Salzburg.
Krebs, W., 1981. The geology of the Meggen ore deposit. In: K.H. Wolf (Editor) (This volume).
Krebs, W. and Wachendorf, H., 1973. Proterozoic–paläozoic geosynclinal and orogenic evolution of Central Europe. *Geol. Soc. Am. Bull.*, 84: 2611–2630.
Krebs, W. and Wachendorf, H., 1974. Faltungskerne im mitteleuropäischen Grundgebirge. Abbilder eines orogen. Diapirismus. *Neues Jahrb. Geol. Paläontol. Abh.*, 147 (1): 30–60.
Krüger, H., 1973. *Der sedimentologische und paläogeographische Rahmen der Liegendschichten des Schwefelkies–Zinkblende–Schwerspat-Lagers von Meggen.* Thesis, Technische Universität, Braunschweig, 66 pp.
Kullerud, G., 1953. The FeS–ZnS-system. A geological thermometer. *Norsk Geol. Tidskr.*, 32: 61–147.
Lambert, J.B., 1976. The McArthur zinc-lead-silver deposits: features, metallogenesis and comparisons with some other stratiform ores. In: K.H. Wolf (Editor), *Handbook of Strata-bound and Stratiform Ore Deposits, 6.* Elsevier, Amsterdam, pp.535–585.
Lelong, F., Tardy, Y., Grandin, G., Trescases, J.J. and Boulange, B., 1976. Pedogenesis, chemical weathering and processes of formation of some supergene ore deposits. In: K.H. Wolf (Editor), *Handbook of Strata-Bound and Stratiform Ore Deposits, 3.* Elsevier, Amsterdam, pp.93–173.
Lindgreen, W., 1933. *Mineral Deposits.* Wiley, New York, 4th ed., 930 pp.
Lindgreen, W. and Irving, J.D., 1911. The origin of the Rammelsberg ore deposits. *Econ. Geol.*, 6: 303–313.
Love, L.G., 1965. Micro-organic material with diagenetic pyrite from the lower Proterozoic Mount Isa shale and a Carboniferous shale. *Proc. Yorks. Geol. Soc.*, 35 (part 2, no. 9): 187–202.
Love, L.G. and Amstutz, G.C., 1966. Review of microscopic pyrite from Devonian Chattanooga Shale and Rammelsberg Banderz.- *Fortschr. Mineral.*, 43: 273–309.
Lovering, T.S., 1963. Epigenetic, diplogenetic, syngenetic and lithogenetic deposits. *Econ. Geol.*, 58: 315–331.
Lusk, J. and Crockett, J.H., 1969. S-isotope fractionation in coexisting sulfides from the Heath Steele B-1 orebody, New Brunswick, Canada. *Econ. Geol.*, 64: 147–155.
Maucher, A., Schultze-Westrum, H.-H. and Zankl, H., 1962. Geologisch-lagerstättenkundliche Untersuchungen im ostpontischen Gebirge. *Bayer. Akad. Wiss., Math.-Naturwiss. Kl. Abh., N.F.*, 109.

McMahon Moore, J., 1971. Fold styles in the ore bodies of Meggen and Rammelsberg. *Trans. Inst. Min. Metall. Ser. B.*, 80: 108–115.

Paul, D.J., 1975. Sedimentologische und geologische Untersuchungen zur Rekonstruktion des Ablagerungsraumes vor und nach der Bildung der Rammelsberger Blei–Zink-Lager im Oberharz. *Geol. Jahrb.*, D, 12: 3–93.

Quade, H., 1976. Genetic problems and enviromental features of volcano-sedimentary iron-ore deposits of the Lahm-Dill type. In: K.H. Wolf (Editor), *Handbook of Strata-Bound and Stratiform Ore Deposits, 7.* Elsevier, Amsterdam, pp. 255–294.

Ramdohr, P., 1928. Über den Mineralbestand und die Strukturen der Erze des Rammelsberges. *Neues Jahrb. Mineral., A,* 57 (2) Festschrift Mügge, pp.1013–1068.

Ramdohr, P., 1953a. Über Metamorphose und sekundäre Mobilisierung. *Geol. Rundsch.*, 42: 11–19.

Ramdohr, P., 1953b. Mineralbestand, Strukturen und Genesis der Rammelsberg-Lagerstätte. *Geol. Jahrb.*, 67: 367–494.

Ramdohr, P., 1961. One thousand years of Mining at the Rammelsberg, Harz Mountains, Germany. *Proc. Geol. Assoc. Can.*, 13: 13–21.

Roedder, E., 1976. Fluid inclusion evidence on the genesis of ores in sedimentary and volcanic rocks. In: K.H. Wolf (Editor), *Handbook of Strata-Bound and Stratiform Ore Deposits, 2.* Elsevier, Amsterdam, pp.67–110.

Saksela, M., 1957. Die Entstehung der Outokumpu-Erze im Licht der tektonisch-metamorphen Stoffmobilisierung. *Neues Jahrb. Mineral., Abh.*, 91: 278–302.

Sangster, D.T., 1976. Sulphur and lead isotopes in Strata-bound deposits. In: K.H. Wolf (Editor), *Handbook of Strata-Bound and Stratiform Ore Deposits, 2.* Elsevier, Amsterdam, pp.219–266.

Saxby, J.D., 1976. The significance of organic matter in ore genesis. In: K.H. Wolf (Editor), *Handbook of Strata-Bound and Stratiform Ore Deposits, 2.* Elsevier, Amsterdam, pp.111–133.

Scherp, A., 1958. Die Dichterze der Blei-Zink-Erzlagerstätte Ramsbeck im Sauerland. *Erzmetall,* 11: 600–607.

Scherp, A., 1974. Die Herkunft des Baryts in der Pyrit–Zinkblende-Baryt-Lagerstätte Meggen. *Neues Jahrb. Geol. Paläontol. Monatsh.*, 1974, 1: 38–53.

Scherp, A. and Strübel, G., 1974. Zur Barium–Strontium-Mineralisation. *Mineral. Deposita,* 9: 155–158.

Schmidt, W.E., 1918. Über die Entstehung und über die Tektonik des Schwefelkies- und Schwerspatlagers von Meggen a.d. Lenne nach neueren Aufschlüssen. *Jahrb. Preuss. Geol. Landesanst.*, 39: 23–72.

Schmidt, W.E., 1932. Die Stratigraphie des Unterdevons und des Mitteldevons in der Umgebung des Rammelsberger Lagers. *Sitzungsber. Preuss. Geol. Landesanst.*, 7: 39–46.

Schmidt, W.E., 1933a. Tektonik und Genesis des Rammelsberger Erzlagers. *Metall. Erz,* 30: 343–344.

Schmidt, W.E., 1933b. Das Rammelsberger Lager, sein Nebengestein, seine Tektonik und seine Genesis. *Z. Berg.-, Hütten-Salinwes.*, 81: 247–270.

Schneider, H.-J., 1964. Facies differentiation and controlling factors for the depositional lead–zinc concentration in the ladinian geosyncline of the eastern Alps. In: G.C. Amstutz (Editor), *Developments in Sedimentology, 2. Sedimentology and Ore Genesis.* Elsevier, Amsterdam, pp.29–45.

Schot, E.H., 1973. *Erzmikroskopische, petrographische und geochemische Beobachtungen an der Erzlagerstätte Rammelsberg bei Goslar/Harz.* Thesis, Universität Heidelberg.

Schot, E.H. and Ottemann, J., 1969. Elektrum und Kobellit im Meliert-Erz vom Rammelsberg. *Neues Jahrb. Mineral Abh.*, 112: 101–115.

Schott, W., 1976. Mineral (inorganic) resources of the oceans and ocean floors. In: K.H. Wolf (Editor), *Handbook of Strata-Bound and Stratiform Ore Deposits, 3.* Elsevier, Amsterdam, pp. 245–294.

Schouten, C., 1937. *Metasomatische Probleme. Mount Isa, Rammelsberg, Meggen, Mansfeld und künstliche Verdrängung.* Scheltema and Holkema, Amsterdam, 147 pp.

Schulz, 1911. Beiträge zur Kenntnis der Kieslagerstätte des Rammelsberges. *Manuskript, Archiv Preuss. Geol. Landesanst.* Nr. 2099.

Schumann, H., 1952. Ein besonderer Fall von Gefügergelung. *Fortschr. Mineral.*, 29/30: p.79.

Schuster, G., 1867. Über die Kieslagerstätte am Rammelsberg bei Goslar. *Berg.- Hüttenmänn. Z.*, 26: 307–308.

Shadlum, T.N., 1971. Metamorphic textures and structures of sulphide ores. *Soc. Min. Geol. Jpn., Spec. Issue*, 3: 241–250.

Shadlum, T.N., 1973. On the origin of "kies"-ore and Pb–Zn deposits in sediments. In: G.C. Amstutz and A.J. Bernard (Editors), *Ores in Sediments*. Springer, Heidelberg, pp.267–273.

Siemes, H., 1961. *Betrachtungen zur Verformung und zum Rekristallisationsverhalten von Bleiglanz*. Thesis, T.H. Aachen.

Siemes, H., 1967. *Experimentelle Stauchverformung von polykristallinen Bleiglanzen. Bestimmung der Festigkeitseigenschaften unter allseitigem Druck zwischen 1 und 5000 bar sowie röntgenographische Untersuchung der auftretenden Texturen*. Hab.-Schrift T.H. Aachen.

Siemes, H., 1970. Experimental deformation of galena ores. In: P. Paulitsch (Editor), *Experimental and Natural Rock Deformation*. Springer, New York, N.Y., pp. 165–208.

Söhle, H., 1899. Beitrag zur Kenntnis der Erzlagerstätte des Rammelsberges bei Goslar. *Österr. Z. Berg-Hüttenwes.*, 47: 563–568.

Sommerlatte, H., 1959. Die Blei–Zink-Erzlagerstätte von Bawdwin in Nord-Burma. *Z. Dtsch. Geol. Ges.*, 110 (3): 491–504.

Stanton, R.L., 1972. *Ore Petrology*. McGraw-Hill, New York, N.Y., 713 pp.

Stelzner, A.W. and Bergeat, A., 1904. *Die Erzlagerstätten*. Felix, Leipzig, 470 pp. (1st part).

Strauss, G.K., 1970. Sobre la Geologia de la provincia piritifera del Suroeste de la Peninsula Iberica y de sus yacimientos, en especial sobre la mina de pirita de Lousal (Portugal). *Mem. Inst. Geol. Min. España*.

Strauss, G.K. and Madel, J., 1974. Geology of massive sulphide deposits in the Spanish–Portuguese Pyrite Belt. *Geol. Rundsch.*, 63 (1): 191–211.

Taylor, G.H., 1971. Carbonaceous Matter: A Guide to the Genesis and History of Ores. *Soc. Min. Geol. Jpn., Spec. Issue* 3: 283–288.

Trebra, F.W.H. v., 1785. *Erfahrungen vom Innern der Gebirge*. Dessau; Leipzig, 244 pp.

Trudinger, P.A., 1976. Microbiological processes in relation to ore genesis. In: *K.H. Wolf (Editor), Handbook of Strata-Bound and Stratiform Ore Deposits, 2*. Elsevier, Amsterdam, pp. 135–190.

Udubasa, G., 1972. *Syngenese und Epigenese in metamorphen und nicht metamorphen Blei–Zink-Erzlagerstätten aufgezeigt an den Beispielen Blazna-Tal (Ost-Karpaten, Rumänien) und Ramsbeck (Westfalen, Bundesrepublik Deutschland)*. Thesis, Univ., Heidelberg.

Vaughan, D.J., 1976. Sedimentary geochemistry and mineralogy of the sulfides of lead, zinc, copper and iron and their occurrence in sedimentary ore deposits. In: K.H. Wolf (Editor), *Handbook of Strata-Bound and Stratiform Ore Deposits, 2*. Elsevier, Amsterdam, 317–363.

Violo, M., 1969. Experimental studies on galena and sphalerite deposition and on galena remobilization. In: *Remobilization of Ores and Minerals – Symposium*, Instituto di Giacimenti Minerai, Univ. Cagliari, pp. 59–80.

Vogt, J.H.L., 1894. Über die Kieslagerstätten vom Typus Röros. Vigsnäs, Sulitelma in Norwegen und Rammelsberg in Deutschland. *Z. Prakt. Geol.*, 2: 41–50, 117–134, 173–181.

Wedepohl, K.H., Delevaux, M.H. and Doe, B.R., 1978. The potential source of lead in the Permian Kupferschiefer bed of Europe and some selected Palaeozoic mineral deposits in the Federal Republic of Germany. *Contrib. Mineral. Petrol.*, 65: 273–281.

White, D.E., 1968. Environments of generation of some base-metal ore deposits. *Econ. Geol.*, 63: 301–335.

White, D.E., Muffler, L.J.P. and Truesdell, A.H., 1971. Vapor-dominated hydrothermal systems compared with hot water systems. *Econ. Geol.*, 66: 75–97.

Whitten, E.H.T., 1966. *Structural Geology of Folded Rocks*. Rand McNally, Chicago, 678 pp.

Wiechelt, W., 1904. Die Beziehungen des Rammelsberger Erzlagers zu seinen Nebengesteinen. *Berg-Hüttenm. Z.*, 63: 285–288, 297–301, 313–316, 329–333, 341–345, 357–361.

Wimmer, F., 1877. Vorkommen und Gewinnung der Rammelsberger Erze. *Z. Berg-, Hütten-Salinenwes.*, 25: 119–131.

Wolf, K.H., 1976. Ore genesis influenced by compaction. In: G.V. Chilingar and K.H. Wolf (Editors), *Compaction of Coarse-Grained Sediments, 6*. Elsevier, Amsterdam, pp.475–676.

Wolf, K.H. and Chilingar, G.V., 1974. Diagenesis of sandstones and compaction. In: G.V. Chilingar and K.H. Wolf (Editors), *Compaction of Coarse-Grained Sediments, 2*. Elsevier, Amsterdam, pp.69–444.

Wolff, L., 1913. Die Erzlagerstätte des Rammelsberges.- *Z. Berg-, Hütten-Salinenwes.*, 61: 457–513.

Chapter 11

GEOLOGY OF THE PINE POINT LEAD–ZINC DISTRICT

J. RICHARD KYLE

INTRODUCTION

The Pine Point mining district is located on the south shore of the Great Slave Lake in the District of Mackenzie, Northwest Territories, Canada, about 800 km north of Edmonton, Alberta, and 180 km south of Yellowknife, N.W.T. (Fig. 1). The district consists of about 40 lead–zinc orebodies in a Middle Devonian carbonate barrier complex. The area has had a long and colorful history since the lead–zinc occurrences were examined by prospectors en route to the gold fields of the Yukon and Alaska in 1898. The low precious-metal content, remote location, and extensive alluvial and vegetative cover hampered exploration and development of the deposits. The exploration program beginning

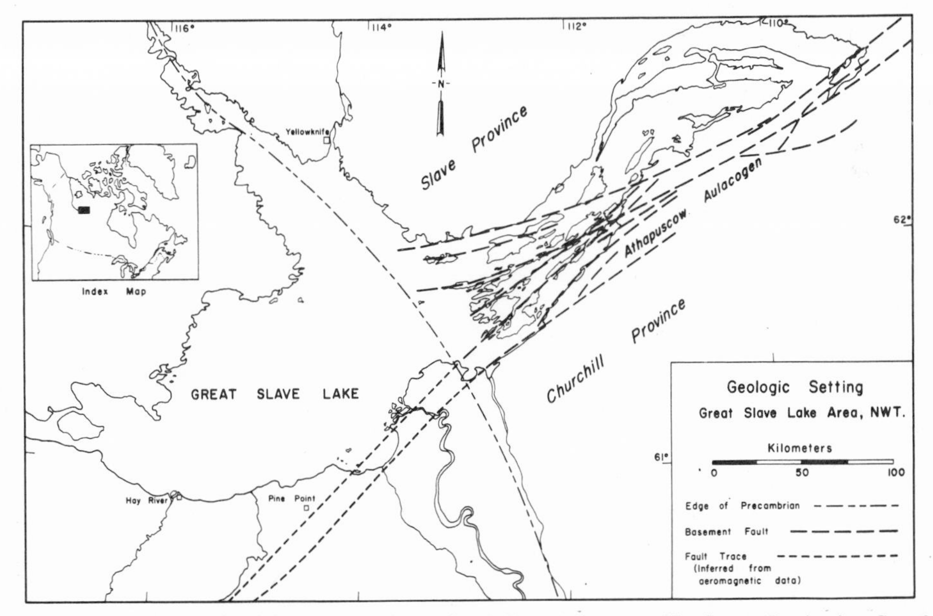

Fig. 1. Location and geologic setting of the Great Slave Lake area, Northwest Territories, Canada. (Modified after Norris, 1965 and Hoffman et al., 1974.)

about 1940 was based on the concept that the Pine Point sulfide deposits were related to the MacDonald Fault which could be traced southwestward from the Precambrian Shield in the East Arm of the Great Slave Lake and projected beneath Paleozoic cover to the Pine Point area. It was reasoned that if numerous concealed orebodies lay along a linear trend in the Devonian dolostones overlying the projection of the fault zone, the aggregate reserves from many isolated near-surface deposits might make a major mining operation feasible in spite of the remote location. This concept resulted in the discovery of

TABLE I

Pine Point district production and reserves 1964–1979

	Production			Reserves		
	Ore (tons)	Pb (%)	Zn (%)	Ore (10^6 tons)	Pb (%)	Zn (%)
1964	14,070	18.6	25.8			
1965	75,356	4.3	7.6	21.5	4.0	7.2
	364,168	22.5	29.1			
1966	1,457,990	4.9	10.5	37.8	2.9	6.8
	282,309	18.8	26.3			
1967	1,521,000	4.7	9.7	40.5	2.6	6.8
	333,000	18.0	27.9			
1968	2,138,000	3.5	6.6	39.3	2.6	6.8
	353,000	19.0	25.0			
1969	3,605,000	3.2	7.4	41.8	2.4	6.3
1970	3,860,000	3.0	7.1	43.5	2.5	6.0
	92,600	14.5	21.5			
1971	3,892,000	2.6	6.5	41.9	2.4	6.0
1972	3,810,000	2.7	6.2	40.9	2.4	6.0
1973	3,896,000	2.9	6.0	38.3	2.3	5.7
1974	4,135,000	2.5	5.3	39.5	2.2	5.7
1975	3,905,000	2.4	4.9	39.2	2.0	5.4
1976	3,773,000	1.7	5.3	36.2	2.0	5.4
1977	3,443,000	2.1	5.3	37.5	2.1	5.3
1978	3,290,000	2.6	5.9	37.3	1.9	5.1
1979	3,291,000	1.9	5.5	38.0	1.9	5.0
Total milled	46,381,158	2.7	6.2			
Total direct shipping	1,439,147	19.3	26.7			
Total production	47,820,305	3.2	6.7			

Compiled from Pine Point Mines Limited Annual Reports 1964–1979.

orebodies with no surface expression, and mining operations became economically viable in 1961 with the decision of the Federal Government to build a railroad to the Great Slave Lake. Increased exploration drilling, aided by induced-polarization techniques, resulted in the establishment of substantial reserves, and the first ore shipment was made in late 1964. Total production of the district to the end of 1979 is 47.8 million tons of ore averaging 6.7% zinc and 3.27% lead (Table I). Present official ore reserves are 38.0 million tons of ore grading 5.0% zinc and 1.9% lead (Pine Point Mines Limited, Annual Report 1979).

In addition to the economic importance of the district, its geological setting has been used to define the widely accepted model of basinal evolution for the genesis of carbonate-hosted lead–zinc deposits of the so-called Mississippi Valley-type (Beales and Jackson, 1966; Jackson and Beales, 1967). The Pine Point deposits have been the subject of numerous investigations, and it is appropriate that this extensive data be reviewed to evaluate the basinal evolution model for the origin of Pine Point orebodies in particular and carbonate-hosted lead–zinc deposits in general.

GEOLOGICAL SETTING

Middle Devonian (Givetian) stratigraphic relationships in western Canada have received considerable attention because of the economic significance of these units not only in the lead–zinc district at Pine Point but also in the subsurface where equivalent strata form important hydrocarbon reservoirs. The Pine Point barrier complex dips gently to the southwest from the outcrop belt near the Great Slave Lake and hosts several major oil fields in northwestern Alberta and scattered smaller gas fields in Alberta, British Columbia, and adjacent Northwest Territories (Dunsmore, 1973; De Wit et al., 1973).

Stratigraphy

Pre-Givetian stratigraphy. Three diamond drill holes extend to the Precambrian basement in the Pine Point area. Pre-Givetian stratigraphy in these holes and for the southern Great Slave Lake region as defined by Norris (1965) from subsurface and outcrop data is summarized in Fig.2. The easternmost mineralized zone (N204) in the Pine Point district is about 75 km southwest of the first exposures of Precambrian rocks in the Churchill Province (Fig. 1). The nature of the Precambrian underlying the immediate Pine Point area is little known; two drill holes penetrate short sections of micaceous quartzite and biotite granodiorite.

The Old Fort Island Formation is the oldest Paleozoic unit in the area and unconformably overlies the Precambrian (Fig. 2). It is typically a friable, fine- to medium-grained quartz sandstone (Plate I: 1) with minor variegated siltstone and shale. The formation may be as much as 40 m thick, but the thickness is quite variable because the unit

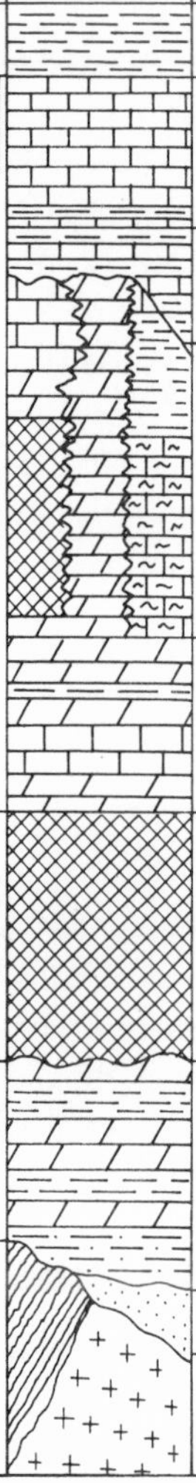

AGE	STRATIGRAPHY	FORMATION	THICKNESS (meters)	DESCRIPTION	MINERALIZATION
U. DEVONIAN (Frasnian)		HAY RIVER		Calcareous shale, minor limestone	
U. DEVONIAN (Frasnian)		SLAVE POINT	50-70	Argillaceous limestone, minor dolostone, calcareous mudstone	
M. DEVONIAN (Givetian)		WATT MOUNTAIN	15-45	Limestone and dolostone, waxy green mudstone interbeds	
M. DEVONIAN (Givetian)		PINE POINT GROUP	75-150	Upper — Limestone of reefal and associated depositional facies; extensive coarse-crystalline dolostone (Presqu'ile); transitional into calcareous shale (Buffalo River Fm.) to NW Lower — Fine-crystalline dolostones of reefal and associated depositial facies; transitional into bituminous limestone to NW and evaporites (Muskeg Fm.) to SE	
M. DEVONIAN (Givetian)		KEG RIVER	65-75	Argillaceous dolostone and limestone	
M. DEVONIAN (Eifelian)		CHINCHAGA	90-110	Anhydrite and gypsum, minor dolostone, limestone, and mudstone	
ORDOVICIAN or older		MIRAGE POINT	60-90	Dolostone, mudstone, siltstone, anhydrite, and gypsum	
PRECAMBRIAN		OLD FORT ISLAND	0-30	Friable, fine to medium-grained sandstone	
PRECAMBRIAN				Micaceous quartzite and granodiorite ?	

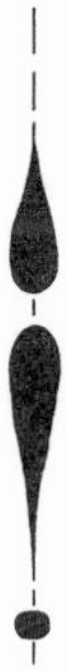

Fig. 2. Stratigraphic column of Paleozoic formations, southern Great Slave Lake area. (After Norris, 1965 and Skall, 1975.)

is apparently preserved in depressions on the Precambrian erosional surface and may be completely absent where it pinches out against high areas. The Old Fort Island Formation is considered to be Middle Ordovician or older (Norris, 1965).

The Mirage Point Formation transitionally overlies the Old Fort Island Formation or unconformably overlies the Precambrian basement where the basal sandstone unit is not present. In outcrop it consists of a thinly interbedded sequence of dark red, purple, and orange-red dolostone, argillaceous dolostone, sandy dolostone, gypsiferous dolostone, dolomitic mudstone, very fine-grained dolomitic sandstone, green and red shale, and gypsum (Plate I: 2,3). The formation may be as much as 178 m in thickness, but is much thinner in the Pine Point area presumably because this region was on the depositional shelf closer to the edge of the Precambrian Shield. The Mirage Point Formation is considered to be Upper to Middle Ordovician or older in age (Norris, 1965).

The Chinchaga Formation unconformably overlies the Mirage Point Formation and typically consists of light gray to brown anhydrite with minor brown to brown-gray, dense, fine-crystalline dolostone (Plate I: 4). Brecciated limestone and dolostone and contorted and brecciated anhydrite are present in some areas. The Chinchaga Formation ranges from about 90 to 115 m in thickness and is lower Middle Devonian (Eifelian) in age (Norris, 1965).

Keg River Formation. Middle Givetian stratigraphic relationships regionally consist of a narrow carbonate barrier, the Pine Point barrier complex, which separates carbonate and shale strata deposited in the deep-water environment of the Mackenzie Basin from extensive back-reef evaporite deposits of the Elk Point Basin (Fig. 3). The Middle Givetian units are of particular economic importance because they are host to the vast majority of the lead–zinc ore in the Pine Point district (Fig. 2). The following discussion of Givetian stratigraphy is summarized (Table II) largely from Skall (1975), and the reader is referred to this excellent work for details of the stratigraphic relationships. Generalized stratigraphic features of the Pine Point area are shown in plan in Fig. 4 and in section in Fig. 5; stratigraphic position of some key orebodies is illustrated schematically in Fig. 6.

The Keg River Formation (Facies A) conformably overlies the Eifelian Chinchaga Formation and is early Givetian in age. The unit consists of medium to dark gray-brown, dense to sucrosic dolostone with varying amounts of argillaceous and carbonaceous material and occasional chert nodules (Plate I: 5). In the lower part of the unit, remnants of medium gray-brown micritic limestone are preserved. The formation maintains a fairly consistent thickness of about 70 m throughout the Pine Point area. One to three widely distributed bluish-gray, pyritic, calcareous shale beds occur 35–50 m above the base (Plate I: 6). The individual units, termed E-shale markers (Campbell, 1950), vary from a few centimeters to as much as a meter in thickness, and the beds occur over a vertical distance of 5–11 m. The faunal assemblage of crinoids, brachiopods, and corals, the ubiquitous argillaceous wisps, and the uniform lithology of this facies indicate the moderately agitated depositional environment of a marine platform (Skall, 1975).

Pine Point Group. Skall (1975) has shown that previous subdivisions of Givetian stratigraphy in the Pine Point area are inadequate to represent the highly variable strata of the Pine Point barrier complex. Therefore, the Pine Point Group has been defined to include all lithologies which form integral parts of the barrier complex. It includes the Sulphur Point Formation, the Presqu'ile Formation, and most of the Pine Point Formation of Norris (1965); facies designations B through K replace formational ranks (Table II). These facies are clearly defined by composition, fabric, and paleoecology, although the

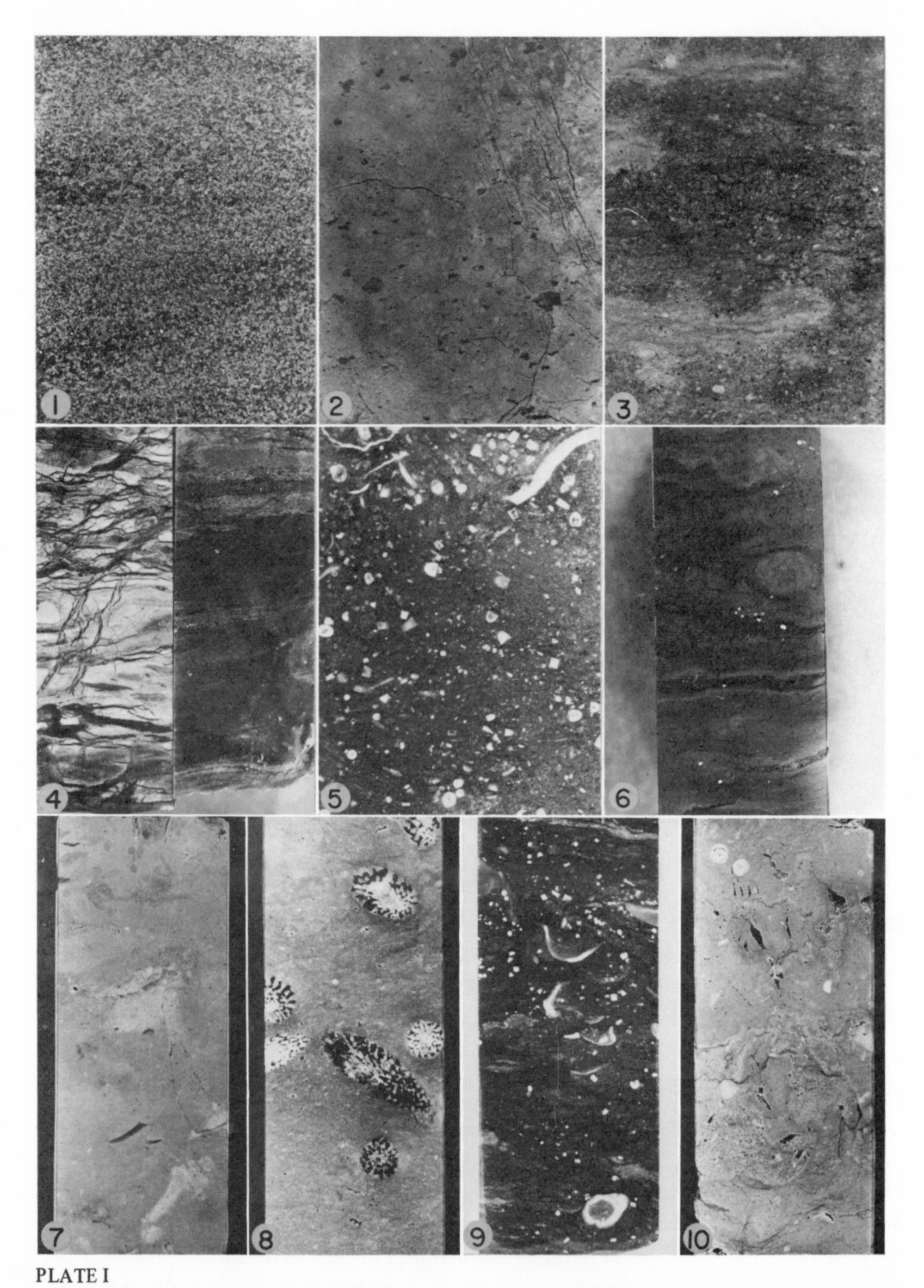

PLATE I

Lower Paleozoic and lower Pine Point barrier rock types, (samples are polished cores, 35 mm in horizontal dimension).

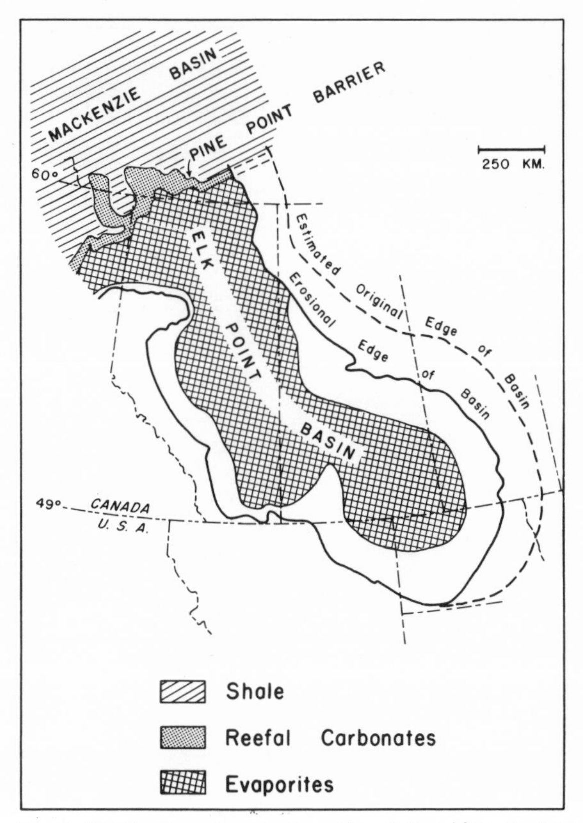

Fig. 3. Middle Devonian stratigraphic relationships, western Canada. (Modified after Maiklem, 1971 and Skall, 1975.)

---

1. Old Fort Island Formation – reddish, fine-grained, friable quartz sandstone.
2. Mirage Point Formation – orange, fine-crystalline dolostone with blebs and veins of anhydrite.
3. Mirage Point Formation – reddish- and greenish-gray dolomitic mudstone with abundant scattered sand grains.
4. Chinchaga Formation – light to medium brown anhydrite with thin dolostone laminae and anhydrite-filled fractures.
5. Keg River Formation – Facies A – medium brown, argillaceous limestone with abundant crinoid ossicles and scattered brachiopod fragments.
6. Keg River Formation – E-shale – bluish-gray, dolomitic shale with scattered crinoid ossicles.
7. Pine Point Group – Facies B-1 – light gray-brown, dense dolostone with scattered stromatoporoid fragments.
8. Facies B-1 – medium brown, sandy dolostone with *Thamnopora* impregnated with bitumen.
9. Facies B-2 – dark gray-brown, dense dolostone with abundant crinoid ossicles and thin-shelled brachiopods (partly silicified).
10. Facies B-3 – light brown, dense dolostone with fine fractures and areas of good intergranular ("spongy") porosity after leached fossils and burrows.

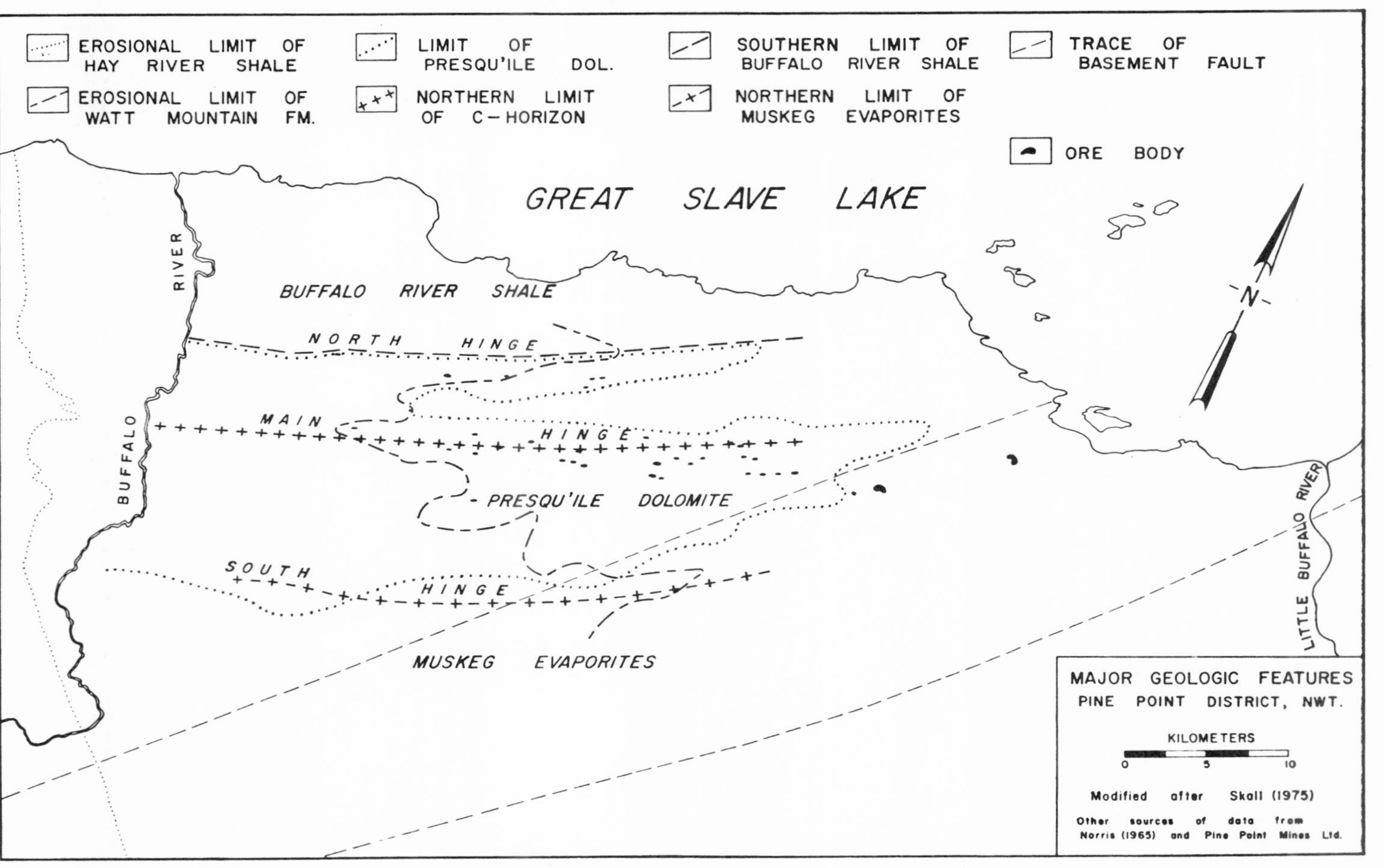

Fig. 4. Major geologic features of the Pine Point mining district.

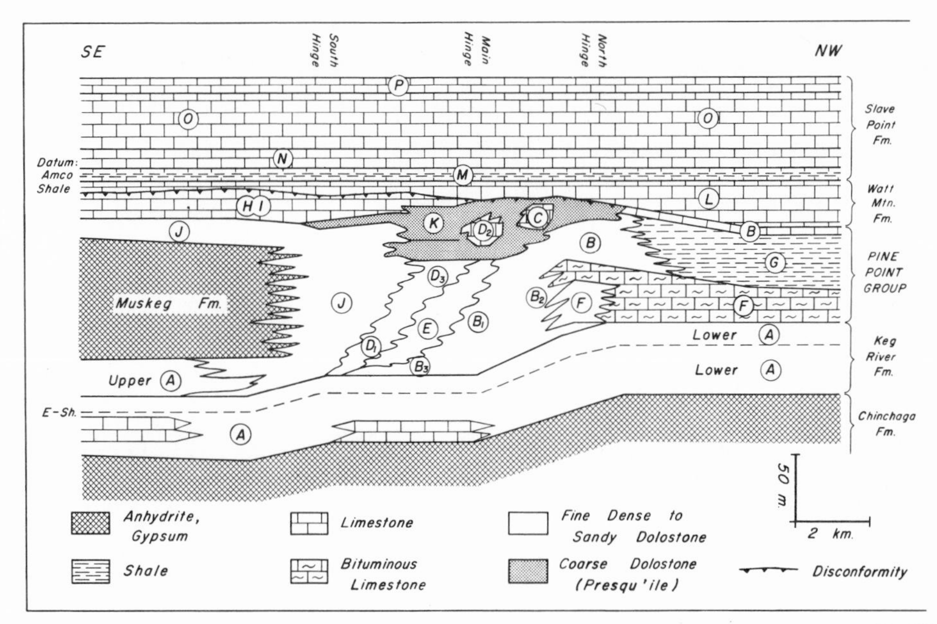

Fig. 5. Generalized Middle Devonian stratigraphic relationships, Pine Point mining district. (Modified after Skall, 1975.)

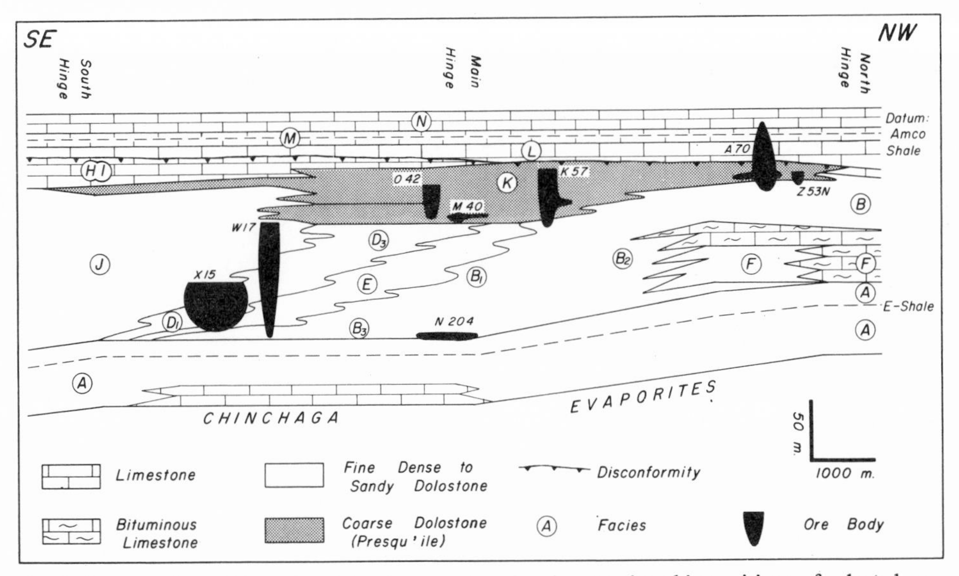

Fig. 6. Cross-section of Pine Point barrier complex showing stratigraphic positions of selected ore-bodies.

TABLE II

Givetian stratigraphy, Pine Point district (After Skall, 1975 and Adams, 1975).

AGE	FORMATION	FACIES	MAXIMUM THICKNESS (METERS)	DESCRIPTION	FAUNA*	DEPOSITIONAL ENVIRONMENT
LATE GIVETIAN	SLAVE POINT	P	12	Limestone, dark brown, dense micrite, slightly argillaceous, interclasts common	Massive corals and stromatoporoids(+++), Amphipora (+), thin-shelled brachiopods (+), Stachyodes (+), crinoids (++)	Deep Marine Platform
		O	50	Limestone, light to dark brown, micrite or sand, large interclasts common, slightly argillaceous to common shaley laminae, burrowed	Massive stromatoporoids (+), Amphipora (+ to +++), thin-shelled brachiopods (++)	Shallow Marine Platform
		N	18	Limestone, often dolomitic, light gray, dense micrite; blotchy bedding, stromatolite, and fenestrate structures common	Amphipora (+), calcispheres (+), Charophyta (+), brachiopods (−), ostracods (−)	Tidal Flat
		M	10	1. Shale, gray to blue-gray, usually calcareous and mottled with disseminated sulfides	Crinoids (−), thick-shelled brachiopods (−)	Marine
				2. Limestone or dolostone, light to medium brown, sand or micrite, argillaceous	Brachiopods (+ to +++), massive stromatoporoids (++), crinoids (+)	Marine Shoal
	WATT MOUNTAIN	L	45	Limestone, may be dolostone or dolomitic; very light gray, light gray, buff, light green gray; micrite with minor sandy beds; laminite, stromatolite, blotchy bedding and fenestrate structures common; minor gypsum; waxy green shale interbeds	Charophyta (+ to ++), calcispheres (+), gastropods (+), ostracods (+), Amphipora (+)	Predominately Tidal Flat, Lacustrine
DISCFM.		K	65	Dolostone, buff to light gray, coarse-crystalline, often with large vugs; uniform granular, mottled, zebra, and breccia-moldic textures common; white dolomite may be abundant	Rarely recognizable; variable depending on original lithology	Secondary effect imposed on B,C,D,E,H,I, and J facies
		Muskeg	120	Anhydrite and gypsum, white to light brown, dense; massive, laminated, and nodular bedding; may be interbedded with J facies dolostone		Restricted Evaporitic
		J	145	1. Dolostone, light gray to blue-gray, dense to sucrosic; laminite, blotchy bedding, stromatolite and fenestrate structures common	Charophyta (+)	Tidal Flat
				2. Dolostone, light to medium brown, sucrosic with intergranular porosity; slightly argillaceous to laminated		Subtidal to Tidal Flat
				3. Dolostone, light brown to buff, sucrosic to sandy with good intergranular porosity; some faint argillaceous wisps	Amphipora (+), corals (+), stromatoporoids (+), brachiopods (+)	Backreef Subtidal
				4. Dolostone, buff to medium brown, sucrosic with some argillaceous wisps, may be vuggy	Amphipora (++ to +++)	Lagoonal
				5. Dolostone, very light gray, dense micrite	Corals (+), dendroid stromatoporoids (+)	Lagoonal

MIDDLE GIVETIAN	PINE POINT GROUP	I	30	Limestone, very light gray, pelleted micrite and sand; fenestrate, laminite, and stromatolite structures common	Amphipora (++ to +++), gastropods (+), corals (+), stromatoporoids (+), Stachyodes (+), brachiopods (+), calcispheres (+),	Backreef Platform Lagoonal and Tidal Flat
		H		Limestone, very light gray, micrite with bioclastic debris	Gastropods (+ to +++), stromatoporoids (++), Stachyodes (++), corals (++), brachiopods (++), Amphipora (+)	Backreef Platform Tidal Flat and Lagoonal
		G	60	Shale, dark gray to blue-gray, fissile, calcareous, disseminated iron sulfides	Thin-shelled brachiopods (−)	Marine Pelagic
		F	35	Limestone, occasionally dolomitic, dark brown to black, dense micrite, very bituminous and argillaceous	Crinoids (+), Tentaculites and Styliolina (+ to +++), thin-shelled brachiopods (+ to +++)	Basinal Deep Marine
		E	45	Dolostone, buff to light brown, sucrosic to sandy with good intergranular porosity, often friable; lacks argillaceous material	Stromatoporoids (−), corals (−), thick-shelled brachiopods (−), Amphipora (−)	Forereef
		D	30	1. Dolostone, light brown to buff, sucrosic to sandy matrix with abundant fossils; may have large vugs and good intergranular porosity	Tabular stromatoporoids (++ to +++), massive stromatoporoids (− to ++), Stachyodes (++), dendroid corals (++ to +++), Alveolites (+)	Organic Barrier Reef in Part
				2. Limestone, very light gray, skeletal sand and clasts cemented by sparry calcite	Massive stromatoporoids (++ to +++), Stachyodes (++ to +++), massive corals (++), thick-shelled brachiopods (++), gastropods (++), cephalopods (−)	
				3. Dolostone, buff to blue-gray, dense, fragmental (often vague due to diagenetic modifications)	Massive and dendroid stromatoporoids (+++), tabular stromatoporoids (+ to ++), dendroid corals (+ to ++)	
		C	20	Limestone, buff to very light gray, micrite and bioclastic sand; lacks argillaceous material	Crinoids (+ to +++), thick-shelled brachiopods (++ to +++), corals (+), dendroid and massive stromatoporoids (+)	Shallow Forereef
		B	60	1. Dolostone, rarely limestone, light to medium gray-brown, sandy to sucrosic, slightly argillaceous	Crinoids (+ to ++), dendroid corals (+ to ++)	Offreef
				2. Dolostone, rarely limestone, medium to dark gray-brown to black, dense to sucrosic, very argillaceous, common chert nodules and secondary silicification	Thin-shelled brachiopods (+ to +++), crinoids (++ to +++), dendroid corals (+ to ++)	Marine
				3. Dolostone, light to medium brown, sandy to dense matrix with zones of good intergranular porosity; many large vugs after leached fossils	Massive stromatoporoids (+ to +++), crinoids (+), brachiopods (+)	Offreef to Shallow Platform
EARLY GIVETIAN	KEG RIVER	A	70	Dolostone and limestone, medium to dark brown, dense to sucrosic, varying amounts of argillaceous and carbonaceous wisps, minor chert nodules, E-shale marker beds (1 to 3) occur 35 to 50 meters above base	Crinoids (+ to +++), brachiopods (+ to ++), gastropods (+), tabular stromatoporoids (+), cephalopods (−)	Marine to Shallow Platform

* Faunal Abundance   Abundant (+++), Common (++), Minor (+), Rare (−)

PLATE II

Pine Point Group rock types (samples are polished cores, 35 mm in horizontal dimension, unless another scale is shown).

contacts are generally gradational and interdigitated (Skall, 1975).

Facies B can be divided into three subfacies which can be distinguished by the amount of argillaceous material and the faunal constituents. Total thickness of the facies is variable due to the diachronous development of the barrier and intercalations with other facies but is a maximum of 60 m. Facies B grades northward into Facies F (basinal deep marine) and southward into Facies E (shallow fore-reef) (Fig. 5). B-1 is slightly argillaceous, light to medium brown, sandy to sucrosic dolostone with only a few carbonaceous wisps and partings (Plate I: 7,8). The subfacies is occasionally preserved as limestone or calcareous dolostone. Crinoid ossicles and dendroid corals are the most abundant faunal components. B-1 represents an off-reef depositional environment with moderately agitated conditions. B-2 is distinctly argillaceous, medium to dark brown, dense dolostone with many carbonaceous wisps (Plate I: 9). The subfacies is occasionally preserved as limestone. Crinoids, dendroid corals, and thin-shelled brachiopods are often abundant, and chart nodules and secondary silicification may be common. B-2 was deposited in a slightly agitated, open-marine environment. It represents a similar environment to that which prevailed during deposition of the Keg River Formation. However, it occupies a definite position between the organic barrier facies in the south and the basinal facies in the north (Fig. 5) and represents sedimentation on a gently sloping distal fore-reef area rather than an open-marine platform (Skall, 1975). Lithologies intermediate between B-1 and B-2 are common. B-3 is light to medium brown, sandy to dense dolostone with minor argillaceous wisps and partings. Zones of intergranular porosity and large vugs after leached fossils are common, resulting in its designation as the "spongy horizon" (Plate I: 10). Small, bulbous stromatoporoids, crinoids, and brachiopods may still be recognizable. The subfacies is best developed immediately above the Keg River Formation and is host to the N204 orebody (Fig. 6). B-3 represents an off-reef to marine shoal depositional environment (Adams, 1975).

Facies C is tan to very light gray, micritic to sandy limestone often with intercalated layers of skeletal fragments (Plate II: 1). Faunal components include crinoids,

---

1. Facies C – very light gray limestone composed of uniform skeletal sand.
2. Facies D-1.
   a. Medium brown dolostone with tabular stromatoporoid fragments.
   b. Tan sandy dolostone with vague "micritized" tabular stromatoporoids.
3. Facies D-3.
   a. Light brown to gray fragmental dolostone.
   b. Blue-gray, dense, vuggy dolostone with vague fossil forms.
4. Facies D-1 – medium brown dolostone with abundant leached stromatoporoids. X15 orebody, north side, fourth bench.
5. Facies D-2.
   a. Very light gray limestone composed of massive stromatoporoids. K62 orebody, south side, first bench.
   b. Very light gray limestone composed of large fragments of massive stromatoporoids.
6. Facies E – light brown, sandy, friable dolostone with good intergranular porosity.
7. Facies F – black, very bituminous, micritic limestone with *Tentaculites* and *Styliolina*.
8. Facies F – photomicrograph of *Tentaculites* in dense, bituminous limestone. Scale is 0.4 mm.
9. Facies G – Buffalo River Shale – bluish-gray, fissile shale with disseminated iron sulfides.

thick brachiopods, dendroid and massive stromatoporoids, and corals. The facies is only developed in the upper part of the barrier complex (Fig. 5) and is generally preserved as a limestone remnant within Facies K where it was sheltered from secondary dolomitization. Thickness of the unit is widely variable but is generally less than 20 m. Facies C is the result of comminution of reefal debris under agitated, shallow-water conditions in the proximal fore-reef area (Skall, 1975).

Facies D may be divided into three subfacies which can be distinguished by lithology, color, and fabric. Fossils are extremely abundant and include dendroid, tabular, and massive stromatoporoids, dendroid and massive corals, and thick brachiopods. Gastropods, cephalopods, and crinoids occur sporadically. Total thickness of the facies may be as much as 30 m, but it may be completely absent in some areas. D-1 generally consists of light brown to tan, dense to sandy dolostone with good intergranular porosity and abundant fossils (Plate II: 2,4). D-1 is typical of the early development of the barrier and in combination with Facies J is host to the X15 and W17 orebodies (Fig. 6). D-2 is very light gray, bioclastic limestone cemented by sparry calcite (Plate II: 5). It is developed in the upper part of the barrier and is preserved as limestone remnants within Facies K which escaped secondary dolomitization. The facies is very fossiliferous and may satisfy the characteristics of a "true reef". D-3 is tan to blue-gray, dense fragmental dolostone (Plate II: 3) and is the most frequently intersected reefal lithology. The unit occurs most often below Facies K. Diagenesis has obliterated the original fabric often making it difficult to distinguish fragments from matrix, or organic from nonorganic fragments. The faunal assemblage and lack of argillaceous material of Facies D indicate that it was deposited under very agitated water conditions. Skall (1975) has clearly shown that Facies D was an organic barrier which controlled Middle Devonian sedimentation throughout much of western Canada, but which only rarely developed into a "wave-resistant erect rigid structure" by the actions of "sediment-binding biotic constituents" as defined by Lowenstam (1950) and Klovan (1964). Therefore, Facies D is designated the "Organic Barrier Facies" and includes a variety of fossiliferous, reefal lithologic types of which a true reef is an exceptional development. Skall (1975) states that "The Organic Barrier Facies, in comparison with contemporaneous sediments of the barrier complex, is formed in the most agitated zone of wave action and separates sheltered deposits on the leeward side from sediments that formed in more open marine realms."

Facies E is tan to light brown, sucrosic to sandy dolostone with good intergranular porosity (Plate II: 6). It may be quite friable, and the lack of argillaceous and carbonaceous material is characteristic. The fauna is identical to that of Facies D but generally consists only of minor amounts of fine fossil debris. These clean sands are developed only in the lower part of the barrier complex and reach a maximum thickness of 45 m (Fig. 5). The massive uniform lithology, the scarcity of argillaceous material, and the faunal assemblage indicate that Facies E was deposited under agitated water conditions and resulted from the attrition of Facies D skeletal material in the immediate fore-reef area (Skall, 1975).

Facies F is dark brown to black, dense, micritic, argillaceous, and very bituminous limestone with minor dolostone (Plate II: 7). Fauna consists of abundant *Tentaculites* sp., *Styliolina* sp., and thin-shelled, articulated brachiopods, and minor crinoids and corals. Facies F is identical to the Bituminous Shale and Limestone Facies of the Pine Point Formation as defined by Norris (1965). The unit attains a maximum thickness of 35 m near the shore of the Great Slave Lake. The ecology, argillaceous and bituminous content, and the position of the lithology within the barrier complex (Fig. 5) indicate that Facies F was deposited under very quiet-water conditions and is representative of deep-marine basinal sedimentation (Skall, 1975).

The name Buffalo River Shale (Facies G) was proposed by Campbell (1950) for the dark green to bluish gray, fissile, calcareous shale with disseminated iron sulfides (Plate II: 9) that was originally recognized in drill holes just west of the mouth of the Buffalo River. Fossils are rare and are generally restricted to thin-shelled brachiopods. Facies G is as much as 60 m thick near the shore of Great Slave Lake; to the south, it interfingers with and is eventually replaced by Facies B (Fig. 5). Stratigraphic relationships, fossil assemblage, and lithology classify Facies G as a deep-marine basinal deposit. The unit, at least in part, represents strata which are time equivalent to the upper part of the carbonate barrier platform (Skall, 1975).

Although Facies H and I were originally defined separately (Skall, 1970), now they usually are not differentiated because of their lithologic similarities and lack of clear-cut boundaries. The very light gray limestone of these facies consists of skeletal sand and clasts and alternating micrite and pelleted micrite layers. Sand predominates over micrite in Facies H, and micrite is dominant in Facies I. Small calcite blades after gypsum molds are occasionally present, but no evidence exists for significant amounts of evaporites in these facies. Fossils may be abundant and include *Amphipora, Stachyodes,* massive stromatoporoids, corals, gastropods, brachiopods, and traces of calcispheres and crinoids. The abundance of gastropods is diagnostic of Facies H (Plate III: 1), whereas the dominance of *Amphipora* is characteristic of Facies I (Plate III: 2). The combined facies attain a maximum thickness of about 50 m south of the barrier core and become thinner northward where they interfinger with and are altered to dolostones of the Presqu'ile (Fig. 5). Skall (1975) abandoned the term Sulphur Point Formation as defined by Norris (1965) because the type section has been shown to include both Upper Givetian Watt Mountain Formation (Facies L) and only part of the Middle Givetian Facies H and I. Facies I was deposited in a shallow lagoonal basin which periodically became a tidal flat area resulting in the formation of carbonates under very restricted conditions, represented by the stromatolite, laminite, and fenestrate structures. Gradual incorporation of more reef-derived detritus resulted in the formation of Facies H (Skall, 1975).

Facies J may be divided into five subfacies which are generally interbedded. J-1 and J-3 are the most abundant lithologic types, followed by J-2; J-4 and J-5 are uncommon. Facies J attains a maximum total thickness of about 145 m; individual subfacies units usually vary from a few centimeters to a few meters, but seldom exceed 8 m in thickness.

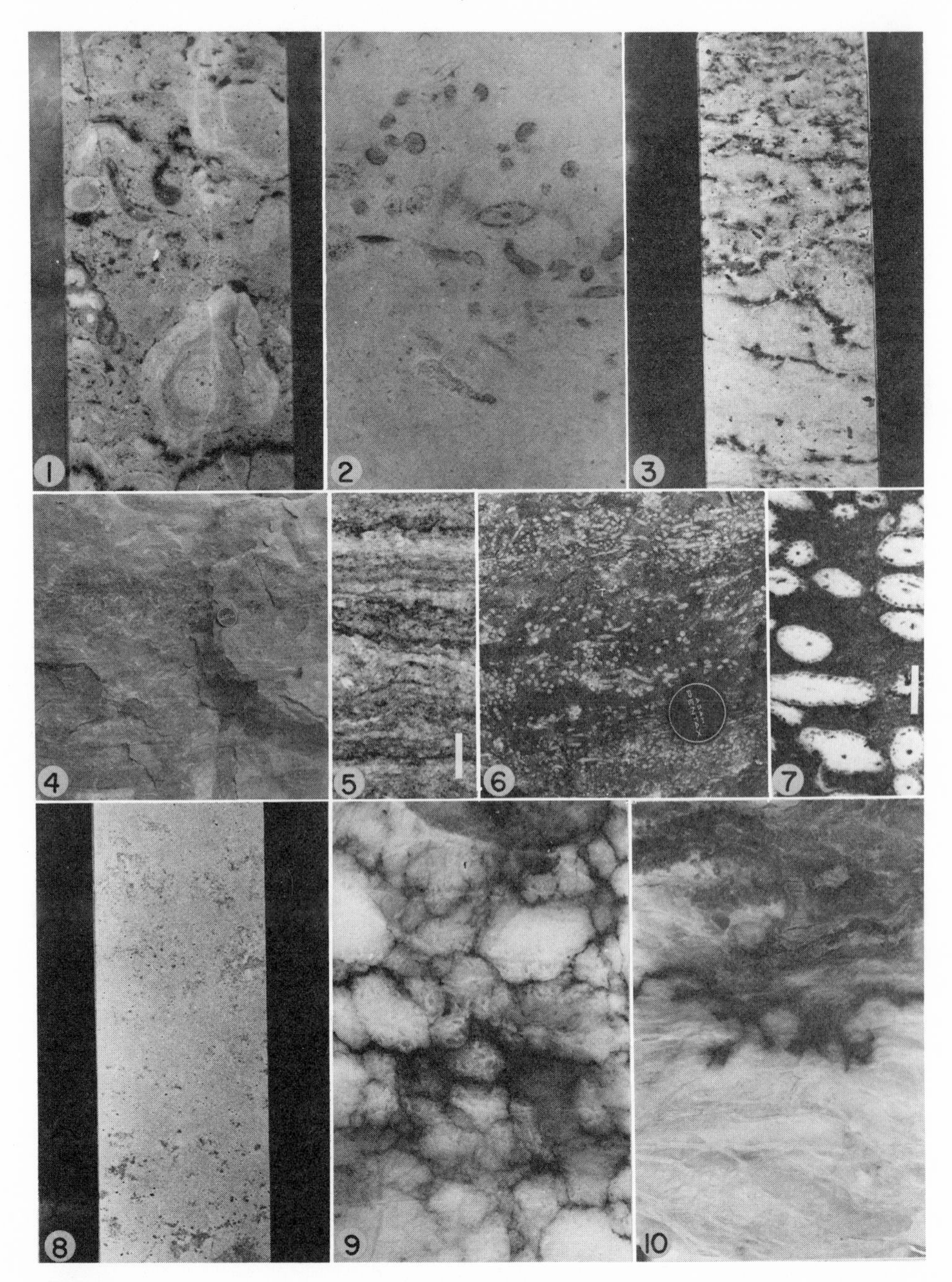

PLATE III

Pine Point Group rock types (samples are polished cores, 35 mm in horizontal dimension, unless another scale is shown).

1. Facies H – very light gray limestone with gastropods and algal-coated skeletal grains.

To the south Facies J becomes interbedded with the evaporites of the Muskeg Formation; to the north the upper part of Facies J interdigitates with the coarse-crystalline dolostone of the diagenetic Facies K (Fig. 5). One to three thin Facies J beds, designated the C-Horizon by Campbell (1966), extend north as much as 5 km from the main South Flank body into the Presqu'ile. The subunits of Facies J are the dominant host rocks of the X15 and W17 orebodies (Fig. 6). J-1 is light gray to blue-gray, dense to sucrosic dolostone which commonly exhibits laminite, stromatolite, blotchy bedding, and fenestrate structures (Plate III: 3). These sedimentary structures and minor spores of *Charophyta* indicate that J-1 was deposited in a tidal flat environment. J-2 is light to medium brown, sucrosic dolostone with intergranular porosity. Minor amounts of argillaceous and carbonaceous material are concentrated in thin laminae and partings (Plate III: 4,5). Macrofossils are not present; intraclasts, stromatolites, and fenestrate structures are occasionally preserved. J-2 is believed to have been deposited in a tidal flat to subtidal environment. J-3 is light brown to tan, sucrosic to sandy dolostone with intergranular porosity (Plate III: 8). The unit is less argillaceous than J-2 with only faint shaley wisps. The faunal assemblage includes minor *Amphipora,* brachiopods, corals, and bulbous stromatoporoids. J-3 is lithologically similar to off-reef Facies B-1 and fore-reef Facies E, but the association with distinct back-reef lithologies and the presence of *Amphipora* indicate that J-3 was deposited in a back-reef, slightly agitated subtidal environment. J-4 is tan to medium brown, sucrosic dolostone which may be slightly argillaceous and vuggy. The numerous to abundant *Amphipora* are diagnostic (Plate III: 6,7) and indicate that J-4 is the result of lagoonal sedimentation (Skall, 1975). J-5 is very light gray, dense, micritic dolostone which may contain minor corals and dendroid stromatoporoids. This common lithology is believed to be representative of back-reef lagoonal sedimentation (Adams, 1975).

Skall (1975) did not include the evaporite strata of the Muskeg Formation as a facies of the Pine Point Group, and this unit does not extend into the Pine Point mining district (Fig. 4). However, it represents an important aspect of Middle Givetian sedimentation in western Canada and is discussed here and shown in Table II within the Pine Point Group. Middle Devonian evaporite strata in the Elk Point Basin extend as far south as North Dakota (Fig. 3) and contain the economically important potash deposits in the Prairie Evaporite of south-central Saskatchewan. The Muskeg Formation consists of white

---

2. Facies I – very light gray, micritic limestone with common *Amphipora.*
3. Facies J-1 – light blue-gray, dense dolostone with blotchy mottling, probably due to bioturbation.
4. Facies J-2 – medium brown, faintly laminated, dense dolostone; lens cap is 54 mm in diameter. W17 orebody, west side, third bench.
5. Facies J-2 – polished core showing crenulated laminae, probably of algal origin; scale is 5 mm.
6. Facies J-4 – medium brown, dense dolostone with abundant *Amphipora;* lens cap is 54 mm in diameter. X15 orebody, west side, third bench.
7. Facies J-4 – polished core with abundant *Amphipora;* scale is 5 mm.
8. Facies J-3 – light brown, sandy dolostone with good intergranular porosity.
9. Muskeg Formation – white, nodular gypsum.
10. Muskeg Formation – light to medium brown, dense anhydrite.

PLATE IV

Presqu'ile Facies K.

1. Bedded Facies K in west wall of N38A open pit, first bench. Note abundant zones of white dolomite above thin bed of Facies J-2 dolostone (C-Horizon). Bench height is about 8 m.

to light brown, bedded anhydrite which may be nodular, mosaic, laminated, or massive (Plate III: 9,10). Evaporites interbedded with Facies J dolostone extend to within 10 km of the southernmost zone of orebodies (Fig. 4). Fifteen kilometers further south the Muskeg Formation is 120 m in thickness and largely evaporitic (Skall, 1975). The nodular and mosaic types that are intimately associated with thin-bedded, algal-laminated dolostones are believed to be the result of shallow lagoonal, intertidal, and supratidal sedimentation (Bebout and Maiklem, 1973). The bedded massive and laminated anhydrite with only scattered dolomite rhombs and minor argillaceous material are thought to be the result of sedimentation in a shallow-water restricted lagoon (Bebout and Maiklem, 1973; Skall, 1975).

Facies K or Presqu'ile Facies is tan to light gray, vuggy, coarse-crystalline dolostone (Plates IV and V). Mottled, uniform granular, zebra, and breccia-moldic textures are common, and white dolomite often is abundant. Skall (1975) restricts Facies K to the coarse-crystalline dolostone which occurs between the Watt Mountain Formation and Facies B, D, E, and J (Fig. 5). Coarse-crystalline dolostone lithologies which can be identified by color, fabric, relic fossil content, and stratigraphic position as belonging to Facies B, D, E, and J (Plate V: 9) may be present within the lower part of Facies K. Skall (1975) mentions the rare occurrence of coarse-crystalline dolostone in the Watt Mountain and Slave Point Formations. The contact between Facies K and the overlying Watt Mountain Formation is marked by a zone of green clay and rubble. Facies K varies greatly in thickness but reaches a maximum of about 65 m along the Main Hinge. Along the North Hinge, it is about 20 m thick. The northern limit of the Presqu'ile coincides with the abrupt transition of the barrier carbonate facies into the Buffalo River Shale (Facies G) (Figs. 4 and 5). To the south the unit forms a distinct interfingering contact with the limestone strata of Facies H, I, C, and D-2 and with the dolostones of Facies J. The limestone facies also may be preserved as isolated remnants within Facies K which are of great benefit in interpreting the original nature of the strata now so greatly altered (Plate V: 1,2). The contact between limestone and coarse-crystalline dolostone may either follow bedding planes or cut irregularly across stratigraphic contacts. Because of the gentle dip of the barrier to the southwest, Facies K is eroded in the eastern part of the area (Fig. 4). The Presqu'ile continues in the subsurface to the west and is intersected by many drill holes in northeastern British Columbia (Skall, 1975).

---

2. Well-bedded back-reef Facies K in south wall of N42 open pit. Height of exposed section is about 25 m.
3. White dolomite-healed "breccia" in upper Facies K in south wall of K57 open pit, fourth bench.
4. Pervasive white dolomite in upper Facies K in southeast wall of K57 open pit, fourth bench.
5. Boxwork white dolomite in west wall of K57 open pit, fourth bench.
6. Boxwork white dolomite with porosity-occluding coarse calcite. K57 orobody, west side, fourth bench.
7. Pervasive white dolomite in upper Facies K; core specimens from K57 area.
8. Buff, coarse-crystalline dolostone in lower Facies K; K57 area.
   a, b. Massive, vuggy dolostone with relic massive stromatoporoids.
   c, d. Uniform granular dolostone, probably relic skeletal debris; pervasive iron sulfides account for darker color of d.

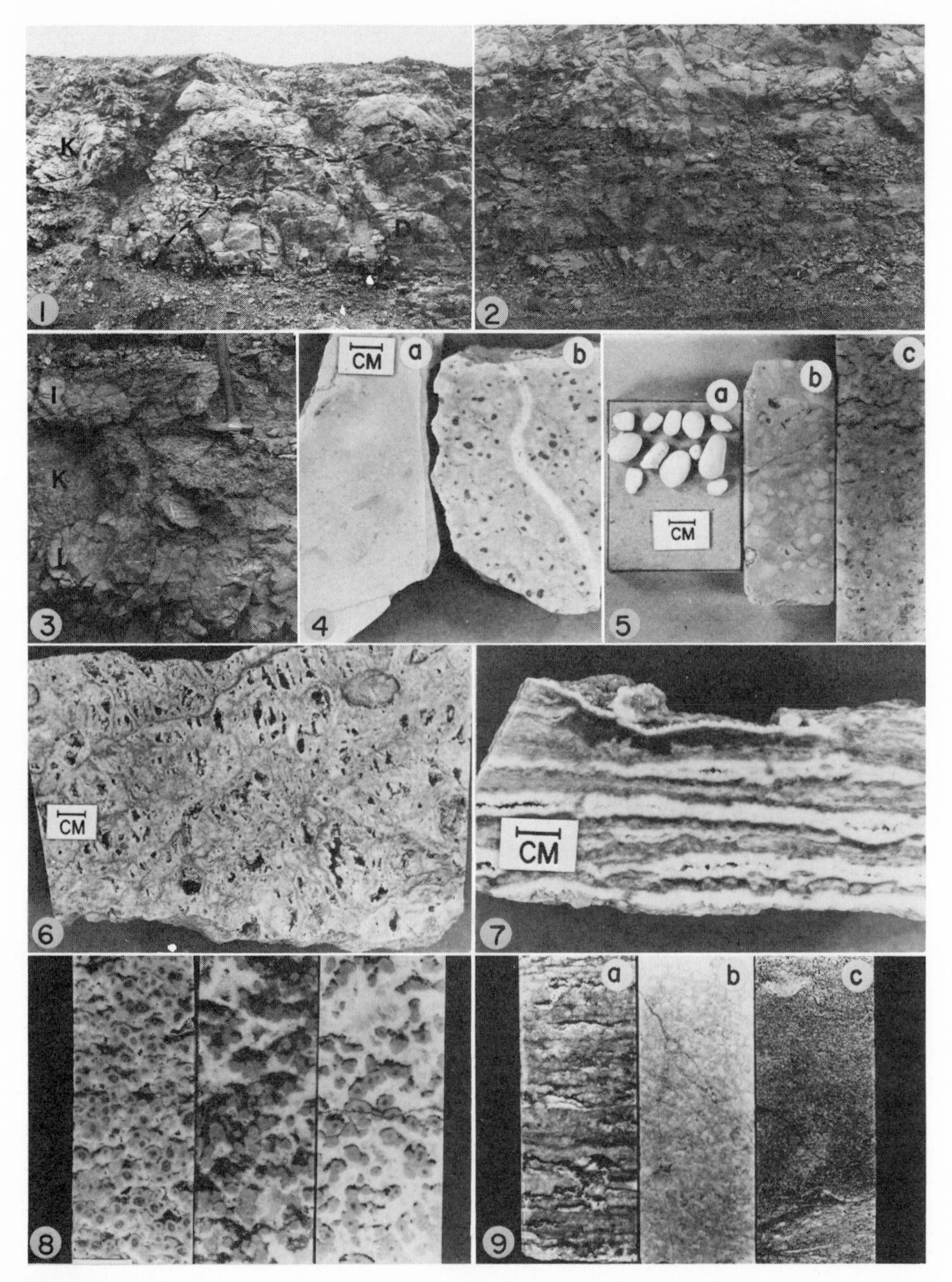

PLATE V

Presqu'ile Facies K.

1. Irregular contact of coarse-crystalline Facies K dolostone with unaffected near-reef Facies D-2 limestone in J44 ramp; steeply-dipping fracture is filled with gray-green clay.

South of the Main Hinge, the thin beds of Facies J dolostone termed C-Horizons may be used to divide Facies K into an upper and lower unit (Plate IV: 1). Although not precisely corresponding to the C-Horizon break, the nature of the upper part and the lower part of Facies K may be quite different. The lower part tends to be tan to light gray, coarse-crystalline, vuggy dolostone in which the original gross nature of the strata may be recognizable. Although greatly modified, it is sometimes possible to ascertain former clastic carbonate, lagoonal, or reefal textures (Plate IV: 8; V: 6; VIII). If white dolomite is present, it occurs as an open-space filling of pores or discrete fractures. The original nature of the upper part of Facies K is often completely obliterated by the pervasive introduction of white dolomite (Plate IV: 7). The percentage of white dolomite and of open space is usually much greater in the upper part than the lower part of Facies K. The contact between these types is not distinct but is highly irregular and variable. The contact of Facies K with the underlying facies is usually gradational over a few centimeters to a few meters.

Skall (1975) has shown that the Presqu'ile Facies does not represent a "coarsely recrystallized reef core" as previously assumed by Norris (1965) and Campbell (1967) but is the result of alteration which affected various facies of the barrier. Although the Organic Barrier Facies D-2 largely has been converted into coarse-crystalline dolostone, the back-reef (Facies H and I), shallow fore-reef (Facies C), and off-reef (Facies B) also were affected, and they undoubtedly represent a volumetrically larger amount of altered strata than the reef core. Although most of the rocks now comprising the Presqu'ile Facies were deposited under back-reef conditions, there is no evidence to suggest that significant

---

2. Interbedded coarse-crystalline dolostone of Facies K and micritic limestone of Facies I on right passing to left into section of entirely coarse-crystalline dolostone, N42 ramp.
3. Close-up of part V: 2.
4. Specimens of interbedded lithologic types, N42 ramp.
   a. Very light gray Facies I limestone with skeletal sand and gravel.
   b. Light tan, coarse-crystalline Facies K dolostone with granular texture.
5. Transformation of carbonate sediment into Facies K; interbedded and gradational lithologies from drill core.
   a. Skeletal gravel, largely dendroid stromatoporoid fragments.
   b. Lithified skeletal gravel (Facies D-2).
   c. Coarse-crystalline Facies K dolostone derived from D-2 limestone.
6. Light gray, coarse-crystalline Facies K dolostone consisting of massive stromatoporoids with some corals.
7. Light brown, coarse-crystalline dolostone with parallel layers of white dolomite ("zebra rock").
8. Coarse-crystalline, granular Facies K dolostone types with varying amounts of white dolomite.
9. Coarse-crystalline Facies K dolostones, possibly derived from pre-existing fine-crystalline dolostones.
   a. Light gray, laminated coarse-crystalline dolostone associated with C-Horizon; possibly former Facies J dolostone.
   b. Light brown, coarse-crystalline dolostone occurring in lower part of Facies K; possibly former Facies E dolostone.
   c. Medium brown, argillaceous, coarse-crystalline dolostone with relic thin-shelled brachiopods and crinoid ossicles from lower part of Facies K; probably former Facies B dolostone.

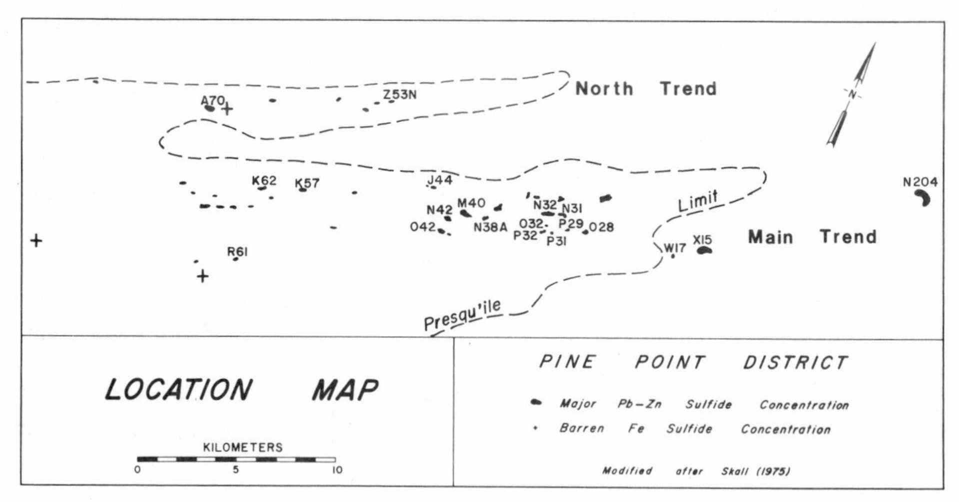

Fig. 7. Distribution of major sulfide concentrations relative to coarse-crystalline dolostone (Presqu'ile).

amounts of evaporites were present (Skall, 1975). Most orebodies in the Pine Point district occur in the Presqu'ile Facies (Figs. 6 and 7). The development of the hosting structures, the coarse-crystalline dolostone, and the orebodies will be considered in detail in later sections.

Watt Mountain Formation. Facies L consists of tan, very light gray, light gray, and light green-gray, micritic limestone and rare dolostone (Plate VI: 1,2,3). Wiley (1970) recognized fourteen basic lithologic types and four faunal assemblages which can be used to divide the Watt Mountain Formation into eight microfacies indicative of slightly different depositional environments. Argillaceous and sandy carbonate lithologies are occasionally present. Laminites, stromatolites, and blotchy bedding are common. Minor amounts of anhydrite and secondary gypsum occur as thin lenses, and discontinuous interbeds of waxy green shale, a few centimeters to a meter or more in thickness, are common. Minor thin brachiopods are the only marine fossils of any consequence; gastropods, calcispheres, and ostracods occur sporadically. Oogonia of *Charophyta* may be common, especially in some green shaley layers. In the southern and central part of the area, Facies L disconformably overlies the facies of the upper Pine Point Group with a pronounced rubble contact and is usually less than 15 m thick (Fig. 5). To the north Facies L directly overlies Facies B and attains a maximum thickness of about 45 m. In this area the disconformable surface is not present, and the off-reef Facies B grades over a few meters into Facies L through a transition zone with a mixed biota. The deposition of Facies L took place in a variety of restricted environments, primarily tidal flat and lacustrine but to a lesser extent lagoonal or restricted marine. The waxy green shales are considered to be the result of resedimentation of residual material that developed on the exposed barrier (Skall, 1975).

Slave Point Formation. The Slave Point Formation conformably overlies the Watt Mountain Formation and may be divided into four facies in the Pine Point area. Facies M is the oldest member and consists of two lithologic types which comprise three units. Total thickness of the facies is usually less than 10 m. M-1 is gray to blue-gray, calcareous, mottled shale which contains disseminated iron sulfides and is commonly referred to as the Amco shale or the Amco marker. The shale is usually about 3 m thick and contains traces of crinoids and brachiopods. M-1 represents open-marine sedimentation. Skall (1975) used the Amco Shale in conjunction with the E-Shales to document development of the Pine Point barrier complex. M-2 consists of light to medium brown, sandy micritic limestone, or rarely dolostone, units which occur both above and below the M-1 shale unit. The limestones are slightly argillaceous and contain numerous brachiopods and massive stromatoporoids and some crinoid ossicles (Plate VI: 4). The lower M-2 biostromal unit represents the open-marine shoal environment which replaced the tidal-flat sedimentation of Facies L. The upper M-2 unit is the result of marine regression following the deposition of the M-1 shale and preceding shallow-water deposition of Facies N (Skall, 1975).

Facies N is light gray, dense, micritic limestone, or occasionally dolostone, which generally has laminite, stromatolite, blotchy bedding, or fenestrate structures (Plate VI: 5). Evaporite crystal molds and gypsum nodules occur rarely. It is lithologically similar to Facies L but lacks the green shale layers. Fauna consists of minor *Amphipora, Charophyta,* calcispheres, and rare brachiopods and ostracods. Facies N ranges from 10 to 18 m in thickness and grades upward into Facies O. The environment of deposition was predominantly that of tidal-flat sedimentation (Skall, 1975).

Facies O is light to dark brown limestone that ranges from sandy micrite to micritic sand (Plate VI: 6). The sand is generally nonskeletal and consists of sand-size intraclasts; larger intraclasts are common. The limestone is slightly argillaceous with shaley wisps and thin shale layers. Lithologies more typical of facies N are occasionally intercalated. Thin-shelled brachiopods are the most common fossil, followed by *Amphipora* and massive stromatoporoids. Thickness of Facies O ranges from 25 to 50 m depending on the nature of the contact with Facies N (Adams, 1975). Facies O represents sedimentation under slightly agitated conditions of a shallow-marine platform (Skall, 1975).

Facies P is dark brown, dense, micritic limestone which contains argillaceous wisps and numerous large intraclasts (Plate VI: 7). It is lithologically similar to Facies O but contains numerous crinoids and brachiopods instead of an *Amphipora* assemblage. The unit ranges in thickness from 6 to 12 m. Facies P represents a deepening marine platform and foreshadows marine sedimentation of the Upper Devonian (Frasnian) Hay River Shale (Skall, 1975). Braun (1967) considers the contact between the Slave Point and Hay River Formations to be disconformable in the Hay River area, but Skall (1975) found no evidence of a disconformable surface in drill core. From the standpoint of paleoenviron-

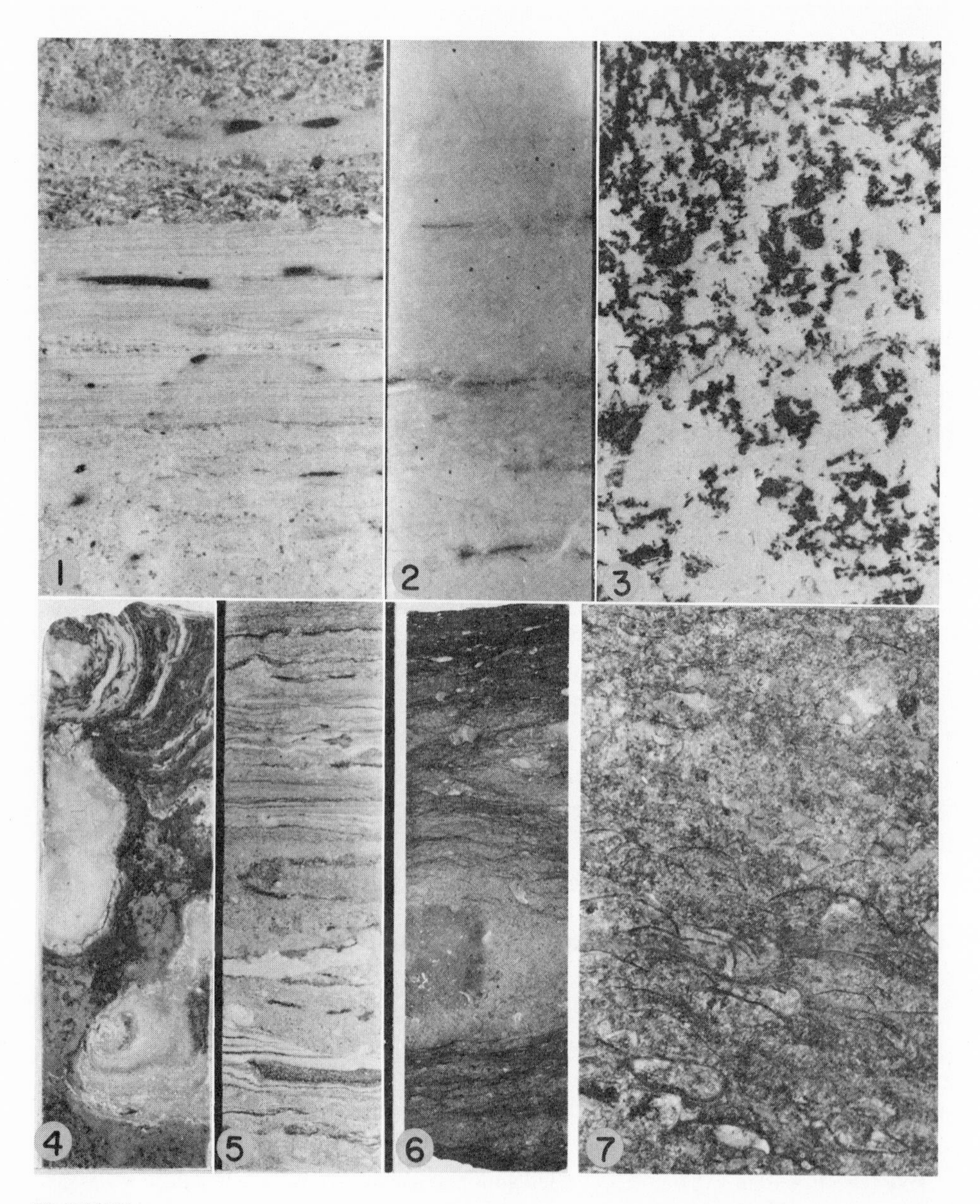

PLATE VI

Watt Mountain and Slave Point Formations (samples are polished cores, 35 mm in horizontal dimension).

1. Watt Mountain Formation – Facies L – light gray, dense limestone with algal and skeletal sand laminae.
2. Facies L – light green-gray, argillaceous dolostone with scattered blotchy mottling and shale chip conglomerate.
3. Facies L – very light gray, micritic limestone impregnated with bitumen.

ments, there is no reason to anticipate a major break in the sedimentation pattern of gradual deepening water conditions represented by Facies N, O, and P, and the Hay River Formation.

Structural evolution

Pre-Givetian tectonics. The fault zone trending N45°E can be traced for over 500 km in Precambrian rocks of the Churchill Province of the Canadian Shield before it is covered by Paleozoic rocks northeast of Pine Point (Fig. 1); a similar structural trend is present along its projection in the subsurface of northeast British Columbia (Sikabonyi and Rogers, 1959). This trend marks a major zone of prolonged tectonic disturbance which first became active in early Proterozoic time. Sedimentologic and structural investigations by Hoffman (1969) have revealed that the Proterozoic succession in the East Arm of Great Slave Lake was deposited in a long-lived, deeply subsiding linear trough, and Hoffman et al. (1974) have interpreted the zone to be an aulacogen. It is not known how far this feature extends to the west of the exposed Precambrian rocks under Phanerozoic cover and the Great Slave Lake. Hoffman et al. (1974) suggest that aulacogens are especially susceptible to reactivation and may control sedimentation even after long periods of dormancy.

Following the late Proterozoic and early Paleozoic erosional period, the Old Fort Island and Mirage Point formations were deposited on a gently sloping shelf. Silurian or Lower Devonian strata have not been recognized in the area, suggesting that the region was emergent and undergoing erosion during this time. The Chinchaga Formation was deposited during early Middle Devonian on the eroded surface. Surface sections and limited subsurface information suggest that the Chinchaga is a relatively tabular unit in the southern Great Slave Lake area (Norris, 1965).

Development of the Pine Point barrier complex. The detailed stratigraphic analysis of Skall (1975) has permitted documentation of the development of the carbonate sediments of the Pine Point barrier. His interpretation of the sequence of events is shown in Fig. 8. Following deposition of the platform carbonates of the Keg River Formation (Facies A), gentle tectonic arching of the area between early and middle Givetian was responsible for establishment of shoal conditions and initiation of barrier development. During this first stage (Fig. 8A), the depositional environments of the Organic Barrier Facies (D), the Clean Arenite Facies (E), the South Flank Facies (J), and probably the

---

4. Slave Point Formation – Facies M-2 – light brown limestone with bulbous stromatoporoids impregnated with bitumen.
5. Slave Point Formation – Facies N – light gray brown, laminated limestone.
6. Slave Point Formation – Facies O – dark brown, micritic limestone with abundant irregular argillaceous laminae.
7. Slave Point Formation – Facies P – medium brown limestone with common thin-shelled brachiopods and small intraclasts.

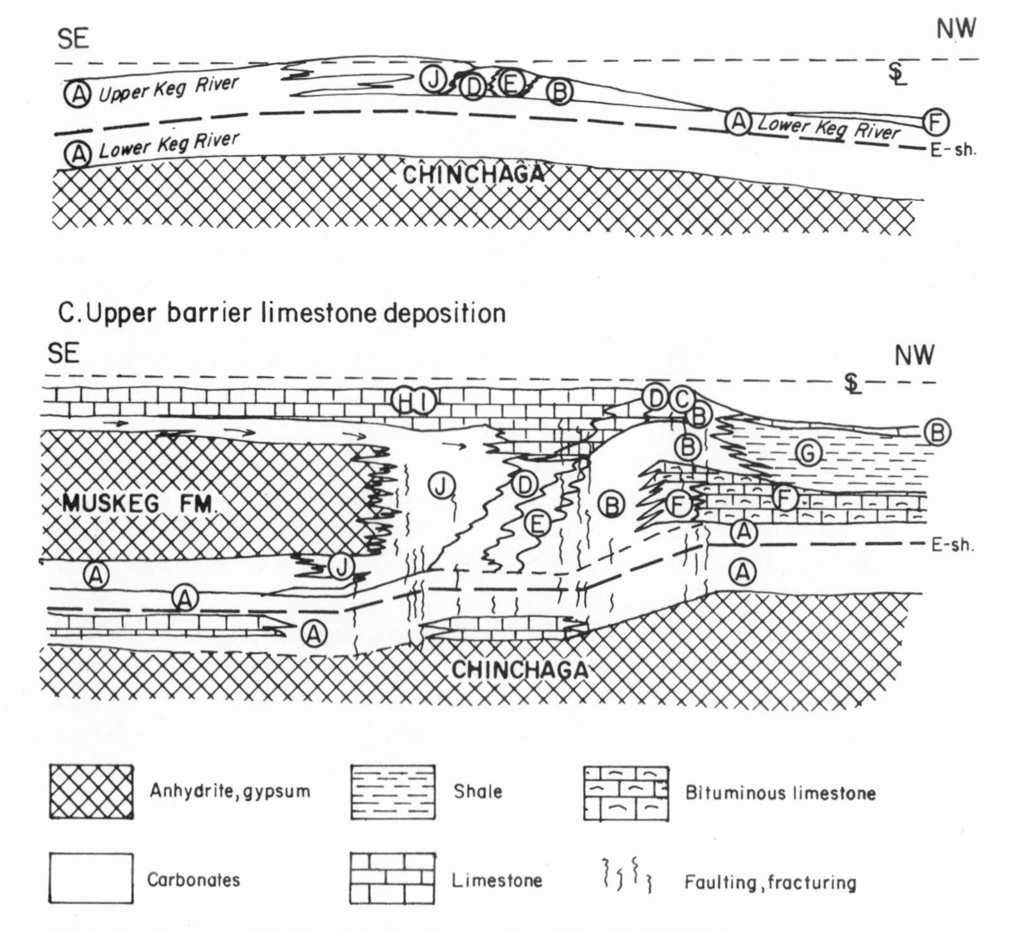

Fig. 8. Development of the Pine Point barrier complex. (After Skall, 1975.)

Tentaculites Facies (F) came into existence. The barrier apparently emerged hundreds of kilometers from the nearest coast and exerted only minor influence on regional sedimentation during initial development. With additional tectonic adjustments and concomitant stabilization of the organic framework of the barrier, more restricted sedimentary environments were created.

A rather abrupt increase in thickness of sediments between the Amco and E shales is present in the southern part of the Pine Point area (Figs. 2 and 6) and is interpreted to be the result of tectonic adjustments along what is referred to as the South Hinge (Skall, 1975). This subtle tectonic movement caused a higher rate of subsidence to the south and resulted in the establishment and maintenance of restricted conditions which promoted

B. Barrier stabilization and Muskeg evaporite deposition

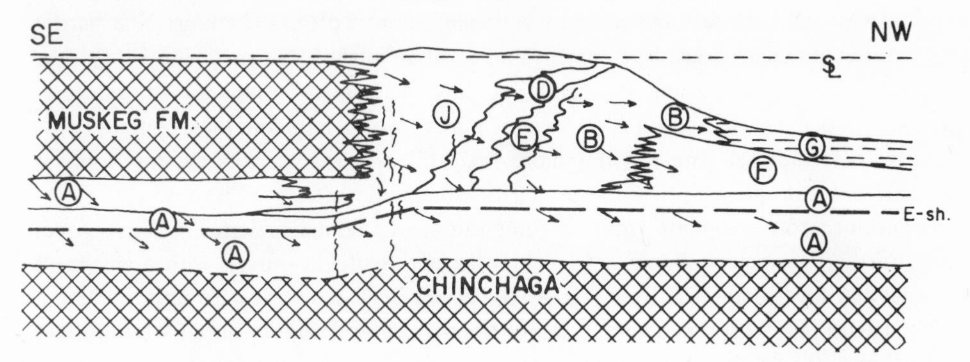

D. Barrier karstification

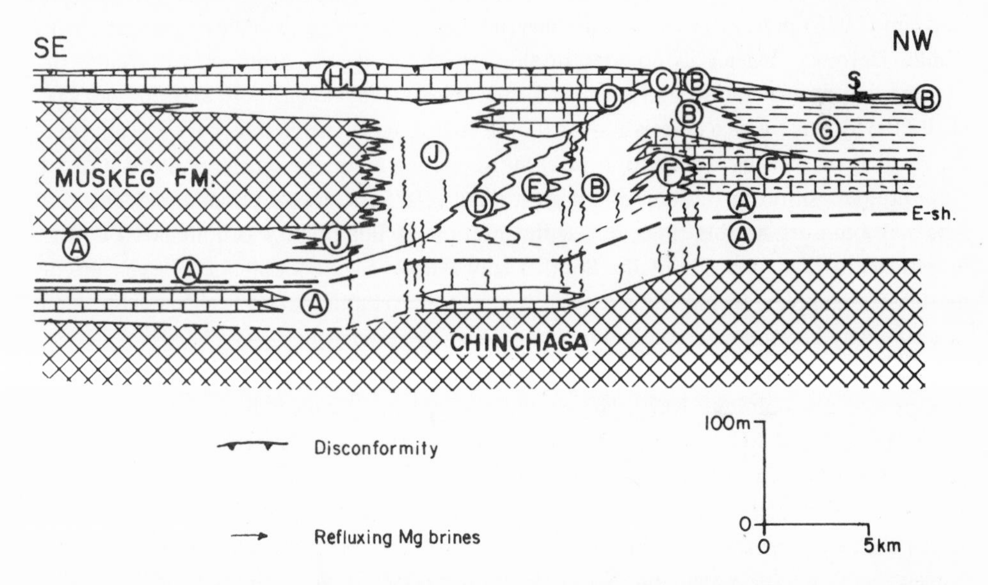

precipitation of calcium sulfates in vast evaporite pans in the back-reef area. The barrier became firmly established during this time and was effectively separated from the evaporite area by the extensive tidal-flat zone of the South Flank Facies (Fig. 8B). Precipitation of evaporites continued as long as the higher rate of subsidence was maintained in the south and resulted in the deposition of 120 m of interbedded evaporite and dolostone strata of the Muskeg Formation. Displacement along the South Hinge amounted to only 20 m over a prolonged period of time but was reponsible for the rapid facies change between the South Flank Facies and the Muskeg Formation (Skall, 1975).

While evaporites were being deposited south of the South Hinge, the barrier continued to grow on the more stable north slope. The Organic Barrier Facies (D) grew seaward in time over its own fore-reef deposits because of the very slow rate of subsidence

(Fig. 8B). Skall (1975) recognized that during its total development, the barrier migrated horizontally about 10 km and climbed vertically about 165 m. Therefore, the barrier units represent diachronous development and form an angle between 1° and 2° with the time units of the Amco and E Shales. Intricate facies relationships are the result of sporadic barrier growth, presumably due to minor sea-level fluctuations. Wiley (1970) indicates that the reef core was not more than 450 m wide and that the reef top was probably not more than 15 m above the surrounding sediments. The Clean Arenite Facies (E) continued to accumulate from the diminution of skeletal material in the fore-reef area. Further downslope the skeletal debris became gradually mixed with argillaceous material and formed the Off-Reef Facies (B). Facies B graded basinward into the *Tentaculites* Facies (F); with time Facies F was overlain and eventually replaced by the Buffalo River Facies (G). During this early stage of barrier development, the limestones were converted to fine dense to sandy dolostones (Skall, 1975). Maiklem (1971) and Bebout and Maiklem (1973) present evidence for subaerial exposure during early development of the Middle Devonian barrier carbonates in the Elk Point Basin. Although differences in stratigraphic nomenclature and scale and area of study between these and the work of Skall (1975) make exact correlations difficult, it appears that this time period is approximately that marked by termination of early development of the barrier and cessation of evaporite deposition in the Pine Point area. Skall (1975) does not recognize evidence for subaerial exposure at this time, but sedimentation at this point was dominated by the supratidal tidal-flat deposits of the South Flank Facies (Fig.8B). Ephemerial exposure of the barrier complex would only require minor sea-level fluctuation; evidence for such exposure may be somewhat masked by later diagenetic effects.

Growth of the barrier was interrupted by a second phase of tectonic adjustments that created the Main and North Hinges and resulted in a marked change in sedimentary facies (Figs. 4 and 8C). Again, the exact time of tectonic adjustment is not known; it occurred after deposition of the E Shales but prior to deposition of the Watt Mountain Formation. Skall (1975) suggests that the second tectonic phase began at a late stage of barrier development and probably lasted throughout the late middle Givetian. Pre-Amco displacement along the Main and North Hinges was responsible for thinning of the barrier by 45 m between the E Shales and Amco markers. A higher rate of subsidence was maintained in the south and now affected the previous tidal-flat area as well (Fig. 8C). Evaporite precipitation was terminated, and tidal-flat conditions were only occasionally affected in the back-reef. Instead, the extensive lagoonal deposits of the Gastropod and *Amphipora* Facies (H and I) dominated the back-reef area. The Organic Barrier Facies was represented by subfacies D-2. The Clean Arenite Facies (E) was replaced by the Shallow Fore-Reef Facies (C), and sedimentation of the Off-Reef Facies (B) and the Buffalo River Facies (G) continued (Skall, 1975). The upper part of the barrier was not affected by the processes that converted the lower barrier lithologies into fine, dense to sandy dolostones, and the facies of the upper part of the barrier remained limestones (Fig. 8C).

Growth of the Pine Point barrier was terminated by marine regression that resulted in a partial disconformity and the development of a karst surface. Subaerial exposure and diagenesis affected the topographically higher parts of the barrier (Facies C, D, H, and I), but shallow-water sedimentation continued in the lower-lying off-reef area (Facies B) (Fig. 8D). Skall (1975) reconstructed the post-erosional topography of the barrier using the overlying Amco Shale as the datum. The higher areas of the barrier are within 12 m of this marker, but the Off-Reef Facies is only within 42 m. Therefore, most of the barrier was at least 30 m above sea level and was probably more, as it cannot be determined how much of the barrier was eroded during this time (Fig. 8D). The erosional period has been timed by ostracod data to mark the boundary between the middle and late Givetian (Skall, 1975).

Following the erosional period, the late Givetian Watt Mountain Formation (Facies L) was deposited during initial marine transgression. The irregular surface on the eroded barrier and the sporadic nature of marine encroachment resulted in the several ephemeral shallow-water depositional environments that are represented by the varied lithologies of Facies L. Wiley (1970) recognized five phases of marine transgression separated by periods of minor regression. Marine transgression enveloped the eroded barrier from the north; basinward the Watt Mountain Formation is thicker and conformably overlies the Off-Reef Facies B (Wiley, 1970). Facies M was the result of short-lived, deepening marine conditions; this unit was followed by the tidal flat sedimentation of Facies N. Facies O and P and the Hay River Shale reflect slow but steady increase in subsidence which was prevalent over the entire Pine Point area (Skall, 1975).

Therefore, the subtle tectonic adjustments along a N65°E trend were responsible for pronounced facies development of the Pine Point barrier complex during middle Givetian time. Displacements along the South, Main, and North Hinge zones were compensated for by faulting in consolidated strata, slumping of unconsolidated sediments, and shifts in depositional environments. The South Hinge coincides with the northernmost occurrence of evaporites of the Muskeg Formation, and the Main Hinge marks the northernmost extension of the C-Horizon. The North Hinge is slightly south and parallel to the southern limit of the Buffalo River Facies (Figs. 4 and 5). Middle Devonian faulting and fracturing is concentrated along the hinge zones but is not restricted to them (Skall, 1975). Numerous gentle folds trending parallel to the hinge zones are also present. Although successful exploration of the Pine Point district was based on the concept of relationship of mineralization to basement faults projected from the East Arm of Great Slave Lake, Norris (1965) pointed out that the N65°E structures do not parallel the trace of the basement faults based on aeromagnetic data (Fig. 1) and suggested that the two trends are tectonically unrelated. Although both sets of structures are tensional features, the exact relationship between basement faults and the Middle Devonian tectonic adjustments must be regarded as uncertain.

Post-Devonian tectonics. Little is known about post-Devonian tectonics in the southern Great Slave Lake area largely because of the absence of late Paleozoic and younger strata in most of the region. Skall (1969) has recognized faults with post-Amco displacement in the Pine Point area. Late Paleozoic and Cretaceous strata are preserved in some areas west of Great Slave Lake (De Wit et al., 1973) and probably once extended much further to the east of the present outcrop areas. De Wit et al. (1973) feel that the most important uplifts in the region took place during the late Paleozoic and early Mesozoic and during the Late Cretaceous and Tertiary (?). These tectonic movements may have caused further displacement along older fault zones and are responsible for the present southwesterly dip of the Pine Point barrier complex.

DIAGENESIS AND DEVELOPMENT OF SULFIDE-HOSTING STRUCTURES

Diagenesis is used here in the broad sense of Murray and Pray (1965) to include "... those natural changes which occur in sediments or sedimentary rocks between the time of initial deposition and the time – if ever – when the changes created by elevated temperature, or pressure, or by other conditions can be considered to have crossed the threshold into the realm of metamorphism." It appears that all of the processes that have affected the Pine Point barrier, including sulfide mineralization, are within the realm of diagenesis (Jackson and Beales, 1967; Dunsmore, 1973). Dolomitization and karstification are major aspects of carbonate diagenesis at Pine Point which were of great importance in the preparation of the fluid-transporting and sulfide-hosting structures. (For a general summary of limestone and dolomite diagenesis, see Chilingar et al., 1979a, b.)

Dolomitization

All of the Pine Point orebodies occur in dolostones of several types in several depositional facies (Fig. 6). Although geochemical and isotopic data are available for some dolostone types (Fritz and Jackson, 1972), the lack of such data for the extensive dolostones of the lower Pine Point Group (Fig. 5) precludes an effective classification based on these parameters. Therefore, the descriptive subdivisions of "fine, dense to sandy dolostone" and "coarse dolostone" will be used (Skall, 1975).

Fine-crystalline dolostones. The "fine, dense to sandy" dolostones of Skall (1975) may be divided into two types. One type includes all fine-crystalline dolostones for which field and petrographic relationships suggest that deposition took place in a supratidal to intertidal environment. These dolostones consist of very uniform, very fine-crystalline (generally less than 20 μm), dense rocks with laminite, stromatolite, intraclast, bioturbation, and fenestrate structures and without megafossils. Subfacies J-1 and J-2 are probably of this type (Table II), and the dolostones of the Watt Mountain and Slave Point Formations probably formed under similar conditions (Fritz and Jackson, 1972).

Other units consist of fine-crystalline dolostones for which stratigraphic, petrologic, and faunal evidence indicates that these units were deposited as calcium carbonate sediments in a subtidal normal marine to slightly restricted environment. These lithologies generally are fine-crystalline (20–150 μm), dense to sandy dolostones with varying amounts of coarser skeletal material. The dolostones of Facies B-1, B-2, D-1, D-3, E, J-3, J-4, and J-5 of the Pine Point Group are of this type (Table II). With the exception of a minor amount of Facies B limestone, these dolostones in the Pine Point Group have no preserved calcium carbonate precursors (Fig. 5). Thus, it appears that dolomitization was a relatively early process which equally affected lithologies of many differing depositional environments within the barrier complex. Strata deposited in supratidal, intertidal, and subtidal environments are completely dolomitized.

Skall (1975) proposed that all of the fine, dense to sandy dolostones of the lower Pine Point Group were the result of the migration of dense, high Mg/Ca brines from the vast evaporite pans represented by the Muskeg Formation through the lower barrier (Fig. 8B), the so-called reflux mechanism for dolomitization popularized by Adams and Rhodes (1960). There are hydrologic objections to the reflux model, and Hsu and Siegenthaler (1969) have proposed that dolomitization in this type of geologic setting is the result of evaporative pumping of seawater through the carbonate sediments. Maiklem (1971) has documented regional subaerial exposure of the Pine Point barrier at the period of time corresponding to the cessation of tectonic adjustments along the South Hinge in the Pine Point area (Fig. 8B). Therefore, meteoric water also may have influenced the formation of the fine-crystalline dolostones of the lower barrier.

Coarse-crystalline dolostone. Skall (1975) defined Facies K (Presqu'ile) as the coarse-crystalline dolostone which occurs between the lower barrier units and the Watt Mountain Formation (Fig. 5). The distribution of the coarse-crystalline dolostone is further restricted to the area between the Hinge Zones and below the post-middle Givetian disconformable surface (Figs. 4 and 5). This lithology consists of dolomite crystals greater than 200 μm in size, and the unit is a diagenetic facies superimposed on back-reef, reef, and fore-reef strata (Skall, 1975). In contrast to the ubiquitous fine-crystalline dolostones of the lower barrier, the original limestone lithologies are preserved as isolated remnants within the coarse-crystalline dolostone (Plate V: 1–5) and as extensive back-reef strata of Facies H and I (Fig. 5). Contacts between coarse-crystalline dolostone and limestone commonly transect bedding, are irregular, and relatively sharp. Relic depositional textures are preserved locally in the coarse-crystalline dolostone, particularly in the lower units (Plates IV and V). White dolomite is a ubiquitous accessory, and its pervasive introduction has obliterated much of the original nature of the upper part of Facies K in some areas (Plate IV). The contact of the coarse-crystalline dolostone with the underlying fine-crystalline dolostones of the lower barrier is commonly abrupt but locally is gradational over a few meters. These relationships indicate that the coarse-crystalline dolostone is distinctly different from the fine-crystalline dolostone and originated by a different mechanism.

Four mechanisms have been proposed for the origin of the coarse-crystalline Facies K dolostone: (1) dolomitization of coarse-grained reefal limestones (Norris, 1965), (2) recrystallization of fine-crystalline dolostones (Campbell, 1967), (3) dolomitization of barrier limestones by pre-sulfide heated brines circulating along the Hinge Zones during post-Givetian times, and (4) dolomitization of barrier limestone related to the mixing of meteoric water with seawater during the post-middle Givetian erosional period (Kyle, 1977). The coarse-crystalline nature of Facies K does not reflect an original coarse-grained sediment because it is not restricted to the reefal facies and because the coarse-grained reefal lithologies of the lower barrier have not been converted to coarse-crystalline dolostone. The extensive recrystallization of fine-crystalline dolostone into Facies K is unlikely because of the thin beds of fine-crystalline Facies J dolostone (C-Horizons) preserved within the coarse-crystalline dolostone. The coarse-crystalline dolostone is not an alteration effect directly related to sulfide mineralization because it is much more extensive than sulfide mineralization, is not associated with the orebodies in the lower barrier, and is not present above the disconformity in those orebodies that extend into the Watt Mountain Formation.

Both the reflux and evaporative pumping mechanisms can be eliminated as possible methods of generating the coarse-crystalline dolostone because the evaporitic conditions responsible for the deposition of the Muskeg evaporites and tidal-flat lithologies of Facies J had ceased to exist in the Pine Point area by the time of upper barrier sedimentation. Instead, predominantly lagoonal sedimentation resulted in Facies H and I (Fig. 8C), and the existence of bedded evaporites in the upper Pine Point Group cannot be demonstrated. Skall (1975) acknowledged that the distribution of the coarse-crystalline dolostone below the Watt Mountain disconformity suggests an origin related to the erosional surface. He chose to relate its development to post-Givetian circulation of warm, magnesium-rich fluids along the Hinge Zones. Skall suggested that the erosional period represented by the disconformity served to increase the permeability of the barrier and that the Watt Mountain shale beds overlying the disconformity restricted the upward migration of dolomitizing fluids. This interpretation is influenced greatly by fluid-inclusion data (Roedder, 1968a) which indicate that the distinctive white dolomite associated with sulfide ore minerals was formed at temperatures of 90–100°C from brines which contained 15–20 wt.% total salts, that is, brines similar to those which deposited the sulfides. Not only is the development of the coarse-crystalline dolostone genetically unrelated to sulfide mineralization, but also as Skall (1975) suggested, white dolomite is the product of late alteration primarily of the coarse-crystalline dolostone by fluids of the sulfide-depositing system. In fact, a great deal of the white dolomite may postdate the period of major sulfide deposition. Consequently, there is no evidence to indicate that relatively hot, strongly saline fluids were responsible for the development of the coarse-crystalline dolostone. Further, the influence of non-saline water in the formation of the coarse-crystalline dolostones is suggested by their sodium contents of generally less than 100 ppm; in contrast, the fine-crystalline dolostones contain as much as 600 ppm Na (Fritz and Jackson, 1972).

The strongest evidence concerning the origin of the coarse-crystalline Facies K dolostone is its distribution below the Watt Mountain disconformity and immediately "landward" of the most "seaward" extent of the partial erosional surface (Fig. 5). The disconformity is not exposed in the Pine Point area, but the abundant drill-hole information indicates that the paleoerosional surface is one of relatively low relief with a characteristic karst topography which gently slopes "seaward", generally at less than 1 m/km. Thus, the situation provides a classic model for the dynamic mixing of meteoric water with normal marine water in the subsurface during the post-middle Givetian exposure of the barrier complex. The low relief and lack of major surface drainage on the erosional surface and the apparent lack of stratigraphic restrictions to vadose water flow in the upper barrier suggest that the water table was not much above sea level. A representative figure of 1 m above mean sea level approximately 10 km from the point of termination of the partial disconformable surface (i.e., the middle Givetian "shore") is reasonable in comparison with the analogous modern hydrologic system of the northern Yucatan Peninsula (Back and Hanshaw, 1970). Assuming that the permeability of the upper Pine Point barrier complex was relatively homogeneous during the erosional period, the Ghyben–Herzberg principle can be applied (Back and Hanshaw, 1970). This principle states that for every unit of fresh water above the mean sea level, the thickness of the freshwater lens floating on salt water of ocean-water density is about 40 units. With a hydraulic head of 1 m, the freshwater lens would have extended within the upper barrier to a depth of about 40 m below the mean level of the post-middle Givetian sea. A zone of meteoric water and seawater mixing would have existed below this depth and could account for the position of the maximum thickness of coarse-crystalline Facies K dolostone about 10 km "landward" from "shore" and sloping gently upward to mean sea level (Figs. 5 and 8D).

Kyle (1977) has suggested that the development of the coarse-crystalline Facies K dolostone was related to the dynamic mixing of meteoric water with seawater as an integral part of barrier diagenesis during the post-middle Givetian erosional period. This mixing model has been proposed for dolomitization in several varied carbonate environments (e.g. Badiozamani, 1973; Land, 1973; Folk and Land, 1975); at Pine Point it is consistent with evidence of paleogeography and other effects of subaerial exposure. Magnesium required for dolomitization could have been supplied by seawater and by magnesium expulsion during mineralogic stabilization of upper barrier carbonates in the freshwater phreatic zone. The mixing zone was probably rather thin at any particular time, but short- and long-term fluctuations in sea level, including the sporadic advance of the Watt Mountain sea, resulted in migration of the mixing zone and concomitant dolomitization of the entire area now represented by Facies K (Fig. 5).

Karstification and development of sulfide-hosting porosity

Although Beales and Jackson (1968) and Jackson and Folinsbee (1969) suggested that subaerial exposure and karstification of the Pine Point barrier could have been important in the preparation of the host rock for sulfide concentration, it was not until the

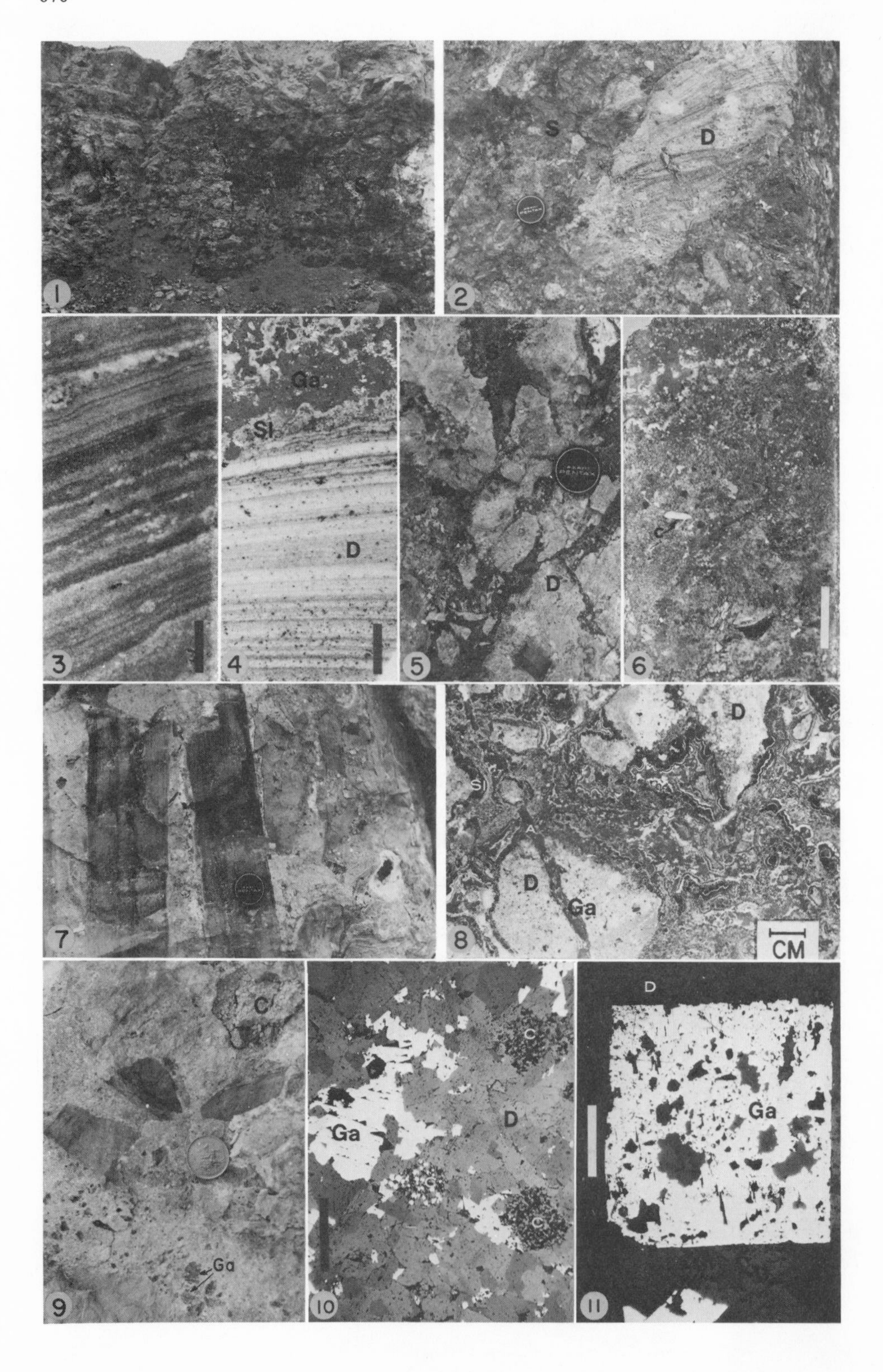
1
K
S
2
S
D
3
4
Ga
Sl
D
5
D
6
C
7
8
D
Sl
D
Ga
CM
9
C
Ga
10
C
Ga
D
C
11
D
Ga

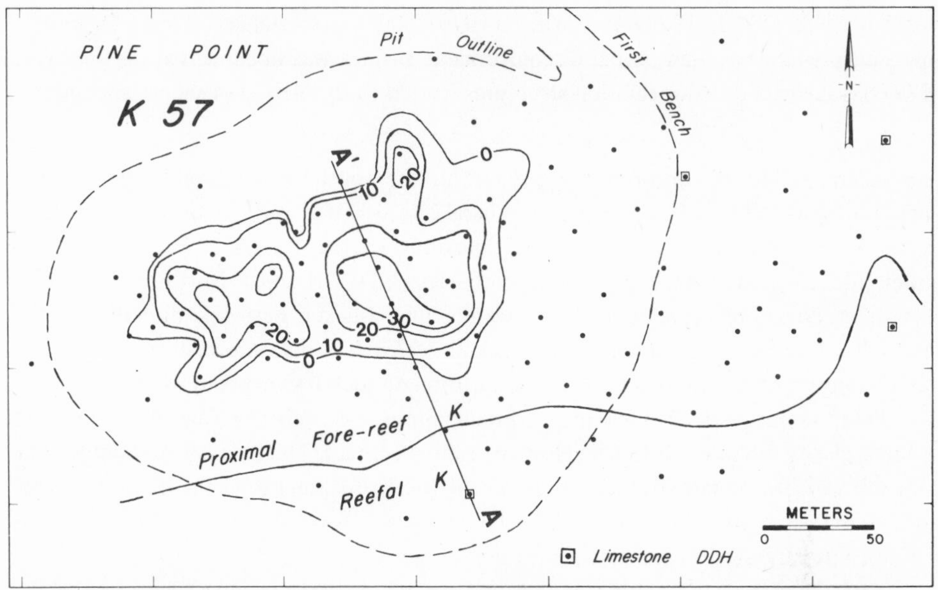

Fig. 9. General geologic features of the K57 area. Dots indicate location of diamond drill holes; limestone drill holes indicate areas of upper Pine Point Group limestone preserved within Facies K. Isopach is detritus thickness in meters.

PLATE VII

Sulfide-hosting features associated with prismatic orebodies in Facies K.

1. Abrupt transition (dashed line) from the iron-rich perimeter (*S*) of N38A orebody to sulfide-deficient coarse-crystalline dolostone of upper Facies K. N38A open pit, southeast side, first bench; width of scene is about 8 m.
2. Isolated remnant block of laminated detritus (*D*) within massive sphalerite and galena (*S*); K62 open pit, first bench.
3. Medium gray, laminated detritus with scattered green clay blebs and white dolomite. Sample K57-29-80; scale is 1 cm.
4. Light gray, laminated detritus (*D*) with introduced sphalerite (*Sl*) and galena (*Ga*). Sample K62-25; scale is 1 cm.
5. Detritus (*D*) with scattered tidal-flat dolostone fragments; irregular fractures cemented by sphalerite and galena (*S*). K62 open pit, first bench; lens cap is 54 mm in diameter.
6. Dolostone detritus with pervasive introduction of about 75% sphalerite and galena; note preserved blebs of green clay (*C*). Sample A70-4-193; scale is 1 cm.
7. Large tidal-flat dolostone fragments in detritus; K62 open pit, center, first bench.
8. Detritus "fragments" (*D*) cemented by colloform sphalerite (*Sl*) and galena (*Ga*); note isolated colloform crusts without substrates and galena veinlet transecting detritus fragment and its colloform sphalerite rim (*A*). Sample N38A-7.
9. Fragments of tidal flat dolostone and green clay (*C*) in detritus; galena crystals (*Ga*) in matrix. K62 open pit; coin is 23 mm in diameter.
10. Photomicrograph of dolostone detritus (*D*) with intercrystalline galena (*Ga*) and clay (*C*). Sample Z53N-1-52, polished surface 6320, reflected light; scale is 0.4 mm.
11. Photomicrograph of galena in dolostone detritus; irregular crystal edges and remnants of detritus within the single crystal suggest that the crystal formed by successive stages of detritus dissolution and galena precipitation. Sample K62-5, polished surface K-5, reflected light; scale is 0.4 mm.

work of Skall (1970, 1975) and Wiley (1970) that the existence of the post-middle Givetian, pre-late Givetian partial disconformable surface was documented. Kyle (1977, 1979) has identified major solution structures related to this period of subaerial exposure and has documented their relationship to sulfide concentrations.

Karstification. The disconformity is present in the southern and central part of the Pine Point area (Figs. 4 and 5) where the basal late Givetian Watt Mountain Formation, Facies L, overlies the coarse-crystalline dolostone of the Presqu'ile Facies K or the undolomitized upper-barrier limestones of Facies C, D-2, H and I. North of the coarse-crystalline dolostone development, a considerably thicker Facies L directly overlies Facies B without apparent disconformity (Skall, 1975). The disconformable contact commonly consists of a zone of green clay and carbonate rock fragments overlying the upper Pine Point Group strata which may contain minor green clay in vugs, fractures, and bedding planes for some distance below the rubble contact. Thickness of the rubble zone is highly variable, apparently depending on the local topography on the disconformable surface, but generally is less than 5 m. In the A70 area, a waxy green shale bed as much as 4 m thick usually overlies and grades into the rubble zone.

Stratigraphic studies within and around several mineralized zones have revealed the presence of major solution features (Kyle, 1977). These features appear to control the distribution of sulfides and to account for the geometry of the "prismatic" orebodies with restricted horizontal dimensions relative to vertical extent of the "tabular" orebodies with restricted vertical extent relative to the horizontal dimensions (Skall, 1972). Within many prismatic orebodies in the upper Pine Point barrier are strata which are not present in standard Facies K sections, even immediately adjacent to the mineralized zones (Plate VII: 1). Generally, this material is light gray, fine- to coarse-crystalline, friable dolostone (Plate VII), commonly with a detrital texture. This detritus may be laminated (Plate VII: 1,2,4) and may contain fragments of green clay and fine-crystalline tidal-flat dolostone, particularly in the upper portion (Plate VII: 3,5,6). Generally, these fragments are only a few centimeters in the longest dimension but are occasionally as much as a meter (Plate VII: 5).

The K57 area illustrates well the nature of the detritus and its relationship to adjacent strata and to sulfide concentrations of both the prismatic and tabular types. The K57 orebody and the geologically similar K62 orebody occur within the coarse-crystalline Presqu'ile dolostone along the Main Hinge Zone in the western part of the district (Fig. 7). In this area, Facies K has a maximum preserved thickness of about 45 m and may be divided into upper and lower units. The lower unit consists of tan to light gray, coarse-crystalline dolostone which contains megafossil (Plate IV: 8a,b) and uniform granular (Plate IV: 8c,d) lithologic types. These rocks are interpreted to represent reefal and proximal fore-reef depositional environments, respectively. A small remnant of Facies D-2 limestone was exposed within the coarse-crystalline dolostone in the south side of the K57 open pit. An irregular, interfingering facies boundary between the two lithologic types of the lower Facies K is present just to the south of the mineralized zone (Figs. 9 and 10).

Direct evidence concerning the original nature of the upper Facies K unit has been largely obliterated by the pervasive introduction of white dolomite (Plate IV: 4-7). Isolated limestone remnants lateral to this unit (Figs. 9 and 10) suggest that this material was originally the lagoonal Facies H/I. This relationship would be anticipated due to the general northwestward progradation of depositional environments in the area.

Within the zone of coarse-crystalline dolostone is an irregularly elliptical area containing light gray, fine- to coarse-crystalline, friable dolostone detritus (Figs. 9 and 10). The detritus zone is about 190 m long and 100 m wide and has a maximum detritus thickness of about 35 m in two areas. Although the top of most of the detritus zone is truncated at its subcrop, Facies L units are present in four drill holes, thus suggesting that the present detritus thickness represents most of the original material. The thickest section of detritus overlies Facies B (Fig. 10). Near the detritus zone, the contact between Facies B and Facies K is indistinct because the upper part of Facies B is medium- to coarse-crystalline with poorly preserved faunal components. Three stratigraphic units have been recognized in the K57 detritus zone (Fig. 10; Kyle, 1977). The stratigraphically lowest unit consists of relatively clean, fine- to coarse-crystalline dolostone which contains blocks of coarse-crystalline Facies K dolostone, particularly near the base. Minor amounts of detritus may be present in fractures and along bedding planes for some distance in the units underlying the detritus zone. Thickness of the lower unit is highly variable and attains a maximum of 18 m. A thinner unit of laminated detritus, consisting of alternating bands of light and medium gray dolostone with rare green clay blebs (Plate VII: 2–4), overlies the lower detritus unit (Fig. 10). Thickness of this unit is also variable, largely due to disruption by sulfide introduction, and is a maximum of 5 m. The upper unit consists of nonlaminated detritus with common small fragments of green clay and fine-crystalline dolostone, particularly in the upper part (Plate VII: 5–10), and is a maximum of 15 m in thickness.

A major detritus zone is also present in the A70 area, a zone of extensive sulfide concentration within Facies K along the North Hinge about 6 km northwest of K57 (Fig. 7). The middle Givetian stratigraphic sequence has not been truncated by recent erosion in this area. The coarse-crystalline Presqu'ile dolostone reaches a maximum of about 25 m in thickness but is entirely absent locally; the lower part of Facies K often retains the textural features of Facies B. A northward prograding depositional sequence of fore-reef, reef and back-reef limestone facies is common away from the zone of maximum coarse-crystalline dolostone development (Figs. 11 and 12); these limestones are interbedded with and laterally equivalent to Facies K. The major elliptical zone of detritus about 400 m in length and 175 m in width is surrounded by standard coarse-crystalline Facies K dolostone. Maximum detritus thickness is about 20 m, and two depocenters are indicated by structural contours on the base of the detritus zone and by thickness contours of the detritus interval (Fig. 11). The waxy green shale bed which overlies the disconformity rubble contact also is present on top of the detritus zone. In cross-section the detritus zone is broad and shallow and is overlain by late Givetian units (Fig. 12). As in

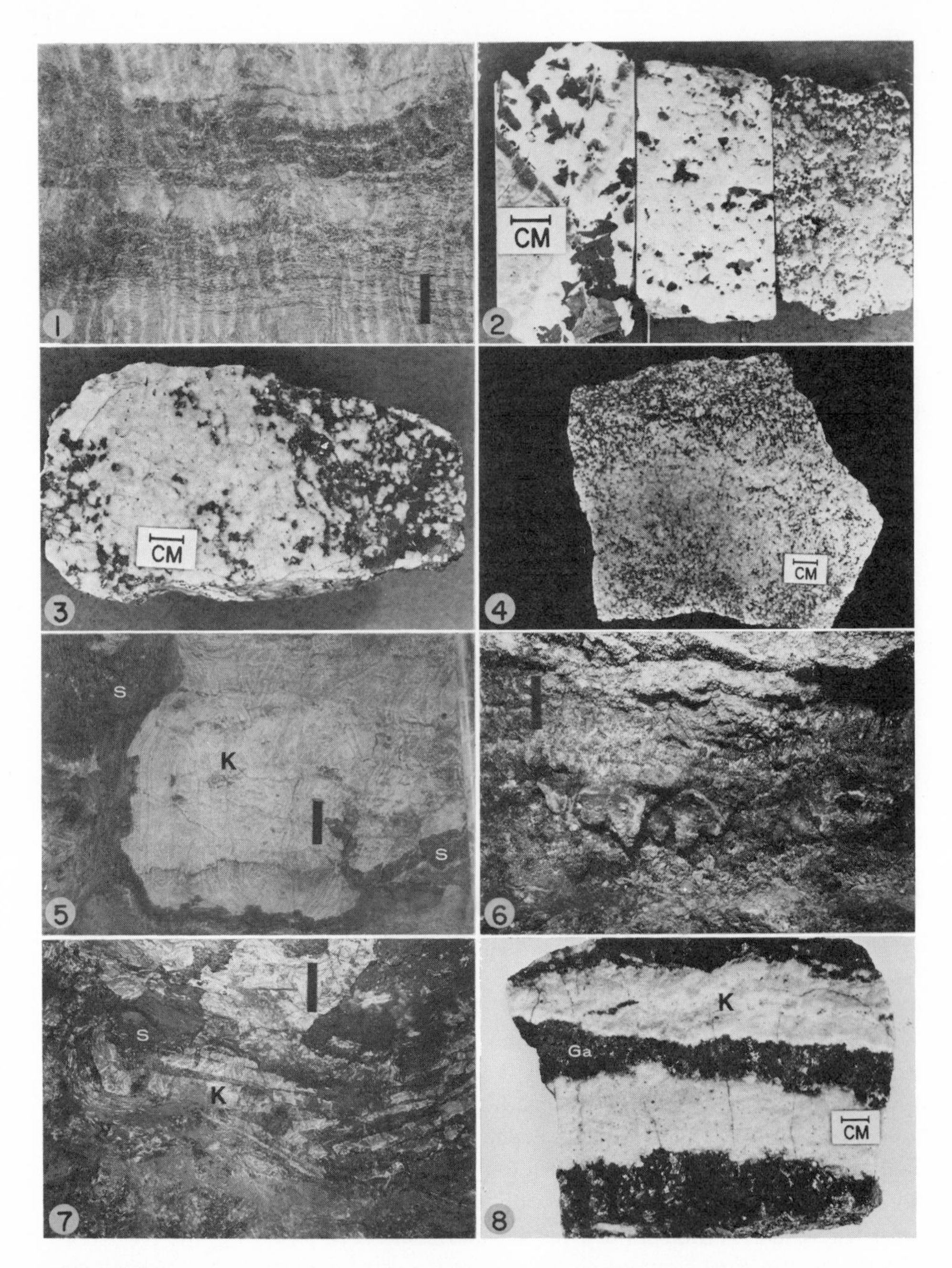

PLATE VIII

Sulfide-hosting features associated with tabular orebodies in Facies K.

1. Coarse-crystalline sphalerite in stratabound porosity zones in lower Facies K dolostone. M40 orebody, stope 3336-7; scale is 30 cm.

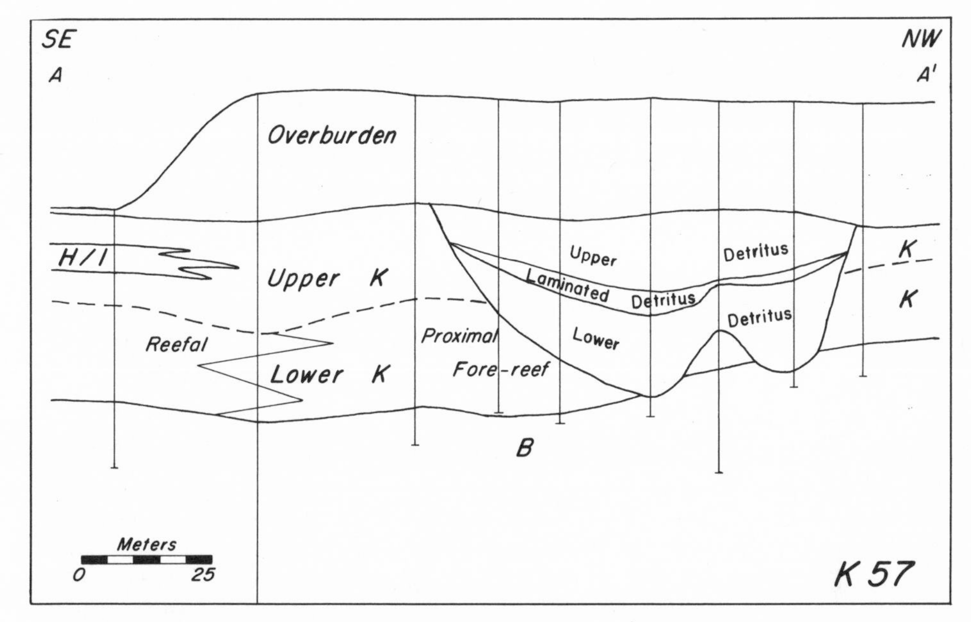

Fig. 10. Geologic cross-section of the K57 area. Vertical lines represent diamond drill holes.

the case of K57, the detritus zone overlies Facies B where thickest and Facies K around its margins.

Dolostone detritus has been documented to occur in at least eleven of the prismatic orebodies within the coarse-crystalline Facies K dolostone (Kyle, 1977). The lithologic types of detritus recognized in K57 have been observed in other detritus zones, but additional work is needed to establish the consistency of the sequence. It is also apparent that some sulfide-bearing zones lack major amounts of associated detritus. These zones correspond to the tabular orebodies in the upper barrier complex (Skall, 1972). Although there are small areas of massive sulfide concentration in these zones, most of the ore consists of stratabound sulfides confined to a relatively narrow stratigraphic interval (Plate

---

2. Uniform granular Facies K dolostone with fine- to very coarse-crystalline dark reddish-brown sphalerite. Cores from tabular sulfide zone east of K57 open pit, samples, left to right, K57-142-101, K57-128-78, K57-128-92.
3. Uniform granular Facies K dolostone with coarse-crystalline galena. Sample K57-20.
4. Uniform granular Facies K dolostone with fine-crystalline dark reddish-brown sphalerite. Sample N38A-25.
5. Facies K dolostone surrounded by massive sphalerite and galena (*S*) in a macropore. M40 orebody; scale is 30 cm.
6. Stalactitic sphalerite in tabular macropore. M40 orebody, 3134 drift east; scale is 10 cm.
7. Collapse breccia in lower Facies K dolostone cemented by sphalerite and galena (*S*). M40 orebody, stope 3337-16; scale is 30 cm.
8. Breccia of uniform granular Facies K dolostone fragments cemented by massive coarse-crystalline galena (*Ga*). Sample K62-9.

VIII). Tabular sulfide zones are common in the lower part of the coarse-crystalline Facies K dolostone in some areas; many of the prismatic orebodies have contiguous tabular sulfide concentrations.

M40 is a tabular orebody located in the central portion of the Main Hinge Zone within the cluster of N38A, N42, and O42 prismatic orebodies (Fig. 7). The sulfide concentration in M40 is stratabound in the lower part of the coarse-crystalline Facies K dolostone and averages about 3 m in thickness, although it reaches a maximum of 15 m (Fig. 6). Most of the ore occurs as open-space filling of vuggy porosity (Plate VIII: 1–4), and local collapse breccias also host sulfides (Plate VIII: 7,8). Some massive sulfide concentrations are present (Plate VIII: 5); sulfide textures indicate growth from the walls and suggest that these zones were large voids prior to mineralization. Some of these zones are thin and tabular, and contain gravity-controlled stalactitic sulfide forms (Plate VIII: 6). Major amounts of dolostone detritus of the type common in prismatic orebodies have not been recognized in M40.

Solution features associated with sulfide concentrations within the coarse-crystalline dolostone of the upper barrier differ somewhat from the sulfide-hosting features within the fine-crystalline dolostones of the lower barrier (Plate IX). The latter sulfide-bearing zones occur northeast of the present-day erosional limit of the Presqu'ile (Fig. 7), but presumably once were covered by the coarse-crystalline dolostone. Both the W17 and X15 orebodies are largely within Facies D-1 and the J subfacies (Fig. 6). Sulfide concentrations may be either massive without lithic components or distributed in intercrystalline porosity in reefal and back-reef lithologies. In addition to intraformational reefal breccias, two other types are present in W17. One type consists of fragments of Facies J fine-crystalline dolostone in a matrix of light gray, sandy, fine-crystalline dolostone; appreciable amounts of sulfides occur within breccia matrix and fragments (Plate IX: 2,3). The second type consists of Facies J fragments in a dark gray, dense micritic, calcareous matrix (Plate IX: 4). Fragments are largely dolostone, but calcareous dolostone and limestone fragments are also present; some blocks have bleached rims (Plate IX: 4). Sulfides are uncommon in this type. In addition, some breccias consist of Facies J dolostone blocks cemented by sulfides without apparent rock matrix (Plate IX: 5), and some irregular areas of sulfides occur in Facies J dolostone lacking evidence of pre-mineralization disruption by major fracturing or brecciation (Plate IX: 6). These breccias are local features and are largely coextensive with the mineralized zone. A local zone of green clay occurs in the upper part of W17 and marks the central part of a major collapse zone (Figs. 13 and 14; Kyle, 1977). The green clay has a maximum preserved thickness of about 10 m and contains pyrite and sphalerite. The green clay is similar to the waxy green shales of the disconformity and the Watt Mountain Formation, but the exact relationship between these units is indeterminable.

N204 is the most northeasterly and the lowest stratigraphically of the presently known mineralized zones (Fig. 7). Sulfides occur in the upper part of a laterally extensive zone of vuggy and intercrystalline porosity and incipient breccia (Plate IX: 8–10), known

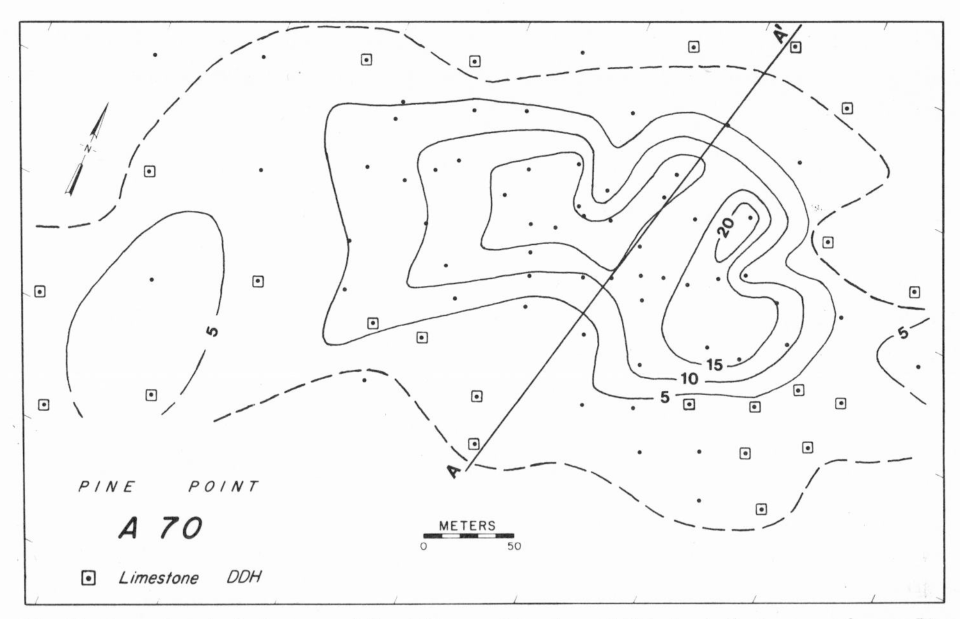

Fig. 11. General geologic features of the A70 area. Limestone drill holes indicate areas of upper Pine Point Group limestone preserved within Facies K; dashed line outlines area of Facies K thickness greater than 10 m. Isopach is detritus thickness in meters.

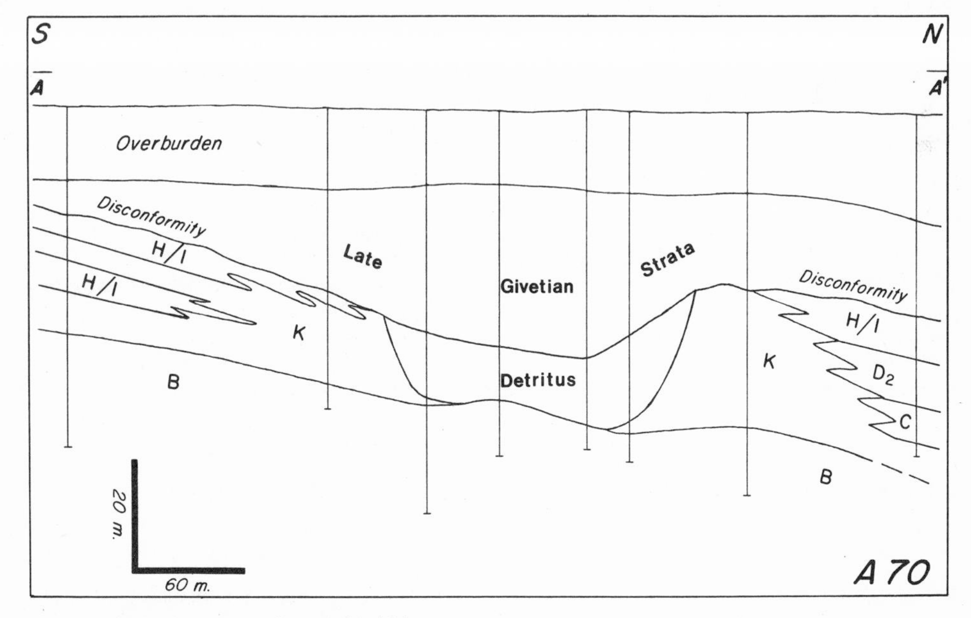

Fig. 12. Geologic cross-section of the A70 area.

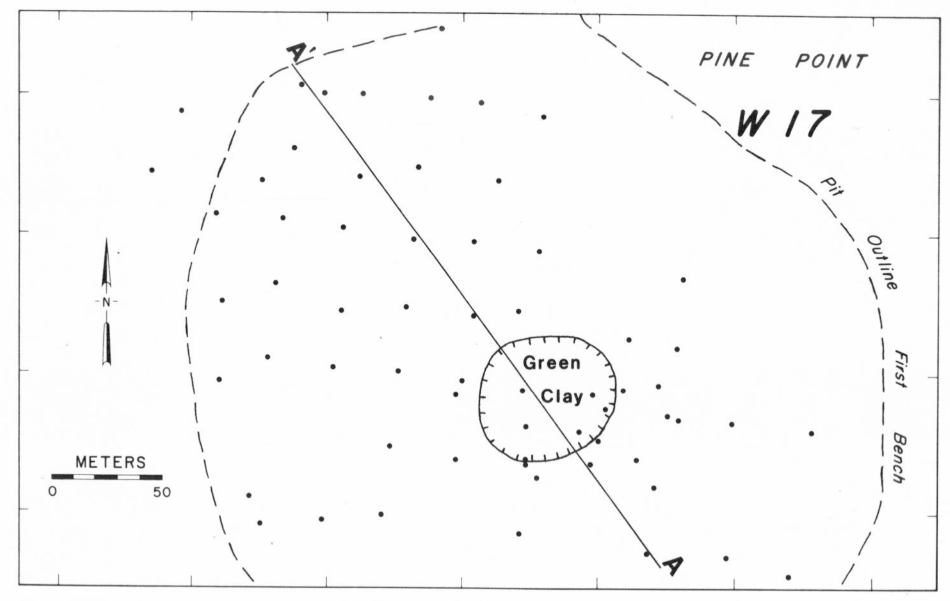

Fig. 13. General geologic features of the W17 area.

PLATE IX

Sulfide-hosting features in the fine-crystalline dolostones of the lower Pine Point Group.

1. Depression in Facies J dolostone exposed in west wall of W17 open pit; note boundary "fault" on right side. Height of exposed section is about 45 m.
2. Rock matrix breccia with pervasive massive sulfides (*S*); note extremely irregular edges on fragment of J-4 dolostone. W17 open pit, sixth bench; scale is 15 cm.
3. Rock matrix breccia (*BR*) consisting of J-2 and J-3? dolostone fragments in fine-crystalline, light gray dolostone matrix; note irregular contact with massive sulfides (*S*). The absence of sulfide encrusting growth forms on the irregular contact suggests that the sulfides formed by successive stages of carbonate dissolution and sulfide precipitation. Sample W17-17.
4. Fragments of dolostone and limestone in micritic, calcareous matrix; note bleached rims on some fragments. W17 open pit; lower scale in cm.
5. Breccia of J-2 dolostone fragments cemented by encrusting sulfides. W17 open pit; lens cap is 54 mm in diameter.
6. J-2 dolostone with irregular areas of massive sulfides (*S*); note that tidal flat laminations are not displaced adjacent to the sulfides. W17 open pit; scale is 20 cm.
7. Massive ore composed of intimate mixture of marcasite, pyrite, sphalerite, and galena. W17 open pit, third bench.
8. Facies B dolostone with sulfides in intercrystalline porosity and galena in thin fractures. Core, N204 orebody; scale is 1 cm.
9. Facies B dolostone with dark brown coarse-crystalline sphalerite in vugs and in intercrystalline porosity in burrows. Core, N204 orebody; scale is 1 cm.
10. Pervasive sulfides in intercrystalline porosity in Facies B dolostone. Core, N204 orebody; scale is 1 cm.

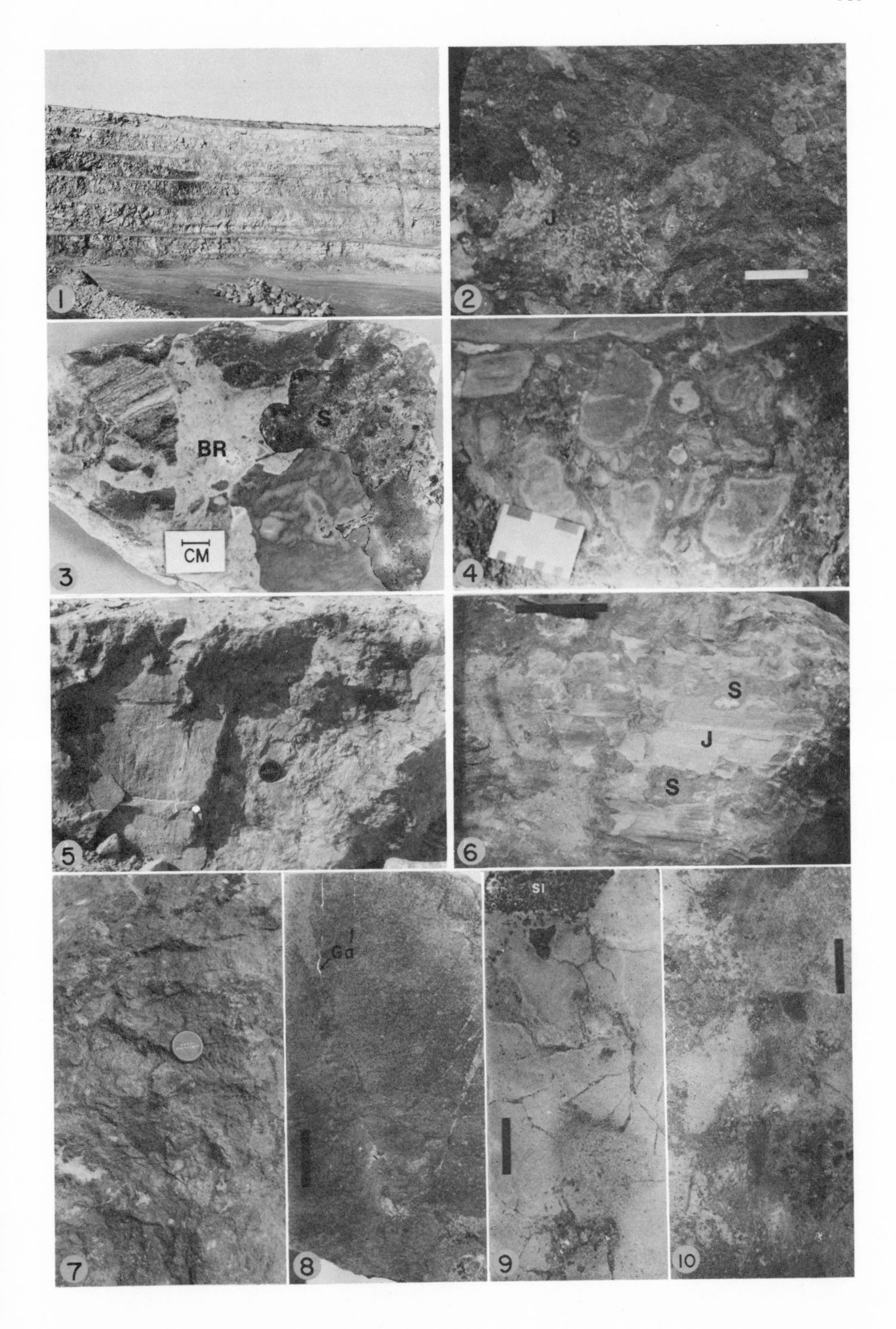

1
S
J
2
BR
S
CM
3
4
5
S
J
S
6
Gd
Sl
7
8
9
10

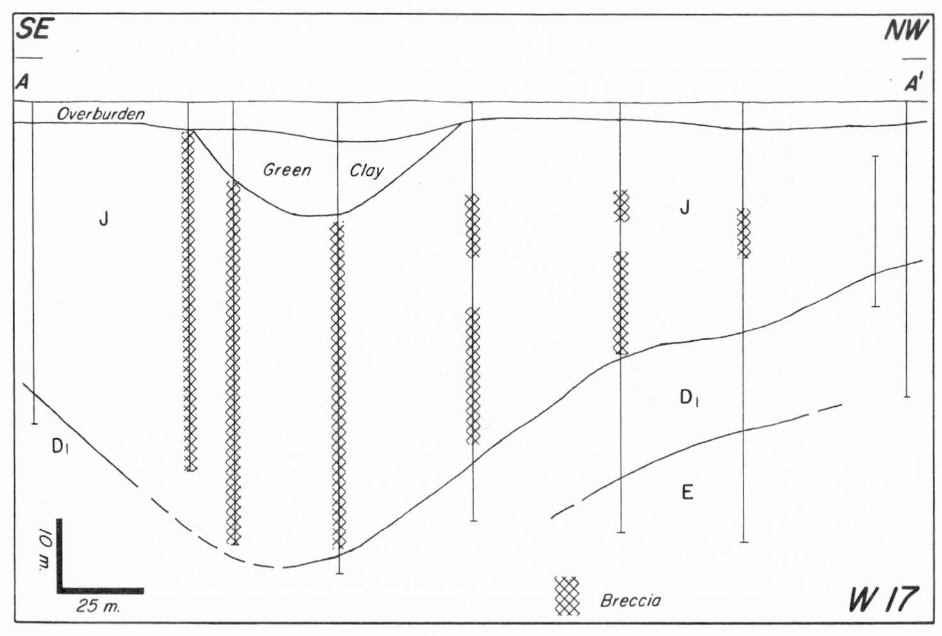

Fig. 14. Geologic cross-section of the W17 area.

as the Spongy Horizon, in the fine-crystalline dolostones of subfacies B-3 just above the Keg River Formation (Fig. 6).

Relationship of orebodies to solution features. Solution features in the Pine Point barrier complex were of considerable importance in the localization of sulfide concentrations. The detritus zones appear to have been the loci for the deposition of the prismatic orebodies and contiguous sulfide concentrations in the upper barrier. This association can be demonstrated readily by comparing the distribution of detritus with the total volumetric amount of sulfides.

K57 has a maximum total sulfide concentration of about 50 vol.% over a 53-m interval (Fig. 15). The greatest sulfide concentration is within the detritus zone (Fig. 9), but an irregular stratabound zone of lower sulfide content is present to the east of the massive zone (Figs. 15 and 16). Most of the sulfides in the stratabound zone occur as pore-filling of the uniform granular lithology (proximal fore-reef) in the lower unit of the coarse-crystalline Facies K dolostone (Fig. 10; Plate VIII: 2,3). Sulfide minerals within the detritus zone fill fractures and breccia interstices, but large intervals are massive. Despite extensive vuggy porosity within the pervasive white dolomite of the upper Facies K (Plate IV), the boundary of the prismatic orebody is sharp, and the upper Presqu'ile does not contain megascopic sulfides. Minor sulfides occur in intercrystalline porosity in the Facies B dolostone which underlies the Presqu'ile (Figs. 10 and 16).

A70 has a maximum of over 40 vol.% total sulfides over a 53-m interval (Fig. 17).

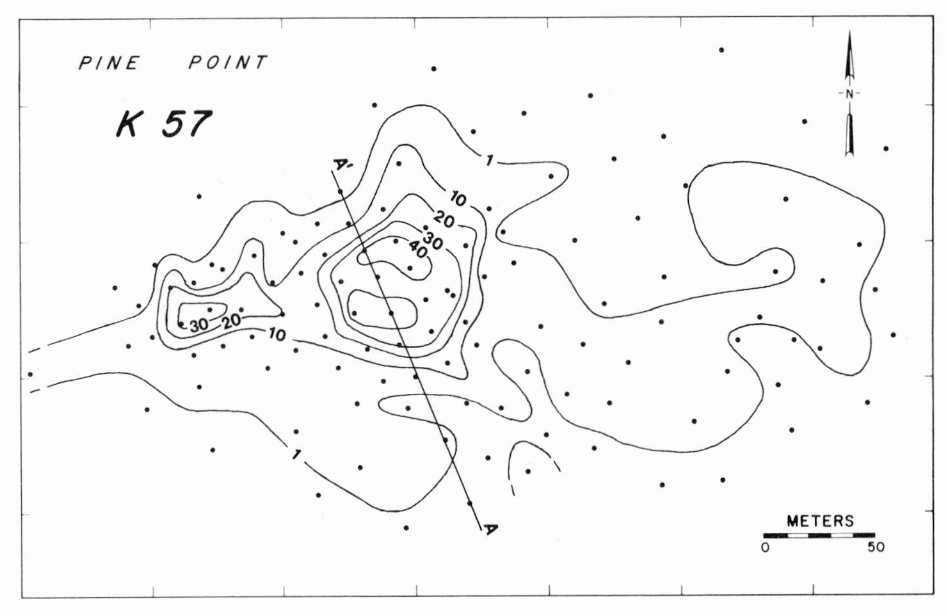

Fig. 15. Volume percent total sulfides in K57 orebody, 149.4–202.7 m.

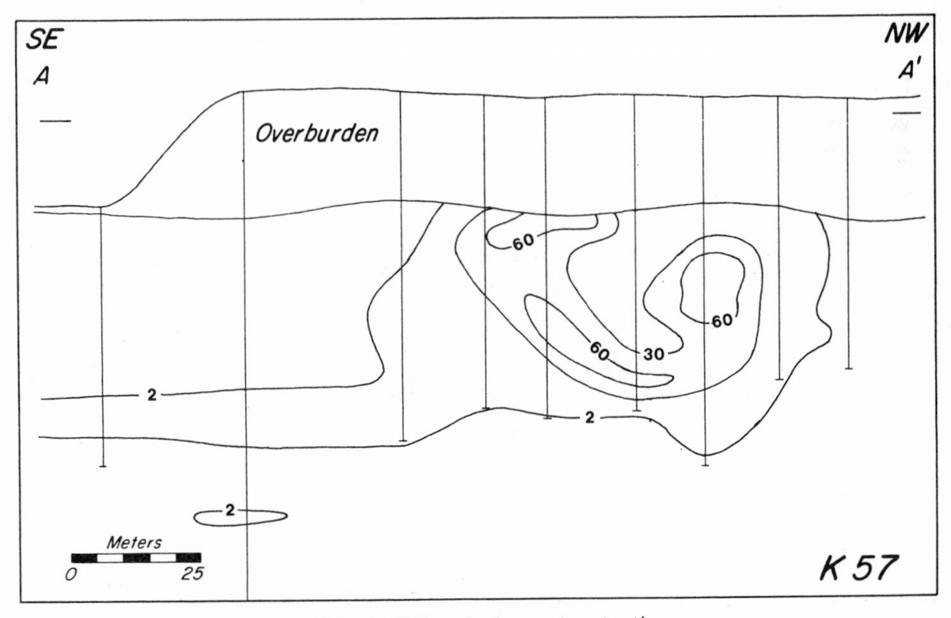

Fig. 16. Volume percent total sulfides in K57 orebody, section A–A'.

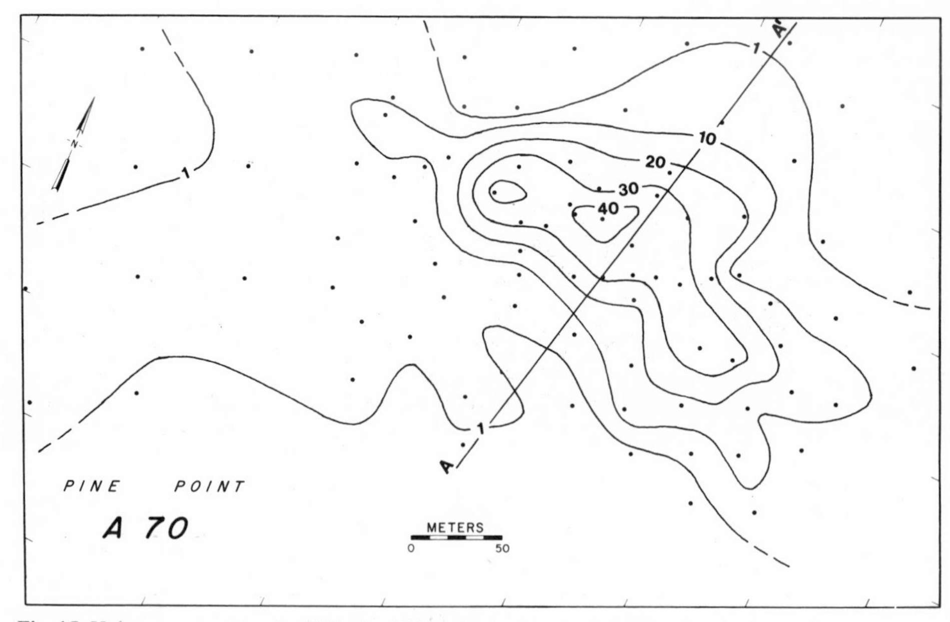

Fig. 17. Volume percent total sulfides in A70 orebody, 114.3–167.6 m.

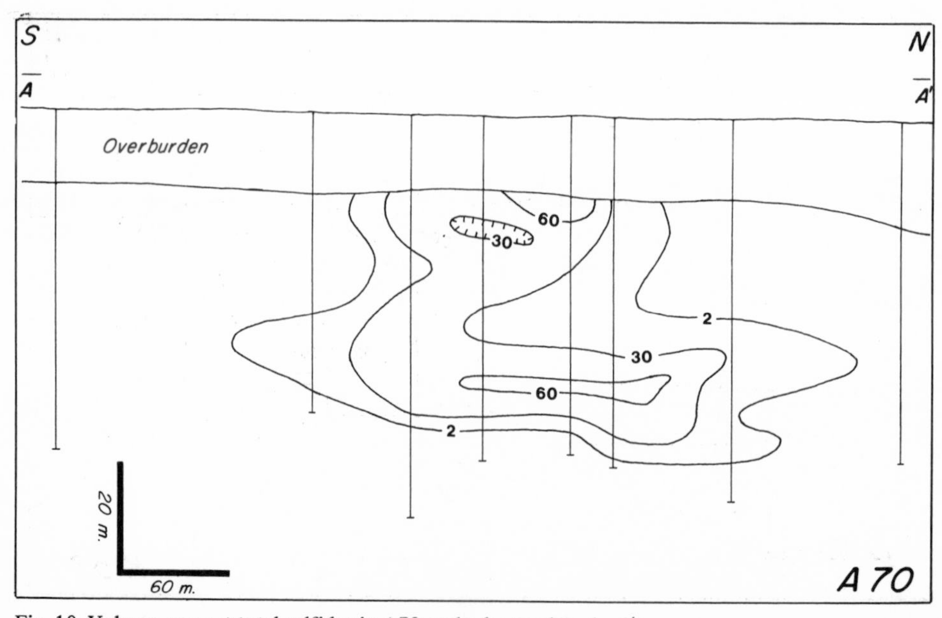

Fig. 18. Volume percent total sulfides in A70 orebody, section A–A'.

The prismatic portion of the sulfide concentration is coincident with the major detritus zone (Figs. 11 and 12), but a broad stratabound zone of lower sulfide concentration occurs within the coarse-crystalline Presqu'ile dolostone, particularly northeast and southwest of the detritus zone (Fig. 17). Downward displacement of late Givetian strata has taken place in an irregularly elliptical area overlying the detritus zone (Fig. 12). The late Givetian strata and the detritus zone are fractured and brecciated. Sulfides fill fractures and form concentrations at the expense of pre-existing materials, particularly the carbonate detritus (Fig. 18; Plate VII: 6,10,11).

A definite relationship also exists between areas of collapse (Plate IX: 1) and the sulfide concentrations in the fine-crystalline dolostones of the lower barrier. W17 has a maximum of over 50 vol.% total sulfides over a 99-m interval (Fig. 19), and some 25-m intervals contain over 85% sulfides (Fig. 20). The area of greatest sulfide concentration is largely coincident with the green clay zone which represents the central area of collapse (Fig. 13). A correlation exists between sulfide concentrations and the various breccia types (Figs. 14 and 20). Sulfides fill fractures and replace breccia matrix and fragments (Plate IX: 2–6). Most sulfides appear to be present in intervals of brecciated Facies J dolostone (Fig. 14), but in high sulfide intervals, there is little lithic material remaining (Plate IX: 7).

Detritus or collapse breccias do not appear to have been of major importance in the localization of the N204 sulfide concentrations. However, local downward displacement

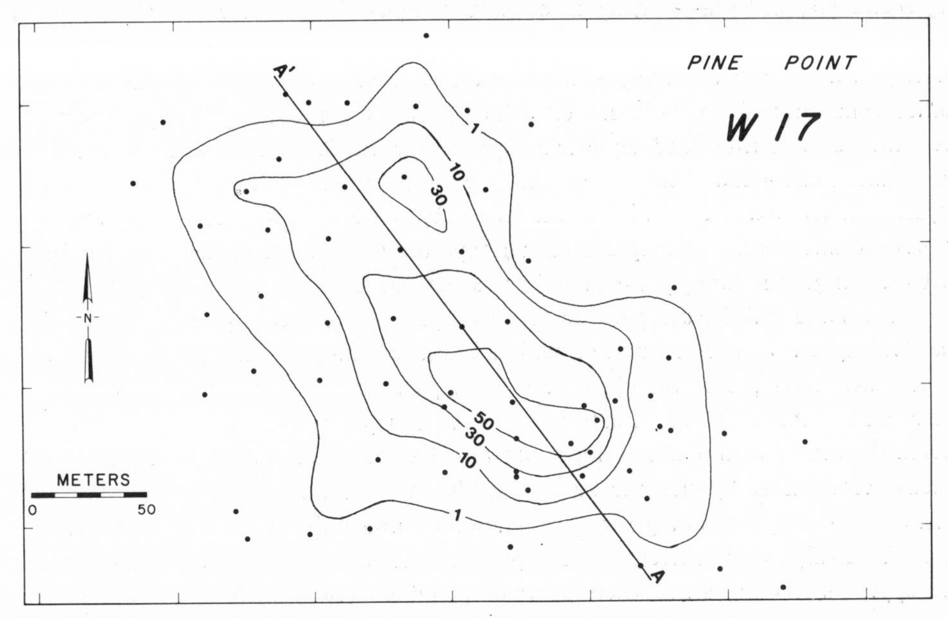

Fig. 19. Volume percent total sulfides in W17 orebody, 102.1–201.2 m.

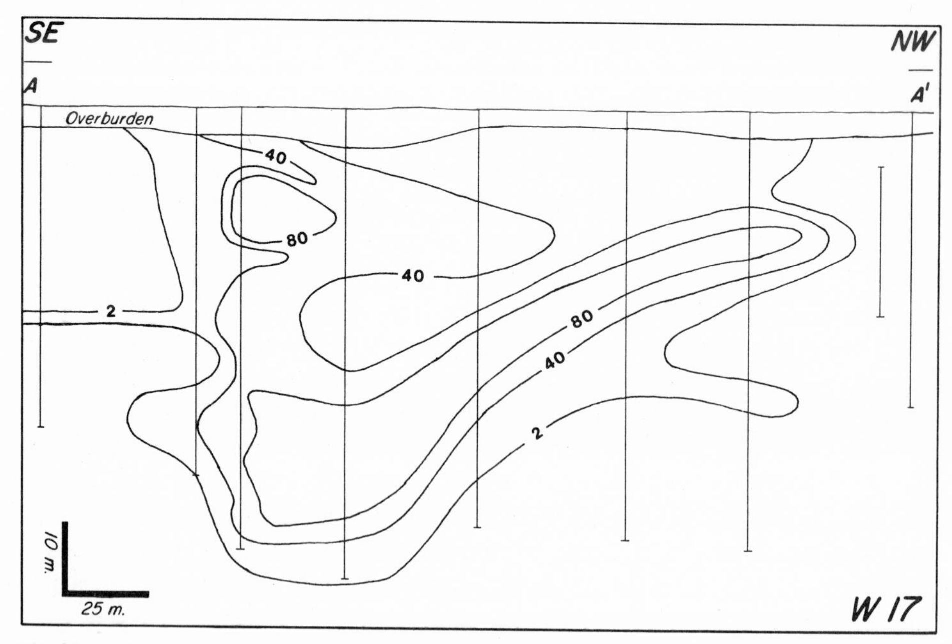

Fig. 20. Volume percent total sulfides in W17 orebody, section $A-A'$.

affects strata as deep as the underlying E-Shales. Highest sulfide concentrations occur in the "Spongy Horizon" adjacent to these local slump areas, but sulfides also occur at higher stratigraphic levels overlying the downward displaced zones (Adams, 1973).

Development of sulfide-bearing solution features. Solution features associated with sulfide concentrations in the coarse-crystalline Facies K dolostone of the upper barrier are either large detritus-filled depressions open to the disconformable surface or macropores and stratabound zones of increased porosity in the lower Presqu'ile without apparent direct connection with the disconformable surface. These features are thought to have formed during prolonged subaerial exposure with attendant karstification and carbonate diagenesis during post-middle Givetian emergence (Kyle, 1977, 1979).

The sulfide-filled macropores in the lower part of Facies K (e.g. M40; Plate VIII) that lack apparent direct connection with the disconformity are interpreted as former caves which developed in the upper part of the phreatic zone during the period of the stable water table (Fig. 21, situation A). Local collapse in the cave system created breccias which now also host sulfide minerals. Extensive dissolution of limestone is readily accomplished by meteoric water which has not become saturated with respect to calcite due to CO_2 loss during seepage of rainwater through the vadose zone to the water table (Thraikill, 1968). Therefore, macropores and caves developed in the limestone of the upper Pine Point barrier where a bypass permitted vadose flow instead of vadose seepage to the water table (Fig. 21). The vadose bypasses did not remain open enough to

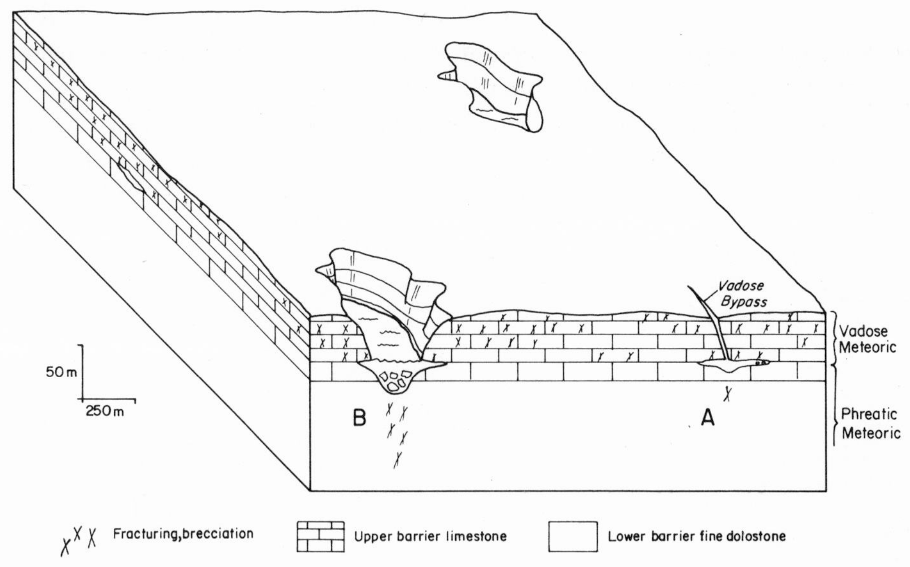

Fig. 21. Vadose and phreatic zones in the upper Pine Point barrier during the post-middle Givetian erosional period.

permit complete filling of the caves by detritus during the close of the erosional period.

Horizons of increased porosity and permeability apparently were created in the upper part of the phreatic zone adjacent to the zones of major dissolution. This process was responsible for creation of most of the hosting porosity for the tabular sulfide concentrations, such as M40, which are common in the lower Presqu'ile (Plate VIII), particularly along the Main Hinge. The development of horizons of increased permeability probably was controlled by the local supply of meteoric water with the chemical capacity to produce increased porosity. Once meteoric water became saturated with respect to calcite, effective porosity enlargement ceased. This restriction may account for the apparent limited lateral extent of these horizons in the lower Presqu'ile rather than development of a continuous zone of increased permeability along the position of the upper part of the paleo-phreatic zone.

The detritus zones are interpreted as filled compound dolines which developed largely by solution activity in the limestone Facies C, D-2, H, and I of the upper barrier (Fig. 21, situation B). Original stratigraphic facies was not the major factor controlling development of the dolines, because they occur in back-reef (e.g. N38A, O42), reefal (e.g. K57, K62), and intercalated facies (e.g. A70). The dolines are elongate in a northeasterly direction parallel to the Hinge Zones, suggesting that karstification was controlled by the same structural factors that governed carbonate sedimentation. The dolines are as much as 400 m in length and have length-to-width ratios ranging from about 2 to 3.5. Doline walls were relatively steep and irregular as suggested by the abrupt transition from detritus to wall rock, and depths of at least as much as 35 m are indicated. In addition to detritus-filled dolines, green clay and carbonate detritus occur in solution-enlarged joints, bedding planes, and vugs near the disconformity.

The developmental sequence envisaged for the formation of the sulfide-bearing,

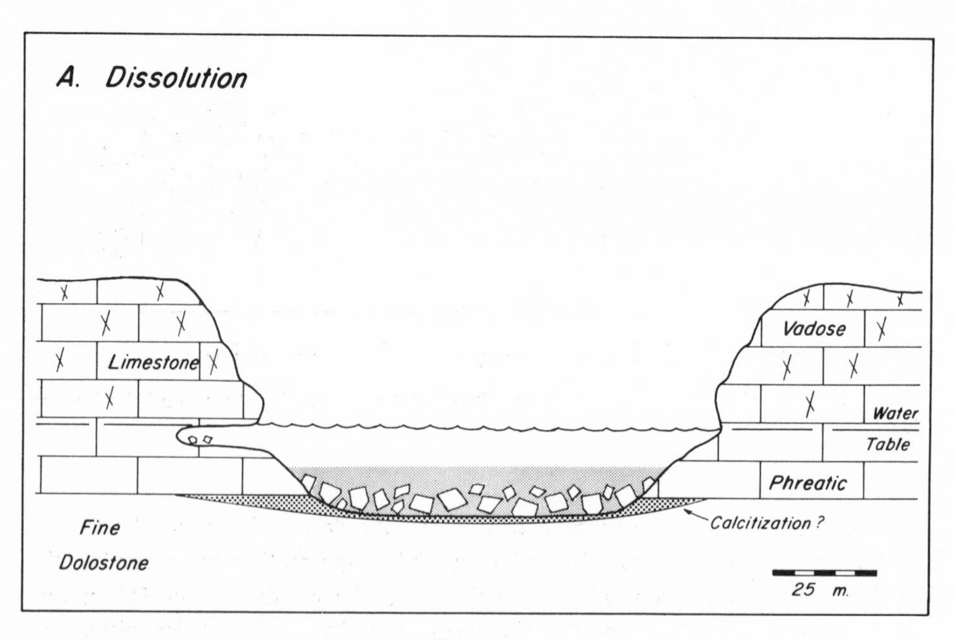

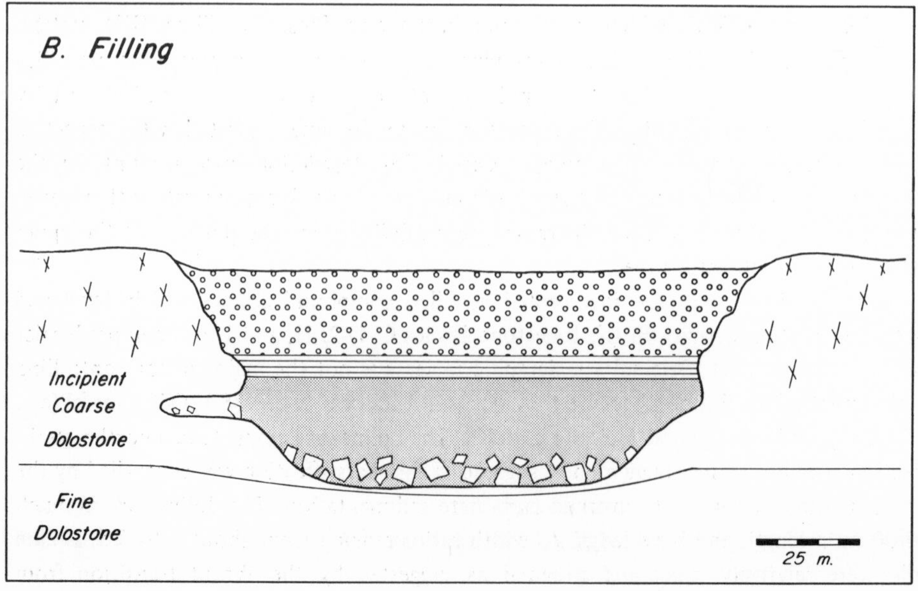

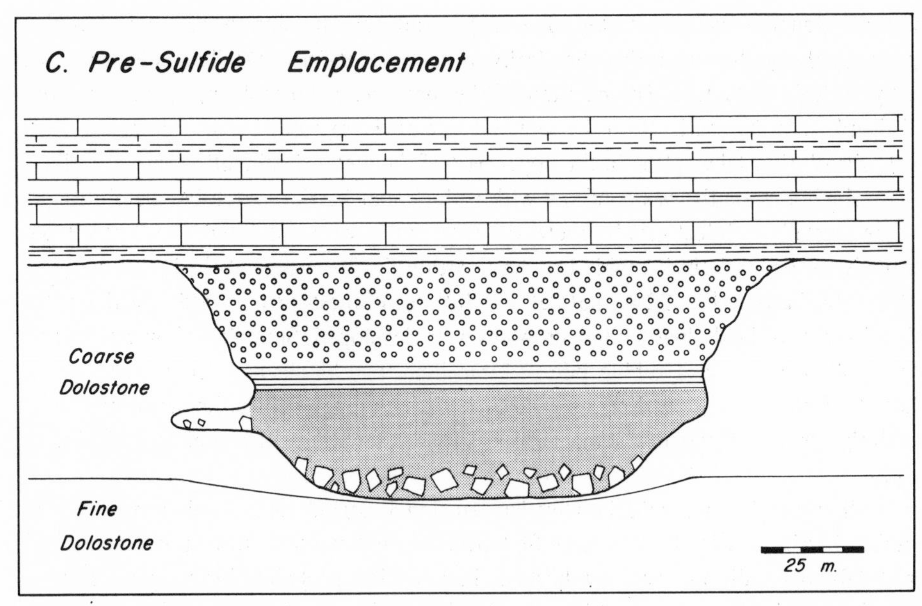

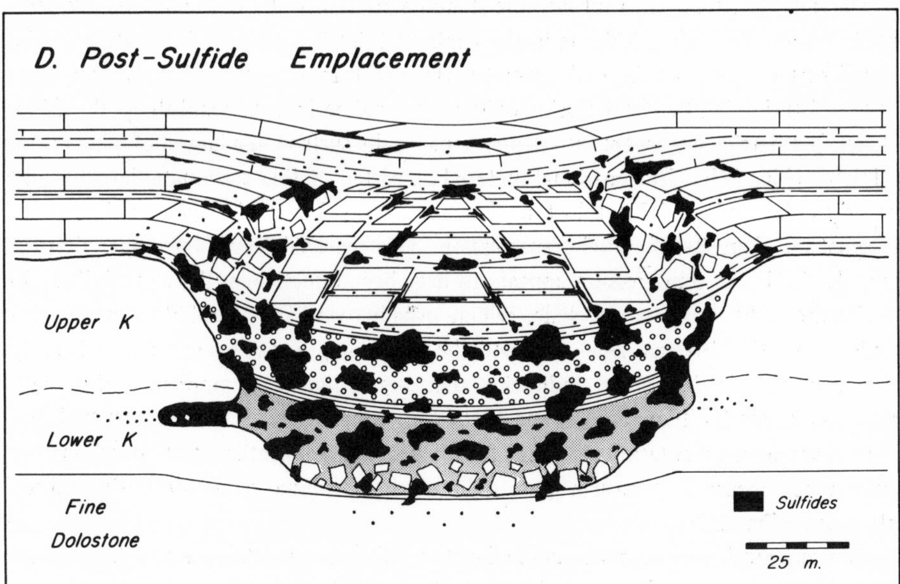

Fig. 22. Developmental sequence for sulfide-bearing, detritus-filled dolines.

detritus-filled dolines is shown in Fig. 22. During subaerial exposure, the upper barrier limestones were subject to chemical attack by rainwater, which is undersaturated with respect to calcite (Thrailkill, 1968). Dissolution was concentrated at joint intersections, generally along the dominant northeasterly trend. Dissolution proceeded until the entire limestone sequence was penetrated (Fig. 21). Additional dissolution was slowed at this stage, not only by the underlying chemically more resistant fine-crystalline dolostone, but also by hydrologic factors. The position of the meteoric water table may have been controlled indirectly by sea level on the fore-reef side of the exposed barrier. Since most carbonate dissolution takes place in the upper part of the meteoric phreatic zone (Thrailkill, 1968), which apparently was confined within the upper barrier, extensive dissolution of the fine-crystalline dolostones by meteoric water was greatly inhibited. The water table eventually became relatively stable at a level within the lower part of the limestone units of the upper barrier (Fig. 22A). Lateral dissolution of limestone occurred near the water table adjacent to the dolines, and macropores were developed. Slumping of oversteepened doline walls resulted in a rubble pile of large limestone blocks and finer debris on the doline floor as in the lower detritus zone in K57. Slumping of the doline walls may have effectively sealed off part of the macropores so that they were not completely filled with detritus.

Following the post-middle Givetian erosional period and development of the karstified barrier, marine transgression first affected the low-lying "near-shore" areas, and local tidal-flat conditions were established. Periodically, probably during major storms, fine carbonate detritus was swept into the standing body of water in the dolines and sedimented in thin layers (Fig. 22B) as in the laminated detritus in K57. Inundation of the erosional surface was sporadic with several periods of transgression and minor regression (Wiley, 1970). During the final closing of the erosional period, the regolith developed on the karstified barrier was swept into depressions on the surface, the largest of which were the dolines (Fig. 22B). This erosional material consisted of fine carbonate detritus with some green clay and tidal-flat dolostone fragments eroded from the ephemeral tidal-flat and lacustrine deposits of initial transgression. This material forms the upper detritus zone in K57. Marine transgression appears to have been effective in filling the dolines. In the A70 area, the waxy green shale, which usually overlies the rubble contact, also extends across the detritus zone. This and the succeeding intervals are not consistently thicker over the detritus zone, indicating that the doline was completely filled with detritus during marine transgression and that little compaction of detritus occurred during late Givetian sedimentation (Fig. 22C). Collapse of late Givetian strata overlying the detritus zones occurred at least in part as the result of volume reduction during sulfide emplacement (Fig. 22D).

The relationship of the post-middle Givetian episode of karstification to the sulfide-hosting structures in the fine-crystalline dolostones of the lower barrier complex is unclear. The collapse zone and local rock matrix breccias in W17 bear some resemblance to sulfide-hosting features in the upper barrier, but the upper part of the W17 zone has been eroded, and it is impossible to determine if it originally extended into the upper barrier. None of the Facies K mineralized zones extend for any appreciable distance into the fine-crystalline dolostones. The vuggy and intercrystalline porosity which hosts sulfide minerals in N204 is the result of leaching of mineralogically unstable skeletal components and burrows; such horizons result from vadose leaching in modern carbonates (e.g. Matthews, 1974), but intrastratal dissolution could produce similar effects.

It seems unlikely that the zone of most intensive carbonate dissolution, the upper phreatic, could have extended into the lower levels of the barrier complex during the post-middle Givetian erosional period. However, carbonate dissolution may have occurred in the zone of mixing of fresh water and seawater which existed within the lower barrier during this period. Another possibility is that the sulfide-hosting structures in the fine-crystalline dolostones resulted from an earlier period of subaerial exposure and carbonate diagenesis, as Maiklem (1971) has proposed for the middle Givetian of the Pine Point barrier complex.

ORE DEPOSITS

General characteristics

The Pine Point district consists of about 50 known Pb–Zn–Fe sulfide bodies in an area of about 1000 km^2 (Fig. 7); these bodies vary considerably in size, geometry, metal percentages and ratios, sulfide textures, and host-rock relationships. Sulfide bodies range in size from less than 100,000 to as much as 15 million tons and are elongated in a north-

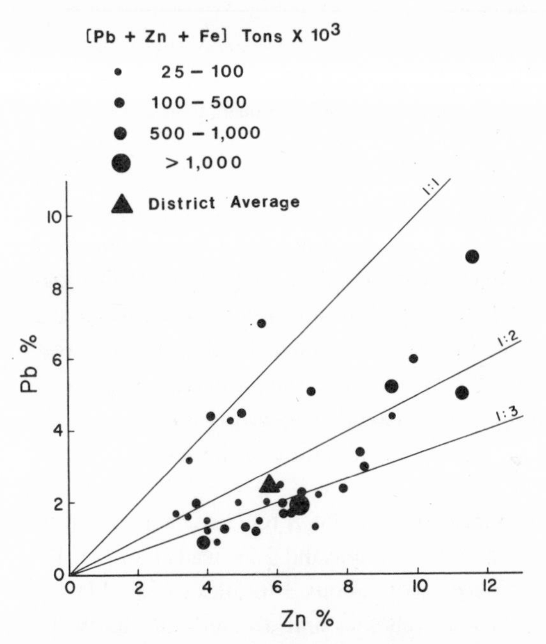

Fig. 23. Weight percent of Pb and Zn in major orebodies.

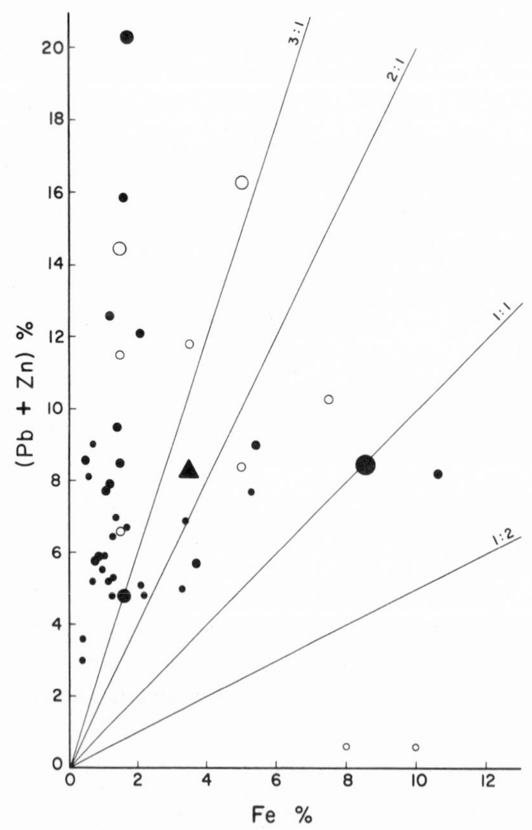

Fig. 24. Weight percent of combined Pb–Zn and Fe in major sulfide bodies. Open circles indicate deposits for which one or more metal percentages, usually Fe, are poorly defined by present data.

easterly direction generally parallel to the Hinge Zones. Individual bodies are stratabound in relatively narrow intervals, but major sulfide bodies occur in several stratigraphic positions in a 200-m section of the Pine Point barrier complex (Fig. 6). Some orebodies may be classified as either tabular or prismatic (Skall, 1972), but this division is somewhat artificial because some sulfide zones have both prismatic and tabular concentrations (e.g. K57, A70), and because it separates orebodies of otherwise similar characteristics (e.g. X15, W17). Most of the sulfide concentrations are incomplete because of erosional truncation, and the following discussion will reflect the current nature of the sulfide bodies at different levels of preservation.

The metal content of the orebodies ranges from about 3 to 11.5% zinc and from about 0.8 to 9% lead; the district average is about 5.8% zinc and 2.2% lead (Fig. 23). The combined lead–zinc content of the bodies ranges from about 3 to 20.5% (Fig. 24). Significant tonnages of massive ore exist in some orebodies and account for almost 1.5 million tons of direct shipping material averaging over 45% combined lead–zinc (Table I).

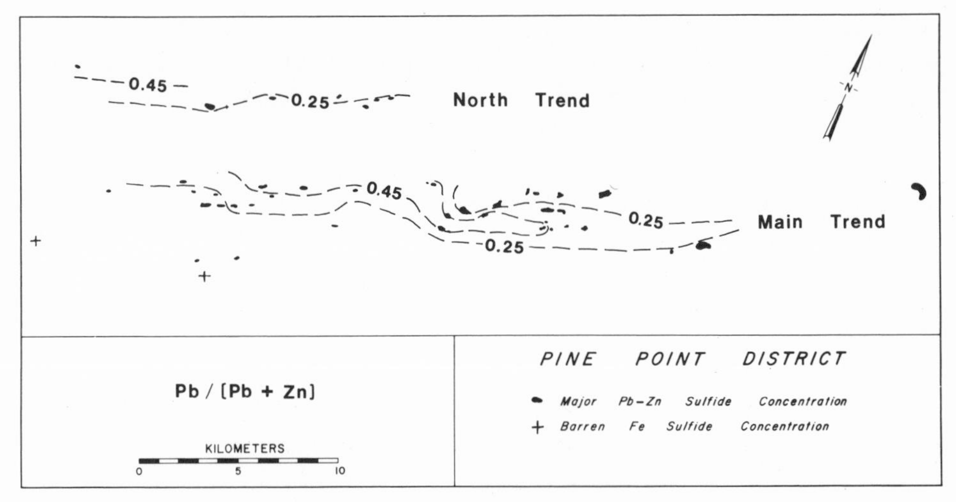

Fig. 25. District contour plot of Pb/(Pb + Zn) ratios of orebodies.

The iron content of the orebodies varies from less than 0.5 to about 10.5% with a district average of about 3.5% (Fig. 24). Some relatively small "barren" sulfide bodies consist of iron sulfides with only minor lead and zinc.

The Pb/(Pb + Zn) ratios of the Pine Point sulfide bodies range from 0.6 to 0.1 with a district average of about 0.3. The strongest peak at about 0.25 is greatly influenced by the large X15 orebody with this ratio. The Fe/(Pb + Zn + Fe) ratios of the lead–zinc orebodies range from 0.6 to less than 0.1. The district average is about 0.3, but several distinct peaks are evident. Again, the strongest peak at about 0.5 is dominated by the X15 orebody. The small "barren" iron sulfide bodies have Fe/(Pb + Zn + Fe) ratios greater than

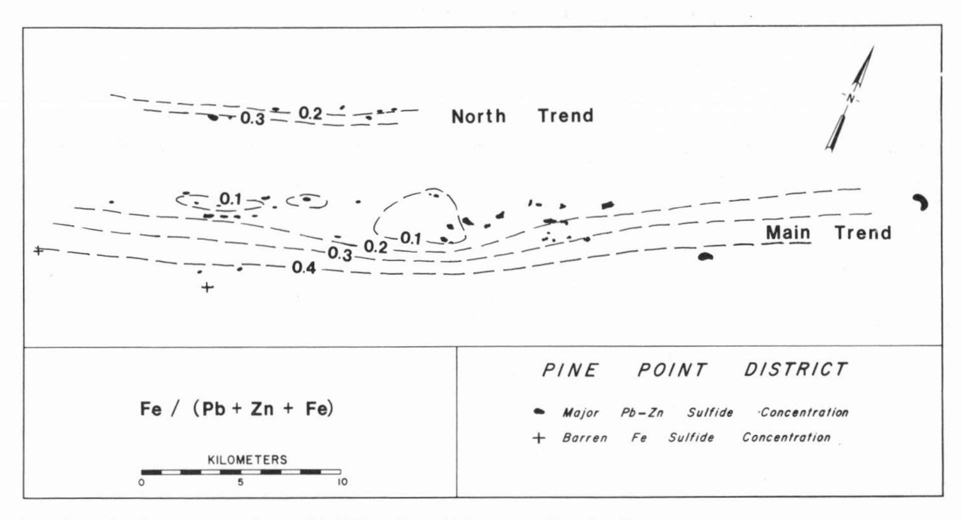

Fig. 26. District contour plot of Fe/(Pb + Zn + Fe) ratios of orebodies.

0.9. Orebodies in which the greatest sulfide content is in a prismatic zone, particularly those within the upper barrier, are usually lead-rich with Pb/(Pb + Zn) averaging about 0.4. Conversely, tabular orebodies are relatively zinc-rich with Pb/(Pb + Zn) averaging less than 0.3. There does not appear to be a consistent relationship between geometry and iron content of the sulfide bodies. Although these data are for incompletely preserved sulfide bodies, the apparent absence of vertical metal zoning within individual bodies and the presence of district-wide metal zoning indicate that these ratios reflect real variations in the Pb–Zn–Fe ratios of individual sulfide bodies (Kyle, 1977).

Metal zoning

District. A district-wide pattern of major metal zoning can be demonstrated using metal ratios determined from ore-reserve data (Kyle, 1977, 1979). The Pb/(Pb + Zn) ratios of orebodies increase from about 0.2 in the southeast to about 0.5 in the northwest in zones parallel to the Main Trend (Fig. 25). A similar pattern, less well defined by present data, exists along the North Trend of orebodies. The Fe/(Pb + Zn + Fe) ratios decrease strikingly from about 0.5 in the southeast to about 0.1 in the northwest in zones parallel to the Main Trend (Fig. 26). Again, a similar pattern exists along the North Trend. Three barren iron sulfide bodies fit well in this pattern as determined by the metal ratios of all orebodies (Fig. 26).

The weighted averages of lead, zinc, and iron from sulfide bodies in various stratigraphic positions within the Pine Point barrier suggest weak vertical metal zoning within the district (Table III). Although the stratigraphic distribution of major sulfide bodies does not permit statistical verification of metal trends, sulfide bodies in the upper barrier (Facies K, L, M) have relatively higher lead and lower iron than those in the lower barrier (Facies A, D, J). However, there is considerable variation in the metal ratios of the upper barrier orebodies, and some (e.g. A70, R61) have metal ratios similar to the sulfide bodies of the lower barrier.

Individual orebodies. K57 is a prismatic sulfide body with an adjacent discontinuous tabular sulfide zone (Figs. 27, 28); some intervals of essentially massive sphalerite and

TABLE III

Metal ratios of sulfide bodies relative to hosting facies

Dominant host rock facies	Total number of sulfide bodies	Number of sulfide bodies >100,000 tons (Pb + Zn)	Cumulative tons (Pb + Zn)	Pb/(Pb + Zn)	Fe/(Pb + Zn + Fe)
K, L, M	50	17	$5.6 \cdot 10^6$	0.33	0.17
J, D	2	2	$2.1 \cdot 10^6$	0.23	0.49
A	1	1	$0.8 \cdot 10^6$	0.20	0.27

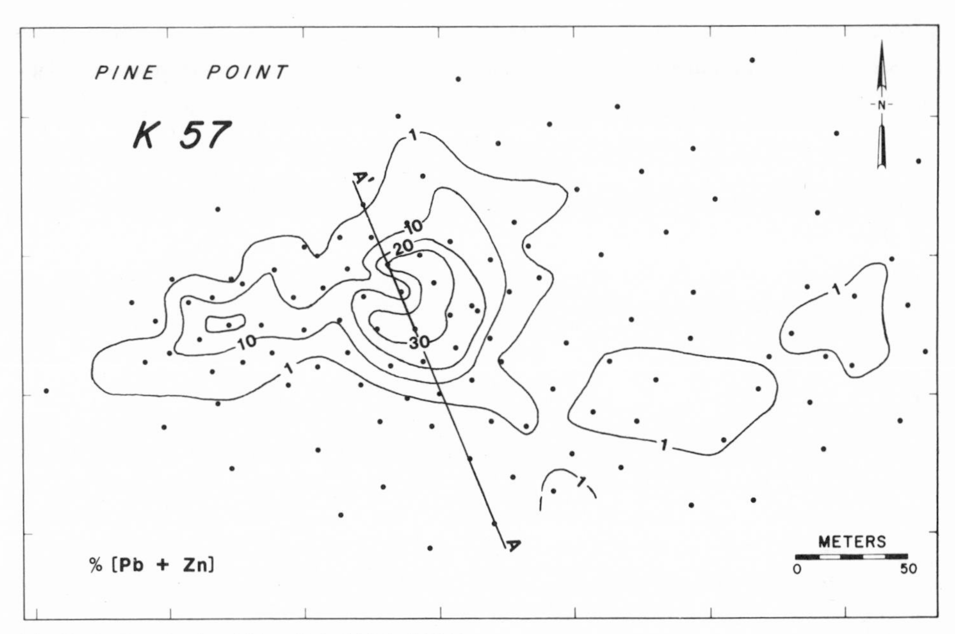

Fig. 27. K57 orebody, %(Pb + Zn); 149.4–202.7 m.

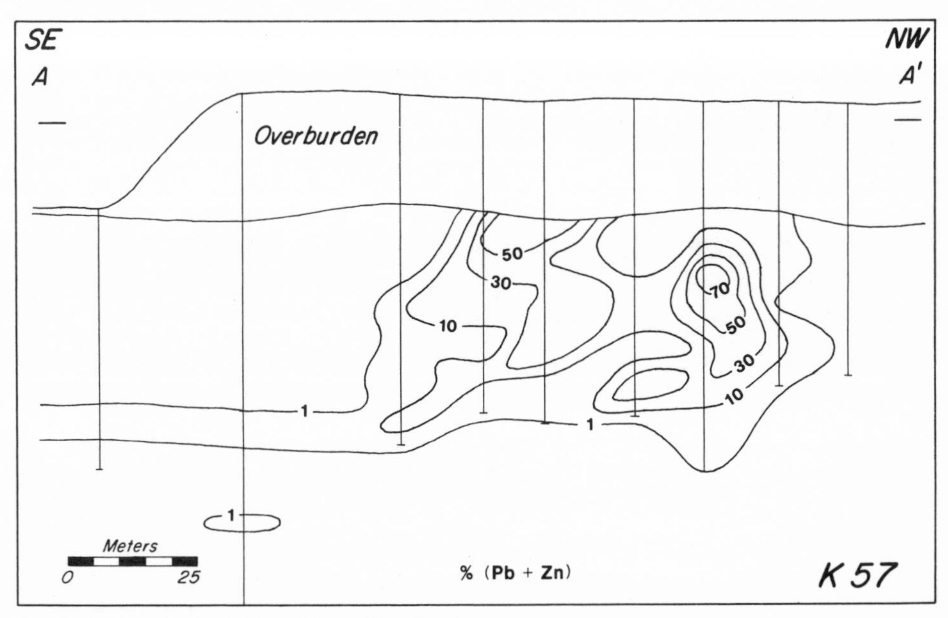

Fig. 28. K57 orebody, %(Pb + Zn), section A–A'.

galena are present with over 75 wt.% combined lead–zinc (Fig. 28). Preproduction ore reserves were 1.8 million tons averaging 7.0% lead, 5.6% zinc, and 1.2% iron. The high-grade portion is about 150 m long and a maximum of about 75 m wide; it contains a maximum of 38.7 wt.% combined lead–zinc over a 53-m interval (Fig. 27). Major amounts of lead are restricted to the prismatic zone and reach a maximum of 32.9% in the northeastern part. Maxima of 14.3% zinc and 3.7% iron are reached in the prismatic zone. The Pb/(Pb + Zn) ratio increases from 0.3 on the periphery of the sulfide body to 0.7 in the center (Fig. 29), thus reflecting the zinc-rich zone surrounding the lead-rich core. The relative abundance of iron in the perimeter of the sulfide body is shown by outward increase of Fe/(Pb + Zn + Fe) from less than 0.1 to 0.3 (Fig. 30). There does not appear to be a simple vertical metal zonation (Figs. 31, 32); instead, zoning appears to consist of a three-dimensional iron envelope around the lead-rich center.

A70 is a prismatic sulfide body (Figs. 33, 34) containing 2.8 million tons of ore averaging 5.0% lead, 11.3% zinc, and about 5% iron. An extensive contiguous tabular sulfide zone containing primarily Fe sulfides with only minor lead and zinc is present northeast and southwest of the orebody (Fig. 17). The high-grade core is about 200 m long and 80 m wide and contains a maximum of 27.8% combined lead–zinc over a 53-m interval (Fig. 33). Major amounts of lead and zinc are restricted to the prismatic body and reach maxima of 8.8 and 19.6%, respectively. Iron content of the orebody is quite variable and attains a maximum of about 7% around the perimeter of the prismatic body. The relative abundance of lead in the high-grade core is shown by the increase of Pb/

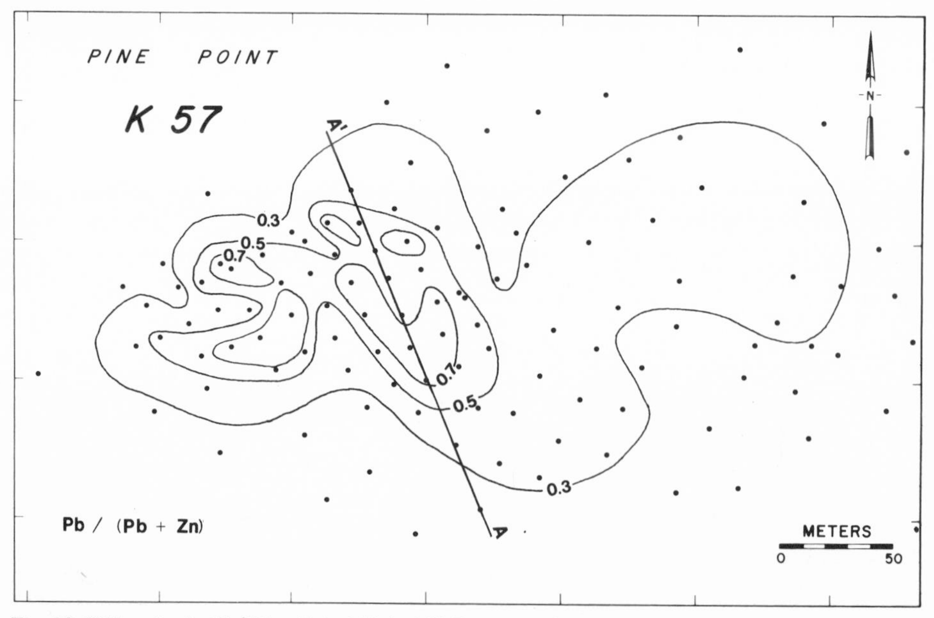

Fig. 29. K57 orebody, Pb/(Pb + Zn), 149.4–202.7 m.

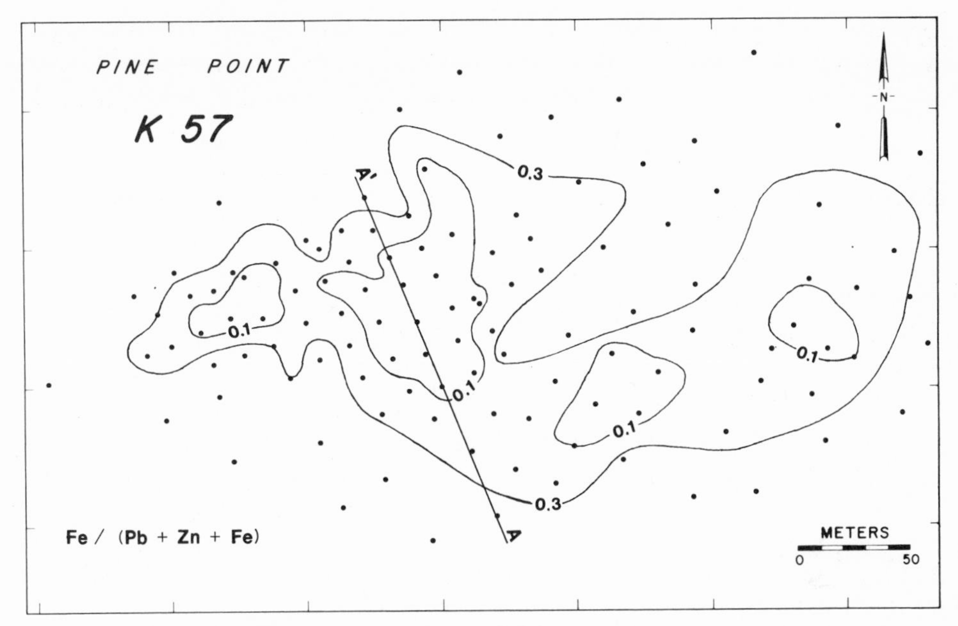

Fig. 30. K57 orebody, Fe/(Pb + Zn + Fe), 149.4–202.7 m.

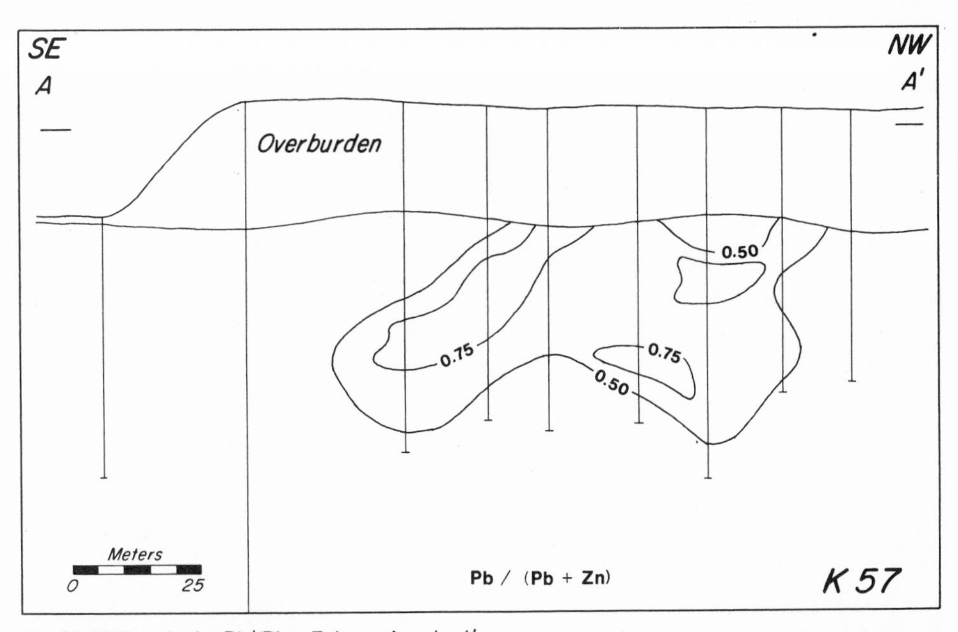

Fig. 31. K57 orebody, Pb/(Pb + Zn), section A–A'.

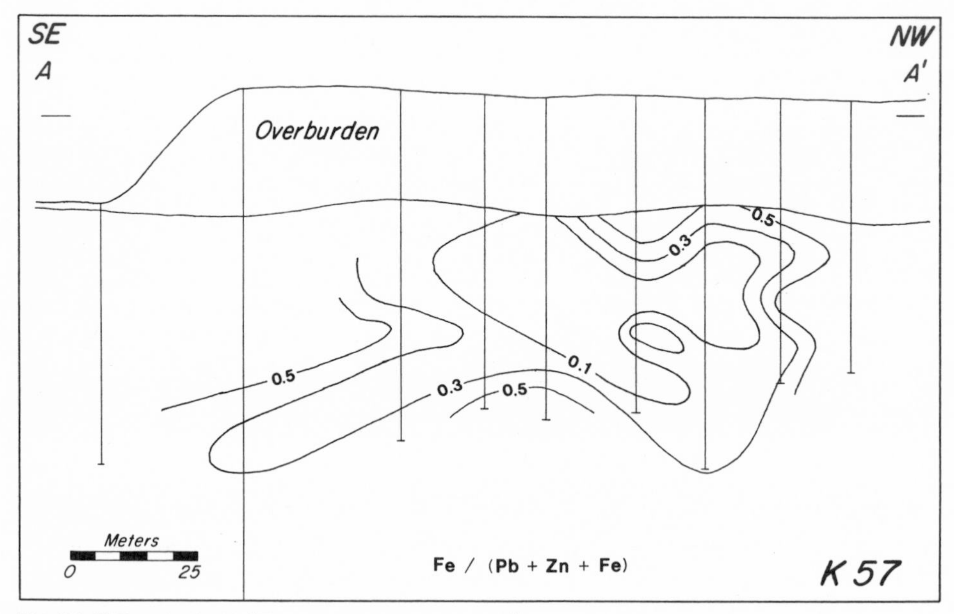

Fig. 32. K57 orebody, Fe/(Pb + Zn + Fe), section A–A'.

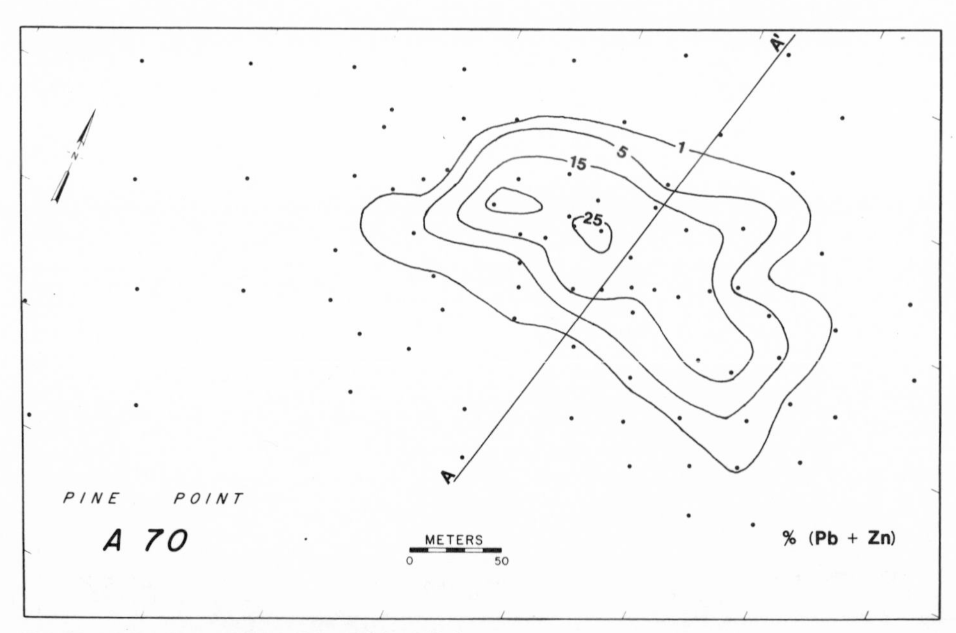

Fig. 33. A70 orebody, %(Pb + Zn), 114.3–167.6 m.

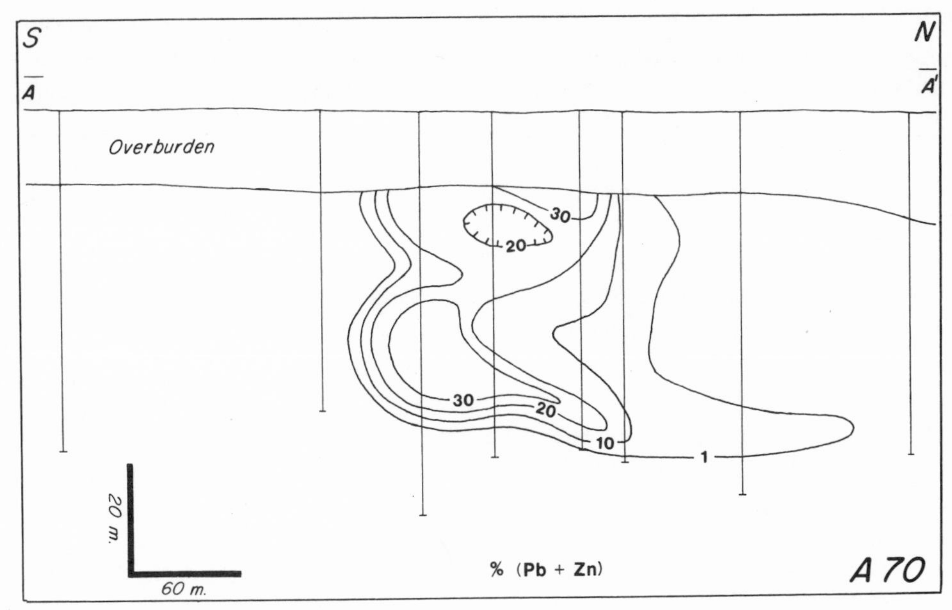

Fig. 34. A70 orebody, %(Pb + Zn), section A–A'.

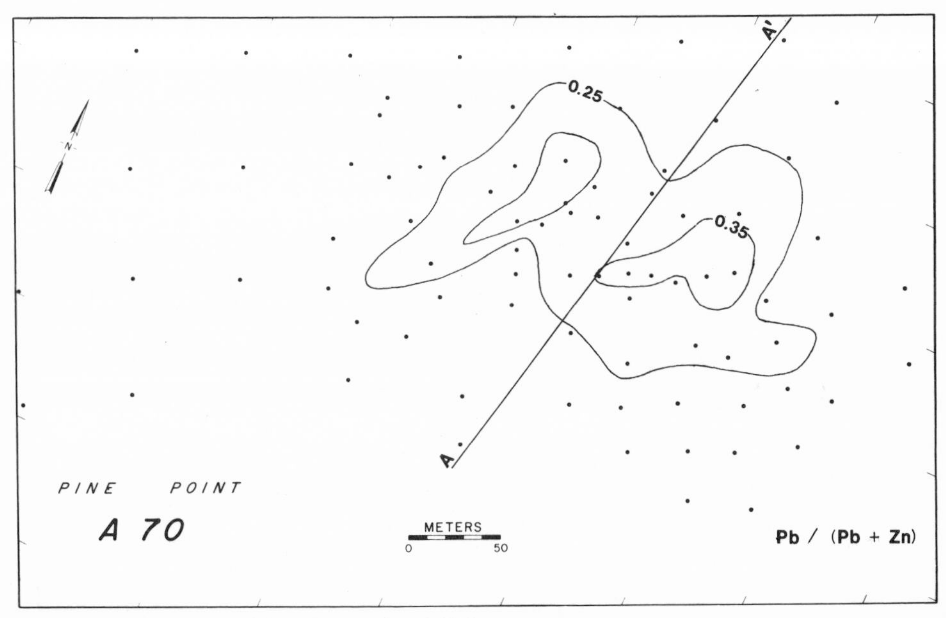

Fig. 35. A70 orebody, Pb/(Pb + Zn), 114.3–167.6 m.

(Pb + Zn) from 0.25 around the periphery to 0.35 in the center (Fig. 35). There appears to be a decrease in relative lead content from south to north as reflected by the decrease of Pb/(Pb + Zn) from more than 0.3 to less than 0.2. The abundance of iron on the perimeter of the prismatic body is shown by the marked increase of Fe/(Pb + Zn + Fe) from more than 0.7 to less than 0.1 (Fig. 36). Again, metal distribution consists of a three-dimensional iron envelope around the zinc-rich zone which surrounds the lead-rich center.

W17 is a prismatic sulfide body (Figs. 37, 38) containing preproduction ore reserves of 3.9 million tons averaging 2.0% lead, 6.2% zinc, and 10.6% iron. The high-grade portion is about 140 m long and 85 m wide and contains a maximum of about 18% combined lead–zinc over a 99-m interval. Although major concentrations of lead, zinc, and iron are largely coincident and reach maxima of 4.9, 12.9, and 23.1%, respectively, the high-iron zone is more extensive. Lead appears to decrease slightly from southwest to northeast as indicated by the decrease of Pb/(Pb + Zn) from 0.3 to 0.2 (Fig. 39). In section, Pb/(Pb + Zn) varies irregularly between 0.4 and 0.2. The three-dimensional increase in relative iron content is shown by the increase of Fe/(Pb + Zn + Fe) from less than 0.5 in the center to more than 0.7 on the periphery (Fig. 40).

N204 is an extensive low-grade tabular sulfide body with a Pb/(Pb + Zn) ratio varying irregularly between 0.25 and 0.15 in the principal mineralized zone. The absolute iron content is greatest on the east side and is part of a discontinuous relatively high iron envelope with Fe/(Pb + Zn + Fe) greater than 0.4 (Fig. 41).

Although individual sulfide bodies differ greatly in metal percentages and ratios,

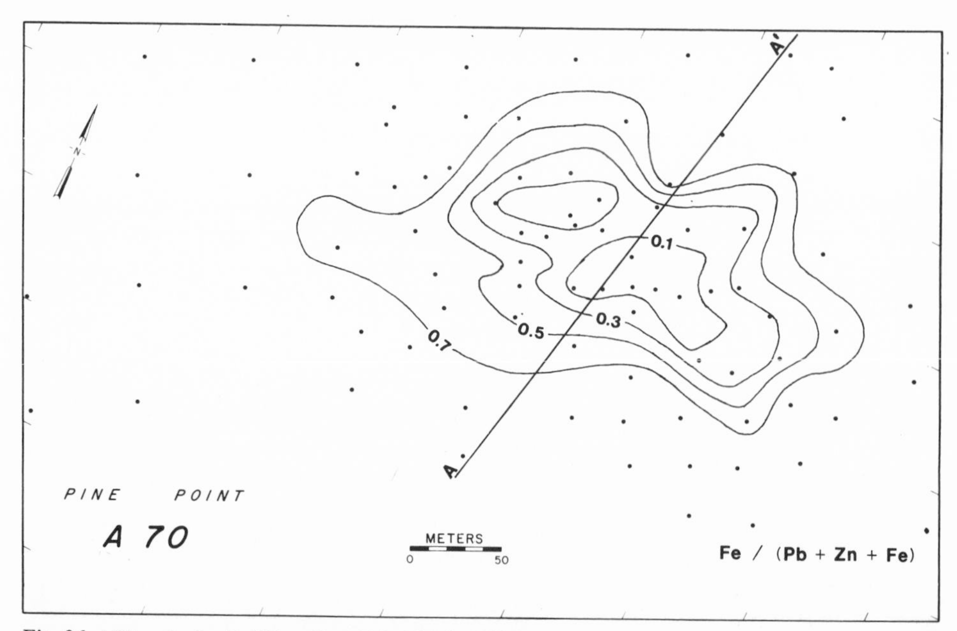

Fig. 36. A70 orebody, Fe/(Pb + Zn + Fe), 114.3–167.6 m.

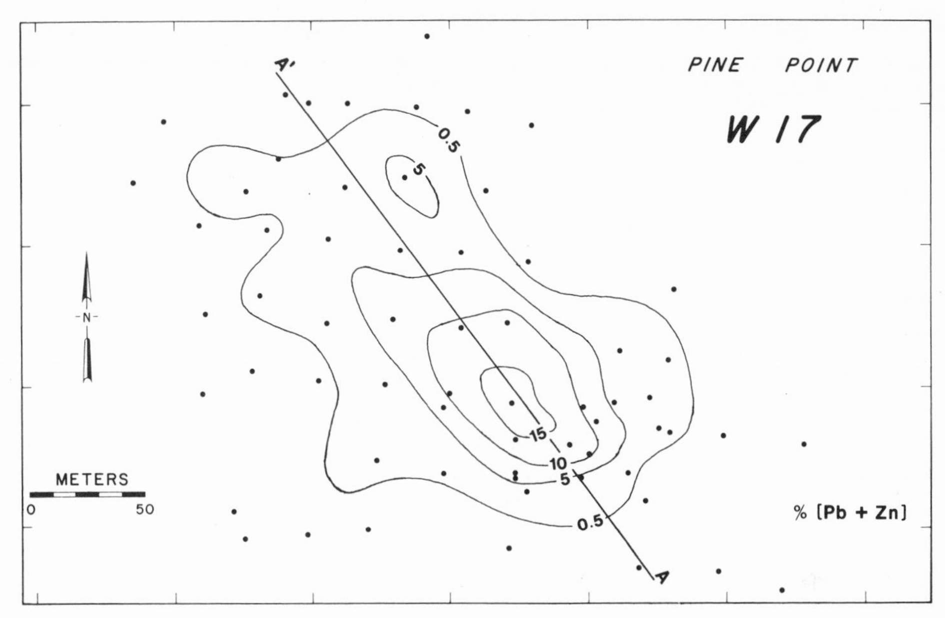

Fig. 37. W17 orebody, %(Pb + Zn), 102.1–201.2 m.

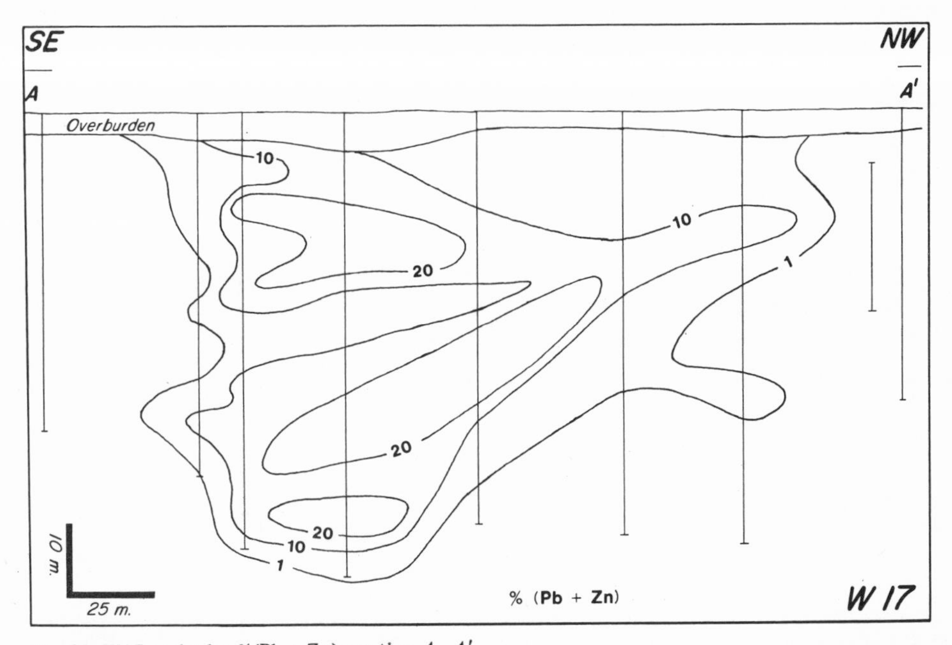

Fig. 38. W17 orebody, %(Pb + Zn), section $A–A'$.

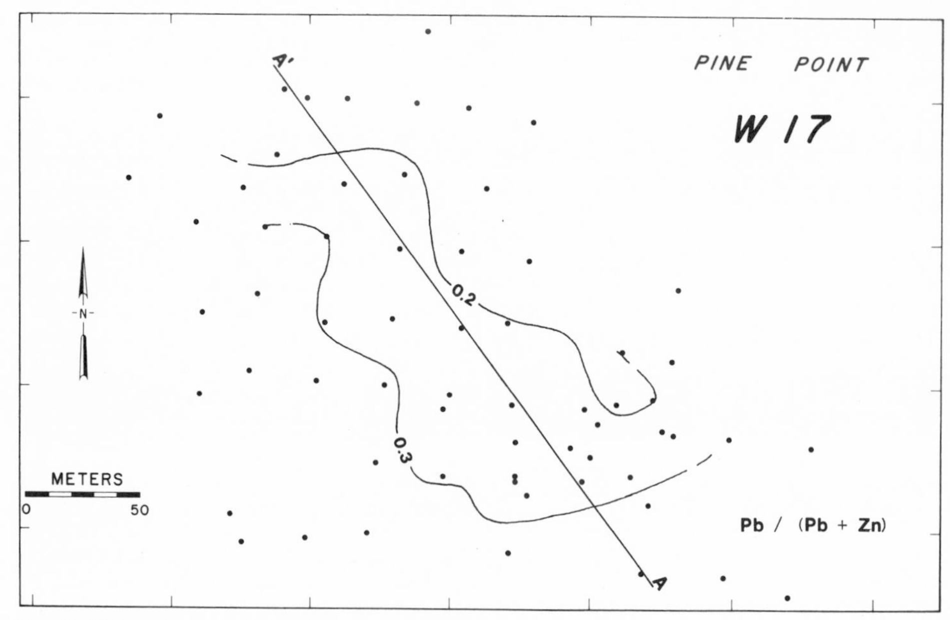

Fig. 39. W17 orebody, Pb/(Pb + Zn), 102.1–201.2 m.

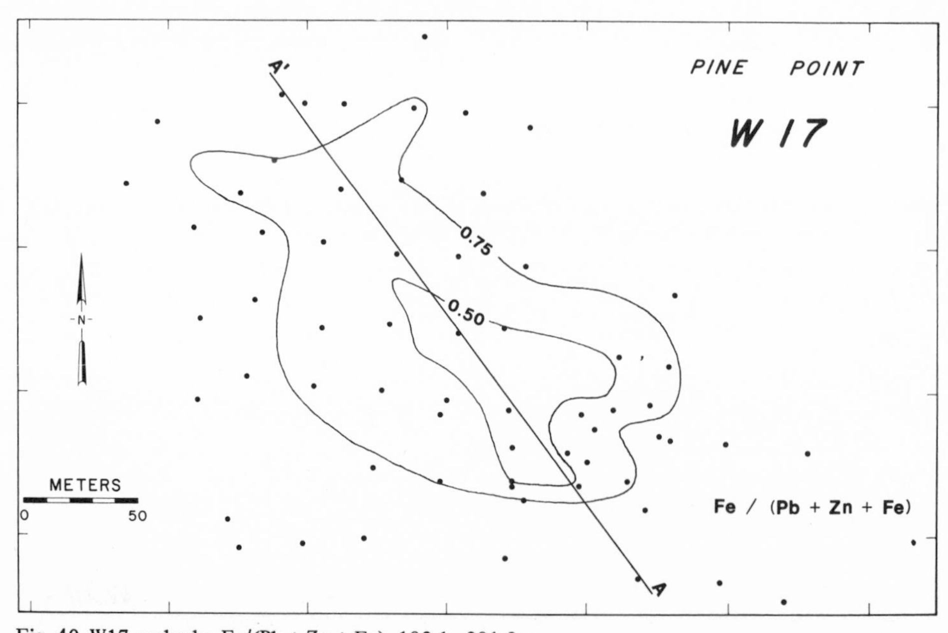

Fig. 40. W17 orebody, Fe/(Pb + Zn + Fe), 102.1–201.2 m.

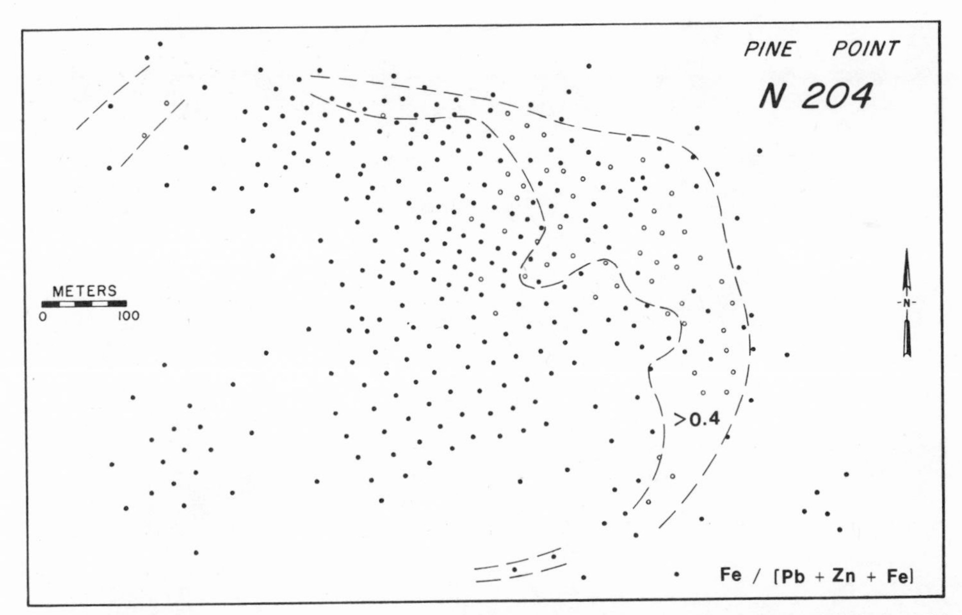

Fig. 41. N204 orebody, Fe/(Pb + Zn + Fe), principal zone. Open circles indicate drill holes with greater than 4% Fe.

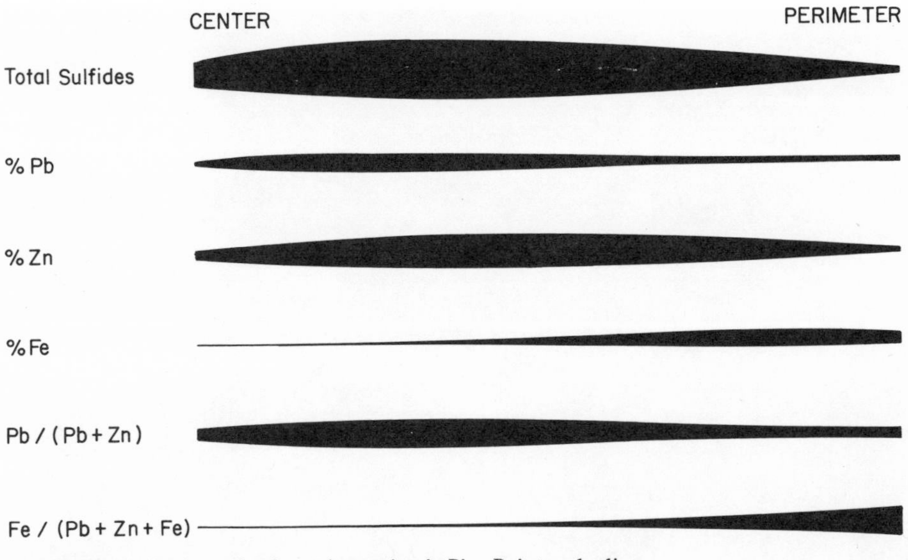

Fig. 42. Generalized trends of metal zonation in Pine Point orebodies.

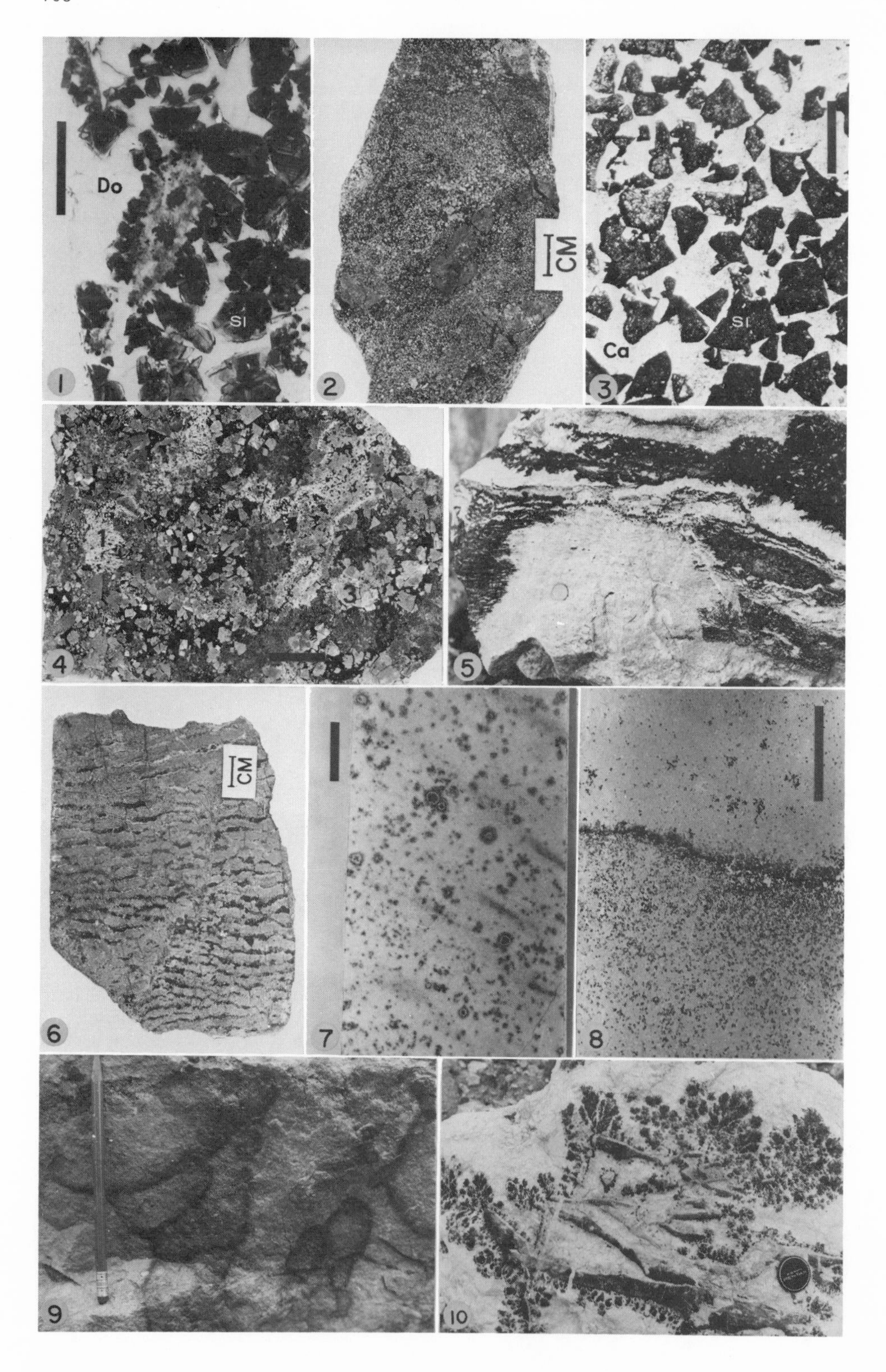
Do
Si
CM
Ca
Si
1
3
CM
1
2
3
4
5
6
7
8
9
10

some consistent trends in relative metal distribution are apparent (Kyle, 1977, 1979). The generalized pattern of metal distribution in the Pine Point orebodies is a lead–zinc, high-grade center with high Pb/(Pb + Zn) and low Fe/(Pb + Zn + Fe) passing outward into a zinc-rich, high-grade zone with lower Pb/(Pb + Zn) and low Fe/(Pb + Zn + Fe) which grades into an iron-rich, low-grade envelope with low Pb/(Pb + Zn) and high Fe/(Pb + Zn + Fe) (Fig. 42). This zonation appears to be applicable to both prismatic and tabular sulfide bodies throughout the barrier complex. Low-grade sections exist locally in the high-grade center, and tabular sulfide concentrations may be present adjacent to the prismatic zone. Transition between high total sulfide material and barren host rock is commonly abrupt.

Mineralogy

The Pine Point sulfide bodies are composed almost exclusively of sphalerite, galena, pyrite, and marcasite with a non-metallic gangue of dolomite and calcite. Minor amounts of pyrrhotite, celestite, barite, gypsum, anhydrite, fluorite, sulfur, and bitumen are present within the host-rocks.

Sphalerite. Sphalerite, ZnS, is the most common sulfide mineral in most of the ore-bodies and occurs as individual tetrahedral crystals, as "colloform" and banded crusts, as

---

PLATE X

Sulfide textures.

1. Zoned tetrahedral crystals of reddish-brown sphalerite (*Sl*) with encrusting white dolomite (*Do*). Sample J44-20; scale is 1 cm.
2. Yellow sphalerite crystals, fragments, and broken crusts in a matrix of calcite; calcite has been stained with Alizarin red-S and appears dark. Sample X15-15.
3. Photomicrograph of X15-15; sphalerite (*Sl*) appears dark in the coarse-crystalline calcite (*Ca*). Transmitted light; scale is 0.4 mm.
4. Three textural types of sphalerite in X15 and W17 orebodies. Type 1 consists of coalescing fine-crystalline spherules of dark reddish-brown sphalerite, often with yellow rims. Type 2 is the yellow variety of X: 2, and type 3 is greenish-brown coarse-crystalline sphalerite with associated cubic galena. Interstitial calcite has been stained with Alizarin red-S and appears dark. Sample X15-20; scale is 1 cm.
5. Irregularly banded coarse-crystalline Facies K dolostone with introduced banded fine-crystalline greenish-brown sphalerite. N38A open pit, first bench, northeast side; coin is 23 mm in diameter.
6. Fine-crystalline sphalerite bands with interlayer galena and marcasite from location in X: 5. Sample N38A-8.
7. Zoned reddish-brown sphalerite crystals in dense very fine-crystalline dolostone of Facies L. Sample A70-1-80; scale is 1 cm.
8. Concentration "front" of sulfides, largely sphalerite, in fine-crystalline Facies J-3 dolostone. Sample X15-38; transmitted light; scale is 1 cm.
9. Anastomosing network of concentration "fronts" in Facies J-3 dolostone; note apparent bleached areas adjacent to sulfide concentrations. X15 open pit, fourth bench, north side.
10. Dendritic and ramose sphalerite forms with galena growing on substrates of dolostone breccia fragments and encrusted by coarse calcite. J44 stockpile, exact location unknown; lens cap is 54 mm in diameter.

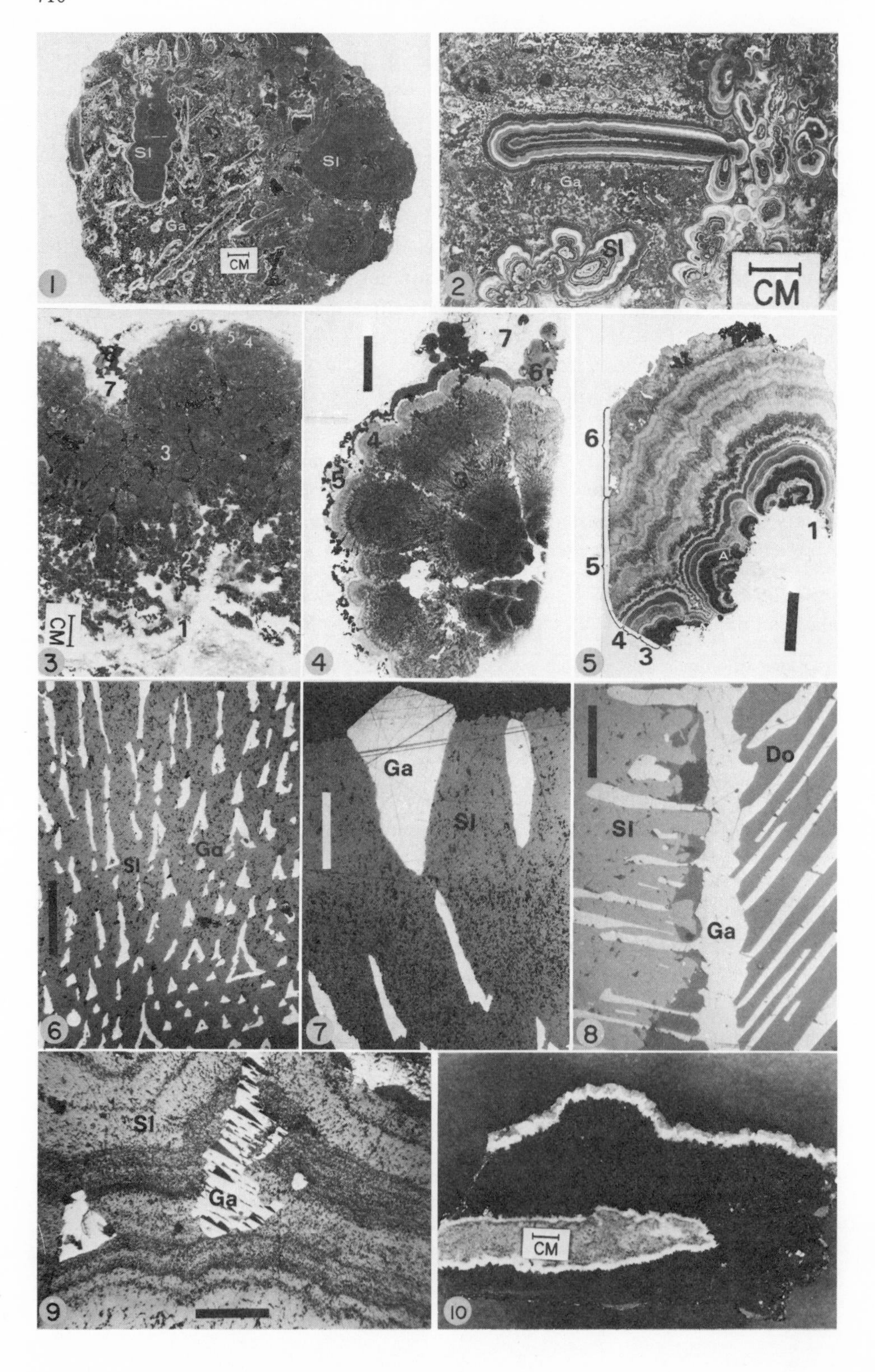

1
Sl
Sl
Ga
CM
2
Ga
Sl
CM
3
6
5
4
7
3
2
1
CM
4
7
6
4
5
3
5
6
5
4
3
1
A
6
Sl
Ga
7
Ga
Sl
8
Do
Sl
Ga
9
Sl
Ga
10
CM

ramose and dendritic forms, and as intimate intergrowths with other sulfides (Plates X–XIII). Individual crystals range from less than 1 mm to as much as 4 cm; many project into open spaces, but some occur without associated porosity in fine-crystalline sulfide aggregates and within dense host dolostone (Plate X: 7–9). Sphalerite occurs in yellow, tan, orange, light reddish-brown, dark reddish-brown, and dark brown varieties; some irregular laminae are purple in transmitted light. Commonly sphalerite occurs in "colloform" crusts of varying thicknesses composed of differently colored layers (Roedder, 1968b) (Plate XI); crystals are optically continuous across many color bands and often extend across the entire crust (Plate XI: 5). Some colloform sphalerite appears to be stalactitic in form (Plate XI: 1,2), but this material is generally broken and is not in growth position. All ZnS appears to be in the form of sphalerite, although local birefringent areas suggest some wurtzite polytype layers.

A preliminary study of "stratigraphy" in colloform sphalerites from a number of orebodies has shown that sphalerite bands ("strata") can be traced consistently among samples from individual orebodies, using megascopic criteria of color, habit, relative thickness, associated minerals, and stratigraphic sequence (Plate XI: 4,5). Certain sphalerite strata and sequences of strata are absent from individual samples; their absence

PLATE XI

Sulfide textures.

1. Broken colloform sphalerite (*Sl*) in a matrix of skeletal galena (*Ga*). Sample O42-1.
2. Dark and light reddish-brown and tan colloform sphalerite in a matrix of sphalerite and galena. Sample 11,676; O42 open pit.
3. Thick sulfide crust consisting of the following units, in order of deposition: (*1*) dolomite; (*2*) dendritic dark reddish-brown sphalerite with minor galena; (*3*) ramose orange sphalerite with about 30% skeletal galena and rare pyrite; (*4*) orange sphalerite with minor skeletal galena; (*5*) yellow sphalerite; (*6*) reddish-brown sphalerite with cubic galena; (*7*) dolomite; and (*8*) dark brown sphalerite crystals with octahedral galena. Sample M40-7.
4. Layers 3–7, sample M40-7 (XI: 3). Transmitted light; scale is 1 cm.
5. Thick sulfide crust consisting of the following units: (*1*) dolomite; (*2*) alternating light and dark reddish-brown sphalerite with minor skeletal galena; (*3*) reddish-brown sphalerite with about 10% skeletal galena; (*4*) alternating yellow and tan microcrystalline sphalerite layers; initial yellow sphalerite layer fills skeletal galena molds transecting several layers of earlier strata (*A*); (*5*) alternating yellow and orange sphalerite bands; unit crystals extend across the stratigraphic sequence; and (*6*) yellow and orange sphalerite bands with octahedral galena crystals. Sample N42-1; scale is 1 cm.
6. Photomicrograph of skeletal galena in sphalerite, sample M40-7 (XI: 3). Reflected light; scale is 0.4 mm.
7. Photomicrograph of "molar tooth" of galena with crown extending beyond growth limit of sphalerite stalactite. Sample M40-27, reflected light; scale is 0.4 mm.
8. Photomicrograph of skeletal galena with sphalerite and dolomite. Sample M40-8, reflected light; scale is 0.4 mm.
9. Photomicrograph of galena in colloform sphalerite showing the response of sphalerite precipitation to the presence of the galena crystal. Sample O42-1, reflected light, partly crossed nicols; scale is 0.4 mm.
10. Thick crust of colloform sphalerite with galena; note white dolomite underlying and overlying sulfides. Sample O28-2.

is apparently due to non-deposition rather than chemical erosion. Gross sequences of strata ("groups") can be correlated between some orebodies, but the textural details of the stratigraphic sequence are commonly quite different, even among adjacent orebodies (Plate XI: 4,5). These features suggest that local conditions greatly influenced sulfide precipitation.

Microprobe analyses of sphalerites from several Pine Point orebodies reveal that iron is the most common trace element and ranges from 0.15 to 10.3 wt.% (Table IV). Although Roedder and Dwornik (1968) did not find a direct correlation between color and iron content of colloform sphalerite layers, analyses of sphalerite of several colors indicate a general pattern of greater iron content with darker color (Fig. 43). Since iron is the only common trace element, there also appears to be a complementary correlation between lower zinc content and darker color. In general, these analyses are in agreement with the findings of Scott and Barnes (1972) which suggested that dark color in sphalerite is indicative of decreased metal and increased sulfur content. There does not appear to be a positive relationship between iron sulfide content of individual orebodies and iron content of sphalerite from those bodies. For instance, sphalerite from the iron-rich X15 orebody is generally lower in iron than the average of all sphalerite samples (Kyle, 1977). Pb, Cu, Cd, and Mn are highly erratic in distribution and reach maxima of 1.05, 0.16, 0.32, and 0.02%, respectively (Table IV). Pb, Cu, and Mn do not appear to vary systematically with sphalerite color; Cd, on the other hand, appears to be most abundant in the tan, pulverulent microcrystalline variety (Fig. 43; Plate XI: 1,2,5). Ni, Co, and Ag were not detected in any sample. These limited analyses do not reveal a consistent trend in the composition of sphalerite relative to paragenetic stage, associated minerals, host rock type, or location within the district (Kyle, 1977).

Galena. Galena, PbS, is present in varying amounts in all orebodies and occurs as indi-

TABLE IV

Composition of Pine Point sulfide minerals

Mineral		Zn	S	Pb	Fe	Mn	Cu	Cd	Total
Sphalerite	maximum	66.64	33.61	1.05	10.30	0.02	0.16	0.32	100.04
	minimum	55.47	31.48	0.00	0.15	0.00	0.00	0.00	98.07
	mean (69)	64.16	33.38	0.21	2.23	0.01	0.02	0.05	99.67
Galena	maximum	1.30	13.66	86.60	0.15	n.a.	0.24	n.a.	99.98
	minimum	0.00	13.09	84.47	0.00	n.a.	0.00	n.a.	98.44
	mean (24)	0.23	13.34	85.76	0.01	n.a.	0.08	n.a.	99.44
Pyrite/marcasite	maximum	n.a.	53.95	n.a.	46.82	n.d.	1.27	n.a.	100.04
	minimum	n.a.	51.93	n.a.	45.00	n.d.	0.00	n.a.	98.63
	mean (31)	n.a.	53.26	n.a.	46.10	n.d.	0.16	n.a.	99.54

n.a. not analyzed; n.d. not detected.

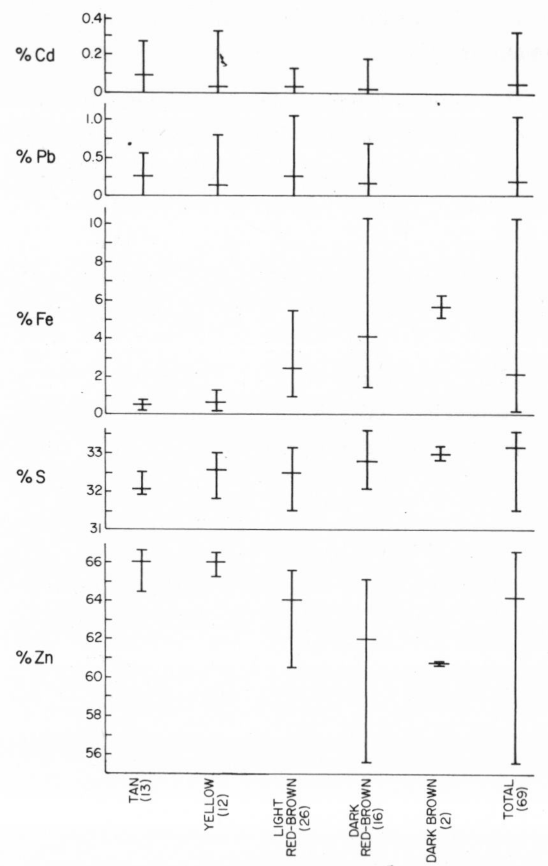

Fig. 43. Composition of Pine Point sphalerite relative to color. Vertical bar is the range; horizontal bar is the mean.

vidual cubic and cubo-octahedral crystals, as skeletal hopper, bladed, and ramose forms, and as intimate intergrowths with other sulfides (Plates X–XIII). Individual crystals range from less than 1 mm to as much as 5 cm (Plate XII), and galena has essentially the same modes of occurrence as sphalerite. Galena is present within colloform and stalactitic sphalerite, generally oriented subperpendicular to sphalerite growth bands (Plate XI: 7). These and other forms are crystallographic entities which often extend for several centimeters in the direction of growth and commonly transect sphalerite stratigraphy. Galena co-precipitation with sphalerite is indicated by intimate intergrowths (Plate XI: 3,4,6,7) and by the apparent response of sphalerite growth bands to galena crystal interference (Plate XI: 9).

Microprobe analyses of galena reveal little variation in composition relative to orebody, associated minerals, paragenetic position, or crystal habit (Kyle, 1977). Zn, Cu, and

Fe are the only trace elements present and reach maxima of 1.30, 0.24, and 0.15% (Table IV), respectively; Ag was not detected in any sample.

Pyrite, marcasite, and pyrrhotite. The polytypes of FeS_2, pyrite and marcasite, are present in varying amounts in all Pine Point sulfide bodies (Plates XII and XIII); pyrrhotite, $Fe_{1-x}S$, is rare, and its distribution is not well known (Plate XIII: 10). Although the pyrite/marcasite ratio varies greatly among specimens, marcasite appears to be the dominant iron sulfide, particularly in those sulfide bodies which contain abundant iron. The decomposition of iron-rich ore upon exposure further suggests the abundance of easily oxidized marcasite. Marcasite and pyrite are often intimately intergrown with other sulfides (Plate XII: 10), and the iron sulfides may be brecciated and enclosed in a matrix of later sulfides (Plate XIII: 9). Iron sulfides are commonly massive, but locally have vuggy, boxwork, radiating, and ramose forms (Plate XIII). Marcasite generally occurs as the characteristic bladed orthorhombic crystals, sometimes twinned on (101), ranging from less than 0.1 mm to as much as 1 cm (Plate XIII: 4,5). Pyrite is usually in poorly formed cubes (Plate XII: 10), occasionally with octahedral faces, and individual crystals up to 3 cm are present in the Green Clay Zone of W17 (Plate XIII: 3).

Microprobe analyses of iron sulfides from several orebodies indicate little compositional variation between marcasite and pyrite or between the same mineral from different orebodies, mineral assemblages, or crystal habits (Kyle, 1977). Cu is erratic in distribution and reaches a maximum of 1.25% (Table IV); Co, Ni, Mn, and Ag were not detected in any sample.

PLATE XII

Sulfide textures.

1. Skeletal galena aggregate containing 3 cm cubic galena crystals. Sample K62-34.
2. Large cubic galena crystals from clay-filled vug in X15 orebody.
3. Photomicrograph of galena crystals with colloform sphalerite; massive galena associated with colloform sphalerite (e.g. XI: 1, XII: 1) commonly is composed of skeletal crystal aggregates. Sample 8341 (O42 open pit), reflected light; scale is 0.4 mm.
4. Coarse reddish-brown sphalerite (*Sl*) overgrown by late cubo-octahedral galena (*Ga*) and white dolomite. Sample N42-15; scale is 1 cm.
5. Hopper galena crystals overgrown by white dolomite; arrow shows relative direction of mineralizing fluid movement as indicated by crystal asymmetry (Kesler et al., 1972). Sample M40-36; scale is 1 cm.
6. Bladed galena crystals in center of banded reddish-brown sphalerite. Sample K57-22.
7. Coarse cubic galena crystals in a matrix of fine-crystalline, pulverulent, tan sphalerite and skeletal galena. Sample K62-10.
8. High-grade ore consisting of sulfide and Facies D-1 dolostone fragments cemented by calcite; note large iron sulfide fragments (*A*) and marcasite rind on one lithic fragment (*B*). Sample X15-26.
9. Pulverulent, tan colloform sphalerite formed around irregular masses of iron sulfides (*FS*); colloform sphalerite is uncommon in the orebodies in the fine-crystalline dolostones of the lower Pine Point Group. Sample X15-6.
10. Photomicrograph of fine-crystalline ramose sphalerite enclosing marcasite and pyrite within massive iron sulfides (*FS*); large euhedral crystals on periphery are pyrite (*Py*). Sample X15-9, reflected light, scale is 0.4 mm.

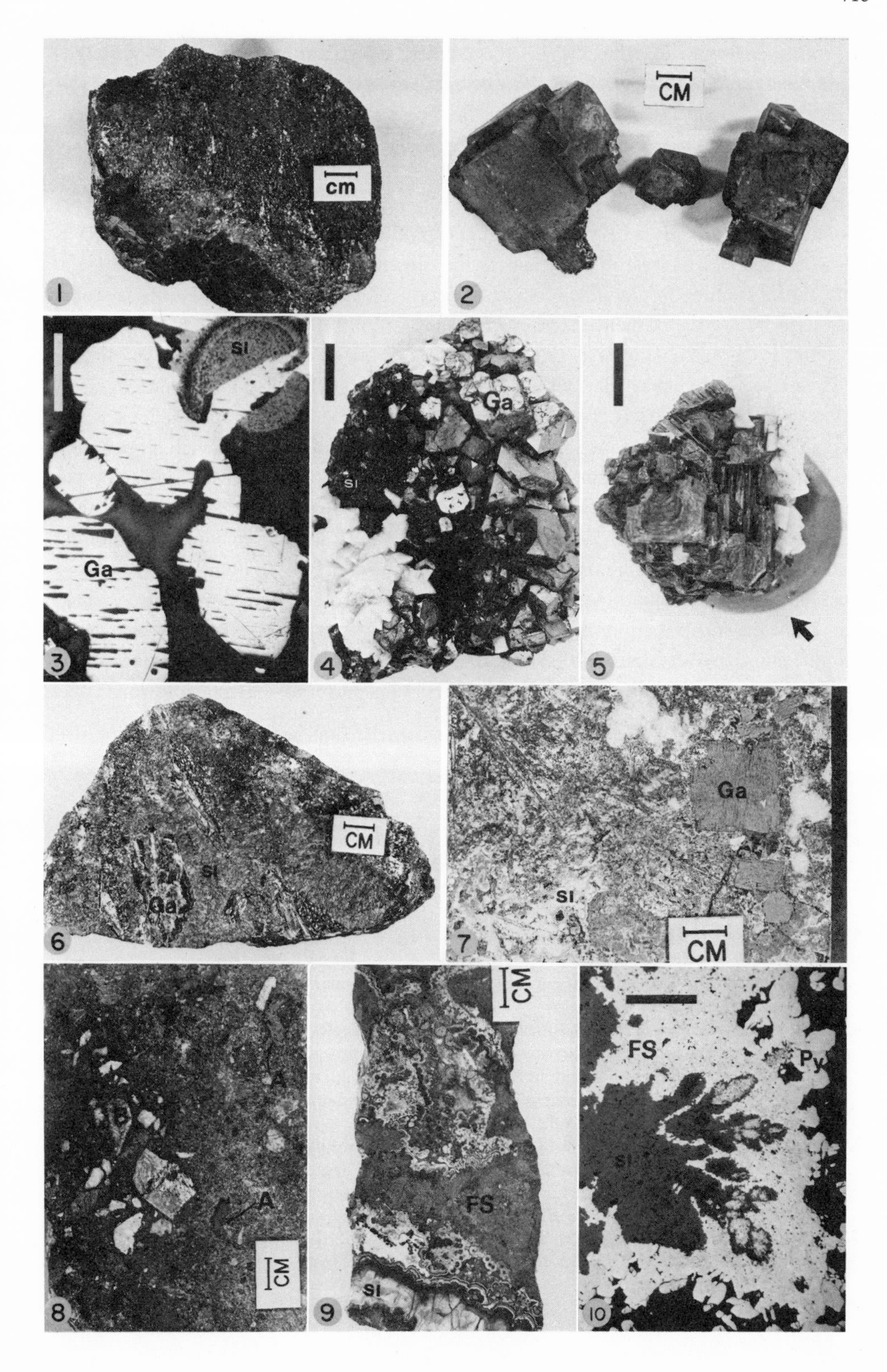
cm
1
CM
2
si
Ga
3
Ga
si
4
5
CM
si
Ga
6
Ga
si
CM
7
A
A
CM
8
CM
FS
si
9
FS
Py
si
10

Dolomite. Dolomite, $CaMg(CO_3)_2$, is extremely abundant in the coarse-crystalline Facies K dolostone, particularly in the upper part (Plate XIV), and minor amounts occur locally throughout the Givetian strata in the Pine Point area. Dolomite occurs as single crystals, thick banded crusts, fracture and breccia cement, breccia-moldic forms, and pervasive replacement of coarse dolostone (Plates IV and V). Individual rhombohedral crystals commonly reach 1 cm in size and are generally curved (baroque). Most of the dolomite is white, some is light to medium gray, and a minor amount is pink (Plate XIV). The white color is, at least in part, due to the presence of a vast number of fluid inclusions; fine-crystalline iron sulfides impart the gray color. The pinkish dolomite is probably caused by a trace element, perhaps Mn. The relationship between the white and gray dolomites is complex. They may be interlayered, one may pass abruptly into the other in the same layer, or one may fill fractures in the other (Plate XIV: 1–4). Dolomite in direct contact with iron sulfides is commonly pale gray, but most gray dolomite is not directly associated with megascopic iron sulfides. White and gray dolomite are much more extensive than sulfide mineralization; commonly, there is less dolomite in the sulfide bodies than in the adjacent unmineralized Facies K. Dolomite is both earlier and later than sulfide minerals (Plate XI: 3,5,10; XIV: 1,2). Brecciated sulfides cemented by dolomite (Plate XIV: 2–4) and the absence of sulfides with the white dolomite in the very porous upper Facies K dolostone suggest that a great deal of white dolomite may have formed later than the main period of sulfide deposition. Vague boundaries between white dolomite and coarse-crystalline dolostone (Plates IV: 7; V: 8; XIV: 5) further suggest that white dolomite largely formed by isochemical replacement of pre-existing dolostone.

Calcite. Calcite, $CaCO_3$, is rather erratic in its distribution and abundance in the dis-

---

PLATE XIII

Sulfide textures.

1. Rare example of acicular crust of marcasite on massive iron sulfides. Sample P31-1.
2. Vuggy boxwork of pyrite with marcasite and minor colloform sphalerite. Sample X15-12.
3. Pyrite crystal groups and nodules with minor encrusting coarse-crystalline, dark reddish-brown sphalerite from Green Clay Zone of W17.
4. Photomicrograph of twinned orthorhombic marcasite crystals. Sample X15-6, reflected light, partly crossed nicols; scale is 0.2 mm.
5. Acicular crystals of marcasite with intercrystal filling of galena. Sample X15-26, reflected light, scale is 0.2 mm.
6. Pervasive introduction of iron sulfides along fracture in Facies L tidal-flat dolostone. Sample R61-1.
7. Photomicrograph of iron sulfide rim in R61-1 (XIII: 6); introduction of acicular marcasite crystals along dolostone crystal contacts has resulted in irregular rhombic outlines, particularly toward fracture (extreme top). Reflected light; scale is 0.2 mm.
8. Photomicrograph illustrating pervasive introduction of iron sulfides (*FS*) along irregular rhombic crystal boundaries in dolostone. Sample R61-38-64, reflected light; scale is 0.2 mm.
9. Fragments of massive iron sulfides in a matrix of fine-crystalline marcasite, pyrite, sphalerite, and galena. Sample W17-2.
10. Photomicrograph showing rare occurrence of interstitial pyrrhotite (*Po*) along grain boundaries in massive pyrite (*Py*). Sample X15-12, reflected light; scale is 0.2 mm.

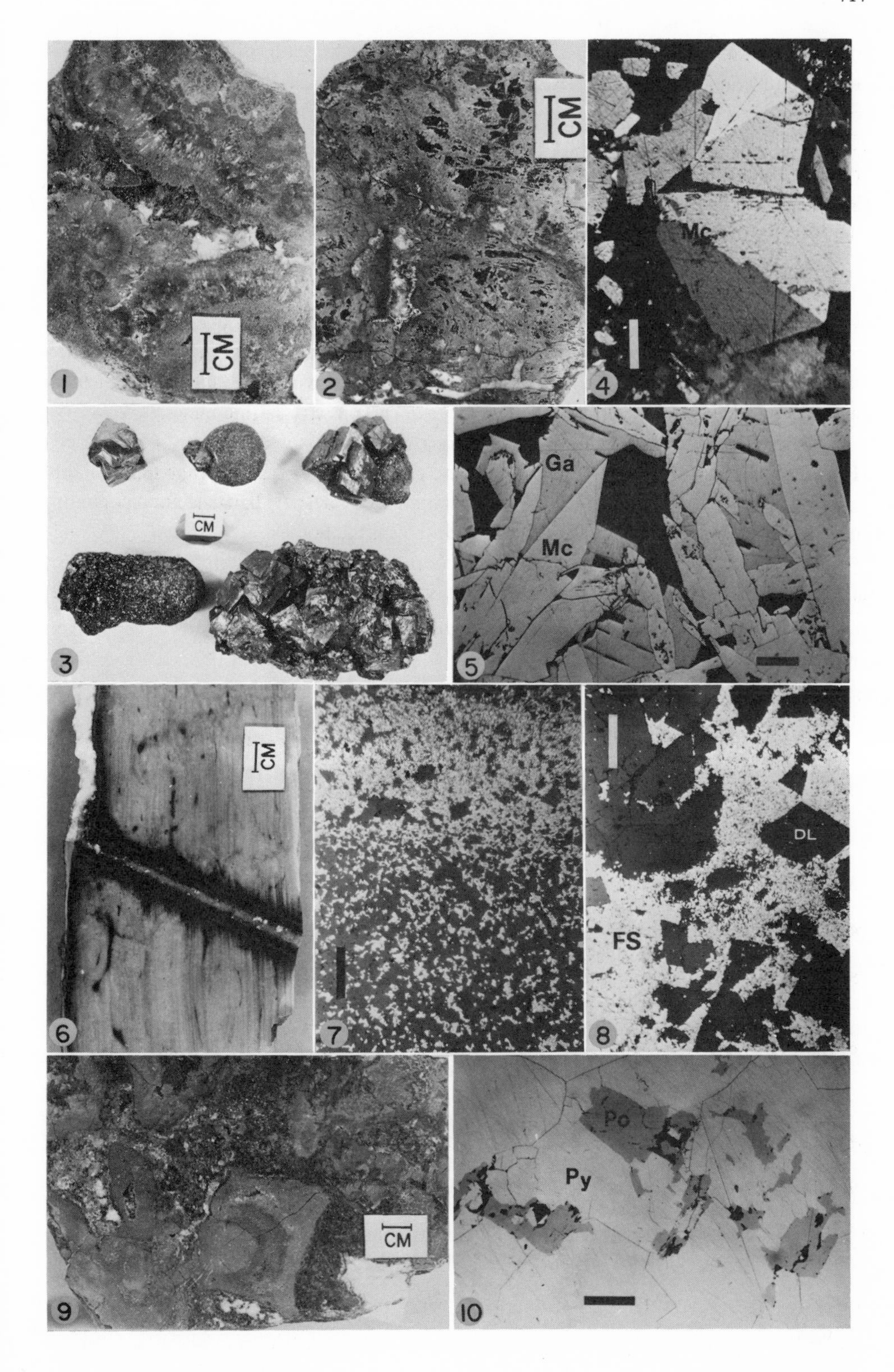
CM
CM
Mc
CM
Ga
Mc
CM
DL
FS
Po
Py
CM
1
2
3
4
5
6
7
8
9
10

trict. It occurs as single crystals, as thick crusts, as vug- and fracture-fillings, and as breccia cement and is commonly colorless, white, or amber (Plate XIV). Simple and complex scalenohedral crystals are the most common forms; individual crystals reach as much as 25 cm in the longest dimension. Calcite occurs also as medium gray, radiating, acicular crystals within the fine-crystalline reefal lithologies in the lower Pine Point Group. Except for calcite interlayered with white dolomite, calcite was one of the last minerals to form, and its distribution reflects the distribution of pre-existing porosity.

Celestite, barite, and fluorite. Colorless, orthorhombic celestite, $SrSO_4$, is present within the coarse-crystalline Facies K dolostone and is usually associated with dolomite, calcite, and native sulfur (Plate XIV: 8). Coarse-crystalline celestite associated with calcite usually is not recognized in routine core examination and is more abundant than generally acknowledged. It has not been observed in association with sulfide minerals.

Barite, $BaSO_4$, is commonly listed among the minerals which occur at Pine Point (Beales and Jackson, 1968). W.E. Wiley (pers. comm., 1974) suggests that some of the reported barite may have been misidentified dolomite or celestite. Barite is not a common mineral at Pine Point and is not associated with the sulfide bodies.

Fluorite, CaF_2, is commonly listed as occurring in the Pine Point district (Beales and Jackson, 1968). It has been observed only in one short section of drill core in the

---

PLATE XIV

Gangue mineral textures.

1. White dolomite overlain by gray dolomite, dark reddish-brown coarse-crystalline sphalerite, and white dolomite; the gray dolomite passes abruptly to white dolomite with a faint pink tint on the right. Sample M40-31; scale is 1 cm.
2. Central area (*1*) of white dolomite containing coarse-crystalline dolostone with relic fossil outlines overlain by gray dolomite (*2*) with dark reddish-brown sphalerite, grayish white dolomite (*3*), gray dolomite (*4*), and a thick rim of white dolomite (*5*). Units *3* and *4* fill fracture in gray dolomite and sphalerite of unit *2* at *A*. Sample J44-26.
3. Fragments of massive iron sulfides (*FS*) with coarse-crystalline reddish-brown sphalerite cemented by white dolomite. Compare with XII: 6. Sample N38A-30.
4. Fragments and bands of fine-crystalline, light reddish-brown sphalerite cemented by light gray and white dolomite; compare with XII: 6. Sample N38A-31.
5. Coarse-crystalline orange sphalerite with white dolomite in uniform granular dolostone of lower Facies K. Vague boundaries between dolostone and white dolomite suggest that some white dolomite grew by cannibalization of the coarse dolostone. Sample K57-25-158; scale is 1 cm.
6. Thick carbonate crust with massive white calcite and white dolomite on gray dolomite. Right half of core has been stained with Alizarin red-S so that calcite appears dark and dolomite remains unchanged. Sample K57-95-131.
7. Large calcite crystals from vug in J44 orebody; coin is 23 mm in diameter.
8. Tabular crystals of colorless celestite (*Ce*) overgrown by coarse-crystalline yellow native sulfur (*S*) from vug in Facies K. Sample 3301-421; scale is 1 cm.
9. Banded light reddish-brown sphalerite (*Sl*) with bitumen sphere (*Bi*) and calcite (*Ca*). Sample K62-36.
10. White dolomite crust on gray dolomite with bitumen. Bitumen sphere has white dolomite around it and may have been present during white dolomite growth. Sample K57-32.

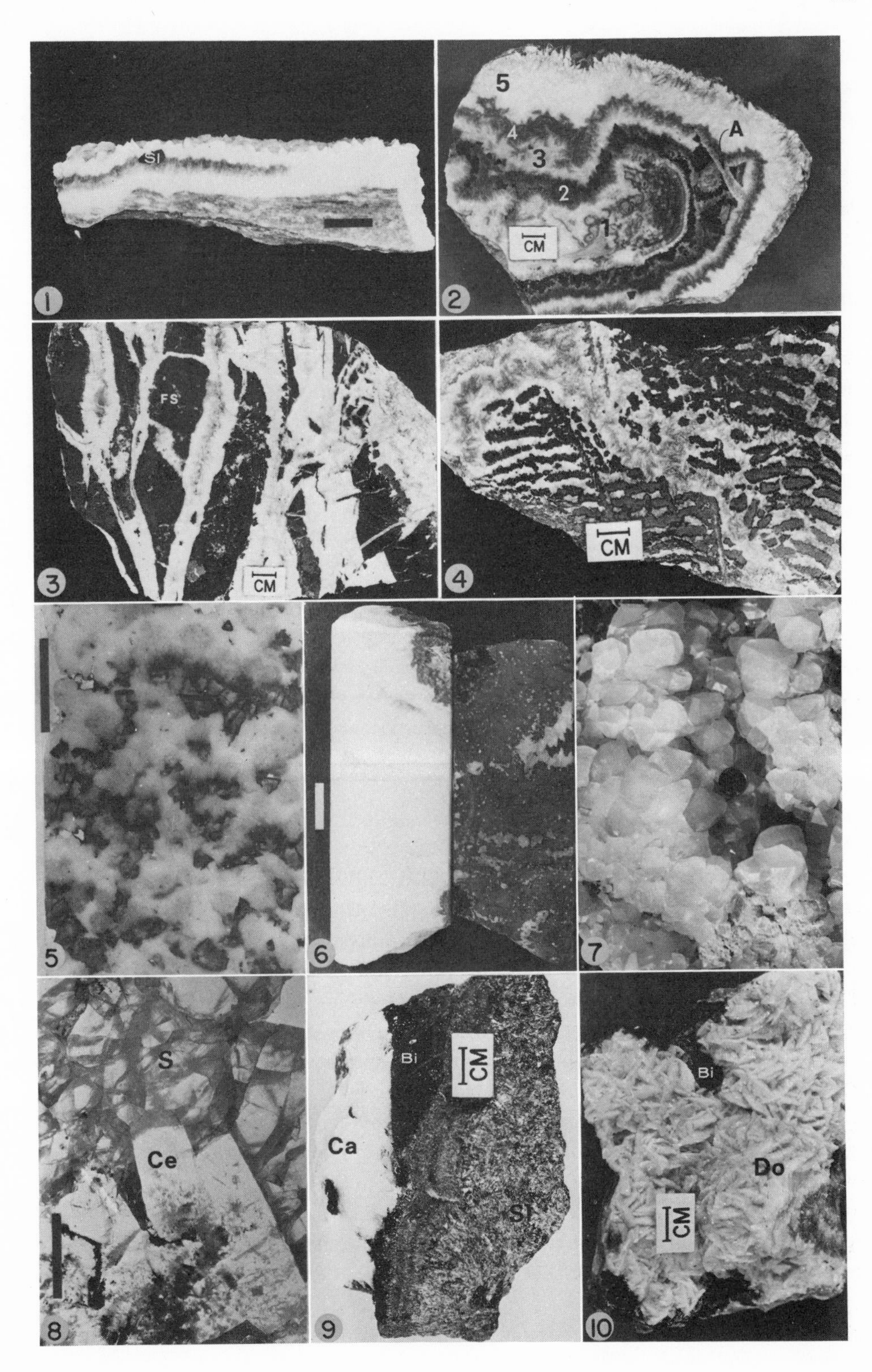
Sl
5
4
3
2
1
A
CM
FS
CM
CM
S
Ce
Bi
CM
Ca
Sl
Bi
Do
CM

coarse-crystalline Facies K dolostone (W.E. Wiley, pers. comm., 1975) and must be regarded as very rare.

Anhydrite and gypsum. Anhydrite, $CaSO_4$, and its hydrated analogue, gypsum, $CaSO_4 \cdot 2\,H_2O$, occur as extensive back-reef evaporite strata of the Muskeg Formation. They also fill porosity in and locally replace barrier lithologies, particularly in the fine-crystalline reefal rocks of the lower Pine Point Group. Gypsum and anhydrite occasionally have been observed in spatial proximity to sulfide bodies (Jackson, 1971), but the relationship between sulfides and sulfates has not been clearly established.

Sulfur. Coarse-crystalline, yellow native sulfur, S, is relatively common in the district and is associated with dolomite, calcite, celestite, and bitumen (Plate XIV: 8). It has been observed throughout the Givetian carbonate facies but is abundant only in Facies K. This relationship may be due in part to the greater availability of late porosity in the coarse-crystalline dolostone. Sulfur occurs also as the result of present-day decomposition of marcasite. Native sulfur occurs occasionally in spatial proximity to sulfide minerals, but a genetic association in inconclusive.

Bitumen. Bitumen, a complex mixture of devolatized hydrocarbons, is present in vuggy and intergranular porosity in the carbonate rocks (Plate XIV: 9,10). It forms spheres as much as 2 cm in diameter when present in open vugs and is most abundant in Facies K. White dolomite appears to have formed around bitumen in some instances (Plate XIV: 10). Bitumen occurs within the sulfide bodies, but in all of the observed examples bitumen is later than the sulfide minerals (Plate XIV: 9).

Sulfide textures and sulfide-carbonate relationships

Upper barrier orebodies. Colloform and banded sulfide textures are common in the orebodies in Facies K and younger strata (Plate XI). These textures and some well-formed coarse sulfide crystals undoubtedly were formed in open space, but in view of other textures, some of this porosity may not have existed prior to sulfide emplacement. That is, fluids of the sulfide-depositing system may have significantly enlarged pre-existing porosity by dissolution of carbonates, particularly within the detritus and breccia zones, and collapse of overlying strata may have created additional porosity (Fig. 22D). Dissolution enlargement is suggested by fragments of paragenetically early sulfides within later sulfides (Plate XI: 1), remnants of carbonate detritus within coarse-crystalline sulfides (Plate VII), corroded edges of dolomite rhombs in dolostone (Plate XIII: 7,8), and sulfides within dense dolostone (Plate X: 7). Beales (1975) has proposed that sulfides formed in equilibrium with the white dolomite of the Pine Point orebodies because of apparent distinct contacts between the two. However, the contact of the white dolomite and sulfides is irregular locally. Evidence for carbonate dissolution is associated more commonly with the prismatic orebodies (Plate VII). The tabular sulfide bodies generally appear to be simple open-space filling of pre-existing porosity (Plate VIII).

Lower barrier orebodies. Colloform sphalerite textures are present (Plate XII: 9) in the orebodies in the fine-crystalline dolostones of the lower barrier but are not as common as in the upper barrier orebodies. Instead, the dominant early generation sphalerite consists of dark reddish-brown coalesced spherules, often with yellow rims and generally less than 1 mm in diameter (Plate X: 4). Massive ore consists of a complex aggregate of fine and coarse sphalerite, marcasite, pyrite, and galena (Plates IX: 7; X: 4; XII: 10); fragments of sulfides in a matrix of later sulfides of coarse-crystalline calcite are common (Plates X: 2; XII: 8; XIII: 9). Sulfide breccia textures and highly irregular contacts between massive sulfides and dolostone or dolostone breccia suggest local conditions of disequilibrium between the fluids of the sulfide-depositing system and dolomite (Plate IX: 2,3,6).

Paragenesis

A detailed paragenetic sequence applicable throughout the district is difficult to establish because of the highly variable sulfide textures and modes of occurrence and the rather individual nature of the orebodies. The generalized sequence is shown schematically in Fig. 44. White and gray dolomite deposition generally precedes sulfide precipitation in the orebodies within Facies K. Marcasite is the earliest sulfide mineral in most orebodies, followed by, and often intergrown with, pyrite. Sphalerite deposition overlaps the iron sulfide stage, and the early textures may be fine-crystalline, colloform, dendritic,

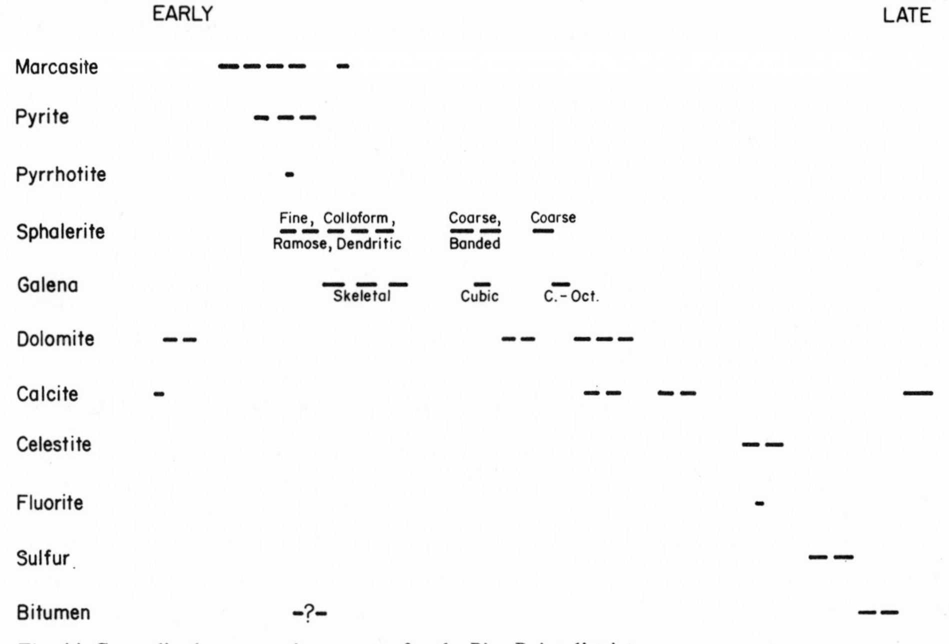

Fig. 44. Generalized paragenetic sequence for the Pine Point district.

or ramose. Galena may be intergrown with this sphalerite and is commonly skeletal. Coarse bands and crystals of late sphalerite associated with cubic and cubo-octahedral galena are generally the last sulfides deposited. A stage of white and gray dolomite and massive white calcite deposition follows sulfide precipitation. This generation of dolomite is believed to be more extensive than the pre-sulfide generation. Celestite, fluorite, and native sulfur belong to a later period of deposition. Coarse crystals of calcite fill remaining porosity. Most bitumen is a late feature, but some may be of an earlier generation.

It is possible to define more detailed sequences of mineral deposition for some individual orebodies. The orderly paragenesis of the tabular M40 orebody in the coarse-crystalline dolostone of the lower Facies K begins with dark reddish-brown dendritic sphalerite with some skeletal galena and minor iron sulfides on a white dolomite substrate (Plate XI: 3). The dendritic zone passes into a thick ramose network of orange sphalerite and abundant intergrown skeletal galena; thin bands of orange and yellow sphalerite with minor skeletal galena cap this zone. Yellow sphalerite is overlain by reddish-brown sphalerite with common cubic galena. White dolomite encloses the latest sulfide stage of coarse-crystalline dark brown sphalerite and cubo-octahedral galena. Coarse-crystalline calcite has filled much of the remaining open space. The sequence is best developed in the thick zone of massive sulfides in the central area of tabular mineralization (Plate VIII: 5).

The paragenesis of the prismatic W17 orebody in the fine-crystalline dolostone of the lower Pine Point Group apparently is not as sequential as it is in M40. Abundant marcasite and pyrite are the first sulfides to form on the dolostone substrate; rare pyrrhotite occurs along fractures and grain boundaries of massive pyrite. Dark reddish-brown fine-crystalline sphalerite spherules are intergrown with the iron sulfides. Tan, pulverulent colloform sphalerite is another paragenetically early form and may contain some skeletal galena. Coarse yellow and greenish-brown sphalerite and cubic galena are a later sulfide stage. Calcite may cement fragments of these sulfides and fill porosity. Dolomite rarely occurs in the sulfide body.

MINERALIZATION: DISCUSSION

The origin of carbonate-hosted lead–zinc deposits of the so-called Mississippi Valley-type has been the subject of considerable debate for over 50 years. The basinal evolution model developed at Pine Point (Beales and Jackson, 1966) has been applied widely to explain the origin of other carbonate-hosted deposits. Only now that development and diagenetic modification of the Pine Point barrier complex and the characteristics of the sulfide bodies have been determined is it possible to consider fully the complex problem of ore genesis. Genetic aspects to be considered for any mineral deposit are the nature of the transporting fluid, source of the components in the deposit, primarily metals and sulfur in this case, direction of fluid movement, reasons for mineral precipitation and concentration, and timing of mineralization. (For general discussions on related topics, see review by Wolf, 1976.)

Nature of the transporting fluid

Based on available production and reserve information, the Pine Point district contains about 85 million tons of ore averaging about 2.5% Pb, 6.0% Zn, and 3.5% Fe. A reasonable addition to this figure based on subeconomic sulfide occurrences and undiscovered orebodies indicates a total metal concentration of over 2 million tons of Pb, 5 million tons of Zn, and 3 million tons of Fe. These sulfide concentrations also contain 15 million tons of S, not including the native S throughout the barrier. Assuming that the present mineral assemblages and the trace-element content of the sulfides represent the

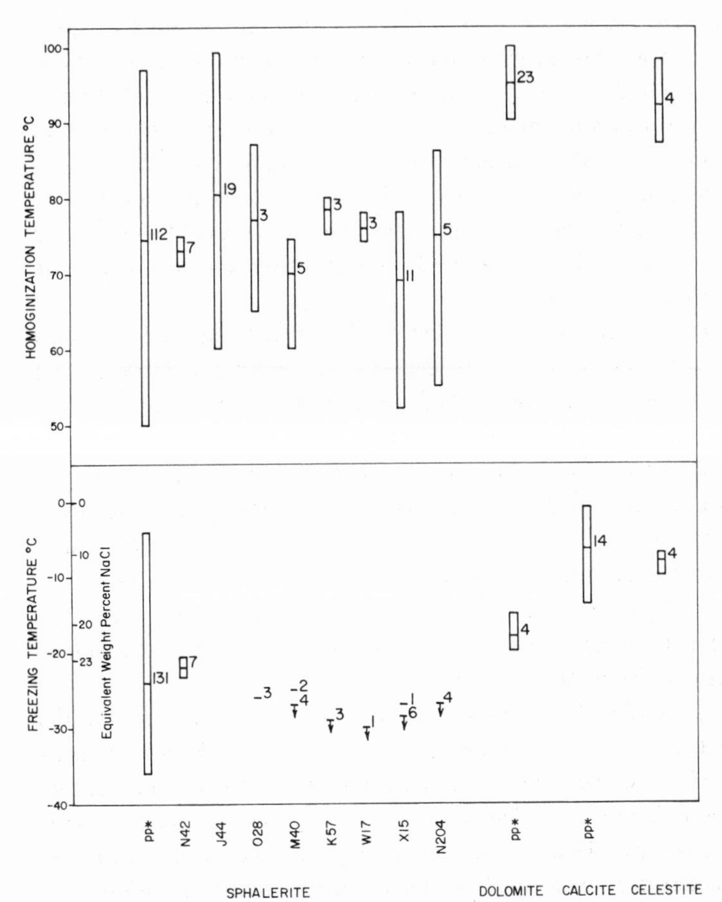

Fig. 45. Freezing and homogenization temperatures of fluid inclusions, Pine Point orebodies. (Data from Roedder, 1968a and Kyle, 1977.) PP inclusions are those of Roedder (1968a) from unspecified locations during early production from the district, probably the N42, O42, and J44 orebodies. Horizontal bar is the mean; arrows with the freezing temperatures indicate that these inclusions remained unfrozen after several hours at these temperatures.

content of the mineralizing fluid, then Pb, Zn, and Fe were the only common metals and were present in a ratio of 2 : 5 : 3. Cu, Mn, and Cd are present locally in minor quantities, but separate minerals have not been recognized. Ca-, Mg-, and Sr-bearing minerals are present, but their relationship to the sulfide minerals is unclear because of paragenetic and fluid inclusion differences; Ca, Mg, and Sr are common components of carbonate rocks.

Homogenization temperature determinations for primary fluid inclusions in sphalerite by Roedder (1968a) and Kyle (1977) indicate that the depositing fluids ranged in temperature from 51 to 99°C (Fig. 45). Freezing temperatures commonly ranged from −10 to −35°C; many inclusions remained as unfrozen, supercooled liquid at −78°C after several hours (Roedder, 1968a). These freezing temperatures indicate that the sphalerite-depositing solutions were highly saline, ranging from about 15 to over 23% total salts, calculated on the basis of pure NaCl. No daughter minerals are present, but Roedder (1968a) recognized NaCl · 2 H_2O (?) hydrate crystals in many inclusions as the last solid phase to melt on warming. Inclusions in white dolomite crystals associated with sulfide minerals had similar homogenization temperatures ranging from 90° to 100°C, but salinities from 15 to 20% NaCl are indicated (Roedder, 1968a). Similar results were obtained for celestite from the Facies K dolostone with homogenization temperatures of 86–98°C and salinities of 10–15% NaCl (Fig. 45). Coarse calcite crystals from the orebodies have salinities less than 10% (Roedder, 1968a).

The relatively low temperatures and high salinities of the Pine Point fluid inclusions are characteristic of the carbonate-hosted Pb–Zn deposits. The distinct temperature and composition differences between inclusion fluids from magmatic mineral deposits and from carbonate-hosted deposits (Roedder, 1976) and the general absence of evidence for magmatic or volcanogenic processes in the carbonate-hosted mineral districts (Ohle, 1959) suggest that the carbonate-hosted deposits originated by non-magmatic processes. This discussion will develop further the concept of a "sedimentogenic" origin (Sangster, 1976a). Chemical analyses of fluid inclusions from carbonate-hosted deposits show that the contained solutions are dense brines with Cl, Na, Ca, K, and Mg in order of decreasing weight percent (Roedder, 1976). These brines are remarkably similar to the subsurface brines present in many oil fields (White, 1968).

Modern metal-rich subsurface brines have been discovered in the Salton Sea, the Red Sea (White, 1968), the Cheleken Peninsula (Lebedev, 1972), the Western Canada basin (Billings et al., 1969), and the Gulf Coast (Carpenter et al., 1974). The last two occurrences are far removed from areas of igneous activity and high geothermal gradients and have geologic settings compatible with carbonate-hosted Pb–Zn deposits. These subsurface brines have characteristics similar to those determined from fluid inclusions. The metal source for these brines and for carbonate-hosted deposits is unclear but can be considered relative to the processes which have affected sedimentary basins.

Source of metals

Similarities between oil-field brines and inclusion fluids and the spatial association of some hydrocarbon and sulfide accumulations have prompted some authors to suggest genetic affiliations between hydrocarbons and carbonate-hosted Pb–Zn deposits (Beales and Jackson, 1966, 1968; Jackson and Beales, 1967; Skinner, 1967; Dozy, 1970; Dunsmore, 1973, 1975; Macqueen, 1976, 1979; and others). Although there are different opinions among these authors concerning metal sources and causes of sulfide precipitation, all suggest that carbonate-hosted Pb–Zn sulfide deposits, like petroleum and natural gas, are the result of normal processes active during the evolution of sedimentary basins.

Inherent in the basinal evolution model for development of carbonate-hosted Pb–Zn deposits is the concept of an original source "bed" from which metals were extracted prior to concentration as metal sulfides in a suitable "trap". Source beds for hydrocarbons can be demonstrated to have been organic-rich, fine-grained clastic sediments (e.g. Erdman and Morris, 1974); such sediments have been considered by some authors to have been a likely source of metals as well. Other authors have proposed that the immediate metal source for the deposits was either the host carbonate strata, associated evaporites, or sandstone aquifers through which the brines passed.

Fine-grained clastic sediments. The basinal evolution model for Pine Point (Beales and Jackson, 1966) invokes metal supply from compacting shales in the Mackenzie Basin and sulfur supply from the evaporites in the Elk Point Basin (Fig. 3). Favorable sites for sulfide precipitation were provided by the porous rocks of the Pine Point barrier complex. Jackson and Beales (1967) emphasize that shales generally contain more Pb and Zn than do carbonate rocks and that the large amount of water originally present in argillaceous sediments would have to be expelled during compaction and could transport released metals. However, argillaceous-sediment dewatering is a complex process involving not only compactional release of interpore water but also the dehydrational release of interlattice water (Burst, 1976). Salinity of pore water increases with depth due to ionic filtration by clays, thus indicating that relatively "fresh" water is expelled during compaction (Magara, 1974; Burst, 1976). Petroleum is highly soluble in and could be readily transported by "fresh" water, whereas increases in salinity cause hydrocarbon exsolution (Price, 1976). On the other hand, base-metal solubility in the 100°C region under consideration increases radically with increasing salinity because of metal-chloride complexing (Nriagu and Anderson, 1971). Therefore, the relationship between hydrocarbon accumulations and carbonate-hosted Pb–Zn deposits cannot be simple because the fluids which can effectively transport petroleum and base metals are of greatly different salinities. Transport of metals as organic-metal complexes is a possibility (Saxby, 1976) but introduces other problems. If metal-bearing brines are derived from basinal shales, metal extraction must have taken place during a later phase of clay-mineral dehydration as suggested by Macqueen (1976). These reactions could take place at depths of a few kilometers (Magara, 1974) which, assuming normal geothermal gradients, would be in the

75–175°C range generally applying to carbonate-hosted mineral deposits (Macqueen, 1976, 1979).

Shales, particularly organic-rich ones, do contain more Pb and Zn than most other sedimentary rocks; this fact does not necessarily mean that metals were released at an earlier stage in their diagenetic history. Also, shales are enriched in many metals, including uranium (e.g. Macqueen et al., 1975), and thus they seem too complex a source for the simple element suite present in the carbonate-hosted deposits. In addition, shale sequences which could serve as potential source rocks are absent in the vicinity of many of the important carbonate-hosted deposits of the Mississippi Valley.

Coarse-grained clastic sediments. The Pb-isotope composition of the Pb-rich ores of Southeast Missouri suggest that the metals were derived from the carbonate cement of the Lamotte Sandstone, the aquifer which overlies the Precambrian basement and underlies the host Bonneterre Dolostone (Doe and Delevaux, 1972). Sangster (1976b) points out that the few analyses presented do not eliminate the potash-feldspar component of the basal arkosic Lamotte as the immediate Pb source with the granitic basement constituting the original source. Helgeson (1967) and Banaszak (1979) support derivation of metals by reaction of heated, acidic brines with silicate sediments. The geologic setting of the Pine Point district is similar to Southeast Missouri because the host dolostones are not more than 300 m above the Precambrian basement and sandstone unit, the Old Fort Island Formation, overlies the basement (Fig. 2). Pine Point ores also are Pb-rich relative to most of the other major carbonate-hosted districts. On the other hand, Pine Point lead isotopes lack the distinct anomalous character of the Southeast Missouri deposits (Sangster, 1976b). Also, circulation of the metal-bearing brine from the basal sandstone to the host dolostone would be greatly inhibited by the intervening impermeable Chinchaga evaporite sequence.

Carbonate sediments. Because of their ubiquitous carbonate host rocks, it is reasonable to consider the possibility that trace amounts of Pb and Zn from these rocks were concentrated into economic sulfide deposits. Collins and Smith (1972) and Bernard (1973) suggest that metal concentration is accomplished during the weathering cycle that creates the "karstic" porosity which hosts much of the ore. Metals are released by mechanical weathering and chemical leaching in the vadose zone; deposition of sulfides takes place in areas of inhibited circulation below the layer of active circulation at the top of the phreatic zone (Bernard, 1973). Paterson (1975) emphasizes the textural similarities between some Pine Point ores and carbonate speleothems from modern caves. However, meteoric water at near-surface temperature is markedly different from the highly saline, 50–100°C mineralizing solutions indicated for Pine Point by fluid inclusion evidence. The explanation of Bernard (1973) that these hot, saline fluids are the result of recrystallization of "karstic" sulfides in the deep subsurface was reviewed and judged invalid by Roedder (1976). The differences in timing of karstification and sulfide deposition as shown for Pine Point are also difficult to reconcile by the model of Bernard (1973). Johnson (1972) advocates that subsurface brines are enriched in metals during dolomitization of reefal complexes.

Although the mechanism is not clear, a strong case can be developed for metal supply from carbonate rocks. Probably the most supportive evidence is the common association of galena, sphalerite, fluorite, and barite with carbonate rocks, particularly dolostones, not only in major and minor mining districts (Heyl, 1968), but also in open spaces in many Phanerozoic shelf carbonate sequences. These minor occurrences have fluid inclusion characteristics which are indistinguishable from the major districts (E. Roedder, pers. comm., 1975). Economic concentrations thus appear to be the result of a favorable combination of hosting structures, hydrology, and timing.

Evaporites. Dunsmore (1975) emphasizes the evaporitic nature of many of the host carbonate rocks, an association previously noted by Davidson (1966). He shows that the residual brines from modern seawater evaporitic pans when adjusted for diagenetic changes, have chemical compositions essentially identical to typical fluid-inclusion brines from carbonate-hosted deposits. Thiede and Cameron (1978) provided the first systematic study of the base-metal content of an evaporitic sequence. Although there is considerable variation among evaporite lithologies, lead and zinc are concentrated in the Elk Point evaporites of Saskatchewan by factors of 10^2–10^6 over that of seawater. Considerable tonnages of metals were available for release during gypsum dehydration at depths of 600–900 m (Burst, 1976) or during evaporite solution in the deep subsurface (Thiede and Cameron, 1978). One serious objection previously noted for shale metal sources is that evaporite sequences generally are not as spatially related to Pb–Zn deposits as at Pine Point, and certainly there are many major evaporite basins without known Pb–Zn deposits. However, supply of metals and sulfur in a single solution eliminates the problem of having two solutions arrive at the same site in order for sulfide precipitation to occur (Dunsmore, 1975; Banaszak, 1979).

Source of sulfur

Approximately 15 million tons of sulfur are present in sulfide form in the Pine Point orebodies, not considering the native sulfur and sulfates within the host rocks which have an uncertain relationship to the sulfides. Since the role of sulfur is that of chemically "trapping" metals to form an economic sulfide accumulation, its ultimate source and subsequent history is a vital aspect of ore genesis. Sulfur-isotope compositions provide some limitations as to source.

Sasaki and Krouse (1969) provide extensive sulfur-isotope data for sulfides from P29, N32, N42, O42, and X15 orebodies at Pine Point (Fig. 46). The average of 118 sulfide samples is +20.1‰ with a standard deviation of 2.6. The means for galena, marcasite, pyrite, and sphalerite are +18.4, +19.3, +19.7, and +21.6‰, respectively. These values are very similar to the ratios (+19 to +20‰) determined for Middle Devonian evaporites. The most reasonable source for the sulfide sulfur is Middle Devonian seawater sulfate, perhaps supplied in connate brines from the Elk Point evaporites

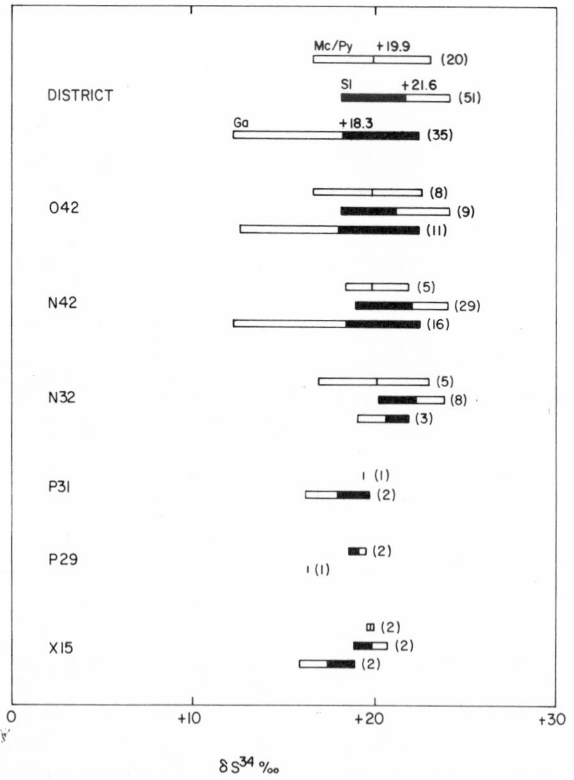

Fig. 46. Sulfur-isotope composition of Pine Point sulfides. (Data from Sasaki and Krouse, 1969).

(Sasaki and Krouse, 1969). The small range of sulfide sulfur isotopes could be the result of isotopic homogenization of H_2S before fixation as sulfides (Sangster, 1976b).

Fluid movement

Consideration of the evidence for fluid movement is pertinent to evaluation of the concept of sulfide precipitation by fluid mixing. Skall (1975) demonstrates conclusively the importance of Middle Devonian tectonic adjustments along the N65°E trend of the Hinge Zones to the establishment of depositional environments; subsequent adjustments along the Hinge Zones were responsible for the creation of the diagenetic environments during the post-middle Givetian erosional period. The combined effect of depositional and superimposed diagenetic facies formed permeability zones subparallel to the N65°E trend. The uniform crystalline dolostones of Facies E and to a lesser extent Facies D and B appear to be the most permeable units in the lower barrier. The lower part of Facies K is highly permeable, whereas the upper, highly vuggy Facies K section is probably not an effective aquifer. The thick, impermeable Chinchaga evaporite section is a barrier to vertical fluid movement, as is the intercalated shale and dense carbonate sequence of the late Givetian Facies L, M, and N. These relationships suggest that fluid movement was channeled along the Hinge Zones and within the rocks of the Pine Point Group.

Kesler et al. (1972) show that horizontal asymmetry of single sulfide crystals and groups in ten Pine Point orebodies suggest that the principal flow directions of the mineralizing fluid were both northeast and southwest along the barrier trend. A strong flow vector is directed towards the southeast, and other divergences of flow from the barrier trend are apparent. This pattern can be interpreted to represent sulfide deposition spreading from point sources along joint patterns.

Speculation concerning the direction of mineralizing fluid movement is permitted also by metal zoning in the district and in individual orebodies. Based on the decreasing relative solubilities of Fe, Zn, and Pb sulfides (Barnes and Czamanske, 1967), the metal distribution pattern for the Main and North Trends (Figs. 25, 26) suggests overall fluid movement from southeast to northwest; similarly, the ratios of the cumulative metal tonnages relative to stratigraphic position (Table III) suggest a vertical component to fluid movement. In addition, the concentric metal zonation of individual orebodies is most readily explained by local vertical movement of the mineralizing fluid.

Sulfide precipitation and concentration

Certain mineral assemblages and textures in carbonate-hosted Pb–Zn ores may aid in defining geochemical characteristics of the depositing fluids. The relationship between sulfides and carbonates is important particularly in indicating the cause of precipitation; if carbonates are stable during sulfide deposition, increase in reduced sulfur is the most likely reason for precipitation (Anderson, 1973, 1975, 1978). Beales (1975) points out instances of apparent carbonate–sulfide equilibrium, yet Kyle (1977) recognized numerous examples suggesting carbonate instability (dissolution) *during* the sulfide-depositing episode at Pine Point (e.g. Plates VII: 2,6,11; IX: 2,3,6). This evidence does not necessarily indicate that carbonates were undergoing active dissolution during sulfide precipitation; rather, it may indicate alternating episodes of carbonate dissolution and sulfide precipitation. Similarly, reduced sulfur may have been supplied periodically; as Anderson (1975) emphasizes, regular replenishment of reduced sulfur is a requirement for large sulfide concentrations because of the small amount of H_2S which can occupy available pores.

Solubility data for galena at 100°C indicate that the transporting brine must be extremely acid if both lead and reduced sulfur are carried in the same solution (Anderson, 1975); yet, it is certain that any solution within a thick carbonate sequence would have been quickly buffered by calcite and dolomite. Therefore, the most reasonable mechanism for producing local conditions of carbonate disequilibrium is through mixing of solutions of differing character (Runnells, 1969; Plummer, 1975), probably one containing metals and another reduced sulfur. The mixing of these solutions is likely to cause rapid crystallization resulting from the relatively large degree of supersaturation (Anderson, 1975). Such a phenomenon is suggested by the early stage of colloform sphalerite and skeletal galena at Pine Point. Skeletal and other incomplete forms of galena have been produced experimentally by mixing aqueous lead chloride with a solution containing H_2S (Leleu and Goni, 1974). The mixing of fluids is reflected by the $\delta\ ^{18}O$ composition of carbonate gangue minerals in the Pine Point deposits (Fritz, 1976). Banaszak (1979) contends that moderately acid solutions can carry reduced sulfur and metals complexed by chloride if reduced sulfur and oxidized sulfur are not in equilibrium. Acidity is most readily maintained if the fluid is transported largely in non-carbonate aquifers and is locally introduced into the carbonate host rocks. According to this model, sulfide precipitation will occur as the result of cooling, neutralization, and dilution of the transporting brine.

The Cheleken Peninsula is an excellent example of a modern stratigraphic section with separate aquifers which contain fluids of greatly differing salinities, metal percentages, sulfur oxidation states, and temperatures (Lebedev, 1972). Artificial mixing of these fluids produces spectacular growth of sulfides and sulfates. Kyle (1977, 1979) has proposed that natural mixing of fluids from separate aquifers in paleo-dolines and breccia zones in the Pine Point barrier complex resulted in slower but just as spectacular growth of sulfides. This mechanism is supported by the association of the prismatic orebodies with dolines and breccia zones within an otherwise undisturbed stratigraphic section, by the sharp boundaries and individual nature of these orebodies, by the concentric metal zoning pattern in individual orebodies, and by the divergences in fluid flow directions which appear to indicate point sources for the orebodies.

Timing of mineralization

Probably the single most enigmatic aspect of carbonate-hosted Pb–Zn deposits is the general problem of determining the age of mineralization. Yet, timing is of paramount importance in any genetic model for these deposits (for general discussion see Wolf, 1976). It is clear from geologic evidence that many of these sulfide accumulations postdate the deposition of their host rocks, i.e., they are "epigenetic" rather than "syngenetic". In the case of Pine Point, sulfides occur in strata as young as the late Givetian Facies N, but there is no geologic evidence to indicate a precise age of mineralization during the 375 million years from the Givetian to the present.

Most of the classic carbonate-hosted Pb–Zn deposits of the Mississippi Valley are characterized by markedly radiogenic lead isotopes of the so-called J-type with $^{206}Pb/^{204}Pb$ ratios of 20 or greater (Heyl et al., 1974). Isotopic zonation in some districts has been interpreted to represent solution flow vectors; some individual galena crystals are zoned with more radiogenic isotopes on the perimeter (Heyl et al., 1974). The lead isotopic composition of 40 galena samples from the Pine Point district has been determined (Cumming and Robertson, 1969; Cumming et al., 1971; Kyle, 1977). Collectively, these samples represent sixteen orebodies and additional mineralized drill intersections; the analyses reported by Kyle (1977) are for thirteen samples selected to supplement previous data and to represent variables of geographic location, hosting facies, galena habit, associated minerals, and paragenetic position (Fig. 47). Although some variation is shown by these data, the analyses are spread along the "204-error" line, suggesting that analytical error of the least abundant isotope is the probable cause of most variation. Thus, the analyses can be considered as a single population, and a single-stage model for their evolution can be applied (D.F. Sangster, pers. comm., 1976). Using the methods of Cumming and Richards (1975), a model age of 310–320 million years can be calculated from the mean isotopic composition. The mid-Carboniferous age thus determined can be considered as the valid age of mineralization in the absence of other evidence. The non-radiogenic character of the Pb may be interpreted to indicate either a juvenile source, or a source thoroughly mixed during sedimentation or transport so as to approximate single-stage conditions (Sangster, 1976b). District zonation or variation in lead-isotopic percentages or ratios due to stratigraphic position, associated minerals, galena habit, or paragenetic stage is not indicated by present data.

The study by Beales et al. (1974) demonstrates great potential for the use of natural remnant magnetism to determine age of mineralization. Paleomagnetic pole positions computed for ores and their host rocks from two carbonate-hosted Pb–Zn districts are identical within statistical uncertainty. These results suggest that the ores, although clearly epigenetic, are of approximately the same age as the host rock (within 25 million

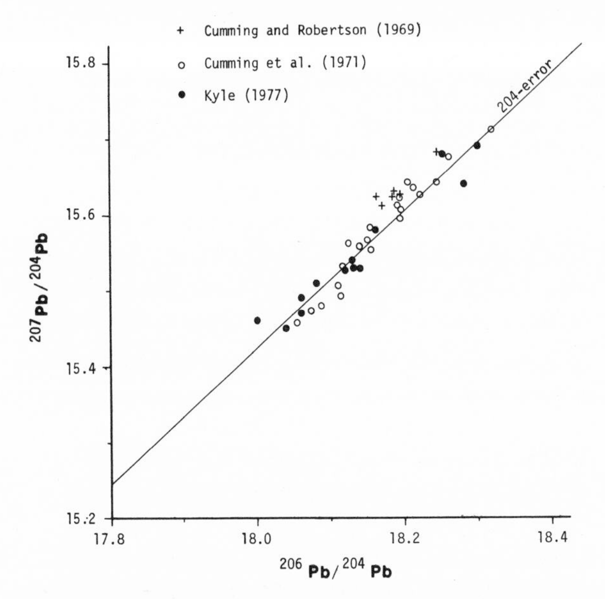

Fig. 47. Lead-isotope composition of Pine Point galenas. (Data from Cumming and Robertson, 1969, Cumming et al., 1971, and Kyle, 1977.)

years at their limit of accuracy), which is in agreement with geologic evidence. This method of age determination should become increasingly beneficial as apparent polar wandering curves are better defined by future work and as more sensitive techniques are developed to measure low levels of remnant magnetism. Beales et al. (1974) were unable to detect the remnant magnetism for a sample of Pine Point ore, but Beales and Wu (1979) indicate that paleomagnetic data suggest a late Devonian age for mineralization at Pine Point.

In addition to the geologic age of mineralization, the length of time involved in sulfide deposition is also of interest. For example, as Roedder (1976) shows, 100 million tons of 10% ore could be deposited within a time span of 1000 to 10 million years with a precipitation change of 1000 to 10 ppm, respectively, using geologically reasonable values for fluid flow rates, quantity of flow, fluid density, and bulk porosity. Roedder (1968a) recognized regular ("periodic") color bands in Pine Point colloform sphalerites and interpreted them to be annual varves, perhaps resulting from seasonal mixing of fluids of different compositions. If the total ore deposition consists of about 7 cm thick colloform crusts on many substrates, as Roedder assumes, and each varve has a thickness of 7 μm, then 100 million tons of 10% ore could be deposited in about 350,000 years with a precipitation change of 1000 ppm and flow rates of 10 gallons per minute. Extremely rapid sulfide growth has been demonstrated for the Cheleken Peninsula where the artificial mixing of natural metal- and sulfur-bearing brines results in colloform sphalerite precipitation at the rate of 1 cm per year (Lebedev, 1972). Periods of rapid precipitation are suggested also by Pine Point sulfide textures such as colloform sphalerite, skeletal galena, and intimate intergrowths (Plates XI: 3–8; XII: 5); also periods of slower growth are indicated by large crystals (Plates XII: 2,4; XIII: 3). Interruptions in sulfide deposition are indicated by intergrown dolomite (Plate XI: 3).

COMPARISON WITH OTHER CARBONATE-HOSTED DISTRICTS [1]

There have been many attempts to define the characteristics of the carbonate-hosted lead–zinc–fluorite–barite deposits of the Mississippi Valley-type (e.g. Ohle, 1959; Snyder, 1968; Brown, 1970; Hagni, 1976; Sangster, 1976a) and to classify this diverse group of deposits (e.g. Brown, 1967) into geologically meaningful subdivisions. It is apparent from these classifications that there is a great deal of variation among the geologic features of these districts. Classifications that place particular weight on the presence or absence of a particular characteristic may exclude districts of otherwise compatible characteristics; the Pine Point deposits illustrate this case. Although the Pine Point district has many of the same general characteristics of the other carbonate-hosted districts (Table V) and has been used to model the genesis of this type of deposit, it lacks

[1] See Laznicka's Chapter 3, Vol. 8, on "The concept of ore types".

TABLE V

Comparison of the characteristics of Pine Point with major Mississippi Valley-type deposits (After Hagni, 1976; Brown, 1970; Snyder, 1968; and Ohle, 1959)

Mississippi Valley-type deposits	Pine Point deposits
I. *Host-rock character*	
A. Major districts occur in shallow-water carbonates of late Proterozoic to Triassic age.	A. Middle Devonian (Givetian) reef and associated carbonate barrier lithologies.
B. Certain stratigraphic horizons are the principal producers, but stratabound ore extends over a larger stratigraphic section.	B. Stratabound orebodies at several intervals over 200-m stratigraphic section.
C. Unconformities underlying and within the mineralized sequence influenced depositional and diagenetic facies.	C. Post-middle Givetian, pre-late Givetian partial disconformity influenced depositional and diagenetic facies.
II. *Diagenetic modification*	
A. Dolomitization; most ore deposits are confined to dolostones of several types.	A. Orebodies confined to fine-crystalline dolostones of lower barrier complex or coarse-crystalline dolostones of upper barrier complex.
B. Recrystallization; local recrystallization of host rocks adjacent to ore deposits.	B. Neomorphic growth of dolomite, probably unrelated to mineralization.
C. Silicification; developed to varying degrees.	C. No silica associated with ore zones.
D. Karstification; often important in preparation of ore host.	D. Karstification important in preparation of hosting structures.
III. *Regional mineralization controls*	
A. Districts occur on the margins of major sedimentary basins or above major positive structures.	A. District located at the margin of Elk Point Basin and Mackenzie Basin near flank of Precambrian Shield.
B. Major structural features important in ore localization.	B. N65°E Hinge Zones important in controlling sedimentation, diagenesis, and mineralization; relation to N45°E basement structures is unclear.
C. Districts located near facies boundaries of sedimentary or diagenetic origin.	C. District located in carbonate barrier complex between evaporite and shale facies; some orebodies near diagenetic facies boundary between limestone and coarse-crystalline dolostone.
IV. *Local mineralization controls*	
A. Sedimentary structures; reefs, bars, slump breccias, and local facies were important in ore localization.	A. Sedimentary facies, modified by diagenetic processes, were important particularly in localization of tabular orebodies.
B. Tectonic structures; folding and minor faulting were important ore controls in some districts.	B. Some orebodies are associated with low-amplitude fold trends and minor faults.
C. Diagenetic structures; solution collapse breccias and zones of secondary porosity are important ore hosts in most districts.	C. Filled dolines, collapse breccias, and secondary porosity zones were important controls of mineralization.

(continued)

TABLE V (continued)

Mississippi Valley-type deposits	Pine Point deposits
V. *Ore character*	
A. Coarse crystals of ore and gangue minerals resulting from the open-space filling of vugs, fractures, and breccias form the most common ore type.	A. Colloform, skeletal, and coarse sulfide crystals resulting from open-space filling comprise the majority of the ore.
B. Disseminated "replacement" ore is associated with cavity-filling ore in some districts.	B. Disseminated sulfides, developed by local removal of carbonate grains, are common in some ore zones.
C. Fissure veins are associated with some strata-bound deposits.	C. No associated veins.
D. Size of orebodies is limited by size of the hosting structure	D. Enlargement of pre-existing structures during mineralization was important in some orebodies.
E. Most deposits are relatively low grade.	E. District averages about 8% combined Pb–Zn, with significant amounts of higher grade ore.
VI. *Ore mineralogy and paragenesis*	
A. Principal minerals are restricted to a small number with simple chemical compositions.	A. Common minerals include only sphalerite, galena, marcasite, pyrite, dolomite, and calcite.
B. Minor minerals are limited.	B. Pyrrhotite, celestite, gypsum, anhydrite, fluorite, sulfur, and bitumen are present in minor amounts.
C. Minerals were deposited in a well-defined sequence in which most of one mineral tends to be deposited before most of another.	C. Principal minerals were deposited in a general sequence of dolomite, marcasite/pyrite, colloform sphalerite and skeletal galena, coarse crystalline sphalerite and galena, dolomite, and calcite.
VII. *Chemical constituents of the ore*	
A. Major elements include variable amounts of Zn, Pb, Fe, Cu, Ba, F, Ca, Mg, Si, S, C, and O.	A. Major elements include Zn, Pb, Fe, Ca, Mg, S, C, O.
B. Minor elements present in principal minerals: Fe, Ag, Cd, Ge, In, Co, and Hg in sphalerite; Ag, Sb, Bi, and As in galena; Co, Ni, Ag, and Cu in pyrite and marcasite; Sr, Y, Ba, and Mn in calcite.	B. Minor elements present in principal minerals: Fe, Pb, Cu, Mn, and Cd in sphalerite; Zn, Cu, and Fe in galena; Cu in pyrite and marcasite.
C. Wide range in lead isotope composition of galena; generally enriched in the radiogenic lead isotopes (J-type) in the midcontinent deposits with $^{206}Pb/^{204}Pb$ of 20 or greater.	C. Galena is not enriched in radiogenic lead isotopes with an average $^{206}Pb/^{204}Pb$ of 18.2.
D. Wide range in sulfur-isotope composition of sulfides; generally sulfides are depleted in ^{34}S relative to ^{32}S in comparison with evaporite sulfate of the same geologic age as the carbonate host rocks.	D. Average of sulfide sulfur isotopes is +20.1‰ in comparison with Middle Devonian evaporite sulfate of +19 to +20‰.

TABLE V (continued)

Mississippi Valley-type deposits	Pine Point deposits
VIII. *Mineralizing fluid character*	
A. Fluid inclusions contain concentrated Na–Ca–Cl brines; younger minerals may contain more dilute solutions.	A. Brine salinity in sphalerite generally greater than 20 wt.% total salts; younger dolomite, celestite, and calcite have salinities ranging from 20% to less than 10%.
B. Homogenization temperatures of fluid inclusions generally range from 50 to 175°C.	B. Homogenization temperatures range from 50 to 100°C.
C. Textures generally suggest equilibrium deposition of sulfides on carbonates.	C. Evidence is present for stages of both equilibrium and disequilibrium deposition of sulfides on carbonates.

the anomalous "J-type" lead isotope signature of the classic lead–zinc districts in the midcontinent of the United States, and therefore, it would not be considered by some authors (e.g. Brown, 1970; Heyl et al., 1974) to be of Mississippi Valley-type. However, the presence of many similar features (Table V) and the existence of other carbonate-hosted districts without J-type lead isotopes (e.g. Silesia; Sangster, 1976b) suggests that these deposits are genetically related, albeit variations in genetic mechanisms are required to account for these differences.

The Pine Point deposits have a number of other features that differ from the classic carbonate-hosted districts. The virtual absence of silica associated with the ore zones probably reflects the paucity of siliciclastic sediments in the Pine Point area. Pine Point is a premier example of facies control of mineralization with orebodies confined to a carbonate barrier complex less than 20 km wide that separates coeval shale and evaporite strata; this close association of sedimentary facies is not apparent in many districts. The rapid facies change reflects the nature of the tectonic hinge zones that controlled sedimentation; tectonic control of sedimentation and mineralization in most districts is very subtle. The Pine Point ores are generally higher grade than those of most of the Mississippi Valley-type deposits. In part, this reflects economic factors, but the high-grade nature of the ore may be a result of the apparent ability of the mineralizing fluids to enlarge pre-existing ore-hosting structures at Pine Point. The colloform sphalerite and skeletal galena of the paragenetically early sulfide generations are much more common than in most districts. The extremely simple mineralogic and chemical composition of Pine Point ore is somewhat atypical. The virtual absence of fluorite and barite, as well as the presence of celestite, native sulfur, and pyrrhotite is atypical for the Mississippi Valley-type districts but is not unique. The Pine Point orebodies tend to have rather individual natures, i.e., variations in mineral and metal ratios, textures, paragenesis, hosting structures, etc. are typical, even among nearby orebodies. This discontinuity is believed to reflect the nature of the ore-hosting structures and fluid movement; the stratabound "paleoaquifer" systems of many of the Mississippi Valley-type districts (Hagni, 1976) has resulted in relatively continuous and regular sulfide deposition over large areas. The difference between sulfide and contemporaneous seawater sulfate sulfur isotopes is much less than for most carbonate-hosted districts (Sangster, 1976b). The mineralization temperatures for the Pine Point deposits are lower than for most of the Mississippi Valley-type deposits (Roedder, 1976).

SUMMARY

The Pine Point district contains about 40 Pb–Zn orebodies within a Middle Devonian carbonate barrier complex that developed along tectonic hinge zones separating contemporaneous evaporite and shale basins. The orebodies occur in two trends parallel to the hinge zones and are stratabound in dolostones of several depositional facies over a 200-m section of the barrier. The sulfide bodies range in size from 100,000 tons to more than 15 million tons and contain up to 20% combined Pb–Zn as sphalerite and galena. Marcasite and pyrite are ubiquitous accessories, and the Fe content of orebodies ranges from less than 1% to greater than 10%. The overall Pb : Zn : Fe ratio for the district is about 2 : 5 : 3. Orebodies are prismatic with large vertical relative to horizontal dimensions, and tabular with large horizontal relative to vertical dimensions. Prismatic orebodies are generally Pb-rich with Pb/(Pb + Zn) ratios averaging about 0.4; tabular orebodies are Zn-rich with ratios averaging less than 0.3. Individual orebodies are zoned with a Pb-rich, high-grade core passing outward into a Zn-rich, high-grade zone which grades into an Fe-rich, low grade envelope. The sulfide concentrations have abrupt contacts with barren host rocks. Orebodies in both the Main and North Trends are more Pb-rich and less Fe-rich from southeast to northwest in zones parallel to the hinge zones.

Paleo-solution structures related to post-middle Givetian subaerial exposure of the barrier complex are important ore hosts. Numerous irregularly elliptical dolines as much as 400 m long and 35 m deep developed through dissolution of limestones by meteoric water. These dolines subsequently were filled with erosional detritus. Caves and tabular zones of increased permeability were formed in the upper part of the phreatic zone. The coarse-crystalline Facies K dolostone may have been created by mixing of fresh water and seawater during this period of subaerial exposure. These paleo-solution features were aquifers and loci for sulfide deposition in the coarse-crystalline Facies K dolostone. Dolines host prismatic orebodies, whereas caves and tabular permeable zones contain tabular orebodies. Origin of the breccias which host sulfides in the fine-crystalline dolostone of the lower barrier is less apparent, but these are believed to be solution features as well.

Ore textures are complex and exhibit paragenetically early marcasite and pyrite with rare pyrrhotite followed by colloform and banded sphalerite with skeletal galena. Large, well-formed sphalerite and galena crystals are late. Dolomite is both earlier and later than sulfides; calcite, celestite, gypsum, fluorite, sulfur, and bitumen are later and generally not associated with the sulfides. Fluid-inclusion evidence indicates that the Pine Point sulfides were deposited by highly saline, 50–100°C brines. These brines appear to have originated within the sedimentary sequence, but the immediate metal source cannot be defined by present data. The most likely source of sulfur is the Middle Devonian evaporites. The Pb-isotope compositions of district galenas are nonradiogenic with a mean $^{206}Pb/^{204}Pb$ ratio of about 18.2. These data suggest a mid-Carboniferous age of mineralization, considerably younger than the middle Givetian dolostone host. Colloform sphalerite and skeletal galena indicate rapid early sulfide deposition, while coarse crystals suggest slower late sulfide growth. Evidence for both sulfide–carbonate equilibrium and disequilibrium conditions is apparent, perhaps relating to periodic fluctuations in the supply of reduced sulfur. High-grade sulfide concentrations are localized in paleodolines and breccia zones because these transgressive features were the bypasses between different aquifers and acted as natural mixing sites for mineralizing fluids. This model is supported by the occurrence of prismatic orebodies in dolines and breccia zones within an otherwise undisturbed stratigraphic section, the sharp boundaries and individual nature of the ore zones, the concentric metal zoning pattern in orebodies, and apparent divergences in fluid flow directions.

Many questions remain unanswered for the Pine Point district, including the source of metals and the timing of mineralization. The concept of multiple phases of metal enrichment should be considered. Exhalative supply of metals to the sea floor along the N65°E Hinge Zones perhaps resulted in metal concentration in shales, carbonates, or evaporites. Subsequent metal enrichment of sedimentary brines occurred, and metals were fixed as sulfides probably upon encountering reduced sulfur. Mechanical (and chemical?) reworking of sulfides as a result of carbonate dissolution accompanying sulfide deposition appears to have resulted in further sulfide concentration in some orebodies.

ACKNOWLEDGEMENTS

I am grateful to Pine Point Mines Ltd. and Cominco Ltd. for providing summer employment at Pine Point and for permission to publish this material. This paper represents part of a doctoral research project at the University of Western Ontario under the supervision of R.W. Hutchinson; D.F. Sangster provided much valuable guidance while serving as an external advisor. R.W. Macqueen and Karl H. Wolf reviewed the manuscript and made valuable suggestions for its improvement. I am grateful to Ms. Linda Davis Kyle and Mrs. Betty Kurtz for assistance in manuscript preparation.

REFERENCES

Adams, D.H., 1973. N204 study. In: *Pine Point Mines 1974 Exploration Proposal.*

Adams, D.H., 1975. Stratigraphic correlations over the Pine Point area. Pine Point Mines report (unpublished).

Adams, J.E. and Rhodes, M.L., 1960. Dolomitization by seepage refluxion. *Am. Assoc. Pet. Geol. Bull.*, 44: 1912–1920.

Anderson, G.M., 1973. The hydrothermal transport and deposition of galena and sphalerite near 100°C. *Econ. Geol.*, 68: 480–492.

Anderson, G.M., 1975. Precipitation of Mississippi Valley-type ores. *Econ. Geol.*, 70: 937–942.

Anderson, G.M., 1978. Basinal brines and Mississippi Valley-type ore deposits. *Episodes*, 2: 15–19.

Back, W. and Hanshaw, B.B., 1970. Comparison of chemical hydrogeology of the carbonate peninsulas of Florida and Yucatan. *J. Hydrol.*, 10: 330–368.

Badiozamani, K., 1973. The Dorag dolomitization model-application to the Middle Ordovician of Wisconsin. *J. Sediment. Petrol.*, 43: 965–984.

Banaszak, K.J., 1979. A coherent basinal-brine model of the genesis of Mississippi Valley Pb–Zn ores based in part on absent phases. *Soc. Min. Eng. AIME Prepr.*, *79-94*: 9 pp.

Barnes, H.L. and Czamanske, G.K., 1967. Solubilities and transport of ore minerals. In: H.L. Barnes (Editor), *Geochemistry of Hydrothermal Ore Deposits.* Holt, Rinehart, and Winston, New York, N.Y. pp. 334–381.

Beales, F.W., 1975. Precipitation mechanisms for Mississippi Valley-type ore deposits. *Econ. Geol.*, 70: 943–948.

Beales, F.W. and Jackson, S.A., 1966. Precipitation of lead–zinc ores in carbonate reservoirs as illustrated by Pine Point ore field, Canada. *Trans. Inst. Min. Metall.*, 75: B278–B285.

Beales, F.W. and Jackson, S.A., 1968. Pine Point – a stratigraphical approach. *Can. Inst. Min. Metall. Bull.*, 61: 867–878.

Beales, F.W. and Wu, Y., 1979. Paleomagnetics applied to the study of Mississippi Valley-type ore deposits. *Soc. Min. Eng. AIME Program Abstr.*, 33.

Beales, F.W., Carracedo, J.C. and Strangway, D.W., 1974. Paleomagnetism and the origin of Mississippi Valley-type ore deposits. *Can. J. Earth Sci.*, 11: 211–223.

Bebout, D.G. and Maiklem, W.R., 1973. Ancient anhydrite facies and environments, Middle Devonian Elk Point Basin, Alberta. *Bull. Can. Pet. Geol.*, 21: 278–343.

Bernard, A.J., 1973. Metallogenic processes in intra-karstic sedimentation. In: G.C. Amstutz and A.J. Bernard (Editors), *Ores in Sediments.* Springer, New York, N.Y., pp. 43–57.

Billings, G.K., Kesler, S.E. and Jackson, S.A., 1969. Relation of zinc-rich formation waters, northern Alberta to the Pine Point ore deposit. *Econ. Geol.*, 64: 385–391.

Braun, W.A., 1967. Upper Devonian ostracod faunas of Great Slave Lake and northern Alberta. *International Symposium on Devonian Systems,* 2: 617–652.

Brown, J.S. (Editor), 1967. Genesis of Stratiform Lead–Zinc–Barite–Fluorite Deposits (Mississippi Valley Type Deposits). *Econ. Geol. Monogr.*, 3: 443 pp.

Brown, J.S., 1970. Mississippi Valley type lead–zinc ores. *Miner. Deposita,* 5: 103–119.

Burst, J.F., 1976. Argillaceous sediment dewatering. In: F.A. Donath et al. (Editors), *Annu. Rev. Earth Planet. Sci.*, 4: 293–318.

Campbell, N., 1950. The Middle Devonian in the Pine Point area, N.W.T. *Proc. Geol. Assoc. Can.*, 3: 87–96.

Campbell, N., 1966. The lead–zinc deposits of Pine Point. *Can. Inst. Min. Metall. Bull.*, 59: 953–960.

Campbell, N., 1967. Tectonics, reefs, and stratiform lead–zinc deposits of the Pine Point area, Canada. In: J.S. Brown (Editor), *Genesis of Stratiform Lead–Zinc–Barite–Fluorite Deposits in Carbonate Rocks. Econ. Geol. Monogr.*, 3: 59–70.

Carpenter, A.B., Trout, M.L. and Pickett, E.E., 1974. Preliminary report on the origin and chemical evolution of lead- and zinc-rich oil field brines in central Mississippi. *Econ. Geol.*, 69: 1191–1206.

Chilingar, G.V., Bissell, H.J. and Wolf, K.H., 1979a. Diagenesis of carbonate sediments and epigenesis (or categenesis) of limestones. In: G. Larsen and G.V. Chilingar (Editors), *Diagenesis in Sediments and Sedimentary Rocks.* Elsevier, Amsterdam, pp. 249–422.

Chilingar, G.V., Zenger, D.H., Bissell, H.J. and Wolf, K.H., 1979b. Dolomites and dolomitization. In: G. Larsen and G.V. Chilingar (Editors), *Diagenesis in Sediments and Sedimentary Rocks.* Elsevier, Amsterdam, 423–563.

Collins, J.A. and Smith, L., 1972. Sphalerite as related to the tectonic movements, deposition, diagenesis and karstification of a carbonate platform. *24th Int. Geol. Congr., Proc. Sec.*, 6: 208–215.

Cumming, G.L. and Richards, J.R., 1975. Ore lead isotope ratios in a continuously changing earth. *Earth Planet. Sci. Lett.*, 28: 155–171.

Cumming, G.L. and Robertson, D.K., 1969. Isotopic composition of lead from the Pine Point deposit. *Econ. Geol.*, 64: 731–732.

Cumming, G.L., Burke, M.D., Tsong, F. and McCullough, H., 1971. A digital mass spectrometer. *Can. J. Phys.*, 49: 956–965.

Davidson, C.F., 1966. Some genetic relationships between ore deposits and evaporites. *Trans. Inst. Min. Metall.*, 75: B216–B225.

De Wit, R., Gronberg, E.C., Richards, W.B. and Richmond, W.O., 1973. Tathlina area, District of Mackenzie. In: *The Future Petroleum Provinces of Canada. Can. Soc. Pet. Geol. Mem.*, 1: 187–212.

Doe, B.R. and Delevaux, M.H., 1972. Source of lead in Southeast Missouri galena ores. *Econ. Geol.*, 67: 409–435.

Dozy, J.J., 1970. A geological model for the genesis of the lead–zinc ores of the Mississippi Valley, U.S.A.. *Trans. Inst. Min. Metall.*, 79: B163–B170.

Dunsmore, H.E., 1973. Diagenetic processes of lead–zinc emplacement in carbonates. *Inst. Min. Metall. Trans.*, 82: B168–B173.

Dunsmore, H.E., 1975. *Origin of Lead–Zinc Ores in Carbonate Rocks: a Sedimentary-Diagenetic Model.* Thesis, Royal School of Mines, Imperial College, London (unpublished).

Erdman, J.G. and Morris, D.A., 1974. Geochemical correlation of petroleum. *Am. Assoc. Pet. Geol., Bull.*, 58: 2326–2337.

Folk, R.L. and Land, L.S., 1975. Mg/Ca ratio and salinity: two controls over crystallization of dolomite. *Am. Assoc. Pet. Geol. Bull.*, 59: 60–68.

Fritz, P., 1976. Oxygen and carbon isotopes in ore deposits in sedimentary rocks. In: K.H. Wolf (Editor), *Handbook of Strata-Bound and Stratiform Ore Deposits, 2.* Elsevier, Amsterdam, pp. 191–218.

Fritz, P. and Jackson, S.A., 1972. Geochemical and isotopic characteristics of Middle Devonian dolomites from Pine Point, Northern Canada. *24th Int. Geol. Congr., Proc. Sec. 6*: 230–243.

Hagni, R.D., 1976. Tri-state ore deposits: the character of their host rocks and their genesis. In: K.H. Wolf (Editor), *Handbook of Strata-Bound and Stratiform Ore Deposits, 6*. Elsevier, Amsterdam, pp. 457–494.

Helgeson, H.C., 1967. Silicate metamorphism in sediments and the genesis of hydrothermal ore solutions. In: J.S. Brown (Editor), *Genesis of Stratiform Lead–Zinc–Barite–Fluorite Deposits in Carbonate Rocks. Econ. Geol. Monogr.*, 3: 333–342.

Heyl, A.V., 1968. Minor epigenetic, diagenetic, and syngenetic sulfide, fluorite, and barite occurrences in the central United States. *Econ. Geol.*, 63: 585–594.

Heyl, A.V., Landis, G.P. and Zartman, R.E., 1974. Isotopic evidence for the origin of Mississippi Valley-type mineral deposits: a review. *Econ. Geol.*, 69: 992–1006.

Hoffman, P.F., 1969. Proterozoic paleocurrents and depositional history of the east arm fold belt, Great Slave Lake. *Can. J. Earth Sci.*, 6: 441–462.

Hoffman, P., Dewey, J.F. and Burke, K., 1974. Aulacogens and their genetic relations to geosynclines. In: R.H. Dott Jr. and R.H. Shaver (Editors), *Modern and Ancient Geosynclinal Sedimentation. Soc. Econ. Paleontol. Mineral. Spec. Publ.*, 19: 38–55.

Hsu, K.J. and Siegenthaler, C., 1969. Preliminary experiments on hydrodynamic movement induced by evaporation and their bearing on the dolomite problem. *Sedimentology,* 12: 11–26.

Jackson, S.A., 1971. *The Carbonate Complex and Lead–Zinc Orebodies, Pine Point, Northwest Territories, Canada.* Thesis, Univ. of Alberta, Edmonton, Alta. (unpublished).

Jackson, S.A. and Beales, F.W., 1967. An aspect of sedimentary basin evolution: the concentration of Mississippi Valley-type ores during late stages of diagenesis. *Bull. Can. Pet. Geol.*, 15: 383–433.

Jackson, S.A. and Folinsbee, R.E., 1969. The Pine Point lead–zinc deposits, N.W.T., Canada – Introduction and paleoecology of the Presqu'ile reef. *Econ. Geol.*, 64: 711–717.

Johnson, A.M., 1972. Metal-bearing brines from reef complexes (abs.): *Geol. Soc. Am. Abstr. Programs,* 4: 553.

Kesler, S.E., Stoiber, R.E. and Billings, G.K., 1972. Direction of flow of mineralizing solutions at Pine Point, N.W.T., *Econ. Geol.*, 67: 19–24.

Klovan, J.E., 1964. Facies analysis of the Redwater Reef Complex, Alberta, Canada, *Can. Pet. Geol. Bull.*, 12: 1–100.

Kyle, J.R., 1977. *Development of Sulfide-Hosting Structures and Mineralization, Pine Point, Northwest Territories.* Thesis, Univ. Western Ontario, London (unpublished).

Kyle, J.R., 1979. Mineralization controls, Pine Point district, Northwest Territories. *Soc. Min. Eng. AIME Preprint* 79-50: 19 pp.

Land, L.S., 1973. Contemporaneous dolomitization of middle Pleistocene reefs by meteoric water, North Jamaica. *Bull. Mar. Sci.*, 23: 64–92.

Lebedev, L.M., 1972. Minerals of contemporary hydrotherms of Cheleken. *Geochem. Int.*, 9: 485–504.

Leleu, M. and Goni, J., 1974. Sur la formation biogeochemique de stalactites de galene. *Miner. Deposita.*, 9: 27–32.

Lowenstam, H.A., 1950. Niagaran reefs of the Great Lakes area. *J. Geol.*, 58: 430–487.

Macqueen, R.W., 1976. Sediments, zinc and lead, Rocky Mountain Belt, Canadian Cordillera. *Geosci. Can.*, 3: 71–81.

Macqueen, R.W., 1979. Base metal deposits in sedimentary rocks: some approaches. *Geosci. Can.*, 6: 3–9.

Macqueen, R.W., Williams, G.K., Barefoot, R.R. and Foscolos, A.E., 1975. Devonian metalliferous shales, Pine Point region, District of Mackenzie. *Geol. Surv. Can. Pap.*, 75-1 (Part A); 553–556.

Magara, K., 1974. Compaction, ion filtration, and osmosis in shale and their significance in primary migration. *Am. Assoc. Pet. Geol. Bull.*, 58: 283–290.

Maiklem, W.R., 1971. Evaporative drawdown – a mechanism for water-level lowering and diagenesis in the Elk Point Basin. *Bull. Can. Pet. Geol.*, 17: 194–233.

Matthews, R.K., 1974. A process approach to diagenesis of reefs and reef associated limestones. In: L.F. Laporte (Editor), *Reefs in Time and Space. Soc. Econ. Paleontol. Mineral. Spec. Publ.*, 18: 234–256.

Murray, R.C. and Pray, L.C., 1965. Dolomitization and limestone diagenesis – an introduction. In: *Dolomitization and Limestone Diagenesis: a Symposium. Soc. Econ. Paleontol. Mineral. Spec. Publ.*, 13: 1–2.

Norris, A.W., 1965. Stratigraphy of Middle Devonian and older Paleozoic rocks of the Great Slave Region, Northwest Territories. *Geol. Surv. Can. Mem.*, 322.

Nriagu, J.O. and Anderson, G.M., 1971. Stability of the lead (II) chloride complexes at elevated temperatures. *Chem. Geol.*, 7: 171–183.

Ohle, E.L., 1959. Some considerations in determining the origin of ore deposits of the Mississippi Valley type. *Econ. Geol.*, 54: 769–789.

Paterson, D.M., 1975. *A Mineralographic Investigation of Pine Point Ores.* Thesis, Univ. British Columbia, Vancouver, B.C. (unpublished).

Plummer, L.N., 1975. Mixing of seawater with calcium carbonate ground water. In: E.H.T. Whitten (Editor), *Quantitative Studies in the Geological Sciences. Geol. Soc. Am. Mem.*, 142: 219–236.

Price, L.C., 1976. Aqueous solubility of petroleum as applied to its origin and primary migration. *Am. Assoc. Pet. Geol. Bull.*, 60: 213–244.

Roedder, E., 1968a. Temperature, salinity, and origin of the ore-forming fluids at Pine Point, Northwest Territories, Canada, from fluid inclusion studies. *Econ. Geol.*, 63: 439–450.

Roedder, E., 1968b. The noncolloidal origin of "colloform" textures in sphalerite ores. *Econ. Geol.*, 63: 451–471.

Roedder, E., 1976. Fluid-inclusion evidence on the genesis of ores in sedimentary and volcanic rocks.

In: K.H. Wolf (Editor), *Handbook of Strata-Bound and Stratiform Ore Deposits, 2.* Elsevier, Amsterdam, pp. 67–110.

Roedder, E. and Dwornik, E.J., 1968. Sphalerite color banding: lack of correlation with iron content, Pine Point, Northwest Territories, Canada. *Am. Mineral.,* 53: 1523–1529.

Runnels, D.D., 1969. Diagenesis, chemical sediments, and mixing of natural waters, *J. Sediment. Petrol.,* 39: 1188–1201.

Sangster, D.F., 1976a. Carbonate-hosted lead–zinc deposits. In: K.H. Wolf (Editor), *Handbook of Strata-Bound and Stratiform Ore Deposits, 6.* Elsevier, Amsterdam, pp. 447–465.

Sangster, D.F., 1976b. Sulphur and lead isotopes in stratabound deposits. In: K.H. Wolf (Editor), *Handbook of Strata-Bound and Stratiform Ore Deposits, 2.* Elsevier, Amsterdam, pp. 219–266.

Sasaki, A. and Krouse, H.R., 1969. Sulfur isotopes and the Pine Point lead–zinc mineralization. *Econ. Geol.,* 64: 718–730.

Saxby, J.D., 1976. The significance of organic matter in ore genesis. In K.H. Wolf (Editor), *Handbook of Strata-Bound and Stratiform Ore Deposits, 2*. Elsevier, Amsterdam, pp. 111–133.

Scott, S.D. and Barnes, H.L., 1972. Sphalerite–wurtzite equilibria and stoichiometry. *Geochim. Cosmochim. Acta,* 36: 1275–1295.

Sikabonyi, L.A. and Rodgers, W.J., 1959. Paleozoic tectonics and sedimentation in the northern half of the West Canadian Basin. *J. Alberta Soc. Pet. Geol.,* 7: 193–216.

Skall, H., 1969. Special study 1969-8. Cominco Ltd. report (unpublished).

Skall, H., 1970. Geology of the Pine Point Barrier Complex. Cominco Ltd. report (unpublished).

Skall, H., 1972. Geological setting and mineralization of the Pine Point lead–zinc deposits. In: *Major lead–zinc deposits of western Canada. 24th Int. Geol. Congr. Guideb. A24-C24:* 3–18.

Skall, H., 1975. The paleoenvironment of the Pine Point lead–zinc district. *Econ. Geol., 70*: 22–45.

Skinner, B.J., 1967. Precipitation of Mississippi Valley type ores: a possible mechanism. In: J.S. Brown (Editor), *Genesis of Stratiform Lead–Zinc–Barite–Fluorite Deposits. Econ. Geol. Monogr.*, 3: 363–369.

Snyder, F.G., 1968. Geology and mineral deposits, midcontinent, United States. In J.D. Ridge (Editor), *Ore deposits of the United States, 1933–1967.* Am. Inst. Min. Metall. Pet. Eng., New York, N.Y., pp. 257–286.

Thiede, D.S. and Cameron, E.N., 1978. Concentration of heavy metals in the Elk Point evaporite sequence, Saskatchewan. *Econ. Geol.,* 73: 405–415.

Thrailkill, J.V., 1968. Chemical and hydrologic factors in the excavation of limestone caves. *Geol. Soc. Am. Bull.,* 79: 19–46.

White, D.E., 1968. Environments of generation of some base-metal ore deposits. *Econ. Geol.,* 63: 301–335.

Wiley, W.E., 1970. *Middle Devonian Watt Mountain Formation, Pine Point, N.W.T.* Thesis. Univ. Saskatchewan (unpublished).

Wolf, K.H., 1976. Ore genesis influenced by compaction. In: G.V. Chilingar and K.H. Wolf (Editors), *Compaction of Coarse-Grained Sediments, 2.* Elsevier, Amsterdam, pp. 475–676.

Chapter 12

STRATA-BOUND TIN DEPOSITS

BERND LEHMANN and HANS-JOCHEN SCHNEIDER

INTRODUCTION

According to the common textbook classification of ore deposits, the occurrence of tin is genetically related to the latest stage of granitic intrusions or their subsequent vein mineralization. This latter group represents the major part of "primary" tin deposits.

Only recently has it been recognized that there are Precambrian synsedimentary tin mineralizations which are part of the family of massive sulfide deposits (Baumann and Weinhold, 1963; Mulligan, 1978). Although this type has presently no economic significance in tin mining, it bears great metallogenic importance in general. The prevailing assumption that primary tin mineralization is necessarily related to granitic rocks needs review in light of these current findings.

Besides the massive sulfide-related type there is a wide variety of hardrock tin deposits which display strata-bound features. This group of ore deposits differs from the common magmatic-hydrothermal tin deposits not only in shape and structural relationships, but also reflects a distinct and in some cases very different genetic evolution.

Information on strata-bound tin deposits is limited, and the scarcity of detailed petrographic and geochemical investigations does not currently permit a comprehensive evaluation. This chapter, therefore, must be considered as a first attempt to delimit the strata-bound tin occurrences as an individual group of ore deposits; an approach which probably cannot overcome some subjectivity, but hopefully will stimulate the somewhat stagnant discussion on tin deposits.

The spectrum of strata-bound tin deposits is classified below into two groups according to the age of the host rocks. The subdivision within these groups follows partly non-genetic descriptive terms. As a preliminary approach we prefer this stratigraphic and geometric grouping, in view of the sometimes conflicting genetic evidence for these ore deposits, although ultimately a more genetic classification will be preferable. The following types of strata-bound tin deposits are discussed by one example each:

(1) Deposits in Precambrian host rocks:

(a) Stratiform volcanogenic–sedimentary deposits (e.g. Halsbrücke and Gierczyn districts, GDR and Poland);

(b) Stratiform sedimentary deposits with hydrothermal-exhalative influence (e.g.

Sullivan Mine, Canada);

(c) Strata-bound pocket- and pipe-shaped replacement deposits in arkosites (e.g. Rooiberg district, South Africa);

(2) Deposits in Phanerozoic host rocks:

(a) Stratiform replacement (?) deposits in volcano-sedimentary rocks (e.g. Cleveland Mine, Australia);

(b) "Bed veins" in clastic rocks (e.g. Belitung district, Indonesia);

(c) Network and disseminated mineralization in quartzitic rocks (e.g. Kellhuani district, Bolivia);

(d) Stratiform skarn-type deposits (e.g. Okehampton district, England).

Moreover, there are some tin deposits of strata-bound affinity which are not treated in this paper. These are: Cenozoic placer formations (see summary in Taylor, 1979), wood-tin veinlets and stockworks in rhyolitic lava sheets (Mexico, Bolivia, Nevada; see Knopf, 1916; Ahlfeld, 1945; Gonzalez, 1956), and pegmatitic lenses (Central and South Africa; see Pelletier, 1964; Taupitz, 1978).

DEPOSITS IN PRECAMBRIAN HOST ROCKS

Halsbrücke district, Erzgebrige (G.D.R.) and Gierczyn area, Izera Mountains (Poland)

The well-known tin deposits of the Erzgebirge are bound to the endo- and exocontacts of highly differentiated granitic intrusions of Carboniferous age. Syngenetic Sn-enrichments in Proterozoic strata of the crystalline basement have been discovered only recently (Baumann and Weinhold, 1963; Baumann, 1965; Baumann et al., 1976; Weinhold, 1977). From the neighboring region of the Izera Mountains (Western Sudetes), Jaskolski (1960, 1962) has reported analogous synsedimentary Sn-occurrences of the same lithostratigraphic position. The Precambrian stratiform tin mineralization throws a new light on the metallogenesis of the Erzgebirge province (G.D.R., C.S.S.R.).

The crystalline basement of the Erzgebirge is made up of monotonous biotite gneisses and biotite–muscovite gneisses of the Graugneis Series, which are exposed in the northwestern part of the mountain range (Fig.1). Following these oldest rocks of miogeosynclinal character (meta-graywackes), there are widespread gneisses and schists of very variable lithology known as the Pressnitz Series (Lorenz and Hoth, 1964). An intense Assyntian initial magmatism is indicated by frequent intercalations of meta-volcanics of the spilite–keratophyre–quartz keratophyre association. Alternating deposits of volcanics, cherts, lime- and mudstones and conglomerates is common (Fig.2). These sediments have been transformed by polymetamorphism to two-mica gneisses, amphibolites, and aphanitic rocks of skarn-type paragenesis (misleadingly named "felsites" by the local investigators). The Pressnitz Series contains several mineralized horizons in lateral toothing with spilitic meta-volcanics and cherts. There are oxide (magnetite, cassiterite),

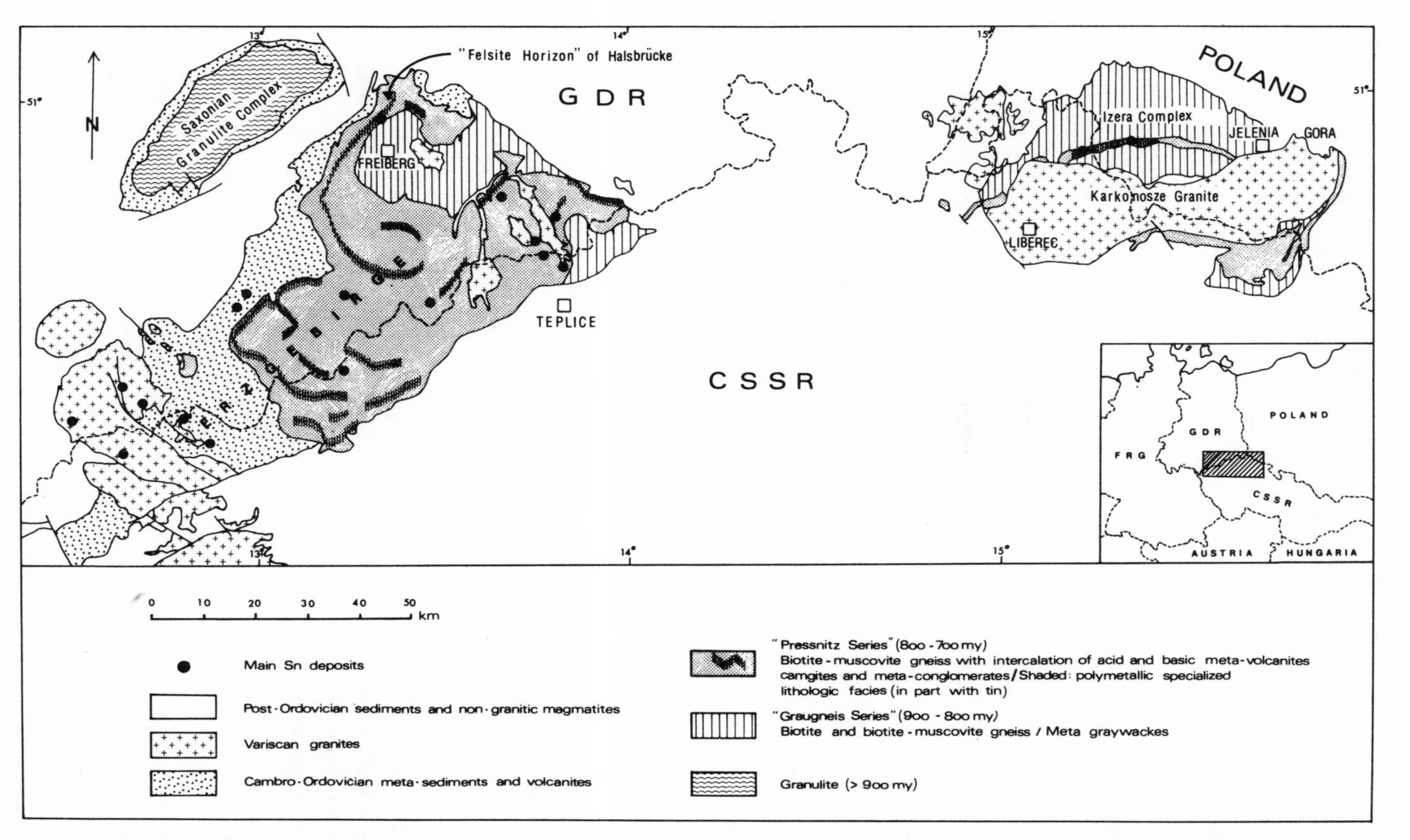

Fig. 1. Geologic sketch map of the Erzgebirge and Izera Mountains (Western Sudetes). Except for the Halsbrücke and Izera Mountain tin deposits, all indicated ore deposits are in close spatial relationship to granitic intrusions of Variscan age which often cannot be represented on the scale mapped. (Compiled from various regional geologic maps and from Baumann and Tischendorf, 1978.)

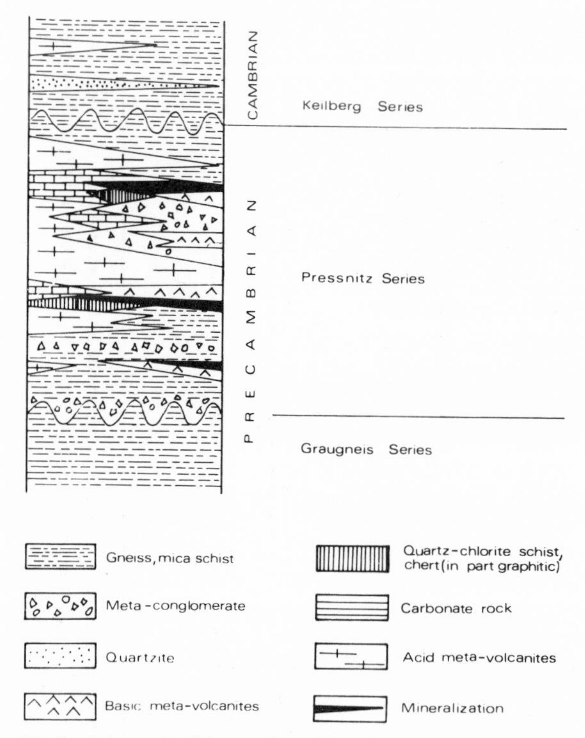

Fig.2. Schematic lithostratigraphic column of the Pressnitz Series northwest of Freiberg (G.D.R.) ("Felsite Horizon of Halsbrücke"). (After Weinhold, 1977.)

sulfide (Fe, Cu, Zn, Pb) and silicate (chamosite) parageneses. The syngenetic character of the mineralization is displayed by:

(1) the occurrence of the mineralization over large areas in the same lithostratigraphic horizon in association with a distinct lithofacies; and

(2) sedimentary fabrics on various scales (from outcrop dimensions down to microscopic scale), slump structures, relics of primary ore precipitation textures.

An exhalative-volcanogenic origin of the ores in a submarine eugeosynclinal environment seems probable and is further substantiated by clastic enrichment-horizons of cassiterite (fossil placers) in Cambrian meta-graywackes (Baumann and Tischendorf, 1978).

Cassiterite–sulfide beds of economic potential of this type are known from the "Felsite Horizon" of Halsbrücke in the area of Freiberg (Fig. 3) (G.D.R.), as well as from the area between Nove Mesto P. Smrk (C.S.S.R.) and Gierczyn (Poland) in the Izera Mountains. The latter ore occurrences are about 140 km east of the Halsbrücke deposits, but belong to the same lithostratigraphic complex and were mined from the 16th to 18th centuries for Sn, Co, and Cu.

The equivalents of the Pressnitz Series in the Izera Mountains form an east–west

Fig.3. Layered pyrite-cassiterite ore (gray) in graphitic quartz–chlorite schist (black). Lager 2, Halsbrücke near Freiberg, G.D.R. (Photograph by courtesy of Dr. G. Weinhold.)

trending arc about 35 km long that extends from Lazne Lipverda (C.S.S.R.) in the west to Wojcieszyce, near Jelenia Gora (Poland), in the east. At both ends, the 700–800 m thick series borders the Variscan Karkonosze Granite (Fig.4). Minor tin occurrences in the Karkonosze Granite have been reported by Kozlowski and Karwowski (1975).

The schists display katazonal polymetamorphism and are of very variable lithology. Main mineral constituents are chlorite (thuringite), muscovite, quartz, biotite, garnet, kyanite, with minor amounts of tourmaline, fluorite, and scheelite. Mineralization is bound to a prominent horizon of garnet–mica schist of less than 1 m in thickness and can be traced over 14 km from Mata Kamienica up to Nove Mesto p. Smrk. Intercalations with small lenses of amphibolites are frequent. The main ore minerals are pyrite, pyrrhotite, sphalerite, chalcopyrite, cassiterite, and galena. In some localities Bi-, Ag- and Co-minerals are abundant. The presence of cassiterite before garnet blastesis is indicated

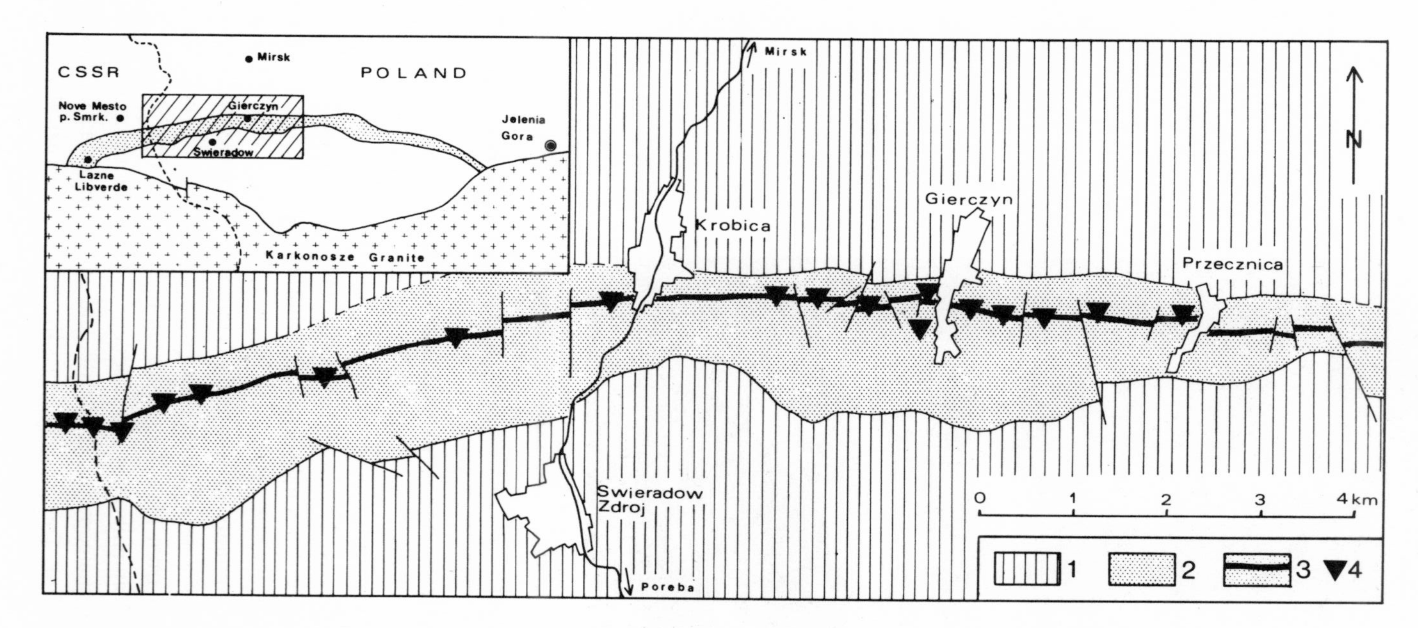

Fig. 4. Generalized geologic map of the Gierczyn area, Izera Mountains, Poland. (After Berg, 1925, and Berg and Ahrens, 1925.) *1* = augen and flaser gneiss (meta-graywacke); *2* = chlorite–mice–quartz schist (+ garnet, kyanite, chloritoid); *3* = garnet "felsite" with amphibolite lenses; *4* = ancient mines (Sn–Cu–Co mineralization).

by pre-kinematic inclusions of cassiterite in almandine (Szalamacha and Szalamacha, 1974).

Both in the Halsbrücke and in the Gierczyn deposits, the cassiterite is of very pale, yellowish to reddish color, fine-grained and often needle-shaped. Relics of gel-cassiterite are corroded by recrystallized needle tin. Very fine-grained globular aggregates of needle tin indicate recrystallization from colloidal cassiterite masses.

The composition of the cassiterites is distinctly different from the common "pneumatolytic" cassiterites of the Erzgebirge. Ti, Zr, Ga, and Bi contents are about one order of magnitude lower, the "temperature-diagnostic" elements Nb, Ta, and Sc are very low (<20 ppm). Also the cassiterites show a deficiency in Li, W, Mn, and In contents, which points to special conditions of formation. Weinhold (1977) proposed a genetic model of ore genesis with precipitation of hydrocassiterite gel from hydrothermal vents on the sea floor and metamorphic dehydration to botryoidal–reniform aggregates.

The primary tin concentrations of the crystalline basement were remobilized in part by an exogene cycle in the Cambro-Ordovician. In some conglomeratic and quartzitic strata of the Caledonian cover complex, elevated Zr and Sn contents are recorded (residual enrichment of heavy minerals). Also Caledonian meta-volcanics display a geochemical enrichment in tin (Baumann and Tischendorf, 1978).

The regional Sn-content of the Pressnitz Series is in the order of 100 ppm. Based on analyses of 1725 samples of pre-Variscan rocks of the Erzgebirge, Weinhold (1977) calculated a regional crustal abundance of 68 ppm Sn (Table I). The crystalline basement of the Erzgebirge can thus be regarded as a possible metal source of the tin in granites generated by the Variscan grantitization.

Besides the syngenetic strata-bound tin occurrences, there are several stratiform tin

TABLE I

Comparison of crustal clarkes with regional element abundances of the crystalline basement of the Erzgebirge (1725 samples; after Weinhold, 1977).

	Global crustal abundance (from Rösler and Lange, 1976) (ppm)	Regional crustal abundance (ppm)	Ratio of regional to global crustal abundance
Ti	4500	2330	0.5
V	90	81	0.9
Cr	83	104	1.3
Mn	1000	1140	1.1
Cu	47	86	1.8
Zr	170	222	1.3
Sn	2.5	68	27
Ba	650	636	1.0
Pb	16	81	5.1

deposits of skarn-type with epigenetic characteristics in the Erzgebirge (Hösel and Pfeiffer, 1965; Lorenz and Hoth, 1968; Hösel, 1968, 1973).
Their distinguishing features are:

(1) strong hydrothermal alteration and metasomatic transformation processes;

(2) association with granitic cupolas, ore distribution control by the granite surface (i.e. distance to the granite contact);

(3) impregnations of ore metals emanate from disconformable structures; and

(4) cassiterite composition correspond to that of hydrothermal and "pneumatolytic" vein-type formations.

Sullivan Mine, British Columbia (Canada)

The Sullivan Mine in southern British Columbia is one of the largest base-metal sulfide deposits in the world and since 1910 has produced more than 100 million m.t. [1] of lead, zinc, and silver ore with substantial by-products of Cd, Sb, Au, and Sn.

The deposit is of stratiform type, in Middle Proterozoic metasediments, and, like other stratiform base-metal deposits, it has been the subject of considerable debate regarding ore genesis. In previous volumes of this Handbook, the Sullivan Mine has been discussed by Gilmour (Vol. 1), Thompson and Panteleyev (Vol. 5) and Sangster and Scott (Vol. 6) in the context of global or regional studies of massive sulfide deposits. The evidence presented by these authors in favor of an essentially synsedimentary origin of the orebody has been supported by recent sulfur-isotope studies by Campbell et al. (1978a, b).

What makes the Sullivan polymetallic orebody highly relevant to the present discussion is its anomalously high tin content, which was estimated by Pentland (1943) at about 0.05%. On the basis of recent tonnage-recovery figures, the tin content is considerably lower because of the shifting of the bulk of production to areas remote from the tin-rich zones, which are centered around the pyrrhotite core of the ore lens. Since the beginning of recovery of tin in 1941, the production so far has amounted to roughly 10,000 m.t. of Sn (Mulligan, 1975).

Although the Sullivan orebody has been widely discussed, there is little reference in the literature to its tin mineralization, which will be briefly reviewed here, based on the papers by Freeze (1966), Mulligan (1975), Ethier et al. (1976) and Campbell et al. (1978a, b).

The Sullivan orebody is enclosed in beds of the Middle Proterozoic Aldridge Formation, a sequence of mainly clastic meta-sediments of low regional metamorphic grade. Although the orebody is stratiform, the degree to which layering is apparent varies widely. In the western sector of the mine, the ore tends to be massive, and stratigraphic correlations are for the most part not obvious. Furthermore, there are indications of post-

[1] m.t. = metric tons.

sedimentary introduction of hydrothermal fluids which caused tourmalinization and albitization in the footwalls and hanging walls, respectively.

In contrast, throughout the eastern sector, the sulfide ore occurs in five distinct horizons of decimeter to meter scale, separated by barren beds of argillaceous meta-siltstones and quartzites. Individual laminae within stratigraphic units can be traced for hundreds of meters. Bedding, slump structures, and local erosional truncations of laminae within the sulfide units argue strongly for a synsedimentary origin.

The ore-bearing sequence thins toward the eastern perimeter of the ore zone, and the sulfide minerals become progressively more disseminated. This thinning is reflected proportionately in both clastic sedimentary beds and ore beds and indicates subsidence in the central part of the basin as deposition took place.

The footwall of the orebody consists of intraformational conglomerates and breccias of local extent, as well as of laminated argillaceous to arenaceous rocks. In a zone restricted to the central and western parts of the orebody the footwall complex has been intensely brecciated, mainly prior to the deposition of the ore-bearing horizon. Tectonic activity continued to some extent during ore deposition and waned during sedimentation of the hanging-wall complex.

The ore mineralogy consists predominantly of pyrrhotite, pyrite, galena, and sphalerite. Minor constituents are chalcopyrite, arsenopyrite, magnetite, cassiterite, and boulangerite. Most of the cassiterite occurs as rounded, microcrystalline grains. Metal distribution throughout the orebody is roughly concentric in plan. A central iron zone (pyrrhotite) is enclosed by an irregular zone characterized by elevated Pb, Ag, Sn and As values. Toward the margins of the orebody, the Zn-content increases, and Sb is concentrated at the periphery of the deposit. Vertical zonation is displayed by the upward succession pyrrhotite → pyrrhotite + sphalerite + galena → sphalerite + galena (Ethier et al., 1976). The entire orebody contains some cassiterite, but the main concentrations are near the outer part of the central pyrrhotite zone. High Sn-values are associated with the pyrrhotite horizon just above the footwall, and the Sn-content tends to diminish toward the hanging wall.

Some small, Sn-rich pockets are associated with faults and fractures, of which the "tin zone fracture" is the most prominent (Fig.5). Samples of pyrrhotite ore from this zone commonly contain 1–2%, Sn and the large-scale, pervasive tourmalinization of the footwall complex is accompanied by Sn dissemination at the 100 ppm level (Mulligan, 1975). Fig. 5 demonstrates the penecontemporaneous relation between sedimentation and tectonic activity, which – under a then prevailing epigenetic replacement concept – was observed by Freeze (1966, p.270): "Where the [tin zone] structure has been explored intensively, it was found to pass upward through the main ore zone, with diminishing intensity, becoming a relatively insignificant structure as it approaches the hanging wall".

Sedimentation and tectonism were accompanied by hydrothermal activity, and, in accord with recent sulfur-isotope studies by Campbell et al. (1978a, b), the resulting

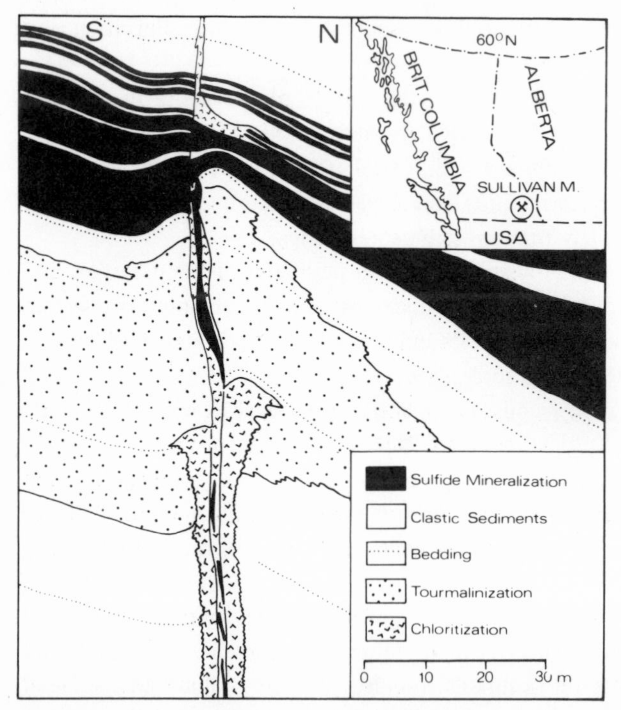

Fig.5. The "tin zone fracture" of the Sullivan Mine, Canada, demonstrating the synchronous relationship between tectonism, sedimentation, and hydrothermal activity. The fracture is regarded as one of the feeder zones for tourmalinization of the footwall complex, for stratiform sulfide deposition and, in its waning stage, for albitization of the hanging-wall complex. (Modified after Freeze, 1966.)

interaction is described in the following model of ore genesis:

(1) Tectonically induced local basin development with deep fracturing and local brecciation (collapse structures);

(2) Hydrothermal discharge at the sediment/sea-water interface, sulfide precipitation from the metal-bearing brines by biologically reduced sulfate from sea-water and sedimentation of laterally extensive sulfide horizons; and

(3) Continuation of small-scale tectonism and thermal brine activity during and after deposition of the hanging-wall unit, thereby producing albitization halos in the hanging wall.

This is mainly associated with the early phase of hydrothermal activity and is concentrated in feeder zones and in basal sulfide units around the loci of submarine exhalations.

There is no evidence for volcanic influence in the Sullivan deposit, and in terms of tectonic and compositional environment its formation might be compared with the Red Sea deeps, although the sedimentation of the Sullivan ore complex took place under somewhat more euxinic conditions.

Rooiberg district, Transvaal (South Africa)

The cassiterite mineralization of the Rooiberg and Leeuwpoort mining area occurs in a volcano-sedimentary sequence of more than 2 b.y. in age, belonging to the upper part of the Precambrian Transvaal System. In the Rooiberg district, this sequence is preserved as an isolated roof-block of triangular shape with a diameter of about 35 km, known as the "Rooiberg Fragment", enclosed within and on top of the late granitic phase of the western Bushveld Igneous Complex (Fig.6).

All the economic tin deposits in the Rooiberg Fragment are located near or in the

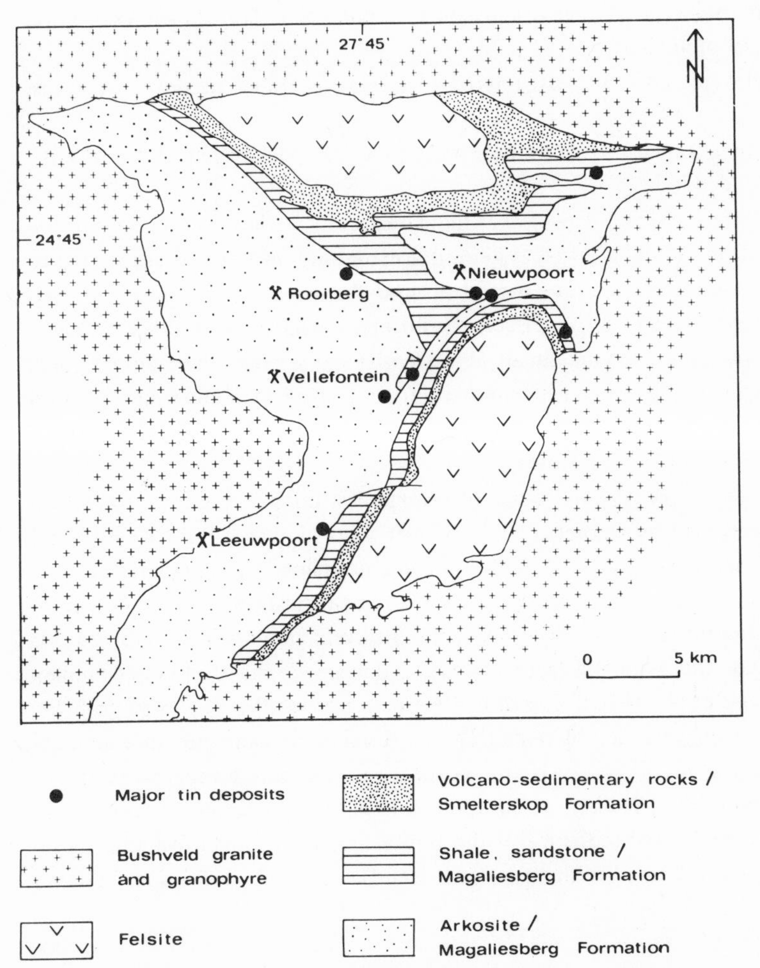

Fig.6. Geologic setting of the Rooiberg district, Transvaal, South Africa. (Generalized after Stear, 1977.)

transitional sedimentary contact between massive, cross-bedded arkosites and overlying rhythmically interbedded sandstones, siltstones and shales. These clastic sediments of the "Magaliesberg Stage", with a thickness of about 900 m, are followed by the approximately 700 m thick "Smelterskop Stage" of volcano-sedimentary series which grade into felsitic lavas and tuffs (stratigraphic classification following Leube and Stumpfl, 1963). An outstanding feature of the Rooiberg tin district is the confinement of mineralization to a defined stratigraphic horizon of about 250 m in thickness over about 20 km of strike. This horizon consists of orthoclase-rich arkosites with an upwardly increasing shale component.

The prevailing paragenesis of the ore deposits is cassiterite, magnetite, tourmaline and carbonates, associated with varying amounts of sulfides (pyrrhotite, pyrite, chalcopyrite, bismuthinite, sphalerite, galena, and some Ni-minerals). The tin mineralization occurs in two types of orebodies:

(1) discordant veins and breccias, related to strata-controlled tensional fractures; and

(2) strata-bound replacement bodies, occurring as pocket-shaped ore-shoots, annular pipes and bedded lodes.

The structural analysis by Stear (1977) indicated a close relationship between the strata-bound mineralization and a thrust system of horizontally disposed forces related to the intrusion of the Bushveld granites in and around the Rooiberg Fragment. In this model the strata-bound mineralization reflects the different mechanical response between cross-bedded arkosites and overlying shales at a transitional stratigraphic contact zone. The detailed petrographic, geochemical and structural studies by Leube and Stumpfl (1963) provided good evidence for the genesis of the tin deposits by hydrothermal fluids in genetic relation to the Bushveld granites.

However, the metallogenic importance of the granitic intrusives as an ultimate tin source has been questioned by Garnett (1967, p.144), who suggested that the Bushveld granites "acted only as an agent of transportation and conversion" by assimilation of pre-existing "tin-bearing sediments". With respect to this transformistic view it may be interesting to note that the initial $^{87}Sr/^{86}Sr$ ratios of the Bushveld granites (0.715–0.724) are distinctly higher than those of the layered mafic sequence of the Bushveld Complex (0.706–0.709) (Davies et al., 1970). The Sr-isotope data suggest that the most probable origin of the granitic magma lies in remelting of crustal rocks, and not in fractional crystallization from the Mafic Phase. If this is the case, the heat source necessary to induce melting of sialic crust is likely to be the intruding mafic magma that ultimately crystallized to form the layered sequence of the Bushveld Complex. The available radiometric age data are in agreement with this model (see Davies et al., 1970; Lenthall and Hunter, 1977).

DEPOSITS IN PHANEROZOIC HOST ROCKS

Cleveland Mine, Tasmania (Australia)

The Cleveland cassiterite–sulfide deposit is situated about 60 km southwest of Burnie in western Tasmania. Although known since the beginning of this century, it came into production only in 1968 and currently produces about 300,000 m.t./year of ore at a grade of 0.75% Sn and 0.25% Cu (Palmer, 1976).

The Cleveland orebodies occur in a steeply dipping, unfossiliferous sedimentary sequence of probable Cambrian age, which consists of a basal graywacke unit with thin horizons of shale and chert (Crescent Spur Mica Sandstone Formation); an intermediate suite of marine shales, cherts, carbonaceous material and lenses of spilitic rocks (Hall's Formation); and a conformably overlying thick pile of spilites with interbedded pyroclastics, shales and cherts (Deep Creek Basic Volcanics Formation) (Table II). The lavas within the Deep Creek Basic Volcanics are typical albite–actinolite–chlorite rocks with accessory quartz, epidote, Fe- and Ti-oxides, and relict clinopyroxene.

The sequence is affected by low-grade regional metamorphism. About 500 m west of the mine, ultrabasic intrusions are present, and about 5 km to the east, the 350 m.y. old Meredith Granite crops out.

Mineralization is confined to the Hall's Formation (Fig.7) and is conformable with the bedding on a megascopic as well as microscopic scale. The ore lenses attain a maximum thickness of about 30 m and a maximum strike length of 600 m and are usually intimately associated with cherts. They display a fine compositional layering, with laminae ranging down to 0.05 mm in width. The laminae consist of microcrystalline quartz, quartz–tourmaline, carbonate–chlorite, tourmaline–fluorite, and sulfides, together with gradational intermediate types (Cox and Glasson, 1971).

The sulfide-rich horizons and lenses are composed mainly of pyrrhotite (partly altered to pyrite and marcasite) with chlorite, carbonates, cassiterite, quartz, fluorite,

TABLE II

Stratigraphic column of the Cleveland Mine area, Tasmania (according to Palmer, 1976)

Deep Creek Basic Volcanics	Spilites with intercalations of shales, cherts, limestones	>200 m
	Gray shales, cherts, graywackes, ore lenses	5–100 m
Hall's Formation	Spilites and pyroclastics	0–15 m
	Gray shales, cherts, graywackes, ore lenses	0–50 m
	local unconformities	
Crescent Spur Mica Sandstone	Graywackes with intercalations of shales and cherts	>200 m

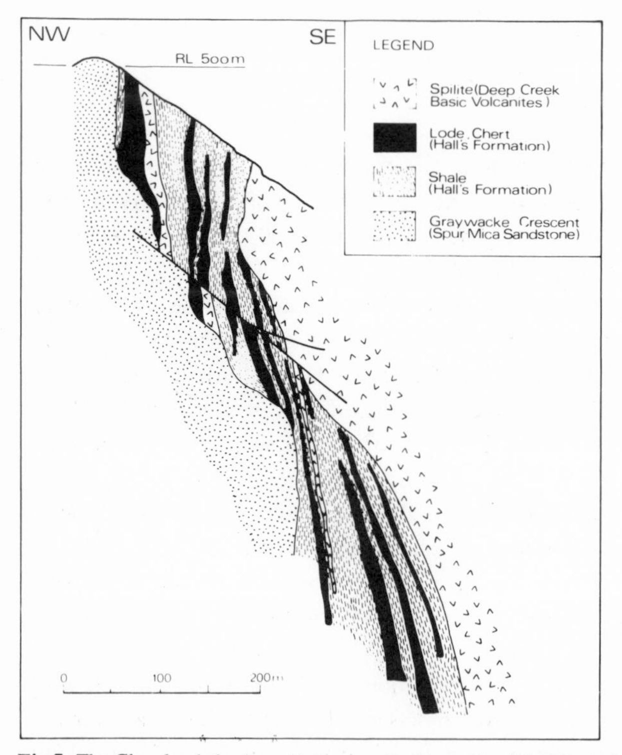

Fig.7. The Cleveland tin deposit, Tasmania, Australia. (Modified after Palmer, 1976.)

actinolite and tourmaline. The most common sulfide is pyrrhotite (average in ore zone: 35 vol%). Chalcopyrite averages about 1 vol%, and sphalerite less than 0.1 vol%. Total tin is of the order of 1 vol%, of which 5–15% is stannite. Cassiterite occurs as euhedral to subhedral grains of 0.1–1.00 mm in diameter. Arsenopyrite, tetrahedrite, wolframite, and stilpnomelane are present locally (Palmer, 1976).

The restriction of mineralization to discrete layers within a well-defined volcanically influenced shale–chert sequence indicates an almost entirely stratigraphic control. The hitherto prevailing interpretation of the ore formation of the Cleveland deposit by replacement of finely layered calcareous shales by magmatogenic-hydrothermal solutions (Cox and Glasson, 1971; Palmer, 1976) is essentially based on the relative proximity (5 km) of the tin-bearing Meredith Granite of Devonian age. However, there are no convincing observations of hydrothermal wall-rock alteration. The sericitization halo around the Cleveland deposit, reported by Cox and Glasson (1971) has been proven to be of very widespread extent and can be attributed to the low-grade regional metamorphism (Ransom and Hunt, 1975).

The nearby Sn deposits of Mount Bischoff and Renison Bell have been regarded as a further argument for an epigenetic origin of the Cleveland deposit (Cox and Glasson,

1971). These orebodies display strata-bound and minor fault-controlled characteristics, but belong to different stratigraphic units. The host rock of the Mount Bischoff cassiterite–sulfide deposit is a dolomitic horizon of probable Late Precambrian age, overlain by tuffaceous graywackes. A zone of interconnecting quartz porphyry dykes embraces the mineralized area. The dykes are of Devonian age and are partly altered in quartz–topaz–cassiterite greisens (Knight, 1975).

The Renison Bell tin deposits lie within a sequence of Lower Cambrian sediments and volcanics (Newnham, 1975). Cassiterite–sulfide mineralization is confined to carbonate beds of several meter thickness, interlayered with and overlain by shales, cherts and spilitic rocks. The presence of a hidden magmatic body is inferred from weak "contact-metamorphism" (Patterson, 1976).

Although there are some reasons to see a genetic relation of the tin deposits of the Cleveland–Mount Bischoff–Renison Bell area to the Devonian granites, their strata-bound character and association with spilite-type volcanism argue against an epigenetic interpretation. Knight (1974, 1975) decided for a synsedimentary origin of the orebodies with remobilization during granite intrusions, an interpretation, which needs verification by geochemical data for the spilitic rock units.

Belitung (Billiton) district (Indonesia)

The thick Permo-Carboniferous sedimentary complex of Belitung comprises a considerable number of strata-bound tin deposits. They are scattered in the northern part of the island over a distance of more than 45 km in a W–E direction, the regional strike of the sedimentary formation. Granite intrusions of Mesozoic age are cutting through the older structures (Fig.8).

According to Adam (1960) the "bedding plane lodes" of the Klappa Kampit Mine, the most important one of this group, are intercalated into a sequence of alternating layers of sandstone and shale ("clay shale") of more than 500 m in thickness. Mine maps exhibit the spatial (stratigraphic) linkage of the mineralization to radiolaritic layers. Younger faulting has split up the stratiform ores (Fig.9).

The mineral association of the ore deposits consists of predominantly pyrite, pyrrhotite, and magnetite, with cassiterite, chalcopyrite, sphalerite, arsenopyrite, galena, and minor amounts of bismuth minerals, as well as fluorite, tourmaline, biotite, garnet, pyroxene, siderite, etc. Wolframite is absent throughout the stratiform tin ore lodes (Adam, 1960). Although the paragenesis has been considered to indicate a "pneumatolytic replacement origin", Adam (1960) points out the fact that the two nearest outcrops of granitic rocks are situated about 10 km away from the Klappa Kampit Mine. Geologic data about the Belitung ore deposits are very scarce, and their genesis is "enigmatical", as stated by Hosking (1970, p.42) who surveyed the strata-bound ores at Klappa Kampit.

Very recently the Adit 22 orebody has been discovered adjacent to the now paralyzed Klappa Kampit Mine. There is only the brief description by Taylor (1979,

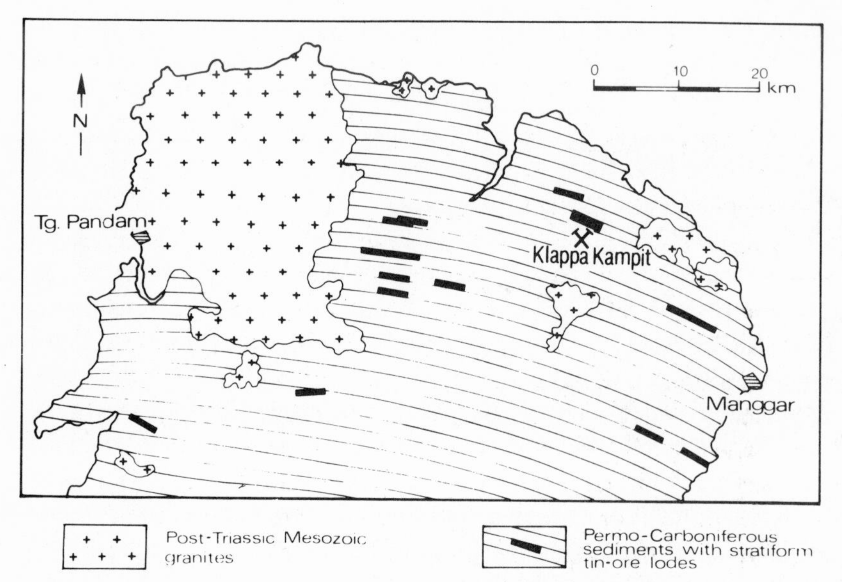

Fig.8. Strata-bound tin ore lodes in the sedimentary complex of Belitung, Indonesia. (Modified after Adam, 1960.)

p.248) available at present, which makes some genetic similarities to the massive sulfide-type tin occurrences evident, as discussed above (Fig.10): "The orebody consists of a dark green-black phlogopite shale with sulphides occurring parallel to the bedding in the form of discontinuous layers producing an overall effect very similar to a typical banded volcanogenic ore. The sulphides are principally pyrrhotite although occasionally pyrite predominates. Cassiterite occurs as very fine grains or lensoid aggregates in association with the pyrrhotite (pyrite) layers. The zone is currently being extracted via large-scale underground stoping techniques. The origin of this deposit is uncertain (Personal communication–M. Jones, University of Adelaide, Australia) and although premature to

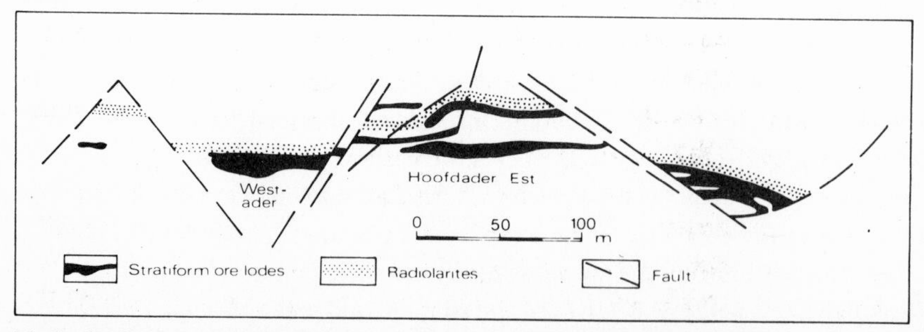

Fig.9. Stratiform mineralizations in stratigraphic relationship to radiolaritic layers. Klappa Kampit Mine, Main Lode, 3rd level; Belitung, Indonesia. (Redrawn after Adam, 1960.)

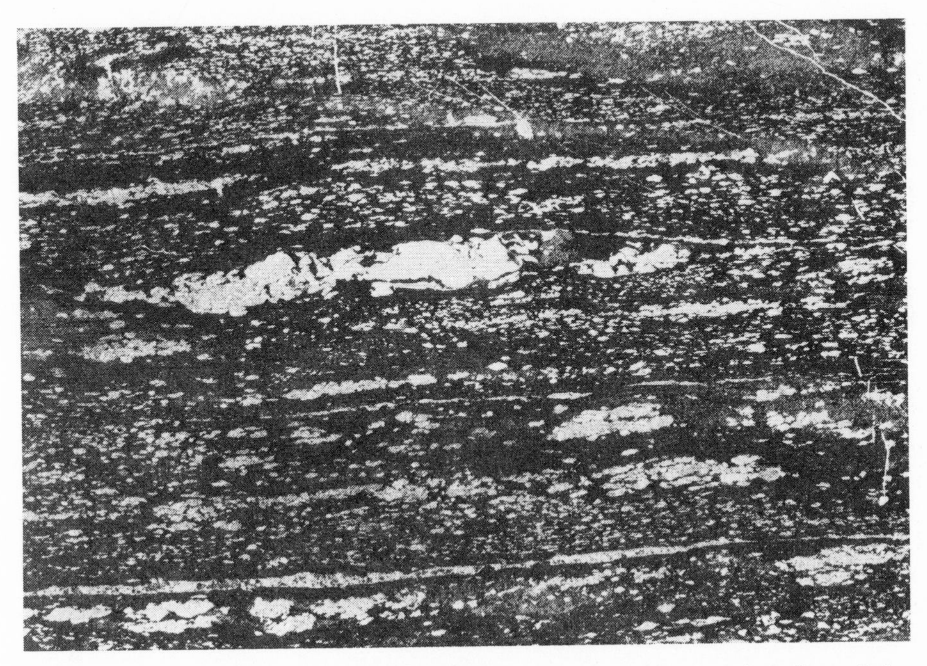

Fig.10. Layered/banded (?) ore from the Adit 22 tin deposit, Kelapa Kampit, Belitung, Indonesia. Elongate pyrrhotite aggregates (pale) associated with cassiterite (gray–fine-grained) in a matrix of phlogopite (dark). Long axis of photograph = 5.0 cm. (After Taylor, 1979, fig.7.2; photograph by courtesy of the author.)

assume a volcanogenic association, it can certainly be regarded as a new species of tin deposit with considerable exploration implications".

Kellhuani district (Bolivia)

The Eastern Cordillera of Bolivia consists of a Lower Paleozoic clastic sequence of more than 10,000 m in thickness. In a defined lithostratigraphic series of Upper Ordovician–Silurian age, strata-bound tin mineralization occurs which is linked to certain quartzitic layers – the so-called "mantos". Deposits of this type are known over more than 500 km of strike in the East Andean chains and reach from San José/Amarete in the north of Ocuri (near Potosi) in the south (Fig.11) (Schneider and Lehmann, 1977).

Due to their low-grade tin mineralization in the order of about 0.X wt%, they generally have been of minor economic interest in the past. Therefore, they have not been studied in any detail as compared with the famous Sn-Ag deposits of magmatic/subvolcanic origin. Modern information on the manto-type tin deposits is available only for the northern district of the Bolivian Andes.

The Kellhuani-Huallatani mining district is situated about 20 km north of La Paz in

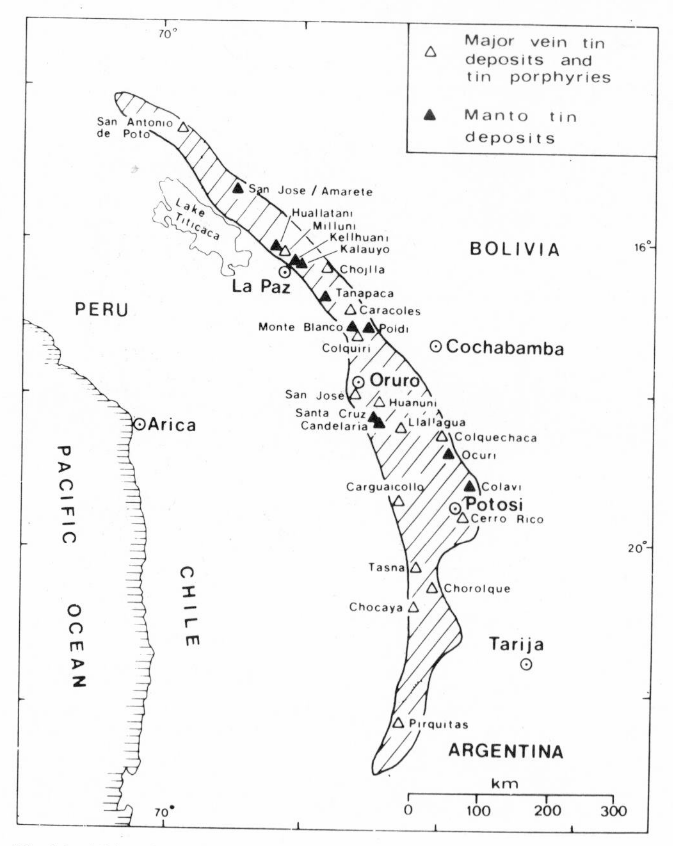

Fig.11. Main tin deposits of the Bolivian tin province with special reference to the strata-bound manto deposits. (Modified after Schneider and Lehmann, 1977.)

the Lower Paleozoic sedimentary complex of the Cordillera Real. The area has been studied recently by Lehmann (1979).

Tin mineralization is confined to the quartzitic layers of the Silurian Catavi Formation, an alternating series of quartzites and black shales, under- and overlain by predominantly argillaceous metasediments of very low to low-grade regional metamorphism. The strata-bound mineralization displays a remarkable congruency with the quartzitic beds (Fig.12) and can be traced by hundreds of small galleries and adits over at least 20 km of strike along the Catavi Formation. The quantitatively dominating black shales are barren. Although stratigraphically conformable on a regional and local scale, the mineralization shows discornformable relations on the outcrop scale. Mineralization is essentially confined to a network of veinlets scattered through the quartzitic country rock, indicating intense shattering prior to and during ore deposition. The ore mineralogy is simple in paragenesis and consists of quartz, tourmaline and/or chlorite, and cassiterite. Sulfides are very sub-

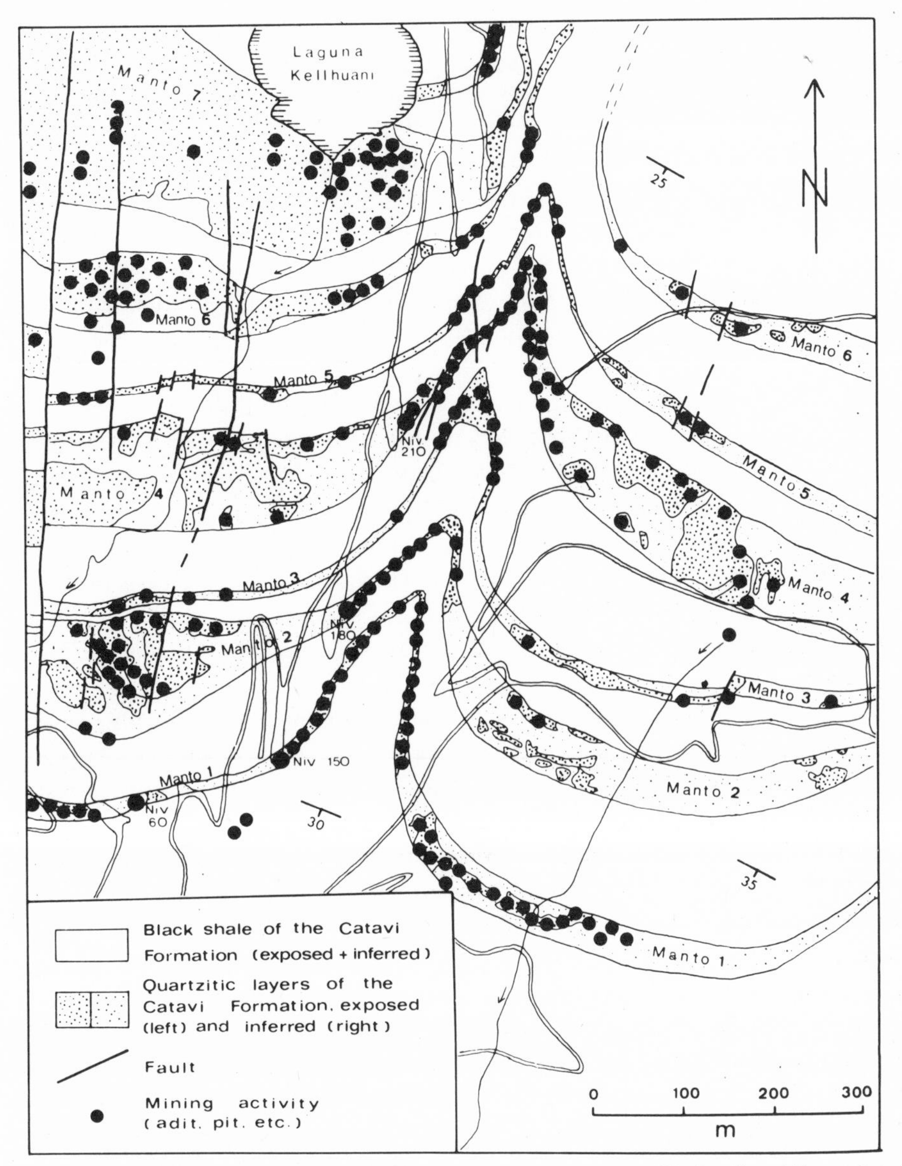

Fig.12. Congruency of tin mineralization and quartzitic layers of the Silurian Catavi Formation. Manto deposit Kellhuani, Bolivia. (After Lehmann, 1979.)

ordinate; fluorite, albite and muscovite are abundant locally.

The mineralized quartzites ("mantos") show distinct recrystallization textures and display features of pervasive hydrothermal alteration. Tourmalinization and/or chloritization, sericitization, and granoblastic disseminations of cassiterite and fluorite are preferentially located in argillaceous parts of the quartzites, in which typical replacement textures on a millimeter-scale are developed.

The disseminated Sn of the quartzites is in the order of 100 ppm, but bulk content of larger rock volumes with network mineralization is up to 1 wt% Sn. At present, the explored reserves of Kellhuani Mine are about 9 million m.t. of ore with a grade of 0.5% Sn (Lehmann and Schneider, 1980).

This figure is composed only of the exploration results of two of the more than fifteen mantos with thicknesses ranging between 0.8 and 30 m. However, taking into account the large lateral extent of the tin mantos far beyond the concession limits of Kellhuani Mine, the ore potential of this strata-bound mining district is exceptional. The previously adopted, selective small-scale production is actually being transformed into modern large-scale mining, and open-pit production is scheduled for the next few years.

Besides the strata-bound network tin mineralization, there are some vein-type deposits cross-cutting in the Catavi Formation. The biggest of these vein deposits is the Sn–Zn Milluni Mine. Ore mineralogy of the vein deposits is characterized by the superimposition of an extensive base-metal sulfide stage on the tin oxide mineralization. The main minerals of the sulfidic stage are pyrite, arsenopyrite, sphalerite, chalcopyrite, stannite, bismuthinite and galena. In the scope of paragenesis there is a striking difference between the base-metal-rich variety of the vein tin deposits and the predominantly oxidic mineralization of the tin mantos.

The strata-bound tin deposits of the Kellhuani area can be associated with a large hydrothermal aureole centered on a small stock of granite porphyry of Mesozoic age ("Chacaltaya granite porphyry") (Fig.13). This stock shows strong geochemical specialization in Sn and B and is partly greisenized. A second, smaller hydrothermal aureole in the Huallatani area points to a hidden magmatic body. Large-scale hydrothermal alteration is confined principally to the quartzitic strata and displays a zonal distribution, with the granoblastic mineral assemblage quartz–chlorite–siderite–cassiterite in the outer zone, and quartz–tourmaline–chlorite–sericite/muscovite–siderite–fluorite–cassiterite in the inner zone. Primary dispersion halos of Sn, B, F, Li and Cs follow the same geometric pattern, but with element-specific halo extent.

The lithologic control of hydrothermal circulation and associated tin mineralization is taken to be the result of the specific mechanical response of the host rocks to tectonic and hydraulic stress generated by the emplacement and cooling of the Chacaltaya granite porphyry system. The quartzitic rocks underwent intense fracturing to local brecciation, whereas the incompetent black shales suffered essentially plastic deformation. The network-fractured, highly permeable quartzite beds provided ideal channelways for hydrothermal fluids in an early stage of mineralization, but with increasing ore deposition permeability was rapidly reduced. Later stages of mineralization were localized in longer-lived structures such as vein systems, which explains the paragenetic complexity of the vein deposits relative to the "mantos".

A preliminary survey of the Kellhuani-Huallatani manto deposits by Schneider and

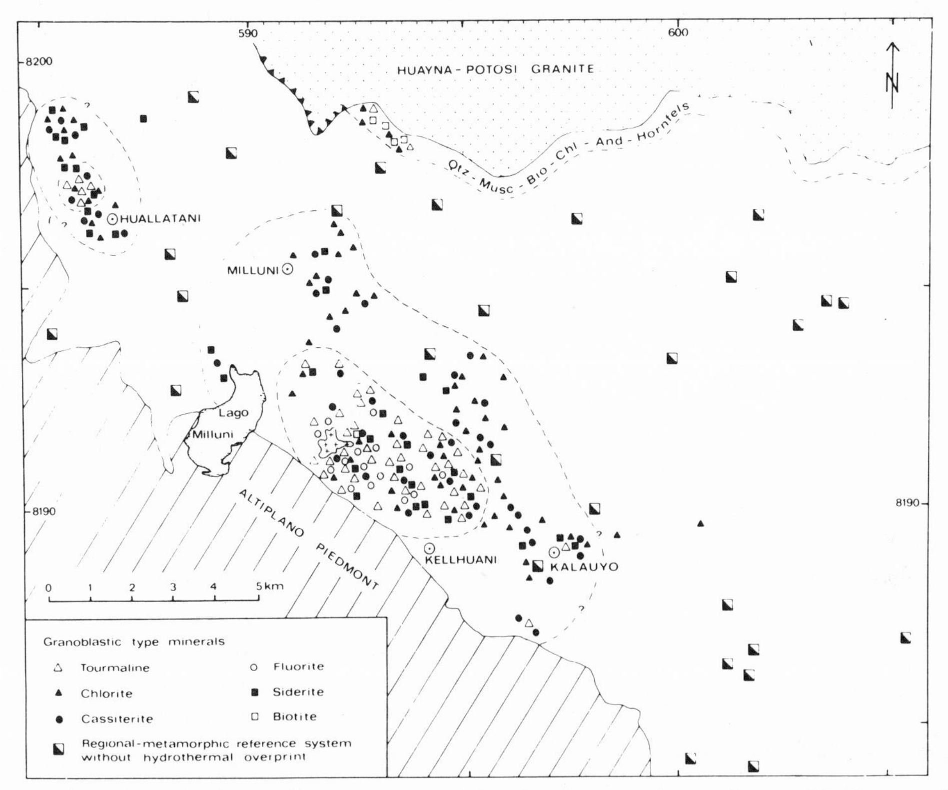

Fig.13. Hydrothermal alteration zoning around the Chacaltaya granite porphyry and in the area of Huallatani, Bolivia. (After Lehmann, 1979.)

Lehmann (1977) gathered various arguments for a syngenetic origin of the tin mineralization which was thought to have formed by remobilization of fossil cassiterite placers. However, the following geochemical study by Lehmann (1979) showed that the Paleozoic country-rocks reveal no tin anomaly relative to average sedimentary rocks. On the other hand, a distinctive boron anomaly related to detrital tourmaline has been discovered which seems to characterize the Lower Paleozoic meta-sediments of the Bolivian tin province. The possibility exists therefore that there might still be undiscovered sedimentary cassiterite accumulations which by virtue of the high specific weight of cassiterite are to be expected to be limited to a much smaller aerial extent than the tourmaline dispersion pattern. As proposed for the Rooiberg district, therefore, the granitic intrusion might have acted essentially as an agent of redistribution by leaching and assimilation of detrital cassiterite and reprecipitation in cupola structures.

Okehampton district (England)

The ancient mining district of Okehampton lies along the northern edge of the Dartmoor Granite in the Cornwall tin province, and has been worked for Cu, Pb, As and Sn (Fig.14). The mineralization is restricted to a lithologically distinct group of the Lower Carboniferous Meldon Formation. This group of about 100–150 m in thickness is clearly differentiated from the monotonous clastic sediments below and above by the presence of black limestones, siliceous mudstones, cherts and volcanic rocks of the spilite suite. Rapid lateral variations in lithology are common. The volcanic–sedimentary sequence is repeated tectonically by intense folding and is overturned to the southeast (Edmonds et al., 1969).

Emplacement of the Permo-Carboniferous Dartmoor Granite was accompanied by formation of a thermal aureole in the surrounding country rocks, and the especially reactive Ca–Mg-rich volcanic–sedimentary complex has been transformed into calc-silicate hornfelses of variable structure and composition. Characteristic rock-forming minerals are andradite, grossularite, wollastonite, hedenbergite, vesuvianite, datolite, axinite, quartz, carbonates, and actinolite. Impregnations of sulfides occur predominantly in massive garnet–actinolite rock horizons, and the perfectly conformable nature of mineralization was recognized by the miners at an early date. Smith (1878) points out the continuity of the ore horizon, which extends over more than 15 km from the Fanny Mine in the west to the Throwleigh Mine to the east, and varies in thickness from several decimeters to several tens of meters. Disseminated sulfide mineralization in the mostly coarse-grained

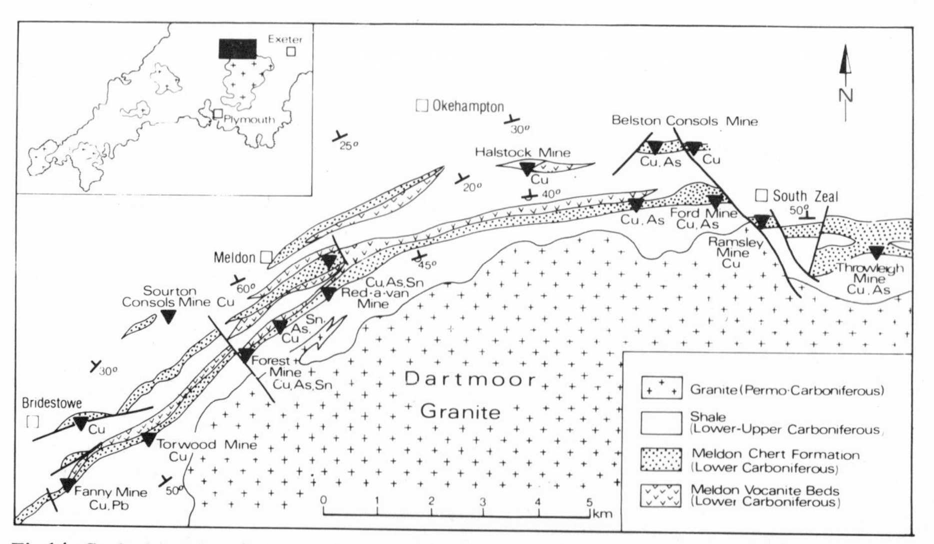

Fig.14. Geologic setting of the strata-bound skarn-type deposits of the Okehampton district, Cornwall, England. (Generalized after Dunham, 1969.)

TABLE III

Tin in some stratiform and strata-bound massive sulfide deposits

Deposit	Economic metals	Host rocks	Age of host rocks	Tin mineral	Sn (%)	Reference
Gierczyn (Poland)	Cu, Pb, Zn, Co, Sn	garnet–mica schist with amphibolite lenses	Precambrian (Proterozoic)	cassiterite	0.7–0.8 *	Putzer (1940)
Halsbrücke (G.D.R.)	Cu, Zn, Pb, Sn	spilitic meta-volcanics, cherts, conglomerates	Precambrian (Proterozoic)	cassiterite (relics of gel-cassit.)	0.01–0.1	Baumann and Weinhold (1963)
Kidd Creek (Canada)	Zn, Cu, Pb, Ag, Sn, Cd	rhyolitic meta-volcanics, cherts, black shale, consanguineous quartz feldspar porphyry stock	Precambrian (Archaean)	cassiterite	0.14	Mulligan (1975)
South Bay (Canada)	Zn, Cu, Ag, Sn	rhyolitic meta-volcanics, consanguineous quartz feldspar porphyry stock	Precambrian (Archaean)	cassiterite	0.25	Mulligan (1975)
Brunswick No. 6 + 12 (Canada)	Zn, Pb, Cu, Ag, Sn	acid to basic meta-volcanics, cherts	Orodovician	cassiterite, stannite	0.05–0.1	Mulligan (1975), Luff (1977)
Sullivan (Canada)	Zn, Pb, Ag, Sn	shales, meta-siltstones, quartzites, conglomerates	Precambrian (Proterozoic)	cassiterite	0.05	Pentland (1943)
Bleikvassli (Norway)	Zn, Pb, Cu	mica schists and gneisses	Lower Paleozoic	cassiterite, stannite	0.03	Vokes (1960)
Boliden (Sweden)	Cu, As, Au, Ag Co, Zn, Pb	acid meta-volcanics, phyllites	Precambrian	stannite	0.006	Grip and Wirstam (1970)

* Mined out in the 16th–18th centuries.

garnet beds consists of pyrrhotite, pyrite, chalcopyrite, sphalerite, löllingite, arsenopyrite, and bornite. Minor constituents are molybdenite, fluorite, and manganese oxides; secondary copper minerals are locally abundant.

Most mining has been done for copper, though records of output are unknown or incomplete. Tin is reported to have been produced from the Forest and Red-a-van Mines (De la Beche, 1839), which prompted a recent survey of tin distribution by El-Sharkawi and Dearman (1966). According to the latter study, the Sn-contents are generally in the range of 0.1–1.0 wt% and have been found along the mining district. The tin mineralization is confined to garnet–sulfide beds, and a maximum with up to 6.8 wt% SnO_2 is localized at the Red-a-van Mine. At high whole-rock Sn-contents, the tin occurs as malayaite, $CaSnSiO_5$, in association with löllingite, and wollastonite or grossularite/andradite. Elsewhere, stanniferous grossularite and andradite with up to 0.2 wt% SnO_2-content are the main Sn-bearing phases. Cassiterite has not been observed in recent investigations, which are hindered by poor outcrop exposure. Nevertheless, its presence seems probable according to ancient tin mining reports (De la Beche, 1839; Smith, 1878).

The formation of the strata-bound Cu–Zn–As–Sn–B skarns may be generally explained by metasomatic introduction of ore fluids in genetic relation to the Dartmoor Granite, which, however, in the sector covered by Fig.14 bears no indications of Sn-mineralization. The calcic environment of ore deposition is reflected by the malayaite and stanniferous garnet paragenesis together with axinite, datolite, and danburite, corresponding to cassiterite and tourmaline respectively, in a noncalcic environment (for information on phase relations between cassiterite and malayaite in the skarn environment, see Burt, 1978).

Thermal transformation of Ca–Mg-rich volcanic–sedimentary beds into calc-silicate horizons of high density is thought to have provided the necessary permeability for effective strata-bound channelways for hydrothermal ore solutions.

CONCLUSIONS

The foregoing examples treat only some of the most characteristic types of strata-bound tin occurrences without attempting complete coverage of this topic. On the basis of degree of stratigraphic control, these various types can be arranged in a continuous series, ranging from epigenetic network, skarn and replacement deposits to syngenetic massive sulfide deposits, and the individual occurrences often display transitional features.

The epigenetic end-member of this series is represented by stratiform skarn/replacement deposits in a calcareous environment, and quartzitic "mantos" in a non-calcareous environment, respectively. These deposits are closely related to granitic intrusives, although ore deposition is controlled chemically and/or structurally by specific litho-stratigraphic units. The stratigraphic control reflects high permeability of the host rock unit during mineralization, which is provided either by chemical reactivity or by mechanical properties of specific units of a given rock sequence. The former case is realized in skarn horizons, where permeability is increased principally by the transformation of calcareous material into denser calc-silicate phases, the latter by intercalation of competent and incompetent strata, where either fracture deformation or plastic flow is predominant.

Strata-bound skarn Sn-mineralization is rather widespread and major deposits have been reported from Cornwall (Dines, 1956), Malaya and Indonesia (Hosking, 1969), China (Hsieh, 1963), U.S.S.R. (Smirnov, 1977), and the Erzgebirge (Hösel, 1968). Strata-bound Sn-mineralization in fractured quartzite/sandstone beds is known from Bolivia (Ahlfeld and Schneider-Scherbina, 1964; Schneider and Lehmann, 1977), the Erzgebirge province (Hösel, 1973), and Yukon, Canada (Ed Spooner, pers. comm., currently under investigation by CCH Resources Ltd., Vancouver). A combined case of chemical reactivity with specific mechanical response of a defined litho-stratigraphic unit seems to be represented by the Rooiberg tin district, where the mineralized arkosites display replacement phenomena superimposed on a local fracture pattern.

The syngenetic end-member of the series of strata-bound tin occurrences under discussion are tin-bearing massive sulfide deposits (Table III). This type is fairly abundant in Canada, where the Kidd Creek Mine is currently the foremost tin producer in North America (Mulligan, 1975; Walker et al., 1975).

A characteristic feature of the Canadian examples is that the stratiform deposits grade stratigraphically downward into zones of disseminated or stringer ore, obviously not conformable. These zones are commonly marked by extensive fracturing, brecciation, and hydrothermal alteration, and they are thought to represent the channelways through which metal-bearing solutions reached the surface, i.e. sea bottom (cf. Walker et al., 1975; Campbell et al., 1978).

In general, the tin-bearing masive sulfide deposits are associated with volcanic rocks and cherty horizons, the Sullivan Mine being a major exception. In the Archaean greenstone belts of the Canadian Shield, the deposits are typically in rhyolitic rocks, which mark the end-stage of basic–intermediate–acid volcanic cycles. The volcanic host rocks seem to indicate slightly anomalous Sn-contents, but geochemical data are sparse (Mulligan, 1975). Tin and base-metal mineralization is often found in areas where the volcanic pile is intruded by consanguineous acid porphyries, which suggests some genetic similarities to the formation of Kuroko deposits (Sato, 1974).

A second type of volcanic association with tin-bearing massive sulfide deposits is the spilite–keratophyre–quartz keratophyre group. The spilitic suite is part of the characteristic volcanic–sedimentary sequence, which comprises basic volcanics of highly variable composition, pyroclastics, cherts, and graphitic and Ca-Mg-rich sediments. Examples of this association are found in the Upper Proterozoic basement of the Erzgebirge tin province, but there are probably further occurrences in Paleozoic eugeosynclinal environments of Asia and Australia (Hosking, 1969, 1970; Knight, 1974).

The Lower Paleozoic spilitic rocks of the Eastern Alps, which are genetically related to major strata-bound scheelite deposits, display a geochemical specialization in Sn, besides anomalies in other lithophile elements and a large number of base-metals (Höll and Maucher, 1976).

In general, spilitic rocks are very often associated with a wide variety of ore deposits (Amstutz, 1958, 1974), though their metallogenic significance needs review in the light of current understanding of sub-seafloor hydrothermal circulation/metamorphism. The possible tin potential of basic magmas has been stressed by Dmitriev et al. (1971), and Barsukov and Dmitriev (1972).

Most strata-bound tin deposits show transitional features between the epigenetic and syngenetic end-members of the continuous series discussed above. A genetic classification therefore is difficult without detailed geochemical investigations, which mostly are lacking. It is still common to prejudice the discussion of the formation of tin deposits by the a priori postulation of a granitic source of the hydrothermal fluids. Taking into account the conspicuous lithostratigraphic position of many skarn-type or replacement(?) deposits in sequences of spilitic rocks and cherts, this assumption seems questionable. The specific host rock type indicates a volcanic-sedimentary control of mineralization which is far more pronounced than the sometimes doubtful linkage to a granite body.

In view of the abundance of Cenozoic cassiterite placers, located especially in Southeast Asia, the very sparse data on fossil exogene Sn-accumulations is surprising.

Baumann and Tischendorf (1978) report an average detrital SnO_2-content of 150 ppm in the Cambrian meta-graywackes of the Erzgebirge. Cassiterite occurs in the heavy mineral spectrum together with zircon and rutile, and is thought by them to have been derived from the degradation of the Precambrian stratiform ore deposits of the Pressnitz Series.

A similar situation has been found in Hunan, South China, where tillites and basal black shales of Lower Cambrian age are enriched in Sn, which is taken as an indication of a Precambrian tin mineralization of still unknown extent (Hsieh, 1963).

Quartzitic horizons with clastic cassiterite at the 100 ppm level have been described by Samama (1971) from the Cambrian of the Massif Central, France.

These fossil placer Sn-occurrences bear no known economic potential, but they provide important hints to the metallogenic evolution of tin provinces, which commonly seem to be characterized by the persistence of Sn far back in their geologic evolution.

ACKNOWLEDGEMENTS

Reviews and valuable criticisms were kindly given by Philip Candela, Robert Loucks and Ulrich Petersen. The first author wishes to acknowledge "Deutscher Akademischer Austauschdienst" for a NATO Research Fellowship at Harvard University during which this paper has been finished.

REFERENCES

Adam, J.W.H., 1960. On the geology of the primary tin-ore deposits in the sedimentary formation of Billiton. *Geol. Mijnbouw*, 39: 405–426.

Ahlfeld, F., 1945. Los Yacimientos de "Estano Madera" de Macha (Bolivia) y Yacimientos similares del Noroeste Argentino. *Notas Mus. Univ. Nac. La Plata Geol.*, 10(36): 35–54.

Ahlfeld, F. and Schneider-Scherbina, A., 1964. Los yacimientos minerales y de hidrocarburos de Bolivia. *Dep. Nac. Geol., Bol., 5 (Espec.)*: 1–388.

Amstutz, G.C., 1958. Spilitic rocks and mineral deposits. *Mo. Sch. Mines Tech. Ser.*, 96: 1–11.

Amstutz, G.C., 1974. *Spilites and spilitic rocks.* Springer, Berlin, 482 pp.

Barsukov, V.L. and Dmitriev, L.V., 1972. The earth's upper mantle as a possible source of ore material. *Geochem. Int.*, 1972: 1017–1040.

Baumann, L., 1965. Zur Erzführung und regionalen Verbreitung des "Felsithorizontes" von Halsbrücke. *Freiberg. Froschungsh.*, C, 186: 63–81.

Baumann, L. and Tischendorf, G., 1978. The metallogeny of tin in the Erzgebirge. In: M. Stemprok, L. Burnol and G. Tischendorf (Editors), *Metallization Associated with Acid Magmatism, 3.* Czech. Geol. Surv., Praha, pp.17–28.

Baumann, L. and Weinhold, G., 1963. Zum Neuaufshluss des sog. "Felsithorizontes" von Halsbrucke. *Z. Angew. Geol.*, 1963: 338–345.

Baumann, L., Tischendorf, G., Schmidt, K. and Gubitz, K.B., 1976. Zur minerogenetischen Rayonierung des Territoriums der Deutschen Demokratischen Republik. *Z. Geol. Wiss.*, 4: 955–973.

Berg, C., 1925. *Geologische Karte von Preussen und benachbarten deutschen Ländern. Blatt Wigandsthal-Tafelfichte, 1 : 25.000, Nr. 2944 and 3006.* Preuss. Geol. Landesanstalt, Berlin.

Berg, G. and Ahrens, W., 1925. *Geologische Karte von Preussen und benachbarten deutschen Ländern. Blatt Friedeberg am Queiss, 1 : 25.000, Nr. 2945.* Preuss. Geol. Landesanstalt, Berlin.

Burt, D.M., 1978. The silicate-borate-oxide equilibria in skarns and greisens. The system $CaO-SnO_2-SiO_2-H_2O-B_2O_3-CO_2-Fe_2O_3$. *Econ. Geol.*, 73: 269–282.

Campbell, F.A., Ethier, V.G. and Krouse, R.H., 1978a. Geochemistry of the massive pyrrhotite lens of the Sullivan orebody. *Econ. Geol.*, 73: 1389–1390.

Campbell, F.A., Ethier, V.G., Krouse, H.R. and Both, R.A., 1978b. Isotopic composition of sulfur in the Sullivan orebody, British Columbia. *Econ. Geol.*, 73: 246–268.

Cissarz, A. and Baum, F., 1960. Vorkommen und Mineralinhalt der Zinnerzlagerstätten von Bangka (Indonesien). *Geol. Jahrb.*, 77: 541–580.

Cox, R. and Glasson, K.R., 1971. Economic geology of the Cleveland Mine, Tasmania. *Econ. Geol.*, 66: 861–878.

Davies, R.D., Allsopp, H.L., Erlank, A.J. and Manton, W.I., 1970. Sr-isotope studies on various layered mafic intrusions in southern Africa. *Geol. Soc. S. Afr., Spec. Publ.* 1: 576–593 (Symposium on the Bushveld Igneous Complex and other layered intrusions).

De la Beche, H.T., 1839. *Report on the Geology of Cornwall, Devon, and West Somerset.* Longman, Orme, Brown, Green and Longmans, London, 648 pp.

Dines, H.G., 1956. *The Metalliferous Mining Region of South-West England, 2.* H.M.S.O., London, 795 pp.

Dmitriev, L., Barsukov, V. and Udintsev, G., 1971. Rift-zones of the ocean and the problem of ore-formation. *Soc. Min. Geol. Jpn., Spec. Issue* 3: 65–69.

Dunham, K.C. (Editor), 1969. *Okehampton, Sheet 324, Geol. map 1 : 63,360.* Geol. Surv. Great Britain. Engl. and Wales, Southampton.

Edmonds, E.A., McKeown, M.C. and Williams, M., 1969. *British Regional Geology. South-West England.* H.M.S.O., London, 3rd ed., 130 pp.

El-Sharkawi, M.A.H. and Dearman, W.R., 1966. Tin-bearing skarns from the north-west border of the Dartmoor Granite, Devonshire, England. *Econ. Geol.*, 61: 362–369.

Ethier, V.G., Campbell, F.A., Both, R.A. and Krouse, H.R., 1976. Geological setting of the Sullivan orebody and estimates of temperatures and pressure of metamorphism. *Econ. Geol.*, 71: 1570–1588.

Freeze, A.C., 1966. On the origin of the Sullivan orebody, Kimberley, B.C. *CIM Spec. Vol.*, 8: 263–294.

Garnett, R.H.T., 1967. The underground pursuit and development of tin lodes. In: W. Fox (Editor), *A Technical Conference on Tin.* Int. Tin Council, London, pp.137–200.

Gilmour, P., 1976. Some transitional types of mineral deposits in volcanic and sedimentary rocks. In: K.H. Wolf (Editor), *Handbook of Strata-Bound and Stratiform Ore Deposits, 1.* Elsevier, Amsterdam, pp.111–160.

Gonzales, R.J., 1956. Riqueza minera y Yacimientos minerales de Mexico. *Int. Geol. Congr. Mexico,* 20: 1–498.

Grip, E. and Wirstam, A., 1970. The Boliden sulphide deposit. A review of geo-investigations carried out during the lifetime of the Boliden Mine, Sweden (1924–1967). *Sver. Geol. Unders.,* 64: 1–68.

Höll, R. and Maucher, A., 1976. The strata-bound ore deposits in the Eastern Alps. In: K.H. Wolf (Editor), *Handbook of Strata-Bound and Stratiform Ore Deposits, 5.* Elsevier, Amsterdam, pp.1–36.

Hösel, G., 1968. Die Skarnlager im Raum Schwarzenberg (Erzgebirge). *Ber. Dtsch. Ges. Geol. Wiss., B,* 13: 469–477.

Hösel, G., 1973. Zur Genese der sogenannten Zinnlager von Aue und Bockau im Erzgebirge. *Z. Angew. Geol.,* 19: 4–8.

Hösel, G. and Pfeiffer, L., 1965. Geologie, Petrographie und Genese der Skarnlagerstätte Pöhla (Erzgebirge). *Z. Angew. Geol.,* 11: 169–180.

Hosking, K.F.G., 1969. Aspects of the geology of the tin fields of South-East Asia. In: W. Fox (Editor), *A Second Technical Conference on Tin, 1.* Int. Tin Council, Bangkok, pp.39–80.

Hosking, K.F.G., 1970. The primary tin deposits of South-East Asia. *Miner. Sci. Eng.,* 2: 24–50.

Hsieh, C.Y., 1963. A study of tin deposits in China. *Sci. Sin.,* 12: 373–390.

Jaskolski, S., 1960. Beitrag zur Kenntnis über die Herkunft der Zinnlagerstätten von Gierczyn (Giehren) im Iser-Gebirge, Niederschlesien. *Neues Jahrb. Mineral. Abh.,* 94: 181–190.

Jaskolski, S., 1962. Erwägungen über die Genese zinnführender Schiefer im Isergebirge (Niederschlesien). *Pol. Akad. Nauk, Pr. Geol.,* 12: 33–53.

Knight, C.L., 1974. Metallogenesis in the Tasman geosyncline. In: A.K. Denmead, G.W. Tweedale and A.F. Wilson (Editors), *The Tasman Geosyncline – A Symposium.* Geol. Soc. Aust. Queensland Div., pp.247–256.

Knight, C.L., 1975. Mount Bischoff tin orebody. In: C.L. Knight (Editor), *Economic Geology of Australia and Papua New Guinea, 1. Metals. Australas. Inst. Min. Metall., Monogr. Ser.,* 5: 591–592.

Knopf, A., 1916. Tin ore in northern Lander County, Nevada. *U.S. Geol. Surv. Bull.,* 640-G: 125–138.

Kozlowski, A. and Karwowski, L., 1975. Genetyczne wakazniki mineralizacji W-Sn-Mo na obszarze karkonosko-izerkim. *Kwart. Geol.,* 19: 67–73.

Lehmann, B., 1979. Schichtgebundene Sn-Lagerstätten in der Cordillera Real/Bolivien. *Berliner Geowiss. Abh., A,* 14: 1–135.

Lehmann, B. and Schneider, H.-J., 1980. Schichtgebundene Zinnlagerstätten als mögliche Zukunftsreserve. *GDMB Schriftenreihe,* 15 pp., in press.

Lenthall, D.H. and Hunter, D.R., 1977. The geochemistry of the Bushveld granites in the Potgietersrus tin field. *Precambrian Res.,* 5: 359–400.

Leube, A. and Stumpfl, E.F., 1963. The Rooiberg and Leeuwpoort tin mines, Transvaal, South Africa. *Econ. Geol.,* 58: 391–418; 527–557.

Lorenz, W. and Hoth, K., 1964. Die lithostratigraphische Gliederung des kristallinen Vorsilurs in der fichtelgebirgisch-erzgebirgischen Antiklinalzone. *Beih. Geol.,* 44: 1–44.

Lorenz, W. and Hoth, Kl, 1968. Die geologische Stellung der erzgebirgischen Skarnlager. Ber. deutsch. *Ges. Geol. Wiss., B,* 13: 497–503.

Luff, W.M., 1977. Geology of Brunswick No. 12 Mine. *CIM Bull.,* 70 (782): 109–119.

Mulligan, R., 1975. Geology of Canadian tin occurrences. *Geol. Surv. Can. Econ. Geol. Rep.,* 28: 1–155.

Mulligan, R., 1978. Tin in stratabound massive sulphide deposits. In: M. Stemprok, L. Burnol, and G. Tischendorf (Editors), *Metallization Associated with Acid Magmatism, 3.* Czech. Geol. Surv., Praha, pp.439–446.

Newham, L.A., 1975. Renison Bell tinfield. In: C.L. Knight (Editor), *Economic Geology of Australia and Papua New Guinea, 1. Metals. Australas. Inst. Min. Metall., Monogr. Ser.,* 5: 581–583.

Palmer, K.G., 1976. The Cleveland tin deposit. *Excursion Guide no. 31 AC, 25th IGC,* pp.27–31.

Patterson, D.J., 1976. The Renison tin deposit. *Excursion Guide no. 31 AC, 25th IGC,* pp.36–41.

Pelletier, R.A., 1964. *Mineral Resources of South-Central Africa.* Oxford Univ. Press, Cape Town, 277 pp.

Pentland, A.G., 1943. Occurrence of tin in the Sullivan Mine. *CIM Trans.,* 46: 17–22.

Putzer, H., 1940. Die zinnführende Fahlbandlagerstätte von Giehren am Isergebirge. *Z. Dtsch. Geol. Ges.,* 92: 137–158.

Ransom, D.M. and Hunt, F.L., 1975. Cleveland tin mine. In: C.L. Knight (Editor), *Economic Geology of Australia and Papua New Guinea, 1. Metals. Australas. Inst. Min. Metall., Monogr. Ser.,* 5: 584–591.

Rösler, H.J. and Lange, H., 1976. *Geochemische Tabellen.* VEB Deutsch. Verlag Grundstoffindustrie, Leipzig, 2nd ed., 674 pp.

Samama, J.-C., 1971. Paléoplacers à tungstène-étain dans les séries métamorphiques du Vivarais oriental. *C.R. Acad. Sci. Ser., D,* 272: 516–518.

Sangster, D.F. and Scott, S.D., 1976. Precambrian, strata-bound, massive Cu–Zn–Pb sulfide ores of North America. In: K.H. Wolf (Editor), *Handbook of Strata-Bound and Stratiform Ore Deposits, 6.* Elsevier, Amsterdam, pp.129–222.

Sato, T., 1974. Distribution and geological setting of the Kuroko deposits. *Min. Geol., Spec. Issue,* 6: 1–9.

Schneider, H.-J. and Lehmann, B., 1977. Contribution to a new genetical concept on the Bolivian Tin Province. In: D.D. Klemm and H.-J. Schneider (Editors), *Time- and Strata-Bound Ore Deposits.* Springer, Berlin, pp.153–168.

Smirnov, V.I. (Editor), 1977. *Ore Deposits of the USSR, 3.* Pitman, London, 492 pp.

Smith, W.W., 1878. On the occurrence of metallic ores with garnet rock. *Trans. R. Geol. Soc. Cornwall,* 9: 38–45.

Stear, W.M., 1977. The strata-bound tin deposits and structures of the Rooiberg fragment. *Trans. Geol. Soc. S. Afr.,* 80: 67–78.

Szalamacha, M. and Szalamacha, J., 1974. Geologiczna i petrograficzna charakterystyka lupkow zmineralizowanych kasyterytem na przykladzie kamieniolomu w Krobicy (Geological and petrographic characteristics of schists mineralized with cassiterite on the basis of material from the quarry at Krobica). *Inst. Geol.,* 279 (23): 59–90.

Taupitz, K.C., 1978. Die erzführenden Pegmatite des "Zambezi Mobile Belt". *Erzmetall,* 31: 61–70; 133–137.

Taylor, R.G., 1979. *Geology of Tin Deposits.* Elsevier, Amsterdam, 543 pp.

Thompson, R.I. and Panteleyev, A., 1976. Stratabound mineral deposits of the Canadian Cordillera. In: K.H. Wolf (Editor), *Handbook of Strata-Bound and Stratiform Ore Deposits, 5.* Elsevier, Amsterdam, pp.37–108.

Vokes, F.M., 1960. Contributions to the mineralogy of Norway. No. 7. Cassiterite in the Bleikvassli ore. *Nor. Geol. Tidsskr.,* 40: 193–201.

Walker, R.R., Matulich, A., Amos, A.C., Watkins, J.J. and Mannard, G.W., 1975. The geology of the Kidd Creek Mine. *Econ. Geol.,* 70: 80–89.

Weinhold, G., 1977. Zur prävaristischen Vererzung im Erzgebirgskristallin aus der Sicht seiner lithofaziellen und geotektonisch-magmatischen Entwicklung während des assyntisch-kaledonischen Ära. *Freiberg. Froschungsh.,* C, 320: 1–53.